P9-CED-424

# PRENTICE HALL GENERAL SCIENCE

'OR GRADES 6-9

## SOLID SCIENCE, SIMPLY UNDERSTOOD

**The key to unlocking the world of tomorrow...*today*!**

# BROAD AND IN-DEPTH COVERAGE ORGANIZED FOR EASY LEARNING AND TOTAL TEACHING FLEXIBILITY

**Comprehensive and Balanced Coverage** integrated through two units of life science, earth science, and physical science in each text provides total flexibility for teaching one-, two-, or three-year general science programs.

**Relevant and Up-to-date Content** lets students relate science to the ever-changing world around them.

**Dynamic Unit Openers** with exciting visuals, vividly written introductory paragraphs, and lists of chapter titles motivate the student at the beginning of each unit.

UNIT TWO

Classification of Living Things

CHAPTERS
6 Classification
7 Viruses, Bacteria, and Protists
8 Nonvascular Plants and Plantlike Organisms
9 Vascular Plants
10 Animals: Invertebrates
11 Animals: Vertebrates
12 Mammals

**Exciting Chapter Openers** arouse students' attention through full-color pictures and compelling stories that relate science to their everyday world.

Physical and Chemical Changes

**A List of Chapter Sections and Chapter Objectives** provides students with clear directions for study—organizing chapter content and pinpointing learning goals at a glance.

# NOT JUST READABLE... BUT UNDERSTANDABLE!

**Clear and Flowing Narrative** explains concepts through interesting examples and anecdotes at the reading level of your students.

**Short, Digestible Paragraphs** help students grasp concepts quickly.

**Dynamic Photographs and Large Detailed Illustrations** on nearly every page promote understanding through visual reinforcement.

Figure 19-1 *This photograph shows hot, molten lava erupting from a volcano on Hawaii.*

Figure 19-2 *A pattern of roughly six-sided rock columns forms where mud flats dry out. This pattern is similar to the skin of solid rock that shrunk and split to form the Giant's Causeway.*

## 19-1 Rocks of Liquid and Fire: Igneous Rocks

The strange six-sided rocks of the Giant's Causeway were forged in fire. But not by giants, and not in an ordinary fire.

Deep within the earth it is so hot that some of the minerals that make up rocks melt and flow like liquid. The liquid rock is less dense than solid rock, so the liquid rock tends to rise—just as bubbles rise through water. This hot, molten rock deep inside the earth is called **magma.**

Sometimes the magma reaches the surface. It may erupt from a volcano. Or it may seep through weak spots in the ground. Magma that has reached the earth's surface is called **lava.**

At the site of the Giant's Causeway, melted rock seeped toward the surface, forming large pools of melted rock material. As the rock cooled and solidified, it shrank and cracked. The cracks formed a six-sided pattern. You can see a pattern something like this where mud flats dry out in places like the Painted Desert in Arizona. As the lower levels of the pool of melted rock material cooled and hardened, the pattern of cracks spread downward. Each pool broke up into rows of six-sided rock columns.

The rocks of the Giant's Causeway are basalt. They belong to one group of rocks called **igneous** (IG-nee-us) **rocks.** The word "igneous" comes from the Latin word "ignis," meaning coming from fire. **All rocks formed from the cooling and hardening of magma are igneous rocks.**

### Extrusive Rocks

Basalt is only one kind of igneous rock. It is a kind known as **extrusive** (ek-STROO-siv) **rock.** Extrusive rocks form from melted rock or lava that cools and hardens at or near the earth's surface. Some extrusive rocks originally were pushed slowly onto the earth's surface, much as toothpaste is pushed out of a tube. Other extrusive rocks formed as melted rock material erupted rapidly from places such as volcanoes.

434

Figure 19-3 *These photographs show that basaltic rock can have very different shapes. Devil's Postpile National Monument (left) is a spectacular mass of blue-gray basalt columns. Shiprock (right) is the remains of the basaltic inner core of a volcano.*

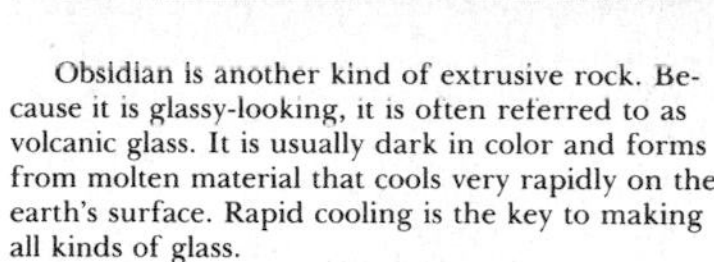

Obsidian is another kind of extrusive rock. Because it is glassy-looking, it is often referred to as volcanic glass. It is usually dark in color and forms from molten material that cools very rapidly on the earth's surface. Rapid cooling is the key to making all kinds of glass.

Sometimes igneous rocks form so quickly that bubbles of volcanic gases are trapped in them. Rocks formed in this way contain many holes and tiny needlelike slivers of volcanic glass. Pumice (PUH-mis) is an example of this kind of rock. Because pumice is filled with bubbles, it is a rock that actually floats on water! The particles of volcanic glass in pumice make it a good polishing agent. Have you had your teeth cleaned at a dentist's office? The gritty stuff in the cleaning powder was probably crushed pumice.

### Intrusive Rocks

Another kind of igneous rock is granite, one of the earth's most common types of rock. Chemically, granite is similar to obsidian. However, granite is usually white or gray, rather than black. Also granite contains large crystals, while obsidian contains none.

Rocks that form from melted rock or magma that cools and hardens deep below the earth's surface are called **intrusive** (in-TROO-siv) **rocks.** They intrude, or push between surrounding rocks, while in a molten state. Huge masses of granite form within the earth

Figure 19-4 *This obsidian dome is an example of an extrusive rock that cools rapidly on the earth's surface.*

435

**Large, Easy-to-read Type Size** facilitates reading and learning.

**Vocabulary Words** are introduced in boldface type, defined in context the first time they appear in a unit, and are accompanied by phonetic respellings to assure proper pronunciation.

**Key Ideas** appear in boldface to guide study efforts and assure concept mastery.

# SKILLS-ORIENTED FOR EXPERIENCING THE EXCITEMENT OF SCIENCE

**Activities** include quick and easy hands-on activities, computational exercises, library assignments, and science writing activities found in the margins of each chapter. Great fun-filled ways for reinforcing and extending content in addition to developing every student's science skills.

directly on the retina. In certain cases, the image falls in front of the retina or behind it. If the eyeball is too long, the image forms in front of the retina. A person has difficulty seeing objects at a distance but no trouble seeing objects nearby. This condition is called **nearsightedness.** A correcting lens would have to make the light rays diverge before they enter the eye. What kind of lens does this? Right, a concave lens. Figure 5-27 will help you understand this idea.

If the eyeball is too short, the image is focused behind the retina. The person can see distant objects clearly but has difficulty with nearby objects. This problem is called **farsightedness.** A convex lens, which makes light rays converge, is used to correct farsightedness. See Figure 5-27.

**Activity**

*Hole in Your Hand*

Here is an optical illusion for you to try.

1. Roll a sheet of notebook paper into a tube and hold one end of it up to your right eye.
2. Place the side of your left hand against the tube with palm toward you at a distance of 15 centimeters from your eye.
3. Keep both eyes open and look at a distant object. Describe what you see.

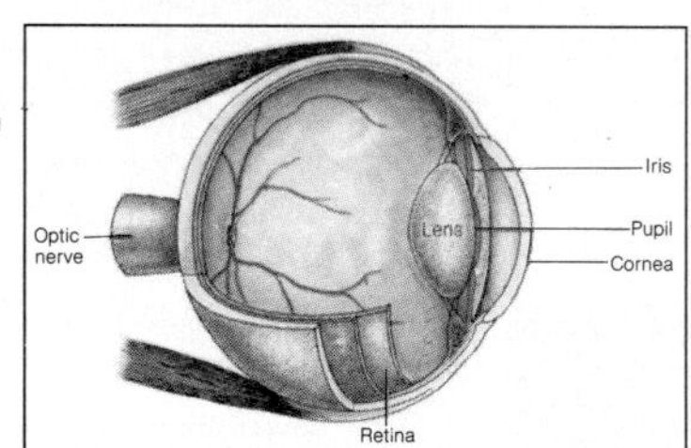


**Figure 5-26** *The various parts of the eye all work together to enable you to see.*

126

**LABORATORY ACTIVITY**

***The Effect of Water Depth on Sediments***

**Purpose**

In this activity, you will find out what effect differences in water depth have on the settling of sediments containing mixed particles.

**Materials** *(per group)*
Plastic tubes of different lengths containing sediment samples and water

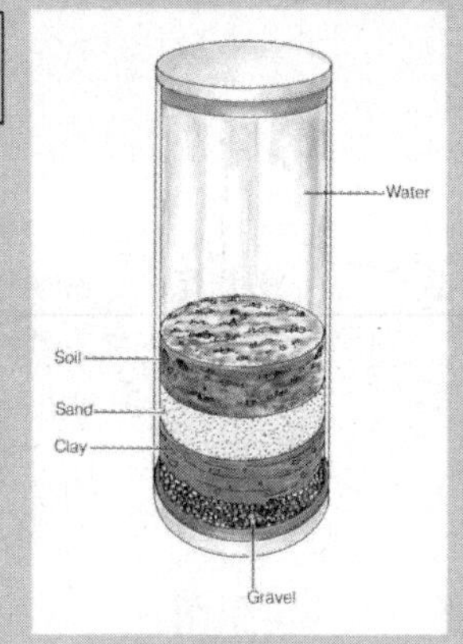


**Procedure**

1. Select one tube from those provided by your teacher.
2. Check to see that each end of the tube is securely capped.
3. Hold the tube by both ends and gently tip it back and forth until the sediment is *thoroughly* mixed throughout the water.
4. Set the tube in an upright position in a place where it will not be disturbed.
5. Repeat steps 1 through 4 for each of the tubes remaining.
6. Carefully observe which type of sediment settles first.

**Observations and Conclusions**

1. What general statement can you make about the effect that the size of sediment particles has on the order in which sediment settles in a tube?
2. Observe each column carefully. Make a detailed sketch to illustrate the heights of the different layers formed by the settling sediment in the tube containing the shortest column of water.
3. Describe the effect that the length of a water column has on the number and height of layers formed in a sediment sample containing mixed particles.

308

**Laboratory Investigations** in every chapter reinforce concepts, give students hands-on experience using the scientific method, and develop manipulative laboratory and critical thinking skills. All experiments use inexpensive materials and include clear step-by-step directions, useful visuals, and questions that guide students to drawing conclusions.

...ic activity and m... building.

6. Scientists have estimated the age of the earth by the radioactive dating of rock fragments called meteorites.

10. At one time, all the lands on the earth were a single continent called Uinta.

**CONCEPT REVIEW: SKILL BUILDING**

*Use the skills you have developed in the chapter to complete each activity.*

1. **Making charts** Make a chart of the geologic eras. Include when each era began and ended and the life forms that characterize each era.
2. **Making calculations** A radioactive substance has a half-life of 500 million years. After 2 billion years, how many half-lives have passed? How many kilograms of a 10-kg sample would be left at this time? If the half-life were 4 billion years?
3. **Interpreting diagrams** Use the diagram to answer the following questions.
   a. According to the way in which layers C, D, E, and F lie, what might have happened in the past?
   b. Which letter shows the unconformity?
   c. List the events that occurred from oldest to youngest. Include the order in which each layer was deposited and when the fault, intrusion, and unconformity were formed. Explain why you chose this order.

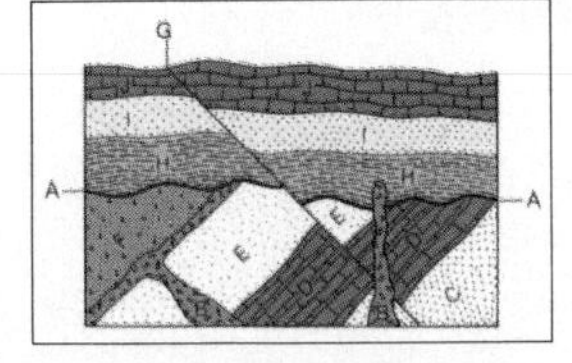


4. **Relating facts** Explain why present-day clams are so similar to 500-million-year-old clam fossils.
5. **Applying concepts** Why have rocks that account for the first 600,000 years of the earth's existence never been found on the earth? Explain your answer.

**CONCEPT REVIEW: ESSAY**

*Discuss each of the following in a brief paragraph.*

**New Skill-building Exercises** at the end of every chapter provide more interesting and challenging questions to further develop critical thinking skills.

# RELATES SCIENCE TO OUR EVERYDAY LIVES

**Science Gazettes** are exciting mini-magazines at the end of each unit that present highly motivational and thought-provoking articles on current discoveries, issues, and imaginative, but fact-based, future scenarios related to science. Plus special science reading skills activities for each Gazette article in the corresponding Teacher's Resource Book.

SCIENCE GAZETTE

Adventures in Science

The Search for SUPERCONDUCTORS

Karl Mueller and Johannes Bednorz

Imagine trains that fly above their tracks at airplane speeds and powerful computers that fit in the palm of your hand. Picture unlocking the secrets of the atom, or skiing on slopes made of air. Purely imagination? Not really! All of these things—and more—have been brought closer to reality by the work of Dr. Karl Alex Mueller and Dr. Johannes Georg Bednorz. These two dedicated scientists have changed fantasy to fact through their work with superconductors.

SCIENCE GAZETTE

Issues in Science

Are we DESTROYING the GREATEST CREATURES of the SEA?

Futures in Science

Hypersonic Planes: FLYING FASTER THAN THE SPEED OF SOUND

dow may resonate to a loud noise outside. A vase resting on a radio may "ring" if a certain note is played. And a singer standing near a piano can make a piano string vibrate if she sings a note with the same frequency as the piano string.

Although you may not be aware of it, you are applying the idea of resonance every time you tune your radio to a particular station. Each radio station broadcasts at a specific frequency. When you "tune in" to your favorite station, you are matching the frequency of that station.

**Making Music**

Musical instruments produce sounds in different ways, depending upon the instrument. Percussion instruments, such as drums and cymbals, are set vibrating by being struck. In wind and brass instruments, such as flutes, clarinets, trumpets, and trombones, columns of air within the instrument are made to vibrate at various frequencies.

**Activity**

*Making Music*

The piano, guitar, and trombone are examples of the three different kinds of musical instruments. The piano is a percussion instrument; the guitar is a stringed instrument; and the trombone is a wind instrument. Each of these instruments produces sound vibrations in a different way.

Examine each instrument and describe how it works. A diagram would be helpful in explaining this information.

**Career:** *Sound Mixer*

**HELP WANTED: SOUND MIXER** To regulate sound levels and sound quality in recording studio. High school diploma and commercial radio operator's license required. Some training in operating electronic equipment needed. Experience in broadcasting desirable.

The scene is set for the train robbery. The television cameras carefully follow the bandits as they chase the speeding train, while a technician sits at a table busily turning knobs and dials. The technician adjusts several microphones so that the many sounds that will be heard are correctly balanced. The sounds of the bandits' voices, galloping hoofs, chugging train, and background music all have to be combined in the most striking and exciting way.

The technician who controls volume level and sound quality during a radio, television, or motion picture production is called a **sound mixer**. Sound mixers control the output of individual sources of sound so that a balance of music, dialogue, and sound effects is obtained.

Sound mixers also direct the installation of microphones and amplifiers. Using special testing equipment, they locate defects in the sound equipment and make repairs.

Anyone interested in a career as a sound mixer can obtain more information by writing to the Federal Communications Commission, 1919 M Street N.W., Washington, DC 20554.

99

**Careers** in every chapter highlight realistic opportunities in science for students at every ability level.

**Real-life Applications** and up-to-date content show how science affects our everyday experiences and lives.

**2-3 Heating and Refrigeration Methods**

Controlling the temperature of the environment is an important application of scientific knowledge about heat. Warming and cooling homes and workplaces involves designing and constructing various types of heating and refrigeration systems. **The methods used in all heating and cooling systems apply the knowledge of heat transfer and energy conversion.**

**Figure 2-15** *HEATING SYSTEMS*

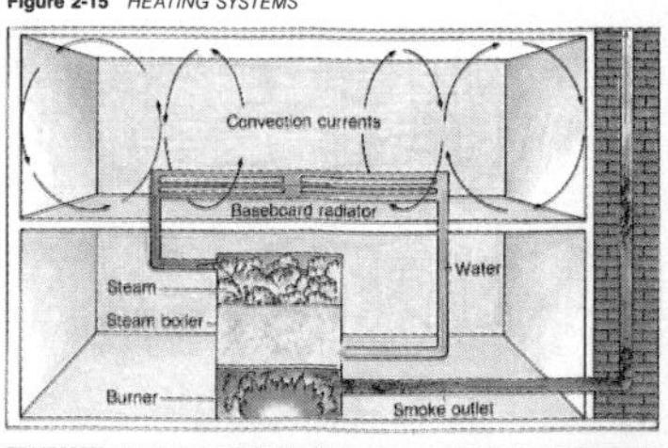


*Steam Heat* In a boiler above the furnace, water changes to steam. Steam has higher potential energy than liquid water. It stores more heat. The steam passes through pipes to radiators. Metal radiators have large surface areas that permit rapid heat transfer to the surrounding air. The heat transfer is by conduction and radiation. Heated air near the radiator rises in convection currents and circulates through the room. As the steam releases its stored energy in the form of heat, it cools and changes back to liquid water. The liquid water returns to the boiler, and the cycle begins again.

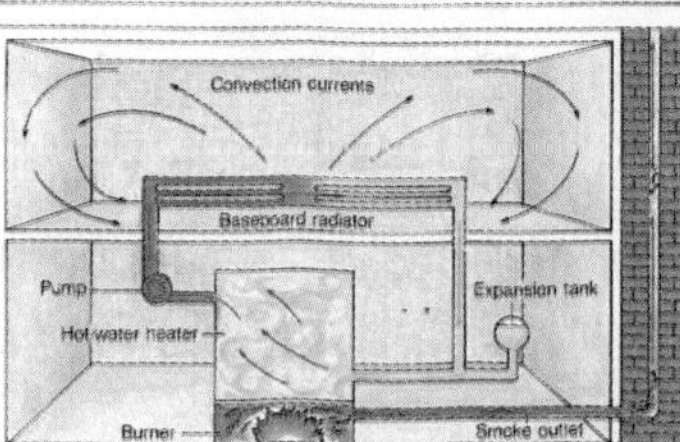


*Hot Water* Heat released by burning fuel or by electric coils in a furnace is absorbed by water in a boiler. The hot water is piped through the rooms, usually in radiators. Heat flows from the hot water to the cooler air in the room. As the air near the radiator warms and rises, convection currents are formed. The cooled water returns to the boiler, and the process begins anew.

44

# INVOLVES STUDENTS ON EVERY PAGE

## 15-1 Nutritional Substances

Throughout the world, more than five billion chemical factories operate both day and night. These factories produce a great variety of chemicals from raw materials. Strangely, in order for these factories to work properly, their internal temperature must be kept at almost exactly 37° C. For the most part, the chemical changes carried out in these factories could not be done anywhere else. Moreover, each of these factories can discover its own faulty work. They can also reproduce and repair some of their parts, and

**Figure 15-1** *The body needs and uses more than 50 nutrients. These nutrients are grouped into six categories. Which nutrient is contained in all foods?*

***THE SIX BASIC NUTRIENTS***

| Substances | Sources | Need For |
|---|---|---|
| Proteins | Soybeans, milk, eggs, lean meats, fish, beans, peas, cheese | Growth, maintenance, and repair of tissues<br>Manufacture of enzymes, hormones and antibodies |
| Carbohydrates | Cereals, breads, fruits, vegetables | Energy source<br>Fiber or bulk in diet |
| Fats | Nuts, butter, vegetable oils, fatty meats, bacon, cheese | Energy source |
| Vitamins | Milk, butter, lean meats, leafy vegetables, fruits | Prevention of deficiency diseases<br>Regulation of body processes<br>Growth<br>Efficient biochemical reactions |
| Mineral salts | | |
| Calcium and phosphorus compounds | Whole-grain cereals, meats, milk, green leafy vegetables, vegetables, table salt | Strong bones and teeth<br>Blood and other tissues |
| Iron compounds | Meats, liver, nuts, cereals | Hemoglobin formation |
| Iodine | Iodized salt, seafoods | Secretion by thyroid gland |
| Water | All foods | Dissolving substances<br>Blood<br>Tissue fluid<br>Biochemical reactions |

366

## 19-4 The Rock Cycle

Rocks go through many changes. Igneous and sedimentary rocks can change to metamorphic rocks. Metamorphic rocks, too, may be remelted and become igneous rocks again. **The continuous changing of rocks from one kind to another over long periods of time is called the rock cycle.**

**Figure 19-16** *This diagram of the rock cycle shows the many ways that rocks are changed. What changes sedimentary rocks to metamorphic?*

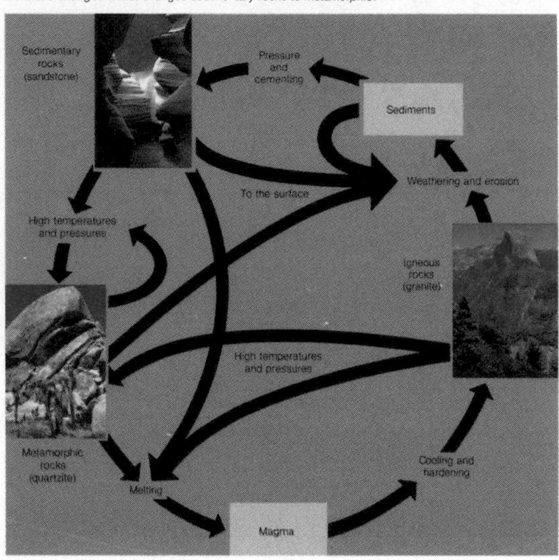


444

**Charts, Maps, and Graphs** reinforce chapter content and help students organize, interpret, and apply scientific data.

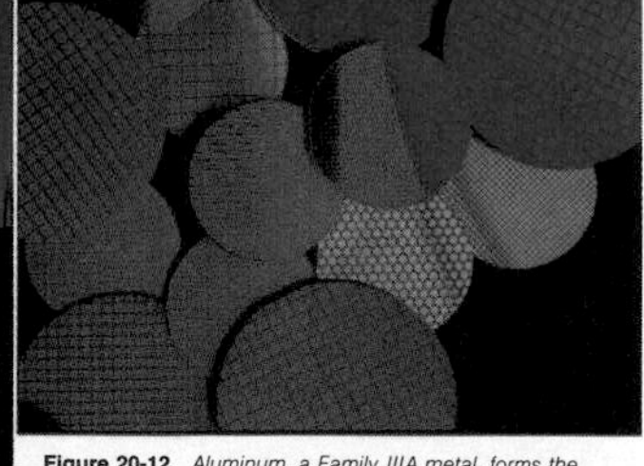

**Figure 20-12** *Aluminum, a Family IIIA metal, forms the outer walls of the World Trade Center towers in New York (left). These wafers, on which computer circuits will be built, are made of silicon, a metalloid of Family IVA (right).*

gasoline. Carbon compounds are so numerous that a whole branch of chemistry is devoted to their study. This branch is called organic chemistry.

The **nitrogen family,** Family VA, contains an element that makes up about 78 percent of the air around you—nitrogen. The atoms of elements in this family have five valence electrons. These atoms tend to share electrons when they bond with other atoms. Can you identify the metalloids in this family?

**Figure 20-13** *Family VA is the nitrogen family and Family VIA is the oxygen family. Which elements in these families are gases?*

| VA | VIA |
|---|---|
| 7 **N** Nitrogen 14.007 | 8 **O** Oxygen 15.999 |
| 15 **P** Phosphorus 30.974 | 16 **S** Sulfur 32.06 |
| 33 **As** Arsenic 74.922 | 34 **Se** Selenium 78.96 |
| 51 **Sb** Antimony 121.75 | 52 **Te** Tellurium 127.60 |
| 83 **Bi** Bismuth 208.98 | 84 **Po** Polonium (209) |

**Figure 20-14** *This explosion is caused by TNT, which is an organic compound containing nitrogen, a Family VA nonmetal.*

472

**Interactive Questions** in the text and figure captions actively involve students in the lesson and make concepts more understandable and easier to learn. Great for promoting class discussions and sharpening critical thinking skills.

# THOROUGH REVIEW AND REINFORCEMENT TO MASTER SCIENCE CONCEPTS AND SKILLS

**Figure 10-39** *Insects defend themselves in many ways. The wasp (left) uses its stinger, the assassin bug (top right) is camouflaged, the bombardier beetle (center right) sprays a foul-smelling chemical, and the eye spot coloring on this moth (bottom right) startles its predators.*

thorns of plants. Some insects have the ability to spray foul-smelling chemicals at an enemy. Other insects have markings that frighten birds and other animals that might eat them. In Figure 10-39, you can see two large "eyespots" on the wings of the moth. These spots startle predatory animals and may confuse a predator long enough for the insect to make an escape.

***SECTION REVIEW***

1. What are the three main sections of an insect's body?
2. What are the four stages of metamorphosis?
3. List three kinds of social insects.

247

## Section Reviews
check recall and basic understanding

**CHAPTER REVIEW**

SUMMARY

**7-1 Viruses**

- Viruses cannot carry out any life processes unless within a living cell.
- Viruses are tiny particles that contain a center of either DNA or RNA surrounded by a protein coat.
- Reproduction of viruses occurs only when the virus invades a living cell.
- Bacteriophages are viruses that infect bacterial cells.
- When a virus's hereditary material enters a living cell, the virus takes control of the cell. The cell begins to produce new viruses, which are released from the cell.

**7-2 Bacteria**

- Bacteria are simple unicellular monerans.
- Bacteria have three basic shapes: cocci, bacilli, or spirilla.
- Bacterial cells have a cell wall and a cell membrane surrounding the cytoplasm. The hereditary material of a bacterial cell is spread throughout the cytoplasm.
- Some bacteria have flagella, or whiplike structures, that enable them to move.
- Some bacteria called autotrophs use energy from the sun or chemicals to make food.
- Bacteria that must obtain their food from other organisms are called heterotrophs.
- Bacteria usually reproduce by binary fission, or splitting in two.

**7-3 Protozoans**

- Protozoans are unicellular animallike microorganisms. Their hereditary material is confined in a nucleus.
- Amoebas move and obtain food by the movement of their pseudopods. Amoebas reproduce by binary fission.
- Ciliates, such as paramecia, have small hairlike cilia used for obtaining food and for movement.
- Paramecia reproduce by binary fission and conjugation.
- Flagellates use whiplike flagella to move. There are two groups of flagellates. The members of one group contain chlorophyll and can make their own food. The second group of flagellates do not contain chlorophyll and cannot make their own food.
- Sporozoans cannot move from place to place. Sporozoans are parasites and have complex life cycles.

VOCABULARY

*Define each term in a complete sentence.*

| | | | |
|---|---|---|---|
| **anal pore** | **cilium** | **flagellum** | **pseudopod** |
| **antibiotic** | **coccus** | **food vacuole** | **RNA** |
| **bacillus** | **conjugation** | **gullet** | **saprophyte** |
| **bacteriophage** | **contractile vacuole** | **microbiology** | **spirillum** |
| **bacterium** | **cytoplasm** | **nucleus** | **toxin** |

## Chapter Summaries
state key concepts taught in each section.

## Vocabulary Exercises
reinforce important scientific terms used in the chapter.

## Comprehensive Chapter Reviews
check factual recall, concept understanding, and critical-thinking skills development through a variety of question formats—including multiple-choice, completions, true or false, short answers, and essay. Plus *NEW* skill-building questions that sharpen higher order thinking skills for students at every ability level.

CONTENT REVIEW: MULTIPLE CHOICE

*Choose the letter of the answer that best completes each statement.*

1. Viruses are
   a. shaped like spheres. b. shaped like tadpoles.
   c. shaped like cylinders. d. any of these shapes.
2. What is used to observe viruses?
   a. the unaided eye b. an electron microscope
   c. a light microscope d. all of these
3. The scientist who gave viruses their name was
   a. Martinus Beijerinck. b. Wendell Stanley.
   c. Dimitri Iwanowski. d. Anton van Leeuwenhoek.
4. The type of hereditary material in a virus may be
   a. DNA. b. RNA.
   c. either DNA or RNA. d. neither DNA nor RNA.
5. An organism that feeds on dead matter is called a(n)
   a. autotroph. b. saprophyte. c. parasite. d. flagellum.
6. An amoeba moves by means of
   a. flagella. b. cilia. c. pseudopods. d. conjugation.
7. Which organism reproduces by binary fission and conjugation?
   a. amoeba b. paramecium c. sporozoan d. *Euglena*
8. *Euglena* is an example of a(n)
   a. flagellate. b. ciliate. c. sporozoan. d. amoeba.
9. Which group of protozoans cannot move by themselves?
   a. amoebas b. ciliates c. flagellates d. sporozoans
10. The most familiar type of sporozoan causes
    a. pneumonia. b. tuberculosis.
    c. malaria. d. African sleeping sickness.

CONTENT REVIEW: COMPLETION

*Fill in the word or words that best complete each statement.*

1. A virus that infects a bacterial cell is called a(n) ________.
2. The bacteria that look like rods are called ________.
3. Bacteria usually reproduce through ________.
4. Bacteria that make nitrogen compounds from nitrogen gas are called ________.
5. Some bacteria produce poisons called ________.
6. Paramecia have ________ that enable them to move.
7. The spherical structure in a cell that releases enzymes to digest food is the ________.
8. In paramecia, food particles are swept by cilia into an indented structure called a(n) ________.
9. A flagellate that is both an autotroph and a heterotroph is the ________.
10. The structure in *Euglena* that responds to changes in light is the ________.

166

CONTENT REVIEW: TRUE OR FALSE

*Determine whether each statement is true or false. If it is true, write "true." If it is false, change the underlined word or words to make the statement true.*

1. A virus is made up of hereditary material surrounded by a protein coat.
2. The outermost boundary of a bacterial cell is the cell membrane.
3. Cilia are long, thin, whiplike structures that are used to propel some microorganisms.
4. Saprophytes are organisms that feed on living organisms.
5. Endospores allow bacteria to survive unfavorable conditions.
6. Protozoan means false foot.
7. In a paramecium, the oral groove is the funnel-shaped structure that ends in a food vacuole.
8. The contractile vacuole pumps excess water out of an amoeba.
9. In a paramecium, undigested material is eliminated through an opening called the anal pore.
10. *Euglena* is an example of a flagellate.

CONCEPT REVIEW: SKILL BUILDING

*Use the skills you have developed in the chapter to complete each activity.*

1. **Interpreting a graph** The following graph shows the growth of bacteria. Describe what is happening at each of the numbered growth stages. Suggest a hypothesis to explain the growth pattern in stage 3 and in stage 4.

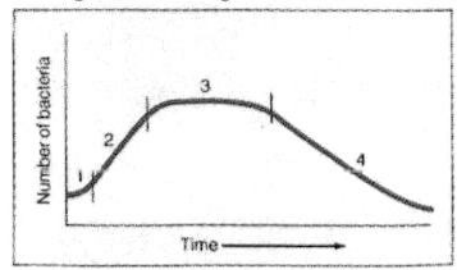


2. **Applying definitions** Suppose a biologist reported the discovery of a sporozoan that moves by means of pseudopods and has a thick cell wall made of cellulose. Could the report be true? Explain.
3. **Applying concepts** The growth of bacteria is slowed by cooler temperatures. How can you apply this information to control food spoilage?
4. **Designing an experiment** Penicillin is an antibiotic that kills some types of bacteria. Suppose you have five different bacteria cultures. Design an experiment to determine which of the cultures could be destroyed by penicillin. Include a control and a variable in your experiment.

CONCEPT REVIEW: ESSAY

*Discuss each of the following in a brief paragraph.*

1. Compare viruses to cells.
2. Explain why some bacteria are harmful.
3. What contributions did Dimitri Iwanowski and Martinus Beijerinck make to the study of viruses?
4. What is the difference between autotrophic and heterotrophic bacteria?
5. Compare the methods of movement in an amoeba, a paramecium, and a *Euglena*.
6. Describe how viruses reproduce.

167

# NEW WRAPAROUND TEACHER'S EDITION PROVIDES ALL THE TEACHING SUPPORT YOU NEED...

**Section Previews** highlight the key concepts presented in each section.

**Performance Objectives** identify what the student should learn from the lesson.

**Science Terms** list all the boldfaced vocabulary terms introduced in the section along with the page numbers on which each term is defined.

**Highlighted Sentences** on the reduced student page draw your attention to the topic sentences and key concepts found in each lesson.

## 18-1 WHAT IS A MINERAL?

### SECTION PREVIEW 18-1

Minerals are composed of one or more elements. The large number of naturally occurring elements form more than 2000 different minerals. For a substance to be classified as a mineral, it must be an inorganic solid that occurs naturally in the earth. It must also have a definite chemical composition and an orderly atomic structure. Minerals are the building blocks of rocks on the earth. About 12 minerals make up most of the earth's crust.

The atoms of the elements that make up minerals combine in a fixed proportion and in a definite geometric pattern. The orderly arrangement of the atoms determines the crystal shape of a mineral. There are six basic crystal shapes: isometric, tetragonal, orthorhombic, monoclinic, triclinic, and hexagonal.

### SECTION OBJECTIVES 18-1

1. **Define the term mineral.**
2. **Identify minerals by their physical properties.**
3. **Describe some uses of minerals.**

### SCIENCE TERMS 18-1

**mineral p. 416**
**inorganic p. 416**
**crystal p. 416**
**magma p. 417**
**luster p. 418**
**hardness p. 419**
**Mohs hardness scale p. 419**
**streak p. 421**
**cleavage p. 422**
**fracture p. 422**

**Figure 18-1** *Chalk (top), aragonite (center), and pearl (bottom) are all made of calcium carbonate or limestone. But only aragonite is a mineral. Why?* ❶

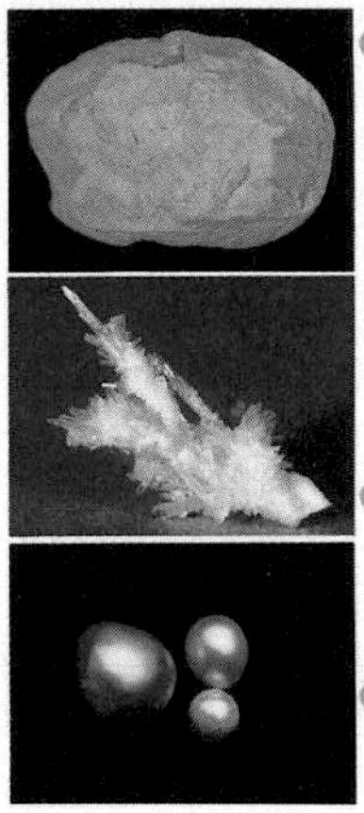

416

### 18-1 What Is a Mineral?

The substance you just read about—diamond—is a mineral. A **mineral** is a naturally occurring substance formed in the earth. A mineral may be made of a single element, such as copper, gold, or sulfur. Or a mineral may be made of two or more elements chemically combined to form a compound. For example, the mineral halite is the compound sodium chloride, which is made of the elements sodium and chlorine.

In order for a substance to be called a mineral, it must have five special properties. The first property of a mineral is that it is an **inorganic** substance. organic substances are not formed from living things or the remains of living things. Calcite, a compound made of calcium, carbon, and oxygen, is a mineral. ❶ It is formed underground from water containing these dissolved elements. Coal and oil, although found underground, are not minerals because they are formed from decayed plant and animal life.

The second property of a mineral is that it occurs naturally in the earth. Steel and cement are manufactured substances. So they are not minerals. But gold, silver, and asbestos, all of which occur naturally, are minerals.

The third property of a mineral is that it is always a solid. The minerals you just read about—halite, calcite, gold, silver, and asbestos—are solids.

The fourth property of a mineral is that, whether it is made of a single element or a compound, it has a definite chemical composition. The mineral silver is ❷ made of only silver atoms. The mineral quartz is made entirely of a compound formed from the elements silicon and oxygen. So even though a sample of quartz may contain billions of atoms, its atoms can only be silicon and oxygen joined in a definite way.

The fifth property of a mineral is that its atoms are arranged in a definite pattern repeated over and ❸ over again. This repeating pattern of atoms forms a solid called a **crystal.** A crystal has flat sides, or faces that meet in sharp edges and corners. All minerals have a characteristic crystal shape.

### TEACHING STRATEGY 18-1

**Motivation**

Display samples of such common minerals as salt, mica, sulfur, quartz, gold, silver, copper, and graphite. Have students observe the samples and note the characteristics of each one. Also have students note ways in which some of the substances are similar. For example, gold, silver, and copper are all metals; salt and sulfur are powdery; mica and quartz are glassy and have an obvious crystal structure.

**Content Development**

Emphasize the definition of a mineral as a substance that is not live and was never part of a living thing. Some students may be familiar with the terms *organic* and *inorganic*. Explain that a mineral is an inorganic substance, because it was not formed from a living thing, or the remains of a living thing.

Point out that although certain types of minerals such as precious metals and gems have similar properties, the entire group of substances known as minerals is amazingly diverse. The diversity of minerals is evidenced by the fact that minerals are found in almost every aspect of our lives.

**Content Development**

Review the five characteristics that all minerals must meet. Point out that substances such as coal and oil are

416

## TEACHING STRATEGIES

boxed in red, are useful classroom-proven teaching suggestions categorized to meet every kind of teaching need found in the science classroom:

- **MOTIVATION**—stimulating activities, demonstrations and thought questions for introducing each lesson.
- **CONTENT DEVELOPMENT**—teaching tips for helping students better understand the concepts.
- **SKILLS DEVELOPMENT**—ideas for developing content through skills-oriented activities and questions.

# RIGHT ON THE LESSON PAGE!

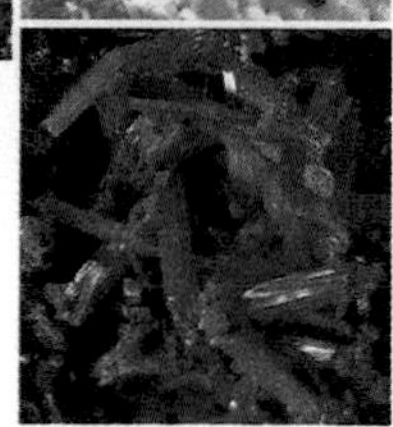

There are more than 2000 different kinds of minerals. But all minerals have the five special properties you just read about. Using these five properties, you can now define a mineral in a more scientific way. **A mineral is a naturally occurring, inorganic solid that has a definite chemical composition and crystal shape.**

### *Formation and Identification of Minerals*

Almost all minerals come from the material deep inside the earth. This material is hot liquid rock called **magma.** When magma cools, mineral crystals are formed. How magma cools and where it cools determine the size of the mineral crystals.

When magma cools slowly beneath the earth's crust, large crystals form. When magma cools rapidly beneath the earth's crust, small crystals form. Sometimes the magma reaches the surface of the earth and cools so quickly that crystals do not form at all. ❹

Crystals may also form from a mineral dissolved in a liquid. When the liquid evaporates, or changes to a gas, it leaves behind the mineral as crystals. Here too, the size of the crystals depends on the speed of evaporation.

Because there are so many different kinds of minerals, it is not an easy task to tell one from another. However, minerals have certain physical properties that can be used to identify them.

**Figure 18-2** *Crystals are found in many different shapes, such as six-sided cubes of fluorite (top right), radiating crystals of wavelite (top left), and needlelike structures of croccolite (bottom).*

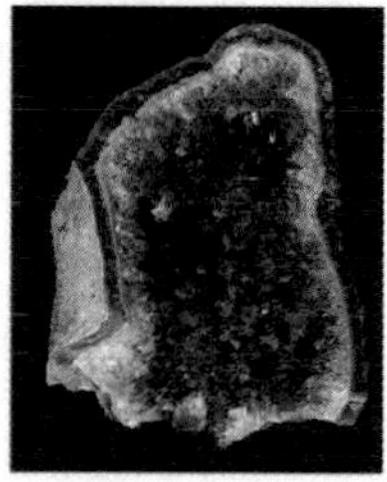

**Figure 18-3** *A geode is a rock whose hollow interior is lined with mineral crystals, usually quartz, formed by evaporation.*

417

**ANNOTATION KEY**

❶ Chalk and pearl are formed from living things. A mineral, such as aragonite, must be inorganic. (Applying concepts)

❶ Thinking Skill: Applying definitions

❷ Thinking Skill: Relating properties

❸ Thinking Skill: Identifying patterns

❹ Thinking Skill: Identifying processes

**TEACHER DEMONSTRATION**

Heat enough water to fill 3 Styrofoam cups. Fill each cup with hot water and place a thermometer in each. Record the starting temperature. Pour the heated water in cup #1 into a flat tray. On cup #2, place a lid. On cup #3, place a lid and wrap it with a towel. Wait 10 minutes and check the temperatures. Have students predict the results.

- **Which cup will have cooled the most?** (cup #1)
- **Why did it cool the fastest?** (exposure to air)
- **How might this compare to minerals being formed in and on the earth's surface?** (The more insulation, the longer it takes to cool. The longer it takes to cool, the larger the crystal of the same mineral.)

found naturally occurring on the earth but that they are not minerals since they do not have a crystal shape and are organic substances made from the remains of once-living things.

**Content Development**

Explain to students that most minerals come from hot molten rock called magma. Magma that makes it to the earth's surface is called lava. Not all magma makes it to the earth's surface—most remains below. Magma that remains deep within the earth's surface cools slowly, and this allows the crystal time to grow larger than it would if it cooled quickly.

**Motivation**

Have students add salt to a beaker of water until no more salt will dissolve. Gently heat the solution or leave the beaker uncovered for a couple of days. As the water evaporates, crystals will form. Have students look at the crystal shape under a microscope or other magnifier.

**Reinforcement**

Show students various types of minerals that exhibit crystal shapes. Review the basic characteristics of minerals as students examine your samples.

417

**The Annotation Key** supplies answers to all in-text and figure caption questions in addition to identifying thinking skills developed throughout the narrative in the student text.

**Historical Notes** instantly place science topics within a historical context.

**Background Information** provides additional information for making science topics more meaningful to students.

**Teacher Demonstrations,** complete with inquiry type questions, add new excitement to teaching science.

**Facts and Figures** offers informative tidbits to perk up the lesson.

**"Tie-ins"** link the topic at hand to other areas of science, as well as to mathematics, the social sciences, and the humanities.

- **REINFORCEMENT**—useful tips for reteaching basic concepts and facts to students who require extra assistance.
- **ENRICHMENT**—more in-depth information and challenging activities for stimulating your advanced students.
- **ANSWERS TO SECTION REVIEW QUESTIONS**—short answers printed in red to make grading easier and faster.
- **COMMON ERRORS**—mistakes that some students might make while solving mathematics related problems.

# TOTAL TEACHER SUPPORT FOR

**Laboratory Manuals** for each of the three student texts include over 60 laboratory investigations that reinforce concepts, develop laboratory and critical-thinking skills, and illustrate everyday life applications. The **Annotated Teacher's Editions** include safety tips, materials and suppliers lists, answers to all questions, and much more.

**Computer Test Banks** provide an abundance of test items, with a state-of-the-art test generator for use with Apple and IBM personal computers. Teachers can create tests, prepare alternate tests, modify questions, add questions, and make use of many other options.

Dial-A-Test—Prentice Hall's exclusive service allows teachers to call our toll-free number to request customized tests and answer keys mailed within 48 hours of the call.

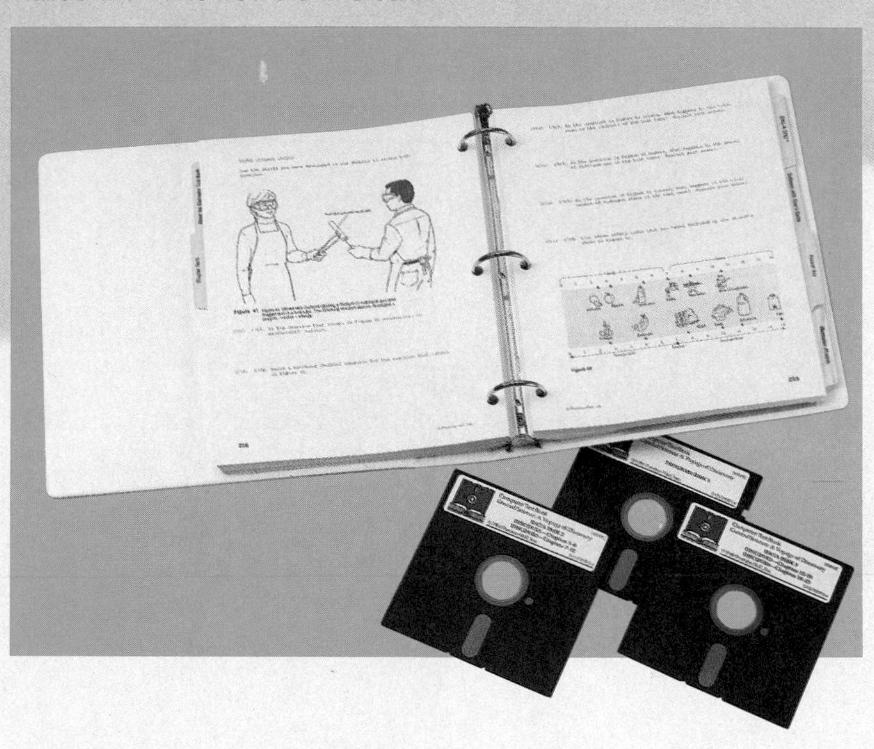

**General Science Transparencies** include 70 full-color overhead transparencies that span the content of the entire Prentice Hall General Science series. Great for explaining difficult concepts and bringing new excitement into the classroom. Also includes a Teacher's Guide with teaching strategies and questions and answers for each transparency.

# TOTAL TEACHING SUCCESS!

**Teacher's Resource Books** for each student text supply a wealth of worksheets, activities, tests, and planning material—all to give you more options for selecting materials that work best with your teaching style, time limitations, and students' needs. Inside you'll find:

- **DAILY LESSON PLANS** featuring content and skill objectives, key terms and concepts, and teaching suggestions for remedial, average, or advanced student ability levels—all organized onto one handy two-page chart cross-referencing all the components of the program.
- **VOCABULARY WORKSHEETS** that reinforce new science terms through fun-filled word puzzles.
- **ACTIVITIES WORKSHEETS** that include remedial, average, and enriched activities for every chapter so all your students better understand science.
- **MASTER FORMS,** ideal as handouts or overhead transparencies, that challenge students to fill in missing information.
- **LABORATORY WORKSHEETS** that can be used with the laboratory investigations in the student text.
- **CHAPTER AND UNIT TESTS** for evaluating mastery of general science.
- **SCIENCE READING SKILLS** sections that include reading activities corresponding to Science Gazettes in the student text. Each exercise sharpens skills such as Identifying a Main Idea, Using Context Clues, Making Predictions, and Making Inferences.
- **SCIENCE GRAPHING SKILLS** section for teaching graphing in a fun-filled manner.
- **SAFETY AND LABORATORY SKILLS** section that includes a safety contract, safety test and laboratory practicals.

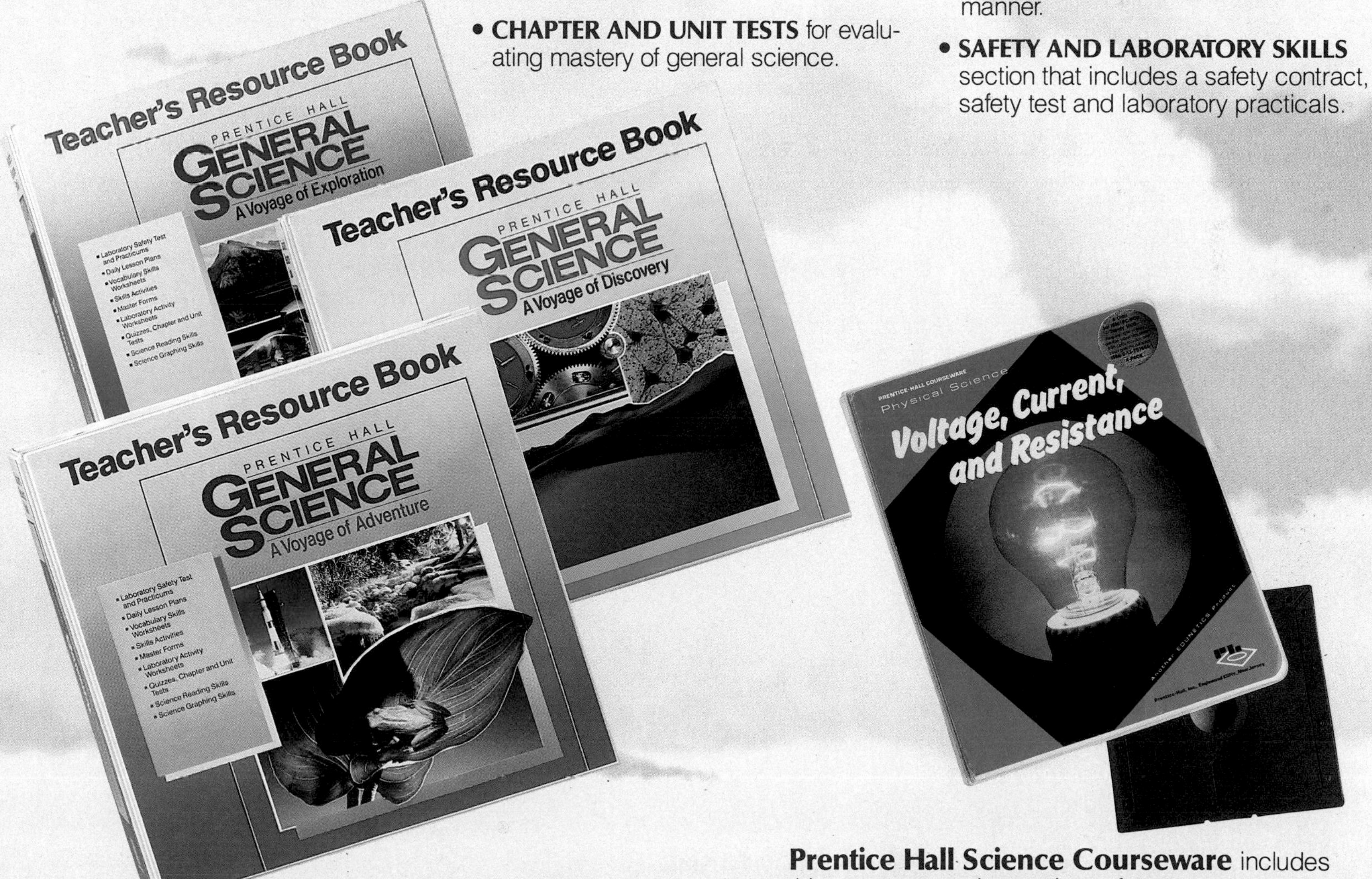


**Prentice Hall Science Courseware** includes thirty easy-to-use interactive software programs that reinforce and extend life, earth, and physical science concepts and skills through dynamic graphics and a sound educational design.

# SOLID SCIENCE CONTENT THAT MEETS ALL YOUR CURRICULUM NEEDS

## FLEXIBLE • READABLE • SKILLS-ORIENTED

## PRENTICE HALL GENERAL SCIENCE

### A Voyage of Adventure

**UNIT 1 CHARACTERISTICS OF LIVING THINGS**
1 Exploring Living Things
2 The Nature of Life
3 Cells
4 Tissues, Organs, and Organ Systems
5 Interactions Among Living Things

**UNIT 2 CLASSIFICATION OF LIVING THINGS**
6 Classification
7 Viruses, Bacteria and Protists
8 Nonvascular Plants and Plantlike Organisms
9 Vascular Plants
10 Animals: Invertebrates
11 Animals: Vertebrates
12 Mammals

**UNIT 3 MATTER**
13 General Properties of Matter
14 Physical and Chemical Changes

**UNIT 4 STRUCTURE OF MATTER**
15 Atoms and Molecules
16 Elements and Compounds
17 Mixtures and Solutions

**UNIT 5 COMPOSITION OF THE EARTH**
18 Minerals
19 Rocks
20 Soils

**UNIT 6 STRUCTURE OF THE EARTH**
21 Internal Structure of the Earth
22 Surface Features of the Earth
23 Structure of the Atmosphere

### A Voyage of Discovery

**UNIT 1 ASTRONOMY**
1 Exploring the Universe
2 Stars and Galaxies
3 The Solar System
4 Earth and Its Moon

**UNIT 2 THE CHANGING EARTH**
5 History of the Earth
6 Surface Changes on the Earth
7 Subsurface Changes on the Earth

**UNIT 3 HUMAN BIOLOGY**
8 Skeletal and Muscular Systems
9 Digestive System
10 Circulatory System
11 Respiratory and Excretory Systems
12 Nervous System
13 Endocrine System
14 Reproduction and Development

**UNIT 4 HUMAN HEALTH**
15 Nutrition and Health
16 Infectious Diseases
17 Chronic Disorders
18 Drugs, Alcohol, and Tobacco

**UNIT 5 CHEMISTRY OF MATTER**
19 Atoms and Bonding
20 The Periodic Table
21 Chemical Reactions

**UNIT 6 MECHANICS**
22 Force and Work
23 Motion and Gravity

### A Voyage of Exploration

**UNIT 1 FORMS OF ENERGY**
1 Exploring Energy
2 Heat Energy
3 Electricity and Magnetism
4 Sound
5 Light
6 Nuclear Energy

**UNIT 2 SCIENCE AND TECHNOLOGY**
7 Energy Resources
8 Chemical Technology
9 Space Technology
10 The Computer Revolution
11 Pollution

**UNIT 3 OCEANOGRAPHY**
12 Currents and Waves
13 The Ocean Floor
14 Ocean Life

**UNIT 4 WEATHER AND CLIMATE**
15 The Atmosphere
16 Weather
17 Climate

**UNIT 5 HEREDITY AND ADAPTATION**
18 Genetics
19 Human Genetics
20 Changes in Living Things Over Time
21 The Path to Modern Humans

**UNIT 6 ECOLOGY**
22 Biomes
23 Pathways in Nature
24 Conservation of Natural Resources

## TOTAL TEACHING SUPPORT FOR TOTAL TEACHING SUCCESS!
## Each volume of PRENTICE HALL GENERAL SCIENCE includes:

- Student Edition
- Annotated Teacher's Edition
- Teacher's Resource Book
- Laboratory Manual
- Laboratory Manual, Annotated Teacher's Edition
- Transparencies
- Software
- Computer Test Bank

**For more information, call Toll-free: 1-800-848-9500**

**PRENTICE HALL**
School Division of Simon & Schuster
Englewood Cliffs, NJ 07632

# PRENTICE HALL GENERAL SCIENCE
## A Voyage of Adventure

**Dean Hurd**
Physical Science Instructor
Carlsbad High School
Carlsbad, California

**Charles William McLaughlin**
Chemistry Instructor
Central High School
St. Joseph, Missouri

**Susan M. Johnson**
Associate Professor of Biology
Ball State University
Muncie, Indiana

**Edward Benjamin Snyder**
Earth Science Instructor
Yorktown High School
Yorktown Heights, New York

**George F. Matthias**
Earth Science Instructor
Croton-Harmon High School
Croton-on-Hudson, New York

**Jill D. Wright**
Professor of Science Education
University of Tennessee
Knoxville, Tennessee

**Prentice Hall**
A Division of Simon & Schuster
Englewood Cliffs, New Jersey 07632



0–13–717877–8
10 9 8 7 6 5 4 3 2 1

Prentice-Hall of Australia, Pty. Ltd., Sydney
Prentice-Hall of Canada Inc., Toronto
Prentice-Hall Hispanoamericana, S.A., Mexico
Prentice Hall of India Private Ltd., New Delhi
Prentice-Hall International (UK) Limited, London
Prentice-Hall of Japan, Inc., Tokyo
Prentice-Hall of Southeast Asia Pte. Ltd., Singapore
Editora Prentice-Hall do Brasil Ltda., Rio de Janeiro

# Contents of Annotated Teacher's Edition

# Other Components of *Prentice Hall General Science*

*Prentice Hall General Science* has been designed as a complete program for use with junior high or middle school general science students. The student text contains the topics most widely covered by general science courses. The auxiliary materials described below are available to augment the program.

### ANNOTATED TEACHER'S EDITION

This comprehensive teacher's edition includes all the information needed to teach *Prentice Hall General Science*. It features pedagogical aids for science instruction; teaching strategies for units, chapters, and chapter sections; background information; historical notes; interesting facts and figures; suggested teacher demonstrations; science, technology, and society suggestions; answers to all text and end-of-chapter questions; additional end-of-chapter questions; section previews; section objectives; and shading in of key topic sentences. A comprehensive list of laboratory materials, science suppliers, teacher's bibliography, and audiovisual and software suggestions are also included.

### LABORATORY MANUAL—STUDENT EDITION

Each textbook has an accompanying Laboratory Manual. The manuals measure 8½ inches by 11 inches, are five-hole punched, and consumable. Each laboratory activity is keyed to a particular chapter in the student text. There is at least one lab per chapter. These laboratory investigations provide a traditional laboratory approach to the topics presented in the text and employ standard laboratory equipment.

Each investigation begins with background information for the student, a comprehensive list of materials, a step-by-step procedure, safety caution symbols keyed to specific steps in the procedure, an observations section, a conclusions section, a section that poses additional critical thinking and application questions, and also a "going further" section that allows industrious students to do further experiments.

### LABORATORY MANUAL—ANNOTATED TEACHER'S EDITION

Each Annotated Teacher's Edition provides answers to questions posed in the investigations, includes setup instructions, and suggests places to purchase special equipment, if any is required.

### TEACHER'S RESOURCE BOOK

Each Teacher's Resource Book, or TRB, contains a wide variety of skills-related material for the teacher, most of which is clearly keyed to various academic levels. Materials in the TRB are presented on a chapter-by-chapter basis; that is, correlated to the chapters in the student text and presented in the same sequence. Materials for each chapter include

1. Daily lesson plans integrating all components of *Prentice Hall General Science*
2. Vocabulary Skill Worksheets
3. At least 3 blackline master activities—each keyed to academic level
4. Laboratory Activity Worksheets
5. Transparency Master Forms
6. Chapter and Unit Tests

In addition to the chapter-by-chapter materials, the TRB contains a section on science reading skills, a booklet on science graphing skills, a section on laboratory safety and assessment that includes a safety contract and laboratory practicals, a section on running a science fair, general science wall posters, and Prentice Hall's exclusive activity exchange.

### SCIENCE COURSEWARE

Prentice Hall Computer Software Programs provide a vast multiprogram project under the general title Science Courseware. Such courseware employs state-of-the-art graphics in order to present on the computer screen events, concepts, and data that cannot be presented in the traditional classroom setting. The goal of Prentice Hall Courseware is to allow the student—through a maximum of student participation and interaction—to master science material that cannot be easily grasped through the printed word, through laboratory experiences, or by using other traditional classroom materials and techniques.

All Prentice Hall science courseware is correlated to chapters in the textbook and offers review, reteaching, glossaries, basic skills instruction, simulations, tests, and a complete management system.

### GENERAL SCIENCE COLOR TRANSPARENCIES

Packaged in a convenient three-ring binder, the four-color general science transparencies allow the teacher to use visual aids to demonstrate difficult general science concepts and processes.

# To the Teacher

### RATIONALE OF THE PROGRAM

Science education is a vital force in helping your students recognize the critical importance of scientific developments in today's world—and tomorrow's. The authors and editors of *Prentice Hall General Science* have designed this textbook and its auxiliary materials to meet this educational goal. This program provides your students with the basic knowledge of science as it relates to them and to their own range of experiences. However, this program goes even further. It makes it possible for young people to use their abilities to develop an appreciation of the basic concepts in science. Historical achievements in the field of science, career paths, and some thoughts on the future all will contribute to the growth and development of your students.

### BALANCE IN THE GENERAL SCIENCE PROGRAM

The *Prentice Hall General Science* program has been developed in accordance with basic principles of science education. The program offers a balance between textual and investigative material, with enough flexibility to suit individual teaching styles and classroom needs.

The textual material presents relevant and recent facts that are used to build science concepts. Illustrations, teaching captions, and in-text questions encourage students to participate, to draw on previously learned information, to make judgments, and to inquire, thereby forming a basis for conceptual learning.

### SKILLS IN THE GENERAL SCIENCE PROGRAM

The *Prentice Hall General Science* program provides for the full development of science skills that are inherent in science. The skills package within the textbook includes Laboratory Activities, marginal activities, and Skill Building activities at the end of each chapter. Through these investigations and activities, students gain firsthand experience with such learning skills and processes as *observing, classifying, identifying, measuring, inferring, hypothesizing, interpreting,* and *predicting*.

By utilizing the skills package provided in the textbook, teachers can be assured that their students will receive a comprehensive program that reinforces and extends all the skills that are applicable to general science. Furthermore, since most skills investigations and activities require report writing and the recording of data, teachers can easily assess whether their students have developed proficiency in skills development.

### CRITICAL THINKING IN THE GENERAL SCIENCE PROGRAM

Aside from reinforcing basic science skills, the Prentice Hall General Science program provides the teacher with a wealth of ways in which to reinforce critical thinking skill development. The student text is written in such a manner as to present students with the same types of problems often faced by scientists, calling upon students to think critically as scientists do. Many marginal activities and end-of-chapter activities call upon students to use basic critical thinking skills, ranging from low-order skills such as observing to high-order skills such as inferring and synthesizing. Furthermore, activities suggested in the Annotated Teacher's Edition, the Laboratory Manuals, and the Teacher's Resource Book provide a wide variety of skill building, critical thinking materials to enable teachers to pick and choose those materials most appropriate for their students.

### READABILITY AND STUDENT COMPREHENSION AND INTEREST

The authors and editors of *Prentice Hall General Science* have taken many steps to improve the readability of this text as a means of facilitating student comprehension and interest.

*Prentice Hall General Science* is designed with an open, clean format that uses exciting and relevant visuals to enhance student interest. The text is written in a style that appeals to and accommodates the wide range of comprehension levels found in the junior high or middle school. The readability of the text has been carefully controlled so that most students will be able to understand the science concepts and content presented.

Studies have shown that students often feel hindered and frustrated by scientific terms because the words are alien to them, are usually difficult to pronounce, and are hard to remember. To remedy this situation, *Prentice Hall General Science* introduces scientific terms in boldfaced type. Definitions accompany the new terms, and phonetic pronunciation guides are given where necessary.

In addition, photographs and illustrations, most of which are in full color, visually reinforce the text material, thereby aiding readability and comprehension. Most photographs and illustrations include an explanatory caption and frequently an inductive question. The answers to these caption questions can be discerned by the student in three ways:

1. Referring back to text material
2. Basing the response directly on the photograph or illustration
3. Using previously acquired knowledge

Other aids to readability and comprehension include single-concept paragraphs, frequent in-text questions, section checkup questions, chapter summaries, and end-of-chapter questions. These features help to make the text more readable by providing immediate learning evaluations and by continually eliciting student participation and response.

Since a clear guide to what students will be expected to learn in each chapter is a proven readability aid, *Prentice Hall General Science* includes an outline of chapter sections with objectives keyed to each section on chapter-opening pages. In each section, a key idea statement tying together the most important concepts in the section is included in boldfaced type. These boldfaced key ideas are extremely helpful for students who need some guidance as to what is the most important information in each section.

# Overview of Science Education

### READING AND LANGUAGE DEVELOPMENT IN THE SCIENCE CLASSROOM

The type of reading required of secondary students is different from that required of elementary students. The vocabulary is more difficult, the sentence structure is more complex, the writing is denser, and the concepts presented are more challenging. In addition, reading in the science area presents difficulties of its own. Students must be able to isolate important facts and organize them, form hypotheses based on these facts, test alternatives, and draw conclusions.

In order for students to utilize any text, appropriate readability is essential. The Dale-Chall formula indicates that this text is written on the reading level of most junior high school students. The introduction of new terms in boldfaced type, the pronunciation guides and definitions, and the review questions at the end of each section and chapter improve the readability of this text. However, you will still need to help your students develop the specialized skills they need to read in the science area.

### *Vocabulary in the Science Area*

Reading in the science area presents students with a largely unfamiliar technical vocabulary. Words such as covalent and metamorphosis do not come up in everyday conversation. In addition, terms with which studens may be familiar in an everyday context, such as electricity, have a specialized meaning in a scientific context. Students may require some assistance in dealing with these difficult terms. Try doing the following before giving a reading assignment from this text:

1. Identify the key concepts and vocabulary words in each chapter.
2. Pronounce all new words. Each new scientific term in each unit is introduced in boldfaced type and followed by a phonetic pronounciation guide where necessary. It is important for you to pronounce these terms and any others with which your students may have trouble. The pronunciation key is located at the beginning of the Glossary.
3. Define all new and potentially difficult terms. New terms are defined in text and, for most terms, again in the Glossary.
4. Draw students' attention to the charts, drawings, and photographs that will help them understand new words and concepts. All photographs and artwork have been carefully selected to help students visualize concepts presented throughout the text.

### *Reading in the Science Area*

Besides helping students develop a science vocabulary, a science course should also help them develop skills in the following:

1. Reading for exact meaning
2. Identifying the main ideas expressed in chapter sections
3. Classifying information and organizing ideas obtained from reading the text
4. Noting cause-and-effect, relationships
5. Gaining accurate information from the visual representations throughout the text
6. Understanding the scientific formulas and symbols in the text
7. Reading directions accurately, especially for carrying out Laboratory Investigations and Activities suggested in the text
8. Locating and using different sources of information

Science teachers sometimes overlook the development of these specialized reading skills. These skills are necessary for students to develop into mature and independent readers both inside and outside of the science classroom.

### *Writing in the Science Area*

Developing good writing skills goes hand in hand with developing good reading skills. Both are important for effective communication.

Because our society relies heavily on written expression and printed material, it is important for students to possess effective writing skills. Your students will have frequent opportunities to display and develop their writing skills while using the *Prentice Hall General Science* text. When using this textbook, you should expect students to

1. Keep records of all investigations and activities. Many investigations and activities require students to write a short report detailing observations and conclusions made during an investigation or activity.
2. Record notes from text material and your lectures and discussions. For students to fully comprehend the material presented by such sources, they need to learn to organize and summarize information. Writing the information will help students retain and understand the material better.
3. Write the answers to Section Review questions, as well as the questions at the end of each chapter. Essay questions at the end of each chapter require students to write brief essays on topics related to the chapter. All such exercises should be used not only for reinforcement and as an evaluative and diagnostic aid, but also as an opportunity to develop writing skills.

### *Speaking in the Science Area*

The *Prentice Hall General Science* text offers many opportunities to develop communication skills. When students have a chance to express their own thoughts or to interpret the thoughts of others, oral and written communication improve and become a means for learning. Thus, speaking is another important language skill that can and should be further developed in the science classroom. Oral communication is the most effective and common means of communicating and is the basis for a sound program in reading and writing.

1. **Presentations.** Students of this age are curious about their surroundings and the natural events that occur in their environment. Have students make oral presentations to their classmates about any general science topics that interest them, or results obtained from investigations and activities. Encourage students in the audience to participate by asking relevant questions.
2. **Discussion.** Encourage students to participate in class discussions. Techniques for encouraging student involvement in discussions are presented in a later section, Questioning in the Science Area.
3. **Dramatizations.** Have students act out concepts presented in this text, such as important discoveries of scientists or the impact pollution may have on a town or an individual.

## *Listening in the Science Area*

Even though speaking is the most common form of communication, it is ineffective without a listener. Listening skills should be developed concurrently with speaking skills. Students need to learn to respect each other's viewpoints.

Encourage good listening habits in your students by being a good listener yourself. Pause after asking a question. If no student offers an answer, rephrase the question. Teaching effective listening is important in helping all your students become better science students.

Listening constitutes a major portion of the communication process. Listening skills can be developed by encouraging students to

1. Listen to questions posed by the teacher and other students. For example, before proceeding with a Laboratory Activity, you may ask a question such as "What conditions do you think bring about an earthquake?" A question of this nature would involve the entire class in a discussion and would require students to listen to responses from their classmates.
2. Listen to directions for carrying out Activities. It is important for students to be able to follow directions, especially safety precautions, in a science class.

   Have your students practice listening skills by giving them oral instructions on how to make something—a paper airplane, for instance. Repeat the directions frequently and ask your students to repeat the instructions as they follow them. This will prepare students for Laboratory Activities by helping them follow a thought sequence and by emphasizing the need for accuracy in communication and interpretation.
3. Listen to explanations or descriptions of natural phenomena provided by you, other students, or guest speakers. You can help your students listen properly by telling them what to listen for, how to listen to the material presented, and how to mentally organize or write down what they hear.

## *Questioning in the Science Area*

Developing communication in your science classroom involves your participation. Much of a teacher's class time is spent asking questions. In fact, research has shown that teachers use questioning more frequently than any other single teaching technique. You ask questions to develop creative learning situations, evaluate your students' progress, give directions, correct behavior, and initiate instructions.

It is important to understand and use good questioning techniques and strategies during the instructional process. Thought-provoking questions and improved questioning techniques can help you develop and sustain student interest, provide new ways to deal with subject matter, and give purpose to your student evaluations.

You should design your objectives at a variety of cognitive levels. Your questions should also reflect the various levels of your performance objectives. Research indicates that many teachers unconsciously concentrate most of their questions at the lowest level in the cognitive domain—knowledge. Answering questions at this level requires a simple recall of facts. For instance,

1. What does the word element mean?
2. In what year was the planet Pluto discovered?

Notice that each question requires only a short response or the recall of a definition or fact. These questions could be rephrased to require your students to operate at a higher thinking level. For instance,

1. Explain the difference between an element and a compound.
2. How did the wobbling of Neptune's orbit lead to the discovery of Pluto?

These questions would result in longer student responses, would require answers that go beyond the simple recall of facts, and would probably encourage greater discussion in your classroom. Questions should, therefore, be asked at a variety of cognitive levels.

Besides the type of question you ask, the number of questions you ask affects student response. It is easy to ask too many questions during an instruction period. Studies have found that some teachers ask questions at a rate of 180 questions per science lesson! This rapid-fire method of questioning and calling on a student to respond immediately after the question is asked leads to brief student responses.

Thus, wait-time serves a twofold purpose in the classroom: (1) It provides an atmosphere more conducive to discussion and learning and (2) students learn to use wait-time to organize a more complete answer.

Along with waiting 3 to 5 seconds after asking a question, pausing after a student response is also helpful. This second pause, or silent-time, increases the chance that the student will add to his or her response or that other students will add to the initial response. If you follow these simple techniques of waiting before and after a student's response, more students may becomes involved, you may not need to ask as many questions, and the questions you do ask will probably be of a higher cognitive level.

## TEACHING HETEROGENEOUS CLASSES

*Prentice Hall General Science* has been designed to meet the needs of students of all ability levels. Through careful analysis of readability, the text has been monitored to ensure that students in the junior high or middle school can read and comprehend the material presented in the text. Moreover, large photographs and illustrations, which reinforce and extend the material in the text, are important tools in helping students comprehend facts and concepts in science.

In order to help teachers meet the needs of students of varying ability levels in a heterogeneous classroom, all of the marginal activities have been keyed to an ability level in the Annotated Teacher's Edition. These ability levels are remedial, average, and enriched. In this way, the teacher can assign activities on an individual basis, depending on the ability level of different students in the classroom.

## TEACHING "SPECIAL" STUDENTS

Certain state and federal laws have mandated that all students are to have access to the least restrictive learning environment possible. Thus, many "special" students, those with physical and mental disabilities, are being mainstreamed into nonspecialized classes. This action challenges the teacher to accommodate a much wider range of student abilities, needs, and interests.

### *Students with Learning Problems*

Learning processes that include inferences and abstract reasoning are often more difficult for students who have learning problems. Such students include those who have some degree of mental retardation. In order to better help such students grasp facts and concepts in science, it is important to provide daily learning goals at a pace that will allow the goals to be achieved. These students will benefit greatly from the use of concrete examples in the classroom that relate back to daily life. The need to reinforce lessons is also important to such students. Furthermore, since many of these students will have experienced failure in their studies, it is vital to provide as much positive reinforcement as possible. Emphasize success and minimize failure whenever you can.

### *Students with Visual Problems*

Students who are blind, as well as those with limited sight, are more dependent on senses such as hearing than are other students. As a result, such students should always be seated where they will be able to hear the teacher and their classmates most easily. Tape recording lessons will help these students study and go over material at their own pace. Also, classmates can be a great aid by providing descriptions of photographs and illustrations in the text.

### *Students with Hearing Problems*

Students with hearing problems are far more dependent on the written word than other students. Usually, these students should be seated near the front of the room so that they can read the teacher's lips. The teacher should enunciate every word and avoid talking too quickly. All instructions and assignments should be written down for these students. Allow students who cannot hear well to copy the notes taken in class by classmates.

### *Students with Other Physical Problems*

Students who have physical problems that require crutches or wheelchairs will need extra room to get around in the classroom. Take care to make sure such students do not try to stretch their limits beyond their physical capabilities, but do not treat them any more differently than necessary so that they will feel an integral part of the class.

Students who have physical problems due to disorders such as muscular dystrophy, or other disorders that deter motor coordination, will often have trouble in the laboratory setting. Holding flasks, pouring liquids, and using other equipment may be beyond their capabilities. If these students can write, it is often best to assign them the task of recording during Laboratory Activities, while their lab partners carry out the more physical aspects of the activity.

Some students may have illnesses such as diabetes or epilepsy. In general, such students will not need any special care. However, the teacher should be aware of any special problems or symptoms these illnesses might present, in order to obtain prompt medical attention when necessary.

# Features of the Student Text

*Prentice Hall General Science* has been set up to provide a flexible and varied approach to teaching science. The text is divided into units, sufficiently self-contained to be taught separately and in any order. Since science cannot be compartmentalized into discrete packets of information, there are areas in which topics on one unit overlap with topics in other units. However, in order to retain the flexibility of the program and allow the teacher to begin with any unit he or she desires, any concept of definition introduced for the first time in any unit is considered unfamiliar and taught as if the topic is a new one.

One exception to this flexibility is Chapter 1. This chapter introduces students to the scientific method, various branches of science, the metric system, tools of measurement, and the need for safety in the laboratory. It is recommended that all students complete Chapter 1 first. At that point, the teacher may jump to the unit in the text that best fits his or her curriculum needs.

#### UNIT OPENERS

Each unit begins with a two-page spread that includes a large dramatic photograph or illustration. Accompanying the visual is a short overview that both introduces the topics to be discussed in the unit and provides motivational text to capture student interest. A listing of the chapters in the unit is also included in the unit-opening spread.

#### CHAPTER OPENERS

Each chapter begins with a two-page spread. Like the unit openers, large photographs and illustrations are employed to grab student attention immediately. A short, concise caption informs the reader as to the nature of the visual. The visual can also be readily identified through the chapter opening text. This text, often written in an anecdotal style, serves to entice the student to read further. Intriguing questions and unusual data are employed to hold the student's attention and to motivate the student to find out more about the topic.

Also included in the chapter-opening spread is a list of the main sections in each chapter. This listing serves as an instructional outline for both student and teacher. Moreover, chapter objectives are given in the chapter-opening spread. Thus, the chapter opener serves the dual purpose of initiating the student's desire to learn more about the material and of alerting the student to the specific objectives, or goals, that he or she is expected to grasp when the chapter is completed.

#### CHAPTER SECTIONS

As noted, each main chapter section is listed in the chapter opening. These sections are numbered consecutively on both the chapter-opening page and in the text itself.

Numbering the main sections helps distinguish the main topics in the chapter from the subtopics. Subtopics in each main section are set apart and boldfaced.

In most sections, the students will find in-text questions based on the material just presented. Some in-text questions require simple factual recall. Other in-text questions employ more advanced critical thinking skills such as predicting and relating.

Within each main section are numerous photographs and illustrations that help teach and reinforce the topics found in the section. Most visuals are large and in color, to further hold student interest. Data charts and graphs are interspersed in the text as well, to provide further information. Each numbered section also contains one sentence that is set apart in boldfaced type. This sentence alerts the student to the key idea of that section.

**SECTION REVIEW**

At the end of every numbered section is a Section Review. These review questions, usually short-answer type questions, allow the teacher to quickly verify whether the important topics in that section have been grasped by the student. In general, at least one review question relates back to the Section Objective listed at the beginning of the chapter.

**MARGINAL ACTIVITIES**

In most cases, each main numbered section of the text contains at least one marginal activity. These activities are placed in the margin to avoid disrupting the student's reading of the text. The Activities can be used in a variety of ways, including homework, extra credit, and class projects. Each Activity is keyed to an academic level—remedial, average, or enriched—in the Annotated Teacher's Edition. In this way, teachers can assign those they think are most appropriate for a particular student's academic needs.

The marginal Activities appear with the generic term Activity. However, on pages T-12–T-39 of the Annotated Teacher's Edition is a detailed skills scope and sequence for all marginal Activities, as well as for Laboratory Activities and skill-building questions found at the end of each chapter. There are five types of marginal activities: hands-on, computational, reading comprehension/vocabulary, library, and field Activities. All activities require the student to utilize science skills applicable to all scientific endeavors. These skills range from simple skills such as observing and comparing to more high-level skills such as relating, predicting, and applying. In many activities, the student is required to write a report, prepare a visual presentation in the form of charts and diagrams, or graph any observations and data.

**CAREERS**

Each chapter contains a Career feature that introduces students to a possible career choice in science related to that particular chapter. Careers, it should be noted, range from those requiring a high school diploma to those requiring a doctorate. Each Career feature consists of a full-color photograph depicting a person in this career at work, a description of the work done by people in this career, and an address to which interested students may write for further information. In order to increase career awareness and encourage students to contemplate and begin planning their own futures, a Help Wanted ad is included in each Career to help relate such careers to real-life career possibilities.

**LABORATORY ACTIVITIES**

Each chapter contains one full-page Laboratory Activity just prior to the chapter summary page. Such activities provide students with the opportunity to work in the laboratory and actively participate in investigating science problems. Most of the activities are designed to reinforce concepts presented in the text, but a few are designed to supplement the material in the text. Easy-to-obtain materials are used in such activities.

Each Laboratory Activity clearly outlines the Purpose of the activity, the Materials needed, and the Procedure to follow. A section called Observations and Conclusions alerts students to any observations or data they are to collect and tells students how to organize their data. In general, data is organized in the form of charts or graphs. Conclusions questions tie up the Laboratory Activity, calling upon students to analyze their data and draw various conclusions. Often the Conclusions questions ask students to use their data to reinforce or further establish a scientific concept or theory. In addition, special safety symbols alert students when important safety precautions must be observed.

**CHAPTER REVIEW**

*Chapter Summary and Vocabulary*

At the end of every chapter is a chapter summary section. The summary is divided into groupings based on the main numbered sections in the chapter. Under each grouping is a list of key sentences that describe the most important concepts presented in the chapter. The chapter summary might be considered a detailed outline of the chapter content.

Following the summary is a listing of the vocabulary words included in the chapter. The vocabulary words include all the boldfaced terms from the chapter.

### *End-of-Chapter Questions*

A wide variety of questions end each chapter. The first three question sections are Content Review questions. These questions are listed in the order in which the material is introduced in the chapter. Content Review questions are broken down into ten multiple choice questions, ten completion questions, and ten true-or-false questions. In the true-or-false questions, students must not only identify an incorrect statement but they must make it correct by substituting the correct word or phrase into the statement.

Also included in the end-of-chapter questions are two Concept Review sections. The first is titled Skill Building. Skill-building questions require the student to draw upon science skills and are based on material presented in the chapter. Skill-building questions range from simple computational questions to more difficult questions utilizing higher order skills such as inferring and relating. Often the student is asked to examine illustrations in the chapter and then analyze the illustrations in the form of charts and graphs.

The second type of Concept Review questions are essay questions. Essay questions reinforce and extend concepts, both scientific and societal. Since they require the student to write his or her answer in paragraph form, essay questions enhance the student's ability to write and report on science topics.

## END-OF-UNIT MATERIAL

### *Science Gazettes*

At the end of every unit, is a section entitled Science Gazette. Each Science Gazette is a Science Reader, which consists of three different types of articles. One type, called Adventures in Science, profiles a particular scientist and the path the scientist took, or is taking, to make a significant discovery. The second type of article, called Issues in Science, presents a nonbiased discussion of a contemporary scientific issue. The third type of article, called Futures in Science, presents a possible future scenario based on current scientific theory, experimentation, or data.

The articles of the Science Gazette have been carefully chosen and designed to maximize the motivation of students of all levels in their study of science. Each article is related to topics presented in the unit. The articles can be used to stimulate class discussions, as the basis for individual short essay homework assignments, to introduce or wrap up lessons on related topics, and to develop science reading skills.

In the Annotated Teacher's Edition, additional background material is included on each Science Gazette article. In addition, a teaching strategy for each article is presented, as well as Additional Questions and Topic Suggestions. Moreover, in the Teacher's Resource Book, specific activities and handouts related to each Gazette article are included in the Science Reading Skills section. These activities test basic science reading skills and provide the teacher with an excellent framework for assessing whether students have grasped the concepts in the Science Gazette articles.

## END-OF-TEXT FEATURES

Following the text are several features designed to aid students in their understanding of science.

### *For Further Reading*

A Bibliography for each chapter in the textbook is provided in a section called For Further Reading. The books listed in the Bibliography will help students who wish to do further research on a topic they find interesting, as well as help students reinforce the material they have learned.

### *Appendices*

A variety of appendices are located in the back of the textbook.

### *Glossary*

Scientific terms introduced in the text are listed in the Glossary in alphabetical order. Each term is clearly defined.

### *Index*

The Index provides students with an easy-to-use reference listing of subjects covered in the text.

# *Prentice Hall General Science* Skills Scope and Sequence

The following charts provide the basic skills framework presented in *Prentice Hall General Science.* Basic skills through more complex critical thinking skills are included in these charts. The Skills Scope and Sequence keys in all marginal Activities, all Laboratory Activities, and all end-of-chapter Skill Building questions.

## CHAPTER 1 EXPLORING LIVING THINGS pp. 12–33

| TEXT REFERENCE | HANDS-ON | COMPUTATIONAL | VOCABULARY/WRITING | LIBRARY | FIELD | OBSERVING | COMPARING | MANIPULATIVE | RELATING | MEASURING | INFERRING | RECORDING | SAFETY | APPLYING | DIAGRAMING | DESIGNING | CLASSIFYING | HYPOTHESIZING | PREDICTING |
|---|---|---|---|---|---|---|---|---|---|---|---|---|---|---|---|---|---|---|---|
| Activity: Theories and Laws, p. 15 | | | X | X | | | X | | X | | | | | X | | | | | |
| Activity: Making Observations, p. 16 | X | | | | | X | X | X | | | X | | | | X | | | | |
| Activity: Early Scientists, p. 17 | | | X | X | | | X | | | | | | | X | | | | | |
| Activity: To Grow or Not to Grow, p. 18 | X | X | | | | X | X | X | | X | X | | | X | | X | | | |
| Activity: Practicing Metric Measurements, p. 22 | X | X | | | | X | | X | X | X | | X | | X | | | | | |
| Activity: Metric Conversion, p. 23 | | X | | | | | | | | | | | | | | | | | |
| Activity: Using Métric Measurements, p. 24 | X | X | | | | X | X | | | X | | X | | | | | | | |

## CHAPTER 1 EXPLORING LIVING THINGS pp. 12–33 (continued)

| TEXT REFERENCE | HANDS-ON | COMPUTATIONAL | VOCABULARY/WRITING | LIBRARY | FIELD | OBSERVING | COMPARING | MANIPULATIVE | RELATING | MEASURING | INFERRING | RECORDING | SAFETY | APPLYING | DIAGRAMING | DESIGNING | CLASSIFYING | HYPOTHESIZING | PREDICTING |
|---|---|---|---|---|---|---|---|---|---|---|---|---|---|---|---|---|---|---|---|
| Laboratory Activity: A Moldy Question, p. 30 | X | | | | | X | X | X | X | X | X | X | | X | | | | | X |
| Skill Building 1, p. 33 | | | | | | | | | X | X | X | | | X | | | | | |
| Skill Building 2, p. 33 | | X | | | | | | | X | X | | X | | X | | | | | |
| Skill Building 3, p. 33 | | | | | | | | | X | | | | X | X | | | | | |
| Skill Building 4, p. 33 | | | X | | | | | | X | | X | | | X | | | | | |
| Skill Building 5, p. 33 | | X | | | | | X | | X | | X | | | X | | | | | X |

## CHAPTER 2 THE NATURE OF LIFE pp. 34–55

| TEXT REFERENCE | HANDS-ON | COMPUTATIONAL | VOCABULARY/WRITING | LIBRARY | FIELD | OBSERVING | COMPARING | MANIPULATIVE | RELATING | MEASURING | INFERRING | RECORDING | SAFETY | APPLYING | DIAGRAMING | DESIGNING | CLASSIFYING | HYPOTHESIZING | PREDICTING |
|---|---|---|---|---|---|---|---|---|---|---|---|---|---|---|---|---|---|---|---|
| Activity: Disproving Spontaneous Generation, p. 37 | | | X | X | | | | | X | | | | | X | | | | | |
| Activity: Catch Those Rays, p. 38 | X | | | | | X | | X | | | X | | | | | | | X | |
| Activity: Counting Calories, p. 40 | X | X | | | | | X | | X | X | | X | | X | X | | | | |
| Activity: Observing Stimulus-Response Reactions, p. 43 | X | | | | | X | | | X | | | X | | | | | | | |
| Activity: You're All Wet, p. 45 | | X | | | | | | | | | | | | | | | | | |
| Activity: Breathe Easy, p. 46 | X | | | | | | | X | X | | X | | | X | | X | | | |
| Activity: Temperature Range, p. 47 | | X | | | | | | | | | | | | | | | | | |
| Activity: A Starchy Question, p. 48 | X | | | | | | X | X | X | | | X | | X | | | | | X |
| Laboratory Activity: You Are What You Eat, p. 52 | | | | | | X | X | | X | | X | X | | X | | | | | |
| Skill Building 1, p. 55 | | | | | | | | | | | X | | | X | | | | X | |
| Skill Building 2, p. 55 | | | | | | | | | X | | X | | | X | | | | | |
| Skill Building 3, p. 55 | | | | | | | X | | | | X | | | X | | | | | |

**CHAPTER 2 THE NATURE OF LIFE pp. 34–55** (continued)

| TEXT REFERENCE | HANDS-ON | COMPUTATIONAL | VOCABULARY/WRITING | LIBRARY | FIELD | OBSERVING | COMPARING | MANIPULATIVE | RELATING | MEASURING | INFERRING | RECORDING | SAFETY | APPLYING | DIAGRAMING | DESIGNING | CLASSIFYING | HYPOTHESIZING | PREDICTING |
|---|---|---|---|---|---|---|---|---|---|---|---|---|---|---|---|---|---|---|---|
| Skill Building 4, p. 55 | | | | | | | | | X | | | | | X | | | | | |
| Skill Building 5, p. 55 | | | | | | | | | X | | X | | | X | | | | | |
| Skill Building 6, p. 55 | | | | | | | X | | | | X | | | X | | | | | |
| Skill Building 7, p. 55 | | | | | | | | | X | | X | | | X | | | | X | |

## CHAPTER 3 CELLS pp. 56–77

| TEXT REFERENCE | HANDS-ON | COMPUTATIONAL | VOCABULARY/WRITING | LIBRARY | FIELD | OBSERVING | COMPARING | MANIPULATIVE | RELATING | MEASURING | INFERRING | RECORDING | SAFETY | APPLYING | DIAGRAMING | DESIGNING | CLASSIFYING | HYPOTHESIZING | PREDICTING |
|---|---|---|---|---|---|---|---|---|---|---|---|---|---|---|---|---|---|---|---|
| Activity: Word Clues, p. 65 | | | X | | | | | | X | | | | | X | | | | | |
| Activity: Plant and Animal Cells, p. 66 | X | | | | | X | X | X | X | | | | | | | | | | |
| Activity: The Big Egg, p. 70 | X | | | | | X | X | X | | | X | | | | | | | X | |
| Activity: Making Cell Models, p. 71 | X | | | | | X | | X | | | | | | X | | | | | |
| Laboratory Activity: Things Look Different Under a Microscope, p. 74 | X | | | | | X | X | X | | | | | X | X | | | | X | |
| Skill Building 1, p. 77 | | | | | | X | | | X | | | | | | | | X | | |
| Skill Building 2, p. 77 | | | | | | | | | X | | | | | X | | | | | |
| Skill Building 3, p. 77 | | | | | | | | | X | | | | | X | | | | X | |
| Skill Building 4, p. 77 | | | | | | X | | | | X | | | | X | | | | | |

## CHAPTER 4 TISSUES, ORGANS, ORGAN SYSTEMS pp. 78–91

| TEXT REFERENCE | HANDS-ON | COMPUTATIONAL | VOCABULARY/WRITING | LIBRARY | FIELD | OBSERVING | COMPARING | MANIPULATIVE | RELATING | MEASURING | INFERRING | RECORDING | SAFETY | APPLYING | DIAGRAMING | DESIGNING | CLASSIFYING | HYPOTHESIZING | PREDICTING |
|---|---|---|---|---|---|---|---|---|---|---|---|---|---|---|---|---|---|---|---|
| Activity: Division of Labor, p. 81 | | | | | X | X | X | | | | | | | X | | | | | |
| Activity: An Amazing Feat, p. 87 | | X | | | | | | | | | | | | | | | | | |
| Laboratory Activity: Have You Tasted These?, p. 88 | | | | X | | X | | | | | | | | X | | | X | | |
| Skill Building 1, p. 91 | | | | | | | | | X | | | | | X | | | X | | |
| Skill Building 2, p. 91 | | | | | | | | | X | | | | | X | | | X | | |
| Skill Building 3, p. 91 | | | | | | | | | | | | | | X | X | | X | | |
| Skill Building 4, p. 91 | | | | | | | | | X | | | | | X | | | X | | |
| Skill Building 5, p. 91 | | | | | | | | | | | | | | X | | | X | | |

## CHAPTER 5 INTERACTIONS AMONG LIVING THINGS pp. 92–119

| TEXT REFERENCE | HANDS-ON | COMPUTATIONAL | VOCABULARY/WRITING | LIBRARY | FIELD | OBSERVING | COMPARING | MANIPULATIVE | RELATING | MEASURING | INFERRING | RECORDING | SAFETY | APPLYING | DIAGRAMING | DESIGNING | CLASSIFYING | HYPOTHESIZING | PREDICTING |
|---|---|---|---|---|---|---|---|---|---|---|---|---|---|---|---|---|---|---|---|
| Activity: Identifying Interactions, p. 97 | X | | | | | X | X | | X | | | | | X | X | | | | |
| Activity: Home Sweet Home, p. 99 | X | | | | | | X | X | X | | | | | X | X | | | | X |
| Activity: Drawing a Food Web, p. 105 | X | | | | | | | | X | | | | | X | X | | X | | |
| Activity: Your Place in a Food Chain, p. 107 | X | | | | | | X | | X | | | | | X | X | | | | |
| Activity: Shady Survival, p. 109 | | X | | | | | | | | | | | | X | | | | | X |
| Activity: Changes in Populations, p. 115 | | | X | X | | | | | X | | | | | X | | | | | |
| Laboratory Activity: A Little Off Balance, p. 116 | X | | | | | X | X | X | X | | | | X | X | | | | | X |
| Skill Building 1, p. 119 | | | | | | | | | X | | | | | X | | | X | | |
| Skill Building 2, p. 119 | | | | | | | X | | X | | | | | X | X | | X | | |
| Skill Building 3, p. 119 | | | | | | | | | X | | X | | | X | | | | X | X |
| Skill Building 4, p. 119 | | | | | | | X | | X | | | | | X | | | | | |
| Skill Building 5, p. 119 | | | | | | | | | X | | X | | | X | | X | | | |
| Skill Building 6, p. 119 | | | | | | | | | X | | X | | | X | | | | X | X |

## CHAPTER 6 CLASSIFICATION pp. 130–145

| TEXT REFERENCE | HANDS-ON | COMPUTATIONAL | VOCABULARY/WRITING | LIBRARY | FIELD | OBSERVING | COMPARING | MANIPULATIVE | RELATING | MEASURING | INFERRING | RECORDING | SAFETY | APPLYING | DIAGRAMING | DESIGNING | CLASSIFYING | HYPOTHESIZING | PREDICTING |
|---|---|---|---|---|---|---|---|---|---|---|---|---|---|---|---|---|---|---|---|
| Activity: What's in a Name?, p. 134 | | | X | | | | | | X | | X | | | X | | | | | |
| Activity: A Secret Code, p. 137 | | | X | | | | | | X | | X | | | X | | | | | |
| Activity: Classification of Plants, p. 141 | X | | | | | | X | X | X | | | X | | | | | X | | |
| Laboratory Activity: Whose Shoe Is That?, p. 142 | X | | | | | X | X | | X | | X | X | | X | | | X | | |
| Skill Building 1, p. 145 | | | | | | | X | | X | | | | | X | X | | X | | |
| Skill Building 2, p. 145 | | | | | | X | X | | X | | | | | X | X | X | X | | |
| Skill Building 3, p. 145 | | | | | | | X | | X | | X | X | | X | X | | X | | |
| Skill Building 4, p. 145 | | | X | | | | | | X | X | | | | X | | | X | | |
| Skill Building 5, p. 145 | | | | | | | X | | X | | X | | | X | | | X | X | X |
| Skill Building 6, p. 145 | | | | | | | X | | X | | X | | | X | | | X | | |
| Skill Building 7, p. 145 | | | | | | | | | X | | X | | | X | | | | X | X |
| Skill Building 8, p. 145 | | | | | | | | | | | X | | | X | | | | X | |

## CHAPTER 7 VIRUSES, BACTERIA, AND PROTISTS pp. 146–167

| TEXT REFERENCE | HANDS-ON | COMPUTATIONAL | VOCABULARY/WRITING | LIBRARY | FIELD | OBSERVING | COMPARING | MANIPULATIVE | RELATING | MEASURING | INFERRING | RECORDING | SAFETY | APPLYING | DIAGRAMING | DESIGNING | CLASSIFYING | HYPOTHESIZING | PREDICTING |
|---|---|---|---|---|---|---|---|---|---|---|---|---|---|---|---|---|---|---|---|
| Activity: Building a Bacteriophage, p. 150 | X | | | | | | | X | X | | | | | X | X | | | | |
| Activity: Shapes of Bacteria, p. 154 | X | | | | | X | X | | | | | | | | X | | X | | |
| Activity: Observing Bacteria, p. 157 | X | | | | | X | X | X | X | | | | | X | | | | X | |
| Activity: A Protozoan Population, p. 159 | | X | | | | | | | | | | | | | | | | | |
| Activity: Harmful Microorganisms, p. 160 | | | X | | | | | | X | | | | | X | | | | | |
| Activity: Capturing Food, p. 163 | X | | | | | X | X | X | X | | | | | X | | | | | |
| Laboratory Activity: Where Are Bacteria Found?, p. 164 | | | | | | X | X | X | | | X | | X | | X | | | X | |
| Skill Building 1, p. 167 | | X | | | | | X | | X | | | | | X | | | | X | |
| Skill Building 2, p. 167 | | | | | | | X | | X | | | | | X | | | | | X |
| Skill Building 3, p. 167 | | | | | | | | | X | | | | | X | | | | | X |
| Skill Building 4, p. 167 | | | | | | | | | | | X | | | X | | X | | X | X |

## CHAPTER 8 NONVASCULAR PLANTS AND PLANTLIKE ORGANISMS pp. 168–191

| TEXT REFERENCE | HANDS-ON | COMPUTATIONAL | VOCABULARY/WRITING | LIBRARY | FIELD | OBSERVING | COMPARING | MANIPULATIVE | RELATING | MEASURING | INFERRING | RECORDING | SAFETY | APPLYING | DIAGRAMING | DESIGNING | CLASSIFYING | HYPOTHESIZING | PREDICTING |
|---|---|---|---|---|---|---|---|---|---|---|---|---|---|---|---|---|---|---|---|
| Activity: Algae Life Cycles, p. 172 | | | | X | | | X | | | | | | | | X | | X | | |
| Activity: Collecting Algae, p. 176 | X | | | | | X | X | X | | | | | | | | | | | |
| Activity: Making Spore Prints, p. 181 | X | | | | | X | | X | X | | | | | | | | | | |
| Activity: Growing Mold, p. 183 | X | | | | | X | | X | X | | X | | | X | | | | | |
| Activity: Observing a Slime Mold, p. 180 | X | | | | | X | | X | X | | | | | X | | | | | |
| Laboratory Activity: Examining a Slime Mold, p. 188 | X | | | | | X | | X | X | | | | X | | | X | | | |
| Skill Building 1, p. 191 | | | | | | | | | X | | | | | X | | | | | X |
| Skill Building 2, p. 191 | | | | | | | | | X | | | | | X | | | | X | |
| Skill Building 3, p. 191 | | | | | | | | | X | | | | | X | | | | | |
| Skill Building 4, p. 191 | | | | | | | | | | | | | | | | | | X | |
| Skill Building 5, p. 191 | | | | | | | | | | | | | | X | | X | | X | |
| Skill Building 6, p. 191 | | | | | | | | | X | | | | | | | | | X | X |
| Skill Building 7, p. 191 | | | | | | | | | X | X | X | | | X | | | | | |

## CHAPTER 9 VASCULAR PLANTS pp. 192–215

| TEXT REFERENCE | HANDS-ON | COMPUTATIONAL | VOCABULARY/WRITING | LIBRARY | FIELD | OBSERVING | COMPARING | MANIPULATIVE | RELATING | MEASURING | INFERRING | RECORDING | SAFETY | APPLYING | DIAGRAMING | DESIGNING | CLASSIFYING | HYPOTHESIZING | PREDICTING |
|---|---|---|---|---|---|---|---|---|---|---|---|---|---|---|---|---|---|---|---|
| Activity: Observing Ferns, p. 196 | X | | | | X | X | | | X | | | | | X | X | | | | |
| Activity: Plant Transport, p. 198 | X | | | | | X | X | X | | | | | | X | | | | | |
| Activity: Annual Rings, p. 200 | X | | | | X | X | X | | | | | | | X | | | | | |
| Activity: Plant Responses, p. 202 | | | X | | | | | | X | | | | | X | | | X | | |
| Activity: Perfect or Imperfect?, p. 208 | X | | | | | X | X | | X | | | | | X | X | | X | | |
| Activity: Seed Germination, p. 211 | X | | | | | X | | X | | | | | | X | | X | | X | |
| Laboratory Activity: Geotropism, p. 212 | X | | | | | X | X | X | X | | X | | | X | X | | | X | X |
| Skill Building 1, p. 215 | | | | | | | X | | X | | | | | | | | | | |
| Skill Building 2, p. 215 | | | | | | | X | | | | | | | | X | | | | |
| Skill Building 3, p. 215 | | | | | | | | | | | | | | | | | | X | X |
| Skill Building 4, p. 215 | | | | | | | | | | | | | | X | | | | | |
| Skill Building 5, p. 215 | | | | | | | | | X | | X | | | | | | | | |
| Skill Building 6, p. 215 | | | | | | | | | | | | | | X | | X | | X | X |

## CHAPTER 9 VASCULAR PLANTS pp. 192–215 (continued)

| TEXT REFERENCE | HANDS-ON | COMPUTATIONAL | VOCABULARY/WRITING | LIBRARY | FIELD | OBSERVING | COMPARING | MANIPULATIVE | RELATING | MEASURING | INFERRING | RECORDING | SAFETY | APPLYING | DIAGRAMING | DESIGNING | CLASSIFYING | HYPOTHESIZING | PREDICTING |
|---|---|---|---|---|---|---|---|---|---|---|---|---|---|---|---|---|---|---|---|
| Skill Building 7, p. 215 | | | | | | | | | X | | | | | | | | | | |
| Skill Building 8, p. 215 | | | | | | | | | | | | | | X | | X | | | |
| Skill Building 9, p. 215 | | | | | | | | | X | | | | | | | | | | X |

## CHAPTER 10 INVERTEBRATES pp. 216–251

| TEXT REFERENCE | HANDS-ON | COMPUTATIONAL | VOCABULARY/WRITING | LIBRARY | FIELD | OBSERVING | COMPARING | MANIPULATIVE | RELATING | MEASURING | INFERRING | RECORDING | SAFETY | APPLYING | DIAGRAMING | DESIGNING | CLASSIFYING | HYPOTHESIZING | PREDICTING |
|---|---|---|---|---|---|---|---|---|---|---|---|---|---|---|---|---|---|---|---|
| Activity: Observing a Sponge, p. 219 | X | | | X | | X | X | X | X | | | | | | | | | | |
| Activity: Observing Hydra, p. 224 | X | | | | | X | | X | | | | | | | X | | | X | |
| Activity: A Worm Farm, p. 227 | X | | | | | X | | X | X | | | | | | | | | | |
| Activity: Mollusks in the Supermarket, p. 230 | X | | | | X | | X | | X | | | | | X | X | | | | |
| Activity: Symmetry, p. 237 | | | X | | | | X | | X | | | | | X | | | | | |
| Activity: The Life of a Mealworm, p. 233 | X | | | | | X | | X | X | | | X | | X | | | | | |
| Laboratory Activity: Which Way Did That Isopod Go?, p. 248 | X | | | | | X | | X | X | | X | X | | X | | X | | X | |
| Skill Building 1, p. 251 | | | | | | | | | X | | X | | | X | | | | X | X |
| Skill Building 2, p. 251 | | | | | | | X | | X | | | | | X | X | | X | | |
| Skill Building 3, p. 251 | | | | | | | X | | | | X | | | X | | | | | |
| Skill Building 4, p. 251 | | | | | | | | | X | | X | | | X | | | | X | |
| Skill Building 5, p. 251 | | | | | | | | | X | | X | | | X | | | | X | |
| Skill Building 6, p. 251 | | | | | | | | | X | | | | | X | | | | | |
| Skill Building 7, p. 251 | | | | | | | | | X | | | | | X | | | | X | |

**CHAPTER 10 INVERTEBRATES pp. 216–251** (continued)

| TEXT REFERENCE | HANDS-ON | COMPUTATIONAL | VOCABULARY/WRITING | LIBRARY | FIELD | OBSERVING | COMPARING | MANIPULATIVE | RELATING | MEASURING | INFERRING | RECORDING | SAFETY | APPLYING | DIAGRAMING | DESIGNING | CLASSIFYING | HYPOTHESIZING | PREDICTING |
|---|---|---|---|---|---|---|---|---|---|---|---|---|---|---|---|---|---|---|---|
| Skill Building 8, p. 251 | | | | | | | | | X | | | | | X | | | | X | X |
| Skill Building 9, p. 251 | | | | | | | | | X | | X | | | X | | | | | |
| Skill Building 10, p. 251 | | | | | | | | | X | | X | | | X | | | | X | |

## CHAPTER 11 ANIMALS: VERTEBRATES pp. 252–279

| TEXT REFERENCE | HANDS-ON | COMPUTATIONAL | VOCABULARY/WRITING | LIBRARY | FIELD | OBSERVING | COMPARING | MANIPULATIVE | RELATING | MEASURING | INFERRING | RECORDING | SAFETY | APPLYING | DIAGRAMING | DESIGNING | CLASSIFYING | HYPOTHESIZING | PREDICTING |
|---|---|---|---|---|---|---|---|---|---|---|---|---|---|---|---|---|---|---|---|
| Activity: Comparing Sizes, p. 257 | | X | | | | | X | | | | | | | | | | | | |
| Activity: Observing a Fish, p. 259 | X | X | | | | X | | X | X | | | | | | X | | | | |
| Activity: The Truth About Snakes, p. 265 | | | X | X | | | X | | X | | | | | X | | | | | |
| Activity: A Symbiotic Relationship, p. 270 | | | X | X | | | | | X | | | | | X | | | | | |
| Activity: Comparing Feathers, p. 273 | X | | | | | X | X | | X | | | | | | X | | | | |
| Activity: Bird Watching, p. 275 | | | | | X | X | X | | | | | | | X | | | | | |
| Laboratory Activity: Classifying Vertebrate Bones in Owl Pellets, p. 276 | X | | | | | X | X | X | X | | X | | X | X | | | X | X | |
| Skill Building 1, p. 279 | | | | | | | | | X | | X | | | X | | | X | | |
| Skill Building 2, p. 279 | | | | | | | X | | X | | X | | | X | | | X | X | X |
| Skill Building 3, p. 279 | | | | | | | | | X | | X | | | X | | | | X | |
| Skill Building 4, p. 279 | | | | | | | | | X | | X | | | X | | | | X | |
| Skill Building 5, p. 279 | | | | | | | X | | X | | X | | | X | | | | X | |
| Skill Building 6, p. 279 | | | | | | | | | X | | X | | | X | | | | X | X |

## CHAPTER 11 ANIMALS: VERTEBRATES pp. 252–279 (continued)

| TEXT REFERENCE | HANDS-ON | COMPUTATIONAL | VOCABULARY/WRITING | LIBRARY | FIELD | OBSERVING | COMPARING | MANIPULATIVE | RELATING | MEASURING | INFERRING | RECORDING | SAFETY | APPLYING | DIAGRAMING | DESIGNING | CLASSIFYING | HYPOTHESIZING | PREDICTING |
|---|---|---|---|---|---|---|---|---|---|---|---|---|---|---|---|---|---|---|---|
| Skill Building 7, p. 279 | | | | | | | | | X | | X | | | X | | | X | X | |
| Skill Building 8, p. 279 | | | | | | | | | X | | X | | | X | | | | X | |
| Skill Building 9, p. 279 | | | | | | | | | X | | | | | X | | X | | | |

## CHAPTER 12 MAMMALS pp. 280–301

| TEXT REFERENCE | HANDS-ON | COMPUTATIONAL | VOCABULARY/WRITING | LIBRARY | FIELD | OBSERVING | COMPARING | MANIPULATIVE | RELATING | MEASURING | INFERRING | RECORDING | SAFETY | APPLYING | DIAGRAMING | DESIGNING | CLASSIFYING | HYPOTHESIZING | PREDICTING |
|---|---|---|---|---|---|---|---|---|---|---|---|---|---|---|---|---|---|---|---|
| Activity: Pet Mammals, p. 283 | X | | | | | X | X | | X | | | X | | | | | | | |
| Activity: Vertebrate Body Systems, p. 284 | X | | | X | | | X | | X | | | | | X | | | | | |
| Activity: The Fastest Runner, p. 290 | | X | | | | | | | | | | | | | | | | | |
| Activity: Useful Mammals, p. 292 | | | | X | | | X | | X | | | X | | | | | | | |
| Activity: Migration of Mammals, p. 296 | X | | | X | | | X | | X | | | X | | | X | | | | |
| Laboratory Activity: Classifying and Comparing Mammals, p. 298 | X | | | X | | | X | | X | | X | | | X | | | X | | |
| Skill Building 1, p. 301 | | | | | | | | | X | | | | | | | | | | |
| Skill Building 2, p. 301 | | | | | | | | | X | | | | | X | X | | X | | |
| Skill Building 3, p. 301 | | | | | | | X | | | | | | | | | | X | | |
| Skill Building 4, p. 301 | | | | | | | | | | | X | | | X | | | | X | |
| Skill Building 5, p. 301 | | | | | | | | | X | | X | | | X | | | | | |
| Skill Building 6, p. 301 | | | | | | | X | | X | | | | | X | | | | X | |
| Skill Building 7, p. 301 | | | | | | | | | X | | | | | X | | X | | | |

## CHAPTER 13 GENERAL PROPERTIES OF MATTER pp. 312–327

| TEXT REFERENCE | HANDS-ON | COMPUTATIONAL | VOCABULARY/WRITING | LIBRARY | FIELD | OBSERVING | COMPARING | MANIPULATIVE | RELATING | MEASURING | INFERRING | RECORDING | SAFETY | APPLYING | DIAGRAMING | DESIGNING | CLASSIFYING | HYPOTHESIZING | PREDICTING |
|---|---|---|---|---|---|---|---|---|---|---|---|---|---|---|---|---|---|---|---|
| Activity: Demonstrating Inertia, p. 316 | X | | | | | X | X | X | X | | | | | X | | | | | |
| Activity: A Quick Weight Change, p. 318 | | X | | | | | | | | | X | | | | | | | | |
| Activity: Volume of a Solid, p. 319 | X | | | | | | | X | | X | | | | | | | | | |
| Activity: Describing Properties, p. 320 | X | | X | | | X | | | | | | | | X | | | X | | |
| Activity: Archimedes and the Crown, p. 322 | | | X | X | | | | | | | | | | X | | X | | | X |
| Laboratory Activity: Inertia, p. 324 | | | | | | X | X | X | X | X | X | | | X | | | | X | |
| Skill Building 1, p. 327 | | | | | | | X | | | | | | | | | | | | |
| Skill Building 2, p. 327 | | X | | | | | | | | | | | | | | | | | |
| Skill Building 3, p. 327 | | | | | | | | | | | | | | | | X | | | |
| Skill Building 4, p. 327 | | | | | | | | | | | | | | X | | | | | |
| Skill Building 5, p. 327 | | | | | | | | | X | | | | | | | | | | |
| Skill Building 6, p. 327 | | | | | | | | | | | X | | | | | | | | |

## CHAPTER 14 PHYSICAL AND CHEMICAL CHANGES pp. 328–349

| TEXT REFERENCE | HANDS-ON | COMPUTATIONAL | VOCABULARY/WRITING | LIBRARY | FIELD | OBSERVING | COMPARING | MANIPULATIVE | RELATING | MEASURING | INFERRING | RECORDING | SAFETY | APPLYING | DIAGRAMING | DESIGNING | CLASSIFYING | HYPOTHESIZING | PREDICTING |
|---|---|---|---|---|---|---|---|---|---|---|---|---|---|---|---|---|---|---|---|
| Activity: Observing Viscosity, p. 333 | X | | | | | X | X | X | | X | X | | | X | | | | | |
| Activity: Determining Particle Space, p. 335 | X | | | | | X | X | X | X | X | X | | | | | | | X | |
| Activity: Charles's Law, p. 336 | X | | | | | X | X | X | X | X | | | | X | | | | | |
| Activity: Melting Ice and Freezing Water, p. 338 | X | | | | | X | X | X | | | X | X | | | | | | X | |
| Activity: Holding Particles Together, p. 341 | | | | X | | | X | | X | | | | | X | | | | | |
| Activity: An Almost Ruined Day, p. 342 | | | X | | | | | | X | | | | | X | | | | | |
| Activity: Physical and Chemical Changes, p. 345 | X | | | | | X | | X | X | | X | | | X | | | | | |
| Laboratory Activity: Conservation of Mass, p. 346 | X | X | | | | X | X | X | X | X | | X | X | X | | | | | |
| Skill Building 1, p. 349 | | | | | | | | | | | | | | X | | | | | |
| Skill Building 2, p. 349 | | | | | | | | | | | | | | | | | X | | |
| Skill Building 3, p. 349 | | | | | | | | | | | | | | | | | X | | |
| Skill Building 4, p. 349 | | | | | | | | | X | | | | | | | | | | |
| Skill Building 5, p. 349 | | | | | | | | | | | X | | | | | | | | |

## CHAPTER 15 ATOMS AND MOLECULES pp. 360–375

| TEXT REFERENCE | HANDS-ON | COMPUTATIONAL | VOCABULARY/WRITING | LIBRARY | FIELD | OBSERVING | COMPARING | MANIPULATIVE | RELATING | MEASURING | INFERRING | RECORDING | SAFETY | APPLYING | DIAGRAMING | DESIGNING | CLASSIFYING | HYPOTHESIZING | PREDICTING |
|---|---|---|---|---|---|---|---|---|---|---|---|---|---|---|---|---|---|---|---|
| Activity: A Mental Model, p. 363 | X | | | | | | | | X | | X | | | X | | | | X | X |
| Activity: Making Indirect Observations, p. 366 | X | | | | | X | X | X | | | X | | | | | | | | |
| Activity: Electron Arrangement, p. 369 | | X | | | | | | | X | | | | | X | | | | | |
| Laboratory Activity: Flame Tests, p. 372 | X | | | | | X | X | X | X | | | X | X | X | | | | | X |
| Skill Building 1, p. 375 | | X | | | | | | | | | | | | X | | | | | |
| Skill Building 2, p. 375 | | | | | | | | | X | | X | | | X | X | | | X | X |
| Skill Building 3, p. 375 | | | | | | | X | | X | | | | | | | | | | |
| Skill Building 4, p. 375 | | | | | | | | | X | | X | | | X | | | | | |

## CHAPTER 16 ELEMENTS AND COMPOUNDS pp. 376–389

| TEXT REFERENCE | HANDS-ON | COMPUTATIONAL | VOCABULARY/WRITING | LIBRARY | FIELD | OBSERVING | COMPARING | MANIPULATIVE | RELATING | MEASURING | INFERRING | RECORDING | SAFETY | APPLYING | DIAGRAMING | DESIGNING | CLASSIFYING | HYPOTHESIZING | PREDICTING |
|---|---|---|---|---|---|---|---|---|---|---|---|---|---|---|---|---|---|---|---|
| Activity: Getting to Know the Elements, p. 378 | X | | | | X | X | | | X | | | | | X | | | | | |
| Activity: Naming and Counting Atoms, p. 385 | | X | | | | | | | | | | | | X | | | | | |
| Laboratory Activity: Marshmallow Molecules, p. 386 | | | | | | | X | X | X | | X | | | X | | X | | | |
| Skill Building 1, p. 389 | | | | | | | | | | | | | | | | | X | | |
| Skill Building 2, p. 389 | | | | | | | | | | | X | | | | | | | | |
| Skill Building 3, p. 389 | | | | | | | | | X | | | | | | | | | | |
| Skill Building 4, p. 389 | | X | | | | | | | | | | | | | | | | | |
| Skill Building 5, p. 389 | | | | | | | | | | | | | | | | | | | X |
| Skill Building 6, p. 389 | | | | | | | | | | | | | | | | X | | | |
| Skill Building 7, p. 389 | | | | | | | | | | | | | | X | | | | | |
| Skill Building 8, p. 389 | | | | | | | | | | | | | | X | | | | | |

## CHAPTER 17 MIXTURES AND SOLUTIONS pp. 390–403

| TEXT REFERENCE | HANDS-ON | COMPUTATIONAL | VOCABULARY/WRITING | LIBRARY | FIELD | OBSERVING | COMPARING | MANIPULATIVE | RELATING | MEASURING | INFERRING | RECORDING | SAFETY | APPLYING | DIAGRAMING | DESIGNING | CLASSIFYING | HYPOTHESIZING | PREDICTING |
|---|---|---|---|---|---|---|---|---|---|---|---|---|---|---|---|---|---|---|---|
| Activity: A Mixture Collection, p. 393 | X | | X | | | | | | X | | | | | X | X | | | | |
| Activity: Expressing Solubility, p. 397 | | | X | X | | | | | | | | | | X | | | | | |
| Activity: Water Dissolves Most Substances, p. 398 | X | | | | | X | X | X | X | | | | | X | X | | | | |
| Activity: Solubility of a Gas in a Liquid, p. 399 | X | | | | | X | X | X | X | | | | | | | | | X | |
| Laboratory Activity: Examination of Freezing Point Depression, p. 400 | X | | | | | X | X | X | X | X | | | X | X | | | | X | |
| Skill Building 1, p. 403 | | | | | | | X | | X | | | | | X | | | X | | |
| Skill Building 2, p. 403 | | | | | | | X | | X | | | | | X | | | | | |
| Skill Building 3, p. 403 | | | | | | | | | X | | | | | X | | | X | | |
| Skill Building 4, p. 403 | | | | | | | X | | X | | | | | X | | | | | |
| Skill Building 5, p. 403 | | | | | | | | | | | X | | | X | | | | X | |
| Skill Building 6, p. 403 | | | | | | | | | X | | X | | | X | | | | X | X |
| Skill Building 7, p. 403 | | | | | | | | | X | | X | | | X | | | | X | |
| Skill Building 8, p. 403 | | | | | | | | | X | | X | | | X | | | | X | |
| Skill Building 9, p. 403 | | | | | | | | | X | | | | | | | | | X | X |
| Skill Building 10, p. 403 | | | | | | | X | | | | | | | | | X | | | |

## CHAPTER 18 MINERALS pp. 414–431

| TEXT REFERENCE | HANDS-ON | COMPUTATIONAL | VOCABULARY/WRITING | LIBRARY | FIELD | OBSERVING | COMPARING | MANIPULATIVE | RELATING | MEASURING | INFERRING | RECORDING | SAFETY | APPLYING | DIAGRAMING | DESIGNING | CLASSIFYING | HYPOTHESIZING | PREDICTING |
|---|---|---|---|---|---|---|---|---|---|---|---|---|---|---|---|---|---|---|---|
| Activity: Mineral Hardness, p. 421 | X | | | | | X | X | | X | | | | | X | X | | | | |
| Activity: Crystal-System Models, p. 422 | X | | | | | | | X | X | | | | | X | | | | | |
| Activity: Mineral Deposits, p. 426 | X | | | X | | | | | X | | | | | X | X | | | | |
| Laboratory Activity: Forming Mineral Crystals, p. 428 | X | | | | | X | X | X | X | X | | | X | X | | | | | |
| Skill Building 1, p. 431 | | | | | | | X | | X | | | | | X | | | | | |
| Skill Building 2, p. 431 | | | | | | | | | X | | | | | X | | | | | |
| Skill Building 3, p. 431 | | | | | | | X | | X | | | | | X | | | | | |
| Skill Building 4, p. 431 | | | | | | | X | | | | | | | | | | X | | |
| Skill Building 5, p. 431 | | | | | | | X | | X | | | | | X | | | | | |
| Skill Building 6, p. 431 | | | | | | | X | | X | | | | | X | | | | | |

## CHAPTER 19 ROCKS pp. 432–449

| TEXT REFERENCE | HANDS-ON | COMPUTATIONAL | VOCABULARY/WRITING | LIBRARY | FIELD | OBSERVING | COMPARING | MANIPULATIVE | RELATING | MEASURING | INFERRING | RECORDING | SAFETY | APPLYING | DIAGRAMING | DESIGNING | CLASSIFYING | HYPOTHESIZING | PREDICTING |
|---|---|---|---|---|---|---|---|---|---|---|---|---|---|---|---|---|---|---|---|
| Activity: Sedimentation, p. 437 | X | | | | | X | | X | X | | X | | | X | | | | X | |
| Activity: Coral Conversions, p. 440 | | X | | | | | | | | | | | | | | | | | |
| Activity: Rock Quarries, p. 441 | | | X | X | | | | | X | | | | | X | | | | | |
| Activity: A Rock Walk, p. 443 | | | X | | X | X | | | | | | | | X | | | | | |
| Activity: Rock Samples, p. 445 | X | | | | X | X | X | | | | X | | | X | | | X | | |
| Laboratory Activity: Making a Sedimentary Rock, p. 446 | X | | | | | X | X | X | X | X | | | X | X | | | | | |
| Skill Building 1, p. 449 | | | | | | | X | | X | | | | | X | | | | | |
| Skill Building 2, p. 449 | | | | | | | | | X | | | | | X | | | | | |
| Skill Building 3, p. 449 | | | | | | | | | | | | | | X | | | X | | |
| Skill Building 4, p. 449 | | | | | | | | | X | | | | | X | | | | | |
| Skill Building 5, p. 449 | | | | | | | | | X | | | | | X | | | | | |
| Skill Building 6, p. 449 | | | | | | | | | | | | | | | X | | | | |
| Skill Building 7, p. 449 | | | | | | | | | X | | | | | X | | | | | X |

## CHAPTER 20 SOILS pp. 450–467

| TEXT REFERENCE | HANDS-ON | COMPUTATIONAL | VOCABULARY/WRITING | LIBRARY | FIELD | OBSERVING | COMPARING | MANIPULATIVE | RELATING | MEASURING | INFERRING | RECORDING | SAFETY | APPLYING | DIAGRAMING | DESIGNING | CLASSIFYING | HYPOTHESIZING | PREDICTING |
|---|---|---|---|---|---|---|---|---|---|---|---|---|---|---|---|---|---|---|---|
| Activity: Humus, p. 453 | X | | | | | X | X | X | X | | | | | | | | X | | |
| Activity: Studying Soil Layers, p. 454 | X | | | | | X | X | X | X | | | | | X | | | | | |
| Activity: Leaching, p. 459 | | | X | | | | X | | X | | | | | X | | | | | |
| Activity: The Use of Fertilizers, p. 460 | | | X | X | | | X | | X | | | | | X | | | | | |
| Laboratory Activity: Determining Rates of Weathering, p. 464 | X | | | | | X | X | X | X | X | | X | X | X | | | | | |
| Skill Building 1, p. 467 | | | | | | | X | | X | | | | | X | | | | | |
| Skill Building 2, p. 467 | | | | | | | X | | X | | | | | X | | | | | |
| Skill Building 3, p. 467 | | | | | | | X | | X | | X | | | X | | | | | |
| Skill Building 4, p. 467 | | | | | | | | | X | | | | | | | | | | |
| Skill Building 5, p. 467 | | | | | | X | X | | | | | | | X | | | | | |

## CHAPTER 21 INTERNAL STRUCTURE OF THE EARTH pp. 478–499

| TEXT REFERENCE | HANDS-ON | COMPUTATIONAL | VOCABULARY/WRITING | LIBRARY | FIELD | OBSERVING | COMPARING | MANIPULATIVE | RELATING | MEASURING | INFERRING | RECORDING | SAFETY | APPLYING | DIAGRAMING | DESIGNING | CLASSIFYING | HYPOTHESIZING | PREDICTING |
|---|---|---|---|---|---|---|---|---|---|---|---|---|---|---|---|---|---|---|---|
| Activity: Speed of Seismic Waves, p. 483 | | X | | | | | X | | | | | | | | | | | | |
| Activity: How Many Earths?, p. 485 | | X | | | | | X | | | | | | | | | | | | |
| Activity: Model of the Earth's Interior, p. 487 | | | | | | | | | X | | | | | X | | | | | |
| Activity: Visiting the Earth's Core, p. 488 | | | X | X | | | X | | X | | | | | X | | | | | |
| Activity: Simulating Plasticity, p. 491 | X | | | | | X | X | X | X | X | | | | X | | | | X | |
| Activity: The Earth's Crust, p. 493 | X | | | | | X | X | X | X | | | | | X | | X | | | |
| Activity: Mohorovicic's Discovery, p. 495 | | | X | X | | | | | | | | | | | | | | | |
| Laboratory Activity: Building an Active Geyser, p. 496 | X | | | | | X | | | X | | | | X | X | | | | | |
| Skill Building 1, p. 499 | | | | | | | | | X | | X | | | X | | | | X | X |
| Skill Building 2, p. 499 | | | | | | | | | X | | | | | X | | | | X | |
| Skill Building 3, p. 499 | | X | | | | | X | | X | | | | | | X | | | | |

## CHAPTER 22 SURFACE FEATURES OF THE EARTH pp. 500–521

| TEXT REFERENCE | HANDS-ON | COMPUTATIONAL | VOCABULARY/WRITING | LIBRARY | FIELD | OBSERVING | COMPARING | MANIPULATIVE | RELATING | MEASURING | INFERRING | RECORDING | SAFETY | APPLYING | DIAGRAMING | DESIGNING | CLASSIFYING | HYPOTHESIZING | PREDICTING |
|---|---|---|---|---|---|---|---|---|---|---|---|---|---|---|---|---|---|---|---|
| Activity: Earth's Mountains, p. 506 | | X | | | | | | | | | | | | | | | | | |
| Activity: Continental Sizes, p. 508 | X | X | | | | X | | X | X | | | | | X | | | | | |
| Activity: Surface Features, p. 510 | | | X | | | | X | | X | | | | | X | | | | | |
| Activity: Reservoirs, p. 512 | | | X | X | | | | | X | | | | | X | | | | | |
| Activity: Simulating the Water Cycle, p. 513 | X | | | | | X | | X | X | | X | | | X | | | | | |
| Activity: Drought and the Water Table, p. 516 | X | | | | | X | | X | X | | | | | X | | | | X | X |
| Laboratory Activity: Examining Differences Between Fresh and Salt Water, p. 578 | X | | | | | X | X | X | X | X | X | X | | X | | | | X | X |
| Skill Building 1, p. 521 | | | | | | | | | X | | | | | X | | | | | |
| Skill Building 2, p. 521 | | X | | | | | | | | | | | | | | | | | |
| Skill Building 3, p. 521 | | | | | | | | | X | | | | | X | | | | | X |
| Skill Building 4, p. 521 | | | | | | | | | | | | | | X | | | | | |
| Skill Building 5, p. 521 | | | | | | | | | | | X | | | | | | | X | |
| Skill Building 6, p. 521 | | | | | | | | | X | | | | | | | | | | |
| Skill Building 7, p. 521 | | | | | | | | | X | | | | | X | | X | | | |

## CHAPTER 23 STRUCTURE OF THE ATMOSPHERE pp. 522–541

| TEXT REFERENCE | HANDS-ON | COMPUTATIONAL | VOCABULARY/WRITING | LIBRARY | FIELD | OBSERVING | COMPARING | MANIPULATIVE | RELATING | MEASURING | INFERRING | RECORDING | SAFETY | APPLYING | DIAGRAMING | DESIGNING | CLASSIFYING | HYPOTHESIZING | PREDICTING |
|---|---|---|---|---|---|---|---|---|---|---|---|---|---|---|---|---|---|---|---|
| Activity: Air Pressure, p. 527 | X | | | | | X | | | X | | X | | | X | | | | | |
| Activity: Temperature Changes in the Troposphere, p. 528 | X | X | | | | X | | X | X | X | | X | | X | | X | | | |
| Activity: Meteor Hunt, p. 532 | | X | | | X | X | | | | | | X | | | | | | | |
| Activity: Listening to the Radio, p. 535 | | | | | X | X | X | | X | | X | | | | | | | | |
| Laboratory Activity: Using Atmospheric Pressure to Crush a Can, p. 538 | X | | | | | X | X | | X | | X | | X | X | | | | X | X |
| Skill Building 1, p. 541 | | | X | | | | X | | X | | | | | X | | | | | |
| Skill Building 2, p. 541 | | X | | | | | | | | | X | | | | | | | | |
| Skill Building 3, p. 541 | | | | | | | | | X | | X | | | X | | | | | |
| Skill Building 4, p. 541 | | | | | | | X | | | X | | | | | X | | X | | |

# Features of the Annotated Teacher's Edition

The *Prentice Hall General Science* Annotated Teacher's Edition is the most complete, comprehensive, and pedagogically sound teacher's edition available for the junior high/middle school science teacher. The basic structure of this teacher's edition provides reduced student pages. This reduction allows a wide variety of teaching materials to be wrapped around the student page, but the reduction is not so great that the teacher cannot easily read all student material.

You will immediately note that on most pages the right and left margin columns have a blue-tinted background color. All materials found in these margin columns pertain to background information for the teacher, interesting facts and figures the teacher may want to present to the class, answers and additional information relating to all marginal Activities, teacher demonstration ideas, and other extremely useful information. The exact nature of the components of these margin columns will be explained in the pages that follow.

You will also immediately notice that an area boxed in red is found at the bottom of most teacher pages. This boxed-in area has been designed to provide teaching strategies specifically geared to the information presented in the text. Thus, by separating teaching strategies in one box and including other relevant information in the side columns only, the teacher can use this Annotated Teacher's Edition with a minimum of training or experience. The kinds of teaching strategies included for each lesson will be discussed in detail in the pages that follow.

One important aspect of this wrap-around Annotated Teacher's Edition is that all relevant material is right there along with the student pages for easy reference. Ease of use was an important criterion in the development of this teacher's edition. Of equal importance is the fact that the teaching material and marginal column material have been carefully controlled so that all materials relate directly to the two-page spread in the student edition. That is, all materials on the teacher's edition pages refer directly to those same student pages. Teachers need not try to decipher which teaching instructions or background information is applicable to which student page. All applicable materials are included directly along with the student pages so no flipping of pages is required to look for answers or teaching strategies. Again this makes the *Prentice Hall General Science* Annotated Teacher's Edition the most comprehensive and functional teacher's edition available.

#### UNIT OPENERS

Wrapped around the two-page unit openers in *Prentice Hall General Science* are a wide variety of instructional materials and teaching information. In the blue-tinted side columns you will find a Unit Overview, which provides a short overview of the facts and important concepts covered in a particular unit. Following the overview are Unit Objectives, which tie together in objective form the facts and concepts that are employed in the entire unit. Thus, through unit objectives, the teacher is presented with a basic understanding of the broad objectives students should meet during their study of a particular unit. Also included in the side columns is a list of each chapter in the unit and a brief description of the topics covered in each chapter.

At the bottom of each unit opener, located in the red-bordered strategy box, is a teaching strategy for introducing the unit. In general, the strategy calls upon the teacher to first have students observe the unit-opening photograph. Questions based on the photograph are often provided. Many are open-ended and require some degree of critical thinking. A basic design feature of all questions posed in this teacher's edition is that a small bullet is placed before each question. The question is set in boldfaced type so that all suggested questions are immediately obvious at a glance. The answers to such questions are placed in parentheses immediately after the question.

After students observe the photograph, they are called upon to read the text material that accompanies the unit-opening visual. Again, questions and teaching strategies are included in order to motivate and provide general interest to the student.

**CHAPTER OPENERS**

The blue-tinted columns that wrap around the chapter-opening pages begin with a Chapter Overview. Like the Unit Overview, the Chapter Overview provides the teacher with the basic facts and concepts that will be presented in the chapter. Following the overview is a Teacher Demonstration, which has been written to motivate students and provide a conceptual framework for the information that is to come in the chapter. Also included in the marginal columns of the chapter opener are a listing of appropriate audiovisual materials, a bibliography for teachers and students, and appropriate software materials for the chapter.

The bottom of the chapter-opening spread is reserved for a chapter-opening teaching strategy and is called Introducing the Chapter. The format for the chapter introductory teacher materials is similar in scope and design to the unit-opening introductory teacher materials.

**CHAPTER SECTION MATERIALS**

Each major section in *Prentice Hall General Science* is numbered for easy reference. The column in the margin directly next to the numbered section begins with a Section Preview. The Section Preview alerts the teacher to the basic facts and concepts that will be presented in that specific section of the chapter. Following the Section Preview is a numbered listing of Section Objectives for that particular section. Following the Section Objectives are all the bold-faced vocabulary words found in that section, which are listed under the title Science Terms. In addition, the page on which each term is introduced is included along with the term.

The blue-tinted columns on the right and left that follow each numbered section contain a wide variety of instructional materials. These instructional materials appear, whenever applicable, only in the marginal columns. Background Information provides the teacher with extra information about the topics being covered. Historical Notes provide a historical framework for the concepts being covered. Aside from the Teacher Demonstration that is used to introduce each chapter, at least one more Teacher Demonstration, and often more than one, is included within the margin columns of each chapter. Again, these demonstrations are meant to be motivational and often include questions to be asked before and after the demonstration. A feature called Facts and Figures may also be found in the marginal columns of each chapter. This feature provides interesting information that the teacher may want to present to the class. Science, Technology, and Society materials are also included in every chapter.

Because science cannot be taught in a vacuum, *Prentice Hall General Science* provides numerous Tie-Ins to other areas of science and to other curriculum areas, such as history, government, and art. These Tie-Ins help relate general science to other areas of science, as well as to the students' everyday life.

Also included in the marginal columns of the Annotated Teacher's Edition are the instructional materials related to the marginal Activities in the text. These teaching notes are always found directly on the same two-page spread on which they are located in the student text. Usually, they are found right beside the accompanying student Activity. The teacher is provided with the skills to be employed with each Activity, the type of activity that is being assigned, the level of each Activity (remedial, average, or enriched), any materials that are required, and the answers or suggested answers to the Activity.

Finally, the column materials that pertain to the student text pages include an Annotation Key. The Annotation Key is a key to all in-text questions, as well as a key to all thinking skills employed on a particular text page. You will note that the student pages in the Annotated Teacher's Edition contain small numbers in red beside each in-text or caption question. The same number, in red, is found in the Annotation Key with the correct answer. In parentheses after the answer is the science skill that the student must employ to answer the question. You will also note small numbers in blue next to paragraphs on the student pages of the Annotated Teacher's Edition. These blue numbers are repeated in the Annotation Key and alert the teacher to the types of thinking skills that the student must use to read, grasp, and understand the facts and concepts presented in the chapter. Each two-page spread in the text material for each chapter contains an Annotation Key. In this way, all answers and thinking skills employed on the spread in the student book are immediately visible and answered on the accompanying teacher page. No flipping between pages is necessary to find all the answers and thinking skills.

Below the student pages in the Annotated Teacher's Edition is an area boxed in red. This area is reserved for teaching strategies. Each numbered section in the text begins with a teaching strategy called Motivation. The strategy employed in Motivation is to present students with an activity, a thought question, a demonstration, or some type of teaching tool to help interest and motivate students. Motivation strategies are also employed throughout the section. Following the motivational ideas that begin each section is a teaching strategy called Content Development. Content Development information is designed to help the teacher teach the basic facts and concepts presented in the chapter. Like Motivation strategies, Content Development ideas are interspersed throughout each section. Another feature that is located in the teaching strategy box is called Skills Development. Under this title is a listing of the skills that will be developed, followed by an activity or a set of questions to help test basic science skills. Yet another strategy employed in the teaching strategy box is called Reinforcement. These features help the teacher reinforce basic concepts and facts for students who may not always grasp the facts and concepts without some extra help. For students who are highly academic, a feature called Enrichment is also interspersed throughout each chapter. These enrichment ideas provide the teacher with a way to help academic students go beyond the text material. Finally, at the end of each numbered section in the text are the answers to Section Review questions presented in the textbook. Section Review answers are set in red type for easy reference.

As you look over the student pages in your Annotated Teacher's Edition, you will notice that certain sentences have been shaded. This shading provides teachers with the most important topic sentences and concepts that are being taught in a particular chapter. The shading basically provides an outline of key facts and figures in each chapter.

## LABORATORY ACTIVITY ANNOTATIONS

The Laboratory Activities in the *Prentice Hall General Science* textbook are always located at the back of the chapter, just prior to the chapter summary materials. In the Annotated Teacher's Edition, a great deal of extra information is provided for the teacher on each activity. All such information is provided on blue-tinted pages. The first feature for all lab activities is called Before the Lab. This tells the teacher when to prepare any materials for the activity and any special instructions that may be necessary prior to the activity. The next feature is called Pre-lab Discussion. This section tells the teacher which concepts presented in the chapter should be reviewed prior to the activity, as well as other general information to be discussed with students prior to their doing the activity. Variables and hypotheses are brought out in the Pre-lab Discussion, whenever applicable.

Following the Pre-lab Discussion is a section called Skills Development. This section alerts the teacher to which skills students will employ while completing the activity. After the Skills Development feature is a section called Safety Tips. While the student text includes safety symbol alerts, this Safety Tips section alerts the teacher to any potential safety problems that may occur if students do not follow accepted safety practices in the laboratory.

A section called Teaching Strategy for Lab Procedure is next. It provides the teacher, when necessary, with a strategy to be employed while students complete the activity. Next comes a section called Observations and Conclusions, in which answers to questions in the activity are provided.

Finally, for each Activity there is a section called Going Further: Enrichment. Based on the lab work performed, this section may include additional activities, critical thinking questions, and application questions, as well as other enrichment ideas.

## END-OF-CHAPTER QUESTIONS

The last two pages in a chapter contain end-of-chapter questions. Answers to all multiple choice, completion, true or false, skill building, and essay questions are provided in the Annotated Teacher's Edition. Also included in the Annotated Teacher's Edition are Additional Questions and Topic Suggestions. These may be additional activities or questions teachers may want to assign their students. Finally, a section called Issues in Science is included in the Annotated Teacher's Edition. This section provides one or more issues that can be discussed, debated, or assigned for homework.

## SCIENCE GAZETTES

At the end of every unit are three Science Gazette articles. The first is an Adventures in Science profile of a scientist or group of scientists. The second is an Issues in Science article presenting a current scientific issue in a nonbiased manner. The third article is a Futures in Science article in which a future scenario is described. All such future scenarios are based on current scientific data on some of the advances and technologies in science that will be commonplace in the future. In the Annotated Teacher's

Edition, extra background information is presented on each Gazette article. Also included are questions, many of which are critical thinking in nature, that are based on the Gazette article. A teaching strategy for each Science Gazette as well as suggested debate topics are also included in the Annotated Teacher's Edition.

# Science Safety Guidelines

### SCIENCE SAFETY CLASSROOM DO'S AND DON'T'S

It is essential that students follow safety guidelines whenever performing a Laboratory Activity. Make sure your students read the safety section in Chapter 1. You may also want to read the following do's and don't's to your class.

#### *Do*

1. Wear protective goggles when working with chemicals, burners, or any substance that might get into your eyes. Many materials in a lab can cause injury to the eyes and even blindness.
2. Learn what to do in case of specific accidents such as getting acid in your eyes or on your skin. (Rinse acids on your body with lots of water.)
3. Before starting any Laboratory Activity or other experiment, make a list of the things that could go wrong that might hurt you. Then make a list of what you should do if the accident happens.
4. Make sure you have a fire extinguisher nearby to put out a fire.
5. Work with a friend, when you can, under the supervision of a science teacher or adult who understands lab safety rules.
6. Work in a well-ventilated area.
7. Learn how to use first aid to quickly treat burns, cuts, and bruises. Seek help if you are injured.
8. Read directions for an experiment carefully two or more times. Follow the directions exactly as they are written.

#### *Don't*

1. Mix chemicals "for the fun of it." You might produce an explosive reaction that could seriously injure you.
2. Taste, touch, or smell any chemical that you don't know for a fact is harmless. Many chemicals are poisonous.
3. Heat any chemical that you are not instructed to heat. A chemical that is harmless when cool can be dangerous when heated.
4. Heat a liquid in a closed container. The expanding gases produced may blow the container apart, injuring you.
5. Perform an experiment in which you must connect wires to house current. You could electrocute yourself. (Use dry cells instead.)
6. Tilt a test tube toward yourself or anyone else (or hold it upright) when you are heating its contents. (Always tilt the tube away from yourself and others.)
7. Perform any experiment for which you do not have written instructions (in a text or from your teacher).

### FIELD TRIP SAFETY

This text offers a variety of field investigations, or field trips, that can be used to extend your students' studies and interests in science. Although the program outlined in this book can be successfully carried out within the confines of a classroom, the suggested field investigations have been included to provide an opportunity for your students to have firsthand evidence and experiences outside of the classroom.

1. **Site Selection:** The text suggests possible field trip sites, but it is up to you to select an appropriate field trip site for your students, depending upon your locale. To aid you in making the necessary arrangements, we suggest that you make a school file of available sites in your area. The file should contain specific instructions concerning who to contact at the site, directions or a map to the site, fees (if any), the hours the site is open to the public, and availability of meal facilities and restrooms.

It is suggested that you make a visit to the site prior to the field trip to inspect the facilities. While you are there, locate and inventory work-study areas. Make a list of specific equipment your students will need and note the site restrictions and danger spots. Also note facilities for the handicapped if any of your students have physical limitations.

2. **Planning the Trip:** Be sure that the field trip is justified in view of the school's educational program and your individual lesson objectives. Request written permission for the trip from school personnel and keep this written permission on file. After being granted permission, send a written statement of your destination, departure and arrival times, mode of transportation, and necessary expenditures to each student's parent or guardian.

Meanwhile, provide time in class for advance research on the site. Correlate the projected field trip with your lessons and text material. Tell your students the why, where, and when of the field trip. Be sure to inform them of any special equipment or clothing they will need—special shoes, shorts, hand lenses, notebooks, and so on.

3. **The Actual Field Trip:** Make a head count of your students at each boarding and departure and periodically during the trip. Each adult should be provided with a list of the students he or she is to supervise and should remain with that group throughout the entire trip. While you are on the way to the site, discuss the investigation with your students. When you arrive at the site, keep the group together unless you have planned otherwise.

Make certain the students understand the purpose of the field trip. Have them make sketches, drawings, plans, or maps or take notes. Do not allow students to remove anything from its natural setting unless carefully selected items are taken for observation and returned to their natural habitats.

Most importantly, be enthusiastic but don't rush. Don't try to crowd too much into one field trip. Keep in contact with the individuals in the group and be alert for the "teachable moment."

4. **Follow-up Activities:** A good field trip provides a base experience for other class activities. While you are returning to the school, have students exchange ideas and discuss their experiences and observations at the site. Encourage students to ask questions and propose future activities related to the field trip. Schedule individual or group reports and have the students evaluate the trip.

Later, you may want to have your students prepare exhibits or displays using their sketches, maps, photographs, or other materials from the trip. Have them use the library to investigate questions arising from the trip. A number of library investigations can usually be proposed after a successful field trip. Remember that the learning value of a field trip depends largely on you and the type of follow-up activities you provide or encourage.

# List of Suppliers

**LABORATORY MATERIALS**

Carolina Biological Supply Company
Burlington, NC 27215

DAMON/Educational Division
115 Fourth Avenue
Needham, MA 02194

Edmund Scientific Company
103 Gloucester Pike
Barrington, NJ 08007

Fisher Scientific Company
Stansi Educational Materials Division
1259 Wood Street
Chicago, IL 60622

Hubbard Scientific Company
2855 Shermer Road
Northbrook, IL 60062

La Pine Scientific Company
6001 Knox Avenue
Chicago, IL 60018

Prentice-Hall Equipment Division
10 Oriskany Drive
Tonawanda, NY 14150

Sargeant-Welch Scientific Company
7300 North Linder Avenue
Skokie, IL 60076

Science Kit, Inc.
777 East Park Drive
Tonawanda, NY 14150

Scientific Glass Apparatus Company
737 Broad Street
Bloomfield, NJ 07003

Turtox/Cambosco
Macmillan Science Company, Inc.
8200 South Hoyne Avenue
Chicago, IL 60620

## A-V SUPPLIERS

**BFA**
Phoenix/BFA Films
468 Park Avenue South
New York, NY 10016

**Cor**
Coronet/MTI Film and Video
108 Wilmot Road
Deerfield, IL 60015

**EBE**
Encyclopedia Britannica
Educational Corporation
425 N. Michigan Avenue
Chicago, IL 60611

**Eye Gate**
Eye Gate Media
146-01 Archer Avenue
Jamaica, NY 11435

**NGS**
National Geographic Society
Educational Services
Dept. 79
Washington, DC 20036

Guidance Associates/Center for the Humanities
90 South Bedford Street
Mt. Kisco, NY 10546
*Note: Materials from* **PH Media** *can now be ordered from this supplier.*

**Walt Disney**
Walt Disney Educational Media Company
500 South Buena Vista Street
Burbank, CA 91521

## COMPUTER SOFTWARE SUPPLIERS

Carolina Biological Supply Company
2700 York Road
Burlington, NC 27215

Datatech Software Systems
19312 East Eldorado Drive
Aurora, CA 80013

Educational Dimensions Group
P.O. Box 126
Stamford, CT 06904

Focus Media, Inc.
839 Stewart Avenue
Garden City, NY 11530

Prentice-Hall, Inc.
Prentice Hall School Division
Englewood Cliffs, NJ 07632

# Comprehensive List of Laboratory Materials

| Item | Quantities per group | Chapter |
|---|---|---|
| Alum | 50 g | 18, 19 |
| Aluminum foil | 100 m sheet | 10 |
| Antacid, or seltzer, tablet | 1 | 20 |
| Balance, triple-beam | 1 | 14, 17, 19 |
| Ball bearings | 4 | 22 |
| Barium nitrate solution, 0.5M | 4 mL | 15 |
| Beaker | | |
| 100 mL | 4 | 20 |
| 250 mL | 3 | 14, 19, 21, 22 |
| 500 mL | 1 | 17 |
| Borax | 25 g | 18 |
| Bread | 2 slices | 1 |
| Broom | 1 | 13 |
| Bunsen burner | 1 | 15, 21, 23 |
| Calcium nitrate solution, 0.5M | 4 mL | 15 |
| Can, metal, with secure seal | 1 | 23 |
| Clay, modeling | small piece | 9, 22 |
| Cookbooks | 3 | 4 |
| Copper sulfate | 25 g | 18 |
| Cover slip | 1 | 8 |
| | 4 | 3 |
| Dental floss | 50 cm | 18 |
| Dictionary or encyclopedia | 1 | 4 |
| Dissecting needle | 1 | 8, 11 |
| *Elodea* or other aquatic plant | 8 | 5 |
| Ferric nitrate, $Fe(NO_3)_3$ | 2.4 g | 14 |
| Fertilizer or plant food | small amount | 5 |
| Filter paper containing slime mold | 1 | 8 |
| Flask, or peanut butter jar with screw top | 1 | 14 |
| Food coloring | | |
| blue, green, red, yellow | 1 bottle of each | 16 |
| Forceps | 1 | 8 |
| Funnel, glass | 1 | 21 |
| Glucose | 5.4 g | 17 |
| Graduated cylinders | | |
| 50 mL | 1 | 15 |
| 100 mL | 1 | 19, 20 |
| 250 mL | 1 | 15, 17 |
| Hydrochloric acid, 6M | 50 mL | 15 |
| Ice, crushed | 150 mL | 17 |
| Isopods | 10 | 10 |

| Item | Quantities per group | Chapter |
|---|---|---|
| Jar | 1 | 10 |
| with lid | 2 | 1 |
| 2-L wide-mouthed | 2 | 5 |
| Lab coat | 1 per student | 15 |
| Lithium nitrate solution, 0.5M | 4 mL | 15 |
| Magnifying glass | 1 | 11, 18 |
| Mammal photographs | 2 to 3 sets (14 classes) | 12 |
| Marshmallows, large | 25 | 16 |
| Medicine dropper | 1 | 8, 9 |
| | 4 | 1 |
| Menu, school lunch | 1 | 2 |
| Metric ruler | 1 | 11, 13, 19, 22 |
| Microscope | 1 | 3, 8 |
| Newspaper, small piece | 1 | 3 |
| Oatmeal flakes | small box | 8 |
| Objects of various masses | 1 set of 3 | 13 |
| Oven mitten | 1 | 23 |
| Owl pellet | 1 | 11 |
| Paper | | |
| construction, white (20 cm × 28 cm) | 14 sheets | 12 |
| white (20 cm × 28 cm) | 13 sheets | 2, 11, 20 |
| Paper towel | 1 | 9 |
| | few sheets | 10 |
| Pencil | | |
| glass-marking | 1 | 7, 9, 17, 18, 19, 20 |
| Petri dish | 5 | 9, 18 |
| with sterile nutrient agar | 1 | 8 |
| | 5 | 7 |
| Photograph, magazine | 1 | 3 |
| Plastic container | 1 | 19 |
| Potassium nitrate solution, 0.5M | 4 mL | 15 |
| Potassium thiocyanate, KSCN | 1 g | 14 |
| Reference book | | |
| on mammals | 1 | 12 |
| on nutrition | 1 | 2 |
| Rock salt | few crystals | 14 |
| Rubber gloves | 1 pair per student | 15 |
| Safety goggles | 1 pair per student | 15, 21, 23 |

| Item | Quantities per group | Chapter |
|---|---|---|
| Salt | | |
| coarse | 100 mL | 17 |
| table | 25 g | 18 |
| Salt water, 3 different concentrations | 200 mL of each | 22 |
| Sand | 1 small bag | 19 |
| Scissors | 1 | 9 |
| Seeds, corn | 4 | 9 |
| Shoe box with lid | 1 | 10 |
| | 3 | 13 |
| Slide, microscope | 1 | 8 |
| | 4 | 3 |
| Soap | 1 bar | 7 |
| Sodium chloride | 1.8 g | 17 |
| Sodium nitrate solution, 0.5M | 4 mL | 15 |
| Stirring rod | 1 | 18, 19 |
| Straw, plastic drinking | 1 | 22 |
| Strontium nitrate solution, 0.5M | 4 mL | 15 |
| Sucrose (table sugar) | 35.2 g | 17, 18 |
| Tape | | |
| masking | 1 small roll | 8, 9, 10 |
| transparent | 1 small roll | 12 |
| Teaspoon | 1 | 5 |
| Test tubes | | |
| small | 1 | 14 |
| medium with stoppers | 7 | 15 |
| large with stoppers | 8 | 14, 17 |
| Test tube brush | 1 | 15 |
| Test tube clamp | 1 | 15 |
| Test tube holder | 1 | 15 |
| Test tube rack | 1 | 15, 17 |
| Thermometer, $-10°C$ to $100°C$ | 1 | 17 |
| Toothpicks | 1 small box | 16 |
| Tripod | 1 | 21, 23 |
| Water | | |
| distilled | 1 L | 15 |
| pond | 4 L | 5 |
| tap | 3.015 L | 14, 15, 17, 18, 20, 21, 22, 23 |
| Wire gauze | 1 | 21, 23 |
| Wire loop with wooden handle | 1 | 15 |

PRENTICE HALL

# GENERAL SCIENCE

## A Voyage of Adventure

PRENTICE HALL

# GENERAL SCIENCE

## A Voyage of Adventure

**Dean Hurd**
Physical Science Instructor
Carlsbad High School
Carlsbad, California

**Charles William McLaughlin**
Chemistry Instructor
Central High School
St. Joseph, Missouri

**Susan M. Johnson**
Associate Professor of Biology
Ball State University
Muncie, Indiana

**Edward Benjamin Snyder**
Earth Science Instructor
Yorktown High School
Yorktown Heights, New York

**George F. Matthias**
Earth Science Instructor
Croton-Harmon High School
Croton-on-Hudson, New York

**Jill D. Wright**
Professor of Science Education
University of Tennessee
Knoxville, Tennessee

PRENTICE HALL
Englewood Cliffs, New Jersey
Needham, Massachusetts

## Prentice Hall General Science Program: Third Edition

Student Text and Annotated Teacher's Edition
Laboratory Manual and Annotated Teacher's Edition
Teacher's Resource Book
General Science Color Transparencies
General Science Courseware
Computer Test Bank with DIAL-A-TEST™ Service

## Other programs in this series

Prentice Hall General Science *A Voyage of Discovery* © 1992
Prentice Hall General Science *A Voyage of Exploration* © 1992

### General Science Reviewers


John K. Bennett
Science Specialist
State Department of Tennessee
Cookeville, Tennessee

Sue Teachey Bowden
Department of Science Education
East Carolina University
Greenville, North Carolina

Edith H. Gladden
Curriculum Specialist
Division of Science Education
Philadelphia, Pennsylvania

Gordon Neal Hopp
Carmel Junior High School
Carmel, Indiana

Stanley Mulak
Science Supervisor
Springfield School District
Springfield, Massachusetts

Richard Myers
Science Instructor
Cleveland High School
Portland, Oregon


### Reading Consultant


Patricia N. Schwab
Chairman, Department of Education
Guilford College
Greensboro, North Carolina





ISBN 0-13-717869-7

10 9 8 7 6 5 4 3 2 1

Prentice-Hall of Australia, Pty. Ltd., Sydney
Prentice-Hall Canada Inc., Toronto
Prentice-Hall Hispanoamericana, S.A., Mexico
Prentice-Hall of India Private Ltd., New Delhi
Prentice-Hall International (UK) Limited, London
Prentice-Hall of Japan, Inc., Tokyo
Prentice-Hall of Southeast Asia Pte. Ltd., Singapore
Editora Prentice Hall do Brasil Ltda., Rio de Janeiro



**Photograph credits begin on page 573**


Cover Photographs

The three main branches of science studied in a General Science course are illustrated on the cover. The Green frog on the leaf represents Life Science. *(Michel Tcherevkoff, Image Bank)* The launch of *Apollo 15* from the Kennedy Space Center represents Physical Science. *(Photri, The Stock Market)* The stream running through a winter woodland represents Earth Science. *(Sorensen/Bohmer Olse, Tony Stone Worldwide)*

Back Cover Photographs

Top center, Bryon Crader/*Tom Stack & Associates;* Top right, O.S. Pettingill, Jr./*Photo Researchers;* Bottom left, Center for Astrophysics; Bottom center, Breck P. Kent/*Earth Scenes;* Bottom right, Nicholas Devore/*dpi*

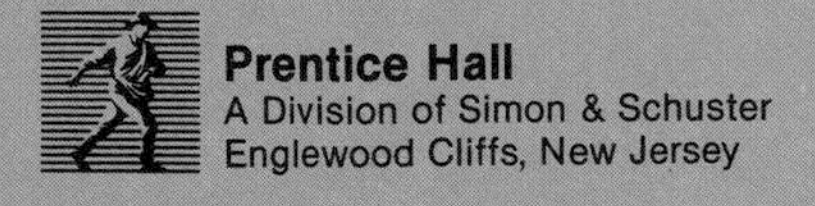


**Prentice Hall**
A Division of Simon & Schuster
Englewood Cliffs, New Jersey

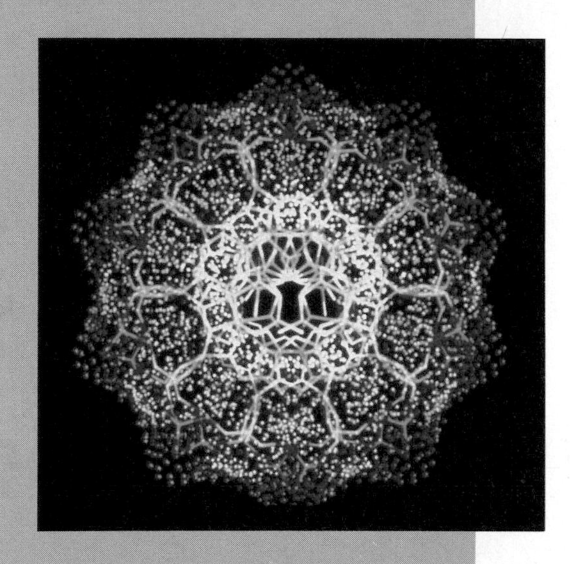

# Contents

## UNIT TWO
## Classification of Living Things 128–309



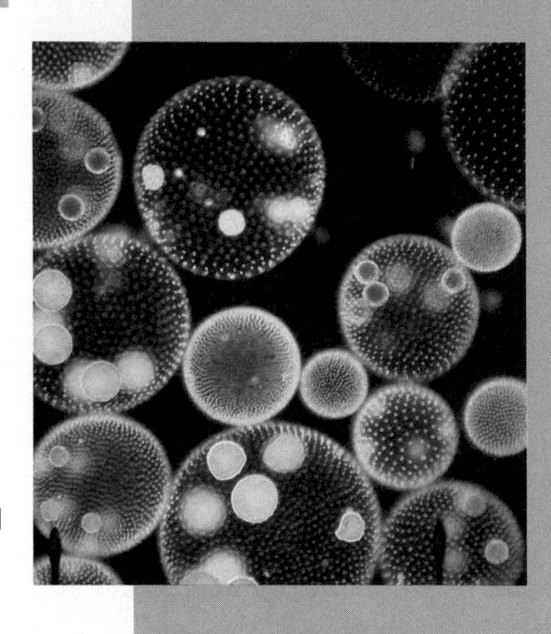

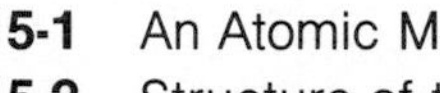

## UNIT SIX

## Structure of the Earth 476–549

# Unit One

# CHARACTERISTICS OF LIVING THINGS

## UNIT OVERVIEW

In Unit One, students are introduced first to the nature of science and to scientific method and measurement, as well as to the tools used by scientists. They next study the characteristics, needs, and chemistry of living things. They find out about the structures and functions of cells, the differences between plant and animal cells, and the reproductive processes of mitosis and meiosis. They read about division of labor and the five levels of organization in living things. Finally, they gain an understanding of the interactions among living things, as they study ecosystems, food and energy relationships, competition, and symbiosis.

## UNIT OBJECTIVES

1. **Describe the scientific method and explain and make use of the metric system of scientific measurement.**
2. **Discuss the basic characteristics, needs, metabolism, and chemistry of living things.**
3. **Describe the structures, functions, and processes characteristic of cells.**
4. **Explain the concept of division of labor in living things.**
5. **Describe the five levels of organization of living things.**
6. **Describe the basic interactions and relationships among living things.**

10

## INTRODUCING UNIT ONE

Begin your teaching of the unit by having students examine the unit-opening illustration, which is an artist's rendering of the ocean floor at great depths. You may wish to see whether the students can infer the answers to the following questions before they read the unit introduction.

- **What is the vessel shown in the picture?** (It is a special submarine.)
- **What is the submarine doing?** (It is gathering specimens of organisms from the ocean floor.)
- **What are these organisms?** (They include clams, mussels, and crabs.)
- **Why is it necessary for the submarine to direct beams of light at the ocean floor?** (No sunlight is able to penetrate to such depths.)
- **What pressure conditions do you think exist in such a place?** (Pressure is extremely high.)
- **What would happen to an unprotected human diver at such depths?** (He or she would be crushed to death because of the water pressure.)
- **What is the red material on the right side of the picture?** (It is magma from within the earth.)
- **What temperature conditions does it create nearby?** (Temperatures are very high.)

Now have students read the unit

# Characteristics of Living Things

The tiny submarine slips deeper and deeper into the ocean. Inside, three scientists prepare for their arrival on the sea floor. Outside the submarine, there is nothing but darkness. With a soft bump, the submarine touches bottom. The depth gauge shows 2590 meters. The water pressure is a crushing 264 kilograms per square centimeter. Lava oozes from cracks in the ocean floor. The lava heats the water to a blistering 371° C, hot enough to melt lead.

Suddenly the floodlights of the submarine flash on. The view is astounding. Giant clams and pale yellow mussels lie on the sand. A white crab walks by. The scientists wonder how living things can exist in such a harsh place. And they realize with excitement that finding out will be a great adventure. This is an adventure made possible by thousands of years of curiosity and learning about the nature of living things.

CHAPTERS



*Amazing life forms of the ocean bottom*



## CHAPTER DESCRIPTIONS

**1 Exploring Living Things** Chapter 1 begins with a discussion of the scientific method. The metric system is explained next. Microscopes and other instruments used by biologists are described. Finally, the importance of proper laboratory safety procedures is discussed.

**2 The Nature of Life** In Chapter 2, the basic characteristics of living things are described. Metabolism and metabolic activities are then explained, as are the basic needs of organisms. The chapter closes with a treatment of the chemistry of life, the distinction between elements and compounds, and organic substances.

**3 Cells** The various living and nonliving structures in plant and animal cells are discussed in Chapter 3. Diffusion and osmosis are compared. The role of reproduction is examined, and the processes of mitosis and meiosis are contrasted.

**4 Tissues, Organs, and Organ Systems** Chapter 4 introduces the concept of division of labor within living things. The five levels of organization—cells, tissues, organs, systems, and organisms—are explained next. The dependence among these levels of organization is also analyzed.

**5 Interactions Among Living Things** In Chapter 5, living things are discussed in terms of their environment and relationships. Ecosystems, communities, populations, habitats, and niches are explained, as are food chains, food webs, and energy pyramids. Competition, together with the symbiotic relationships of commensalism, mutualism, and parasitism are also described.

introduction, and then ask them the following questions.

• **How are the living things able to withstand these conditions?** (They possess various special adaptations that enable them to do so—for example, their cell pressures are very great, to counteract outside pressures.)

• **What would happen if these living things were brought up to the surface without protection?** (They would probably not survive under the greatly changed conditions—for example, cells within them might burst, since the pressure they exert outward is no longer balanced by the external pressure.)

## Chapter 1

# EXPLORING LIVING THINGS

## CHAPTER OVERVIEW

In this chapter, students will be introduced to the study of science. They will learn how scientists work and how scientific problems are solved.

Students will discover that scientists solve problems in a systematic way. The series of steps that scientists use is called a scientific method.

Students will learn that the common language of measurement in science is the metric system. They will be introduced to the metric units for length, volume, mass, weight, and temperature.

Students will discover that scientists use various tools in their work. Among these are the microscope, the computer, and X-rays.

In the last section of this chapter, students will be introduced to the basic rules of laboratory safety. They will learn how to identify symbols of laboratory safety used in this text and lab manual, and how to cope with laboratory emergencies.

## INTRODUCING CHAPTER 1

Begin by having students observe the chapter-opener photo, then ask,

- **What do you see in this picture?** (a strange, lizardlike creature, very large, with striped markings on its back, and back legs that are much longer than its front legs; the creature also appears to have many sharp teeth)
- **What do you see in the landscape surrounding the animal?** (palm trees, other trees and bushes, a small waterfall, mountains in the far background, rocks, sand or dirt, and small weeds or grasses)

Direct students' attention to the chapter-opener text and ask,

- **What did scientists at Rocky Hill find beneath the fallen building?** (unusual scratches on the exposed rock)
- **What did the scientists suspect about these scratches?** (that they might be related to dinosaurs)

Point out to students that these scratches represent a type of fossil. A fossil is the remains or evidence of a living thing. Fossils that are formed from the footprints or other imprint of an organism—without the actual organism being embedded in the rock—are called trace fossils.

Continue by asking students,

- **What did Dr. Coombs know about the scratches right away?** (that they were dinosaur footprints)

Emphasize to students that Dr. Coombs also observed many other things about the scratches. Have students assist you in listing his observations on the chalkboard:

*Observations:*

1. Scratches in groups of 3
2. Scratches 130 cm apart
3. Only tips of dinosaur's toes touched the rock

# 1 Exploring Living Things

**CHAPTER SECTIONS**

**1-1 What Is Science?**
**1-2 Scientific Measurements**
**1-3 Tools of a Scientist**
**1-4 Science Safety in the Laboratory**

**CHAPTER OBJECTIVES**

*After completing this chapter, you will be able to:*

**1-1** Describe the various steps of a scientific method.
**1-2** Make and understand metric measurements.
**1-3** Describe several tools used by scientists.
**1-3** Compare different types of microscopes.
**1-4** Apply safety procedures in the classroom laboratory.

A howling wind shook the walls of the wooden temporary building at Rocky Hill, Connecticut. Torrents of rain swept over the building's roof. With a groan, the building suddenly fell apart.

Unknown to the scientists at Rocky Hill, where the Connecticut State Dinosaur Park is located, the collapse of the building was soon to lead to an unexpected discovery. For later, when the earth was moved to make room for the new building, strange scratches were found on the exposed rock. An expert on dinosaurs, Dr. W. P. Coombs, Jr., of Western New England College in Massachusetts, was contacted.

Dr. Coombs took one look at the mysterious scratches and immediately knew what they were—dinosaur footprints. But there was more to the story. The scratches appeared in groups of three. Dr. Coombs concluded the scratches were made by an animal having three toes with sharp claws. They were clearly the marks of a meat-eating dinosaur.

Carefully, the scientist measured the distance between the footprints—130 centimeters. Dr. Coombs reasoned that only a very large animal, perhaps the 7-meter-long *Megalosaurus,* could take such long strides. There was something very peculiar about the footprints, however. Only the tips of the dinosaur's toes seemed to have touched the rock. But *Megalosaurus* did not run on its toes, at least not on land.

Dr. Coombs could draw only one conclusion. The prints were made under water, which kept most of the animal's weight off the rock. Dr. Coombs had discovered the first evidence of swimming, meat-eating dinosaurs. A simple accident, a sharp eye, and some smart detective work had led to a scientific discovery.

*Megalosaurus—the first swimming, meat-eating dinosaur*



## TEACHER DEMONSTRATION

Begin by asking students,

• **How far apart were the dinosaur footprints described in the chapter opener text?** (130 cm)

Have one or two students begin at one end of the classroom and measure a distance of 130 cm. Have the students make a chalk mark. Then have the students continue to make marks 130 cm apart across the room. Point out that these chalk marks represent dinosaur footprints.

• **How does the distance between dinosaur footprints compare with the size of the footsteps you usually take?** (The distance between dinosaur prints is much greater.)

Have students team up with a partner and measure the size of each other's normal footsteps. Have several students record their results on the chalkboard.

• **On the average, how does the size of your footstep compare with the size of a dinosaur's footstep?** (It should be 1/5–1/4 as great.)

Now have students rejoin their partners and measure and record each other's heights.

• **How long was the dinosaur described in the chapter opener?** (7 meters long)

• **On the average, how do your heights compare with the length of the dinosaur?** (They should be about 1/5–1/4 as great.)

• **What do you notice about the height and footstep ratios that you just calculated?** (They are quite similar.)

• **Can you propose a relationship between the size of an organism and the length of its footstep?** (The longer or taller an organism is, the longer will be its footstep.)

For each observation, have students note the conclusions drawn:

*Observation 1:* The dinosaur had 3 toes and sharp claws. Therefore, the dinosaur must have been a meat-eating dinosaur.

*Observation 2:* Animal must have been large, probably a Megalosaurus.

*Observation 3:* Animal could have walked on its toes, but the Megalosaurus did not do this on land. Therefore, the footprints must have been made under water.

• **Based on his observations and conclusions, what did Dr. Coombs finally conclude about the origin of the footprints?** (They were made by a swimming, meat-eating dinosaur—the first evidence of any such animal to be discovered.)

## TEACHER RESOURCES

**Audiovisuals**

*The Scientific Method,* film, EBE
*What Is Science?,* film, Cor

**Books**

Dixon, B., *What Is Science For?* Collins
Samuel, E., *Order in Life,* Prentice-Hall

## 1-1 WHAT IS SCIENCE?

### SECTION PREVIEW 1-1

In this section, students will be introduced to the study of science. They will learn that scientists uncover truths about nature called facts. Then scientists use facts to solve larger mysteries of nature.

When scientists work, they use a logical series of steps called a scientific method. Students will follow an example in the text to see how a typical problem in life science is solved using a scientific method.

In the last part of the section, students will be introduced to the three main branches of science—life science, earth science, and physical science. They will learn that within each branch are many smaller branches.

### SECTION OBJECTIVES 1-1

1. **Define the term fact.**
2. **Describe the various steps in a scientific method.**
3. **Explain how a scientific experiment is designed.**
4. **Discuss the main branches of science.**

### SCIENCE TERMS 1-1

**scientific method p. 15**
**hypothesis p. 16**
**variable p. 17**
**control p. 18**
**data p. 18**
**zoology p. 19**
**botany p. 19**
**astronomy p. 19**
**chemistry p. 19**
**physics p. 19**

## 1-1 What Is Science?

The universe around you and inside of you is really a collection of countless mysteries. It is the job of scientists to solve those mysteries. And, like any good detective, a scientist uses special methods to find truths about nature.

These truths are called facts. An example of a fact is that the earth is populated with millions of different kinds of living things. But science is more than a list of facts. Jules Henri Poincaré, a famous nineteenth-century French scientist, put it this way: "Science is built up with facts, as a house is with stones. But a collection of facts is no more a science than a heap of stones is a house."

So scientists go further than simply discovering facts. Scientists try to use facts to solve larger mysteries of nature. In this sense, you might think of facts as clues to scientific mysteries. An example of one of these larger mysteries is how the relatively few and simple organisms of three billion years ago gave rise to the many complex organisms that inhabit the earth today.

**Figure 1-1** *It is a fact that this red diamondback rattlesnake injects poison into its prey. It is a hypothesis that the rattler locates its injured victim by following the smell of its own venom.*

### TEACHING STRATEGY 1-1

#### Motivation

Have students observe the classroom and its contents for several minutes, then ask,

- **Based on your observations, can you state 10 facts about our classroom?** (Answers will vary; possible observations include: color of walls, number of windows, dimensions, number of chairs and desks, shape of the room, and so on.)
- **How do you know that these things are true?** (You can see, touch, measure, or count them.)

#### Content Development

Use the Motivation activity to introduce the idea that a fact is a piece of information that can be verified. We know, for example, that the classroom is x meters wide because we can take a meterstick and check it.

Point out that science is concerned with facts about nature. These facts may relate to living things, or they may relate to matter that is not part of any living thing. Scientific facts may also describe energy and the interactions that occur between matter and energy.

Emphasize that scientists do more than uncover facts; they seek to explain why and how things happen as they do. For example, a scientist may wonder why two children in the same family have different color hair and eyes, or why a candle goes out in a closed container.

- **Which do you think is harder to**

### *Scientific Methods*

To uncover scientific facts and solve scientific mysteries, scientists can use any one of a number of **scientific methods.** There are various basic steps in these methods. But these steps need not be followed in any particular order, although some orders make more sense than others. Sometimes the order in which a scientist tries to solve a mystery depends on the nature of the mystery. **The basic parts of any scientific method are the following:**

**Stating the problem**

**Gathering information**

**Suggesting an answer for the problem**

**Performing an experiment to see whether the suggested answer makes sense** ❶

**Recording and analyzing the results of experiments or other observations**

**Stating conclusions**

The following example shows how a scientific method was used to solve an actual problem. As you will see, the basic steps of a scientific method often overlap.

### *Stating the Problem*

Most people know enough to walk the other way if they should run into a rattlesnake. However, if you could safely observe a rattler, you would discover a rather curious kind of behavior.

With fangs flashing and body arching, the deadly rattler strikes. The snake's fangs quickly inject poisonous venom into its victim. Then, in a surprise move, the rattler allows the wounded animal to run away! But the rattlesnake will not miss its intended meal. After waiting for its poison to take effect, the rattler follows the trail of the injured animal. ❷

Although the rattler cannot see well, somehow it manages to find its victim on the dense, dark, forest floor. Clearly, something leads the snake to its prey. What invisible trail does the snake follow in tracking down its bitten prey? This is a *problem* that scientists recently tried to solve.

**Activity**

*Theories and Laws*

Scientists sometimes use observations and experiments to develop a theory. A theory is a broad scientific explanation for things that happen in nature. If the theory is tested and confirmed, it may become a law.

Using books and reference materials in the library, find out about a theory in life science that became a law.



## TIE-IN/LANGUAGE ARTS

The word *science* comes from the Latin word *scire,* which means "to know."

## ANNOTATION KEY

❶ Thinking Skill: Sequencing events

❷ Thinking Skill: Interpreting observations

## Activity

**Theories and Laws**

**Skills: Applying definitions, writing reports**

**Level: Average**

**Type: Library/vocabulary comprehension**

This simple activity will help reinforce the concept that a theory or law may be accepted as true by the scientific community for a time, but that theories and laws can change as more information is revealed. Written or oral reports should be well organized and present findings in a logical, consistent manner.

**discover—facts, or the answers to questions involving "how and why"?** (Answers may vary; many may feel that facts are easier to discover, yet certain facts may be very difficult to discover.)

- **What types of facts might be very hard to discover?** (Facts about things that we cannot see or touch, such as atoms, the interior of the earth, or heavenly bodies in distant galaxies.)

### Reinforcement

Challenge students to see how many scientific facts they can list in 10 minutes. Some possible examples include: radius of the earth, distance to the moon, rising and setting of the sun, shape of a snowflake, normal temperature of the human body, composition of water.

### Content Development

Point out that when a person wonders why or how something happens and decides to investigate, he or she is a scientist. Tell the students that scientists are detectives. Explain that scientists are continually involved in solving problems. Their activity, or work, is directed toward one goal, the solution to the problem.

Explain that scientists must collect all possible evidence related to the problem. Point out that some of the evidence might be quite strange, seem confusing, and appear unrelated. It is then the job of scientists to "think" through the information and decide what the next step should be. Explain that scientists attempt to match and correlate the evidence.

## Activity

**Making Observations**
**Skills: Observing, comparing, diagraming, inferring**
**Level: Remedial**
**Type: Hands-on**
**Materials: Ten similar objects, such as tree leaves**

This activity will help students understand the importance of careful observations. Only if students write down their observations carefully will other students be able to discover the object they selected based on their observations. Make sure you check all observations and drawings carefully. Once the activity has been completed, relate it to the importance of observations in the scientific method.

**Figure 1-2** *The rattler does not have an especially keen sense of sight. Its pit organs detect the body heat given off by this field mouse.*

### Activity

*Making Observations*

**1.** Collect ten similar objects, such as maple leaves, string beans, or blades of grass.

**2.** Choose one object from your collection and observe it carefully.

**3.** Write down all of your observations and draw a diagram of the object.

**4.** Put all ten objects together and mix them up.

**5.** Now see if a classmate can pick out your object using your observations.

Which observations were helpful in identifying your object?

### Gathering Information on the Problem

The first step in solving a scientific problem is to find out or review everything important related to it. For example, the scientists trying to solve the rattlesnake mystery knew that a rattlesnake's eyes are only sensitive to visible light. However, they also knew that a pair of organs located under the animal's eyes detects invisible light in the form of heat. These heat-sensing pits pick up signals from warmblooded animals. The signals help the snake to locate its intended prey. But the heat-sensing pits cannot help the snake find a wounded victim that has run many meters away. Some other process must be responsible for that.

The scientists knew that a rattler's tongue "smells" certain odors in the air. The rattler's tongue picks up these odors on an outward flick. The odors enter the snake's mouth on an inward flick. The scientists also knew that the sight or smell of an unbitten animal did not trigger the rattler's tracking action. Using all this information, the scientists were able to suggest a solution to the problem.

### Forming a Hypothesis

A suggested solution is called a **hypothesis** (high-PAH-thuh-sis). A hypothesis is almost always formed after the information related to the problem is carefully studied. But sometimes a hypothesis is the

16

## 1-1 (continued)

### Motivation

Show the class a picture of a plant or animal that is not common to your area. Discuss the picture.

- **What do you know about this living thing?** (Accept all answers.)
- **What would you like to know about this living thing?** (Accept all answers.)
- **What do you think scientists know about this living thing?** (Accept all answers.)
- **What do you think scientists might *not* know about this living thing?** (Accept all answers.)
- **How might a scientist find out the information that was not known?** (Accept all answers.)

### Content Development

Emphasize to students that a scientific problem is often stated in the form of a question. You can illustrate this concept by posing the following situation:

Suppose that you have three gardens in your yard–one on the east side, one on the south side, and one on the west side. The plants are all of the same type, but you notice that the plants on the south side are at least 5 cm taller on the average than plants on the east side or the west side.

- **Can you recognize the scientific problem in this situation?** (the observation that plants on the south side are taller than plants on the east side or the west side)
- **Can you state this problem in the form of a question?** (Why are plants on the south side taller on the average than plants on the west side or the east side?)

### Skills Development

***Skills: Observing, relating facts***

Have students work in small groups. Ask each group to choose a familiar outdoor area such as a part of the school grounds, their own backyard, or a portion of a park or woods. Have students observe the area they

result of creative thinking that often involves bold, original guesses about the problem. In this regard, forming a hypothesis is like good detective work, which involves not only logic, but hunches, intuition, and the taking of chances.

To the problem, "What invisible trail does a rattler follow in tracking down its prey?" the scientists suggested a hypothesis. The scientists suggested that *after the snake wounds its victim, the snake follows the smell of its own venom to locate the animal.* ❷

### Experimenting

The scientists next had to test their hypothesis by performing certain activities and recording the results. These activities are called experiments. Whenever scientists test a hypothesis using an experiment, they must make sure that the results of the experiment clearly support or do not support the hypothesis. That is, they must make sure that one, and only one, factor affects the results of the experiment. The factor being tested in an experiment is called the **variable.** In any experiment, only one variable is tested at a time. Otherwise it would not be clear which variable had caused the results of the experiment.

**Activity**

*Early Scientists*

The science of biology owes a great deal to the work of many early scientists. Using books and reference materials in the library, look up the following names.

Anton van Leeuwenhoek

Robert Hooke

Leonardo da Vinci

Gerty Radnitz Cori

Write a brief report about the scientists, including information about the contribution of each.

**Figure 1-3** *Sinking its fangs into its victim, the rattler injects the mouse with deadly poison.*

**Activity**

**Early Scientists**
**Skills: Researching, report writing**
**Level: Average**
**Type: Vocabulary/Writing**

This activity helps students to develop library research skills, which are also important in science. Encourage them to bring out in their reports the aspects of scientific method used by the scientists whose contributions they are researching.

**ANNOTATION KEY**

❶ Thinking Skill: Applying facts
❷ Thinking Skill: Hypothesizing

**TIE-IN/GOVERNMENT**

Laws change in government as well as in science. For example, certain drugs that were once legally obtainable at drugstores are now illegal drugs that cannot be used under any circumstances. An interesting example is heroin, which at the turn of the century could be bought as tablets in many drugstores. Today, of course, heroin is illegal.

have chosen and list as many facts as they can. Then challenge each group to state at least one scientific problem suggested by the facts. For example, if students notice moss growing on rocks, they might ask, "How can moss grow without soil?" This activity can be extended by having students suggest hypotheses and possible experiments for the problems they have stated.

**Content Development**

Explain that after all of the clues and facts are found, a scientist can then suggest a solution. Point out that a suggested solution to a problem is called a hypothesis.

Point out that a hypothesis involves not only logic, but hunches, intuition, and the taking of chances. Tell students that a hypothesis should not only consist of all of the known facts but should also predict other things or events that are likely to occur. There are two main parts to a hypothesis: an explanation and a prediction.

**Reinforcement**

Emphasize that a hypothesis must be tested by careful experiments.

- **What do you think should be done if a hypothesis is tested and the experiments prove it false?** (Another hypothesis should be formed and also tested by experiment.)
- **Suppose an experiment yields inconclusive results—that "maybe" the hypothesis is true? What should be done?** (Answers may vary; perhaps the hypothesis needs to be stated in a more precise way, or perhaps the experiment needs to be redesigned.)

## ANNOTATION KEY

❶ Thinking Skill: Relating cause and effect

❷ Thinking Skill: Applying data

❸ Thinking Skill: Classifying

### Activity

**To Grow or Not to Grow**
**Skills: Hypothesizing, computational, designing, measuring**
**Level: Enriched**
**Type: Hands-on**
**Materials: Several potted plants, salt**

This activity will help students understand the importance of designing and conducting an experiment under careful laboratory conditions. It will also help them understand the need for a control setup to test for a particular variable. Check observations and experimental setups to be certain that students have correctly followed the scientific method. Also check their graphs for accuracy in terms of the data they obtain. Although this activity is actually intended to reinforce the scientific method rather than to obtain the correct answer, students should find that adding salt to the soil decreases plant growth and may even kill some plants.

In the rattlesnake experiments, the scientists tested whether the snake's venom formed an invisible trail that the snake followed. The venom was the variable, or single factor, that the scientists wanted to test. The scientists performed the experiment to test this variable.

First, the scientists dragged a dead mouse that had been struck and poisoned by a rattlesnake along a curving path on the bottom of the snake's empty cage. When the snake was placed in its cage, its tongue flicked rapidly, its head moved slowly from side to side, and it followed the exact trail the scientists had laid out. The results seemed clear, but the scientists had one more experiment to perform.

To be sure it was the scent of the venom and no other odor that the snake followed, the scientists ran a **control** experiment. ❶ A control experiment is run in exactly the same way as the experiment with the variable, but the variable is left out. So the scientists dragged an unbitten dead mouse along a path in the cage. The experiment was exactly the same, except this mouse had not been poisoned. This time the snake seemed disinterested. Its tongue flicked very slowly and it did not follow the path.

### Activity

*To Grow or Not to Grow*

Your friend tells you that plants grow slower when salt is added to their soil.

1. Design and conduct an experiment to find out if this is true. Make sure your experiment has a control and a variable.
2. Record all data.
3. Then make a graph of your results. You can label the horizontal axis of your graph "Time in Days" and the vertical axis "Height in Centimeters."
4. Judging from the results of your experiment, can you conclude that your friend was correct? Explain your answer.

### *Recording and Analyzing Data*

The rattlesnake experiments were repeated many times, and the scientists carefully recorded the **data** from the experiments. Data include observations such as measurements.

### *Stating a Conclusion*

After analyzing the recorded data, the scientists concluded that the scent of venom was the only factor that could cause a rattlesnake to follow its bitten victim. Rattlesnake venom is made up of many ❷ different substances. Exactly which ones are responsible for the snake's behavior are as yet unknown. As is often the case in science, a solution to one mystery brings to light another mystery. Using scientific methods similar to those described here, scientists hope to follow a path that leads to the solution to this new mystery.



## 1-1 (continued)

### Content Development

Point out that once a hypothesis is developed, the scientist must continue research and perform activities that will prove the hypothesis to be "right" or "wrong." Tell students that these activities are called experiments. Explain that all activities and/or observations must be "for" or "against" the hypothesis to be of any service. Point out that if the experiment supports the hypothesis, the hypothesis is strengthened. If the experiment contradicts the hypothesis, the hypothesis must be reviewed and possibly changed.

Explain that the experiment to test any hypothesis must check only one thing, or factor, at a time. Point out that the factor being tested in an experiment is called the variable.

### Skills Development

***Skill: Applying concepts***

Tell the class that one student had a hypothesis that a pet gerbil "preferred" to eat sunflower seeds rather than small processed animal food pellets. Discuss possible experiments that might test the hypothesis.

- **What experiments might be done to test the hypothesis?** (Accept all logical answers.)
- **What is the experimental variable?** (sunflower seeds)
- **What is the control?** (small processed animal food pellets)

### Branches of Science

Science is divided into many branches according to subject matter. Each branch is made up of a small or large topic in science. There are three main branches of science.

*Life science* Life science deals with living things and their parts and actions. Smaller branches of life science include **zoology,** the study of animals, and **botany,** the study of plants.

*Earth science* Earth science is the study of the earth and its rocks, oceans, volcanoes, earthquakes, atmosphere, and other features. Usually, the earth sciences also include **astronomy.** Astronomers explore nature beyond the earth. They study such objects as stars, planets, and moons. ③

*Physical science* Physical science is the study of matter and energy. For example, some physical scientists explore what substances are made of and how they change and combine. This branch of physical science is called **chemistry.** Other physical scientists study forms of energy such as heat and light. This is the science of **physics.**

**Figure 1-4** *Life science includes the study of animals such as these penguins who nest in Antarctica.*

**Figure 1-5** *The earth-orbiting Space Shuttle* (left) *helps gather important information about the nature of the earth and the universe beyond. The laws of motion, a part of the science of physics, explain how Olympic cyclists speed along their circular track* (right).

19

- **What other hypothesis might be suggested?** (Lead students to suggest different types of food.)
- **How might the hypothesis be tested?** (Accept all logical answers. Lead students to suggest setting up an experiment with a *single* variable and a control.)

## Skills Development

***Skill: Designing an experiment***

Have students work in small groups. Challenge each group to design an experiment to test one of the following statements:

**1.** Plants grow faster in direct sunlight than in indirect sunlight.
**2.** Water evaporates faster in warm temperatures than in cool temperatures.
**3.** Lilies open and close because of the light and darkness caused by the rising and setting sun.

Stress to students that each experiment must have only one variable; students should also design a control experiment. If time and available materials permit, you may wish to have groups attempt to carry out their experiments.

## Reinforcement

Write each of the steps of the scientific method on a set of index cards. Place the cards face down on a desk or table. Have each student pick a card at random. Challenge the student to discuss how this particular step of the scientific method was carried out in the sample problem about the rattlesnake. For example, if a student draws, "form a hypothesis," he or she might say, "The scientists suggested that after the rattlesnake wounds its victim, the snake follows the smell of its own venom to locate the animal."

## Enrichment

Point out to students that data from experiments is often presented in the form of a graph. Have students find examples of graphs in books, newspapers, and magazines. Then discuss with students how these graphs make data easier to understand and analyze. Also discuss with students different types of graphs, such as line, bar, and circle.

## Skills Development

***Skill: Making a diagram***

Have students work in small groups. Challenge each group to make a creative tree diagram that shows the various main and smaller branches of science. Encourage students to use reference materials to find additional smaller branches—such as oceanography and geology—as part of earth science.

## 1-2 SCIENTIFIC MEASUREMENTS

### SECTION PREVIEW 1-2

In this section, students will be introduced to the metric system of measurement. They will learn that the metric system is used by scientists all over the world.

Students will learn that the metric system is based on powers of ten. They will discover that calculations with metric measurements are convenient to make since each unit in the metric system is ten times larger than the previous unit.

Students will learn to distinguish between mass and weight. They will also learn about the relationship between an object's mass and its volume, which is called density.

### SECTION OBJECTIVES 1-2

1. **Discuss the importance of scientific measurements.**
2. **Identify and compare the basic units of the metric system of measurement.**
3. **Distinguish between an object's mass and its weight.**
4. **Explain what is meant by the density of an object.**

### SCIENCE TERMS 1-2

**metric system p. 20**
**meter p. 21**
**liter p. 22**
**kilogram p. 22**
**mass p. 22**
**weight p. 22**
**gravity p. 22**
**density p. 23**
**Celsius p. 24**

*SECTION REVIEW*

1. What is a hypothesis?
2. Why does an experiment have only one variable?
3. What are the three main branches of science?

## 1-2 Scientific Measurements

As part of the process of experimenting and gathering information, scientists must make accurate measurements. They must also be able to share their information with other scientists. To do this, scientists must speak the same measurement "language." **The common language of measurement in science used all over the world is the metric system.** Scientists often refer to the **metric system** as the International System of Units, or SI.

The metric system is a decimal system, or a system based on ten. That is, each unit in the metric system is ten times larger or ten times smaller than the next unit. Metric calculations are easy to do because they involve multiplying or dividing by ten. Some frequently used metric units and their abbreviations are listed in Figure 1-6.

**Figure 1-6** *The metric system is easy to use because it is based on units of ten. What is the basic unit of length in the metric system?* ❶

*COMMONLY USED METRIC UNITS*

| Length | Mass |
|---|---|
| Length is the distance from one point to another.<br>A meter is slightly longer than a yard.<br>1 meter (m) = 100 centimeters (cm)<br>1 meter = 1000 millimeters (mm)<br>1 meter = 1,000,000 micrometers (μm)<br>1 meter = 1,000,000,000 nanometers (nm)<br>1 meter = 10,000,000,000 angstroms (Å)<br>1000 meters = 1 kilometer (km) | Mass is the amount of matter in an object.<br>A gram has a mass equal to about one paper clip.<br>1 kilogram (kg) = 1000 grams (g)<br>1 gram = 1000 milligrams (mg)<br>1000 kilograms = 1 metric ton (t) |
| **Volume** | **Temperature** |
| Volume is the amount of space an object takes up.<br>A liter is slightly larger than a quart.<br>1 liter (L) = 1000 milliliters (mL) or 1000 cubic centimeters ($cm^3$) | Temperature is the measure of hotness or coldness in degrees Celsius (°C).<br>0°C = freezing point of water<br>100°C = boiling point of water |

### 1-1 (continued)

#### Section Review 1-1

**1.** A suggested solution to a problem.
**2.** To make it clear which factor caused the results of the experiment.
**3.** Life, earth, and physical science.

### TEACHING STRATEGY 1-2

#### Motivation

Write the following "recipe" on the chalkboard.

YUMMY CHOCOLATE BROWNIES
Combine together:
some butter
some sugar
a little salt
a whole lot of flour
a bunch of nuts
a lot of chocolate
a little baking powder
Mix well, heat up the oven, and bake for a long time.

• **What is wrong with this recipe?** (You can't tell how much of each ingredient to use, how long to bake the brownies, or what temperature oven to use.)

What is wrong with using such words as "some," "a lot," and "a little"? (These amounts mean differ-

### Length

The **meter** is the basic unit of length in the metric system. One meter is equal to 39.4 inches. Sometimes scientists must measure distances much longer than a meter, such as the distance a bird may fly across a continent. To do this, scientists use a unit called a kilometer. The prefix *kilo-* means 1000. So a kilometer is 1000 meters, or about the length of five city blocks.

A hummingbird, on the other hand, is too small to be measured in meters or kilometers. So scientists use the centimeter. The prefix *centi-* means that 100 of these units make a meter. Thus there are 100 centimeters in a meter. The hummingbird is only about 5 centimeters long. For objects even smaller, another division of the meter is used. The millimeter is one thousandth of a meter. The prefix *milli-* means 1/1000. One meter equals 1000 millimeters. In

**Figure 1-7** *Scientists use many tools to study their world. A metric ruler* (top, left) *measures length. A graduated cylinder* (top, right) *helps measure the volume of liquids. A triple-beam balance* (bottom, left) *may be used to measure mass. And a Celsius thermometer is used to measure temperature.*

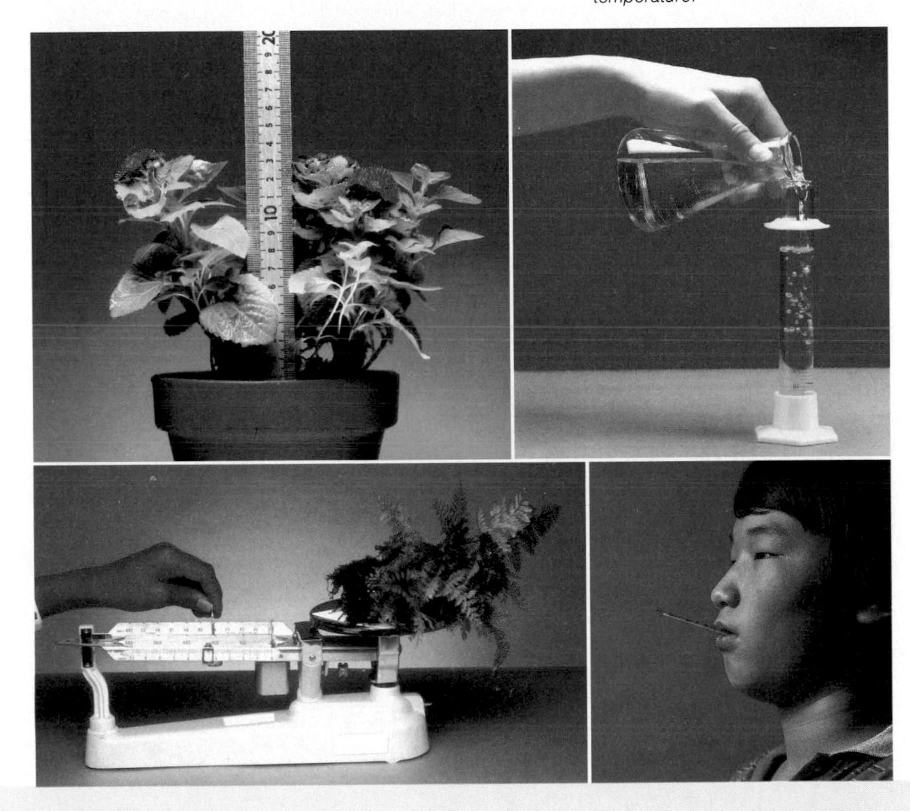

**ANNOTATION KEY**

❶ **Meter (Relating facts)**

❶ **Thinking Skill: Making comparisons**

## TIE-IN/MATH

Review with students the concept of base in a number system. Point out that the number system we use is a base-10 system, meaning that each decimal place has a value ten times greater than the previous decimal place.

ent things to different people. One might ask, "A lot relative to how much?")

### Content Development

Use the Motivation activity to emphasize the importance of standard measurements. Point out that measurements are important because we often need to know exact quantities or dimensions. We also need to communicate these exact amounts to other people.

Stress that measurements are meaningful to us because they express the sizes of things relative to a known standard. For example, if someone says that a table is two meters long, we can take a meterstick and see exactly how long that table is.

### Content Development

Point out that in order to communicate effectively, scientists must speak the same measurement language.

Tell students that the common language of measurement in science all over the world is the metric system.

Explain that the metric system is a system based on ten. Each unit in the metric system is ten times larger or ten times smaller than the next unit.

### Content Development

Write "meter (symbol m)" on the chalkboard. Point out that the meter is the basic unit of length in the metric system.

Show the students a meterstick.

Explain that to measure distances much longer than a meter, scientists use kilometers. *Kilo* means 1000. So, a kilometer is 1000 meters. Write "1000 m = 1 km" on the chalkboard.

Point out that to measure distances smaller than a meter, scientists use the centimeter. Explain that a centimeter is 1/100 of a meter. Explain that it takes 100 centimeters to make a meter. Write "1 m = 100 cm" and "1 cm = 0.01 m" on the chalkboard.

### Skills Development

***Skill: Comparing***

Point out to students that they can use decimals to compare units listed within each category in Figure 1-6. For example:

1 cm = 0.01 m
1 mm = 0.001 m
1 m = 0.001 km

## Activity

**Practicing Metric Measurements**
**Skills: Making measurements, observing, recording, relating, applying**
**Level: Average**
**Type: Computational, manipulative**
**Materials: Meterstick**

In this activity, students apply their knowledge of metric units of length to the measurement of familiar objects, whose dimensions they may be used to thinking of in terms of English units. They should also be able to find a number of products (in particular, food items) whose masses and volumes are expressed, on their containers, in metric as well as English units.

## HISTORICAL NOTES

The French Revolution brought a new government to power that decided to produce a rational set of units for all measurements—the everyday as well as those used in science and technology. They called their system the metric system. The base unit of length was the meter (from the Greek word *metron,* meaning "to measure"). They chose the meter to be one ten-millionth of the distance between the equator and the North Pole.

### Activity

*Practicing Metric Measurements*

Using a meterstick, measure the length of the following objects. Choose the proper metric prefixes for the measurements.

height of a door

diameter of a dime

length of your foot

thickness of a slice of bread

Look around your home for products whose volume is measured in liters or whose mass is measured in grams. Make a list of at least ten items. For each example, give the product and the measurement.

**Figure 1-8** *In the metric system, height or length is measured in units called meters. As you can see from this photograph, height varies from one person to another.*

bright light, the diameter of the pupil of your eye is about 1 millimeter. Even this unit is too large to use when describing the sizes of the smallest germs, such as the bacteria that cause sore throats. These germs are measured in micrometers, or millionths of a meter, and nanometers, or billionths of a meter.

### Volume

The amount of space an object takes up is called its volume. The **liter** is the basic unit of volume in the metric system. The liter is slightly larger than the quart. Here again scientists use divisions of the liter to measure smaller volumes. The milliliter, or cubic centimeter, is 1/1000 of a liter. There are 1000 milliliters, or cubic centimeters, in a liter. The volumes of both liquids and gases are measured in liters and milliliters. For example, a scientist may remove a few milliliters of blood from an animal in order to study the types of substances the blood contains.

### Mass and Weight

The **kilogram** is the basic unit of mass in the metric system. **Mass** is a measure of the amount of matter in an object. Your experience tells you that there is more matter in a tree trunk than in a leaf. Therefore, a tree trunk has more mass than a leaf.

One kilogram is slightly more than two pounds. For smaller units of mass, the gram is used. Remember that the prefix *kilo-* means 1000. There are 1000 grams in a kilogram. One milligram measures an even smaller mass, 1/1000 of a gram. How many milligrams are there in one kilogram? ❶

Mass is not the same as **weight.** Weight is a measure of the force of attraction between objects due to **gravity.** All objects exert a force of gravity on other objects. The strength of the force depends on the mass of the objects and the distance between them. You may not be aware of it, but the gravity of your body pulls the earth toward you. At the same time, the gravity of the earth pulls you toward its center. The mass of the earth is, however, much greater than your mass. As a result, the force of gravity you exert is very small compared to that of the earth. You remain on the surface of the earth because of



## 1-2 (continued)

### Content Development

Emphasize to students that one milliliter (mL) is equal to one cubic centimeter (cc). Point out that either term may be used in laboratory instructions or scientific literature.

### Reinforcement

Prepare sets of flashcards that students can use to practice converting units of metric measure. Some examples of conversions are given below:

1 meter = _______ centimeters (100)
1 meter = _______ millimeters (1000)
1 kilometer = _______ meters (1000)
1 liter = _______ milliliters (1000)
1 liter = _____ cubic centimeters (1000)
1000 _______ = 1 meter (millimeters)
1000 meters = 1 _______ (kilometer)
1 liter = 1000 _______ (milliliters)
1 kilogram = _______ grams (1000)
1 gram = 1000 _______ (milligrams)

### Skills Development

***Skill: Making a model***

Have teams measure and cut out six paper squares that measure 10 cm by 10 cm. Tell them to make a cube by attaching the squares together with tape.

- **How many centimeters is the cube on each side?** (10 cm)
- **How do we calculate the volume of a rectangular object?** (multiply the

Figure 1-9 *Astronaut Bruce McCandless, weightless in space, uses his jet-powered backpack and foot restraint to maneuver around the Space Shuttle.*

the force of the earth's gravity. And it is the size of this force that is measured by your weight. As the distance between objects decreases, the force of gravity between them increases. Therefore, you would weigh a little more at sea level than on the top of a mountain. At sea level, you are closer to the center of the earth.

It may be apparent to you now that mass is a constant and weight is not. Because weight can change, it is not a constant. For example, a person who weighs a certain amount on Earth would weigh much less on the moon. Can you explain this? The force of gravity of the moon is about one sixth that of the earth. Therefore, a person's weight on the moon would be one sixth that on the earth. The mass of the person, however, does not change. It is the same on the moon or the earth because the amount of matter in the person does not change.

DENSITY The measurement of how much mass is contained in a given volume of an object is called its **density.** Density can be defined as the mass per unit volume of a substance. The following formula shows the relationship between density, mass, and volume.

$$\text{density} = \frac{\text{mass}}{\text{volume}}$$

**Activity**

❶ *Metric Conversion*

Complete the following metric conversions.

a. 21,537 millimeters = ________ meters

b. 425 kilometers = ________ centimeters

c. 6.87 grams = ________ kilograms

d. 96.3 milliliters = ________ liters

e. 11 milliliters = ____ cubic centimeters



## ANNOTATION KEY

❶ **One million (Calculating)**

❶ **Thinking Skill: Applying concepts**

❷ **Thinking Skill: Applying formulas**

## Activity

**Metric Conversion**

**Skills: Making calculations, applying, relating**

**Level: Average**

**Type: Computational**

This activity will help students reinforce their ability to convert metric units. You may want to discuss the concept of dimensional analysis with students prior to the activity. Students answers should be **a.** 21.537 **b.** 42,500,000 **c.** .00687 **d.** .0963 **e.** 11.

Gravity can increase a person's mass. (Mass is constant; gravity can only increase weight.)

length times the width times the height)

- **What unit of volume would we get by multiplying cm times cm times cm?** (cubic centimeters, or $cm^3$)

Point out that the unit $cm^3$ is read as "cubic centimeters."

- **What is the volume of the cube?** (1000 cubic centimeters, or 1000 $cm^3$)

## Content Development

Some students may confuse mass and weight. To help reinforce the distinction, challenge students to find the errors in sentences such as these:

This apple weighs 200 grams. (Grams are a measure of mass, not weight)

Objects of equal mass always weigh the same. (only if they are in the same location with respect to gravity)

## Skills Development

***Skill: Making a model***

Have students work in small groups. Provide each group with two containers of the same volume (small beakers or boxes work well), some sand, some cotton, and a balance scale. Challenge each group to create a demonstration that illustrates the property of density. Then have each group present their demonstration to the class.

## 1-3 TOOLS OF A SCIENTIST

### SECTION PREVIEW 1-3

Scientists use a variety of tools to explore the world around them. In this section, students will learn about the tools that are particularly useful to life scientists.

Students will learn that a microscope enables scientists to see objects or parts of objects that are too small to be seen with the unaided eye. Students will read about two types of microscopes: the compound light microscope, and the electron microscope.

Students will be introduced to lasers, tools that are used in numerous ways by doctors. They will also learn about the use of X-rays, CAT scan, and NMR.

Students will discover that the computer is valuable tool to scientists. With the computer, scientists are able to collect, store, and analyze data.

### SECTION OBJECTIVES 1-3

1. **Discuss the importance of tools to scientists.**
2. **Describe some tools used by life scientists.**
3. **Explain the difference between a light microscope and an electron microscope.**

### SCIENCE TERMS 1-3

**microscope p. 24**
**compound light microscope p. 26**
**lens p. 26**
**electron microscope p. 26**

**Figure 1-10** *Scientists measure temperature in degrees Celsius. Temperature affects all living things. This chameleon must step lightly as it walks across a hot road.*

Density is an important concept because it allows scientists to identify and compare substances. Each substance has its own characteristic density. For example, the density of water is $1 g/cm^3$ while that of butterfat is $0.91 g/cm^3$. Based on this data, will butterfat sink or float in water? ❶

#### Temperature

Scientists measure temperature according to the **Celsius** scale, in degrees Celsius. The fixed points on the scale are the freezing point of water at sea level, 0° C, and the boiling point of water, 100° C. The range between these points is 100 degrees, and each degree is $1/100$ of the difference between the freezing point and boiling point of water. Normal human body temperature is 37° C, while some birds maintain a body temperature of 41° C.

#### SECTION REVIEW

1. What are the basic units of length, volume, and mass in the metric system?
2. What is the difference between mass and weight?
3. If a scientist wanted to identify an unknown substance, would it be more helpful for her to measure its mass or its density? Explain.

**Activity**

*Using Metric Measurements*

Use the appropriate scientific tools and the metric system to measure the following. Construct a chart of your results.

length of the textbook
length of your arm
temperature indoors
temperature outdoors
volume of a glass of water
volume of a bucket of water

### 1-3 Tools of a Scientist

**Scientists use a wide variety of tools, ranging from simple microscopes to complex computers.** The scientist chooses the tools most useful for solving a specific problem.

#### Microscopes

Have you ever looked through a magnifying glass to examine a leaf or the body of an insect more closely? If so, you used a simple **microscope.** A microscope is an instrument that produces an enlarged image of an object. A magnifying glass is a simple



### 1-2 (continued)

#### Skills Development

***Skill: Making calculations***

Have students calculate the densities of the following:

**1.** What is the density of a rock that has a mass of 273 g and a volume of $105 cm^3$? ($273 g/105 cm^3 = 2.6 g/cm^3$)

**2.** What is the density of a $158 cm^3$ liquid that has a mass of 142.2 grams? ($142.2 g/158 cm^3 = 0.9 g/cm^3$)

#### Skills Development

***Skill: Applying formulas, making calculations***

Since many temperature measurements are given in degrees Fahrenheit, it may be helpful for students to know the formulas for Fahrenheit-Celsius conversion:

$$°C = (F° - 32) \times 5/9$$

$$°F = (C° \times 9/5) + 32$$

Provide students with several Celsius and Fahrenheit temperatures to convert. Then challenge students to write their own problems involving Celsius-Fahrenheit conversion. Have students trade problems with a partner and solve.

#### Section Review 1-2

**1.** Meter, liter, kilogram.
**2.** Mass, a measure of the amount of

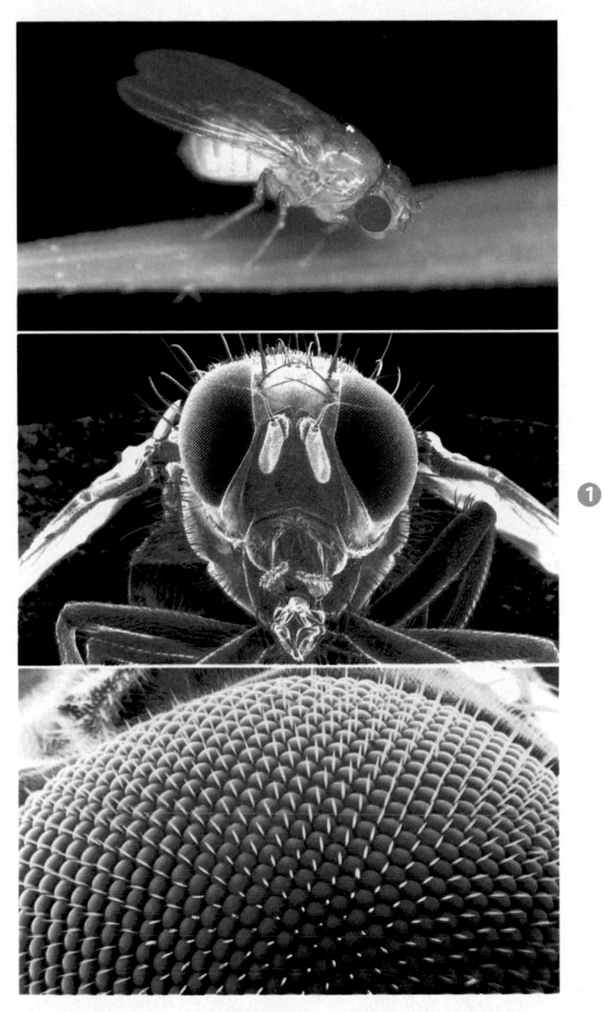

**Figure 1-11** *On this unmagnified fruitfly* (top), *notice the two large, red structures, which are the eyes. Then look at the head of the fruitfly, which is magnified 60 times* (center). *The two rounded structures on each side of the head are the eyes. A scanning electron microscope produced this detailed, three-dimensional image of the eye of the fruitfly* (bottom).

❶

25

## TIE-IN/GOVERNMENT

The Founding Fathers of our American republic used a type of metric system in currency that we still use today. They established our decimal system of coinage, with 100 cents equal to one dollar. All United States monies are decimal multiples or decimal fractions of the base dollar.

### Activity

**Using Metric Measurements**
**Skills: Measuring, recording, comparing**
**Level: Remedial**
**Type: Hands-on/computational**
**Materials: Metric ruler, Celsius thermometer, graduated cylinder**

After students make the necessary measurements, have them compare their measurements with the rest of the class. In this way, you can stress the importance of careful measurements and the possibility of experimental error when measuring.

### ANNOTATION KEY

❶ **Float (Applying concepts)**
❶ **Thinking Skill: Making comparisons**

matter, is constant. Weight, a measure of the force of gravitational attraction, can vary.
**3.** Density, since substances have a characteristic density, but not a characteristic mass.

## TEACHING STRATEGY 1-3

### Motivation

Begin by asking students,

- **What do you think of when you hear the word "tool"?** (Answers will vary; many will probably associate the word tool with such things as a hammer, saw, screwdriver, and so on.)
- **Do you ever think of a fork and spoon as tools?** (Answers may vary.)

Point out that the most general definition of a tool is an instrument used or worked by hand. A more specific definition is an instrument or apparatus used in performing an operation necessary to the practice of one's vocation or profession.

## SCIENCE, TECHNOLOGY, AND SOCIETY

There is a basic difference between science and technology. Science explains what *is* and technology applies this scientific knowledge to create new devices and procedures. We sometimes refer to basic research as "pure science" and to technology as "applied science." Technology makes practical use of the knowledge gained through basic research.

Technological advances without prior advances in basic science will ncessarily be limited. We must not only realize the link between basic research and technology, but we must also realize that science is not just a fix-it enterprise. While technological fixes or cures for problems are important, science discoveries may be even more useful in a preventive mode.

What kinds of basic research might be helpful in solving problems related to energy, population, food supply, air pollution, water pollution, conservation of resources, disease, etc.? What are some of the implications and consequences involved in such dilemmas as greater crop production versus ecological damage, bigger and faster cars versus energy consumption, conquering of diseases versus overpopulation, etc.?

microscope because it has only one **lens.** A lens is a curved piece of glass that bends light rays as they pass through it. In certain lenses, this bending increases the size of an object's image.

**THE COMPOUND LIGHT MICROSCOPE** Optical, or light, microscopes that have more than one lens are called **compound light microscopes.** These microscopes use light to make an object look up to 2000 times larger than it really is. Compound light microscopes can be used to examine the cells in your body.

**ELECTRON MICROSCOPES** An **electron microscope** does not use light to magnify the image of an object. Instead, it uses a beam of tiny particles called electrons. Pictures produced by this beam are focused on a television screen or photographic film. Electron microscopes can magnify objects hundreds of thousands of times. One type, the scanning electron microscope, or SEM, produces a three-dimensional

### Career: *Electron Microscopist*

**HELP WANTED: ELECTRON MICROSCOPIST** Bachelor's degree in histotechnology preferred. Training in the operation of the electron microscope and the interpretation of electron micrographs required.

Until the invention of the electron microscope in the 1930s, scientists examined the inner structures of living things by viewing slices of plant and animal tissues through light microscopes. Light microscopes magnify only large cell structures, not tiny structures. The tiny cell structures can be seen only when viewed through an electron microscope. Users of electron microscopes need special training and extensive practice before they become **electron microscopists.** Once their skill is developed, they can use the electron microscope to bring objects as small as bacteria, viruses, or even atoms into focus.

Electron micrographs are photographs of specimens seen through the electron microscope. By studying these photographs, research scientists learn more about the normal activities of cells. Human tissue micrographs help doctors diagnose patients.

Electron microscopists work in hospitals, universities, and research laboratories. If you wish to know more about a career in electron microscopy, write to the National Society for Histotechnology, P.O. Box 36, Lanham, MD 20706.

### 1-3 (continued)

#### Content Development

Discuss with students the major characteristics of the compound microscope and the electron microscope.

- **Which microscope has the greater magnification?** (the electron microscope)
- **How do the two types of microscopes differ in what they use to magnify objects?** (A compound microscope uses light; an electron microscope uses a beam of electrons.)
- **Why do you think the electron microscope rather than the compound microscope is used to view viruses?** (Viruses are too small to be seen with the light microscope.)

#### Skills Development

***Skills: Observing, comparing***

Have students work in small groups. Provide each group with a magnifying glass and a microscope. Also provide students with several small objects such as a leaf, a swatch of cloth, and a potato chip. Have students view each object first with the unaided eye, then with the magnifying glass, then with the microscope. Challenge students to sketch each view of the object so they have a series of pictures similar to those shown in Figure 1-11.

image. Biologists use electron microscopes to study such things as viruses and parts of cells.

### Lasers

A simple microscope uses light and lenses to magnify an image. The laser uses light in an entirely different way than a microscope does. A laser produces a narrow, intense beam of light. Lasers have many biological uses. A laser beam can be a surgeon's "light scalpel," cutting and sealing off blood vessels. Lasers can also be used to destroy clumps of cancer cells.

**Figure 1-12** *Lasers produce a narrow, concentrated beam of light that can be used in medicine to cut through, destroy, or repair damaged tissue.*

### Computers

Computers are electronic devices that collect, analyze, display, and store data. They have a wide range of uses in the biological sciences. In medicine, for example, computers help doctors diagnose diseases and prescribe treatments. Computers also help researchers gather information about the structure and function of cells and the activities of all living things.

### Seeing Through Barriers

In order for scientists to learn more about living things, they must be able to view the inside of organisms *from the outside.* Certain tools make this investigation possible.

**X-RAYS** Invisible radiations known as X-rays have been used by scientists for almost one hundred years. X-rays easily pass through barriers such as skin and muscle but tend to be blocked by more dense materials such as bone. The result is a picture of the interior of an object. X-rays are most useful for taking pictures of bones inside an organism. ❶

**CAT** A relatively new technique, Computerized Axial Tomography scanning, or CAT scan, provides a two-dimensional picture of an object. A beam of radiation may take as many as 720 separate exposures of the object. A computer then constructs a picture by combining and analyzing each exposure. CAT scans are used to produce pictures of the head and other body parts.

**Figure 1-13** *Top view of a computer-generated image of a DNA molecule.*

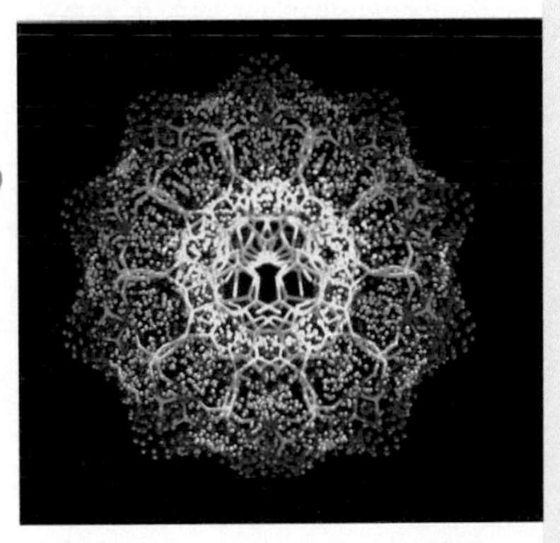



## ANNOTATION KEY

❶ Thinking Skill: Applying technology

## BACKGROUND INFORMATION

The word laser stands for *L*ight *A*mplification *S*timulated *E*mission of *R*adiation. A laser, which produces a powerful highly directional monochromatic beam of light, makes use of the internal energy of atoms and molecules. Lasers have been constructed from ruby crystals, mixtures of inert gases, and cubes of gallium arsenide. Lasers are used in eye surgery, holography, cutting metals, printing, and communications.

## Enrichment

A fascinating application of lasers is holography. Have students find out what holography is and how it is used. Challenge students to find samples of holograms to display to the class.

## Content Development

Remind students of the discussion about facts in section 1-1. Point out that many of the tools that scientists use help them uncover facts. For example, the microscope enables scientists to see things that are too small to be seen with the unaided eye. The X-ray, CAT scan, and NMR enable scientists to discover facts about the inside of objects. These facts also cannot be seen with the unaided eye.

Point out that the computer has a unique role to play with regard to facts—it is able to collect, store, and analyze data. For example, the computers used by the National Weather Service in Washington, DC, collect information about weather from stations all over the world. The computers store and analyze the data, then generate updated weather forecasts every 10 minutes for the entire country! Computers can also display and apply facts in many different ways. For example, when scientific satellites record data about the earth's surface or the ocean floor, the information is relayed to a computer that generates a detailed map of the area.

## Enrichment

Direct students' attention to Figure 1-13. Explain that this is just one example of a computer-generated model of a molecule. Challenge students to find more examples of computer-generated molecules. Also have students find out why these models are so useful to scientists.

## 1-4 SAFETY IN THE SCIENCE LABORATORY

### SECTION PREVIEW 1-4

In this section, students will be introduced to the basic rules of laboratory safety. Students will also be introduced to the laboratory safety symbols that are used in this text and the accompanying lab manual.

Students will discover that the laboratory can be an exciting place for the "hands-on" study of science. They will also learn that laboratory work can be dangerous if directions and safety rules are not followed carefully.

This section provides an excellent opportunity for students to be introduced to the laboratory facilities and equipment that they will be using. It also provides an opportunity for students to learn the emergency procedures that they should follow in case of an accident.

### SECTION OBJECTIVES 1-4

1. **Apply safety rules in the laboratory.**
2. **Identify the laboratory safety symbols and discuss the meaning of each.**
3. **Describe how to respond to an emergency in the laboratory.**

### ANNOTATION KEY

❶ **Heat-resistant glove (Interpreting charts)**

❶ **Thinking Skill: Applying technology**

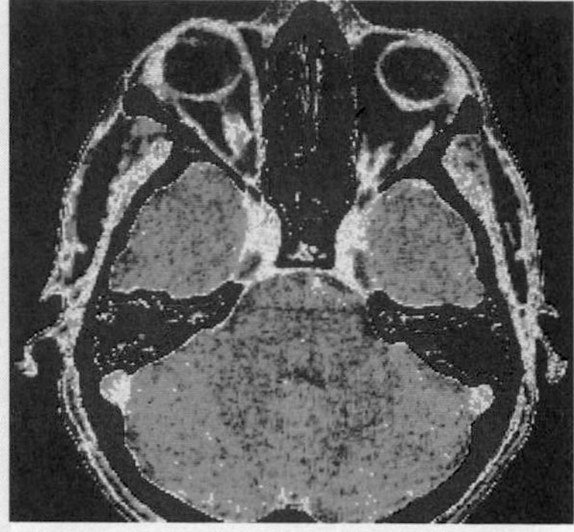

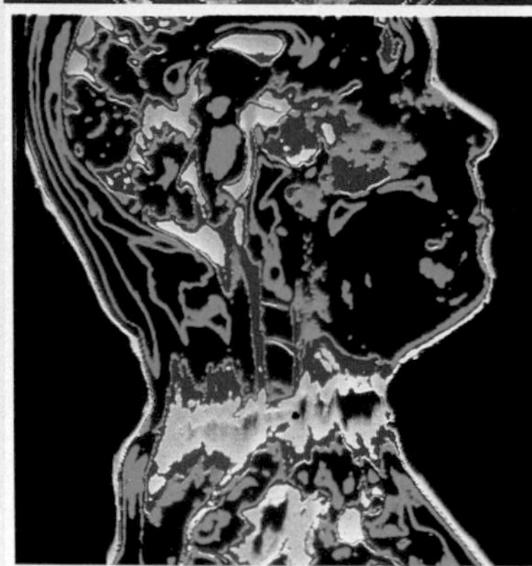

**Figure 1-14** *This image of a person's head* (top) *was produced by CAT scanning. NMR images* (bottom) *help scientists study the internal structure of body parts.*

**NMR** Another tool for seeing and studying the inside of objects is Nuclear Magnetic Resonance, or NMR. This tool uses magnetism and radio waves to ❶ produce images. With no apparent harmful effects on living tissue, NMR promises to be a valuable tool for studying the structure of body cells and how they function.

#### *SECTION REVIEW*

1. Explain the difference between the compound light microscope and the electron microscope.
2. Name three other tools of the scientist and briefly explain their uses.

## 1-4 Safety in the Science Laboratory

The scientific laboratory is a place of adventure and discovery. Some of the most exciting events in scientific history have happened in laboratories. For example, the structure of DNA, the blueprint of life, was discovered by scientists in laboratories.

To better understand the facts and concepts you will read about in science, you may work in the laboratory this year. If you follow instructions and are as careful as a scientist would be, the laboratory will turn out to be an exciting experience for you.

Scientists know that when working in the laboratory, it is very important to follow safety procedures. **The most important safety rule is to always follow your teacher's directions or the directions in your textbook exactly as stated.** You should never try anything on your own without asking your teacher first. And when you are not sure what you should do, always ask first.

As you read the laboratory investigations in the textbook, you will see safety alert symbols. Look at Figure 1-15 to learn the meanings of the safety symbols and all the important safety precautions you should take. If you do not understand a rule, ask your teacher about it. You may even want to suggest further safety rules that apply to your classroom.

28

### 1-3 (continued)

#### Section Review 1-3

**1.** One uses light to magnify objects, the other uses electrons.
**2.** Lasers: to repair or destroy damaged tissue; computers: to collect, analyze, display, and store data; X-rays, CAT, NMR scans: to observe internal structures

### TEACHING STRATEGY 1-4

#### Motivation

Make photocopies or drawings of the seven laboratory safety symbols shown in Figure 1-15. Display the symbols to the class without the titles and descriptions.

Point to each symbol in turn and ask,

• **What does this symbol make you think of?** (Answers may vary; certain symbols such as the poison bottle, eye, and hand are quite obvious; some of the other symbols, such as the flask, razor blade, and safety goggles, may be rather elusive.)

#### Content Development

Continue the Motivation activity by explaining to students that each of the safety symbols represents an aspect of laboratory work that requires caution. These aspects include han-

**LABORATORY SAFETY: RULES AND SYMBOLS**

**Glassware Safety** 

1. Whenever you see this symbol, you will know that you are working with glassware that can be easily broken. Take particular care to handle such glassware safely. And never use broken glassware.
2. Never heat glassware that is not thoroughly dry. Never pick up any glassware unless you are sure it is not hot. If it is hot, use heat-resistant gloves.
3. Always clean glassware thoroughly before putting it away.

**Fire Safety** 

1. Whenever you see this symbol, you will know that you are working with fire. Never use any source of fire without wearing safety goggles.
2. Never heat anything—particularly chemicals—unless instructed to do so.
3. Never heat anything in a closed container.
4. Never reach across a flame.
5. Always use a clamp, tongs, or heat-resistant gloves to handle hot objects.
6. Always maintain a clean work area, particularly when using a flame.

**Heat Safety** 

Whenever you see this symbol, you will know that you should put on heat-resistant gloves to avoid burning your hands.

**Chemical Safety** 

1. Whenever you see this symbol, you will know that you are working with chemicals that could be hazardous.
2. Never smell any chemical directly from its container. Always use your hand to waft some of the odors from the top of the container towards your nose—and only when instructed to do so.
3. Never mix chemicals unless instructed to do so.
4. Never touch or taste any chemical unless instructed to do so.
5. Keep all lids closed when chemicals are not in use. Dispose of all chemicals as instructed by your teacher.
6. Rinse any chemicals, particularly acids, off your skin and clothes with water immediately. Then notify your teacher.

**Eye and Face Safety** 

1. Whenever you see this symbol, you will know that you are performing an experiment in which you must take precautions to protect your eyes and face by wearing safety goggles.
2. Always point a test tube or bottle that is being heated away from you and others. Chemicals can splash or boil out of the heated test tube.

**Sharp Instrument Safety** 

1. Whenever you see this symbol, you will know that you are working with a sharp instrument.
2. Always use single-edged razors; double-edged razors are too dangerous.
3. Handle any sharp instrument with extreme care. Never cut any material towards you; always cut away from you.
4. Notify your teacher immediately if you are cut in the lab.

**Electrical Safety** 

1. Whenever you see this symbol, you will know that you are using electricity in the laboratory.
2. Never use long extension cords to plug in an electrical device. Do not plug too many different appliances into one socket or you may overload the socket and cause a fire.
3. Never touch an electrical appliance or outlet with wet hands.

**Animal Safety** 

1. Whenever you see this symbol, you will know that you are working with live animals.
2. Do not cause pain, discomfort, or injury to an animal.
3. Follow your teacher's directions when handling animals. Wash your hands thoroughly after handling animals or their cages.

**Figure 1-15** *You should become familiar with these safety symbols because you will see them in the laboratory investigations in the textbook. What is the symbol for special safety precautions with heat?* ❶

### SECTION REVIEW

1. What is the most important general rule to keep in mind when working in the laboratory this school year?
2. Explain why the laboratory is important in scientific research.
3. Suppose your teacher asked you to boil some water in a test tube. What precautions would you take so that this activity would be done safely?



## TEACHER DEMONSTRATION

Take this opportunity to point out to students the various safety features of the laboratory that they will be using. Point out the location of fire extinguishers, water and/or showers, first-aid kit, and so on.

Discuss with students the procedure that you want them to follow if an accident should occur. You might want to write this procedure on a large piece of poster board and display it in the lab.

dling glassware, working with fire and extreme heat, working with chemicals, using sharp instruments, and using electricity. Point out that the eye and hand symbols are a reminder that in many situations, special care must be taken to protect the face (especially the eyes) and hands.

• **What two pieces of laboratory equipment are provided to protect the eyes and hands?** (safety goggles and heat-resistant gloves.)

## Enrichment

Have students collect and report on newspaper and magazine articles that describe work in the scientific laboratory. Such articles may relate to work done in industry, commercial research facilities, or universities.

## Reinforcement

Make a set of flash cards with one of the laboratory safety symbols on each card. Have students pick a card at random and name the area of safety that the symbol represents. Then have other students describe the rules that apply to this area of safety.

## Skills Development

***Skills: Inferring, relating cause and effect***

Choose various safety rules from the chart in Figure 1-15. Ask students to explain the "why" of certain rules. For example, you might ask,

• **Why should you never heat anything in a closed container?** (Most substances, including air, expand when heated. An increase in pressure in the container could cause an explosion.)

• **What do you think might happen to chemicals if you left the lids off their containers for any length of time?** (They might evaporate; this could be hazardous if poisonous vapors were given off. Some chemicals might also combine with moisture in the air.)

## Section Review 1-4

**1.** Follow your teacher's directions or the directions in your textbook exactly as stated.
**2.** Exciting discoveries are often made in laboratories.
**3.** Wear safety goggles, heat water in an open container, never reach across the flame, use clamp or tongs, or heat-resistant gloves, keep work area clean, don't wear loose, hanging clothing.

# LABORATORY ACTIVITY
# A MOLDY QUESTION

## BEFORE THE LAB

1. **At least one day prior to the activity, gather enough materials for your class, assuming six students per group. Note that water must be available to each group.**
2. **Make sure that all bread used is of the same type. Ordinary commercial white bread works well.**
3. **Note that the first part of this activity will require about 15 minutes the first day, then about 5 minutes every few days during the next two weeks. The second experiment (Step 4) will require about the same amount of time.**

## PRE-LAB DISCUSSION

The focus of this activity is to give students hands-on experience with the scientific method, particularly in terms of identifying variables and designing an experiment. Before beginning the Pre-Lab discussion, have students read the Laboratory Activity carefully. Then ask,

- **What is the scientific problem presented in this lab?** (What factors affect the growth of bread mold?)
- **Can you offer a hypothesis to the problem?** (Accept all answers.)
- **According to Step 1 of the Procedure, what factors about the two slices of bread will be kept the same during the experiment?** (Both will be in closed jars; both will have the same amount of water added.)
- **What will be different about the two slices of bread?** (One will be placed in a dark closet, while the other will be placed in the sun.)
- **What is the obvious variable of these two environments?** (light)
- **What is a less obvious variable?** (temperature)
- **Do you foresee a problem in carrying out an experiment with two variables?** (yes)
- **Why?** (It may be hard to tell which variable has caused the results of the experiment.)

## LABORATORY ACTIVITY

### *A Moldy Question*

**Purpose**

In this activity, you will investigate variables that may affect the growth of bread mold.

**Materials** *(per group)*
2 jars with lids
2 slices of bread
1 medicine dropper

**Procedure**

1. Put half a slice of bread into each of two jars. Moisten each half slice with ten drops of water. Cap the jars tightly. Keep one jar in sunlight and place the other in a dark closet.

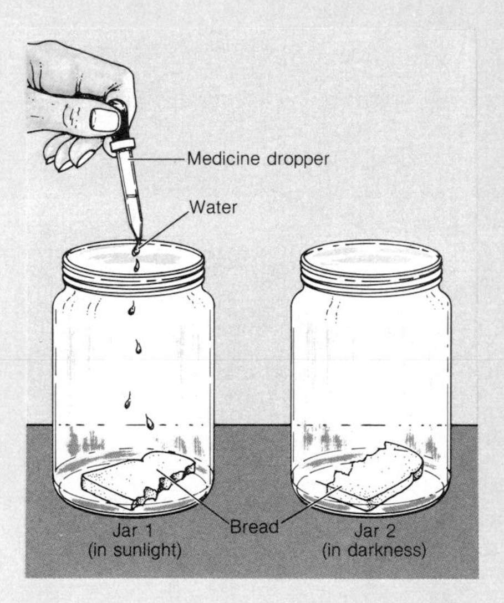


2. Observe the jars every few days for about two weeks. Record your observations. Does light seem to influence mold growth?
3. Ask your teacher what scientists know about the effect of light on mold growth. Was your conclusion correct? Think again. What other conditions might have affected mold growth?
4. Did you think of temperature? How about moisture? Light, temperature, and moisture are all possible variables in this activity. Design a second experiment to retest the effect of light on mold growth—or to test one of the other variables. Test only one variable at a time. Other groups of students will test the other two variables. Then you can pool your results and draw your conclusions together.

**Observations and Conclusions**

1. In your second experiment, what variable were you testing? Did you have a control for your experiment? If so, describe it.
2. Study the class data for this experiment. What variables seem to affect mold growth?
3. Cathy set up the following experiment: She placed a piece of orange peel in each of two jars. She added 3 milliliters of water to jar 1 and placed it in the refrigerator. She added no water to jar 2 and placed it on a windowsill in the kitchen. At the end of a week, she noticed more mold growth in jar 2. Cathy concluded that light, a warm temperature, and no moisture are ideal conditions for mold growth. Discuss the accuracy of Cathy's conclusion.

30

## SKILLS DEVELOPMENT

Students will use the following skills while completing this activity.

1. Observing
2. Manipulative
3. Recording
4. Comparing
5. Inferring
6. Relating
7. Applying
8. Predicting

## OBSERVATIONS AND CONCLUSIONS

**1.** The variable chosen will vary. The control will be a similar jar and piece of bread that does not contain the variable (moisture, for example) but is otherwise identical to the other, experimental setup.
**2.** Moisture and high (but not too high) temperatures have a positive effect on mold growth.

# CHAPTER REVIEW

## SUMMARY

### 1-1 What Is Science?

- A scientific method is a process scientists use to discover facts and truths about nature.
- The basic steps of any scientific method are stating the problem, gathering information, forming a hypothesis, experimenting, recording and analyzing data, and stating a conclusion.
- A suggested solution to a problem is called a hypothesis.
- An experiment should have only one variable, or factor being tested.
- Every experiment should have a control. A control experiment is run in exactly the same way as the experiment with the variable, but the variable is left out.
- The three main branches of science are life science, earth science, and physical science.

### 1-2 Scientific Measurements

- The metric system is the system of measurement used in science. It is a decimal system, or a system based on ten.
- The meter is the basic unit of length in the metric system.
- The liter is the basic unit of volume in the metric system.
- The kilogram is the basic unit of mass in the metric system.
- Mass is a measure of the amount of matter in an object. Weight is a measure of the attraction between objects due to the force of gravity.
- Density is the measurement of the amount of mass that is contained in a given volume of an object.
- The Celsius temperature scale has 100 degrees between the freezing and boiling points of water.

### 1-3 Tools of a Scientist

- The compound light microscope and the electron microscope magnify small objects and produce enlarged images of them.
- Lasers and computers have important applications in biology.
- X-rays, CAT, and NMR can provide pictures of internal body structures.

### 1-4 Science Safety in the Laboratory

- When you are working in the science laboratory, it is important for you to follow correct safety procedures at all times and to be familiar with your safety symbols.
- The most important safety rule is to always follow your teacher's directions or the directions of your textbook exactly as stated.

## VOCABULARY

*Define each term in a complete sentence.*

**astronomy**
**botany**
**Celsius**
**chemistry**
**compound light microscope**
**control**
**data**
**density**
**electron microscope**
**gravity**
**hypothesis**
**kilogram**
**lens**
**liter**
**mass**
**meter**
**metric system**
**microscope**
**physics**
**scientific method**
**variable**
**weight**
**zoology**

31

## GOING FURTHER: ENRICHMENT

**Part 1**

Have students carry out the same activity using different types of bread, particularly those with and without preservatives.

**Part 2**

Mold is a growth produced by a fungus. Have students find out more about the characteristics of fungi and the different types of conditions in which they grow.

**3.** The experiment was not well designed because Cathy varied more than one condition at the same time and therefore could not tell which variable was responsible for the difference in mold growth. The experiment should thus be redesigned, and it should also be repeated several times.

## CHAPTER REVIEW

### MULTIPLE CHOICE

| | | | | |
|---|---|---|---|---|
| **1.** b | **3.** a | **5.** a | **7.** d | **9.** d |
| **2.** c | **4.** b | **6.** c | **8.** b | **10.** c |

### COMPLETION

**1.** hypothesis
**2.** variable
**3.** Botany
**4.** metric
**5.** 100
**6.** one-thousandth
**7.** Mass
**8.** 100
**9.** microscope
**10.** laser

### TRUE OR FALSE

**1.** T
**2.** F data
**3.** T
**4.** F plants
**5.** T
**6.** F liter
**7.** T
**8.** F mass
**9.** F lens
**10.** F closed

### SKILL BUILDING

**1.** a. meter; b. milliliter; c. Celsius thermometer; d. kilometer; e. grams; f. centimeter or millimeter; g. kilogram; h. liter
**2.** a. 13,000; b. 0.476; c. 52; d. 740; e. 12.891; f. 65,000
**3.** a. The flame could easily ignite your clothing. Always place a flame in a position that is convenient so that no student needs to reach across the flame; b. Substances could boil out and spill onto other students. Always make sure you never point a heated test tube towards yourself or others; c. As the substance is heated, it will expand and could cause the container to explode. Never heat a substance in a closed container; d. The substance could be poisonous. Never taste any chemical unless specifically instructed to do so.
**4.** What this statement says is that the mass of an object is determined by the amount of matter in this object. Cut the object in half, and the mass is also cut in half. However, density is a characteristic of a particular material. Each substance has a particular density and doubling or changing the size of the substance will not change the density of that substance.
**5.** a. less than, b. sink, c. 40 g

### ESSAY

**1.** The steps include stating the problem, gathering information on the problem, forming a hypothesis, testing the hypothesis by experimenting, recording and analyzing the data, and stating a conclusion.
**2.** An experimental setup contains the variable that is being tested. The control setup is exactly like the experimental setup, except that it does not contain the variable to be tested.
**3.** Mass is a measure of the amount of matter and is constant for an object. Weight is a measure of the force of gravity acting on an object and can change.
**4.** A simple microscope uses a single lens and a beam of light to magnify objects. A compound microscope also uses a beam of light, but two or more lenses. An electron microscope uses a beam of electrons to magnify objects.
**5.** Lasers can be used as "light scalpels" to cut and seal off blood vessels,

#### CONTENT REVIEW: MULTIPLE CHOICE

*Choose the letter of the answer that best completes each statement.*

**1.** The branch of physical science that deals with the composition of substances, their changes and combinations is
a. astronomy. b. chemistry. c. physics. d. botany.
**2.** The basic unit of length is the
a. liter. b. gram. c. meter. d. kilogram.
**3.** One kilometer is equal to
a. 1000 meters. b. 1/1000 meter. c. 100 meters. d. 10 meters.
**4.** One centimeter is equal to
a. 100 meters. b. 1/100 meter. c. 10 meters. d. 1000 meters.
**5.** The liter is the basic unit of
a. volume. b. mass. c. weight. d. density.
**6.** The basic unit of mass is the
a. liter. b. gram. c. kilogram. d. meter.
**7.** The amount of mass contained in a given volume of an object is called
a. gravity. b. temperature. c. area. d. density.
**8.** The freezing point of water is
a. 32° C. b. 0° C. c. 100° C. d. 212° C.
**9.** A device doctors can use to perform surgery is the
a. X-ray. b. CAT scan. c. NMR. d. laser.
**10.** The most important laboratory safety rule is to
a. wear a lab coat. b. have a partner.
c. follow directions. d. use the metric system.

#### CONTENT REVIEW: COMPLETION

*Fill in the word or words that best complete each statement.*

**1.** A(n) ____________ is a suggested solution to a problem.
**2.** The factor being tested in an experiment is the ____________.
**3.** ____________ is the study of plants.
**4.** The ____________ system is the system of measurement used in science.
**5.** One meter is equal to ____________ centimeters.
**6.** A millimeter is ____________ of a meter.
**7.** ____________ is a measure of the amount of matter in an object.
**8.** Water boils at ____________° C.
**9.** A(n) ____________ magnifies and produces an enlarged image of an object.
**10.** A narrow beam of intense light is produced by a(n) ____________.

#### CONTENT REVIEW: TRUE OR FALSE

*Determine whether each statement is true or false. If it is true, write "true." If it is false, change the underlined word or words to make the statement true.*

**1.** The process used by scientists to discover facts and truths about nature is called a <u>scientific method</u>.
**2.** Recorded observations that often involve measurements are called <u>conclusions</u>.

32

3. The control experiment is the experiment without the variable.
4. Botany is the study of animals.
5. The prefix kilo- means 1000.
6. The basic unit of volume is the meter.
7. The force of attraction between objects is called gravity.
8. Your weight on the moon would be the same as it is on the earth.
9. A mirror is a curved piece of glass that bends light rays as they pass through it.
10. In the laboratory, never heat anything in an open container.

## CONCEPT REVIEW: SKILL BUILDING

*Use the skills you have developed in the chapter to complete each activity.*

1. **Classifying metric units** Which metric units would you use to measure each of the following?
   a. the length of your classroom
   b. the amount of milk you had for breakfast
   c. the temperature outside
   d. the distance from school to your home
   e. the mass of a grasshopper
   f. the length of your big toe
   g. the mass of a great white shark
   h. the amount of water in a swimming pool
2. **Making calculations** Complete the following metric conversions.
   a. 13 L = ____ mL
   b. 476 g = ____ kg
   c. 52 mL = ____ $cm^3$
   d. 74 cm = ____ mm
   e. 12,891 mg = ____ g
   f. 65 km = ____ m
3. **Following safety rules** Explain the potential danger involved in each of the following situations. Describe the safety precautions that should be used to avoid injury to you or your classmates.
   a. reaching across a flame
   b. pointing a test tube that is being heated toward yourself or others
   c. heating a substance in a closed container
   d. tasting an unknown chemical in order to identify it
4. **Applying concepts** Explain why every substance has a characteristic density, but no substance has a characteristic mass.
5. **Making calculations** The density of water is 1 $g/cm^3$. Therefore, objects with densities less than 1 $g/cm^3$ float in water.
   a. Is the density of air greater or less than 1$g/cm^3$?
   b. Will an object with a mass of 49 g and a volume of 21 $cm^3$ float in water?
   c. Sample X sinks in water. Would a 25 $cm^3$ sample of X have a mass of 20 g or a mass of 40 g?

## CONCEPT REVIEW: ESSAY

*Discuss each of the following in a brief paragraph.*

1. Briefly describe the basic steps of a scientific method.
2. Compare an experimental setup with a control setup.
3. Explain the difference between mass and weight.
4. Compare a simple, a compound, and an electron microscope.
5. Describe how lasers and computers are being used in medicine.
6. Explain how chance sometimes plays a role in scientific discoveries.



and they can be used to destroy clumps of cancer cells. Computers can help in diagnosis, prescription, and information gathering.
**6.** Very often a scientific discovery is made through serendipity, or through chance. That is, scientists are not always looking for a particular event or situation, but when such an event occurs, the scientist must be able to recognize that something has occurred that should be investigated.

## ADDITIONAL QUESTIONS AND TOPIC SUGGESTIONS

**1.** When solving a scientific problem, why is it important to form a hypothesis, rather than simply experimenting and gathering data? (A hypothesis gives a focus to the investigation by stating a specific idea or solution to be tested.)
**2.** Suppose that you ran the same experiment three times to test a particular hypothesis, and all three trials yielded different results. What might you conclude about your hypothesis and the design of your experiment? What might you do to get more conclusive results? (Answers may vary. A possible answer is that there is a hidden variable in the experiment. Trying to determine what this variable is and eliminating it could yield better results. Also, the hypothesis might be faulty in that it could be focusing on a solution that is not relevant to the problem, or a solution that is not practical to test. Revising the hypothesis might produce more conclusive results.)
**3.** Would the density of an object be the same in outer space as it is on earth? Why or why not? (According to the definition of density it would be, since mass and volume do not change from one location to another. However, certain properties that we commonly associate with density, such as objects less dense than water floating on water, would be rather meaningless, since all objects "float" in space due to weightlessness.)

## ISSUES IN SCIENCE

The following issue can be used as a springboard for discussion or given as a writing assignment.
**1.** A considerable effort has been made in recent years to use the metric system in all aspects of everyday life. For example, you have probably seen signs on the highway stating the speed limit in km/hr as well as in miles/hr. Some people object to this, saying that it is an unnecessary nuisance, and that some of the metric measurements, such as expressing weight in Newtons, are meaningless to most people. Other people claim that it is important to standardize world measurements, and to increase people's ability to understand science by using the system of measurement that scientists use. What is your opinion? How successful do you think the conversion to the metric system has been? Do you think it is worth the effort?

## Chapter 2

# THE NATURE OF LIFE

## CHAPTER OVERVIEW

Life is all around us. Life exists in tiny ponds, on mountain peaks, in shady meadows, and in busy cities. Forms of life are many and varied. To understand life and living things, we must look for the general characteristics of all living matter.

Living things are able to move, grow, reproduce, respond to a stimulus, and perform certain chemical activities. Living things interact with each other. Their needs for survival depend on the existence of oxygen. Without oxygen, all living things, animals and plants, die. Organisms combine oxygen with other substances to release or store energy. This energy is used in different ways.

This chapter includes detailed discussions of the elements most abundant in living things as well as their common characteristics and needs. In addition, there are sections describing the ways elements combine to form compounds and how the most common compounds in living things are formed.

## INTRODUCING CHAPTER 2

Draw attention to the illustration on page 34. Ask students if the place in the picture resembles any place they have been or seen. Ask if they think there is any life in the place in the picture. Point out that the picture is meant to represent primitive earth, as life was forming. After students have read the introduction to Chapter 2, ask:

- **What chemicals were probably present in the atmosphere of primitive earth?** (hydrogen, hydrogen sulfide, methane, and ammonia)

Point out that hydrogen, hydrogen sulfide, methane, and ammonia are barely present in the atmosphere of today's earth. These chemicals, however, have been detected in the atmosphere of Jupiter.

- **Why would scientists speculate about life evolving on Jupiter?** (because Jupiter's atmosphere resembles the atmosphere of primitive earth)

Stanley Miller filled a flask with four gases. Explain that these four gases were hydrogen, methane, ammonia, and gaseous water. Tell the students that Miller knew these gases made up much of the atmosphere 4 billion years ago. Point out that he wanted to find the missing key to the building-blocks of life.

# The Nature of Life

**CHAPTER SECTIONS**



**CHAPTER OBJECTIVES**

*After completing this chapter, you will be able to:*

**2-1** Discuss the basic characteristics of all living things.

**2-1** Describe metabolism and the activities involved in metabolism.

**2-2** Identify the basic needs of all living things.

**2-3** Distinguish between elements and compounds.

**2-3** Describe the organic compounds that are the building blocks of life.

Slowly, the scientist fills the clear glass flask. First he pours in three colorless gases. The odor is awful—a combination of rotten eggs and swamp gas that stings the scientist's nose and brings tears to his eyes. Now the scientist adds two more gases. Nothing seems to happen. The flask looks empty. But the gases it contains may be changed into something very special, if everything goes right!

The mixture needs a spark to produce the necessary change. The scientist sends a surge of electricity through the flask again and again. Suddenly, a sticky brown coating begins to form on the walls of the flask. The mixture of gases inside is changing—turning into substances that scientists believe may help them solve a key mystery of life.

Magic? It may seem to be, and at times the scientist may seem to be a magician. But this exciting experiment was actually performed, and its results are being used by scientists today as they attempt to study the "stuff of life."

In 1952, the American scientist Stanley Miller mixed together three foul-smelling gases: hydrogen sulfide, methane, and ammonia. To this mixture he added hydrogen, a colorless, odorless gas, and gaseous water. Then he passed an electric current through the colorless mixture. Soon a brown tarlike substance streaked the sides of the container. Dr. Miller analyzed the tarlike substance and found that it contained several of the same substances that make up all living things. From nonliving chemicals, Stanley Miller had made some of the building blocks of life.

*Primitive Earth, on which life evolved*

35

## TEACHER DEMONSTRATION

Draw a timeline or clock indicating a 24-hour cycle, beginning at midnight. Compare the 24-hour cycle to the history of the earth. Point out that the oldest known fossils appeared at about 6 A.M.; the oldest known complex organisms between 8 and 9 P.M.; the oldest plants between 9 and 10 P.M.; and humans in the last 30 seconds. This will help demonstrate that the variety of life on earth is actually a relatively recent phenomenon.

## TEACHER RESOURCES

**Audiovisuals**

*Breathing and Respiration,* filmstrip, BFA

*Excretion,* filmstrip, McGraw-Hill

*Reproductive System,* filmstrip, McGraw-Hill

*The Human Body: Nutrition and Metabolism,* filmstrip, Cor

**Books**

Keeton, W. T., *Biological Science,* Norton

Stare, F. J., and M. McWilliams, *Living Nutrition,* Wiley

Tuckerman, M. M., and S. J. Turco, *Human Nutrition,* Lea and Febiger

Ward, B, *Food and Digestion,* Watts

Hold up an empty flask.

- **What is in this flask?** (Most students will say "nothing.")

Point out the flask is full of air and air is a mixture of gases.

Tell students that after Stanley Miller put the gases in the flask, he sealed it. Then he boiled the gases for a week.

- **What do you guess happened?** (Accept all logical answers.)

Tell students nothing happened. Miller decided he needed another substance in his experiment. He sent a surge of electricity through the flask again and again. Explain that the electricity created lightning bolts, which turned the gases to a brown sticky coating. Point out that after testing this coating, Miller confirmed that the four gases and the electricity had actually created the building blocks of life.

- **What do you think is meant by "the building blocks of life"?** (Accept all logical answers.)
- **What is life?** (Accept all answers.)
- **What is life made of?** (Accept all answers.)
- **What does life need?** (Accept all logical answers.)

Point out that life is all around us. For us to know what life is, we must first learn what it is that makes a thing alive.

## 2-1 CHARACTERISTICS OF LIVING THINGS

### SECTION PREVIEW 2-1

In this section, students are introduced to the variety of living things. They discover that people once believed that life could be produced from nonliving matter. This theory was known as spontaneous generation, and it was not disproved until the seventeenth century. It is also explained that living things are primarily composed of the same elements—carbon, hydrogen, nitrogen, and oxygen. These elements and others interact in ways that are predictable, but it is not understood how they produce the property of life. Finally, there is a discussion of living things and how they share the characteristics of movement, metabolism, growth and development, life span, response to stimuli, and reproduction.

### SECTION OBJECTIVES 2-1

1. **List the elements common to all living things.**
2. **Describe the theory of spontaneous generation.**
3. **Describe the characteristics of living things.**

### SCIENCE TERMS 2-1

**spontaneous generation p. 37**
**metabolism p. 39**
**ingestion p. 39**
**digestion p. 39**
**respiration p. 39**
**excretion p. 40**
**life span p. 41**
**stimulus p. 42**
**response p. 42**
**reproduction p. 43**
**sexual reproduction p. 43**
**asexual reproduction p. 43**

**Figure 2-1** *A great variety of animals and plants inhabit the earth. The llama* (top, left) *may not be as familiar to you as the lovely crocus* (bottom, left). *The baobab tree* (top, right) *and the aardvark* (bottom, right) *may also be unfamiliar. Yet all are made up of the same basic elements.*

## 2-1 Characteristics of Living Things

Take a short walk in the city or the country and you will see an enormous variety of living things. In fact, scientists estimate that there are over five million different types of organisms on the earth, ranging from one-celled bacteria to huge blue whales. Yet all living things are composed mainly of the same basic elements: carbon, hydrogen, nitrogen, and oxygen. These elements make up the gases (1) Miller placed in his flask. These four elements along with iron, calcium, phosphorus, and sulfur all link together to form the stuff of life.

Well-known chemical rules govern the way these elements combine and interact. But less well under-

36

### TEACHING STRATEGY 2-1

#### Motivation

Bring in pictures of living things ranging in size from microscopic to huge. Example are bacteria, moss, flowers, redwood trees, humans, giraffes, and whales.

- **What do these animals and plants have in common?** (They are living things.)
- **What do living things have in common?** (Answers may vary, but possible answers include: they are all composed of the same elements, and they all share the characteristics of movement, metabolism, growth and development, life span, response to stimuli, and reproduction.)

#### Content Development

Point out that all living things are composed almost entirely of the same basic elements: carbon, hydrogen, nitrogen, and oxygen. Tell students that these are the elements that Miller placed in the flask. Explain that methane is a combination of carbon and hydrogen, ammonia is a combination of nitrogen and hydrogen, and water is a combination of hydrogen and oxygen. Point out that these four elements, with small amounts of iron, calcium, phosphorus, and sulfur combine and react with each other to form the matter of life.

stood is what gives this collection of chemicals a very special property—the property of life.

But what are the characteristics that make living things special? That is, what distinguishes even the smallest organism from a lifeless streak of brown tar ❷ on a laboratory flask? **Living things are able to move, grow, reproduce, respond to a stimulus, and perform certain chemical activities.**

### Life from Life

People did not always understand that living matter is so special. Until the 1600s, many people believed in the theory of **spontaneous generation.** According to this theory, life could spring from nonliving matter. For example, people believed that mice came from straw and frogs and turtles developed from rotting wood and mud at the bottom of a pond.

In 1668, an Italian doctor named Francesco Redi disproved this theory. In those days, maggots, a wormlike stage in the life cycle of a fly, often appeared on decaying meat. People believed that the rotten meat had actually turned into maggots, and that flies formed from dead animals. In a series of experiments, Redi proved that the maggots hatched from eggs laid by flies. Today there is no doubt that living things can arise *only* from other living things.

#### Activity

*Disproving Spontaneous Generation*

Redi's conclusions were not well accepted. The controversy over the theory of spontaneous generation continued for many years. Each of the men listed below was involved in either proving or disproving the theory. Using books and reference materials in the library, find out what contribution each man made to a better understanding of living things.

John Needham

Lazzaro Spallanzani

Louis Pasteur

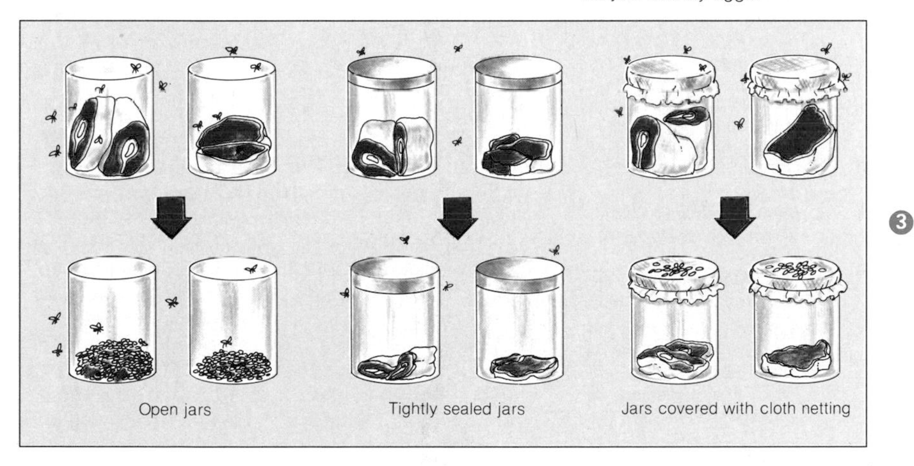


**Figure 2-2** *Redi's experiment helped disprove spontaneous generation. No maggots were found on the meat in jars covered with netting or tightly sealed. Maggots appeared on the meat only when flies were able to enter the jars and lay eggs.*

❸



#### Activity

**Disproving Spontaneous Generation**
**Skills: Report writing, relating, applying**
**Level: Average**
**Type: Library/writing comprehension**

Encourage students to examine the findings of the three scientists in terms of the logic of their scientific thought and methods. This is relatively simple to do in the cases of Spallanzani and Pasteur, who turned out to be correct in their rejection of the theory of spontaneous generation. However, even Needham, who incorrectly held to this theory, formulated hypotheses, conducted experiments, and came to conclusions in a logical, scientific manner. Still, there were certain problems with aspects of his experimental setups and, of course, in his failure to recognize that life forms could have entered the system from the external environment. This activity helps to reinforce the concept that theories change over time.

### ANNOTATION KEY

❶ **Thinking Skill: Identifying patterns**
❷ **Thinking Skill: Synthesizing facts**
❸ **Thinking Skill: Interpreting diagrams**

### Content Development

Point out that many years ago people believed that life could come from nonliving matter. Explain that this theory was called the theory of spontaneous generation.

Tell students that in 1668, an Italian biologist named Francesco Redi disproved this theory. Explain that he proved that livings things can arise only from living things.

### Skills Development

***Skill: Interpreting illustrations***

Have students observe Figure 2-2.

- **How does the meat appear in all of the first jars?** (fresh and new)
- **What do you think is happening in the second jar of each illustration?** (Flies are entering the jars in the first illustration but not the jars in the second and third illustrations.)
- **What is happening in the third jar of each illustration?** (The flies are sitting on the meat in the first illustration. They are not in the jars in the second and third illustrations. But in the third illustration, the flies have laid eggs on top of the cloth.)
- **What has taken place in the fourth jar in each illustration?** (Maggots are on the meat in the first jar. The flies have left the fourth jar in the second illustration. There are maggots on the top of the cloth netting in the third illustration, but there are no maggots on the meat.)

## FACTS AND FIGURES

The Monarch butterfly is one of many insects that travels in yearly migrations. The longest documented Monarch flight was from Ontario to Mexico, a total of about 2400 kilometers. The trip took four months and seven days—not bad for an insect with a mass of only 0.4 grams!

### Activity

**Catch Those Rays**
**Skills: Hypothesizing, manipulative, inferring, relating, observing, drawing conclusions**
**Level: Average**
**Type: Hands-on**
**Materials: Two coleus plants in soil and pots, water**

Answers will vary in this activity, depending on the amount of light and the conditions of the plants. It is most important to make sure students have followed the scientific method while performing this activity. In general, students should find that coleus plants react to light by bending toward the source of the light. The plants that were not directly in front of the window will bend toward the window because that is where the light is coming from.

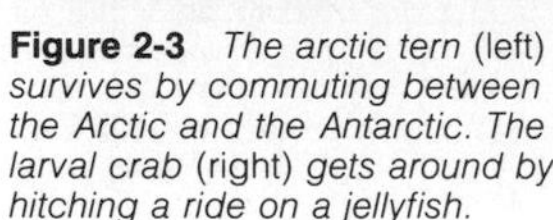

**Figure 2-3** *The arctic tern (left) survives by commuting between the Arctic and the Antarctic. The larval crab (right) gets around by hitching a ride on a jellyfish.*

### Activity

*Catch Those Rays*

1. Obtain two coleus plants of equal size.
2. Place one of the plants on a table or windowsill directly in front of a window.
3. Place the other plant on a table of the same height. Move this table so that it is about 70 cm to the left or the right of the same window.
4. Water both plants once or twice a week or whenever the soil feels dry.
5. Observe the growth of the plants for three weeks.

Write down your observations and the date on which you make them. What conclusion can you draw about the movement of plants in response to light?

### *Movement*

The arctic tern is a small, gull-like bird that at first glance may not appear to be a record breaker. See Figure 2-3. Yet for seven months of every year, the arctic tern is in flight, covering a distance of nearly 32,000 kilometers. The tern begins its journey near the Arctic, travels south to its winter quarters in the Antarctic, and then flies back north again. Its yearly trip gives the tern the "long-distance record" for birds in flight. An ability to move through the environment is an important characteristic of many living things.

Animals must be able to move in order to find food and shelter. In times of danger, swift movement can be the difference between safety and death. Of course, animals move in a great many ways. Fins enable fish such as salmon to swim (1) hundreds of kilometers in search of a place to mate. The kangaroo uses its entire body as a giant pogo stick to bounce along the Australian plains looking for scarce patches of grass upon which to graze.

Most plants do not move in the same way animals do. Only parts of the plants move. The stems of the common houseplant *Pothos* bend toward the sunlight coming through windows. Its leaves turn to catch the sun's rays.



### 2-1 (continued)

#### Content Development

Point out that the ability to move through one's environment is an important characteristic of living things. Explain that living things move in different ways. Tell students that humans move on two legs, while most other mammals move on four legs.

- **How do birds move?** (They fly.) **And fish?** (They swim.)
- **How would you characterize the movement of a snake?** (Accept all logical answers.)
- **What are some living things that are not animals?** (plants)
- **How do plants move?** (Accept all answers.)

Explain that only parts of plants move. Tell students that most plants grow at the top. Point out they also move toward any light source.

#### Content Development

Point out that living things cannot move if they do not receive and have energy. Explain that energy is stored and released by the building up and breaking down of chemical substances in living things. Tell students that these energy-related chemical activities in an organism are called metabolism.

Point out that metabolism is another characteristic of living things.

### *Metabolism*

Building up and breaking down is a good way to describe the chemical activities that are essential to life. During some of these activities, simple substances combine to form complex substances. These substances are needed for growth, to store energy, and to repair or to replace living materials. During other activities, complex substances are broken down, releasing energy and usable food substances. ❷ Together, these chemical activities in an organism are called **metabolism** (muh-TA-buh-li-zuhm). Metabolism, then, is another characteristic of living things.

Metabolism includes many chemical reactions that go on in the body of a living thing. However, metabolism usually begins with eating.

**INGESTION** All living things must either take in food or produce their own food. For most animals, **ingestion,** or eating, is as simple as putting food into their mouths. But some organisms, such as worms that live inside animals, absorb food directly through their skin.

Green plants do not have to ingest food because they make it. Using their roots, green plants absorb water and minerals from the soil. Tiny openings in the underside of their leaves allow carbon dioxide to enter. The green plants use the water and carbon dioxide, along with energy from the sun, to make food.

**DIGESTION** Getting food into the body is the first step in metabolism, but there is a lot more to metabolism than just eating. The food must be digested in order to be used. **Digestion** is the process by which food is broken down into simpler substances. Later, some of these simpler substances are reassembled into more complex materials for use in the growth ❸ and repair of the living thing. Other simple substances store energy that the organism will use for its many activities.

**RESPIRATION** Organisms combine oxygen with the products of digestion to produce energy. The energy is used to do all the work of the organism. The process by which living things take in oxygen and use it to produce energy is called **respiration.**

**Figure 2-4** *Until baby thrushes are old enough to fly, they must rely on their mother for food.*

## TEACHER DEMONSTRATION

Show the class a rock, a living plant, and a small animal such as a gerbil. (If a small animal is not available, use a photograph of a squirrel, dog, or cat.) Discuss the things using questions similar to the following:

- **What do you observe?** (Accept all answers.)
- **How are these things similar?** (Accept all answers.)
- **How are these things different?** (Accept all answers.)
- **Which things can move by themselves?** (Accept all answers.)
- **Which things can produce other things like themselves?** (Accept all logical answers.)

## FACTS AND FIGURES

Some helpful bacteria play an important role in the digestive processes of cows, sheep, goats, and even humans.

## ANNOTATION KEY

❶ Thinking Skill: Relating concepts
❷ Thinking Skill: Identifying processes
❸ Thinking Skill: Sequencing events

Explain that for metabolism to take place in an organism, all living things must either take in food or produce their own food.

### Content Development

Demonstrate some of the properties of respiration by comparing it to combustion. Bring in wooden matches and a glass jar with a lid.

- **What is produced when the match is lit?** (flame and energy)

While the match is burning, place it in the jar, and close the lid of the jar. Have students observe what happens.

- **Why does the flame go out when the jar is closed?** (Because there is no oxygen in the jar, and combustion requires oxygen.)

Point out that combustion and respiration both require oxygen, and that they both produce energy.

### Reinforcement

Explain that biologists call plants "producers" because they produce their own food, as well as provide food for animals. Animals are called "consumers" because they must rely on plants and other animals for food.

### Enrichment

Foods contain fats, carbohydrates, and proteins, in varying combinations. The amount of energy produced by each is measured in calories. Have students find out how much energy is produced by each, in calories per gram, using an encyclopedia or other reference.

## FACTS AND FIGURES

Giant squid can survive only in very cold water. The oxygen-carrying ability of their blood depends on the temperature of the water. In warm water, squid will suffocate because their blood cannot carry enough oxygen.

### Activity

**Counting Calories**
**Skills: Making calculations, measuring, comparing, relating, applying, recording, diagraming**
**Level: Remedial**
**Type: Computational/hands-on**

Students' computational and organization skills are reinforced in this activity. They should have little trouble making a calories chart of their favorite snack foods. You may want to have students compare their charts and find ways to cut calories without cutting enjoyment.

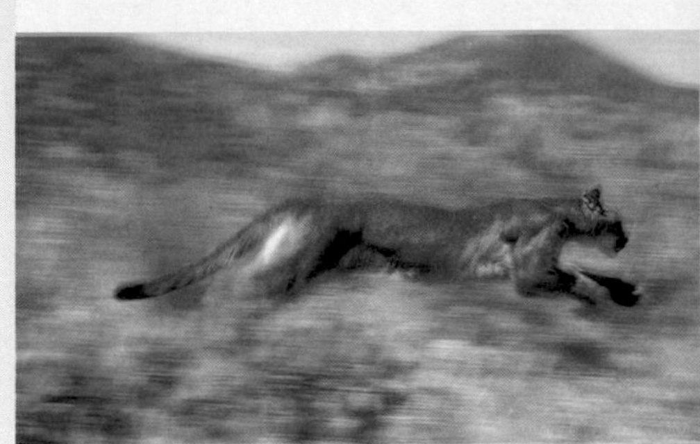
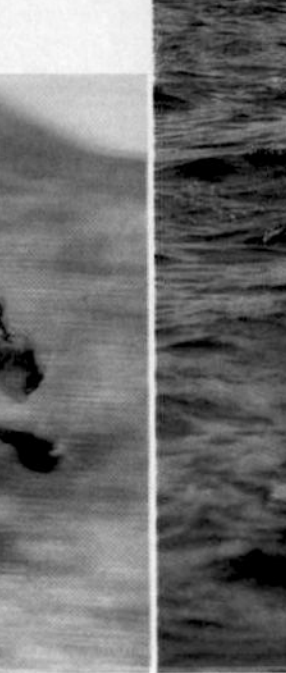

**Figure 2-5** *A cougar* (left) *could not run very fast without the energy produced during respiration. Dolphins* (right) *also need oxygen for respiration, so they continually surface for breaths of air.*

Land animals like yourself have lungs that remove oxygen from the air. Most sea animals have gills that absorb oxygen dissolved in water. Some sea animals, however, come to the surface to breathe with their lungs. Whales, porpoises, and dolphins are examples of air-breathing sea animals. Some of these animals can remain under water for as long as 120 minutes!

Plants, too, need oxygen to stay alive. Most plants absorb oxygen through tiny pores in their leaves. Plants use oxygen for respiration, as do almost all living things. For respiration is the main process that provides energy necessary to living things. You get this energy by combining the foods you eat with the oxygen you breathe. The amount of energy a food can supply is measured in units called Calories.

**Activity**

*Counting Calories*

The amount of energy a food can supply is measured in Calories. In general, people gain weight when they take in more Calories than they use up. Often the extra Calories come from snacks. Ten average-sized potato chips, for example, contain about 110 Calories.

Make a chart listing ten of your favorite snack foods and the number of Calories contained in each. This information may be obtained from food packages or reference materials.

**EXCRETION** Not all the products of digestion and respiration can be used by the organism. Some products are waste materials that must be released. The process of getting rid of waste materials is called **excretion.** If waste products are not removed, they will eventually poison the organism.

### *Growth and Development*

Standing under a giant oak tree, you might marvel at the fact that it grew from a tiny acorn. Within that acorn, metabolic activity produced a supply of energy from stored food. The acorn used this energy to begin to grow into a tree. Without the energy of metabolism, no living thing can grow. And growth is one of the characteristics of living things.



## 2-1 (continued)

### Motivation

Have students observe Figure 2-5. Read the caption.

- **How does the cougar get oxygen into its system?** (Accept all logical answers.)
- **Does the cougar get oxygen into its body the same way that we do?** (yes)
- **How does the dolphin get oxygen into its system?** (Accept all logical answers.)

### Content Development

Point out that living things need oxygen to mix with the products of digestion in order to release energy. Tell students that most land animals breathe with their lungs. Explain that some water animals come to the surface to breathe, while others absorb oxygen directly from the water.

Point out that most plants absorb oxygen through tiny leaf pores.

### Content Development

Point out that not all the products of digestion and respiration can be used by the organism. Explain that the process of getting rid of waste materials is called excretion.

Tell students that metabolic waste products may include such substances as water, carbon dioxide, ammonia, urea, and other indigestible substances.

Point out that plants can excrete waste products directly to their surroundings through their leaves. In most animals, the waste products must be carried out of the system with a liquid. Explain that if the waste products are not excreted, they can become poisonous to the organism.

### Motivation

Growing sprouts is a good way to observe an organism's growth and de-

**Figure 2-6** *A tiger swallowtail caterpillar* (left) *will grow and develop into an adult butterfly* (right).

But growing tnings do more than just increase in size. They also develop and become more complex. Sometimes this development results in dramatic changes. A tadpole, for example, swims for weeks in a summer pond. However, one day that tadpole becomes the frog that sits near the water's edge. And surely the caterpillar creeping through a garden gives little hint of the beautiful butterfly it will soon become. Certainly development must be added to your list of the characteristics of living things. ❷

Different organisms grow at different rates. A person will grow from a newborn baby to an adult in about 18 years. But puppies become adult dogs in only a few years. And a lima bean seed becomes a bean plant in just a few weeks. Some living things, such as certain insects, can change from an egg to an adult within a few days.

### Life Span

One of the important characteristics of living things is **life span,** or the maximum length of time a particular organism can be expected to live. Life span varies greatly from one type of organism to another. For example, an elephant lives for about 60 or 70 years. A bristlecone pine tree may live to be 5500 years old!

In certain organisms, growth and development take up most of the life span. The mayfly, for example, spends two years in lakes or ponds growing and

**Figure 2-7** *According to this chart, what is the maximum life span of a tortoise?* ❶

**MAXIMUM LIFE SPANS**

| Organism | Life Span |
|---|---|
| Adult mayfly | 1 day |
| Marigold | 8 months |
| Mouse | 1–2 years |
| Dog | 13 years |
| Horse | 20–30 years |
| Alligator | 56 years |
| Asiatic elephant | 78 years |
| Blue whale | 100 years |
| Human | 117 years |
| Tortoise | 152 years |
| Bristlecone pine | 5500 years |

**ANNOTATION KEY**

❶ 152 years (Interpreting charts)
❶ Thinking Skill: Making comparisons
❷ Thinking Skill: Sequencing events

## SCIENCE, TECHNOLOGY, AND SOCIETY

Food is necessary for human survival. In order to grow more food, scientists have developed new varieties of seeds, more efficient machinery, and better growing methods. Although such advances have increased food production, they have also driven many smaller, family farms out of existence. The smaller farms cannot compete with the larger, more productive ones. Many farm workers have lost their farmland and even their homes. What possible solutions do you see for the plight of the farmer in the United States?

velopment. Have students add a teaspoon of alfalfa seeds or mung beans to a jar filled with water. Stir gently. Cover the opening of the jar with a screen or a piece of cheesecloth, and let the jar sit overnight. Empty the water the following day, rinse and empty again, and store the jar at an angle. (Storing the jar at an angle allows further drainage.) Keep the jar out of direct sunlight, and rinse twice daily. Have students observe and describe the growth and development of the sprouts for 4 or 5 days. Have them place the jar in sunlight the fourth or fifth day. (This will encourage chlorophyll in the leaves of the sprouts.) When sprouts are about 2 cm, have students empty the jars. Eat with salads or sandwiches!

### Content Development

Tell students that all living things show increases in size during their lifetime. Point out that metabolism supplies living things with the energy they need to grow. Explain that growth is one of the characteristics of living things. Explain that growth is more than just growing in size. As things grow, they become more complex.

Point out that different organisms grow at different rates. Explain that growth varies with the individual type of organism. Point out that humans take about 18 years to grow to adulthood. Tell students that puppies become adult dogs in just a few years. Explain that some organisms such as sponges and many plants continue to grow all of their lives.

Explain that how long organisms live is called their life span. Explain that each type of organism has a different life span. Although an individual organism may live longer or shorter than other individuals, the maximum life span of the group of organisms can be determined over many years of observation.

## ANNOTATION KEY

❶ **Blinking, sneezing, running, flinching, coughing. (Relating facts)**

❶ **Thinking Skill: Relating cause and effect**

❷ **Thinking Skill: Making generalizations**

## TIE-IN/HISTORY

In 1926, Russian physiologist Ivan Pavlov published *Conditioned Reflexes,* which discussed conditioned responses to stimuli, including the response of salivary glands in dogs that were subjected to stimuli associated with food. From this came the notion of behaving like "Pavlov's dog."

**Figure 2-8** *The adult mayfly* (left) *lives for only about one day. The bristlecone pine* (right) *has a life span of about 5500 years.*

developing into an adult. However, the adult lives for only about one day. It finds a mate and starts a new family of mayflies. Then it dies.

### Response

When the alarm clock rings in the morning, you wake up. The smell of eggs and toast makes your mouth water as you hurry to breakfast. On your way to school, you stop at a red light. In just a matter of several hours, you have reacted to signals in your surroundings that determine much of your behavior.

❶ Scientists call each of the signals to which an organism reacts a **stimulus.** A stimulus is any change in the environment, or surroundings, of an organism that produces a **response.** A response is some action or movement of the organism.

Some stimuli come from outside the organism's body. For example, smells and noises are stimuli to which you respond. So is tickling. Light and water are stimuli to which plants respond. Other stimuli come from inside an organism's body. A lack of oxygen in your body is a stimulus that often causes you to yawn.

Some plants have special responses that protect them. For example, when a gypsy moth caterpillar chews on the leaf of an oak tree, the tree responds by producing bad-tasting chemicals in its other leaves. The chemicals discourage all but the hungriest caterpillars from eating these leaves. Can you ❶ think of responses that help you protect yourself?

**Figure 2-9** *Even if they must swim thousands of kilometers, most salmon return to the same stream in which they were hatched to lay their eggs.*

42

## 2-1 (continued)

### Content Development

Point out that scientists call each of the signals to which an organism reacts a stimulus. Explain that a stimulus is any change in the environment that causes the organism to react. Point out that the reaction is called a response.

Point out that responsiveness is of great importance to living things. It allows them to approach food and to avoid harmful things in the environment. Explain that some plants have special responses that protect them. Point out that when an animal or insect eats a leaf, the plant may produce bad-tasting chemicals in its other leaves. Other plants curl up as if they were withering.

Explain that the responses of most animals to a given stimulus are very specific. Point out that a dog has a keen sense of smell. If it smells food, it will run to eat it. If it smells an unfamiliar animal or person it may bark, growl, or become very defensive.

### Skills Development

***Skill: Applying concepts***

Have students make a list of ten different animals or plants and describe their responses to stimuli such as touch, sound, light, darkness, or other stimuli. Have students divide into teams of three to six students per team, share their lists, and make a detailed composite list of the various responses. Have teams share and discuss the various stimuli and responses of the organisms.

- **What did your team list as stimuli–response reactions for the various organisms?** (Accept all suggestions.)
- **What was the stimulus?** (Accept all logical answers.)
- **What was the response?** (Accept all logical answers.)
- **How does a ________ respond to a light stimulus?** (Accept all logical answers.
- **How does a ________ respond to a touch stimulus?** (Accept all logical answers.)

Ask other questions as appropriate.

## Reproduction

You probably know that dinosaurs that lived 230 million years ago are now extinct. Yet crocodiles, which appeared before dinosaurs, are still living today. An organism becomes extinct when it no longer produces other organisms of the same kind. In other words, all living things of a given kind would become extinct if they did not reproduce.

The process by which living things give rise to the same type of living thing is **reproduction.** Crocodiles, for example, do not produce dinosaurs. Crocodiles only produce more crocodiles. You are a human, and not a water buffalo, a duck, or a tomato plant, because your parents are human. An easy way to remember this is *like produces like.*

There are two different types of reproduction: **sexual reproduction** and **asexual reproduction.** Sexual reproduction usually requires two parents. Most higher forms of plants and animals reproduce sexually. ❷

Some living things reproduce from only one parent. This is asexual reproduction. This type of reproduction can be as simple as an organism dividing into two parts. Bacteria reproduce this way. Yeast form growths called buds, which break off and then form new yeast plants. Geraniums and African violets can grow new plants from part of a stem, root, or leaf of the parent plant. All of these examples demonstrate asexual reproduction.

Sexual and asexual reproduction have an important function in common. In each case, the offspring receive a set of very special chemical "blueprints," or plans. These blueprints determine the characteristics of that type of living thing and are passed from one generation to the next.

### SECTION REVIEW

1. What is the theory of spontaneous generation, and what did Redi's experiments reveal about the theory?
2. What is metabolism? What are some processes that are part of metabolism?
3. What are the six basic characteristics of living things?

**Figure 2-10** *Strawberries reproduce asexually when their stems produce bundles of roots, each of which forms a new plant.*

### Activity

*Observing Stimulus-Response Reactions*

Listed below are a few stimuli that produce responses known as reflexes. Observe your reaction, or that of your classmates or family members, to each stimulus.

**1.** Hold your hands close to a friend's face. Quickly clap your hands. Watch your friend's eyes as you do this. What happens?

**2.** Cover one of your eyes for a minute. Then remove your hand and at once look in a mirror. Describe the changes in your eye.

**3.** While looking in a mirror, shine a flashlight into your eyes. What happens to the size of your pupils?

**4.** Cut a slice of lemon. Bring it close to your mouth or, if you choose, put it in your mouth. What is your reaction?



### Activity

**Observing Stimulus–Response Reactions**
**Skills: Relating cause and effect, observing, recording**
**Level: Remedial**
**Type: Hands-on**
**Materials: Slice of lemon, flashlight, mirror**

Students will gain a better understanding of the concept of stimulus–response after completing this activity. They will learn that many responses to stimuli are involuntary. Students should discern that the response to clapping hands is blinking of the eyes. The response to covering their eyes and then removing the cover is dilation of the pupils. The response to shining a light in the eye is the contraction of the pupil. The response to the sour lemon taste is puckering of the lips and mouth.

### Motivation

Bring in pictures of common plants that reproduce asexually. These include spider plants, geraniums, coleus plants, strawberries, and potatoes. Point out that each of these plants uses different parts to reproduce asexually. For example, the coleus can reproduce via a leaf-cutting, and strawberries can reproduce via runners, or roots can grow on the surface of the soil. Ask students if anyone has a plant at home grown from cuttings. Have them make more cuttings, and bring the cuttings into the classroom. Grow some plants!

### Content Development

Point out that reproduction in nearly all living things is either sexual or asexual.

Write "sexual" and "asexual" on the chalkboard.

Explain that sexual reproduction results from the union of two cells. Write "2 cells" on the chalkboard under the word sexual. Explain that these cells come from two distinct sexes—male and female. Write "male and female" under "2 cells."

Point out that in asexual reproduction a single parent cell splits. Write "1 cell" on the chalkboard under asexual. Explain that this type of reproduction can be as simple as an organism dividing into two parts. Write "divides" on the chalkboard under "1 cell."

### Section Review 2-1

**1.** That life could spring from nonlife; that it was false.
**2.** Chemical building-up and breaking-down activities in an organism; ingestion, digestion, respiration, excretion.
**3.** Movement, metabolism, growth and development, life span, response, reproduction.

## 2-2 NEEDS OF LIVING THINGS

### SECTION PREVIEW 2-2

For a living organism to survive, it needs energy, food, water, oxygen, living space, and the ability to maintain a constant body temperature. The primary source of energy for a living organism is the sun. The sun allows plants to produce their own food and grow. Animals receive the sun's energy *indirectly* through the plants or other animals they eat as food. Water makes up the largest part of any living organism. The absence of water leads to death. Living organisms receive oxygen either directly through the air they breathe or from the air that is dissolved in water. Homeostasis is the ability of an organism to keep conditions inside its body the same even though conditions in its external environment change.

### SECTION OBJECTIVES 2-2

1. **Identify the needs of living things.**
2. **Explain how the sun is the primary source of energy for all living things.**

### SCIENCE TERMS 2-2

**competition** p. 47
**homeostasis** p. 47
**warmblooded** p. 48
**coldblooded** p. 48

**Figure 2-11** *This empty shell is home to a hermit crab, as well as to the sea anemones growing on the shell.*

## 2-2 Needs of Living Things

Living things interact with one another and with their environment. These interactions are as varied as the living things themselves. Birds, for example, use dead twigs to build nests, but they eat live worms and insects. Crayfish build their homes in the sand or mud of streams and swamps. They absorb a chemical called lime from these waters to build a hard body covering. Crayfish rely, however, on living snails and tadpoles for their food. And some people rely on crayfish for a tasty meal!

It seems clear, then, that living things depend on both the living and nonliving parts of their environment. **In order for a living organism to survive, it needs energy, food, water, oxygen, living space, and the ability to maintain a fairly constant body temperature.**

### Energy

All living things need energy. The energy can be used in different ways. A lion uses energy to chase and capture its prey. The electric eel defends itself by shocking its attackers with electric energy.

What is the source of this energy so necessary to living things? The primary source of energy for most living things is the sun. Plants use the sun's light energy to make food. Some animals feed on plants and in that way obtain the energy stored in the plants. Other animals may then eat the plant eaters. In this way, the energy from the sun is passed on from one living thing to another.

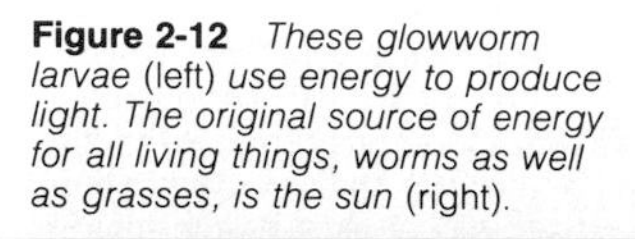

**Figure 2-12** *These glowworm larvae* (left) *use energy to produce light. The original source of energy for all living things, worms as well as grasses, is the sun* (right).

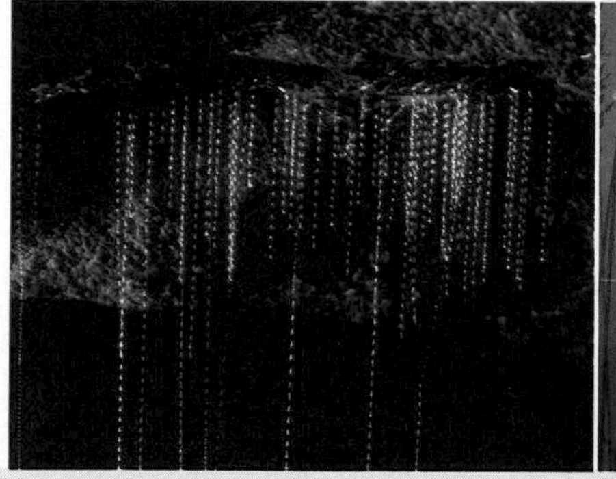

### TEACHING STRATEGY 2-2

#### Motivation

Point out that clothing, food, and shelter allow humans to meet most of their needs as organisms. Have students note what they're wearing, what they've eaten for breakfast, and what type of house or apartment they live in. Ask students to describe what their clothing, food, and houses are made of, or where their clothing, food, and houses come from. Do they come from living or nonliving things?

#### Content Development

Energy is transferred from one organism to another in food. Ecologists refer to this transfer as a food chain. An example of a food chain is a cow feeding on grass, and humans eating meat from a cow. Ask students to give other examples of food chains. Use outside references, if necessary.

#### Content Development

Point out that living things interact with other living things and their environment. Explain that it would be difficult, if not impossible, to live if they did not eat plants or animals.

• **How could you live without eating plants or animals?** (Accept all logical answers. Lead students to realize they could not live without eating plants or animals.)

Explain that in order for a living organism to survive, it needs energy, food, water, oxygen, living space, and the ability to maintain a constant body temperature.

**Figure 2-13** *Many animals and plants have unusual ways of obtaining food. The woodpecker finch* (left) *uses a twig to get at insects in the bark of trees. The Venus' flytrap snares a grasshopper with special leaves* (right).

### *Food and Water*

Have you ever heard the saying "You are what you eat"? Certainly this saying does not mean you will become a carrot or a hamburger if you eat these foods. But it does tell you that what you eat is very important. Food is a need of living things.

**FOOD** The kind of food organisms eat varies considerably. You would probably not want to eat eucalyptus leaves, yet that is the only food a koala eats. A diet of wood may not seem tempting, but for ❸ the termite it is a source of energy and necessary chemical substances. And surely a green plant's diet of simple sugar would not suit your taste.

Although all plants and animals must obtain food to live, some do it in very unusual ways. For example, the Venus' flytrap has special leaves that attract insects. When an insect lands on these leaves, they immediately close and trap the insect. Then the plant slowly digests the insect.

**WATER** Although you would probably not enjoy it, you could live a week or more without food. But you would die in a few days without water. It may surprise you to learn that 65 percent or more of your body is water. Other living things are also made up mainly of water.

Aside from making up much of your body, water serves many other purposes. Most substances dissolve in water. In this way, important chemicals can be transported easily throughout an organism. The blood of animals and the sap of trees, for example, are mainly water.

**Activity**

*You're All Wet*

About 65% of your body mass is water.

**1.** Find your mass in kilograms.

**2.** Use that number to determine how many kilograms of water you contain.

45

## FACTS AND FIGURES

Perhaps the laziest of all animals that hunt for food is the angler fish. An angler fish has a "living lantern" at the back of its mouth. Smaller fish are attracted to the light the angler fish produces. Once the smaller fish enter the angler's mouth, it snaps its mouth closed and captures a tasty meal.

## ANNOTATION KEY

❶ **Thinking Skill: Making generalizations**
❷ **Thinking Skill: Relating concepts**
❸ **Thinking Skill: Identifying relationships**

## Activity

**You're All Wet**
**Skill: Making calculations**
**Level: Enriched**
**Type: Computational**

This activity will help students reinforce their mathematical skills while they discover just how much of their body is water. Students' results will vary, depending on their mass.

## Skills Development

***Skill: Showing relationships***

Make signs saying "plant" and "animal." Place them on the bulletin board. Have each student bring in a picture of an animal. Give them several pieces of string or yarn. Tell them to place the picture on the bulletin board and connect a piece of yarn to what the animals eat.

- **Which animals eat only plants?** (Accept all logical answers.)
- **Which animals eat only animals?** (Accept all logical answers.)
- **Which animals eat both plants and animals?** (Accept all logical answers.)

## Content Development

Point out that all living things need energy. Tell students that the major source of energy is the sun. Explain that the light of the sun allows plants to produce their own food to grow.

- **What happens to the plant after it grows?** (Accept all answers. Lead students to suggest that the plant uses the food, and animals might eat the plant.)

Point out that after an animal has eaten the plant, it has gained the energy in the plant. Explain that some animals eat other animals.

- **Can you think of any animals that eat other animals?** (Accept all logical answers. Answers might be tigers, bears, dogs, humans, etc.)

Point out that the energy from the sun is passed on from one living thing to another.

## Activity

**Breathe Easy**
**Skills: Designing an experiment, inferring, applying, relating, manipulative**
**Level: Enriched**
**Type: Hands-on**
**Materials: Bromthymol blue**

This activity will help reinforce students' ability to design and carry out an experiment. Most students will be able to infer that if they breathe into water and carbon dioxide enters the water to form carbonic acid, then the bromthymol blue will turn the water blue. Thus, exhaled air must contain carbon dioxide.

### Activity

*Breathe Easy*

The air you inhale contains oxygen. But what is in the air you exhale? Is there carbon dioxide in this air?

**1.** Use the following facts to devise an experiment that will determine whether carbon dioxide is present in exhaled air:

- The indicator bromthymol blue turns blue when an acid is present.
- Carbon dioxide, when added to water, forms carbonic acid.

**2.** After you have devised your experiment, show it to your teacher. With your teacher's approval, try the experiment. What were your results? What conclusions can you draw?

Most chemical reactions in living things cannot take place without water. Metabolism would come to a grinding halt without water. And it is water that carries away many of the metabolic waste products produced by living things. For green plants, water is also a raw material needed to make food.

### Air

You already know that oxygen is necessary for the process of respiration. Most living things get oxygen directly from the air or from air that has dissolved in water. When you breathe out, you release the waste gas called carbon dioxide. However, this gas is not wasted. The green plants around you use carbon dioxide in the air to make food.

### Living Space

Do you enjoy hearing the chirping of birds on a lovely spring morning? Surprisingly, the birds are not simply singing. Rather, they are staking out their territory and warning intruders to stay away.

Often there is only a limited amount of food, water, and energy in an environment. As a result, only a limited number of the same kind of living thing can survive there. That is why many animals defend a certain area that they consider to be their living

**Figure 2-14** *Here two bighorn sheep are engaged in battle.*

46

## 2-2 (continued)

### Reinforcement

Remind students that living things need oxygen to combine with the products of digestion in order to release energy during respiration. Oxygen is received from inhaled air or from air dissolved in water.

### Content Development

Point out that living things release waste gas when they exhale. Explain that this gas is carbon dioxide. Tell students that plants use carbon dioxide from the air to make food.

### Motivation

Make a bulletin board of various animals and their shelters. Show various plants in a garden and trees in woods.

Have students observe the bulletin board. Discuss the locations and types of possible animal shelters with some of the following questions.

- **Why would you find it strange to see a rabbit build a shelter in a tree?** (Rabbits do not climb trees.)
- **Why do animals build their shelters where they do?** (Accept all logical answers.)
- **What does the type of shelter an animal builds have to do with the type of life it leads?** (Accept all answers.)

### Content Development

Point out that animals build their shelters in areas that are protected and have access to food. Explain that all areas do not have the same amount of protection nor the same amount of food. Therefore, there is competition for the better areas.

- **Is this fact true of humans?** (Accept all logical answers. Lead students to suggest that people move or live where they can find work.)

space. The male sunfish, for example, defends its territory in ponds by flashing its colorful fins at other sunfish and darting toward any sunfish that comes too close. You might think of this behavior as a kind of **competition** for living space. ❶

Competition is the struggle among living things to get the proper amount of food, water, and energy. Animals are not the only competitors. Plants compete for sunlight and water. Smaller, weaker plants often die in the shadow of larger plants.

### *Proper Temperature*

During the summer, temperatures as high as 58°C have been recorded on the earth. Winter temperatures can dip as low as −89°C. Most organisms cannot survive at such temperature extremes because many metabolic activities cannot occur at these temperatures. Without metabolism, an organism dies.

Actually, most organisms would quickly die at far less severe temperature extremes if it were not for **homeostasis** (ho-mee-o-STAY-sis). Homeostasis is the ability of an organism to keep conditions inside its body the same even though conditions in its external environment change. ❷ Maintaining a constant body temperature, no matter what the temperature of the surroundings, is part of homeostasis. Birds and certain other animals, such as dogs and horses, produce enough heat to keep themselves warm at low temperatures. Trapped air in the feathers of birds keeps

**Activity**

*Temperature Range*

Calculate the range in temperature between this highest recorded temperature, 58°C, and this lowest, −89°C.

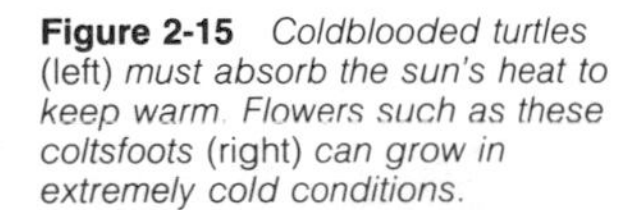

**Figure 2-15** *Coldblooded turtles* (left) *must absorb the sun's heat to keep warm. Flowers such as these* coltsfoots (right) *can grow in extremely cold conditions.*

## Activity

**Temperature Range**
**Skills: Making calculations**
**Level: Remedial**
**Type: Computational**

This exercise illustrates the wide range of temperature extremes on the earth and also reinforces subtraction skills. Since a negative number is involved, the exercise may be difficult for some students. If so, you may suggest that these students "count" from the lower to the higher temperature. The answer is 147° C.

## ANNOTATION KEY

❶ **Thinking Skill: Applying concepts**
❷ **Thinking Skill: Applying definitions**

## SCIENCE, TECHNOLOGY, AND SOCIETY

A list of basic human needs would include energy, food and water, oxygen, shelter, clothing, living space, and proper temperature. However, you may have heard the saying, "Man does not live by bread alone." Human needs go beyond the purely physical or biological environment. Human needs are found in social and economic environments as well.

Consider each of the following human needs: education, meaningful employment, health and safety, leisure time, respect, care and affection. Are these needs basic to all humans? If people are deprived of these needs, do they suffer just as if they were deprived of food and water? Are there any other human needs that you would add to this list?

## Enrichment

Have students select a habitat, and "design" an animal to live in that habitat. The habitat will have a particular climate. The animal will be warmblooded or coldblooded. Various physical and behavioral adaptations will allow the animal to maintain a constant body temperature. Physical adaptations include varying the size of extremities, and varying coloration. Behavioral adaptations include hibernation and lying in the sun. Encourage students to be creative. They should draw or make models of their designs (with such materials as modeling clay or papier maché). They should also describe their animals' physical and behavioral adaptations. References such as an encyclopedia or a biology textbook may be useful.

## 2-3 CHEMISTRY OF LIVING THINGS

### SECTION PREVIEW 2-3

In this section, students are introduced to elements that form compounds when they are chemically joined. Compounds are classified according to whether or not they contain the element carbon. Inorganic compounds usually do not contain carbon, and organic compounds do contain carbon. Families of organic compounds that form the basis of life are discussed. They are: carbohydrates, fats and oils, proteins, enzymes, and nucleic acids.

### SECTION OBJECTIVES 2-3

1. **Compare elements and compounds.**
2. **Compare organic and inorganic compounds.**
3. **Name and describe the families of compounds necessary for life.**

### SCIENCE TERMS 2-3

| | |
|---|---|
| **element p. 48** | **fat p. 49** |
| **compound p. 48** | **oil p. 49** |
| **inorganic compound p. 49** | **protein p. 50** |
| **organic compound p. 49** | **amino acid p. 50** |
| **carbohydrate p. 49** | **enzyme p. 51** |
| **glucose p. 49** | **catalyst p. 51** |
| | **nucleic acid p. 51** |
| | **DNA p. 51** |
| | **RNA p. 51** |

them cool when temperatures get too high. Panting and sweating do the same for dogs and horses. Animals that maintain a constant body temperature are called **warmblooded** animals. Warmblooded animals can be active during both day and night, in hot weather and in cold.

(1) Animals such as reptiles and fish have body temperatures that can change somewhat with changes in the temperature of the environment. These animals are called **coldblooded** animals. To keep warm, a coldblooded reptile, such as a crocodile, must spend part of each day lying in the sun. At night, when the air temperature drops, so does the crocodile's body temperature. The crocodile becomes lazy and inactive. Coldblooded animals do not move around much at relatively high or low temperatures.

#### *SECTION REVIEW*

1. Name five basic needs of living things.
2. What is the relationship between the sun and the energy needed by all living things?
3. What is the difference between warmblooded organisms and coldblooded organisms?

#### Activity

*A Starchy Question*

What are some foods that contain starch?

**1.** Choose four food samples that you think contain starch and four that you think do not.

**2.** Place a few drops of iodine on each sample. Record your observations in a data table.

**3.** Iodine turns a blue-black color in the presence of starch. According to your results, which foods contain starch? Which do not? How did your results compare with your predictions?

## 2-3 Chemistry of Living Things

What do a diamond, foil wrap, and a light bulb filament have in common? These objects look very different and certainly have very different uses, but all are examples of **elements.** An element is a pure substance that cannot be broken down into any simpler substances by ordinary means. A diamond is the element carbon. Foil wrap is the element aluminum. And the light bulb filament is the element tungsten. Carbon, oxygen, hydrogen, nitrogen, copper, gold, and sulfur are elements.

When two or more elements are chemically joined together, **compounds** are formed. Water is a compound made up of the elements hydrogen and oxygen. Table salt, which you probably use to flavor your food, is a compound made up of sodium and chlorine. Sand and glass are compounds composed

48

### 2-2 (continued)

#### Content Development

Point out that animals that maintain a constant body temperature are called warmblooded. Explain that warmblooded animals can live in almost any environment because they produce enough heat to keep themselves warm.

Explain that animals that have body temperatures that can change with changes in the temperature of the environment are called coldblooded. Point out that seasonal or daily temperature changes for coldblooded animals mean adaptation or death. Explain that when the body temperature lowers, the metabolic rate also lowers. Tell students that reptiles and fish do not move much at relatively high or low temperatures because their metabolic rate changes.

### Section Review 2-2

**1.** Energy, food, water, air, living space, proper temperature
**2.** The sun is the primary source of this energy.
**3.** Warmblooded: maintain constant body temperature; coldblooded: body temperature varies with temperature of environment.

### TEACHING STRATEGY 2-3

#### Motivation

Of the 92 elements that occur in nature, only 6 make up about 99% of all life. These are carbon, hydrogen, nitrogen, oxygen, phosphorus, and

of the elements silicon and oxygen. There are thousands of different compounds all around you. In fact, you are made up of many compounds. Scientists classify compounds in two groups.

### Inorganic Compounds

Compounds that may or may not contain the element carbon are called **inorganic compounds.** Most inorganic compounds do not contain carbon. However, carbon dioxide is an exception. Table salt, ammonia, rust, and water are examples of inorganic compounds.

### Organic Compounds

Many compounds contain carbon, which is usually combined with other elements such as hydrogen and oxygen. These compounds are called **organic compounds.** There are more than three million different organic compounds. Some of these compounds are the basic building blocks of life. **Organic compounds that are basic to life include carbohydrates, fats and oils, proteins, enzymes, and nucleic acids.** ❷

**CARBOHYDRATES** The main source of energy for living things is **carbohydrates.** Carbohydrates are made up of the elements carbon, hydrogen, and oxygen. Sugar and starch are two important carbohydrates. Fruit and candy contain sugar. Potatoes, rice, noodles, and bread are sources of starch. What are some foods that you eat that contain sugar and starch? ❶

All carbohydrates are broken down inside the body into a simple sugar called **glucose.** The body then uses glucose to produce the energy needed for life activities. If an organism has more sugar than it needs for its energy requirements, it will store the sugar for later use. The sugar is stored as starch. Starch is a stored form of energy. ❸

**FATS AND OILS** Energy-rich organic compounds made up of carbon, hydrogen, and oxygen are **fats** and **oils.** Fats are solid at room temperature; oils are liquid at room temperature.

**Figure 2-16** *The active element sodium* (top) *combines chemically with poisonous chlorine gas* (center) *to form crystals of ordinary table salt* (bottom).

## Activity

**A Starchy Question**
**Skills: Manipulative, recording, comparing, relating, applying, predicting**
**Level: Average**
**Type: Hands-on**
**Materials: Iodine**

Students should easily be able to tell which foods contain starch through this simple iodine test. You may want to have students compare their predictions prior to testing for starch.

## ANNOTATION KEY

❶ **Answers will include various kinds of fruits and candies, as well as potatoes and breads. (Making predictions)**

❶ **Thinking Skill: Making comparisons**

❷ **Thinking Skill: Making generalizations**

❸ **Thinking Skill: Relating cause and effect**

## FACTS AND FIGURES

Desert plants must be able to survive drastic temperature changes. Daytime temperatures of over 38°C can drop by more than 25°C at night.

sulfur. Humans are about 63% oxygen, 19% carbon, 9% hydrogen, 5% nitrogen, and less than 1% phosphorus and sulfur. You can demonstrate these percentages by making piles of beans of different sizes, each pile representing each element. (The size of each pile should correspond to the percentage of each element.) Write the names and percentages of each element on the chalkboard. Explain that we are made of these 6 elements, which combine to form compounds, including carbohydrates, fats, and proteins.

### Content Development

Point out that everything is made of matter. Explain that matter is anything that takes up space and has mass. Tell students that the basic building blocks of matter are atoms.

Explain that atoms are composed of three main particles: protons, neutrons, and electrons. Point out that the center of an atom is called a nucleus. Explain that the nucleus contains two different types of particles. The nucleus has positively charged particles known as protons and neutral particles called neutrons.

### Content Development

Explain that atoms also have tiny negative particles orbiting around the nucleus. The negatively charged particle is called an electron. The negative charge of each electron balances the positive charge of each proton to equal a neutral atom. In a neutral atom, the number of electrons is always equal to the number of protons.

## HISTORICAL NOTES

It wasn't until relatively recently that the rules governing the way elements interact were well-understood. In 1869, scientists Lothar Meyer and Dmitri Mendeleev separately developed a system for arranging all the known elements. This system was revised in 1912 by a scientist named Mosely, and is still in use today. The Periodic Table of Elements arranges the 96 naturally occurring elements and the 7 elements created by humans in a way that allows us to understand and predict many of their interactions. (A copy of the Periodic Table may be found in most chemistry textbooks. You may want to show it to your students, and point out the symbols for carbon, hydrogen, nitrogen, and oxygen.)

## ANNOTATION KEY

❶ Thinking Skill: Relating facts

❷ Thinking Skill: Identifying processes

**Figure 2-17** *The foods on this picnic table provide the organic compounds that are the basic building blocks of life.*

Fats can provide twice as much energy as carbohydrates. Foods high in fat, such as butter, cheese, milk, nuts, and some meats, should be included in your diet, but not to excess. Fats are stored by the body as body fat. And although this stored fat helps keep you warm, protects your internal organs from injury, and gives you energy, it can make you overweight.

**PROTEINS** Like carbohydrates and fats, **proteins** are organic compounds made up of carbon, hydrogen, and oxygen. But proteins also contain the element nitrogen and sometimes the elements sulfur and phosphorus. Some important sources of proteins are eggs, meat, fish, beans, nuts, and poultry.

The building blocks of proteins are **amino acids.** There are about 20 different types of amino acids. But because amino acids combine in many ways, they form thousands of different proteins.

### Career: *Enzymologist*

**HELP WANTED: ENZYMOLOGIST** Requires Ph.D. in biochemistry and experience in this field. Needed for research project. Will investigate the action of various poisons on enzymes found in living things.

Two young children enter the hospital emergency room sweating and complaining of headaches and nausea. They have eaten the pits of some purple berries they found growing in a field. Although the berries themselves were harmless, the pits contained poisonous chemicals called cyanides. Scientists know that in the body cyanides quickly combine with an important respiratory enzyme. When this enzyme is bonded with cyanide, the enzyme can no longer perform its normal, vital job. As a result, cells cannot use oxygen in the process of metabolism.

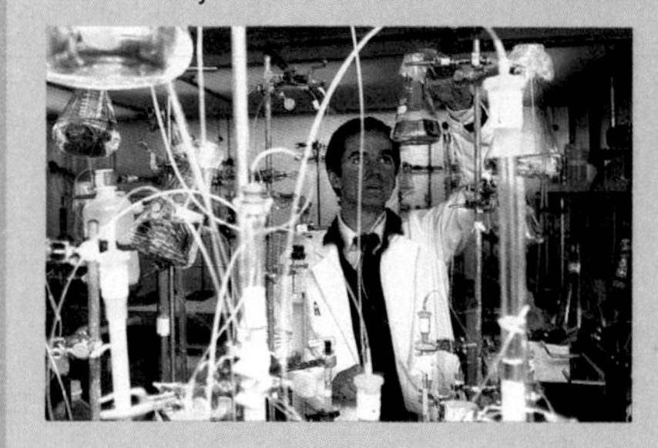

Scientists who do research into the action of enzymes are called **enzymologists.** They isolate, analyze, and identify enzymes. They determine the enzyme's effects on body functions. The human body contains more than one thousand different enzymes. Enzymologists know much about these enzymes.

Enzymes are now being used in the manufacture of detergents, bread, cheese, and coffee. Physicians use enzymes for cleaning wounds, dissolving blood clots, and treating certain diseases. Enzymologists are learning how enzymes can be helpful in cleaning up the environment.

Enzymologists can use their knowledge and skill in industry, in public and private research institutes, and in colleges and universities. To learn more about this career, write to the American Society of Biological Chemists, 9650 Rockville Pike, Bethesda, MD 20014.

50

### 2-3 (continued)

### Reinforcement

A simple activity can demonstrate how carbohydrates are broken down into sugar in the body. Obtain the type of soda cracker made of flour (a starch), water, and baking powder. Have each student chew a cracker well and hold it in his or her mouth for several minutes

- **Does the taste of the soda cracker change? How?** (Yes. It becomes sweeter.)

Explain that there is an enzyme in the mouth that breaks down the starch into sugar.

### Enrichment

By studying crystals with X-rays, geologists in the twentieth century have been able to determine that there are exactly 32 ways for atoms to form crystals.

### Motivation

You can demonstrate that certain foods contain fat. Such foods will leave a translucent spot when rubbed on a brown paper bag. Butter, oil, or any type of nut butter can be used. The spot is created by oil associated with the fat.

### Content Development

Of the 20 amino acids that form proteins, humans can synthesize (or

Proteins perform many jobs for an organism. They are necessary for the growth and repair of body structures. Proteins are used to build body hair and muscle. Proteins provide energy. Some proteins, such as those in blood, carry oxygen throughout the body. Other proteins fight germs that invade the body. Still other proteins make chemical substances that start, stop, and regulate many important body activities.

1

**ENZYMES** A special type of protein that regulates chemical activities within the body is called an **enzyme.** Enzymes act as **catalysts.** A catalyst is a substance that speeds up or slows down chemical reactions, but is not itself changed by the reaction. Without enzymes, the chemical reactions of metabolism could not take place.

**NUCLEIC ACIDS** Do you remember the "blueprints" of life discussed earlier? These blueprints are organic chemicals called **nucleic acids.** Nucleic acids are very large compounds that store information that helps the body make the proteins it needs. The nucleic acids control the way the amino acids are put together so that the correct protein is formed. This process is similar to the way a carpenter uses a blueprint to build a house. Now you understand why nucleic acids are called the blueprints of life.

One nucleic acid is **DNA,** deoxyribonucleic acid. DNA stores the information needed to build a protein. DNA also carries "messages" about an organism that are passed from parent to offspring. Another nucleic acid, **RNA,** ribonucleic acid, "reads" the DNA messages and guides the protein-making process. Together, these two nucleic acids contain the information and carry out the steps that make each organism what it is.

2

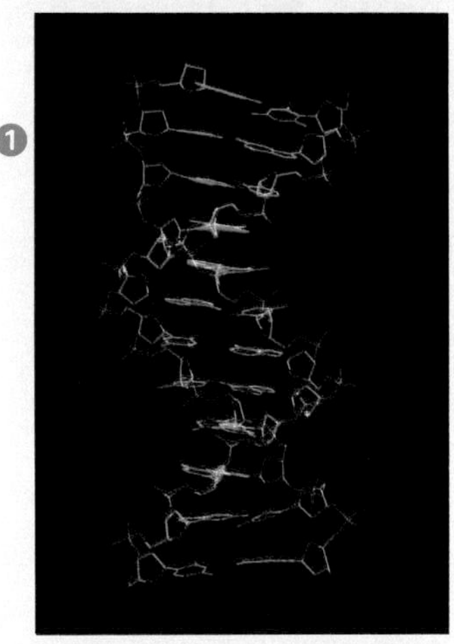

**Figure 2-18** *This computer-generated image shows the structure of the nucleic acid DNA. DNA stores information needed to build proteins and carries "messages" from one generation to the next.*

### *SECTION REVIEW*

1. What is the basic difference between inorganic and organic compounds?
2. Of what importance are carbohydrates, fats and oils, and proteins to living things?
3. What are two nucleic acids? What is the function of each of these nucleic acids?



## TIE-IN/MEDICINE

Animal fats are usually solid at room temperature, while vegetable fats are usually liquid. Most animal fats are also called saturated, which means they contain more hydrogen than "unsaturated" fats. As the body ages, saturated fats contribute to the development of fatty deposits called cholesterol. Cholesterol forms on the insides of blood vessels, and can cause blood to clot. If blood clots in one of the arteries that carry blood to the heart or brain, death can result. For this reason, doctors and nutritionists now recommend that people reduce the amount of foods in their diets that contain saturated fats.

make) 12. The other 8 must be obtained from food, and are called essential amino acids.

## Reinforcement

Review the concept of a catalyst with students. Point out that a catalyst can speed up or slow down a reaction, but that the catalyst itself is not changed after the chemical reaction.

• **If 2 mg of a catalyst are required during a chemical reaction, how much of the catalyst will be left after the reaction?** (2 mg)

## Enrichment

Have students research what happens when DNA passes messages from parent to offspring that produce abnormal results, as with genetic diseases. Have groups or individuals prepare reports that describe the causes and characteristics of different genetic diseases. An encyclopedia or biology textbook may be useful.

## Section Review 2-3

**1.** All organic compounds contain carbon; most inorganic compounds do not.
**2.** Carbohydrates provide energy. Fats and oils also produce energy and provide warmth and protection. Proteins are used for growth and repair, to provide energy, and to regulate body activities.
**3.** DNA stores information needed to build proteins and carries "messages" that are passed from parent to offspring. RNA "reads" DNA messages and guides protein making.

## LABORATORY ACTIVITY
## YOU ARE WHAT YOU EAT

### BEFORE THE LAB

**Provide enough copies of the school lunch menus for each group to have a copy. Also, obtain a reference book on food or nutrition to help students determine whether foods contain large amounts of carbohydrates, fats, or proteins, and whether they come from plant or animal sources.**

### PRE-LAB DISCUSSION

Have students hypothesize about whether the foods on the menus contain large amounts of carbohydrates, fats, or proteins, and whether they come from plant or animal sources. You may want to review the properties of carbohydrates, fats, and proteins before beginning the lab.

### SKILLS DEVELOPMENT

Students will use the following skills while completing this activity.

1. Recording
2. Comparing
3. Inferring
4. Applying
5. Relating
6. Observing

### TEACHING STRATEGY FOR LAB PROCEDURE

You may want to remind students of the tests for starches and fats, using iodine and brown paper bags, respectively. These tests may be useful in determining whether a food contains large amounts of carbohydrates or fats. You may also want to point out that foods such as meat, nut butters, dairy products, and beans contain large amounts of proteins.

## LABORATORY ACTIVITY

### *You Are What You Eat*

**Purpose**

In this activity, you will study your school's lunch menus for one week.

**Materials** *(per group)*

School lunch menus for the current week
Pencil
Paper
Reference book or textbook on foods and nutrition

**Procedure**

1. Obtain a copy of your school's lunch menus for one week.
2. Make a table listing the four basic food groups: meat group, vegetable-fruit group, milk group, bread-cereal group. For each day of the week, place the items from the menu in the appropriate food group.

| Meat Group | Vegetable-Fruit Group | Milk Group | Bread-Cereal Group |
|---|---|---|---|
| Hamburger | | | Roll |

3. Make a second table listing the three major nutrients: carbohydrates, fats, and proteins. List those foods containing large amounts of these nutrients under the proper heading.
4. On a third table, identify those foods that are plants or plant products and those that are animals or animal products.

| Carbohydrates | Fats | Proteins |
|---|---|---|
| Roll | | Hamburger |

| Plants or Plant Products | Animals or Animal Products |
|---|---|
| Roll | Hamburger |

**Observations and Conclusions**

1. Study the data you have collected and organized. What conclusions can you draw regarding your school's lunch program?
2. According to your data, do the foods represent a balanced diet? Do foods in certain categories appear much more often than foods in some other categories?
3. What changes, if any, would you make in the menus?

52

### OBSERVATIONS AND CONCLUSIONS

**1.** In most cases, students will determine that school-lunch menus provide a balanced diet.
**2.** Again, students will determine that their school lunch provides a balanced diet. However, they will probably find that some foods appear on the menus far more frequently than others.
**3.** Accept all reasonable answers, assuming that they still provide a balanced diet.

### GOING FURTHER: ENRICHMENT

**Part 1**

Have students make a list of the foods they eat at home in the course of one day or one week. Enter these foods in each of the tables from steps 2 through 4 of the lab. Ask students if they think their diet shows a balance between the 4 food groups, and between carbohydrates, fats, and proteins.

# CHAPTER REVIEW

## SUMMARY

### 2-1 Characteristics of Living Things

- The elements most abundant in living things are carbon, hydrogen, nitrogen, and oxygen.
- Metabolism is the sum of all chemical activities essential to life. Ingestion, digestion, respiration, and excretion are metabolic activities that occur in all organisms.
- Life span is the maximum length of time a particular organism can be expected to live.
- A living thing reacts to a stimulus, which is a change in the environment, by producing a response.
- Reproduction is the process by which organisms produce offspring.

### 2-2 Needs of Living Things

- Living things need energy for metabolism. The primary source of energy for all living things is the sun.
- All living things need food and water.
- Oxygen in the air or dissolved in water is used by all organisms during respiration. Carbon dioxide is used by plants to make food.
- Competition is a struggle for food, water, and energy.
- Homeostasis is the ability of an organism to keep conditions constant inside its body when the outside environment changes.
- Warmblooded animals maintain a constant body temperature. Coldblooded animals have body temperatures that change with changes in the external temperature.

### 2-3 Chemistry of Living Things

- Elements are pure substances that cannot be broken down by ordinary chemical means. Compounds are formed when two or more elements are chemically joined.
- Most inorganic compounds do not contain the element carbon. Organic compounds do contain carbon. The organic compounds important to living things are carbohydrates, fats and oils, proteins, enzymes, and nucleic acids.
- DNA and RNA are two important nucleic acids. They carry the information for and control the building of proteins that make each organism what it is.

## VOCABULARY

*Define each term in a complete sentence.*

**amino acid**
**asexual reproduction**
**carbohydrate**
**catalyst**
**coldblooded**
**competition**
**compound**
**digestion**
**DNA**
**element**
**enzyme**
**excretion**
**fat**
**glucose**
**homeostasis**
**ingestion**
**inorganic compound**
**life span**
**metabolism**
**nucleic acid**
**oil**
**organic compound**
**protein**
**reproduction**
**respiration**
**response**
**RNA**
**sexual reproduction**
**spontaneous generation**
**stimulus**
**warmblooded**



**Part 2**

Have students research federal guidelines for school lunch programs. They should try to determine how these guidelines insure a balanced diet. Have students report their findings.

## CHAPTER REVIEW

### MULTIPLE CHOICE

| | | | | |
|---|---|---|---|---|
| 1. b | 3. d | 5. c | 7. b | 9. b |
| 2. d | 4. c | 6. a | 8. a | 10. d |

### COMPLETION

| | |
|---|---|
| 1. Metabolism | 6. water |
| 2. digestion | 7. warmblooded |
| 3. reproduction | 8. compound |
| 4. stimulus | 9. carbon |
| 5. sun | 10. catalyst |

### TRUE OR FALSE

| | |
|---|---|
| 1. T | 6. T |
| 2. F spontaneous generation | 7. F warmblooded |
| 3. F ingestion | 8. T |
| 4. T | 9. T |
| 5. F excretion | 10. T |

### SKILL BUILDING

**1.** A peach pit would be considered living since it exhibits the basic characteristics of living things such as growth and development.

**2.** The second set of jars was tightly sealed. As such, a reasonable conclusion could have been that without sufficient oxygen, maggots could not develop on the meat through spontaneous generation. Thus, this set of jars did not disprove the theory of spontaneous generation. Nor did it provide any evidence in favor of the theory.

**3.** A sand dune may "grow" as wind and other factors pile up more and more sand in the dune. An organism, on the other hand, grows as a result of chemical activity that causes more of the organism's living tissue to be produced. This process requires food, energy, and often oxygen. It is not simply a change in mass as a result of increasing the amount of a particular substance, like sand piled into a sand dune.

**4.** Biological organisms are made up of chemical substances. Also, many chemical reactions occur within a living organism. To fully understand the processes that occur in living things, a knowledge of chemistry is required. The study of biochemistry deals specifically with the chemistry of living things.

**5.** Reproduction is vital for the survival of a species because through reproduction more organisms of that species form. Reproduction is not vital to the survival of an organism. That is, an organism will not die if it does not reproduce.

**6.** Sexual reproduction is more likely to result in species diversity because the offspring receive traits from both parents. Asexual reproduction generally results in an organism exactly the same as the parent in most details.

**7.** This recipe may have worked because the grains could well have attracted mice into the open pot. An observer might suggest that the mice developed through spontaneous generation. The easiest way to disprove this was a result of spontaneous generation would be to observe the pots carefully to make sure mice did not enter. Another way would be to seal the jars so that air could enter, but organisms such as mice could not.

### CONTENT REVIEW: MULTIPLE CHOICE

*Choose the letter of the answer that best completes each statement.*

**1.** The theory that life could spring from nonliving matter is called
a. spontaneous respiration. b. spontaneous generation.
c. asexual reproduction. d. homeostasis.

**2.** The building up and breaking down of chemical substances necessary for life is called
a. reproduction. b. excretion. c. competition. d. metabolism.

**3.** The process of combining oxygen with the products of digestion to produce energy is called
a. reproduction. b. ingestion. c. excretion. d. respiration.

**4.** Which organism cannot produce offspring by asexual reproduction?
a. yeast b. bacterium c. human being d. geranium plant

**5.** The gas in air used by plants to make food is
a. oxygen. b. nitrogen. c. carbon dioxide. d. hydrogen.

**6.** The struggle among living things to obtain the resources needed to live is called
a. competition. b. homeostasis.
c. metabolism. d. spontaneous generation.

**7.** Which of these is a warmblooded animal?
a. fish b. bird c. reptile d. crocodile

**8.** Which of these is a substance made entirely of a single element?
a. diamond b. table salt c. water d. glass

**9.** Carbohydrate is to glucose as
a. fat is to oil. b. protein is to amino acid.
c. DNA is to RNA. d. hydrogen is to oxygen.

**10.** The complex compound that carries the information needed to make protein from amino acids is
a. carbohydrate. b. homeostasis. c. glucose. d. DNA.

### CONTENT REVIEW: COMPLETION

*Fill in the word or words that best complete each statement.*

**1.** __________ is the sum of all the chemical activities essential to life.

**2.** The process by which food is broken down into simpler substances is known as __________.

**3.** The process by which living things produce offspring is called __________.

**4.** A response is the reaction of an organism to a(n) __________.

**5.** All organisms directly or indirectly receive energy from the __________.

**6.** About 65 percent of your body mass is composed of __________.

**7.** Animals that maintain a constant body temperature are __________.

**8.** Two or more elements joined together form a(n) __________.

**9.** Most inorganic compounds do not contain the element __________.

**10.** Because it can influence the rate of chemical reactions without itself changing, an enzyme is a(n) __________.

## CONTENT REVIEW: TRUE OR FALSE

*Determine whether each statement is true or false. If it is true, write "true." If it is false, change the underlined word or words to make the statement true.*

1. Most forms of life are composed almost entirely of carbon, hydrogen, nitrogen, and oxygen.
2. The theory of spontaneous combustion states that life can spring from nonliving matter.
3. The process by which an organism puts food into its body is digestion.
4. Green plants use water and carbon dioxide to make food.
5. The process of getting rid of waste material is called reproduction.
6. The maximum length of time a particular organism can be expected to live is called its life span.
7. Animals that maintain a constant body temperature are coldblooded animals.
8. Ammonia, carbon dioxide, and water are examples of inorganic compounds.
9. Stored fat helps keep you warm and protects your internal organs from injury.
10. DNA and RNA guide the protein-making process in living things.

## CONCEPT REVIEW: SKILL BUILDING

*Use the skills you have developed in the chapter to complete each activity.*

1. **Making inferences** Is a peach pit living or nonliving? Explain.
2. **Identifying relationships** Figure 2-2 shows three sets of jars that illustrate Redi's experiments. Explain why the second set of jars did not provide enough evidence to disprove the theory of spontaneous generation.
3. **Making comparisons** How is the growth of a sand dune different from the growth of a living organism?
4. **Relating concepts** Why is the study of chemistry important to the understanding of life science?
5. **Drawing conclusions** Which life function is necessary for the survival of a species but not necessary for the survival of an individual?
6. **Applying concepts** Which is more likely to result in increased variety among organisms, sexual reproduction or asexual reproduction?
7. **Relating cause and effect** During the time people believed in spontaneous generation, a scientist developed this recipe for producing mice: Place a few wheat grains and a dirty shirt in an open pot; wait three weeks. Suggest a reason why this recipe may have worked. How could you prove that the appearance of mice was not due to spontaneous generation?

## CONCEPT REVIEW: ESSAY

*Discuss each of the following in a brief paragraph.*

1. List five characteristics of living things.
2. Your friend believes that because they do not move, plants cannot be alive. Use specific examples to explain why your friend is wrong.
3. Describe four metabolic processes.
4. Describe four organic compounds that are the building blocks of life.
5. Explain this statement: The primary source of energy for all organisms is the sun.

55

## ESSAY

**1.** Living things generally are able to move, grow, reproduce, respond to stimuli, and perform chemical activities.
**2.** One example students might cite is the fact that plants will move toward sunlight. Also, plants such as the Venus' flytrap can move in response to the stimulus of insects touching their leaf hairs.
**3.** One example of metabolism is ingestion, in which organisms take in food materials. Another example is digestion, in which food materials are broken down into simpler substances. A third example is respiration, in which organisms combine oxygen and food materials to produce energy the organism can use. A fourth example is excretion, in which waste products are produced and removed from the body. Point out to students who query you on this fourth factor that excretion involves numerous chemical reactions prior to the wastes being produced and excreted.
**4.** Carbohydrates are one example of organic compounds, and include starches and sugars. Fats and oils are also organic compounds. At room temperature, fats are usually solid, whereas oils are usually liquid. Fats and oils are energy-rich organic compounds. Another organic compound is protein. Proteins are made up of chains of amino acids and are the compounds from which the body builds most of its tissue. Nucleic acids, both DNA and RNA, are organic compounds which carry the blueprints for protein-building, as well as the traits that are passed on from parents to offspring.
**5.** Plants produce food by using raw material obtained from the air, soil, and water in combination with energy from the sun. The plants then obtain energy by breaking down these foods through respiration. Animals either eat plants for food or eat other animals that eat plants. In either case, the animal produces energy through respiration by combining the food with oxygen. Because the source of an animal's food is either other plants or plant-eating animals, all plants and animals get their energy either directly or indirectly from the sun.

## ISSUES IN SCIENCE

The following issues can be used as springboards for discussion or given as writing assignments.
**1.** Some people choose to include little or no meat in their diets, for various reasons. Yet others wonder whether a vegetarian diet provides the body with enough protein, since foods that aren't derived from animal sources tend to contain less protein. Also, foods that are not derived from animal sources are deficient in 2 of the 8 "essential amino acids"—the amino acids that aren't made by the body. Do you think a vegetarian diet can provide the body with enough protein?
**2.** The life span of humans is being lengthened through medical technology. Do you believe it is advantageous for people to live longer? Explain.

# Chapter 3

# CELLS

## CHAPTER OVERVIEW

This chapter deals with the cell, the basic unit of life. As the chapter introduction indicates, cells have existed for billions of years and have developed the most amazing adaptations and the most complex interrelationships within multicellular organisms.

Although the shape and function of individual cells may vary greatly, the basic internal structures of cells are remarkably similar in all organisms. Cellular structures, such as the nucleus, mitochondria, and chloroplasts, are much the same in appearance and function regardless of the kind of organism in which they are found.

The first section of this chapter deals with the basic structure and functions of the living and nonliving parts of the cell, and with the differences between plant and animal cells. The second section describes the processes of metabolism, diffusion, osmosis, and reproduction (including mitosis and meiosis.)

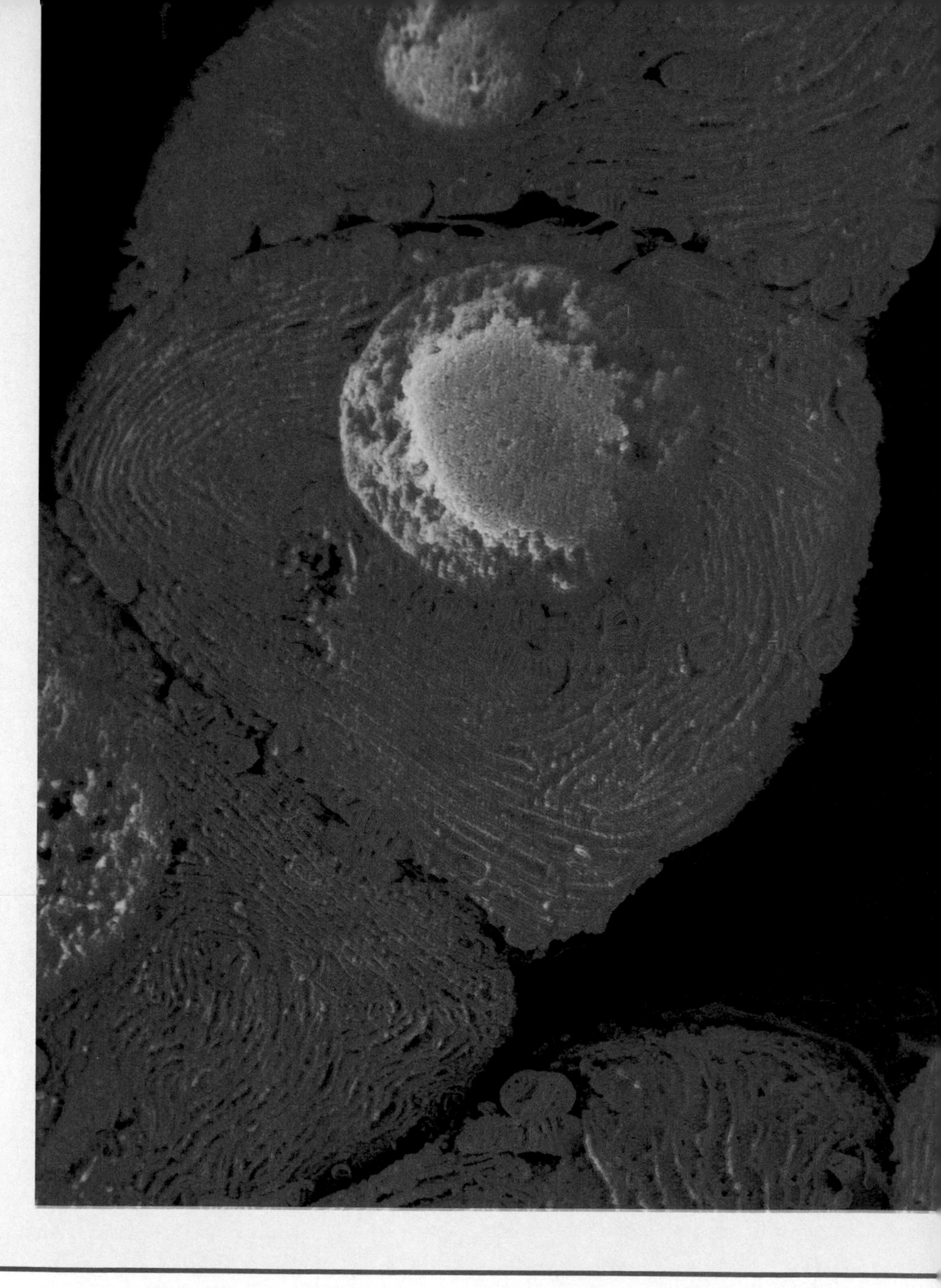

## INTRODUCING CHAPTER 3

Have students examine the photograph on page 56. Explain to them that these are human cells magnified many times.

- **How do you think this photograph was made?** (It was taken through an electron microscope.)
- **How many times larger than the actual cells are the cells in this photograph?** (10,000 ×)

Display one of your school's light microscopes.

- **What is the highest magnification that can be obtained using this microscope?** (Most classroom microscopes will have a 40 × high-power objective. Multiply this by the magnification of the eyepiece (usually 10 ×) to get the total magnification: 40 × × 10 × = 400 ×.)
- **How much greater is the magnification obtained by the electron microscope used to photograph these cells?** (10,000 × ÷ 400 × = 25 times greater)

Although this photograph shows a magnification of 10,000 ×, electron microscopes can magnify an object even more—up to 300,000 ×.

Read the chapter introduction on page 57 and the caption for the photograph.

- **What kind of cells are these?** (human skin cells)

# 3 Cells

**CHAPTER SECTIONS**

**3-1 Cell Structure and Function**
**3-2 Cell Activities**

**CHAPTER OBJECTIVES**

*After completing this chapter, you will be able to:*

**3-1** Describe the structure and function of the parts of a cell.

**3-1** Compare a plant with an animal cell.

**3-2** Describe the activities of a cell.

**3-2** Distinguish between diffusion and osmosis.

**3-2** Explain the importance of reproduction to a living organism.

**3-2** Describe the processes of mitosis and meiosis.

When it first appeared on earth about 3.5 billion years ago, it was a tiny structure made up of tinier parts alive with activity. As the years passed, it changed a bit here and a bit there. New parts developed through unknown processes. And it became able to do increasingly different and fascinating jobs—it could build complex chemicals, it could release bursts of energy, and eventually it could even move by itself.

Millions of years later it would join not only with others of its kind, but with others not quite like it. Together they would form the most amazing inhabitants of the universe—living things with many harmonious parts, such as plants that grow from the soil and animals that race over it.

Today these tiny structures still exist. They are a bridge to the distant past. They are the building blocks of all living things. They are cells! Cells are fascinating and, in many ways, very mysterious objects. Scientists continue to probe the secrets of cells like explorers journeying through parts of an uncharted world.

*Human skin cells magnified 10,000 times*



## TEACHER DEMONSTRATION

Fill the bottom of a petri dish about 5 mm deep with a dilute (10 percent) nitric acid solution. After darkening the classroom, place the dish on an overhead projector and turn it on. Being careful not to let students identify the materials being used, carefully add a large drop of mercury to the acid. **CAUTION:** *Be careful when handling mercury; it is poisonous.* After students have observed the activity of the drop of mercury, add a few crystals of potassium dichromate. The potassium dichromate will seem to be eaten by the mercury. If another drop of mercury is added to the dish, the two drops appear to compete for the food source. If this action stops, break up the mercury drops with a stirring rod.

- **Are these drops living or nonliving? Explain** (Accept all answers.)

Introduce several more items, such as a goldfish, a plastic spoon, an insect, a paper clip, and an earthworm, and have students decide on the appropriate category—living or nonliving. Then introduce several "trickier" items, such as a dead leaf, an egg, a preserved animal specimen, an onion, a seed, and a piece of wood.

- **Are these items living or nonliving?** (After some debate, you will probably want to expand the living category to "living or once-living" so that it will encompass all items.)
- **Why did you group these items as living/once-living? What do they have in common?** Focus on the concept that the items are all composed of cells.)

- **If you had an unknown sample of cells, how could you identify the organism that produced them?** (Several answers may be correct, and you may want to return to this question after students have read Chapter 3. There would be several clues. You could decide whether they were plant or animal cells by the presence or absence of certain organelles. For example, the cell wall and chloroplasts would be found only in plant cells. An analysis of the chromosomes would tell you the species from which the cells came.)

## TEACHER RESOURCES

### Audiovisuals

*Cells: A First Film,* film, Phoenix/BFA
*Cell Biology: Life Functions,* film, Cor
*Learning About Cells,* film, EBE
*The Living Cell: An Introduction,* film, EBE

### Books

Dowben, R., *Cell Biology,* Harper & Row
Swanson, C. P., *The Cell,* Prentice-Hall

### Software

*Cell Reproduction,* Prentice-Hall

## 3-1 CELL STRUCTURE AND FUNCTION

### SECTION PREVIEW 3-1

Most cells share similar characteristics. Plant cells have rigid cell walls made of cellulose that protect the cell's contents and help support the plant as it grows. Just inside the cell wall is the cell membrane, which regulates the passage of materials into and out of the cell.

Cell activities are controlled by the nucleus and its chromosomes. Chromosomes are composed of the nucleic acids DNA and RNA, which function in the building of proteins and in the passing of traits from generation to generation.

Floating in the jellylike cytoplasm are several other cell structures. The endoplasmic reticulum is a maze of passageways involved in the manufacture and transport of proteins. Granular bodies called ribosomes are found on the inner surface of some endoplasmic reticuli and also floating freely in the cytoplasm. Ribosomes are involved in protein production. Rod-shaped structures called mitochondria store energy produced from the breakdown of simple foods. Round structures called lysosomes aid in digestion.

Vacuoles serve as storage areas for food, water, or waste products. Chloroplasts, found only in plant cells, contain chlorophyll for capturing the sun's energy and making food.

## 3-1 Cell Structure and Function

You are about to take an imaginary journey. It will be quite an unusual trip because you will be traveling inside a living organism, visiting its tiny cells.

All living things are made up of one or more **cells,** which are the basic units of structure and function. Most cells are much too small to be seen without the aid of a microscope. In fact, most cells are smaller than the period at the end of this sentence. Yet believe it or not, within these tiny cells are even smaller structures. **The structures within the cells function in storing and releasing energy,** (1) **building and repairing cell parts, getting rid of waste materials, responding to the environment, and increasing in number.**

There are many types of cells. For example, human cells include muscle cells, bone cells, and nerve cells. In plants, there are leaf cells, stem cells, and root cells. However, whether in an animal or a plant, most cells share certain similar characteristics. So hop aboard your imaginary ship and prepare to enter a

**Figure 3-1** *A typical animal cell consists of many different structures, each having a characteristic shape and function. What are some of the functions carried out by cells?* (1)

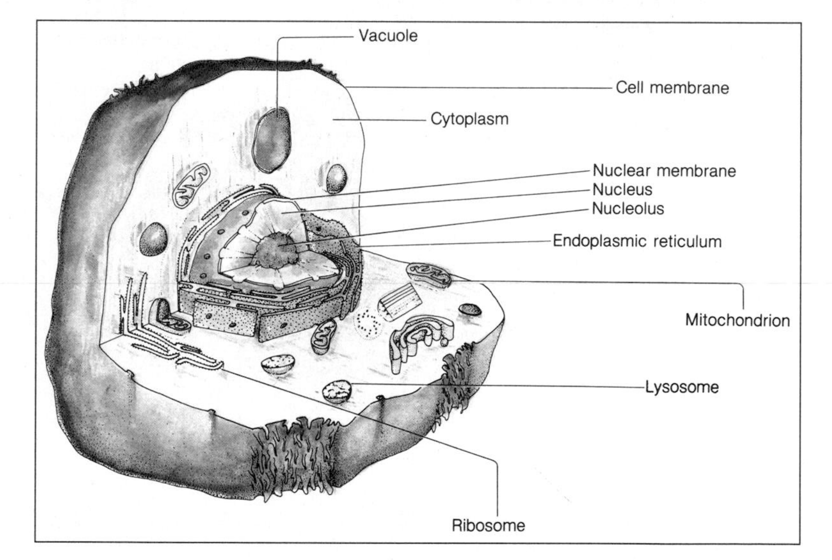


58

### TEACHING STRATEGY 3-1

#### Motivation

Prepare two demonstration slides of onion-skin cells or cheek cells. Leave one specimen unstained and stain the other with Lugol's iodine or methylene blue. Have students take a quick look at both slides. If a microprojector is available, the slides can be projected for the entire class to view together.

- **What similarities and differences exist between the two specimens?** (Same kind of cells, but the coloring was different; in the colored preparation, cell structures were easier to see.)
- **What structures could you identify?** (cell membrane—or cell wall in onion cells—nucleus, and cytoplasm)

Explain to students that many of the cells and other objects that they view with the microscope have been dyed or stained for easier viewing. The colors that they see are artificial; they are not the "real" color of the object. In some cases, two or three dyes may be used. Different cell structures react differently to these dyes—some absorb one kind of dye more readily than another. Thus, the end result is a multicolored cell with different structures stained with different colors such as a dark blue nucleus, red vacuoles, and light blue cytoplasm.

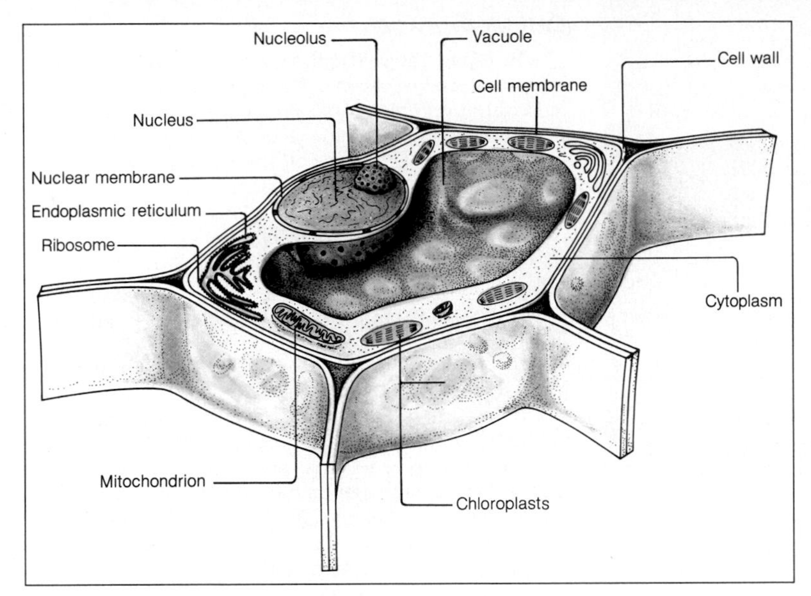


**Figure 3-2** *A plant cell has many of the same structures as an animal cell. What structures do you see in this diagram that are not found in animal cells?* ❷

typical plant cell. You will begin by sailing up through the trunk of an oak tree. Your destination is that box-shaped structure directly ahead. See Figure 3-2.

### *Cell Wall*

Entering the cell of an oak tree is a bit difficult. First you must pass through the **cell wall.** Strong and stiff, the cell wall is made of cellulose, a nonliving material. Cellulose is a long chain of sugar molecules that the cell manufactures. Although the cell wall is hard, it does allow water, oxygen, carbon dioxide, and certain dissolved materials to pass through it. So sail on through.

The rigid cell wall is found only in plant cells. This structure helps give protection and support so the oak tree can grow tall. Think for a moment of grasses and flowers that support themselves upright.  No doubt you can appreciate the important role the cell wall plays for the individual cell and for the entire plant.



## SECTION OBJECTIVES 3-1

1. **Name seven cell structures and describe the function of each.**
2. **Describe the composition of the nucleus.**
3. **Compare the structures of plant cells and animal cells.**

## SCIENCE TERMS 3-1



## ANNOTATION KEY

❶ **Storing and releasing energy, building and repairing cell parts, getting rid of waste materials, responding to the environment, and increasing in number (Making generalizations)**

❷ **Cell wall, chloroplasts, and lysosomes (Interpreting diagrams)**

❶ **Thinking Skill: Making generalizations**

❷ **Thinking Skill: Relating cause and effect**

Show students illustrations of cells or other specimens stained in this manner or have additional prepared slides set up for them to view. Remind students that these are not the true colors of cells but that dyes make viewing and identification easier.

### Content Development

This section introduces students to the structures of cells by taking them on an imaginary journey through a cell. Follow the same approach in your teaching to help hold students' attention. The imaginary-journey approach can make learning about the cell exciting and interesting.

### Content Development

Cell structures carry out all cell functions, such as storage and release of energy, construction and repair of cell parts, removal of wastes, responses to the environment, and reproduction. The outermost covering of a plant cell is the cell wall, a structure that is not found in animal cells.

- **What is the nonliving material that makes up the cell wall?** (cellulose)
- **Name several foods you eat that contain a large amount of cellulose.** (celery, rhubarb, asparagus, cucumber)
- **How do rigid cell walls of plants make them look and feel different from animals?** (Plants have stiff, rigid outer parts such as tree trunks, whereas most animals are quite soft and flexible.)

## BACKGROUND INFORMATION

The outer covering of a plant cell, the cell wall, is not elastic and flexible as is the cell membrane. Thus, the cell does not change shape easily under pressure.

The cell membrane, or plasma membrane, is made up of protein molecules scattered within and throughout a double layer of fat molecules. As these molecules move, they aid in the transport of materials through the membrane.

The nucleus, which is the largest and most prominent structure in many cells, is enclosed by a double membrane, the nuclear membrane. This membrane is perforated by fairly large pores, which serve as passageways for the movement of material between the nucleus and the cytoplasm of the cell.

Most of the nucleus is occupied by the threadlike chromosomes, which are not easily seen except during cell division. The material of the chromosomes stains readily with various basic dyes, and is therefore called chromatin. *Chromatin* is a term that comes from the Greek word "chromos" meaning color.

## FACTS AND FIGURES

Not all cells have nuclei. Mammalian red blood cells have nuclei when they are first formed in the bone marrow. However, they lose their nuclei before they enter the bloodstream.

**Figure 3-3** *The cell wall gives support and protection to plant cells, enabling grasses, flowering plants, and trees to grow upright.*

### *Cell Membrane*

Once past the cell wall, you prepare to enter the living material of the cell. Biologists call all the living material in *both* plant and animal cells **protoplasm.** Protoplasm is not a single structure or substance. It is a term used to describe all the living materials in a cell.

The first structure of protoplasm you encounter is a thin, flexible envelope called the **cell membrane.** In a plant cell, the cell membrane is just inside the (1) cell wall. In an animal cell, the cell membrane forms the outer covering of the cell. Look again at Figures 3-1 and 3-2.

The cell membrane has several important jobs. You can discover one of its jobs on your own. Push down on your skin with your thumb. Your skin does not break, does it? Now lift your thumb. The skin bounces back to its original position. Your skin can do this because the cell membrane around each skin cell is elastic and flexible. It allows the cell to change shape under pressure. The flexible cell membrane also keeps the protoplasm of the cell separated from the environment outside the cell.

As your ship nears the edge of the cell membrane, you notice that there are tiny openings in the membrane. You steer toward an opening. Suddenly, your ship narrowly misses being struck by a chunk of floating material passing out of the cell. You have discovered another job of the cell membrane. This membrane helps control the movement of materials into and out of the cell.

Everything that the cell needs, from food to oxygen, enters the cell through this membrane. And harmful waste products exit through the cell membrane. In this way, the cell stays in smooth-running order, keeping conditions inside the cell the same even though conditions outside the cell may change. (2) This ability of a cell to maintain a stable internal environment is called homeostasis. Now sail on to an important structure inside the living cell.

### *Nucleus*

A large oval structure comes into view just ahead of you. This structure is the control center of the cell—the **nucleus** (NOO-klee-uhs), which acts as the



## 3-1 (continued)

### Content Development

The cell membrane, found in both plant and animal cells, is a very important structure. It controls the passage of materials into and out of the cell. Cell requirements, such as food and oxygen, enter through the cell membrane. Waste products exit through the cell membrane.

### Reinforcement

Have students look again at Figure 3-2, a typical plant cell. Locate the cell wall and the cell membrane. Now have students look at Figure 3-1.

- **Can you locate the cell wall in this diagram?** (No. Animal cells do not have cell walls. They do have cell membranes.)
- **What general shape does the plant cell have?** (rectangular or boxlike)

The diagram of the animal cell in Figure 3-4 has been drawn in a "cutaway" fashion to show the inside of the cell.

- **If this cell were whole, what general shape would it have?** (spherical, like a ball)

You can demonstrate this by cutting a wedge out of an apple.

Reinforce the concept that all cells have thickness. They are three-dimensional, not flat, although they may appear to be flat when viewed through the microscope. Although cells come in many different sizes and shapes, plant cells are more rigid

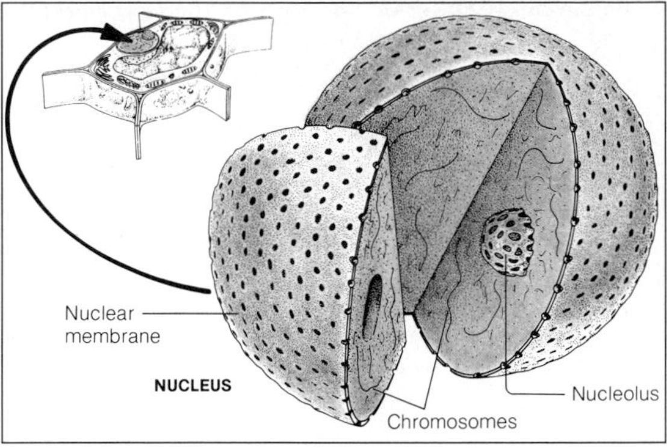


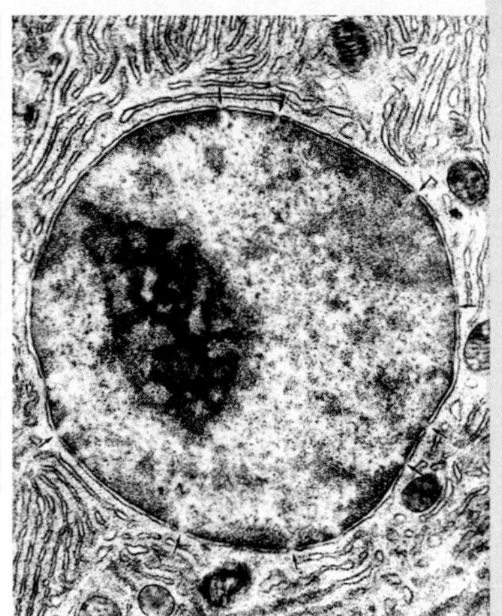

**Figure 3-4** *This nucleus of a typical cell* (left) *shows the nuclear membrane, chromosomes, and nucleolus. In this photograph of the nucleus* (right), *the arrows point to tiny pores in the nuclear membrane.*

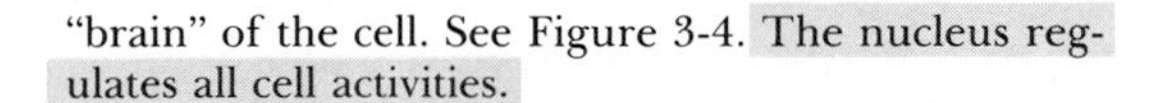

"brain" of the cell. See Figure 3-4. The nucleus regulates all cell activities.

**NUCLEAR MEMBRANE** The thin membrane that separates the nucleus from the protoplasm of the cell is called the **nuclear membrane.** This membrane is similar to the cell membrane in that it allows materials to pass into and out of the nucleus. Small openings, or pores, are spaced regularly around the nuclear membrane. Each pore acts as a passageway into and out of the nucleus. Set your sights for that pore ahead and carefully glide into the nucleus. ❸

**CHROMOSOMES** Those thick, rodlike objects floating in the nucleus are called **chromosomes.** Steer very carefully to avoid colliding with and damaging the delicate chromosomes. For the chromosomes direct the activities of the cell and pass on the traits of the cell to new cells.

The large, complex molecules that make up the chromosomes are compounds called nucleic acids. You may recall from Chapter 2 that nucleic acids store the information that helps a cell make the proteins it needs. And proteins are necessary for life. Some proteins, for example, are used to form structural parts of the cell such as the cell membrane. Other proteins make up different enzymes and



## ANNOTATION KEY

❶ **Thinking Skill: Comparing cells**
❷ **Thinking Skill: Applying concepts**
❸ **Thinking Skill: Making comparisons**

## FACTS AND FIGURES

An organism's size depends on the number, not the size, of its cells. Thus, whale cells are no larger than mosquito cells. The whale simply has more cells.

and boxlike, whereas animal cells are more flexible and rounded.

## Motivation

To demonstrate that living cells come in a variety of sizes and shapes, obtain a few prepared slides of cells from various sources. Have students examine the slides under a microscope. Explain that all cells—regardless of shape, size, or function—are made up of the same basic parts.

## Content Development

The nucleus is the director of activities for the cell. It is surrounded by a nuclear membrane that functions much like the cell membrane. A nucleolus and chromosomes are found inside the nucleus.

## Skills Development

***Skill: Interpreting illustrations***

Have students locate the nucleus in the plant cell (Figure 3-2) and in the animal cell (Figure 3-1). Now have them observe Figure 3-4, which is a closer look at this structure. Point out to students that a miniature version of Figure 3-2 appears in the upper left-hand corner of Figure 3-4. This is done so that the students can see the location of the nucleus in the cell. Next, the nucleus is "blown up" to given students a more detailed look. Finally, an electron photomicrograph of the nucleus is included to show what an actual nucleus looks like.

Have students read the caption for Figure 3-4.

- **What nuclear structures are shown in the diagram?** (nuclear membrane with pores, nucleolus, chromosomes)
- **What do the chromosomes look like?** (threads, worms, etc.)

Explain to students that the electron microscope is used to view cellular structures too small to be seen with a light microscope.

## ANNOTATION KEY

❶ **Thinking Skill: Comparing nucleic acids**

❷ **Thinking Skill: Sequencing events**

## FACTS AND FIGURES

A typical cell is composed of 70% water, 15% proteins, 10% fats, 4% DNA and RNA, and 1% carbohydrates.

hormones used inside and outside the cell. Enzymes and hormones regulate cell activities.

The two nucleic acids that make up the chromosomes are DNA and RNA. You can think of these nucleic acids as the "blueprints" of life. Working together, DNA and RNA store the information and ❶ carry out the steps in the protein-making process—a process necessary to life. The DNA remains in the nucleus. But the RNA, carrying its protein-building instructions, leaves the nucleus through a nuclear pore. So hitch a ride on the RNA leaving the nucleus and continue your exploration of the cell.

**NUCLEOLUS** As you prepare to leave the nucleus, you see a small, dense object float past. It is the **nucleolus** (noo-KLEE-uh-luhs), or "little nucleus." You are looking at a cell structure whose function remains something of a mystery. Biologists know that

### Career: *Medical Artist*

**HELP WANTED: MEDICAL ARTIST** Advanced degree in medical illustration required. College background in biology or zoology desirable. Needed to prepare accurate and creative drawings for medical textbooks.

Have you ever looked at a health book or a magazine about health that did not have diagrams? Probably not, because many concepts and objects related to health cannot be easily understood by reading only words. For example, cell structures and functions, body parts, and organism development can best be "described" through the use of drawings. Making such drawings is the job of a special group of artists.

Artists who create visual materials dealing with health and medicine are **medical artists.** Sometimes their work involves viewing a specimen through various kinds of microscopes in order to draw it. Or a medical artist may dissect and study the parts of animals and plants. Some medical artists work closely with surgeons or other kinds of doctors in order to prepare accurate drawings of medical conditions.

Most medical artists are employed by publishers, medical, veterinary, or dental schools, or by hospitals with programs in teaching and research. To receive more information about this field, write to the Association of Medical Illustrators, Route 5, Box 311F, Midlothian, VA 23113.

### 3-1 (continued)

### Content Development

Explain to students that chromosomes are composed of protein and two nucleic acids, DNA and RNA. DNA is the molecule of heredity. It stores all of the information that makes a human with his or her own set of unique characteristics.

- **Under what circumstances is the DNA of one person exactly like the DNA of another person?** (identical twins)

Inform students that each cell in an organism, except for sex cells, which will be discussed later, has the same number of chromosomes.

- **How many chromosomes are there in each cell in a human being?** (46)

Explain to students that chromosomes and DNA have two very important functions: (1) They pass on characteristics to new cells. In the case of egg and sperm cells, they pass on characteristics from parent to offspring. (2) They control the making of all of the proteins that are necessary for life.

Tell students that RNA also plays a role in this protein synthesis. RNA copies the protein-making codes from DNA and carries them out into the cytoplasm where the proteins are assembled. The nucleolus, a smaller structure within the nucleus, produces and stores RNA.

### Skills Development

***Skill: Making inferences***

Have students refer to Figure 3-5 for a look at the next structure they will study, the endoplasmic reticulum.

- **What does the endoplasmic reticulum look like?** (a series of canals or tubelike passageways)
- **Based on the shape of the endoplasmic reticulum, what do you think its function might be?** (transportation of materials)

### Content Development

Explain to students that the endoplasmic reticulum is indeed involved in transportation. The endoplasmic reticulum is involved in both the manufacture and transport of proteins. These proteins, as well as other materials, are transported to other locations within the cell or to the outside of the cell through this system of passageways, the endoplasmic reticulum.

the nucleolus is made up of RNA and protein. And they believe this tiny structure may play an important role in making proteins for the cell.

### *Endoplasmic Reticulum*

Outside the nucleus, floating in a clear, thick fluid, your ship needs no propulsion. For here the jellylike material called **cytoplasm** is constantly moving. Cytoplasm is the term for all the protoplasm, or living material of the cell, *outside* the nucleus.

Steering will be a bit difficult here because many particles and tubelike structures are scattered throughout the cytoplasm. You cannot help noticing a maze of tubular passageways that leads out from the nuclear membrane. These clear tubes form the **endoplasmic reticulum** (en-doh-PLAZ-mik ri-TIK-yuh-luhm). See Figure 3-5.

The endoplasmic reticulum is involved in the manufacture and transport of proteins. You can see that it is well suited for its job. Its network of passageways spreads throughout the cell. Proteins made in one part of the cell can pass through the endoplasmic reticulum to other parts of the cell. Other materials can be transported to the outside of the cell through this system. ❷

**Figure 3-5** *The endoplasmic reticulum* (left) *manufactures and transports proteins. In this photograph of the endoplasmic reticulum* (right), *the dark dots lining the tubelike passageways are ribosomes.*

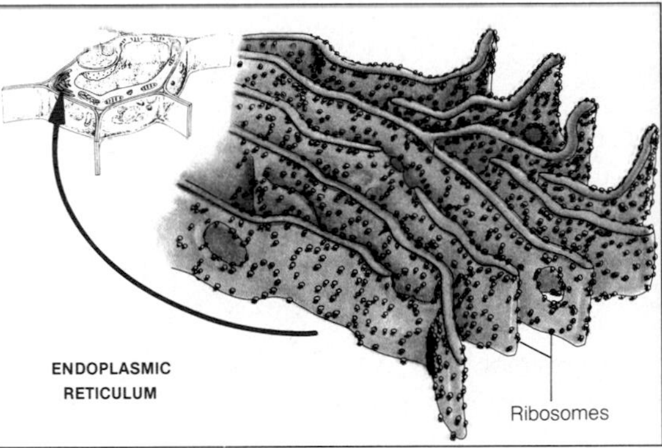


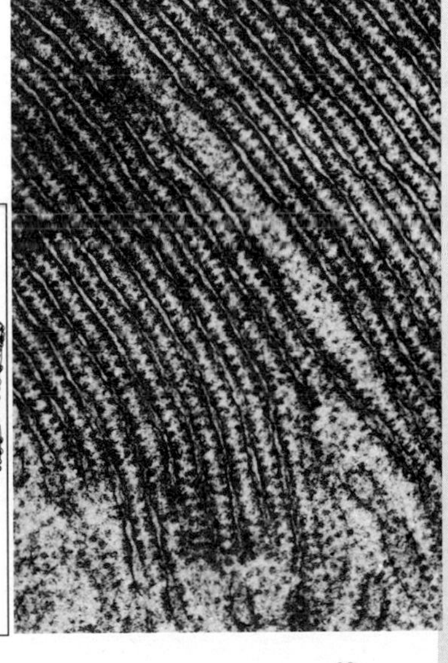



## BACKGROUND INFORMATION

To continue growing and dividing, cells must form additional cell membranes and duplicate cellular structures. Those cells that are not growing or dividing need to replace cellular structures and to synthesize proteins that are needed for normal cell functioning. The site of many of these processes is the endoplasmic reticulum (ER).

In an area of a cell that is actively synthesizing protein, the outer surfaces of the ER are covered with rounded bodies called ribosomes. This type of ER has a rough appearance and is thus called rough ER. ER that lacks ribosomes is called smooth ER.

## SCIENCE, TECHNOLOGY, AND SOCIETY

The study of cells was greatly advanced by the invention of the electron microscope in the 1930s. More recent technological developments have enabled scientists to look even more closely at cell structures.

In a method known as cell fractionation, cell parts can now be separated and studied individually. The cells are first ground up in a blender and then spun at high speeds (centrifuged). Nuclei, which are larger and denser than other cell parts, settle to the bottom. The isolated nuclei can then be broken down further.

Another recent development is the invention of a high-voltage electron microscope that permits biologists to view much thicker sections of biological materials. These huge microscopes stand more than 10 meters high and have a mass greater than 20 tons. They can produce images of the cell's internal structures as X-rays can produce photographs of your body's internal structures.

Students can conduct library research to find out about other biological tools and methods that have contributed to recent cell studies. They can also find out what kinds of discoveries have been made through the use of these tools and methods.

## Reinforcement

- **How many cellular structures that are involved in making proteins have been studied so far?** (nucleus, chromosomes, nucleolus, and endoplasmic reticulum)
- **What role does each of these have in the protein-making process?** (nucleus: control center for all cell activities; chromosomes: activity directors within the nucleus, contain DNA that has the code for all proteins; nucleolus: produces RNA; endoplasmic reticulum: transports proteins to other locations within and outside of the cell)

## Motivation

On a bulletin board, display photomicrographs of parts of a plant cell and an animal cell. Label all parts.

## BACKGROUND INFORMATION

Ribosomes are the sites of protein synthesis and are therefore found in almost all cells. Some ribosomes are attached to the endoplasmic reticulum, some are free in the cytoplasm, and some are inside other cellular structures.

Mitochondria have two membranes. The inner membrane is folded inward at various points. These folds, called cristae, increase the surface area of the membrane. Greater surface area means greater "working space."

## ANNOTATION KEY

❶ **It produces a great deal of protein. (Inferring)**

❷ **Muscle cells. They require a great deal of energy. (Inferring)**

❶ **Thinking Skill: Sequencing events.**

❷ **Thinking Skill: Hypothesizing**

## FACTS AND FIGURES

A cell may contain half a million ribosomes. This number varies with the amount of protein manufactured by the cell.

## FACTS AND FIGURES

Some cells, such as human liver cells, contain more than 1000 mitochondria.

### *Ribosomes*

Look closely at the inner surface of one of the endoplasmic passageways. Attached to the surface are grainlike bodies called **ribosomes,** which are made up mainly of the nucleic acid RNA. Ribosomes are the protein-making sites of the cell. The RNA in the ribosomes, along with the RNA sent out from the nucleus, directs the production of proteins. Ribosomes are well positioned as they not only help make ❶ proteins, they "drop" them directly into the endoplasmic reticulum. From there the proteins go to any part of the cell that needs them.

Not all ribosomes are attached to the endoplasmic reticulum. Some float freely in the cytoplasm. Watch out! There go a few passing by. The cell you are in has many ribosomes. What might this tell you about its protein-making activity? ❶

### *Mitochondria*

As you pass by the ribosomes, you see other structures looming ahead. These structures are **mitochondria** (migh-toh-KON-dree-uh; singular: mitochondrion), which are the main source of energy for the cell. Somewhat larger than the ribosomes, these rod-shaped structures are often referred to as the "powerhouses" of the cell. See Figure 3-6.

**Figure 3-6** *The mitochondrion is the "powerhouse" of the cell.*

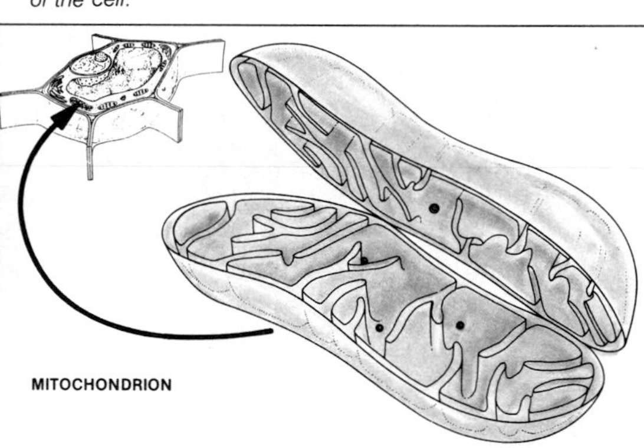




## 3-1 (continued)

### Content Development

Have students refer to Figure 3-5 and explain that the dots attached to the inner surfaces of the endoplasmic reticulum are structures called ribosomes. Remind students that RNA copies the protein-making codes from DNA and carries them out into the cytoplasm. The ribosomes are the destinations for this RNA; they are the sites of the protein making. The RNA from the nucleus works with the RNA in the ribosomes to assemble the proteins. Some ribosomes float freely in the cytoplasm, but most of them are attached to the endoplasmic reticulum.

- **Why are most of the ribosomes located on the endoplasmic reticulum?** (It is convenient and efficient to have a manufacturing site close to a means of transportation. Proteins made by the ribosomes are "dropped" into the endoplasmic reticulum, which carries them where they are needed in the cell.)

### Content Development

Explain to students that protein making and other cell activities require energy. Most of this energy is supplied by the mitochondria within the cell. This is a good example of how independent cell structures work for the good of the entire cell.

### Content Development

- **What kinds of cells would be expected to contain large numbers of**

Inside the mitochondria, simple food substances such as sugars are broken down. Large amounts of energy are released during the breakdown of sugars. The mitochondria gather this energy and store it in special energy-rich molecules. These molecules are convenient energy packages that the cell uses to do all its work. The more active a cell is, the more mitochondria it has. Some cells, such as a human liver cell, contain more than 1000 mitochondria. Would you expect your muscle cells or your bone cells to have more mitochondria? ❷

Because mitochondria have a small amount of their own DNA, some scientists believe that they were once tiny organisms that invaded living cells millions of years ago. The DNA molecules in the mitochondria were passed from one generation of cells to the next as simple organisms evolved into complex ones. Now all living cells contain mitochondria. No longer invaders, mitochondria are an essential part of living cells. ❷

### Vacuoles

That large, round, water-filled sac you see floating in the cytoplasm of this plant cell is called a **vacuole** (VA-kyoo-ohl). Both plant and animal cells have vacuoles. However, plant cells often have one very large vacuole while animal cells have a few small vacuoles.

Vacuoles act like storage tanks. Food, enzymes, and other materials needed by the cell are stored inside the vacuole sacs. Vacuoles also can store waste products. In plant cells, vacuoles are the main water storage areas. When water vacuoles in plant cells are full, they swell and make the cell plump. This plumpness keeps a plant firm. During its lifetime, a plant cell may increase to 500 times its original size because of an increase in the amount of water in its expanding vacuoles.

### Lysosomes

If you carefully swing your ship around the lakelike vacuole, you may be lucky enough to see a **lysosome.** Lucky because lysosomes are common in animal cells but not often observed in plant cells.

#### Activity

*Word Clues*

The definition of a word can often be determined by knowing the meaning of its prefix. Look up the prefix for each of the following words, and tell how it relates to the definition.

chloroplast
protoplasm
mitochondria
vacuole
chromosome
lysosome
cytoplasm



## Activity

**Word Clues**
**Skills: Applying definitions, identifying patterns, relating roots and word parts**
**Level: Average**
**Type: Vocabulary/writing**

This simple activity will reinforce how helpful understanding word parts can be in determining scientific definitions. Students should find that the prefix *chloro-* means "green" and chloroplasts are structures within cells of green plants that contain the pigment chorophyll. The prefix *proto-* means "first in time" or "first in importance" and protoplasm is a term used to describe all of the living materials in a cell. The first living thing, of course, contained protoplasm. The prefix *mito-* means "thread" and mitochondria are small rodlike (or threadlike) structures in a cell that supply most of the energy for the cell. The word *vacuole* comes from the Latin word *vacuus,* which means "empty." A vacuole is a relatively clear, fluid-filled sac within a cell. *Lyso-* comes from a Greek word pertaining to dissolving. Lysosomes are the structures within cells that contain enzymes for digestion. The prefix *cyto-* means "of the cell," and cytoplasm is the protoplasm outside of the nucleus of a cell.

**mitochondria?** (active cells such as muscle and liver cells)

- **How do mitochondria obtain the energy that is needed for the cell to do work?** (Simple food substances, such as sugars, are broken down. This breakdown releases energy.)
- **What kind of energy is released by the breakdown of food?** (chemical energy)

Explain to students that mitochondria have been a controversial issue in the development, or evolution, of cells. Mitochondria contain a small amount of their own DNA and can manufacture some of their own proteins. Therefore, some scientists think that mitochondria originated as separate organisms. These tiny organisms invaded living cells and resided within them. Eventually, over millions of years, they became essential parts of living cells rather than separate organisms.

## Content Development

Have students locate vacuoles in Figure 3-2 (plant cell) and Figure 3-1 (animal cell).

- **How do the vacuoles in plant cells differ from those found in animal cells?** (Plant cells generally contain one large vacuole, whereas animal cells may have several smaller ones.)

## BACKGROUND INFORMATION

Like mitochondria, chloroplasts contain DNA, are separated from the cytoplasm by two outer membranes, and reproduce independently of the nucleus. Chloroplasts are the site of photosynthesis, the process that uses the energy of sunlight to synthesize food molecules.

### Activity

**Plant and Animal Cells**
**Skills: Observing, comparing, manipulative, relating**
**Level: Remedial**
**Type: Hands-on**
**Materials: Onion skin, glass slide, iodine stain, medicine dropper, cover slip, microscope, toothpick, methylene blue**

Have students complete this activity after they have read and discussed the section on plant and animal cells. Point out the similarities of the cells: Both have mitrochondria, vacuoles, ribosomes, nuclei, nuclear membranes, chromosomes, cell membranes, and cytoplasm. The differences are that plant cells have chloroplasts and a cell wall. However, students will not see chloroplasts in onion cells since the onion is the part of the plant's stem that grows underground. To view chloroplasts, have students observe cells from an *Elodea* leaf.

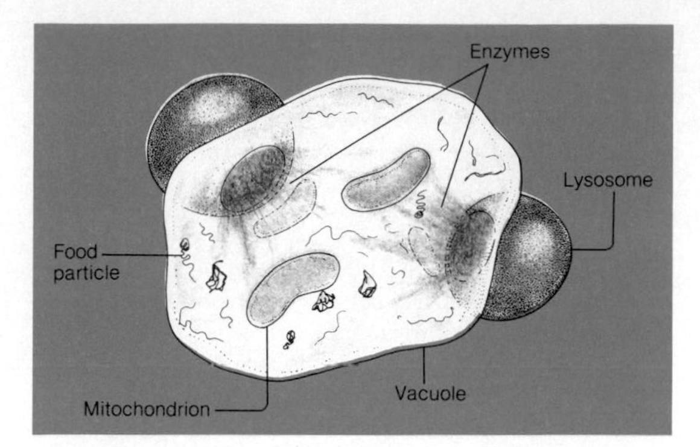


**Figure 3-7** *Enzymes from a lysosome digest the contents of a vacuole.*

### Activity

*Plant and Animal Cells*

**1.** To view a plant cell, remove a very thin transparent piece of tissue from an onion.
**2.** Place the onion tissue on a glass slide.
**3.** Add a drop of iodine stain to the tissue and cover with a cover slip.
**4.** Observe the onion under the low and high power of your microscope. Draw a diagram of one onion cell and label its parts.
**5.** To view an animal cell, gently scrape the inside of your cheek with a toothpick.
**6.** Gently tap the toothpick on the center of another glass slide.
**7.** Add a drop of methylene blue and cover with a cover slip.
**8.** Observe the cheek cells under the high and low power of your microscope. Draw a diagram of one cell and label its parts.

Compare the two cells.

Lysosomes are small, round structures involved with the digestive activities of the cell. See Figure 3-7. Lysosomes contain enzymes that break down large food molecules into smaller ones. Then these smaller food molecules are passed on to the mitochondria, where they are "burned" to provide energy.

Although lysosomes digest substances in the cytoplasm, you need not worry about your ship's safety! The membrane surrounding a lysosome keeps the enzymes from escaping and digesting the entire cell. However, lysosomes do digest whole cells when the cells are injured or dead. In an interesting process in the growth of a tadpole into a frog, lysosomes in the tadpole's tail cells digest the tail. Then this protoplasmic material is reused to make new body parts for the frog. (1)

### Chloroplasts

Your journey is just about over, and you will soon be leaving the cell. But first look around you. Floating in the cytoplasm are large, irregularly shaped structures that are easily sighted. They are green! These structures are **chloroplasts,** and they contain a green pigment called chlorophyll. Chlorophyll captures the energy of sunlight and uses it to make food for the plant cell.

Chloroplasts are found *only* in plant cells. However, the sea slug, an animal, often eats plants that

66

## 3-1 (continued)

### Content Development

Have students examine Figure 3-8. Explain that the large green structures are called chloroplasts.

- **In what kind of cells are chloroplasts found?** (cells of green plants)
- **Why are chloroplasts green?** (They contain a green pigment called chlorophyll.)
- **What is the function of chlorophyll?** (It captures the sun's energy and uses it to make food for the plant.

### Reinforcement

Using flashcards that have the names of each of the cell structures on one side and location, function, and the type of cell in which each is found on the other side, ask any of the following questions that are appropriate for each flashcard.

- **Where is this structure located?** (Answers will vary.)
- **What is the function of this structure?** (Answers will vary.)
- **Is this structure found in plant cells, animal cells, or both?** (Answers will vary.)

As students answer the questions, turn the cards over to display the correct answers. Suggest that students may wish to make their own set of flashcards for study purposes.

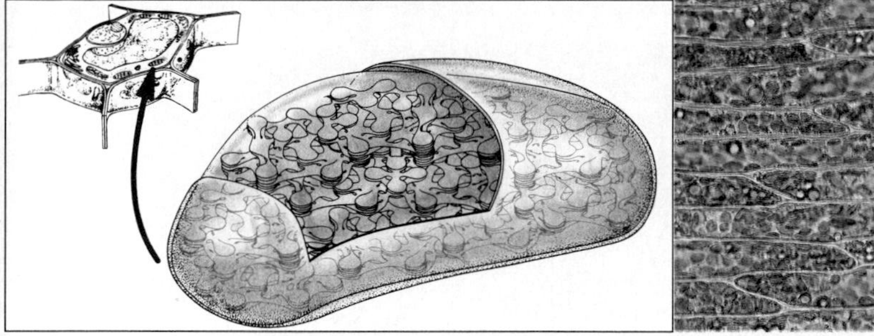
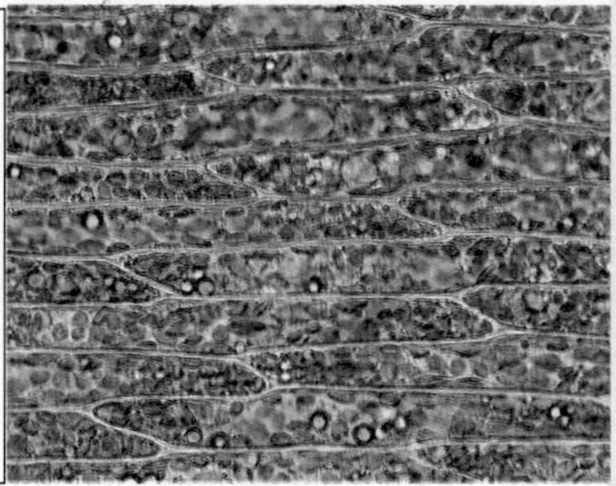

**Figure 3-8** *Chloroplasts* (left), *the food-making sites, are found only in the cells of green plants* (right).

contain chloroplasts. After the sea slug digests the plant, some of the chloroplasts get into the cells of its digestive system. There the chloroplasts continue to make food just as they do in a plant. The process goes on for a week or so and provides the sea slug's cells with food for energy. ❷

### *SECTION REVIEW*

1. What does protoplasm mean?
2. What two nucleic acids are found in the nucleus?
3. Name the structures in the cytoplasm needed for the manufacture and transport of proteins.
4. Describe three differences between plant cells and animal cells.

## 3-2 Cell Activities

Each structure within a cell performs a vital activity. And the cell as a whole carries out the chemical processes necessary to life. **Life activities performed by cells include metabolism, diffusion, osmosis, and reproduction.** The tiny cell is like a miniature factory that produces many kinds of chemicals. Like any factory, a cell must have energy to do work. Working day and night, a cell traps, converts, stores, and uses energy.



## ANNOTATION KEY

❶ **Thinking Skill: Applying concepts**
❷ **Thinking Skill: Relating concepts**

## Section Review 3-1

1. All the living material in a cell.
2. DNA and RNA.
3. Ribosomes and endoplasmic reticulum.
4. Plants cells have cell walls, chloroplasts, few lysosomes, and usually one large vacuole. Animal cells do not have cell walls or chloroplasts and have many lysosomes and vacuoles.

## TEACHING STRATEGY 3-2

### Motivation

Explain to students that cells are somewhat like miniature factories. Refer them to Figure 3-9 and ask them to complete a chart that compares cells to factories. You may wish to allow them to work in groups of three to four students. The three column headings for their charts should be "Cell Part," "Factory Counterpart," and "Function."

### Content Development

• **Where is the energy needed to operate factories obtained?** (Most factories burn fossil fuels—oil, coal, or natural gas; others use hydroelectric or nuclear power. A few may burn wood or other types of biomass, use wind, or utilize geothermal energy.)

Explain to students that cells, like factories, must have energy to do work. Even while they are sleeping, their bodies need energy to keep them alive. Many body activities are still going on while they sleep.

• **What are some of these activities?** (Accept all reasonable responses such as the heart pumping blood through the body, breathing, cells growing and reproducing, food being broken down, the brain sending out signals, etc.)

Stress that all of these activities require energy.

• **Can cells make energy?** (No. Energy cannot be created or destroyed. However, cells can change energy from one form into another.)

• **Where do cells get energy?** (They obtain it from food.)

Tell students that in order to release the stored chemical energy in food, it must undergo many chemical reactions. The sum of these reactions, which involves the building up and breaking down of molecules, is called metabolism.

• **Which releases energy: putting together new molecules or breaking down existing molecules?** (Breaking down existing molecules releases energy. Putting together new molecules uses energy.)

## 3-2 CELL ACTIVITIES

### SECTION PREVIEW 3-2

The vital functions of metabolism, diffusion, osmosis, and reproduction are carried out by specific cell structures. Metabolism is the general term used to describe the total of the chemical activities carried out by an organism.

Diffusion is the process by which materials enter and leave the cell through pores in the cell membrane. The net movement of molecules is always from areas of higher concentration to areas where there are fewer of that type of molecule. Osmosis is a special kind of diffusion that involves the movement of water molecules.

Cells must divide, or reproduce, in order for an organism to grow. This cell reproduction is accomplished through a precise series of steps that ensures that the new cell is exactly like the parent cell.

### SECTION OBJECTIVES 3-2

1. **Explain the similarities and differences between diffusion and osmosis.**
2. **Compare the processes of mitosis and meiosis.**

### SCIENCE TERMS 3-2

**metabolism p. 68**
**diffusion p. 69**
**osmosis p. 70**
**mitosis p. 71**
**meiosis p. 73**

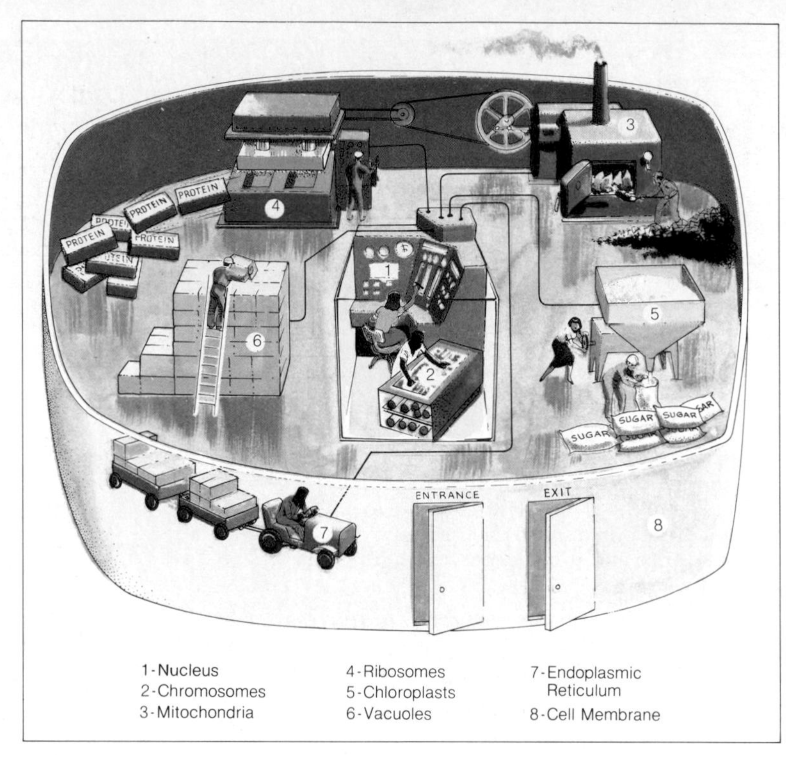


**Figure 3-9** *A cell, such as this plant cell, is like a miniature factory that carries out all the activities necessary to life. What is the function of structure 7?* ❶

### *Metabolism*

Even while you sleep, you need energy to keep you alive. But where does this energy come from? Cells provide it. Although cells do not *make* energy, they do *change* energy from one form to another. Cells obtain energy from their environment and convert it into a usable form.

This conversion process is very complex. And it involves many chemical reactions. Some reactions break down molecules. Other reactions build new molecules. ❶ The sum of all the building-up and breaking-down activities that occur in a living cell is called **metabolism.**

68

### 3-2 (continued)

#### Content Development

Explain to students that metabolism is a continuous process that begins in the digestive tract and the lungs and goes on in every cell of the body. The raw materials of food, water, and oxygen are converted into living tissue, energy, and waste products.

#### Motivation

Open a bottle of perfume or other strong-smelling substance. Ask students to raise their hands when they can smell it. Have them note the pattern of diffusion of the substance through the room.

- **Are the perfume molecules more concentrated inside or outside the bottle?** (inside)
- **In which direction do they tend to move?** (out of the bottle)
- **After a minute or so, are the perfume molecules more concentrated in the front of the room (near the open bottle) or in the back of the room?** (The front of the room; students nearest the bottle smelled it first.)
- **In which direction do the perfume molecules tend to move?** (from the front to the back of the room—outward in all directions from the bottle)

Think for a moment of all the things cells do: grow, repair structures, absorb food, manufacture proteins, get rid of wastes, and reproduce. The energy for these activities is locked up in the molecules in food. As a result of metabolism, the stored energy in food is set free so it can be used to do work.

### *Diffusion*

Remember how you sailed through the cell membrane to enter the cell? Well, the substances that get into and out of the cell do the same thing. Food molecules, oxygen, water, and other materials enter and leave the cell through openings in the cell membrane by a process called **diffusion.**

Why does diffusion occur? Molecules of all substances are in constant motion, continuously colliding with one another. This motion causes the molecules ❷ to spread out. The molecules move from an area where there are more of them to an area where there are fewer of them. See Figure 3-10.

If there are many food molecules outside the cell, for example, some diffuse through the membrane into the cell. At the same time, waste materials built up in the cell pass out of the cell by diffusion.

You can observe the process of diffusion for yourself. Drop some ink into a glass of water and watch what happens. The drop of ink spreads throughout the water, getting lighter in color as the ink molecules move through the liquid.

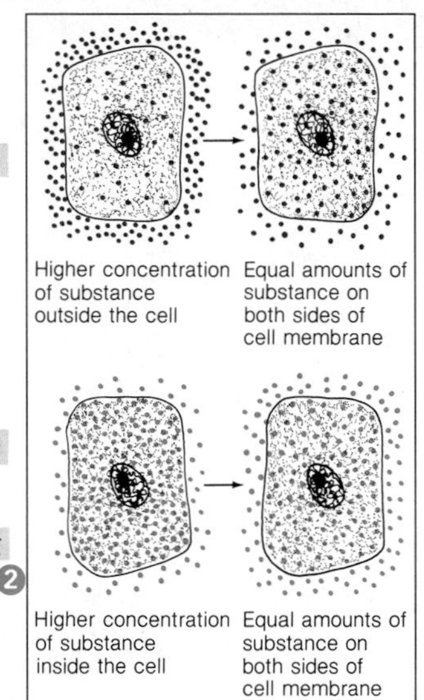


**Figure 3-10** *Diffusion is the movement of molecules of a substance into a cell* (top) *or out of a cell* (bottom). *Substances move from places where they are more concentrated to places where they* ❸ *are less concentrated.*

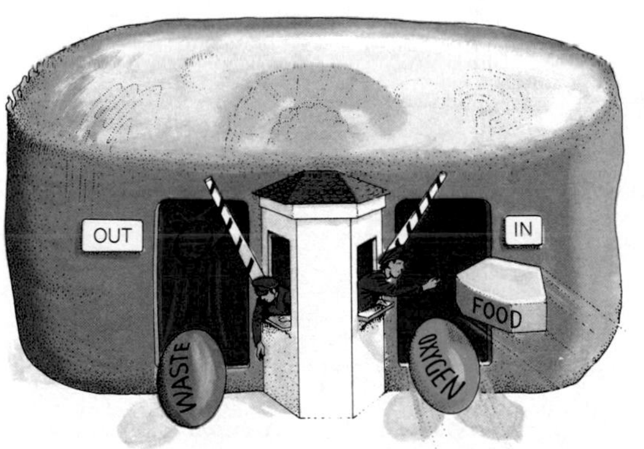


**Figure 3-11** *The cell membrane is selective, permitting oxygen and food molecules to enter and waste materials to leave.*



## TEACHER DEMONSTRATION

To illustrate the general process of diffusion, add a few drops of vegetable coloring to a beaker of water. Students can observe diffusion as the vegetable coloring molecules spread out through the water. In time, the water will become a uniform color.

- **Why do the molecules of vegetable coloring spread out?** (Molecules are in constant motion and move from areas where they are highly concentrated to areas where there are fewer of them.)
- **If there were two beakers of water—one containing hot water and the other cold water—how would diffusion be affected?** (Accept all predictions, then demonstrate. The vegetable coloring will diffuse faster in hot water than in cold water.)
- **What general conclusions can be drawn from this demonstration?** (Molecules diffuse more rapidly at higher temperatures or the rate of diffusion is affected by temperature.)
- **What other factors might affect the rate of diffusion?** (initial concentration of the substances and pressure)

## ANNOTATION KEY

❶ **It is involved in the manufacture and transport of proteins. (Applying definitions)**

❶ **Thinking Skill: Applying definitions**

❷ **Thinking Skill: Relating cause and effect**

❸ **Thinking Skill: Making observations**

Point out that in all cases, the movement of molecules in diffusion is from an area of higher concentration to areas of lower concentration.

## Content Development

Remind students that molecules of all substances, even solids, are in constant motion. Such collisions cause the molecules to spread out. The movement of molecules will be from an area where there are more of them to an area where there are less of them.

Have students refer to Figure 3-10.

- **When does the diffusion process stop?** (when equal concentrations of the substance are on both sides of the cell membrane)
- **Do molecules stop moving when diffusion is complete?** (No. Molecules are still in constant, random motion.)
- **What prevents harmful materials from diffusing into a cell or needed materials from "oozing" out of the cell?** (The cell membrane acts as a gatekeeper. It is selectively permeable—only allowing certain materials to pass in or out.)
- **What is the most important substance that passes through the cell membrane?** (water)

## Activity

**The Big Egg**
**Skills: Inferring, observing, comparing, analyzing data, measuring, hypothesizing**
**Level: Remedial**
**Type: Hands-on**
**Materials: Hardboiled egg, string, metric ruler, beaker of water**

In this activity, students observe the process of diffusion by watching a hardboiled egg increase in circumference when it is placed in water. Point out to students that because water is passing through the egg membranes, this process is called osmosis. You may want to have students observe water diffuse out of the egg in a similar manner by placing a hardboiled egg in a concentrated salt solution.

## FACTS AND FIGURES

There are more than 26 billion cells in a newborn human infant. This number increases to approximately 100 trillion by adulthood.

## ANNOTATION KEY

❶ Thinking Skill: Relating cause and effect
❷ Thinking Skill: Sequencing events
❸ Thinking Skill: Applying concepts

### Activity

*The Big Egg*

Try this activity and you will see osmosis for yourself!

1. Prepare a hardboiled egg.
2. Peel it very carefully.
3. Using a piece of string, measure the egg's circumference in millimeters. Record this information in a chart.
4. Place the egg in a beaker of water for five days. Measure the circumference of the egg every day and record this in the chart.

How does the circumference of the egg vary? Can you observe any other changes in the egg's appearance? What process is demonstrated in this activity?

If substances can move into and out of the cell through the membrane, what keeps the protoplasm from oozing out? What keeps harmful materials from moving in? The cell membrane safeguards the contents of the cell because it is selective. That is, it permits only certain substances, mainly oxygen and food molecules, to diffuse into the cell. Only waste products such as carbon dioxide are allowed to diffuse out of the cell.

### Osmosis

Water is the most important substance that passes through the cell membrane. In fact, about 80 percent of protoplasm is water. Water passes through the cell membrane by a special type of diffusion called **osmosis.** During osmosis, water molecules move from a place of greater concentration to a place of lesser concentration. This movement keeps the cell from drying out.

Suppose you put a cell in a glass of salt water. The concentration of water outside the cell is lower than the concentration of water inside the cell. So water leaves the cell, and the cell starts to shrink. If too much water leaves the cell, the cell dries up and dies.

❶ Now, if the cell was put in a glass of fresh water instead of salt water, just the opposite occurs. Water enters the cell, and the cell swells. This happens because the concentration of water is lower inside the cell than outside the cell. As you might imagine, if too much water enters the cell, the cell bursts.

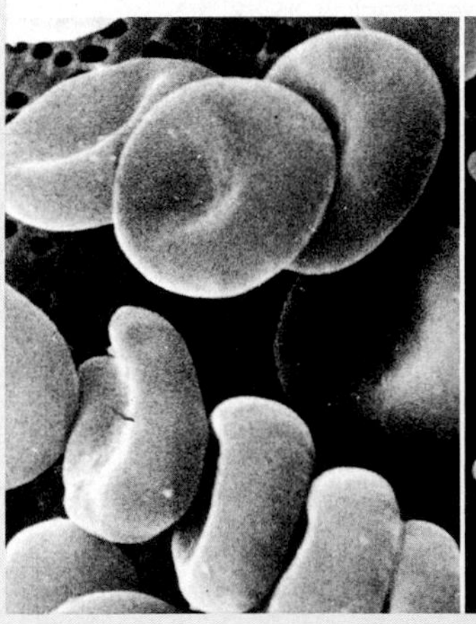
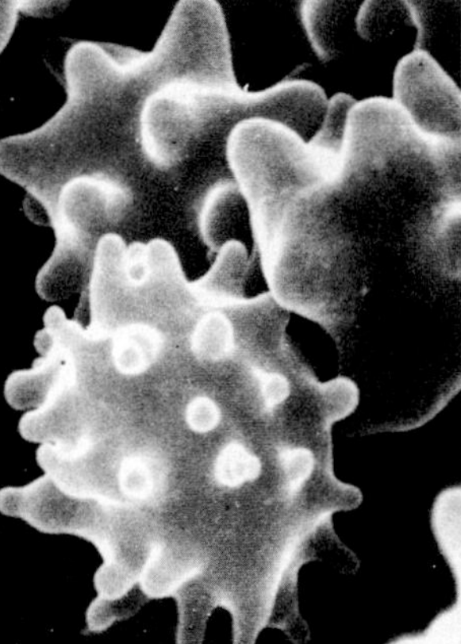

**Figure 3-12** *Normal red blood cells* (left) *will shrink* (center) *if too much water leaves the cells. If too much water enters the cells, they will swell* (right).

## 3-2 (continued)

### Motivation

Show students several slices of raw potato and celery that have been immersed in fresh water for several hours and several other slices that have been immersed in salt water for the same length of time.

• **How do those potato and celery slices differ?** (The ones from the salt water are limp and more flexible. They can be bent without breaking. The ones from the fresh water are stiffer and crisper.)

• **How do you explain this difference?** (When placed in a saltwater environment, the concentration of the water molecules inside the potato and celery cells is higher than it is outside the cells. Therefore, water moves out of the potato and celery cells, causing them to shrink. In fresh water, the concentration of water molecules is greater outside of the cells. Therefore, water moves into the potato and celery cells. The vacuoles in the cells become filled with water and push against the cell walls making the potato and celery slices stiff.)

### Content Development

Explain to students that the process in which water molecules move into and out of cells is called osmosis. Osmosis is a special kind of diffusion that is limited to the movement of water. Osmosis operates in the same manner as diffusion; the water mole-

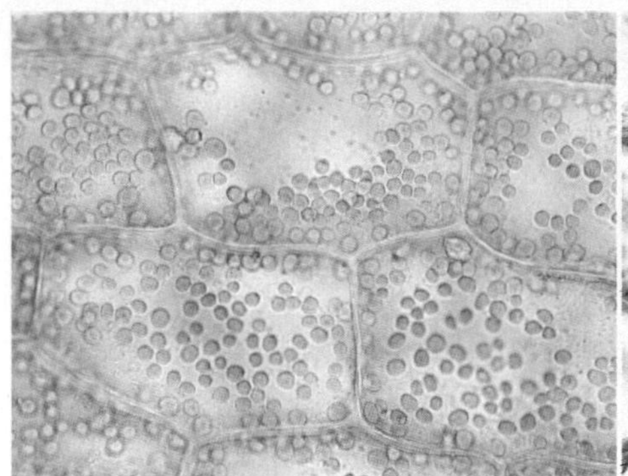
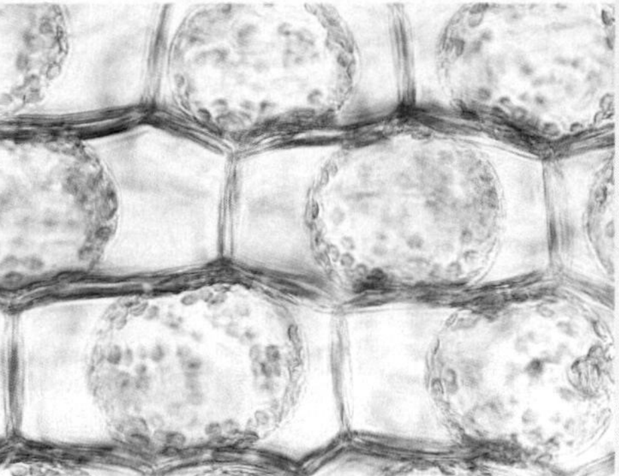

**Figure 3-13** *In these normal plant leaf cells (left), the movement of too much water from the cell vacuoles causes the cell contents to shrink away from the cell wall (right).*

## Reproduction

How many cells do you think you are made of? The number of cells in an organism depends on its size, and so no specific number can be given. But you may be surprised to learn that as an adult, you will consist of about 50 thousand billion cells! All of these cells come from just one original cell.

**GROWTH AND DEVELOPMENT** Body cells must undergo reproduction in order for the total number of cells to increase and for the organism to grow. These cells reproduce by dividing into two new cells. Each new cell, called a daughter cell, is identical to the other and to the parent cell. How does this process occur?

If a parent cell—a skin cell, leaf cell, or bone cell, for example—is to produce two identical daughter cells, then the exact contents of its nucleus must go into the nucleus of each new cell. In other words, the "blueprints" of life in the parent cell must be passed on to each daughter cell. Now, if the parent cell simply splits in half, each daughter cell would get only half the contents of the nucleus—only half the "blueprints." It would no longer be the same kind of cell—a skin cell, leaf cell, or bone cell. ❷

Fortunately, this is not what happens. For just before a cell divides, all the material in the nucleus is duplicated, or copied. The duplication and division of the nucleus and of the chromosomes is called **mitosis** (migh-TOH-sis). ❸

### Activity

*Making Cell Models*

Make an original cell model using food products.

1. Dissolve some colorless gelatin in warm water.
2. Pour the gelatin in a rectangular pan (for a plant cell) or a round pan (for an animal cell).
3. Find edible materials that resemble cell structures. Place these materials in the gelatin before it begins to gel.
4. On a sheet of paper, identify each cell structure.

71

### Activity

**Making Cell Models**
**Skills: Identifying, observing, applying**
**Level: Remedial**
**Type: Hands-on**
**Materials: Colorless gelatin, warm water, rectangular or round pan, various edible materials**

This activity will help students become familiar with the shapes and structures of plant and animal cells. Not only will the students use their creative skills in determining which edible materials they will use, but also they will use their critical thinking skills in identifying the structures. After the students have constructed their "cells," you may wish to have the students display them or have a "cell-eating party."

cules always move from an area where they are more concentrated to areas where there are fewer water molecules.

### Content Development

Explain to students that all cells do not reproduce at the same time. If they did, there would soon be far too many cells, as illustrated by the story about doubling. Cells reproduce more often during the growing years,

Tell students that throughout their lives some cells are reproducing every day to replace worn out or dead cells. New cells must be produced in a precise manner so that they contain all of the information that the old cells carried.

- **In what part of the cell is this important (the blueprints for your body) carried?** (in the DNA of the chromosomes, which are in the nucleus)

Explain to students that if a parent cell simply split in half to form two new daughter cells, each new daughter cell would receive only one-half of the DNA (or blueprints). At the next division, each of the four new cells would contain only one-fourth of the original DNA information, and so on. Students should see that this process would not work effectively.

To guard against this loss of information, the DNA in the nucleus is duplicated just prior to cell division. Then, the nucleus and cytoplasm divide, forming two new daughter cells. Each new cell has all of the DNA information that the parent cell contained. This process is called mitosis, or cell division.

## BACKGROUND INFORMATION

In mitosis, a parent cell divides to form two new daughter cells. During mitosis, the cell passes through five stages: interphase, prophase, metaphase, anaphase, and telophase. The new cells are smaller, but they are identical to the original cell, having the same number of chromosomes as the original cell.

Meiosis results in the production of gametes, or reproductive cells. The original cell divides twice, thus forming either egg or sperm cells that have only half the number of chromosomes as the original cell, or all other body cells, have. Upon fertilization, then, the resulting zygote, or fertilized egg, has a full set of chromosomes, half contributed from the egg and the other half from the sperm.

## FACTS AND FIGURES

At any given time, about 35% of the cells in the human body are undergoing mitosis. Some cells, such as nerve and muscle cells, never divide once they are formed.

## FACTS AND FIGURES

Mitosis takes from 45 minutes to 5 hours, depending on the cell. Prophase is generally the longest of the stages.

**Figure 3-14** *During mitosis, each chromosome is duplicated before the nucleus divides. In this drawing, the magician represents a nucleus. And the balloons represent chromosomes.*

Mitosis is a very complex process that occurs in several stages. During mitosis, each chromosome in the nucleus makes a copy of itself. See Figure 3-14. When the nuclear membrane disappears, one set of chromosomes moves to one end of the parent cell, and the other set of chromosomes moves to the opposite end. A new nuclear membrane forms around each set of chromosomes. The parent cell then divides into two. Each daughter cell contains an exact copy of the chromosomes of the parent cell. The complete "blueprint" of life is passed from one cell to another. The two new skin cells are exactly like the parent skin cell and so are able to do the same job. And the two new nerve cells are identical in form and function to their parent nerve cell.

Mitosis, then, is the process by which the cells of an organism reproduce to form exact copies of one another. Through mitosis, both plants and animals grow larger, repair damaged parts, and replace dead

**Figure 3-15** *In one stage of mitosis, the chromosomes in the nucleus are duplicated* (left). *In another stage, one of each pair of chromosomes moves to the opposite ends of the nucleus* (center). *Finally, the nucleus splits and two new cells with identical chromosomes are formed* (right).

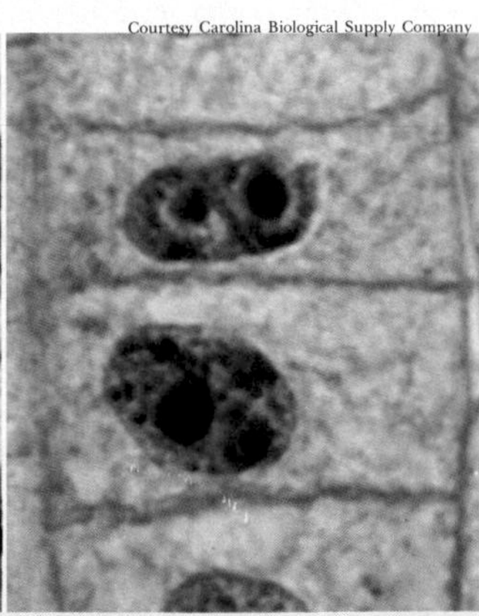

### 3-2 (continued)

#### Content Development

You may wish to go into greater detail concerning the process of cell division. Point out the stages that are involved in this process and have students view prepared slides to observe each of these stages. Prepared onion root-tip slides work well for this activity and can be ordered from a biological supply house or borrowed from the high school biology department.

#### Content Development

Have students follow the stages of mitosis pictured in Figure 3-15 as you explain them.

Stress that during the first stage (photograph at left), the chromosome material is duplicated so that each new cell will have the same amount of DNA as the parent cell. In another stage (center photograph), the chromosome pairs are pulled apart and move to opposite ends of the nucleus. In the final stage of mitosis (photograph at right), two new cells are formed, and the cytoplasm divides in half with a new cell membrane forming between the two new cells. In plant cells, the cytoplasm does not divide in this manner. Rather, a cell plate forms between the two new cells and divides the cytoplasm in half. The cell plate then becomes part of the new cell wall between the plant cells.

Reinforce the concept that the stages of cell division ensure that each new cell receives all of the DNA information carried by the parent cells. For example, the root tip of a plant contains many cells undergoing mitosis. The formation of new cells makes the root longer and allows it to push through the ground toward water. ❷

**PRODUCTION OF OFFSPRING** Mitosis is a method to reproduce cells inside an organism. However, there is another kind of reproduction in which a new organism comes from two parents. This is called sexual reproduction. Sexual reproduction involves the joining of two special types of cells—the male sex cell and the female sex cell—in a process known as fertilization. The offspring of fertilization contains a *combination* of the chromosomes of the male sex cell and the female sex cell.

Sex cells reproduce by a special type of cell division called **meiosis** (migh-OH-sis). Meiosis is a form of cell division that *halves* the number of chromosomes in a male or female sex cell as it forms. When the sex cells later join, the offspring gets half its chromosomes from the male and the other half from the female. The offspring's cells now have the same number of chromosomes as the cells of the parents. ❸

### Activity

*How Many Cells?*

Suppose a cell splits in half every 10 minutes. How many cells will there be after 1½ hours?

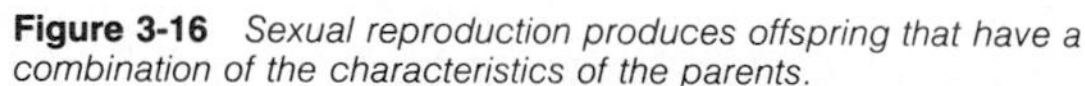

Figure 3-16 *Sexual reproduction produces offspring that have a combination of the characteristics of the parents.*

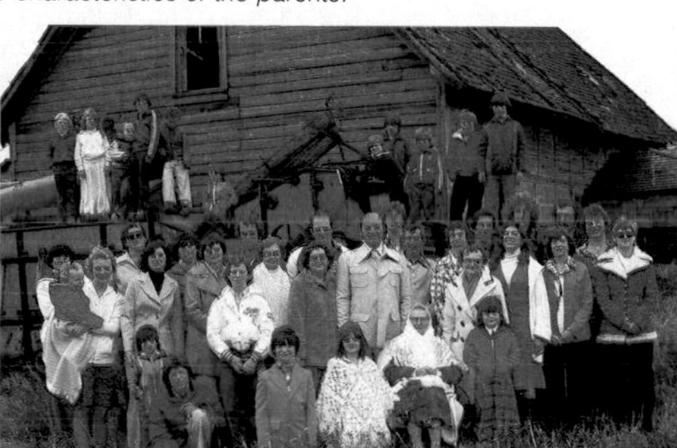

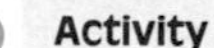

### SECTION REVIEW

1. What is metabolism?
2. Describe the process of osmosis.
3. What process causes an organism to grow?

73

## ANNOTATION KEY

❶ Thinking Skill: Sequencing events
❷ Thinking Skill: Applying concepts
❸ Thinking Skill: Making comparisons

## FACTS AND FIGURES

In some algae and fungi, mitosis is not followed by the division of the cytoplasm. This results in the formation of multinucleate cells.

cell. As a result, each new daughter cell has exactly the same number and kind of chromosomes as the parent cell.

## Content Development

Explain to students that mitosis is responsible for creating more cells within an existing organism. In order to make a new organism, sexual reproduction must occur between the male sex cell, or sperm cell, and the female sex cell, or egg cell. This process is called fertilization, and results in an offspring that has a combination of chromosomes from both sex cells.

Explain that sex cells are created by a process called meiosis. In this process, the number of chromosomes in a sex cell is halved so when two sex cells unite to form a new organism, that organism has the normal number of chromosomes.

Remind students that a normal human being has 46 chromosomes in its body cells.

- **After undergoing meiosis, how many chromosomes will a human sex cell have?** (23)
- **If each sex cell (male and female) contributes all of its chromosomes when joining together, how many chromosomes will the new human being have?** (46)
- **How many chromosomes would the new human being have if its parents' sex cells did not undergo meiosis?** (92)

To help avoid confusion, prepare overhead transparencies ahead of time or draw each stage of meiosis on acetate as it is explained.

## Reinforcement

Have students make models of the various stages of mitosis and meiosis. This can be done as a small group project or as a homework assignment. Posters and clay models are two of the simplest kinds of models to make. The posters can be made three-dimensional by using pipe cleaners or yarn for chromosomes, string for the cell membrane, jellybeans for nuclei, and so on.

## Section Review 3-2

1. Sum of building-up and breaking-down activities in a cell.
2. Water passes from a place of greater concentration to a place of lesser concentration.
3. Mitosis.

# LABORATORY ACTIVITY THINGS LOOK DIFFERENT UNDER A MICROSCOPE

## BEFORE THE LAB

1. **Cut out small pieces (5 mm × 5 mm) of newspaper print and colorful photographs from old magazines and newspapers.**
2. **Gather all other materials. You will need as many sets of materials as you have microscopes.**
3. **Divide the class into groups of two to four students, depending on the number of microscopes available.**

## PRE-LAB DISCUSSION

If this is the first time that students will be using compound microscopes, you will need to go over the parts and care of the microscope. (See Appendix C on page 555.)

Instruct students to read through the complete laboratory procedure first, and then to reread and follow it step-by-step. Discuss the procedure by asking questions like the following:

- **What is the purpose of this laboratory activity?** (to learn how to observe objects using a microscope)
- **Why is it necessary to place a cover slip over the material that is being observed?** (The cover slip protects the lens in the objective from coming into contact with the specimen.)
- **When using a microscope with several objectives, which objective should be used first when focusing on an object?** (Always use the lowest-power objective first.)
- **Why should the fine adjustment knob only be used when focusing under high power?** (The coarse adjustment knob moves the objective much too quickly (in greater increments). There is danger of crashing into the slide and causing damage to both the lens objective and the slide.)

LABORATORY ACTIVITY

### *Things Look Different Under a Microscope*

**Purpose**

In this activity, you will examine newspaper and magazine paper under a microscope.

**Materials** *(per group)*
Microscope
Microscope slide
Cover slip
Small pieces of newspaper print
Small pieces of colorful magazine photographs

**Procedure**

1. Obtain a small piece of newspaper print and place it on a clean microscope slide. Cover the slide with a cover slip.
2. Place the slide on the stage of the microscope. The newspaper should be facing up and be in the normal reading position. **CAUTION:** *Always use a cover slip to protect the objective lens from coming into contact with the material being observed.*
3. With the low-power objective in place, follow your teacher's directions for focusing on a letter.
4. While focusing on a letter, move the slide to the left. Which way does the letter seem to move? Now move the slide to the right. Which way does the letter seem to move this time?
5. While looking through the eyepiece, adjust the slide so that a letter is in the center of your field of view. Now, looking at the stage and objectives from the side, revolve the nosepiece until the high-power objective clicks into place. Using *only* the fine adjustment knob, bring the letter into focus. Describe what you see. Now follow the same steps using magazine paper.

**MICROSCOPE PARTS AND THEIR FUNCTIONS**

1. ***Arm*** Supports the body tube
2. ***Eyepiece*** Contains the magnifying lens you look through
3. ***Body tube*** Maintains the proper distance between the eyepiece and objective lenses
4. ***Nosepiece*** Holds high- and low-power objective lenses and can be rotated to change magnification
5. ***Objective lenses*** Low-power lens usually magnifies 10 times; high-power lens usually magnifies 40 times
6. ***Stage clips*** Hold the slide in place
7. ***Stage*** Supports the slide
8. ***Diaphragm*** Regulates the amount of light let into the body tube
9. ***Mirror*** Reflects the light upward through the diaphragm, the specimen, and the lenses
10. ***Base*** Supports the microscope
11. ***Coarse-adjustment knob*** Moves the body tube up and down for focusing
12. ***Fine-adjustment knob*** Moves the body tube slightly to sharpen the image

**Observations and Conclusions**

1. What letter did you choose? How does this letter appear when viewed through the microscope? Focus on another letter. How does it appear? What conclusion can you draw about the way objects appear when viewed under the microscope?

74

## SKILLS DEVELOPMENT

Students will use the following skills while completing this activity.

1. Safety
2. Observing
3. Manipulative
4. Comparing
5. Applying
6. Hypothesizing

## SAFETY TIPS

Make sure students understand the proper procedure for handling a microscope. Caution them that glass microscope slides and cover slips are very fragile. Alert students to notify you immediately if they should break any glassware, particularly if they are cut by the broken glass.

# CHAPTER REVIEW

## SUMMARY

### 3-1 Cell Structure and Function

- The cell wall, found only in plant cells, is made of nonliving cellulose and gives protection and support to the cell.
- Protoplasm is all the living material of a cell.
- The cell membrane is a thin, flexible membrane that regulates the movement of materials into and out of the cell.
- The control center of the cell is the nucleus. It is surrounded by the nuclear membrane.
- Chromosomes found in the nucleus are made of nucleic acids, which are the "blueprints" of life.
- Cytoplasm is all the protoplasm outside the nucleus.
- The endoplasmic reticulum, a network of tubelike passageways, is the site of the manufacture and transport of proteins.
- Ribosomes, often attached to the endoplasmic reticulum, are the protein-making sites.
- Mitochondria are the "powerhouses" of the cell.
- Vacuoles store food, water, enzymes, and waste products.
- Lysosomes play a role in the digestive activities of the cell.
- Chloroplasts, found only in green plant cells, capture the energy of the sun and use it to make food for the cell.

### 3-2 Cell Activities

- The sum of all the activities that occur in a living cell is called metabolism.
- Food, oxygen, water, and other materials enter and leave the cell by a process called diffusion.
- Water passes through the cell membrane by a type of diffusion called osmosis.
- Mitosis is the duplication and division of the nucleus and of the chromosomes during cell division, which results in daughter cells identical to the parent cell.
- During the formation of the male and female sex cells, meiosis takes place and the chromosome number is halved.

## VOCABULARY

*Define each term in a complete sentence.*

| | | | |
|---|---|---|---|
| **cell** | **cytoplasm** | **metabolism** | **nucleus** |
| **cell membrane** | **diffusion** | **mitochondria** | **osmosis** |
| **cell wall** | **endoplasmic reticulum** | **mitosis** | **protoplasm** |
| **chloroplast** | **lysosome** | **nuclear membrane** | **ribosome** |
| **chromosome** | **meiosis** | **nucleolus** | **vacuole** |

## CONTENT REVIEW: MULTIPLE CHOICE

*Choose the letter of the answer that best completes each statement.*

**1.** The cell wall is made of a nonliving material called
a. protoplasm. b. nucleic acid. c. cellulose. d. chromosomes.

75

# GOING FURTHER: ENRICHMENT

**Part 1**

To practice focusing procedures, have students examine a variety of prepared slides. You might have several "mystery slides" (labeled by number only) with which students attempt to identify the material being observed.

**Part 2**

To illustrate the concept "depth of focus," have students prepare a wet-mount slide of two pieces of thread in the form of an X. They will find that when the top thread is clearly in focus, the bottom thread will be slightly out of focus, and vice versa. Explain that the microscope focuses on a single plane at a time. By moving the adjustment knobs, they are focusing on different levels, or planes, through the specimen.

# TEACHING STRATEGY FOR LAB PROCEDURE

You will probably want to circulate around the room, instructing students who are having trouble observing objects through the microscope. A common complaint you may need to address is that of students who remark that all they can see are the hairs of their eyelashes.

# OBSERVATIONS AND CONCLUSIONS

**1.** The letter chosen will be enlarged and will appear reversed and upside down.

## CHAPTER REVIEW

### MULTIPLE CHOICE

**1.** c **2.** d **3.** a **4.** b **5.** b **6.** d **7.** a **8.** c **9.** b **10.** a

### COMPLETION

**1.** protoplasm
**2.** homeostasis
**3.** DNA, RNA
**4.** mitochondria
**5.** vacuoles
**6.** lysosomes
**7.** diffusion
**8.** osmosis
**9.** mitosis
**10.** meiosis

### TRUE OR FALSE

**1.** F plant
**2.** F nucleus
**3.** T
**4.** F cytoplasm
**5.** T
**6.** F chlorophyll
**7.** T
**8.** F leave
**9.** T
**10.** F mitosis

### SKILL BUILDING

**1.** Nucleus: controls all cellular activities; chromosomes: control the production of proteins and the passing on of traits from one cell to another in cell division; mitochondria: produce energy for the cell; ribosomes: help in protein production; chloroplasts: perform photosynthesis in plant cells in order to produce food for the cell; vacuoles: storage areas for food, water, etc.; endoplasmic reticulum: provides a passageway for the movement of cellular materials; cell membrane: provides a selective barrier keeping various substances in the cell, while allowing other substances to either enter or leave the cell.
**2.** The best site to implant chloroplasts in human cells would be the skin cells. In order for photosynthesis to take place, the energy from the sun is needed, and the skin cells are the closest cells to the sun.
**3.** A plant that has lost a lot of water will contain shrunken vacuoles in its cells. There will be no internal pressure in the cells, and they will begin to sag. This causes the plant to begin to wilt. However, when water is added, it will move into the area of lesser concentration in the vacuoles. This will cause an increase in the internal pressure in the cells, which pushes on all of the cell walls. Soon the plant appears fresh and healthy again.
**4.** Cell A reproduces most quickly at 35°C. As the temperature increases between 10°C and 35°C, the rate of reproduction increases. As the temperature increases between 35°C and 50°C, the rate of reproduction decreases.

### ESSAY

**1.** If the cell contains a cell wall and chloroplasts, then it is a plant cell. If it does not have these structures, then it is an animal cell.
**2.** Control of the cell: nucleus; produces energy for the cell: mitochondria; manufactures proteins: ribosomes; performs photosynthesis in plant cells in order to produce food for cells: chloroplasts; stores materials;

---

**2.** The cell membrane
a. provides protection for the cell.
b. keeps the cell's inner contents together.
c. regulates the movement of materials into and out of the cell.
d. does all of these.

**3.** The "brain" of the cell is its
a. nucleus. b. mitochondria. c. ribosomes. d. cytoplasm.

**4.** The rodlike structures that direct the activities of the cell and pass on traits to new cells are the
a. chloroplasts. b. chromosomes. c. ribosomes. d. lysosomes.

**5.** The network of passageways that transports proteins throughout the cell is known as the
a. nuclear membrane. b. endoplasmic reticulum.
c. mitochondria. d. vacuole.

**6.** Food, water, and wastes are stored in
a. lysosomes. b. chromosomes. c. ribosomes. d. vacuoles.

**7.** Food-making structures found in cells of green plants are called
a. chloroplasts. b. chromosomes.
c. chromoplasts. d. mitochondria.

**8.** Water usually moves through a cell membrane from an area of
a. lesser concentration to greater concentration.
b. equal concentration to equal concentration.
c. greater concentration to lesser concentration.
d. none of these.

**9.** The total number of cells in an organism increases as a result of
a. respiration. b. reproduction. c. homeostasis. d. diffusion.

**10.** Each daughter cell gets an exact copy of the chromosomes of the parent cell through the process of
a. mitosis. b. diffusion. c. homeostasis. d. osmosis.

#### CONTENT REVIEW: COMPLETION

*Fill in the word or words that best complete each statement.*

**1.** All the living material in a cell is called ______.
**2.** The ability of a cell to maintain a stable internal environment even if external conditions change is called ______.
**3.** ______ and ______ are two important nucleic acids.
**4.** Simple food substances are broken down and the energy they release is stored in structures called ______.
**5.** Food, enzymes, and other materials needed by cells are stored inside sacs called ______.
**6.** Structures involved in the digestive activities of the cell are called ______.
**7.** The passage of food, oxygen, water, and other materials in and out of a cell is called ______.
**8.** When only water enters or leaves a cell, the process is called ______.
**9.** During ______, the cells of an organism reproduce to form exact copies of one another.
**10.** During ______, the chromosome number in the male sex cell and in the female sex cell is halved.

76

## CONTENT REVIEW: TRUE OR FALSE

*Determine whether each statement is true or false. If it is true, write "true." If it is false, change the underlined word or words to make the statement true.*

1. The outer covering of the animal cell is the cell wall.
2. The cytoplasm controls all cell activities.
3. Chromosomes are made of nucleic acids.
4. All the living material outside the nucleus is called protoplasm.
5. Ribosomes are the protein-making sites of the cell.
6. The green pigment found in special structures of a plant cell is called chromatin.
7. The sum of all the activities that occur in a living cell is called metabolism.
8. If a cell is placed in a glass of salt water, water will enter the cell.
9. The duplication and division of the nucleus and of the chromosomes is called mitosis.
10. The chromosome number in the male sex cell and in the female sex cell is halved during fertilization.

## CONCEPT REVIEW: SKILL BUILDING

*Use the skills you have developed in the chapter to complete each activity.*

1. **Classifying organelles** On a sheet of paper, list the various activities of a cell. Then, next to each activity, give the organelle that is involved in that activity.
2. **Making inferences** If it were possible to implant chloroplasts in human cells, which cells would be the best sites? Explain.
3. **Relating facts** Your favorite plant has begun to wilt. You feel the soil in its pot and find that it is very dry. You water your plant and, about 20 minutes later, you discover that the plant is standing up straight again. How does this observation relate to osmosis?
4. **Interpreting graphs** At different temperatures, cells reproduce at different rates. According to the accompanying graph, at what temperature does cell A reproduce most quickly? What happens to the rate of reproduction as the temperature increases between 10° C and 35° C? As the temperature increases between 35° C and 50° C?

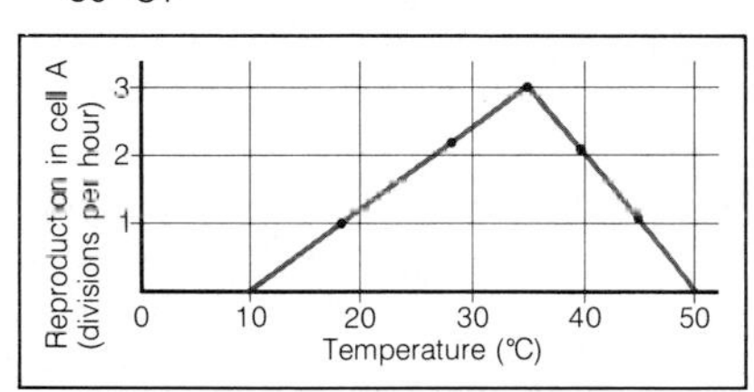


## CONCEPT REVIEW: ESSAY

*Discuss each of the following in a brief paragraph.*

1. Explain how you can distinguish between a plant cell and an animal cell.
2. Describe five functions of a cell and name the structure involved in each function.
3. Why is it important that a cell membrane be selective in allowing materials into and out of the cell?
4. Compare the processes of mitosis and meiosis in terms of introducing new characteristics into the offspring.



vacuoles; provides a passageway for the movement of material throughout the cell: endoplasmic reticulum; provides a selective barrier that allows certain materials to pass into and out of the cell: cell membrane

**3.** If the cell membrane were not selective, important substances in the cell might diffuse out of the cell causing the cell to die. Also, wastes and poisons might diffuse into the cell, again causing the cell to die.

**4.** Mitosis, in which the exact duplicates of cells are formed, cannot introduce new characteristics into the offspring. Meiosis, which involves chromosome halving, together with fertilization (the combination of chromosomes from two parents that occurs later), typically gives an offspring characteristics that may be different from those of either parent.

## ADDITIONAL QUESTIONS AND TOPIC SUGGESTIONS

**1.** A new object has been discovered! How would you decide whether it is living or nonliving? (Duplicate the object's environment as closely as possible. Then look for obvious signs of life such as movement. Examine it more closely for signs of life processes such as growth, breathing, and so on. Examine it microscopically for the presence of cells.)

**2.** Why is the recommended Calorie intake for a 13-year-old higher than for an adult of the same weight? (Teenagers have higher energy needs because they are undergoing periods of rapid growth and development.)

**3.** Most jellyfish live in salt water. If a marine jellyfish were placed in a freshwater pond, what do you think would happen to it? (The jellyfish would die. Its cells would swell and burst as water diffused from an area of higher concentration outside the cells to an area of lower concentration inside the jellyfish cells.)

**4.** Using what you already know about diffusion and osmosis, explain why you often get thirsty after eating salty foods, such as potato chips or popcorn? (Eating salty foods adds more salt to your body fluids. Because the water concentration is less outside the cells than inside them, water moves out of the cells by osmosis, creating thirst.)

**5.** Would meiosis be a suitable type of cell division to produce more skin cells? Explain your answer. (Meiosis would produce cells with half the number of chromosomes instead of the full number. As such, it would not be appropriate for the division of skin cells.)

## ISSUES IN SCIENCE

The following issue can be used as a springboard for discussion or given as a writing assignment.

Some of today's scientists are studying the process of aging. Imagine that the aging process of cells could be slowed, thus increasing the average life span of humans to 200 years. What changes might this bring about in the world as it is known today?

## Chapter 4

# TISSUES, ORGANS, AND ORGAN SYSTEMS

## CHAPTER OVERVIEW

The preceding chapters dealt with the nature of life and with the basic unit of life, the cell. This chapter deals with the higher levels of organization in multicellular organisms: the tissues, organs, and organ systems. There must be close coordination between structures within a given level—for example, between cells within a tissue or between organs within an organ system. Such coordination is essential to the performance of complex activities—for example, the hunting activity of the cheetah described in the chapter introduction.

The first section of the chapter describes in a general way the division of labor, or allotment of functions, that are characteristic of healthy multicellular organisms. The second section of the chapter describes the cells, tissues, organs, and organ systems that represent the levels of organization within an organism.

## INTRODUCING CHAPTER 4

Begin your introduction of Chapter 4 by having the students examine the photograph on page 78.

- **Identify this animal.** (cheetah)
- **To what group of animals is the cheetah related?** (catlike animals)
- **In what part of the world are cheetahs found?** (Africa and Southern Asia)
- **Are there any other living things in the photograph? Identify them.** (yes—trees)

Then have students read the chapter introduction on page 79.

- **What characteristics of the cheetah make it such an excellent hunter?** (Accept all logical responses such as its speed, size, strength, sharp claws and teeth, etc.)
- **What characteristics might an antelope be able to use to avoid being captured by the cheetah?** (Characteristics such as speed, good eyesight and hearing, a sensitive sense of smell, etc. will help the antelope.)

Stress to students that the success of a predator or the ability of an animal to escape a predator depends on how well the parts of the animal's body will work together. A wounded animal will probably not be successful as a predator or an escaping prey because not all of its body parts would be functioning properly.

# Tissues, Organs, and Organ Systems

CHAPTER SECTIONS

4-1 A Division of Labor

4-2 Organization of Living Things

CHAPTER OBJECTIVES

*After completing this chapter, you will be able to:*

**4-1** Explain what is meant by division of labor in living things.

**4-2** Describe the five levels of organization of living things.

**4-2** Explain how the levels of organization of living things are dependent upon one another.

Poised high upon the edge of a jagged rock, the cheetah seems almost motionless. Yet every muscle is tensed and ready to spring into action.

Its nostrils quiver as the familiar scent of an antelope reaches its nose. The sleek animal's back arches, its ears prick up to catch even the faintest sound in the bushes below, and its tongue slowly and smoothly licks its lips.

In anticipation, the cheetah's heart pounds faster and its breathing quickens. A low, menacing sound escapes from its throat. Its paws curl more tightly around the rock as its limbs flex and tighten, flex and tighten. Its flanks ripple with excitement, and juices pour into its mouth as it prepares to descend upon its prey.

The unsuspecting antelope calmly grazes on the plain just below. A young antelope lies near, secure in the protection of its parent. Above them, the cheetah silently readies itself for attack. Every part of the animal—from the smallest cell to its entire body—is working together to ensure that the cheetah will eat well this night.

Cheetah in search of prey



## TEACHER DEMONSTRATION

Obtain a photograph of a single-celled organism such as a bacterium, paramecium, or amoeba.

- **How would you describe this organism?** (Accept all logical responses. Students should note that this is a single-celled organism.)

Then display a photograph of any complex, multicellular organism.

- **How would you describe this organism?** (Accept all logical responses.)

Now display both photographs.

- **How are these organisms alike?** (Both are living.)
- **How are they different?** (Accept all logical responses. Student should understand that while one of the organisms consists of only a single cell, the other contains many cells.)
- **Which organism is more complex?** (The multicellular organism)

Explain to students that in this chapter they will learn how the body parts of multicellular organisms are able to work together for the benefit of the organism.

## TEACHER RESOURCES

### Audiovisuals

*Man: The Incredible Machine,* film, NGS

*The Heart and Circulatory System,* film, EBE

*The Human Nervous System,* 8 filmstrips, Guidance Associates

### Books

Elson, L. M. *It's Your Body,* New York: McGraw-Hill.

Harrison, R. J., and W. Montagna. *Man,* Englewood Cliffs, N. J.: Prentice-Hall.

Nourse, A. E., et al. *The Body,* New York: Time, Inc.

## 4-1 A DIVISION OF LABOR

### SECTION PREVIEW 4-1

From one-celled bacteria to multicellular mammals, the cell is the basic unit of structure and function. In complex organisms, however, there is a division of labor among cells. Different cells are assigned to different jobs. Cells combine to form specialized parts or levels of organization. As each part of the organism does its special job, a harmony is produced within the organism which keeps it alive and healthy.

### SECTION OBJECTIVES 4-1

1. **Describe the division of labor.**
2. **Relate the division of labor to its effect on the life processes of a complex, multicellular organism.**

### SCIENCE TERMS 4-1

**division of labor _ p. 81**

**Figure 4-1** *Multicellular organisms, such as this beetle and the green plant upon which it feeds, carry on a wide variety of complex activities.*

**Figure 4-2** *Different parts of this bat's body work in harmony with each other as the bat prepares to attack. If the tree frog's body parts work well together, the frog may escape.*

## 4-1 A Division of Labor

Bacteria lead very simple lives. They do not walk, run, or listen to music. They do not see, smell, talk, or shake hands. They just grow, divide, and then grow some more. A cheetah, on the other hand, leads a life that is not that simple. Its life includes activities like roaming the plains, climbing trees, hunting for food, and playing with other cheetahs.

The whole body of a bacterium is made up of one cell. On its own, that single cell must do all the jobs that keep a bacterium alive. This accounts, in part, for the simple lives bacteria live. A cheetah, however, is multicellular—it is made of a great many cells. And these numerous cells are organized into the different parts of the cheetah's body, enabling the animal to carry out more complex activities than a single-celled bacterium can.

Some of the cheetah's body parts, such as muscle cells, are relatively small and simple. Other parts, like strands of muscle fibers, are larger and more complex. Still other parts, the cheetah's heart and circulatory system, for example, are even larger and more complex. However, all the different parts, whether small or large, interact with each other, perform their special functions, and keep the cheetah alive.

### TEACHING STRATEGY 4-1

#### Motivation

Explain to students that in nature specialization is the key to progress. In order for organisms to perform more and more complex tasks, finer and finer divisions of labor among cells is necessary. But specialization is achieved only at the sacrifice of versatility.

Make an analogy between specialization in living things and specialization in society. Show photographs of the organisms as they are discussed. Use the following as an example.

**1.** Robinson Crusoe level—one-celled bacterium

A single individual living alone on a deserted island must perform all functions of survival by himself. The single-celled bacterium must perform all of the functions for the survival of the bacterium.

**2.** Loose cooperative such as a nomadic tribe—slime-mold colony

Members of a nomadic tribe help each other in major tasks such as food gathering. A slime-mold colony may be composed of 50,000 single cells. Each retains its individuality, yet all work together to forage for food.

**3.** Small town—hydra

In a small town, everyone knows about everyone else's business and occasionally does it. A hydra has specialized cells and body parts; but if chopped up and sieved through a fine cloth, one or more complete animals can be regenerated from the pulpy mass. If a hydra is cut in half, the bottom half grows a new mouth and tentacles, and the top half grows a new foot or stalk. In both cases, the cells that previously performed one function may assume a new function and location.

**4.** Busy metropolis—human

In a big city, everyone goes about his or her own business. Yet, all work together for proper functioning of the society. Human body parts have different functions, yet all

There is a **division of labor** within the body of the cheetah, as within most living things. **Division of labor means that the work of keeping the organism alive is divided among the various parts of its body.** Each part has a specific job to do. And as the part does its special job, it works in harmony with all the other parts to keep the organism healthy and active.

### *SECTION REVIEW*

1. What is the basic difference between a bacterium and an organism such as the cheetah?
2. What is meant by the term "division of labor" in living things?

**Activity**

② *Division of Labor*

The levels of organization in an organism represent a division of labor among the various parts so that the work of the living thing gets done.

Examine your community or an aspect of your community and explain how division of labor works there. Describe its effectiveness. How does it compare with the division of labor in an organism?

## 4-2 Organization of Living Things

The arrangement of specialized parts within a living thing can be referred to as levels of organization. **The five basic levels of organization, arranged from the smallest, least complex structure to the largest, most complex structure are cells, tissues, organs, organ systems, and organisms.**

### *Level One: Cells*

In Chapter 3, you learned that all living things are made up of **cells.** These microscopic units of structure and function are the building blocks of life. Some living things, such as a bacterium, are made of only one cell. For these organisms, the single cell exists as a free-living organism. It does all the jobs that keep the organism alive.

In multicellular organisms, different cells perform specialized tasks. Muscle cells, for example, help the cheetah move through its environment, climb rocks, and spring to attack. Nerve cells receive and send messages throughout the cheetah's body. Every kind of cell, however, depends on other cells for its survival and for the survival of the entire organism. Those muscle cells move only when set into action by nerve cells. And both muscle cells and nerve cells rely on blood cells for oxygen.



### Activity

**Division of Labor**
**Skills: Observing, comparing, applying**
**Level: Average**
**Type: Field**

This activity introduces students to the concept of biological coordination by means of an analogy to the cooperation within a community of human beings. Point out later that such groups of like organisms, or populations, are often referred to as a sixth level of biological organization beyond the five discussed.

### ANNOTATION KEY

① Thinking Skill: Making comparisons
② Thinking Skill: Applying definitions

### FACTS AND FIGURES

The largest cell is the yolk of an ostrich egg. It is about the size of an orange.

must work together for survival. Human tissues have limited regenerative powers.

### Content Development

Explain that the work of keeping an organism alive is often divided among the different parts of its body. Each part performs a specific job. This is called division of labor.

- **Describe how division of labor works in an automobile manufacturing plant?** (Different individuals are assigned to different jobs such as producing the various auto parts, assembling them, painting the finished product, etc.)
- **What would happen if one part of the production or assembly line breaks down?** (Other parts are affected and eventually cease to function.)
- **Describe how division of labor works in your body.** (Different parts of the body are responsible for different jobs—heart pumps blood, brain thinks, stomach helps digest food, lungs take in oxygen, etc.)
- **What would happen if one part of your body breaks down?** (The individual becomes ill, ceases to function properly, and may eventually die.)
- **What can be done to restore a body part to its proper working order?** (medication, surgery, rest, organ transplants)

### Section Review 4-1

1. A bacterium is one-celled and leads a simple life. A cheetah is multicellular and leads a complex life.
2. Work of keeping the organism alive is divided among different parts of its body.

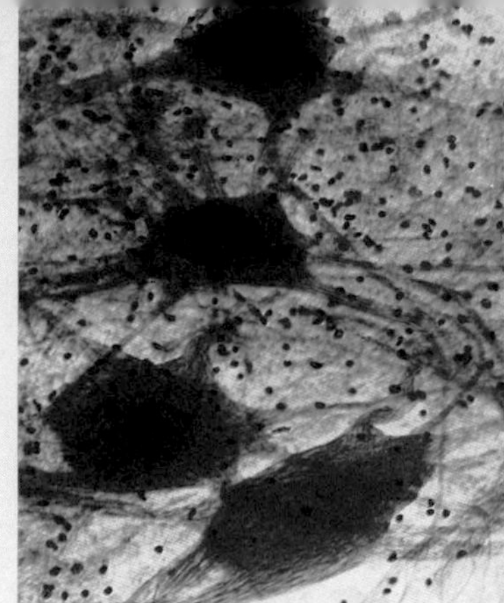
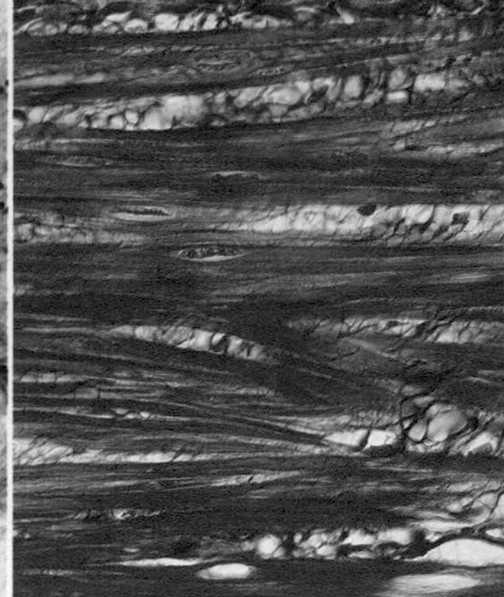
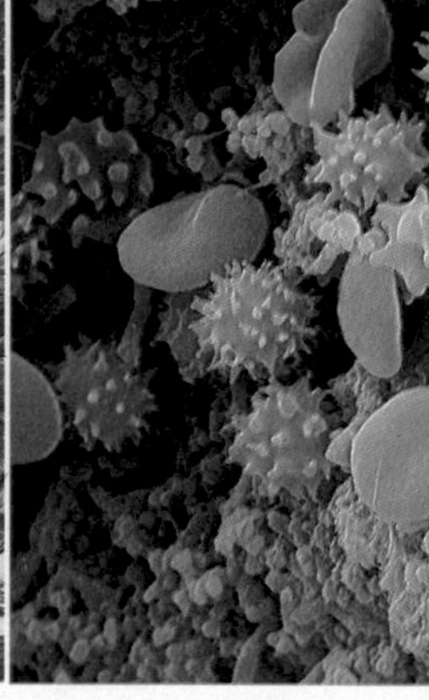

**Figure 4-3** *Cells come in different shapes and sizes, as you can see from these nerve* (left), *muscle* (center), *and blood cells* (right).

## 4-2 ORGANIZATION OF LIVING THINGS

### SECTION PREVIEW 4-2

In multicellular organisms, cells generally do not work alone. Cells that are similar in structure and function are joined to form tissues, such as bone tissue, blood tissue, xylem, and phloem. Organs are groups of tissues working together, such as the heart, eye, root, and leaf. Organs work together as organ systems. Examples of organ systems include the nervous system, digestive system, and excretory system. Finally, organ systems work together to form a functioning organism.

### SECTION OBJECTIVES 4-2

1. **List and give examples of each of the five levels of organization found in complex living things.**
2. **Explain how the various levels of organization interact with one another.**

### SCIENCE TERMS 4-2

**cell** p. 81
**tissue** p. 82
**organ** p. 83
**system** p. 84
**organism** p. 86

### *Level Two: Tissues*

In a multicellular organism, cells usually do not work alone. Cells that are similar in structure and function are joined together to form **tissues.** Tissues are the second level of organization.

Like the single cell of the bacterium, each tissue cell must carry on all the activities needed to keep that cell alive. But at the same time, tissues perform one or more specialized functions in an organism's body. In other words, tissues work for themselves as well as for the good of the entire living thing.

For example, bone cells in the cheetah form bone tissue, a strong, solid tissue that gives shape and support to the bodies of animals. Blood cells are part of blood tissue, a liquid tissue responsible for carrying food, oxygen, and wastes throughout the cheetah's body. Some muscle tissue helps move the cheetah's legs, neck, and other body parts. Another kind of muscle tissue enables its heart to pump blood. A third type of muscle tissue lines its stomach and helps the cheetah digest its meal.

Plants have tissues too. The leaves and stems of a plant are covered by a type of tissue called epidermis, which protects the plant and prevents it from losing water. Tissue known as xylem (ZIGH-luhm) conducts water and dissolved minerals up through

**Figure 4-4** *The sundew plant has special hairlike tissues that produce a sticky fluid that traps insects.*

### TEACHING STRATEGY 4-2

#### Motivation

Have students name four activities that they perform daily. Write these activities on the chalkboard, leaving space beneath each. Ask students what body structures help them complete these activities. List these structures below the activities.

Impress upon students the idea that even the simplest of tasks usually involves the coordination of two or more body parts.

#### Content Development

Explain to students that a tissue consists of cells that are similar in structure and function and are joined together.

- **Name several kinds of tissues found in animals.** (muscle tissue, nerve tissue, connective tissue—cartilage, bone, blood, fat; epithelial tissue—skin)
- **Name several kinds of tissues found in plants.** (vascular tissue—xylem, phloem, and cambium; epidermis, parenchyma tissue—leaf)

Explain to students that the next level of organization after tissues is organs. Tissues work together to form organs. An organ may consist of several kinds of tissues. For example, your heart contains nerve tissue. muscle tissue, and blood tissue.

- **Name some other organs in the human body.** (eyes, ears, nose, lungs, brain, kidneys, skin, etc.)
- **What different kinds of tissues are found in your foot?** (muscle tissue, bone tissue, blood tissue, nerve tissue, skin tissue)

the stems to the leaves. And another special tissue called phloem (FLOH-em) brings food made in the leaves back down to the stems and roots.

### Level Three: Organs

The cheetah is a clever hunter. Stalking about on its padded feet, it uses its keen senses of smell, sight, and hearing to locate its prey. Then, as its lungs fill with oxygen, it moves with incredible speed and strikes down its victim with a blow of its front paws.

The cheetah's eyes, ears, nose, and lungs are several of the **organs** that help the animal find food and stay alive. Organs are groups of different tissues working together. The cheetah's heart, for example, is an organ composed of nerve tissue, muscle tissue, ❷ and blood tissue. Each tissue does its special job. Nerve tissue signals the muscle tissue to contract. Muscle tissue contracts and causes the heart to pump the blood tissue. Blood tissue carries oxygen and wastes to and from every cell in the cheetah's heart—and eventually to cells throughout the entire body.

**Figure 4-5** *Various organs in the cheetah's body work together so that the animal can run as fast as 115 kilometers per hour.*

83

## BACKGROUND INFORMATION

There are levels of organization below the cellular level. Organelles, molecules, and atoms are subcellular levels of organization. There are also levels of organization above that of the organism. Organisms are grouped together according to their relationship to one another to form populations, communities, biomes, and the biosphere.

## FACTS AND FIGURES

Cells are shaped like rods, spirals, circles, daisies on a stalk, snowflakes, stringbeans, and even blobs of jelly.

## ANNOTATION KEY

❶ Thinking Skill: Relating concepts

❷ Thinking Skill: Applying concepts

## Reinforcement

One of the simplest levels of organization in a multicellular organism is the cellular level. Make certain that students are familiar with the general characteristics of cells presented in Chapter 3 before studying about the specializations of different kinds of cells.

- **What is a cell?** (basic unit of structure and function in all living organisms)
- **What part of the cell controls its activities?** (the nucleus)
- **What cell structure separates one cell from another?** (in plant cells, the cell wall; in animal cells, the cell membrane)
- **What is metabolism?** (the sum of all activities that occur in a living cell)
- **What process is responsible for forming new body cells?** (mitosis)

## Skills Development

***Skill: Interpreting illustrations***

Have students observe the three photographs in Figure 4-3.

- **Describe the shapes of the cells in each of the photographs.** (Nerve cells have a dark center portion, cell body, and branching structures; muscle cells are threadlike; blood cells are round, some are smooth and some have projections.)
- **What is the function of each of the cells?** (Nerve cells receive and send messages throughout the organism's body; muscle cells help the organism move; and blood cells carry oxygen, food, and wastes.)

## Enrichment

Show students drawings or photographs of the following: bone cell, bone tissue, a bone (organ), skeletal system, and human. Have students arrange these in the proper order from cell to organism.

## BACKGROUND INFORMATION

Though they are bonded together, tissue cells, except those of certain muscles, do not fuse or even come into actual contact with one another. A kind of cementing substance appears to fill the spaces between them. In some cases, spikelike structures bridge the gaps between cells and bind the cell membranes together.

In the case of cancer cells, cell bonding seems to be weakened. Cells break loose. These loosened cells can migrate and spread.

Many of the processes of normal tissue development are duplicated on a small scale as a wound heals. The tissues not only grow as needed; they also stop growing as soon as the healing process is finished. The capacity for cells to grow and reproduce diminishes with age. Thus, wounds in older people heal less quickly than those of the young.

## ANNOTATION KEY

❶ **Regulates various body functions (Applying definitions)**

❶ Thinking Skill: Relating concepts

❷ Thinking Skill: Making comparisons

**Figure 4-6** *Each organ system is made up of groups of organs that work together. What is the function of the endocrine system?* ❶

**ORGAN SYSTEMS**

| System | Function |
|---|---|
| Skeletal | Protects and supports the body |
| Muscular | Supports the body and enables it to move |
| Skin | Protects the body |
| Digestive | Receives, transports, breaks down, and absorbs food throughout the body |
| Circulatory | Transports oxygen, wastes, and digested food throughout the body |
| Respiratory | Permits the exchange of gases in the body |
| Excretory | Removes liquid and solid wastes from the body |
| Endocrine | Regulates various body functions |
| Nervous | Conducts messages throughout the body to aid in coordination of body functions |
| Reproductive | Produces male and female sex cells |

You are probably familiar with the names of many of the organs that make up your body. Your brain, stomach, kidneys, and skin are some examples. Can you name some others?

Plants have organs too. Roots, stems, and leaves are common plant organs. Like animal organs, plant organs are made up of groups of tissues performing the same function. For example, various tissues in the leaf help this organ make food for the plant.

### Level Four: Systems

Like cells and tissues, organs seldom work alone. They "cooperate" with each other to form specific **systems.** A system, then, is a group of organs that work together to perform certain functions.

The cheetah is able to spot its prey because its sense organs—its eyes, ears, and nose—do their job. They receive messages in the form of sights, sounds, and smells. The cheetah's brain processes this information. ❶ And its spinal cord and nerves send out impulses to all parts of its body so that the animal is able to prepare to attack. Each of these organs, doing its individual job as well as working together with other organs, forms part of the cheetah's nervous system.

A living thing cannot survive unless all its systems are "go." For example, its digestive system must receive, carry, break down, and absorb food. Its excretory system must remove certain waste materials from its body. And its endocrine system must make chemical substances called hormones. Hormones speed up or slow down the activities of many organs. They also play a part in the growth and development of the organism.

The organs that make up a system vary in number and complexity from one kind of living thing to another. For example, a very simple animal called the hydra has a nervous system that is a simple net ❷ of nerves spread throughout the organism. More complex animals, such as the earthworm, have a more highly organized system of nerves and a primitive brain. The most highly developed animals, people, for example, have nervous systems consisting of a complex brain, a spinal cord, and nerves.



## 4-2 (continued)

### Motivation

As you teach students about various kinds of cells and tissues, make use of prepared slides to make clear the special features of each. Slides can be purchased from biological supply houses or borrowed from the high school biology department.

### Content Development

Explain to students that an organ system is a group of organs that work together to perform certain activities.

- **As you are sitting in your seats listening to me, what body systems are working for you now? What functions are they performing?** (Accept all logical responses such as respiratory system—breathing; nervous system—sending messages to the brain; muscular system—body movement; skeletal system—erect posture, etc.)
- **Do any of your body systems stop functioning completely when you sleep? Explain your answer.** (No. Although many body systems will slow down when the body is at rest, none will stop functioning completely.)
- **Which of your body systems would work the hardest when you exercise?** (circulatory and respiratory)

### Skills Development

***Skill: Interpreting charts***

Explain to students that organs seldom work alone. They cooperate to form organ systems. Have students refer to Figure 4-6.

- **Name at least one organ in each of**

**Figure 4-7** *The skin of the camel* (left), *part of its excretory system, helps it get rid of waste materials and excess heat. The circulatory system of the eucalyptus tree* (right) *brings water and other substances to the very top of the tree.*

## Career: *Histologic Technician*

**HELP WANTED: HISTOLOGIC TECHNICIAN** Position involves preparation of tissue for microscopic study and detection of cell features that suggest disease or damage. Completion of a formal training program or an associate degree program in histotechnology required.

New drugs and chemical food additives are constantly being manufactured and introduced into the marketplace. Before these substances can be used, however, they must be carefully tested to ensure that they are safe for human use. The United States Food and Drug Administration requires these tests. Controlled amounts of a new drug or additive are given to research animals over a period of time. Scientists then study the tissues of the animals, looking for possible damage or disease.

People who prepare animal or plant tissue for microscopic study are called **histologic technicians** or histotechnologists. They prepare tissue by first treating it with special chemicals that prevent decomposition of the tissue. They then remove water from the tissue so that they can place it in liquid wax. The wax quickly hardens and gives support to the tissue so that the tissue can be sliced into extremely thin sections with a very sharp knife. The sections are placed on microscope slides and are stained with various dyes that help to distinguish tissue structures. A specially trained technician or a doctor examines the slide.

Opportunities for histologic technicians exist in various fields of medicine, in marine biology, and in certain areas of industrial and university research. For further information, write to the National Society for Histotechnology, 5900 Princess Garden Parkway, Suite 805, Lanham, MD 20706.

## SCIENCE, TECHNOLOGY, AND SOCIETY

Recently, scientists developed a substitute for blood that is chemically similar to Teflon.™ This blood substitute is called Fluosol and has an oxygen-carrying capacity that is 20 times greater than human plasma. Fluosol also can be stored for three years, whereas whole blood can only be stored for three weeks. Blood groups do not have to be cross matched in order to use Fluosol and the recipient does not have to worry about the presence of disease-causing organisms in the donated human blood. Fluosol has another feature—its ability to carry oxygen increases as the temperature decreases.

Are there any disadvantages in using Fluosol? One disadvantage is that Fluosol can not carry out all of the functions of human blood because it does not contain such things as clotting agents, antibodies, or immunoglobins. Another and perhaps the greatest disadvantage is that, in less than 12 hours, Fluosol gradually breaks down into waste gases. These gases are exhaled out of the body through the lungs. A third disadvantage requires that a person who has been given Fluosol also must be given supplemental oxygen to breathe.

Have students examine the advantages and disadvantages of using Fluosol. Then have them design situations or scenarios in which the use of Fluosol, or other blood substitutes, might be preferable to the use of plasma or whole blood.

**these organ systems.** (Answers will vary. Skeletal—any bone such as rib, humerus, sternum; muscular—any muscle such as biceps, triceps, pectoralis major; digestive—stomach, pancreas, large intestine; circulatory—heart, artery, vein; respiratory—lung, nose, throat; excretory—skin, kidneys; endocrine—thyroid, adrenal glands, pituitary; nervous—brain, spinal cord; reproductive—ovary, testis)

- **Which system produces male and female sex cells?** (reproductive)
- **What is the function of the circulatory system?** (to transport oxygen, wastes, and digested food throughout the body)

You may wish to discuss how respiratory (breathing) systems vary in complexity from one organism to another. Amoebas take in oxygen directly through cell membranes. Earthworms breathe through their moist skin. Fish have gills that remove oxygen from the water. Land animals have the most complex systems consisting of the lungs.

### Enrichment

You may wish students to do library research on the anatomy and physiology of various organs or organ systems of human beings or other organisms. Students should then report their findings to the class.

## TEACHER DEMONSTRATION

Begin this demonstration by explaining that the highest level of organization of living things is the organism. Tell students that an organism is an entire living thing that carries out all of its basic life functions.

• **What are some examples of organisms?** (Accept all logical answers.)

Write the phrase "Organism X" on the chalkboard and have students list the characteristics they think an organism (plant or animal) should have to be well-suited for the area in which they are living. As students generate ideas, list them on the chalkboard. For example, in warm areas students might list the need for little hair or fur, an efficient body cooling system, and so forth.

Next have students list the organ systems that contribute to these characteristics. For example, the circulatory system, respiratory system, and excretory system all contribute to the cooling of an organism.

After the students have completed this activity, show photographs of organisms that live in their area. Have students compare their list of ideal characteristics with those displayed by the organisms in the photographs.

## ANNOTATION KEY

**❶ Cells, tissues, organs, systems, organism (Sequencing events)**

❶ Thinking Skill: Applying definitions

❷ Thinking Skill: Synthesizing information

### *Level Five: Organisms*

Throughout this chapter, you have been reading the word "organism." Perhaps you have already determined what an **organism** is. An organism is an entire living thing that carries out *all* the basic life functions. You, like the cheetah, are an organism. ❶ A buttercup is an organism, as are a cactus, a caribou, and a whale. Even a one-celled bacterium is an organism. The organism is the highest level of organization of living things.

A complex organism is a combination of organ systems. Each system performs its particular function, but all the systems work together to keep the organism alive. Without the smooth operation of any one of its systems, a living thing could not survive.

The behavior of the cheetah you read about earlier is the result of many systems performing their specific functions and interacting with other systems. For example, when the cheetah's nervous system senses the antelope's presence, it signals the muscular system. The cheetah's spine arches, its tail goes

**Figure 4-8** *The diagram below shows the levels of organization in an organism. What is the order of organization from least to most complex?* ❶

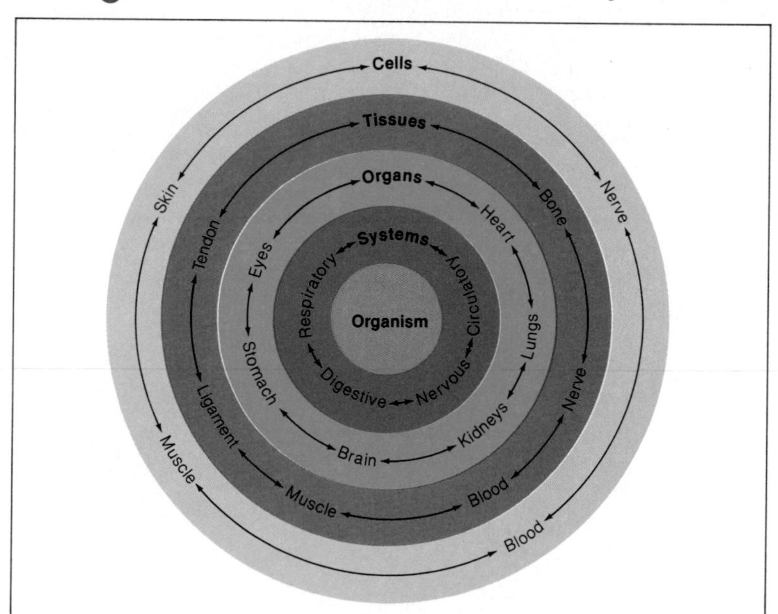


86

### 4-2 (continued)

### Content Development

Explain to students that the fifth level of organization in living things is the organism. An organism is a combination of organ systems.

• **A bacterium is an organism. Does it contain tissues, organs, and organ systems?** (No. Bacteria are one-celled organisms. The five levels of organization are found only in multicellular organisms.)

• **Is an apple tree an organism? Explain your answer.** (Yes. An apple tree contains cells (root tip), tissues (xylem and phloem), organs (roots), and organ systems (root system), all working together.)

Reinforce the concept that as organisms become more complex, they become more specialized. However, each level must interact with every other level in order to maintain smooth functioning and survival of the organism.

### Skills Development

***Skill: Interpreting diagrams***

Have students refer to Figure 4-8.

• **In which level of organization would a ligament be found?** (tissue)

• **What are three examples of organs?** (heart, lungs, kidneys, brain, stomach, eyes)

Figure 4-9 *Survival often depends on the cooperation of body parts—from simple cells to complex systems.*

straight back, and its hind legs extend forward. In this position, the cheetah is able to accelerate to speeds of 75 to 115 kilometers per hour. The nervous system also alerts organs of the endocrine system, which increase their production of hormones. As a result, the cheetah's heart rate increases and more oxygen than usual is delivered to the muscle cells. As you can see, no one process or one system is independent of the others.

Cells, tissues, organs, systems, organisms. By now one thing should be clear to you: Each level of organization interacts with every other level. And the smooth functioning of a complex organism is the result of all its various parts working together.

The antelope, of course, is also an organism. The parts of its body must work together as well. In fact, once the antelope spots the cheetah lying in wait, its body parts work fast and furiously to spare it an untimely death at the paws of the cheetah. Often the cheetah wins this deadly battle, but this time the antelope and its young escape into the bush—just as the cheetah is about to jump and pounce. Tonight, at least, the cheetah will go hungry.

### SECTION REVIEW

1. List and define the five levels of organization of living things.
2. Give an example of two organ systems working together within an organism.

**Activity**

2 *An Amazing Feat*

Bone and muscle work together to perform amazing feats. A hummingbird can stay in one place in midair by beating its wings 80 times per second. How many beats is this in 5 minutes?



- **In which level of organization would blood be found?** (cell and tissue)

## Section Review 4-2

1. Cells: basic unit of structure and function in living things; tissues: cells that are similar in structure and function; organs: groups of different tissues working together; organ systems: groups of organs that work together to perform certain functions; organism: entire living thing that carries out the basic functions of life.
2. Accept all logical answers. For example, the respiratory system permits the exchange of oxygen and carbon dioxide in the body and the circulatory system carries oxygen to cells and removes the waste carbon dioxide.

## Activity

**An Amazing Feat**
**Skill: Making calculations**
**Level: Average**
**Type: Computational**

Students should find that the hummingbird beats its wings about 48,000 times in ten minutes.

## SCIENCE, TECHNOLOGY, AND SOCIETY

Many people in the United States have made the decision to donate certain body organs and tissues to science when they die. These people carry organ-donor cards, which indicate their wishes. Among the tissues and organs used for research and transplants are bones, skin, corneas, parathyroid and pituitary glands, blood vessels, kidneys, livers, lungs, and hearts.

Each of the 50 states has passed legislation that governs the donating of organs. Under this legislation, mentally competent people can donate any or all of their bodies to hospitals, medical schools, dental schools, organ storage facilities, or scientists engaged in research, therapy, transplantation, or education.

The intention to donate an organ can be placed in a person's will, or in a written statement signed in the presence of two witnesses. In most states, a person's driver's license also can give permission for an organ donation in case of a fatal accident. People under the age of 18 must have a parent or legal guardian countersign their donor card.

Most laws also permit the next of kin to authorize donations unless it is shown that the person would have objected to the donation, if he or she were alive. If a person changes his or her mind after obtaining a donor card, he or she can destroy it and/or notify the motor vehicle bureau.

Have students list the advantages and the disadvantages of donating organs or tissues. Have them discuss and compare their lists.

## LABORATORY ACTIVITY
## HAVE YOU TASTED THESE?

### BEFORE THE LAB

1. **Divide the class into groups of three or four students.**
2. **Gather all materials at least one day prior to the investigation. You should purchase enough food to meet your class needs, assuming three to four students per group.**

### PRE-LAB DISCUSSION

Have students read the complete laboratory procedure. Discuss the activity by asking questions similar to the following:

- **What is the purpose of this laboratory activity?** (to taste some foods which come from the cells, tissues, and organs of animals)
- **Did you eat a breakfast this morning?** If so, did any of the foods that you ate come from the cells, tissues, or organs of animals? (Accept all logical responses.)
- **What common foods consist of the cells, tissues, or organs of plants?** (Accept all logical responses.)

### SKILLS DEVELOPMENT

Students will use the following skills while completing this activity.

1. Applying
2. Observing
3. Classifying
4. Relating

### TEACHING STRATEGY FOR LAB PROCEDURE

You may wish to circulate around the room during this activity to ensure that students are performing the activity and to aid students who may be having difficulty classifying a particular food.

## LABORATORY ACTIVITY

### *Have You Tasted These?*

#### Purpose

In this activity, you will be introduced to some unusual foods that are derived from cells, tissues, and organs of various animals.

**Materials** *(per group)*
Dictionary or encyclopedia
Cookbooks

#### Procedure

1. Examine the following list of foods:

| | |
|---|---|
| tripe | jerky |
| sweetbreads | suet |
| chitterlings | marrow |
| escargot | pâté de foie gras |
| sushi | caviar |

How many of these foods can you identify? How many of them have you eaten?

2. Using a dictionary, encyclopedia, or cookbook, look up the foods that no one in the class was able to identify. Name the animal that is the source of each of these foods. From what part of the animal is each of these foods derived?

#### Observations and Conclusions

1. Classify each of these foods using the levels of organization you learned in this chapter as categories: cells, tissues, organs, organ systems, organisms.
2. List other animal organs that are used as foods. How many of these have you tasted? How were these foods prepared?

### ANSWERS TO QUESTIONS IN PROCEDURE

1. Answers will vary.
2. Tripe: muscular lining of stomachs of ruminants; sweet-breads: pancreas or thymus of calf; chitterlings: intestine of pig; escargot: snail; sushi: muscle of fish; jerky: dried muscle of cattle; suet: fat of sheep or cattle; marrow: tissue that fills cavities of bones of cattle; pâté de foie gras: liver of goose; caviar: fish eggs.

### OBSERVATIONS AND CONCLUSIONS

1. Tripe: tissue; sweetbreads: organ; chitterlings: organ; escargot: organism; sushi: tissue; jerky: tissue; suet: tissue; marrow: tissue; pâté de foie gras: organ; caviar: cells
2. Answers will vary.

## CHAPTER REVIEW

### SUMMARY

**4-1 A Division of Labor**

- In a one-celled organism, the single cell performs all the activities necessary to keep the organism alive.
- In multicellular organisms, different parts of the body have specific jobs to do. This division of labor keeps the organism alive.

**4-2 Organization of Living Things**

- The least complex level of organization of living things is the cell. The cell is the basic unit of structure and function of all organisms.
- In multicellular organisms, different cells perform specialized tasks. Every kind of cell, however, depends on other cells for its survival and for the survival of the entire organism.
- The five basic levels of organization—arranged from the smallest, least complex structure to the largest, most complex structure—are cells, tissues, organs, organ systems, and organisms.
- Cells that are similar in structure and function are joined together to form tissue. For example, bone cells in a cheetah form bone tissue, a strong, solid tissue that gives shape and support to the body of the animal.
- Plants as well as animals have tissue. For example, the leaves and stems of a plant are covered by a type of tissue called epidermis, which protects the plant and prevents it from losing water.
- Groups of different tissues work together as organs. For example, the cheetah's heart is an organ composed of nerve tissue, muscle tissue, and blood tissue. Each tissue does its special job.
- Roots, stems, and leaves are common organs in plants.
- A group of different organs working together to perform certain functions is known as a system.
- The organs that make up a system vary in number from one kind of living thing to another. The most highly developed animals, for example, have nervous systems consisting of a complex brain, a spinal cord, and nerves. However, a simple animal such as the hydra has a nervous system that is only a simple net of nerves spread throughout the organism.
- An organism is an entire living thing that carries out all the life functions. Complex organisms are made up of systems. The organism is the highest level of organization of living things.
- In a complex organism, each system performs its particular function, but all systems work together to keep the organism alive.

### VOCABULARY

*Define each term in a complete sentence.*

| | | |
|---|---|---|
| **cell** | **organ** | **system** |
| **division of labor** | **organism** | **tissue** |

### CONTENT REVIEW: MULTIPLE CHOICE

*Choose the letter of the answer that best completes each statement.*

**1.** In a bacterium, the single cell carries on
a. digestion only. b. respiration and digestion only. c. all life functions. d. excretion only.



## GOING FURTHER: ENRICHMENT

**Part 1**

If possible, speak to the home economics department about the preparation of a few of these foods for sampling by your students. If this is not feasible, suggest to students that they or other members of their families prepare and sample these foods.

**Part 2**

Have students do research on other unusual plant and animal cells, tissues, and organs that are used as food. Impress upon students the idea that the unusual nature and limited availability of many of these foods cause them to be considered gourmet delicacies.

## CHAPTER REVIEW

### MULTIPLE CHOICE

| | | | | |
|---|---|---|---|---|
| **1.** c | **3.** c | **5.** c | **7.** a | **9.** a |
| **2.** a | **4.** b | **6.** b | **8.** c | **10.** d |

### COMPLETION

| | |
|---|---|
| **1.** cell | **6.** organs |
| **2.** nerve | **7.** organs |
| **3.** muscle | **8.** organs |
| **4.** tissue | **9.** hormones |
| **5.** blood | **10.** organism |

### TRUE OR FALSE

| | |
|---|---|
| **1.** F one-celled | **6.** F tissues |
| **2.** T | **7.** T |
| **3.** F five | **8.** F organs |
| **4.** F cell | **9.** F systems |
| **5.** T | **10.** F cannot |

### SKILL BUILDING

**1.** a. tissue f. tissue
b. organ g. organ
c. organism h. organ
d. organism i. organism
e. organ system j. tissue

**2.** a. circulatory e. excretory
b. nervous f. nervous
c. respiratory g. skin
d. digestive h. respiratory

**3.** From top to bottom: cell, tissue, organ, organ system, organism.

**4.** Answers will vary. For example, changes in light intensity will cause the pupils in the eyes to become larger or smaller, depending upon whether there is too much or too little light. The eyes and the brain are involved in this response.

**5.** Answers will vary. On hearing a wakeup alarm, sense organs receive information that they convey via the nervous system. On rising, muscles are moved due to nervous stimuli. Eating and digesting breakfast requires the use of muscles and of the organs of ingestion and digestion.

### ESSAY

**1.** A simple organism is typically unicellular; a complex one contains many cells that function together according to the various levels of organization.

**2.** See Figure 4-6 on page 84.

**3.** Cells: building blocks of life—bacterium, amoeba, muscle cell, etc.; tissues: cells that are similar in structure and function—bone tissue, blood tissues, xylem, phloem, etc.; organs: groups of different tissues working together—heart, eye, root, leaf, etc.; organ systems: groups of organs that work together to perform certain functions—nervous, digestive, excretory systems, etc.; organisms: living things that carry out all the basic life functions—bacterium, amoeba, cheetah, buttercup, cactus, etc.

**4.** Because a one-celled organism must perform all the life functions in a single cell, its individual cell is often more complex than cells in multicellular organisms, which often divide up the work that must be done to keep the organism alive.

**2.** The concept of various body parts each doing a specific job is called
a. division of labor. b. reduction division.
c. differentiation. d. specialization.

**3.** The basic units of structure and function of all organisms are the
a. bones. b. systems. c. cells. d. tissues.

**4.** Multicellular organisms have
a. one cell. b. many cells. c. solar cells. d. no cells.

**5.** Cells that are similar in structure and function are joined together to form
a. organs. b. systems. c. tissues. d. organisms.

**6.** Xylem, phloem, and epidermis are examples of plant
a. organs. b. tissues. c. cells. d. systems.

**7.** Roots, stems, and leaves are examples of plant
a. organs. b. cells. c. tissues. d. bones.

**8.** The highest level of organization in living things is the
a. cell. b. organ. c. organism. d. system.

**9.** From the most complex to the least complex, the levels of organization of most living things are
a. organism, systems, organs, tissues, cells.
b. systems, cells, tissues, organs, organism.
c. organism, systems, organs, cells, tissues.
d. cells, tissues, organs, systems, organism.

**10.** In order for any multicellular organism to stay alive,
a. each level of organization must perform its specific job.
b. all levels of organization must interact with one another.
c. neither of these.
d. both of these.

### CONTENT REVIEW: COMPLETION

*Fill in the word or words that best complete each statement.*

**1.** The least complex level of organization in a living thing is the basic unit of structure called the ______________.

**2.** Messages are received and transmitted throughout a complex organism's body by ______________ cells.

**3.** The type of tissue that helps an organism move is ______________ tissue.

**4.** Xylem and phloem are examples of plant ______________.

**5.** The type of tissue that carries oxygen and wastes to and from cells of the heart is ______________ tissue.

**6.** Eyes, kidneys, and skin are examples of animal ______________.

**7.** Roots, stems, and leaves are common plant ______________.

**8.** A system is a group of ______________ that work together to perform a certain activity.

**9.** Chemical substances called ______________ are produced by the endocrine system and regulate the activities of many organs of the body.

**10.** A(n) ______________ is an entire living thing made up of systems that carry out all the basic life functions.

## CONTENT REVIEW: TRUE OR FALSE

*Determine whether each statement is true or false. If it is true, write "true." If it is false, change the underlined word or words to make the statement true.*

1. A bacterium is a multicellular organism.
2. In a complex organism, there is a division of labor among the different parts of the body.
3. There are seven basic levels of organization in living things.
4. The unit of structure and function of living things is the nucleus.
5. Groups of cells similar in structure and activity are called tissues.
6. Organs are made up of groups of different systems, each performing its particular job.
7. The heart, lungs, and eyes are examples of animal organs.
8. Leaves are important plant tissues.
9. Most organisms are composed of many tissues, each carrying out a specific life function.
10. An organism can survive without the smooth operation of any one of its basic systems.

## CONCEPT REVIEW: SKILL BUILDING

*Use the skills you have developed in the chapter to complete each activity.*

1. **Classifying objects** To which level of organization does each of the following belong?

   a. xylem
   b. ear
   c. bacterium
   d. geranium
   e. skeleton
   f. muscle
   g. root
   h. bone
   i. cheetah
   j. blood

2. **Relating facts** Name the organ system to which each of the following organs belongs.

   a. heart
   b. eye
   c. lung
   d. stomach
   e. kidney
   f. brain
   g. skin
   h. nose

3. **Making diagrams** Draw a pyramid to illustrate the five basic levels of organization within an organism. The most complex level should be at the bottom of the pyramid, the least complex at the top.
4. **Relating cause and effect** Describe two changes in your environment that your nervous system might respond to. In each case, describe the response. Which organs are involved in each response?
5. **Applying concepts** Use the sequence of activities that you do from the time you wake up in the morning to the time you leave for school to explain how your body systems work together.

## CONCEPT REVIEW: ESSAY

*Discuss each of the following in a brief paragraph.*

1. Explain the difference between a simple organism and a complex organism.
2. Identify five different organ systems. Then explain how each of these systems is necessary for the normal functioning of an entire organism.
3. Describe the five levels of organization in an organism and give two examples of each level.
4. Defend this statement: One-celled organisms are often more complex than the individual cells of multicellular organisms.



## ADDITIONAL QUESTIONS AND TOPIC SUGGESTIONS

**1.** What level of organization is each of the following:
a. elephant b. liver c. cactus
d. leaf e. amoeba
(elephant—organism; liver—organ; cactus—organism; leaf—organ; amoeba—cell)

**2.** How do the various cell structures illustrate a division of labor? (Each of the structures of a cell are associated with a specific function or activity. Ribosomes act to make proteins; mitochondria produce and store chemical energy; lysosomes aid in digestion, etc.)

**3.** Why would a single-celled living thing such as a bacterium be able to be placed in both the first and fifth levels of organization? (Although a bacterium consists of only one cell, level 1, it carries out all of its life functions by itself, level 5.

**4.** What is meant by the statement: "When a person's cells are sick, he or she is also sick"? (Because the cells work together to form tissues, organs, and organ systems, their improper functioning will have a negative effect on the organism as a whole.)

## ISSUES IN SCIENCE

The following issue can be used as a springboard for class debate or given as a written homework assignment.

You have an identical twin brother or sister whose kidneys have failed and needs a kidney transplant. After a detailed analysis, a kidney specialist determines that your kidneys most closely match those of your brother or sister and that you would be the best donor. What considerations will you have to take into account before deciding whether or not to become a donor?

## Chapter 5

# INTERACTIONS AMONG LIVING THINGS

## CHAPTER OVERVIEW

Ecology teaches that everything is connected to everything else. In this chapter, students will learn about these interconnections beginning with the concept of environment and ecosystem. The first section includes the structure of ecosystems and the role of organisms in communities.

Energy flow through ecosystems includes concrete references to food chains and food webs. The functional classification of producers, consumers, and decomposers is introduced and examples are provided. An energy pyramid is used to illustrate energy losses at successive feeding levels.

Relationships among organisms are given an entire section. The concept of limiting factors is introduced: competition, predation, commensalism, mutualism, and parasitism are described as possible limiting factors. Human impacts on ecosystems are described using Mono Lake as an example.

### INTRODUCING CHAPTER 5

Begin your introduction of Chapter 5 by having students examine the photograph on page 92 and read the chapter introduction on page 93. Explain to students that a tern is a type of sea gull which is smaller and more slender in body and bill size than a typical sea gull. If possible, obtain a photograph of a tern from a nature or wildlife magazine.

After students have read the chapter introduction, ask

- **What natural occurrence was responsible for the decrease in the tern population?** (the growth of grass on the beaches)
- **Why was this not a problem in the past?** (Small field mice called voles ate the grass.)
- **What do you think are some possible reasons for the disappearance of the voles from Great Gull Island?** (Accept all logical responses such as predation by another organism, a decrease in food supply, human interference, environmental changes, etc.)
- **What other populations might be disturbed by the disappearance of the terns from Great Gull Island?** (Accept all logical responses such as fish and other small marine organisms, etc.)

Stress to students the importance of the relationships that exist between

# Interactions Among Living Things

**CHAPTER SECTIONS**

**5-1 Living Things and Their Environment**

**5-2 Food and Energy in the Environment**

**5-3 Relationships in an Ecosystem**

**CHAPTER OBJECTIVES**

*After completing this chapter, you will be able to:*

**5-1** Describe the interdependence between living and nonliving things in an environment.

**5-2** Explain the relationships that exist among producers, consumers, and decomposers in an ecosystem.

**5-2** Describe a food chain, food web, and energy pyramid.

**5-3** Explain why competition among organisms occurs in every ecosystem.

**5-3** Compare the symbiotic relationships of commensalism, mutualism, and parasitism.

**5-3** Predict how a disturbance in the balance in one part of an ecosystem can affect the entire ecosystem.

For more than 30 years, Great Gull Island in Long Island Sound, New York, has been an exciting outdoor laboratory for ornithologists, or people who study birds. The ornithologists have been studying two graceful sea birds—the common tern and the roseate tern.

Gradually, a dangerous inhabitant of the island began to take control and threaten the lives of the terns. This enemy was grass! Scientists know that terns need bare, sandy beaches on which to build nests and raise their young. By 1981, the beaches of Great Gull Island were covered with grass. Human efforts to remove it were useless. Were the terns doomed?

A hundred years ago, Great Gull Island had a large population of field mice called voles. These voles fed on the roots and stems of the island's grasses. But by 1981 voles had long been gone from the island. Could voles now be brought back to the island in an effort to save the terns?

Scientists captured 36 voles and set them loose on the island. Over the next three years, the tern population soared. So did the vole population at first. Then it suddenly shrank. The little voles had eaten up much of their food supply. Will the vole population bounce back? Will a natural balance of living things return to Great Gull Island? Scientists hope so, but only time will tell.

*Vole on Great Gull Island*

93

## TEACHER DEMONSTRATION

If possible, take students on a five-minute field trip on the grounds of the school or in a nearby park. Instruct students that they will be asked to take notes on the kinds of organisms they observe (both plant and animal) and the areas in which these organisms live.

- **What organisms did you observe?** (Answers will vary. Compile a class list of all organisms noted.)
- **Where in the environment did each organism live?** (Add a column to your list of organisms detailing where they were seen.)
- **What was the energy source for these organisms?** (Students will reply that the sun is the energy source for green plants. They should be able to note that some of the animals eat green plants while other animals eat the plant-eaters. Lead students to the idea that in either case the sun is the original source of energy for most living things on earth.
- **In what ways were the organisms dependent on one another for survival?** (The obvious answer will be those organisms that eat other organisms. Lead students to the idea that a robin depends on a tree for its nest, etc.)
- **In what ways were the organisms dependent on the nonliving environment for their survival?** (Answers will likely include shelter, living space, air to breathe, soil to grow, etc.)

Point out that in this chapter students will be looking into the ways living things interact with one another and with their environment.

different types of organisms. All organisms are in some way dependent upon other organisms for life.

- **What are some other examples of one type of organism having an impact on another organism in its environment?** (Accept all logical responses.)

## TEACHER RESOURCES

**Audiovisuals**

*Ants: Backyard Science,* film, BFA

*Interrelationships of Living Things,* filmstrips, Eye Gate

*Nature's Half Acre,* film, Walt Disney

**Books**

Boughey A. S., *Ecology and Population,* Macmillan

Street, P., *Animal Partners and Parasites,* Taplinger

**Software**

*Food Chains and Webs,* Prentice Hall

## 5-1 LIVING THINGS AND THEIR ENVIRONMENT

### SECTION PREVIEW 5-1

This section introduces students to some of the fundamental concepts in ecology. It defines the term *environment* and, using a spider web as a model, describes how biotic and abiotic factors interact. The section also carefully defines the terms *ecosystem, community,* and *population.* The section concludes by explaining the difference between habitat, the place where an organism is usually found and niche, the role or "occupation" an organism plays in its habitat.

### SECTION OBJECTIVES 5-1

1. **Define environment.**
2. **Relate ecology to the relationships of living things with their environment.**
3. **Distinguish between biotic and abiotic factors in the environment.**
4. **Compare ecosystems, communities, and populations.**
5. **Describe the difference between habitat and niche.**

### SCIENCE TERMS 5-1

**environment p. 95**
**ecology p. 95**
**ecosystem p. 95**
**community p. 97**
**population p. 98**
**habitat p. 99**
**niche p. 99**

## 5-1 Living Things and Their Environment

Deep within the cool, dark waters of the earth's oceans lives an unusual variety of animals and other living things. Some animals glow like streetlights in the dark, attracting their prey with their light. Other animals swim very slowly, using huge eyes to find their way in a nighttime world. The only other living things that can survive in these waters are the fungi. Fungi do not need light to live.

On land, in a lush tropical rain forest, tall trees with clinging vines form a giant umbrella. This umbrella blocks the sun and casts a permanent shadow on the forest floor. In the treetops, brightly colored

**Figure 5-1** *An environment includes all the living and nonliving things that interact with one another.*

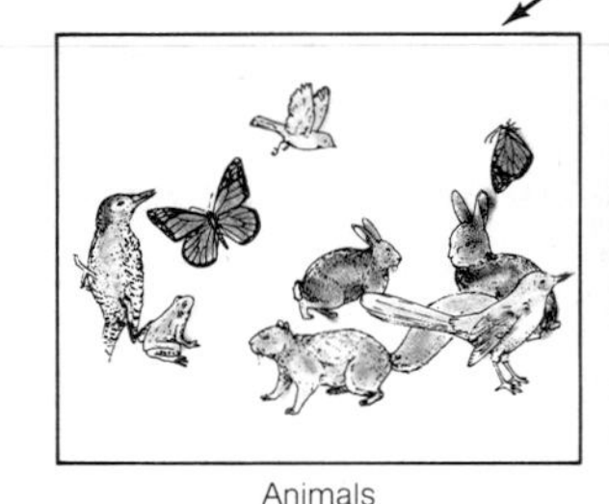


Plants

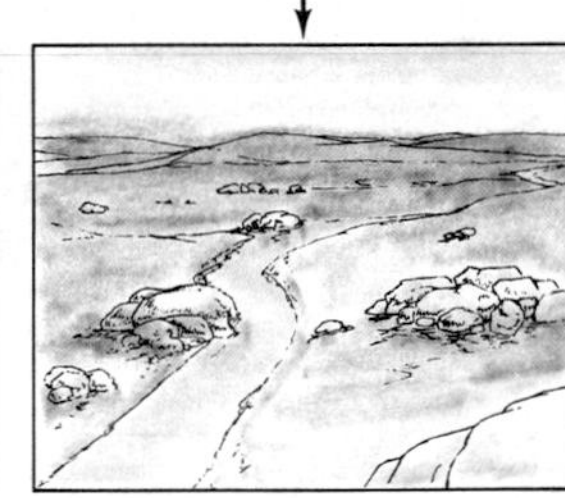


94

### TEACHING STRATEGY 5-1

Begin this section by reviewing the important terms and being certain that students understand their meanings. The term *ecology* comes from the Greek word *oikos* meaning "house" or "place to live." Ecology is the study of organisms in their surroundings or homes. Point out that groups of similar organisms make up populations; groups of populations make up communities; and communities plus abiotic factors, or nonliving factors, make up ecosystems. The terms *population* and *community* refer to the biotic, or living, components of the ecosystem. Explain that populations have different niches because they have different food needs, hunting methods, and ways of protecting themselves. Niche is often described as an organism's "occupation" or the way to makes a living. Habitat is likened to an "address" or place where an organism lives.

#### Motivation

Ask students to describe their environment twenty years from now. What kind of efforts are they willing to make today to ensure their dreams come true?

#### Content Development

Emphasize the idea of interdepen-

parrots munch on seeds while monkeys chatter to one another. Snakes and lizards climb up and down the tree trunks in search of food. And long-nosed anteaters calmly make their way on the ground.

The deep sea and a rain forest are only two of many different **environments** found on the earth. An environment includes all the living and nonliving things with which an organism may interact. **All of the living and nonliving things in an environment are interdependent.** ❶ You can think of each kind of environment as being like a giant spider web. However, the threads of this web are not spun from silk. The threads of an environmental web are the plants, animals, soil, water, temperature, and light found there.

Consider for a moment what happens when an insect gets caught in a spider's web. As one thread of the web is disturbed, the shaking motion is transferred to all the threads that are part of the web. So a spider resting some distance from the trapped insect suddenly receives a signal that dinner is nearby!

In an environmental web, all the living and nonliving things make up different threads. Some threads are delicate. Others are hardy. But all have one thing in common: Like a real spider web, changes in one thread may be transmitted to and have an effect on other threads in the environment.

In order to understand the changes that can occur in an environment, and how they can affect the environment, you must study the science called **ecology.** Ecology is the study of the relationships and interactions of living things with one another and with their environment. ❷ Scientists who study these interactions are called ecologists.

*Figure 5-2 Like the threads of a spider web, the various parts of an environment are interconnected. Changes in one part of the environment can affect other parts.*

### *Ecosystems*

Living things inhabit many different environments. From the poles to the equator, life can be found underground, in air, in water, and on land. A group of organisms in an area that interact with each other, together with their nonliving environment, is often called an **ecosystem.** An ecosystem can be as tiny as a drop of pond water or a square meter of a garden. Or it can be as large as an ocean or an entire forest. ❸

95

## ANNOTATION KEY

❶ Thinking Skill: Making generalizations

❷ Thinking Skill: Applying definitions

❸ Thinking Skill: Relating concepts

## FACTS AND FIGURES

The word *ecology* comes from the Greek words meaning "the study of the home." The "home" includes the environment in which organisms live and the interactions of organisms with both the living and nonliving parts of the environment.

dence throughout the chapter and especially in this section. Ask how all the people in your community are related. Have students describe how various occupations (or niches) in your community function together while they share the same habitat. Ask them to describe similar occupations in forest or beach habitats.

- **What are some ways biotic (living) and abiotic (nonliving) factors affect each other?** (Rocks weather and produce soil. When plants get established in the soil they reduce erosion and hold the soil in place. When the plants die, their bodies add organic matter and nutrients to the soil. Thus the soil helps the plant and the plant helps the soil—this is just one example.)
- **How can ecosystems range in size from a drop of water to an ocean?** (The ecologist defines the ecosystem as the area of study; as long as both biotic and abiotic factors are defined and included, the study area is considered an ecosystem.)

## Skills Development

***Skill: Making a model***

Have students develop a model of their school as an ecosystem. They will need to identify the biotic and abiotic factors in this ecosystem and decide how they interact.

- **Where does the energy, water, and food come from?** (Accept all logical responses.)
- **What are the niches of the students, teachers, parents, administration, and support people?** (Accept all logical responses.)
- **Does this ecosystem have a product?** (Accept all logical responses.)

## Reinforcement

Have students refer to Figure 5-1.

- **What are the biotic factors in this illustration?** (plants and animals)
- **What are the abiotic factors?** (all nonliving objects)

## Motivation

Have students describe the environment in which they live versus the environment that they would consider perfect to live in.

## ANNOTATION KEY

1. Thinking Skill: Identifying relationships
2. Thinking Skill: Relating concepts
3. Thinking Skill: Applying concepts

Figure 5-3 *These copepods are among the animals that live in pond water.*

If, for example, you look at a drop of pond water through a microscope, you will see a world of living things. Tiny plants, such as diatoms and green algae, float through the water in this microscopic ecosystem. Rotifers and copepods, very small animals, chase after these plants and eat them.

In the forest ecosystem, other microscopic organisms can be found. Some live in the soil and feed off dead animals and plants. Much larger living organisms such as foxes roam through the forest looking for food, while birds fly above. The foxes eat earthworms, birds, fruits, and small animals such as mice. [1] Fox families live in burrows, which are holes dug into the ground. The trees of the forest supply shelter for birds, which eat insects and earthworms. From this example you can see that the various threads or factors in an ecosystem are interdependent. In other words, the plants, animals, and nonliving parts of the ecosystem all interact with one another.

Figure 5-4 *The deciduous forest is the ecosystem for many plants and animals that interact with one another.*

96

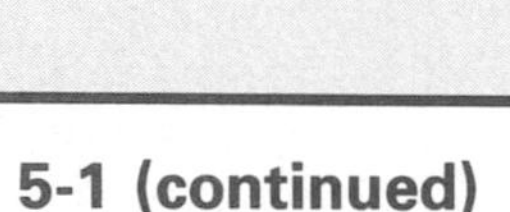

## 5-1 (continued)

### Content Development

Compare the terms *community* and *population.* Point out that a population is made up of only one type of organism. All organisms of a particular type that live in the same general area make up a population.

- **What types of populations live in your area?** (people, squirrels, grasses, etc.)
- **Why do you think these particular populations live where they do?** (They are able to find the factors necessary for life such as food, water, and shelter.)
- **Which populations would be most likely to move to another area and survive?** (People, of course, move all the time. Students may also point to various populations of birds that migrate.)

Point out that a community includes all the living things that live in a particular area.

- **Is it likely that an entire community might move to another area?** (No, although much of a community may move when the environment is damaged by pollution or by events such as a forest fire or volcano explosion.)

You may wish to expand the concept of communities by telling your students that although communities are orderly, changes do occur within them. Use the changes in a forest as an example: leaves change color, die, and fall off the trees; birds may leave in search of better living conditions, and so forth.

### Skills Development

***Skill: Interpreting diagrams***

Have students refer to Figure 5-4.

- **What populations are represented in this diagram?** (Accept any organism visible in the illustration.)
- **What other populations might be present which are not shown in this diagram?** (Accept all logical responses such as fish, worms, insects, moss, etc.)
- **How might the populations be affected by a forest fire?** (Accept all logical responses.

Show students slides or photographs of different communities and have them identify the populations that would likely be found there.

### Reinforcement

Discuss with students other examples of populations, such as lilies in a meadow, earthworms in a lawn,

### Communities

An ecosystem is composed of both living and nonliving things. The living part of any ecosystem—all the different organisms that live together in that area—is called a **community.** Fish, insects, frogs, and plants are members of a pond community. The insects, birds, small animals, and trees make up a community in a forest ecosystem. Are these living things interdependent? If you have ever observed pond life, you may know the answer to this question. Frogs eat insects, and fish eat tadpoles. Insects eat plants, as do some of the fish.

In a desert ecosystem's community, insects and birds feed on plants. Other animals, such as lizards, feed on insects and bird eggs. And larger animals eat lizards. Can you see a pattern to these communities? The plants are a source of food to plant-eating animals. And the rest of the organisms feed on the plant eaters and on other animals. So although not all the animals in these communities eat plants, they could not exist without them! Perhaps you are asking yourself what plants eat. Most plants make their own food, as you will soon learn in the next section.

### Populations

You probably live in a community. And you know from your experience that your community is made up of many different kinds of living things: people, dogs, cats, birds, fish, roses, and daisies.

**Activity**

*Identifying Interactions*

Ecosystems are all around you. Choose one particular ecosystem—for example, an aquarium, swamp, lake, park, or city block—and study the interactions that occur among the living and nonliving parts of the ecosystem. Include drawings and diagrams in your observations.

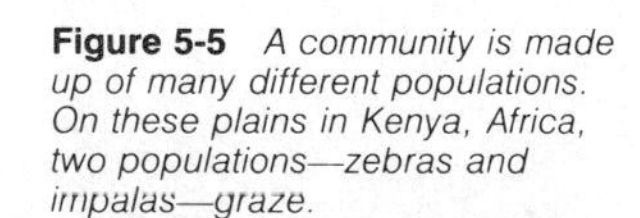

**Figure 5-5** *A community is made up of many different populations. On these plains in Kenya, Africa, two populations—zebras and impalas—graze.*

## Activity

**Identifying Interactions**
**Skills: Classifying, diagraming, observing, comparing, relating, applying**
**Level: Average**
**Type: Hands-on**

Encourage students to make on-site observations of the ecosystem they choose. Remind they that, as in all field trips, they should not go alone and they should inform an adult of their whereabouts. Explain that their observations should be analyzed and presented in terms of the ecological concepts covered in this chapter. Ask them to share their drawings and diagrams with the class. Use their drawings to illustrate some of the points you will make in teaching this chapter.

## FACTS AND FIGURES

Lemmings are tiny mouselike animals that live in Norway, Sweden, and Canada. Every three or four years, when their populations get very large, lemmings swim across rivers and lakes to find new food sources. But because they cannot tell a lake from a sea, thousands of lemmings plunge into the ocean and drown.

spruce trees in a forest, bees in a hive, and goldfish in an aquarium.

### Content Development

Explain to students that communities are often named by using the dominant plants found in the community. Oak-hickory forests, beech-maple, long-leaf pine, and aspen-spruce are examples of common communities. In other communities, such as lakes, oceans, and tropical rain forests, there is no dominant organism.

When discussing communities and populations, ask,

- **Give some examples of how foxes, snakes, and mice are interdependent in a grassland community.** (Foxes eat both snakes and mice; snakes eat mice and are eaten by foxes; mice are eaten by both foxes and snakes, but they usually eat seeds from the plants.)
- **What is another name for the kinds of living things that make up populations?** (species)
- **What are some organisms you would expect to see in a park near your home? On the plains of Africa?** (Accept all logical responses.)
- **What are some of the adaptations animals have to help them survive in their environment?** (Accept all logical responses.)

## ANNOTATION KEY

1. Thinking Skill: Applying definitions
2. Thinking Skill: Making comparisons
3. Thinking Skill: Relating cause and effect

Each kind of living thing makes up a **population** in the community. A population is a group of the *same* type of organism living together in the same area. (1) For example, all the trout living in a lake are a population of trout. All the mesquite bushes surviving in the desert make up another population.

Different ecosystems support different populations of plants and animals. You might expect to see squirrels in a park near your home. The park is the squirrel's natural environment. But, certainly, you would be quite surprised to see a group of zebras grazing there! For the natural environment of zebras is the African plain.

In any ecosystem, the most successful organisms are those that are best adapted to their environment. Perhaps the finest example of an animal population that is well adapted to its environment is the tortoise of the Galapagos Islands. Located in the Pacific Ocean off the western coast of South America, the Galapagos Islands are the home of two different kinds of tortoises. On some of the islands, the tortoises have short necks and shells that hang over the backs of their necks. These shells act like stiff collars and prevent the tortoises from raising their heads. On other islands of the Galapagos, the tortoises have long necks and the part of the shells above their necks is high. These tortoises can raise their heads.

The tortoises with the short necks and low shells live on islands where food is close to the ground. So these tortoises do not have to raise their heads to find food. The tortoises with long necks and high (2) shells live on islands where the only vegetation these animals can eat is high off the ground. So the tortoises must raise their heads high in order to reach food. In other words, each kind of tortoise is adapted to its own environment in ways that help it obtain food.

**Figure 5-6** *The yellow crab spider is well adapted to its environment. Looking much like a yellow mustard flower, it easily attracts insects and then traps them.*

### *Habitats and Niches*

Where would you look for deer in a forest? Would you look for them in the same places you would look for birds or snakes? Probably not. You would expect to find deer among the trees and bushes of the forest; the birds in the branches of the trees; and the snakes in the soil.

98

## 5-1 (continued)

### Enrichment

Arrange for a wildlife specialist to speak to your students about some of the techniques used to determine the populations of certain wildlife species.

### Skills Development

***Skill: Applying concepts***

Point out that populations have different niches because they have different food needs, hunting habits, and ways of protecting themselves. Have students make a list of animals native to their area and ask them to describe the habitats and niches of these animals.

### Motivation

Take a short walk around your school's campus and identify the habitats of some of the organisms that live there. Insects are particularly well adapted to sharing habitats with humans. When you return to the classroom, try to identify the niches of the various organisms you discovered.

### Reinforcement

Make sure that students understand the difference between a habitat and a niche. Often students incorrectly understand a niche as a spatial rather than a role-related concept. Point out that a habitat is an area of the environment. To be successful, an organism's habitat must provide it with food and shelter. A niche is the particular role an organism plays in its habitat. That is, a niche includes only the part of the habitat that a particular organism needs or uses.

### Content Development

Explain to students that habitats sometimes overlap. If the organisms do not eat the same food, however, there is usually no problem. For example, the squirrel inhabits the inside of a hollow tree trunk and gets its food from among the leaves on the ground. These leaves may be the habitat of a land snail. But since the snail is not looking for the same food as the squirrel, the overlapping of habitats causes little interaction between the two animals.

### Reinforcement

Show students slides or photographs of different animal and plant species and have them identify their habitats and niches.

Figure 5-7 *The habitat of these condors is the high, barren peaks of the Andes Mountains of South America.*

The forest contains a community of living things. And each member of that community has a certain place where it lives. The place in which an organism lives is called its **habitat.** A burrow, a tree, or a cave is a habitat. A habitat provides food and shelter for an organism. Plants have habitats too. The beautiful orchid, for example, grows from the branches of tall trees in the rain forest. The branches of these trees are the orchid's habitat.

In addition to a habitat, each organism in a community plays a particular role. The organism has a **niche,** or "occupation." A niche is everything the organism does and everything the organism needs within its habitat. And an organism's niche can affect the lives of other organisms.

Beavers, for example, build dams across streams, creating small lakes. The waters of the stream no longer rush along, wearing down the stream banks. Plants can now grow alongside the lake. Fish, such as trout, can make their home in the calm waters of the lake, and songbirds can nest along the quiet banks. Meanwhile, the beaver lives in an underwater home that it makes from sticks and mud. So the beaver's niche not only makes a home for the beaver, it also makes a habitat in which other animals and plants can thrive.

Organisms in an ecosystem can share the same habitat without any problems, but they *cannot* occupy the same niche. This fact seems obvious since sharing the niche would mean competing for the same

### Activity

*Home Sweet Home*

What does it take to build an animal home?

**1.** Choose one of the following animals and find out what kind of shelter it builds: beaver, trapdoor spider, mud dauber wasp, prairie dog, cliff swallow, termite.

**2.** On a sheet of paper, draw a picture of the animal's shelter and make a list of the materials needed to build it.

**3.** Build a model of your animal's shelter, using the same materials the animal would use whenever possible.

Predict what would happen if the materials that an animal uses to build its shelter were not available.



### Activity

**Home Sweet Home**
**Skills: Making models, diagraming, relating, applying, comparing, predicting**
**Level: Average**
**Type: Hands-on**

Students will gain a better knowledge of the habitats of various organisms through this activity. Encourage students to use everyday materials from home to construct their habitat. Place the best habitats on display when the activity has been completed.

## Content Development

Explain to students that plants that live on the branches of trees or other plants are called epiphytes. Orchids are one example. Spanish moss and a variety of other air plants are also epiphytes. Epiphytes are more common in tropical and subtropical regions than in temperate regions. Let students develop various hypotheses as to why this observation is true.

## Content Development

Stress to students that the niche of one organism can affect the lives and niches of other organisms. Beavers are animals that are capable of modifying the environment so that it suits their needs. When they dam up a fast-moving stream, they destroy the habitat of many other organisms but create habitats for others.

- **What habitats might be destroyed by the beaver's dam building?** (Accept all logical responses.)
- **What habitats might be created by the beaver's dam building?** (Accept all logical responses.)

Stress to students that humans are similar to beavers in that they create habitats for themselves and in doing so destroy the habitats of other organisms.

- **Besides other people, what kind of organisms commonly live in human habitats?** (a variety of insects, especially roaches, spiders, and ants, rats and mice, pigeons, starlings, and other birds, etc.)

Point out that although two organisms in the same habitat cannot occupy the same niche, species in different habitats often occupy similar niches. Such similar species are often called ecological equivalents. Grassland herbivores with similar niches are found on different continents; large fish hawks are distributed around the world.

## 5-2 FOOD AND ENERGY IN THE ENVIRONMENT

### SECTION PREVIEW 5-2

Energy flow through the environment is discussed in this section. Again, interconnectness is a good theme to emphasize throughout this section. Energy from the sun is what connects all living things to each other. Photosynthesis is briefly reviewed and the term *autotrophs* (green plants or producers) is introduced. Consumers are classified as herbivores, carnivores, and omnivores. The roles of scavengers and decomposers are explained and examples are provided.

Food chains, food webs, and energy pyramids are discussed and illustrated. The loss of energy as it moves through each step of the food chain is carefully explained.

### SECTION OBJECTIVES 5-2

1. **Trace the path of solar energy through a simple food chain that includes humans.**
2. **Describe how plants capture and store solar energy.**
3. **Compare autotrophs and heterotrophs.**
4. **Describe a food web.**

### SCIENCE TERMS 5-2

**photosynthesis p. 101**
**autotroph p. 101**
**producer p. 101**
**heterotroph p. 102**
**consumer p. 102**
**scavenger p. 102**
**decomposer p. 102**
**food chain p. 104**
**food web p. 105**

**Figure 5-8** *The dam that this beaver is building will become its underwater home, as well as a habitat for other animals and plants.*

food and space. Here is an example of how similar organisms in a habitat occupy different niches.

Three types of birds are found on the Galapagos Islands. The three types of birds are the blue-footed booby, the red-footed booby, and the white booby. These three birds all feed on the same kind of fish in the nearby waters. However, the blue-footed booby feeds close to shore, while the red-footed booby flies further out to sea to catch its food. And the white booby dives for its meals in the water in between. Can you think of another example of organisms sharing habitats but not niches? (1)

#### *SECTION REVIEW*

1. What is ecology?
2. What is the difference between a community and a population?
3. What is an organism's habitat? Its niche?

### 5-2 Food and Energy in the Environment

All living things need energy in order to live. They use energy to carry on all the basic life functions. For most living organisms, the immediate source of energy is the food they eat. But this food



### 5-1 (continued)

#### Section Review 5-1

**1.** Study of relationships and interactions of living things with one another and with their environment
**2.** Community includes all organisms in an ecosystem; population includes only organisms of same type
**3.** Habitat: place in which organism lives; niche: everything organism does and needs in its habitat

### TEACHING STRATEGY 5-2

#### Motivation

Have you thanked the sun today? Have students design bumper stickers and/or billboards that thank the sun for energizing the earth.

#### Content Development

Point out that the ultimate source of energy for all living things is the sun. Aside from keeping our external temperature in the environment somewhat constant (within a range that living things can survive), the sun's energy is used by green plants to produce food in a process called photosynthesis.

- **What does the term *photo* mean?** (light)
- **What does the term *synthesis* mean?** (to build or construct)
- **What does photosynthesis mean?**

can be traced all the way back to green plants, which use the sun's energy. Thus, all the energy used by plants and animals comes directly or indirectly from the sun.

### Green Plants: Food Factories

Green plants have one very important advantage over other living things. They can make their own food, while most other living things cannot. Plants make food by the process of **photosynthesis.** During photosynthesis the green parts of plants, especially the leaves, capture the energy of sunlight and use it to make glucose, a type of sugar. The glucose is formed when the plants chemically combine water and carbon dioxide. The reason green plants can carry on photosynthesis is that they contain the green pigment chlorophyll. It is the chlorophyll that captures the energy of sunlight. In addition to glucose, oxygen is also produced during photosynthesis. In fact, photosynthesis is an important source of oxygen on the earth. If it were not for green plants, animals would quickly use up all the oxygen available to them.

Because green plants use the energy of the sun directly to make their own food, they are called **autotrophs** (AWT-uh-trohfs), or "self-feeders." Plants use glucose to provide energy for carrying out life functions. But plants also combine the glucose with other chemicals to make starches, fats, and proteins. As you can see, photosynthesis is a very important process to green plants. But photosynthesis is also very important to animals. Can you guess why this process is so important? ❷

### Interactions Among Organisms

As you know, all organisms in an ecosystem are interdependent. **Organisms can be classified into three main groups based on how they obtain energy: producers, consumers, and decomposers.** ③

PRODUCERS Green plants make their own food and are thus the food **producers** of the ecosystem. Animals cannot make their own food. They must eat either plants or other animals that eat plants.

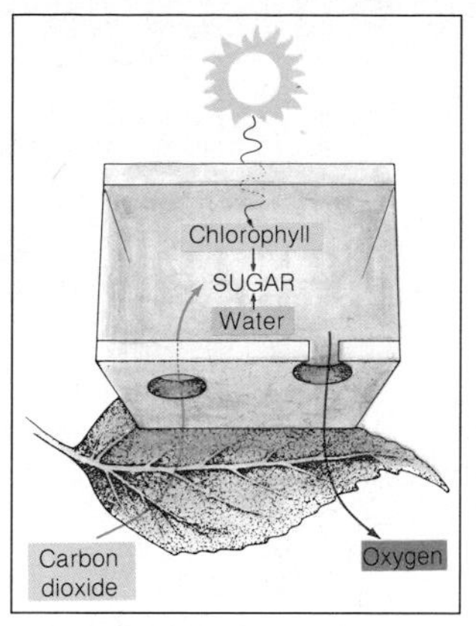


**Figure 5-9** *The leaves of green plants are food factories that make glucose during photosynthesis.* ②

**Figure 5-10** *Indian pipe plants lack chlorophyll and cannot make their own food. But a certain fungus that lives on nearby green plants forms a bridge to the Indian pipe plants. The fungus takes in glucose from the green plants. And the Indian pipes take the glucose from the fungus.*

## FACTS AND FIGURES

Scientists estimate that the total production of glucose by the earth's plants, both on land and in the sea, is about 91 billion metric tons a year.

## FACTS AND FIGURES

The usual equation for photosynthesis shows the overall reaction; in fact, there are more than 80 chemical reactions acting in sequence to complete the photosynthesis process.

## ANNOTATION KEY

❶ **Foxes and wolves in a forest, trout and bass in a lake, etc. (Applying concepts)**

❷ **Through photosynthesis, plants provide food and oxygen for animals. (Making inferences)**

① **Thinking Skill: Identifying relationships**

② **Thinking Skill: Identifying processes**

③ **Thinking Skill: Making generalizations**

(to use light—sun's energy—to build or construct food)

Point out that the main product of photosynthesis for plants is the sugar glucose. However, a byproduct of photosynthesis that is very important to life on earth is the production of oxygen during the process.

## Skills Development

***Skill: Interpreting illustrations***

Use Figure 5-9 to reinforce the process of photosynthesis. Explain that although less than 1 percent of the available solar energy is captured by green plants, this tiny amount is responsible for almost all life on earth.

• **Where does the carbon dioxide needed for the reaction come from?** (mostly from the air)

• **What parts of plants, besides leaves, can capture solar energy?** (any part that contains chlorophyll, including stems, branches, and even roots)

## Reinforcement

Remind students that plants take in carbon dioxide and give off oxygen as a part of photosynthesis.

## Content Development

Explain to students that autotrophs are organisms that produce their own food.

• **What are some examples of autotrophs?** (Accept all logical responses.)

• **Why are autotrophs so important to animals?** (They make their own food and act as a food source.)

Stress to students that the words *food* and *energy* are used interchangeably in the text because food is just a form of chemical energy that is stored as tissue and used to perform various body function.

## FACTS AND FIGURES

In order to survive, a 50-kilogram wolf needs to eat about 2700 kilograms of moose per year. A moose, in turn, needs to eat about 35,000 kilograms of plants per year.

## ANNOTATION KEY

❶ Thinking Skill: Relating definitions
❷ Thinking Skill: Interpreting diagrams
❸ Thinking Skill: Applying concepts

**CONSUMERS** An organism that feeds directly or indirectly on producers is called a **consumer.** Consumers are also called **heterotrophs** (HET-uhr-uh-trohfs), or "other-feeders."

❶ There are many kinds of consumers. Some organisms such as mice, insects, and rabbits are plant eaters. Snakes, frogs, and wolves are flesh eaters. They consume animals that are plant eaters or animals that feed on plant eaters.

In the northern parts of Norway, Sweden, and Finland, a plant called the reindeer moss carries on photosynthesis. Reindeer feed on the reindeer moss. And people who live in this area eat the reindeer. Can you identify the producers and consumers?

A special type of animal consumer feeds on the bodies of dead animals. These organisms are called **scavengers.** Jackals and vultures are examples of scavengers. So are crayfish and snails, who "clean up" lake waters by eating dead organisms.

**DECOMPOSERS** After plants and animals die, organisms called **decomposers** use the dead organic matter as a food source. Unlike scavengers, decom-

❷ **Figure 5-11** *Organisms are classified as producers, consumers, or decomposers, depending on how they get their food. Producers are autotrophs. Consumers and decomposers are heterotrophs.*



### 5-2 (continued)

#### Content Development

Explain that *autotrophs* literally means "self-feeders." Green plants or producers are capable of feeding themselves. Animals that eat green plants or other animals are called "other-feeders" or heterotrophs.

#### Content Development

Explain to students the terms *producer* and *consumer.* Point out that consumers are also called heterotrophs. Heterotrophs are organisms that eat plants or plant-eating animals.

- **What organisms in your home are producers?** (green plants)
- **What organisms in your home are consumers?** (students, parents, and perhaps pets)

#### Content Development

Help students master the terms *producer, consumer,* and *decomposer* by asking for examples. Point out that this classification scheme is based on the way organisms get their food. When discussing consumers, you may wish to introduce the terms *herbivore* (plant-eating animal), *carnivore* (animal-eating animal), and *omnivore* (plant- and animal-eating animal).

- **What are some consumers that live in fresh water?** (Answers will vary but may include aquatic insects, minnows, tadpoles, and fish.)
- **What are some producers that live in fresh water?** (Answers will vary but may include algae and water

**Figure 5-12** *The small stickleback fish, which feeds on water fleas, is about to become food for the larger pike fish* (left). *Spotted hyenas are often scavengers* (right).

posers break down dead plants and animals into simpler substances. In the process, they return important materials to the soil and water. You may be familiar with the term "decay," which is often used to describe this process. Bacteria and mushrooms are examples of decomposers.

## Career: *Game Warden*

**HELP WANTED: GAME WARDEN** College degree in zoology, ecology, or wildlife management desired. Must be familiar with animal behavior and habitats within this state.

The crack of the rifle broke the peaceful quiet of the forest, turning the game warden's attention from the nest full of screeching young robins to more serious matters. This was not the hunting season so there was no reason for anyone to use a gun in the forest. The warden walked carefully to the spot from where the sound seemed to come, calling out "hello" so as not to be mistaken for a target.

This is a scene from the duties of the **game warden.** Each state has rules and regulations protecting its wildlife. The game warden's job is to enforce these rules. They may warn offenders, fine them, arrest them, and confiscate guns, equipment, and any illegally killed animals if necessary.

Game wardens have other duties, too. They keep accurate records of the wildlife in their areas. They are responsible for planning controlled hunts. Wardens issue hunting permits and conduct safety programs to educate hunters. During emergency situations, they may take part in rescue operations.

If you are interested in this career, write to the National Park Service, U.S. Department of the Interior, 18th & C Streets, NW, Washington, DC 20240.

103

## SCIENCE, TECHNOLOGY, AND SOCIETY

To illustrate the complexity of food webs and the unforeseen consequences of human interference, consider the following true incident. In the 1950s, the World Health Organization attempted to eliminate malaria from Borneo. Since they knew that malaria was spread by mosquitoes, the environment was sprayed with DDT to kill the mosquitoes. Other animals, however, were also affected by the DDT.

Cockroaches ingested the DDT and they, in turn, were consumed by geckoes (insect-eating lizards). Cats caught and ate a much greater number of geckoes than usual since the geckoes' movements were slowed down due to DDT-caused nerve damage. As geckoes are natural predators of caterpillars, the diminished gecko population led to an increase in the caterpillar population. The caterpillars began eating the thatched roofs of the houses in Borneo—causing many of the roofs to collapse. Even worse, rats moved in from the forest since most of the cats had died from DDT poisoning. With the rats, came the rat fleas which carry the bacteria that causes plague. And plague can be more immediately fatal than malaria!

In attempts to solve the problems they had created, the World Health Organization stopped spraying DDT and parachuted a large number of cats into Borneo. Eventually, the situation returned to normal.

Have students construct the food web involved in this scenario. Have them cite other instances of the consequences of human intervention in established ecosystems.

plants such as *Elodea* or hydrilla.)

• **What are some scavengers that live in fresh water?** (In addition to crayfish, common freshwater scavengers are turtles and catfish.)

### Reinforcement

Distinguish between scavengers and decomposers by pointing out that scavengers use the energy in dead organisms to carry on their life processes. They store the nutrients as complex molecules. Decomposers also use the energy for life processes, but they release most of the nutrients into the environment as simple chemicals.

### Enrichment

Have students do library research on carnivorous plants such as the sundew, Venus' flytrap, and pitcher plant.

## TEACHER DEMONSTRATION

A food web can be constructed by using a long piece of string and running it back and forth among students "named" for animals and plants. Be sure to include the sun. After twenty or so "organisms" are included in the food web (include all of the students in your class if you wish), have one "organism" drop out, releasing the string. Everyone who is connected to the loose piece must drop their string. Continue until the entire web collapses.

- **How could the loss of one organism in a food web cause the entire web to collapse?** (All organisms in a food web are interrelated; the loss of one organism affects the lives of other organisms that depend on it for life.)

**Figure 5-13** *In this Antarctic food chain, microscopic plants known as diatoms* (top) *are the food of zooplankton* (top, center), *tiny water animals. The zooplankton are then eaten by krill* (bottom, center), *shrimplike animals. Krill feed many creatures, including popeyed squid* (bottom). *The right whale* (bottom, right) *feeds on both krill and squid.*

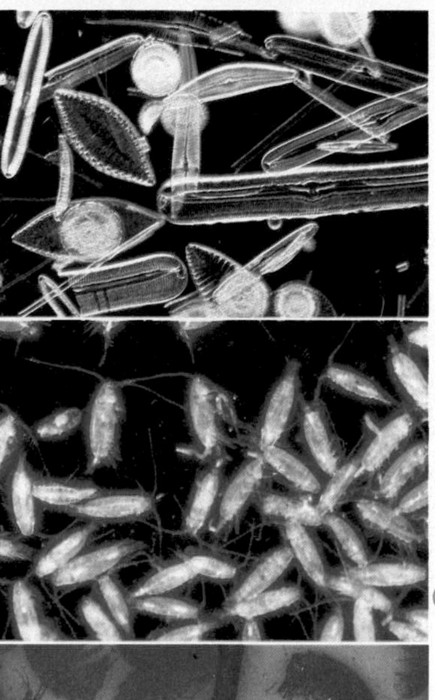

Decomposers are essential to the ecosystem for two reasons. They rid the environment of dead plant and animal matter. But even more important, decomposers return substances such as nitrogen, carbon, phosphorus, sulfur, and magnesium to the environment. These substances then are used by other plants to make food, and the cycle begins again. If the nutrients were not returned to the environment, organisms within that ecosystem could not survive for long.

### *Food Chains and Food Webs*

Within an ecosystem, there are food and energy links between the different plants and animals living there. A **food chain** illustrates how groups of organisms within an ecosystem get their food and energy.

Green plants are autotrophs, or producers. So they are the first link in a food chain. Animals that eat the green plants are the second link. And animals that eat animals that eat plants are the third link.

Here are two examples of food chains. On land, a grasshopper eats a plant. A bird then eats the grasshopper. Finally, a snake eats the bird. In a freshwater pond, microscopic green plants are eaten by water fleas. In turn, the water fleas become the food of minnows, or small fish. Larger fish such as perch and bass feed on the minnows. If no animal eats the perch or bass, it will eventually die and become a source of food for a scavenger like the crayfish. As the perch or bass decays, important substances are returned to the lake water. The microscopic green plants use these substances to make their food. The

## 5-2 (continued)

### Content Development

Have students trace (on paper) the food energy from their last meal back to the sun.

- **Is this an example of a food chain or a food web?** (Unless students ate only one thing, they should draw a food web.)

### Motivation

Encourage students to give examples of food chains and food webs from other habitats. Most students are familiar with animals that live in Africa and other exotic places.

### Enrichment

Not all organisms remain on the same place in food chains throughout their lives. Tadpoles are herbivores, but frogs and toads are carnivores. Many insects are herbivores during their larval stages but become carnivores when they mature. Challenge students to find other examples of animals that change their feeding or trophic level during different phases of their life cycle.

### Content Development

Explain to students that in nature a food web is a far more accurate way of depicting the relationships among various organisms than a food chain. A food web shows all of the various food pathways in a particular ecosystem.

- **Why is a food web a better way of describing the path of food in an ecosystem?** (Because it is rare that a particular organism eats only one type of organism or is eaten by only one type of organism. A food chain actually shows only a portion of the interactions in nature depicted by a food web.)

### Skills Development

***Skill: Interpreting diagrams***

Have students refer to Figure 5-14.

- **What are the producers in this food web?** (seaweed, phytoplankton, and zooplankton)

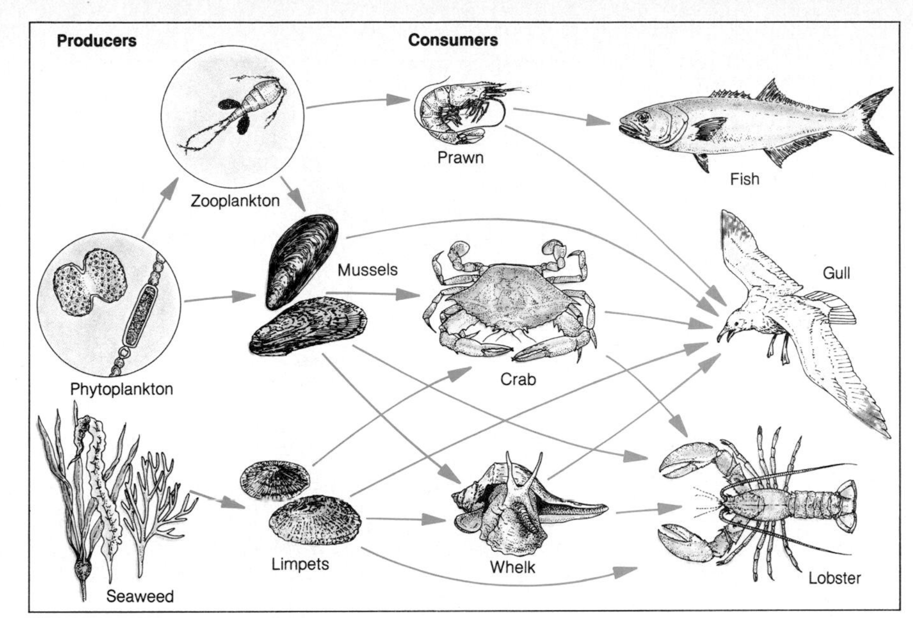


**Figure 5-14** *This ocean food web is a complex overlapping of many individual food chains. How many food chains can you find within the food web?* ❷

food chain goes on and on. This food chain could have been extended beyond the perch or bass. Can you think of who the next consumer might be? ❶

Food chains are good descriptions of how food energy is passed from one organism to another. But interactions among organisms often are more complex because most animals eat more than one type of food.

For example, in a saltwater environment, sea snails eat various types of seaweed. But seaweed is also eaten by other small fish. And these small fish may become food for larger fish, birds, and octopuses. There are several different food chains involved here, and they overlap to form a **food web.** ❸ A food web includes all the food chains in an ecosystem that are connected together. See Figure 5-14.

### *Feeding and Energy Levels*

A feeding level is the location of a plant or an animal along a food chain. Because green plants produce their own food, they form the first feeding level. Consumers that eat plants form the second

**Activity**

*Drawing a Food Web*

Using the following plants and animals, draw the food web as you think it might exist: grass, mice, deer, mountain lions, vultures, bacteria.



## ANNOTATION KEY

❶ **Bear or person (Applying concepts)**
❷ **Eighteen (Interpreting diagrams)**
❶ **Thinking Skill: Identifying processes**
❷ **Thinking Skill: Applying concepts**
❸ **Thinking Skill: Relating concepts**

## Activity

**Drawing a Food Web**
**Skills: Diagraming, applying, relating, classifying**
**Level: Average**
**Type: Hands-on**

This activity provides students with practice in analyzing feeding relationships. They might draw arrows that point from grass to mice and deer, from deer to mountain lions, from mountain lions to vultures, and from all of these organisms to bacteria, which serve as decomposers.

- **What are the consumers of this food web?** (prawn, mussels, limpets, crab, whelk, fish, gull, and lobster)
- **Which consumers feed directly on the producers?** (prawn, mussels, and limpets)
- **What organisms are "missing" from this food chain?** (decomposers such as bacteria)
- **What organism has the most varied diet?** (the gull)

### Reinforcement

Have students construct a food web of the organisms found in the environment around your school.

### Motivation

Have interested students form a large food web on a bulletin board in the classroom, using string and pictures cut from magazines to indicate the food links that are formed between organisms.

### Content Development

Emphasize to students that all of the energy coming from green plants originally came from the sun.

- **Why are so many plants needed to support herb-eating consumers?** (Most herb-eating consumers (herbivores) are not very efficient at converting the energy stored in plants into animal tissue, so they need large quantities of plants.)
- **Why are so many herb-eating consumers needed to support those organisms that eat them?** (Again, animal-eating animals [carnivores] are not very efficient users of the energy stored in the tissues of herb-eating animals. In addition, carnivores use a great deal of energy to hunt and capture prey.)

## ANNOTATION KEY

❶ Thinking Skill: Interpreting diagrams
❷ Thinking Skill: Relating concepts

### Activity

**Your Place in a Food Chain**
**Skills: Identifying patterns, applying, relating, comparing, diagraming**
**Level: Enriched**
**Type: Hands-on**

Students should be able to derive a food chain for each of the major components of each meal. Certain of the foods may contain a large number of ingredients, whose separate food chains may be difficult to determine. Students should be encouraged to do library research on the organisms from which the various ingredients and foods were derived and to report their findings to the class.

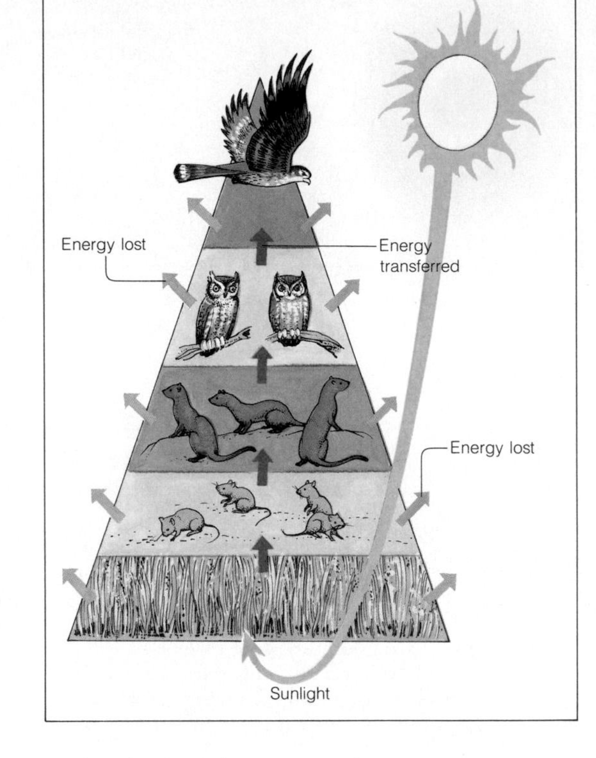


**Figure 5-15** *In this food chain, grasses and other plants make their own food by photosynthesis. Mice feed on the grasses and are consumed by weasels. In turn, owls feed on weasels. Owls also become a source of food for the hawks. At each point along the food chain, some energy is lost.*

feeding level. And consumers that eat animals that eat plants form a third feeding level.

At each feeding level, however, much of the energy in food is lost. Each organism at a particular feeding level uses up some of the energy locked in food to carry out its life activities. In addition, a great deal of energy is lost in the form of heat as it is transferred from one feeding level to another. What does this mean for living things at higher energy levels? There is less energy available to organisms at each higher level. The loss of energy as it moves through a food chain can be pictured as an energy ❶ pyramid. See Figure 5-15. The energy pyramid represents the decreasing amount of energy available at each feeding level.



## 5-2 (continued)

### Reinforcement

Make sure that students understand the concept of an energy pyramid and energy feeding levels. Point out that the amount of energy lost at each level of the food pyramid is substantial and that the number of organisms at each level decreases drastically as the amount of available energy that can be used per organism decreases.

Have students refer to Figure 5-15.

- **What are the most plentiful organisms in this pyramid?** (grasses)
- **What are the least plentiful organisms in this pyramid?** (hawks)
- **Which organisms feed primarily on organisms that eat producers?** (weasels)
- **In most cases, what happens to the size of the organisms involved as you go up the energy pyramid?** (The organisms usually get larger.)

### Skills Development

***Skill: Applying concepts***

Explain to students that about 10 percent of the energy is transformed from one feeding (or trophic) level to the next. Starting with a bass that has a mass of 5 kg, have students calculate how many kilograms of bream (small freshwater fish) it would take to support the bass. Point out that bass eat bream.

- **How many kilograms of bream would it take to support a 5 kg bass?** (50 kg)
- **How many kilograms of aquatic insects would it take to support the bream?** (500 kg)
- **How many kilograms of grass would it take to support the aquatic insects?** (5000 kg)

Since the autotrophs, or producers such as green plants, support all other living organisms on the earth, they form the base of the energy pyramid. This first level is very broad. Each level above the first level is made up of consumers. And each successive level of consumers in the pyramid is narrower than the one beneath it. As energy moves through the pyramid, from the first feeding level up through all the other levels, much of it is lost.

### SECTION REVIEW

1. What is photosynthesis?
2. What is the difference between autotrophs and heterotrophs?
3. What is a food chain? A food web?

**Activity**

*Your Place in a Food Chain*

Every time you eat, you are assuming a particular place in a food chain. You can determine your place in a food chain through the following activity.

Make a list of all the foods you ate for breakfast, lunch, and dinner. For each food, determine the food chain, using a pyramid to represent your findings. Be sure to answer the following questions.

1. Who are the producers?
2. Who are the consumers?
3. At what level are you a consumer?

## 5-3 Relationships in an Ecosystem

Whether an ecosystem is a huge patch of ocean bottom or just a dark corner of a forest floor, all the living and nonliving things in it must interact successfully in order for the ecosystem to survive. From

**Figure 5-16** *Living things, such as this adult and young chameleon, must interact with other living and nonliving things in the ecosystem. If conditions are right and organisms can find food, water, and shelter, they will reproduce and increase their populations.*



## 5-3 RELATIONSHIPS IN AN ECOSYSTEM

### SECTION PREVIEW 5-3

This section covers the major relationships among organisms and other factors in the environment. The idea of limiting factors is introduced, but most of the section deals with relationships between organisms.

Students learn about competition and why it is important. The role of predators is carefully explained. Three types of symbiotic relationships—commensalism, mutualism and parasitism—are described and illustrated.

Finally, the important idea of balance in ecosystems is introduced. Human interference with ecosystems is illustrated using California's Mono Lake as an example.

### SECTION OBJECTIVES 5-3

1. **Give examples of abiotic limiting factors.**
2. **Describe how competition can serve as a limiting factor.**
3. **Define a symbiotic relationship.**
4. **Distinguish between and give examples of commensalism, mutualism, and parasitism.**
5. **Describe the effects of natural phenomena and human interference on ecological balance.**

### SCIENCE TERMS 5-3

**competition p. 108**
**symbiosis p. 109**
**commensalism p. 100**
**mutualism p. 110**
**parasitism p. 112**

### Enrichment

Challenge students to research and develop a class report on the relative efficiency of animals commonly used by people as food. Chickens, turkeys, hogs, and cattle are good starters.

- **Which organisms seem to be the most efficient converters of energy?** (Have students base their answers on their research.)
- **Which organisms seem to be the least efficient converters of energy?** (Have students base their answers on their research.)
- **Because the world's population is growing so rapidly, why does efficiency play an important role?** (World supplies of animal protein are limited, and the more efficient converters will feed more people.)

Have students complete their research by investigating unusual animals that are used as food sources in other countries.

### Section Review 5-2

1. Process by which green plants make food
2. Autotrophs make their own food; heterotrophs do not.
3. Food chain shows food and energy links between organisms in an ecosystem. Food web includes all the overlapping food chains in an ecosystem.

## ANNOTATION KEY

❶ Thinking Skill: Making generalizations

❷ Thinking Skill: Relating cause and effect

❸ Thinking Skill: Applying concepts

Figure 5-17 *The snowshoe hare shares its habitat with moose. During the cold winter, the hare must successfully compete with the moose for food, or else the hare will die.*

their environment, all the living organisms must be able to obtain food, water, and shelter. Given the right conditions, the plant and animal populations can reproduce themselves and increase their number. However, if certain living and nonliving factors in the environment interfere with the growth of the population, the size of the population will be limited.

### Competition

Ecosystems often cannot satisfy the needs of all the living things in a particular habitat. When there is only a limited amount of food, water, shelter, or even light in an environment, **competition** occurs. ❶ **Competition is a type of relationship in which organisms struggle with one another and their environment in order to obtain the materials they need to live.**

Most of the relationships among living things in an ecosystem are based on competition. The moose and the snowshoe hare, for example, share the same habitat and compete for the same food. Birch trees are the most important food source for these two animals. However, the small hares can hardly compete with the large, hungry moose. So the hares may face a shortage of food, and possibly death, during the cold winter.

Competition often has a positive effect on an ecosystem. Through natural competition, different populations of animals usually are prevented from growing so large that they disrupt the ecosystem. But sometimes outside forces can unbalance natural competition. For example, people interfered with such a natural plan when they introduced rabbits to Australia in 1788. The rabbits had few natural enemies, so the rabbit population grew very quickly. ❷ By the middle of the 1800s, these harmless-looking animals were stripping the grasslands bare of vegetation. Cattle herds that also grazed on this vegetation were threatened with starvation. Finally, a disease-causing virus was given to many of the rabbits by scientists. These rabbits died, and so the cattle were saved from starvation.

Plants, too, compete with one another for light, carbon dioxide, minerals, and water. In the forest, plants usually compete for light. Tall trees soak up

108

## TEACHING STRATEGY 5-3

### Motivation

Have students bring in photographs of various nature scenes. Ask them to predict which limiting factors would be most important in each scene in terms of limiting the populations of organisms in that scene.

### Content Development

Introduce the idea of limiting factors along with biotic potential. All populations have a tremendous biotic potential or ability to reproduce. A variety of factors prevent populations from exceeding the ability of the environment to support them. A limiting factor is the one that is in the shortest supply in an ecosystem. It might be nutrients that limit plant growth. If additional nutrients are added, water might be the next limiting factor. If the area is irrigated, space or overcrowding might be the next limiting factor. The sum of all limiting factors is often called environmental resistance.

### Skills Development

***Skill: Predicting***

After a discussion of limiting factors, have students predict what factors will limit the number of people on earth. After you have listed several on the chalkboard, discuss each one. Have students rank them from the most important to the least important.

### Content Development

Competition is often described as a "struggle," but most organisms evolve ways to reduce competition. Small differences in the places where organisms feed, what they feed on, and even the time they feed all help reduce competition between organisms of different species.

- **What does the idea of niche tell us about competition?** (Because two species cannot occupy the same niche, they cannot directly compete with each other for all their needs. This is called the competitive exclusion principle.)
- **What about organisms of the same species?** (Most competition is be-

plenty of sunlight. Smaller trees that also need sunlight may not survive in the shade of these big trees. However, shade can be an advantage to some trees. The red maple tree is hardier than the sugar maple tree, and it usually outnumbers the sugar maple tree in a forest. But the sugar maple tree can grow in the shade, and the red maple tree cannot. So if a young red maple tree and a young sugar maple tree are competing for space and food in the shade of a forest, the sugar maple tree will be more likely to survive.

### Symbiotic Relationships

Instead of competing with one another, some plants and animals survive by "living together" and helping one another. Such a partnership is called **symbiosis** (sim-bee-OH-sis). Symbiosis is a relation-hip in which one organism lives on, near, or even nside another organism. **A symbiotic relationship nay benefit either one partner or both partners in he relationship.** Some symbiotic relationships may ound rather strange to you. But no matter how odd hey may seem, all symbiotic relationships have one hing in common. They provide the means by which one or both of the organisms can survive. 3

COMMENSALISM High in the branches of a tree, a arge, fierce hawk called the osprey builds a big plat-orm-shaped nest for its eggs. Smaller birds such as parrows and wrens set up their homes beneath the sprey's nest. Because the osprey eats mostly fish,

#### Activity

*Shady Survival*

Suppose 140 trees are growing in a shaded area of a forest. Of those trees, 65 percent are sugar maples and 35 percent are red maples. How many trees of each kind are there?

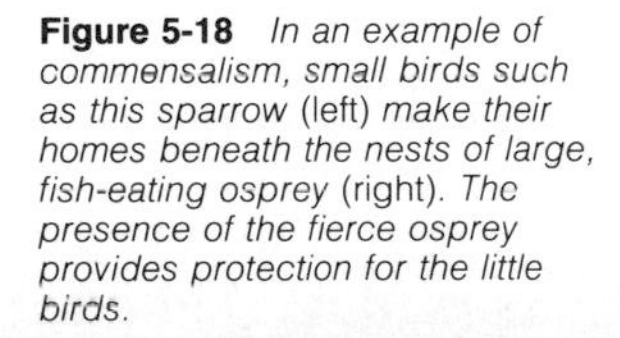

**Figure 5-18** *In an example of commensalism, small birds such as this sparrow* (left) *make their homes beneath the nests of large, fish-eating osprey* (right). *The presence of the fierce osprey provides protection for the little birds.*

#### Activity

**Shady Survival**
**Skills: Making calculations, applying, predicting**
**Level: Remedial**
**Type: Computational**

This activity reinforces mathematical skills. Students must multiply 65% (or 0.65) by 140 and then multiply 35% (or 0.35) by 140. The answer is 91 sugar maples and 49 red maples.

#### FACTS AND FIGURES

A tiny bird called the plover plays dentist to a crocodile in another example of mutualism. The plover hops inside the crocodile's mouth and feeds on the bloodsucking leeches found there. After its tasty meal, the bird hops out, leaving the crocodile free of pests.

tween members of the same species because they have the same niche and the same adaptations to use the niche. This is called intraspecific competition.)

- **Give some example of plant populations competing with each other.** (Answers will vary but should include competition for light, water, and nutrients.)

Point out that plants also have developed ways to avoid competition. Vines and air plants have developed adaptations to help them reduce competition for light.

#### Reinforcement

To reinforce the idea that competition need not always occur when different species of animals or plants live near each other, explain to students that three different types of warblers all live in the same forest and have similar diets. However, each type of warbler feeds at a different height in the trees, thus avoiding competition for food.

#### Content Development

Write "symbiosis" on the chalkboard. Explain to students that it means living in close association with another species. Point out that symbiosis is an ecological association involving some energy transfer or some other ecological benefit.

## ANNOTATION KEY

1. Thinking Skill: Identifying relationships
2. Thinking Skill: Identifying relationships
3. Thinking Skill: Relating processes
4. Thinking Skill: Applying facts

these smaller birds are in no danger from the osprey. In fact, the little birds obtain protection from their enemies by living close to the fierce hawk.

Beautiful tropical orchids survive in dense jungles by growing high above the shade. There among the branches of trees, the flowers get a great deal of sunlight. The roots of these flowers are exposed, so they can take water right out of the air. The tree does not benefit from this relationship. It simply plays the role of "good neighbor."

Both of these symbiotic relationships are examples of **commensalism.** In commensalism, one organism in the partnership benefits. The other organism is neither helped nor harmed.

In another example of commensalism, whales make good neighbors for tiny, crusty-looking animals called barnacles. If you could swim alongside a ❶ whale, you would see these barnacles attached to the whale's sides. There the barnacles get a free ride through vast areas of the ocean, which greatly increases their chances of finding food. The whale, unharmed by the barnacles, simply ignores its tiny passengers!

**MUTUALISM** Some relationships are necessary for the survival of both organisms. This type of symbiosis is called **mutualism** and is helpful to both organisms. Food and protection are two common reasons organisms share such a partnership.

**Figure 5-19** *This furry ratel* (left) *works with the honey guide bird* (right) *so that both can find their favorite foods—honey and beeswax.*

110

### 5-3 (continued)

### Content Development

List the three types of symbiosis—commensalism, mutualism, and parasitism on the chalkboard.

- **In what relationship do both organisms benefit?** (mutualism)
- **In what relationship does one organism benefit and the second is harmed?** (parasitism—the host is harmed or even killed)
- **In what relationship does one organism benefit and the second is neither aided nor harmed?** (commensalism)
- **What kind of symbiotic relationship do flowering plants and bees have?** (mutualism)

Have students provide examples of each type of symbiotic relationship.

### Motivation

Have students give examples of commensalism, mutualism, and parasitism as they occur within human society.

### Content Development

Use the discussion of the osprey and sparrow to illustrate the concept of commensalism. Point out that commensalism is a form of symbiosis in which one partner benefits and the other partner is neither benefited nor harmed. Ask students,

- **In what ways are the sparrows helped by their relationship with the osprey?** (They are protected from their natural enemies by the hawk.)
- **In what ways are the osprey helped or hurt by their relationship with the sparrows?** (They are neither helped nor hurt.)
- **What do you think would happen to the sparrow population if there**

In certain areas of Africa, a small bird known as the honey guide bird lives in a mutualistic relationship with a furry little animal called a ratel. The honey guide bird loves to eat beeswax, but it is too small to break into a bee's nest easily. The ratel likes honey, but it cannot always find a supply on its own. So the two animals work together. The honey guide locates a bee's nest and chirps loudly for the ratel. The ratel moves toward the sound. This chirping and following continues until the ratel reaches the bee's nest. With its sharp claws, the ratel rips the nest open and enjoys a fine feast. The honey guide bird then gets its chance to finish the beeswax.

Strange, undersea creatures called tubeworms create an unusual team with bacteria. At the bottom of the Pacific Ocean in an area near the Galapagos Islands, cracks in the ocean floor leak hot water. This water contains a foul-smelling substance called hydrogen sulfide. Nearby, tubeworms that are three meters long grow together in small bunches. The tubeworms, which lack intestines for digesting food, attach themselves to the ocean bottom, looking very much like rubbery flagpoles bending in the breeze. Each worm has a red, blood-filled organ sticking out of the top of its white tube. These red organs filter the hydrogen sulfide out of the water. Inside the tubeworms, bacteria use the hydrogen sulfide to carry on chemical reactions that release energy. The bacteria use this energy to make food for themselves. The tubeworms also can use the food made by the bacteria. They do not need to digest it, so the absence of intestines is not a problem. This strange relationship between tubeworms and bacteria allows these two organisms to survive deep in the inky darkness of the ocean bottom.

In some cases of mutualism, an animal will help feed another in return for protection. Large fish-eating birds called herons and ibises make their nests in the trees of the Florida swamps. At the base of these trees live poisonous snakes, who gather there to catch pieces of fish the birds may drop. At the same time these snakes are getting a free meal, they are providing protection for the herons and ibises. For the poisonous snakes keep away raccoons and other animals that would feast on the birds' eggs and baby chicks.

2

3

4

**Figure 5-20** *Poisonous snakes live at the base of the tree in which this white ibis makes its nest. They keep raccoons and other animals away from the ibis's eggs and baby chicks.*

**was a decrease in the osprey population?** (This would most likely cause a decrease in the sparrow population because the sparrows would lose one of their protective mechanisms.)

- **What do you think would happen to the osprey population if there was a decrease in the sparrow population?** (The osprey population would most likely be unaffected.)

## Content Development

Explain to students that mutualism is a type of symbiosis that occurs when both organisms benefit from the relationship. Use the discussion of the ratel and the honey guide bird to illustrate the concept of mutualism.

- **In what way is the ratel helped by its relationship with the honey guide bird?** (The ratel is led to beeswax by the honey guide bird.)
- **In what way is the honey guide bird helped by its relationship with the ratel?** (The ratel breaks into the bees' hive to expose the beeswax.)
- **What do you think might happen to the ratel population if there was a decrease in the honey guide population?** (The ratel population might decrease if there was not a suitable food substitute.)
- **What do you think might happen to the honey guide bird population if there was a decrease in the ratel population?** (The honey guide bird population might decrease if there was not a suitable food substitute.)

Discuss other examples of mutualism with the students such as the shark and the remora fish, and the termite and its intestinal protozoa.

## ANNOTATION KEY

❶ **If it did, it would lose its food source. (Inferring)**

❶ **Thinking Skill: Identifying relationships**

❷ **Thinking Skill: Making generalizations**

❸ **Thinking Skill: Relating cause and effect**

**Figure 5-21** *The relationship between these water buffaloes and tickbirds is an example of mutualism. The tickbirds feed on tiny insects they pluck from the buffaloes's hides, and the buffaloes get a good cleaning.*

As you can see, mutualism can occur in just about any ecosystem. All that is necessary is the need for two organisms to cooperate with each other in order that they both may survive.

**PARASITISM** So far you have seen that in a relationship between two organisms, both may benefit or just one may benefit. In the relationship called **parasitism,** however, one partner not only does *not* benefit but is actually harmed by the other organism. The organism that benefits is the parasite. The organism that the parasite lives off and harms is called the host. A parasite has special adaptations that help it to take advantage of its host.

The sea lamprey, a very primitive fish that lacks jaws, is a blood-sucking parasite. This strange-looking fish resembles a swimming tube that has fins on its back. The lamprey's round mouth acts as a suction cup by which the lamprey attaches itself to ❶ other fish. Even as the victim tries to shake the lamprey loose, the parasite's toothed tongue carves a hole in the fish and sucks out its blood. Small fish often die from loss of blood. However, a successful parasite does not usually kill its victim. Can you explain why? ❶

Plants, as well as animals, can be parasites. The dodder plant lives by obtaining all of its food from host plants, such as clover and alfalfa. Wrapping its pale stem around the host plant, the dodder pushes its "suckers" onto its host. Then it releases itself completely from the soil and stays attached to the host plant for support and food.

**Figure 5-22** *The vinelike dodder plant is a parasite. It benefits by getting all its food from its host plant, which is harmed by the relationship.*

### 5-3 (continued)

### Content Development

When discussing parasitism, stress to students that a successful parasite lives off of its host, but rarely kills it. If the parasite were to kill its host it would soon die as well unless it quickly found another host.

After discussing the examples of parasitism cited in the text, ask students,

- **What are some examples of parasitism in which humans acts as hosts?** (Accept all logical responses such as lice, fleas, ticks, and tapeworms.)
- **Why do you think humans would make such excellent hosts for parasites?** (Human hosts are very unlikely to die because of predation or competition for food from other humans. They are unfortunately a very stable and reliable food source.)

### Skills Development

***Skill: Applying concepts***

- **What is the difference between parasitism and commensalism?** (The host is harmed in a parasitic relationship; it is not in a commensalistic relationship.)
- **How does a predator differ from a parasite?** (Parasites live off their host for long periods of time; predators kill and quickly eat their prey; parasites are usually much smaller than their host; predators are usually much larger than their prey.)
- **How are they similar?** (Both get their energy from other organisms.)

### Reinforcement

There are many parasites associated with humans. Assign students to develop reports on parasitic organisms that cause these diseases: typhoid fever, tuberculosis, cholera, syphilis, African sleeping sickness, dysentery, and malaria.

### Enrichment

Parasitism is an effective method of controlling populations. Have students conduct library research and develop brief reports on instances where scientists have introduced parasites to control populations of certain organisms.

**Figure 5-23** *In May of 1980, the lush, green forests of Mount St. Helens* (top, left) *were destroyed by a volcanic eruption. Volcanic ash covered the soil* (top, right). *But within a month, life began to appear amid the fallen trees and ash* (center). *A year later, flowers began to bloom and birds and insects returned to the area* (bottom). *An ecosystem had been reborn.*

### *Balance in the Ecosystem*

An ecosystem is a finely balanced environment in which living things successfully interact in order to survive. **A disturbance in the balance in one part of an ecosystem can cause problems in another part of the ecosystem.** Such disturbances can be the result of nature or of human interference. ②

A natural disaster regularly strikes the giant panda, which lives in the bamboo forests of China and Tibet. The bamboo plants, which make up most of the panda's diet, produce seeds about once every 100 years. Then the bamboo plants die. It takes the new bamboo plants several years to grow large enough for the pandas to eat them. During this time, the pandas face a serious shortage of food. Many of them die. ③

Ecologists do not always have to study the death of living things in order to understand the importance of interdependence among organisms in an ecosystem. Sometimes such interdependence can be studied by observing the rebirth of an ecological system.

In May of 1980, Mount St. Helens, a volcano in the state of Washington, exploded. Thousands of trees, shrubs, flowers, and animals were destroyed by the eruption. Volcanic ash covered the soil as far away as 14 kilometers from the volcano. What had once been a beautiful, green forest soon looked like

113

## Content Development

The relationships in ecosystems evolved over thousands of years. The interconnectedness of these relationships was demonstrated with string during the food web activity. Symbiotic relationships are good examples of how closely "things" in natural ecosystems are tied together. Natural systems are very diverse and have many complex relationships.

## Content Development

Explain to students that natural phenomena are sometimes responsible for creating an imbalance in an ecosystem. The eruption of Mount St. Helens in May 1980 is an excellent example of this.

When discussing the example of the giant panda and the bamboo plant, explain that although this phenomenon is not completely understood, the century bamboo plants that grow in the forests of China and Tibet all produce seeds and die at the same time, thus creating the serious food shortage for the pandas.

You may wish to add that the problem of natural depletion of the panda population is further complicated by the fact that animal specialists have had great difficulty in getting pandas to mate and produce healthy offspring in captivity.

## Enrichment

Assign students to research the development and invasion of a volcanic island by plants and animals.

- **Are these ecosystems simple or complex when compared to ecosystems found on the continents?** (Simple because of the lack of diversity.)
- **What future problems might be encountered by the species that invade a volcanic island?** (Accept all logical responses, such as isolation from new or abundant food sources or destruction of the ecosystem by another volcanic eruption.)

## BACKGROUND INFORMATION

The final word on Mono Lake has yet to be written. The Committee to Preserve Mono Lake is working hard with scientists and environmentalists to gradually replenish the water supplies of the lake. However, the needs of society are also being taken into consideration in any environmental plans. Whether or not the lake will be saved is an open question, although most environmentalists believe it can be saved.

## ANNOTATION KEY

1. Thinking Skill: Sequencing events
2. Thinking Skill: Relating cause and effect

the barren surface of the moon. Within a month, however, life began to return to the area. Roots of the red-flowered fireweed bush and other plants that had survived pushed growing stems up through the ash. These plants attracted insects such as aphids, which feed on the juices of the fireweed. In turn, birds came into the area to feed on insects. Spiders crawled on the ash-covered surface. When these animals died, their bodies fertilized the ash, returning important nutrients to the soil. Hoofed animals such as elk wandered through the area, breaking up the ash cover and leaving holes through which more seeds could sprout. (1)

Once plant and animal life started again, relationships among organisms were reestablished. What was once a bleak, lifeless landscape is now an area filled with colorful flowers, green shrubs, and growing trees. Scientists are carefully studying this area for clues to the secret of how living things can turn a barren land into a lively ecosystem inhabited by interdependent organisms.

People can often be the cause of the destruction of an ecological system. Mono Lake, in eastern California, is a beautiful saltwater lake fed by streams of melting snow from the Sierra Nevada Mountains. Two small islands within Mono Lake attract thou-

**Figure 5-24** *The ecosystem of Mono Lake (left), once disturbed by human interference, is now returning to its delicate balance (right).*



### 5-3 (continued)

### Content Development

Explain to students that other natural disasters that disturb ecosystems include wild fire, hurricanes, tornadoes, and floods. Discuss these examples with your students, pointing out that many organisms have evolved special ways of surviving natural disasters. For example, where fires are common, species may develop deep roots and can sprout new growth from the protected root system. Most tropical trees are flexible and bend during strong winds produced by storms.

### Content Development

Stress to students that human activity tends to simplify ecosystems. Managed forests tend to have one or two primary species; large farms usually grow one or two crops for market; even cities tend to be less complicated than natural ecosystems. Mono Lake is a good example of how humans often fail to see the long-term effects of their actions.

- **Why were the islands in Mono Lake a good nesting place for birds?** (They were protected from predators and the lake provided lots of food.)
- **Why did the water levels in the lake drop?** (People began to use the water from streams that fed the lake.)
- **What happened when the supply of water was reduced?** (The water got too salty and the shrimp, a major food source for the birds, died. In addition, land bridges formed and predators were able to reach the islands.)
- **How might the Mono Lake ecosystem be saved?** (Find a new water supply for Los Angeles; start a campaign to conserve water.)
- **What have you learned from the Mono Lake story?** (Answers will vary but should include the idea of interconnectedness.)

sands of birds, especially sea gulls. More than 80 species of birds nest on these islands and feed on the lake's shrimp, flies, and algae. That is, until 1981. During the summer of that year, many baby sea gulls were found dead. How strange this seemed to be for an ecosystem that had long provided food, water, and shelter for the sea gull population. As people investigated the situation, they discovered that the ecosystem had been disturbed by actions taken far away from the lake many years before.

About 40 years ago, the city of Los Angeles began to use water from the major streams that feed into Mono Lake. Less and less water emptied into Mono Lake, and the lake began to dry up. Thousands of acres of dust formed where there was once water.

As the amount of water in the lake decreased, the concentration of salt dissolved in the water increased. The shrimp that sea gulls fed on could not survive in water so salty. As the shrimp died, less food was available for the sea gulls. Baby sea gulls starved to death.

To make matters worse, as the water level in Mono Lake dropped, a land bridge that connected the shore to the nesting islands formed. Coyotes crossed this bridge, killed many sea gulls, and invaded the gulls' nests.

Many people want to save Mono Lake. They realize that human actions have damaged the lake and disturbed an ecosystem. Without a water conservation plan, Mono Lake will continue to dry up. And the delicate balance between living and nonliving things in this ecosystem will be seriously altered.

Mono Lake illustrates how important it is to understand how an ecosystem operates and how it can be damaged by human activity. With this sort of knowledge, people can save the environment, and perhaps even improve it!

### *SECTION REVIEW*

1. What is competition? Why does it occur?
2. Compare the relationships of commensalism, mutualism, and parasitism.
3. What are the two main causes of disturbances in the balance within an ecosystem?

### Activity

*Changes in Populations*

Isle Royale is a long, narrow island in Lake Superior, 25 kilometers from the shore of Canada. In the early 1900s, a few moose reached the island and within 20 years increased their population to more than 2000.

During the next 40 years, the moose population underwent several changes. Using books and other materials in the library, find out what changes occurred on Isle Royale and what caused the changes. Include answers to the following questions in your report.

**1.** What are limiting factors and in what ways do they affect a population?

**2.** How do birth and death rates affect a population?

**3.** What is the definition of a population cycle?



## Activity

**Changes in Population**
**Skills: Report writing, researching, applying, relating**
**Level: Enriched**
**Type: Library/Vocabulary comprehension**

By doing this library activity, students will learn about factors that limit population size. Have students explain the concept of population cycle, using this and other examples, taken both from their experience and from further library research.

## Skills Development

***Skill: Predicting***

Based on what students have learned about Mono Lake, challenge them to predict what will happen to the ecosystem if water is restored and if water is not restored.

## Enrichment

Have students do library research on other disturbances to ecosystems. This research should include examples of damage caused by human activity and interference.

## Enrichment

Arrange for an environmentalist from an organization such as the Sierra Club or the Environmental Protection Agency to speak to your students about ecological conservation. If you are unable to schedule a speaker, have your students do research on the efforts being made by these and other organizations to preserve the wildlife and natural land resources of the United States and other countries.

## Section Reviews 5-3

**1.** Competition is the struggle by organisms to obtain all of the materials necessary for survival such as food, shelter, energy, and so on. Competition occurs when there is not enough of these resources to fill the needs of all of the organisms in the environment.
**2.** Commensalism: one organism benefits, the other is unaffected; mutualism: both organisms benefit; parasitism: parasite benefits, host is harmed.
**3.** Natural changes and human interference

# LABORATORY ACTIVITY
# A LITTLE OFF BALANCE

## BEFORE THE LAB

Gather all equipment at least one day prior to the investigation. The size of the jars is not particularly important as long as they are the same size. The large jars used in school cafeterias are perfect. If *Elodea* is not available, other aquarium plants can be used. You might want to get students to measure the length of the plants or get a rough idea of the mass of the individual strands in baggies. Challenge students to develop a way to mark each individual plant so that it can be identified later.

## PRE-LAB DISCUSSION

Have students develop data tables in which they can record their observations of jars A and B. Encourage them to quantify their observations and include observations of things that might affect the experimental results. Examples might be weather or a particularly warm spell during the experiment. You might wish to assign one student to record the temperature at the beginning and end of each school day. The water with the plant food may grow cloudy if there is algae present. This is an important observation because the algae are competing with the *Elodea* for nutrients.

If you do not cover the jars, water will evaporate. If you cover them tightly, gases can accumulate at the top and interfere with normal gas exchanges that occur in natural systems. Ask students how they can solve this problem. (Cover the jars and put two or three small holes in each lid.)

## LABORATORY ACTIVITY

### *A Little Off Balance*

**Purpose**

In this activity, you will observe how the addition of lawn fertilizer can affect the balance of an aquatic ecosystem.

**Materials** *(per group)*
2 2-L wide-mouthed jars
Pond water
8 *Elodea* (or other aquatic plant)
Lawn fertilizer (or house plant food)
Teaspoon

**Procedure**

1. Label the jars A and B.
2. Fill each jar about three-fourths full with pond water.
3. Place four *Elodea* in each jar.
4. Add one-half teaspoon of lawn fertilizer to jar B.
5. Place the jars next to each other in a lighted area.
6. Observe the jars daily for three weeks. Record your observations.

**Observations and Conclusions**

1. Were there any differences between jars A and B? If so, when did you observe these differences?

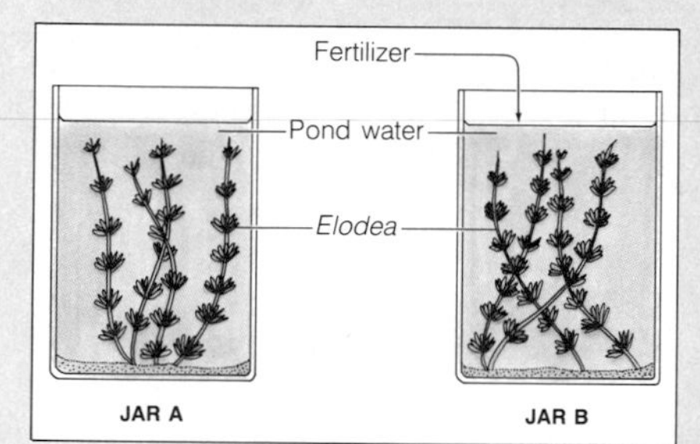


2. What was the control in this experiment? The variable?
3. Why did you place the jars next to each other? Why did you place them in the light?
4. What effect did the fertilizer have on the *Elodea*?
5. Lawn fertilizer contains nitrogen, phosphorus, and potassium. These nutrients are often present in sewage as well. Predict the effects of dumping untreated sewage into ponds and lakes.

116

## SKILLS DEVELOPMENT

Students will use the following skills while completing this investigation.

1. Observing
2. Manipulative
3. Comparing
4. Safety
5. Relating
6. Applying
7. Predicting

## SAFETY TIPS

Once the jars are put in place, only the teacher should be allowed to move them. If possible, place the jars so students can walk around them and observe the entire jar.

## OBSERVATIONS AND CONCLUSIONS

**1.** Students should observe more growth of *Elodea* in jar B than in jar A.
**2.** Jar A was the control, and fertilizer was the variable.
**3.** The jars were placed next to each other so that both would be in the same environment in order to avoid any hidden variables. They were

## CHAPTER REVIEW

### SUMMARY

**5-1 Living Things and Their Environment**

- Ecology is the study of the relationships and interactions of living things with one another and with their environment.
- An ecosystem is a group of organisms in an area that interact with each other, together with their nonliving environment.
- The living part of an ecosystem is called the community.
- A population is a group of the same type of organism living together in the same area.
- Different ecosystems support different living things.
- The place in a community in which an organism lives is its habitat. Its niche is its role in the community. Habitats can overlap, but organisms cannot share the same niche.

**5-2 Food and Energy in the Environment**

- Green plants are autotrophs because they make their own food by photosynthesis.
- Green plants are the producers in an ecosystem; heterotrophs are the consumers.
- Scavengers feed on dead animals.
- Decomposers are organisms that break down dead plants and animals into simpler substances during the decay process. As a result, important substances are returned to the soil and water.
- A food chain illustrates how groups of organisms within an ecosystem get their food and energy.
- Food chains often overlap or connect to form food webs.
- Each plant or animal in a food chain has a particular location called a feeding level.
- At each successive level in a food chain, energy is lost. The loss of energy as it moves through a food chain can be pictured as an energy pyramid.

**5-3 Relationships in an Ecosystem**

- Competition among organisms for food, water, and shelter exists in all ecosystems.
- Symbiosis is a relationship in which two organisms live together for the benefit of either one or both of the partners.
- In commensalism, one organism in the partnership benefits.
- A relationship in which both partners benefit is called mutualism.
- In parasitism, one organism benefits but only by harming the other organism.
- An ecosystem is a finely balanced environment in which living things successfully interact in order to survive.
- Disturbances in an ecosystem can be caused by nature or by human interference.

### VOCABULARY

*Define each term in a complete sentence.*

| | | | |
|---|---|---|---|
| **autotroph** | **ecology** | **heterotroph** | **producer** |
| **commensalism** | **ecosystem** | **mutualism** | **scavenger** |
| **community** | **environment** | **niche** | **symbiosis** |
| **competition** | **food chain** | **parasitism** | |
| **consumer** | **food web** | **photosynthesis** | |
| **decomposer** | **habitat** | **population** | |



## GOING FURTHER: ENRICHMENT

**Part 1**

You may want to repeat this experiment by doing the following:

**1.** Add measured amounts of plant food every week.
**2.** Use a different kind of aquatic plant.
**3.** Put the jars in a shady corner.
**4.** Compare various kinds of fertilizers.
**5.** Use a dish detergent with phosphate in place of the plant food.

**Part 2**

After students have completed the lab, have them investigate cultural eutrophication and the impacts of human activities on freshwater lakes and ponds.

**Part 3**

What would happen if you used tap water instead of pond water? Ask students to write a hypothesis and then design an experiment to test the hypothesis. Encourage them to conduct the experiment and report on their results.

placed in the light so that the *Elodea* could perform photosynthesis.
**4.** Fertilizer probably caused an increase in the growth of *Elodea.*
**5.** Untreated sewage, when dumped into ponds and lakes, often causes increased growth of algae and other plants. This increased growth, in turn, upsets the natural balance of the lake or pond, and increases the process of eutrophication. At this point, you may want to discuss eutrophication, the nutrient enrichment of aquatic systems, with your class. You should point out that eutrophication is a natural process but is often accelerated by the activities of people.

## CHAPTER REVIEW

### MULTIPLE CHOICE

1. c 3. d 5. d 7. c 9. d
2. a 4. a 6. a 8. b 10. c

### COMPLETION

1. ecosystem
2. habitat
3. photosynthesis
4. scavengers
5. decomposers
6. food web
7. food pyramid
8. competition
9. mutualism
10. parasite

### TRUE OR FALSE

1. F ecology
2. T
3. T
4. T
5. F heterotroph
6. T
7. F Producers
8. F less
9. T
10. F parasitism

### SKILL BUILDING

**1.** This is a mutualistic relationship because both organisms benefit. The tickbird obtains food, and the rhinoceros has a parasite removed.
**2.** Check food webs for accuracy. Students should conclude that the grass and tree are producers and all other organisms are consumers.
**3.** The insecticide could destroy insects that are among the lowest-level consumers in the ecosystem. If enough insects were destroyed, organisms that feed primarily on insects would find their food supplies depleted. These organisms might die of starvation, causing organisms that eat them to die off, and so on. Also, since insects often play a role in plant pollination, their destruction could have a negative effect on the plant life in the area.
**4.** Grasses are the main producers in the ecosystem. They are an important food source for zebras. Zebras, in turn, are an important food source for lions.
**5.** Answers will vary, but should include the idea that the natural habitats or organisms living in the area cannot be destroyed; nor can existing supplies of water and shelter space.
**6.** The initial effect would be a severe depletion of the prey on the island. If the new predator was successful, the other predators might die off. In time, with the depletion of prey, the new predator might die off as well.

### ESSAY

**1.** Environment: all the living and nonliving things with which an organism interacts; ecosystem: all of the living and nonliving things in a particular area; community: all the living things in a particular area; population: all the living things of the same kind in a particular area; habitat: the area in which a particular organism lives; niche: the particular "role" an organism plays in an ecosystem. From these definitions, students should have little trouble demonstrating that all of these things are interconnected.
**2.** A food chain shows the relationship between a producer in an ecosystem and various consumers that feed off the producer or off of consumers that feed off the producer. A food web shows all of the food chains that are

#### CONTENT REVIEW: MULTIPLE CHOICE

*Choose the letter of the answer that best completes each statement.*

**1.** The study of the relationships of living things and their environment is called
a. commensalism. b. parasitism. c. ecology. d. botany.
**2.** A forest is an example of a(n)
a. ecosystem. b. population. c. food chain. d. niche.
**3.** A group of the same type of organism living together in the same area is called a(n)
a. ecosystem. b. habitat. c. niche. d. population.
**4.** An organism's "occupation," or role, in an ecosystem is called its
a. niche. b. habitat. c. community. d. level.
**5.** Green plants make their own food by
a. mutualism. b. adaptation. c. population. d. photosynthesis.
**6.** In order to make food, green plants need
a. water, carbon dioxide, and sunlight.
b. oxygen, carbon dioxide, and sunlight.
c. glucose, oxygen, and carbon dioxide.
d. glucose, nitrogen, and phosphorus.
**7.** A whale is an example of a(n)
a. producer. b. autotroph. c. consumer. d. scavenger.
**8.** How organisms in an ecosystem get their food can be described by a(n)
a. energy web. b. food chain.
c. population pyramid. d. symbiotic relationship.
**9.** Two organisms living together for the benefit of both is an example of
a. commensalism. b. parasitism.
c. decay. d. mutualism.
**10.** The delicate balance of an ecosystem can be disturbed by
a. nature. b. human interference.
c. both of these. d. neither of these.

#### CONTENT REVIEW: COMPLETION

*Fill in the word or words that best complete each statement.*

**1.** A group of organisms and their nonliving environment is a(n) ______.
**2.** The place in a community in which an organism lives is called its ______.
**3.** The process by which a green plant makes its own food is called ______.
**4.** Organisms that feed on the bodies of dead animals are called ______.
**5.** Organisms known as ______ break down dead plants and animals.
**6.** When food chains overlap or are connected together, a(n) ______ is formed.
**7.** Energy loss through a food chain can be pictured as a(n) ______.
**8.** There is ______ among organisms for food, water, light, and shelter.
**9.** In ______ each animal benefits from the partnership.
**10.** An organism that lives off and harms its host is called a(n) ______.

118

## CONTENT REVIEW: TRUE OR FALSE

*Determine whether each statement is true or false. If it is true, write "true." If it is false, change the underlined word or words to make the statement true.*

1. The study of the relationships of living things to one another and to their environment is called photosynthesis.
2. The living part of an ecosystem is called a community.
3. A tree, a cave, or a pile of leaves is an example of an organism's habitat.
4. Green plants can make food because they contain chlorophyll.
5. An organism that cannot make its own food is called a(n) autotroph.
6. Decomposers help return important substances to the environment.
7. Consumers, such as green plants, are the first link in a food chain.
8. Within a food chain, there is more energy available at each higher feeding level.
9. Most relationships among living things in an ecosystem are based on competition.
10. A type of symbiosis in which one organism benefits and the other organism is harmed is commensalism.

## CONCEPT REVIEW: SKILL BUILDING

*Use the skills you have developed in the chapter to complete each activity.*

1. **Applying definitions** The African tickbird lives on the back of the rhinoceros, where it picks bloodsucking ticks off the back of the huge animal. What type of symbiosis is this? Explain your answer.
2. **Making diagrams** Draw a food web that includes the following organisms: deer, snake, owl, mouse, grasshopper, wolf, hawk, rabbit, grass, frog, tree. Then identify each organism as a producer or a consumer.
3. **Making predictions** How could the spraying of an insecticide interfere with the balance in an ecosystem?
4. **Relating concepts** Lions, zebras, and grasses are three populations that live on the plains of Africa. How are these populations interdependent?
5. **Developing a model** Pretend you are in charge of building a city. What design features would you use to make the city part of the existing ecosystem?
6. **Applying concepts** Suppose a new predator was introduced into an island ecosystem, and it reproduced successfully. What could happen to the prey and to some of the predators on the island?

## CONCEPT REVIEW: ESSAY

*Discuss each of the following in a brief paragraph.*

1. How is each of the following terms related to the others: environment, ecosystem, community, population, habitat, niche.
2. What is the difference between a food chain and a food web? Describe a food chain and a food web that include you.
3. Why are decomposers an important part of an ecosystem?
4. Compare three types of symbiosis and give an example of each.
5. What is the role of green plants in an ecosystem?



interconnected in a particular ecosystem. Students' examples that include themselves should be checked for accuracy.
**3.** Without decomposers, all of the matter in living things would be lost from the ecosystem when an organism died. If that were the case, all of the matter needed by living things would have been used up long ago and the earth would be covered with the remains of dead organisms. Because of decomposers, matter can be cycled through the ecosystem.
**4.** Commensalism: a relationship in which one organism benefits and the other is neither helped or harmed. Examples include orchids and the trees upon which they grow. Mutualism: a relationship in which each organism benefits. Examples include honey guide bird and furry ratel. Parasitism: a relationship in which one organism benefits and the other is harmed. Examples include the lamprey and the fish upon which it feeds.
**5.** Green plants are the primary producers in an ecosystem. Through photosynthesis, they use the energy from the sun and materials from the environment to produce food.

## ADDITIONAL QUESTIONS AND TOPIC SUGGESTIONS

**1.** Have students observe a 1-meter plot of land for several months. They should take notes on all the organisms they see, when they see them, and any changes they observe during the time period. Students should write an ecological report on their observations, using the terms they have learned from this chapter.
**2.** Have students predict what would happen if a sudden plague or catastrophe somehow destroyed all of the earth's decomposers.
**3.** Study the following food chain: plants → insects → frogs → birds. In order for the bird population to increase, which population will have to be increased first? Explain your answer. (The plant population must first be greatly increased. An increase in the plant population will produce an increase in the insect population. This will increase the frog population, which will eventually result in an increase in the bird population.)

## ISSUES IN SCIENCE

The following issue can be used as a springboard for class debate or can be assigned as a writing homework.

As human populations grow, the amount of land available for wildlife is dwindling. Should people be more concerned with their own needs or the needs of the environment. Relate your answer to the interdependedness of biotic and abiotic factors in the environment.

## Unit One

## CHARACTERISTICS OF LIVING THINGS

### ADVENTURES IN SCIENCE: KATHARINE PAYNE AND THE "LANGUAGE" OF ELEPHANTS

#### BACKGROUND INFORMATION

Katharine Payne has been conducting research in animal communication for many years. With her husband Roger, she spent 20 years studying whale behavior and communication. Mrs. Payne discovered that all humpback whales in the same group change the melodies of their songs in exactly the same way.

In the area of elephant communication, it has been known for years that elephants communicate by roars and bellows as well as by smell and touch. However, recent evidence gathered by Payne and others suggests that elephants may also communicate by low-frequency sounds (14 to 24 hertz). These vibrations are too low for detection by humans, whose average hearing threshold is about 30 hertz. It is believed that these sounds are produced by vibrations of the elephants' vocal cords, which set up sympathetic vibrations in the forehead, causing it to "flutter." As in wolves, the voices of different elephants have different harmonic structures. Some of their infrasonic calls have audible tones that sound to us like rumbling noises.

Scientists are not certain that these long-distance, low-frequency sounds are used for communication among elephants. If they are, then certain puzzling aspects of elephant behavior might be explained. Among these behaviors is the ability of males to locate females during their brief periods of male "must" (heightened sexual activity) and female fertility (or estrus). These periods of about two days a month may occur when the males and females are several kilometers apart. Somehow the males are able to "home in" on the females.



Adventures in Science

# katharine payne & the "Language" of elephants

Sometimes Mrs. Payne follows the travels of an elephant herd in her land rover. Rumbling noises picked up by microphones near a water hole tell her that elephants are on the way.

Soon the elephants appear—roaring, trumpeting, snorting. Humans can easily hear these elephant sounds. However, Mrs. Payne, a biologist working for Cornell University, has found that elephants make some sounds that human ears cannot hear.

#### TEACHING STRATEGY: ADVENTURE

##### Motivation

Play records or tape recordings of animal sounds. These may include wolf calls, dog barks, bird calls, humpback whale songs, or porpoise squeals. Explain to students that many animals produce sounds that we can hear naturally. Other animals emit sounds whose frequencies are either too high or too low for us to hear.

##### Content Development

- **What chance event "turned on" Katharine Payne to the idea that elephants might emit low-frequency sounds that human ears cannot detect?** (While observing elephants at the zoo, she felt the air around her throb.)

These newly discovered elephant sounds can travel over a much greater distance than an elephant's trumpet or roar. Mrs. Payne wondered whether elephants use these sounds to locate each other in the vast plains and forests of Africa. In 1986, Mrs. Payne went to Africa to find out.

A chance discovery in the fall of 1984 gave her the idea for her trip. While observing elephants in a zoo, Mrs. Payne felt the air around her throb, as if thunder were rolling in from some far-off place. "Only after a week," Mrs. Payne recalls, "did I think that these might actually be very low frequency sounds that I couldn't hear."

Later, she returned to the zoo with two other researchers from Cornell. Using sensitive recording equipment, they discovered that the elephants were indeed making sounds with a frequency too low for human ears to hear.

You may know that sounds have different frequencies. The sounds made by a flute have a very high frequency. The sounds made by a tuba have a very low frequency. Human ears can hear sounds over a wide range of frequencies. But they cannot hear sounds of a very high frequency, such as the squeaks made by a bat flying through the night air. And, Mrs. Payne found out, humans cannot hear the low-frequency sounds made by elephants. Suddenly, researchers had a clue to explain one long-standing mystery of elephant behavior.

Wild elephants constantly move from place to place in well-organized herds and family groups. But sometimes individual elephants become separated from their group as they wander over many kilometers of grassland or forest. Yet, without a signal that can be detected by humans, the separated elephants come together. "It's always been mysterious," says Mrs. Payne. "African elephant researchers have always said that there must be some kind of ESP between the animals. And there is. It's extrasensory as far as human beings are concerned, but ordinary as far as elephants are concerned."

**Elephants have few natural enemies. Why are they in danger of becoming extinct?**

Mrs. Payne and the Cornell team think that elephant calls are the key to elephant communication. Most of the elephants' calls are too low for human ears to hear. But these low-frequency calls can travel over great distances, and elephants have no trouble hearing them.

While in Africa, Mrs. Payne observed that elephants' foreheads flutter and their ears flap when they make low-frequency calls as well as calls humans can hear. The forehead flutterings most likely occur when sounds made by the elephants' vocal cords cause the animal's forehead to vibrate.

Some of the elephant calls bring calves running to their mothers. Other calls cause elephants to respond over a great distance. Sometimes the sounds can be heard by human observers, says Mrs. Payne, as "soft, puttering, furry rumbles." But, Mrs. Payne points out that the sounds humans can hear are only a small part of the calls made by elephants.

During the second part of Mrs. Payne's research, recorded calls will be played back to the elephants to see how they react. In the past, researchers have placed electronic collars on some wild elephants. The collars are used to keep track of the animals as they wander over great distances. Mrs. Payne would like to put microphones on some of the collared elephants so that the sounds they make as they move can be recorded. "Our ultimate hope is that our work may increase the elephant's chances of survival," says Mrs. Payne.



- **How did she test the hypothesis that the elephants were producing low-frequency sounds?** (She returned to the zoo with recording equipment that was sensitive enough to detect sounds too low for humans to hear.)
- **If you were standing near an elephant, how could you tell when it was uttering low-frequency calls?** (Although you could not hear them, you could observe the fluttering of the elephant's forehead combined with the flapping of its ears. You could probably feel a throbbing sensation in the air, too.)

## Teacher's Resource Book Reference

After students have read the Science Gazette article, you may want to hand out the Reading Skills worksheet based on the article in your Teacher's Resource Book.

## ADDITIONAL QUESTIONS AND TOPIC SUGGESTIONS

**1.** We know that elephants make at least two different kinds of sounds. Describe the similarities and differences between these two types of sounds. (An elephant's trumpet or roar can be heard by humans. Lower frequency sounds that cannot be heard by humans are also emitted. These sounds travel over much greater distances than the "roars" do. Both types of sounds are probably used for communication.)

**2.** Why are Mrs. Payne and others interested in learning more about elephants' communication systems? (They hope that their work will increase the elephants' chances of survival. Students may also mention such things as satisfaction of scientific curiosity and increasing our basic knowledge of animal behavior.)

**3.** Several other animals are known to emit sounds that humans cannot detect with the unaided ear. The high-frequency sonar used by bats, the songs of humpback whales, and the "talking" of porpoises are interesting topics for further research.

## CRITICAL THINKING QUESTIONS

**1.** Why would hunters and poachers be interested in killing elephants? (Elephants are killed primarily for their ivory tusks. They are also sometimes hunted for sport.)

**2.** In addition to using sounds, what are some other methods by which animals communicate? (Answers will vary. Some methods include body posture or position of certain body parts such as head, tail, or ears; ritualistic movements including dances; facial expressions; touch; smell; heat detection; detection of electrical currents or water movements, etc.)

You may wish to have students differentiate between communication among members of the same species or family group and communication between one species and another.

## Unit One

# CHARACTERISTICS OF LIVING THINGS

## ISSUES IN SCIENCE: PESTS OR PESTICIDES: WHICH WILL IT BE?

### BACKGROUND INFORMATION

Because of the use of pesticides, approximately one-third less food is lost due to pests in the United States than is lost in the rest of the world. Because of the use of pesticides, food prices in the United States are 30% to 50% lower than they would be otherwise. Pesticides have increased the amount of food production per hectare of land, making it possible to grow more food in less space.

So much for the good news. The bad news is that pesticides can and do damage the environment. Because it is difficult to make a pesticide that is totally specific, organisms that are helpful to human beings and food production may be accidentally killed. Pesticides can upset the ecological balance in an area by accidentally destroying the natural enemies of pests. If a pesticide is released into the environment carelessly or in excessive amounts, it can pollute land, air, and water, and cause illness or death to animals, fish, plants, birds, and human beings. In addition, pesticides can actually cause an increase in the population of certain pests by forcing the natural selection of the hardiest members of the species. These pests ultimately produce a generation that is resistant to the pesticide, and is stronger than ever.

Another problem with some pesticides is that they break down slowly and, therefore, remain in the environment for many years. Most hazardous in this way are the chlorinated hydrocarbons, which include such insecticides as DDT, aldrin, dieldrin, and heptachlor. In the 1970s, the U.S. Environmental Protection Agency (EPA) banned the widespread use of these insecticides, although their use is permitted in an emergency.

SCIENCE GAZETTE

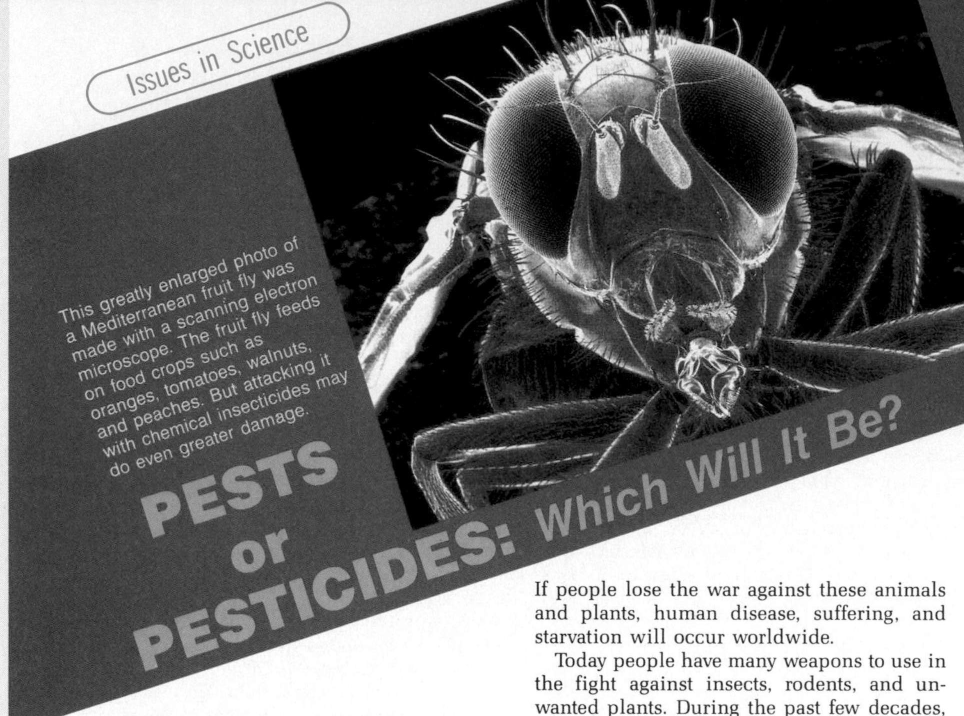
Issues in Science
PESTS or PESTICIDES: Which Will It Be?

This greatly enlarged photo of a Mediterranean fruit fly was made with a scanning electron microscope. The fruit fly feeds on food crops such as oranges, tomatoes, walnuts, and peaches. But attacking it with chemical insecticides may do even greater damage.

There is a war being fought right now that has been going on for thousands of years. It is a fight for survival against armies that far outnumber the earth's entire human population. It is the war people wage against insects, rats, mice, weeds, and fungi.

Some insects, rats, and mice carry deadly diseases such as malaria and typhoid fever. Throughout history, diseases carried by pests have killed hundreds of millions of people. These animals also threaten food supplies around the world.

Other threats to world food supplies are weeds and fungi. Weeds compete with crops for water and nutrients. Some fungi cause plant diseases that result in huge crop losses.

If people lose the war against these animals and plants, human disease, suffering, and starvation will occur worldwide.

Today people have many weapons to use in the fight against insects, rodents, and unwanted plants. During the past few decades, scientists have made or discovered many chemicals that kill pests. These chemicals, which are called pesticides, have helped increase food production by killing the pests that destroy crops and eat stored food. Pesticides have also saved many human lives by killing disease-carrying pests.

Unfortunately, people have not always used pesticides wisely. Farmers and others who use pesticides have accidentally killed useful animals and plants. Also, incorrect use of certain pesticides has made it harder to kill some pests. That is because many pests are now able to withstand assaults of deadly chemicals. In other words, the pests have become resistant to the pesticides.

122

### TEACHING STRATEGY: ISSUE

#### Motivation

Display items such as the following: a mousetrap, a can of aerosol insecticide, flypaper, weed killer, and mothballs.

- **What do all of these items have in common?** (Accept all answers.)

#### Content Development

Point out that all of the displayed items represent methods of pest control. Not all the items are pesticides, however. Pesticides are, by definition, chemicals used to kill pests.

Pesticides can be classified according to the type of pest they seek to kill. Insecticides are directed at insects; rodenticides, at rodents such as mice or rats; fungicides, at fungi; and herbicides, at weeds.

In addition, pesticides are sometimes washed into rivers and streams, where they kill fish and other animals. Winds can spread pesticides over hundreds or thousands of miles. This pollutes areas far from where the chemical was used to kill pests. Pesticides have injured and killed people in all parts of the world.

So now we face a difficult problem. How can we save crops and kill disease-carrying insects without harming the environment?

### The Other Side of Pesticides

There are now about 35,000 different pesticides on the market. Each year in the United States alone, about one-half billion kilograms of these chemicals are used by farmers, homeowners, and industry.

Specific pesticides called herbicides and fungicides have helped farmers to reduce crop losses due to weeds and fungi. Yet during the past 30 years, the amount of crops destroyed by insects has nearly doubled. This is true despite the fact that farmers have been using more powerful insecticides in greater quantities than ever before!

Incorrect use of pesticides may account for the alarming comeback of malaria in countries where it had been practically wiped out. More than 200 million people in Asia, Africa, and Latin America suffer from this disease, which is spread from person to person by mosquitoes. Perhaps because of overspraying of crops, mosquitoes are becoming resistant to insecticides that used to control them.

This begins when a few resistant insects survive after being sprayed with insecticide. These survivors then produce more mosquitoes that are resistant to the insecticide. In a short time, most of the population of insects in the sprayed area is resistant.

Insecticides can also kill the natural enemies of insect pests. For example, insecticides have killed ladybugs in some apple orchards. Now there are no more ladybugs in these orchards to keep apple-eating insects, such as mites, under control.

Because of the harmful effects of pesticides, some people believe that these chemicals are

One way to combat insect pests is to spray them from the air with chemical insecticides. But this method, called crop dusting, also affects living things other than pests.



Since the banning of the chlorinated hydrocarbons, major changes have taken place in pest-control research. Today, nearly 70% of the U.S. Department of Agriculture's budget for pest-control research is devoted to finding alternatives to pesticides. These alternatives include: biological control, which introduces the natural enemies of pests into the environment; the use of natural pesticides, which can be extracted from plants or animals that produce substances to ward off their natural enemies; the genetic engineering of pest-resistant crops; and the use of cultivation methods that discourage the buildup of certain pests. Also under consideration is the use of integrated pest management. Integrated pest management seeks to combine many techniques of pest control (including limited use of pesticides when necessary) in as environmentally compatible a manner as possible.

## ADDITIONAL QUESTIONS AND TOPIC SUGGESTIONS

**1.** Have you had any personal experiences with pest control in which there were unwanted side effects? Describe what happened and your feelings about the incident. (Accept all answers. Students may have had experiences such as using an insect repellent, only to find that the fumes made them feel sick, or caused illness in a pet cat or dog.)

**2.** Go to the supermarket, hardware store, or local garden center and find out what kinds of pesticides are being sold. Read the labels and note the contents of each product, how the product is to be used, and the kind of pest it is intended to kill. Also note any cautions or warnings that might be included about using the pesticide.

**3.** Get together with several classmates and research the kinds of pests most often found in your state. Then make a map to show where these pests are concentrated.

### Content Development

Emphasize to students the need for some form of pest control. Point out that approximately *half* of the food produced in the world each year is lost as a result of the damage caused by pests. The variety of pests is amazing—in the United States alone, there are approximately 19,000 different species of agricultural pests.

Another problem with pests is that they can bring disease to human beings. For example, malaria is a serious, sometimes fatal disease that is transmitted from person to person by a type of mosquito.

### Reinforcement

Review with students the alternatives to pesticides discussed in the text. One of these alternatives is understanding how crops and pests interact with their environment. Another alternative involves cultivation tech-

## CRITICAL THINKING QUESTIONS

**1.** Although the EPA has banned the use of extremely hazardous pesticides such as DDT, U.S. chemical companies still manufacture these insecticides in large quantities for sale abroad. Do you think this practice is ethical? Explain your answer. (Accept all well-supported answers. You may wish to point out to students that many European countries have also banned the use of DDT, but that it is still used in large quantities in South America, Central America, Asia, and Africa.)

**2.** Suppose that a farmer uses a pesticide to kill Bug X, which is damaging the corn crop. The corn yields increase dramatically for two years. Then, in the third year, large amounts of corn are destroyed, although the farmer continues to use the pesticide as before. Can you offer an explanation? (Answers may vary. The two most likely explanations are that the pesticide killed natural enemies of other pests who began to devour the corn; or that Bug X became resistant to the use of the pesticide.)

## CLASS DEBATE

Divide the class into teams of 4–6 students. Challenge each team to imagine that they represent members of the U.S. Environmental Protection Agency (EPA). Ask students to dramatize a meeting of the agency in which they debate the pros and cons of new laws governing the use of pesticides. You may wish to have students base their discussion on actual laws that have been recently enacted by the EPA, or you may wish to have students write their own laws based on the issues discussed in this article.



too dangerous to use. But those in favor of using pesticides disagree. They say that pesticides would not be so dangerous if people knew how to use them properly.

For example, pesticides should not be sprayed in fields when the wind is blowing. When the wind is blowing, much of the pesticide is blown away. The pesticide then travels to pollute the air people breathe, the water they drink, and the rivers and lakes in which fish swim. People should also be taught to use only as much pesticide as they need for a particular job. This would reduce the overspraying that can lead to resistant pests.

### New Weapons

One way to lessen our dependence on pesticides is to understand how crops and insects interact with their environments. The study of how living things interact with their environments is called ecology. Ecology provides clues to how to control pests by changing the environment of either the pest or its victim.

Farmers use pesticides against such destructive insects as the tobacco hornworm. The hornworm attacks many kinds of crops, including these tomato plants. However, pesticides can also harm helpful insects, animals, and even people.

For example, a plant called barberry often carries a fungus that causes a disease called black stem rust. When barberry grows near wheat, the rust spreads from the barberry to the wheat. Some farmers have protected their wheat from black stem rust by destroying nearby barberry plants rather than by spraying fungicide.

Other diseases can be controlled by crop rotation. Periodically planting crops other than cabbage in a cabbage field is an example. This method prevents disease-causing organisms that attack only cabbage from building up in the soil.

Also, scientists might discover new chemicals that are harmful only to pests. For example, scientists at the University of Georgia discovered that the oil in orange peels kills fire ants, wasps, and fleas but is harmless to other animals.

Not all new types of pest control are so down-to-earth, however. By studying how the crop-eating desert locust interacts with its environment, scientists have been able to use satellites to fight this pest.

The desert locust is a flying insect related to the grasshopper. From time to time, millions of these locusts gather and sweep across Africa and India, eating every crop and blade of grass in their path. No weapons have been able to stop these insects once they take flight. But in recent years, satellites orbiting the earth have been taking pictures of areas in Africa and India where locusts might breed. Scientists can identify possible breeding areas by the amount of moisture they contain. If satellite photos show that an area is moist enough for locust eggs to survive there, scientists warn the people that may be threatened. The people can then concentrate their pesticide spraying in these areas.

The better we understand how living things interact with their environments, the more clues we will find for controlling pests without using large amounts of chemicals.



## ISSUE (continued)

niques, such as crop rotation. A third alternative is the use of natural chemicals, such as the oil found in oranges.

## Teacher's Resource Book Reference

After students have read the Science Gazette article, you may want to hand out the reading skills worksheet based on the article in your Teacher's Resource Book.

Futures in Science

Ronda and her family lived in a space settlement on Pluto. One day, a strong radiation storm swept across the Purple Mountains of their planet. There had been many such storms on the planet in the year, 2101. The module in which Ronda and her family lived had been directly in the path of the storm. Somehow, radioactive dust penetrated the sealed glass that served as windows in the module. As a result, Ronda was blind. Radiation had destroyed the nerves that carried the electrical signals from Ronda's eyes to her brain. Her brain could no longer interpret what her eyes were "seeing."

Months after the storm, Ronda sat nervously in a plush armchair in the waiting room of Venus General Hospital. Today was the day the bandages would be removed from her eyes. Ronda was terrified that the operation to restore her sight might have been a failure. She did not want to rely on a seeing-eye robot for the rest of her life.

As the doctors removed the bandages, Ronda thought about the computer that had been implanted in her brain. No larger than a grain of rice, the computer was programmed to record all the images Ronda's eyes picked up and then translate them into messages her brain could understand. The computer was designed to work exactly like the eye nerves that had been destroyed.

The bandages fell from Ronda's eyes. She could see! The living computer inside her head had restored her sight.

Today scientists believe that living computers will be a reality in the not-so-distant

THE COMPUTER THAT LIVES!

## Unit One

## CHARACTERISTICS OF LIVING THINGS

### FUTURES IN SCIENCE: THE COMPUTER THAT LIVES!

### BACKGROUND INFORMATION

Human sight is not made possible by the eyes alone; sight also depends on the vision center in the brain and on the optic nerves that relay images from the eye to the brain. Light enters the pupil of the eye and passes through the lens. An image then forms on the retina, which is at the back of the eye. The retina contains light-sensitive cells that produce nerve impulses that travel along the optic nerves to the vision center of the brain. In the vision center, the images are interpreted and the person "sees."

When any part of the three components of sight—eye, optic nerves, or vision center—are impaired, the ability to see is lost. The living computer described in the article is taking the place of optic nerves that have been destroyed by radiation.

Relevant to the concept of a living computer is the present-day use of a computer to help a paralyzed girl walk again. In 1983, Dr. Jerrold Petrofsky devised a system in which a high-speed computer delivered tiny shocks to the leg muscles of a girl who had been paralyzed for five years. The computer controlled the firing of impulses in just the right order to make the girl's muscles work; tiny sensing devices on her legs told the computer what the muscles were doing. Dr. Petrofsky hopes someday to make the system small enough to be put inside the human body.

### TEACHING STRATEGY: FUTURE

#### Motivation

Begin by asking students the following question.

- **What makes it possible for you to see?** (Answers may vary. Most students will probably say their eyes.)

#### Content Development

Continue the discussion by explaining to students that sight is not made possible by the eyes alone; the eyes can only pick up images. In order for a person to see, the images formed at the back of the eye (on a part of the eye called the retina) must be relayed to the vision center of the brain via optic nerves. It is in the vision center that the images are interpreted and the person "sees."

- **According to the article, what part of the sight process was disrupted**

## ADDITIONAL QUESTIONS AND TOPIC SUGGESTIONS

**1.** Space colonies may be a life-style option in the future. Does this story make you want to live in a space colony? Why or why not? (Accept all answers. Some students may be wary of radiation storms that could cause blindness; others may be intrigued by the possibility of traveling to other planets and living in a strange world.)
**2.** Find out more about ways in which bacteria break down substances. Two topics that you might research include the importance of bacteria in soil and the role of bacteria in foods such as yogurt.
**3.** Why might creative thinking and reasoning ability be possible with a "living computer"? (Accept all answers. A logical answer is that since reasoning and creative thinking are functions of human beings, a computer made of living substances might more closely resemble the human brain.)

Vacuum tubes like these made the first computers possible.

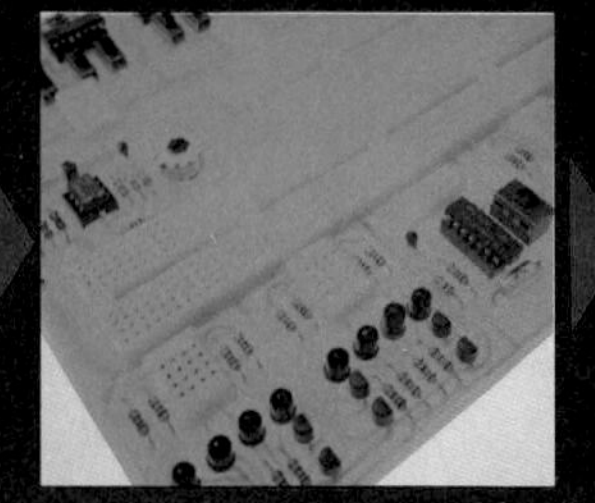

The computers of the 1950s used transistors instead of vacuum tubes.

Today, thousands of transistors are packed onto a silicon chip.

future. Living computers, like the one in Ronda's brain, require no outside power source and never need to be replaced. To understand how living computers may be possible, let's look briefly at how computers have evolved since the 1800s.

The first computers, made up of clunky gears and wheels, were turned by hand. They had only the simplest ability to answer questions based on the information stored inside them. By 1950, computers were run by electricity and operated with switches instead of gears. Information was stored when thousands of switches turned on and off in certain ways. While a lot of information could be fed into electrical memory banks, the computers of the 1950s were still very crude. In fact, the typical 1950s computer took up an entire room and could do less than many video-arcade games of the 1980s.

### Time Marches On

The computers of the 1980s contain thousands of switches and can be placed on a tabletop. The reason for the compactness of these computers is the silicon chip. Engineers can put hundreds of switches on a tiny piece of the chemical silicon. These tiny pieces of silicon, known as chips, are manufactured using laser beams and microscopes. It is these chips that record the information when, for example, you tell a computer your name.

But even with the silicon chip, modern computers cannot really think creatively or reason. And a computer has less sense than an ordinary garden snail. Experts feel that in order for computers to "graduate" to higher-level tasks, a whole new way must be developed of storing information in them. The key to developing a new system of information storage may lie in molecules of certain chemicals.

Why molecules? Scientists know that when some molecules are brought together, interesting changes take place. For example, electricity can jump from one molecule to another almost as if tiny switches were being shut on and off between them. Can we learn how to work these tiny switches? If so, then perhaps a whole new type of computer could be built!

This new computer might be able to hold more information in a single drop of liquid than today's computer could store in an entire roomful of chips. As you can imagine, the molecules in this computer of the future would have to be pretty special. And they would have to be produced in a new way.

### Leave It to Bacteria

One of the most popular current ideas concerning how these molecules could be produced is: Let bacteria do it for us! Bacteria



## FUTURE (continued)

for Ronda? (the part in which images are relayed to the brain)
- **Why could images from Ronda's eyes no longer be transmitted to the brain?** (Her optic nerves had been destroyed.)
- **What made it possible for Ronda to see again?** (A tiny computer that could function like optic nerves was implanted in her brain.)
- **Suppose parts of Ronda's eyes or vision center had been destroyed. Do you think these could have been replaced by computers?** (Accept all answers.)

### Enrichment
- **What aspect of this article lets you know immediately that the story takes place in the future?** (the fact that Ronda and her family live in a space settlement on Pluto)

Point out to students that someday travel to outer space may be quite commonplace, and that people may actually be able to live in space or on other planets.

Divide the class into small groups. Challenge each group to imagine what life would be like in a space settlement such as the one described in the article. Have the groups translate their ideas into dramatizations to share with the class.

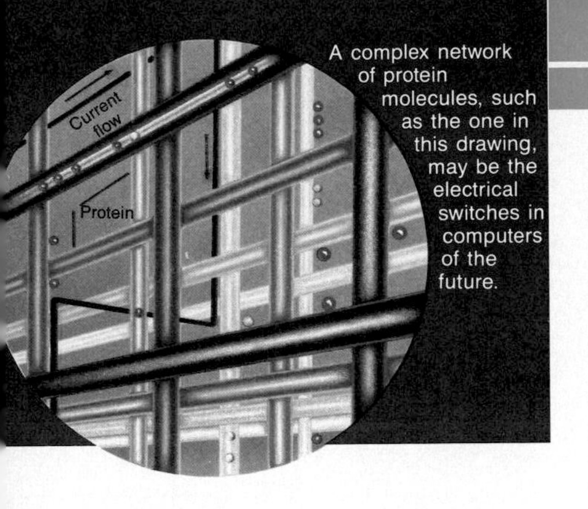

A complex network of protein molecules, such as the one in this drawing, may be the electrical switches in computers of the future.

are all around us. They constantly break down very complicated chemicals into molecules. Bacteria are at work in our bodies, in our food, and in our environment every minute of every day. Some scientists feel that bacteria could be "taught" to make special molecules. These molecules, when mixed together, could produce the flow of electricity needed to make a computer.

Of course, bacteria cannot be taught in the same sense that people can. Bacteria are not able to "learn." However, scientists can now *control* bacteria in many unusual ways. There are new techniques available that allow scientists to combine two different types of bacteria to produce a third, totally different type. In the future, bacteria may produce chemicals that have never been seen before.

In terms of a living computer, imagine that some bacteria have been taught to make special molecules. These bacteria could be grown in a special container and fed a particular substance to produce certain molecules. If the molecules could be told, or programmed, to do the right things, you would have a computer. And the computer would actually be alive because the bacteria live, grow, and produce molecules inside their container.

Think about the living computer implanted in Ronda's brain that allowed her to see again. Remember that bacteria need food to make molecules. Suppose that the computer in Ronda's brain was fed by her own blood, like all the other cells in her body. If this were the case, Ronda's computer would live as long as Ronda herself.

It may be many years before the living computer becomes a reality. Scientists must learn more about such things as how molecules react together, and how they can be programmed. But many scientists await the day when they can look at a computer and say, "It's alive."

When nerves connecting the eye to the brain are destroyed, no electrical signals can be carried. A person cannot see. By implanting a computer the size of a grain of rice, the person's sight is restored. The computer is designed to work exactly like the eye nerves.

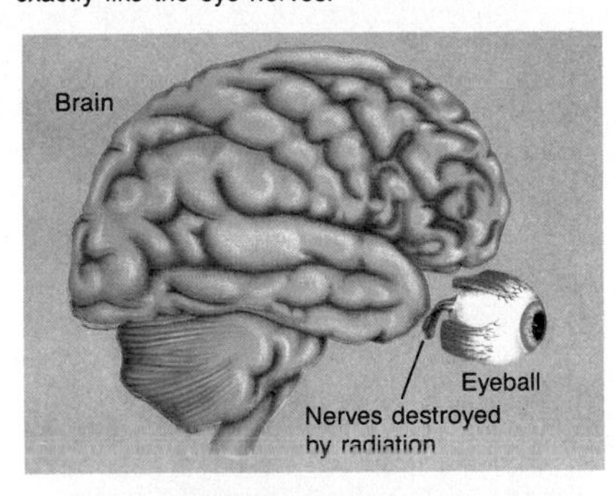


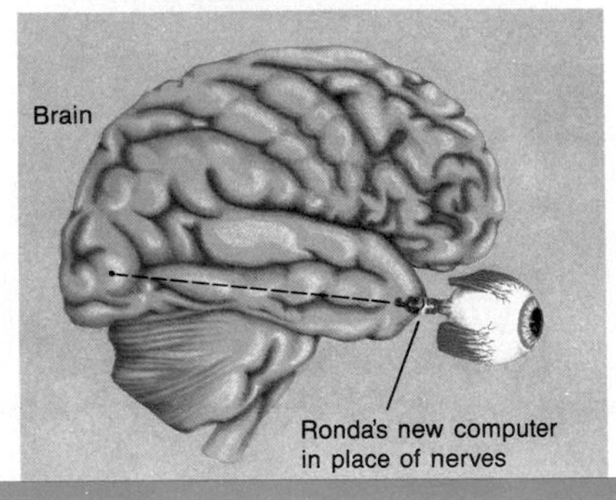


## CRITICAL THINKING QUESTIONS

**1.** What aspect of this story lets you know that in the future, interplanetary travel will be a rather normal activity? (Ronda, who lives on Pluto, goes to a hospital on Venus for treatment.)

**2.** Do you think the author of this story was being totally realistic in letting Ronda's blindness be the only result of the radiation storm that swept across the space colony? Why or why not? (Answers may vary. Some students may realize that radiation contamination usually causes extensive and long-term damage; it can cause many other physical impairments and even death. It is also somewhat unrealistic that no mention is made of the damage caused to other members of Ronda's family; or of the long-term effects of radiation contamination on the space colony.)

**3.** What possible hazards can be associated with the idea of implanting a computer in the human brain? (Answers may vary. An obvious answer is the possibility of the body's rejecting a foreign substance.)

### Skills Development

***Skill: Making diagrams***

Have students make creative diagrams to show how computers have changed in size since the 1800s. Have students include the tiny computer described in this article as part of the diagram. Students may wish to do additional research to find out more about the actual appearance and dimensions of the computers that are described in this article.

### Enrichment

Encourage interested students to study the mechanisms of other physiological functions, such as digestion, hearing, and locomotion. Challenge students to consider how living computers might be able to restore these functions if they should become impaired.

### Teacher's Resource Book Reference

After students have read the Science Gazette article, you may want to hand out the reading skills worksheet based on the article in your Teacher's Resource Book.

## Unit Two

# CLASSIFICATION OF LIVING THINGS

## UNIT OVERVIEW

In Unit Two, students are introduced to the concept of taxonomic classification. They learn about classification history and about modern classification systems and groups. They then study some of the general characteristics of the biological kingdoms (this book uses a five-kingdom system). Then they take up, in turn, the characteristics of different kinds of organisms. These include viruses, bacteria, and protists; nonvascular plants and plantlike organisms; vascular plants; invertebrate and nonmammalian vertebrate animals; and mammals.

## UNIT OBJECTIVES

1. **Explain the use of classification systems, and list the seven major classification groups.**
2. **State the general characteristics of organisms in the plant, animal, protist, monera, and fungi kingdoms.**
3. **Describe the characteristics of viruses, bacteria, and protists.**
4. **Describe the characteristics of plants and plantlike organisms.**
5. **Describe the characteristics of invertebrates and nonmammalian vertebrates.**
6. **Describe the characteristics of mammals.**

## INTRODUCING UNIT TWO

Begin your teaching of the unit by having students examine the unit-opening photograph, which shows a pond scene in Minnesota. Before students read the unit introduction, you may wish to ask them the following questions.

- **What living things can you see in this photograph?** (A deer, several birds, and various coniferous and deciduous plants are visible.)
- **What other living things are probably present in such a scene but are not visible in the photograph?** (The pond probably contains various types of fish and other aquatic organisms. Numerous small mammals probably inhabit the woods. Also, microscopic organisms must be present.)
- **Would you say that these life forms are very similar?** (No. They display great diversity.)
- **Do they all have anything in common?** (They share the general characteristics of living things.)
- **Are some of the organisms more closely related than others? Give examples.** (Yes. The trees and grasses, for example, are more closely related to each other than to the deer.)
- **How might you attempt to set up some sort of classification system that would reflect both the general similarities and specific differences characteristic of organisms?**

# Classification of Living Things

On a small pond in northern Minnesota, tiny water-striding insects glide along the slate-blue surface, paying little attention to the deer standing near the pond. Beneath the water, a hungry pop-eyed minnow follows the rapid movement of the insects. The pleasant odors of pond grasses drift to the twitching nostrils of a cottontail rabbit standing beside the edge of the pond. Suddenly, without warning, the croak of a bullfrog breaks the calm silence.

Not far away, a camper with binoculars observes this tiny slice of nature. "There must be some kind of order to this," she thinks to herself. The woman tries to remember the ways in which living things are grouped. Then she smiles as she remembers a strange sentence: "Kings play cards on fat green stools." This sentence is a kind of code that reveals the way living things are grouped. See page 137 for the secret of the code.

CHAPTERS



Deer in pond ecosystem



(Answers will vary. Some students are likely to hit upon the idea of levels of grouping, from broadest to most specific.)

Now have students read the unit introduction. Ask them the following question.

• **What do you think is the secret of the code sentence that is described?** (The initial letter in the words is the same as those of the classification groups: k = kingdoms, p = phyla, c = classes, o = orders, f = families, g = genera, and s = species. Many students will already be familiar with some of these terms.)

## CHAPTER DESCRIPTIONS

**6 Classification** Chapter 6 opens with a historical treatment of biological classification, including the system of binomial nomenclature introduced by Linnaeus. Then it treats the seven major classification groups: kingdoms, phyla, classes, orders, families, genera, and species. Finally, the general characteristics of the plant, animal, protist, monera, and fungi kingdoms are explored.

**7 Viruses, Bacteria, and Protists** The chapter opens with a discussion of the structure and reproduction of viruses, the most important of the "life-borderline" entities. Next, the structures and shapes of autotrophic and heterotrophic bacteria are described. The chapter closes with a treatment of the parts and life functions of the amoeba, paramecium, *Euglena,* and sporozoans.

**8 Nonvascular Plants and Plantlike Organisms** Chapter 8 focuses first upon the characteristics of algae. Fungi, lichens, slime molds, mosses, and liverwarts are discussed next.

**9 Vascular Plants** Chapter 9 opens with a description of the characteristics of ferns. Following this is an explanation of the functions of vascular-plant parts, including roots, stems, and leaves. Photosynthesis and transpiration are also explained. Gymnosperms and angiosperms are compared, and the structure and function of flower parts are described.

**10 Animals: Invertebrates** Chapter 10 deals with invertebrates, animals that do not have a backbone. Various invertebrates are described.

**11 Animals: Vertebrates** Nonmammalian animals that have backbones are discussed in Chapter 11. These include fishes, amphibians, reptiles, and birds. The special adaptations of these organisms are also examined.

**12 Mammals** Chapter 12 opens with an analysis of the five main characteristics of mammals. Next, egg-laying and pouched mammals are described. Finally, the general characteristics of placental mammals and the more specific attributes of ten groups of these mammals are treated.

## Chapter 6

# CLASSIFICATION

## CHAPTER OVERVIEW

From earliest times people have attempted to better cope with their environment by naming and classifying the organisms about them. A system devised by the ancient Greek philosopher Aristotle for classifying animals on the basis of their means of locomotion was used for almost 2000 years. But the foundation for our present system was laid by an English biologist, John Ray, in the seventeenth century. Ray based his classification scheme on anatomical features, and he also established the concept of species as basic life forms capable of interbreeding. In the eighteenth century, the Swedish botanist Carolus Linnaeus established a two-name system known as binomial nomenclature.

The hierarchical classification system devised by Linnaeus is still used today though it has undergone expansion and refinement. Linnaeus focused primarily on structural similarities when classifying organisms, but modern classification systems make use of several other criteria as well. Today, taxonomists consider such additional features as microscopic structures, biochemical makeup, and genetic similarities. In the time of Linnaeus, all organisms were placed either in the plant or animal kingdoms, but now most biologists recognize five kingdoms. Kingdoms Protista, Monera, and Fungi have been created for organisms that fit poorly in the animal or plant kingdoms.

## INTRODUCING CHAPTER 6

Tell the class to look at the picture of the fish on page 130. Because of its fearsome appearance, some students may mistake it for a shark. Call attention to the fish's name, *coelacanth,* in the caption, and point out that coelacanths are not closely related to sharks.

• **Do you notice anything unusual about this fish?** (Accept any answers, but call attention to the fleshy fins.)

Point out that a coelacanth's fins are modified into fleshy flippers which it uses to walk on the ocean floor and pounce on its prey. These fins are unusual in another way in that the spines are made up of cartilage rather than bone, as in other bony fishes.

• **What are the fins of most other fish like?** (Answers will vary, but the fish familiar to most students have nonmuscular fins containing many parallel or radiating thin bones.)

Now ask students to read page 131.

• **What is another reason that coelacanths are of special interest to scientists?** (They were thought to have become extinct more than 60 million years ago.)

Tell the class that the coelacanth described in the text was caught in 1938. Since then over 100 other spec-

# Classification

**CHAPTER SECTIONS**

**6-1 History of Classification**
**6-2 Modern Classification Systems**
**6-3 The Five Kingdoms**

**CHAPTER OBJECTIVES**

*After completing this chapter, you will be able to:*

**6-1** Trace the history of classification systems.
**6-1** Explain how binomial nomenclature is used to classify living things.
**6-2** Identify the seven major classification groups.
**6-3** Give some general characteristics of the plant, animal, protist, monera, and fungi kingdoms.

The people fishing from the boat could not believe their eyes. Lying in the boat's net was a fish none of them ever had seen before. It stretched more than 1.5 meters from the tip of its ugly nose to the end of its fan-shaped tail. Large steel-blue scales covered its body. A powerful lower jaw hung down from a frightening face. But most peculiar of all, its fins were attached to what appeared to be stubby legs!

The unusual fish, caught at the mouth of the Chalumna River in South Africa, was taken to a local museum. There M. Courtenay-Latimer, a South African museum curator, happened to see the fish. After searching through many books, she had found no description of the strange fish. However, she knew of someone who might be able to solve the riddle. That someone was Professor J. L. B. Smith, an African fish expert. So M. Courtenay-Latimer preserved and sent the fish to Professor Smith.

The scientist was shocked. He later wrote, "I would hardly have been more surprised if I met a dinosaur on the street." Professor Smith was looking at an animal thought to have become extinct more than 60 million years ago. Yet, in a flash, Professor Smith had been able to identify the fish as a coelacanth (SEE-luh-kanth). The discovery was exciting because coelacanths are thought to be closely related to fish that evolved into four-footed land animals.

What led Professor Smith to make his startling identification? A knowledge of biological classification, a special system that helps scientists to identify and name organisms.

*Rare photograph of a coelacanth*



## TEACHER DEMONSTRATION

Place these pieces of laboratory equipment on a demonstration table or other location where they can be seen by the class:

| | |
|---|---|
| beaker | graduated cylinder |
| forceps | scissors |
| flask | test tube |
| funnel | test tube clamp. |

• **Is there some way we could separate these pieces of laboratory equipment into two different groups?** (Accept all logical answers. Several criteria could be used to group the objects. For example, the glass items might go in one group and the metal ones in another.)

After the class has suggested a way to separate the objects, discuss the characteristics selected for grouping them. Point out that though the characteristic selected was a good one, other characteristics could have been used as well.

• **Could we now take the objects in each of these groups and subdivide them into two more groups?** (Accept all logical suggestions for subgrouping the objects.)

Explain that when we select certain characteristics to group objects together we are classifying them. Emphasize that humans classify almost everything about them, in some fashion. Ask the class to suggest some everyday examples of classification.

Point out that classification is also important in science, and in this chapter, they will learn about the systems that scientists use to classify living things.

• **Why do you think scientists attempt to classify all living things?** (Accept all answers.)

imens have been caught in the Indian Ocean near the Comoro Islands.

• **Even though Professor Smith had never seen a coelacanth, how was he able to identify it?** (Smith recognized the fish was a coelacanth because he was familiar with biological classification.)

Point out that biologists can use their knowledge of scientific classification to identify unknown plant and animal specimens. Through the centuries as scientists have studied organisms, they have used a standard system to name and classify them. Since biologists everywhere use the same classification system, an unfamiliar specimen can be correctly identified anywhere in the world.

## TEACHER RESOURCES

**Audiovisuals**
*Carolus Linnaeus,* film, EBE
*Classifying Plants and Animals,* film, Cor

**Books**
Dixon, B., *What Is Science For?,* Collins
Samuel, E., *Order of Life,* Prentice-Hall

**Software**
*Classifying Mammals,* Prentice Hall

## 6-1 HISTORY OF CLASSIFICATION

### SECTION PREVIEW 6-1

Though the classification system we currently use was devised over 200 years ago, other systems were used before this. In the fourth century B.C., Aristotle developed a system in which he placed animals into three groups according to the way they moved. In the seventeenth century, John Ray devised a system of classification based on the internal anatomy of plants and animals. The system used today had its beginnings in the eighteenth century with the work of Carolus Linnaeus. Linnaeus placed organisms into large categories, and then subdivided them into smaller categories according to similarities in form and structure. He also developed a standard naming system in which every organism was given two names, a genus and species.

### SECTION OBJECTIVES 6-1

1. **Define taxonomy, and state its purposes.**
2. **Describe the classification systems of Aristotle and John Ray.**
3. **Identify the features of the classification system devised by Linnaeus.**
4. **Define binomial nomenclature and name the two names given to all organisms.**

### SCIENCE TERMS 6-1

**taxonomy p. 133**
**species p. 134**
**binomial nomenclature p. 134**
**genus p. 134**

**Figure 6-1** *Thousands of years ago, people drew this painting on a cave wall in Spain. What message does the painting communicate?* ❶

## 6-1 History of Classification

Thousands of years ago, people began to recognize that there were different groups of living things in the world. Some animals had claws and sharp teeth and roamed the land. Others had feathers and beaks and flew in the air. Still others had scales and fins and swam in the water.

People also made observations about plants. Not only did plants vary in shape, size, and color, but some were good to eat while others were poisonous. In a similar sense, some animals, such as those with sharp teeth, were very dangerous, while others, such as those with feathers, were relatively harmless.

Without knowing exactly what they were doing, these people developed simple systems of classification. **Classification is the grouping of living things according to similar characteristics.** ❶ Knowledge of these characteristics helped people to survive in their environment. For example, they quickly learned to fear the sharp-toothed animals and to hunt the feathered ones.

Perhaps these people painted pictures on the walls of caves to communicate this knowledge. See Figure 6-1. Today, similar but much more complex



### TEACHING STRATEGY 6-1

#### Motivation

Collect and display pictures of these animals: dog, wolf, fox, tiger, lion, house cat. Ask the class to compare these animals.

• **In what ways are all of these organisms similar?** (Answers will vary, but students will likely suggest that all are flesh-eating animals, or carnivores.

• **How might these pictured animals be placed in two different groups?** (Lead students to suggest that the dog, wolf, and fox go in one group, and the tiger, lion, and cat in another.)

• **Why did you group the animals in this way?** (Lead the class to suggest that similar characteristics were used to group the animals. Dogs are more similar to wolves and foxes than they are to lions, tigers, and cats, etc.)

Explain that classification systems based on structural similarities have been used since early times to group and name plants and animals.

• **Are there any other ways in which we could have grouped these animals?** (Answers will vary, but lead students to suggest that other criteria could have been used. For example, the dog and house cat might be placed together since they are common domesticated pets.)

information is found in the drawings, photographs, and words in scientific books. This information is organized into the science of classification, which is called **taxonomy** (tak-SAH-nuh-mee). The scientists who work in this field are called taxonomists. The purpose of taxonomy is to group all the plants and animals on earth in an orderly system. This system helps to provide a better understanding of the relationships among living things. Meaningful names are given to newly discovered animals and plants based on taxonomy. Sometimes taxonomy even helps solve biological riddles of the past.

### *Early Classification Systems*

In the fourth century B.C., the Greek philosopher Aristotle first proposed a system to classify life. He placed the animals into three groups. One group included all animals that flew, another group included those that swam, and a third group included those that walked. ❷

Aristotle classified animals according to the way they moved. Although this system was useful, it caused problems. According to Aristotle, both a bird

**Figure 6-2** *Because of its unusual characteristics and behavior, the giant panda* (right) *was once thought to be more closely related to the raccoon* (left) *than to the bear* (top right). *Today, giant pandas are classified as bears.*

## ANNOTATION KEY

❶ **The painting shows the types of animals people hunted. (Making inferences)**

❶ **Thinking Skill: Classifying organisms**

❷ **Thinking Skill: Making generalizations**

## TIE-IN/MUSIC

Every day students encounter numerous classification systems. Without them our lives would be chaotic. Ask the class to think about the classification systems they encounter as they look for an album in a music store. For example, records, cassette tapes, and compact discs are shelved in different sections of the store. Within each section, the albums are grouped by such music types as rock, jazz, classical, and country. The recordings of the various artists and groups may then be arranged alphabetically by type, and so forth.

Point out that any system for classifying life is arbitrary. It was invented by humans as a convenience to help them better keep track of living things. Different classification schemes are the result of different criteria being selected when grouping organisms.

## Content Development

Write the word "taxonomy" on the chalkboard. Beside the word, write the root words, *taxis* and *nomy.* Explain that taxonomy is derived from the Greek terms *taxis,* meaning arrangement and *nomy,* meaning law.

• **Why do you think that the science of classification is called taxonomy?** (The science of taxonomy provides an orderly set of rules, or laws, for arranging organisms into groups.)

Mention that a scientist who specializes in taxonomy is called a taxonomist. After students have read the introduction to Section 6-1, ask,

• **Why are classification systems needed?** (Taxonomy provides a better understanding of relationships among living things; it establishes a standard system for naming newly discovered organisms; and helps solve biological riddles of the past.)

## Reinforcement

Give students a list of the following food products:

| | |
|---|---|
| margarine | English muffins |
| flour | raisin bran |
| rye bread | grapes |
| sugar | cheddar cheese |
| bacon | chicken legs |
| sirloin steak | donuts |
| corn flakes | yogurt |
| lettuce | potatoes |
| chocolate cake mix | milk |
| hamburger | oranges |

Working in small groups, ask students to classify these items into the sections where they would be found in a supermarket.

## TIE-IN/EARTH SCIENCE

Like life science, other branches of science use classification systems. In earth science, rocks are classified into three main groups: sedimentary, igneous, and metamorphic.

**ANNOTATION KEY**

1 Thinking Skill: Applying definitions

2 Thinking Skill: Identifying groupings

### Activity

**What's in a Name?**
**Skills: Inferring, relating, applying**
**Level: Average**
**Type: Vocabulary/Writing**

Students should enjoy searching for the clues in the chapter introduction that will tell them why the scientific name for the coelacanth is *Latimeria chalumnae*. After being caught at the mouth of the Chalumna River, the fish was seen by M. Courtenay-Latimer. She was responsible for having the fish identified. Hence the coelacanth is named for her and for the river in which it was caught.

and a bat would fall into the same flying group. Yet in some basic respects, birds and bats are very different. Birds, for example, are covered with feathers. Bats, on the other hand, are covered with hair.

Although the system devised by Aristotle would not satisfy today's taxonomists, it was the first attempt to develop a scientific and orderly system of classification. Aristotle's classification system was used for almost 2000 years.

In the seventeenth century, John Ray, an English biologist, set an enormous goal for himself. He decided to collect, name, and classify all the plants and animals in England. But unlike Aristotle, Ray based his system of classification on the internal anatomy of plants and animals. He examined how they behaved and what they looked like as well. Ray was the first person to scientifically use the term **species** (SPEE-sheez). 1 A species is a group of organisms that are able to interbreed, or produce young.

John Ray achieved his goal. Moreover his work aided in the development of more complete and accurate classification systems.

### *Binomial Nomenclature*

The eighteenth-century Swedish scientist Carolus Linnaeus spent the major part of his life developing a new system of classification. The system placed all living things into plant and animal groups according to similarities in form or structure.

Linnaeus also developed a simple system for naming organisms. Before Linnaeus developed his naming system, plants and animals had been identified by a series of Latin words. These words described the physical features of the organism. Sometimes five or six Latin words were used to describe a single organism!

Linnaeus devised a simpler naming system of **binomial nomenclature** (bigh-NOH-mee-uhl NOH-muhn-klay-cher). In this system, each plant and animal are given two names, a **genus** (plural: genera) name and a species name. 2 For example, you could think of the genus name as your family name. The species name could be thought of as your first name. Like your family, a genus may consist of several closely related organisms or only one species.

**Activity**

*What's In a Name?*

The scientific names of living things are derived from different sources. The coelacanth described at the beginning of this chapter was given the name *Latimeria chalumnae*. Find clues in the opening chapter page that will tell you why this name was given to the coelacanth.

134

### 6-1 (continued)

**Content Development**

Point out that for organisms to be considered members of a single species, they must not only be capable of interbreeding, but their young must also be fertile. Although horses and donkeys can interbreed to produce mules, the parent animals are considered two different species since mules are not fertile. On the other hand, organisms may be quite dissimilar in appearance and yet be members of the same species. For example, it is possible for a boxer and collie to produce fertile offspring. Even though boxers and collies look quite different they are classified as different breeds, or varieties, within the same species.

### Section Review 6-1

**1.** Science of classification
**2.** Unlike Aristotle, Ray based his system of classification on the internal anatomy of plants and animals.
**3.** System of giving each plant and animal two names: a genus and a species name

## TEACHING STRATEGY 6-2

**Motivation**

Have students look at Figure 6-3 and read the caption.

- **What is the function of the forelimb of each of these animals?** (A human's forearm is used for grasping and throwing objects, a bat's wing is for flying, and a whale uses its forelimb for swimming.)
- **How are the forelimbs of these**

### *SECTION REVIEW*

1. What is taxonomy?
2. How did the classification systems of Aristotle and Ray differ?
3. What is meant by binomial nomenclature?

## 6-2 Modern Classification Systems

Today, 200 years after Linnaeus completed his work, scientists consider many additional factors when classifying organisms. Of course, scientists still examine large internal and external structures, but they also examine microscopic structures. Scientists even analyze the chemical makeup of an organism.

Although such a classification system may at first seem complicated, it is really quite simple. The modern system of classification does two jobs. First, it groups organisms according to their *basic* characteristics. Second, it gives a *unique name* to an organism that scientists all over the world can use and understand. For example, in North and South America

**Figure 6-3** *Scientists often use skeletons of animals to help in classification. Notice how the similarity of forelimb structures of these animals helps to classify them as being part of the same group, in this case, mammals.*

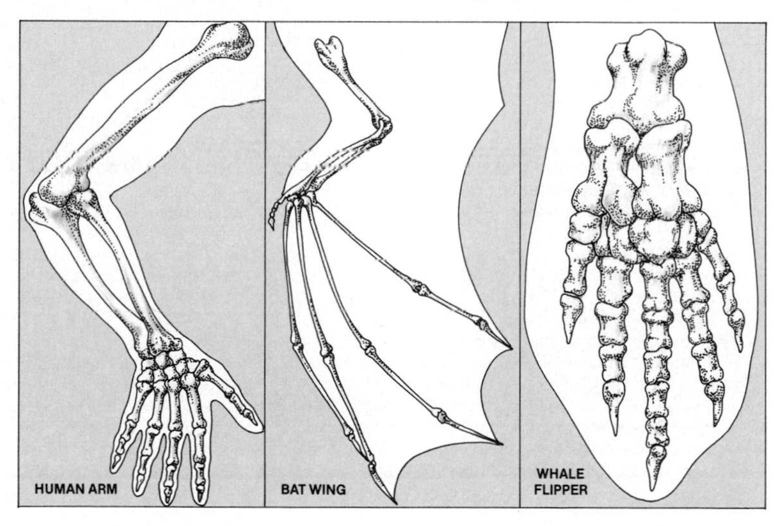


135

## 6-2 MODERN CLASSIFICATION SYSTEMS

### SECTION PREVIEW 6-2

Modern classification systems do two things—they group organisms according to their basic characteristics, and they give organisms a unique name that is used worldwide. All organisms that have been identified are grouped within a hierarchical system consisting of these seven levels: kingdom, phylum, class, order, family, genus, and species. Every level within this hierarchical system indicates something about an organism's characteristics. The organism's genus and species make up its scientific name.

### SECTION OBJECTIVES 6-2

1. **State the two major functions of modern classification systems.**
2. **List in the correct sequence the seven major classification groups.**
3. **Compare and contrast the classification of several well known organisms.**

### SCIENCE TERMS 6-2

**kingdom p. 136**
**phylum p. 136**
**class p. 137**
**order p. 137**
**family p. 137**

**animals similar?** (Though the function is different, each has five digits, and the basic bone structure is similar.)

• **Why would the structure of these animal's forelimbs be useful in classification?** (Similar internal structures can be used to show relationships among animals whose external appearances may be quite different.)

Point out that similar structures that perform different functions are called homologous structures. Homologous structures also suggest descent from a common ancestor.

### Content Development

Emphasize that today taxonomists consider several factors other than structural similarities when classifying organisms. A few factors that might be mentioned are these:

**1.** Fossil evidence: The fossil record often reveals that certain organisms are related because of common ancestors.

**2.** Early development: Scientists believe that similarities in the early embryos of organisms indicate relationships.

**3.** Chemical makeup of an organism: Studies have revealed that certain proteins have a similar chemical makeup in similar species. By studying the chemical structure of these proteins, scientists can get an idea of how closely related various organisms may be.

**4.** Genetic makeup of an organism: Relationships among species can be found by comparing the structure of DNA, the hereditary material of their cells. The more closely related they are, the more similar their DNA will be.

## TIE-IN/LANGUAGE ARTS

Explain that most of the Latin names within the classification of living things from the kingdom name on down are very close to the English names. In most cases, only the endings are different. Point out the following similarities:

| English | Latin |
|---|---|
| animal | Animalia |
| chordate | Chordata |
| mammal | Mammalia |
| carnivore | carnivora |
| felid | Felidae |
| panther | Panthera |

## HISTORICAL NOTES

Linnaeus published a system for classifying plants in 1753, and one for classifying animals in 1758. Initially, his grouping system consisted of just three levels: kingdom, genus, and species. The phylum, class, family, and order levels were added later. Jean Baptiste Lamarck, who is best known for his pre-Darwinian theory of evolution, was largely responsible for expanding and refining the Linnaean system.

one large cat is called a mountain lion by some people, a cougar by others, and a puma by still others. If these people were to talk to one another about the animal, they might think they were talking about three different animals. But scientists throughout the world use only one name for this large cat, *Felis concolor.* This name easily identifies the cat to all scientists, no matter what language they speak.

### *Classification Groupings*

**All living things are classified into seven major** (1) **groups: kingdom, phylum, class, order, family, genus, and species.** The largest group is called a **kingdom.** All animals, for example, belong to the animal kingdom, and all plants belong to the plant kingdom. The second largest group is called a

### Career: *Zoo Keeper*

**HELP WANTED: ZOO KEEPER** High school diploma required. Minimum one year experience as an apprentice zoo keeper needed. Must like animals and be tolerant of both animal and human behavior. Apply in person.

Today is a big day at the zoo. The arrival of a male and female giraffe is expected from Africa. For months, everyone has prepared for the new exhibit. Gardeners planted fruit trees bearing the giraffe's favorite food. In addition, zoo architects designed and built a new home, while tour guides learned the habits of the newcomers. Everyone wants to make sure the giraffes will be safe and comfortable. If all goes well, the pair will breed and produce young giraffes.

At first, the giraffes will be kept away from other animals. Once tests show they are strong and healthy, the **zoo keeper** will help move them to their new home. The zoo keeper will also be responsible for giving food and water to the giraffes. Other responsibilities of the zoo keeper include keeping the animals' area clean and observing and recording their behavior.

The zoo keeper usually is the first to notice any medical problems as well as to treat minor injuries or ailments. However, serious problems must be reported to the senior zoo keeper.

Of the many people working at the zoo, zoo keepers come into the most direct contact with the animals. In fact, zoo keepers sometimes bathe and groom the animals. Occasionally, zoo keepers answer visitors' questions about the inhabitants of the zoo. For further career information, write to the American Association of Zoo Keepers National Headquarters, 635 Gage Boulevard, Topeka, KS 66606.



## 6-2 (continued)

### Content Development

Review the seven major groupings of living things with students. Point out that kingdoms are the largest groupings and that species are the smallest groupings. You may want to tell students that the term *division* is often used in place of phylum when discussing plants.

### Reinforcement

Point out that like all other organisms, humans have a unique genus and species name. Encourage students to find out what our scientific name is. They will find that our genus name is *Homo,* meaning human, and our species name is *sapiens,* meaning wise.

### Motivation

Compare the seven levels of scientific classification to the categories used to keep track of time. Develop the analogy by placing this outline on the chalkboard or on an overhead transparency:

Kingdom . . . . . . . . . . . . . years
Phylum . . . . . . . . . . . . . . months
Class . . . . . . . . . . . . . . . weeks
Order . . . . . . . . . . . . . days
Family . . . . . . . . . . . hours
Genus . . . . . . . . . . . minutes
species. . . . . . . . . . . seconds

Point out that as we go from years to seconds, we go from general to more specific units for identifying time intervals. In a similar way, we go from general categories to specific organisms as we go from kingdom to species. Ask the class to think of other examples of hierarchical classification groupings. For example, in most school districts some of these administrative levels can be identified:

**Figure 6-4** *The variety of living things in a rain forest, such as this one in Uganda, Africa, is greater than in any other place on the surface of the earth. Here gorillas share their territory with countless different plants and animals.*

**phylum** (FIGH-luhm; plural: phyla). Each phylum is made up of **classes.** Within each class are **orders.** In turn, each order is divided into **families** that consist of many related genera. Each genus usually is divided into one or more species.

The words for an organism's genus and species make up its scientific name. The genus name is capitalized, but the species name begins with a small letter. For example, the genus and species name for a wolf is *Canis lupus*. These two names identify the organism. Although most of these names are in Latin, some are in Greek. Scientists estimate that there are at least three million and perhaps as many as ten million different species alive today. Many of them have yet to be identified and named.

### Classifying an Organism

Figure 6-5 shows the classification groupings of several organisms. Each grouping indicates something about the organism's characteristics. For example, the lion belongs to the order Carnivora. ❷

#### Activity

*A Secret Code*

You can remember the correct classification sequence from the largest to the smallest group by remembering this sentence: *Kings play cards on fat green stools*. The first letter of each word is the same as the first letter of each classification group.

1. Think of another sentence you can use to remember the classification sequence.
2. The colors of the rainbow are red, orange, yellow, green, blue, and violet. Create a sentence you can use to remember the colors in the correct order.
3. Think of a sentence you can use to remember the order of the planets from the sun.

137

## ANNOTATION KEY

❶ Thinking Skill: Classifying

❷ Thinking Skill: Interpreting charts

## Activity

**A Secret Code**
**Skills: Applying concepts, relating, inferring**
**Level: Remedial**
**Type: Vocabulary/Writing**

This simple activity will help students gain an understanding of the usefulness of mnemonic devices. Accept all reasonable sentences for each of the suggested activities.

School board
Superintendent
Associate superintendents
Principals
Assistant principals
Teachers
Students

## Content Development

Emphasize that Latin and sometimes Greek is used when writing scientific names. This practice follows a tradition established by Linnaeus. In his day, Latin was used extensively for scholarly writing because it was understood by learned people in various countries. Although Latin is no longer a commonly used language, it is appropriate for scientific nomenclature because it is not subject to change as are some words in modern languages.

Be sure students understand that the genus name is usually a noun, and it is always capitalized. The species name is usually an adjective, and it is not capitalized. Further, be sure they understand that many organisms can have the same genus name, but the species name always identifies only one specific kind of organism. For example, dogs, wolves, and coyotes all belong to the genus *Canis*, for their species names are *familiaris*, *lupus*, and *latrans*, respectively. Sometimes a second name beginning with a lower case letter follows the species name to designate a particular variety of breed within that species. *Ursus americanus cinnamomum* is a cinnamon black bear and *Ursus americanus hamiltoni* is a Newfoundland black bear.

## Enrichment

The text identifies *Canis lupus* as the genus and species names for a wolf. Encourage students to use appropriate references to find out the scientific names of other members of the canid family, such as dogs, coyotes, and jackals. They may also want to prepare reports comparing and contrasting these various types of canids. Some of them are among the most misunderstood animals.

### CLASSIFICATION OF FIVE DIFFERENT ORGANISMS

| | Lion | Onion | Paramecium | Yogurt-making Bacterium | Edible Mushroom |
|---|---|---|---|---|---|
| **Kingdom** | Animalia | Plantae | Protista | Monera | Fungi |
| **Phylum** | Chordata | Tracheophyta | Ciliophora | Eubacteriacea | Basidiomycetes |
| **Class** | Mammalia | Angiospermae | Ciliatea | Schizomycetes | Homobasidiomycetes |
| **Order** | Carnivora | Liliales | Hymenostomatida | Eubacteriales | Agaricales |
| **Family** | Felidae | Liliaceae | Paramecidae | Lactobacillaceae | Agaricaceae |
| **Genus** | *Panthera* | *Allium* | *Paramecium* | *Lactobacillus* | *Agaricus* |
| **Species** | *leo* | *cepa* | *caudatum* | *bulgarius* | *campestris* |

**Figure 6-5** *The classification of five different organisms is shown in this chart. Which organism is a member of the family Liliaceae?* ❶

"Carnivora" is the Latin word for flesh eater. Many other familiar organisms also belong to this order such as dogs, raccoons, and bears. Look at Figure 6-5 again and notice that the lion is in the family Felidae. This family contains not only the lion but other cats, including ordinary house cats.

This knowledge allows you to have a very good idea of how a lion acts and looks even if you have never seen one in person. You can do this because lions and house cats are in the same family. But lions and house cats are not in the same genus and species, which indicates they are somewhat different. For example, taxonomists include the lion in the genus *Panthera* along with other large cats. All cats in the genus *Panthera* roar; they do not purr. Other cats, including house cats, are in the genus *Felis*. They, of course, purr and do not roar. Finally, the species name for the lion is *leo*. In this case, the lion, *Panthera leo*, is the animal that you see jumping through hoops in the circus or lounging lazily in a cage at your local zoo. *Felis domesticus* is the scientific name for your playful pet cat. ❶

### SECTION REVIEW

1. What are the seven major classification groups?
2. What is the scientific name for an onion?



## 6-2 (continued)

### Skills Development

***Skill: Interpreting Charts***

Direct the class to study Figure 6-5.

- **Are the same kinds of classification groupings used for all kinds of organisms?** (Yes. Most organisms can be classified within the seven major classification groups.)
- **To how many different kingdoms do the organisms classified in Figure 6-5 belong?** (Each organism belongs to a different kingdom; therefore these organisms represent five different kingdoms.)
- **At which classification level—kingdom, class, genus, etc.—would you find the greatest number of organisms?** (The greatest number of kinds as well as the greatest number of individuals occurs at the kingdom level.)
- **In which group would the plants be more alike—in the Order *Liliales*** or the Family *Liliacae*? (Plants would be more alike at the family level.)
- **What is the scientific name for one type of edible mushroom?** (*Agaricus campestris*)

If any students ask why the suffix *mycetes* is used for organisms in two different kingdoms, point out that at one time bacteria were commonly classified as fungi. Mycetes refers to fungi.

### Enrichment

Point out that house cats are thought to be descendants of the African wild cat. They may have first been domesticated in Egypt as early as 2000 B.C.

Suggest that interested students prepare reports on the varieties of domestic cats kept by humans. The reports should include information on variations such as coloration, anatomical differences, behavioral patterns, and ancestry.

### Skills Development

***Skill: Interpreting illustrations***

Tell the class to look at Figure 6-6. Call on someone to read the caption and answer the question. Then use some of the following questions to lead a discussion on the figure:

- **What are some common features of the animals at the class level?** (Answers will vary. Lead the class to suggest that as mammals, these animals have hair and they nurse their young.)
- **Why are animals like skunks, wolves, walruses, and lions classified in the same order?** (These ani-

CLASSIFICATION OF THE LION

**Kingdom** Animalia

**Phylum** Chordata

**Class** Mammalia

**Order** Carnivora

**Family** Felidae

**Genus** *Panthera*

**Species** *leo*

**Figure 6-6** *Examine each row from top to bottom. What pattern do you discover?* ❷



## ANNOTATION KEY

❶ **Onion (Interpreting charts)**

❷ **The similarities among organisms increase as the animals are classified from kingdom to species. (Interpreting charts)**

❶ **Thinking Skill: Making comparisons**

mals are similar in that they are carnivores, or flesh eaters.)

- **Is a tiger more similar to a house cat or to a leopard. How do you know?** (A tiger and house cat are classified in the same family, but they are in different *genera*. Therefore the tiger and leopard are more similar.)
- **In which row are the animals most closely related?** (The animals are most closely related at the species level.)

### Enrichment

Have students develop a series of pictures similar to "Classification of the Lion" for another animal.

### Section Review 6-2

1. Kingdom, phylum, class, order, family, genus, species.
2. *Allium cepa.*

## SCIENCE, TECHNOLOGY, AND SOCIETY

In Greek mythology, many creatures existed that were hybrids of real-life animals. For example, Pegasus was a horse with wings, a griffin had the body of a lion and the head and wings of an eagle, and a centaur was part man and part horse. Although these creatures were mythical, today's scientists are producing such animals as goat-sheep composites and Angus-Hereford cattle combinations.

As early as 1982, a goat-sheep "chimera" was born at Cambridge University. The animal, produced from spliced embryos, had the body of a goat, the legs of a sheep, and various combination body parts. Composite Angus-Hereford calves have been born at Louisiana State University Agricultural Experiment Station. They were also produced from a microsurgical combination of pairs of embryos. Other animals have been produced from four or more parent stocks.

The goals of such research are to improve farm production, to preserve endangered species, and to help solve economic problems on a global scale. However, some people, including some scientists, are opposed to this type of research because of the ethical questions that are involved. What do you think? How might our classification system change if large-scale production of composite animals became a reality?

## 6-3 THE FIVE KINGDOMS

### SECTION PREVIEW 6-3

Classification systems are imperfect and they are subject to change. At one time all organisms were placed in either the plant or animal kingdoms. Later, a third kingdom called *Protista* was invented for organisms that did not fit well in either the plant or animal kingdom. More recently, two other kingdoms called *Monera* and *Fungi* have been added to classification systems.

Members of the plant kingdom are mostly multicellular autotrophs. Animals are always multicellular heterotrophs, and, unlike plants, most of them are motile. Both protists and monerans are unicellular, but monerans lack a well-defined nucleus. Fungi are nonmotile like plants, but unlike plants, they do not contain chlorophyll. Thus, they are heterotrophs.

### SECTION OBJECTIVES 6-3

1. **Distinguish between multicellular and unicellular organisms.**
2. **Distinguish between organisms that are autotrophs and heterotrophs.**
3. **List characteristics of each kingdom in the five kingdom classification system.**

### SCIENCE TERMS 6-3

**multicellular p. 140**
**unicellular p. 140**
**autotroph p. 140**
**heterotroph p. 140**

## 6-3 The Five Kingdoms

In Figure 6-5, you may have noticed that some organisms were not in the plant or animal kingdoms. The paramecium, for example, is usually placed in a kingdom called Protista. Scientists invented this kingdom to include organisms that did not seem to fit into either the plant or animal kingdoms. Later, two more kingdoms called Monera (muh-NIHR-uh) and Fungi (FUHN-jigh) were added to many modern classification systems. **Today, most scientists use a system of classification that includes five kingdoms. These five kingdoms are plants, animals, protists, monerans, and fungi.**

**PLANTS** Most plants are **multicellular,** or many-celled, organisms that contain specialized tissues and organs. A few plants are **unicellular,** or one-celled. Some algae are examples of unicellular plants. In general, plants are **autotrophs** (AWT-uh-trohfs). Autotrophs are organisms that can make their own food from simple substances. Most plants contain chlorophyll, a green pigment necessary for making food.

**Figure 6-7** *The ability to move from place to place is one way scientists distinguish the wasp, a member of the animal kingdom, from the berry bush, a member of the plant kingdom.*

140

### TEACHING STRATEGY 6-3

#### Motivation

Call on a student to look up the definition of the term *motile* in a dictionary. Have him/her read the definition aloud and then ask someone to explain what the opposite term, *nonmotile,* must mean.

- **In Figure 6-7, which organism is motile and which one is nonmotile?** (The wasp is motile and the berry bush is nonmotile.)
- **In which of the five kingdoms are most of the organisms motile?** (Most animals and protists are motile.)
- **Which kingdoms are made up of mostly nonmotile organisms?** (Most plants, monerans and fungi are nonmotile. However, some monerans are capable of locomotion.)

#### Content Development

After students have read page 141 ask,

- **Are fungi autotrophs or are they heterotrophs?** (Fungi are heterotrophs because they cannot make their own food.)

Explain that even though fungi and animals are both heterotrophs, their means of taking in and digesting food are quite different. Animals ingest food and digest it inside their bodies. Fungi produce enzymes that digest their food externally. The digested food is then absorbed into their cells.

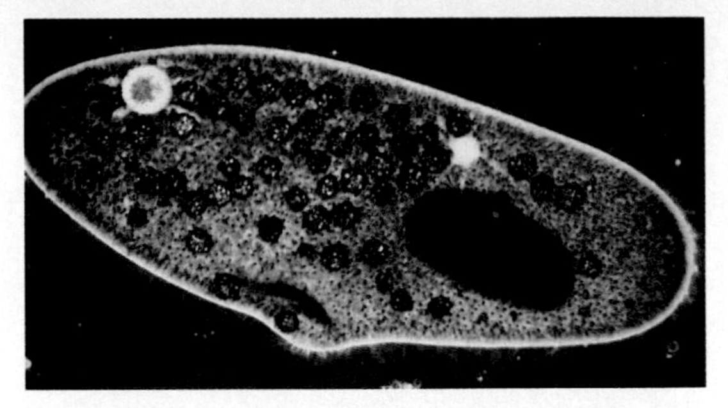

**Figure 6–8** *The unicellular paramecium is a member of the Protista kingdom. This paramecium is magnified 200 times.*

**ANIMALS** Only multicellular organisms are in the animal kingdom. These organisms have tissues and most have organs and organ systems. Unlike plants, animals do not contain chlorophyll and are not able to make their own food. They depend upon autotrophs. These organisms are called **heterotrophs** (HET-uh-roh-trohfs).

**PROTISTS** The Protista kingdom includes most unicellular organisms. Microscopic protozoans such as paramecia are found in this kingdom.

**MONERANS** All of the earth's bacteria are found in the Monera kingdom. One kind of algae, the blue-green algae, are also in the Monera kingdom. Like protists, monerans are unicellular. However, the cells of monerans do not contain a well-defined nucleus. The nucleus is the control center of a cell.

**FUNGI** As you might expect, the world's wide variety of fungi make up the Fungi kingdom. The kinds of fungi you are probably most familiar with are mushrooms and molds. Fungi share many characteristics with plants. However, fungi do not contain the green pigment chlorophyll. So, unlike plants, fungi cannot make their own food.

### SECTION REVIEW

1. In what kingdom are most protozoans found?
2. Define multicellular and unicellular.

### Activity

*Classification of Plants*

1. Obtain 15 different leaves from local plants. **CAUTION:** *Some plants are poisonous. Before you touch any plants, check with a field guide or knowledgeable adult to make sure the plants you choose are not poisonous.*
2. Place each leaf between two sheets of wax paper and glue the edges of the wax paper together.
3. Press the leaves between the pages of a heavy book.
4. Use a key or field guide to find the scientific names of the plants from which each leaf came.
5. Create a poster in which you identify each leaf. Design the poster so that the most closely related leaves are grouped together.



### Activity

**Classification of Plants**
**Skills: Manipulative, comparing, recording, classifying, relating**
**Level: Average**
**Type: Hands-on**
**Materials: 15 different leaves, wax paper, glue, leaf field guide**

This simple activity will reinforce student ability to classify organisms by their basic characteristics, in this case the leaves of various trees. Students should have little trouble determining the identity of each leaf with a field guide.

## TEACHER DEMONSTRATION

Collect some easily obtained specimens and/or pictures of organisms from each of the five kingdoms. If possible find photographs of monerans that show their cell structures. Place the specimens or photographs at several stations around the classroom, and label those that might be unfamiliar to students. Allow ample time for students to move from station to station to observe the displays. As they do so, have them collect this information about each:

1. Is it multicellular or unicellular?
2. If unicellular, does it have a nucleus?
3. Can it move about?
4. Is it an autotroph or heterotroph?
5. To what kingdom does it belong?

When all students have returned to their seats, discuss their observations with them.

- **What kingdoms are made up of mostly or entirely multicellular organisms?** (Most plants and fungi and all animals are multicellular.)
- **How can you distinguish between the two kingdoms of unicellular organisms?** (Monerans lack a well defined nucleus, but a nucleus can usually be observed in the cells of protists.)
- **In what ways are plants and fungi alike? How are they different?** (Both plants and fungi are nonmotile, but fungi lack chlorophyll.)

## Reinforcement

Ask students to gather pictures from sources such as old nature magazines and science journals, and prepare bulletin board displays of organisms representing the five kingdoms. Suitable pictures of protists and monerans may be hard to find. If so, provide references such as biology textbooks from which drawings can be made. In addition to pictures, have students include brief descriptions of kingdom characteristics in the display.

## Section Review 6-3

1. Protist
2. Multicellular means many-celled. Unicellular means one-celled.

## LABORATORY ACTIVITY WHOSE SHOE IS THAT?

### BEFORE THE LAB

1. **Divide the class into groups of six students. Because of class size it may be necessary to set up some groups of five or seven rather than six.**
2. **If desks or tables in your classroom are too small, have the groups do the activity on the floor instead.**

### PRE-LAB DISCUSSION

Have students read the complete laboratory procedure. Discuss the procedure by asking questions similar to the following:

- **What is the purpose of this laboratory activity?** (To develop a system for classifying shoes.)
- **Name some characteristics that can be used to divide and subdivide shoes into groups.** (Answers will vary, but lead students to discuss appropriate characteristics. A brief list of appropriate grouping criteria might be placed on the chalkboard.)
- **How will we know when the activity is finished?** (At the end of the activity each person's shoe will be separate from all of the others.)

### SKILLS DEVELOPMENT

Students will use the following skills while completing this activity.

1. Classifying
2. Recording
3. Comparing
4. Inferring
5. Relating
6. Observing
7. Applying

## LABORATORY ACTIVITY

### *Whose Shoe Is That?*

**Purpose**

In this activity, you will develop a classification system for shoes.

**Materials** *(per group of 6)*
Students' shoes
Pencil and paper

**Procedure**

1. At your teacher's direction, remove your right shoe and place it on a work table.
2. As a group, think of a characteristic that will divide all six shoes into two kingdoms. For example, you may first divide the shoes by the characteristic of color into the brown shoe kingdom and the nonbrown-shoe kingdom.
3. Place the shoes into two separate piles based on the characteristic your group has selected.
4. Next, working only with those shoes in one kingdom, divide that kingdom into two groups based on a new characteristic. The brown shoe kingdom, for example, may be divided into a shoelace group and a non-shoelace group.
5. Further divide these groups into subgroups. For example, the shoes in the shoelace group may be again separated into shoes with rubber soles and shoes without rubber soles.
6. Continue to divide the shoes by choosing new characteristics until you have only one shoe left in each group. Identify the person who owns this shoe.
7. Now repeat this process working with the nonbrown shoes.

**Observations and Conclusions**

1. Test your shoe classification system for accuracy by doing the following: Have a member of your group hold up a shoe. Starting with the kingdom, classify it according to the system you have just developed. The system should lead you to the correct person's name.

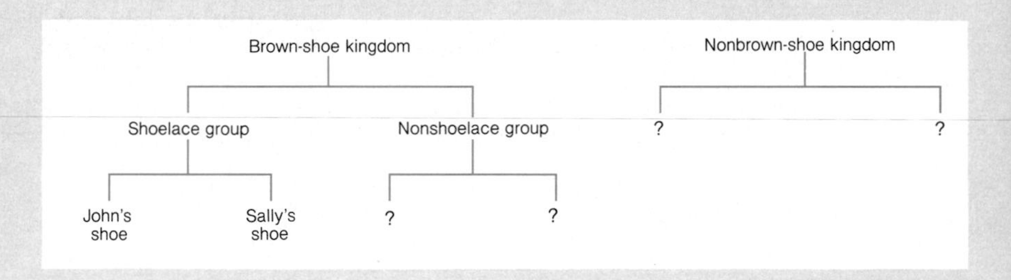


142

### TEACHING STRATEGY FOR LAB PROCEDURE

**1.** As students work, circulate among the groups to ensure that they stay on task. Make it clear that any horseplay, such as throwing or hiding shoes, will not be tolerated.
**2.** As students develop their classification system, have them write out a branching chart similar to that shown on page 142. In some classes it might be desirable to assign one student to serve as recorder to prepare one branching chart for the entire group.
**3.** At the conclusion of the activity have each group copy its branching chart on the chalkboard. Discuss similarities and differences in the various classification systems.

# CHAPTER REVIEW

## SUMMARY


### 6-1 History of Classification

- Classification is the grouping of living things according to similar characteristics.
- The science of classification is known as taxonomy.
- Aristotle was the first person to develop a classification system for living things. His system was used for almost 2000 years.
- In the seventeenth century, John Ray, an English biologist, introduced a classification system that became the foundation of modern systems.
- In the eighteenth century, Carolus Linnaeus, a Swedish scientist, devised a system for naming organisms called binomial nomenclature. Linnaeus also developed a classification system based on the structural similarities of organisms.

### 6-2 Modern Classification Systems

- Today, most scientists use a system of classification that includes five kingdoms. These kingdoms are plants, animals, protists, and fungi.
- Organisms are classified into seven groups. In order of decreasing size, these groups are kingdom, phylum, class, order, family, genus, and species.

### 6-3 The Five Kingdoms

- A five-kingdom system of classification consists of plants, animals, protists, monerans, and fungi.
- Autotrophs are organisms that are able to make their own food.
- Heterotrophs are organisms that cannot make their own food.


## VOCABULARY

*Define each term in a complete sentence.*

**autotroph**
**binomial nomenclature**
**class**
**family**
**genus**
**heterotroph**
**kingdom**
**multicellular**
**order**
**phylum**
**species**
**taxonomy**
**unicellular**

## CONTENT REVIEW: MULTIPLE CHOICE

*Choose the letter of the answer that best completes each statement.*

1. The first person to propose a classification system was
   a. Ray. b. Linnaeus. c. Aristotle. d. Haeckel.
2. A group of organisms that are able to interbreed, or produce young, is a
   a. species. b. family. c. genus. d. kingdom.
3. What language is used most often to name organisms in modern classification systems?
   a. French b. Latin c. English d. German



# GOING FURTHER: ENRICHMENT

**Part 1**

Assign the students to different groups, and have them repeat the activity. At the conclusion, compare the second classification system with the first. Establish the idea that even though all groups were working on the same task, many classification systems can result if different criteria are used when grouping and regrouping the shoes.

**Part 2**

Have students make up a branching chart to classify six or so things or organisms with which they are familiar. For example, have them try to classify:

a. their classmates
b. pets
c. band or orchestra instruments
d. different kinds of balls

# OBSERVATIONS AND CONCLUSIONS

**1.** If a proper and complete classification system has been devised, any given shoe out of the group of six should be uniquely identifiable.

## CHAPTER REVIEW

### MULTIPLE CHOICE

**1.** c **2.** a **3.** b **4.** d **5.** c **6.** d **7.** c **8.** b **9.** a **10.** c

### COMPLETION

**1.** taxonomy
**2.** binominal nomenclature
**3.** genus
**4.** kingdom
**5.** species
**6.** phylum
**7.** Carnivora
**8.** protist
**9.** autotroph
**10.** heterotroph

### TRUE OR FALSE

**1.** F Ray
**2.** F species
**3.** F genus, species
**4.** F
**5.** F Carnivora
**6.** F unicellular
**7.** F monerans
**8.** F autotrophs
**9.** F Plant
**10.** T

### SKILL BUILDING

**1.** Comparisons should be as follows: kingdom: continent; phyla: country; class: county; order: city; family: street; genus: house; species: number.
**2.** Check systems to ensure that they are logical and that each item in the closet can be separated from all other items by following down the classification system
**3.** a. plant, b. animal, c. fungi, d. fungi, e. animal, f. monera, g. plant, h. monera, i. fungi, j. protist.
**4.** a. A sea lion is not a lion, but does live in the sea, b. a starfish is not a fish, c. a horse chestnut is a plant seed not a horse, d. a sea horse is a fish not a horse, e. a jellyfish is not a fish nor is it composed of jelly, f. reindeer moss is a plant not a reindeer, g. a sea cucumber is an animal, not a plant, h. a horseshoe crab is not a crab.
**5.** The *Pinus nigra* and *Pinus strobus* are more closely related since they are in the same genus and are both pine trees.
**6.** A cocker spaniel and a poodle are among the same species since they can interbreed and produce fertile young. A fox and a wolf cannot interbreed and produce fertile young.
**7.** With the advent of the microscope, scientists discovered microscopic organisms that did not seem to fit into the Plant or Animal kingdom. This invention, then, helped spur on the development of the Protist Kingdom.
**8.** Such a system would not help delineate all living things from one another because many livings things live in similar areas and eat similar food substances. For example, both cows and horses eat grass and may live in pastures, but they are different organisms.

### ESSAY

**1.** To organize all living things into one classification scheme and to be able to communicate to different scientists by using a universal language; also, to help determine where new organisms that are discovered fit into the scheme of living things and to better understand the relationship among living things.
**2.** Similarities in structure and form
**3.** Both systems use binomial nomen-

**4.** The largest classification group is the
a. species. b. order. c. phylum. d. kingdom.
**5.** A classification group that is smaller than an order is
a. kingdom. b. phylum. c. family. d. class.
**6.** A classification group that is larger than an order is
a. genus. b. species. c. family. d. class.
**7.** A genus can be divided into
a. phyla. b. orders. c. species. d. families.
**8.** The term "Protista" refers to a
a. class. b. kingdom. c. genus. d. species.
**9.** Which organism is a heterotroph?
a. frog b. maple tree c. seaweed d. spinach
**10.** The animal kingdom is made up of only
a. autotrophs. b. unicellular organisms.
c. multicellular organisms. d. families.

#### CONTENT REVIEW: COMPLETION

*Fill in the word or words that best complete each statement.*

**1.** The science of classification is known as ______________.
**2.** In classification, the two-word naming system is known as ______________.
**3.** The first word in a scientific name is the ______________.
**4.** In classification, the largest group is called a(n) ______________.
**5.** The smallest classification group is a(n) ______________.
**6.** The second largest classification group is a(n) ______________.
**7.** ______________ is the Latin word for flesh eater.
**8.** A paramecium can be classified as a member of the ______________ kingdom.
**9.** An organism that can make its own food is called a(n) ______________.
**10.** An organism that cannot make its own food is called a(n) ______________.

#### CONTENT REVIEW: TRUE OR FALSE

*Determine whether each statement is true or false. If it is true, write "true." If it is false, change the underlined word or words to make the statement true.*

**1.** Aristotle introduced the term "species."
**2.** In Linnaeus's classification system, the smallest group was the genus.
**3.** The correct classification order from largest to smallest groups is kingdom, phylum, class, order, family, species, genus.
**4.** The scientific name for a lion is *Panthera leo*.
**5.** *Felis* is the Latin word for flesh eater.
**6.** Multicellular organisms are composed of only one cell.
**7.** In a five-kingdom classification system, bacteria are classified as plants.
**8.** Green plants are heterotrophs.
**9.** Animal cells contain chlorophyll.
**10.** Heterotrophs are organisms that cannot make their own food.

## CONCEPT REVIEW: SKILL BUILDING

*Use the skills you have developed in the chapter to complete each activity.*

1. **Making comparisons** Use the following words to classify the place in which you live: street, county, continent, city, number, country, house. Then compare each word with one of the seven groups used to classify organisms.
2. **Developing a model** Design a classification system for objects that might be found in your closet. Then draw a diagram that will illustrate your classification system.
3. **Making charts** Create a chart in which you classify each of the following into its correct kingdom:
   a. tulip
   b. wasp
   c. bread mold
   d. straw mushroom
   e. zebra
   f. blue-green alga
   g. oak tree
   h. bacterium
   i. toadstool
   j. paramecium
4. **Making generalizations** Explain why the common name for each of these organisms may be confusing:
   a. sea lion
   b. starfish
   c. horse chestnut
   d. sea horse
   e. jellyfish
   f. reindeer moss
   g. sea cucumber
   h. horseshoe crab
5. **Identifying relationships** Which two of the following three organisms are most closely related: *Morus nigra, Pinus nigra, Pinus strobus?* Explain your answer.
6. **Relating concepts** Explain why a cocker spaniel and a poodle can interbreed and produce young, while a fox and a wolf cannot.
7. **Relating cause and effect** How do you think the invention of the compound microscope affected the classification of living things?
8. **Applying concepts** Why is it that scientists do not classify animals by what they eat or where they live?

## CONCEPT REVIEW: ESSAY

*Discuss each of the following in a brief paragraph.*

1. Explain why it is important for scientists to classify organisms.
2. What type of characteristics did Linnaeus use to develop his classification system of living things?
3. How is the classification system used by scientists today different from the classification system developed by Linnaeus? How is it similar?
4. Describe each of the kingdoms used in the five-kingdom classification system. Give an example of an organism in each kingdom.
5. How do an autotroph and a heterotroph differ? Give an example of each.
6. List the seven major groups in scientific classification in order from the largest to the smallest.



clature. However, scientists today classify organisms by more than structure and form. Also, today scientists group organisms into five kingdoms.
**4.** Plant: most are multicellular and autotrophic. Animal: all multicellular and heterotrophic. Protist: all unicellular with well-defined nuclei. Monera: all unicellular without nucleus. Fungi: similar to plants but do not contain chlorophyll so they are not autotrophic. Examples include: animal (lion), plant (tulip), protist (paramecium), monera (bacteria), fungi (mushroom).
**5.** An autotroph can make its own food, usually by using the energy from the sun in a process called photosynthesis. A heterotroph cannot make its own food and must eat plants or planteaters. Examples of autotrophs are a tulip and a maple tree. Examples of heterotrophs are a whale and a grasshopper.
**6.** Kingdom, phylum, class, order, family, genus, species.

## ADDITIONAL QUESTIONS AND TOPIC SUGGESTIONS

**1.** Why was the binomial nomenclature system developed by Linnaeus an improvement over earlier systems for naming organisms? (Before Linnaeus, five or six Latin names were used to describe an organism. Linnaeus devised a simpler system in which all organisms were given two names.)
**2.** Which of these bears is most closely related? Explain your answer.
American black bear *(Ursus americanus)*
Asian black bear *(Selenarctos thibetanus)*
Polar bear *(Ursus maritimus)*
(The American black bear and polar bear are most closely related since they are both members of the genus *Ursus.*)
**3.** Which kingdom best fits the following descriptions:
(a) Unicellular, no nucleus
(b) Multicellular, heterotroph, do not move about
(c) Multicellular, autotrophs, do not move about
(a—Monera, b—Fungi, c—Plant)

## ISSUES IN SCIENCE

The following issue can be used as a springboard for class discussion, or it can be given as a writing assignment.

Recently scientists have discovered that certain kinds of bacteria are quite different in their chemical makeup from other monerans. They have proposed that a sixth kingdom called *Archaebacteria* be created for these bacteria since they are so different. Other scientists oppose this idea. Some argue that even five kingdoms are too many. If we keep inventing new kingdoms, classification systems may become too complex. What is your opinion? Should a sixth kingdom be established?

## Chapter 7

# VIRUSES, BACTERIA, AND PROTISTS

## CHAPTER OVERVIEW

When viruses invade a host's cell, they first attach themselves to the exterior cell membrane. Then, the genetic material of the virus—either DNA or RNA—is injected into the host cell. The outer coat, made up of protein, is left behind. Viruses use the genetic machinery and supplies in the host cell to cause the host cell to construct more viruses. At some time, the host cell bursts and the viruses are freed to invade other nearby cells.

Bacteria, members of the Monera Kingdom, are definitely living organisms. There are three basic shapes of bacteria: rod-shaped bacilli, sphere-shaped cocci, and spiral-shaped spirilla. Bacteria need food, air, and water for survival. Most bacteria are heterotrophs. A few bacteria can make their own food and are autotrophs, but most are not capable of photosynthesis.

Protozoans constitute the Protista Kingdom. Protists differ from members of the Monera Kingdom by the absence of a cell wall and the presence of a distinct cell nucleus. All are unicellular or colonial. The protists are classified further into four groups—amoebas, ciliates, flagellates, and sporozoans—according to structures by which they move from place to place.

## INTRODUCING CHAPTER 7

Have students observe the photograph on page 146. Because of the photograph's surrealistic qualities, some students may offer imaginative descriptions of what they think it represents. Tell them the photograph shows crystals of erythromycin, an antibiotic, that have been magnified thousands of times. Explain that antibiotics are chemicals that are used to inhibit, or stop, the growth of harmful bacteria.

- **List some other antibiotics.** (penicillin, streptomycin, ampicillin, neomycin, tetracycline)

Point out that antibiotics are produced by living microorganisms, and are capable of destroying other microorganisms or preventing their reproduction. The term antibiotic is derived from the Greek words meaning "against life."

- **What kinds of organisms produce antibiotics?** (Most antibiotics are produced by molds and bacteria.)

Now have students read the chapter introduction on page 147. Explain that Legionnaires' disease was not recognized at first as a new disease because its symptoms were similar to those of pneumonia—high fever, chills, headache, muscle pain, lung congestion. Eventually, 29 people who attended the convention and

# 7 Viruses, Bacteria, and Protists

**CHAPTER SECTIONS**

**7-1 Viruses**
**7-2 Bacteria**
**7-3 Protozoans**

**CHAPTER OBJECTIVES**

*After completing this chapter, you will be able to:*

**7-1** Describe the structure of virus.
**7-2** Classify bacteria according to shape.
**7-2** Describe the structure of a bacterial cell.
**7-3** Define protozoan and describe the structure and behavior of three typical protozoans.
**7-3** Describe a sporozoan.

In July of 1976, the city of Philadelphia was full of people wearing navy blue military caps. These people were members of the American Legion, a veterans' organization that was holding its yearly convention in what is known as "the city of brotherly love." However, before the month was out, Philadelphia might well have been called "the city of brotherly fear." For shortly after the close of the convention, a serious disease struck many of the legionnaires.

What caused the frightening illness, which was given the name "Legionnaires' Disease"? One possible cause was a microorganism, a living thing too small to be seen without the aid of instruments such as microscopes. But what particular microorganism? There were millions.

Dr. Joseph McDade of the Center for Disease Control at Atlanta, Georgia, was one of the medical detectives trying to solve the mystery of Legionnaires' Disease. If a microorganism was responsible for Legionnaires' Disease, McDade would find it. Early in 1977 he discovered the microorganism responsible.

The microorganism McDade identified was a new bacterium, a kind of germ. Fortunately, the new bacterium could be treated with an old kind of medicine called erythromycin, which adds another twist to this strange story. Erythromycin is made by bacteria. So the illness caused by one kind of bacteria was cured by a substance produced by another kind of bacteria.

*Erythromycin crystals*



developed the disease died. Investigators discovered the cause of the disease, an unknown bacterium, living in the water that is used in the air-conditioning system in the hotel in which the convention was held. Since 1977, several other outbreaks of Legionnaires' disease have been reported in the United States and Europe.

• **What kind of organism produces erythromycin?** (bacteria)

Point out that erythromycin, a product of bacteria that live in the soil, is used to treat scarlet fever and some kinds of pneumonia. Mention that this chapter will explain how bacteria and other kinds of microorganisms can be both beneficial and harmful.

## TEACHER DEMONSTRATION

**1.** Obtain four disposable plastic petri dishes that contain nutrient agar.
**2.** Using a source of nonpathogenic bacteria, such as yogurt or buttermilk, dip a sterile cotton swab into the bacteria source, open the top of one petri dish, and gently rub the swab over the entire surface of the agar. Repeat this step, using a new sterile cotton swab for each of the remaining petri dishes.
**3.** Number the petri dishes.
**4.** Cut out a few disks of filter paper that are 1 cm in diameter.
**5.** Dissolve antibiotic tablets or capsules in water. Soak the disks in the antibiotic solution.
**6.** Using forceps that have been dipped in alcohol, pick up one antibiotic disk, open petri dish 1, and place the disk in the center of the petri dish. Repeat this step with petri dish 2.
**7.** Set petri dishes 3 and 4 aside. They will act as the controls.
**8.** Place all of the petri dishes in a warm, dark place for a few days.
**9.** After a few days, examine the petri dishes for evidence of bacterial growth. If bacteria are present, they will appear as white dots or clumps.

• **Where did the bacteria that are growing in the petri dishes obtain their food and water?** (They obtained it from the nutrient agar.)
• **What evidence is there that the antibiotics stopped the growth of bacteria?** (the clear zone around each antibiotic disk)
• **What was the purpose of petri dishes 3 and 4?** (They served as controls.)

## TEACHER RESOURCES

**Audiovisuals**

*Germs and Your Body,* film, Cor
*Infectious Diseases and Man-Made Defenses,* film, Cor
*The Protist Kingdom,* film, BFA

**Books**

Dixon, *Magnificent Microbes,* Atheneum
Jahn, T. L., *How to Know the Protozoa,* Brown

**Software**

*Body Defenses,* Prentice Hall
*Agents of Infection,* Prentice Hall

# 7-1 VIRUSES

## SECTION PREVIEW 7-1

Viruses are tiny crystalline structures that are composed of a single protein and a core of nucleic acid (either DNA or RNA). Although it is open to debate, many scientists feel that viruses are not living things. The reason for this is that viruses lack many of the features of living cells.

Viruses were discovered in the late nineteenth century, when it was found that the infectious substance that causes tobacco mosaic disease were too small to be seen with a light microscope and would pass through a filter that stopped all known bacteria. Furthermore, it was discovered that viruses reproduced only inside living cells. Once inside a host cell, the virus takes over the host cell's control mechanism, and the host cell then produces the virus's proteins and nucleic acids.

## SECTION OBJECTIVES 7-1

1. **Describe the events leading up to the discovery of the tobacco mosaic virus.**
2. **Compare a virus and a living cell.**
3. **Identify the parts of a virus.**
4. **Describe the sequence of events in the reproduction of a bacteriophage.**

## SCIENCE TERMS 7-1

**bacterium p. 148**
**microbiology p. 149**
**virus p. 149**
**DNA p. 150**
**RNA p. 151**
**bacteriophage p. 151**

## 7-1 Viruses

In 1892, a terrible disease swept through certain areas of Russia. Victims of the disease were not people or even animals. They were plants—tobacco plants. Because tobacco was a very important crop, many people were interested in discovering the cause and, perhaps, a cure for the disease. One of these people was a 28-year-old Russian botanist, or plant expert, named Dimitri Iwanowski.

By the late 1800s, scientists had discovered that many diseases of human beings, animals, and plants were caused by unicellular microorganisms called **bacteria** (singular: bacterium). These discoveries were made possible by hard work and the invention of powerful light microscopes. These microscopes made visible microorganisms never seen before. Based on these discoveries Iwanowski assumed the tobacco disease was also caused by bacteria. So he set out to devise an experiment to identify them.

Iwanowski's procedure was simple. He would gather some infected tobacco leaves, crush the leaves, and collect the juice. Then he would pass the juices (1) through a filter whose holes were so small that bacteria could not slip through. Finally, Iwanowski would place the pure juices on healthy tobacco plants. He

**Figure 7-1** *Tobacco mosaic disease causes healthy tobacco plant leaves* (left) *to become spotted* (right).

148

## TEACHING STRATEGY 7-1

### Motivation

Introduce this section by talking with students to determine what they already may know about viruses and some of the diseases they cause.

• **What do you think of when you hear the word virus?** (Accept all answers. Most students probably will see viruses in an unfavorable light because they cause disease.)

• **What are some diseases caused by viruses?** (Accept all answers. At this time, students will probably mention some nonviral diseases.)

Some of human diseases that are caused by viruses include the common cold, influenza, poliomyelitis, measles, mumps, chicken pox, hepatitis, genital herpes, and AIDS. There is also evidence that certain types of cancer may be caused by viruses.

• **Are viruses living?** (Answers will vary with student's previous knowledge of viruses. Do not attempt to answer the question at this time.)

Inform students that as they read Section 7-1, they will learn what a virus is. They will also learn whether viruses are living and how they reproduce.

### Content Development

Be sure students understand that be-

reasoned the plants would remain healthy and this would be proof that bacteria were the cause of the disease.

All this reasoning made very good sense. And Iwanowski performed his experiment perfectly. The only problem was that the healthy plants caught the disease. Iwanowski thought his experiment was a failure. He thought his filters were not well made and that cracks in them must have let through the disease-causing bacteria. However, Iwanowski was wrong. He had made a great discovery without recognizing it. The recognition came six years later from Delft, Holland, about 1600 kilometers to the east.

### *Viruses—Smaller than Bacteria*

In 1898, the Dutch microbiologist Martinus Beijerinck, was also working on the devastating disease of tobacco plants. A microbiologist is a scientist who studies **microbiology,** or the science of microorganisms. Using the best microscopes, Beijerinck searched for bacteria in the diseased, spotted leaves of tobacco plants. He also searched in their filtered juices and in the material trapped by filters that should have contained bacteria. Like Iwanowski's, Beijerinck's experiments seemed to end in failure. He could find no bacteria. Nevertheless, the filtered juices from diseased plants continued to cause healthy plants to become infected.

These repeated observations led Beijerinck to draw a momentous conclusion that had escaped Iwanowski. The filtered juices from diseased plants caused infection, but the juices did not contain bacteria. Therefore, they contained something else that caused infection, a germ much smaller than bacteria. This germ was too small to be visible even under the lenses of the most powerful light microscope. Beijerinck named this germ a **virus,** which comes from the Latin word for poison. ❷

Almost 40 years passed before Wendell Stanley, an American scientist, isolated the virus that caused the disease in tobacco leaves. He called it the tobacco mosaic virus. The word "mosaic" refers to the pattern of spots found on the leaves of diseased plants. See Figure 7-1.

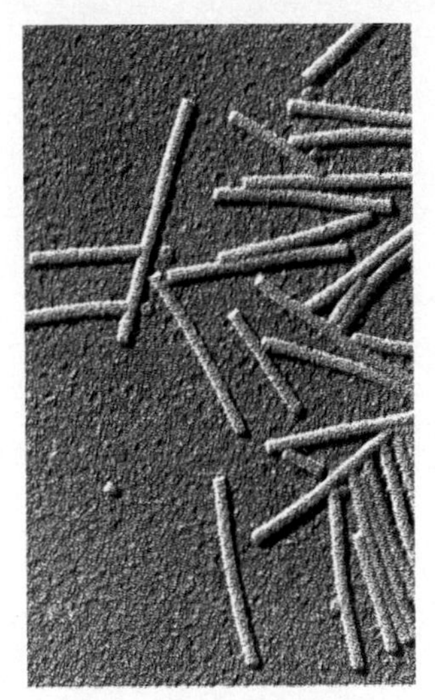

**Figure 7-2** *These rod-shaped objects are tobacco mosaic viruses.*



## HISTORICAL NOTES

In 1935, Wendell Stanley was the first to isolate a virus, which caused tobacco mosaic disease. To obtain an extract containing the tobacco mosaic virus, Stanley ground up over 1000 kilograms of infected tobacco leaves. From this extract, he was able to obtain crystals of the disease-causing virus. When the crystals were placed in water and then spread on healthy tobacco leaves, the plants became infected with the tobacco mosaic disease. For his work, Stanley was awarded the Nobel Prize in Chemistry in 1946.

## ANNOTATION KEY

❶ **Thinking Skill: Problem solving**

❷ **Thinking Skill: Relating cause and effect**

cause of their small size, viruses cannot be seen under a light microscope. For this reason, their structure can be studied only with an electron microscope. Point out that the viruses shown in Figure 7-2 were photographed with an electron microscope.

## Content Development

The word *virus* comes from the Greek meaning "poison." Ask students to recall their most recent bout with a virus, particularly one involving stomach upset—an unpleasant recollection but an easy way to remember word-root association.

## Skills Development

***Skill: Sequencing events***

After students have read about the discovery of viruses, place these dates in a column on the chalkboard: 1892, 1898, 1935. In a second column, alphabetically list these scientists: Martinus Beijerinck, Dimitri Iwanowski, Wendell Stanley. In a third column, list the following events in this order: American scientist who isolated a virus that caused tobacco mosaic disease in plants; Dutch scientist who concluded that germs smaller than bacteria caused tobacco mosaic disease in plants; Russian scientist who tried to prove that bacteria caused the tobacco mosaic disease in plants.

Have students arrange the dates, scientists, and events in the correct sequence. Discuss the significance of each scientist's contribution to the discovery of viruses.

- **Which scientist made a great discovery without realizing it?** (Dimitri Iwanowski did not realize that tobacco mosaic disease was caused by germs too small to be trapped by his filters.)
- **Which scientist first used the word virus?** (The word was first used by Martinus Beijerinck.)
- **Which scientist named the germ that infected tobacco leaves?** (Wendell Stanley)

## ANNOTATION KEY

❶ Thinking Skill: Applying concepts

❷ Thinking Skill: Applying technology

### Activity

**Building a Bacteriophage**
**Skills: Developing a model, making diagrams, applying, relating**
**Level: Remedial**
**Type: Hands-on**
**Materials: Pipe cleaners, construction paper, screws, nuts, bolts, scissors, tape, glue, crayons, screw driver, floral wire**

Students should be able to identify the virus's core, containing the hereditary material, and the tail fibers as the part that the bacteriophage uses to attach itself to a bacterial cell. This activity helps student identify the parts of a bacteriophage. Check student models and diagrams for accuracy. Use the best and most creative models in a classroom display.

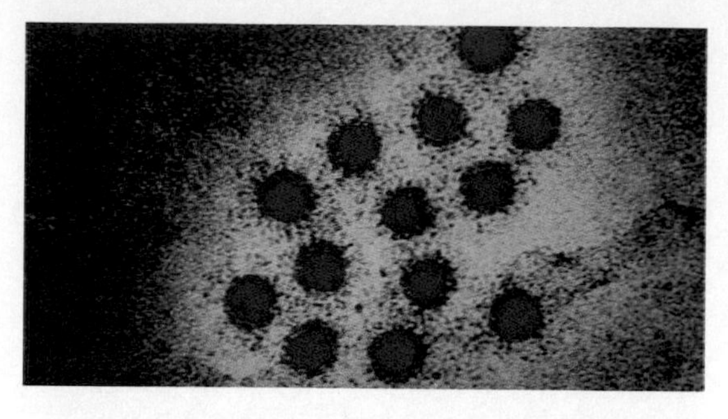

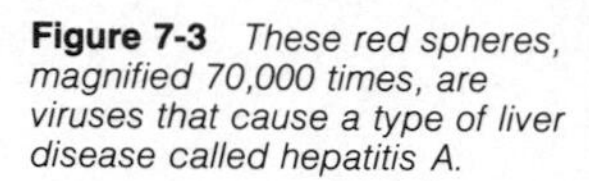

**Figure 7-3** *These red spheres, magnified 70,000 times, are viruses that cause a type of liver disease called hepatitis A.*

Since the discovery of the tobacco mosaic virus, scientists have found that there are many different kinds of viruses. By studying many of these viruses, scientists have learned a great deal about the function and structure of viruses.

### *What Is a Virus?*

Living things come in many shapes and sizes. However, all living things seem to have one thing in common. They all contain cells. Some organisms consist of only a single cell. Other organisms, such as yourself, are made up of many cells. Unlike these organisms, viruses are not cells and do not contain cells. Viruses are tiny particles that contain hereditary material.

Because viruses are not cells, they cannot perform the life functions of living cells. They cannot, for example, take in food or get rid of wastes. In fact, about the only similarity that viruses share with cells is that viruses are able to reproduce. However, viruses cannot reproduce on their own. They need the help of other living cells. For this reason, many scientists consider viruses an unusual form of life. Other scientists strongly disagree and do not classify viruses as living things. Thus, it might help if you think of viruses as being on the threshold of life.

#### Activity

*Building a Bacteriophage*

You can better understand the structure of a bacteriophage by building your own model.

**1.** Use the diagram on page 151 and any of the following materials to build your own model bacteriophage: pipe cleaners, construction paper, screws, nuts, bolts, scissors, tape, glue, crayons, screw driver.

**2.** Make a drawing of your model and label the parts. Which part contains the hereditary material of the bacteriophage? Which part would the bacteriophage use to attach itself to a bacterial cell?

### *Structure of Viruses*

❶ **A virus has two basic parts, a core of hereditary material and an outer coat of protein.** The hereditary material in the virus's core may be either **DNA,**



## 7-1 (continued)

### Motivation

Give each student a peanut in a shell.

• **In what way is the structure of a virus like the structure of a peanut?** (Accept all logical answers.)

• **What part of a virus is similar to the part of the peanut that you eat?** (the hereditary material in the virus's core)

• **What kinds of hereditary material make up the core of the virus?** (DNA or RNA)

• **What is the function of the hereditary material?** (It controls the production of new viruses.)

• **What virus structure is similar to the shell of the peanut?** (the protein coat surrounding the viral nucleic acid)

### Content Development

Explain that because scientists do not agree on whether viruses are living things, they are not classified as living things are. Traditionally, viruses have been classified according to size, shape, and the host they infect. Generally, the outer protein coats of viruses may be manysided, spiral-shaped, or a combination of these two shapes. Viruses are also classified according to whether they infect the cells of plants, bacteria, or animals including humans. Recently, scientists have been using chemical and structural characteristics to classify viruses. For example, those viruses that contain DNA are placed in one group and those containing RNA in another. Structural characteristics are then used to subdivide them into smaller groups.

*deoxyribonucleic acid,* or **RNA,** *ribonucleic acid.* DNA and RNA are substances that control the production of new viruses.

Surrounding the DNA or RNA in the virus is a protein coat. The coat is much like the shell of a turtle. Like the turtle's shell, the protein coat encloses and offers some protection to the virus. In fact, the protein coat is so protective that some viruses survive after being dried and frozen for many years.

With the invention of the electron microscope in the 1930s, scientists were able to see and to study the shapes and sizes of certain viruses. Some viruses, such as those that cause the common cold, have 20 surfaces. Each of these surfaces is in the shape of a triangle that has equal sides. Other viruses look like small threads, while still others resemble small spheres. Some even resemble small spaceships. See Figure 7-5.

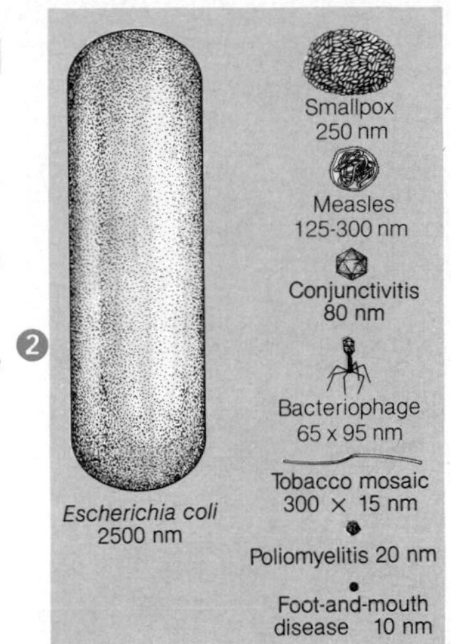


**Figure 7-4** *Viruses vary in size from about 10 to 250 nanometers, or nm. A nanometer is equal to one millionth of a millimeter. Compare the sizes of the viruses on the right to the large bacterium on the left.*

### Reproduction of Viruses

In order to understand how a virus reproduces, causing disease, you can examine the activities of one kind of virus called a **bacteriophage** (bak-TEE-ree-uh-fayj). A bacteriophage is a virus that infects bacterial cells. The word "bacteriophage" means "bacteria eater."

**Figure 7-5** *This drawing shows the structure of a bacteriophage, which attacks bacterial cells. The photograph is of the same virus magnified 190,000 times.*

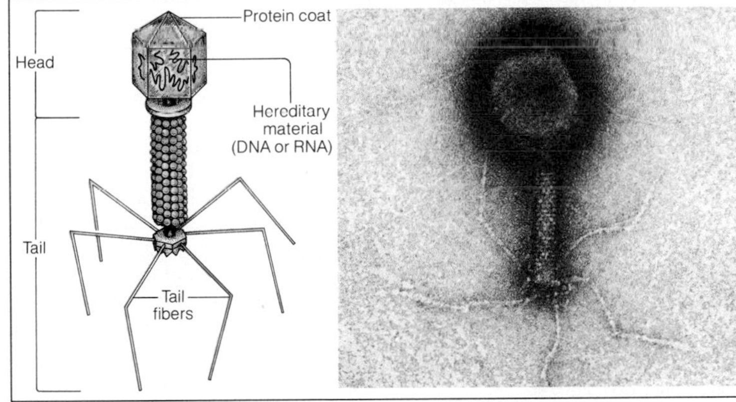




## TEACHER DEMONSTRATION

Have students study Figure 7-4 and read the caption.

- **What unit is used to measure the size of viruses?** (Viruses are measured in nanometers [nm].)

Arrange students in groups of four. Distribute metersticks or metric rules to each group.

- **What are the smallest markings on the meterstick (metric ruler) called?** (The smallest markings are millimeters.)
- **How many nanometers are there in one millimeter?** (1 mm = 1,000,000 nm.)
- **How many smallpox viruses could be placed side-by-side on a line that measured 1,000,000 nm?** (1,000,000 nm ÷ 250 nm = 4000. Approximately 4000 smallpox viruses could be placed side-by-side on a line that measured 1 mm.)

Now show the class a penny and tell them to imagine that it is a smallpox virus. Point out that a distance of 80 meters would be needed to place 4000 pennies side-by-side on a straight line. To help students conceptualize this size, take them to a location such as the schoolyard football field, or sidewalk where they can measure a distance of 80 m. Place a penny on the ground or paved surface and have students use metersticks or metric tape to measure 80 m.

- **If this penny represents a 250 nm smallpox virus, what microscopic distance is represented by 80 m?** (The 80 m distance represents 1 mm or 1,000,000 nm.)

Conclude by mentioning that a smallpox virus is one of the larger viruses. Have students refer to Figure 7-4 to see viruses that are smaller in size than the smallpox virus.

## Content Development

Review the basic parts of the virus as presented in the text. Make sure students can use Figure 7-5 to describe the outer protein coat and inner core of a typical virus. Point out that the virus in this illustration is called a bacteriophage because it infects bacteria. Many students may be surprised to learn that even bacteria can "get sick" due to a viral infection.

## Skills Development

***Skill: Making calculations***

Have students determine the relative sizes of various viruses as compared to the bacterial cell *(Escherichia coli)* shown in Figure 7-4.

## Reinforcement

Review some basic ideas about nucleic acids, DNA and RNA, and the mechanism viruses use in the process of replication.

## FACTS AND FIGURES

In the 1960s and 1970s, several kinds of infectious particles even smaller than viruses were discovered. These particles are called viroids and consist of short, single strands of RNA without any protective protein coat.

## 7-2 BACTERIA

### SECTION PREVIEW 7-2

Bacteria, the simplest of living organisms, are classified in the Kingdom Monera. Bacteria are unicellular organisms that are classified into three basic groups, according to their shape. As monerans, the cells of bacteria have their hereditary material spread throughout the cytoplasm rather than enclosed in a nucleus. Like plants, bacteria have cell walls that form the cells' outermost boundaries. Although some bacteria have flagella that aid in locomotion, many are unable to move on their own.

Although bacteria are simple organisms, they carry on the same life processes as more advanced organisms. Most bacteria are heterotrophs that feed on dead things. Some bacteria are autotrophs that use light energy or chemical energy to produce food. Binary fission, a type of cell division, is the most common means of bacterial reproduction. If environmental conditions become unfavorable, some bacteria produce a structure called an endospore, which protects the bacteria until conditions become favorable again.

### SECTION OBJECTIVES 7-2

1. **List the three basic shapes of bacteria.**
2. **Identify the major structures in a bacteria cell.**
3. **Describe the ways in which bacteria obtain their nutrition.**
4. **Describe the method by which bacteria reproduce.**
5. **List several ways in which bacteria can be helpful and in which they can be harmful.**

### SCIENCE TERMS 7-2

**cell wall p. 152**
**coccus p. 153**
**bacillus p. 153**
**spirillum p. 153**
**cell membrane p. 153**
**nucleus p. 153**
**cytoplasm p. 153**
**flagellum p. 154**
**parasite p. 154**
**saprophyte p. 155**
**endospore p. 155**
**binary fission p. 155**
**antibiotics p. 156**
**toxin p. 157**

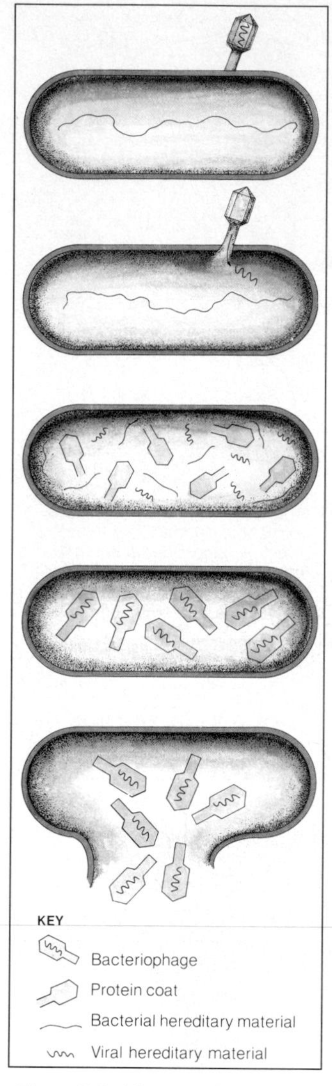


**Figure 7-6** *Viruses reproduce in cells. A virus attaches to the cell and injects its heredity material. The cell then makes more viruses, which burst from the cell.*

In Figure 7-6, you can see how a bacteriophage attaches its tail to the outside of a bacterial cell. The virus quickly injects its hereditary material directly into the living cell. The protein coat is left behind and discarded by the virus. Once inside the cell, the virus's hereditary material takes control of all of the bacterial cell's activities. As a result, the bacterial cell is no longer in control. The cell begins to produce new viruses rather than its own chemicals. (1)

Soon the bacterial cell fills up with new viruses, perhaps as many as several hundred. Eventually, the bacterial cell bursts open. The new viruses are released and infect nearby bacterial cells. This process continues until all living bacterial cells have been infected by the virus.

The viruses that attack plants, animals, and people may vary in size and shape. However, all viruses act in much the same way as bacteriophages.

#### *SECTION REVIEW*

1. What was the first virus to be isolated?
2. Name the two parts of a virus.
3. What is a bacteriophage?

## 7-2 Bacteria

Unlike viruses, bacteria clearly are living organisms. At one time, scientists placed bacteria in the plant kingdom because, like plants, bacteria have **cell walls.** The cell wall is the outermost boundary of plant and bacterial cells. The cell wall helps the cells of these organisms to keep their shape. Today, most classification systems consider bacteria as monerans. You may recall from Chapter 6 that monerans are members of the Monera kingdom. Like all other monerans, bacteria are unicellular. (2)

Bacteria are considered the simplest organisms. However, bacteria are more complex than they may appear. Each cell performs the same basic functions that more complex organisms, including you, perform.



### 7-1 (continued)

#### Skills Development

***Skill: Interpreting diagrams***

Have students refer to Figure 7-6.

• **What process is shown in this set of diagrams?** (the stages by which a bacteriophage takes over a bacterial cell causing it to produce more viruses)

Have students cut out 5 cm × 10 cm pieces of paper. Then have students number the papers from 1 to 5 so that they correspond to each step shown in Figure 7-6. After each number, have students describe, in one or two sentences, what is happening in each stage. When they have finished, call on students to read their descriptions.

### Section Review 7-1

1. Tobacco mosaic virus.
2. Core of hereditary material (DNA

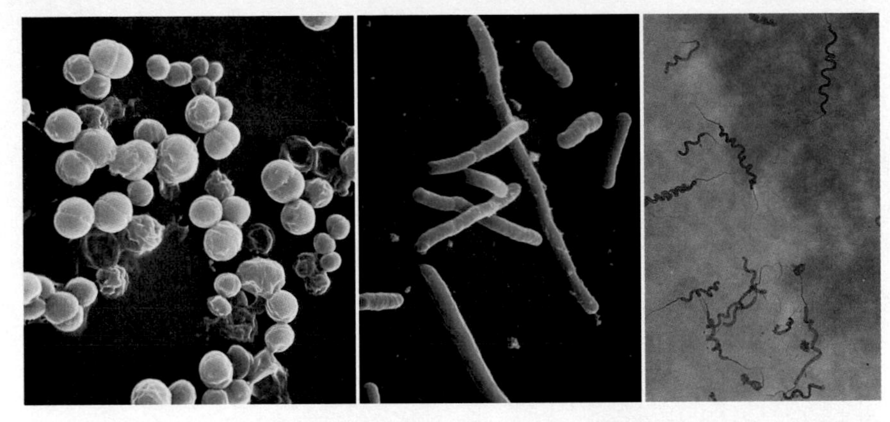

Figure 7-7 *Bacteria are grouped and named according to their three basic shapes. Spherical bacteria are called cocci (left), rodlike bacteria are called bacilli (center), and spiral-shaped bacteria are called spirilla (right).*

## ANNOTATION KEY

❶ Thinking Skill: Applying concepts
❷ Thinking Skill: Classifying organisms
❸ Thinking Skill: Comparing bacterial shapes

Bacteria are among the most numerous organisms on earth. Scientists estimate there are about 2.5 billion bacteria in a gram of garden soil. And the total number of bacteria living in your mouth is greater than the number of people who have ever lived.

**Bacteria are classified according to shape as either cocci, bacilli, or spirilla.** Bacteria that resemble spheres are called **cocci** (KAHK-sigh; singular: coccus). Bacteria that look like rods are called **bacilli** (buh-SIL-igh; singular: bacillus). And spiral-shaped bacteria are called **spirilla** (spigh-RIL-uh; singular: spirillum). See Figure 7-7. ❸

### Structure of Bacteria

All bacteria have a **cell membrane** on the inside of the cell wall. The cell membrane controls which substances enter and leave the bacterial cell. Unlike most other cells, the hereditary material of bacteria is not confined in a **nucleus** (NOO-klee-uhs; plural: nuclei). The nucleus is a spherical structure within a cell that directs all the activities of the cell. Instead, the hereditary material is spread throughout the **cytoplasm** in the cell. The cytoplasm is the jellylike material found inside the cell membrane.

Figure 7-8 *This drawing shows the typical structure of a bacterium.*

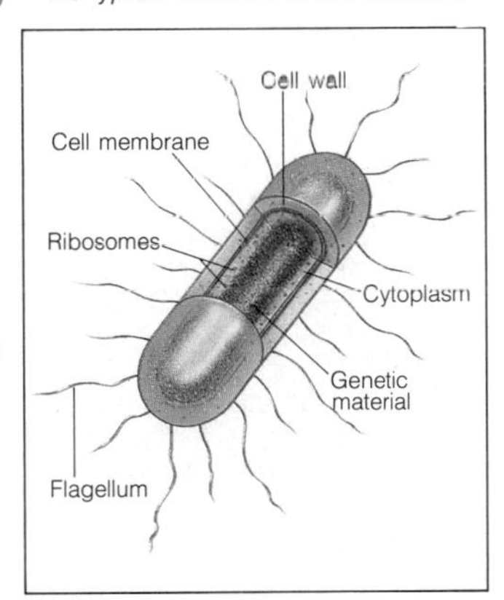




## Content Development

- **How are bacterial cells different from your cells?** (They have a cell wall.)
- **How is the division of labor in a bacterial cell different from your cells?** (Bacterial cells must perform all of the life functions necessary to stay alive. In more complex organisms, such as humans, the cells perform different functions for the whole organism.)

Point out that bacteria are found everywhere and in great numbers.

- **Do you think all bacteria cause disease?** (Students should infer that if that were the case, we would all be sick most of the time owing to the enormous numbers of bacteria in the world.)

## Content Development

Mention that in addition to the structures that are shown in Figure 7-8, some bacteria also have outer layers called capsules surrounding their cells. The capsule is made of a slimy, sticky material that allows the bacteria to stick to surfaces. The capsules of disease-causing bacteria may also protect them from being engulfed by white blood cells.

Also point out that some bacteria remain together to form pairs, chains, or clumps of cells after they reproduce. Those bacteria that form groups are called colonial forms.

- **Which of the bacteria shown in Figure 7-7 appear to form colonies?** (Cocci and bacilli form colonies.)

or RNA) and an outer coat of protein.
3. Virus that infects bacterial cells.

## TEACHING STRATEGY 7-2

### Motivation

Collect some common items that can be used to model the three basic shapes of bacteria. For example, to illustrate cocci, use beads, malted milk balls, or dried peas. Short pieces of dowel rods or frankfurters can be used to represent bacilli. To make a model of spirilla, twist a pipe cleaner into a coil around a pencil. Show each model, one at a time, and have students identify the type of bacteria that each model represents. As each type is identified, write both the singular and plural forms of the bacterium's name on the chalkboard. Call attention to the differences in the singular and plural spellings.

## BACKGROUND INFORMATION

The term endospore is derived from the spore's site of formation. It is formed inside the cell of certain bacilli. When conditions are unfavorable, the bacterial DNA replicates and splits into two parts. A thick coat then forms around one portion of the DNA and the rest of the bacterial cell breaks down, releasing the endospore. Because endospores can withstand extreme temperatures, chemicals, and drying, special methods must be used to kill them—especially in hospitals and laboratories. One method is to subject the endospores to pressurized steam at about 125°C for 15 to 45 minutes. This method is effective against most endospores.

### Activity

**Shapes of Bacteria**
**Skills: Classifying, observing, diagraming, comparing**
**Level: Average**
**Type: Hands-on**
**Materials: Prepared slides, microscope**

Students should be able to easily distinguish the three shapes of bacteria. Check their diagrams to ensure that they have identified the shapes of the bacteria correctly.

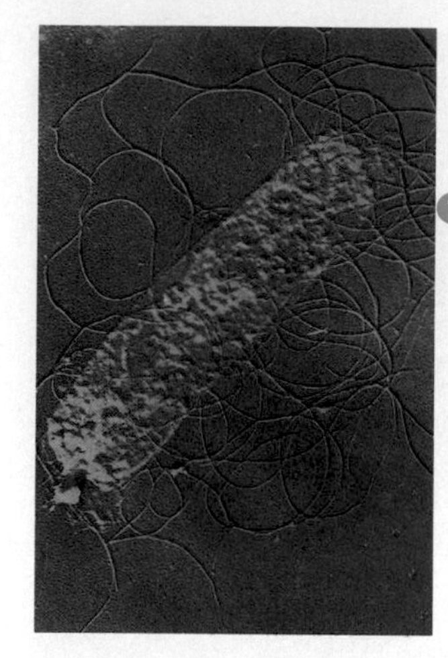

**Figure 7-9** *The bacterium,* Proteus vulgaris, *uses its many whiplike flagella to move through water.*

Many bacteria are not able to move on their own. However, they can be carried from one place to another by air and water currents, clothing, and other objects. Other bacteria have special structures that help them move in watery surroundings. One type ① of structure is a **flagellum** (fla-JEL-um; plural: flagella). A flagellum is a long, thin, whiplike structure that propels the bacterial cell. Some bacteria may have many flagella.

### *Life Functions of Bacteria*

Bacteria are in water, air, and the upper layers of soil. In fact, bacteria live almost everywhere, even in places where other living things cannot survive. For example, some bacteria were discovered living in volcanic vents at the bottom of the Pacific Ocean south of Baja, California. Temperatures here are as high as 250° C. Unlike most living things, some bacteria can thrive without oxygen. However, all bacteria need water or they eventually die.

② Most bacteria are heterotrophs. These bacteria feed on other living things or on dead things. Bacteria that feed on living organisms are called **parasites** (PA-ruh-sights). These bacteria cause infections in

**Figure 7-10** *Bacteria that live in Morning Glory Pool, a very hot spring in Yellowstone National Park, make up the blue-green growth.*

#### Activity

*Shapes of Bacteria*

How do the shapes of bacteria differ?

**1.** Obtain three prepared slides containing bacteria.

**2.** Observe each slide under low and high power of the microscope.

Draw and label a few cells from each slide. Identify each slide as containing the following bacteria shapes: cocci, bacilli, or spirilla.

154

## 7-2 (continued)

### Motivation

Begin by asking students what happens to the leaves when they rake them in the fall. Some will say they are placed in plastic bags and thrown away with the trash. Others might say that they are piled up in a far corner of the property. A few might even say they compost theirs!

- **In a forest, where do the leaves go?** (Some might say they blow away; others might say that they rot away.)
- **How do the leaves decompose?** (with aid of bacteria and fungi)
- **Why does a leaf decompose?** (As heterotrophs, bacteria and fungi use the dead leaves and other dead material as food and return many other chemical compounds to the soil.)
- **What would happen if all of the bacteria suddenly disappeared?** (We would be buried in the carcasses of dead plant and animal material.)

### Content Development

Mention that although some bacteria thrive at very high temperatures, others can live where temperatures are extremely low. Bacteria have been found living in Antarctica at −7°C. However, most bacteria grow best at temperatures between 20°C and 40°C.

### Motivation

Have students look up the words *binary* and *fission* in a dictionary. They will find that binary refers to some-

**Figure 7-11** *Saprophytic bacteria of decay break down material in a dead tree and return important substances to the soil.*

people, animals, and plants. Bacteria that feed on dead things are called **saprophytes** (SA-pruh-fights). Why are saprophytes among the most important organisms found on earth? ❶

Some bacteria are autotrophs. Like green plants, some of these food-making bacteria use the energy of sunlight to produce food. Other bacteria use the energy in substances such as sulfur and iron to make food.

Conditions may become unfavorable for bacteria. If the food, water, or air supply of bacteria becomes used up, some bacteria form microscopic **endospores.** An endospore is shaped like a ball or oval and is surrounded by a thick protective membrane. The protective membrane enables the endospores to survive long periods of boiling and disinfectant chemicals. When food, air, and water again become available, the endospores develop into active bacteria. ③

Most bacteria reproduce by **binary fission.** In binary fission a cell divides into two cells. Under the best conditions, most bacteria reproduce quickly. Some types can double in number every 20 minutes. At this rate, after about 24 hours the offspring of a single bacterium would have a mass greater than 2 million kilograms, or as much as 2000 small cars. In a few more days, their mass would be greater than that of the earth. Obviously, this does not happen. ④

**Figure 7-12** *These bacteria are splitting in two, a process of reproduction called binary fission. Which of the three basic groups of bacteria are reproducing here?* ❷

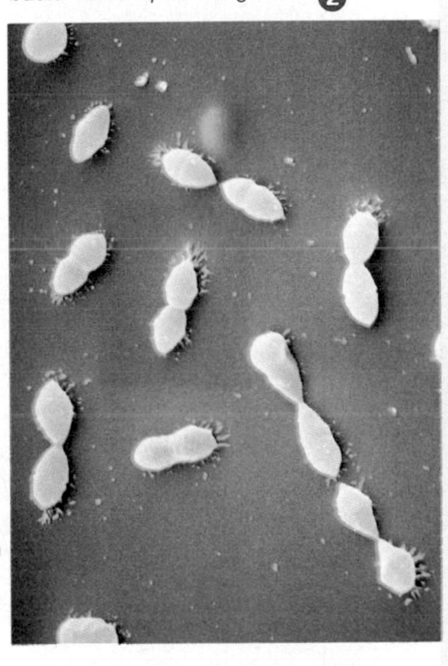

155

## ANNOTATION KEY

❶ **They break down the dead remains of plants and animals into substances that can be used again by other organisms. (Inferring)**

❷ **Cocci (Relating concepts)**

① Thinking Skill: Relating facts

② Thinking Skill: Making comparisons

③ Thinking Skill: Relating cause and effect

④ Thinking Skill: Making calculations

## FACTS AND FIGURES

The bacterium *Sulfolobus acidocaldarius* changes sulfur into sulfuric acid in order to get energy to live.

## SCIENCE, TECHNOLOGY, AND SOCIETY

Acquired Immune Deficiency Syndrome (AIDS) is a communicable disease caused by a virus. The most widely used name for the AIDS virus is HIV. The AIDS virus attacks and destroys certain white blood cells called T cells or T lymphocytes, which aid in the body's immune response. This creates a shortage of T cells and lessens the body's ability to produce antibodies to fight diseases. Victims of AIDS generally die from diseases that a healthy immune system would be able to fight. Two common causes of death are a particularly deadly type of pneumonia (pneumocystis carinii pneumonia) and a rare type of skin cancer (Kaposi's sarcoma).

The AIDS virus can be spread by sexual contact, by sharing intravenous-drug needles, or by receiving infected blood through transfusion. Women infected with the AIDS virus also may transmit the disease to their developing babies.

Have students research current statistics regarding the spread of AIDS and the kinds of treatment available. Review the manner in which viruses multiply in the body. Discuss ways of preventing the spread of AIDS.

thing that has two distinctive parts. Fission means a process of splitting. As the students read their definitions, write them on the chalkboard.

• **Why is reproduction in bacteria called binary fission?** (Students should conclude that during binary fission a cell divides into two cells.)

Be sure to point out that binary fission is not the same process as mitosis (Section 3-2). In mitosis, chromosomes are duplicated and distributed equally to two daughter cells. However, bacterial DNA is not enclosed within a nucleus. The hereditary material is contained in an unpaired, circular DNA molecule, having only about one-thousandth the DNA found in cells that reproduce by mitosis. A bacterial cell duplicates its DNA well in advance of the cell division, and may contain two or three copies at once. This DNA is attached to the cell membrane.

## TIE-IN/ HISTORY

Perhaps the most devastating bacterial disease ever to have afflicted humans is the bubonic plague, also known as the black death. Millions of people have died from this disease, which was caused by a bacterium found in fleas. The disease is transmitted to humans by rats that carry the fleas. Over the centuries, many epidemics of bubonic plague have been recorded. Bubonic plague is thought to have been a major factor that contributed to the decline of the Roman Empire. From 1334 to 1351, the disease was widespread over much of Europe and Asia. In one year alone, 25 million people, which was one-fourth of the population of Europe, died. Another great outbreak occurred in India between 1894 and 1914, in which over 10 million people died. Massive programs to exterminate rats, and clean up unsanitary conditions have reduced the incidence of bubonic plague in recent times. Yet, occasional epidemics are still reported, especially in parts of Asia and Africa.

## FACTS AND FIGURES

Some types of bacteria have a guidance system that enables them to distinguish north from south. Within these bacteria are tiny sacs that contain magnetite, which helps the bacterium sense the earth's magnetic field.

In the real world, the rate of reproduction slows down because the bacteria soon use up their food, water, and space. They also produce wastes that eventually poison them.

### Helpful Bacteria

Most types of bacteria are not harmful and do not cause disease. For example, many food products, especially dairy products, are produced with the help of bacteria. Bacteria are needed to make cheeses, butter, yogurt, and sour cream. Some species of bacteria are used in the process of tanning leather.

Some helpful bacteria are used to fight other, harmful bacteria. These helpful bacteria, for example, can help produce **antibiotics.** Antibiotics destroy or weaken disease-causing bacteria. Scientists also have found a way to turn certain types of bacteria (1) into "chemical factories." The DNA within these bacteria is changed. The bacteria then produce large

### Career: *Bacteriologist*

**HELP WANTED: BACTERIOLOGIST**
College degree required; master's or Ph.D. desirable. Strong background in science necessary. Applicant should be creative and innovative. Position involves testing and research. Past experience as lab helper/technician an asset.

The specially equipped plane landed at the Atlanta airport. Within minutes, white-coated technicians rushed the test tube the plane was carrying to the Center for Disease Control. There, under strict laboratory conditions, the contents of the test tube were analyzed.

Through the powerful lens of the microscope, a tiny rod-shaped object became visible. Was it a living organism? The answer soon became obvious. Twenty minutes later a second rod-shaped object became visible under the lens of the microscope. The organism had reproduced.

Now a more important question had to be answered. Could this rod-shaped bacteria cause disease in human beings? Was it related to a type of bacteria that had caused diseases such as diphtheria, typhoid fever, and plague in the past? A **bacteriologist** would determine the type of bacteria it was and whether or not it was dangerous. In San Francisco, a sick patient and her doctor waited eagerly for the results.

Many bacteriologists specialize in identifying unknown microorganisms from the 2000 or so known types of bacteria. Others try to devise methods to combat harmful bacteria. Still others study how disease-causing bacteria may be spread in our environment. For further information on this career, write to the American Society of Microbiology, 1913 I Street N.W., Washington, DC 20006.



### 7-2 (continued)

#### Motivation

- **Aren't germs really bacteria and viruses?** (Yes, but only a small minority are harmful.)
- **List some of the good things or products derived from bacteria.** (yogurt, sour cream, vinegar, and cheese)
- **What do you notice about the flavor of some of these products?** (They are sour.)
- **Why do you think these products are sour?** (because the action of the bacteria in the fermentation process yields an acid product that gives the characteristic sour taste)

Briefly discuss the role of bacteria in producing some of these foods. Point out that the first step in producing such dairy products as cheese is to introduce certain types of bacteria into milk or cream. Through their metabolic activities, bacteria form lactic acid, which produces the desired flavors associated with the products. Hard cheeses, such as cheddar, are further processed by the action of other bacteria and molds during ripening. In manufacturing vinegar, bacteria convert alcohol in apple cider or wine to acetic acid. In the making of sauerkraut, bacteria convert sugars in cabbage leaves to lactic acid.

Next show students a leather product, such as a belt or a glove. Discuss the use of bacteria in tanning leather.

- **How are bacteria used to produce this product?** (Students should sug-

amounts of important substances such as insulin. Insulin helps to control the rate at which the human body breaks down sugar. People who do not produce enough of their own insulin can use the insulin produced by the bacteria.

Certain bacteria, called nitrogen-fixing bacteria, can make nitrogen compounds from nitrogen gas in the atmosphere. Nitrogen-fixing bacteria turn nitrogen gas, which plants cannot use for food, into nitrogen compounds that plants can use for food. These nitrogen-fixing bacteria often appear as lumps on the roots of plants such as alfalfa. See Figure 7-13. Nitrogen-fixing bacteria also help replace the nitrogen compounds in the soil. Without such nitrogen-fixing bacteria, most nitrogen compounds in the soil would be quickly used up and plants could no longer grow. ❷

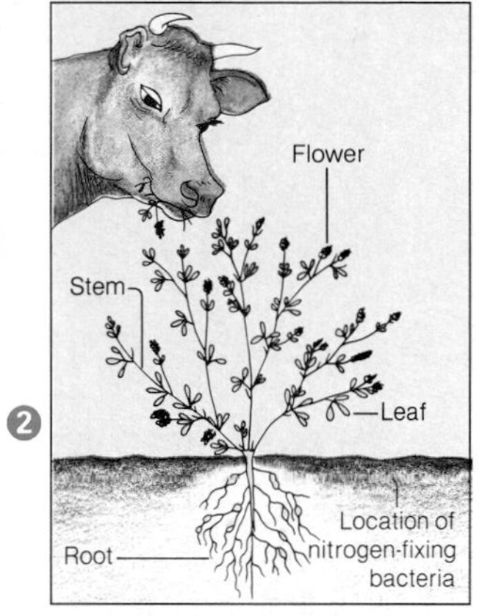


**Figure 7-13** *Nitrogen-fixing bacteria on the roots of alfalfa plants convert nitrogen gas in the air to nitrogen compounds that enter the soil. These compounds are needed by plants for growth.*

### *Harmful Bacteria*

Although most bacteria are harmless, and some are helpful, a few can cause trouble. The trouble comes in a number of forms—food spoilage, diseases of people, diseases of farm animals and pets, and diseases of food crops. Fortunately, there are a number of defenses against attacks by harmful bacteria.

**FOOD SPOILAGE** Spoilage can be prevented or slowed down by heating, drying, salting, or smoking foods. Each one of these processes prevents or slows down the growth of bacteria. Milk, for example, is heated to 71° C for 15 seconds before it is placed in containers and shipped to the grocery or supermarket. This process, called pasteurization, destroys most of the bacteria that would cause the milk to spoil quickly. Heating and then canning foods such as vegetables, fruits, meat, and fish are also used to prevent bacterial growth. But if the foods are not sufficiently heated before canning, bacteria can grow inside the can and produce poisons called **toxins.** ❸

**HUMAN DISEASES** These diseases caused by bacteria include strep throat, certain kinds of pneumonia, diphtheria, tuberculosis, and whooping cough. Some of these diseases can be prevented by an immunization shot. Others can be treated with antibiotics.

**Activity**

*Observing Bacteria*

**1.** Obtain some plain yogurt. Add water to the yogurt to make a very thin mixture.

**2.** With a medicine dropper, place a drop of the yogurt mixture on a glass slide.

**3.** With another dropper, add one drop of methylene blue to the slide.

**4.** Carefully place a cover slip over the slide.

**5.** Observe the slide under the low and high power of the microscope.

Describe your observations. What are the shapes of the bacteria? Why do you think you had to use methylene blue?



## ANNOTATION KEY

❶ Thinking Skill: Applying technology

❷ Thinking Skill: Applying concepts

❸ Thinking Skill: Applying technology

**Activity**

**Observing Bacteria**

**Skills: Making observations, comparing, manipulative, relating, applying, hypothesizing**

**Level: Remedial**

**Type: Hands-on**

**Materials: Plain yogurt, water, medicine dropper, glass slide, methylene blue, cover slip, microscope**

Students should be able to see rod-shaped bacteria, or bacilli, under the microscope. The methylene blue stain enables them to see the bacteria better. Yogurt is made from milk that has been fermented by bacteria.

gest that leathers are tanned by bacterial action.)

Explain that tanning is a process in which preserved animal hides are made into leather. These hides are soaked in solutions containing bacteria that secrete chemicals that soften the hides.

## Content Development

Students often confuse the words antibiotics and antibodies. Make a special effort to help them distinguish between these terms. Have a student read the definition of antibiotics as given in the glossary. Point out that "anti" means against and "biotic" refers to living things. Thus, antibiotics are substances produced by one living thing that work against other living things. Antibodies, on the other hand, are chemicals that are produced by white blood cells in response to foreign substances, or bodies, that have invaded the body.

## Skills Development

***Skill: Drawing a conclusion***

Refer students to Figure 7-13 and have them read the caption. Emphasize that although 78 percent of the atmosphere is gaseous nitrogen, it cannot be taken in by plants. When nitrogen is converted to nitrates by nitrogen-fixing bacteria, it can then be used by plants.

- **Many farmers plant crops such as wheat or corn in a field for one year. Next year they will plant other crops such as alfalfa or clover in the same field. Why do they follow this practice of alternating crops from year to year?** (Students should reason that different crops have different needs. Corn and wheat use large amounts of nitrates from the soil. Nitrates can be replaced by planting legumes, such as clover, alfalfa, or soybean.)

## 7-3 PROTOZOANS

### SECTION PREVIEW 7-3

Protozoans are unicellular, animal-like organisms that are generally found in watery environments. Unlike bacteria, their hereditary material is enclosed in a nucleus. Protozoans are classified in the Protista kingdom, and they are subdivided into smaller groups, primarily on the basis of their means of locomotion.

### SECTION OBJECTIVES 7-3

1. **Compare protozoans and bacteria.**
2. **Describe the methods of locomotion of amoebas, ciliates, flagellates, and sporozoans.**
3. **Identify the cell structures of the amoeba, the paramecium, and *Euglena*.**
4. **Compare the methods of feeding and reproduction in the amoeba, the paramecium, and *Euglena*.**
5. **Name and describe a human disease caused by sporozoans.**

### SCIENCE TERMS 7-3

**protozoan p. 159**
**pseudopod p. 159**
**food vacuole p. 160**
**contractile vacuole p. 160**
**cilium p. 160**
**oral groove p. 161**
**gullet p. 161**
**anal pore p. 161**
**conjugation p. 161**
**eyespot p. 163**

**Figure 7-14** *Leaves on this pear tree have been attacked by a bacterial disease called fire blight.*

**ANIMAL DISEASES** Bacteria cause diseases in animals. These diseases include anthrax, which attacks sheep, horses, cattle, and other large farm animals, and fowl cholera, which attacks chickens. Vaccinations can prevent both diseases.

**PLANT DISEASES** These diseases caused by bacteria include fire blight of apples and pears and soft rot of vegetables. There are many ways of treating plant diseases, including the use of antibiotics and various chemicals.

#### *SECTION REVIEW*

1. What are the three basic shapes of bacteria?
2. What are two ways in which bacteria are helpful? Harmful?

## 7-3 Protozoans

Two billion years ago, the earth was a strange and barren place. No animals roamed the land, swam in the sea, or flew in the air. No trees, shrubs, or grasses grew from the soil. From the air, the earth

**Figure 7-15** *The radolarian, a protozoan, is surrounded by a hard silicon shell.*

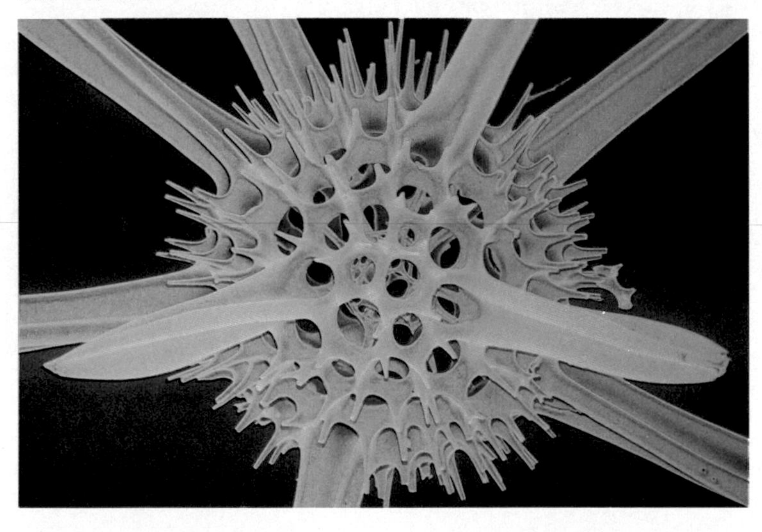

### 7-2 (continued)

#### Reinforcement

To control bacteria, various antiseptic solutions can be found to clean everything from your breath to the countertop to the diaper pail. Let's compare how some of these substances work. Have students bring in various antiseptics or disinfectants such as mouthwashes, cleaning products, and other materials associated with killing bacteria. Ahead of time, you will need bacterial field plates you have prepared or purchased. Have students soak dots of filter paper made by a paper punch in the solutions they provide. Then, using sterile technique, transfer these dots to the bacterial fields, noting the cleaning agent being added.

#### Section Review 7-2

**1.** Spheres—cocci; rods—bacilli; spirilla—spirals.
**2.** Helpful bacteria—aid in production of dairy products and antibiotics and used in the tanning of leather. Harmful bacteria—can cause food spoilage and diseases of people, farm animals, pets, and food crops.

### TEACHING STRATEGY 7-3

#### Motivation

Introduce protozoans by presenting some large pictures, wall charts, overhead transparencies or slides that show the structure of the amoeba, paramecium, and *Euglena*. If such visuals are not available, have students refer to Figures 7-16, 7-18, and 7-20.
• **What are some ways in which**

looked gray, brown, and blue. But even though it looked lifeless, an unseen new form of life was taking hold in its blue waters. This form of life, the **protozoans,** represented a giant step in the parade of living things that would follow. **Protozoans are unicellular animal-like organisms.** The protozoans still inhabit the world's seas, lakes, rivers, and ponds.

Protozoans differ in many ways from the bacteria that first inhabited the earth. For one thing, protozoans do not have cell walls. They do contain a distinct cell nucleus, which contains hereditary material. Protozoans are, in fact, the first animal-like organisms to evolve on the earth. For this reason, they were given their name, which comes from the Greek words for "first animal." However, protozoans are not true animals and are classified in the Protista kingdom.

### Amoebas

Amoebas (uh-MEE-buhz) make up one of the simplest groups of protozoans. Although some amoebas are visible to the unaided eye, most are microscopic. Under the microscope, an amoeba looks like a blob of jelly.

The amoeba moves slowly in a watery environment by extending its cell membrane and cytoplasm. This extension is called a **pseudopod** (soo-duh-pahd), or "false foot." When the pseudopod is fully extended, it pulls the amoeba along with it.

**Activity**

*A Protozoan Population*

The largest existing protozoan is *Pelomyxa palustrius.* It may grow to a length of up to 1.5 centimeters. If all the *Pelomyxa palustrius* in a 30-meter-long pond were placed end-to-end and equaled the pond's length, how many would be present in the pond?

**Figure 7-16** *This drawing shows the structure of a typical amoeba. The photograph of the amoeba was magnified 160 times.*

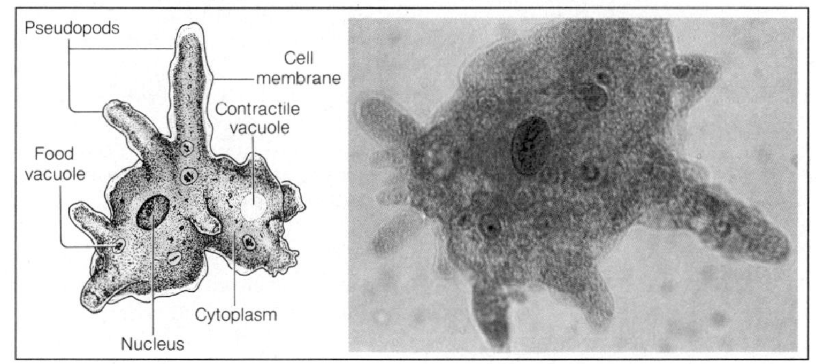


159

**Activity**

**A Protozoan Population**
**Skill: Making calculations**
**Level: Average**
**Type: Computational**

Students must first convert 30 meters into centimeters by multiplying by 100. Then students should divide 3000 by 1.5 to determine the answer. There would be 2000 *Pelomyxa palustrius* protozoans in the pond.

different ways in which protozoans move. Explain that the way in which they move provides a basis for classifying them.

## Content Development

Students may not see much similarity between the radiolarian shown in Figure 7-15 and the cells of the other protozoans. Point out that Figure 7-15 shows only the radiolarian's outer silicon shell. The cell that was located near the center of the radiating projections is no longer present. After the cells of radiolarians die, their shells settle on the ocean floor to form sandlike deposits that pile one upon the other. When radiolarians are alive, they float about and trap other floating microorganisms with their stiff, sticky projections.

## Skills Development

***Skill: Making comparisons***

To compare characteristics of an amoeba, paramecium, and *Euglena,* have students prepare a chart with these column headings: Protozoan; Sketch of cell; Method of movement; How food is obtained; Method of reproduction. As students read about these representative protozoans, have them complete their charts.

**these organisms are similar to bacteria?** (Accept all logical answers. Some students may mention, that like bacteria, they are unicellular and have a cell membrane and cytoplasm.)

- **Do you notice anything about the cells of these organisms that makes them different from the cells of bacteria?** (Accept all logical answers. Many specialized structures may be mentioned but be certain that students notice that these protozoans have well-defined nuclei, but lack cell walls.)
- **How do protozoans move?** (At this time allow students to speculate as to how protozoa move.)

Mention that protozoans have several animal-like characteristics. For this reason, protozoans were once classified as one-celled animals. However, most biologists now classify them in the Protista kingdom. Have students pay special attention to the

## Activity

**Harmful Microorganisms**
**Skills: Relating cause and effect, researching, applying**
**Level: Average**
**Type: Library/writing**

Have students report their findings to the class, possibly assigning each student a particular disease to research.

## BACKGROUND INFORMATION

Amoebas (sarcodines) and ciliates include a huge number of organisms, which are represented here by amoebas and paramecia. Among the Protists, the structures for locomotion of these two groups are quite elaborate. These organisms are active feeders that digest food in vacuoles. Binary fission produces new individuals in both groups. Conjugation is described between paramecia.

Ciliates are the most complex protozoa because they possess many specialized structures for feeding and response. Ciliates are found in both fresh and marine water where there is an abundance of decaying organic matter to support the bacteria on which most paramecia feed. Most ciliates are free-living, but some are parasites of mollusks and crustaceans.

### Activity

*Harmful Microorganisms*

Using reference materials in the library, prepare a report on the following diseases.

- measles
- smallpox
- amoebic dysentery
- poliomyelitis
- pneumonia
- tetanus

In your report, include information about how these diseases are transmitted to human beings. Also include what types of treatment are used to prevent or cure the diseases.

The amoeba also uses pseudopods to obtain food. As the amoeba nears a small piece of food, such as a smaller protozoan, the amoeba extends its pseudopod around the food. Soon the food particle is engulfed, or surrounded, by the pseudopod. A spherical **food vacuole** (VAK-yoo-ohl) in the pseudopod forms around the engulfed food particle. The food vacuole releases special digestive chemicals, called enzymes, in order to break down the captured food. (1) The digested food can then be taken in and used by the amoeba. The waste products left behind in the food vacuole are eliminated through the cell membrane.

To supply itself with energy, the amoeba requires oxygen as well as food. Oxygen passes from the watery environment into the amoeba through the cell membrane. At the same time, waste products, such as carbon dioxide, pass out of the amoeba through the cell membrane. Excess water in the amoeba is (2) pumped out through the cell membrane by a **contractile vacuole.**

Like bacteria, amoebas reproduce by binary fission. A parent cell divides into two new identical cells. Each of these new cells has the same amount and kind of hereditary material as the parent cell.

Amoebas respond simply to changes in their environments. They are sensitive to bright light and move to areas of dim light. Amoebas are also sensitive to certain chemicals, moving away from some and toward others.

### *Ciliates*

The protist kingdom also includes the ciliates (SIHL-ee-ayts). These protozoans have **cilia** (SIHL-ee-uh; singular: cilium). Cilia are small hairlike projections on the outside of the cells. Sweeping through the water like tiny oars, the cilia enable these organisms to move, and help them to obtain food. In addition, the cilia function as sensors through which these organisms receive information about their environment.

One of the most familiar ciliates is the paramecium (pa-ruh-MEE-shi-uhm; plural: paramecia). A hard membrane covers the outer surface of the paramecium. This membrane gives the paramecium its

**Figure 7-17** *The drawing shows the wavelike motion of a cilium.*

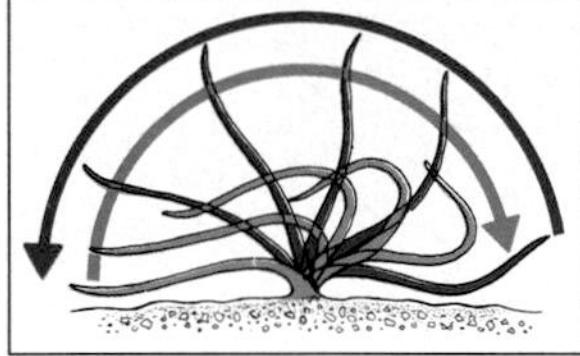

160

## 7-3 (continued)

### Content Development

Tell students that amoebas are usually found living on submerged vegetation at the edge of freshwater ponds. It is often difficult to identify amoebas collected in this way, because they blend in with the debris on which they live. **CAUTION:** *Advise students not to go near a pond or any body of water to collect specimens unless they are accompanied by an adult.*

### Motivation

Distribute drawings of a paramecium and have students draw arrows that show the pathway of food as it is taken in, digested, and distributed through the paramecium. Have students use a different-colored pencil to indicate where undigested materials are eliminated. Have them label each of the following structures: oral groove, gullet, food-vacuole forming, food vacuole, and anal pore. Check drawings.

### Content Development

Point out that the paramecium's large nucleus contains multiple copies of the hereditary material found in the small nucleus. Experiments have shown that if the large nucleus is removed, the paramecium will die. A paramecium, on the other hand, can live without its small nucleus, but it cannot reproduce by conjugation.

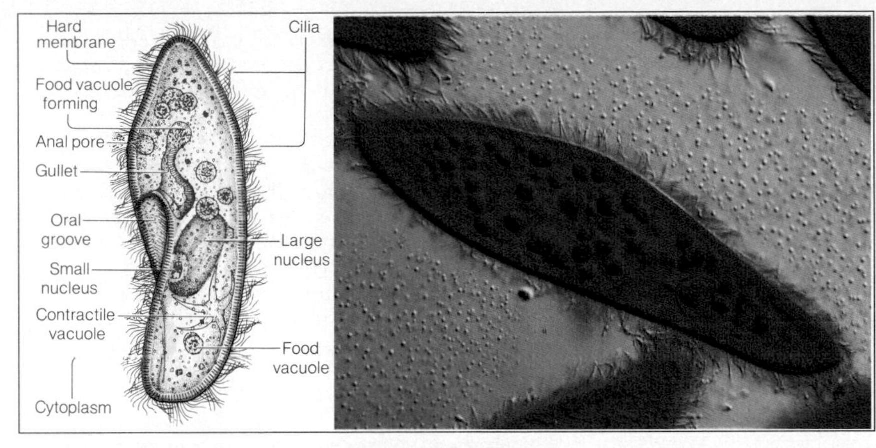


**Figure 7-18** *This drawing shows the structure of a typical paramecium. The photograph of the paramecium was magnified 140 times.*

slipper shape. The cilia of the paramecium move food particles in the water into its **oral groove.** The oral groove is an indentation in the paramecium. Food goes from the oral groove into the **gullet.** The gullet is a funnel-shaped structure that ends at the food vacuole. The food vacuole distributes food particles as it moves through the organism. Undigested material is eliminated through the **anal pore.**

Located in the cytoplasm of paramecia are a small nucleus and a large nucleus. The small nucleus controls reproduction. The large nucleus controls all other life functions.

Reproduction in paramecia occurs in two ways. Like an amoeba, paramecia reproduce by binary fission. A paramecium may also share its hereditary material with another paramecium through a process called **conjugation** (kahn-joo-GAY-shuhn). During conjugation, two paramecia join together. Soon the larger nuclei disappear and the smaller nuclei divide in two. Then, one of the smaller nuclei from each pair passes into a special tube joining both paramecia. After exchanging the hereditary material, the two paramecia move away from each other. A new small nucleus and a new large nucleus form in each of the two paramecia. Conjugation is beneficial ❸

161

## FACTS AND FIGURES

The sarcodines consist of four groups: amoebas, foraminiferans, radiolarians, and heliozoans. Although most amoebas do not have shells, many of them do. All members of the other three groups have shells.

## ANNOTATION KEY

❶ Thinking Skill: Relating facts
❷ Thinking Skill: Identifying processes
❸ Thinking Skill: Identifying processes

## Reinforcement

Obtain photographs or diagrams that show paramecia reproducing by binary fission and by conjugation. Such photographs or diagrams are commonly found in high school biology textbooks. These visuals can be used to make simple chalkboard drawings that illustrate the processes. If a microprojector is available, obtain prepared microscope slides that show these forms of reproduction. Prepared slides of paramecia undergoing fission and conjugation are available from biological supply companies. Have students draw each process.

- **When paramecia reproduce by binary fission, does the cell divide lengthwise or crosswise?** (The cell divides crosswise.)
- **Which nuclei are exchanged during conjugation?** (micronuclei)

(Emphasize that the exchange of hereditary material during conjugation better adapts paramecia to changing environmental conditions than does binary fission.)

## Skills Development

***Skill: Applying definitions***

Have students recall the definitions of asexual and sexual reproduction (Section 2-1).

- **In what two ways does reproduction occur in paramecia? Which process is sexual and which is asexual?** (Binary fission is asexual and conjugation is sexual.)
- **Explain why binary fission is an example of asexual reproduction.** (In binary fission, there is only one parent.)
- **How is conjugation an example of sexual reproduction?** (Conjugation involves an exchange of hereditary material between two parent cells.)

## Skills Development

***Skill: Interpreting diagrams***

Refer to students to Figure 7-17.

- **What does the outer arrow show about the movement of this one cilium?** (For the power stroke, the cilium is almost straight as it moves from right to left.)
- **What does the inner arrow show?** (As it recovers to the starting position, the cilium bends to reduce drag.)
- **If there were other cilia beating in the same direction, would the animal move to the right or left on this page?** (to the right)

## BACKGROUND INFORMATION

In contrast to the complex locomotion structures of the sarcodines and ciliates, those of many flagellates seem less developed. Sporozoans have no special means for moving and are all parasitic. *Euglena,* an autotroph that contains chloroplasts, represents the flagellates. The complicated life cycle of many sporozoans is illustrated by *Plasmodium,* which causes malaria in humans.

The spores produced by sporozoans are not to be confused with the endospores produced by certain bacteria. Unlike endospores, which are protective structures formed within bacterial cells, the spores of sporozoans are cells that result from multiple cell divisions following sexual reproduction. It is during the spore stage that a sporozoan is transferred from one host to another.

## TIE-IN/ SOCIAL STUDIES

Efforts to control parasitic diseases require careful attention to both the scientific and human behavioral aspects. For instance, during the early 1960s in Sri Lanka, a determined effort of spraying for mosquitoes and caring for sick people resulted in essentially eradicating malaria. But the human nature to become complacent was not taken into consideration. That, plus the development of resistance in the mosquitoes and the *Plasmodium* resulted in more than a million reported cases of malaria in Sri Lanka in 1983.

**Figure 7-19** *Volvox are flagellates that form colonies. Volvox are autotrophs because they contain the green substance chlorophyll, which enables them to make their own food.*

because it allows paramecia to share hereditary characteristics that may help the paramecia become better able to survive a changing environment.

### *Flagellates*

The flagellates include protozoans that have flagella. Most of the flagellates, such as *Euglena,* are unicellular. However, some flagellates, such as *Volvox,* form colonies, or clusters of cells.

There are two groups of flagellates. One group contains the green substance chlorophyll. These flagellates are autotrophs. *Volvox* is an example of this type of flagellate.

The second group of flagellates do not contain chlorophyll. They are heterotrophs. Most of these flagellates are parasites. Some of them cause disease such as African sleeping sickness. This group of disease-causing flagellates is transmitted to people and animals by the tsetse fly.

**Figure 7-20** *This drawing shows the structure of the autotrophic flagellate called* Euglena. *The photograph is of a typical* Euglena.

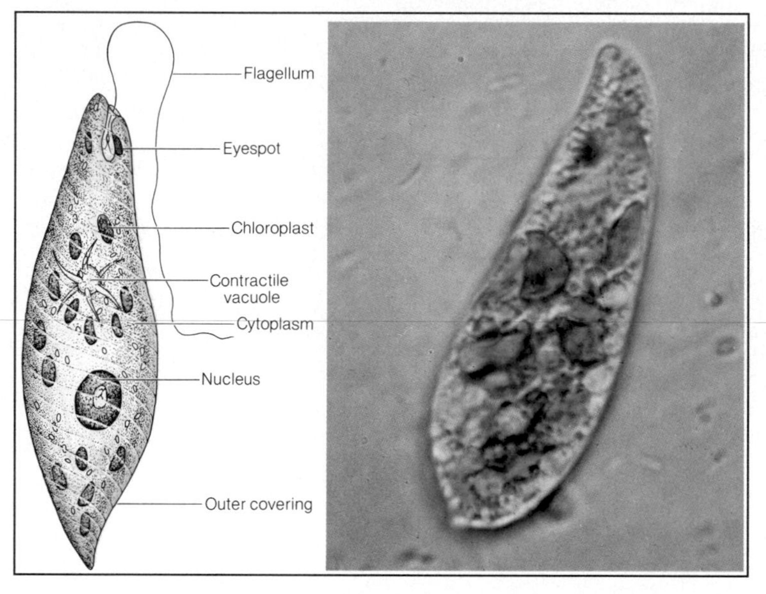


162

### 7-3 (continued)

#### Motivation

Have the class look at Figure 7-19 and read the caption. Point out that although *Volvox* appears to be multicellular, it is actually a colony of unicellular organisms. A colony is a cluster of individual cells living together. A *Volvox* colony is a hollow ball-shaped colony that consists of 500 to 60,000 individual cells that are held together by cytoplasmic threads.

• **Why are *Volvox* green?** (They contain chlorophyll.)

• **Are *Volvox* heterotrophs or autotrophs? Explain why.** (*Volvox* are autotrophs because they contain chlorophyll.)

Have students locate the smaller, more numerous cells. Each of these small cells has two flagella that beat in a coordinated manner, thus causing the entire colony to spin like a ball. The flagellated cells are incapable of reproduction.

Now have students locate the larger cells. Explain that these cells are nonmotile, but they function in reproduction. Both types of cells can perform photosynthesis.

• **The small, flagellated cells depend on the large cells to keep the colony moving. How might the large reproductive cells depend on the smaller cells?** (Accept all logical answers. The larger cells might be dependent on the smaller cells to keep them near the surface of the pond to obtain light for photosynthesis.)

A flagellate that is both an autotroph and a heterotroph is *Euglena*. *Euglena* is an oval-shaped organism with one pointed end and one rounded end. The rounded end contains the flagellum. Also located in this area in the cytoplasm is the **eyespot.** The eyespot is a reddish structure that is sensitive to light. *Euglena* tends to move toward areas where there is enough light. *Euglena* uses light energy to make its own food.

Like the amoeba and paramecium, *Euglena* reproduces by binary fission. However, unlike most protozoans, *Euglena* splits in two lengthwise. This type of division produces two cells that are mirror images of each other.

### *Sporozoans*

Sporozoans are protozoans that cannot move from place to place. These organisms are parasites whose foods are the cells and body fluids of animals.

Many of these sporozoans have complicated life cycles. During their life cycles, sporozoans form spores. Each spore contains hereditary material and a small amount of cytoplasm. Eventually, the sporozoans release these spores. Each spore can develop into a mature sporozoan.

Perhaps the most familiar type of sporozoan is the organism that causes the disease malaria. This sporozoan is carried by the *Anopheles* mosquito. ❷ When an infected mosquito bites a human being, the mosquito injects its saliva, and the spores enter the person's body. The infected person soon begins to experience chills and fever, which are two symptoms of malaria.

### *SECTION REVIEW*

1. What is a pseudopod?
2. What are cilia?
3. How do amoebas, paramecia, and *Euglena* reproduce?
4. What is the function of the eyespot in *Euglena*?
5. How do sporozoans differ from other protozoans?

**Activity**

*Capturing Food*

How does a paramecium capture its food?

**1.** Place one drop of paramecia from a paramecium culture on a slide.

**2.** Add one drop of *Chlorella*, a green algae, to the slide.

**3.** To a second slide, add another drop of paramecia. Then, add a drop of India ink.

**4.** Observe the two slides under the low and high power of the microscope.

What happened to the *Chlorella?* What did the addition of India ink enable you to see? Describe the movement of the India ink into a paramecium.



## ANNOTATION KEY

❶ Thinking Skill: Making comparisons

❷ Thinking Skill: Relating cause and effect

## FACTS AND FIGURES

The largest existing protozoan is *Pelomyxa palustrius,* which may grow to a length of up to 1.5 centimeters.

## FACTS AND FIGURES

One of the most complex flagellates lives in the digestive system of termites. It is covered by hundreds of long flagella. Termites are able to eat wood, only because these flagellates digest the wood particles. Termites do not have the enzyme needed to digest wood on their own.

## Activity

**Capturing Food**
**Skills: Making observations, manipulative, comparing, relating, applying**
**Level: Average**
**Type: Hands-on**
**Materials: Paramecia, slides, *Chlorella*, India ink, microscope**

The *Chlorella* is ingested by the paramecia. As a result, green-colored food vacuoles form. By adding India ink to the paramecia on the slide, the action of the cilia in the paramecia's oral grooves can be observed. Also the formation of a black-colored food vacuole can be seen. The movement of the cilia creates a current of water, which moves the India ink into the oral groove and then eventually into the food vacuole. The undigested India ink will pass out of the paramecia through the anal pores.

## Content Development

When discussing *Euglena,* explain that ordinarily these organisms are autotrophs. However, experiments have shown that if they are placed in darkness in nutrient-rich solutions their chloroplasts will disintegrate. The *Euglena* will then become heterotrophs, absorbing nutrients through their cell membranes. If returned to a lighted area, they will continue to be heterotrophic.

## Section Review 7-3

**1.** Extension of an amoeba's cell membrane and cytoplasm.
**2.** Hairlike projections.
**3.** All reproduce by binary fission; paramecia also reproduce by conjugation.
**4.** Enables *Euglena* to move toward light.
**5.** Sporozoans cannot move from place to place; have complicated life cycles.

# LABORATORY ACTIVITY WHERE ARE BACTERIA FOUND?

## BEFORE THE LAB

1. **If you are using glass petri dishes be sure they are scrubbed clean and dried before pouring your agar. This agar need not be sterile, as the poured plates should be autoclaved as a unit. If you are using sterile plastic dishes, your nutrient agar must be sterile, as these dishes cannot be autoclaved.**
2. **Provide a clean environment for the students' work and stress aseptic procedure. Most importantly, stress to students the fact that bacteria are everywhere, and our body does a good job of preventing them from infecting us. However, they are going to be dealing with concentrated colonies of potentially hazardous organisms so be sure to note that once the dishes have been contaminated, according to the instructions, they are *never* to be opened again until they are autoclaved. Even turning the dishes over several times could release drops of contaminated condensation from the dish, so handling should be kept to a minimum. Taping the dishes closed might be the deterrent needed to keep out the defiantly curious student.**
3. **Be sure to provide a germicidal soap for all students to use after handling the incubated dishes.**

## PRE-LAB DISCUSSION

You have seen individual bacteria through the microscope and are aware of their shapes.

- **Can you see any bacteria on your hands?** (No)
- **What do you think bacteria will look like in the petri dish?** (dots)
- **How will you be able to identify what kinds of bacteria are present?** (You do not really have to, but various colors will be your first indicators, and after that microscopic analysis will have to be done.)
- **Will there be things other than bacteria in the dish?** (You will more than likely have growths of mold.)
- **Predict which dish will show the most colonies?** (This will vary based on conditions.)
- **How does your contamination procedure illustrate good scientific method?** (Only one variable is being considered for each dish.)
- **Why is one dish left uncontaminated?** (It is the control to show that there is nothing in the dishes that will cause anything to grow.)

## SKILLS DEVELOPMENT

Students will use the following skills while completing this investigation.

1. Observing
2. Comparing
3. Inferring
4. Manipulative
5. Safety
6. Designing
7. Hypothesizing

## LABORATORY ACTIVITY

### *Where Are Bacteria Found?*

**Purpose**

In this activity, you will determine some common sources of bacteria.

**Materials** *(per group)*

5 petri dishes with sterile nutrient agar
Glass-marking pencil
Pencil with eraser
Soap and water

**Procedure**

1. Turn each petri dish bottom side up on the table. **Note:** *Be careful not to open the petri dish.*
2. With the glass-marking pencil, label the bottom of the petri dishes containing the sterile nutrient agar A to E. Turn the petri dishes right side up.
3. Remove the lid of dish A and lightly rub a pencil eraser across the petri dish. Close the dish immediately.
4. Remove the lid of dish B and leave it open to the air until the class period ends. Then close the lid.
5. Remove the lid of dish C and lightly rub your index finger over the surface of the agar. Then close the lid.
6. Wash your hands thoroughly. Remove the lid of dish D and lightly rub the same index finger over the surface of the agar. Then close the lid.
7. Do not open dish E.
8. Place all five dishes upside-down in a warm, dark place for three or four days.
9. After three or four days, examine each dish. **CAUTION:** *Do not open dishes.*
10. On a sheet of paper, construct a table similar to the one on this page. Then fill in the table.
11. Return the petri dishes to your teacher. Your teacher will properly dispose of the petri dishes.

**Observations and Conclusions**

1. How many colonies, or similar types of bacteria, appear to be growing on each petri dish? How can you distinguish between different bacterial colonies?

| Petri Dish | Source | Description of Bacterial Colonies |
|---|---|---|
| A | | |
| B | | |
| C | | |
| D | | |
| E | | |

2. Which petri dish has the most bacterial growth? Which has the least?
3. Which petri dish was the control?
4. Did the dish that you touched with your unwashed finger contain more or less bacteria than the one that you touched with your washed finger? Explain.
5. Explain why the agar was sterilized before the investigation.
6. Design an experiment to show if a particular antibiotic will inhibit bacterial growth.
7. Suggest some methods that might stop the growth of bacteria.
8. What kinds of environmental conditions seem to influence where bacteria are found?

## CHAPTER REVIEW

### SUMMARY

**7-1 Viruses**

- Viruses cannot carry out any life processes unless within a living cell.
- Viruses are tiny particles that contain a center of either DNA or RNA surrounded by a protein coat.
- Reproduction of viruses occurs only when the virus invades a living cell.
- Bacteriophages are viruses that infect bacterial cells.
- When a virus's hereditary material enters a living cell, the virus takes control of the cell. The cell begins to produce new viruses, which are released from the cell.

**7-2 Bacteria**

- Bacteria are simple unicellular monerans.
- Bacteria have three basic shapes: cocci, bacilli, or spirilla.
- Bacterial cells have a cell wall and a cell membrane surrounding the cytoplasm. The hereditary material of a bacterial cell is spread throughout the cytoplasm.
- Some bacteria have flagella, or whiplike structures, that enable them to move.
- Some bacteria called autotrophs use energy from the sun or chemicals to make food.
- Bacteria that must obtain their food from other organisms are called heterotrophs.
- Bacteria usually reproduce by binary fission, or splitting in two.

**7-3 Protozoans**

- Protozoans are unicellular animallike microorganisms. Their hereditary material is confined in a nucleus.
- Amoebas move and obtain food by the movement of their pseudopods. Amoebas reproduce by binary fission.
- Ciliates, such as paramecia, have small hairlike cilia used for obtaining food and for movement.
- Paramecia reproduce by binary fission and conjugation.
- Flagellates use whiplike flagella to move. There are two groups of flagellates. The members of one group contain chlorophyll and can make their own food. The second group of flagellates do not contain chlorophyll and cannot make their own food.
- Sporozoans cannot move from place to place. Sporozoans are parasites and have complex life cycles.

### VOCABULARY

*Define each term in a complete sentence.*

| | | | |
|---|---|---|---|
| **anal pore** | **cilium** | **flagellum** | **pseudopod** |
| **antibiotic** | **coccus** | **food vacuole** | **RNA** |
| **bacillus** | **conjugation** | **gullet** | **saprophyte** |
| **bacteriophage** | **contractile vacuole** | **microbiology** | **spirillum** |
| **bacterium** | **cytoplasm** | **nucleus** | **toxin** |
| **binary fission** | **DNA** | **oral groove** | **virus** |
| **cell membrane** | **endospore** | **parasite** | |
| **cell wall** | **eyespot** | **protozoan** | |



## SAFETY TIPS

Caution students not to open the petri dishes once they are contaminated. Make sure students wash their hands after the investigation.

## TEACHING STRATEGY FOR LAB PROCEDURE

Hopefully students will not haphazardly contaminate their dishes before the appointed time. This could influence results but is not hazardous. It is after the incubation time that we do not want any mishaps. Any dish with live bacteria must be treated with caution, especially because you have no control over what was introduced. Dropped and broken or otherwise opened dishes should be cordoned off and cleaned up by you or another responsible individual—*not* by the students. Be sure to have water and germicidal soap available.

## OBSERVATIONS AND CONCLUSIONS

**1.** Answers will vary, but students should be able to distinguish between the clumps of bacterial colonies. Some may be shaped or colored differently.
**2.** There should be more colonies on dishes A, B, and C than on the others. Dish E should have the least growth.
**3.** Dish E
**4.** Because washing removed bacteria, the dish touched by the unwashed finger should show more growth.
**5.** To kill any bacteria already on the agar.
**6.** Experiments should include a control in which no antibiotic was used and an experimental setup in which an antibiotic disk was used.
**7.** Washing, sterilization, etc.
**8.** Bacteria can be found wherever they can find food and a suitable temperature. Most bacteria require oxygen as well.

## GOING FURTHER: ENRICHMENT

Students should be directed to come up with other ideas for contaminating sterile dishes. They have been told since they were young children that this, that, and the other thing is dirty or covered with germs. Develop a list of contamination sources, both contact and exposure. Suggestions for contact: comb, money, hair, a sterile swab that has been rubbed over someone's braces, and the always popular "kiss of death." (Have a student volunteer to kiss the agar, reminding them that it is perfectly pure and clean.) Besides classroom air, other locations could be suitable for exposure: gym locker, cafeteria, inside a gym shoe or may be inside a lunch box. Imagination is your only limitation.

You can then pose a similar line of questioning as proposed by the lab text. You have simply provided more variables to consider and have provided a good lesson about the distribution of germs in our environment.

## CHAPTER REVIEW

### MULTIPLE CHOICE

| | | | | |
|---|---|---|---|---|
| **1.** d | **3.** a | **5.** b | **7.** b | **9.** d |
| **2.** b | **4.** c | **6.** c | **8.** a | **10.** c |

### COMPLETION

**1.** bacteriophage
**2.** bacilli
**3.** binary fission
**4.** nitrogen-fixing bacteria
**5.** toxins
**6.** cilia
**7.** food vacuole
**8.** oral groove
**9.** *Euglena*
**10.** eyespot

### TRUE OR FALSE

**1.** T
**2.** F cell wall
**3.** F Flagella
**4.** F Parasites
**5.** T
**6.** F Pseudopod
**7.** F gullet
**8.** T
**9.** T
**10.** T

### SKILL BUILDING

**1.** In 1 and 2, the bacteria are reproducing and growing because conditions for growth are favorable. In 3, the growth of bacteria has leveled off. The bacteria are not reproducing as rapidly, because the conditions for growth are becoming unfavorable. In 4, there is a decrease in the growth of bacteria because conditions have become unfavorable. As a result, some of the bacterial population is beginning to die off.
**2.** This report could not be true. Sporozoans do not have any means of locomotion.
**3.** By placing food in cooler places, such as in a refrigerator or a freezer, the growth of bacteria will be slowed down. As a result, less food will spoil and can be used for a longer period of time.
**4.** Students should suggest placing the penicillin in each of the five bacteria cultures. If the penicillin inhibits the growth of some of the bacterial cultures, there will be clear areas where there is no growth of bacteria. As a control, students should set up the same types of bacterial cultures but without the penicillin.

### ESSAY

**1.** Viruses are not cells. Unlike cells, viruses cannot reproduce on their own, although they do contain hereditary material. Viruses cannot perform the other functions of cells either.
**2.** Some bacteria can spoil food and cause disease in people, animals, and food crops.
**3.** Dimitri Iwanowski and Martinus Beijerinck attempted to discover the bacteria that they believed were the cause of the tobacco mosaic disease. But even plant juices from which bacteria had been filtered out were able to spread the disease. Beijerinck concluded correctly that the juices contained viruses, which are smaller than bacteria.
**4.** Autotrophic bacteria contain chlorophyll and can make their own food. Heterotrophic bacteria cannot make their own food.

## CONTENT REVIEW: MULTIPLE CHOICE

*Choose the letter of the answer that best completes each statement.*

**1.** Viruses are
a. shaped like spheres. b. shaped like tadpoles.
c. shaped like cylinders. d. any of these shapes.
**2.** What is used to observe viruses?
a. the unaided eye b. an electron microscope
c. a light microscope d. all of these
**3.** The scientist who gave viruses their name was
a. Martinus Beijerinck. b. Wendell Stanley.
c. Dimitri Iwanowski. d. Anton van Leeuwenhoek.
**4.** The type of hereditary material in a virus may be
a. DNA. b. RNA.
c. either DNA or RNA. d. neither DNA nor RNA.
**5.** An organism that feeds on dead matter is called a(n)
a. autotroph. b. saprophyte. c. parasite. d. flagellum.
**6.** An amoeba moves by means of
a. flagella. b. cilia. c. pseudopods. d. conjugation.
**7.** Which organism reproduces by binary fission and conjugation?
a. amoeba b. paramecium c. sporozoan d. *Euglena*
**8.** *Euglena* is an example of a(n)
a. flagellate. b. ciliate. c. sporozoan. d. amoeba.
**9.** Which group of protozoans cannot move by themselves?
a. amoebas b. ciliates c. flagellates d. sporozoans
**10.** The most familiar type of sporozoan causes
a. pneumonia. b. tuberculosis.
c. malaria. d. African sleeping sickness.

## CONTENT REVIEW: COMPLETION

*Fill in the word or words that best complete each statement.*

**1.** A virus that infects a bacterial cell is called a(n) __________.
**2.** The bacteria that look like rods are called __________.
**3.** Bacteria usually reproduce through __________.
**4.** Bacteria that make nitrogen compounds from nitrogen gas are called __________.
**5.** Some bacteria produce poisons called __________.
**6.** Paramecia have __________ that enable them to move.
**7.** The spherical structure in a cell that releases enzymes to digest food is the __________.
**8.** In paramecia, food particles are swept by cilia into an indented structure called a(n) __________.
**9.** A flagellate that is both an autotroph and a heterotroph is the __________.
**10.** The structure in *Euglena* that responds to changes in light is the __________.

166

## CONTENT REVIEW: TRUE OR FALSE

*Determine whether each statement is true or false. If it is true, write "true." If it is false, change the underlined word or words to make the statement true.*

1. A virus is made up of hereditary material surrounded by a protein coat.
2. The outermost boundary of a bacterial cell is the cell membrane.
3. Cilia are long, thin, whiplike structures that are used to propel some microorganisms.
4. Saprophytes are organisms that feed on living organisms.
5. Endospores allow bacteria to survive unfavorable conditions.
6. Protozoan means false foot.
7. In a paramecium, the oral groove is the funnel-shaped structure that ends in a food vacuole.
8. The contractile vacuole pumps excess water out of an amoeba.
9. In a paramecium, undigested material is eliminated through an opening called the anal pore.
10. *Euglena* is an example of a flagellate.

## CONCEPT REVIEW: SKILL BUILDING

*Use the skills you have developed in the chapter to complete each activity.*

1. **Interpreting a graph** The following graph shows the growth of bacteria. Describe what is happening at each of the numbered growth stages. Suggest a hypothesis to explain the growth pattern in stage 3 and in stage 4.

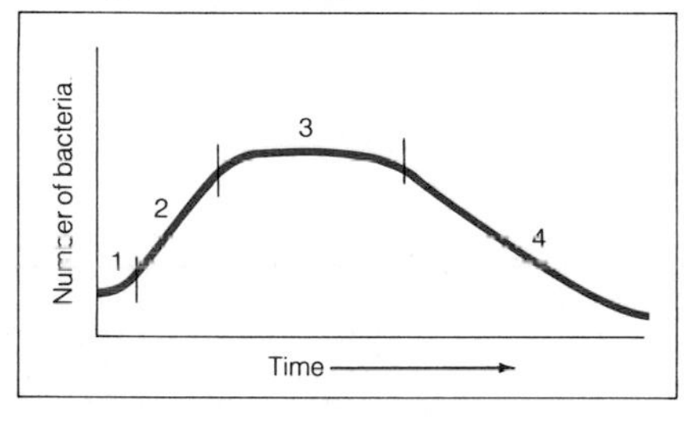


2. **Applying definitions** Suppose a biologist reported the discovery of a sporozoan that moves by means of pseudopods and has a thick cell wall made of cellulose. Could the report be true? Explain.
3. **Applying concepts** The growth of bacteria is slowed by cooler temperatures. How can you apply this information to control food spoilage?
4. **Designing an experiment** Penicillin is an antibiotic that kills some types of bacteria. Suppose you have five different bacteria cultures. Design an experiment to determine which of the cultures could be destroyed by penicillin. Include a control and a variable in your experiment.

## CONCEPT REVIEW: ESSAY

*Discuss each of the following in a brief paragraph.*

1. Compare viruses to cells.
2. Explain why some bacteria are harmful.
3. What contributions did Dimitri Iwanowski and Martinus Beijerinck make to the study of viruses?
4. What is the difference between autotrophic and heterotrophic bacteria?
5. Compare the methods of movement in an amoeba, a paramecium, and a *Euglena*.
6. Describe how viruses reproduce.

167

**5.** An amoeba moves by means of pseudopods; a paramecium moves by means of cilia; and a *Euglena* moves by means of flagellum.
**6.** Viruses reproduce in living cells by attaching themselves to the cell and injecting their hereditary material. The cell then makes more viruses, which burst from the cell.

## ADDITIONAL QUESTIONS AND TOPIC SUGGESTIONS

**1.** Why don't scientists agree about whether viruses are living things? (Viruses are composed of nucleic acids and proteins, which are substances found in all living things. Yet viruses do not carry on any life processes of organisms, and they can reproduce only with the help of living cells.)
**2.** Bacteria that are heterotrophs may be either parasites or saprophytes. What is the difference? (Parasitic heterotrophs feed on living organisms, while saprophytic heterotrophs feed on dead things.)
**3.** How is a *Euglena* similar to a plant and an animal? (*Euglena* has chloroplasts and carries on photosynthesis like plants. Like most animals, *Euglena* is capable of locomotion.)
**4.** Go to an encyclopedia and look up the work of Louis Pasteur and his development of the rabies vaccine. How is is produced? How is it administered? How does it work?
**5.** One common amoeba is about 0.25 mm long. How many amoebas this size could fit end to end in 1 mm? (4) In 1 cm? (40) In 1 inch? (102)

## ISSUES IN SCIENCE

The following issues can be used as a springboard for discussion or given as writing assignments.
**1.** Some health professionals feel that one goal of the World Health Organization should be the elimination of malaria from the earth. Others say that disease helps to keep the world population from growing too large, and diseases should not be wiped out. Others say the goal would take too much money to reach. Others think that another disease would quickly replace one that was ended. Explain how you feel.
**2.** What do you think would happen to life on earth if all protozoans were killed? What do you think would happen if all living things except protozoans were killed?
**3.** What limitations should be placed on turning bacteria into "chemical factories"(genetic engineering)? Human genes for the production of various hormones have been incorporated in bacterial DNA and made to function. Are we going too far? What are the dangers?

## Chapter 8

# NONVASCULAR PLANTS AND PLANTLIKE ORGANISMS

## CHAPTER OVERVIEW

Those plants and plantlike organisms that lack specialized tissues for carrying food and water throughout their bodies are studied together in this chapter. All of these organisms lack true roots, stems, and leaves. They are usually aquatic or found in wet habitats, and most of them are much older in origin than the vascular plants.

Algae are photosynthetic autotrophs containing chlorophyll and often a variety of other pigments. Almost all algae are aquatic, and they exhibit great diversity in size, cell arrangement, and specialization. Some algae are the likely ancestors of the more advanced land plants.

Like many algae, most fungi have bodies composed of simple threadlike filaments. But unlike algae, all fungi are heterotrophs. Further, their style of heterotrophism is unique in that they grow on their food source and absorb nutrients that are digested externally. Fungi vary greatly in form, but most of them reproduce by production of spores.

Lichens and slime molds are difficult to include in any classification scheme. Though they appear to be a single organism, lichens are actually a symbiotic association of certain fungal filaments and unicellular algae. The slime molds undergo a life cycle in which at times they resemble fungi, but at other times they have characteristics of protozoans.

Mosses and liverworts are the only group of nonvascular, photosynthetic organisms that are adapted for life on land. Even so, the lack of a vascular system limits these plants to damp habitats, and they cannot achieve the large sizes of some of the vascular plants.

## INTRODUCING CHAPTER 8

Ask students to look at the photograph on page 168. Point out that the pillar-shaped mounds are not ordinary rocks. Instead, they were produced by a simple form of living organism in the water. Write the word *stromatolite* on the chalkboard, and explain that the organisms that produce the stromatolites are blue-green algae.

- **How do you think that simple organisms, like blue-green algae, could have produced stromatolites?** (Accept all answers, but lead students to suggest they could be the remains of dead blue-green algae or the remains of something given off by their cells.)
- **Do you think the stromatolites were built rapidly or over long periods of time?** (Accept all answers, but lead students to suggest that long

# 8 Nonvascular Plants and Plantlike Organisms

**CHAPTER SECTIONS**



**CHAPTER OBJECTIVES**

*After completing this chapter, you will be able to:*

**8-1** Define nonvascular plants.
**8-1** Identify the characteristics of algae.
**8-2** Identify the characteristics of fungi.
**8-3** Describe lichens and slime molds.
**8-4** Compare mosses and liverworts.

The sun blazes in a cloudless sky. Far to the east, a tall pillar of orange smoke billows from an erupting volcano. The air, almost empty of oxygen, blows swiftly over the ocean, whipping up waves that crash against a huge sand bar. But there is no one here to witness or record these happenings. No people sail on this ocean or wade in the protected tidal pool, which is dotted with countless peculiar pillows of rock. No fish swim below the ocean's surface, nor do any birds fly above it. This is the world of 2.5 billion years ago. Here, there are no signs of life—except for those strange rocky pillows. One day in the future scientists will recognize these softly rounded rocks as the first structures built by living things.

Scientists call these softly rounded rocks stromatolites, which comes from Greek words that roughly mean "mattress of stone." But what living thing could build these 2.5-billion-year-old stony mattresses? The answer is blue-green algae. Undisturbed in protected pools of very salty water, the blue-green algae grew even larger and larger. As they did so, they produced a rocky mineral called lime. The lime piled up—year after year, century after century—and built a seascape of gray-brown cushions. And in certain parts of the world, such as in Shark Bay on the northwest coast of Australia, the process still goes on.

*Stromatolites in Shark Bay, Australia*



periods of time are required.)

Now ask students to read page 169 for an explanation of how stromatolites are formed today, as well as in the past. Explain that blue-green algae often live in places like hot springs or extremely salty water where few other organisms can survive. The sand bar at the mouth of Shark Bay keeps ocean water from flowing freely in and out. As water evaporates from the pool, considerable salt is left behind, making the bay uninhabitable for most organisms except blue-green algae. Thus the lime secreted by their cells piles up over long periods of time to form the pillarlike stromatolites.

• **What can stromatolites tell us about life in the past?** (Accept all answers, but lead students to conclude that simple organisms like blue-green algae have been living on the earth for a very long time.)

## TEACHER DEMONSTRATION

If possible, prepare a display of some of the organisms representing those to be studied in this chapter. Some possibilities include:

**1.** Floating algae from a pond or the seashore, or growing on the sides of an aquarium;
**2.** Bracket fungi, mushrooms, or a moldy piece of food;
**3.** A clump of moss;
**4.** Lichens on a rock or tree bark;
**5.** A slime mold on a rotting piece of wood.

Provide ample time for students to examine the specimens and/or photographs. After students return to their seats, announce that they have just completed some observations of organisms like those they will read about in Chapter 8.

• **Were some of these organisms familiar to you? How many could you identify?** (Answers will depend on the specimens displayed and prior experiences, but allow time for students to tell about their observations.)

• **Did the organisms appear to have anything in common?** (Accept all logical answers, but lead students to suggest that all were plants or they appeared to be more like plants than animals.)

Some students may also notice that though the specimens are plantlike, they lack familiar plant structures like leaves and flowers.

• **In what kind of environment might most of these organisms live?** (Accept all logical answers, but lead students to suggest that they probably live in water or in shady, damp places.)

## TEACHER RESOURCES

### Audiovisuals

*Algae, Fungi, and Lichens,* filmstrip, Eye Gate
*Kingdom of Plants,* filmstrip, National Geographic Society

### Books

Bold, C., and C. L. Hundell, *The Plant Kingdom,* Prentice-Hall
Brodie, J., *Fungi: Delight of Curiosity,* University of Toronto Press
Burns, *The Plant Kingdom,* Macmillan

## 8-1 ALGAE

### SECTION PREVIEW 8-1

Algae are simple nonvascular plants or plantlike organisms that possess chlorophyll. Lacking a transport system, they are generally found living in water or in moist habitats. The bodies of algae vary from microscopic, unicellular forms through colonial and filamentous forms to large multicellular seaweeds. Since all algae contain chlorophyll, they are photosynthetic. Most reproduce by binary fission, though many of them have complicated life cycles involving sexual and asexual stages. The chlorophyll of many algae is often masked by a variety of pigments. Thus algae may exhibit a variety of colors. Though not all scientists agree on the classification of algae, color is frequently used as a criterion for grouping them.

In this chapter the algae are classified in six groups according to the colors of the predominant pigments. The simplest algae, the blue-greens, are placed in the Kingdom Monera with the bacteria because of their simple cell structure. The other five groups of algae are classified as plants. Green algae vary from planktonic forms to terrestrial forms living on the surface of rocks and trees. Golden algae include the diatoms that live within an overlapping glassy cell wall. The largest and most complex algae are the brown and red algae. Most forms are found in the oceans. The fire algae are unicellular, and their cell walls resemble plates of armor.

### SECTION OBJECTIVES 8-1

1. **Define nonvascular plants.**
2. **Describe the characteristics of algae.**
3. **Relate chlorophyll to the process of photosynthesis.**
4. **List traits and give example of blue-green algae, green algae, golden algae, brown algae, red algae, and fire algae.**
5. **Recognize the ecological and economic importance of algae.**

## 8-1 Algae

In a northern California forest, a giant redwood tree over 200 years old stands through a violent rainstorm. Deep in the soil the tree's roots soak up water. Soon thin tubes in the tree's trunk will carry the water nearly 60 meters to the top of the tree. Without these tubes, the redwood could not bring water to its millions of living cells.

If you were walking through this forest, you would probably marvel at the height of the redwood tree. However, unless you looked closely, you might not notice a smaller plant growing on the bark near the bottom of the tree. Unlike the redwood, this (1) plant cannot grow to majestic heights. It is one of the earth's **nonvascular** (nahn-VA-skyuh-ler) **plants.** Nonvascular plants do not have transportation tubes to carry water and food throughout the plant.

The plant living on the bark of the redwood tree is a type of **alga** (AL-guh; plural: algae—AL-jee). **Algae are nonvascular plants that contain chlorophyll.** Because algae have no way of transporting water and food over long distances, they must live close to a (2) source of water.

**Figure 8-1** *The green patches floating on the surface of this lake are colonies of threadlike algae.*

### TEACHING STRATEGY 8-1

#### Motivation

Begin this lesson by showing the class two or more common vascular plants. Potted plants, such as geranium or coleus, might be used, or trees and shrubs that can be viewed from classroom windows might be observed. It is not necessary that the names of the plants be known.

- **Do all of you agree that each of these organisms is a plant?** (Accept all logical answers, but students will likely be in agreement.)
- **How do you know that these are plants?** (Call attention to such plant characteristics as their green color due to chlorophyll, and the presence of roots, stems, leaves, and possibly flowers.)

Now show an example of some nonvascular plant. Some easily ob-

Nonvascular plants such as the algae are often called simple plants. Simple plants lack true stems, leaves, and roots. However, do not let the term "simple" fool you. Nonvascular plants have managed to survive for hundreds of millions of years while other forms of life have come into being and then died off. In fact, through their long history nonvascular plants have become well adapted to many different environments on the earth.

### Where Algae Live

No doubt you enjoy a visit to the zoo. Walking past one of the cages, you might notice a greenish stump hanging from a large branch. You look more closely and you see that the stump is breathing. The stump is alive! Looking at the small plaque on the cage, you discover that the "stump" is an animal called the three-toed sloth. Interestingly, the hair of this and many other three-toed sloths is the home for certain green algae. Because the sloth is such a slow-moving animal, it is unable to escape if there is danger. The algae's green color helps the sloth to blend in with its surroundings until the danger has passed. In return, the sloth gives the algae a place to live.

Most algae, of course, do not live on sloths. Algae live in many different places. The majority of algae live in watery environments such as oceans, ponds, and lakes. Other algae grow in the soil, on the sides of houses, and at the base of trees. Certain species of algae thrive in the near-boiling water of hot springs (3) such as those in Yellowstone National Park. Other algae grow in snow. Some even grow in the icy waters of such places as the Antarctic continent, in which lies the South Pole.

**Figure 8-2** *Algae live in many different places. Some grow on the bark of trees* (top). *Others are at home in the hair of a three-toed sloth* (bottom).

### Structure of Algae

Some species of algae are unicellular, or one-celled, and can only be seen with a microscope. Quite often unicellular algae group together to form colonies. These colonies may even attach themselves to one another and form chains of algae.

Beaches often become littered with seaweed after a heavy storm. Seaweed is one type of multicellular



## SCIENCE TERMS 8-1

**nonvascular plant p. 170**
**alga p. 170**
**chlorophyll p. 172**
**photosynthesis p. 172**
**air bladder p. 176**
**bioluminescence p. 178**

## ANNOTATION KEY

(1) **Thinking Skill: Making comparisons**
(2) **Thinking Skill: Making generalizations**
(3) **Thinking Skill: Making comparisons**

## FACTS AND FIGURES

About 650 million years ago, some 70 percent of the different kinds of algae on the earth died out in the earliest known great extinction of living things.

Point out that in this chapter we will read about a wide variety of nonvascular plants, as well as some other organisms that are similar to nonvascular plants.

### Content Development

Be sure that students understand that without a transport system, nonvascular plants will be limited in size and confined to watery or damp environments. Plants can achieve large sizes or live in dry locations only if they have specialized tissues for carrying water to those cells some distance from a water source. Emphasize that all nonvascular plants have a simple structure. Most cells are near the surface where they make direct contact with their watery environ ment. Thus water enters the cells directly from the environment. Point out that the algae are one type of nonvascular plant, and they live in a watery environment.

tained specimens include *Spirogyra,* which is found floating on ponds during warm weather months. Other possibilities are a clump of moss or a piece of bark covered with *Protococcus.* This powdery appearing green alga can be collected throughout the year.

- **This is also a plant. How is it similar to the other plants?** (Focus attention on the fact that simple plants also possess chlorophyll.)
- **How is it different from the other plants?** (Lead students to suggest that it has a simpler structure. True roots, stems, and leaves are not present.)

Direct everyone's attention to the term nonvascular plant in the text. Discuss the meaning of this term, and then ask:

- **Which of these plants that we observed is nonvascular?** (Accept all correct answers.)

### Activity

**Algae Life Cycles**
**Skills: Comparing, diagraming, classifying**
**Level: Average**
**Type: Library**

The information gathered by students should be of use and interest to other students. Display their labeled diagrams of the rather complicated life cycles involved, and refer to them as you describe the various types of algae.

### Activity

*Algae Life Cycles*

Using reference materials in the library, find out more about the life cycles of the different groups of algae. Summarize your findings by drawing labeled diagrams of each type of life cycle.

algae. Multicellular algae can grow quite large. The giant kelp, for example, may stretch over 30 meters.

Whether unicellular or multicellular, all algae are autotrophs, or organisms that can make their own food. All algae contain **chlorophyll,** a green substance found in green plant cells. Chlorophyll is used in the process of **photosynthesis** (foh-tuh-SIN-thuh-sis). In photosynthesis, plants use the energy in sunlight to make their own food. Plants make food by combining carbon dioxide from the air with water and minerals from the soil. During the process of photosynthesis, oxygen is released from the plant.

Early in the earth's history there was very little oxygen in the atmosphere. In time, algae and other simple green plants, along with certain bacteria, released vast amounts of oxygen into the air. Eventually there was enough oxygen in the air to allow other forms of life to develop. If it were not for these simple green plants, no animals or people would now exist on the earth. And, as a matter of fact, if green plants were to suddenly vanish from the earth, oxygen in the air would soon become so rare that most living things would vanish.

Algae have several similarities and differences. Most algae reproduce by binary fission. Others have very complicated life cycles. Regardless of these differences, scientists have placed the algae into six groups. The groups are arranged according to the color of the algae.

**Figure 8-3** *Several kinds of seaweeds, which are types of algae, washed up on a New England beach.*

#### Blue-Green Algae

In 1883, on the island of Krakatoa in Indonesia, a great volcano exploded. The sound of the explosion was heard by people thousands of kilometers away. In an instant, all living things on the island were destroyed, leaving behind only bare rock. Yet a few years later, one form of life had begun to grow on the barren island. This living thing was a blue-green alga.

Soon a layer of blue-green algae carpeted much of the island. The layer of algae eventually became so thick that plants were able to grow on top of it. Today, as they did before the volcanic eruption, many kinds of plants live and flourish on the island of Krakatoa.



### 8-1 (continued)

#### Content Development

Ask students to recall the process of photosynthesis that was introduced in Section 5-2. Point out that this important process involves an interaction between plants and their environment.

- **What green substance within their cells do plants need for making their own food?** (Chlorophyll is needed for photosynthesis.)
- **What is the name of this food-making process?** (Photosynthesis is the process by which plants make food.)
- **What is the energy source for photosynthesis?** (Energy in sunlight is used for photosynthesis.)
- **What materials from the environment does a plant use to make food?** (The material requirements for food synthesis are carbon dioxide, water, and minerals from the soil.)
- **What substance does the plant release during photosynthesis?** (Oxygen is released to the environment.)

#### Skills Development

***Skill: Designing an experiment***

Tell the class to imagine that a pond snail and an aquarium plant have been placed in pond water in a test tube. A cork seals the tube so that no outside air can get it. After five days the snail and plant are both alive, and they appear to be healthy.

- **How did the snail remain alive without any air getting into the tube?** (Students will likely indicate that the plant produced something, perhaps oxygen, needed by the snail.)

Explain that they have formed a hypothesis—a hunch or explanation that can be tested. Divide the class into groups of three or four and ask them to design a simple experiment to test their hypothesis. Ask each team to write out a plan and/or diagrams showing the experimental setup. Bring the class together for a debriefing and sharing of plans. Lead students to conclude that their hypothesis might be tested by comparing sealed tubes containing plants and snails with sealed tubes containing snails only. A class demonstration

Figure 8-4 *Algae are found in many environments on the earth. These blue-green algae* (left) *are in a hot spring in Yellowstone National Park. Some blue-green algae* (right) *grow below the frozen surface of Lake Hoare in Antarctica.*

Blue-green algae are considered to be the simplest of all algae. In fact, unlike all other algae, blue-green algae are not considered plants. They are placed in the Monera kingdom, along with bacteria. ❷ All blue-green algae need to live is sunlight, nitrogen and carbon dioxide gas from the air, and a few minerals in their water supply. It is likely that they were the first organisms to grow on land.

Blue-green algae are special for other reasons. Some blue-green algae can remove nitrogen gas from the air and combine the nitrogen with other substances to make nitrogen compounds. In this way, blue-green algae help provide the nitrogen compounds that plants need to live. Normally, farmers must use fertilizers to replace nitrogen compounds used by green plants. However, in places like Asia where blue-green algae live in the water, farmers can plant rice year after year without the need of fertilizers. ❸ The blue-green algae fertilize the rice paddies naturally.

Not all blue-green algae are blue-green in color. In some algae, the blue-green substance is masked by another substance. For example, one species of blue-green algae is actually red. It lives in the Red Sea, which accounts for the sea's name.



## BACKGROUND INFORMATION

The blue-green algae are so named because in addition to chlorophyll, their cells contain a bluish pigment called phycocyanin. They may also contain other pigments that give them a black, brown, yellow, or red appearance. The blue-green algae occur as unicellular, filamentous, or colonial forms. They have no specialized structures for locomotion, though some species move slightly by swaying back and forth. Though found most often in freshwater ponds, they also live in salt water, on damp rocks, and in moist soil. In lakes that are polluted with organic matter, they sometimes multiply rapidly causing great algal blooms.

## ANNOTATION KEY

❶ **Thinking Skill: Making generalizations**
❷ **Thinking Skill: Applying concepts**
❸ **Thinking Skill: Applying technology**

of the experiment might then be set up. *Elodea* is an appropriate aquarium plant to use. Be sure to relate the proposed experiment to the discussion in the text of oxygen as a by-product of photosynthesis.

### Motivation

Have the class look at Figure 8-4.

• **What is unusual about the algae shown here?** (Some blue-green algae thrive in such places as hot springs or under Antarctic ice where few other types of life can survive.)

Emphasize that despite their simple cell structures, some blue-green algae are remarkably well adapted to live in habitats devoid of other organisms. They can grow almost any place where moisture is present.

• **What other evidence is given in your text to show that blue-green algae can live where other types of life are missing?** (Lead students into a discussion of the colonization by blue-greens of bare rock on Krakatoa. Point out that in a similar way, these algae may have been the first plantlike organisms to move onto land from the earth's ancient oceans.)

### Content Development

Explain that blue-green algae were at one time classified as plants. After it was discovered that their cells lack nuclei and other complex organelles, such as chloroplasts, they were reclassified with the bacteria in the Kingdom Monera. Many biologists now consider the blue-green algae to be a type of bacteria, and they refer to them as cyanobacteria.

## BACKGROUND INFORMATION

There are over 7000 species of green algae. Most species live in freshwater, but others live in salt water, in soil, on rocks and trees, and even in snow. The green algae are the most diverse type of algal plants. They vary from unicellular, microscopic forms through colonial forms to large multicellular marine forms over 8 meters long. Many move by means of flagella. Reproduction occurs in several ways. Some species reproduce only asexually, but many others have complex life cycles that alternate between asexual and sexual forms.

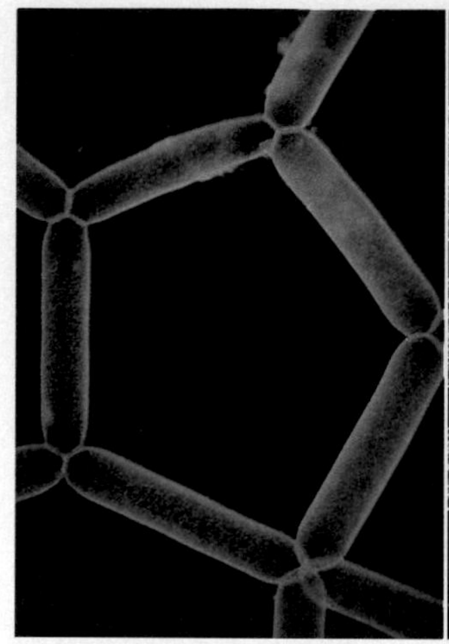

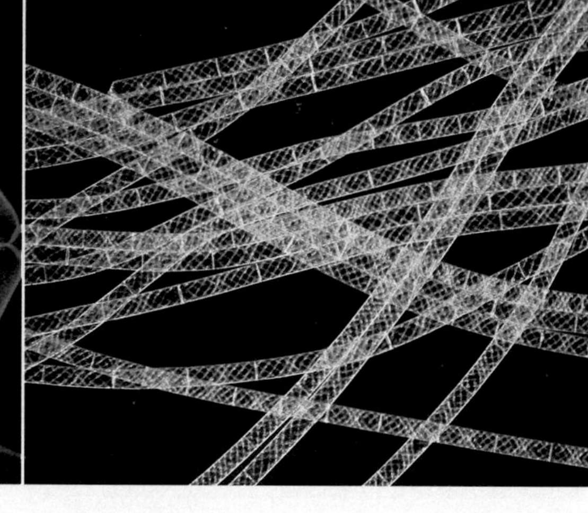

**Figure 8-5** *The green alga* (left), *magnified 50 times, forms "water nets" on ponds.* Spirogyra (right) *is a green alga that forms filaments and is found in pools of fresh water.*

### *Green Algae*

The year is 2001. Deep in space, a silvery ship is on its second year in a four-year journey to study the planets. On board, the two astronauts are about to finish dinner. Although they brought along enough food to last the entire journey, they could not carry enough oxygen to last several years. Are the astronauts doomed to suffocate in space? Of course not.

In a tiny room near the back of the ship lies a tub of water filled with green algae. The green algae use the carbon dioxide wastes exhaled by the astronauts to make food during photosynthesis. In return, the algae release their own waste, oxygen, for the astronauts to breathe. If you looked closely at the tub, you could see bubbles containing oxygen floating toward the surface. In this way, the astronauts and the green algae support each other's lives.

This scene is one that will not be played out for some time. However, scientists today are hard at work developing methods of growing green algae in (1) closed environments. These methods will allow the algae to produce the oxygen future travelers will use in their explorations of space.

Right now you can find green algae only at home on the earth. Most green algae live in the water, but

174

## 8-1 (continued)

### Skills Development

***Skill: Interpreting illustrations***

Direct the class to look at Figure 8-5 and read the caption.

- **How are these two types of algae similar?** (Both are types of green algae.)
- **Where are these green algae found?** (Both live in fresh water.)
- **Do these green algae appear to be unicellular or multicellular?** (In the figure, both appear to be multicellular.)

Point out that the alga in the left figure is *Hydrodictyon.* Strictly speaking, *Hydrodictyon* is neither unicellular nor multicellular. Each of the five cells making up the closed filament contains many nuclei. These nuclei result from multiple nuclear divisions without a corresponding division of the cytoplasm and cell walls. Green algae made up of such multinucleate cells are said to be *coenocytic.* On the other hand, *Spirogyra* is truly multicellular because the division of its nucleus is followed by division of the cytoplasm and the formation of cell walls. After division, the cells remain together to form long filaments.

- **Why is *Hydrodictyon* nicknamed "water net?"** (Accept all answers.)
- **The root word *spira* means "a coil" and the root word *gyro* means "to whirl or turn." Explain why *Spirogyra* is a good name.** (*Spirogyra* is characterized by a ribbonlike, spiraling chloroplast.)

### Enrichment

Since many green algae are a rich source of nutrients, research has been conducted to determine the feasibility of using them as a source of human food. One type that has shown some promise of being raised for such purposes is unicellular *Chlorella.* Encourage students to conduct research and prepare reports on the possibility of producing human food from *Chlorella.* Some references give recipes for making cookies from *Chlorella.* If such a reference is found, students might try to obtain some *Chlorella* and bake cookies to be shared with the class.

### Motivation

Begin the discussion of golden algae by displaying a few of the following

some plants may grow on the branches, stems, and leaves of trees. A few types of green algae may anchor themselves to rocks and pieces of wood.

### Golden Algae

Every morning and evening you probably brush your teeth. Chances are that the last thing you would consider brushing with is algae. Yet a part of many toothpastes is made of golden algae. Do not let their name fool you, however. Golden algae all contain the green substance chlorophyll. Their golden color comes from a mixture of orange, yellow, and brown substances in their cells.

The most common, and certainly the most attractive, golden algae are the diatoms (DIGH-uh-tuhms). If you looked at diatoms through a microscope, they would resemble tiny glass boxes. See Figure 8-6. The appearance of diatoms is due to a glassy material in their cell walls. When diatoms die, their tough glass-like walls remain. In time, the walls collect in layers and form deposits of diatomaceous (digh-uh-tuh-MAY-shuhs) earth. Diatomaceous earth is a coarse, powdery material. For this reason, it is often added to toothpastes to help polish teeth. And, because ❷

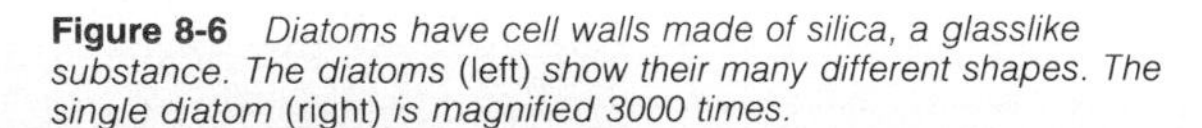

**Figure 8-6** *Diatoms have cell walls made of silica, a glasslike substance. The diatoms* (left) *show their many different shapes. The single diatom* (right) *is magnified 3000 times.*

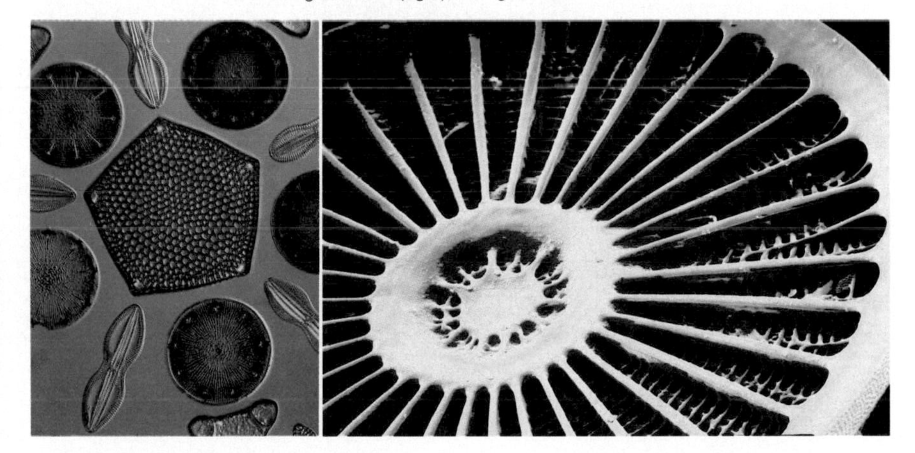



## BACKGROUND INFORMATION

In addition to diatoms, the golden algae also include about 1500 species of golden-brown algae and 600 species of yellow-green algae. All members of this group are unicellular, and they contain accessory pigments along with chlorophyll, causing their colors to range from brown to yellow-green. Although most algae store their food as some form of carbohydrate, the golden algae generally store theirs as oils. Because of their oil reserves some biologists have suggested that these algae might be a potential source of oil for fuel.

## FACTS AND FIGURES

Diatoms are the most numerous of all algae. Biologists believe that nearly 10,000 species exist. Though small, they are nevertheless a major provider of oxygen because of their vast numbers.

## ANNOTATION KEY

❶ **Thinking Skill: Applying technology**
❷ **Thinking Skill: Sequencing events**

items in which diatomaceous earth may be an ingredient, or is used in its production:

Toothpaste, car polish, silver polish (used as an abrasive);

Paint (used as a filler);

Acoustical tile (used as a soundproofing material).

• **What do these products have to do with algae?** (Diatomaceous earth, derived from the glassy cell walls of diatoms, is used in making these products. Diatoms are a type of golden algae.)

Discuss the specific use of diatomaceous earth in each product displayed. Then demonstrate the structure of a diatom cell wall with a petri dish. A crumpled tissue or paper towel can be placed in the petri dish to represent the cell itself. Explain that the top and bottom halves of the petri dish are like the overlapping glassy cell wall. Point out that the cell walls contain silica, a compound used in making glass. It is the silica that produces diatomaceous earth.

### Content Development

Point out that diatoms are mostly marine, but some species are found in freshwater and terrestrial habitats. Most of the plankton—the microscopic organisms floating at the ocean's surface—is composed of diatoms. Diatoms are probably the major food-producing organisms of the oceans.

## Activity

**Collecting Algae**
**Skills: Observing, comparing, manipulative**
**Level: Enriched**
**Type: Hands-on**
**Materials: Bark with algae, glass dish, slide and cover slip, microscope**

Students may succeed in collecting samples of a number of different species of algae. The drawings of the various samples seen through a microscope should be displayed, and students should be encouraged to conduct library research to attempt to identify the algae.

## ANNOTATION KEY

1 Thinking Skill: Relating concepts
2 Thinking Skill: Relating concepts

### Activity

*Collecting Algae*

Some algae appear as green masses on the bark of trees.

**1.** Find such algae and remove it from the tree along with some bark.
**2.** Place the algae in a dish containing a small amount of water. Cover the dish.
**3.** After a few days, scrape off some of the green material and place it on a glass slide containing a drop of water. Cover with a cover slip.
**4.** Examine the slide under the low and high powers of a microscope.

How does this alga compare with the photographs in Figure 8-5? Draw and label this alga under low and high power.

diatomaceous earth reflects light, it is also added to the paint used to indicate separate traffic lanes on highways.

### Brown Algae

For centuries, sailors whispered tales of a huge sea within a sea in the Atlantic Ocean southeast of Bermuda. This sea, the Sargasso Sea, the sailors said was filled with endless tangles of brown seaweed. The name of the sea comes from the brown *Sargassum* algae that grow in great amounts in this part of the Atlantic Ocean. Legend had it that ships trapped in this thick growth of seaweed would remain here forever, forming a fleet of dead ships guarded by skeletons. Today, some people believe that the sailors traveling with Christopher Columbus mutinied not out of fear of falling off the side of the earth, but out of fear of being trapped in the Sargasso Sea.

The legends about the Sargasso Sea are merely a product of people's strong imagination. However, the brown algae that float there are very real. The *Sargassum* lie on or near the ocean's surface. There they are able to receive enough sunlight to perform photosynthesis. They stay near the surface because they have **air bladders.** Air bladders are tiny grape-

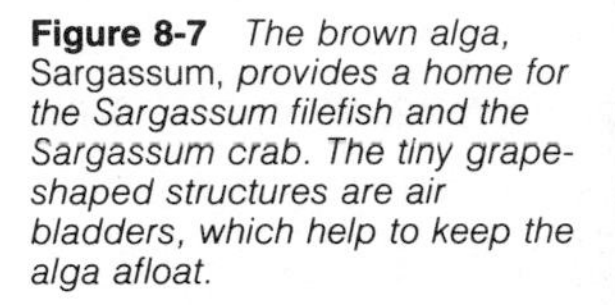

**Figure 8-7** *The brown alga,* Sargassum, *provides a home for the Sargassum filefish and the Sargassum crab. The tiny grape-shaped structures are air bladders, which help to keep the alga afloat.*

176

## 8-1 (continued)

### Content Development

The bodies of some brown algae consist of simple branching filaments. Other are quite large with a variety of specialized tissues. Some of them, such as *Sargassum* seen in Figure 8-7, have a branched structure called a holdfast at the base to secure them to rocks or other objects on the sea floor. A stemlike structure called the stipe extends from the holdfast. Above the stipe is a leaflike blade with air bladders to keep it afloat. Students may mistakenly refer to the holdfast, stipe, and blade of some large brown algae as roots, stems, and leaves, respectively. Explain that though structures resemble advanced plant organs, their internal organization is different. Brown algae do not have true vascular tissues composed of xylem and phloem.

Call attention to the air bladders shown in Figure 8-7 and described in the text.

• **Why are air bladders so important to *Sargassum* and other large brown algae?** (The air bladders keep the blades afloat so they can receive maximum sunlight for photosynthesis.)

### Skills Development

***Skill: Making inferences***

Ask the class to look at Figure 8-7 and read the caption. Emphasize the fact that many organisms make their homes within beds of large brown algae.

• **Can you think of a reason why the filefish and crab make their home among strands of *Sargassum?*** (Accept all logical answers, but lead students to infer that their color enables them to blend in with their environment. Thus, they are protected from predators.)

shaped structures. These air bladders act like inflatable life preservers.

Brown algae, such as *Sargassum,* are the largest and most complex of the algae. Most brown algae live in salt water and are called seaweeds or kelps. Some brown algae float freely while others remain attached to the sea floor.

### Red Algae

Red algae, named for a red substance found in their cells, make up another large group of algae. Some red algae produce a jellylike material called agar (AH-guhr). Agar is a substance on which scientists grow bacteria.

Red algae can grow to be several meters long, but they never reach the size of brown algae such as *Sargassum.* Red algae usually grow attached to rocks on the ocean floor. Their red color allows them to absorb a part of sunlight that can penetrate deep into the ocean. For this reason, some species of red algae grow as far as 200 meters below the ocean's surface. ❷

**Figure 8-8** *Unlike brown algae which float free, red algae usually are attached to rocks on the ocean floor. Red algae also are smaller and more delicate than brown algae.*

### Fire Algae

In the winter and spring of 1974, the waters near the west coast of Florida suddenly turned brown and red. Soon thousands of dead fish washed up on shore. One young boy became very ill after eating clams taken from the area. This was not the first time a "red tide" had struck a Florida beach, or, for that matter, other beaches in the United States. Red tides have swept onto beaches in such places as New England and Los Angeles.

The red tide is not a tide at all. It is a large group of unicellular fire algae. Fire algae range in color from yellow-green to orange-brown. Some of these fire algae produce poisons. These poisons can injure or kill living things.

If you observed fire algae through a microscope, you would notice something interesting. Fire algae have cell walls that look like plates of armor. In addition, fire algae have two flagella that help propel them forward. So the algae might look like tiny submarines spinning through the water like tops.

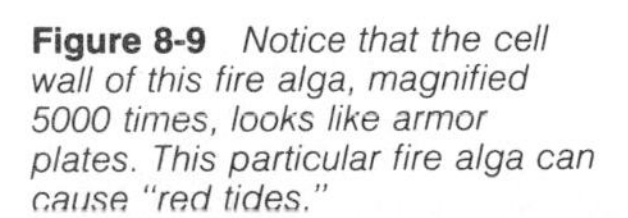

**Figure 8-9** *Notice that the cell wall of this fire alga, magnified 5000 times, looks like armor plates. This particular fire alga can cause "red tides."*

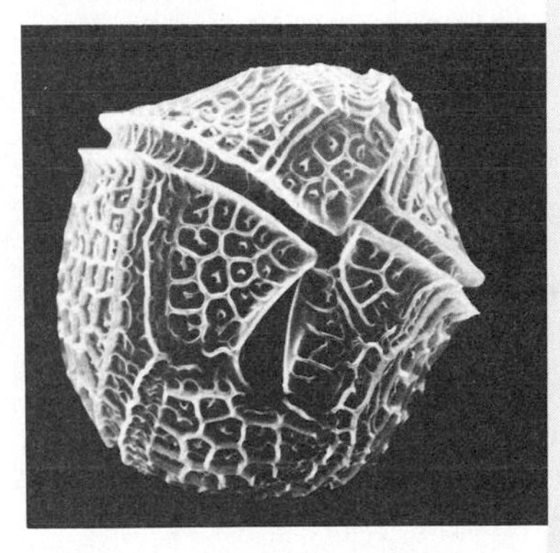

177

• **Can you think of a term that describes this characteristic of the filefish and crab?** (Students may be aware of the terms *camouflage* or *protective coloration.*)

Ask the class to name other examples of organisms that are protected by their coloration.

## Content Development

Explain that biologists are not sure of the cause of outbreaks of red tides.

Shellfish, such as oysters and clams, are not killed by the powerful toxins produced by the red-tide fire algae. However, shellfish can concentrate the toxins in their tissues, making them unfit for human consumption. The seafood industry often loses millions of dollars due to outbreaks of red tides.

## SCIENCE, TECHNOLOGY, AND SOCIETY

Several new breakthroughs in fungal research may prove to be beneficial for humans in a variety of ways. Many of our most common antibiotics are produced by filamentous fungi. However, some of the older antibiotics (such as penicillin) have become less effective against infectious diseases due to the development of resistant strains of bacteria. Years of scientific research have resulted in a new generation of antibiotics, which are also produced from fungi. Cephalosporins, from the fungus *Cephalosporium,* and new varieties of penicillin are among these new antibiotics.

A new type of pesticide may be developed from the spores of a predatory fungus found in the rain forests of Costa Rica. When a spore from this fungus falls on an insect or spider, it germinates and the hyphae enter the animal's body. Growth of the hyphae interfere with the animal's nervous system, causing death. The spores germinate into mushrooms, which grow on the dead carcass and soon release more spores. Scientists are considering the idea of using these fungal spores to control crop pests. The spores would be suspended in a liquid and sprayed on crops. Insects attacking the crops would become infected with the predatory fungus and die; the plants would be unharmed.

Other recent experiments with fungi have led to the discovery of a common wood rot fungus that transforms lignite coal into a liquid. Further research with this fungus has enabled scientists to isolate an enzyme that dramatically speeds up the reaction. Perhaps this enzyme can be used to liquefy coal underground so that it can be pumped to the surface like oil.

These are just three ways in which fungi have played key roles in the advancement of science and technology. Which of these three discoveries do you think will be most beneficial to humankind? For which of these discoveries would you advocate financial support for continued research?

## 8-2 FUNGI

### SECTION PREVIEW 8-2

Like algae, the bodies of fungi are nonvascular. But unlike algae, they lack chlorophyll, so they are always heterotrophs. Since they cannot make food, nor move around like animals to find it, the fungi always live in or on a food source. Some live as parasites, taking their food from a living host. Others are saprophytes, obtaining their food from dead organisms or the products of organisms.

Some fungi, such as yeasts, are unicellular. Most fungi have multicellular bodies composed of threadlike filaments called hyphae. The hyphae produce digestive enzymes that are released into the substrate on which the fungus grows. The food, which is digested externally, is then absorbed by the hyphae. Most fungi reproduce by producing spores in structures called fruiting bodies. In many fungi the spores can by produced by either a sexual or asexual process.

### SECTION OBJECTIVES 8-2

1. **Distinguish between fungi that are parasites and those that are saprophytes.**
2. **Describe the food absorbing and reproductive structures common to most fungi.**
3. **Identify the structures of a mushroom's fruiting body.**
4. **Describe fermentation and reproduction in yeasts.**
5. **Name two examples of molds.**

Some fire algae have a characteristic that has amazed sailors since they first set sail on the world's oceans. These sailors often saw a strange glow at night in the water behind their ships. What the sailors did not know was that the glow was produced by fire algae. This glow is called **bioluminescence** (bigh-oh-loo-muh-NE-suhns), which means "having living light." The glow is similar to that produced by (1) fireflies, or "lightning bugs."

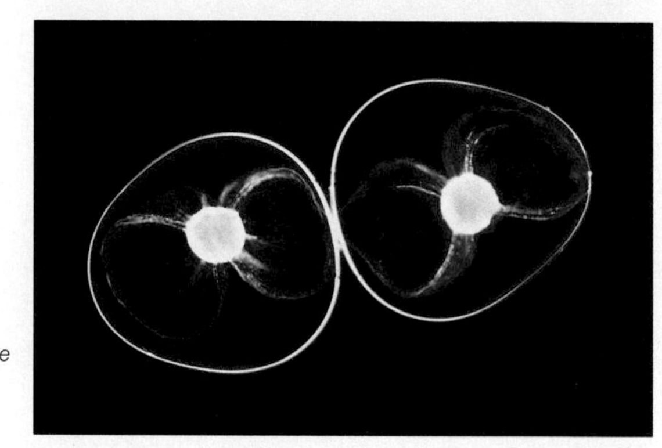

**Figure 8-10** *These two fire algae cells produce a cold light, or a glow, called bioluminescence.*

### *SECTION REVIEW*

1. What are nonvascular plants?
2. List six groups of algae.
3. What is bioluminescence?

## 8-2 Fungi

In 1927, scientists discovered a deadly elm tree disease in England. The disease, called Dutch elm disease, caused the leaves of the trees to wither. Eventually, the trees died. You would think that elm trees in the United States, separated from England by the Atlantic Ocean, would be safe. However, despite repeated warnings, wood of the English elm trees was shipped to this country to make furniture. By 1930, Dutch elm disease had struck trees in Ohio.



### 8-1 (continued)

#### Skills Development

***Skill: Classifying algae***

Prepare a worksheet or overhead transparency to help students identify characteristics and examples of the six kinds of algae.

Classifying Algae

| | |
|---|---|
| blue-green algae | brown algae |
| green algae | red algae |
| golden algae | fire algae |

(fire) Cause of red tides

(brown) *Sargassum*

(green) Possible source of oxygen for astronauts

(blue-green) Classified in Kingdom Monera

(fire) Two flagella produce spinning motion

(red) Produce agar for growing bacteria

(golden) Diatoms

(brown) Largest and most complex algae

(green) *Spirogyra*

(golden) Resemble tiny glass boxes

(fire) Some produce bioluminescence

(blue-green) Provide nitrogen for rice plants

**Figure 8-11** *The elm tree (left) has been attacked by a fungus that causes Dutch elm disease. This disease is carried by the bark beetle whose wormlike larvae (right) burrow through the bark of the tree.*

Today, Dutch elm disease has spread to almost every state in the country. The disease is carried from tree to tree by insects called bark beetles. However, as scientists learned, Dutch elm disease actually is caused by a type of **fungus,** (FUHNG-guhs; plural: fungi) carried by the beetles.

**Fungi are nonvascular plantlike organisms that have no chlorophyll.** Therefore, fungi cannot perform photosynthesis. Because fungi must obtain their food from living or once-living organisms, they are heterotrophs. ❷

Fungi that live and grow on other living things are called parasites. The fungus that kills elm trees is an example of a parasite. Other parasitic fungi grow on and destroy crop plants, which people depend upon for food. Fungi can also attack people directly, as anyone who has ever suffered from the itching of athlete's foot can tell you. Animals, too, can be victims of fungi. Some fungi, for example, infect fishes, including those that are raised in home aquariums. ❸

Other species of fungi get their food from once-living organisms. These fungi are called saprophytes. They, along with many types of bacteria, are the earth's "clean-up crew." These fungi decompose, or

**Figure 8-12** *The four shelflike structures growing on this tree are shelf fungi. These fungi are parasites and get their food from the living tree.*

179

## SCIENCE TERMS 8-2

**fungus p. 179**
**hyphae p. 180**
**fruiting bodies p. 180**
**cap p. 181**
**stalk p. 181**
**gills p. 182**
**fermentation p. 182**
**budding p. 182**

## ANNOTATION KEY

❶ Thinking Skill: Relating facts
❷ Thinking Skill: Relating concepts
❸ Thinking Skill: Applying concepts

## TEACHING STRATEGY 8-2

### Motivation

Use a free-writing activity to introduce the study of fungi. Write the term *fungi* and its singular form *fungus* on the chalkboard. Point out that these plantlike organisms will be studied next. Now ask the class to write for about five minutes on what they already know or think they know about fungi. As this activity is diagnostic in nature, accept any honest ideas, even though many inaccuracies and misconceptions will be presented at this time. When finished, group students into teams of three or four and have them share writings. Ask each team to prepare a list of questions that they would like to have answered about fungi.

Do not criticize students for their lack of knowledge about the topic or attempt to grade their papers on the basis of accuracy. Instead, identify problem areas, and clarify misinformation as you teach Section 8-2. The questions generated by each group can further serve as a guide to content development.

### Content Development

Explain that the fungus that causes Dutch elm disease is spread by two kinds of bark beetles—the elm leaf beetle and the longhorned beetle. The fungus grows in the new ring of wood adjacent to the bark where it interferes with the flow of sap.

(golden) Source of diatomaceous earth
(brown) Have air bladders to keep them afloat
(blue-green) Some types live in hot springs
(red) Capable of living in deep ocean waters

Ask students to identify the group of algae that matches each example or description. Answers are given in parentheses.

### Section Review 8-1

1. Nonvascular plants are plants without transportation tubes to carry food and water throughout the plant.
2. Blue-green, green, golden, brown, red, and fire.
3. Bioluminescence is light produced by living things such as fire algae.

## BACKGROUND INFORMATION

Hyphae consist of tubelike structures containing cytoplasm and hundreds of nuclei. In some fungi the hyphae are divided into sections by cross walls called septa, but there are perforations in these walls. Thus the cytoplasm and nuclei are able to move freely throughout the hyphae. Other fungi lack septa altogether. As hyphae grow they form a branching tangled mass called a mycelium. It is the mycelium that forms the body of a fungus.

As a fungus matures a part of the mycelium develops into a reproductive structure, or fruiting body. Spores are produced within fruiting bodies by either a sexual or asexual process. Fungi can also reproduce asexually by fragmentation. In this process, fragments of broken hyphae are carried by wind or water to new locations. If conditions there are favorable, the fragments will develop into new hyphae.

**Figure 8-13** *Fungi have many shapes and can be very colorful. The bird's nest* (top left), *morel* (top right), *scarlet cup* (bottom left), *and death cup* (bottom right) *are different kinds of fungi.*

break down, dead plant and animal matter. These broken-down products become the foods of other living things. Without saprophytes, dead plants and animals would soon litter and pollute the earth.

### Structure of Fungi

Fungi have several types of structures. Most members of the Fungi kingdom have threadlike structures called **hyphae** (HIGH-fee; singular: hypha). These hyphae produce special chemicals called enzymes. The enzymes digest or break down the cells of living or dead organisms. The broken-down cells and their chemicals are used by fungi as food. Fungi absorb food through the walls of the hyphae.

Most fungi reproduce by means of spores. These (1) spores are contained in special **fruiting bodies.**



## 8-2 (continued)

### Skills Development

***Skill: Making comparisons***

Tell the class that at one time, algae and fungi were placed together in a single group called the *thallophytes* within the plant kingdom. Now fungi are placed in their own kingdom.

- **Name some ways in which fungi are like algae.** (The bodies of fungi are made up of threadlike filaments like those of some algae. Further, the cells of both are surrounded by cell walls. Both algae and fungi are nonvascular organisms lacking true roots, stems, and leaves.)
- **What are some ways in which fungi are different from algae?** (Unlike algae, fungi have no chlorophyll for carrying out photosynthesis. Therefore, they are either saprophytes or parasites, obtaining food made by other organisms.)
- **Compare parasites and saprophytes. How do these kinds of organisms obtain their food?** (Parasites obtain their food from living organisms, while saprophytes feed on dead remains of other organisms.)
- **Name some parasitic fungi.** (Accept all plausible answers. Several examples are given in Section 8-2.)
- **Name some fungi that are saprophytes.** (Several examples are given in Section 8-2).

### Content Development

Use visuals to help students distinguish between the hyphae and fruiting body. A chalkboard, overhead transparency, or wall chart illustrating the hyphae and reproductive structures of either a mushroom or bread mold would be appropriate.

Point out that in many fungi, such as mushrooms, the fruiting body is the most conspicuous structure. Call attention to the various forms of fruiting bodies shown in Figure 8-13, and emphasize that these colorful structures produce reproductive spores. Be sure that students understand that hyphae are

Fruiting bodies are structures that form from the closely packed hyphae. When the spores are released and land in moist areas, such as on the forest floor, they grow into new fungi.

### *Mushrooms*

Have you ever ordered a pizza with all the trimmings? If so, you probably ate one type of fungi called the mushroom. All you may know about the mushrooms on pizza is that they taste good. But the pizza maker has to know a lot more because certain mushrooms are poisonous, and poisonous mushrooms on a pizza certainly would not improve sales.

Fortunately, the mushrooms on pizzas and stocked on grocery shelves come from special farms and are safe to eat. However, mushrooms that grow in the wild often are not safe to eat. One of the most poisonous wild mushrooms is the death cup. This well-named mushroom produces a poison that acts much like the venom of a rattlesnake. Since most people cannot tell the difference between poisonous and nonpoisonous mushrooms, wild mushrooms should never be picked or eaten. ❷

The umbrella-shaped part of the mushroom you probably are most familiar with is the mushroom's **cap.** It is actually the mushroom's fruiting body, which contains spores. See Figure 8-14. The cap can be any color, depending on the species of mushroom. The cap is at the top of a **stalk,** a stemlike

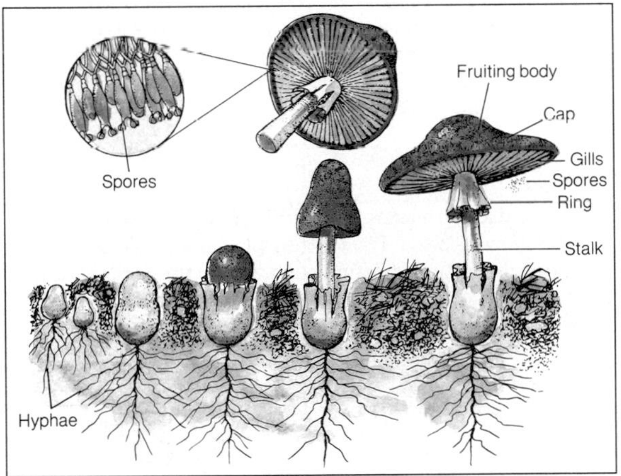


**Figure 8-14** *As a mushroom develops, the fruiting body emerges from under the ground and releases spores. These spores then develop into new mushrooms.*

#### Activity

*Making Spore Prints*

How are spores involved in the reproduction of mushrooms?

**1.** Place a fresh mushroom cap, gill side down, on a sheet of white paper. Cover the cap with a large beaker, open end down.

**2.** After several days, carefully remove the beaker and lift off the mushroom cap. You should find a spore print on the paper.

**3.** Spray the paper with clear varnish or hair spray to make a permanent spore print.

**4.** Examine the print with a magnifying glass.

Explain your print in terms of the reproduction of mushrooms.



the growing, food-absorbing structures that spread over the food surface on which the fungus grows. The hyphae are often hidden from view, growing under soil or leaf litter, or within the bark of a decaying log.

## Enrichment

Some students may be interested in doing library research and preparing reports on mushrooms and other fungi that are used for food. Reports could also include information on fungi, such as yeasts, that are used in preparing certain foods and beverages. Provide an opportunity for reports to be shared with the class.

## Activity

**Making Spore Prints**
**Skills: Observing, manipulative, relating**
**Level: Average**
**Type: Hands-on**
**Materials: Fresh mushroom cap, white paper, large beaker, clear varnish or hair spray, magnifying glass**

This activity requires part of two class periods several days apart. It provides an interesting hands-on introduction to the study of mushrooms. The observation of the spore print is best done after students have completed the section on mushrooms. Students should obtain a spore print consisting of lines radiating outward like spokes. Examination with a magnifying glass will reveal the tiny, dotlike spores that make up the print. These spores are produced by small reproductive structures within the mushroom's gills. Under the correct conditions, the spores grow into new mushrooms.

## ANNOTATION KEY

❶ Thinking Skill: Classifying fungi parts
❷ Thinking Skill: Applying facts

## BACKGROUND INFORMATION

The club fungi include some parasitic varieties. Among these are the rusts and smuts, which cause considerable damage to grain crops. One family of rusts attacks wheat, oats, and barley, causing millions of dollars of damage each year. Rusts often have complex life cycles that involve different hosts. For example, wheat rust infects wheat and barberry plants at different stages of the life cycle. Smuts produce black, slimy swellings on ears of corn.

## TEACHER DEMONSTRATION

Show the class a package of dry yeast. Point out that if conditions for growth and reproduction become unfavorable, yeasts form spores within their cells and become dormant. The dry yeast within this package is a collection of dormant yeast spores.

- **What do we need to do to cause the yeast to become active?** (Most students will be aware that the yeast can be activated by adding water to the package contents.)

Point out that in addition to moisture, yeasts also need a source of food.

- **What do yeasts use as a source of food energy?** (The source of energy for most yeasts is sugar.)

Prepare a culture of yeast cells by providing the conditions needed to activate the packaged yeast. To culture the yeast, mix about 5 mL of molasses and 500 mL of warm tap water in a beaker, and and 1/2 package of dry yeast. Stir the mixture, cover the top loosely with aluminum foil, and allow it to stand for about 30 minutes in a warm place.

- **Why was molasses added to the dry yeast mixture?** (The molasses contains the sugar needed for food.)

Have the class use microscopes to view yeast cells under low and high power. The cells might be seen better if a drop of methylene blue stain is added to the culture under high power. Everyone should make a drawing to record his or her observations.

**Figure 8-15** *When a water droplet hits this puff ball fungus, the fungus responds by releasing a cloud of spores.*

**Figure 8-16** *Yeasts reproduce by budding, as shown in this photograph. The cells that result can form new yeast colonies.*

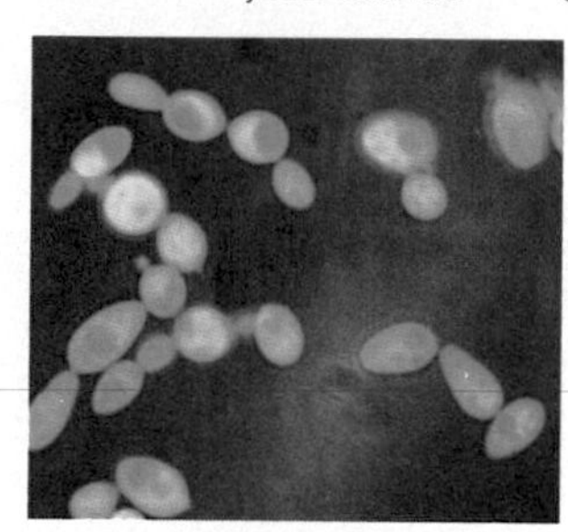

structure that has a ring near its top. If you were to turn the mushroom upside down, you would see its (1) **gills.** The gills are the mushroom's spore factories.

Not all mushrooms resemble the types that you see in the grocery store. The puffball, for example, looks like a giant softball and can grow up to 60 centimeters in diameter. When a raindrop hits the puffball, a tiny puff of "smoke" is given off. The puff of smoke contains thousands of spores. See Figure 8-15.

### Yeasts

Most people cannot help but stop and take a deep breath when they pass a bakery. There is something about the aroma of fresh bread that excites our senses. The next time you pass a bakery you might whisper a soft thank-you to another type of fungi, the yeasts.

Yeasts, like other fungi, have no chlorophyll. They obtain their energy through a process called **fermentation.** During fermentation, sugars and starches are changed into alcohol and carbon dioxide gas. At the same time, energy is released.

Bakers add yeast to bread dough. As the dough bakes, the yeast produces carbon dioxide. This causes the bread to rise. The carbon dioxide also (2) produces millions of tiny bubbles in the bread, which you can see as holes in a slice of bread.

Although yeasts are unicellular, they can clump together to form chains of cells. Yeasts reproduce by forming spores or by **budding.** See Figure 8-16. During budding, a portion of the yeast cell pushes out of the cell wall and forms a tiny bud. In time, the bud forms a new yeast.

### Molds

Centuries ago people treated infections in a rather curious way. They often placed decaying breads, cheeses, and fruits such as oranges on the infection. Although the people did not have a scientific reason to do this, every once in a while the infection was cured. What these people did not and could not know was that the cure was due to a type of fungus, called mold, which grows on certain foods.



### 8-2 (continued)

### Content Development

Mention that yeasts and some molds are classified as sac fungi, fungi that produce their spores in a saclike cell called an ascus. With about 30,000 species, the sac fungi comprise the largest and most economically significant group of fungi. Among the sac fungi are plant parasites that cause diseases such as powdery mildew; chestnut blight, which has eliminated the American chestnut; and Dutch elm disease, which is now destroying the American elm. The fungus that causes ergot in rye also causes the disease known as "St. Anthony's Fire." Epidemics of this disease in medieval Europe crippled or killed many people and livestock.

On the positive side, the antibiotic penicillin is produced by a sac fungus. Small doses of ergot are also used medically, to lower blood pressure and ease migraine headaches. Other sac fungi are used in the production of blue cheeses, such as Roquefort. Morels and truffles are prized as food by many gourmets. Yeasts are of enormous commercial value in the baking process and as a source of alcoholic beverages and industrial alcohol.

In 1928, the Scottish scientist, Sir Alexander Fleming found out why this treatment worked. Fleming discovered that a substance produced by the mold *Penicillium* could kill certain bacteria that caused infections. Fleming named the substance penicillin. Since that time penicillin, an antibiotic, has saved millions of lives.

You have probably seen the common mold that grows on bread. This mold looks like tiny fluffs of cotton. Actually, these fluffs are groups of long hyphae that grow over the surface of bread. Shorter hyphae grow down into the bread and resemble tiny roots. The shorter hyphae release the enzymes that break down chemicals in the bread. The broken-down chemicals are food for the mold and are absorbed by the hyphae. See Figure 8-17.

Perhaps you have noticed tiny black spheres on bread mold. These spheres are spore cases, which produce spores. The spores are carried from one place to another through the air. When the spores land on food, they begin to develop into a new mold.

**Activity**

*Growing Mold*

③ Here is a way to grow your own mold.

④ **1.** Obtain a large covered container. Line the container with moist blotting paper.

**2.** Place an orange in the container and put the cover on.

**3.** Store the container in a dark place for a few days.

Describe the color and appearance of the mold. What conditions are necessary for the growth of mold? Are molds autotrophs or heterotrophs? Explain your answer.

**Figure 8-17** *Bread mold reproduces by producing spores. They are released from structures called spore cases. Long hyphae anchor the mold to bread. Short hyphae absorb food from bread.*

⑤

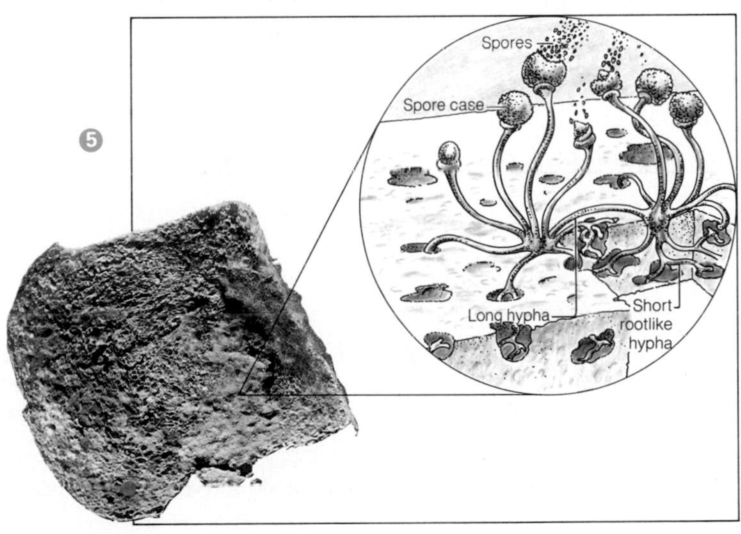


183

## Activity

**Growing Mold**
**Skills: Manipulative, observing, inferring, relating, applying**
**Level: Average**
**Type: Hands-on**
**Materials: Covered container, blotting paper, orange**

Students will likely see some mold growth when they complete this activity. The mold will probably be either green, black, or white. Students should infer that moisture and a food source are necessary to promote mold growth and that molds are heterotrophs.

## ANNOTATION KEY

❶ **Thinking Skill: Classifying mushroom parts**
❷ **Thinking Skill: Relating cause and effect**
❸ **Thinking Skill: Applying technology**
❹ **Thinking Skill: Making observations**
❺ **Thinking Skill: Interpreting illustrations**

## Motivation

Introduce the discussion of the discovery of the antibiotic penicillin by showing the class some examples of common molds of the genus *Penicillium*. Bluish-green molds of this genus are often seen growing on citrus fruits, in Roquefort cheese, and sometimes on bread. Ask the class to recall that antibiotics are chemicals produced by helpful bacteria that weaken or destroy disease-causing bacteria (Section 7-2). Explain that some molds also produce antibiotics. Be sure to point out that penicillin is not actually derived from the mold on an orange or Roquefort cheese, but from a closely related species.

## Skills Development

***Skill: Making comparisons***

Tell students to reexamine Figures 8-14 and 8-17.

- **How are mushroom and bread mold similar in the way they obtain their food?** (Both of these fungi have threadlike hyphae which penetrate into the surface on which they grow. The hyphae release enzymes in the food surface and the broken-down food is absorbed.)
- **What similarities are there in the fruiting bodies of these two fungi?** (In both fungi the spores are produced in special reproductive structures sitting atop a stalk.)
- **What differences do you see in the fruiting bodies of mushroom and bread mold?** (In a mushroom the spores are produced and released by the gills under the cap. In bread mold the spores are produced inside spore cases. When the spore cases open, the spores are released.

## 8-3 LICHENS AND SLIME MOLDS

### SECTION PREVIEW 8-3

Lichens are not a single organism but a combination of a fungus and an alga living in a symbiotic relationship. In this association the fungus provides the alga with water and minerals absorbed from the surface on which the lichen grows. In turn, the alga makes food for the fungus and itself. Lichens are capable of living in a wide variety of environments ranging from frozen arctic regions to dry hot deserts. They also can survive on bare rocks, tree bark, and on mountain tops. Lichens are called pioneer plants because they are often the first plants to colonize a rocky barren area. Acids released as a metabolic by-product begin the process of soil formation, making it possible for other organisms to follow.

Slime molds are organisms having stages in their life cycle that are like those of both fungi and protozoans. In one of its stages, it resembles a giant amoeba that spreads slowly over the surface on which its grows. In another stage, it produces reproductive spores within colorful fruiting bodies like those of some fungi.

### SECTION OBJECTIVES 8-3

1. **Describe the structure and role of the organisms making up a lichen.**
2. **Define *symbiosis*.**

### SCIENCE TERMS 8-3

**lichen p. 185**
**symbiosis p. 185**
**slime mold p. 186**

### Career: *Mushroom Grower*

**HELP WANTED: MUSHROOM GROWER** Must know basics of mushroom growing and have an interest in botany. High school diploma needed. Business and marketing courses desirable.

People have been using mushrooms to flavor food for thousands of years. Usually the mushrooms were found growing wild in fields. It was not until the late 1800s that mushrooms were grown on a large scale in the United States for sale in the marketplace.

**Mushroom growers** must grow their crop indoors, where conditions can be controlled. The growers keep careful watch over the special devices that control temperature, humidity, and ventilation. In addition, mushroom growers must know how to identify and treat diseases of mushrooms.

Growers work in all stages of mushroom production. Because mushrooms live on dead or decayed matter, growers prepare mixtures of corn and hay. This mixture is placed in a shallow bed and then steam heated to destroy harmful insects, worms, and fungi. Then the hyphae, or cottony masses of material from which the mushrooms will develop is planted in the mixture of corn and hay and covered with a layer of soil.

Two to three months later, the mushrooms are picked just before the gills appear under the caps. The mushrooms are sent to large canning companies or supermarkets the same day they are picked. Some mushroom growers sell directly to consumers. If becoming a mushroom grower appeals to you, you can write to the American Mushroom Institute, P.O. Box 373, Kennett Square, PA 19348.

### SECTION REVIEW

1. How are fungi different from algae?
2. What are hyphae?
3. How do yeast cells reproduce?
4. What is penicillin?

## 8-3 Lichens and Slime Molds

Suppose someone asked you what kind of organism can live in the hot, dry desert as well as the cold, wet Arctic. What if the person added that this organism can also survive on bare rocks, wooden



### 8-2 (continued)

#### Section Review 8-2

1. Fungi are heterotrophic parasites and saprophytes. Algae are autotrophs.
2. Hyphae are threadlike structures in fungi that produce enzymes. These enzymes are used to digest food, which fungi then absorb through the hyphae.
3. Yeast cells reproduce by budding.
4. Penicillin is an antibiotic produced by a mold.

### TEACHING STRATEGY 8-3

#### Motivation

Bring in some specimens of lichens and allow students to examine them. In most areas they can be collected from tree bark or the surface of rocks. If possible bring in some of the substrate along with the lichen specimen. The lichens may be brittle, but they can be sprayed with water to soften them.

- **Do any of you know what this is?** (If the specimen is a lichen on a rock or piece of bark, students may incorrectly identify it as moss or algae. Be sure that the class is aware of the specimen's correct identification before proceeding.)

poles, the sides of trees, and even the tops of mountains? You might reply that no one organism can survive in so many different environments. In a way, ❶ your response would be right. For although **lichens** (LIGH-kuhnz) can actually live in all of these environments, they are not one organism but two. **A lichen is made up of a fungus and an alga that live together.** Combined, these two organisms can live in many places that neither could survive in alone.

The fungus part of the lichen provides the alga with water and minerals that the fungus absorbs from whatever the lichen is growing on. The alga part of the lichen uses the minerals and water to make food for the fungus and itself. This type of relationship, in which two different organisms live together, is called **symbiosis** (sim-bigh-OH-sis).

Lichens are called pioneer plants because they are often the first plants to appear in rocky, barren areas. Lichens release acids that break down rock and cause it to crack. Dust and dead lichens fill the cracks, which eventually become fertile places for other organisms to grow. In time, the rocky area may become a lush, green forest. ❷

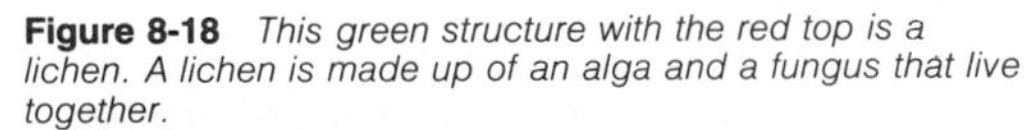

**Figure 8-18** *This green structure with the red top is a lichen. A lichen is made up of an alga and a fungus that live together.*

## BACKGROUND INFORMATION

Though lichens are found in a wide variety of natural habitats, they are usually absent from areas severely affected by air pollution. Lichens absorb their needed moisture and minerals from rain water, and they are extremely susceptible to toxic materials that may be borne by rain. Thus the presence or absence of lichens in an area can serve as an indicator of that area's air quality.

## TIE-IN/ PHYSICAL SCIENCE

One type of lichen is used in the manufacture of litmus, a substance that is used to determine whether a solution is an acid or a base.

## ANNOTATION KEY

❶ **Thinking Skill: Relating facts**
❷ **Thinking Skill: Sequencing events**

## FACTS AND FIGURES

Lichens grow extremely slowly. Some crustose lichens grow as little as 0.1 to 10 mm a year. A large patch of crustose lichen on a rock may represent more than 4000 years of growth.

Explain that the lichen is a combination of two organisms living together.

- **What two kinds of organisms make up a lichen?** (A lichen is made up of an alga and a fungus living together.)
- **In what kinds of places can lichens live?** (Lichens live in a wide variety of environments. Some examples of their wide range are given in the text.)

### Content Development

Write the word "symbiosis" on the chalkboard, and call on someone to read its definition from the text. Expand on the definition by asking students to recall that in a symbiotic relationship at least one of the organisms benefits. The second organism may also benefit, or it may be harmed.

- **Why is a lichen a good example of symbiosis?** (Two different kinds of organisms, an alga and a fungus, live in a close relationship.)
- **How does the fungus benefit from the symbiosis?** (The fungus obtains food from the alga.)
- **How does the alga benefit from the symbiosis?** (The alga obtains water and minerals that are absorbed by the fungus.)

## Activity

**Observing a slime mold**
**Skills: Observing, relating, applying, manipulative**
**Level: Enriched**
**Type: Hands-on/Field**
**Materials: Slime mold, magnifying glass**

Slime molds are slow moving and difficult to find, so this investigation should probably be carried out only by patient and observant students. Such students may even be able to bring in living samples of slime molds. They should also collect the moist bark or wood to which the slime molds are clinging, and they should keep the samples moist. Slime molds may also be obtained from biological supply houses.

**Figure 8-19** *The lavender bloblike organism growing on this dead leaf is a slime mold.*

While walking through a forest, you may have seen a bloblike organism growing in moist areas, on dead leaves, rotting logs, and other types of decaying material. This organism is called a **slime mold** because of its appearances during its life cycle. **The slime mold resembles a protozoan and a fungus during the two stages of its life cycle.**

In one of the slime mold's stages, it resembles a jellylike mass of protoplasm, engulfing and digesting food like a giant amoeba. In the other stage, the slime mold develops fruiting bodies like its cousin the mushroom. The fruiting bodies produce spores. The spores develop into new slime molds.

### *SECTION REVIEW*

1. What is a lichen?
2. Describe a slime mold's life cycle.

#### Activity

*Observing a Slime Mold*

Slime molds may be found as yellow, red, or white masses under the bark of fallen logs or under boards in moist places. With a hand lens, carefully observe a slime mold. How does it move from place to place? Describe the shape of the slime mold.

## 8-4 Mosses and Liverworts

In a barren cold part of northern Europe, above the Arctic Circle, is the region of Lapland. Its rocky landscape has a few small trees and beds of lichens and **mosses.** The people who live in this icy place are



## 8-3 (continued)

### Section Review 8-3

**1.** A lichen is a fungus and an alga that live together.
**2.** In one stage of a slime mold's life cycle, it is an amoebalike mass of protoplasm. In the other stage, it resembles a fungus, and develops fruiting bodies that produce spores. These spores develop into new slime molds.

## TEACHING STRATEGY 8-4

### Motivation

Provide specimens of mosses for students to examine. These may be collected locally or ordered from a biological supply company. If collected, they should be wrapped in newspaper or placed in plastic bags with a small amount of the soil in which they were growing to be transported to class. Provide hand lenses for students to use in their examination. Call attention to the arrangement of the leaflike and stemlike structures.

• **What is the evidence that mosses can carry on photosynthesis?** (Their green color indicates they contain chlorophyll.)

Tell the class that the green leafy part of the moss plant is called a *gametophyte.*

• **Why are the structures making up the gametophyte not really true stems and leaves?** (They do not contain vascular tissues.)

Have students use their lens to examine the rhizoids.

• **What is the function of these root-like structures?** (They anchor the plant to the earth.)

• **Why are mosses always so small in size?** (Without vascular tissue to transport materials throughout their bodies they cannot grow large.)

### Content Development

Call attention to the capsules shown in Figure 8-20. Point out that mosses have a life cycle involving both sexual and asexual reproduction. Male and

Figure 8-20 *The rounded structures at the top of these haircap mosses are capsules. Spores are produced within these capsules.*

called Lapps. For them, everything in their environment is important, including the mosses. Because the mosses are soft and keep in warmth, the Lapps use them to line their baby cradles.

**Mosses are small green nonvascular plants that have stemlike and leaflike parts.** They live almost everywhere in the world. They are found near or on the ground, close to sources of water. Mosses absorb water through rootlike structures called **rhizoids** (RIGH-zoids). Because mosses contain chlorophyll, they make their own food through photosynthesis. They live in damp and cool places such as the shaded surfaces of trees and rocks.

Although **liverworts** share characteristics with mosses, they are different plants. **Liverworts are small green nonvascular plants that have flat leaf-like parts.** Each liverwort looks like a tiny green leaf, although it is not a real leaf. Some liverworts look like miniature livers, which explains how these simple plants got their name. Like mosses, liverworts grow in moist places and have rhizoids. Certain liverworts have a special way of reproducing. New plants grow from pieces broken off older plants.

Figure 8-21 *The small, flat, green plants growing on the rock are liverworts.*

### *SECTION REVIEW*

1. What is the function of a rhizoid?
2. How are liverworts like mosses? Unlike mosses?



## 8-4 MOSSES AND LIVERWORTS

### SECTION PREVIEW 8-4

Mosses and liverworts are photosynthetic, nonvascular plants that are adapted for life on land. However, they are always found in moist locations. Lacking vascular tissues, the cells in their simple bodies must be near the surface in order to permit the absorption of water and nutrients directly from the environment. Thus, mosses and liverworts are always small plants found at or near ground level.

Mosses and liverworts lack true roots, stems, and leaves, but they have rootlike structures called rhizoids that anchor them to the ground. Mosses have a leaflike appearance and grow more upright than liverworts, which grow flat along a surface.

### SECTION OBJECTIVES 8-4

1. **Describe characteristics of mosses and liverworts.**

### SCIENCE TERMS 8-4

**mosses p. 186** **liverworts p. 187**
**rhizoids p. 187**

### FACTS AND FIGURES

Certain plants that are called mosses are not true mosses. For example, Spanish moss, which grows on trees in the south, is a flowering plant related to the pineapple. Reindeer moss, found in cold regions, is a lichen.

female organs on the green leafy part of the moss produce sperm and egg cells during the sexual part of the cycle. After fertilization, the spore capsules seen in the figure develop on top of the green plant. Spores that are carried by the wind to a new location then develop into more green moss plants.

### Reinforcement

Suggest that students try to collect some mosses and liverworts and grow them in a terrarium. They will need to be collected with some soil remaining about the rhizoids and transported to school in damp newspapers or paper towels. A terrarium can be prepared by placing some coarse gravel about 3 cm deep on the bottom of an aquarium or large glass container. Spread about 2 cm of sand over the gravel and then about 3 cm of garden soil on top of the sand. The mosses or liverworts can be placed on top of the soil. Water should be added about half way up the gravel layer, and the top should be covered with a glass cover.

### Section Review 8-4

1. Rhizoids absorb water and anchor certain plants to the ground.
2. Liverworts and mosses live in similar places and have rhizoids. However, liverworts are smaller than and look different from mosses.

# LABORATORY ACTIVITY EXAMINING A SLIME MOLD

## BEFORE THE LAB

1. **At least one day prior to the activity, gather enough materials for students to work in teams of two to four.**
2. **Disposal plastic petri dishes containing agar can be purchased from a biological supply company. You can also purchase dehydrated agar and prepare the petri dishes yourself. To culture slime molds, it is not necessary to sterilize the agar.**
3. **Cultures of the slime mold *Physarum polycephalum* in petri dishes can be obtained from biological supply companies. Prepare and dispense the filter paper squares containing the slime mold at the time they are needed. To do this, use a dissecting needle or cotton swab to transfer a small amount of mold to a wet 2 cm × 2 cm square cut from filter paper.**
4. **The oatmeal flakes can be pulverized with a mortar and pestle.**

## PRE-LAB DISCUSSION

Ask the class to read the problem, materials, and procedures sections of the activity.

• **What is the purpose of this activity?** (The purpose of this activity is to observe some of the characteristics of a living slime mold.)

Go over the list of materials and discuss any of the items that may not be familiar.

• **What do you think is the purpose of the agar and the oatmeal flakes?** (They are a source of food for the slime mold.)

• **What changes do you think will take place in the slime mold during this investigation?** (Accept all logical answers, but students will likely hypothesize that the mold will spread over the agar.)

## LABORATORY ACTIVITY

### *Examining a Slime Mold*

**Purpose**

In this activity, you will observe the characteristics of a slime mold.

**Materials** *(per group)*

Petri dish containing agar
Filter paper containing slime mold
Crushed oatmeal flakes

| | |
|---|---|
| Water | Forceps |
| Glass slide | Microscope |
| Masking tape | Medicine dropper |
| Cover slip | Dissecting needle |

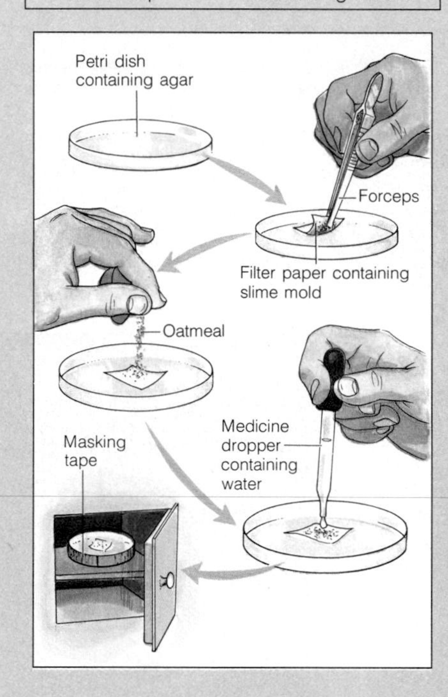


**Procedure** 

1. With a forceps, place the small piece of filter paper containing the slime mold in the center of the petri dish.
2. Sprinkle some crushed oatmeal flakes next to the piece of filter paper.
3. Add two to three drops of water to the slime mold and oatmeal flakes.
4. Cover the petri dish. Seal the dish with masking tape.
5. Place the sealed petri dish in a cool, dark place.
6. Examine the petri dish each day for three days. Record your observations.
7. After three days, remove a small amount of the slime mold from the petri dish and place it on a glass slide.
8. Examine the slime mold under the low power of a microscope.
9. With a dissecting needle, puncture a branch of the slime mold. **CAUTION:** *Be careful when using a dissecting needle.* Observe the slime mold for a few minutes.

**Observations and Conclusions**

1. Describe the changes that took place in the slime mold during the three-day observation period.
2. What activity did you observe in the slime mold after placing it on the glass slide?
3. Describe what happened to the puncture that you made in the slime mold.
4. Why was oatmeal added to the petri dish?
5. Is the slime mold a heterotroph or an autotroph? Explain.
6. Based on your observations, describe the characteristics of a slime mold.
7. Describe an experiment to determine the response of the slime mold to certain substances, such as salt or sugar.



## SKILLS DEVELOPMENT

Students will use the following skills while completing this activity.

1. Observing
2. Manipulative
3. Safety
4. Designing an experiment
5. Relating

## SAFETY TIPS

Make sure students understand the proper procedure for handling a microscope and glassware. Caution them that glass microscope slides, cover slips, and petri dishes are very fragile. Alert students to notify you immediately if they should break any glassware, particularly if they are cut by the broken glass. Caution students to be very careful when working with

# CHAPTER REVIEW

## SUMMARY

**8-1 Algae**

- Algae are nonvascular organisms that lack true stems, roots, and leaves.
- Algae contain chlorophyll, a green substance needed for photosynthesis.
- The majority of algae live in watery environments such as oceans, ponds, and lakes. Other algae grow in soil, on the sides of trees, or even near hot springs.
- Some species of algae are unicellular, or one-celled. These algae often group together to form colonies.
- Algae are placed into six groups according to their color. They are grouped as blue-green, green, golden, brown, red, and fire algae. All but the blue-green are plants.

**8-2 Fungi**

- Fungi are nonvascular organisms that lack true stems, roots, and leaves.
- Unlike algae, fungi do not contain chlorophyll and cannot perform photosynthesis.
- Fungi are heterotrophs. They cannot make their own food. Some fungi are parasites, organisms that feed on other living things. Others are saprophytes, organisms that feed on once-living organisms.
- Most members of the Fungi kingdom have threadlike structures called hyphae, which produce chemicals called enzymes.
- Most fungi reproduce by means of spores, which are contained in special structures called fruiting bodies.
- Yeasts obtain energy through a process called fermentation.

**8-3 Lichens and Slime Molds**

- Lichens are made up of two organisms, a fungus and an alga, that live together. This type of relationship is called symbiosis.
- Slime molds are found in moist areas on once-living leaves, rotting logs, and other types of decaying material.

**8-4 Mosses and Liverworts**

- Mosses are nonvascular plants that have rootlike structures called rhizoids. Rhizoids help to absorb water from the soil.
- Liverworts are smaller than mosses. Some liverworts reproduce when pieces of the plant break off older plants.

## VOCABULARY

*Define each term in a complete sentence.*

**air bladder**
**alga**
**bioluminescence**
**budding**
**cap**
**chlorophyll**
**fermentation**
**fruiting body**
**fungus**
**gill**
**hypha**
**lichen**
**liverwort**
**moss**
**nonvascular plant**
**photosynthesis**
**rhizoid**
**slime mold**
**stalk**
**symbiosis**



a dissecting needle. Students should alert the teacher immediately if they should cut themselves with the needle. Have students wash their hands after the investigation.

## TEACHING STRATEGY FOR LAB PROCEDURE

**1.** Point out that microscopes, slides, cover slips, and dissecting needles are not needed for the first day's procedure.
**2.** Remind students to place a label on their petri dishes for identification. Glass-marking pens or masking tape labels can be used.
**3.** Explain that the cover slip will have to be removed from the slide before the slime mold branch can be punctured. You might demonstrate the procedure before students try it.

## OBSERVATIONS AND CONCLUSIONS

**1.** The slime mold should resemble an amoeba and spread across and eat the oatmeal flakes. The slime mold should also increase in size.
**2.** The slime mold moves across the glass slide and will engulf anything in its path.
**3.** After a short time, the puncture will disappear.
**4.** The oatmeal is a food source for the slime mold.
**5.** The slime is a heterotroph because it cannot make its own food through photosynthesis.
**6.** A slime mold is a heterotroph that acts like an amoeba, engulfing its food.
**7.** Place a small amount of the slime mold on one end of a glass slide and a small amount of salt or sugar on the other end. Observe whether the slime mold engulfs the salt or sugar.

## GOING FURTHER: ENRICHMENT

**Part 1**

Have students try the experiments that are described in Observations and Conclusions Question 7. Some groups can check the effects of salt and others the effect of sugar on the growth of the molds. Suggested procedure: Students might add a small amount of sugar or salt to the crushed oats before placing them in the petri dish.

**Part 2**

Some students may want to perpetuate their slime mold cultures. If so, they can transfer the mold to a new petri dish by cutting a small square of agar from their present petri dish and placing the mold on the agar surface in a new dish. Some crushed oats could be sprinkled around the agar block. The mold will spread to the new agar surface in about three days.

## CHAPTER REVIEW

### MULTIPLE CHOICE

| | | | | |
|---|---|---|---|---|
| 1. c | 3. d | 5. b | 7. d | 9. d |
| 2. c | 4. c | 6. b | 8. b | 10. c |

### COMPLETION

1. nonvascular
2. photosynthesis
3. diatom
4. fruiting bodies
5. cap
6. stalk
7. mushroom
8. budding
9. symbiosis
10. rhizoids

### TRUE OR FALSE

1. T
2. T
3. F brown
4. F golden
5. T
6. F fire
7. T
8. F heterotrophs
9. F fungi
10. F lichen

### SKILL BUILDING

**1.** You would expect to find more fungi in a shady forest than in a sunny field. Fungi live in moist, shady areas.
**2.** Most algae live in shallow water or float on the surface of the water because they need to get as much sunlight as possible for photosynthesis.
**3.** Your friend is mistaken. Mosses are nonvascular plants and thus cannot grow very large.
**4.** A logical hypothesis is that mushroom spores lying on the forest floor become "activated" to grow due to the presence of water from the rainfall.
**5.** Experiments should be logical and well thought out. They should include a control and a variable setup.
**6.** The most immediate effect would be on the organisms that eat the fungi. These organisms would have to find an alternate food source to survive. Since fungi also have a role as decomposers, one might expect a reduction in the amount of decomposition of dead material such as rotting logs and leaves on the forest floor.
**7.** The experiment was designed to show the rate of photosynthesis in algae under high- and low-intensity light at varying temperatures. Algae that received high-intensity light showed a marked increase in photosynthetic rates as the temperature increased until the temperature rose above 30°C, at which point the rate dropped dramatically. Algae under low-intensity light did not increase in photosynthetic rate regardless of an increase in temperature, but did drop off in rate at about 30°C.

### ESSAY

**1.** Algae, like large green plants, contain chlorophyll, perform photosynthesis, and are autotrophs. Fungi lack chlorophyll and are heterotrophs.
**2.** Algae may be used in food, as a polish in toothpastes, and as oxygen sources in spaceships. Students may suggest additional answers.
**3.** Fungi are called heterotrophs be-

## CONTENT REVIEW: MULTIPLE CHOICE

*Choose the letter of the answer that best completes each statement.*

1. The green substance that is found in the cells of green plants and is needed for photosynthesis is
   a. diatomaceous earth. b. hypha. c. chlorophyll. d. rhizoid.
2. Algae whose gritty, glasslike skeletons form a substance often used in toothpaste are
   a. blue-green algae. b. green algae.
   c. golden algae. d. red algae.
3. Tiny structures that enable brown algae to float near the surface of the water are called
   a. rhizoids. b. hyphae. c. fruiting bodies. d. air bladders.
4. The structures that fire algae use to propel themselves through the water are called
   a. rhizoids. b. cilia. c. flagella. d. pseudopodia.
5. Most fungi have threadlike structures called
   a. caps. b. hyphae. c. fruiting bodies. d. stalks.
6. Yeasts change sugars into alcohol and carbon dioxide during a process known as
   a. budding. b. fermentation. c. symbiosis. d. conjugation.
7. An important product produced by the *Penicillium* mold is
   a. chocolate milk. b. toothpaste. c. fertilizer. d. an antibiotic.
8. The black, rounded structures in bread mold produce
   a. rhizoids. b. spores. c. fruiting bodies. d. hyphae.
9. Mosses have rootlike structures that are called
   a. hyphae. b. stalks. c. caps. d. rhizoids.
10. A small green nonvascular plant that has flat leaflike structures is known as a(n)
    a. mushroom. b. alga. c. liverwort. d. fungus.

## CONTENT REVIEW: COMPLETION

*Fill in the word or words that best complete each statement.*

1. All plants that lack true roots, stems, and leaves are ______________.
2. The process by which green plants make their own food is ______________.
3. An example of a golden alga is a(n) ______________.
4. In most fungi, spores are contained in special ______________.
5. The umbrella-shaped part of a mushroom is the ______________.
6. In a mushroom, the cap is found at the top of a(n) ______________.
7. A puffball is an example of a(n) ______________.
8. Yeasts reproduce by forming spores or by ______________.
9. A relationship in which two different organisms live together is ______________.
10. A moss absorbs water from its environment with its rootlike ______________.

## CONTENT REVIEW: TRUE OR FALSE

*Determine whether each statement is true or false. If it is true, write "true." If it is false, change the underlined word or words to make the statement true.*

1. Most algae live in a watery environment.
2. Algae contain chlorophyll.
3. Blue-green algae are considered to be the most complex of all algae.
4. Diatoms are examples of red algae.
5. *Sargassum* is an example of a brown alga.
6. Some types of brown algae are capable of bioluminescence.
7. Dutch elm disease is caused by a fungus.
8. Fungi are autotrophs.
9. Yeasts are unicellular algae.
10. A slime mold is made up of an alga and a fungus.

## CONCEPT REVIEW: SKILL BUILDING

*Use the skills you have developed in the chapter to complete each activity.*

1. **Making predictions** Would you expect to find more fungi in a sunny field or in a shady forest? Explain your answer.
2. **Relating facts** Why do most algae live in shallow water or float on the surface of the water?
3. **Applying concepts** A friend tells you that he has seen mosses that were two meters tall. Is your friend mistaken? Why or why not?
4. **Developing a hypothesis** Develop a hypothesis to explain why in many forests, mushrooms suddenly spring up out of the soil after a rainstorm.
5. **Designing an experiment** Design an experiment to show how light affects the growth of bread mold. Be sure to include a control in your experiment.
6. **Relating concepts** How would a forest be affected by the removal of all fungi?
7. **Interpreting a graph** The graph shows the results of a photosynthesis experiment in which algae were used.

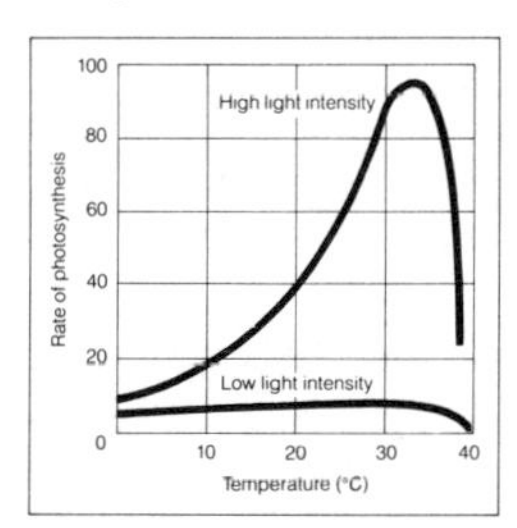


What do you think the purpose of the experiment was? What conclusion can you draw from the graph?

## CONCEPT REVIEW: ESSAY

*Discuss each of the following in a brief paragraph.*

1. Explain why algae are classified as plants and fungi are not.
2. Describe three uses of algae.
3. Why are fungi called heterotrophs?
4. Describe the six groups of algae.
5. What is fermentation? Why are yeast cells used in baking bread?
6. Describe the structure of the mushroom and explain the function of each structure.



cause they cannot make their own food.

**4.** The six groups of algae are blue-green, green, golden, fire, brown, and red. Descriptions should be similar to those presented in the chapter.

**5.** Fermentation is an energy-releasing process in which sugars are changed into alcohol and carbon dioxide. Yeast is used in baking because the carbon dioxide it releases through fermentation causes dough to rise.

**6.** Cap: mushroom's fruiting body that contains reproductive spores; stalk: stemlike structure that holds up the cap and attaches the mushroom to the ground or to the substrate on which it lives; gills: the structures that produce spores

## ADDITIONAL QUESTIONS AND TOPIC SUGGESTIONS

**1.** Explain why algae that live in water do not need a vascular system. (The cells of these algae are in direct contact with the environment. Water and minerals are taken in directly.)

**2.** In some classification systems, fungi are placed in the plant kingdom, but most biologists now place them in a kingdom by themselves. In what ways are fungi like plants? In what ways are they different? (Fungi and plants are alike in that both are generally nonmotile and have rigid cell walls. The chief difference is that fungi have no chlorophyll and must obtain food from another source.)

**3.** Why are lichens able to grow in places where few other plants can grow? (The symbiotic relationship between the alga and fungus making up the lichens enables them to live in locations where these organisms could not survive separately.)

**4.** Use a reference book to find out why a type of moss known as sphagnum is commercially important. (Because of its absorptive qualities, dried sphagnum is bagged and sold as peat moss. It is widely used by gardeners as a mulch and soil conditioner, and by nurserymen as a packing material for fragile plants.)

## ISSUES IN SCIENCE

The following issue can be used as a springboard for discussion or given as a writing assignment.

Certain antibiotics, similar to penicillin, are added to the feed of farm animals that are being raised to provide meat for humans. The antibiotics keep the animals free from certain diseases and increase their growth rate. However, some scientists believe that the organisms that cause human and animal diseases may develop a resistance against antibiotics if antibiotics are used too freely. They want to stop the practice of using antibiotics in the food of meat animals. What is your opinion and why?

# Chapter 9

# VASCULAR PLANTS

## CHAPTER OVERVIEW

Plants that contain transporting tubes are called vascular plants. In this chapter, students will learn the characteristics of vascular plants. They will learn that all vascular plants have leaves, stems, and roots, which contain tubes that carry materials throughout the plant.

Students will learn that ferns are among the oldest vascular plants on the earth. They will learn that ferns go through two stages in their life cycle. Depending upon what stage it is in, the plant reproduces by either asexual or sexual reproduction.

Students will learn that a seed contains a young plant, stored food, and a seed coat. Students will learn that gymnosperms are seed plants that produce uncovered seeds. They will learn that the largest group of gymnosperms are the conifers or cone-bearing plants.

Students will also learn that angiosperms are seed plants that produce covered seeds. They will learn that angiosperms produce flowers that contain the reproductive organs of a plant.

Students will learn that photosynthesis is the process by which green plants use carbon dioxide, water, and light energy to produce glucose and oxygen. They will learn that photosynthesis occurs in the leaves and stems of green plants or wherever chlorophyll-containing chloroplasts are located in the plant.

## INTRODUCING CHAPTER 9

Have students observe the photograph on page 192. Mention that this plant is rare and grows only in one place in the United States.

• **In what kind of environment does this plant appear to live?** (Accept all answers. Students will probably notice the rocks in the vicinity and suggest a harsh environment.)

Explain that the plant is a dwarf plant that typically grows approximately 1.5 to 5 cm in height. Its leaflets are about 6 mm long and its flowers have a diameter of about 8 mm. Although the plant is colorful, its small size can make it relatively inconspicuous among the rocks.

• **Describe the structures that appear to be surrounding the flowers.** (Students should see fuzzy, hairlike tufts around the flowers.)

Have students read the chapter introduction on page 193. As they read, have them look for reasons why this plant is endangered.

• **What is this plant's name?** (Robbins cinquefoil)

• **Where is the only place that Robbins cinquefoil is found?** (on Mt. Washington in New Hampshire)

• **What should be done to protect the Robbins cinquefoil?** (Accept all reasonable answers.)

Point out that although the

# 9 Vascular Plants

CHAPTER SECTIONS



CHAPTER OBJECTIVES

*After completing this chapter, you will be able to:*

- **9-1** Compare vascular and nonvascular plants.
- **9-2** Identify the functions of roots, stems, and leaves in seed plants.
- **9-2** Explain the processes of photosynthesis and transpiration.
- **9-3** Describe the characteristics of gymnosperms.
- **9-4** Describe the characteristics of angiosperms.
- **9-4** Discuss the processes of pollination, fertilization, seed germination, and seed dispersal.

Life at the top of Mount Washington in New Hampshire is not easy. Winds often whip through the air at more than 160 kilometers an hour. Winter temperatures can dive to −40°C. Yet even in this harsh environment, living things can survive. As a matter of fact, one of the rarest plants in the world lives here. The tiny plant, less than 5 centimeters tall, is called the Robbins cinquefoil (SINK-foil). This plant is able to survive the wind and cold because it grows under patches of snow and ice. The snow and ice act as a blanket that protects the plant from low temperatures and from blowing away.

Today, there are fewer than 4000 Robbins cinquefoil plants in the world. However, it is not the cold environment that threatens them, but rather people. Hikers enjoying the pleasures of Mount Washington often walk near the plant. As they do, they kick away some of the soil around the plant. Without the soil, new seeds cannot grow, which endangers the future of the plant.

Can the Robbins cinquefoil be saved? The federal government has gone to great lengths to protect the plants. But, in the end, it will be up to people to save the Robbins cinquefoil. Through education and understanding, people will know next time to step around a tiny plant growing at the top of a tall mountain.

*Robbins cinquefoil on Mount Washington*



Robbins cinquefoil is tiny and rare, it possesses all the features of vascular plants that will be studied in this chapter. Vascular plants are the most diverse, widespread, and economically important plants on the earth.

## TEACHER DEMONSTRATION

Prepare a display of some plants or plant parts representing those that will be studied in this unit. If actual specimens are not available, photographs of each will suffice.

• **Can you identify any of these plants or plant structures?** (Answers will vary depending upon the display.)

• **In what ways are these plants or plant structures different from the plants that we've read about earlier?** (Accept all logical answers.)

Call attention to such structures as roots, stems, leaves, flowers, and seeds that are associated with vascular plants. Emphasize that although nonvascular plants have many of the same features as vascular plants, nonvascular plants lack true roots, stems, and leaves.

• **What evidence is there to indicate that these plants carry on photosynthesis?** (Accept all answers. Students should suggest that the leaves and stems of plants that perform photosynthesis are green because they contain chlorophyll, a substance needed for photosynthesis.)

• **Although these plants share many common characteristics, they also have many differences. What are some of these differences** (Accept ail logical answers.)

Conclude by pointing out that despite their obvious differences, these plants have one characteristic in common. They are all vascular plants or are parts of vascular plants. In this chapter, students will learn the features and diversity of vascular plants.

## TEACHER RESOURCES

### Audiovisuals

*Flowering Plants,* filmstrip with cassette, Eye Gate
*Flowers: Structure and Function,* film, Cor
*Green Plants and Sunlight,* film, EBE

### Books

Devlin, R., *Plant Physiology,* D. Van Nostrand
Galston, A. W., et al, *Life of the Green Plant,* Prentice-Hall

### Software

*Plant Processes,* Prentice Hall

## 9-1 FERNS

### SECTION PREVIEW 9-1

Although they are often found in the same habitats as mosses, ferns grow more upright. The reason for this is that ferns have conducting tubes, which carry materials throughout the plants. This enables ferns to grow up and away from the ground, which contains their source of water. Because ferns have true roots, stems, and leaves, they are vascular plants. The most noticeable part of the fern is its leaves, called fronds. The stem, or rhizome, grows along the surface of the soil or just beneath the ground. Roots grow from the rhizomes.

A fern's life cycle includes an asexual stage and a sexual stage. These stages alternate with each other. During the asexual stage, spores are released from the spore cases on the underside of the fronds. The spores develop into small heart-shaped structures that produce sperm and egg cells. The mature fern develops from this structure after fertilization.

### SECTION OBJECTIVES 9-1

1. **Compare vascular and nonvascular plants.**
2. **Identify the parts of a fern.**
3. **Describe the two stages in a fern's life cycle.**

### SCIENCE TERMS 9-1

**vascular plant p. 194**
**frond p. 195**
**rhizome p. 195**
**asexual reproduction p. 196**
**sexual reproduction p. 196**

## 9-1 Ferns

Imagine walking through a rain forest, such as Washington State's Olympic National Park. As you enter, you immediately feel the cool dampness of the air. Tall trees, such as firs, spruces, and cedars, are everywhere. Mosses drape the branches and trunks of trees and hold moisture like a sponge.

As you continue your walk, you notice feathery green leaves above some moss plants. Many of these leaves belong to ferns. Unlike mosses, which are nonvascular plants, ferns are **vascular plants.** Vascular plants contain transporting tubes that carry material throughout the plant. Remember that mosses must live close to the ground. Mosses have no transporting tubes to carry water and nutrients to their parts.

Ferns are one of the oldest plants on the earth. They appeared more than 300 million years ago and soon became the most abundant type of plant. These ferns were as large as trees. Today, some ferns the size of trees are still found in tropical rain forests.

Most ferns disappeared as the earth's climate changed. All that remained of some ferns was dead

**Figure 9-1** *The photograph on the left shows tree ferns as they appear today in a rain forest in Costa Rica. The photograph on the right shows a model of a fern forest during the Carboniferous period.*

### TEACHING STRATEGY 9-1

#### Motivation

After students have read the opening paragraphs of Section 9-1 and studied Figure 9-1, display a wall map of the Western Hemisphere. Point out the location of Olympic National Park in the state of Washington and the country of Costa Rica.

- **What kinds of forests are found in Olympic National Park and in Costa Rica?** (rain forests)
- **What kinds of plants grow in rain forests?** (tall trees, mosses, and ferns)
- **Are the plants in the rain forest in Washington likely to be exactly the same as those in the rain forests of Costa Rica? Explain your answer.** (Students should mention that Washington has a temperate climate, while Costa Rica has a tropical climate. As a result, the plants that are growing in these two areas will be somewhat different.)

Refer students to the tree ferns in Figure 9-1. Mention that some varieties of tree fern grow up to a height of 25 cm.

#### Content Development

Emphasize that mosses and ferns are often found growing in the same locations.

- **Why do ferns grow more upright**

plant material. This material later formed great coal deposits. Most living or once-living material contains carbon. Because coal is made up mostly of carbon, the geological period in which ferns were abundant is called the Carboniferous Period.

### Structure of Ferns

**Like all vascular plants, ferns have true leaves, stems, and roots.** In fact, a fern's **fronds,** or leaves, are the part of the plant you usually notice. See Figure 9-2. For the most part, developing fern leaves are curled at the top and resemble the top of a violin. Because of their appearance, these developing leaves are called fiddleheads. As they mature, the fiddleheads uncurl until they reach their full size. ❷

Following the leaves down toward the ground, you find the **rhizome** (RIGH-zohm), or stem of the fern. In some ferns, the rhizome looks like a large fuzzy, brown caterpillar with leaves. The rhizomes of many ferns grow along the surface of the soil. In other types of ferns, the rhizomes grow beneath the ground. Roots grow from the rhizomes and anchor the ferns to the ground. The roots also absorb water and minerals for the plant.

Transporting tubes travel throughout the fern's leaves, stems, and roots. These tubes carry materials throughout the plant. These special tubes allow the fern to grow taller than nonvascular plants.

### Reproduction of Ferns

As the fern grows, it goes through two stages in its life cycle. If you were to look at both stages of the same plant, you would think that each was a different plant.

The fern uses a different type of reproduction in each stage. In the first stage of a fern's life cycle, small brown structures appear on the underside of the fronds. These structures are spore cases and contain spores. See Figure 9-3. Once the spore cases ripen, they open and release spores. Spores are carried through the air and eventually fall to the ground. If growth conditions are favorable, the spores begin to grow into new fern plants. The development of new organisms from spores is an ❸

**Figure 9-2** *The curled structures at the tops of these ferns are called fiddleheads because they resemble the tops of violins. Fiddleheads are the developing fronds, or leaves, of ferns.*

**Figure 9-3** *The tiny brown spots on the underside of this fern frond are spore cases. Spore cases contain spores.*

## TEACHER DEMONSTRATION

Obtain one or more potted ferns to call attention to some of their special characteristics. If you do not already have ferns on hand, inexpensive ones can be bought at a florist, greenhouse, or plant department of a discount store.

- **What type of plant is this?** (Most students will probably identify the plant as a fern.)
- **What is the stem of a fern called?** (rhizome)
- **Where is the rhizome of a fern generally found?** (Along the surface of the soil or beneath the ground.)
- **What is the job of the roots of a fern?** (The roots anchor the fern to the ground. Roots also absorb water and minerals for the plant.)
- **What is the job of the transporting tubes?** (to carry materials throughout the plants)

## ANNOTATION KEY

❶ **Thinking Skill: Making comparisons**
❷ **Thinking Skill: Making generalizations**
❸ **Thinking Skill: Identifying processes**

## FACTS AND FIGURES

Because ferns were the most abundant plants during the Carboniferous Period, this period is also called the Age of Ferns.

**than mosses?** (Unlike mosses, ferns are vascular plants. Because ferns have transport tubes that carry materials throughout the plants, they can grow away from the soil. Mosses, on the other hand, must grow close to the soil because they do not have transport tubes.)

Point out that, as vascular plants, ferns are well-adapted for life on land. Most species of ferns grow in damp tropical habitats. Many other species can withstand harsh environments.

### Skills Development

***Skill: Making Comparisons***
Have students look at Figure 9-3 and read the caption. Compare reproduction in ferns and fungi by posing the following questions:

- **What are the brown structures on the underside of the fern frond?** (These structures are spore cases bearing reproductive spores.)
- **How do the spores usually reach a new location after they are released?** (Spores are generally carried by air.)
- **What are some other plantlike organisms that also reproduce by producing spores?** (fungi)

Review spore production in such fungi as mushrooms, puffballs, and bread molds. Discuss similarities and differences in the ways in which spores are produced and dispersed in fungi and ferns.

## 9-2 SEED PLANTS

### SECTION PREVIEW 9-2

The dominant plants on earth are the seed plants. Like ferns, seed plants have vascular tissues and asexual and sexual stages in their life cycles. Seed plants are divided into two groups, gymnosperms and angiosperms, on the basis of whether or not their seeds are covered by a protective wall.

Seed plants have true roots, stems, and leaves. Roots anchor the plant in the ground and absorb water and minerals from the soil. The stem supports other plant parts and displays the leaves to sunlight. Xylem tissue forms tubes within the stem through which water and minerals are carried upward to the leaves. Other transport tubes form from phloem tissue and carry food that was manufactured in the leaves to other parts of the plant. The leaf is the main organ of photosynthesis, and it also releases excess water into the air. Although roots, stems, and leaves are specialized for certain functions, they are composed of similar tissues and function together in carrying out the plant's life processes.

### SECTION OBJECTIVES 9-2

1. **Describe the characteristics of seed plants.**
2. **List the functions of roots, stems, and leaves.**
3. **Compare herbaceous and woody stems.**
4. **Describe photosynthesis and transportation.**

### SCIENCE TERMS 9-2

**seed** p. 196
**gymnosperm** p. 197
**angiosperm** p. 197
**xylem** p. 198
**phloem** p. 198
**cambium** p. 199
**annual ring** p. 200
**herbaceous stem** p. 201
**woody stem** p. 201
**photosynthesis** p. 202
**stoma** p. 203
**transpiration** p. 204
**epidermis** p. 204
**guard cells** p. 204

**Activity**

*Observing Ferns*

Visit an area, such as a park or nature center, in your neighborhood where ferns grow. Such areas usually have lots of shade and moisture. Examine a fern plant. Observe the brown spots on the underside of the fronds. These spots are the spore cases and contain the spores.

Carefully remove a frond and place it in a small plastic bag. Plant the frond on top of a pot of moist soil. Be sure that the spore cases are facing down. Make sure that the soil is always moist.

In about four weeks, you will notice new ferns growing. What type of reproduction has occurred? Using reference material, draw the life cycle of a fern.

example of **asexual reproduction.** Asexual reproduction is the formation of an organism from a single parent.

The next stage of a fern's life cycle begins as the spores grow into structures that do not resemble their parents. These structures look like tiny heart-shaped green plants. The plants develop special tissues that produce male and female sex cells. The male sex cell is called the sperm and the female sex cell is called the egg. Both are located on the plant. The sperm unites with the eggs by swimming across the plant through dew and rainwater.

The united sperm and egg grow into a new fern plant. This plant is similar in appearance to the plant with fronds that produced spores. The new fern plant was produced by **sexual reproduction.** In sexual reproduction, an organism develops from the uniting of two different sex cells.

Although the fern's life cycle seems complicated, it is successful. Ferns have survived, with little change, for millions of years.

### SECTION REVIEW

1. How are vascular plants different from nonvascular plants?
2. What is a frond? A rhizome?
3. What is the difference between asexual and sexual reproduction?

**Figure 9-4** *The sticklike structure growing out of the avocado fruit is a developing plant. This plant grew from the seed within the avocado.*

## 9-2 Seed Plants

If you wanted to plant a vegetable garden, you would begin by loosening the soil. Also you would have to make sure that your garden had a source of water. Only then would you plant your **seeds.** A seed contains a young plant, stored food, and a seed coat. With proper care, the seeds in your garden would begin to grow and, after a few months, produce adult plants loaded with vegetables.

(1) **Seed plants are vascular plants that produce seeds and have true roots, stems, and leaves.** Seed plants also go through the same two reproductive



### 9-1 (continued)
### Section Review 9-1

**1.** Vascular plants contain transporting tubes, while nonvascular plants do not.
**2.** A frond is a fern's leaf, while a rhizome is a fern's stem.
**3.** In asexual reproduction, an organism develops from a single parent. In sexual reproduction, an organism develops from the uniting of two different sex cells.

### TEACHING STRATEGY 9-2

#### Motivation

Begin the discussion of seed plants by displaying some examples of gymnosperm seeds and angiosperm seeds. For example, a pine cone might be shown to illustrate gymnosperm seeds. Any familiar, seed-bearing fleshy fruit or pod, opened to reveal the seeds, might be used as a representative angiosperm seed.

**Figure 9-5** *This zucchini squash plant is an example of an angiosperm. Angiosperms produce seeds covered by a protective wall and flowers.*

stages as ferns. Seed plants are divided into two groups based on the structure of their seeds. The seeds of **gymnosperms** (JIM-nuh-spermz), *are not* covered by a protective wall. ❷ In **angiosperms** (AN-jee-uh-spermz), the seeds are covered by a protective wall.

Seed plants are among the most numerous plants on the earth. Also, they are the plants with which most people are familiar. The trees in the forest, the vegetables in a garden, and the cotton plant are all examples of seed plants. What other seed plants can you name? ❶

### Roots

Anyone who lives near trees no doubt has seen what a violent storm can do to them. The tremendous force of the storm's wind can pull a tree right out of the ground. A large hole marks the spot where the tree once was anchored to the ground. If you were to look carefully at the exposed base of the tree, you would notice small and large structures that resemble tentacles. These structures are the tree's roots, which anchor the plant in the ground. Roots also absorb water and minerals from the soil. In addition, the roots of some plants, such as those of the



## Activity

**Observing Ferns**
**Skills: Observing, relating, applying, diagraming**
**Level: Average**
**Type: Field/Hands-on**
**Materials: fern frond, small plastic bag, pot of moist soil**

The two reproductive stages of ferns are dramatically different in appearance. This activity should allow students the opportunity of observing the large frond-leafed sporophyte generation in nature. They will also be able to grow the tiny prothallus gametophyte generation in the classroom and to observe the resultant formation of new sporophytes that develop by means of sexual reproduction. Have students display their drawings, both of the actual ferns and of the fern life cycle as pictured in reference books.

## ANNOTATION KEY

❶ **Answers will vary. (Relating concepts)**
❶ **Thinking Skill: Applying concepts**
❷ **Thinking Skill: Making comparisons**

Discuss the differences between gymnosperms and angiosperms by asking the following:

- **How are the seed-bearing parts of these plants different?** (Accept all logical answers. Students should mention that in one example the seeds are exposed, while in the other they are enclosed.)
- **What group of seed plants produces uncovered seeds?** (gymnosperm)
- **What are the names of some familiar gymnosperms?** (pines, cedars, firs, spruces, and hemlocks)
- **What group of seed plants produces seeds that are covered by a protective wall?** (angiosperms)

Next have students look at Figure 9-5 and read the caption.

- **What plant structure forms a protective wall around the seeds of angiosperms?** (The protective wall, or ripened ovary.)
- **Name some familiar angiosperms.** (Answers will vary.)

### Content Development

As roots are discussed, list their functions on the chalkboard. Because a root grows underground on most seed plants, it: 1. anchors the plant in place and 2. absorbs water and minerals from the soil. In some plants, roots store food.

## FACTS AND FIGURES

Scientists calculated that the root system of a rye plant measures more than 609,000 meters in length.

### Activity

**Plant Transport**
**Skills: Making comparisons, manipulative, observing, applying definitions**
**Level: Remedial**
**Type: Hands-on**
**Materials: Medium-sized jar, water, vegetable coloring, twig, celery stalk, knife**

In this activity, students will observe the transport of water containing vegetable coloring up the stem of the plants to the leaves. Students should report that the celery has a herbaceous stem, while the twig has a woody stem. The xylem tissue, which carries the liquid, can be seen as dots in the cross section and, in particular, as rings of dots in the woody twigs. Students should determine that the xylem is the structure that transported the colored water up the stem.

**Figure 9-6** *Although all roots absorb materials from the soil and anchor a plant to the soil, they do not all look the same. Notice the differences in the roots of the banyan trees* (left), *grasses* (top right) *and carrots* (bottom right).

### Activity

*Plant Transport*

**1.** Fill a medium-sized jar one-fourth full of water. Then add a few drops of vegetable coloring and stir.

**2.** Place a freshly cut twig from a tree and a stalk of celery in the jar so that only the cut part of each is underwater.

**3.** After 24 hours, remove the twig and the stalk.

**4.** With a knife cut off the portion of each plant that was underwater. **CAUTION:** *Be very careful when using a knife.* Discard these portions.

**5.** Cut another small section across the bottom of each remaining portion. Examine.

How do the two sections differ? How are they alike? What structure transports the colored water up the plants?

carrot, store food for the plant. The roots of some plants, such as those of turnips, go straight down. Other plants, such as the grasses, have slender roots that spread out in many directions. See Figure 9-6.

Tiny root hairs cover the surface of many roots. Root hairs are microscopic extensions of single cells. The root hairs greatly increase the surface area of the root. They allow the plant to take in water and minerals from the soil. The minerals pass through the root hairs into tissue called **xylem** (ZIGH-luhm). The xylem tissue forms tubes that carry water and minerals from the roots up through the plant. The roots also contain another kind of tissue called **phloem** (FLOH-uhm). Phloem tissue forms tubes that carry food substances down from the plant's leaves. 1

### Stems

Have you ever seen a group of maple trees with buckets attached to their trunks and wondered what was going on? Remember that the stem, or trunk, of

198

## 9-2 (continued)

### Reinforcement

Have students think about some roots that humans use for food. Then have them make a list of as many of these roots as they can. Or suggest that small groups visit a produce market or the fresh fruits and vegetables section of a supermarket and list all of the root foods they see on display. Point out that not all plant parts that grow underground are roots. For example, an onion bulb is an extension of a plant's leaves, and a white potato, shown in Figure 9-10, is actually an underground stem.

### Content Development

Have students look at the different types of roots shown in Figure 9-6.

• **How are the roots of carrots and grasses different?** (Carrots have one large main root, while grasses have slender, branched roots.)

Write the words *taproot* and *fibrous root* on the chalkboard. Explain that plants with one main root, such as the carrot, have a taproot system. Those with branching roots, like grasses, have a fibrous root system.

• **What are some other plants that have taproots?** (Some examples include beets, turnips, radishes, and trees.)

• **Which kind of root, taproot or fibrous root, extends deeper into the ground?** (Taproots generally extend deeper than fibrous roots.)

• **Which type of root could better hold soil in place?** (Due to their branching nature, fibrous roots are more effective in holding soil in place.)

Point out that the banyan tree, in

the tree contains xylem and phloem tissue. Within these tissues, there flows a liquid called sap. The sap in the xylem tissue contains water and minerals, while the phloem tissue's sap contains sugary plant food. Usually, the sugary sap moves downward. At certain times of the year in some trees, such as the sugar maple, the sap moves both up and down. The sap is collected from sugar maple trees and used to produce the maple syrup that you pour over your breakfast pancakes!

Stems also contain a tissue called **cambium** (KAM-bee-uhm). Cambium is the growth tissue of the stem. New xylem and phloem cells are produced in the cambium, causing the wood in the stem to become

**Figure 9-7** *Maple syrup comes from the sap of certain types of maple trees. This tree is being tapped for its sugary sap.*

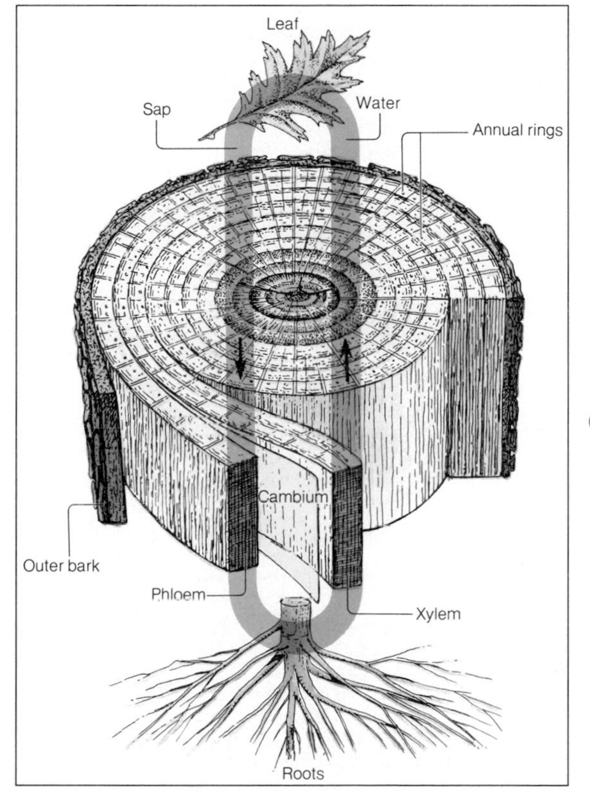


**Figure 9-8** *In a stem, the xylem transports water and other materials from the roots up to the leaves. The phloem conducts sap from the leaves down to the roots. Cambium is the growth tissue of the plant and produces new xylem and phloem cells.*

❷



## ANNOTATION KEY

❶ **Thinking Skill: Comparing vascular tissue**

❷ **Thinking Skill: Interpreting diagrams**

Figure 9-6, has roots that extend from the base of its stem. These roots are called prop roots because they give added support to the plant. If possible, show the class a photograph of a corn stalk with prop roots growing out of its stem at the base.

## Motivation

To enable students to examine root hairs, germinate some soaked radish seeds on pieces of damp filter paper or paper toweling in a covered petri dish. Root hairs will develop in a few days. Provide hand lenses or stereomicroscopes for small groups of students. Emphasize that the root hairs greatly increase the root's absorbing surface.

## Skills Development

***Skill: Making Generalizations***

Write the following on the chalkboard or an overhead transparency: xylem (xylon), phloem (phloos). Point out that the word *xylem* is derived from the Greek word *xylon,* which means wood. *Phloem* is derived from the Greek word *phloos,* meaning bark. Next have students examine Figure 9-8.

- **Why were the terms xylem and phloem chosen for the plant's transport tissues?** (Students should conclude that the wood of trees consists of xylem, and phloem cells are found within the inner bark.)

## Enrichment

Encourage interested students to conduct research and write reports on the production of maple syrup. If maple syrup is produced in your area, perhaps a field trip can be arranged for students to view the tapping of trees and the processing of the sugar and syrup.

## Skills Development

***Skill: Making calculations***

- **If water moves up the xylem of a 200-m-tall tree at the rate of 5 cm/sec, how many seconds does it take for the water to reach the tree's top?** (Students must convert centimeters to meters in order to divide. So 5 cm/sec becomes .05 m/sec. Students should then divide 200 m by .05 m/sec. The answer is 4000 sec, or 66.6 min., or 1.1 h. This is how long it takes the water to reach the top of the tree.)

## BACKGROUND INFORMATION

A tree's approximate age can be determined by counting its annual rings. However, to be accurate, the count must be taken on a cut surface as near the soil level as possible. Counts taken higher up the stem will indicate only the age from the time that part of the tree developed.

## HISTORICAL NOTES

Dendrochronologists are scientists who use the annual rings of trees to conduct historical research. By matching annual rings of living trees with wood samples from ancient structures, dendrochronologists are often able to determine the age of these structures. For example, beams from Indian pueblos in Arizona and New Mexico have been found to be close to 2000 years old.

### Activity

**Annual Rings**
**Skills: Observing, comparing, applying**
**Level: Average**
**Type: Field/Hands-on**

The growth rings of most severed tree trunks should be quite easily visible. You might caution students that the number of rings is not an exact indicator of the age of a tree, because in some years more than one growth layer can be added.

**Figure 9-9** *The age of a tree can be determined by counting the number of rings in its stem. Each ring represents one year's growth.*

### Activity

*Annual Rings*

Find a sawed-off tree stump or a fallen tree. Count the number of annual rings to determine the age of the tree. Notice the widths of different rings. A wide ring is produced during a year of heavy rainfall, while a narrow ring is produced during a year of little rain. Closely examine one ring. Within each ring, you should be able to see a wide growth area and a narrow growth area. Which of these areas do you think took place in the spring? In the summer? Explain why.

thicker. Each year, as a new layer of xylem cells grows, it wraps itself around the layer before it. Because each layer is one year's growth of xylem cells, these layers are called **annual rings.** Scientists estimate the age of a tree by counting its annual rings.

Plant stems vary in size, shape, and function. The trunk, branches, and twigs of a tree are all stems. Some plants, such as the cabbage, have stems so short that you probably would not recognize them.

Stems provide the means for the transportation of water, minerals, and food. Another function of the stems is to support the other parts of the plant. In addition, the stems hold the leaves up in the air so that they can receive sunlight and make their own food.

Although most stems grow vertically and above the ground, some grow like roots. In some plants, such as the iris and lily of the valley, the stems grow horizontally and under the ground. Like those of (1) ferns, these stems are called rhizomes. Other underground stems called tubers store food. The potato, for example, is a tuber. The "eyes" of the potato are buds. Each bud is capable of developing into a new potato plant. Certain spring plants, such as onions and tulips, produce a short, thick stem called a bulb.

**Figure 9-10** *A potato is actually a special kind of stem called a tuber that grows under the ground. The fingerlike structures growing on the potato are actually new potato plants.*

## 9-2 (continued)

### Skills Development

***Skill: Drawing a Conclusion***

Place a drawing that shows annual growth rings of varying widths on the chalkboard, overhead transparency, or on a handout sheet. This drawing may be similar to Figure 9-9. Notice in this figure that the rings nearest the center are wider than those nearest the outside. Point out that the width of annual rings provides a guide to past conditions of rainfall and temperature.

- **What is the approximate age of this tree?** (Answers will vary depending upon the number of annual rings.)
- **What evidence indicates that a tree may have gone through a period of dryness?** (Sections where the annual rings are the narrowest.)
- **Suppose that several years ago other trees once growing near a tree were cut down. What effect might be observed in the size of annual rings of the tree?** (Accept all logical answers. Students may suggest that the annual rings probably would become wider. The tree would be able to obtain more water and minerals from the soil if competing trees were removed.)

### Content Development

Explain that stems have several im-

**Figure 9-11** *Stems are either herbaceous or woody. The soft, green herbaceous stems on the left are of a daisylike plant. On the right are the rigid woody stems of trees.*

Bulbs have thick leaves that contain food for the plant. When planted, the bulbs develop into new plants.

Stems may be placed into two groups based on their hardness. A **herbaceous** (her-BAY-shuhs) **stem** is green and soft. Sunflowers, eggplants, peas, beans, and tomatoes are examples of plants with herbaceous stems. Most plants that have herbaceous stems are annuals. An annual is a plant that completes its life cycle within one growing season. A plant's life cycle begins when the plant starts to grow from a seed and ends when the plant produces its own seeds. Wheat, rye, and tobacco plants are examples of annuals.

Unlike herbaceous stems, **woody stems** are rigid. ❷ These stems have large amounts of woody xylem tissue. Plants that have woody stems are called perennials (puh-REN-ee-uhls). A perennial is a plant that lives for more than two growing seasons. Included in the group of perennial plants are trees and woody shrubs. Plants that live for two growing seasons are called biennials (bigh-EN-ee-uhls). Biennials



## TEACHER DEMONSTRATION

Obtain two paper cups, a plastic knife, blue and red vegetable coloring, and fresh white carnations. Place some red vegetable coloring into one paper cup and some blue vegetable coloring into another paper cup. Carefully, slit the stem of a carnation into equal halves. Place one half of the stem into the red vegetable coloring and the other half of the stem into the blue vegetable coloring. Have the students observe the results.

• **What has happened to the carnation?** (The tips of the flower are now red and blue.)

• **What caused the color change?** (the vegetable coloring)

• **How did the vegetable coloring get to the petals of the flower?** (It was transported up through the xylem tissue in the stem.)

Tell students that water and other minerals are transported through the flower in a similar manner.

## ANNOTATION KEY

❶ Thinking Skill: Relating structures

❷ Thinking Skill: Comparing stems

portant functions: 1. conduct water and minerals absorbed by the roots upward through xylem tissue to the leaves; 2. conduct food produced in the leaves downward to the roots through the phloem; 3. display leaves to the sun so they receive maximum sunlight; and 4. support plant structures. In some plants, food may be stored in the stem.

• **What are some plant stems that are eaten by people?** (asparagus, broccoli, and white potatoes. Some students may mention celery. However, a celery stalk is an enlarged leaf petiole rather than a stem.)

### Skills Development

***Skill: Classifying herbaceous and woody stems***

After discussing the differences between herbaceous and woody stems, have students look at 10 photographs of plants and classify them as either herbaceous or woody.

### Content Development

Write the words *annual, perennial,* and *biennial* at the top of the chalkboard. As these types of plants are discussed, list examples given in the text in the appropriate column. Students may be able to offer additional examples. Mention that many people are not aware that certain plants are biennials. Many of them are harvested for food during their first growing season. Thus they have no opportunity to grow into the second season.

## TIE-IN/SOCIAL STUDIES

For years, the leaves of the tea plant have been brewed to make a warm beverage. Have students investigate the parts of the world where the tea plant (*Camellia sinensis*) grows. Determine the economic effect of this plant on its native countries.

### Activity

**Plant Responses**
**Skills: Relating, applying, classifying**
**Level: Average**
**Type: Writing**

The tropisms mentioned are responses to light (phototropism), gravity (geotropism), water (hydrotropism), touch (thigmotropism), and chemical substances (chemotropism). The information gathered can be used to plan classroom demonstrations of these tropisms.

**Figure 9-12** *Leaves come in a wide variety of types and shapes. The tulip tree leaf* (left) *is a simple leaf, while the poison ivy leaf* (right) *is a compound leaf.*

### Activity

*Plant Responses*

The growth response of a plant to a stimulus, or change in its surroundings, is called a tropism. If a tropism is positive, the plant grows toward the stimulus. If the tropism is negative, the plant grows away from the stimulus. Using reference materials in the library, look up information on the following tropisms:

phototropism
geotropism
hydrotropism
thigmotropism
chemotropism

Write a brief report in which you describe and give examples of each of these tropisms. Present your report to the class.

may be either herbaceous or woody. Many garden vegetable plants, such as beets, celery, carrots, and cabbage, are biennials.

### Leaves

Have you ever heard of the drug digitalis (di-ji-TAL-is)? It is given to people who have certain types of heart problems. This drug is not made in a laboratory. It is made from the dried leaves of a garden flower called the foxglove. The leaves are found growing along the stem of the plant.

Plant leaves vary greatly in shape and size. For example, leaves of the foxglove are long and oval-shaped. Pines, firs, and balsams have needle-shaped (1) leaves. Other plants have flat, wide leaves. Most leaves have a stalk and a blade. The stalk connects the leaf to the stem, while the blade is the thin, flat part of the leaf. Most of the cells in which food making takes place are in the blade.

In addition, there are two types of leaves. Some leaves have only one blade and are called simple leaves. The silver maple leaf and apple leaf are examples of simple leaves. Leaves such as the black locust and buckeye have two or more blades. These are examples of compound leaves.

Whatever their shape, leaves are very important structures. Within the leaves, the sun's energy is trapped and used to produce food. This process is called **photosynthesis.** It is the largest and most important manufacturing process in the world.

202

### 9-2 (continued)

### Skills Development

***Skill: Classifying Leaves***

Have students refer to Figure 9-12.

• **How is the tulip tree leaf different from the poison ivy leaf?** (The tulip tree leaf is a simple leaf and the poison ivy leaf is a compound leaf.)

As the differences between these leaves are discussed, emphasize that the flat part of the leaf is called the blade. In addition to the blade, most leaves also have a stalk called a petiole that attaches the blade to the stem. Compound leaves have several blades called leaflets arranged along the petiole.

Set up several stations around the classroom where various examples of locally collected simple and compound leaves are displayed. If the identity of the leaves is known, write its name on a card beside each leaf. Then have students visit each station and classify the leaves as either simple or compound. If possible, include both pinnately compound and palmately compound leaves among the specimens. Pinnately compound leaves are those, such as ashes and locusts, in which the leaflets are arranged along opposite sides of the petiole. In palmately compound leaves, such as chestnut or buckeye, the leaflets arise from the tip of the petiole. As an alternative approach to this activity, you might have students collect examples of simple and compound leaves to be brought to class and identified.

### Content Development

Summarize photosynthesis by writing the word and chemical equations for the process on the chalkboard. Compare photosynthesis to a manufacturing process.

In photosynthesis, the sun's light energy is captured by chlorophyll, the green substance in plants. The sun's energy is used to combine water from the soil and carbon dioxide from the air. The food made by this combination is a sugar called glucose. Glucose is used by the plant for growth and to repair its parts. Glucose can be stored in special areas in the roots and stems. Photosynthesis also yields a waste product—oxygen! Of what value is this product of photosynthesis? ❶ ❷

Photosynthesis is a complicated process, although an equation can be used to sum up what occurs. An equation is the scientist's shorthand for describing a reaction. This is the equation for photosynthesis.

**carbon dioxide + water $\xrightarrow[\text{chlorophyll}]{\text{sunlight}}$ glucose + oxygen**

$$6CO_2 + 6H_2O \xrightarrow[\text{chlorophyll}]{\text{sunlight}} C_6H_{12}O_6 + 6O_2$$ ❸

In photosynthesis, carbon dioxide enters the leaves through **stomata** (STOH-muh-tuh; singular: stoma). Stomata are openings in the surface of the leaf. Although most plants have stomata on the lower surface of their leaves, some plants have stomata on the top surface of their leaves. Stomata also permit oxygen to pass out of the leaf into the air.

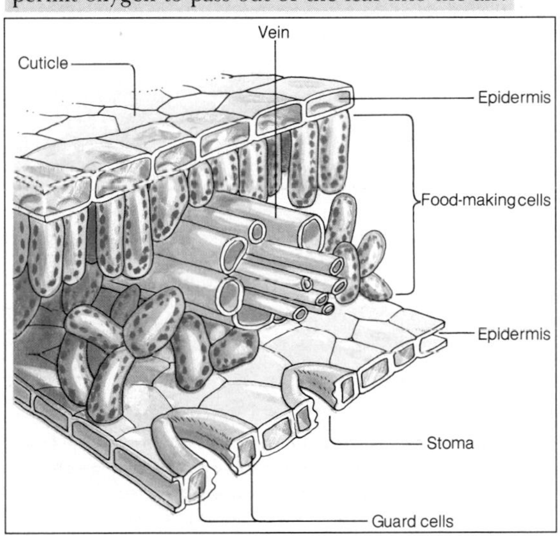


**Figure 9-13** *This section of a leaf shows the tissues and structures found in various parts of the leaf.*



## TEACHER DEMONSTRATION

To demonstrate that oxygen is released during photosynthesis, fill a large beaker or battery jar about three-fourths full of water. Add a spoonful of sodium bicarbonate to the water as a source of carbon dioxide. Obtain about five sprigs of healthy *Elodea,* and cut each sprig diagonally near its base with a sharp scissors or a scalpel. Next place the *Elodea* sprigs inside the wide mouth of a glass funnel. Arrange them in the glass funnel so that their cut stems lie within or near the stem of the funnel. Then carefully place the funnel mouth-side down in the beaker or battery jar. Be sure that the stem of the funnel is completely under the water. Now invert a test tube that has been completely filled with water over the funnel stem. Be sure that no water runs out of the inverted test tube as you do this. Finally, place the entire setup in bright light for several hours. After a short time, bubbles of oxygen should be seen escaping from the *Elodea* as it carries on photosynthesis. As oxygen continues to be released, it will gradually replace the water in the test tube.

## ANNOTATION KEY

❶ **Animals use the oxygen during respiration**

❶ **Thinking Skill: Relating structures**

❷ **Thinking Skill: Identifying processes**

❸ **Thinking Skill: Applying formulas**

- **Where is the "factory" in which this manufacturing process takes place?** (Photosynthesis takes place within the leaves of green plants.)
- **What raw materials are needed for this process?** (Carbon dioxide from the air and water from the soil are the raw materials.)
- **What kind of energy is used within the "plant factory?"** (The energy source is light.)
- **How does chlorophyll help the plant make food?** (Chlorophyll captures light energy.)
- **What product is used by the plant?** (glucose)
- **What waste product is given off?** (oxygen)

## Reinforcement

Encourage those students who enjoy drawing to prepare a large drawing of a leaf cross section. Figure 9-13 might be used as a resource. In addition to labeling the various structures, students may also wish to include brief descriptions of their functions with the drawing.

## Enrichment

Suggest that students conduct re search and report on the factors that cause tree leaves to change colors during autumn in certain parts of the country. Have students give their reports orally.

## 9-3 GYMNOSPERMS

### SECTION PREVIEW 9-3

In this section, students will be introduced to the characteristics of gymnosperms. They will learn that gymnosperms are seed plants that produce uncovered seeds. Students will learn that gymnosperms are divided into three main groups.

One group of gymnosperms is the cycads. Students will learn that cycads also have cones. They will learn that cycad trees are either male or female.

Another group of gymnosperms are gingkoes. Students will learn that gingkoes have male or female cones on separate trees.

The third and largest group of gymnosperms is the conifers. Students will learn conifers are cone-bearing plants. They will learn that conifers produce male and female cones on the same tree.

### SECTION OBJECTIVES 9-3

1. **Identify the characteristics of gymnosperms.**
2. **Compare cycads, gingkoes, and conifers.**

### SCIENCE TERMS 9-3

**pollen p. 205** **ovule p. 205**

**Figure 9-14** *Plants release excess moisture, which can be seen on the inside of the jar. The process by which plants release excess water through the leaves is called transpiration.*

**Figure 9-15** *These palmlike trees are cycads, which grow in tropical areas. The cones at the top of the trees are the reproductive structures.*

The stomata also regulate water loss through the leaves of a plant. This regulating process is called **transpiration.** You might be surprised at the amount of water that a plant can give off. Scientists estimate that a single corn plant can give off more than 200 liters of water during a single growing season. This water is enough to fill a very large bathtub. Of course, this amount is not all the water a plant takes in during its growing season. Although the corn plant takes in much more water, the plant uses much of it to make glucose.

If you closely examined a leaf section, you would be able to see how a living food factory works. The outer protective layer of the leaf is called the **epidermis** (E-puh-DER-mis). The epidermis contains many stomata. Each stoma is surrounded by two sausage-shaped structures called **guard cells.** Guard cells regulate the opening and closing of the stomata. When the guard cells swell, the stoma opens and carbon dioxide enters the leaf. When the guard cells relax, the stoma closes. Most stomata are open during the day and closed at night.

Some leaves have a waxy cuticle (KYOO-ti-kuhl) that covers the epidermis. The cuticle prevents the loss of too much water from the leaf. Within the leaf are two thin layers of food-making cells, which contain chlorophyll. In addition, the leaf has small veins that contain xylem and phloem tissue.

#### *SECTION REVIEW*

1. What are the functions of the roots?
2. What is xylem tissue? Phloem tissue?
3. What are the functions of the stem?
4. Define annual, biennial, and perennial.
5. What is the equation for photosynthesis?

## 9-3 Gymnosperms

The word "gymnosperm" comes from two Greek words meaning uncovered and seed. **Gymnosperms are seed plants that produce uncovered seeds.** The three main groups of gymnosperms are conifers



### 9-2 (continued)

#### Skills Development

***Skill: Making calculations***

The number of stomata on the surfaces of a leaf varies from a few thousand to about 100,000 per square centimeter. For this activity, have students assume that there are 10,000 stomata in each square centimeter on the lower surface of potted plants such as geranium, African violets, or bryophyllum. Working in small groups, instruct students to follow these steps to determine the total number of stomata in a leaf of one of these plants or some other readily available potted plant:

**1.** Outline the shape of the leaf on a sheet of graph paper.
**2.** Count the number of small squares within the outline.
**3.** With a metric ruler, find the number of squares that are contained within one square centimeter.
**4.** Count the number of square centimeters within the leaf. Multiply this number by 10,000 to determine the total number of stomata on the leaf's lower surface.

Although this number may not be accurate for the leaf used, students should still be impressed by the large number of stomata that are present on a leaf's surface.

#### Section Review 9-2

**1.** Anchor the plant in the soil, absorb water and minerals from the soil, and store food for the plant.
**2.** Xylem tissue forms tubes that carry materials up through the plant. Phloem tissue forms tubes that carry materials down through the plant.

(KAHN-uh-fuhrz), cycads (SIGH-kadz), and gingkoes (GING-kohz).

The cycads are tropical trees that look like small palm trees or ferns. Cycads have leaves that look like a feathery crown. In the center of the leaves is a large cone. The trees have either male or female cones. Inside the male cones are **pollen.** Inside the female cones are **ovules** (OH-vyoolz). Pollen contain male sex cells called sperm. Ovules contain female sex cells called eggs. The pollen is carried to the female cones by wind. After a male and female sex cell join, a seed forms. As the seeds mature, part of the cone dries up and releases the seeds, which fall to the ground. These seeds develop into new plants.

④ The second group of gymnosperms, gingkoes, contains only one species, which is commonly called the maidenhair tree. Gingkoes come from China and now grow on many streets in the United States. The gingko can grow as tall as 30 meters and has fan-shaped leaves. As with the cycads, male and female cones are found on separate gingko plants. After the female and male sex cells join, yellowish fruits form. These fruits on the female gingko have an awful odor. In fact, the odor is so bad that only male gingko trees are planted on city streets.

The largest group of gymnosperms is the conifers. The word "conifer" means cone-bearing. Like the gingkoes and cycads, they have cones that produce pollen or ovules. Conifers are woody plants that live in the termperate areas of the world.

**Figure 9-16** *Gingko, or maidenhair, trees are often planted along city streets.*

**Figure 9-17** *The conifers, or cone-bearing plants, grow in forests in many cool areas of the world.*

205

## ANNOTATION KEY

❶ **Thinking Skill: Identifying processes**

❷ **Thinking Skill: Relating cause and effect**

❸ **Thinking Skill: Making generalizations**

❹ **Thinking Skill: Comparing gymnosperms**

## FACTS AND FIGURES

The longest-lived evergreen leaves are the needles of the bristlecone pine, which live for 30 years.

3. Transports materials up and down the plant, supports plant parts, and holds leaves up to the sun.

4. An annual lives only one season; a biennial lives two seasons; and a perennial lives for more than two seasons

5. $6CO_2 + 6H_2O \xrightarrow[\text{chlorophyll}]{\text{sunlight}} C_6H_{12}O_6 + 6CO_2$

## TEACHING STRATEGY 9-3

### Motivation

Begin by showing examples of the male and female cones produced by pine trees. If you cannot obtain actual specimens of these cones, use photographs, or an overhead transparency showing these cones. Point out that all gymnosperms produce sperm and egg cells within different structures. In conifers, emphasize that the male cones produce pollen and the female cones produce ovules.

- **How do these two pine cones differ?** (One type of cone is larger than the other and the larger cone appears to be more woody than the smaller cone.)

Point out that the male cones are smaller than the female cones. Male cones appear in clusters at the tips of branches. These cones produce pollen, which in turn, contains sperm cells. Call attention to the woody scales of the female cone. Explain that winged seeds develop at the base of these scales. The seeds developed from reproductive structures within these cones are called ovules.

- **How does the pollen from the male cones reach the female cones?** (Pollen is carried by the wind.)

### Content Development

Point out that about 100 species of cycads are found growing in tropical regions. Only one species, *Zamia,* is found in the southern part of the United States. Although cycads have been known to grow up to a height of 9 m, most are usually about 2 m tall. Cycads are often grown as ornamental plants in conservatories and botanical gardens.

Mention that gingkoes are very easy to recognize because of their fan-shaped leaves that grow in clusters at the tips of branches. Unlike most gymnosperms, gingkoes lose their leaves each autumn.

## 9-4 ANGIOSPERMS

### SECTION PREVIEW 9-4

The largest group of plants is the angiosperms. These plants bear flowers that contain the reproductive organs. The essential parts of a flower are the stamens and pistils, which produce sperm and eggs, respectively.

Pollination is the transfer of pollen containing sperm cells from the stamen to a pistil. This process occurs before fertilization. After fertilization, the fertilized egg becomes a seed and the base of the pistil, the ovary, develops into a fruit. A fruit is a seed-enclosed structure that distinguishes angiosperms from gymnosperms.

Seeds are structures that contain embryo plants and stored food surrounded by protective coats. If conditions are favorable, seeds will germinate and develop into a new plant.

### SECTION OBJECTIVES 9-4

1. **Identify the characteristics of angiosperms.**
2. **List the parts of a flower.**
3. **Describe the processes of pollination and fertilization.**
4. **Identify the structures that make up a seed.**

### SCIENCE TERMS 9-4

**flower p. 206**
**sepal p. 207**
**petal p. 207**
**stamen p. 207**
**pistil p. 207**
**pollination p. 208**
**fertilization p. 208**
**fruit p. 208**
**cotyledon p. 210**
**germination p. 210**

**Figure 9-18** *These wild desert plants produce colorful structures called flowers, a characteristic of angiosperms. Flowers contain the male and female structures of the plant.*

Most conifers, such as pines, cedars, firs, spruces, and hemlocks, are called evergreens. Evergreens are trees that appear to keep their needles or leaves year round. Actually, evergreens lose leaves while adding new ones. The needles or leaves may remain on an evergreen tree for 2 to 12 years.

***SECTION REVIEW***

1. What are the three main groups of gymnosperms?
2. What are evergreens?

## 9-4 Angiosperms

❶ **Angiosperms are seed plants that produce covered seeds.** They make up the largest group of plants in the world. Because they produce **flowers,** angiosperms are also called flowering plants. The flowers of angiosperms fill the earth with beautiful colors and pleasant smells. But more importantly, flowers are the structures that contain the reproductive organs of angiosperms.

Angiosperms differ in size and living environment. Plant sizes range from the tiny duckweed, a water plant less than 1 millimeter long, to a giant redwood tree 100 meters tall. Some angiosperms,

**Figure 9-19** *The seeds of a gymnosperm are not covered by a fleshy fruit, while those of an angiosperm are.*

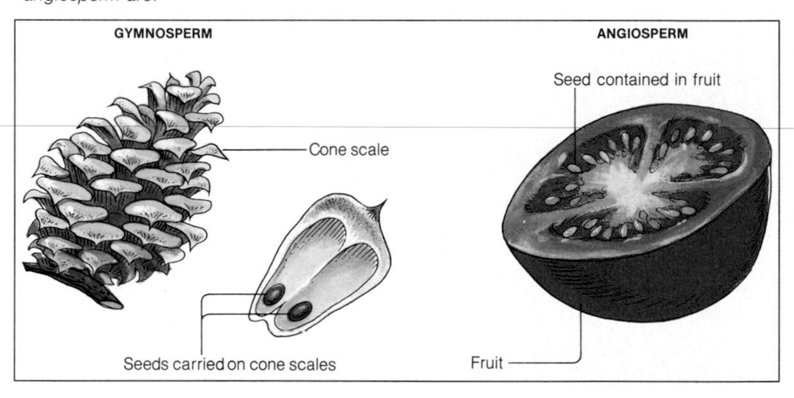


206

### 9-3 (continued)

### Section Review 9-3

1. Cycads, gingkoes, and conifers.
2. Trees that appear to keep their needles or leaves year round.

### TEACHING STRATEGY 9-4

#### Motivation

Display photographs that show some of the modifications in flower design of various plants. First show some photographs of flowers that have conspicuous, brightly colored petals.

• **What is the function of this part of a plant?** (At this time, some students may have misconceptions as to a flower's actual function. So as not to inhibit discussion, accept all answers.)

• **In what ways are the flowers of various plants different?** (Answers will vary, but students probably will mention such characteristics as the colors, shapes, or numbers of the petals.)

Point out that the flower's actual reproductive organs are contained within the ring of petals.

• **Why are many flowers brightly colored?** (Accept all answers.)

Next display a few photographs of flowers that lack brightly colored petals. Here are a few examples.

1. A cauliflower is a cluster of unde-

such as the orchids, live in rain forests where the air and ground are moist. Others, such as the cacti, live in deserts where rain may fall only once a year.

Like gymnosperms, angiosperms form seeds and have leaves, stems, and roots. However, these two groups of plants differ from each other in several ways. The most obvious difference is that only angiosperms produce flowers. ❷

### Flowers

When a flower is still a bud, it is enclosed by leaf-like structures called **sepals** (SEE-puhls). Sepals protect the flower. The sepals open and reveal the flower's **petals.** The colors and shapes of the petals attract insects, which play a vital role in the reproduction of flowering plants. See Figure 9-20. ❸

The petals surround the reproductive organs. The **stamens** (STAY-muhns) are the male organs of the flower. In most flowers, each stamen has two parts. The filament is stalklike and supports the anther. The anther produces the pollen, which contains the sperm.

In the center of the flower are the **pistils.** The pistils are the female organs of the flower. Most flowers have two or more pistils. Some flowers, such as the sweet pea and clover, have only one pistil. The pistils of most flowers have three parts. At the base of the flower is the hollow ovary (OH-vuh-ree). The ovary contains egg cells. A slender tube, called the

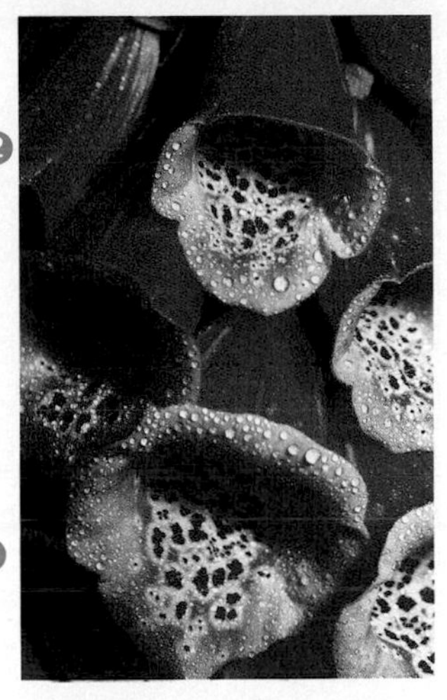

**Figure 9-20** *The patterns inside the petals of this foxglove plant act as a "landing strip" for insects. These patterns help to guide insects to the male and female reproductive structures within the flower.*

**Figure 9-21** *In the photograph, some of the petals have been removed to show the structures inside a flower. The drawing of the flower indicates the names of these structures.*

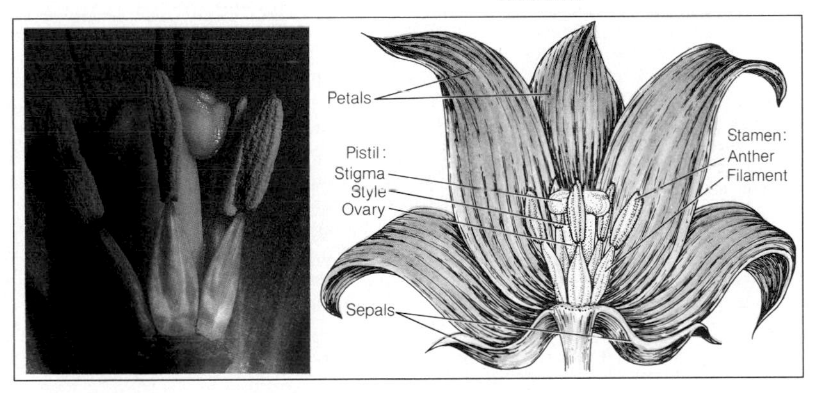




## BACKGROUND INFORMATION

Insects play an important role in the reproduction of flowers. Therefore, in order to reproduce, a flower must be able to attract insects. Plants do this by color, scent, and food such as nectar or pollen. Botanists have found that insects perceive color in a manner different from humans. As a result, blues, mauves, purples, and yellows are the most popular colors for flower-visiting insects.

## ANNOTATION KEY

❶ **Thinking Skill: Making generalizations**
❷ **Thinking Skill: Comparing seed plants**
❸ **Thinking Skill: Identifying flower structures**

veloped flowers along with the branches that support them.
**2.** The tassels and silks of corn are the stamens and pistils of flowers, respectively.
**3.** The flowers of some grasses are composed of clusters of spikelets at the ends of a stalk.
**4.** In poinsettias, the colored structures, often incorrectly called flowers, are modified leaves (bracts). The flowers are the small inconspicuous structures surrounded by the bracts.

Emphasize that although not all flowers are brightly colored, their function is still the same—they produce seeds that give rise to new plants.

## Content Development

Be sure to point out that, like gymnosperms, angiosperms reproduce sexually. The stamens are the male part of the flower that produce pollen grains in which sperm develop. The female part is the pistil. Within the enlarged base of this structure are the ovules, each of which contains an egg. In many angiosperms, a flower contains both stamens and pistils. Such flowers are said to be perfect. In other species, there are separate male and female flowers. These flowers are called imperfect. Imperfect flowers that contain only stamens are called staminate flowers. Those with only pistils are pistillate flowers. For example, in corn the tassels at the top of the plant are the staminate flowers, while the immature ears are the pistillate flowers.

## Skills Development

***Skill: Making Charts***

As students read about flowers and look at Figure 9-21, have them prepare a chart with the heads "Flower Structure" and "Function." Have students complete the chart.

## HISTORICAL NOTES

Records indicate that artificial pollination was practiced as early as 3500 B.C. It is believed that the farmers of Mesopotamia realized that date palm trees were either male or female. By bringing clusters of staminate flowers in contact with pistillate flowers, fruit production was controlled. This ancient method of artificial pollination is still used today.

### Activity

**Perfect or Imperfect?**
**Skills: Classifying flowers, observing, comparing, relating, applying, diagraming**
**Level: Remedial**
**Type: Hands-on**
**Materials: Gladiolus or snapdragon flower, ailanthus flower**

Gladiolus and snapdragon are showy flowers with colorful petals. The petals of ailanthus are small and yellowish green. The gladiolus is a monocot, while the snapdragon and ailanthus are dicots. All are angiosperms. The gladiolus and snapdragon are perfect flowers, while the ailanthus tree is imperfect because it does not have sepals.

**Figure 9-22** *Insects, such as this butterfly, help to transfer pollen from one flower to another.*

style, connects the ovary to a sticky structure at the top of the pistil. This structure is called the stigma (STIG-muh).

In flowers, reproduction occurs in two stages. First, the pollen must be transferred to the stigma. This stage is called **pollination.** In *self*-pollination, pollen is transferred to the stigma of the same flower, or to the pistil of another flower on the same plant.

When pollen is transferred from the flower of one plant to the stigma of another plant, *cross*-pollination occurs. In this kind of pollination, pollen grains are carried from flower to flower by wind, insects, and birds. Usually, cross-pollinated plants have large sweet-smelling flowers. These features help to attract insects and birds.

Because the stigma surfaces of all flowers are sticky, the pollen grains cling to it. Once pollen reaches the stigma, a tube pushes its way down the style to an ovule in the ovary. The sperm from the pollen grain then travels down the tube to the ovule. When the sperm unites with an egg, **fertilization** occurs. Fertilization is the second stage in flower reproduction.

The fertilized egg becomes a seed and the ovary develops into a **fruit.** A fruit is the ripened ovary that encloses and protects the seed. Apples and cherries are examples of fruits.

If it were not for the activities of insects such as bees, moths, and butterflies, many flowering plants

#### Activity

*Perfect or Imperfect?*

A perfect flower contains petals, sepals, stamens, and pistils. An imperfect flower is missing one or more of these structures.

**1.** Obtain either a gladiolus or a snapdragon flower and a flower from an ailanthus, or tree of heaven.
**2.** Examine the reproductive structures within each flower.
**3.** Draw each flower and label its structures.

How are the two flowers different? How are they similar? Classify each flower as perfect or imperfect.



### 9-4 (continued)

#### Motivation

Have students observe Figure 9-22 and read the caption.

- **Name some other insects that also transfer pollen from flower to flower.** (bees, wasps, beetles)
- **Do any animals other than insects transfer pollen?** (Pollinators include birds, such as hummingbirds, and some species of mammals, including bats, a few rodents, and a few small primates.)
- **Why must pollen be transferred from flower to flower in many plants?** (Accept all logical answers. Pollination enables reproduction to occur.)

#### Skills Development

***Skill: Sequencing Events***

After students have read about pollination and fertilization, place the following list of events on the chalkboard, an overhead transparency, or a duplicated worksheet.

Have students number the following in the proper sequence.

1 Pollen is produced by anther of a stamen.
6 Sperm travels through pollen tubes to the ovule.
2 Pollen sticks to body of an insect or bird
8 Fertilized egg becomes a seed; ovary becomes a fruit.
5 Pollen tube enters the ovule within the ovary.
7 Sperm unites with egg.
4 Pollen tube starts to grow down through the style.

would not be able to reproduce. In some cases, only one species of insect can cross-pollinate a single species of plant. For example, in a species of Central American yucca plant, the stamens of the flowers can be reached only by a curved object. That object is the long noselike tube of a particular small moth.

The moth lands on a flower and gathers pollen from the stamen. After rolling the pollen into a ball, the moth flies off to another yucca flower. There the moth crawls deep into the flower, punches a hole in the flower's ovary, and lays its eggs inside the ovary. Then, for some reason not known to scientists, the moth creeps up to the top of the flower's stigma and pushes down the ball of pollen. This action pollinates the yucca. Without this special relationship, both the moth and the yucca would quickly vanish from the earth.

## Career: *Nursery Manager*

**HELP WANTED: NURSERY MANAGER** Requires two- or four-year degree in horticulture. Business background in marketing and management desirable. Nursery or greenhouse experience helpful. Apply in person.

Nearly every community has a garden center that provides its customers with the plants and supplies they need for their homes and gardens. If you pass by a garden center in April, you see pots of tulips and daffodils on display. In July, petunias and other colorful flowers catch your eye. Piles of pumpkins appear in October, and rows of pine trees emerge in December. However, all these plants are not grown at the garden center. They are ordered and delivered by a nursery.

Nurseries are run under the direction of **nursery managers.** The managers supervise the growing of plants and flowers from seeds, seedlings, or cuttings until they are large enough to sell. Nursery managers make sure nursery workers properly fertilize, weed, transplant, and apply pesticides to the plants. In addition, nursery managers make sure that proper inventory records are kept and that the

orders placed by garden centers are filled quickly.

Nursery managers also make many business decisions, such as choosing new equipment, deciding how much money to spend on the growing of various plants, and determining the kinds of plants to sell. Nursery managers also are involved in planning chemical use, in deciding which new plant varieties should be planted, and in directing the building of storage areas. If you enjoy owning a business that involves working with plants, write to the American Society for Horticultural Science, 701 North Saint Asaph Street, Alexandria, VA 22314.



## SCIENCE, TECHNOLOGY, AND SOCIETY

Recently, a new technique has been developed that may enable scientists to produce crops that will grow in desert environments without irrigation. This technique is called electroporation. Electroporation is a process by which plant cells are placed in a special salt solution and are struck with an intense burst of electricity. This action causes the plant cells to develop tiny openings in their cell walls. These openings permit foreign genes to slip through and unite with the plant cell's own genetic material. This method of producing new kinds of plants is much simpler and less time-consuming than older methods of hybridization.

Suppose that genes from a cactus plant could be introduced into a corn or a wheat plant. Then these altered cells could grow into mature plants. If the corn or wheat plant possessed the cactus' ability to grow in a dry environment, deserts could be transformed into corn or wheat fields without the need for expensive irrigation projects.

Although techniques such as electroporation could expand world food production, some people are afraid that altering the genetic makeup of living things could prove dangerous. These people fear that the environment might be endangered because of these altered plants. They also say that because this method is costly, only a few large companies could afford to use it.

Should this type of research continue? What do you think? If you could design a new plant, what features would it have?

3____ An insect or bird brushes pollen from its body onto a stigma.

### Content Development

Explain that angiosperms are unique in that during their reproductive cycles double fertilization takes place. Two male sperm travel downward through the pollen tube into the ovule. Here one sperm unites with an egg to form a zygote. This zygote then develops into the embryo plant. The second sperm fertilizes two structures called polar nuclei, which become food-rich storage tissue called endosperm. The endosperm nourishes the plant as it develops.

### Enrichment

Interested students might try to grow pollen tubes from pollen grains. These are the tubes through which sperm travels to fertilize an egg within the ovule. To do this, have them place a drop of 10% sucrose solution (10 g of sucrose in 90mL of water) on a microscope slide. Then sprinkle some pollen into the solution and cover with a cover slip. Within 30 to 60 minutes, a pollen tube may begin to grow from the pollen grains. The tubes can be observed under the low-power objective of a microscope.

## 9-4 (continued)

### Motivation

Provide samples of some common monocot and dicot seeds for students to examine. Unsalted dry-roasted peanuts and some presoaked kidney beans and corn seeds that have been cut in half from front to back would be appropriate. First distribute some whole peanuts. Then have students split the peanuts in half and compared the halves to Figure 9-23.

• **How many cotyledons does this seed have?** (two cotyledons)

• **What do these cotyledons contain?** (The cotyledons are composed of stored food.)

• **Where is the embryo plant located?** (The embryo plant is located between the halves near one end of the cotyledon.)

Have students examine the embryo with a magnifying glass. If possible, have them identify the epicotyl, hypocotyl, and radicle.

Next distribute the presoaked beans and corn seed halves. Direct the class to carefully remove the seed coat from the bean seed.

• **What is the covering around the bean seed called?** (the seed coat)

• **How does the corn seed differ from the peanut and bean seed?** (Answers will vary. The corn seed has one cotyledon, while the peanut and bean seed have two. In the corn seed, the embryo and stored food are fused.)

### Content Development

Point out that in addition to their seeds, plants that are classified as monocots and dicots also differ in the following ways:

**1.** Monocots have flower parts, such as petals, that are arranged in threes or multiples of three; dicots have flower parts that are arranged in fours or fives or multiples of four or five.

**2.** Monocots have parallel veins in their leaves; dicots have branched leaf veins.

**3.** The bundles of xylem and phloem are scattered throughout a monocot stem. In a dicot stem, the bundles are arranged in rings.

Try to find some wall charts, overhead transparencies, or visuals in reference books that illustrate each of these differences.

### Skills Development

***Skill: Designing an experiment***

Suggest that students work in small groups to design an experiment that tests the effect of temperature on seed germination. Point out that their designs should include a hypothesis that can be tested. Some other considerations include the types and number of seeds that should be used, the variety of temperatures, and the means of recording data.

Allow students to perform their experiments. Be sure that fresh, easily germinated seeds are used. Fresh bean seeds would be appropriate. Because germinating seeds mold easily, have the groups presoak the seeds in

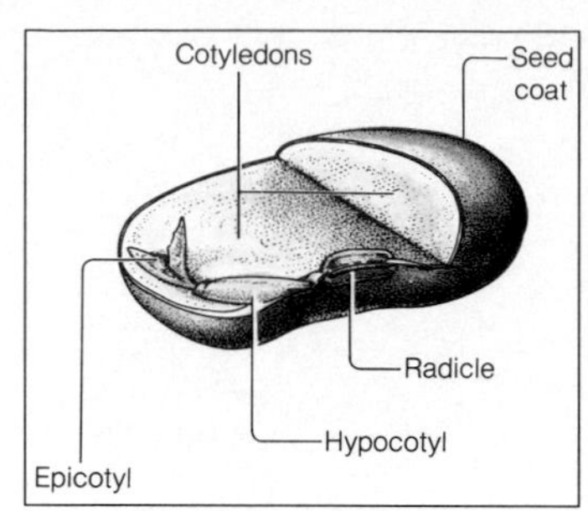


**Figure 9-23** *This bean seed is an example of a dicot, which contains two cotyledons. Within the seed is an embryo plant, which is made up of the epicotyl, radicle, and hypocotyl.*

**Figure 9-24** *The stage in a plant's development when a seed begins to grow is called germination. Notice how the roots and stem grow.*

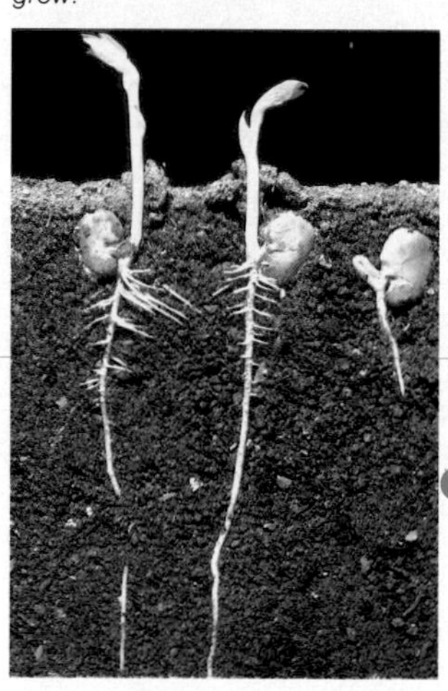

### Seeds

A seed is a tiny package that can develop into a new plant—if conditions are right. Surrounding the seed is a coating that develops from the ovule's wall. Like your winter coat, the seed coat protects what is inside it. Within the seed is a tiny embryo plant and the endosperm. The endosperm is a tissue that contains food the plant will need until it can make its own food.

If you were to remove the seed coat of a bean seed carefully, the seed would divide into two halves. These halves are called **cotyledons** (kaht-uh-LEE-duhnz). A cotyledon is the embryo plant's leaflike structure that stores food. Angiosperms have one or two cotyledons. Those angiosperms that have one cotyledon are called monocotyledons, or monocots. Grasses, orchids, and irises are examples of monocots. Angiosperms with two cotyledons, such as cacti, roses, and peas, are called dicotyledons, or dicots.

Most seeds remain dormant, or inactive, for a short time after being scattered. If there is enough moisture and oxygen, and the temperature is just right, most seeds go through a process called **germination** (jer-muh-NA-shuhn). Germination is the early growth stage of an embryo plant. Some people call this stage "sprouting."

When germination begins, part of the embryo below the cotyledon starts to grow down into the soil. This part of the embryo is called the hypocotyl (high-puh-COT-uhl) and will become the plant's stem. The upper part of the embryo is called the epicotyl (ep-uh-COT-uhl). The epicotyl becomes the leaves, while a part called the radicle gives rise to roots. See Figure 9-24.

### Seed Dispersal

After seeds ripen, they usually are scattered far from where they were made. The scattering of seeds is called seed dispersal (di-SPER-suhl). Seeds are dispersed in many ways. Have you ever picked a dandelion puff and blown away all its tiny fluffy seeds? If you have, you have helped to scatter the seeds. Usually, the wind scatters dandelion seeds.

**Figure 9-25** *Seeds are dispersed, or scattered, in many ways. Seeds of the milkweed (left) and red maple (center) are dispersed by the wind. The seeds of the burdock are carried by people's clothes or animal fur.*

Maple and elm seeds also travel on the wind with the help of their winglike structures. In addition, human beings and animals play a part in seed dispersal. For example, burdock seeds have spines that stick to people's clothing or animal fur. People or animals may pick up the seeds on a walk through a field or forest. At some other place, the seeds may fall off and eventually start a new plant.

The seeds of most water plants are scattered by floating in oceans, rivers, and streams. The coconut, which is the seed of the coconut palm, floats in water. This seed is carried from one piece of land to another by ocean currents.

Other seeds are scattered by a kind of natural explosion, which sends the seeds flying into the air. This is how the dwarf mistletoe disperses its seeds. The popping open of a dwarf mistletoe's seed case can shoot its seeds up to a distance of 15 meters.

### SECTION REVIEW

1. What is a flower?
2. What is pollination? Fertilization?
3. What conditions are necessary for seed germination?

#### Activity

*Seed Germination*

1. Place ten bean seeds in water and let them soak overnight.
2. Remove the seeds from water.
3. In each of ten test tubes, place a piece of blotter (2.5 cm × 5 cm). Place a seed between the blotter and the test tube wall. Moisten the blotters with water.
4. Place five of the test tubes near a sunny window. Place the other five in a warm, dark place such as a closet.
5. Examine the seeds after four days.

What was the control in this experiment? Which seeds germinated? Is light needed for the germination of bean seeds?



## ANNOTATION KEY

1 Thinking Skill: Comparing angiosperms

2 Thinking Skill: Identifying processes

#### Activity

**Seed Germination**
**Skills: Making observations, manipulative, hypothesizing, designing experiments, applying**
**Level: Average**
**Type: Hands-on**
**Materials: 10 bean seeds, water, blotter paper, ten test tubes**

Students should infer that either the seeds in the sun or in darkness could serve as a control. They should find that all seeds germinate, even those in the dark closet. Experimental designs will likely include placing some seeds in a refrigerator and others at room temperature. If the experiment is carried out, students will note that the seeds in the refrigerator will not germinate because warmth is required for germination (about 20°C).

a weak solution of household bleach. The bleach should kill any mold spores on the seed coats. Equal numbers of seeds might then be placed on pieces of blotting paper in petri dishes or other flat bottom containers. Throughout the experiment, keep the blotting paper moist. The groups can expose their containers of seeds to different temperatures by placing some of them in a refrigerator, some at room temperature, and others near a warm location such as a heater.

### Enrichment

People use seeds as a food source. Suggest that interested students conduct a seed survey in which they list every edible seed found in their homes. In addition to such obvious kinds of seeds as nuts, beans, and grains, tell them not to overlook other foods such as flour, coffee, cocoa, caraway, celery seeds, and nutmeg.

### Section Review 9-4

1. Flowers are structures that contain the reproductive organs of angiosperms.
2. Pollination is the transfer of pollen from one flower to another. Fertilization is the joining of the sperm and the egg.
3. Moisture, oxygen, and proper temperature.

## LABORATORY ACTIVITY GEOTROPISM

### BEFORE THE LAB

1. **Gather all the materials at least one day before the activity. Be sure you have enough corn seeds to give four seeds to each group of students.**
2. **At least 24 hours before the activity, soak all the seeds to be used in water.**
3. **Have each group of students decide on a name for their group. This group name will be written on the lid of the petri dish for identification purposes.**

### PRE-LAB DISCUSSION

Explain what a tropism is. Emphasize the idea that a geotropism is a plant's movement due to gravity. Tell students that gravity is a downward pull exerted on all objects on the surface of the earth.

In order to understand this investigation, students will need to know that positive geotropism is a plant's movement toward the source of gravity. Negative geotropism is a plant's movement away from the source of gravity.

### SKILLS DEVELOPMENT

Students will use the following skills while completing this activity.

1. Observing
2. Comparing
3. Relating
4. Applying
5. Hypothesizing
6. Predicting
7. Inferring
8. Diagraming



## *Geotropism*

**Purpose**

In this activity, you will observe how gravity affects the growth of a seed.

**Materials** *(per group)*

4 corn seeds soaked in water for 24 hours

| | |
|---|---|
| Paper towels | Clay |
| Petri dish | Scissors |
| Masking tape | Water |
| Glass-marking pencil | Medicine dropper |

**Procedure**

1. Arrange four soaked corn seeds in a petri dish. The pointed ends of the seeds should all point toward the center of the dish. One of the seeds should be at the 12 o'clock position of the circle, and the other seeds at 3, 6, and 9 o'clock.
2. Place a circle of paper towel over the seeds. Then pack the dish with enough pieces of paper towel so that the seeds will be held firmly in place when the other half of the petri dish is put on.
3. Moisten the paper towels with water. Cover the petri dish and seal the two halves together with a strip of masking tape.
4. With the glass-marking pencil, draw an arrow on the lid pointing toward 12 o'clock. Label the lid with your name and the date.
5. With pieces of clay, prop the dish up so that the arrow is pointing up. Place the dish in a completely dark place.
6. Predict what will happen to the seeds. Then observe the seeds each day for about one week. Make a sketch of them each day. Be sure to return the dish and seeds to their original position when you have finished.

**Observations and Conclusions**

1. Describe your observations. In which direction did the roots and the stems grow?
2. Explain your observations in terms of geotropism, or the effect of gravity on plant growth.
3. What would happen to the corn seeds if the dish was turned so that the arrow was pointing toward the bottom of the dish? To the right or left?
4. Why is it important that the petri dish remain in a stable position throughout the investigation?
5. Suppose you planted all your corn seeds in the soil upside down. In which direction would the stems grow? The roots?
6. Explain why the seeds were placed in the dark rather than near a sunny window.

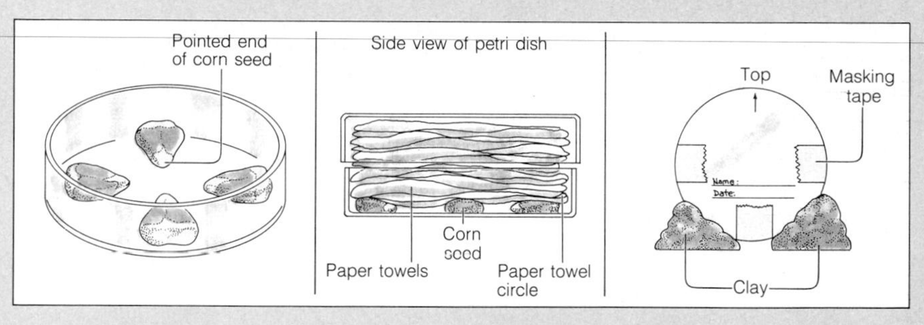




### SAFETY TIPS

Caution students to be very careful when working with scissors and glassware.

### TEACHING STRATEGY FOR LAB PROCEDURE

**1.** Be sure students understand how the corn seeds are to be arranged in the petri dish. They might not be familiar with the concept of a 3, 6, 9, and 12 o'clock position.

**2.** Be sure students understand that the clay should be used to hold the petri dish in a vertical rather than a horizontal position.

### OBSERVATIONS AND CONCLUSIONS

**1.** The corn seeds germinated. The roots will grow down toward the earth and the stems will grow up.

**2.** Roots show positive geotropism, while the stems show negative geotropism.

**3.** The roots would curl around and grow downward, while the stems would curl around and grow upward.

# CHAPTER REVIEW

## SUMMARY

**9-1 Ferns**

- Vascular plants contain tubes that carry material throughout the plant.
- Ferns have true leaves, stems, and roots.
- In asexual reproduction, new fern plants develop from spores.
- In sexual reproduction, fern plants develop from the uniting of a sperm and an egg cell.

**9-2 Seed Plants**

- A seed contains an embryo plant, stored food, and a seed coat.
- Seed plants are divided into two groups: gymnosperms and angiosperms.
- Roots are structures that anchor plants and absorb water and minerals.
- In the roots, minerals and water travel up the plant through the xylem. Food substances travel down the plant through the phloem.
- Stems support plant parts.
- Herbaceous stems are soft and green, while woody stems are rigid.
- During photosynthesis, the sun's energy is used to combine carbon dioxide from the air and water from the soil to produce glucose and oxygen.
- In transpiration, the stomata regulate the amount of water lost through the plant leaves.
- The leaf has an epidermis, two layers of food-making cells, and veins containing the xylem and phloem.

**9-3 Gymnosperms**

- Gymnosperms produce uncovered seeds.
- Cycads, gingkoes, and conifers are the three main groups of gymnosperms.

**9-4 Angiosperms**

- Angiosperms, or flowering plants, produce seeds with a protective covering.
- Flowers contain the male and female reproductive organs. The stamens produce sperm-containing pollen. The pistils contain the egg-producing ovary.
- Fertilization occurs when the sperm unites with the egg in the ovary. The ovary then develops into a fruit that encloses the seed.
- Seeds have one or two cotyledons. A cotyledon is the embryo plant's food storehouse.

## VOCABULARY

*Define each term in a complete sentence.*

| | | | |
|---|---|---|---|
| **angiosperm** | **frond** | **phloem** | **sexual reproduction** |
| **annual ring** | **fruit** | **photosynthesis** | **stamen** |
| **asexual reproduction** | **germination** | **pistil** | **stoma** |
| **cambium** | **guard cell** | **pollen** | **transpiration** |
| **cotyledon** | **gymnosperm** | **pollination** | **vascular plant** |
| **epidermis** | **herbaceous stem** | **rhizome** | **woody stem** |
| **fertilization** | **ovule** | **seed** | **xylem** |
| **flower** | **petal** | **sepal** | |



**4.** Changing the position of the dish would alter the variable being studied.
**5.** Stems would grow up out of soil and the roots down into the soil.
**6.** Seeds do not need light for germination.

## GOING FURTHER: ENRICHMENT

To continue the study of tropisms, students can investigate the effect of light on a plant. Place a seedling on a windowsill. It should be positioned in such a way that the source of light is almost horizontal to the plant. Ask students to predict what may occur. Observe the growth pattern of the seedling to see if their predictions are correct. After performing this experiment, ask students the following questions:

- **Based on your observations, is light a vital need of plants?** (yes)
- **Does a plant try to adapt itself to its surroundings?** (yes)
- **How did this seedling adapt to its particular environment?** (Because it needs light for life, the plant shifted its growth pattern so that it would receive the necessary light.)

## CHAPTER REVIEW

### MULTIPLE CHOICE

1. d 3. d 5. a 7. b 9. d
2. d 4. c 6. b 8. d 10. a

### COMPLETION

1. ferns
2. spores
3. roots
4. cambium
5. compound
6. epidermis
7. gymnosperms
8. stigma
9. germination
10. water

### TRUE OR FALSE

1. T
2. F root hairs
3. F annual ring
4. F perennial
5. F photosynthesis
6. F water
7. F cones
8. F flowers
9. F fertilization
10. T

### SKILL BUILDING

**1.** Only vascular plants can grow tall because they contain transporting tubes that carry material up and down the plants and supporting tissues. Nonvascular plants do not contain transporting tubes, and therefore, they must remain very close to the ground.
**2.** Check drawings. Drawings should resemble the simple and compound leaves shown in Figure 9-12 on page 202.
**3.** The petals would change hue to the color of the colored water as vascular tubes carried the colored water up the stem to the flower.
**4.** The flat, broad shape provides a large surface area for the trapping of light for photosynthesis.
**5.** It might hinder pollination because insects are often the organisms that carry pollen from one flower to another.
**6.** The plant would die because it could not take in carbon dioxide for photosynthesis or release oxygen. Nor could it release water through transpiration.
**7.** This makes it easier for fertilization to occur.
**8.** Check experiments to make sure students have an experimental setup and a control setup.
**9.** Girdling destroys the pathway vascular tubes need to carry food, water, and minerals throughout the tree.

## CONTENT REVIEW: MULTIPLE CHOICE

*Choose the letter of the answer that best completes each statement.*

1. Vascular plants have
   a. true leaves and roots. b. true roots and stems.
   c. true stems and leaves. d. true roots, stems, and leaves.
2. Ferns produce spores during
   a. germination. b. fertilization.
   c. sexual reproduction. d. asexual reproduction.
3. Plant tissue that forms tubes that carry water and minerals up through the plant is called
   a. stoma. b. cambium. c. phloem. d. xylem.
4. Which of these structures is *not* a stem?
   a. bulb b. rhizome c. frond d. tuber
5. During photosynthesis, oxygen leaves the leaf of a plant through structures called
   a. stomata. b. rhizoids. c. guard cells. d. rhizomes.
6. Gymnosperms
   a. produce flowers. b. have seeds with no protective wall.
   c. have seeds with a protective wall. d. are found only in deserts.
7. Which structure produces pollen?
   a. pistil b. anther c. ovary d. style
8. The union of the sperm and the egg is called
   a. germination. b. transpiration. c. pollination. d. fertilization.
9. Seeds can be scattered by
   a. wind. b. water. c. people. d. all of these.
10. How many cotyledons does a monocot contain?
    a. one b. two c. three d. four

## CONTENT REVIEW: COMPLETION

*Fill in the word or words that best complete each statement.*

1. A group of vascular plants that were abundant during the Carboniferous Period were the ______.
2. Small brown structures on the underside of fern fronds produce ______.
3. ______ are fern structures that absorb water and minerals from the soil.
4. In the stem, new xylem and phloem cells are produced by the ______.
5. Leaves that have two or more blades are called ______.
6. The outer, protective cell layer of a leaf is the ______.
7. Conifers belong to a group of seed plants called ______.
8. During pollination, pollen grains cling to the sticky surface at the top of the pistil called the ______.
9. When conditions are just right, a seed may go through a process of growth called ______.
10. The seeds of the coconut palm are dispersed by ______.



### ESSAY

**1.** In the asexual stage, spores are produced on the undersides of fronds. The spores fall and produce the tiny, heart-shaped structures of the sexual stage. These structures produce male and female cells, which unite and produce a new asexual-stage fern.
**2.** Roots anchor plants in the ground, absorb water and minerals from the soil, and sometimes store food. Stems support the plants, hold up leaves to the light, and transport water, minerals, and food.
**3.** Herbaceous stems are green and soft and tend to be found in annual plants. Woody stems are rigid, have large amounts of xylem, and tend to be found in perennial plants.
**4.** Gymnosperms produce uncovered seeds and produce no flowers or fruits. Angiosperms produce flowers and fruits, which contain the seeds.
**5.** In self-pollination, pollen is transferred from a stamen to a pistil of the same flower or of another flower on the same plant. In cross-pollination, the transfer occurs between stamens

## CONTENT REVIEW: TRUE OR FALSE

*Determine whether each statement is true or false. If it is true, write "true." If it is false, change the underlined word or words to make the statement true.*

1. In sexual reproduction, an organism develops from the union of an egg and a sperm.
2. Seed plants absorb water and minerals through microscopic extensions of single cells called guard cells.
3. One year's growth of xylem cells produces a layer called an epidermis.
4. A biennial is a plant that lives for more than two growing seasons.
5. Glucose is the food made by a plant during germination.
6. The cuticle prevents the loss of too much oxygen from the leaf.
7. Conifers produce flowers.
8. Angiosperms produce cones.
9. When a sperm unites with an egg in a flower's ovary, pollination occurs.
10. A fruit is the ripened ovary that encloses and protects the seed.

## CONCEPT REVIEW: SKILL BUILDING

*Use the skills you have developed in the chapter to complete each activity.*

1. **Relating concepts** In your friend's backyard, you find a plant that is 3 meters tall. Is it a vascular or a nonvascular plant? Explain your answer.
2. **Drawing diagrams** Based on what you learned in the chapter, draw a diagram of a simple and a compound leaf. Label the stalk and the blade in each leaf.
3. **Making predictions** Predict what would happen to the petals of a carnation if you placed the stem in a vase filled with colored water.
4. **Applying concepts** Most leaves are broad and flat. How are these characteristics helpful to a plant?
5. **Making inferences** Pesticides are chemicals designed to kill harmful insects. But sometimes these chemicals also kill helpful insects. What effect could this have on angiosperms?
6. **Developing a hypothesis** A scientist performed an experiment in which he covered the leaves of a plant with petroleum jelly. The plant died. Explain why.
7. **Relating facts** What is the advantage for a gymnosperm of having both male and female cones on the same tree?
8. **Designing an experiment** Design an experiment to determine whether or not water is needed for seed germination. Describe your experimental setup. Be sure to include a control.
9. **Relating cause and effect** Girdling is the complete removal of bark from a section of a tree. Explain why girdling can kill a tree.

## CONCEPT REVIEW: ESSAY

*Discuss each of the following in a brief paragraph.*

1. Describe the two stages in the reproductive cycle of a fern.
2. Compare the functions of roots and stems.
3. Distinguish between herbaceous and woody stems.
4. Distinguish between gymnosperms and angiosperms.
5. Discuss the difference between self-pollination and cross-pollination.



and pistils in flowers on different plants.

## ADDITIONAL QUESTIONS AND TOPIC SUGGESTIONS

**1.** Explain why ferns, like mosses, are often found growing in moist environments. (During the fern's life cycle, water must be present for the sperm to swim to the egg for fertilization.)
**2.** Why might a tree die if a beaver or rabbit chews a ring completely around the bark at its base? (In most trees, phloem tubes are part of the inner bark layer. Because phloem carries food that was manufactured in the leaves to the roots, the roots will eventually die if their food supply is cut off.)
**3.** The leaves of most conifers are needle-like. Why might such leaves enable conifers to survive better in drier areas than broad-leafed trees? (Needle-like leaves have fewer stomata on their surface than broad leaves. Therefore, they do not lose as much water through transpiration.)
**4.** Explain why cross-pollinated plants often have larger and sweeter-smelling flowers than plants that self-pollinate. (Large, colorful, sweet-smelling flowers can better attract insects or other animals that transfer the pollen of these plants.)

## ISSUES IN SCIENCE

The following issue can be used as a springboard for discussion or given as a writing assignment.

The great majority of the earth's plants are found in the tropics. Although many of these plants have not even yet been identified, their natural habitat is being cleared for farming, firewood, and manufacturing. Many scientists are alarmed that great numbers of exotic tropical plants will soon become extinct. For this reason, scientists are trying to persuade the governments of tropical nations to preserve the native plants. However, others say that because of the growing human population in these parts of the world, the tropical forests must be cleared to make way for civilization. What is your opinion and why?

## Chapter 10

# ANIMALS: INVERTEBRATES

## CHAPTER OVERVIEW

The simplest invertebrates are the sponges, which are little more than colonies of cells living together. Their bodies remain stationary as their cells filter food and oxygen from the water entering their pores. Somewhat more advanced are the coelenterates. Unlike sponges, their body cells are organized into true tissues.

Worms are grouped into three phyla on the basis of their body plan. Flatworms have ribbonlike bodies and an incomplete digestive tract, whereas roundworms are threadlike and have a digestive tract with two openings. Many flatworms and roundworms are parasites. Annelids have rounded bodies with segments. With segmentation, a greater specialization of body parts is possible.

The majority of mollusks and echinoderms (spiny-skinned animals) are aquatic creatures. Mollusks include such diverse animals as snails, clams, and squids. But despite their diversity, they all have soft bodies covered by a mantle that often secretes a limy shell. The spiny-skinned animals have a unique water vascular system to move about in their marine environment.

Arthropods are characterized by their joint appendages and exoskeleton. In terms of numbers they are the most successful of all groups of animals. Arthropods include crustaceans, centipedes and millipedes, arachnids, and insects.

## INTRODUCING CHAPTER 10

Ask students to look at the picture on page 216.

- **Honeybees live in large groups called colonies. Different members of the colony perform different tasks. What do you think is the job of the bee shown in the picture?** (If students are familiar with honeybees they will be aware that this type of bee gathers nectar from flowers.)
- **What are the bees that gather nectar called?** (Answers may vary. If students do not identify the bee correctly point out that it is a worker.)

Tell students that thousands of bees make up a typical colony. Most of them are workers, which are infertile females. Workers perform all of the jobs in the colony except reproduction. A colony also includes drones and one queen. The only job of the drones is to mate with the queen, the only female capable of reproduction.

Now have the class read page 217. After they have finished, point out that a worker bee will have several jobs during her adult life. After she first emerges from the pupa she serves as a nurse to young larvae and a servant to the drones and queen. In performing this role she feeds these other bees pollen and honey. Later the worker takes on different

# 10 Animals: Invertebrates

**CHAPTER SECTIONS**



**CHAPTER OBJECTIVES**

*After completing this chapter, you will be able to:*

**10-1** Distinguish between vertebrates and invertebrates.

**10-2** Describe the characteristics of sponges.

**10-3** Describe the characteristics of coelenterates.

**10-4** Compare flatworms, roundworms, and segmented worms.

**10-5** Describe the characteristics of mollusks.

**10-6** Describe the characteristics of spiny-skinned invertebrates.

**10-7** Identify the characteristics of arthropods

**10-8** Describe the anatomy and characteristics of insects.

A hard-working person is often described as being "busy as a bee." But, you may wonder, just how busy is a bee? Kirk Visscher, a scientist, wondered the same thing. Visscher is an *entomologist* (en-tuh-MAHL-uh-jist), a person who studies insects.

Visscher knew that one group of bees, worker bees, had the jobs of cleaning up the hive, feeding young bees, finding and processing pollen, and guarding the nest. However, while watching bees in a glass-enclosed colony, Visscher noticed that certain bees also seemed to have the responsibility of removing dead bees from the hive. While most worker bees ignored their dead companions, these undertaker bees grasped the dead bees in their jaws and flew hundreds of meters from the hive before dropping them off.

This special task of beehive undertaking interested Visscher. He wondered what proportion of worker bees acted as undertakers for the hive. To find out, the entomologist painted dots on the bees that removed the dead. He then determined that only about 2 percent of the worker bees actually did the undertaking. And, he discovered another interesting fact. The task of carrying away the dead was only a temporary one. After a few days, the undertakers moved on to other jobs.

*Honeybee flying over a flower*



## TEACHER DEMONSTRATION

Set up a display around the classroom of some common specimens of plants and animals. Some possibilities include

1. algae floating in water
2. clump of moss
3. *Elodea*
4. potted geranium
5. potted ivy
6. potted fern
7. branch from a tree
8. earthworm
9. snail
10. cricket
11. crayfish
12. fish in an aquarium
13. mouse

If you cannot obtain living specimens, then display a variety of pictures of plants and animals or a combination of specimens and pictures. Allow the class ample time to move about the room to examine the specimens or pictures. After they return to their seats, conduct a discussion on some of the similarities and differences between plants and animals.

• **If you were to place the organisms you just looked at into two different groups, how would you do it?** (Answers may vary. Lead students to suggest that some are plants and the others are animals.)

• **How do animals obtain their food?** (Answers will vary. Develop the idea that animals are ingestors of food. Most of them take in food and digest it internally.)

• **Why do you think most animals move about but plants do not?** (Answers will vary. Lead students to suggest that because animals obtain food from their environment, they must move about to find it.)

jobs of building, repairing, and cleaning the wax cells making up the hive. During the final weeks of her life, she becomes a forager searching for pollen and nectar.

• **During the period of their life when workers care for the hive, do all of them perform the same tasks?** (Lead students to suggest that only certain ones among the workers perform such tasks as removing dead bees from the hive.)

Explain that among social insects there is usually a high degree of "division of labor." The insects cooperate with each other by performing specific tasks for the benefit of the entire colony.

• **Do you know of any other kinds of insects in which there is a division of labor?** (Some students may be aware of a division of labor among such other social insects as termites, ants, and wasps.)

## TEACHER RESOURCES

**Audiovisuals**

*Animals Without Backbones,* filmstrip, EBE
*Crustaceans,* film, EBE
*Insects,* film, EBE

**Books**

Blaney, W. M., *How Insects Live,* E. P. Dutton
Buchsbaum, R., *Animals Without Backbones,* University of Chicago Press

## 10-1 CHARACTERISTICS OF INVERTEBRATES

### SECTION PREVIEW 10-1

Those organisms classified in the animal kingdom are usually subgrouped as vertebrates or invertebrates. Vertebrates have a backbone, invertebrates do not. Many people automatically think of vertebrates when hearing the word *animal,* but invertebrates are by far the most numerous. Invertebrates occur worldwide in almost every habitat, and they make up more than 90% of all animal species.

### SECTION OBJECTIVES 10-1

1. **Define vertebrates and give some examples.**
2. **Define invertebrates and give some examples.**

### SCIENCE TERMS 10-1

**vertebrate p. 218**
**invertebrate p. 219**

## 10-1 Characteristics of Invertebrates

Animals! Just hearing the word brings a different image to almost everyone. Some people think of the fierce great cats in Africa. To others, the word brings to mind the friendly porpoise, certainly among the most intelligent of all animals. Of course, for a great many people the word "animal" reminds them of that cuddly puppy they had, or have, or have always wanted to get. All of these animals, no matter how different they may seem, have one thing in common. They are **vertebrates** (VER-tuh-brits). **Vertebrates are animals with a backbone.**

**Figure 10-1** *There are many kinds of invertebrates, or animals without backbones. The skeleton butterfly* (top left), *hercules beetle* (top right), *giant clam* (bottom left), *and flame lobster* (bottom right) *are examples of invertebrates.*

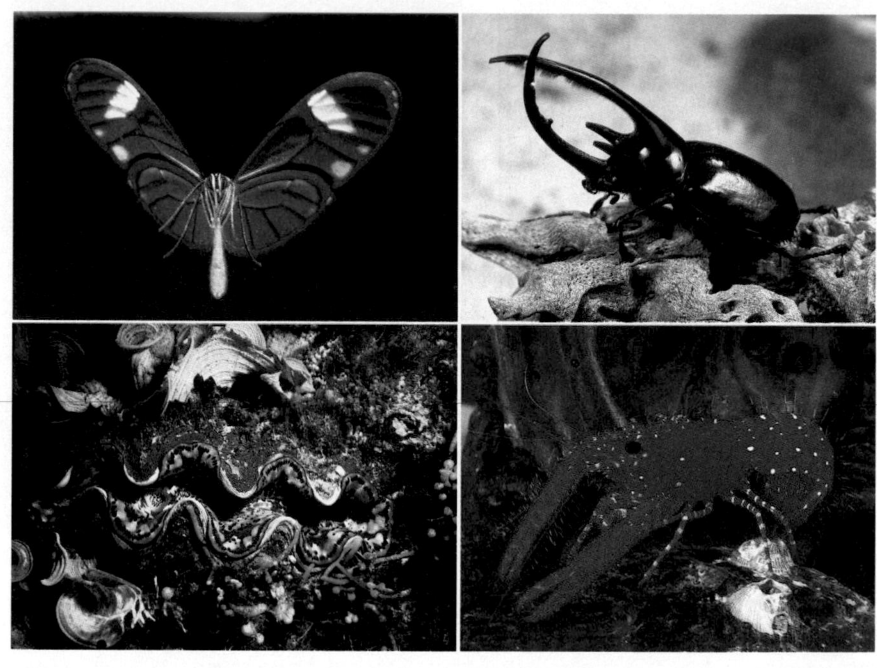

218

### TEACHING STRATEGY 10-1

#### Motivation

Obtain three or more pictures of invertebrates and a similar number of pictures of vertebrates. If you have no bulletin board pictures of animals you might find suitable pictures of animals in books from your school or public library. Colored slides, if available, could also be used. Have students compare the pictures, and then ask,

• **Can you think of a way we can divide these animals into two groups?** (Accept all logical answers, but lead students to suggest that some of the pictured animals have a backbone and some do not.)

• **What do we call animals that have a backbone?** (Animals with a backbone are called vertebrates.)

• **What do we call animals without backbones?** (Animals without backbones are called invertebrates.)

Now ask students to try to write down the name of one invertebrate and one vertebrate. Place two columns headed "invertebrates" and "vertebrates" on the chalkboard or overhead projectual, and then call on students one at a time to name their examples. List them in the appropriate column as they are given. See how many different examples of each the class can generate at this time.

#### Content Development

Emphasize that even though vertebrates are generally larger and more familiar to most people, they are in the minority. No one really knows the number of animal species that exists, but some biologists feel there may be as many as 2 million. Among these, fewer than 50,000 species are vertebrates. Point out that the insects alone, which are just one kind of invertebrate, outnumber all vertebrates.

### Section Review 10-1

1. Vertebrates.
2. Invertebrates.

On the other hand, almost nobody thinks of the earthworm crawling through the soil when the word "animals" is mentioned. Or the mosquito about to raise a bump on another victim. Or the jellyfish with its stinging cells always ready to poison a passing fish. And yet, these organisms are among a group of animals called **invertebrates** that make up more than 90 percent of all animal species. **Invertebrates are animals without a backbone.** They are found in just about every corner of the world.

All animals, including the invertebrates, are multicellular. Most have cells that perform specialized functions. Some cells, for example, are involved with the digestion of food. Other cells help circulate the digested food particles throughout the body. In many animals, these specialized cells are grouped together into tissues. Skin cells, for example, form skin tissue. Tissues that work together can be further organized into organs. For example, the lungs are organs made up of different tissues that work together to help you breathe. Finally, organs that work together often are grouped as organ systems. The mouth, stomach, and intestines all work together as part of the digestive system.

### *SECTION REVIEW*

1. What are animals with backbones called?
2. What are animals without backbones called?

## 10-2 Sponges

Today, the sponge you use to wash the dishes or take a bath is probably made of synthetic material. In the past, however, people used natural, or real, sponges from the sea. A natural sponge is the dried remains of an animal that lives in the ocean, or as in a few cases, in fresh water. You might think a sponge, whether synthetic or natural, looks nothing like any animal you ever saw. In a way you would be right, for sponges seem to be totally unrelated to all other forms of animal life. For many years, in fact, sponges were classified as "plant-animals."

### Activity

*Observing a Sponge*

How is a real sponge different from a synthetic sponge?

1. Obtain a natural sponge from your teacher.
2. Use a hand lens to examine the surface and the pores. Make a drawing of what you see.
3. Tear off a small piece of the sponge and place it on a glass microscope slide. Draw what you observe under the microscope. Spicules are hard, pointed structures that support the body of some sponges. Can you see any spicules? Draw what you observe.
4. Now repeat this activity with a synthetic kitchen sponge. How does it compare with a natural sponge?



### Activity

**Observing a Sponge**
**Skills: Making comparisons, manipulative, observing, relating**
**Level: Enriched**
**Type: Hands-on/library**
**Materials: Synthetic and natural sponge, hand lens**

Synthetic kitchen sponges are made from dry rubber or latex. Latex is a substance found in a wide variety of trees and other plants. During the processing of the sponges, tiny bubbles of air become trapped in the rubber giving the sponge its "natural" look. Most natural sponges have a mineral skeleton made up of tiny, needlelike spicules. The spicules may be either calcium carbonate (limestone) or silica, a glasslike material. In bath sponges, the skeleton consists of fibers of tan-colored protein called spongin. The skeleton of spongin fibers is what remains after a bath sponge dies and its cells are removed. Many sponges have a skeleton of both mineral spicules and spongin fibers.

## 10-2 SPONGES

### SECTION PREVIEW 10-2

Sponges, the simplest invertebrates are commonly called poriferans. Unlike most animals, these aquatic organisms usually remain attached to one spot their entire lives. Their bodies contain numerous pores through which water flows bringing food and oxygen to the cells. Wastes from the cells are eliminated back into the moving water.

### SECTION OBJECTIVES 10-2

1. **Describe the physical appearance of a sponge.**
2. **Describe the methods by which sponges obtain food and oxygen.**

### SCIENCE TERMS 10-2

**pore p. 220**

## TEACHING STRATEGY 10-2

### Motivation

If available, show the class some preserved specimens and/or pictures of sponges. Also have them look at Figure 10-2 on page 220.

- **Why do you think people sometimes mistake sponges for plants?** (Accept all logical answers. Students will likely indicate that sponges are attached to one spot like most plants.)
- **Why are sponges classified as animals?** (They cannot carry on photosynthesis. They catch food and digest it inside their bodies.)
- **Why could sponges be used to take a bath or wash dishes?** (Since their bodies contain numerous pores that absorb water, a sponge's dried remains could be used for cleaning purposes.)

## FACTS AND FIGURES

Sponges grow in a variety of shapes and sizes. The smallest are cuplike and only a few millimeters long, while the largest types are vaselike, branching, or moundlike. Some of these larger species may be up to 1.8 m tall and more than 2 m across.

## ANNOTATION KEY

❶ **Food and oxygen are carried by moving water through the pores into the sponge. (Applying concepts)**

① **Thinking Skill: Identifying processes**

② **Thinking Skill: Applying technology**

③ **Thinking Skill: Making generalizations**

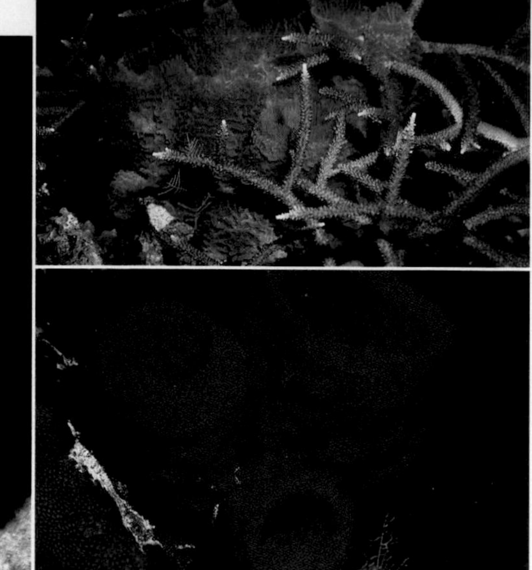

**Figure 10-2** *Although there are different types of sponges, they all share the characteristic that their bodies contain many pores.*

**Figure 10-3** *Notice the pores in this diagram of a sponge. What is the function of the pores?* ❶

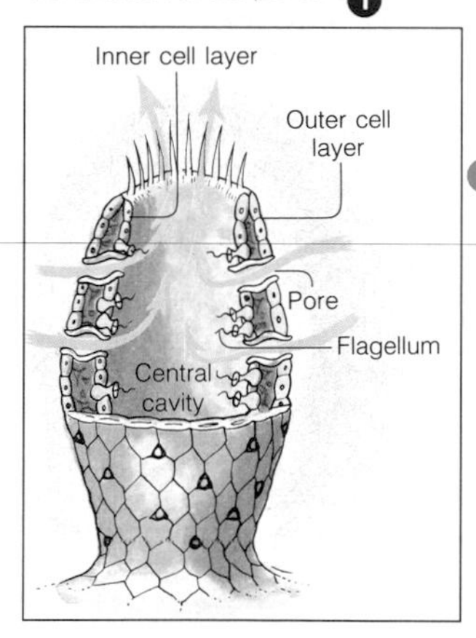


Sponges lead a very simple life. They grow attached to one spot on the ocean floor, usually in shallow water. A sponge stays in that one spot its entire life unless a strong wave or current washes it somewhere else. **The body of a sponge has a great many pores, or holes.** Moving ocean water carries food and oxygen through the **pores** into the sponge. ① The sponge's cells remove food and oxygen from the water and, at the same time, release waste products into the water. Because the pores of a sponge are small, most sponges live in clear water that is free of floating matter that could clog the pores.

Sponge cells are unusual in that they function on their own without any coordination between one another. In fact, some people think of sponges as a colony of cells living together. However, despite their independent functioning, sponge cells have a mysterious attraction to one another. This attraction can be demonstrated easily by passing a sponge through a fine filter so that it is broken into clumps of cells.

220

## 10-2 (continued)

### Content Development

As students look at Figures 10-2 and 10-3, emphasize that even though sponge cells are capable of functioning on their own, there is nevertheless some specialization, or division of labor, among the cells. The cells making up the outer layer of the sponge's body are flat, and they serve as a protective covering. The cells of the inner layer contain flagella, which set up water currents that bring in food particles. These cells, called collar cells, capture and digest food particles that may be in the water. Wastes pass out a large opening at the top of the sponge. In the jellylike mass between these two cell layers are specialized cells called *amoebocytes*. These cells secrete skeletal materials and transport digested food. Some amoebocytes can also form sperm and egg cells that function in sexual reproduction.

### Reinforcement

Ask students to look once again at Figure 10-2.

- **Why are these animals sometimes thought to be plants?** (They remain attached to one place like most plants.)
- **How do these animals take in food and oxygen?** (As ocean water moves into a sponge through its pores, the inner cells remove food and oxygen.)
- **What can a sponge do if it is broken into small clumps of cells?** (The cells can reorganize themselves into the shape of the original sponge.)

## Section Review 10-2

**1.** Sponge cells remove food from the water that has entered through pores.

**2.** Sponge cells release waste products into water.

**3.** To keep their pores free of material.

## TEACHING STRATEGY 10-3

### Motivation

Introduce this section by showing some pictures of the familiar coelen-

Within hours, these cells reform into the shape of the original sponge. No other animal species shares this amazing ability of sponge cells to reorganize themselves.

### *SECTION REVIEW*

1. How do sponges obtain food?
2. How do sponges remove waste products?
3. Why do most sponges live in clear waters that are free of floating matter?

## 10-3 Coelenterates

In the shallow waters off the coast of Puerto Rico, a group of scientists drill into the sea floor. To an ordinary person, the scientists might appear to be hunting for oil and gas, or perhaps valuable minerals. However, these scientists are drilling through layers of coral in order to read a very special "diary"—a diary of the earth's past. The scientists have discovered that corals are extremely sensitive to environmental changes, including changes in temperature and the chemicals in sea water. These changes are recorded in the daily growth rings of coral. By studying the layers of coral that were laid down year after year for more than 500 million years, scientists hope to uncover evidence of past events. For example, many scientists believe the length of the earth's day has increased by several hours or more in the last few hundred million years. Scientists hope to find evidence for this increase in the coral growth rings. ②

Corals are among a group of animals called the coelenterates (si-LEN-tuh-rayts). Also included among the coelenterates are the jellyfish, the hydra, and the sea anemone (uh-NEM-uh-nee). **All coelenterates contain a central cavity with only one opening.** ③ You can think of coelenterates as being cup-shaped animals with an open mouth. Tentacles with stinging cells called **nematocysts** (NEM-uh-toh-sists) surround the mouth of most coelenterates. Coelenterates use nematocysts to stun or kill other animals. It is no

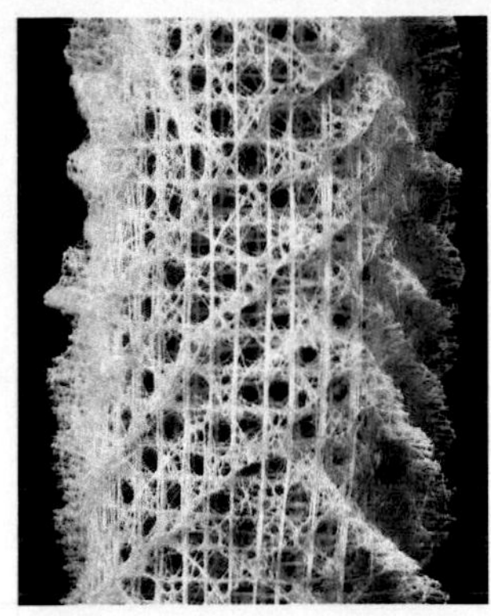

**Figure 10-4** *Venus' flower baskets are sponges having glassy skeletons. These sponges have been on the earth at least 500 million years.*

**Figure 10-5** *Meters below the surface of the ocean, these scientists use a coring machine to remove a sample of coral. Like trees, coral forms annual layers as it grows.*

## 10-3 COELENTERATES

### SECTION PREVIEW 10-3

Phylum Coelenterata, the coelenterates, is a group of aquatic invertebrates having a digestive cavity with only one opening. They are also characterized by tentacles with stinging cells called nematocysts surrounding the mouth. These nematocyst-bearing tentacles are used in capturing prey and forcing it into the mouth. Unlike sponges, coelenterates have true tissues organized into two body layers.

Most coelenterates live in salt water, and many are strikingly beautiful. Familiar coelenterates include the corals, which live together in colonies. Corals produce a hard protective covering of limestone that is left behind when the organisms inside die. Over long periods of time, these limestone layers form coral reefs. Another well-known coelenterate is the sea anemone, an ocean bottom dweller that resembles underwater flowers. Motile coelenterates include the jellyfish, an organism that sometimes annoys swimmers.

### SECTION OBJECTIVES 10-3

1. **Describe characteristics of coelenterates.**
2. **Give examples of organisms classified as coelenterates.**
3. **Compare the methods by which corals, sea anemones, and jellyfish obtain food.**

### SCIENCE TERMS 10-3

**nematocyst p. 221**

terates to be studied. Pictures of coral, sea anemones, jellyfish, and hydras are often shown on biological wall charts. Such pictures may also be found in library reference books.

Explain that even though the next group of animals to be studied appears to be very different, they have several common characteristics.

• **What appears to be around the mouth of these animals?** (Answers may vary, but lead students to indicate that there is a ring of tentacles around the mouth.)

• **What do you think the tentacles are used for?** (Students should indicate that the tentacles are used to capture prey and force it into the mouth.)

Write the term *coelenterates* on the chalkboard. Explain that the name for this group is based on the world *coelenteron* meaning hollow body cavity. Point out that all coelenterates have a hollow body cavity with one opening leading into it. This opening is surrounded by a ring of tentacles in most coelenterates.

### Content Development

Ask a student to read aloud the definition of *nematocysts* from the glossary. Point out that the presence of nematocysts is a trait unique to coelenterates.

## BACKGROUND INFORMATION

Unlike sponges, coelenterates have true tissues—groups of cells specialized for a specific function. The outer layer, *ectoderm,* contains several kinds of cells; the inner layer, or *endoderm,* secretes enzymes to digest food. Between these two layers is a thin jellylike material called *mesoglea.* Inside the body is an opening called the *gastrovascular cavity* where digestion takes place. The digested food is then taken up by the cells lining the cavity. Since there is only one opening to the gastrovascular cavity, any undigested food particles are eliminated through the same opening.

## FACTS AND FIGURES

The Portuguese man-of-war is not a single organism, but a colony of jellyfish kept on top of the water by a gas filled float. This float, which may be up to 30 cm long, looks like a blue or pink plastic bag. The tentacles that dangle from the float may reach a length of 20 m or more, and they are lined with extra large nematocysts whose poison is as potent as that of a cobra. Portuguese man-of-war can produce a sting that is fatal to humans.

**Figure 10-6** *Jellyfish, like this Portuguese man-of-war, use their tentacles to capture prey. Nematocysts, or stinging cells, on the tentacles then paralyze the prey.*

surprise, then, that the stinging cells are near the mouth. For after capturing an animal with their tentacles, coelenterates pull it into their mouth and then into their central body cavity. Once the food is digested, coelenterates then release waste products back out, through their mouth.

Unlike the sponges, coelenterates contain groups of cells that perform special functions. That is, coelenterates have specialized tissues. Some coelenterates like the jellyfish move about in the water using muscle tissues. Others, like the corals, remain in one place. ❶

### Corals

In the warm waters off the coast of eastern Australia lies one of the largest structures ever built by living things—the 2000-kilometer-long Great Barrier Reef. However, the reef was not built by people. Instead, it was built by tiny corals.

Corals, like all coelenterates, are soft-bodied organisms. However, corals use minerals in the water to build a hard protective covering of limestone. When the coral dies, the hard outer covering is left behind. Year after year, for many millions of years, generations of corals live and die, each adding a layer of limestone. In time, a coral reef such as the Great Barrier Reef forms. The outer layer of the reef, then, contains living corals. But underneath this ❷

**Figure 10-7** *When these brightly colored corals die, their skeletons remain. Over the years, the skeletons pile up, creating a coral reef.*

### 10-3 (continued)

#### Content Development

Point out that most reef-forming corals are found only in warm, well-lighted ocean environments where the temperature rarely falls below 23°C. The great reefs of tropical waters are home to a vast variety of marine organisms including fish and such invertebrates as sponges, echinoderms, and crustaceans. Though the great majority of corals flourish in tropical waters, a few reef-building species dwell in deep waters that get as cold as 9°C.

#### Skills Development

***Skill: Relating Concepts***

Review the concept of *symbiosis* that was previously introduced (page 109) by calling attention to the last paragraph in the section on Corals.

- **What kind of organisms live in the bodies of certain corals?** (Algae live inside the coral's body.)
- **Do the algae harm the coral or do they benefit it?** (The algae benefit the coral by making food for them.)
- **What do we call this kind of relationship in which two different kinds of organisms live together?** (The relationship is called symbiosis.)
- **What are some other examples of symbiosis that you studied earlier?** (Accept all logical answers, but lead students to mention the symbiotic relationship between the alga and fungus making up a lichen.)

#### Enrichment

Some students may be interested in conducting research and preparing reports on the Great Barrier Reef, located along the eastern coast of Australia. Encourage students to share their reports with the class.

Figure 10-8 *Surrounding the mouth of this delicate-looking coral are tentacles, which are used to trap food.*

"living stone" are the remains of corals that may have lived when dinosaurs walked the earth.

Corals live together in colonies that can grow into a wide variety of shapes and colors. Some corals look like antlers, others like fans swaying in the water, while still others look like underwater brains.

At first glance, a coral appears to be little more than a mouth surrounded by stinging tentacles. See Figure 10-8. However, there is more to a coral than meets the eye. Algae live inside the coral's body. The algae help make food for the coral. And since algae need sunlight to make food, corals must live in shallow waters where sunlight can reach them. This relationship between a coral animal and an alga plant is among the most unusual in nature. ❸

### *Sea Anemones*

Can you spot the clownfish swimming through the brightly colored plant in Figure 10-9? You might be surprised to discover that the "plant" is actually an animal—a type of coelenterate called a sea anemone. Sea anemones look like underwater flowers. However, the "petals" are really tentacles, and their brilliant coloring helps attract passing fish. When a fish passes over the anemone's stinging cells, the cells poison the fish. The tentacles then pull the fish into the anemone's mouth, and the stunned prey soon is digested.

The clownfish, however, is not harmed by the anemone. It swims safely through the anemone's

Figure 10-9 *The lavender-tipped "petals" in this photograph are actually the tentacles of a sea anemone. Notice the black and white clownfish swimming among the tentacles.*

## TIE-IN/ LITERATURE

Have students do research on the mythological monster, Hydra, for which the freshwater coelenterate is named. In this well-known myth, Hydra had nine heads, and when one was cut off, two grew back in its place. The coelenterate is so-named because of its ability to regenerate body parts.

## ANNOTATION KEY

❶ **Thinking Skill: Making comparisons**

❷ **Thinking Skill: Relating cause and effect**

❸ **Thinking Skill: Identifying relationships**

## SCIENCE, TECHNOLOGY, AND SOCIETY

Marine invertebrates are becoming increasingly important to humans. In recent years, scientists have discovered at least 1500 new compounds from marine organisms. Over 150 of these compounds display medicinal properties such as antimicrobial, anti-inflammatory, antitumor, or cardiotonic activities. Certain sponges and soft corals, for example, contain anti-inflammatory agents that may be useful in treating arthritis. Sea whips have been found to contain lophotoxin, a poisonous protein substance that inhibits muscular activity by its action on the nervous system. Lophotoxin may be useful in treating diseases, such as multiple sclerosis, that affect muscle function. Numerous other marine organisms contain cytotoxins, which inhibit cell development. Cytotoxins may be useful in fighting cancerous cells.

## Motivation

Tell the class to look at Figure 10-9 and then read the text section on sea anemones.

- **How did you distinguish the clownfish from the tentacles of the sea anemone?** (Accept all answers. Students will likely mention that the clownfish is black and white, but the tentacles of the anemone are lavender and white.)
- **In what way is the relationship between the clownfish and sea anemone similar to that of the algae inside the coral's body?** (Students should point out they benefit each other. The relationship is symbiotic.)
- **How do the clownfish and anemone benefit each other?** (The anemone protects the clownfish from predators while the while the clownfish attracts other food fish to the anemone.)

## FACTS AND FIGURES

Jellyfish vary greatly in size. The smallest are less than 2 cm in diameter. The largest one known, *Cyanea arctica,* is more than 1.8 m across, and it has trailing tentacles up to 36 m long.

## ANNOTATION KEY

❶ Thinking skill: Identifying relationships

❷ Thinking Skill: Classifying organisms

## Activity

**Observing Hydra**
**Skills: Making observations, manipulative, diagraming, hypothesizing**
**Level: Enriched**
**Type: Hands-on**
**Materials: Medicine dropper, small dish, hydra, toothpick, *Daphnia,* vinegar**

Students will observe that the hydra contracts all of its tentacles when touched. The hydra uses its tentacles to grab its prey and pull it toward its mouth. The hydra reacts to vinegar by contracting its tentacles.

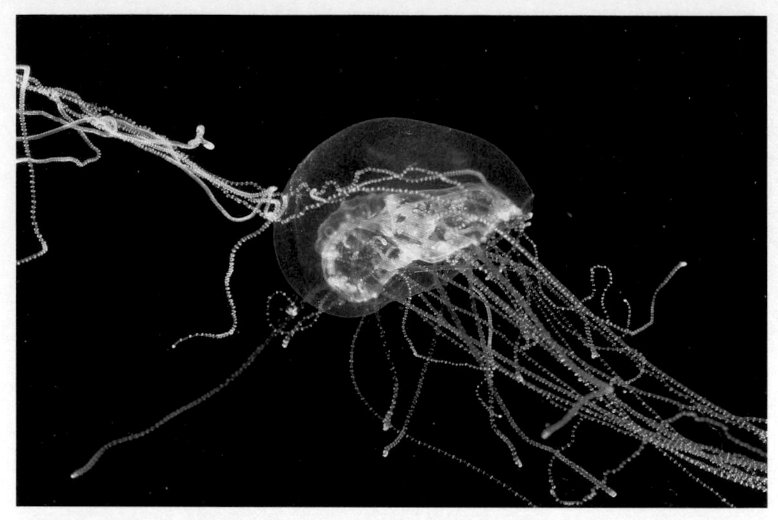

**Figure 10-10** *Jellyfish move with a jetlike motion by contracting their muscle cells. The long tentacles are used to capture food.*

### Activity

*Observing Hydra*

1. Use a medicine dropper to place a hydra in a small dish along with some of the water in which it lives.
2. Observe the hydra. Locate its mouth, body cavity, and tentacles. Draw a diagram in which you label these parts.
3. With a toothpick, gently touch the hydra in three different places. How does the animal respond to each touch?
4. Place some *Daphnia* in the dish. How does the hydra feed?
5. Add one drop of vinegar to the dish. How does the hydra react?

tentacles. In this way, the clownfish is protected from other fish that might try to attack it. At the same time, the clownfish serves as a kind of living bait for ❶ the anemone. Other fish see the clownfish, come closer, and are quickly trapped by the anemone. So the sea anemone and the clownfish live in harmony.

### *Jellyfish*

If you ever swam in the ocean and saw a jellylike cup floating in the water, you probably knew enough to swim away. Most people quickly recognize this coelenterate, the jellyfish. While the jellyfish may look harmless, it can deliver a painful poison through its stinging cells. In fact, even when they are broken up into small pieces, the stinging cells remain active and can sting a passing swimmer who accidentally bumps into them. Of course, jellyfish do not have stinging cells merely to bother passing swimmers. Like the other coelenterates, jellyfish use the stinging cells to capture prey.



## 10-3 (continued)

### Skills Development

***Skill: Making comparisons***

Ask the class to look again at the jellyfish in Figure 10-10 and the sea anemone in Figure 10-9. Point out that even though both of these organisms are coelenterates, their body plans are different.

- **What differences do you see in the shapes of a jellyfish and a sea anemone?** (Lead students to suggest that a jellyfish has a bell-shaped body, and a sea anemone has a cup or tubelike form.)
- **Do both of these coelenterates have tentacles around the mouth?** (Both have tentacles around the mouth.)
- **In jellyfish, do the mouth and tentacles point upward or do they point downward?** (They point downward.)
- **In what direction do the tentacles and mouth point in a sea anemone?** (They point upward in a sea anemone.)

Explain that all coelenterates have a body plan like that of either the jellyfish or sea anemone. The body plan of the jellyfish, with the mouth and tentacles facing downward, is called a *medusa.* The sea anemone's body form with the mouth and tentacles facing upward, is called a *polyp.*

- **Which of these body forms is stationary and which is adapted for swimming around?** (The polyp is usually attached to an underwater

### SECTION REVIEW

1. What are nematocysts?
2. Which of the coelenterates discussed have muscle tissue that can help move them about?
3. What type of coelenterate formed the Great Barrier Reef?

## 10-4 Worms

You might be surprised to find out that a group of invertebrates—the worms—is involved in the formation of many natural pearls. Natural pearls are formed inside oysters. When a particular type of worm gets into an oyster, the worm acts as an irritant. The oyster's response to the worm is to produce a hard material that surrounds the worm. That material is a pearl.

Most people think of a worm as a slimy, squiggly creature. And, in fact, many are. However, there are some worms that look nothing like the worms used to bait fishing hooks. See Figure 10-11. **Worms are classified into three groups: the flatworms, the roundworms, and the segmented worms.** ❷

**Figure 10-11** *Not all worms are slimy and squiggly. The plume worms* (left) *resemble feathered flowers, while the sea mouse* (right) *looks like a furry animal.*

225

## 10-4 WORMS

### SECTION PREVIEW 10-4

The least complex worms are the flatworms, of which there are both parasitic and free-living forms. Among the free-living flatworms are the planarians, which are found living in ponds, lakes, streams, and oceans. These flatworms feed on both dead and living small plants and animals. Other flatworms, such as tapeworms, live as parasites in the bodies of humans and other animals. Roundworms have not only a rounded body, but also a complete digestive system with two openings—a mouth at the anterior end and an anus at the posterior end. Roundworms include many parasitic forms, such as *Trichina,* flukes, and hookworms.

Segmented worms have bodies made up of many sections. These worms are found in oceans and freshwater streams and lakes, but the most familiar segmented worm is the earthworm. Earthworms have many of the organs and systems found in more advanced animals.

### SECTION OBJECTIVES 10-4

1. **Name the three groups of worms and give examples of each.**
2. **Compare the shapes and body plans of the three groups of worms.**
3. **Describe the basic systems of an earthworm.**

### SCIENCE TERMS 10-4

**host p. 226** **gizzard p. 228**
**crop p. 228**

surface, while the medusa is adapted for swimming.)

Conclude by pointing out that some jellyfish go through a life cycle in which they alternate between the medusa and polyp body forms.

### Section Review 10-3

1. Stinging cells.
2. Jellyfish.
3. Corals.

### TEACHING STRATEGY 10-4

#### Motivation

Have students look at the main headings in Section 10-4. Then hold up a piece of ribbon, a short piece of rope or string, and a single strand of some beads on a string.

- **Which of these is similar to those worms called flatworms?** (The flatworms's body is like a ribbon.)
- **Which of the other two groups of worms is most like the piece of rope?** (Since roundworms have a rounded body they are somewhat like the rope.)
- **How are segmented (annelid) worms like this string of beads?** (The bodies of segmented worms are made up of sections like beads on a string.)

## TEACHER DEMONSTRATION

Set up several stations around the classroom for students to observe living flatworms and roundworms. An appropriate flatworm to use is planaria. These nonparasitic flatworms can be collected from under rocks or on submerged vegetation in most streams, ponds, and lakes. They can also be ordered from a biological supply company. Vinegar eels, nonparasitic roundworms, are excellent for student observation.

At some stations place several planarians in a small amount of pond water or aquarium water in a petri dish. Provide hand lenses or a stereomicroscope for students to watch these worms swim about. If possible, set up another station where students can watch planarians feed. To do this, smear some liver on one side of a microscope slide, and invert it on two small wood chips in a petri dish containing some pond water or aquarium water. Place the dish under a stereomicroscope, and focus on the liver smear. Students should be able to see the planarians move to the slide and feed by extending a muscular tube called a pharynx through their mouth opening in the middle of their body.

At other stations, place a few drops of vinegar eel culture in a depression slide or on a plain microscope slide. Do not use a cover slip. Place the slide under low power of a microscope and let students observe. As students watch, they will be able to see the vinegar eels move continuously with a rapid, jerky motion. If the background light is dimmed, they may also be able to pick out a few anatomical details.

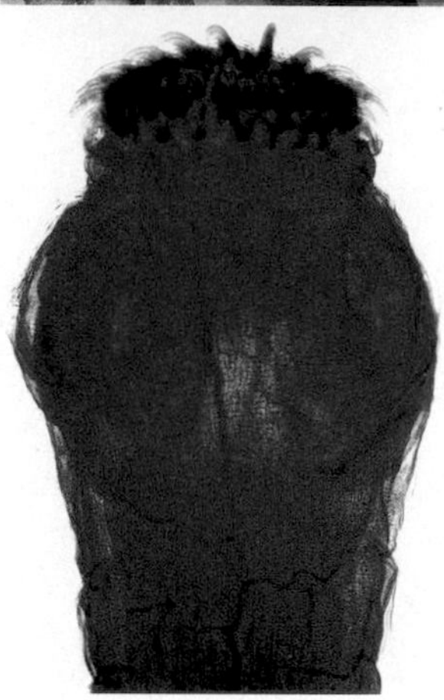

**Figure 10-12** *The planarians* (top) *and the tapeworm* (bottom) *are examples of flatworms.*

### *Flatworms*

You have probably never seen worms that look like those in Figure 10-12. These worms are flatworms. Flatworms, as you might expect, have flat bodies. A planarian is a kind of flatworm. Most planarians live in ponds and streams, often on the bottom of plants or on underwater rocks. Planarians feed on any dead plant or animal matter. However, when there is little food available, some planarians (1) do a rather unusual thing. They break down their own organs and body parts for food. Later, when food is available again, the missing parts regrow. In fact, if a planarian is cut into pieces, each piece eventually grows into a new planarian!

Not all flatworms are found on dead matter. Some are parasites and grow on or in living things. Tapeworms, for example, look like long flat ribbons and live in the bodies of many animals, including human beings.

The head of a tapeworm has special hooks that it uses to attach itself to the tissues of the **host,** or animal in which it lives. The tapeworm then takes food and water from its host. However, the tapeworm gives its host nothing in return.

Tapeworms can grow quite large inside the host. One of the largest tapeworms can live in human beings and grow to be 18 meters long. However, size is not always a good indicator of danger. The most dangerous tapeworm known to humans is only about 8 millimeters long and enters the body through microscopic eggs in food.

### *Roundworms*

You probably have been told time and time again never to eat pork unless it is well cooked. Did you know that the reason has to do with the *Trichina* (trick-EE-nuh) worm? This worm lives in the muscle (2) tissue of pigs. If a piece of pork is cooked improperly, the worm may survive and enter a human's body. In the body, the worm lives in the muscle tissue. The disease it causes, called trichinosis (tri-kuh-NOH-sis), can be very painful and is difficult to cure.

*Trichina* is a kind of roundworm. All roundworms resemble strands of spaghetti with pointed ends.

226

**10-4 (continued)**

**Content Development**

Place the following list on the chalkboard, overhead transparency, or duplicated worksheet:

(1) Grass contaminated with tapeworm eggs eaten by cow

(5) Poorly cooked meat eaten by human

(7) Mature tapeworms produce eggs

(2) Eggs pass to cow's intestine and develop into immature worms

(9) Tapeworm eggs cling to grass and soil

(8) Human eliminates solid wastes containing tapeworm eggs

(4) Blood carries immature worms to cow's muscles (meat)

(6) Immature worms develop into mature tapeworms in human intestine

(3) Immature worms dig out of cow's intestine and enter cow's blood stream

Tell the class that humans can

Roundworms live on land or in water. Many roundworms are animal parasites, although some live on plant tissue. One roundworm, the hookworm, infects more than 600 million people in the world each year. The worms enter the body by burrowing through the skin. Eventually, they end up in the intestines where they live on the blood of the host.

Roundworms, like other worms, have both a head end and a tail end. In fact, worms were the first organisms to develop with distinct head and tail ends. Connecting the two ends is a tube called the digestive tube. Food enters the tube in the head end and waste products leave through the tail end. Although this system is not very complex, it is far more advanced than that of coelenterates. How do coelenterates take in food and release wastes? ❶

### Segmented Worms

The worm you probably are most familiar with is the common earthworm. The earthworm is a type of segmented worm. As you might expect, its body is divided up into numerous segments, usually at least 100. Earthworms, of course, live in the soil. But other segmented worms may live in salty oceans or the fresh waters of lakes and streams. Segmented worms also include sandworms and leeches.

If you ever felt an earthworm, you know it has a slimy outer layer. This layer, made up of slippery mucus, helps the earthworm glide through the soil.

**Activity**

*A Worm Farm*

Here is a way to observe earthworms in soil.

**1.** Fill a clear plastic box with about 2 cm of sand.

**2.** Then place about 7 cm of loosely packed topsoil over the sand.

**3.** Use pond water or tap water that has been standing for a day to slightly moisten the soil. Add more water whenever the soil appears dry.

**4.** Place between six and twelve earthworms in the box.

**5.** Cover the box with clear plastic wrap. Put a few holes in it so that the worms can get air.

**6.** Observe the worms for a week. Keep a record of your observations.

How do the earthworms move through the soil?

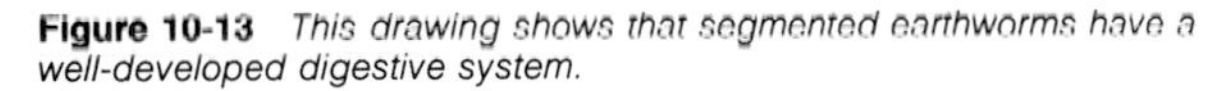

**Figure 10-13** *This drawing shows that segmented earthworms have a well-developed digestive system.*

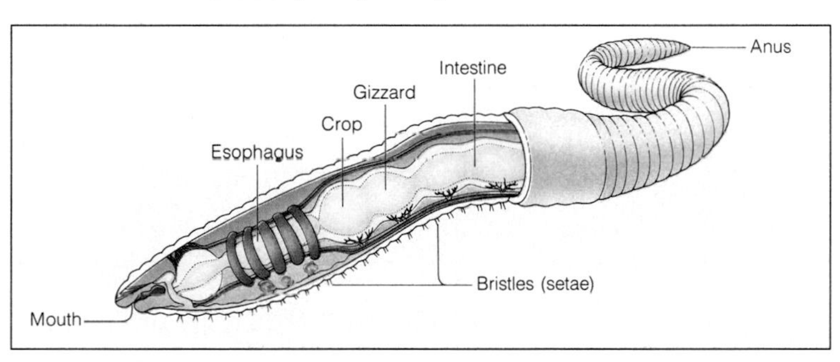


227

**Activity**

**A Worm Farm**
**Skills: Making observations, relating, manipulative**
**Level: Remedial**
**Type: Hands-on**
**Materials: Clear plastic box, sand, pond water, 6 to 12 earthworms, plastic wrap**

In observing an earthworm students should be able to tell the difference between the head and the tail by noticing that the head end is located closer to the clitellum than the tail end is. The clitellum is the light-colored band around the earthworm. Other distinguishing features are a lip at the head end extending over the mouth and a tail that is usually more pointed than the head. Students should notice that the earthworms can move forward or backward by using their setae, or bristles.

Drawings should include the head, tail, setae, and segments.

## ANNOTATION KEY

❶ **Through their mouth (Relating facts)**

❶ **Thinking Skill: Identifying processes**

❷ **Thinking Skill: Relating cause and effect**

become infected by eating contaminated beef that is not cooked well. The adult tapeworm lives in the intestine where it feeds on the host's digested food. Then call attention to the list of events. Point out that the list describes different stages in the life cycle of the beef tapeworm, but the steps are out of order. Tell them to arrange the steps in a logical sequence by placing a number in front of each statement. Notice that the first statement is already numbered, and the correct answers for the others appear in parentheses. Students may work individually, in small groups, or as a whole class.

### Content Development

Emphasize the fact that the roundworms are the first group of animals to be studied that have a complete digestive system.

- **In what important way is the digestive system of roundworms different from that of coelenterates and flatworms?** (Roundworms have a complete digestive system with two openings—a mouth at the head end and an anus at the tail end.)

### Reinforcement

To help students compare the three groups of worms, have them prepare a chart with these headings: group of worms, structure, examples. Instruct them to complete their table by referring to Section 10-4.

## FACTS AND FIGURES

The largest tapeworm that affects humans is the Broadfish tapeworm. This parasite can grow to be about 18 m long with a body composed of up to 4000 sections. In addition to humans, this worm may also infect bears, dogs, foxes, and cats. Among humans it is most common in certain parts of the Baltic region where almost the entire population is infected.

## ANNOTATION KEY

1 Thinking Skill: Identifying processes

2 Thinking Skill: Relating structure and function

Small bristles, or setae, on the segments of an earthworm help it to pull itself along the ground.

Earthworms are good friends to have in the garden. As they pass through the soil, the earthworms feed on dead plant and animal matter. Earthworms create tiny passages as they move. Air enters these passages and improves the quality of the soil. Also, waste products released by earthworms help fertilize the soil. (1)

Earthworms have a well-developed digestive system. After the earthworm eats the food, it passes through several structures before reaching the worm's **crop.** The crop is a saclike organ that stores food. From the crop, the food travels into the worm's muscular **gizzard.** The gizzard grinds up food and then passes it into the worm's intestine. Here, nutrients are removed and enter the worm's circulatory system.

The earthworm has a closed circulatory system. All of its body fluids are contained within small tubes. The fluids are pumped throughout the earthworm's body by a series of special vessels that act like a tiny "heart."

(2) In the respiratory system of the earthworm, oxygen enters through the skin. The earthworm produces carbon dioxide gas, which leaves through the animal's skin. However, the skin must remain moist in order for gases to pass through. If an earthworm remains out in the heat of the sun, the skin dries out and the animal suffocates.

The earthworm's reproductive system has both male and female structures. In one part of the earthworm, sperm cells are produced. In another part,

**Figure 10-14** *Individual earthworms contain both male and female structures. When two earthworms join together, sperm cells are exchanged and eggs are fertilized.*

228

## 10-4 (continued)

### Content Development

- **Into what structures does food go as it travels through an earthworm's digestive system?** (Food taken in through the mouth passes through the esophagus to the crop and gizzard. It then passes to the intestine where nutrients are absorbed. Undigested food particles are eliminated through the anus.)
- **What is the function of the crop and gizzard?** (The crop stores food before it passes into the gizzard where it is ground up.)
- **In what ways are the earthworm's circulatory system like yours?** (The blood is pumped by "hearts" through a system of closed tubes.)
- **How does an earthworm take in oxygen and release carbon dioxide?** (These gases are exchanged through the thin moist skin.)
- **What is the reproductive system of an earthworm like?** (An earthworm has both male and female reproductive structures.)
- **How is fertilization accomplished?** (Earthworms join together and exchange sperm cells that unite with egg cells of the other worm.)

## Section Review 10-4

1. Flatworm.
2. *Trichina* and hookworm.
3. Earthworms reproduce through sexual reproduction. Each worm produces both sperm and eggs. The sperm from one earthworm unites with the egg from another earthworm in a process called fertilization.

## TEACHING STRATEGY 10-5

### Motivation

Many students have a special fascination with squids because of fables and horror movies about their supposed aggressiveness. Ask the class to look at Figure 10-15 and read the open-

egg cells are made. However, despite the fact earthworms contain both types of reproductive cells, reproduction occurs through sexual reproduction. Earthworms mate at night when two worms join together and exchange sperm cells. The sperm of each worm unites with the eggs of the other worm. This process is called fertilization. Soon after, the worms move apart. Each worm then lays a number of fertilized eggs in a slimy shell or cocoon. Eventually, the young worms break out of the shell and crawl away.

Although earthworms have a simple nervous system, they are very sensitive to their environment. Earthworms seem to be able to both sense danger and warn other earthworms as well. They do so by releasing a kind of sweat that helps them glide away much faster and warns other worms in the area at the same time.

### *SECTION REVIEW*

1. A planarian is an example of what type of worm?
2. Name two kinds of parasitic roundworms.
3. How do earthworms reproduce?

## 10-5 Mollusks

In 1873, two men and a young boy were fishing off the coast of Newfoundland in Canada when they spotted a dark object in the water. One of the men poked the object to see what it was. Suddenly, two large green eyes looked up at the men. Before they could move, the strange creature attacked their boat. A huge tentacle wrapped itself around the boat and began to pull it under. The two men were terrified, but the boy thought quickly and chopped through the tentacle with an axe. The water immediately turned jet black as the creature dove beneath the waves. Later, a part of the tentacle left in the boat was fed to hungry dogs on shore. Without knowing it, the boaters had destroyed one of the few parts of a giant squid ever recovered.

Although this story actually happened, most legends about giant squids are false. Giant squids do

**Figure 10-15** *Giant squids can grow up to 17 meters in length. In this historical print, one of the few actual cases of a giant squid attacking people is shown.*



## 10-5 MOLLUSKS

### SECTION REVIEW 10-5

Mollusks are a group of invertebrates having a soft unsegmented body. The body is covered by a thin curtainlike tissue called the mantle. In many species the mantle secretes a hard shell composed of calcium carbonate. Most mollusks also have a thick muscular foot that is used for locomotion. The mollusks include a wide variety of forms that are classified according to the presence of a shell, the type of shell, and the type of foot.

Three major groups of mollusks are discussed in this section. Univalves are those having a one-piece shell. In many univalves, such as snails, the shell is often coiled. Other univalves, such as slugs, have no shell at all. Univalves feed by scraping food with a filelike tongue called a *radula.* Clams, mussels, and scallops are called bivalves because they live in a shell made up of two sections. Bivalves feed by filtering food particles out of the water. The most specialized mollusks are the head-footed mollusks such as the octopus, squid, and nautilus. Most of these mollusks lack an external shell, but in some there is an internal shell. In these mollusks the foot is divided into a number of arms or tentacles surrounding the head. These structures are used for locomotion and capturing prey.

### SECTION OBJECTIVES 10-5

1. **List the characteristics shared by all mollusks.**
2. **Identify the three major groups of mollusks and give examples of each.**
3. **Describe the type of shell in the major groups of mollusks.**
4. **Compare feeding and locomotion among mollusks in the three major groups.**

### SCIENCE TERMS 10-5

**mantle p. 230**
**radula p. 231**

ing paragraphs of Section 10-5. After discussing this incident with the class, allow students to share stories they may have heard about squid attacks on humans.

Point out that even though large squids can be dangerous if provoked, there are no confirmed records of their being naturally aggressive toward humans. Most of them are small shy animals that slip away quietly in the presence of people.

### Content Development

Mention that most squids are small creatures ranging from less than 30 cm to a few meters in length. Yet a few giant ones do live in deep ocean waters. The largest one known, *Architeuthis princeps* can grow up to 9 m long with tentacles of 15 m and a mass of 2000 kg. This giant squid is seldom seen at the surface, but its dead remains have occasionally washed ashore.

**ANNOTATION KEY**

1 Thinking Skill: Making generalizations
2 Thinking Skill: Identifying body parts
3 Thinking Skill: Relating facts

**Activity**

**Mollusks in the Supermarket**
**Skills: Diagraming, relating, applying, comparing**
**Level: Average**
**Type: Field/hands-on**

Answers may include clams, oysters, mussels, squids, octopuses, and conches. The soft bodies of bivalves and univalves are eaten, while only the tentacles of squids and octopuses are usually eaten. These mollusks may be found frozen, fresh, or in cans.

**Figure 10-16** *Many types of mollusks are covered with a hard shell. Notice the different colors and patterns of these mollusk shells.*

not normally attack boats. They usually live in deep ocean waters far from people. In fact, until 1980, when an entire giant squid body was found washed onto a Massachusetts shore, few scientists had ever seen or been able to study the legendary giant squid. Today the remains of this squid are on view in the Smithsonian Institution in Washington, D.C.

Giant squids are included among a group of invertebrates called mollusks (MAHL-uhsks). If you enjoy seafood, you are probably more familiar with other mollusks such as clams, oysters, octopuses, and normal-sized squids. You would not find giant squids on any menu, by the way, even if they could be easily found. They are rumored to taste like ammonia!

The word "mollusk" comes from the Latin word (1) meaning soft. **Mollusks have a soft, fleshy body, which in many species is covered with a hard shell.** See Figure 10-16.

The bodies of most kinds of mollusks having outer shells are similar. Most mollusks have a thick muscular foot. The foot of some mollusks opens and closes their shells. In other species, the foot is used for movement. Some mollusks even use their foot to bury themselves in the sand or mud.

The head region of a mollusk generally contains the mouth and sense organs such as the eyes. The rest of the body contains various organs involved in (2) reproduction, circulation, digestion, and other important processes. Covering much of the body is a soft **mantle.** The mantle is the part of a mollusk that produces the material that makes up the hard shell.

**Activity**

*Mollusks in the Supermarket*

Visit your neighborhood supermarket to look for mollusks that are being sold. Make a chart of all the mollusks you find. Include in your chart the name of the mollusk, the part of the mollusk that is sold, and how it is packaged.



## 10-5 (continued)

### Motivation

Tell students to look at Figures 10-17, 10-18, 10-19, and 10-20. As each figure is examined call on someone to read the caption aloud. Point out that even though these mollusks are quite different, they share some characteristics in common. Next, call on a student to look up and read aloud the derivation of the word *mollusk* in a dictionary. The word is derived from the Latin word *mollis,* meaning soft.

• **Why is the word *mollusk* appropriate for the animals shown in the figures?** (All mollusks have a soft fleshy body. Some students may want to talk about the shell that many mollusks possess, but point out that not all mollusks have this trait in common.)

### Content Development

Emphasize that the soft body of a mollusk is covered by a curtainlike tissue called the *mantle.* In those mollusks with a limy shell, the mantle secretes that shell. But be sure to clarify the fact that not all shelled animals are mollusks.

• **A turtle has a shell, and so do shellfish like crabs and lobsters. Are these animals also mollusks?** (Most students will likely be aware that the origin and composition of turtle and crustacean shells are unlike the calcium carbonate shells of mollusks.)

Also emphasize that mollusks are characterized by a muscular foot. Ask students to look for differences in the structure and function of the foot as they continue to read about the mollusks.

As the animal grows, the mantle enlarges the shell, providing more room for its occupant.

Mollusks are grouped according to certain characteristics. These characteristics include the presence of a shell, the type of shell, and the type of foot.

### *One-Shelled Mollusks*

Many kinds of mollusks have only one shell and are called univalves. "Uni-" is the prefix that means one and "valve" is the word scientists use to mean shell. There are many kinds of univalves, which live in oceans, fresh waters, or on land. However, univalves that live on land still must have a moist environment to survive.

Univalves have an interesting feature in their mouth called a **radula.** The radula resembles a file used by carpenters to file wood. The radula files off bits of plant matter into small pieces that can be swallowed by the univalve. Some species of univalves can inject a poison through the radula that can be dangerous to people.

3

The land univalve you probably are most familiar with is the common garden snail. These snails move slowly along a trail of mucus that they produce. As they release this slippery mucus, they are able to travel easily over rough surfaces because their body does not actually touch the ground. As you can see in Figure 10-17, univalves such as the snail have two eyes located on the end of two stalks sticking out of their head.

**Figure 10-17** *Snails, which are examples of univalves, have eyes on stalks sticking out of their heads and a radula that files off bits of food* (left). *Some snails go into a resting state when the weather becomes too hot* (right).

## TEACHER DEMONSTRATION

To demonstrate the way in which one-shelled mollusks move on their muscular foot, encourage some students to bring some snails or slugs to class. Slugs work best since they have no shell in which to retreat. Place one or two large snails or slugs on a glass plate about 10 cm × 10 cm. Then carefully turn the plate upside down over a culture dish or bowl so that the under side of the foot can be observed. If possible provide hand lenses or a stereomicroscope for students to view the mollusk's movement along the glass plate. After all students have had an opportunity to observe, discuss the movement or the slug or snail.

- **Describe the way in which a one-shelled mollusk moves.** (The slug or snail moves in a gliding motion on its muscular foot.)
- **Why does this animal leave a slimy trail as it moves?** (Lead students to suggest that this organism has mucous glands at the front of its foot that secrete the layer of slime on which it moves.)
- **How could we measure the speed at which this animal moves?** (Accept any logical suggestions. A simple method could involve placing the snail or slug on a glass plate or petri dish over a sheet of graph paper, and measuring the average distance in squares or centimeters that it moves in a minute.)

If students are interested, invite them to try this activity.

## Motivation

Ask the class to look again at Figure 10-17. Point out that the "eyes" at the ends of the stalk function largely to detect changes in light intensity. There is also a shorter pair of stalks on the head, and these can be retracted. It is believed that a snail's organs of smell are located here.

## Enrichment

Many slugs and snails are considered to be harmful because they do great damage to plants on which they feed. Additionally, some snails are dangerous because they serve as intermediate hosts for such animal parasites as the sheep liver fluke and human-blood flukes. Encourage students to prepare reports on some of the ways in which univalve mollusks can be either beneficial or harmful.

## Reinforcement

Some students may be interested in collecting and observing several varieties of snails and slugs. Water species can be collected from submerged vegetation in lakes or ponds. These snails can be maintained in an established aquarium where they will feed on algae and organic debris. Land varieties can be found in moist soil in wooded areas and gardens. These snails can be placed in a terrarium and fed occasionally with lettuce.

## BACKGROUND INFORMATION

Bivalve mollusks have a variety of forms and they vary greatly in size. The smallest clams have shells less than 1 mm long, while some giant clams of tropical ocean waters have shells over 1.5 m across and a mass of more than 230 kg. Since bivalves are filter feeders they are restricted to aquatic habitats.

**Figure 10-18** *Although this sea slug does not have a hard outer shell, it is a mollusk. Its shell, however, is internal.*

## ANNOTATION KEY

1. Thinking Skill: Relating concepts
2. Thinking Skill: Classifying
3. Thinking Skill: Relating cause and effect

It may seem strange to you, but there is one kind of univalve that does not have any visible shell. It is the slug. See Figure 10-18. Many slugs, called sea slugs, live in ocean waters. However, some slugs live on land. Most people are familiar with the simple slugs they often find in moist areas. (1)

### *Two-Shelled Mollusks*

If you have been to the beach, you may have seen clam and mussel shells littering the shore. Clams and mussels are members of a group of mollusks called bivalves, or two-shelled mollusks. Bivalves move (2)

**Figure 10-19** *The bay scallop* (right) *is a bivalve, or two-shelled mollusk. The blue dots are its eyes. The chambered nautilus* (left), *a head-footed mollusk, has a shell that consists of many sections.*

## 10-5 (continued)

### Motivation

Hand a student a preserved clam, and ask him or her to try to open it using only their hands. In all likelihood, they will find it very difficult or impossible to pry open the shell.

- **Why do you think the clam's shell is so hard to open?** (Consider all sensible answers, but lead students to suggest that bivalves, such as clams, have very powerful muscles that open and close the shell.)

Use a chart, overhead transparency, or drawing in a biology textbook to call attention to the two muscles—the *anterior adductor muscle* and the *posterior adductor muscle*—that make a clam's shell difficult to open. Explain that in seafood restaurants the shells are opened by inserting a thin-bladed knife between the two shells and cutting the muscles.

### Content Development

Point out that though scallops move by clapping their two shells together, some of the other mollusks have other means of locomotion. For example, a clam uses its wedge-shaped foot to burrow into sand or mud. Mussels pull themselves by secreting a threadlike substance by which they attach themselves to rocks or other mussels. As adults, oysters do not move at all.

### Skills Development

***Skill: Making predictions***

Tell the class that the nervous system of head-footed mollusks is highly developed. Both squids and octopuses have been used extensively in experiments on the nervous system and intelligence. Discuss the following experiments with your students:

Because an octopus has a well developed brain, scientists have used them in many experiments on intelligence and learning. By giving rewards, such as crabs to eat, and punishments, such as electric shocks,

through the water by clapping their two shells together, which forces water out between the shells. The force of the moving water propels the bivalve.

Bivalves do not have radula. Instead, they feed on small organisms in the water. Bivalves are often called filter feeders since they spend most of their time straining the water for food.

### *Head-Footed Mollusks*

The most highly developed mollusks are the head-footed mollusks. These mollusks include the octopus, the squid, and the nautilus. Most head-footed mollusks do not have an outer shell, but do have some part of a shell within their body. An exception is the chambered nautilus. The nautilus shell consists of many chambers, or rooms. These chambers are small when the animal is young but increase in size as the animal grows. The nautilus constructs a new chamber as it grows. It lives in the outer chamber.

All head-footed mollusks have tentacles that are used to capture food and for movement. However, these mollusks differ in the number and type of tentacles they possess. The octopus, of course, has eight tentacles. Squids have ten tentacles, although two of them differ in shape from the other eight. These two tentacles are shaped like paddles.

Octopuses, squids, and nautiluses use a water propulsion system for movement. They force water out of a tube in one direction, which pushes them along in the opposite direction. Using this "jet" system, these animals can move away rapidly from danger. Squids and octopuses can also produce a purple dye. When they squirt this dye into the water, it hides the animal and confuses predators. The squid or octopus can then make good its escape. In the example earlier in the chapter, the water turned dark when the young boy chopped off the giant squid's tentacle for this reason. ❸

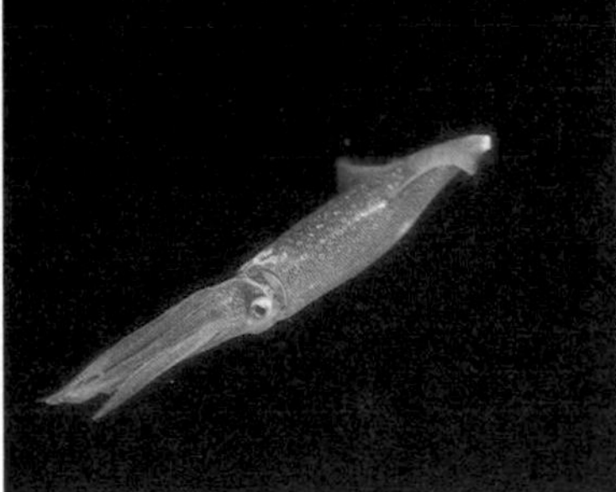

**Figure 10-20** *Most head-footed mollusks, such as the octopus* (top) *and the squid* (bottom), *do not have outer shells.*

### *SECTION REVIEW*

1. List three groups of mollusks and give an example of each.
2. What is the main function of the mantle?
3. How does the radula help univalves eat?

233

## TEACHER DEMONSTRATION

To demonstrate the effect of filter feeding by bivalve mollusks, place a live clam or mussel in a large beaker containing about 200 mL of aquarium water. The mollusks can be collected locally from oceans, streams, or lakes, or they can be obtained from a biological supply company. To another beaker, add 200 mL of aquarium water and an empty mollusk shell as a control. To the water in both beakers, add several drops of India ink to darken the water slightly. Allow the beakers to sit undisturbed for at least 30 minutes. The water in the beaker containing the living mollusk should gradually lighten as the bivalve strains the suspended ink particles from the water.

- **Why did the water in the beaker containing the live clam become clearer?** (Lead students to suggest that the bivalve is a filter feeder. It clarified the water by straining the suspended ink from the water.)
- **What do you think would happen if we repeated the experiment using two live clams in one beaker?** (The water would probably clear more quickly.)
- **What was the purpose of the empty clam shell in the other beaker?** (It served as a control.)

scientists have been able to get an octopus to behave in a certain way. In one experiment, a white plastic square was laid beside a group of crabs placed in a tank with an octopus. When the octopus attempted to eat the crabs it was given an electric shock. At other times, crabs were placed in the tank without a white square beside them. When the octopus attempted to eat these crabs, no shock was given.

- **Can you predict the outcome of this experiment?** (After a while, the octopus had been conditioned to avoid crabs beside the white square and eat only those not near such a square.)
- **What had the octopus apparently learned in this experiment?** (The octopus had learned to associate the crabs beside the square with an electric shock.)
- **What do you predict would be the outcome of an experiment in which an octopus was given a shock when fed crabs that were dyed green, but it was given no shock when fed red crabs?** (The octopus would likely learn to avoid green crabs.)

Point out that the learning process in these experiments is called conditioning. The octopus has learned to respond to a certain stimulus, such as a white square object or green colored crab, in a way that would not be expected under normal circumstances.

## Section Review 10-5

1. Univalve: garden snail; bivalve: clam; head-footed: octopus.
2. The mantle produces the material that makes up the hard shell.
3. It files off bits of plants into small pieces.

## 10-6 SPINY-SKINNED ANIMALS

### SECTION PREVIEW 10-6

The most familiar of the spiny-skinned animals, known also as echinoderms, is the starfish. A unique feature of starfish and their relatives is their water vascular system, a connected system of water-filled canals that serves in locomotion. The internal canals lead to thousands of tube feet on the bottom side of a starfish's arms. The starfish uses its tube feet to move about and also to open the shells of clams on which it feeds. Like other spiny-skinned animals, a starfish has great powers of regeneration.

Other spiny-skinned animals vary widely in appearance. Sea lilies resemble flowers and are found attached to the bottom of the ocean by means of stalks. Sea urchins and sand dollars have rounded bodies without arms. Sea cucumbers are bottom dwellers that resemble vegetables with a fringe of tentacles at one end.

### SECTION OBJECTIVES 10-6

1. **Identify the distinguishing features of the spiny-skinned animals.**
2. **Describe the function of a starfish's tube feet.**
3. **Give examples of several spiny-skinned animals that differ in appearance from starfish.**

**Figure 10-21** *The spiny structures in this photograph are the arms of the crown-of-thorns sea star, a spiny-skinned animal. This animal feeds on the soft bodies of living corals.*

## 10-6 Spiny-Skinned Animals

Earlier you read about the Great Barrier Reef and how it took millions of years to build. Today, the living corals on the surface of the reef are in danger of being destroyed. If they are, the reef will grow no larger. What is damaging the corals? A kind of starfish called the crown-of-thorns sea star is responsible for the damage. See Figure 10-21. The crown-of-thorns eats the soft body parts of the living coral and leaves an empty shell behind.

The crown-of-thorns is a member of a group of animals commonly called the spiny-skinned animals. (1) **Spiny-skinned animals are invertebrates that have a rough, spiny skin.**

### *Starfish*

Starfish are *not* fish. But most *are* shaped like stars. They have five or more arms that extend from a central body. On the bottom of these arms are thousands of tube feet that resemble tiny suction cups. These tube feet not only help the animal to move about, but do much more. For example, when a starfish passes over a clam, one of its favorite foods, the tiny tube feet grasp the clam's shell. The

**Figure 10-22** *This starfish is using its suctionlike tube feet to open a clam's shell.*

### TEACHING STRATEGY 10-6

#### Motivation

Obtain enough preserved starfish for groups of four students to examine. You may be able to borrow a few dissecting specimens from a high school biology lab. As the groups examine the starfish direct their attention to some of these external features:

- **How would you describe the "skin" of a starfish?** (Many coarse spines cover the upper surface.)
- **What do you think is the function of these spines?** (They provide protection against predators.)

Call attention to the small, button-like structure on the upper side of the central disc. Point out that this structure contains many small pores through which water enters a system of internal canals.

- **On which side, top or bottom, do grooves extend down the arms?** (Students should note that these grooves are found on the lower side. Ask the groups to try to locate the small tube feet that line the grooves. Point out that the tube feet resemble tiny medicine droppers, and they help the starfish to move about.)
- **Where is the starfish's mouth located?** (The mouth is in the middle of the central disc on the lower side.)
- **Are there any sense organs on the starfish's body?** (The body does not appear to have any external sense organs; however, you might point out that an eyespot sensitive to light and a tentacle sensitive to touch are lo-

tube feet exert a tremendous force on the clam's shell, and eventually the shell opens. Then the starfish can eat a leisurely meal.

People who harvest clams from the ocean bottom have long been at war with starfish that destroy their clam beds. In the past, when starfish were captured near clam beds, they were cut into pieces and thrown back into the water. However, starfish have an amazing ability to regenerate, or grow back, missing body parts. So by cutting them up, all people did was make sure there would always be more and more starfish than before—exactly the opposite of what was wanted. ❷

### *Other Spiny-Skinned Animals*

Other spiny-skinned animals vary widely in appearance. Some, like the sea cucumber, resemble a vegetable. These animals usually are found lying on the bottom of the sea. Sea cucumbers lack arms. They slowly move along the sea's bottom with tube feet or by wiggling back and forth.

Sea lilies have many arms that look like flower petals. These animals grow on stalks attached to the sea bottom. Most sea lilies remain in one spot for their entire lives.

Sea urchins and sand dollars are round and lack arms. They may be flat like the sand dollar, or dome-shaped like the sea urchin. Some species have long poisonous spines used for protection.

**Figure 10-23** *Like starfish, the sea urchin* (left) *and the sand dollar* (right) *are also spiny-skinned animals.*



## BACKGROUND INFORMATION

Because new starfish can regenerate from a single arm of a starfish with a piece of central disc, clam harvesters no longer cut them in pieces and throw them back in the water. Now they are caught with nets and destroyed. In some cases, the clam beds are treated with powdered lime. The lime kills the starfish, but it is harmless to the clams.

## FACTS AND FIGURES

Sea cucumbers have amazing powers of regeneration, which they use as a means of defense. If they are bothered by an enemy, such as a crab or lobster, they split open near the anus and eject their internal organs. The organs secrete a mass of sticky threads, which entangle the enemy while the sea cucumber slips away. A complete new set of internal organs can be regenerated within a short time.

## ANNOTATION KEY

❶ **Thinking Skill: Making generalizations**
❷ **Thinking Skill: Applying concepts**

cated at the ends of each of the arms.)

### Content Development

Point out that the tube feet of a starfish resemble rows of tiny medicine droppers extending from the grooves on the underside of the arms. The tube feet are connected to a system to water canals inside the arms. Water can be forced from the canals into the tube feet causing them to extend. The suckers of the extended tube feet can then attach to a surface by suction.

- **What are two functions of a starfish's tube feet?** (The tube feet are used for locomotion and for feeding on clams.)
- **How does a starfish use its tube feet to open clam shells?** (The starfish spreads its arms over a clam's shell, and attaches its tube feet to the two halves of the shell. It then uses the suction produced by the tube feet to gradually force the shell open.)

Explain that the starfish then extends its stomach through its mouth opening into the space between the two halves of the clam shell. Digestive juices produced by the starfish partially digest the clam within its own shell. The partially digested clam is taken into the stomach, which is then pulled back into the starfish.

### Reinforcement

Many mollusks and echinoderms are strikingly colorful animals. Suggest that some of the class collect pictures of these organisms and prepare a large collage for the bulletin board. Students with artistic ability might be encouraged to make drawings of representative mollusks and echinoderms to be displayed.

## 10-7 ARTHROPODS

### SECTION PREVIEW 10-7

Arthropods outnumber all other species of animals combined; thus they are the most successful phylum of invertebrates. The arthropods are characterized by having paired jointed appendages and an exoskeleton. The major classes of arthropods include the crustaceans, centipedes, millipedes, arachnids, and insects.

Most crustaceans are found in aquatic environments where they obtain oxygen through gills. Their bodies are segmented, and a highly specialized pair of appendages is attached to each segment. Both centipedes and millipedes appear to be worms with legs. However, centipedes are carnivores with one pair of legs per segment, while millipedes have two pairs of legs per segment, and they feed on plants. The bodies of arachnids are made up of two divisions, and they all have eight legs. This class of arthropods is found in many different environments, and it includes the spiders, scorpions, ticks, and mites.

### SECTION OBJECTIVES 10-7

**1. Name two major characteristics of arthropods.**
**2. State an advantage and a disadvantage of an exoskeleton.**
**3. Identify characteristics of arachnids and crustaceans, and give examples of each.**

### SCIENCE TERMS 10-7

**exoskeleton p. 236**
**gill p. 237**

***SECTION REVIEW***

1. What spiny-skinned animal is causing severe damage to the Great Barrier Reef?
2. What does it mean to regenerate?
3. Name three spiny-skinned animals other than the starfish.

## 10-7 Arthropods

*Figure 10-24 The rigid outer covering of this horseshoe crab is called an exoskeleton. An exoskeleton is characteristic of all arthropods.*

The arthropods (AR-thruh-podz) are the most successful invertebrates on the earth. There are more species of arthropods than all the other animal species combined. Arthropods live in air, on land, and in water. Wherever you happen to live, you can be sure arthropods live there too. In fact, arthropods are our main competitors for food, and if left alone and unchecked, these invertebrates would eventually take over the world.

The name "arthropod" means jointed legs. **Jointed legs and an exoskeleton are the main characteristics of arthropods.** An **exoskeleton** is a rigid outer covering. In some ways, an exoskeleton is similar to the armor once worn by knights as protection in battle. However, there is a drawback to having an exoskeleton. It does not grow as the animal grows. So the arthropod's armor must be shed and replaced

*Figure 10-25 Like all arthropods, these spider crabs have jointed legs and exoskeletons.*

### 10-6 (continued)

### Section Review 10-6

1. Crown-of-thorns sea star.
2. To grow back missing body parts.
3. Sea cucumber, sea lily, sea urchin, and sand dollar.

### TEACHING STRATEGY 10-7

#### Motivation

Introduce this section by displaying some pictures or showing slides of some common insects, spiders, and crustaceans. If pictures or slides are not readily available have the class instead look at the crustaceans, spiders, and insects shown in this chapter. Point out that all of these invertebrates are classified as arthropods. Now, have the class work in teams of three or four to prepare a chart listing as many similarities and differences among crustaceans, insects, and spiders as they can think of. After each group has prepared its chart, have the entire class prepare a single chart on the chalkboard. At this point do not spend too much time discussing inaccurate student perceptions. Such inaccuracies can be clarified as the discussion of this section proceeds. The main purpose now is to get students to focus on what they perceive as similarities and differences among arthropods.

Figure 10-26 *A hermit crab peers out from its borrowed shell atop a fire coral.*

from time to time. While the exoskeleton is being replaced, the arthropod is more vulnerable to attack from other animals.

Arthropods include a variety of animal groups. Among the many types of arthropods are crustaceans, centipedes and millipedes, arachnids, and insects.

### Crustaceans

Do you see the two eyes peering out of the shell in Figure 10-26? These eyes belong to a hermit crab, a crab that uses discarded shells for its home. Crabs, along with lobsters, crayfish, and shrimp, are arthropods that are included among the crustaceans (kruh-STAY-shuhnz). All of these animals live in a watery environment. Crustaceans obtain their oxygen from the water through special structures called **gills.** Even the few land-dwelling crustaceans have gills and must live in damp areas to get oxygen.

The bodies of crustaceans are divided into segments. A pair of limbs or other body parts, such as claws, are attached to each segment. These limbs have many different functions, depending on the organism. The claws of some crabs, for example, are strong enough to enable them to open their favorite food, coconuts. Crabs use their claws to cut through the tough outer husk of a coconut to reach the tender flesh inside. Other limbs are adapted for walking. The female lobster even attaches clusters of eggs to the limbs beneath her tail. She carries the eggs in this way.

#### Activity

*Symmetry*

The body shapes of invertebrates show either radial symmetry, bilateral symmetry, or asymmetry. Visit the reference section of your library. Using a dictionary, look up the meaning of the word "symmetry." Then use a science dictionary or science encyclopedia to define "radial" and "bilateral" symmetry, as well as "asymmetry."

Make a list of the invertebrates discussed in this chapter. Indicate what type of symmetry is shown by each one.



## FACTS AND FIGURES

Despite their name, horseshoe crabs (Figure 10-24) are not true crabs, and they are therefore not classified as crustaceans. They are actually the only living representatives of a group of arthropods whose other members have become extinct. The closest relatives of horseshoe crabs appear to be scorpions and spiders rather than any crustaceans.

## Activity

**Symmetry**
**Skills: Applying definitions, researching, comparing, relating**
**Level: Average**
**Type: Writing/vocabulary**

Symmetry—the arrangement of parts on opposite sides of a plane or around a central axis. Radial symmetry—the arrangement of an organism's parts so that the parts can be separated into mirror-image halves along many lengthwise planes. Bilateral symmetry—the arrangement of an organism's parts so that the parts can be separated into mirror-image halves along only one lengthwise plane. Asymmetry—lack of symmetry. Most invertebrates exhibit bilateral symmetry. Sponges are asymmetrical, while jellyfish, starfish, and sea anemones are radially symmetrical.

## Content Development

Be sure that students appreciate the significance of "jointed legs." Joints in the appendages provide for greater facility in movement. Furthermore, in the arthropods jointed legs are adapted for many different types of motion such as walking, jumping, climbing, and swimming. In addition, the legs may also be modified to aid in food-getting, defense, reproduction, and sensation.

A second major characteristic of arthropods is the exoskeleton made of chitin.

- **How is an exoskeleton beneficial to an arthropod?** (Students will likely answer that the exoskeleton serves as protection against predators.)

Remind students that an exoskeleton is also beneficial in preventing water loss from the body. Thus arthropods are capable of living in dry environments. It also serves as a point for muscle attachment, making efficient movement at the joints possible.

- **What is a disadvantage of an exoskeleton?** (An exoskeleton limits an arthropod's growth. To grow, the exoskeleton must be shed or molted. While molting, an arthropod is more vulnerable to predators.)

## FACTS AND FIGURES

The largest centipede is the Giant Scolpender that lives in Central America. It may reach a length of up to 30 cm and a width of 2.5 cm. Though it feeds mostly on insects, it is also fond of mice and lizards. In the tropics several species of millipedes grow up to 30 cm long with a body width as great as 2 cm.

## TIE-IN/HOME ECONOMICS

Crustaceans are widely used as a source of food for humans. Fine restaurants feature them in numerous specialties, and they are common staples in many households. Ask the class to list as many dishes as they can think of in which crustaceans are used. From the individual lists, compile a class list of crustacean foods on the chalkboard. If any students are interested in cooking they may want to share the recipes for some of their favorite crustacean dishes.

## ANNOTATION KEY

1 Thinking Skill: Making comparisons
2 Thinking Skill: Classifying body parts
3 Thinking Skill: Making comparisons

**Figure 10-27** *Centipedes have one pair of legs in a segment* (top). *Millipedes have two pairs* (bottom). *Centipedes are carnivores. The centipede shown here is injecting poison into its prey—an unlucky toad.*

Crustaceans are able to regrow certain parts of their body. A crab for example, can grow new claws. The stone crab lives in the waters off the coast of Florida. Its claws are considered to be particularly tasty. When a stone crab is caught, one of its claws is broken off and the crab is returned to the water. In about a year's time, the broken claw has regrown. If the crab is unlucky enough to be caught again, that claw may once again be removed.

### Centipedes and Millipedes

Both centipedes and millipedes have been described as worms with legs. In fact, you probably think the main difference between centipedes and millipedes is the number of legs. Actually, both types of arthropods have many legs, and you cannot easily tell them apart by counting. Centipedes have one pair of legs in a segment, while millipedes have two pairs of legs in a segment. (1) However, if you were a tiny earthworm crawling through the soil you would certainly know the difference between the two. Millipedes live on plants and simply would pass you by. Centipedes, however, are carnivorous and are active hunters. The centipede would use its well-developed claws to inject a poison into your body. So, for the earthworm, the difference between the two can mean the difference between life and death.

### Arachnids

In Greek mythology, there is a legend of a young woman named Arachne who challenged the gods to a weaving contest. When the mortal Arachne won, the angered gods tore up her tapestry. Arachne hanged herself in sorrow. The gods, the legend goes, then changed her into a spider. Her tapestry became a spider's web. Today, spiders, along with scorpions, ticks, and mites, are all included in a group of arthropods called arachnids (uh-RACK-nidz). As you can guess, arachnids are named for the young Arachne.

(2) The bodies of arachnids are divided into two main sections: a head-chest section and an abdominal section. Although arachnids vary in size and shape, they all have eight legs. So, if you ever find a small

238

## 10-7 (continued)

### Enrichment

Crustaceans differ from most other arthropods in the number and degree of specialization of their appendages. Encourage interested students to conduct research on a crayfish's appendages and their specialized functions. Have them prepare a large drawing with the appendages and their functions labeled.

### Skills Development

***Skill: Making comparisons***

Mention that many people have difficulty distinguishing between centipedes and millipedes.

- **What does the prefix *centi* mean?** (Many students will be aware that centi means 100.)
- **Does a centipede actually have 100 legs?** (Although they may appear to some people to have 100 legs, the actual number is usually fewer than 30.)
- **How many legs are on each segment of a centipede's body?** (They have one pair of legs per segment.)
- **What does *milli* mean?** (Students generally know that this prefix means 1000.)
- **Do millipedes have 1000 legs?** (No, these arthropods have two pairs of walking legs per segment.)
- **Beside the number of legs, there is another important difference between centipedes and millipedes. On what do these animals feed?** (Centipedes are predators that feed on insects. Millipedes are scavengers that feed on dead vegetation.)

### Motivation

- **What kinds of arthropods are**

**Figure 10-28** *While underwater, the diving spider breathes air it stores in bubbles.*

"bug," one way you can tell if it is an arachnid is to count its legs.

Arachnids live in many environments. Most spiders, for example, live on land. However, one spider lives underwater inside a bubble of air it brings from the surface. Scorpions, for the most part, are found in dry desert areas. Ticks and mites, however, live on organisms. They may live on a plant and stay in one place, or they may live on an animal and go wherever the animal goes. ③

Spiders catch their prey in different ways. Many spiders make webs of fine, yet very strong, strands of silk. Special glands in the abdomens of spiders secrete this silk. Spiders' webs are often constructed in

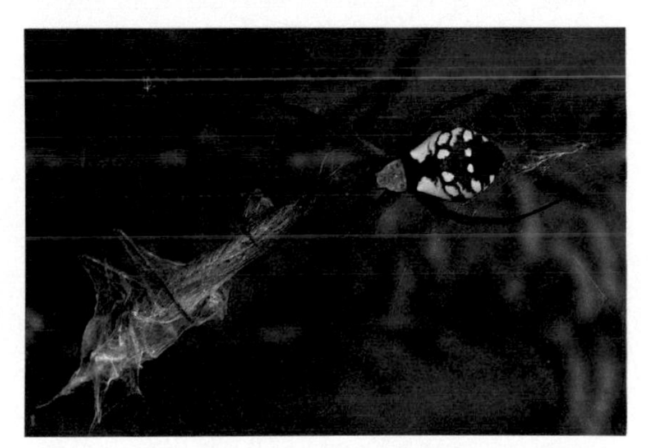

**Figure 10-29** *Like many spiders, this spider traps its prey in a web.*



**classified as arachnids?** (Arachnids include spiders, ticks, mites, scorpions, and daddy long legs.)

Many people look on arachnids with distaste or fear. Since some of them, especially ticks and mites, are parasites, such distaste is justified. Yet, spiders, the most numerous arachnids, feed mainly on insects, and they play an important role in keeping their numbers in check.

To diagnose your students' attitudes, conduct a free-writing activity. Give them five to ten minutes to write individually on the topic, "Many people dislike arachnids because . . . " Then during a follow-up discussion, have students share some of their thoughts. Do not put down students for any misconceptions or biases they may have at this time. But as arachnids, especially spiders, are discussed in class, their beneficial role to humans should be emphasized.

## TEACHER DEMONSTRATION

Obtain a few live crayfish for students to observe. Some students may be able to collect these from local streams, ponds, or marshes. They are also sometimes sold at bait stores as "soft craws," and they can also be purchased from biological supply companies.

Place a few specimens in beakers or battery jars at several locations around the room for small groups of students to examine. Put only a small amount of water in the container. Ask that groups make some specific observations.

- **How many antennae (feelers) does the crayfish have on its head?** (Students should notice it has two long antennae that bend backward over the body. Mention that these are the crayfish's organs of touch, taste, and smell. They should also notice the pair of short branched antennules. These serve as the organs of balance and hearing.)

Provide some small pieces of meat. Ask the groups to place these near the crayfish's head. If it feeds, ask them to note how it obtains and eats the meat.

- **Does the crayfish chew its food or does it swallow it whole?** (The group should notice that it uses the appendages near the mouth to hold the food as it chews it with its mandibles, or jaws.)

Tell the groups to gently place the crayfish on a table top and watch it walk. Caution them to keep fingers away from the claws.

- **How many legs does the crayfish use for walking?** (Unless legs are missing, the crayfish uses its four pairs of walking legs.)

Next, have them place the crayfish in an aquarium or plastic dishpan and observe it as it swims.

- **Describe how the animal moves in water.** (Students should note that the small appendages called *swimmerets* and the flipperlike tail sections are used to move in water. The swimmerets are located on the lower side of the abdomen.)

Summarize by discussing the different kinds of specialized appendages that a crayfish has.

## FACTS AND FIGURES

The poison of only a few species of spiders is harmful to humans. The most dangerous is the bite of the black widow. This poisonous species occurs worldwide in warm areas, including much of the United States. Though deaths from black widow bites are rare, the injected poison can cause severe pain and muscle paralysis. Only the female black widow bites. She is a large glossy-black spider that can be easily identified by a red hourglass marking on the underside of her abdomen.

## BACKGROUND INFORMATION

The arachnids that affect humans most directly are ticks and mites. Ticks, which are generally larger than mites, often live as parasites on humans and their domestic animals. Though a tick bite can be dangerous itself, more often a wide variety of other diseases are transmitted by these arachnids. Among the more dangerous tick-spread diseases are encephalitis, Rocky Mountain spotted fever, and tularemia. Mites are the cause of a number of skin irritations, such as chigger bites and scabies in humans, and mange on dogs and cats. Many human allergies are also thought to be caused by mites borne by dust particles.

## FACTS AND FIGURES

One kind of microscopic mite lives on the eyelashes of all humans.

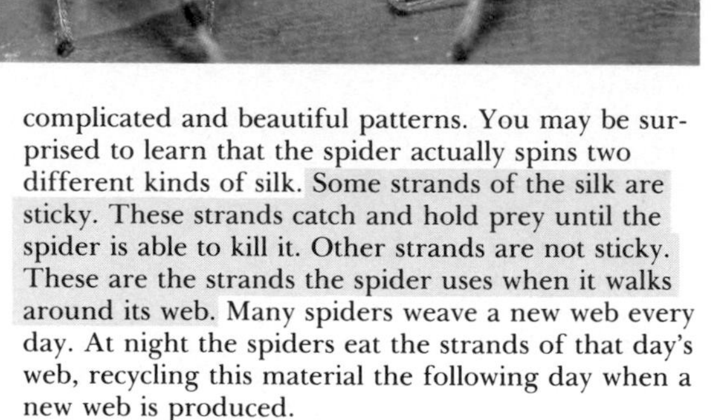

**Figure 10-30** *The fishing spider (left) and the jumping spider (right) do not spin webs to catch their prey. These spiders hunt their prey.*

complicated and beautiful patterns. You may be surprised to learn that the spider actually spins two different kinds of silk. Some strands of the silk are sticky. These strands catch and hold prey until the spider is able to kill it. Other strands are not sticky. These are the strands the spider uses when it walks around its web. Many spiders weave a new web every day. At night the spiders eat the strands of that day's web, recycling this material the following day when a new web is produced.

Some spiders hide and spring out to surprise their prey. The trapdoor spider lives in a hole in the ground covered by a door made of silk and hidden by dirt and pieces of plants. When an unsuspecting insect passes close to the trapdoor, the spider rushes out to catch it.

Once a spider catches its dinner, it uses a pair of fangs to inject venom, or poison, into its prey. Sometimes the venom kills the prey immediately. Other times it paralyzes the prey so that the spider can save living creatures trapped in its web for another day when it needs more food.

Scorpions are active mostly at night. They capture and hold their prey with their large front claws while they inject venom through the stinger in their tail. During the day, scorpions hide under logs, stones, or in holes in the ground. Campers have to be careful when they put their boots on in the morning because a scorpion may have mistaken a boot

**Figure 10-31** *The scorpion, an arachnid, is found in dry environments. Located at the tip of its tail is a stinger, which the scorpion uses to stab and poison its prey.*

## 10-7 (continued)

### Enrichment

Some students may have an interest in making web prints of various species of spiders. To make these prints they will need to obtain either a can of white spray enamel or plastic spray from an art supply store. They will also need a sheet of dark construction paper for each print to be made. To make the print, the students will need to chase the spider from the web, and then coat both sides of the web by spraying it lightly from an angle. The construction paper is then touched to the web, and any part of the web extending beyond the edge of the paper is cut away. The web print should then be allowed to dry. A set of prints can be used to make an unusual bulletin board display.

### Reinforcement

Direct students to make a summary table with three headings at the top: Arthropod Group, Characteristics, Examples. Have them complete the chart with information on Crustaceans, Centipedes, Millipedes, and Arachnids by reviewing Section 10-7

### Section Review 10-7

1. Rigid outer covering.
2. Gills.

lying on the ground for a suitable place to hide to escape the heat of the day.

Ticks and mites live off the body fluids of animals and plants. Many ticks suck blood from larger animals. When they do this they may spread disease. Rocky Mountain spotted fever is spread to people through the bites of ticks. Some ticks and mites live on insects. Other mites live by sucking juices from plant stems and leaves. They may also spread disease.

### *SECTION REVIEW*

1. What is an exoskeleton?
2. Through what structure do crustaceans take in oxygen from the water?
3. Describe the body of an arachnid.

## 10-8 Insects

By now you may have noticed that there is one group of arthropods, or animals with jointed legs, that has not been discussed. This group, of course, is the insects. However, it would be hard to overlook insects for too long since there are more kinds of insects than all other animal species combined. In fact, it has been estimated that there may be as many as

**Figure 10-32** *Insects, like the may beetle* (left) *and the banded woollybear* (right), *vary greatly in appearance.*



3. Arachnid bodies are divided into two sections: a head-chest section and an abdominal section.

## TEACHING STRATEGY 10-8

### Motivation

Consider presenting a graph to help students better conceptualize the vast numbers of insect species inhabiting the earth. Since about 3/4 of all animal species are insects, you could place a circle on the chalkboard, and with colored chalk, section off a wedge comprising 1/4 of the circle to represent all animals other than insects. The remaining 3/4 of the circle, of course, represents the 750,000 or so known species of insects. Point out that of these 750,000 species, at least 275,000 are beetles.

## 10-8 INSECTS

### SECTION PREVIEW 10-8

One group of arthropods, the insects, outnumber all other animal species combined. A reason for this enormous success is the wide variation in mouth parts that adapt different species to a multitude of food sources. Other insect characteristics include a body divided into three sections, three pairs of legs, compound and simple eyes on the head, and an open circulatory system. In most insects there are two pairs of wings enabling them to fly. Because of their exoskeleton, insects must molt to grow. As they develop, many of them pass through a metamorphosis consisting of four distinct stages.

Certain behaviors and defense mechanisms also contribute to the success of insects. Most insects live alone so as to lessen competition for food. At mating time some species release scents called pheromones to attract mates. Other insects, such as bees, live in highly organized societies with a high degree of division of labor. Insects have many defense mechanisms to ensure their survival. Wasps and bees defend themselves with their stingers, while other species rely on camouflage, mimicry, and foul odors to ward off enemies.

### SECTION OBJECTIVES 10-8

1. **Describe the distinguishing characteristics of insects.**
2. **Identify the stages of an insect's metamorphosis.**
3. **Define *pheromone*.**
4. **Describe the social organization of bees.**
5. **List several defense mechanisms of insects.**

### SCIENCE TERMS 10-8

**metamorphosis p. 243**
**larva p. 245**
**pupa p. 245**
**pheromone p. 245**

## ANNOTATION KEY

❶ **Thinking Skill: Making generalizations**
❷ **Thinking Skill: Making observations**
❸ **Thinking Skill: Applying concepts**

## 10-8 (continued)

### Content Development

Call attention to the fact that a major reason for the success of insects is that they have a tremendous variety of food sources. Indeed, foods such as feces, wood, wool clothing, or oil, that are rejected by most animals, are often mainstays in the diets of certain insects. Specialization of the mouthparts makes such diverse feeding habits possible.

- **What are some insects that have chewing mouthparts with jaws that can grind up leaves?** (Answers will vary, but grasshoppers, harvester ants, and potato beetles might be mentioned.)
- **What insects have piercing mouthparts with which they can suck blood or plant juices?** (Mosquitoes have such mouthparts.)
- **What well-known insects have mouthparts adapted as coiled tubes for sucking nectar from flowers?** (Butterflies have mouthparts like this.)
- **Can you think of any insects whose mouthparts are sharp and pointed for biting and piercing?** (Carnivorous insects such as dragonflies have such mouthparts.)

### Content Development

Direct the discussion of the section on insect anatomy by asking questions similar to some of the following.

- **How is an insect's body different from the body of an arachnid?** (An insect's body has three sections, but an arachnid has two body sections.)
- **What are the names of the three main sections of a grasshopper's body?** (The sections are the head, thorax, and abdomen.)
- **How many pairs of legs does an insect have?** (An insect has three pairs of legs.)
- **What structures are located on a grasshopper's head?** (Three simple eyes and a pair of compound eyes are on the head.)

Point out that one pair of antennae are also located on the head. The antennae contain an insect's organs for smelling.

- **How are the antennae of the may beetle in Figure 10-32 different from those of the grasshopper in Figure 10-33?** (A may beetle has featherlike antennae, but they are unbranched in a grasshopper.)
- **How many pairs of wings does a grasshopper have?** (Two pairs of wings can be seen in Figure 10-33. The long, leathery forewings protect the more delicate flying wings underneath.)

Explain that though most insects have two pairs of wings, certain sim-

---

300 million insects for every single person alive on the earth!

Along with birds and bats, certain insects are the only animals that can fly. Insects vary greatly in appearance. If you examine different insect species, for example, you would find that the mouth parts vary greatly. A mosquito's mouth resembles a tiny needle. It can push its needlelike mouth through an animal's skin and remove some of the animal's body fluids. Other insects have mouths that are adapted to eating parts of plants. If you have ever watched a caterpillar eating a leaf, you know how efficient these insects' mouths are.

Many insects compete directly with humans. Insects eat the plants we use for food. Others eat the clothes we wear. If you live in a house made of wood, you may be surprised to find that some species of insects, such as termites, may even be eating your house.

### *Insect Anatomy*

Although both are arthropods, insects are different from arachnids. An arachnid's body is divided into two sections. ❶ **An insect's body is divided into three main sections: a head, a thorax, and an abdomen. And, an insect has three pairs of legs.**

**Figure 10-33** *A rainbow grasshopper investigates a flower. Compare the photograph of the grasshopper with the diagram showing its structure.*

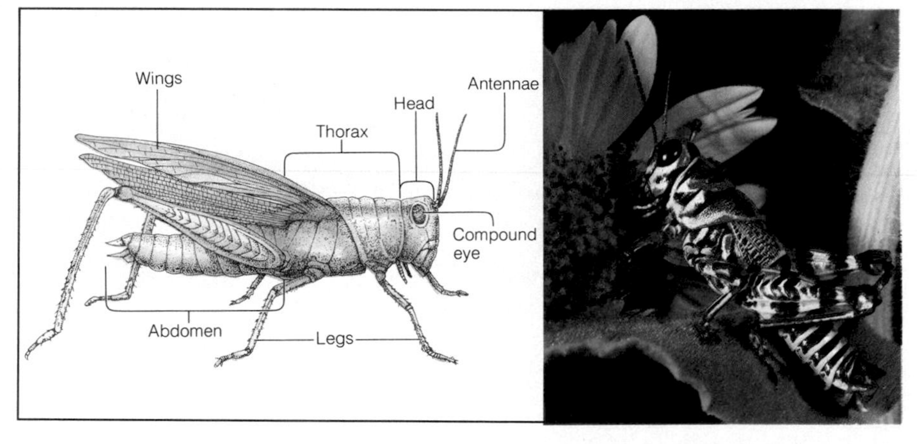


242

The grasshopper is a typical insect in many ways. See Figure 10-33 on page 242. In the grasshopper, the three pairs of legs are not identical. One pair of legs is much larger than the other two. This pair of legs enables the grasshopper to jump away if an enemy gets too close.

If you took a close look at the head of a grasshopper, you would find five eyes peering at you. ❷ The grasshopper has three simple eyes on the front of its head. These eyes can detect only light and dark. However, on the sides of its head are two compound eyes. Compound eyes contain many lenses. See Figure 10-34. These compound eyes can detect some colors, but they are best at detecting movement. The ability to detect movement is important to an animal such as a grasshopper that is hunted daily by other animals for food.

Most insects have wings. The grasshopper has two pairs of wings, and it can fly quite well for short distances. Some kinds of grasshoppers can fly great distances in search of food. Insect flight varies from the gentle fluttering flight of a butterfly to the speedy flight of the hawkmoth, an insect that can fly as fast as 50 kilometers per hour.

Insects have an open circulatory system. Their blood usually is not contained within a system of blood vessels, as it is in humans. The blood moves around the inside of the insect's body, bathing the internal organs. An insect's blood carries food. But ❸ the blood does not carry oxygen. Insects do not have a well-developed system for moving oxygen into the body and waste gases out. Instead, they have a system of tubes that pass through the exoskeleton and into the insect's body. Gases move into and out of the insect's body through these tubes.

### Insect Growth and Development

Insects spend a great deal of time eating, and they grow rapidly. Like other arthropods, insects must shed their exoskeleton as they grow. As insects develop, they pass through several stages. Some species of insects change their appearance completely as they pass through the different stages. This change in appearance due to development is known as **metamorphosis** (met-uh-MAWR-fuh-sis).

**Figure 10-34** *Insects, such as this fly, have compound eyes. Within these eyes are many lenses, which enable the insect to detect the slightest movement of an object.*

#### Activity

*The Life of a Mealworm*

Here is a way to observe metamorphosis for yourself.

**1.** Fill a clean liter jar about one-third full of bran cereal.
**2.** Place four mealworms in the jar.
**3.** Add a few slides of raw potato to the jar.
**4.** Shred some newspaper and place it loosely in the jar.
**5.** Cover the top of the jar with a layer of cheesecloth. Use a rubber band to hold the cheesecloth in place.
**6.** Observe the jar at least once a week. Record all changes that take place in the mealworms.

What is this insect called in its larval stage? In its adult stage? How long was the mealworms' life cycle?



## FACTS AND FIGURES

With its well-developed jumping legs, a grasshopper is capable of jumping a distance of about 20 times the length of its body. A human accomplishing such a feat would have to be able to jump a distance of about 33 m.

## Activity

**The Life of a Mealworm**
**Skills: Making observations, manipulative, recording, applying, relating**
**Level: Average**
**Type: Hands-on**
**Materials: Liter jar, bran cereal, four mealworms, newspaper, raw potato, cheesecloth**

Mealworms are not worms; they are insects. They are the larval stage of the grain beetle. As you can guess, they live in stored meal, grain, or cereal. The complete life cycle of a mealworm may take four to six months.

ple species like lice and fleas are wingless. Insects of the Order Diptera (flies, mosquitoes, gnats, midges, etc.) have only one pair of wings. However, they have a pair of clubbed threads called balancers or halteres in place of the other pair.

## Skills Development

***Skill: Designing an experiment***

It is believed that many pollinating insects identify flowers by their color. Working in small groups, have students design an experiment to see if some common insects have a preference for certain colors. Be sure to remind students to include a hypothesis and both experimental and control variables in their design.

If the groups have difficulty getting started, suggest that they might place insects in two clear bottles, vials, or jars with a passageway between them. The containers could be covered with different colors of cellophane. The experimenters could gauge the insects' color preference, if any, by seeing where they choose to spend most of their time. If some students express a wish to actually do such an experiment, fruit flies might be appropriate insects to use.

## Content Development

Explain that even though an insect has an open circulatory system, it does have a heart and one long blood vessel along the top side of the body. The heart is a long tube closed at the posterior end. When it contracts, blood is forced forward into the blood vessel where it is carried to the vicinity of the head. From there the blood goes out into the body cavity where it bathes the internal organs. The blood reenters the heart through openings located in the abdominal segments.

## BACKGROUND INFORMATION

Pheromones released by some female moths are so potent that a male can be attracted by as little as a ten-billionth of a gram of the substance. The male's antennae bear thousands of sensory hairs that are sensitive to the pheromones released by females of only his own species. In addition to attracting mates, it is believed that pheromones regulate other behavioral features such as development of social castes in termites.

## FACTS AND FIGURES

Some species of termites build gigantic mounds called termitaria (Figure 10-38). These mounds are constructed of soil, fecal matter, and saliva. Termitaria may be 3.5 m in diameter at the base and up to 7.5 m high.

**Figure 10-35** *During metamorphosis, an insect goes through a series of changes. The monarch butterfly begins life as an egg* (top left), *then becomes a larva, also known as a caterpillar* (bottom left), *and finally wraps itself up into a cocoon* (right), *where it becomes a pupa.* 1

**Figure 10-36** *In the last stage of metamorphosis, the monarch has wings and other parts of a typical adult insect. These monarch butterflies are wintering in a tree in southern California.*

### 10-8 (continued)

#### Motivation

Most students are generally familiar with the metamorphosis of frogs from tadpoles to the adult stage. Show the class a picture of tadpoles and adult frogs or perhaps a picture that illustrates the frog life cycle.

- **In what way are the animals in these pictures like those in Figures 10-35 and 10-36 in your text?** (Lead students to suggest that both butterflies and frogs undergo a metamorphosis.)
- **What is meant by the term *metamorphosis*?** (Metamorphosis refers to a series of changes an organism undergoes as it develops from egg to adult.)
- **What are the stages in the metamorphosis of a frog?** (Tadpoles hatch from eggs laid in water. The tadpoles gradually develop into the adult frog.)
- **Describe the metamorphosis of the monarch butterfly.** (A larva, or caterpillar, hatches from an egg. The larva forms a pupa from which the adult butterfly emerges.)
- **Do you know what the larva stage of flies are sometimes called? Of beetles?** (Larval flies are maggots and larval beetles are grubs.)

Mention that metamorphosis in insects as well as amphibians is under the control of hormones.

#### Content Development

Explain that not all insects go through a four-stage metamorphosis like that of the butterfly. In some wingless insects, like springtails, the young are just like the adults except for size. Other insects undergo an incomplete metamorphosis consisting of three stages. An immature wingless nymph hatches from the egg. The nymph undergoes several molts and gradually develops into a winged, sexually mature adult. Grasshoppers, termites, and dragonflies are examples of insects that undergo incomplete metamorphosis.

During metamorphosis, an insect passes through four distinct stages. The first stage is the egg. When the egg hatches, a **larva** emerges. A caterpillar, for example, is the larva of an insect that will one day become a butterfly or a moth. The larva spends almost all of its time eating and can eat all the leaves of a plant in a short time.

Eventually, the caterpillar begins the next phase, the **pupa** (PYOO-puh) phase. In this phase, the caterpillar secretes a covering made of silk or another material. It wraps itself in this cocoon and appears to be sleeping. But inside the cocoon remarkable changes take place. The pupa changes into an adult insect with a completely different appearance. It is often difficult to believe that a beautiful butterfly was once a creeping caterpillar.

### Insect Behavior

Most insects lead solitary lives. They live alone. In this way, they do not compete directly with other members of their species for available food. They come together only to mate and to produce fertilized eggs. Insects attract mates in different ways. One way involves the giving off of a special scent, which you might think of as a kind of perfume. These scents are called **pheromones** (FER-uh-mohnz), extremely powerful chemicals that cannot be smelled by a human. However, even a small amount of a pheromone can be noticed by a potential mate over great distances. Some pheromones produced by female insects can attract a male located more than 11 kilometers away. ❷

Other insects, known as social insects, cannot survive alone. These insects form colonies or hives. Ants, termites, some wasp species, and bees are social insects. They survive as a society of individual insects that perform different jobs. Many of these colonies are highly organized.

A beehive is a marvel of organization. Worker bees, actually infertile females, perform all of the work needed to ensure the survival of the hive. For example, worker bees supply the hive with food. They make the honey and the combs to store it. They feed the queen bee, whose only function is to lay huge numbers of eggs. Worker bees keep the

**Figure 10-37** *Insect behavior often helps other organisms. This moth flies from flower to flower. As it does, it helps pollinate the flowers.*

**Figure 10-38** *This large mound of soil was built by termites. Termites, like bees, are social insects and work and live together in a colony.*

## ANNOTATION KEY

❶ Thinking Skill: Sequencing events

❷ Thinking Skill: Applying concepts

## BACKGROUND INFORMATION

All species of termites are social, but there are fundamental differences between their societies and those of bees. Among bees, only females contribute to activities of the colony because the drones (males) can function only during reproduction. However, in termite societies all young nymphs resemble adults. After their first molt they are capable of caring for themselves, and all of them become workers. These workers are sterile and wingless, but later some of them will develop into soldiers or reproductive males and females. The soldiers have large powerful jaws that they use to defend the nest.

The reproductive males and females of some species are winged, and at times they are produced in enormous numbers. These winged individuals then fly out in swarms to establish new colonies. On finding a new location for a nest, they land, shed their wings, and mate. Together the male (king) and female (queen) share the labor of establishing the new colony. Once the colony is established the king and queen remain together for years and reproduce repeatedly.

## Skills Development

***Skill: Applying concepts***

Tell students that small insects called fruit flies are widely used to study inherited traits. In a certain school, a student asked her science teacher for some fruit flies to use in a project she planned to do. Her teacher gave her a small bottle with some food in it that contained many fruit flies swarming about. Since the food was almost used up, he also gave her another small bottle containing fresh food. The student transferred all of the flies to the new bottle, and then set both the old and new bottles on a shelf in her room. After two days she was surprised to find many small white worms crawling about in the old bottle.

- **Where did the white worms probably come from?** (Most students will likely conclude that the white "worms" are larvae that hatched from eggs that were laid by flies that were in the bottle.)
- **How might the student see if the white worms really are fruit fly larvae?** (Accept all sensible answers. Most students will probably suggest that she could continue to observe the bottle to see if the white worms turn into pupa and then into adult fruit flies.)

## FACTS AND FIGURES

Other insects that undergo metamorphosis are flies, whose larval forms are called maggots, and beetles, whose larval forms are called grubs.

## ANNOTATION KEY

❶ Thinking Skill: Applying concepts
❷ Thinking Skill: Making observations

queen bee clean as well as do the housekeeping for the hive. They also protect the hive. Male bees, whose only function is to fertilize the queen, are unable to feed themselves and so also are dependent upon the workers.

### *Defense Mechanisms*

Insects have many defense methods to ensure their survival. Wasps and bees have stingers that they use to defend them against their enemies. Other insects are masters of camouflage. ❶ These insects survive because their bodies are not easily seen by their enemies. Some insects, for example, resemble sticks and twigs. Other insects resemble leaves or the

### Career: *Beekeeper*

**HELP WANTED: BEEKEEPER** No special education required. Courses in beekeeping, agriculture, and business helpful. Will train to harvest honey from honeycombs and clean, repair, and inspect beehives.

The worker bee hovered over the flower while sucking nectar with its long tongue into its stomach. In the bee's stomach, special chemicals mixed with the nectar. When its stomach was full, it returned to its hive. At the hive it brought the nectar mixture back up through its mouth and placed it in a compartment or cell in the hive. As water evaporated from the mixture, the nectar changed to honey. Other bees placed a wax cap on the honey-filled cell. Honey stored in this way provides food for all the bees during the winter.

People who maintain hives of bees from which they collect honey and beeswax are called **beekeepers.** They sell the honey to bakeries for use in cookies and crackers or package it to sell in markets by the jar. The beeswax is sold for use in making candles, lipsticks, and polishes.

Beekeepers construct wooden boxes with removable frames as hives for the bees. The bees build honeycombs of wax on the frames, which contain many cells to receive the nectar mixtures. Bees are free to come and go so they can bring nectar from flowers back to the hive.

Anyone interested in beekeeping might begin by keeping a hive or two in a backyard or on a roof. Beginner colonies can usually be bought or rented from local beekeepers. For more information on a career working with fascinating insects, write to the New Jersey Beekeepers Association, 157 Five Point Road, Colts Neck, NJ 07722.



## 10-8 (continued)

### Reinforcement

Some students might build an ant colony to house and view ants in the classroom. A simple colony can be built by partially filling a large glass container, such as a battery jar, with some moistened sandy soil. Ants collected from an anthill can then be added to the container. If possible, worker ants and a queen should be placed in the colony. The top of the container can then be covered with a piece of muslin or nylon hose, and it should be set in a pan of water to further prevent the escape of ants. After a while, tunnels made by the ants should be visible through the glass.

### Enrichment

In addition to using pheromones to communicate, honeybees communicate by movements described as dances. Research has shown that when a foraging worker bee locates a source of nectar, she is able to inform the rest of the hive of her find by performing one of two dances. Ask some students to do research on this means of communication, and prepare an oral or written report to be shared with the class. Some visual aids, such as posters, chalkboard drawings, or overhead transparencies, would be useful for the reporter to demonstrate the pattern of the dances.

### Content Development

After the class has studied the section on defense mechanisms, conduct a discussion by asking some of the following questions:

- **What kinds of insects defend themselves by injecting a poisonous secretion with a stinger?** (Bees and wasps sting to defend themselves.)

Mention that some flies and true bugs are also venomous, but they use their mouthparts rather than a

**Figure 10-39** *Insects defend themselves in many ways. The wasp* (left) *uses its stinger, the assassin bug* (top right) *is camouflaged, the bombardier beetle* (center right) *sprays a foul-smelling chemical, and the eye spot coloring on this moth* (bottom right) *startles its predators.*

thorns of plants. Some insects have the ability to spray foul-smelling chemicals at an enemy. Other insects have markings that frighten birds and other animals that might eat them. In Figure 10-39, you can see two large "eyespots" on the wings of the moth. These spots startle predatory animals and may confuse a predator long enough for the insect to make an escape. ❷

***SECTION REVIEW***

1. What are the three main sections of an insect's body?
2. What are the four stages of metamorphosis?
3. List three kinds of social insects.



## FACTS AND FIGURES

In some species of social insects, pheromones are also produced for defensive purposes. If the colony is invaded some individuals release alarm pheromones that trigger aggressive behavior in the others.

Point out that some insects use mimicry to protect themselves. That is, their color patterns resemble those of another insect that a predator might avoid. For example, the wasp beetle has a black and yellow banded body that resembles that of a certain stinging wasp.

### Reinforcement

Provide some mealworms for students to observe their metamorphosis. Live mealworms are often sold at pet stores as food for lizards and amphibians. The mealworms can be cultured in battery jars or other large jars. The jar should be filled about 1/3 full with bran cereal or oatmeal. A few slices of raw potato should be added to provide moisture and food for the adults when they develop. A piece of cheesecloth should be fastened with a rubber band over the top of the jar. Have students observe the jar periodically for changes that take place as the mealworms develop into black grain beetles.

stinger to inject their poison.

• **Name an insect that defends itself by spraying a foul smelling secretion.** (The bombardier beetle can do this.)

Emphasize that many other insects protect themselves from their enemies by means of protective color patterns.

• **Look at the grasshopper in Figure 10-33. How is it protected from its enemies?** (It makes use of camouflage. Its coloration enables it to blend in with the color of its surroundings.)

• **Name an insect that is camouflaged by looking like tree bark.** (The assassin bug resembles bark.)

• **Some insects have color patterns that startle or confuse an enemy. Name an insect that appears to have eye spots on its wings.** (Some moths startle their enemies with eyespot patterns.)

### Section Review 10-8

1. Head, thorax, abdomen.
2. Egg, larva, pupa, adult.
3. Ants, termites, some wasp species, bees.

## LABORATORY ACTIVITY WHICH WAY DID THAT ISOPOD GO?

### BEFORE THE LAB

1. **Several days before the activity, ask students to bring a shoe box to class. The same boxes can be used by different classes, and they can also be saved for use in future years.**
2. **One day before the activity, ask students to collect at least 10 isopods. Students who have extra isopods can share them with students who do not have enough. To keep the isopods healthy until class time, students should be told to place several centimeters of slightly damp soil in their collecting jar. Some leaf litter might also be added to the jar, and the jar should be kept in a cool dark place.**

### PRE-LAB DISCUSSION

Have students read the complete laboratory procedure prior to coming to class. Ask the following questions as the procedure is discussed:

- **What is the purpose of this laboratory activity?** (to determine whether isopods prefer a wet or a dry environment)
- **Based on what you already know about isopods, state a hypothesis.** (Accept any logical hypothesis. A possible hypothesis could be, "If isopods are given a choice of a moist or a dry environment, then they will choose a moist environment.")
- **How will we record our data for the three trials we do?** (Lead students to suggest that a table for recording data is needed.)

Ask for suggestions in designing a data table. After consensus is reached on a suitable format, place a table on the chalkboard as a guide for students to use in setting up their tables.

- **How will we compile class results?** (Each group's average results for each situation should be posted on the class data table.)



### *Which Way Did That Isopod Go?*

**Purpose**

In this activity, you will determine the type of environment isopods prefer.

**Materials** *(per group)*

Collecting jar
10 isopods
Shoe box with a lid
Aluminum foil
2 paper towels
Masking tape
Water

**Procedure**

1. With your collecting jar, gather some isopods. These are usually found under loose bricks or logs. Observe the characteristics of the isopods.
2. Line a shoe box with aluminum foil.
3. Tape down two paper towels side by side in the bottom of the shoe box. Separate them with a strip of masking tape.
4. Moisten the paper towel on the left side of the box only.
5. Place the ten isopods on the masking tape. Put the lid back on the shoe box.
6. Predict what will happen when you open the lid. Wait five minutes. Meanwhile, make a data table.
7. After five minutes, open the lid and quickly count the number of isopods on the dry paper, on the masking tape, and on the moist paper. Record the results in your table.
8. Repeat the procedure two more times. Before each trial, be sure to place the isopods on the masking tape. Record your results in your data table.
9. After you have completed the three trials, find the average result for each column (dry, tape, moist). To do this, add up each column and divide by 3. Record your average results on a class chart.

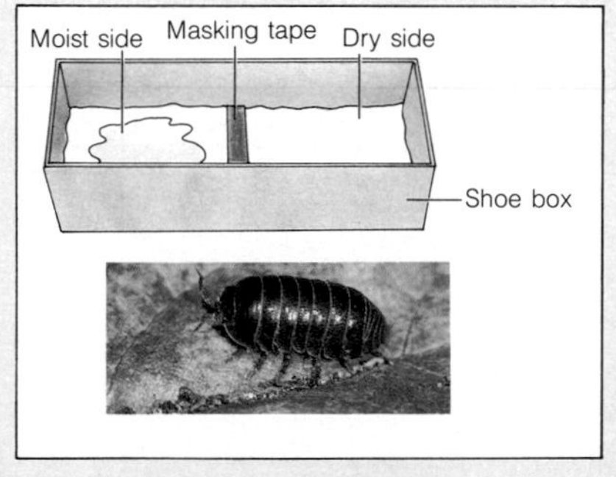


**Observations and Conclusions**

1. How did the isopods react when you opened the lid of the box? How did this reaction compare with your prediction?
2. What was the variable in this experiment?
3. Were there other variables in the experiment that could have affected the outcome? If so, what were they?
4. What was the control in this experiment?
5. How did your results compare to the class results?
6. From the class results, what conclusions can you draw about the habitats isopods prefer?
7. To which group of invertebrates do isopods belong? What characteristics led you to your conclusion?
8. What was the purpose of the masking tape in the experiment?
9. Why did you go through the procedure three times?
10. Design another experiment in which you test the following hypothesis: Isopods prefer dark environments over light environments. Be sure to include a variable and a control in your design.



### SKILLS DEVELOPMENT

Students will use the following skills while completing this activity.

1. Observing
2. Manipulative
3. Recording
4. Hypothesizing
5. Designing
6. Inferring
7. Applying
8. Relating

### SAFETY TIPS

Reminds students that isopods, like all living organisms, must be treated with care and in a humane manner.

### TEACHING STRATEGY FOR LAB PROCEDURE

1. Circulate about the room as students work to ensure that they stay on task.

## CHAPTER REVIEW

### SUMMARY

**10-1 Invertebrates**

- Vertebrates have backbones, while invertebrates have no backbones.
- All animals are multicellular, and most have specialized cells.

**10-2 Sponges**

- The cells of sponges remove food and oxygen from ocean water as the water flows through pores. The out-flowing water carries away waste products.

**10-3 Coelenterates**

- Coelenterates have a cup-shaped body with one opening. Examples include corals, sea anemones, and jellyfish.
- Most coelenterates have stinging cells called nematocysts on their tentacles. Coelenterates sting and capture their prey, which is digested in the central body cavity. Waste products are released through the mouth.

**10-4 Worms**

- Flatworms have flat bodies and live in ponds and streams. Flatworms can regrow missing or cut-off parts. Examples of flatworms are planarians and tapeworms.
- Roundworms resemble strands of spaghetti. Food passes from the mouth end to tail end through a digestive tube. Trichinosis is a disease caused by eating pork containing roundworms.
- Segmented worms, such as earthworms, have segmented bodies and live in soil, salt water, or fresh water.
- Earthworms have a digestive system with a crop and gizzard, a closed circulatory system, a moist skin for gas exchange, a sexual reproductive system, and a simple nervous system.

**10-5 Mollusks**

- Mollusks have soft bodies covered by a shell-producing mantle. Their muscular foot opens and closes the shell and permits movement. Snails, clams, and squids are examples of mollusks.

**10-6 Spiny-Skinned Animals**

- Starfish have tiny tube feet for movement and can regenerate missing parts.
- The spiny-skinned animals also include sea cucumbers, sea lilies, sea urchins, and sand dollars.

**10-7 Arthropods**

- Arthropods are invertebrates that have jointed legs and an exoskeleton. The group includes crustaceans, centipedes and millipedes, arachnids, and insects.

**10-8 Insects**

- Insects have bodies divided into three parts, have three pairs of legs and an open circulatory system.

### VOCABULARY

*Define each term in a complete sentence.*

| | | | |
|---|---|---|---|
| **crop** | **host** | **metamorphosis** | **pupa** |
| **exoskeleton** | **invertebrate** | **nematocyst** | **radula** |
| **gill** | **larva** | **pheromone** | **vertebrate** |
| **gizzard** | **mantle** | **pore** | |

249

**2.** Insist that students enter data in their tables as it is collected rather than trying to remember it to record later. Assist any students who may have difficulty in computing their average results.
**3.** Check to see that all groups enter their data in the class data table.

## OBSERVATIONS AND CONCLUSIONS

During this activity, students will discover that more isopods are found on the moist side of the paper towel and that isopods prefer moist, dark environments. When they are finished, you may want to point out that isopods are usually found in soil under logs or leaves.

**1.** The isopods scattered, looking for a dark environment.
**2.** Variable is moisture. Note: It would not be incorrect for students to infer that the variable was dryness instead of moisture. Either answer is correct.
**3.** If the lid is opened too many times, the variable of light might affect the investigation. Other hidden variables could include heat, time of day, etc.
**4.** Depending on their choice of variables, the control is either moisture or dryness. The control answer should not be the same as the variable answer.
**5.** All groups should get similar results.
**6.** Isopods prefer dark, moist environments.
**7.** Isopods are in the phylum Arthropoda and class Crustacea because they have jointed legs and exoskeletons.
**8.** To separate the moist and dry towels.
**9.** To ensure more accurate results.
**10.** In the student-designed experiment, two shoe boxes could be used. Each box should contain moist towels. Because the variable is lightness and darkness, one box should have a cover and the other should not. Students would find that the isopods prefer dark environments.

## GOING FURTHER: ENRICHMENT

**Part 1**

Have students try the experiment they design in question 10. If time does not permit this to be done in class, then interested students might do it as an out-of-class project.

**Part 2**

If some students would like to maintain a culture of isopods, they can do so by placing them in a container such as a plastic shoe box filled with good garden soil. The soil should be kept slightly moist and a piece of cut potato can be placed on the soil, cut-side-down, for food.

## CHAPTER REVIEW

### MULTIPLE CHOICE

**1.** b **2.** b **3.** b **4.** a **5.** b **6.** d **7.** d **8.** c **9.** b **10.** a

### COMPLETION

**1.** invertebrates
**2.** Nematocysts
**3.** trichinosis
**4.** hookworm
**5.** mollusks
**6.** radula
**7.** spiny-skinned animal
**8.** exoskeleton
**9.** metamorphosis
**10.** pheromones

### TRUE OR FALSE

**1.** T
**2.** F inverte-brates
**3.** T
**4.** F gizzard
**5.** F mantle
**6.** T
**7.** F gills
**8.** F arachnids
**9.** F tubes
**10.** F exoskel-etons

### SKILL BUILDING

**1.** The floating matter might clog up the pores of the sponge and prevent food materials from entering the body of the sponge.
**2.** Charts should be consistent with the information presented in the chapter.
**3.** Coelenterates are more complex because their body parts are more subdivided to perform various functions. Sponges, in general, do not move, whereas most coelenterates do move about and seek out food. Sponges rely on passing currents to carry food into their body.
**4.** No, even if some of the insect's legs had been destroyed, an insect must have three body parts.
**5.** Since water passes through the clam, many of the pollutants could build up in the clam's body and be passed on to whomever ate the clams.
**6.** Answers will vary but should include pollination and the fact that some insects eat other organisms that are harmful to people and plants
**7.** Hookworms burrow through the skin and enter the body, usually through the feet. The best method to prevent this would be to wear shoes in any area that contains hookworms.
**8.** Scientists could use the pheromones to attract insects to insect traps or to attract them away from important food sources.
**9.** The tapeworm uses much of the nutrients from the food people eat, and they do not actually get all the nutrients they need, despite overeating.
**10.** Insects reproduce in large numbers, live in most environments, and eat a wide variety of foods.

### CONTENT REVIEW: MULTIPLE CHOICE

*Choose the letter of the answer that best completes each statement.*

**1.** Tissues that work together are organized into
a. cells. b. organs. c. organ systems. d. tissue systems.
**2.** Which of these invertebrate groups is responsible for building the Great Barrier Reef?
a. sponges b. corals c. sea anemones d. jellyfish
**3.** Scientists hope to learn more about the earth's past by studying the daily growth rings of
a. sponges. b. corals. c. segmented worms. d. mollusks.
**4.** Tapeworms are examples of parasitic
a. flatworms. b. roundworms. c. segmented worms. d. earthworms.
**5.** Which of these is *not* a function of a mollusk's thick muscular foot?
a. movement b. production of a hard outer shell
c. digging d. opening and closing the shell
**6.** Which of these animals can regenerate lost body parts?
a. crabs b. planarians c. starfish d. all of these
**7.** In which group of invertebrates do the animals have jointed legs and exoskeletons?
a. coelenterates b. worms c. mollusks d. arthropods
**8.** Spiders belong to a group of invertebrates called
a. coelenterates. b. crustaceans. c. arthropods. d. insects.
**9.** In which stage of metamorphosis is a caterpillar?
a. egg b. larva c. pupa d. adult
**10.** The only function of the queen bee is to
a. lay large numbers of eggs. b. do the beehive housekeeping.
c. protect the hive. d. make the honey to store in combs.

### CONTENT REVIEW: COMPLETION

*Fill in the word or words that best complete each statement.*

**1.** Animals that do not have backbones are called ____________.
**2.** ____________ are the stinging cells that are found around the mouth of most coelenterates.
**3.** The painful disease caused by eating raw pork containing certain roundworms is called ____________.
**4.** The ____________ is a roundworm that infects over 600 million people each year by burrowing through the skin.
**5.** Giant squids belong to a group of invertebrates called ____________.
**6.** The ____________ is the structure that files off tiny bits of plant matter in most mollusks.
**7.** The crown-of-thorns sea star is an example of a(n) ____________.
**8.** The rigid outer covering of an arthropod is called a(n) ____________.
**9.** The changing appearance of insects as they pass through different stages is called ____________.
**10.** The special scents produced by insects to attract mates are called ____________.

250

### ESSAY

**1.** Some insects, such as flies, eat plants that human beings use for food. Others, such as moths, eat fabrics commonly used for clothing. Still others, such as termites, eat the wood used in constructing houses.
**2.** The sea anemone protects the clownfish against predators by means of stinging cells; the clownfish, in turn, attracts other fish, which are then trapped by the anemone. The earthworm releases a kind of "sweat" that helps it glide away from predators

## CONTENT REVIEW: TRUE OR FALSE

*Determine whether each statement is true or false. If it is true, write "true." If it is false, change the underlined word or words to make the statement true.*

1. All animals are multicellular.
2. Animals that do not have backbones are called vertebrates.
3. Certain worms get into oysters causing the oysters to produce pearls.
4. In an earthworm, the crop is the muscular organ that grinds up food.
5. The soft radula covering the body of a mollusk produces the material making up the mollusk's hard shell.
6. Sea cucumbers and sea lilies are examples of spiny-skinned animals.
7. In crustaceans, the structures that remove oxygen from water are called pores.
8. Spiders are examples of crustaceans.
9. Oxygen and waste gases pass out of the exoskeleton of an insect's body through gills.
10. In order to grow, insects must shed their mantles.

## CONCEPT REVIEW: SKILL BUILDING

*Use the skills you have developed in the chapter to complete each activity.*

1. **Making predictions** Suppose the water in which a sponge lived became polluted with a lot of floating matter. How would the sponge be affected?
2. **Making charts** Construct a chart with three columns. In the first column, list each group of invertebrates you read about in the chapter. In the second column, list the characteristics of each group. In the third column, name two animals from each group.
3. **Making comparisons** Which group of invertebrates is more complex, the sponges or the coelenterates? Explain.
4. **Relating facts** Your friend said she found an insect with four legs and two body parts. Is this possible? Explain.
5. **Making inferences** Why is it unsafe to eat clams from polluted water?
6. **Making generalizations** Explain how insects are helpful to humans.
7. **Relating concepts** Describe a method that would prevent people from being infected by hookworms.
8. **Applying concepts** How might scientists use pheromones to control insect pests?
9. **Relating cause and effect** Explain why people with tapeworm infections eat a lot but still feel hungry and tired.
10. **Applying concepts** Insects are often described as the most successful group of animals. What characteristics of insects could account for this description?

## CONCEPT REVIEW: ESSAY

*Discuss each of the following in a brief paragraph.*

1. Discuss the ways in which insects compete with human beings. Give two examples.
2. Describe the protective method each of the following uses against predators: sea anemone, earthworm, squid, insect.
3. Explain the process of metamorphosis.
4. List the different methods invertebrates use to get food. For each method, give examples of those invertebrates that use it.



more quickly and that also warns other earthworms of danger. The squid releases a dye into the water to hide itself and confuse predators; it then uses its water-propulsion system to move quickly away. Various insects exhibit a wide variety of defensive behaviors and features, including stingers, camouflage, and markings that frighten predators.

**3.** In metamorphosis, insects pass through four stages: egg, larva (which is hatched from an egg), pupa (which is wrapped in a covering secreted by the larva), and adult (which emerges from the pupa).

**4.** Invertebrates get food by the following methods: filtering water (sponges, bivalve mollusks), stinging prey and bringing it into the central body cavity (hydra), breaking down their own body organs, which later regenerate (planarian), absorbing food from hosts in which they live parasitically (tapeworm, hookworm), eating dead plant and animal matter in soil through which they burrow (earthworm), forcing open shells by means of tube feet (starfish), catching prey in silk webs (spider), sucking blood through specialized mouthparts (mosquito), and collecting nectar and making honey (bee).

## ADDITIONAL QUESTIONS AND TOPIC SUGGESTIONS

**1.** A snail and an octopus do not appear to have much in common. Explain why both are classified in the same phylum? (Snails and octopuses can both be classified as mollusks because each has a soft unsegmented body, a muscular foot, and a mantle covering the body organs.)

**2.** A friend brings you an unknown animal he has found. Because it has jointed legs and an exoskeleton, you decide it is an arthropod. You also notice it has two main body sections, eight walking legs, and antennae on its head. In what arthropod group would you classify this animal—arachnid, crustacean, or insect? Give reasons for your choice. (The animal is not an insect because it has two, rather than three, body sections and eight, rather than six, legs. Either an arachnid or a crustacean could have eight walking legs. However, arachnids do not have antennae. The animal must be a crustacean.)

## ISSUES IN SCIENCE

The following issue can be used as a springboard for discussion or given as a writing assignment.

Millions of dollars are spent every year to control insect pests with chemical pesticides. Many people are opposed to this practice because it is costly, and the pesticides can also be harmful to beneficial insects, other animals, and humans. In some cases, insects may create a worse problem because insects can become resistant to them. Others argue that if pesticides were not used to control insects, they would soon take over the earth. Crops would be destroyed resulting in widespread starvation, and diseases spread by insects would cause unbelievable misery and death. What is your opinion and why?

# Chapter 11

# VERTEBRATES

## CHAPTER OVERVIEW

This chapter introduces students to the vertebrate animals, those that have a vertebral column—a backbone of bone or cartilage protecting the nerves of the spinal cord. With their well-adapted brain and body systems, the vertebrates are the most advanced animals living on earth.

Vertebrates are grouped in five classes. In three of the classes—fish, amphibians, and reptiles—the animals have no internal control over their body temperature, and they are said to be coldblooded. Among the cold-blooded vertebrates, the fish are best adapted for a life under water. Amphibians live a double life, for they spend the first part of it as fishlike animals with gills. Later, most of them develop air-breathing lungs that enable them to live on land. Reptiles have dry, scaly skins and eggs protected by a shell, so they are adapted to live their entire lives on land.

Birds and mammals have the ability to maintain a constant internal temperature so they are considered warmblooded. The warmblooded condition, a body-covering of feathers, and forelimbs in the form of wings enable birds to fly. Mammals, the other warmblooded vertebrates, will be presented in the following chapter.

## INTRODUCING CHAPTER 11

Tell the class to observe the picture on page 252 and read the caption.

• **Have you ever seen this kind of fish before?** (Accept all answers.)

Explain that the anglerfish may be unfamiliar to most people since it lives on the ocean floor in deep water.

• **Why do you think this fish is named "anglerfish?"** (Answers may vary. If students cannot answer, point out that the term angler refers to someone who fishes with a hook on a line.)

Now have students read page 253. If they are still not sure why the fish is known as an anglerfish, tell them to see if they can find the answer as they read.

• **What does the anglerfish use to attract smaller fish?** (Students should answer that the structure resembling a worm is part of an organ attached to the anglerfish's head.)

Explain that since it is dark at the depths where some anglerfish live, the tip of the "worm" is capable of giving off light.

• **What are some other adaptations that permit anglerfish to trick its prey?** (Lead students to suggest that it resembles a rock, and it is capable of remaining very still while waiting for prey.)

# 11 Animals: Vertebrates

**CHAPTER SECTIONS**

**11-1 Fish**
**11-2 Amphibians**
**11-3 Reptiles**
**11-4 Birds**

**CHAPTER OBJECTIVES**

*After completing this chapter, you will be able to:*

**11-1** Describe the characteristics of vertebrates.

**11-1** List the three groups of fish and give an example of each.

**11-2** Describe the characteristics of amphibians.

**11-3** Describe the adaptations that allow reptiles to live their entire lives on land.

**11-4** Describe the characteristics of birds.

**11-4** List the four main categories of birds and their adaptations for flight.

A small fish swims past a rock. As it does, it notices a worm wriggling and moves closer to investigate. However, the fish is in for a deadly surprise. No, the worm is not at the end of a fishing line that leads to the surface. But it *is* being used as bait. In fact, the worm is attached to the rock, which is quietly watching the small fish. For the "rock" is not made of minerals; it is actually a fish called an anglerfish. And the "worm" is really an organ attached to the fish's head. In order to catch a meal, the anglerfish must wait patiently, as still as a rock.

The worm lures the fish closer. When the prey is near, the anglerfish takes a big gulp. The small fish ends up in the anglerfish's huge mouth.

To trap its prey, the anglerfish must remain very still and show unlimited patience. However, it can move quickly when it strikes. In fact, the strike of the anglerfish, one of the fastest in the animal kingdom, has been timed at less than 1/1000 of a second!

*Anglerfish luring its prey*



Mention that in some types of anglerfish, only females catch fish. Males, which are much smaller, attach themselves permanently to the body of the female and live as a parasite. Even their circulatory systems fuse. Apparently the only function of the male is to produce sperm for reproduction.

## TEACHER DEMONSTRATION

If possible, display a live representative of each of the five groups of vertebrates. If you do not already have living vertebrates on hand, then ask if some of your students can bring in some of their pets, if they have any of the following:

Goldfish or guppies
Frog or toad
Snake (nonpoisonous) or chameleon
Caged bird (parakeet, canary, etc.)
Mouse, gerbil, hamster, etc.

If it is not possible to display live animals, then show some pictures of representative vertebrates.

- **These animals all look quite different, yet they have one important trait in common. In what way are they alike?** (Accept any sensible answer at this time. Lead students to talk about their internal skeletons, including a backbone.)
- **What are these animals with backbones called?** (These animals are vertebrates.)
- **What are some of the ways in which these animals are different?** (Accept all logical answers, but focus attention on the different types of body coverings.)

Conclude by finding out how many students already know the vertebrate group to which each specimen or pictured animal belongs.

## TEACHER RESOURCES

### Audiovisuals

*Animals with Backbones,* filmstrip with cassette, Eye Gate
*All Things Animal,* film, Barr Productions, Inc.
*Crocodile,* film, Centre Productions, Inc.
*Frogs and Toads—Watch Them Sing!,* film, International Film Bureau, Inc.
*Reptiles,* film, EBE

### Books

Curry J., *Animal Skeletons,* Crane, Russak
Young, J. Z., *The Life of Vertebrates,* 2nd ed., Clarendon Press

### Software

*Life Zones in the Ocean,* Prentice Hall

## 11-1 FISH

### SECTION PREVIEW 11-1

Fish, the first vertebrates on the earth, appeared about 500 million years ago. Like the other vertebrates, fish have a vertebral column, or backbone. This column of bone or cartilage supports and protects the nerve cord running down the fish's back. The vertebral column forms part of the internal skeleton.

There are three groups of fish. The most primitive of the three are the jawless fish, which also lack paired fins and scales covering the body. The most common fish is the lamprey, a parasite of other fish. Sharks, skates, and rays make up a group known as the cartilaginous fish. Though they have jaws and fins, their skeletons are made entirely of cartilage. The most numerous and widespread fish are the bony fish, whose skeletons are composed mainly of bone. An important characteristic of bony fish is their swim bladder, a structure filled with air that helps keep the fish afloat.

### SECTION OBJECTIVES 11-1

1. **Describe the major characteristics of vertebrate animals.**
2. **Name several adaptations of fish for life in water.**
3. **Name the three groups of fish and their characteristics.**

### SCIENCE TERMS 11-1

**vertebra p. 254**
**cartilage p. 257**
**swim bladder p. 259**

## 11-1 Fish

Around 500 million years ago, a tiny animal first appeared in the earth's oceans. This strange animal had no jaws. It did have fins, but they were not like the fins of modern fish. However, there was something very special about this animal—something that would group it with the many kinds of fish that were to come in later years. This early fish was the first animal with a backbone—the first vertebrate.

The main characteristic of all vertebrates, including fish, is a backbone, although you will read about some fish whose backbone is not made of bone at all! The bones that make up a vertebrate's backbone are called **vertebrae** (singular: vertebra). See Figure 11-1. The backbone is part of the vertebrate's internal skeleton, although, again, not all skeletons are made of bone. The skeleton provides support and helps give the body of a vertebrate its shape. One important advantage of an internal skeleton is that it increases in size as the animal grows. It does not have to be shed, as does the exoskeleton of an insect.

1

An important function of the backbone of a vertebrate is to protect the nerves of the spinal cord, which runs down through the center of the backbone. These spinal cord nerves connect the vertebrate's well-developed brain to nerves that carry information to and from every part of the body.

**Figure 11-1** *Like all vertebrates, fish have a series of bones called vertebrae that make up their backbone.*

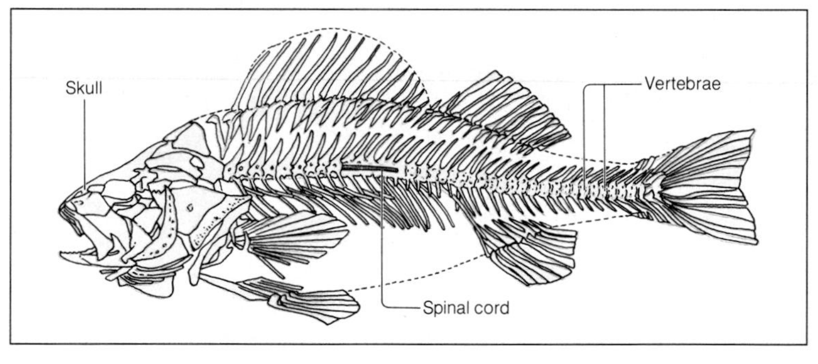


254

### TEACHING STRATEGY 11-1

#### Motivation

Show the class a human skeleton if one is available. If you do not have a skeleton, then try to display a wall chart, overhead transparency, or a picture in a book showing the human skeleton. Ask the class to compare the human skeleton with Figure 11-1 in their book.

- **What are some similarities between your skeleton and a fish's skeleton?** (Accept all sensible answers, but lead students to conclude that both skeletons are composed of bone, and both have a backbone made of vertebrae.)
- **Exactly what are vertebrae?** (Students should answer that the vertebrae are the separate bones making up the entire spinal column or backbone.)
- **Why is a backbone such an important structure?** (Accept all answers at this time, but students should mention that the backbone supports and protects the nerves making up the spinal cord.)

Students often confuse the terms spinal cord and spinal column. Help them to make the distinction by emphasizing that the spinal cord is a nerve cord running through the center of a series of bones called the backbone, or spinal column. The

Figure 11-2 *Some fish, like these in the Coral Sea, travel in schools.*

Fish are the vertebrates that are best adapted to a life under water. Their smooth body, usually covered with scales, is streamlined, allowing them to glide through the water. Most fish have fins. Some fins keep a fish upright so that it does not roll over on its side. Other fins help the fish steer and stop. The side-to-side movement of the large tail fin propels the fish through the water. Rapid movement of its tail fin, for example, allows the pike to travel up to 12 times its body length in less than one second.

Like some invertebrates, fish take oxygen from the water through gills. As a fish swims, water passes through its gills. Oxygen passes from the water into blood vessels in the gills. Carbon dioxide wastes also ❷ pass out of the fish through the gills. Fish have well-developed digestive, circulatory, and nervous systems, as do the other vertebrates you will read about.

Fish live in all of the earth's waters, in both fresh and salt water. As you might expect, however, the same fish usually cannot live in both fresh and salt water. Also, while some fish, like the cod, may live in the cold waters of the Arctic, others, like the ferocious piranha, live in warm, tropical waters. If you take a tropical piranha and place it in arctic waters, it will soon die. Why? Fish are coldblooded. They have no internal control over their body temperature. So changes in their environment can affect their body temperature. If the change is severe, as from tropical to arctic waters, most fish will not survive. However, ❸

Figure 11-3 *Like this emperor angelfish, most fish have fins and streamlined bodies that help them glide through the water.*

255

## FACTS AND FIGURES

Fossil remains indicate that the earliest fish developed over 350,000,000 years ago.

## ANNOTATION KEY

❶ **Thinking Skill: Applying concepts**
❷ **Thinking Skill: Identifying processes**
❸ **Thinking Skill: Relating cause and effect**

(Students should notice in Figure 11-1 that the fins are supported by bones.)

As students look at Figure 11-1, explain that the two fins on the top of the body are called dorsal fins. The fin on the bottom where the body begins to taper is called the anal fin. The dorsal and anal fins function to keep the fish from rolling. The pectoral and pelvic fins are the ones that are paired along the side of the body, and they are used for steering. The tail fin, also known as the caudal fin, provides thrust for movement.

• **What structures enable a fish to breathe in water?** (Fish have gills for breathing.)

### Skills Development

***Skill: Classifying coldblooded and warmblooded animals***

Write the terms "coldblooded" and "warmblooded" atop two columns on the chalkboard. Call on a student to read the definition of each term from the glossary. Then ask the class to name examples of coldblooded and warmblooded vertebrates. List the examples in the proper column as they are named. Coldblooded vertebrates include fish, amphibians, and reptiles. Birds and mammals are warmblooded.

bones are also known as the vertebral column.

• **What is the name given to those animals that have a backbone?** (Vertebrates are animals with backbones.)

### Content Development

Call on a student to read aloud the first sentence on page 255 stating, "Fish are the vertebrates that are best adapted to a life under water." Discuss some of these adaptations by asking some questions similar to the following:

• **How does the shape of a fish's body help adapt it for a life in water?** (The streamlined shape allows it to glide easily through water.)

• **In what way do fins enable a fish to move in water?** (Fins perform such functions as stabilization, steering, stopping, and propulsion.)

• **Are there any bones in the fins?**

## BACKGROUND INFORMATION

Though the sea lamprey is a notorious parasite, most other species of lampreys are small eellike creatures that feed on decayed organic material. Beside lampreys, the only other type of jawless fish is the hagfish. Like lampreys, hagfish lack jaws, scales, and paired fins. Hagfish live only in the sea as bottom-dwelling scavengers.

## TIE-IN/ ECONOMICS

The parasitic sea lamprey is found in both fresh and salt water. The construction of canals to link the Great Lakes with the Atlantic Ocean allowed them to gradually become established in all of the lakes. Their movement into the upper Great Lakes caused a startling decline in the commercial catch of lake trout. For example, in Lake Michigan, the annual catch was reduced from 2,542,500 kg in 1945 to only 180 kg in 1954. Lampreys are now controlled by the use of selective poisons in their larval beds. As a result the lake trout are making a comeback.

**Figure 11-4** *Some fish, such as the potato cod (right), live only in cold water. Other fish, such as* Priacanthus *(left), live only in warm water. Notice the small "cleaner" fish near the tail of* Priacanthus.

since the water temperature in an area remains relatively constant compared to the temperature on land, fish have little trouble maintaining a constant body temperature as long as they remain in their native waters. This results in a saving of energy, for fish do not have to use food energy to keep warm.

❶ **Fish are divided into three groups. The three groups of fish are the jawless fish, the cartilaginous fish, and the bony fish.**

**Figure 11-5** *The lamprey, a jawless fish, attaches to other fish with its suction-cup mouth. Then the lamprey uses its teeth to drill a hole into the other fish.*

### *Jawless Fish*

Jawless fish are the most primitive of all fish. They are so primitive, in fact, that they lack scales and fins as well as jaws. In Figure 11-5, you can see the most common jawless fish—the lamprey. The lamprey looks like a snake with a suction-cup mouth at one end. Even though the fish has no jaws, this suction-cup mouth is very efficient. Using its mouth, the lamprey attaches itself to the soft belly of some ❷ other fish, such as a trout. Then, with its teeth and rough tongue, the lamprey drills a hole into the fish and sucks out its blood and other body fluids. At one time, lampreys in the Great Lakes killed so many



## 11-1 (continued)

### Content Development

Refer students to Figure 11-4. Emphasize to students that each type of fish is suited to a particular environment. Conditions such as whether the water is fresh or salty and the temperature of the water will determine if a particular type of fish can survive there.

- **What do fish found in the Great Salt Lake (Utah) and those found in the Pacific Ocean have in common?** (They both are coldblooded vertebrates requiring a saltwater environment.)
- **Could a fish taken from the Great Salt Lake live in the Pacific Ocean?** (Its survival is dependent upon the water temperature.)
- **Could a fish native to Lake Michigan survive in the waters off the Florida coast?** (It is unlikely because the Florida waters would be warm salt water.)

### Motivation

Tell the class to look at the lamprey in Figure 11-5.

- **How is a lamprey's mouth different from that of other fish?** (The lamprey has a suction-cuplike mouth without jaws.)
- **How is this mouth adapted to obtain its food?** (Answers may vary. Call attention to the suckerlike mouth used to attach to a fish. Students should also notice the rasping teeth that are used to drill into its prey.)

Mention that the lamprey is capable of sucking out a fish's internal organs as well as its blood and body fluids. These destructive predators may grow up to 1 m long. Students are usually fascinated by the grotesque appearance of a lamprey. If possible order a preserved specimen from a biological supply company for them to examine.

other fish that special poisons were used to kill lamprey young.

There is something else that makes jawless fish unusual. To find out what it is, take your ear between your fingers and move it back and forth a few times. Nothing breaks, does it? The reason is that your ear contains a flexible material called **cartilage** (KAHRT-uhl-ij). In fact, before you were born your entire skeleton was made of cartilage. In time, the cartilage was replaced by bone, except in some places like your ear and the tip of your nose. The entire skeleton of a jawless fish is made of cartilage. The skeleton is never replaced by bone tissue. As you might expect, such a fish is very flexible and can be bent so that its head touches its tail without causing any damage to the fish. ③

### *Cartilaginous Fish*

A monstrous figure moves slowly toward a group of skin divers. However, the divers are too intent on their swimming to notice the 18-meter-long fish as it approaches. Then, suddenly, just as it reaches them one of the divers turns and sees the creature. It is a shark! The diver signals her companions. But they do not swim away in fear, even though they are confronted by the largest shark on earth—the whale

**Activity**

*Comparing Sizes*

The great white shark can grow up to 4.5 meters in length and may have a mass of over 700 kilograms. Compare the great white shark's length and mass to your own.

**Figure 11-6** *In spite of their enormous size, whale sharks are harmless to people. Whale sharks only eat microscopic plants and animals.*

## ANNOTATION KEY

❶ Thinking Skill: Making generalizations
❷ Thinking Skill: Relating facts
❸ Thinking Skill: Applying concepts

## Activity

**Comparing Sizes**
**Skills: Comparing, making calculations**
**Level: Average**
**Type: Computational**

Students will have a greater appreciation for the size of sharks after they complete this activity. The best way to have them complete the comparison is for them to draw the sharks to scale on a sheet of graph paper. They can draw themselves to scale over the shark. Naturally, comparisons and computations will vary depending on student size and mass.

## Content Development

Point out that all lampreys lay their eggs in freshwater streams. The young are only about 1 cm long when they hatch, and they burrow into mud at the bottom of the stream where they remain for several years. After they mature they move into open waters where they begin to parasitize fish. The lamprey's reproduction and early development in freshwater streams has made it possible to control them by using selective poisons that kill the lamprey young while leaving other organisms unharmed. Efforts to curtail the lamprey population and restock trout in the Great Lakes have been moderately successful.

## Content Development

Emphasize the idea that the skeletons of both jawless fish and cartilaginous fish are composed of cartilage.

- **What are some traits shared by jawless fish and cartilaginous fish?** (Both are coldblooded vertebrates with skeletons composed of cartilage.)
- **How are jawless fish and cartilaginous fish different?** (Cartilaginous fish have jaws.)
- **Why do you think these fish can survive without a strong skeleton made of bones?** (The body of the fish is supported by the water that surrounds it.)

## Skills Development

***Skill: Making comparisons***

Have students prepare a two-column chart with one column headed "Jawless Fish" and the other headed "Cartilaginous Fish." Down the left side of the chart, have them list these words: mouth, skeleton, fins, scales. As students read about these two groups of fish tell them to compare the structures listed by completing the chart.

## TIE-IN/NUTRITION

For centuries, people have consumed fish. This foodstuff has been found to be an important source of vitamins and minerals. Have interested students research the specific ways the intake of fish aids the human body.

## TEACHER DEMONSTRATION

To help students understand how a fish's swim bladder works, obtain a sausage-shaped balloon, a short piece of rubber or plastic tubing, some string, and an aquarium or clear plastic sweater box. Fill the aquarium or plastic box about half full of water, and use the string to fasten the tubing to the opening of the balloon.

Place the deflated balloon under water and flow through the tubing until the balloon rises in the water. Point out that the balloon is similar to the swim bladder.

• **How does inflating the swim bladder with gas control the depth at which a fish swims?** (Students should answer that increasing the volume of gas in the bladder adjusts the bouyancy so the fish can rise.)

• **How can the fish move to a lower level in the water?** (By releasing gas from the bladder.)

Some students may wonder where the air that enters the swim bladder comes from, and where it goes when it is released. Point out that an organ called the gas gland, or a dense network of blood vessels, pumps the gas in or out of the swim bladder. The gases enter and leave these structures from the fish's blood.

**Figure 11-7** *The great white shark* (left) *and the horn shark* (right) *are examples of cartilaginous fish.*

shark. Instead, they grab the shark's top fin and playfully hitch a short ride. The divers know that the whale shark has little interest in people. It eats only microscopic plants and animals.

Sharks are included in a group of fish called the cartilaginous (kahrt-uhl-AJ-uh-nuhs) fish. Like those of jawless fish, sharks' skeletons are made of cartilage. But unlike the jawless fish, sharks definitely have jaws. See Figure 11-7. In fact, sharks are probably the most successful predators on earth. And they have been so for hundreds of millions of years. When you think of sharks, you probably think of the great white shark, which has a reputation for eating people. Great white sharks, along with a few other types, have been known to attack people occasionally. But for the most part sharks let people alone and prefer to be let alone as well.

Also included among the cartilaginous fish are skates and rays. These fish have two large, broad fins that stick out from their sides. They beat these fins to move through the water, much as a bird beats its wings to fly through air. Rays and skates often lie on the ocean bottom, where they use their "wings" to cover their bodies with sand and hide. When an unsuspecting fish or invertebrate comes near, the hidden skate or ray is ready to attack. Some rays have a poisonous spine at the end of their long, thin tail. However, they use this for defense, not to catch prey. Other rays are able to produce small charges of electricity to stun and capture prey.

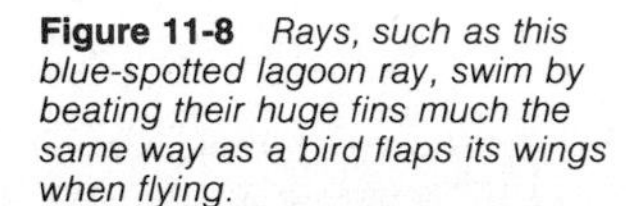

**Figure 11-8** *Rays, such as this blue-spotted lagoon ray, swim by beating their huge fins much the same way as a bird flaps its wings when flying.*

### 11-1 (continued)

### Motivation

Have students look at the shark shown in Figure 11-7, and remind them that sharks may be "the most successful predators on earth."

• **What are some reasons why sharks are such successful predators?** (Accept all sensible answers. From the observations of the figure, students will likely mention such physical characteristics as their razor sharp teeth and powerful fins that allow them to swim swiftly in pursuit of prey.)

You may also want to point out that sharks have a very keen sense of smell and they can pick up a scent more than 400 m away. In addition to their well-developed olfactory sense they have small pits along the sides of the face and body that can sense vibrations in the water caused by other organisms.

### Content Development

Because sharks have a sleek appearance, students may assume that they have no scales. Be sure to emphasize that sharks do have scales, but they are different in structure and appearance from the flat, overlapping scales of body fish. A shark's body is covered with small, enamel-tipped, spinelike scales. These scales feel like sandpaper. The shark's teeth are similar in structure to the scales.

### Bony Fish

Anyone who has eaten a flounder or a trout knows why such fish are called bony fish. Their skeleton is composed of hard bones, many of which are quite small and sharp. That is one reason that many restaurants serve fish with the bones removed. Some of the bony fish, including the tuna and herring you may eat, travel in groups called schools. Because of this schooling behavior, these fish can be caught in large numbers at one time by fishing boats.

One important characteristic of bony fish is their **swim bladder.** The swim bladder is a sac that the fish can empty or fill with air. The swim bladder acts in much the same way as a life preserver that keeps people afloat. However, by letting air in and out of the bladder, a fish can float at any level in the water. That is why a fish can sleep under water and not sink, even though it is not moving its fins.

There are many different kinds of bony fish. Some have made remarkable adaptations to life in the water. The electric eel, for example, can generate considerable amounts of electricity. This eel, found in South American streams, uses the electricity it produces to stun its prey. The remora, a saltwater fish, uses a sucker to attach itself to a shark or other marine animal. It feeds on bits of food the shark leaves behind. Some fish, such as the flounder, are

#### Activity

*Observing a Fish*

**1.** Obtain a preserved fish from your teacher and place it in a tray.

**2.** Observe the fish. Note its size, shape, and color. To which group does your fish belong?

**3.** Draw a diagram of the fish and label as many structures as you can.

**4.** Locate the fish's gill cover. Lift it up and examine the gills with a hand lens. If necessary, use a scissor to cut away the gill cover. **CAUTION:** *Be careful when using sharp instruments.* How many gills do you see?

**5.** Remove one of the scales from the fish.

**6.** Examine the scale under a microscope. Each year a new dark ring is added to the scale. How old is your fish?

**Figure 11-9** *Fish are adapted in many different ways to a life under water. The stonefish* (left) *not only blends in with its surroundings, it also gives off a deadly poison through its spines. The deep sea fish* (right) *has large eyes, a huge mouth, fanglike teeth, and light emitting organs that flash on and off.*

259

#### Activity

**Observing a Fish**
**Skills: Making observations, manipulative, diagraming, calculating, relating**
**Level: Remedial**
**Type: Hands-on**
**Materials: Fish, tray, hand lens, scissors, microscope**

Students should note during this activity that most fish belong to the bony class of fish and have four gills enclosed in a gill chamber at each side of the head. Each gill is composed of two rows of fleshy filaments attached to a gill arch. A flap of bone called a gill cover protects the gills. Naturally, the age of each fish, as shown by its scales, will vary.

#### Skills Development

***Skill: Expressing an opinion***

Though most sharks are predators, the whale shark and the basking shark, which are the two largest, feed only on small plants and animals. Though predatory sharks should not be taken lightly, only a few of them attack humans.

Give students a chance to express their opinions about whether or not they think sharks are vicious killers or merely predators in search of food. Consider using some questions such as these to stimulate the discussion:

- **Do sharks have an undeserved bad reputation?**
- **Should all sharks be treated with extreme caution?**
- **Have some books and movies presented sharks in unrealistic ways?**

Since dialogue and expression of opinions are the goals here, do not put down any sincere ideas that are presented. However, you might tactfully attempt to clarify or correct any obvious misconceptions that are voiced.

#### Content Development

Emphasize the idea that the swim bladder is an important characteristic of bony fish. The swim bladder is a cavity within the fish that can be filled with oxygen and nitrogen taken from the blood. By adjusting the amount of gas in its swim bladder, the bony fish adjusts itself to the water pressure present at various ocean depths.

- **What can a bony fish do because of its swim bladder?** (float at any level in the water)
- **How is a swim bladder similar to a life preserver?** (Both enable an object to float when inflated.)
- **What are some other objects humans use to float in water?** (Answers may include water wings, tire inner tubes, and rafts.)

## 11-2 AMPHIBIANS

### SECTION PREVIEW 11-2

Amphibians are vertebrates that spend the early part of their lives in water and usually live on land as adults. Though amphibians were the first vertebrates to adapt to life on land, they are still closely tied to water. They reproduce in water since their unshelled eggs would dry out if laid on land. Though most develop lungs as adults, their skin must remain moist to permit gas exchange to take place here as well.

The most familiar amphibians are frogs and toads. Both lack tails as adults, and the hind legs, which are larger than the front ones, are adapted for jumping. Though frogs and toads are similar in appearance, frogs generally have a smooth, moist skin, while toads are drier and warty. Both frogs and toads usually go into hibernation during cool months. Another major group of amphibians is comprised of salamanders and newts. They have tails as adults, and both pairs of legs are nearly the same size. As with frogs and toads, salamanders and newts are found in most areas.

### SECTION OBJECTIVES 11-2

**1. List the general characteristics of amphibians.**

**2. Describe the life cycle of amphibians.**

**3. List the identifying characteristics of salamanders and newts.**

### SCIENCE TERMS 11-2

**hibernation p. 261**

**Figure 11-10** *The mudskipper is a "fish out of water." It can crawl along the mud of tropical swamps and breathe in air as well as in water.*

masters of camouflage. They are able to change color to match their surroundings. By changing to the same color as the ocean bottom, the flounder "hides" from predators.

Some fish that live in the dark ocean depths have developed light-producing organs that can flash on and off to attract prey. Other deep-sea fish have huge eyes to help them see in dim light.

You might think that all fish have one thing in common—a need to stay in water at all times. However, the mudskipper and the walking catfish have adaptations that allow them to come out of the water and spend some time on land. And the lungfish can bury itself in mud in order to survive a dry season during which the water of the stream or pond in which it lives evaporates.

#### *SECTION REVIEW*

1. What are vertebrae?
2. What material makes up the skeleton of lampreys and sharks?
3. What structure allows bony fish to float at any depth in the water, even when asleep?

## 11-2 Amphibians

In the forests of Colombia, South America, lives the Choco Indian tribe. Today, as they have done for centuries, these Indians hunt deer, monkeys, and even jaguar with poisoned arrows. To do so, the Indians must first capture a supply of a certain frog that lives in the area. They then collect a milky fluid from the frog's back. The tips of their arrows are dipped into this fluid, a poison so powerful that a few drops can kill several thousand mice. The common name for this frog is no surprise—it is called the arrow-poison frog.

The arrow-poison frog is included among a group of coldblooded animals called the amphibians. The word "amphibian" means double life. And most amphibians do live double lives. **Amphibians spend part of their lives in water and part on land.** All



### 11-1 (continued)

#### Section Review 11-1

1. Bones that make up a vertebrate's backbone.
2. Cartilage.
3. Swim bladder.

### TEACHING STRATEGY 11-2

#### Motivation

Call attention to the arrow poison frog in Figure 11-11.

- **Why is this frog given the name, "arrow poison frog"?** (It produces a poison that is used by Indians to poison their arrows.)
- **In what part of the world are these frogs found?** (They live in tropical forests in South America.)

Explain that the poison in the frog's skin is deadly even to a large predator. However, the Indians can handle the frogs as they extract the poison, since the poison must be taken internally to be effective. The frogs are killed by piercing their skin with a sharp stick. The dead frogs are then held on a stick above a fire. The heat causes the poison to sweat from the frog's skin. It is then scraped from the skin and allowed to ferment. The arrow tips are then coated with this fluid and dried.

amphibians, for example, are born from eggs laid in water. And even those that spend their entire lives on land must return to the water to reproduce. Why? The eggs of amphibians lack a hard outer shell and would dry out if they were not deposited in the water.

There is yet another reason amphibians cannot stray too far from water—or at least from a moist area. Amphibians breathe through their skin, but the skin must remain damp for them to take in oxygen in this way.

**Figure 11-11** *This brilliantly colored frog is called an arrow-poison frog. The poisonous fluid in its skin can kill small animals.*

### *Frogs and Toads*

Have you ever wondered where frogs and toads go in the winter when the temperature drops? Frogs and toads, like all amphibians, are unable to move to warmer climates. But they can survive. Frogs often bury themselves beneath the muddy floor of a lake during the winter. Toads dig through dry ground below the frost line. Then these amphibians go into a winter sleep called **hibernation.** During hibernation, all body activities slow down so that the animal can live on food stored in its body. The small amount of oxygen needed during hibernation passes through the amphibian's skin as it sleeps. Once warmer weather comes, the frog or toad awakes. If you live in the country, you can usually tell when this

③ **Figure 11-12** *Amphibians such as the tree frog* (left) *and toad* (right) *lead a "double life." They spend part of their lives in water and part of them on land.*

261

## BACKGROUND INFORMATION

Although lungfish have gills for gas exchange, they also have primitive lungs. These lungs are connected to the nostrils on the head, making it possible for them to breathe air. By breathing through their lungs, lungfish are able to withstand long periods when their watery environment dries up. One species, the African lungfish, can survive out of water for up to four years. During dry spells it buries itself in mud at the bottom of its dried up river bed, and breathes through a plug made of mud. Other species of lungfish are found in rivers and lakes in Australia and South America.

## ANNOTATION KEY

① **Thinking Skill: Relating concepts**
② **Thinking Skill: Making generalizations**
③ **Thinking Skill: Applying definitions**

## Skills Development

***Skill: Drawing conclusions***

To help students see the advantages of hibernation in winter, ask these questions:

- **Are frogs and toads warmblooded or are they coldblooded?** (Most students are probably aware that they are coldblooded.)
- **What happens to the body activities of coldblooded animals when the temperature drops?** (Lead students to answer that all body activities slow down.)
- **What do most frogs and toads use for food?** (They feed mostly on insects.)
- **Can they obtain this kind of food during the winter months?** (Students should answer that insects are not available during the winter.)
- **What are some advantages of hibernation to frogs and toads during the winter?** (Students should conclude that hibernation is an advantage if low temperatures slow the body activities, and if insect food is scarce.)

## Content Development

Emphasize the fact that though most amphibians can live on land as adults, they must still live in the vicinity of moisture.

- **Why must amphibians return to water to reproduce?** (Because their eggs are unshelled, they would dry out if laid on land.)
- **In addition to using lungs to breathe, how else do amphibians take in oxygen?** (Amphibians also breathe through their skin.)
- **What will happen if an amphibian's skin dries out?** (They will not be able to breathe through their skin, and as a result, they will die.)
- **Why are amphibians not usually found very far from water or moist areas?** (Students should conclude that they need water for their reproduction and to keep their skin moist for gas exchange.)

## FACTS AND FIGURES

The largest amphibians are the giant salamanders of China and Japan. They can grow up to 1.5 m long.

## TEACHER DEMONSTRATION

If possible obtain one or two live frogs and toads. Display the frogs in a glass aquarium containing a small amount of water and a rock or other object on which they can climb out. The toads can be displayed in a similar aquarium, but place loose soil and a small pan of water in it. Be sure to cover the tops of both containers to prevent them from escaping.

If live animals cannot be obtained, then try to display a few pictures or wall charts that illustrate the distinctive features of both animals. After students have had an opportunity to view the living specimens or the pictures, call attention to some of their similarities and differences by asking questions similar to these:

- **In what ways are the skins of frogs and toads different?** (Students should notice that a frog's skin is smooth and thin, but a toad has skin that is bumpy or warty looking.)
- **In what ways are these animals alike?** (Accept all answers. Students will likely mention that the body shapes are similar, and the hind legs are larger than the front ones.)
- **Which of these amphibians is likely to be found farthest away from water?** (Lead students to suggest that toads are better adapted for life on land.)

Mention that toads often bury themselves in moist soil to escape heat and enemies.

happens. The night is suddenly filled with the familiar peeps, squeaks, chirps, and grunts that male frogs and toads use to attract their mates.

Frogs and toads appear similar in shape. But if you touch them, you can tell one difference immediately. Frogs have a smooth, moist skin. Toads are (1) drier and are usually covered with small wartlike bumps. In many toads, the bumps behind the eyes contain a poisonous liquid, which the toad releases when attacked. The attacking animal quickly becomes sick and may even die.

Neither frogs nor toads have tails as adults. However, strange as it may seem, they do have tails when they hatch from eggs in the water. In this stage of their lives they are called tadpoles or polliwogs. A tadpole has gills to breathe under water and feeds on plants. Eventually the tadpole begins to undergo (2) remarkable changes. Its tail begins to disappear. Two pairs of legs take shape. At the same time, its gills begin to close. Inside the tadpole's body, lungs are forming. Soon the tadpole will be an adult toad or frog, ready for its life on land.

If there is one thing most people know about adult toads and frogs, it is that they are excellent jumpers. The main reason for this is that the hind legs of a frog or toad are much larger than the front legs. It is these powerful hind legs that allow the animals to jump so well and help them escape from

**Figure 11-13** *All amphibians lay their eggs* (left) *in a watery environment. Each egg then hatches into a larval form called a tadpole* (right).

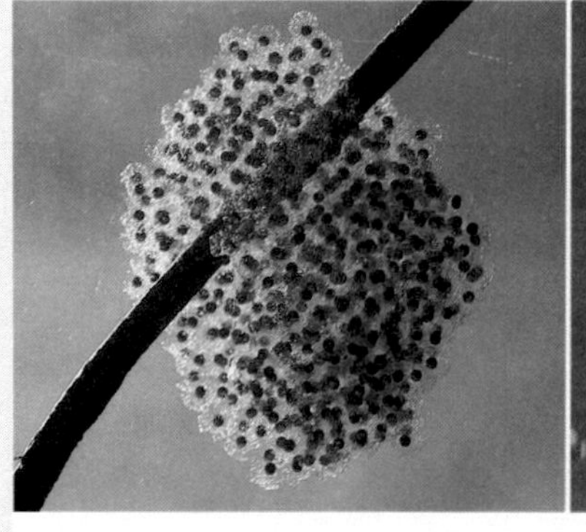

262

### 11-2 (continued)

### Skills Development

***Skill: Sequencing events***

In discussing the life cycle of frogs and toads, ask the class first to recall the stages in the life cycle of an insect. As the names are recalled, write them on the chalkboard connected by arrows:

egg → larva → pupa → adult

- **What do we call this type of development in which organisms change their appearance as they go through different stages?** (This type of development is metamorphosis.)

Now ask students to read carefully the paragraph on amphibian metamorphosis on page 262. As they read, tell them to list the stages in sequence and connect them with arrows.

### Motivation

Tell students that frogs have extraordinary jumping ability. For example, the leopard frog that lives in ponds throughout the United States has a body length of only about 12.5 cm. Yet it is capable of jumping 1.6 m, a distance almost 13 times its length.

Provide some metersticks and have students work in pairs to measure their length from the top of the head to the hips. Then ask each stu-

**Figure 11-14** *The sticky tongue in a frog's mouth is well adapted for catching insects, such as this fly.*

enemies. The largest frog, the goliath frog of West Africa, can jump more than 3 meters with little effort! A frog's large hind legs are also useful in water, where they serve as paddles to propel the animal.

Unlike tadpoles, adult frogs and toads are carnivorous. They catch their prey with a most unusual tongue. The sticky tongue of a frog or toad is attached to the front of its mouth. To trap an insect or other prey, the frog or toad quickly flicks its tongue out of its mouth and catches the insect as it flies by.

### *Salamanders and Newts*

An important ingredient of a witches' brew or magic potion was the eye of a newt. Well, there are no magic potions, but there really are newts. Newts, along with salamanders, are amphibians with tails. Like frogs and toads, these animals have two pairs of legs. However, their legs are not developed like those of a frog. Newts and salamanders cannot jump at all.

Since they are amphibians, salamanders and newts must live in moist areas. Some live in the water all their lives. Others may spend most of their life under a single tree stump. Like frogs and toads, salamanders and newts must lay their eggs in water.

**Figure 11-15** *Like all amphibians, this spotted salamander must live in a moist area. Notice the mass of eggs below the salamander.*

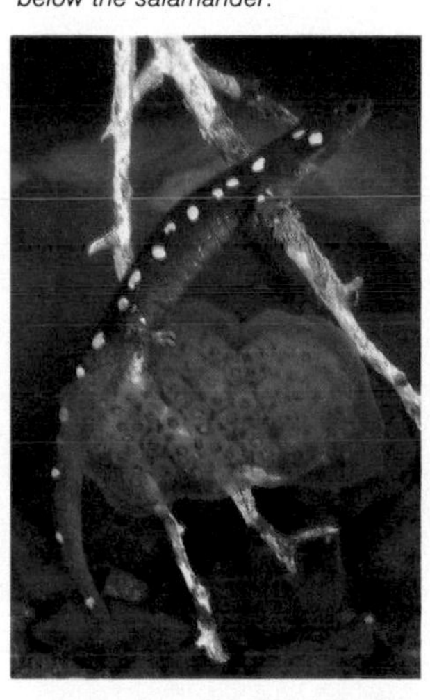

263

dent to continue how far she/he could jump if they were a leopard frog. To do this they need to multiply their measured length by 13.

## Reinforcement

At certain times of the year—usually spring—frog egg masses or tadpoles can be collected from the shallow edges of ponds. At other times they may be available from biological supply companies. If tadpoles can be obtained, some students might enjoy watching their development into adult frogs. It takes about four months for frog eggs to develop into young frogs. If students undertake this activity refer them to an appropriate reference on the care of tadpoles.

## Skills Development

***Skill: Designing an experiment***

If tadpoles can be obtained, have students work in small groups to design an experiment to test the effect of temperature on their development. Tell the groups that their plans must include:

**(1) A hypothesis to be tested**
**(2) Means for establishing different temperature conditions**
**(3) The number of tadpoles to be tested at each temperature**
**(4) Procedures for making observations and recording data**
**(5) Provisions for maintaining all tadpoles in good health throughout the experiment.**

After each group comes up with a plan, bring the class together to share each group's plan. If the plans are adequate, then individual groups or the class as a whole might try the experiment. As a suggested approach, students might keep one group of tadpoles at normal room temperature, and another group at the same stage of development in a cool location, such as an unheated garage. Emphasize that both groups must be fed regularly and their water kept clean.

## Content Development

Point out that almost all salamanders live in temperate regions of the northern hemisphere. Some species are completely aquatic and keep their gills throughout their entire lives. A few species have neither gills nor lungs. All gas exchange takes place through the skin or the roof of the mouth.

## TIE-IN/ LITERATURE

Because most students know that frogs are excellent jumpers, suggest that they read Mark Twain's short story. "The Celebrated Jumping Frog of Calaveras County." If they have already read this tale, then ask someone to summarize it for the class.

## ANNOTATION KEY

❶ **Thinking Skill: Making comparisons**
❷ **Thinking Skill: Sequencing events**

## 11-3 REPTILES

### SECTION PREVIEW 11-3

Reptiles are coldblooded vertebrates that have dry, scaly skin. These vertebrates are fully adapted for a life on land, though many of them live in watery environments. Reptile eggs, unlike those of the amphibians, are enclosed within a shell. Thus, they do not have to return to water to reproduce.

Millions of years ago, reptiles were the dominant animals on earth. Today, three major groups remain: the snakes and lizards, turtles, and crocodilians. The most numerous are the snakes and lizards. Most snakes are nonpoisonous, but some produce venom that is injected through fangs into their prey. Lizards can be distinguished from snakes by the presence of ears and legs. Turtles and tortoises have a bony shell enclosing the body, and their mouths lack teeth. Alligators and crocodiles spend most of their time in water, and they take care of their young.

### SECTION OBJECTIVES 11-3

1. **List the major traits of reptiles.**
2. **Distinguish between snakes and lizards.**
3. **Identify characteristics of turtles and tortoises.**
4. **Distinguish between alligators and crocodiles.**

### SCIENCE TERMS 11-3

**venom p. 265**

***SECTION REVIEW***

1. What does the term "amphibian" mean? How does it apply to amphibians?
2. How do frogs survive the cold winter months?
3. When do frogs and toads have tails and gills? When do they have legs and lungs?

## 11-3 Reptiles

On the barren, windswept shoreline of the Galapagos Islands in the Pacific Ocean, a group of iguanas (i-GWAH-nuhz) cling to the rocks. Wave after wave pounds against the rocks, but the iguanas do not let go. Then, without warning, the iguanas plunge into the cold sea in search of food. The iguanas did not suddenly become hungry. Instead, they basked on the rocks of the Galapagos until the sun warmed their bodies. Only then could these coldblooded animals enter the chilly waters. And once they become cold again in the water, they will have to return to the rocks to warm up.

The iguana, a kind of lizard, belongs to a group of animals called reptiles. (1) **Reptiles are coldblooded vertebrates that have dry, scaly skin and lay eggs on land.** In addition to lizards, reptiles include snakes, turtles, and alligators. Although many rep-

**Figure 11-16** *Sunning themselves on the shore of this Galapagos island, located off the west coast of South America, are marine iguanas. Marine iguanas are a type of reptile. A crab with red-tipped legs keeps them company.*



### 11-2 (continued)

#### Section Review 11-2

1. Double life: part of life spent in water, part spent on land.
2. They go into hibernation.
3. Tails and gills as tadpoles; Legs and lungs as adults.

### TEACHING STRATEGY 11-3

#### Motivation

Ask students to compare the photographs of lizards, snakes, alligators, and turtles found in this section.

- **To what vertebrate group do all of these animals belong?** (Students should identify them as reptiles.)
- **Can you identify any characteristics that these animals share in common?** (Accept all sensible answers, but be sure that their scaly skin is mentioned.)
- **Many fish also have scales. In what way do you think the scales of reptiles are different from those of fish?** (Lead students to answer that a reptile's skin is thicker than that of most fish, and it is dry.)
- **In what ways do some of these reptiles differ from each other?** (Answers will vary, though such distinguishing traits as the turtle's shell

**Figure 11-17** *Some reptiles, such as this gecko, must shed their skin as they grow.*

tiles live in or near water, as do amphibians, they do not have to go through a water-dwelling stage in their lives. And, because their scaly skin is resistant to drying out, reptiles do not have to live in a moist environment. In fact, many reptiles live in deserts. To grow larger, reptiles such as snakes and lizards periodically shed their skins. Snakes actually crawl out of their old skins through holes near their mouths. Lizards shed their skins in strips. Some of them then eat the pieces of old skin. ❷

The eggs of reptiles have a leathery protective shell. This shell prevents the contents of the egg from drying out. So reptiles do not have to return to the water to lay their eggs. Since the eggs are enclosed within a shell, reptiles have developed a system of internal fertilization. The eggs are fertilized within the female's body before the shell forms around the egg. ❸

### Snakes and Lizards

Most people are naturally fearful of snakes. They mistakenly think that all snakes are dangerous. But most snakes are not poisonous and will not harm people. In fact, most snakes are helpful. They eat small animals like mice and rats. However, a good rule to follow is: When in doubt, leave a snake alone!

Those snakes which *are* poisonous have developed special glands that produce their **venom,** or poison. Snakes inject their venom into their prey

#### Activity

*The Truth About Snakes*

Many people hear the word snake and think of a slippery, poisonous animal. But snakes are not slippery. And most of them are not poisonous either.

Use reference books in the library to determine what proportion of the world's snakes actually are poisonous. In addition, find out about the ways in which snakes are helpful to people. Present your findings in a written report.



## ANNOTATION KEY

❶ Thinking Skill: Making generalizations

❷ Thinking Skill: Relating cause and effect

❸ Thinking Skill: Applying concepts

## Activity

**The Truth About Snakes**

**Skills: Report writing, making comparisons, applying, relating, researching**

**Level: Remedial**

**Type: Vocabulary/library**

This activity will help reinforce students' writing and researching skills. Reports should be well written and factually documented. As a class project, you may want to ask students to suggest various ways the numbers of poisonous versus nonpoisonous snakes might be graphically illustrated.

and the absence of legs on snakes will probably be mentioned.)

Also emphasize that reptiles share the trait of coldbloodedness with fish and amphibians.

## Content Development

Ask students to look at Figure 11-17 and read the caption. Emphasize that a reptile's skin is dry and waterproof. Thus their body is protected from drying out in a hot desert.

• **Why are some of the gecko's scales brighter than others?** (The gecko is in the process of shedding.)

• **Why must its skin be shed from time to time?** (The skin is shed as the animal grows.)

Remind the class that the process during which an animal sheds its outer covering is called molting.

• **Do you recall another group of animals that also molts as it grows?** (arthropods)

## Skills Development

***Skill: Making comparisons***

A major advance of reptiles over amphibians is their ability to reproduce on land. Internal fertilization and a shelled egg frees reptiles from the necessity of returning to water to reproduce.

• **In what important way are the eggs of reptiles different from those of amphibians?** (Reptile eggs have a leathery, protective shell.)

• **Why is this covering so important?** (It keeps the egg from drying out; thus they can be laid on land.)

• **Where are the eggs of most female amphibians fertilized?** (Most are fertilized in water, outside the female's body.)

• **How is fertilization among reptiles different from that of amphibians?** (Lead students to answer that reptiles have developed a system of internal fertilization.)

## FACTS AND FIGURES

The world's largest snakes are the anaconda and the reticulated python. Both are reported to grow up to 9 m long. Both of these giant snakes kill their prey by constriction. There are occasional reports of their attacking humans, but such incidents are rare. There is, however, a confirmed case of a reticulated python attacking and swallowing a 90 kg bear.

## TIE-IN/ LITERATURE

For centuries, the image of the snake has been used in literature. Investigate some ancient folklore, myths, or legends to see how the snake has been portrayed. Then, write your own legend about a snake.

## FACTS AND FIGURES

The most venomous land snake is the tiger snake of southern Australia.

## ANNOTATION KEY

❶ **Snakes inject their venom through fangs**

❶ **Thinking Skill: Applying concepts**

❷ **Thinking Skill: Making comparisons**

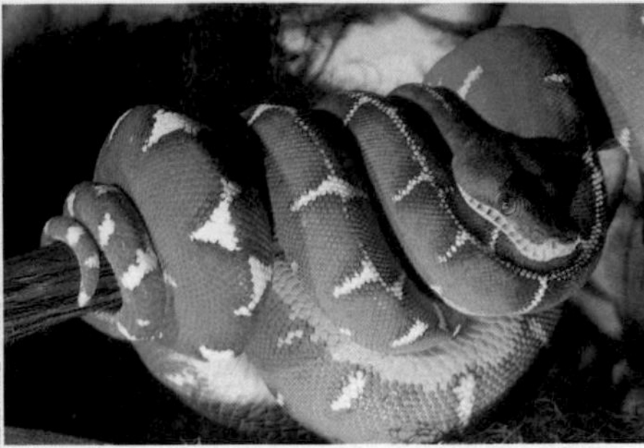

**Figure 11-18** *The king cobra* (left) *is the largest poisonous snake in the world. The emerald tree boa* (right) *kills its prey by squeezing its victim's chest so tightly that it suffocates.*

through special teeth called fangs. Four kinds of poisonous snakes are found in the United States: rattlesnakes, copperheads, water moccasins, and coral snakes. Other poisonous snakes, like the king cobra, the largest poisonous snake in the world, are found on other continents.

Snakes have developed several remarkable ways of finding prey. Many snakes are able to detect the body heat produced by their prey. They have special ❶ pits on the sides of their head that are extremely sensitive to heat. Their tongue is also used as a sense organ. When a snake flicks its tongue in and out of its mouth, it is actually tasting the air, trying to detect those particles in the air that tell the snake that

**Figure 11-19** *The Komodo dragons in this photograph are eating a goat. The largest lizards, Komodo dragons can grow to a length of 3 meters and weigh 113 kilograms.*

266

## 11-3 (continued)

### Content Development

Snakes have several adaptations that permit them to swallow prey larger in width than their own bodies. The lower jaw is not joined directly to the skull. During swallowing it can be drawn down and forward, allowing the snake to take in large prey. Therefore, the body can expand considerably during swallowing. All snakes swallow their food whole, and all digestion takes place in the stomach.

### Content Development

Discuss some of a snake's adaptations for locating prey.

- **Some snakes have a small pit on the head in front of the eyes. What is the function of this pit?** (The pit contains a heat-sensitive organ that helps them detect warmblooded prey.)
- **Why do snake flick their tongues in and out of their mouth?** (The tongue picks up molecules from the air that are given off by nearby prey.)

Explain that the snake transfers the molecules from the air to two cavities called Jacobson's organ located in the roof of its mouth. The Jacobson's organ is extremely sensitive to odors. Also mention that even though snakes lack outer ears, they do have inner ears that can detect vibrations.

food is nearby. Snakes are deaf and have poor eyesight. But their other senses make up for these limitations.

A snake moves by wriggling its muscular body. The scales on its belly help the animal grip the ground and push itself forward. Many snakes, including the water moccasin, are quite at home in the water and can swim along at the surface or submerged.

Lizards differ from snakes in several ways. The most obvious difference is that lizards have legs. Lizards also have ears and can detect sounds. Most lizards are small and eat insects. However, one lizard, the Komodo dragon of Indonesia in Southeast Asia, 2 is very large and looks like a prehistoric dinosaur. The Komodo dragon can swallow small pigs and chickens whole.

Lizards have developed various ways to trap prey. For example, the Gila (HEE-luh) monster, a lizard that lives in the American Southwest, poisons its prey. The Gila monster bites the prey, holds on tight, and rolls over on its side. Meanwhile, poison released from the lizard's lower jaw flows into the wound, aided only by the force of gravity. How does this differ from the way a snake poisons its prey? 1

Some lizards have developed special ways to protect themselves from becoming another animal's dinner. The chameleon is one of several kinds of lizard that can change color to match their surroundings. In this way, the chameleon hides from predators. Another lizard has an even stranger way of escaping.

**Figure 11-20** *The Gila monster* (left) *lives in North American deserts and has a poisonous bite. The anole* (right) *is a lizard that is able to blend in with its surroundings. Can you find the anole in this photograph?*

## SCIENCE, TECHNOLOGY, AND SOCIETY

Hunting is one method of controlling animal populations. We can distinguish among three kinds of hunting: commercial hunting, hunting for food, and recreational hunting. Commercial hunting has been responsible for the extinction of some species, such as the American passenger pigeon, and the near extinction of others.

Hunting for food was once a way of life for many people in the United States and is still a necessity in some parts of today's world. However, in our country today, most individuals hunt for sport. Proponents of recreational hunting say that they enjoy the challenge and sportsmanship, learn about wildlife and gain greater respect for nature. Taxes and license fees from hunters also supply some of the money used for wildlife management. They argue further that hunting is needed to maintain the balance among certain populations. Those opposed to hunting for sport argue that it is a cruel and unnecessary practice that endangers wildlife and their habitats. They are also opposed to hunting because of the dangers involved in the use of firearms.

If wildlife is a public resource, should the public have a right to harvest it? Do the current laws and regulations that govern hunting in your state seem reasonable to you?

### Content Development

Use the following questions to help students distinguish between the way a poisonous snake and a Gila monster poisons its prey:

- **How do snakes release poison into their victims?** (Poisonous snakes have fangs that inject poison.)
- **How does the poisonous Gila monster poison its victim?** (The Gila monster bites its victim and forces poison into the wound from a poison gland in the lower jaw.)

Mention that there are only eight known cases of humans being killed by the bite of a Gila monster. The only other venomous lizard is the beaded lizard of western Mexico.

### Skills Development

***Skill: Relating concepts***

Ask some questions similar to these to aid student understanding of the defense mechanisms of lizards and other animals.

- **How is the way a chameleon protects itself like that of a flounder?** (Both change colors to blend with their surroundings.)
- **What happens if predators attempt to catch certain other lizards by their tail?** (The tail becomes detached and confuses the predator.)

Point out that several species of lizards, especially skinks, have this ability to shed the tail.

- **In what way is this lizard somewhat like a starfish?** (Both animals have the ability to regenerate missing body parts.)

## BACKGROUND INFORMATION

Turtles are the oldest group of living reptiles. The fossil record shows that their ancestors were around even before the dinosaurs, and they have changed very little over time. Turtles are, however, a very small group since there are only a little over 200 species, and several of them are threatened with extinction.

## FACTS AND FIGURES

Turtles and tortoises have a life expectancy greater than most other vertebrates. Most species of turtles live 50 years or more. The record age for any turtle is about 150 years.

If caught by its tail, it will quickly shed the tail. The tail remains behind, wriggling on the ground. This action confuses the predator, and the lizard scampers to safety. Later, the lizard's body regenerates the missing tail.

### Turtles and Tortoises

Turtles and tortoises are two reptiles that look alike but have adapted to different environments. Turtles spend most of their time in the water. Their legs are shaped like paddles and are used for swimming. Turtles can swim quite well; however, on land they move slowly. Tortoises spend most of their time on land. They have solid, stumpy legs used for walking. Tortoises also have claws on their feet that are used for digging.

Covering most of the body of both turtles and tortoises are shells made of plates. Although the shells

### Career: *Fish Farmer*

**HELP WANTED: FISH FARMER** High school diploma desired. Courses in fish behavior, fish physiology, fish diseases, and business helpful. Job involves outdoor work.

In Mark Twain's *Huckleberry Finn,* Huck recalls fixing breakfast for his friend and himself. "I fetched meal and bacon and coffee and coffeepot and frying pan and sugar and tin cups. . . . I catched a good big catfish, too, and Jim cleaned him with his knife and fried him." In early America, when you sat down to a tasty fish dinner, it meant that somebody had had good luck fishing that day. Now, however, chances are that the fish you ate recently was raised and bred on a fish farm. **Fish farmers** throughout America raise catfish, salmon, trout, bass, and other types of fish to be sold at a profit to restaurants and markets.

At a fish farm, farmers begin by stripping eggs from female fish. They place them in moist pans, where they fertilize them with sperm cells from male fish. The fertilized eggs are then kept warm to ensure growth. When the fish hatch and grow to be as long as a person's finger, they are called fingerlings and are moved to rearing ponds until fully grown.

If fish farming interests you and you live in an area that will support some type of fish farming, you can learn more by writing to the Marine Resources Research Institute, South Carolina Wildlife and Marine Resources Department, P.O. Box 12559, Charleston, SC 29412.



## 11-3 (continued)

### Motivation

Obtain a picture or two of a Galapagos tortoise to show the class. Good photos are often seen in encyclopedias or animal reference books. Point out that the largest land tortoises in the world live in the Galapagos. One species may have a shell 1.8 m long and a mass of nearly 275 kg when full grown. Despite their large size, they are quite tame.

- **Other than their gigantic size, in what other way are these tortoises unusual?** (Answer may vary, but lead students to suggest that they may live to be very old.)

Conclude by reviewing the adaptations of the tortoise's necks and shells to their feeding habits.

### Content Development

Write the terms turtle, tortoise, and terrapin on the chalkboard.

- **What is the major difference between turtles and tortoises?** (Turtles spend most of their time in water, while tortoises spend most of their time on land.)
- **Does anyone know the meaning of the term terrapin?** (Some students may be aware that, strictly speaking, turtles live in the sea and terrapins are freshwater turtles. However, even biologists often do not make this distinction.)
- **In what way are the legs of turtles or terrapins adapted for living in water?** (They are shaped like paddles to facilitate swimming.)
- **In what way are the legs of tortoises different?** (They have stumpy legs for walking, and claws on the feet for digging.)

### Content Development

Stress the fact that turtles and tortoises are distinctly different from other reptiles in several ways. The most obvious difference is the bony shell that encloses the body. In most turtles, the shell is composed of a number of hard plates, but in soft-shelled species it is composed of a leathery skin. The inner layer of the shell is usually fused with the vertebrae and ribs. Turtles and tortoises are also different in that they lack teeth. They catch and crush their food with horny beaks.

**Figure 11-21** *The Ridley turtle* (left) *and giant tortoise* (right) *are examples of reptiles with shells. Turtles spend most of their time in water, while tortoises spend most of their time on land.*

offer some protection, they are heavy and slow the animals down. The backbones of these reptiles are fused to their top shell. Turtles and tortoises have no teeth. They have beaks that are similar in structure to the beaks of birds. Many turtles and tortoises eat plants as well as animals. ❷

Turtles and tortoises can live for a very long time. The Galapagos tortoise, for example, may live for as long as 200 years. When sailors first discovered these animals, they captured them and brought them back to their ships. The animals were kept alive and used as a source of fresh meat on long ocean voyages.

A sea turtle called the green turtle is among the most outstanding navigators in the animal kingdom. Soon after the turtle eggs hatch, perhaps on a beach in Brazil, the young turtles head for the ocean. There they wander for many years over thousands of square kilometers. Eventually, the turtles mature and mate. Now they are ready to lay their own eggs. Where do these turtles go? Back to the same beach where they were born! ❸

That beach may be hundreds of kilometers away, across an ocean surface that has no road signs or other markings. Yet the turtles find their way home. How? No one knows. And here is another fascinating mystery. No matter how far they spread out



## BACKGROUND INFORMATION

The shape of a reptile's pupil indicates whether the animal is active at night or during the day. A round pupil indicates the reptile is active during the day. A slitlike pupil indicates the reptile is active at night.

## HISTORICAL NOTES

The great tortoises of the Galapagos Islands have been so greatly reduced in numbers that they are now in danger of extinction. When the islands were first discovered in the seventeenth century, they were reported to be so numerous that a person could walk long distances atop their shells without touching ground. In the 1800s, sailors started capturing them for their meat and oil. Later the dogs and cats of settlers further reduced the population when they started eating the eggs and babies of the tortoises.

## ANNOTATION KEY

❶ Thinking Skill: Making comparisons
❷ Thinking Skill: Relating facts
❸ Thinking Skill: Relating concepts

Point out that actual research has confirmed that when green turtle eggs are incubated at low temperatures, mostly males hatch. Higher temperatures produce mostly females. Similar studies have shown that temperature variations affect sex determination in other reptiles, too.

## Skills Development

***Skill: Drawing conclusions***

After discussing the remarkable navigational abilities of the green turtle, present this hypothetical experiment to the class: "A biologist collected a large quantity of green turtle eggs from a beach and placed them in batches of 25 each at different temperatures. After about 60 days, the following number of eggs had hatched at each temperature:

26° C: 21 males, 2 females
27° C: 18 males, 0 females
28° C: 13 males, 11 females
29° C: 12 males, 12 females
30° C: 1 male, 19 females
31° C: 0 males, 21 females
32° C: 1 male, 20 females."

- **What might be concluded from this biologist's data?** (Students should conclude that temperature is a factor in determining the sex of green turtles.)

## Reinforcement

Students may have difficulty determining common traits of reptiles. Have students identify each of the following statements as true or false.

- **All reptiles hatch from eggs.** (true)
- **Reptiles spend part of their life on land and part in water.** (false)
- **Reptiles breath through gills.** (false)
- **Reptiles reproduce by external fertilization.** (false)

## TIE-IN/SOCIAL STUDIES

Alligators and crocodiles have been hunted for their hides. Investigate the uses of the hides of these reptiles. Include a description of the penalty imposed on a person caught hunting these creatures today.

## ANNOTATION KEY

❶ Thinking Skill: Making comparisons
❷ Thinking Skill: Making generalizations

### Activity

**A Symbiotic Relationship**
**Skills: Report writing, researching, applying, relating**
**Level: Average**
**Type: Library/writing**

Students can easily find out about the relationship between the Egyptian plover and the Nile crocodile. The plover picks leeches and bits of food off the crocodile's teeth. The crocodile gets its teeth cleaned, and the plover gets a free meal. You may encourage students to bring in nature magazines in which this and other symbiotic relationships are pictured.

**Figure 11-22** *How can you tell the difference between an alligator and a crocodile? An alligator has a broad snout* (left). *A crocodile has a narrower snout. When a crocodile closes its mouth, another basic difference can be seen. Some of its bottom teeth are visible* (right).

through the ocean, turtles born on a particular beach find one another in the ocean and return together to that beach. One scientist observed a huge group of green turtles in a line that stretched for more than 100 kilometers. Each turtle was almost exactly 200 meters ahead of the next!

### Alligators and Crocodiles

To most people, alligators and crocodiles are very similar in appearance. Actually there are some differences. The snout of an alligator is broader than the snout of a crocodile, which is narrower and more pointed. ❶ When a crocodile's mouth is closed, some of its bottom teeth are visible. When an alligator closes its mouth, none of its teeth are visible. Most people, of course, do not feel the need to examine the mouth of either of these animals closely enough to tell them apart!

Alligators and crocodiles spend most of the time submerged in water with only their eyes and nostrils showing above the surface. Both kinds of animals eat meat. Their diet consists mainly of fish and other animals that venture too close to their large mouths.

Alligators and crocodiles have unusual reproductive behavior. The female alligator builds a nest of rotting plants in which to lay her eggs. The rotting plants give off heat as they decompose. This heat keeps the eggs warm and helps them to develop. The alligators inside the eggs make low chirping sounds when they are about to hatch. The sounds

### Activity

*A Symbiotic Relationship*

Look for information and write a short paragraph on the relationship between the Egyptian plover and the Nile crocodile.



## 11-3 (continued)

### Content Development

Explain that even though alligators and crocodiles are similar, each has certain distinguishing features.

• **Which of these two thick-skinned reptiles has a broader snout?** (The alligator's snout is broader.)

Mention that when viewed from above, a crocodile's snout has a somewhat triangular shape.

• **In which do the teeth stick out of the mouth even when it is closed?** (The teeth protrude from a crocodile's mouth when closed.)

Students may ask if one is larger than the other, or if the colors are different. Point out that color and size distinctions are not great, and they vary from one species to another. In general, African crocodiles, which sometimes reach lengths of 4.8 meters, are the largest. Yet, the largest American alligator on record was 5.7 meters long.

### Skills Development

***Skill: Drawing a conclusion***

Tell the class that alligators and crocodiles have several traits unique from those of other reptiles. List these crocodilian traits on the chalkboard:

1. The eyes and ears are located on the top of the snout.
2. A fleshy valve at the back of the mouth can tightly cover the passage from the mouth to the lungs.
3. Nostrils on top of the head lead to the air passage behind the throat valve.
4. The nostrils and ears can close.

are heard by the mother, who has remained nearby guarding her eggs. When she hears the hatchlings, she uncovers the eggs. The tiny alligators, looking like small copies of their parents, come out into the world. The female continues to care for the young for some time after they hatch. The male alligator also helps care for the young. This behavior is unlike the behavior of most reptiles, which usually leave their eggs after they are laid. Their young must care for themselves as soon as they hatch.

### SECTION REVIEW

1. What is an important difference between the egg of an amphibian and the egg of a reptile?
2. How are the legs of a turtle different from the legs of a tortoise?

## 11-4 Birds

For many people, birds are the most fascinating and colorful animals on earth. Birds, along with bats and insects, are the only animals with the power of flight, although not all species of birds do fly.

The earliest known bird lived more than 150 million years ago, during a time when reptiles like the dinosaurs ruled the earth. In fact, this early bird had many of the characteristics of reptiles. For example, it had a long bony tail and sharp teeth. Modern birds have neither of these. However, there is no doubt that this early flying creature was a bird. For fossil evidence shows it had feathers, a characteristic that only birds have. If you look at a modern bird, you will notice another characteristic in common with reptiles. Birds have scales on their legs.

**Birds are warmblooded vertebrates that have wings and a body covered with feathers.** Warmblooded animals have the ability to maintain a constant body temperature despite the temperature of their environment. The penguins in Figure 11-23 are able to trap their own body heat under an insulating coat of feathers. Thus, the penguins can withstand very cold temperatures. However,

**Figure 11-23** *Birds have feathers and are warmblooded. These penguins can withstand the cold of Antarctica because they are insulated by a coat of feathers.*

## 11-4 BIRDS

### SECTION PREVIEW 11-4

Birds are warmblooded vertebrates that have wings and a body covering of feathers. The fossil record indicates they have been on earth for more than 150 million years, and they probably evolved from reptiles. On the basis of variations in their beaks, feet, wings, and feeding habits, birds are divided into four main groups: perching birds, water birds, birds of prey, and flightless birds. The bodies of most birds are specially adapted for flying. Their bones are light and hollow, and feathers give them a streamlined shape and provide lift.

Many birds have complicated behaviors related to attracting mates, nesting, and caring for the young. Many of them also migrate to a new environment on a seasonal basis. Different species appear to depend on different mechanisms to navigate while migrating.

### SECTION OBJECTIVES 11-4

1. **List the major traits of birds.**
2. **Describe the adaptations of birds for flight.**
3. **Give examples of several types of bird behaviors that relate to breeding and development.**
4. **List several ways birds may navigate during migration.**

### SCIENCE TERMS 11-4

**talon p. 272**
**contour feather p. 273**
**down feather p. 273**
**migrate p. 275**

• **What do these adaptations tell you about an alligator's or crocodile's life style?** (Lead students to conclude that they are adaptations for lying in water with only the ears, eyes, and nostrils above the surface.)

The ability to close the throat, ears, and nostrils allows them to drag prey below the water to drown it.

### Content Development

Emphasize that unlike other reptiles, some species of alligators and crocodiles care for their eggs and young.

• **Where do the females build their nest?** (The nest is built of rotting plants on the shore.)

• **How are the eggs kept warm?** (Heat is given off as the rotting plants decompose.)

• **Which parent cares for the young after they hatch?** (Both female and male care for the young.)

• **What other vertebrate animals have somewhat similar reproductive and nesting behaviors?** (Answers may vary, but birds should be mentioned by some students.)

### Section Review 11-3

1. The eggs of reptiles are leathery.
2. The turtle's legs are shaped like paddles while the tortoise's legs are solid and stumpy.

## TEACHING STRATEGY 11-4

### Motivation

Call on someone to read aloud the first sentence in Section 11-4. Then conduct an open-ended discussion on some possible reasons for people's fascination with birds.

- **At zoos, large crowds are often seen watching the birds shown in Figure 11-23. Why do you think so many people enjoy these birds?** (Accept all answers.)
- **Why do some people keep birds like the one shown in Figure 11-24 as pets?** (Accept all answers.)
- **Why do you think certain birds are sometimes portrayed as cartoon characters on television?** (Accept all answers.)
- **Do you have a favorite bird? If so, why?** (Accept all answers.)

### Content Development

After students have read page 271, ask them to compile a list of at least five traits of birds. For example: Birds—

1. Are usually capable of flying
2. Have feathers
3. Are warmblooded
4. Lack teeth
5. Have scales on their legs.

After they have compiled their individual lists, compile a master list on the chalkboard and discuss each trait as it is listed.

to maintain a high body temperature birds must consume a great deal of food for energy. For the penguins, that means eating a lot of fish.

### Types of Birds

Birds can be divided into four main groups: perching birds, water birds, birds of prey, and flightless birds. The perching birds are perhaps the most familiar. These birds have feet that are adapted for perching. Their feet can easily grasp a branch. Cardinals, robins, and sparrows are perching birds. The beaks of some of these birds may be long and thin like the beak of the hummingbird. This beak enables (1) hummingbirds to reach deep into flowers for nectar. Other perching birds have beaks that are adapted for cracking seeds or catching insects. The macaw, a kind of parrot, has a beak that is strong enough to crack open tough Brazil nuts. The woodpecker has a beak that can open the bark of trees where insects, the food of woodpeckers, live.

Water birds have feet that are adapted for swimming. Ducks and geese are water birds with feet that resemble paddles. These birds glide across the surface of the water looking for food. All water birds can fly as well as swim.

**Figure 11-24** *Perching birds, such as this macaw, live on land and spend their lives in trees or other high places when they are not flying. Macaws are members of the parrot family.*

Birds of prey are superb fliers and also have keen eyesight. Soaring high in the air, they can spot prey on the ground or in the water far beneath them. Birds of prey eat small animals including fish, reptiles, and other birds. Some, such as a certain eagle, even eat small monkeys.

Birds of prey are able to fly very fast. The peregrine falcon has been clocked at more than 125 kilometers per hour while diving at its prey. Birds of prey have sharp claws called **talons** on their toes. Talons enable the bird to grasp and hold its prey. Some eagles have talons that are longer than the (2) fangs of a lion. Birds of prey also have strong, curved beaks that are used to tear their prey into pieces small enough to be swallowed.

The flightless birds include the ostrich of Africa, the largest bird alive today; the rhea of South America; and the emu and cassowary of Australia. Also included in this group are penguins, flightless birds that are able to "fly" through the water. All of these

272

### Content Development

Mention that birds have several structural and physiological adaptations for flight other than their feathers and light bones. First, the respiratory system is designed to provide large amounts of oxygen needed during flight. A series of air sacs extend from the lungs into the body cavity. These sacs increase the total amount of oxygen available for flying by holding oxygen in reserve. The air sacs further serve to reduce the bird's body weight. Second, birds lack a urinary bladder. By not storing urine in the body, weight is reduced. Another adaptation for reducing weight is the laying of eggs after they are formed rather than storing them inside the body.

Since flying requires an enormous expenditure of energy, birds have a high rate of metabolism to release the energy needed. The body temperature of some birds is as high as 43° C as a result. Birds also have very powerful muscles. As much as 50% of their body weight is muscle.

### Motivation

Ask students to comment on the often used phrase, "He or she eats like a bird." Point out that many people erroneously believe that birds are light eaters. On the contrary, birds eat more for their size than any other vertebrates.

- **Why do you think birds require so much food?** (Answers may vary, but the idea should be developed that birds must consume great

**Figure 11-25** *The paddlelike feet of the wood duck* (top left) *enable it to swim in water. The bald eagle* (bottom left) *has keen eyesight, sharp bill, and talons, or curved claws, for grasping prey. A road runner* (right) *can run as fast as 24 kilometers per hour.*

birds have small wings relative to the size of their bodies. All flightless birds except penguins have strong leg muscles that enable the birds to run quickly. These birds also use their strong legs for defense, kicking viciously at any enemy foolish enough to challenge them.

### Adaptations for Flight

Birds have light, hollow bones—certainly an advantage to any flying animal. Birds are also covered with feathers. They usually have large feathers on their wings. These feathers provide lift, much as an airplane's wings lift it off the ground.

The feathers on the wings and most of the bird's body feathers are called **contour feathers.** These feathers are the largest and most familiar feathers. They give birds their streamlined shape. Other feathers, called **down feathers,** are short and fuzzy ③

**Activity**

*Comparing Feathers*

The body of a bird is covered with feathers.

**1.** Obtain a contour feather and a down feather.

**2.** Examine each feather. Locate the main stem or shaft of the feather. Then locate the vane. The vane is attached to the shaft in the same way that the blade of a leaf is attached to a stem.

**3.** Draw labeled diagrams of the feathers.

How are the feathers alike? Different? How is each adapted for its particular function?



## ANNOTATION KEY

① Thinking Skill: Classifying
② Thinking Skill: Making comparisons
③ Thinking Skill: Applying concepts

**Activity**

**Comparing Feathers**
**Skills: Comparing, diagraming, observing, relating**
**Level: Remedial**
**Type: Hands-on**
**Materials: Contour feather, down feather.**

Most feathers have a stiff central shaft, on each side of which is a flat vane. The vane is made up of hundreds of thin parallel branches from the shaft. These branches are called barbs. The largest feathers are the long contour feathers. These feathers give birds their streamlined shape, which helps them when flying. In addition to feathers with vanes, some birds have down feathers. Down feathers have a short shaft and soft, fuzzy barbs that are not connected into veins. The down feathers provide insulation.

amounts of food to maintain the high metabolic rate needed to fly.)

### Skills Development

***Skill: Interpreting illustrations***

Ask some of these questions to help students identify traits of the four bird groups discussed here:

- **What are the four groups of birds?** (Perching birds, water birds, birds of prey, and flightless birds are the four main groups.)
- **What characteristic identifies the macaw in Figure 11-24 as a perching bird?** (Students should notice the feet are adapted for gripping branches.)
- **What are some other types of perching birds?** (Some familiar examples are cardinals, robins, sparrows, etc.)
- **Which bird in Figure 11-25 is a water bird, and how do you know?** (Students should identify the wood duck as a water bird because its feet are adapted as paddles.)
- **For what are the feet of the eagle in the figure adapted?** (Lead students to answer that its feet have sharp claws called talons for grasping prey.)
- **How are the legs and feet of the road runner different from those of the other birds pictured on these pages?** (They are adapted for rapid running.)
- **To what group does this bird belong?** (The road runner is primarily a flightless bird.)

If anyone asks, road runners do not go "beep-beep" like the popular cartoon character.

## FACTS AND FIGURES

The hoatzin, a bird living in rain forests of South America, has been called a "living fossil." Like the extinct *Archaeopteryx*, it has tiny claws on its wings. Also, unlike most other birds, it leaves the nest on its own for short jaunts soon after hatching. While wandering around on tree branches, it uses the claws to grip small twigs to keep from falling.

## ANNOTATION KEY

❶ Thinking Skill: Applying concepts
❷ Thinking Skill: Relating facts

**Figure 11-26** *Male peacocks are famous for their beautiful tail feathers. They display these feathers to attract females.*

and act as insulation. Most birds have down feathers on their breasts. Down feathers are also found covering young birds after they hatch, but before contour feathers have grown in.

### Breeding Habits and Development

Many birds have interesting behaviors. Bird songs are used to establish a territory, an area where the bird lives. The song chases other birds of the same species away. Establishing a territory is important because it ensures that few birds will compete for food and living space in the same area.

The bright feathers of some male birds are used to attract females. However, brightly colored feathers also make the bird more obvious to predators. So many birds sport bright colors only during the breeding season.

Some birds attract a mate by constructing a large and colorful nesting site. Male bowerbirds use bits of colorful material to call attention to themselves and thereby attract a mate. ❶ The male penguin, on the other hand, does not construct a colorful structure. Instead, he presents his intended mate with a pebble. The pebble indicates that he is ready to breed and care for a youngster.

Most birds build nests. These nests can be little more than a space hollowed out in the ground, or they can be quite elaborate like the nest of the weaver, which is actually more like a crowded apartment house in a big city. All of these structures are designed to protect the eggs and the young birds as they develop.

**Figure 11-27** *Birds are the most famous nest-builders. Weavers use their feet and bills to weave their enclosed nests of grass. These nests hang from tree branches.*

274

## 11-4 (continued)

### Content Development

Point out that feathers help a bird fly. The feathers on the body and wings streamline the bird for flight. Explain that in the flying birds, the wings and flight muscles are larger and more effective than those on the flightless birds.

Point out that the feathers on the wings of flying birds form an airfoil to allow the bird to lift from a surface.

Explain that when a bird flaps its wings, it produces an increased pressure on the underside and slightly reduces the pressure on the upperside of the wing. As a result, air spills around the tip of the wing in an upward direction to reduce this difference. The result is an upward lift.

Point out that airplanes are designed to use an airfoil similar to those of flying birds. Explain that lift and streamlining enable birds and airplanes to fly.

### Content Development

Point out that in addition to the contour feathers and down feathers, birds may have another kind of feather called filoplumes. Filoplumes consist of a long staff with just a few barbs at the tip, and they provide added insulation. Also mention that at least once a year birds molt their feathers and grow new ones. In most cases only a few feathers are shed at one time so the birds continue to fly during molting.

### Content Development

Emphasize the varied courtship behaviors of birds. Point out that in most cases these behaviors are instinctive.

- **What is the function of the songs sung by many male birds?** (These songs are used to claim a territory, and by some species to attract mates.)
- **Why are the feathers of many male birds more attractive than the female's?** (The bright plumage serves to attract females.)

Explain that the courtship displays of many male birds are quite

Unlike a reptile's egg, a bird's egg is encased within a hard, strong shell. The shell protects the developing bird and contains food for the bird. The shell, while seemingly quite solid, allows oxygen to pass into the egg and carbon dioxide to pass out.

Most birds incubate their eggs by sitting on them. This behavior keeps the eggs warm as the young develop. However, the penguin keeps its single egg warm by balancing it on the top of its feet and pressing down on the egg with its abdomen. The egg never rests on the cold ground.

Among birds, sometimes only one parent will incubate the eggs. In other species, both parents take turns incubating the eggs. After the eggs hatch, the young birds are quite helpless. They cannot fly to look for food. Their parents must bring them food and water until they are old enough to fly and take care of themselves.

**Figure 11-28** *Young birds are quite helpless. Their parents must bring them food and water until they are able to fly.*

### Migration

Many birds **migrate,** or move to a new environment, during the course of a year. Some birds migrate over tremendous distances. For example, the American golden plover flies more than 25,000 kilometers when it migrates. Birds migrate for many reasons, but perhaps the most important reason is to follow seasonal food supplies.

Some birds fly south in the winter in search of food. The whooping crane, for example, flies south from its nesting area in Canada to winter on the Texas shore. Birds have developed extremely accurate mechanisms for migrating. Scientists have learned that some birds navigate, or find their way, by observing the sun and other stars. Other birds follow coastlines or other natural formations such as mountain ranges. Still other birds are believed to have magnetic centers in their brain. These centers act like a compass to help the bird find its way. 2

### SECTION REVIEW

1. How do bird legs resemble lizard legs?
2. What are the four main categories of birds?
3. List three possible ways birds may navigate across great distances.

#### Activity

*Bird Watching*

Many people find bird watching a relaxing, enjoyable activity. Obtain a field guide to birds from the library. See if you can borrow a pair of binoculars. Then, early in the morning, see how many kinds of birds you can identify in your neighborhood. Make a list of those you spot. Check your list in different seasons to see if it changes.



### Activity

**Bird Watching**
**Skills: Observing, comparing, applying**
**Level: Enriched**
**Type: Field**
**Materials: Binoculars, bird field guide**

This activity requires a little patience, an observant eye, and an interest in birds. Students can use small field guides to identify birds and larger reference guides to learn about the fascinating behaviors and adaptations of birds. Artistic students may make and display color sketches of the birds they observe. Students interested in photography may take and exhibit snapshots for the class.

elaborate. Most students are probably familiar with the male peacock's brilliant tail plumage and its strutting behavior. Birds of paradise also have brilliant plumage, and the males of some species carry out their courtship rituals while hanging upside down from trees.

• **What kind of birds attract mates by building large colorful nests?** (Bowerbirds build nests to attract mates.)

• **How does a male penguin signal a female he is ready to mate?** (He presents a pebble to a female.)

### Content Development

Explain that the migratory behavior of certain birds is an adaptation for feeding and mating. Migratory birds fly to their winter grounds as food becomes scarce. In spring they fly to other regions to mate. Such behavior ensures the survival of more birds as there are fewer predators in the mating and nesting area. Research has shown that the migratory instinct is controlled by hormones. In turn, environmental conditions, such as length of day and temperature trigger the secretion of these hormones.

### Reinforcement

Explain that all nests are designed to protect the eggs and the young birds as they develop. Point out that the female egg is fertilized by the male sperm "inside" the body. Explain that after the eggs are laid, the female usually incubates, or warms the eggs, by sitting on them. During this time the male feeds the female.

Point out that after the eggs hatch, the birds are too small to hunt for food and water until they are old enough to take care of themselves.

### Section Review 11-4

1. Birds' legs have scales.
2. Perching birds, water birds, birds of prey, and flightless birds.
3. Observing the sun and other stars; following coastlines; or having magnetic centers in their brain.

## LABORATORY ACTIVITY CLASSIFYING VERTEBRATE BONES IN OWL PELLETS

### BEFORE THE LAB

1. **Well in advance of the activity, order a sufficient supply of owl pellets for groups of 2–4 students. Owl pellets are available from biological supply companies.**
2. **One day prior to the lab gather all other necessary materials.**
3. **Set up lab teams of 2–4 students.**

### PRE-LAB DISCUSSION

Have students read through the complete laboratory procedure. Use some of the following questions while discussing the procedure:

- **What is the purpose of this laboratory activity?** (To identify undigested bones of small vertebrates in an owl pellet.)
- **To which group of birds do owls belong?** (Most students will know that owls are birds of prey.)

Explain that owls usually feed on small mammals like those listed in the table. They will also eat small birds, and occasionally rodents. Owls swallow their prey whole, and the soft parts are digested in the stomach. But undigestible parts such as hair, teeth, and bones are regurgitated daily as pellets. An examination of an owl's pellet will provide clues to its feeding habits.

- **As you examine the owl pellet, how will you be able to identify the kinds of animals it has eaten?** (By comparing skulls with the pictures in the table.)

### SKILLS DEVELOPMENT

Students will use the following skills while completing this activity.

1. Observing
2. Applying
3. Relating
4. Manipulative
5. Inferring
6. Hypothesizing
7. Classifying
8. Safety
9. Comparing

## LABORATORY ACTIVITY

### *Classifying Vertebrate Bones in Owl Pellets*

**Purpose**

In this activity, you will identify undigested bones of small vertebrates present in a pellet coughed up by an owl.

**Materials** *(per group)*

Owl pellet
Metal dissecting needle or small spatula
Magnifying glass
Small metric ruler
Pencil and paper

**Procedure**

1. Observe the outside of an owl pellet and record your observations.
2. Gently break the pellet into two pieces.
3. Using a metal dissecting needle or small spatula, separate any bones and fur from the pellet. Also remove all fur from any skulls found in the pellet.
4. Group similar bones together in a pile. For example, put all skull bones in one group. Observe the skulls. Record the length, number, shape, and color of their teeth.
5. Now try to fit together bones from the different piles to form complete skeletons.

**Observations and Conclusions**

1. On the basis of your observations of the skulls, use the chart to identify the kinds of animals eaten by the owl.
2. What are the functions of the different bones you found?

| | |
|---|---|
| **Shrew** | Upper jaw has at least 18 teeth.<br>Skull length is 23 mm or less.<br>Teeth are brown. |
| **House Mouse** | Upper jaw has 2 biting teeth.<br>Upper jaw extends past lower jaw.<br>Skull length is 22 mm or less. |
| **Meadow Vole** | Upper jaw has 2 biting teeth.<br>Upper jaw does not extend past lower jaw.<br>Molar teeth are flat. |
| **Mole** | Upper jaw has at least 18 teeth.<br>Skull length is 23 mm or more. |
| **Rat** | Upper jaw has 2 biting teeth.<br>Upper jaw extends past lower jaw.<br>Skull length is 22 mm or more. |

276

### SAFETY TIPS

Caution students to wash their hands after handling the fur and bones.

### TEACHING STRATEGY FOR LABORATORY PROCEDURE

1. If students seem reluctant to handle the owl pellets, tell them they have been cleaned to eliminate any possibility of contamination. Do not force any squeamish students to handle the pellets. If left alone, such students often join in after watching their teammates work for a while.

## CHAPTER REVIEW

### SUMMARY


**11-1 Fish**

- All vertebrates, including fish, have a backbone made up of bones called vertebrae. The backbone is part of their internal skeleton.
- Jawless fish, such as the lamprey, are the most primitive fish.
- Like that of jawless fish, the skeleton of cartilaginous fish is made of flexible cartilage, rather than hard bone. Examples of cartilaginous fish are sharks, skates, and rays.
- Bony fish have a swim bladder that allows them to float at different levels in the water, even when asleep.

**11-2 Amphibians**

- The term "amphibian" means "double life." Amphibians spend part of their lives on land and part in water.
- The eggs of amphibians do not have hard shells. They must be laid in water or they will dry out.
- Frog and toad eggs hatch into tadpoles. Tadpoles have gills and a tail.
- As adults, frogs and toads develop lungs and legs. The powerful hind legs of these animals allow them to leap great distances.
- Like frogs and toads, salamanders and newts have two pairs of legs.
- Salamanders and newts are examples of amphibians with tails.

**11-3 Reptiles**

- The eggs of reptiles have leathery shells and will not dry out when laid on land.
- Most snakes are nonpoisonous. However, those snakes which are poisonous inject their venom through special teeth called fangs.
- Snakes and lizards are covered with a scaly skin. They must shed their skins to grow larger.
- Turtles are adapted for a life in the water. Tortoises are adapted for a life on land.
- Alligators have a broader snout than do crocodiles.

**11-4 Birds**

- Birds, along with bats and insects, are the only animals that can fly.
- The four main groups of birds are perching birds, water birds, birds of prey, and flightless birds.
- The beaks of many birds are adapted for the kinds of foods they eat.
- Contour feathers give birds their streamlined appearance. Down feathers act as insulation.
- Birds have many ways to attract a mate, including songs, brightly colored feathers, and unusual nests.
- Many birds migrate long distances in search of food and warm climates.


### VOCABULARY

*Define each term in a complete sentence.*

**cartilage** **contour feather** **down feather** **hibernation** **migrate** **swim bladder** **talon** **venom** **vertebra**



## GOING FURTHER: ENRICHMENT

**Part 1**

Tell students that owls regurgitate about 2 pellets a day. Then tell them to use the number of skulls found in their pellet to compute the number of animals that an owl might eat in a year.

**Part 2**

Ask the class to review the concept of food webs on page 105. Then ask them to construct a food web showing the relationship between owl and the animals it feeds on. In order to trace the feeding levels back to plants, students may need to do research on the feeding habits of the prey animals.

**2.** Suggest that as the skulls are separated from the pellets, that each one be placed on a separate square of paper. Data on each skull can be written on the same piece of paper.
**3.** If students are able to fit the bones together to form a skeleton, have them glue these skeletons to a piece of heavy cardboard to be displayed.

## OBSERVATIONS AND CONCLUSIONS

**1.** Answers will vary, depending on the kinds of skulls present.
**2.** The long bones generally are used to support the animal and are used in moving. The bones of the skull are used to protect the brain. Some of the bones, such as the teeth and jawbones, are used in feeding. The ribs protect the internal thoracic organs. The vertebrae protect the spinal cord.

## CHAPTER REVIEW

### MULTIPLE CHOICE

**1.** b **2.** d **3.** a **4.** b **5.** c **6.** d **7.** b **8.** c **9.** d **10.** b

### COMPLETION

**1.** vertebrae
**2.** spinal cord
**3.** swim bladder
**4.** amphibian
**5.** hibernation
**6.** females
**7.** venom
**8.** perching
**9.** Down
**10.** migrate

### TRUE OR FALSE

**1.** F Bony
**2.** F jawless
**3.** F cannot
**4.** F moist
**5.** F hind
**6.** T
**7.** F internal
**8.** F on land
**9.** T
**10.** F Down

### SKILL BUILDING

**1.** Students should be able to infer that they are warmblooded vertebrates since their internal body temperature is usually constant and is not affected by external conditions. Also, people do not have to adjust their behavior patterns during warm or cold weather in order to regulate their body temperature.

**2.** The easiest approach students will suggest is to touch the skin. If it is moist, it is an amphibian. If it is dry, it is a reptile.

**3.** Frogs are coldblooded and cannot adapt to the cold environment of the Antarctic. Their body temperature would fall too low, despite any behavior attempts to keep warm such as basking in the sun.

**4.** Fish must produce a great many eggs to ensure the survival of the species. Although they produce a great many eggs, few of the eggs actually develop and result in an adult fish. The reasons for this include disruptions in the environment of the eggs and the fact that the eggs serve as a food source for other water-dwelling organisms.

**5.** Many organisms mistake the king snake for the poisonous coral snake and do not attempt to attack it. Thus, by its coloring, the king snake is protected from predators because of a case of "mistaken identify."

**6.** The best defense mechanism many fish have is to not be seen. Another fish looking down on the dark coloring of the top of the fish would not spot it because the water appears darker when looking down. On the other hand, when looking up, the water appears lighter and a fish with a light-colored bottom will be less obvious.

**7.** Most students will point out that amphibians must spend much of their lives in or near water. Like fish, amphibians lay their eggs in water. Also, amphibians develop gills during the early stages of their life cycle and are able to breathe underwater.

**8.** The toad may secrete a poisonous liquid that may bother the raccoon. By wiping it on the ground, the raccoon removes this liquid before eating the toad.

**9.** Accept all reasonable designs. Make sure students include a variable and a control setup and that the basic steps of the scientific method are followed.

### CONTENT REVIEW: MULTIPLE CHOICE

*Choose the letter of the answer that best completes each statement.*

**1.** All vertebrates have
a. an exoskeleton. b. an internal skeleton. c. scales. d. a bony skeleton.

**2.** The fins of fish help them
a. steer. b. swim upright. c. move through the water. d. all of these.

**3.** Cartilaginous fish do *not* include
a. cod. b. sharks. c. rays. d. skates.

**4.** Tadpoles are the early stage of
a. reptiles. b. frogs. c. newts. d. salamanders.

**5.** Amphibians with tails include
a. frogs. b. toads. c. newts. d. lizards.

**6.** During a frog's life, it breathes through its
a. skin. b. gills. c. lungs. d. all of these.

**7.** A snake *not* native to the U.S. is the
a. coral snake. b. cobra. c. copperhead. d. rattlesnake.

**8.** One kind of poisonous reptile is the
a. turtle. b. chameleon. c. Gila monster. d. crocodile.

**9.** Modern birds and reptiles have
a. feathers. b. teeth. c. bony tail. d. scales on legs.

**10.** Which are warmblooded?
a. fish b. birds c. reptiles d. amphibians

### CONTENT REVIEW: COMPLETION

*Fill in the word or words that best complete each statement.*

**1.** The bones that make up a vertebrate's backbone are called ______________.

**2.** The backbone helps protect the nerves of the ______________.

**3.** Fish are able to float at different levels in the water because of a structure called the ______________.

**4.** The term ______________ means "double life."

**5.** During the cold winter, frogs can go into a winter sleep called ______________.

**6.** The peeps and grunts of male frogs are used to attract ______________.

**7.** The poison injected by certain snakes is called ______________.

**8.** The sparrow is a kind of ______________ bird.

**9.** ______________ feathers act as insulation for birds.

**10.** Many birds ______________ to warmer climates in the winter.

### CONTENT REVIEW: TRUE OR FALSE

*Determine whether each statement is true or false. If it is true, write "true." If it is false, change the underlined word or words to make the statement true.*

**1.** Jawless fish have a hard, bony skeleton.

**2.** The lamprey is a kind of bony fish.

3. Most fish can live in both fresh and salt water.
4. Amphibians live in a dry environment.
5. The front legs of a toad are very powerful and help the toad jump.
6. The tongue of a frog is attached to the front of its mouth.
7. In reptiles, a system of external fertilization has developed.
8. Tortoises are well adapted to a life in the water.
9. Penguins are a kind of flightless bird.
10. Contour feathers provide insulation for birds.

## CONCEPT REVIEW: SKILL BUILDING

*Use the skills you have developed in the chapter to complete each activity.*

1. **Applying definitions** Are you a warm-blooded or a coldblooded vertebrate? Explain your answer.
2. **Classifying vertebrates** In the woods, you discover a small four-legged, cold-blooded vertebrate. How can you tell whether it is an amphibian or a reptile?
3. **Relating facts** There are no frogs living in Antarctica. Explain why.
4. **Making inferences** The sturgeon, a bony fish, can lay up to 6 million eggs. Why do you think it is necessary for most fish to produce so many eggs?
5. **Relating concepts** The poisonous coral snake has alternating bands of black, bright red, and bright yellow. The harmless scarlet king snake has a very similar color pattern. Why is this distinctive pattern an advantage to the king snake?
6. **Developing a hypothesis** A fish has light coloring on its bottom surface and dark coloring on its top surface. Explain how this coloration could be an advantage to the fish.
7. **Applying concepts** Scientists think that amphibians may have developed from a fishlike ancestor. What characteristics of amphibians provide evidence for such a belief?
8. **Relating facts** When a raccoon catches a toad, it usually wipes the amphibian along the ground before eating it. Suggest a reason for this strange behavior.
9. **Designing an experiment** A scientist wants to know if turtles can detect sound. Design an experiment that she can use. Be sure to include a variable and a control in your experiment.

## CONCEPT REVIEW: ESSAY

*Discuss each of the following in a brief paragraph.*

1. Describe the three types of fish.
2. Because the skins of alligators make excellent leather, these animals have been hunted almost to the point of extinction. Do you think such animals should be protected from hunters? Or do people have a right to hunt all animals, even if they are endangered?
3. Why are reptiles able to live in dry environments when amphibians are not?
4. What are the different mechanisms birds may use to navigate long distances?
5. Describe the life cycle of a frog.
6. What are two adaptations that enable birds to fly?
7. Vertebrates that reproduce by internal fertilization usually produce fewer eggs than do animals that reproduce by external fertilization. Explain why.



## ESSAY

**1.** Jawless fish lack scales, fins, and jaws and have skeletons made of cartilage; some, such as the lamprey, suck out the body fluids of other fish. Cartilaginous fish have jaws and skeletons made of cartilage. Bony fish have skeletons made of hard bone.
**2.** Answers will vary, depending on student opinions.
**3.** Reptiles, unlike amphibians, have dry, scaly skins that are resistant to drying out. Their eggs are protected by a leathery shell, so that they do not dry out and need not be laid in water.
**4.** Some birds navigate long distances by observing the sun and other stars. Others follow coastlines or mountain ranges. Still others may have magnetic centers in their brains that are used as a sort of compass.
**5.** Frogs hatch from eggs in the water as tadpoles. The tadpole breathes through gills and feeds on water plants. In time, the tail disappears and legs begin to form. The gills eventually close as lungs form. Soon the tadpole leaves the water as an adult frog.
**6.** Light, hollow bones help birds by cutting down on their weight and the amount of force required for flight. Strong muscles in their wings help them fly. Feathers also help birds fly by streamlining the bird.
**7.** An organism that is going to carry its young within its body obviously could not carry thousands, even millions, of young at one time. Organisms that have external fertilization deposit their eggs outside their body, where they are then fertilized.

## ADDITIONAL QUESTIONS AND TOPIC SUGGESTIONS

**1.** Bony fish can float at any depth in water, but sharks tend to sink if they stop swimming. Can you think of a reason why this is so? (Bony fish have a swim bladder, but sharks lack this structure.)
**2.** Salamanders and lizards have a similar body form. How is it possible to tell them apart? (A salamander's skin is thin and moist, but a lizard has dry, scaly skin.)
**3.** Compare the way in which a king cobra and an emerald tree boa capture their prey. (A cobra is a poisonous snake that injects venom into its prey. A boa is a constrictor that suffocates its victim by squeezing it.)
**4.** Explain why an eagle's feet would be more like those of a hawk than the feet of a robin. (Eagles and hawks are birds of prey; thus their feet have talons. A robin is a perching bird.)

## ISSUES IN SCIENCE

The following issue can be used as a springboard for discussion or given as a writing assignment.

The only method that has been successful in reducing the number of lampreys in the Great Lakes is to poison their young in the freshwater streams where they develop. A favorite fish of sports fishermen is the coho salmon whose young develop in the same streams. These fishermen believe that the poison may also be responsible for reducing the number of coho. Some have urged that the practice of poisoning lampreys be stopped. What is your opinion?

# Chapter 12

# MAMMALS

## CHAPTER OVERVIEW

Mammals are warmblooded vertebrates that have hair and feed their young with milk produced in mammary glands. In addition to feeding their young with milk, most mammals give their young more care and protection than do other animals. Mammals have the most highly developed brains and are believed to be the most intelligent animals on Earth. Because of their well-developed brains, mammals are capable of complex behavior and learning.

Mammals are divided into three main groups: monotremes, marsupials, and placental mammals. Monotremes lay eggs like birds but feed their young with the fluid secreted by the mammary gland. Marsupials are born premature and finish their development in their mother's pouch. Placental mammals are fully developed at birth. Placental mammals have the ability to nourish their young before birth by means of the placenta, which is present in the female body. The placental mammals are divided into about 20 groups according to how they eat, how they move, or where they live.

In this chapter, students will learn about monotremes, marsupials, and the major groups of placental mammals.

## INTRODUCING CHAPTER 12

Direct students' attention to the photograph of the duckbilled platypus. Ask them to describe what they see. Most students will wonder what a ducklike animal is doing in a chapter on mammals. Some will describe the duckbilled platypus as a cross between a duck and a beaver. Now have them read the chapter introduction.

- **What is this animal called?** (duckbilled platypus)
- **Why is this name appropriate?** (It has a bill like a duck's and flat feet.)
- **Why were scientists confused by this animal?** (Accept all reasonable answers.)
- **Why did scientists decide it was a mammal?** (It has hair and nurses its young.)
- **How is the duckbilled platypus different from other mammals?** (Accept all reasonable answers.)

# 12 Mammals

**CHAPTER SECTIONS**



**CHAPTER OBJECTIVES**

*After completing this chapter, you will be able to:*

**12-1** List the main characteristics of mammals.

**12-2** Describe the characteristics of egg-laying mammals.

**12-3** Describe the characteristics of pouched mammals

**12-4** Describe the characteristics of placental mammals.

**12-4** Name ten groups of placental mammals and give an example of each.

Imagine yourself living in London at the end of the eighteenth century. In the newspaper, the headline "Explorers Discover Strange Creature in Australia" catches your eye. As you read on, you find out that when the first dried skins of the creature arrived at London's Natural History Museum, they were thought to be fakes. The creature had thick fur, webbed feet, and a beak like a duck's. No one had ever seen such a strange animal!

Years later, when English scientists examined the complete body of this creature, they discovered it was not a fake. The creature was real, no matter how strange it looked. But what was it? It had a bill like a bird's, feet like a frog's, and fur like a beaver's. Moreover, it laid eggs and fed its young milk. As hard as it might be for you to believe, scientists decided that the mysterious creature belonged to a certain group of animals called mammals, which includes lions, dogs, and people!

When it came time to decide what to call this weird animal, scientists searched for a name that would describe its strange features. First, the animal was called a platypus (PLAT-uh-puhs), which means "flat-footed." Soon after, scientists discovered that this name had already been given to a flat-footed beetle. Scientists made the name more descriptive and called it duckbilled platypus.

*Duckbilled platypus*



## TEACHER DEMONSTRATION

Obtain a live mammal such as a hamster, guinea pig, or mouse. Show the animal to the class. Allow one or two students to touch or hold the animal gently.

- **What do you notice when you touch this animal?** (It is warm and has fur.)
- **Warmblooded animals that have hair are mammals. How many of you have pet mammals? What kind of mammal do you have?** (Answers will vary. You may wish to write the names of the different mammals on the board.)
- **Has your pet ever had babies? How did your pet take care of its babies?** (Answers will vary. Lead students to note that their pet nursed its young.)

Tell the class that mammals, as well as having hair and being warmblooded, also feed their young with milk from mammary glands.

- **What are some other animals that are mammals?** (Write all appropriate answers on the board.)
- **How are these animals alike?** (They have hair, are warmblooded, and feed their young with milk. Other answers are also appropriate.)
- **How are these animals different?** (Accept all reasonable answers.)

Tell students that they will learn more about the trait of mammals in this chapter.

## TEACHER RESOURCES

### Audiovisuals

*Marine Life Preview,* film, Marineland of Florida

*Rare and Endangered Mammals of the World,* slides, Carolina Biological Supply Co.

### Books

MacDonald, D. (ed.), *The Encyclopedia of Mammals,* Facts on File

Walker, E. P., *Mammals of the World,* 3rd ed., Johns Hopkins University Press

Young, J. Z., *The Life of Vertebrates,* 2nd ed., Clarendon Press

### Software

*Classifying Mammals,* Prentice Hall

## 12-1 CHARACTERISTICS OF MAMMALS

### SECTION PREVIEW 12-1

Mammals are warmblooded vertebrates that have hair, and feed their young with milk produced in mammary glands. Although the 4000 species of mammals look very different from one another, they have several important traits in common. Mammals have brains that are better developed than those of any other group of animals. Mammals are believed to be the most intelligent animals on Earth.

In all mammals, fertilization takes place internally. However, the offspring of mammals develop in several different ways. The differences in mammals' patterns of development are the basis for the classification of mammals into three groups: egg-laying mammals (monotremes), pouched mammals (marsupials), and placental mammals.

### SECTION OBJECTIVES 12-1

1. **Describe the main characteristics of mammals.**
2. **Explain why mammals are thought to be the most intelligent animals on Earth.**
3. **List the three basic groups of mammals.**

### SCIENCE TERMS 12-1

**mammary gland p. 283**

## 12-1 Characteristics of Mammals

A few hundred meters off the coast of California, a small group of animals playfully swim around one another. These whiskered animals are called sea otters. Sea otters spend most of their lives swimming in the ocean. While lying on its back, a sea otter can pound a closed shell against a rock balanced on its chest. The sea otter uses the rock to crack open the shell so that it can eat what is inside.

The sea otter seems rather intelligent. To prevent itself from being swept away by waves, the sea otter wraps itself in strands of giant seaweed growing offshore. The otter uses the seaweed much as an ocean liner uses giant ropes to hold itself close to a pier.

Sea otters belong to a group of vertebrates called mammals. There are about 4000 kinds of mammals on the earth. In addition to sea otters, duckbilled platypuses, lions, dogs, and people, this group also includes such animals as whales, bats, and elephants. Since scientists group together animals that have common characteristics, you might wonder what such different-looking animals could have in common.

Mammals have characteristics that set them apart from all other living things. **Mammals are warmblooded vertebrates that have hair and feed their young with milk produced in mammary glands.** In fact, the word "mammal" comes from the term

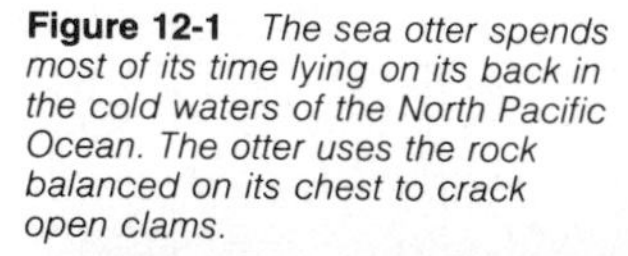

**Figure 12-1** *The sea otter spends most of its time lying on its back in the cold waters of the North Pacific Ocean. The otter uses the rock balanced on its chest to crack open clams.*

### TEACHING STRATEGY 12-1

#### Motivation

Have students rub their hand up and down their arm.

- **What do you feel?** (Accept all logical answers. Lead students to realize they can feel the hair on their arms.)

Divide students into groups of two or three. Have students look at the hair on their arms without a magnifying glass.

- **What do you see?** (Accept all logical answers.)

Give each group a magnifying glass. Have students observe the hair and pores of their arms through the magnifying glass.

- **What does it look like now?** (Accept all logical answers.)
- **Does the amount of hair vary from person to person?** (yes)
- **Do you think humans are unique because of the hair that is on their bodies?** (Accept all logical answers.)

#### Content Development

Point out that humans as well as lions, bears, bats and seals all belong to a class of chordates called mammals. Explain that mammals are warmblooded vertebrates that have hair and feed their young with milk produced in mammary glands. Tell students that the mammary gland is a special gland in a mammal that produces milk.

Figure 12-2 *As with most mammals, a thick coat of hair covers the body and provides warmth for llamas.*

**mammary gland.** Another special characteristic of mammals is that they give their young more care and protection than do other animals. ❷

At one time during their lives, all mammals possess fur or hair. The fur or hair, if it is thick enough, acts as insulation and enables some mammals to survive in very cold parts of the world. Mammals also can survive in harsh climates because, like birds, mammals are warmblooded. The body temperature of mammals remains almost unchanged no matter how the temperature of their surroundings may change. ❸

Mammals are believed to be the most intelligent animals on the earth. This intelligence comes from a brain that is better developed than that of any other group of animals. Human beings, for example, have the most well-developed brain of all mammals and are considered to be the most intelligent of living things.

As in reptiles and birds, fertilization in mammals is internal. But despite the characteristics that all mammals have in common, their young develop in different ways. This difference can be used to place mammals into three basic groups: egg-laying mammals, pouched mammals, and placental mammals. ❹

### SECTION REVIEW

1. List the main characteristics of mammals.
2. What substance is produced by mammary glands?

#### Activity

*Pet Mammals*

If you have a pet cat, dog, mouse, gerbil, guinea pig, horse, or pig, then your pet belongs to the class Mammalia. Make observations of your pet mammal for a few days. On a sheet of paper, make a list of your pet's behaviors. Next to each item, record what the behavior means to you. Do you think other animals of the same species behave the same way? How could you go about finding out if they do?



## ANNOTATION KEY

❶ Thinking Skill: Making comparisons
❷ Thinking Skill: Making generalizations
❸ Thinking Skill: Applying concepts
❹ Thinking Skill: Classifying mammals

### Activity

**Pet Mammals**
**Skills: Observing, applying, relating, comparing, recording**
**Level: Average**
**Type: Hands-on/library**

This activity provides an excellent means of motivating students to learn more about mammals and to extend their powers of observation by studying the nonhuman living things with which they are most familiar. Students will enjoy reporting on the behavior of their pets. By doing library research they can compare this behavior with that of the same types of animals found in nature in the undomesticated state. You may wish to augment this activity with observation of some in-classroom pets, such as gerbils, hamsters, and guinea pigs. Students can care for the animals, keep records of the amounts of food and water they consume, observe their development, and so on.

Point out that the structure of hair changes on some mammals. Point out that the wool of sheep, the quills of porcupines, and the armor of armadillos are all modified types of hair.

## Skills Development

***Skill: Making graphs***

Have students prepare a bar graph comparing the number of species in the five basic vertebrate groups.

| *Group* | *Approximate Number of Species* |
|---|---|
| Fish | 30,600 |
| Amphibians | 3,000 |
| Reptiles | 4,200 |
| Birds | 8,500 |
| Mammals | 4,000 |

Provide graph paper and help students set up an appropriate scale. Remind students to label the axes and bars and to put a title on their graph. Suggest that each bar be colored differently and that a key to the colors or shading be included.

## Section Review 12-1

1. Hair or fur, warmblooded, feeding young with milk from mammary glands, well-developed brains.
2. Milk.

## 12-2 EGG-LAYING MAMMALS

### SECTION PREVIEW 12-2

Monotremes are found only in Australia, New Guinea, and a few nearby islands. There are only three species of monotremes: the duckbilled platypus, the common (or short-beaked) echidna, and the long-beaked echidna. Echidna are more commonly known as spiny anteaters. Monotremes are considered the most primitive mammals because their young develop for a short time within an egg and because they have a cloaca, a chamber into which the excretory, digestive, and reproductive systems empty. The single opening from the cloaca to the outside of the body gave the order its name, Monotremata—"one holed."

Both the spiny anteater and the platypus have special structures for feeding. The spiny anteater has a long, thin snout that is used to probe for food and a sticky tongue that catches insects and termites. The platypus's soft bill is covered with skin that contains many nerves. The platypus uses its bill to locate and find its way around underwater.

### SECTION OBJECTIVES 12-2

1. **Name the two kinds of egg-laying mammals.**
2. **List the major traits of the egg-laying mammals.**
3. **Compare the feeding habits of the spiny anteater and platypus.**

### SCIENCE TERMS 12-2

**monotreme p. 284**

#### Activity

*Vertebrate Body Systems*

Vertebrates have well-developed body systems. However, these systems are not the same in all vertebrates. Choose one of the systems listed below. Use posterboard and colored pencils to draw and label how a particular system changes from fish to amphibians to reptiles to birds to mammals.

nervous system
digestive system
circulatory system
reproductive system
excretory system

## 12-2 Egg-Laying Mammals

A strange-looking animal called the spiny anteater lives in Australia. It has long claws, a tubelike nose, fur, and short, stiff spines like those of a porcupine. See Figure 12-3. What makes the spiny anteater even stranger is that although it is a mammal, it lays eggs! **Mammals that lay eggs are called monotremes.** The spiny anteater and the duckbilled platypus are the world's only **monotremes.**

Just after the female spiny anteater lays her eggs, she places them into a pouch on her abdomen. The eggs hatch in seven to ten days. The young spiny anteaters feed on milk produced by their mother's mammary glands. Milk production, you may recall, is characteristic of mammals.

The female platypus lays her one to three eggs in a burrow that she digs in the side of a stream bank. She stays with the eggs to keep them warm until they hatch. After the eggs hatch, the young platypuses feed on milk produced by their mother.

The unusual body parts of the spiny anteater help it gather its food, which consists of ants and termites. For example, its nose holds a sticky, wormlike

**Figure 12-3** *This porcupinelike animal is actually a spiny anteater. Unlike all other kinds of mammals, the spiny anteater and the duckbilled platypus do not give birth to live young. Instead, they lay eggs that have a leathery shell.*

### TEACHING STRATEGY 12-2

#### Motivation

Obtain photographs and/or pictures of monotremes. Have students examine these photographs and pictures.

• **What kinds of animals do these remind you of?** (Accept all reasonable answers)

• **How are these animals similar to other mammals?** (They have hair. Lead to student to infer that the monotremes also share other mammalian traits.)

• **What sort of things can you tell about these animals just by looking at them? How can you tell?** (Accept all reasonable answers. Students might guess that spiny anteaters are protected by their spines, that platypuses swim because they have webbed feet, and that spiny anteaters can only eat small things because their mouths are small.)

#### Content Development

Monotremes and certain shrews are the only venomous mammals. In the spiny anteater, the structure that produces and delivers the venom is vestigial. However, male platypuses have a bony spur on the back of each ankle. This spur is connected to a venom-producing gland. A jab from the spur can kill a dog and cause severe pain in a human.

tongue that flips out to catch insects. When in danger, the spiny anteater uses its short powerful legs and curved claws to dig a hole in the ground and cover itself until only its spines are showing. These 6-centimeter-long spines scare almost any enemy!

Just as the spiny anteaters have special structures to search for and eat insects on land, the duckbilled platypus has special structures for hunting small animals under water. The platypus is a very good swimmer that has webbed paws and waterproof fur. When underwater, the platypus keeps its eyes closed and uses its soft bill to feel around for snails, mussels, worms, and sometimes small fishes.

### *SECTION REVIEW*

1. What is another name for egg-laying mammals?
2. Name the two egg-laying mammals.

## 12-3 Pouched Mammals

**Mammals with pouches are called marsupials.** Most people think of a **marsupial** (mahr-soo-pee-uhl) as a kangaroo. But kangaroos are not the only marsupials. Some marsupials resemble weasels or

**Figure 12-4** *The bandicoot (left) and wombat (right) are examples of marsupials, or pouched mammals. These mammals live in Australia or nearby islands.*

### Activity

**Vertebrate Body Systems**
**Skills: Comparing, relating, applying**
**Level: Enriched**
**Type: Library/hands-on**
**Materials: Posterboard, colored pencils**

Students can easily find out about each of the vertebrate body systems by examining a biology text. Students will find that as they progress from fish to amphibians to reptiles and finally to birds and mammals that the body systems become more complex. For example, the nervous system of all vertebrates is made up of a spinal cord and brain. In fish, amphibians, reptiles, and birds, the cerebrum's surface is smooth. In mammals, the cerebrum's surface is highly convoluted.

## Section Review 12-2

1. Monotremes.
2. Spiny anteater and duckbilled platypus.

## TEACHING STRATEGY 12-3

### Motivation

Tell students that an adult common opossum is about the same size as a cat. However, a newborn opossum is about 1 cm long, while a newborn kitten is about 7 cm long. If possible, obtain photographs of newborn kittens and newborn opossums to show the class.

- **Which animal is larger at birth?** (cat)
- **Which animal would you expect to be less developed at birth? Why?** (opossum, because it's much smaller)
- **Where do mammals develop before birth?** (inside the mother)
- **Which animal spends less time developing inside the mother?** (opossum)

Point out that a young opossum must undergo much of its development outside the mother's body.

- **How might the opossum care for its newborn offspring?** (Accept all reasonable answers. Lead students to suggest that an opossum needs a place where the young can be kept warm and protected while they develop.)

### Content Development

Point out that animals with pouches are called marsupials. Explain that marsupials are born only partially developed. After birth, the newborn babies are immediately put into this special body pouch near the mother's nippled mammary glands. Tell students that marsupial babies feed on the milk from the mother's mammary glands until they are developed enough to leave the pouch.

## 12-3 POUCHED MAMMALS

### SECTION PREVIEW 12-3

The pouched mammals, or marsupials, give birth to young that are only partially developed. The young marsupial completes its development inside its mother's pouch, where it feeds on milk from the mammary glands. Familiar marsupials include koalas, kangaroos, and opossums.

### SECTION OBJECTIVES 12-3

1. **Identify the major characteristic of marsupials.**
2. **Describe three familiar marsupials.**

### SCIENCE TERMS 12-3

**marsupial p. 285**

### ANNOTATION KEY

1. Thinking Skill: Making comparisons
2. Thinking Skill: Relating cause and effect
3. Thinking Skill: Making comparisons

Figure 12-5 *The young of some marsupials, such as the koala, hitch rides on their mother's back.*

Figure 12-6 *The most familiar marsupial is the kangaroo. A baby kangaroo, called a joey, rests in its mother's pouch.*

(1) teddy bears. There are also marsupials that look like rats and mice.

One of the cuddliest and cutest of marsupials is the koala (koh-AH-luh). See Figure 12-5. The koala's ears are big and round and are covered by thick fur. Unlike most marsupials, the opening of the koala's pouch faces its hind legs rather than its head. Eucalyptus leaves are about the only type of food that koalas eat. Oils from these leaves are an ingredient in cough drops. So you should not be surprised to (2) learn that koalas actually smell like cough drops! Because koalas eat only plant material, they are called **herbivores** (HER-buh-vorz).

#### Kangaroos

"At best it resembles a jumping mouse but it is much larger." These words were spoken in 1770 by the famous English explorer James Cook. He was describing an animal never before seen by Europeans. He later gave the name "kangaroo" to this strange animal.

Kangaroos live in the forests and grasslands of Australia. They have short front legs but long, muscular hind legs and tails. The tail helps the kangaroo to keep its balance and to push itself forward.

When a kangaroo is born, it is only 2 centimeters long and cannot hear or see. Although the kangaroo is only partially developed at birth, its front legs enable it to crawl as much as 30 centimeters to its mother's pouch. How the newborn kangaroo finds its way is a mystery. It would be like a person being blindfolded and earplugged and placed in the center of a strange room the size of half a football field. The person would then be asked to find a single doorway on the first try. After the young kangaroo crawls into the pouch, it stays there for about nine months, feeding on its mother's milk.

#### Opossums

Have you ever heard the phrase "playing possum" and wondered what it meant? When in danger, opossums lie perfectly still, pretending to be dead. In some unknown way, this behavior helps to protect the opossum from its predators.

286

### 12-3 (continued)

#### Enrichment

Although students are probably familiar with koalas and kangaroos, they may not be aware of the many Australian marsupials that resemble commonplace placental mammals. Have students prepare reports on a marsupial such as the Tasmanian wolf, marsupial mole, cuscus, bandicoot, or sugar glider. The report should include information on the similarities and differences between the marsupials and their placental counterparts.

#### Section Review 12-3

1. A pouched mammal; koala, kangaroo, opossum.
2. An animal that eats plants.
3. Opossum.

### TEACHING STRATEGY 12-4

#### Motivation

Some students may have been present at the birth of puppies, kittens, or other animals. Ask them to describe what they observed.

- **How was the offspring attached to its mother?** (by the umbilical cord)

Explain that the other end of the umbilical cord is attached to the placenta, which is expelled after the offspring is born.

- **What is another name for the placenta?** (afterbirth)
- **What does the placenta do?** (Accept all logical answers)

#### Content Development

Explain that placental mammals are classified according to how they eat, how they move, and where they live.

Figure 12-7 *The only marsupial that is native to North America is the opossum. Young opossums cling to their mother as she moves from place to place.*

The opossum is the only marsupial found in North America. It lives in trees, often hanging onto branches with its long tail. Opossums eat fruits, insects, and other small animals.

Female opossums may give birth to as many as 24 opossums at one time. Opossums are partially developed at birth and crawl along their mother's abdomen to her pouch. The newborn opossums are so tiny that all 24 of them could fit into a single teaspoon!

### SECTION REVIEW

1. What is a marsupial? Give two examples.
2. What is an herbivore?
3. Name the only North American marsupial.

## 12-4 Placental Mammals

**Unlike those of egg-laying and pouched mammals, the young of placental mammals develop totally within the female.** The females in this group of mammals have **placentas** (pluh-SEN-tuhz). The placenta is a structure through which the developing young receive food and oxygen while in the mother's body. The placenta also removes wastes from the developing young. After they give birth, female placental mammals, like all mammals, supply their young with milk from mammary glands.

There are more than 20 groups of placental mammals. They are grouped according to how they

Figure 12-8 *Like all mammals, placental mammals feed their young with milk from mammary glands.*

## 12-4 PLACENTAL MAMMALS

### SECTION PREVIEW 12-4

The young of placental mammals develop totally within the female. The placenta is a structure through which the developing young receive food and oxygen while inside the mother. Placental mammals are found all over the world.

There are more than twenty groups of placental mammals. Some are carnivorous; others are herbivorous. They live on land and in the seas. The placental mammals are widely varied in structure from the insect-eating mammals to the primates. They are grouped according to how they eat, how they move, or where they live.

### SECTION OBJECTIVES 12-4

1. **Describe the characteristics of placental mammals.**
2. **Classify ten groups of placental mammals and give an example of each.**
3. **Compare placental mammals to monotremes and marsupials.**
4. **Compare carnivorous and herbivorous placental mammals.**

### SCIENCE TERMS 12-4

**placenta p. 287** **incisor p. 290**
**canine p. 290** **carnivore p. 290**

If students have difficulty answering, suggest they look at some of the figures in Section 12-4.

- **Some marsupials, such as koalas and kangaroos, are herbivores. What are some placental mammals that are herbivores?** (Answer will vary.)
- **What placental mammals are meat eaters, rather than herbivores?** (Accept all sensible answers. Students may mention humans. If so, point out that animals that eat both vegetation and meat are called omnivores.)
- **Why can the leopard seen in Figure 12-14 move more rapidly on land than the walrus in Figure 12-13?** (Students should answer that the limbs of the leopard are adapted for running rapidly to catch prey. Walruses have flippers for moving in water.)
- **What are some of the different places where placental mammals live?** (Accept all answers. Again, if students are unsure, have them look at some of the figures in the text.)

### Content Development

Point out that unlike egg-laying and pouched mammals, the young of placental mammals develop totally within the female. Explain that the placenta is the life-giving organ that is developed within the female mammal. Tell students that a placenta is a thick membranous organ that connects the mother to the baby. Explain that through the membranes, food and oxygen are carried to the baby from the mother and metabolic wastes are removed.

## FACTS AND FIGURES

Some species of shrews are venomous. One shrew may produce enough poison in its salivary glands to kill 200 mice. The shrew's venomous bite allows it to subdue prey considerably larger than itself such as fish and frogs.

## TIE-IN/ LANGUAGE

Because of the feisty nature of shrews, irascible people are sometimes referred to as "shrews." Some other words and expressions can also be traced to shrews. For example, the word "shrewd" originally meant a cunning, villainous person, though its present meaning is somewhat different.

**Figure 12-9** *Mammals that give birth to well-developed young are called placental mammals. Grizzly bears are examples of placental mammals.*

eat, how they move about, or where they live. Of these groups, ten are discussed in the sections that follow.

### *Insect-Eating Mammals*

What has a nose with 22 tentacles and spends half its time in water? If you have given up, the answer is the star-nosed mole. The star-nosed mole gets its name from the ring of 22 tentacles on the end of its nose. See Figure 12-10. No other mammal has this structure. Each tentacle has very sensitive feelers, which enable the mole to find insects to eat and to feel its way around. Even though these moles have eyes, they are too tiny to see anything. The star-nosed mole spends part of its day burrowing beneath the ground and the other part of the day in the water. Star-nosed moles are found in northeastern United States and eastern Canada.

In addition to moles, hedgehogs and shrews are also insect-eating mammals. Because they are covered with spines, hedgehogs look like walking cactuses. When threatened by predators, hedgehogs roll up into a ball with only their spines showing. This action makes the hedgehog's enemy a little less enthusiastic about disturbing the tiny animal.

One type of shrew is the smallest mammal in the world. This animal, called the pygmy shrew, weighs

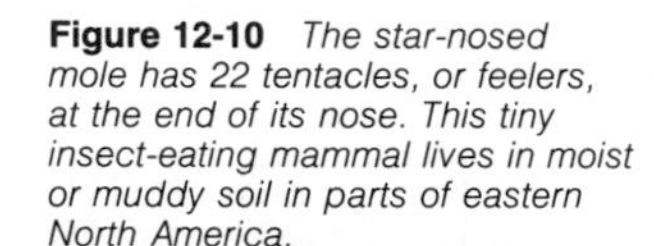

**Figure 12-10** *The star-nosed mole has 22 tentacles, or feelers, at the end of its nose. This tiny insect-eating mammal lives in moist or muddy soil in parts of eastern North America.*

### 12-4 (continued)

#### Motivation

Have students look at the photograph of the star-nosed mole.

- **How did this animal get its name?** (the ring of tentacles at the end of its nose)
- **How does this structure enable the star-nosed mole to catch food, even with poor eyesight?** (The tentacles help the mole find insects.)
- **Where might this mole be found?** (underground)
- **Look at the mole's feet. How are they adapted to the mole's life style?** (The feet are specialized for burrowing or paddling in water)

#### Content Development

Point out that the insect-eating group of placental mammals includes the moles, shrews, and hedgehogs. Explain that moles eat grubs and worms. Shrews eat grubs, worms, mice, and other shrews. Hedgehogs eat almost all available invertebrates, fruit, eggs, and, occasionally, small vertebrates.

#### Motivation

Have students observe Figure 12-12.

- **How many of you have seen a bat, either outside or in a zoo?** (Accept all answers.)
- **What did you notice about the bat?** (Accept all answers.)

#### Content Development

Point out that the skin of the bat is stretched over and between the toe bones of the front legs, as well as attached to the side of the body, the back legs, and the tail. Explain that this stretched skin gives a bat a large wingspread.

Explain that because the skin covers both the front and back legs that a bat cannot walk well. Tell students that a bat's eyesight is very poor. Explain that a bat can maneuver because it listens for the echoes of its own highpitched voice. Point out that as the echo returns to the bat it can adjust its flying pattern so that it doesn't bump into anything.

#### Content Development

Point out that all bats hunt for food at night and rest during the day. However, different species fit into one of three categories in regard to

Figure 12-11 *Insect-eating mammals include the hedgehog* (left) *and shrew* (right). *The hedgehog curls up into a ball when threatened by predators. To keep healthy, shrews must eat twice their weight in insects each day.*

only 1.5 to 2 grams—as an adult! Because shrews are so active, they must eat large amounts of food to maintain their energy. Shrews can eat twice their own weight in insects each day! ①

### *Flying Mammals*

Bats look like mice and are the only flying mammals. In fact, the German word for bat is "Fledermaus," which means flying mouse. Bats are able to fly because they have skin stretched over their arms and fingers, which form wings. Other mammals such as the flying squirrels do not really fly. They simply glide to the ground after leaping from high places. ②

Although a bat's eyesight is poor, its hearing is excellent. While flying, bats give off high-pitched squeaks that people cannot hear. These squeaks bounce off nearby objects and return to the bat as

Figure 12-12 *Bats are the only flying mammals. This fruit-eating bat is carrying a fig in its mouth.*

289

## ANNOTATION KEY

① **Thinking Skill: Relating facts**

② **Thinking Skill: Making comparisons**

## FACTS AND FIGURES

Some shrews exhibit a behavior known as "caravanning." When young shrews leave the nest, they form a line. Each young shrew bites the rump of the one in front of it. The offspring in the front of the line grips the mother's rump. The young shrews hold on so tightly that the entire caravan can be lifted off the ground by picking up the mother.

their social organization and hunting behaviors. One group consists of bats that roost in large groups in caves or trees. At night they all leave together to hunt, and at dawn they return to their roost at the same time. Another group roosts in caves, hollow trees, or buildings with many others of their species. But they go out individually to hunt and they return at different times. A third category is comprised of individualists. They live and hunt alone or in small family groups, and roost in nooks and crannies of all sorts.

## Motivation

Have the class look carefully at the structure of the bat's wing in the photograph.

- **How is the bat's forearm similar to yours?** (Students should notice that a bat has both an upper arm and lower arm, as well as fingers.)

Point out that bats also have a claw on their thumb. They can use the claw to help them move around while roosting. Some also use them to hold and manipulate food.

- **What makes up the bat's wing?** (skin stretched over the legs, arms, and fingers)

Explain that the wings stretch over both the arms and legs, and they are attached to the side of the body as well. In some species, the wing membranes are also attached to the tail. Some bats are unable to use their legs for walking, but others can run and leap very quickly.

- **How are a bat's wings different from those of birds?** (no feathers; different bones supporting the wings)

## Skills Development

***Skill: Making predictions***

Present this hypothetical experiment and have students predict the outcome. "Suppose that in an experiment, some biologists cover the eyes of 100 bats, and place them in a cage filled with moths. They cover the ears of a second group of 100 bats, and place them in another cage of moths."

- **Which group of bats will likely eat the most moths?** (Students should predict that the most moths will be eaten by the bats with covered eyes, since bats depend on sound rather than sight to locate food.)

## Activity

**The Fastest Runner**
**Skill: Making calculations**
**Level: Average**
**Type: Computational**

First, students must calculate how many seconds are in an hour by using the following equation:

$$1 \text{ hr} \times \frac{60 \text{ min}}{\text{hr}} \times \frac{60 \text{ sec}}{\text{min}} = 3600 \text{ sec}$$

Then they must divide 100 kilometers by 3600 seconds to get the answer of .28 kilometers/sec.

**Figure 12-13** *Although most mammals are herbivores, or plant eaters, some are carnivores, or flesh eaters. Flesh eaters have large pointed teeth called canines. Walruses have especially large canines, called tusks.*

echoes. By listening to these echoes, the bat knows where objects are. Bats that hunt insects such as moths also use this method to find their prey.

There are two types of bats: fruit eaters and insect eaters. Fruit-eating bats are found in tropical areas, such as Africa, Australia, India, and the Orient. Insect-eating bats live almost everywhere.

### *Flesh-Eating Mammals*

The ground is frozen solid with ice. Great icebergs move slowly in the sea. The shoreline is almost impossible to see through the ice and snow. The chilling wind makes the air feel colder than it already is. Yet, even in this brutal Arctic environment, there is life. In the distance, a group of walruses pull themselves out of the water onto a sheet of floating ice. As they stretch out on the ice, these large animals resemble sunbathers on a beach.

The frozen sheets of ice and freezing waters of the Arctic are home for the walruses. They are able to live in this freezing environment because they have a layer of fat, called blubber, under their skin that keeps body heat in. In addition to their large size, another very noticeable feature of walruses is their long ivory tusks. These tusks are really special teeth called **canines** (KAY-nighns) that point downward and may grow to 100 centimeters in length.

You also have canines; two in your top set of teeth and two on the bottom. To locate your upper canines, look at your top set of teeth in a mirror. Find your **incisors** (in-SIGH-zers). They are your four front teeth, which are used for biting. On one side of the incisors is a tooth that comes to a point. These teeth are your canines.

Unlike most flesh-eating mammals, walruses do not use their canines for tearing and shredding meat. Instead, the walruses use their tusks to defend themselves from polar bears. The walruses also use their tusks as hooks to help them when climbing onto ice.

All of the mammals in this group, including the walruses, are **carnivores** (KAR-ni-vorz), or flesh eaters. Carnivores are predators. Most carnivores, such as lions, wolves, and bears, have very muscular legs that help them to chase other animals. Carnivores

#### Activity

*The Fastest Runner*

The cheetah is the fastest land animal and can run at speeds of up to 100 kilometers per hour. How far can it run in a second?



## 12-4 (continued)

### Motivation

Display photographs that show the variety among carnivores. Try to include member of all the carnivore groups.

- **What do these animals have in common?** (Accept all logical answers, but lead students to suggest they are all flesh eaters.)
- **What is another name for flesh-eating mammals?** (carnivores)
- **What traits help carnivores catch their prey?** (Accept all logical answers.)

### Skills Development

***Skill: Relating facts***

Tell students that though the terms "carnivore" or "carnivorous" are used to describe flesh-eating mammals, they are also used for nonmammals that eat flesh. Ask students to list five kinds each of fish, amphibians, reptiles, and birds that are carnivores. After the individual lists are completed, discuss the various examples with the class. Students may also be able to recall examples of carnivorous invertebrates.

### Content Development

Explain that many mammals classified as carnivores also include vegetable matter in their diets. In fact, many bears are largely herbivores that eat meat only on occasion. However, they are still classified as carnivores because of the presence of the enlarged canine teeth that makes it possible for them to eat flesh.

### Motivation

Have students rub their tongue across the edge of their upper teeth.

- **What do you feel?** (Accept all logical answers.)

Point out that the center four teeth are called incisors and are used for biting. Explain that the longer pointed teeth on either side are called canines.

Figure 12-14 *Although most mammals run from a predator, this baboon battles a leopard face to face.*

also have sharp claws on their toes to help them hold their prey. Most land-living carnivores include any members of the dog, cat, and bear families. Sea-living carnivores include otters, sea lions, and seals.

## Career: *Animal Technician*

**HELP WANTED: ANIMAL TECHNICIAN** Completion of a two-year animal technology program required. Experience handling animals and knowledge of laboratory procedures desirable. Needed to assist veterinarian.

A veterinarian and assistant are called to a farm to treat a horse. The horse had difficulty chewing and swallowing and now is shaking violently. The veterinarian discovers that the horse was injured by a nail sticking out of a board in the horse's stall.

The vet's assistant prepares the medicine that will make the horse well. Then the vet injects the medicine into the horse. The assistant records the visit.

The vet's assistant is an **animal technician,** someone who has had training in assisting vets and working in laboratories and animal research. They work on farms, in kennels and pounds, in hospitals, and in laboratories.

Animal technicians must work under the supervision and instruction of a veterinarian. An animal technician's duties include record keeping, specimen collection, laboratory work, and wound dressing. They also help with animals and equipment during surgery.

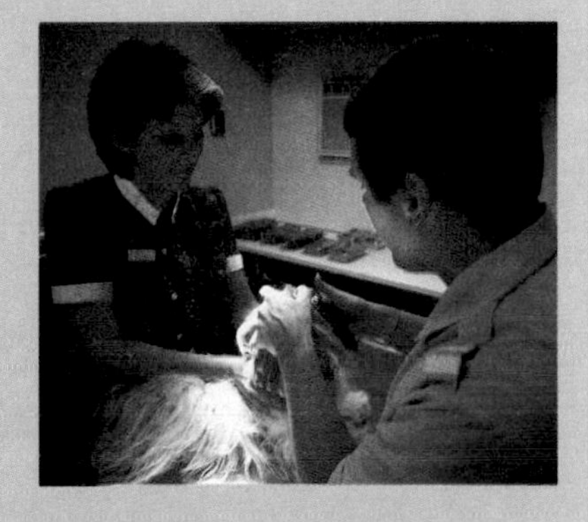

Patience, compassion, and a willingness to be involved with animal health care are important qualities for one interested in becoming an animal technician. To receive more information about this career, write to the American Veterinary Medical Association, 930 North Meacham Road, Schaumburg, IL 60196.



## SCIENCE, TECHNOLOGY, AND SOCIETY

Throughout history, mammal skins have been used for clothing humans. When the last Ice Age ended about 10,000 years ago, people on the North American continent used skins from saber-toothed cats, bears, wolves, and woolly mammoths to keep them warm. Native Americans later used the pelts of elk, deer, buffalo and bear for clothing. Early colonists continued these uses, and the European settlers of North America brought with them a tradition of making clothing out of spun fibers such as sheep's wool.

Modern technology has created synthetic clothing materials such as nylon and polyester. Many of these synthetic materials are derived from nonrenewable natural resources like petroleum. Today, Americans wear clothing made from these synthetic fibers as well as fur coats, leather goods, and woolen products.

Have students analyze the different kinds of clothing they wear according to the natural resources from which they were derived. What kinds of impacts do their clothing preferences have on the environment? In their opinion, which kinds of clothing have the least damaging impact on the environment? Should any products or practices related to the clothing industry be prohibited?

### Content Development

Point out that all flesh-eating mammals, called carnivores, have sharp pointed teeth called canines. Explain that canine teeth are used for tearing and shredding meat.

Tell students that carnivores are predators.

- **What is a predator?** (Accept all answers.)

Explain that a predator is an animal that kills and eats other animals.

Point out that the sea mammal carnivores include the seal, walrus, and sea lion. Explain that the land carnivores are divided into three subgroups. One subgroup includes the bear and raccoon, which walk flat-footed on the soles of their feet. The second subgroup includes carnivores that walk on their toes, such as the cat, dog, wolf, and tiger. The third subgroup walks with a combination, partly on the toes and partly on the soles of the feet. This third subgroup includes the skunk, otter, mink, and weasel.

### Skills Development

***Skill: Identifying patterns***

Have students research and make drawings of the footprints of different carnivorous mammals. Have them identify the mammal and the subgroup of walking style on the back of the drawing. Have other students observe and identify the type of walking style by the drawings.

## FACTS AND FIGURES

Some prehistoric "toothless" mammals, or edentates, were enormous. Certain giant ground sloths grew to the size of modern elephants, and were about 6 m long. An armored, armadillolike edentate was about 5 m long and carried a 3 m shell on its back.

## TIE-IN/ MEDICINE

Armadillos are thought to be the only nonhuman animals that can catch Hansen's disease, or leprosy. Nine-banded armadillos are often used in research of this disease, since the four offspring in a litter are genetically identical.

### Activity

**Useful Mammals**
**Skills: Comparing, relating, recording**
**Level: Average**
**Type: Library**

This is a good activity for slower students as well as average students to work on, because it not only is interesting and relevant to human beings but also provides an opportunity for reviewing the particular characteristics that led to the specific uses of the animals. You can also extend it to bring in the issues of ecological relationships and conservation. Interesting class discussion and debate can be generated as a result of gathering information on this subject.

**Figure 12-15** *The armor-plated armadillo* (left) *and two-toed sloth* (right) *are examples of toothless mammals.*

### *"Toothless" Mammals*

Although "true" anteaters belong to this group of mammals, they are the only members of the group that actually have no teeth. The other members, the armadillos and the sloths, have poorly developed teeth.

Unlike the spiny anteaters mentioned earlier in this chapter, the true anteater does not lay eggs. The young remain inside the female until they are fully (1) developed. However, both types of anteaters have something in common—a long, sticky tongue that is used to catch insects.

The second group of "toothless" mammals are the armadillos. They live in the southern parts of the United States and in Central and South America. These mammals eat plants, insects, and small animals. The most striking feature of the armadillo is its protective, armorlike coat. In fact, the word "armadillo" comes from Spanish, meaning armored.

A type of armadillo called the nine-banded armadillo is the only "toothless" mammal found as far north as the United States. Its special feature is that the female always gives birth to identical quadruplets! This means that the four young are always the same sex and are exactly alike.

Sloths are the third type of "toothless" mammals. There are two kinds of sloths: the two-toed sloth and the three-toed sloth. They eat leaves and fruits and

#### Activity

*Useful Mammals*

Visit your library to find out which mammals have benefited people. On a sheet of paper, make a list of these mammals. Beside each mammal, indicate how it benefited people. Discuss with your class whether the uses listed have been harmful, helpful, or had no effect upon the mammal. Was it necessary for people to use these mammals for their own survival?



### 12-4 (continued)

#### Content Development

Anteaters have large, sickle-shaped claws on their front paws, which they use for defense, and for opening the nests of ants and termites. The largest anteater, the giant anteater, is about 2 m long and has a mass of about 35 kg. The young giant anteater is dependent on its mother until it is about two years old. For the first year of life, the young anteater often rides on its mother's back.

#### Enrichment

Have students use animal encyclopedias and other reference books to find information and pictures of a pangolin, or scaly anteater. Then, have them write a short essay describing the similarities and differences between the scaly anteater and true anteaters.

#### Content Development

Sloths are often distinctly greenish in color. This color is due to symbiotic blue-green algae that grow in the animal's hair. The algae help camouflage the sloth. Sloths also have a symbiotic relationship with cellulose-digesting bacteria, which live in the sloth's stomach.

#### Motivation

Have students examine the photo-

are extremely slow-moving creatures. Sloths spend most of their lives hanging upside down in trees. They can spend up to 19 hours a day resting in this position.

### *Trunk-Nosed Mammals*

Holding its trunk high in the air, the elephant moves clumsily into the deep river. Little by little, the water seems to creep up the elephant's body. Will it drown? The answer comes a few seconds later as the huge animal actually begins to swim!

To an observer on the shore, nothing can be seen of the elephant except its trunk, through which the animal breathes in air. The trunk is the distinguishing feature of all elephants. And it is an amazingly complicated structure whose many kinds of movements are controlled by no fewer than 40,000 muscles!

Elephant trunks are powerful enough to tear large branches from trees. Yet the same trunk is capable of very delicate movements, such as picking up a single peanut thrown by a child at a zoo. These movements are made possible by the work of those 40,000 muscles and the brain that directs them.

Elephants are the largest land animals. There are two kinds of elephants: African and Asiatic. As their names suggest, African elephants live in Africa and ❷ Asiatic elephants in Asia, especially in India and

**Figure 12-16** *Elephants are the only trunk-nosed mammals. The larger ears of the African elephant* (left) *distinguish it from the Asiatic elephant* (right).

## FACTS AND FIGURES

Elephants were at one time found all over the world except in Australia, New Zealand, and Antarctica. The largest one known was the Imperial mammoth that flourished in North America, Europe, Asia, and Africa. During the Ice Age, these elephants were a staple part of the diets of humans, and it is possible they were hunted into extinction. Today only about 1 million African elephants are estimated to live in the wild, and only about 50,000 Asiatic elephants remain.

## ANNOTATION KEY

❶ **Thinking Skill: Making comparisons**
❷ **Thinking Skill: Classifying elephants**

## FACTS AND FIGURES

The heaviest land animal alive today is the African elephant. Males can have a mass of up to 7500 kg, which is equal to that of about eight compact cars. Such elephants can be 7 m long, trunk to tail, and 4 m tall at the shoulder. These animals eat only vegetation, which makes them herbivores.

graphs of the armadillo and sloth in Figure 12-15.

• **Why is the armadillo aptly named?** (Its plates look like armor.)

Explain to students that the unarmored parts of the armadillo are covered with furry skin.

• **What does it mean to be "slothful"? Why is the term "slothful" appropriate?** (lazy; sloths are very slow-moving and apparently lazy animals)

• **How is a sloth adapted for life in the trees?** (camouflage, claws shaped like hooks for hanging from branches)

• **Why are sloths and armadillos placed in the same group of mammals?** (have reduced teeth)

## Content Development

Point out that the trunk is a modification of the upper lip and nose. While it is powerful enough to tear branches from trees, it is also a sensitive organ of smell and touch. Small hairs at the tip enable the elephant to pick up very small objects. Point out that elephants can use their trunks for many purposes. In addition to serving as a snorkel for underwater swimming, it is also used to suck up water when drinking, and squirt it into the mouth. The trunk is also used to spray water over the body to cool off. Elephants also touch trunks to greet each other, and they raise it as a sign of aggression.

## BACKGROUND INFORMATION

The odd-toed and even-toed hoofed mammals, or ungulates, are thought to have diverged from a common ancestor about 60 million years ago. However, the two orders of hoofed mammals share many similarities in form and function.

Most hoofed mammals are adapted for swift running on open ground. Their muscular legs are of equal length and are designed for back and forth movement. The limited movement in the leg joints allows for fast running, but does not permit activities such as climbing and digging. The number of toes is reduced, and the foot is elevated so that all of the body weight is on the tips of the toes. The toes are covered by hard, thick hooves, which are actually enlarged nails.

In addition to being suited to running, most hoofed mammals are suited to low-protein, high-fiber diets of plant matter. A hoofed mammal's teeth have large grinding surfaces which break down cellulose plant-cell walls to release the nutritious cell contents. The digestive system is also modified for handling large quantities of cellulose.

## ANNOTATION KEY

❶ **Thinking Skill: Making comparisons**

❷ **Thinking Skill: Relating cause and effect**

Southeast Asia. Although there are a number of differences between the two kinds of elephants, the most obvious one is ear size. The ears of African elephants are much larger than those of their Asiatic cousins.

### *Hoofed Mammals*

What do pigs, camels, horses, and rhinoceroses have in common? No much at first glance. They ❶ could not look more different. Yet look again—down where these animals meet the ground—and you see a common characteristic. The feet of these animals end in hoofs—two different kinds of hoofs.

One kind of hoof has an even number of toes and belongs to such mammals as pigs, camels, goats, cows, and the tallest of all mammals, the giraffes. The other kind of hoof has an odd number of toes and belongs to mammals such as horses, rhinoceroses, zebras, and tapirs.

Hoofed animals are among the most important "partners" of human beings and have been so for thousands of years. People eat their meat, drink their milk, wear their skins, ride on them, and use them to pull devices used in farming.

**Figure 12-17** *Mammals with hoofs can be even-toed or odd-toed. A giraffe* (left) *has an even number of toes, while the rhinoceros* (top right) *and tapir* (bottom right) *have odd numbers of toes.*

## 12-4 (continued)

### Motivation

Show students several pictures of hoofed mammals such as a horse, cow, deer with horns, giraffe, or bison. Have them look at the hoofed mammals in Figure 12-17.

- **What do you predict these mammals have in common?** (Accept all logical answers.)
- **How are some of these mammals helpful to people?** (Accept all logical answers.)
- **Are these mammals herbivores or are they carnivores?** (herbivores)

Emphasize that great numbers of the hoofed mammals are grazers or browsers on vegetation.

- **How do you think the grazing mammals escape predators?** (Answers will vary. Students will probably mention that some have horns or antlers. Most also have muscular legs for running swiftly.)
- **How are the feet of grazing animals different?** (Students should answer that some are odd-toed and some are even-toed.)
- **Give examples of both odd-toed and even-toed hoofed mammals.** (odd-toed: horse, zebra, rhinoceros; even-toed: pig, cow, sheep, deer, hippopotamus. Accept all correct answers)

**Figure 12-18** *The most numerous of all mammals are the rodents. The harvest mouse* (left) *is the smallest rodent, while the capybara* (right) *is the largest.*

### *Gnawing Mammals*

Hardly a day goes by that people in the country or even in cities do not see a gnawing mammal. There are more gnawing mammals than any others on earth. These mammals are commonly known as rodents.

Among the rodents are such animals as squirrels, beavers, chipmunks, rats, mice, and porcupines. As you might guess, what they have in common has something to do with the way they eat—by gnawing. The common characteristics are four special incisors that are used for gnawing. These teeth are chisel-like and constantly grow for as long as the animal lives. Because rodents gnaw or chew on hard objects such as wood, nuts, and grain, their teeth are worn down as the teeth grow. If this were not the case, a rodent's incisors would grow to be so long that the animal could not open its mouth wide enough to eat. (2)

Some rodents, especially rats and mice, compete with human beings for food. They eat the seeds of plants and many other foods used by people. Rodents also spread serious diseases.

**Figure 12-19** *A pika is a close relative of the rabbit and hare, which are rodentlike mammals. Pikas live on the sides of mountains where they eat the sparse vegetation.*

### *Rodentlike Mammals*

Rabbits, hares, and pikas (PIGH-kuhz) belong to the group of rodentlike mammals. These mammals have gnawing teeth, similar to rodents. But unlike rodents, they have a small pair of grinding teeth behind their gnawing teeth. Another difference



## Content Development

Point out that the even-toed mammals, such as the cow, sheep, pig, and hippopotamus, are subdivided into two groups according to their eating habits. There are cud-chewing and noncud chewing even-toed mammals.

• **Which mammals do you know of that are cud-chewing?** (Most students will say cows.)

Explain that cows, sheep, goats, giraffes, and deer are cud-chewing mammals. Point out that cud-chewing mammals have four divisions to their stomachs. Explain that they eat a large amount of food that quickly passes to the first stomach division. Later it is brought back to the mouth where it is chewed and passed to the second division of the stomach, where division begins.

Point out that in the noncud-chewing mammals the food goes directly to the stomach to start digestion.

Point out that some hoofed placental mammals have horns. Explain that some of the hoofed mammals, such as the moose, deer, and elk, shed their horns every year, while the cow, sheep, and bison never shed their horns.

## Motivation

Have students observe the rodents in Figure 12-18.

• **What traits do these different-sized rodents have in common?** (Lead students to answer that rodents are gnawing mammals.)

• **The gnawing mammals are the most numerous of all. What are some different ones that you know of?** (Numerous answers are possible. As correct answers are given, list them on the chalkboard.)

• **What teeth do these mammals use for gnawing?** (The incisors are the gnawing teeth.)

• **How are a rodent's incisors different from yours?** (They are chisel-shaped and continue to grow throughout life.)

## Content Development

Explain that rabbits, hares, and pikas were once classified as rodents, because they also have chisellike incisors. However, because of other differences they are now placed in another group.

• **How are the teeth of rabbits and hares different from rodent teeth?** (Smaller grinding teeth are behind the gnawing teeth.)

• **In what other way are these animals different from rodents?** (The rodentlike mammals move their jaws from side to side as they chew.)

## FACTS AND FIGURES

The largest mammal of all is the blue whale, whose head to tail length is about 27 m. The largest whale ever caught was 33.58 m long. It can have a mass of up to 140,000 kg. The largest land animal, the African elephant, is tiny by comparison. Its trunk to tail length is about 7 m, and it may have a mass of about 7500 kg.

## ANNOTATION KEY

1. Thinking Skill: Making comparisons
2. Thinking Skill: Classifying water-dwelling mammals
3. Thinking Skill: Relating facts

### Activity

**Migration of Mammals**
**Skills: Comparing, relating, recording, diagraming**
**Level: Average**
**Type: Hands-on/Library**
**Materials: Posterboard, colored pencils, reference books**

Migration studies at this grade level are usually limited to studies of birds; however, the migration patterns of certain mammals are also interesting. The maps drawn by students can be referred to as you teach about the flying, hoofed, and water-dwelling mammals involved.

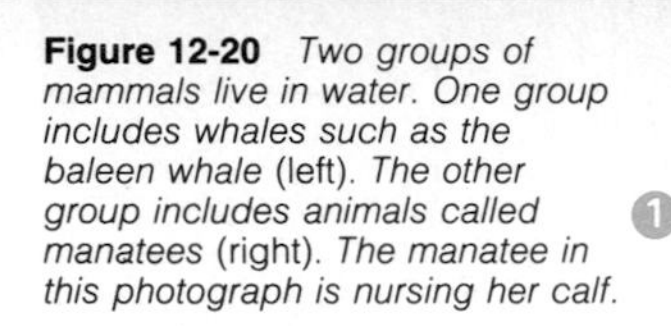

**Figure 12-20** *Two groups of mammals live in water. One group includes whales such as the baleen whale* (left). *The other group includes animals called manatees* (right). *The manatee in this photograph is nursing her calf.*

between these two groups of mammals is that the rodentlike mammals move their jaws from side to side as they chew their food, while rodents move their jaws from front to back as they chew. (1)

### *Water-Dwelling Mammals*

"Thar she blows!" is the traditional cry of a sailor who spots the fountain of water that a whale sends skyward just before it dives. Sailors of the past recognized this sign of a whale, but they did not have the slightest idea that this sea animal was not a fish but a mammal.

Whales, along with dolphins, porpoises, dugongs, and manatees are intelligent. They have hair and feed their young with milk. They are all mammals—mammals that live in water most or all of the time but breathe air. They are also different from one another. Their differences have allowed scientists to place whales, dolphins, and porpoises in one group and manatees and dugongs in a second group.

Animals in the first group spend their entire lives in the ocean. They cannot survive on land. Animals in the second group live in shallower water, often in rivers and canals. They can move around on land but with great difficulty and only for short periods of time when they become stranded. (2)

**Activity**

*Migration of Mammals*

Certain mammals migrate, or move, to places that offer better living conditions. Using posterboard and colored pencils, draw maps that trace the migration of the following mammals: North American bat, African zebra, American elk, and gray whale.

At the bottom of each map, give the reasons why each particular mammal migrates. Display these maps on a bulletin board at school.

### *Primates*

On a visit to your local zoo, you come upon a crowd of people standing and laughing in front of one of the cages. Hurrying over to see what the



## 12-4 (continued)

### Motivation

Direct students' attention to the photographs of the whale and manatees in Figure 12-20.

- **Why were whales once thought to be fish?** (Live in water, streamlined, have fins. Accept all logical answers.)
- **Why are water-dwelling animals classified as mammals?** (warmblooded, feed young with milk, some have hair)
- **How are water-dwelling mammals different from other mammals?** (live in the water, fishlike shape)
- **How do manatees differ from whales?** (size, habitat, manatees eat plants and whales eat other animals)

### Motivation

Tell students that primates evolved from tree-dwelling mammals and that many primates still live in trees. Direct students' attention to the photographs of primates on this page.

- **How do the hands and feet aid a primate in living in trees?** (Accept all logical answers, but be sure that someone mentions five fingers and toes are adaptations for grasping branches.)
- **Where are the eyes located on a primate's face?** (The eyes face forward.)

excitement is about, you hear strange noises. Carefully you make your way to the front of the crowd. What has caused all the excitement and noise? Soon you see a family of chimpanzees entertaining the crowd by running and tumbling around their cage. The baby chimpanzee comes to the front of the cage and extends its hand to you. You are amazed to see how much the chimpanzee's hand looks like yours. It is no wonder they are similar. After all, the chimpanzee along with the gibbon (GIB-uhn), orangutan (oh-RANG-oo-tan), and gorilla are the closest mammals, in structure, to human beings. These mammals, along with baboons, monkeys, and human beings belong to the same group of mammals—the primates. ③

All primates have eyes that face forward, enabling the animal to see depth. The primates also have five fingers on each hand and five toes on each foot. The fingers are capable of very complicated movements, especially grasping objects.

Primates also have large brains and are the most intelligent of all mammals. There is evidence that chimpanzees can be taught to communicate with people by using a kind of sign language. Some scientists have reported that chimpanzees can *use* tools, such as the use of twigs to remove insects from a log. Human beings, on the other hand, are the only primates that can *make* their own tools.

**Figure 12-21** *Chimpanzees* (top left), *gorillas* (top right), *and lemurs* (bottom right) *are all primates. Among the characteristics that primates share are eyes that face forward and hands and feet adapted for grasping.*

### SECTION REVIEW

1. What is the placenta?
2. What are canines? Incisors?
3. Define carnivore.



## TEACHER DEMONSTRATION

Show students the palms of your hands with your fingers and thumb extended. Fold your thumb into the palm and back out several times. Touch each of your fingers in turn with the thumb of that hand. Form your hand into a fist and open it several times. Pick up an object.

- **Can all of you do the exercises I just did?** (yes)
- **Although these exercises are easy for us, they are unique in the world of animals. Why do you think they are unique?** (Accept all answers.)
- **How does being able to move a thumb make us different from most other living things?** (Accept all answers.)

Emphasize that the front-facing eyes permit three-dimensional depth vision.

- **Why do you think depth vision is important when you live in trees?** (Accept all logical answers.)

Mention that even though gorillas are ground-dwellers they often sleep in trees, and young gorillas spend much time in trees.

## Content Development

Point out that only primates stand on two legs, have five fingers on each hand, five toes on each foot, and can move their thumbs in and out (in opposition to the fingers).

Explain that primates include monkeys, gibbons, orangutans, apes, and humans. Point out that by being able to move the fingers and thumb, primates can grasp and hold on to objects. Point out that this grasping allows monkeys to swing from trees.

## Content Development

Mention that primates are adapted to a wide range of habitats and thus show a wide range of postures and methods of locomotion. Some of the primitive primates, like lemurs and tarsiers, resemble squirrels, and like squirrels, they scamper about on tree branches on all fours. Monkeys are somewhat more upright, but they also run along branches on all four limbs. Some apes like gibbons and orangutans spend most of their time in trees, swinging from limb to limb with their arms. Chimpanzees and gorillas spend considerable time on the ground, and they can walk short distances on their hind legs. But most of the time they also use the knuckles of their hands for walking. Humans are the only primates that walk completely upright.

## Section Review 12-4

1. The structure through which developing young receive food and oxygen while in the female.
2. Canines are teeth used for tearing and shredding. Incisors are teeth used for biting.
3. Flesh-eater.

## LABORATORY ACTIVITY CLASSIFYING AND COMPARING MAMMALS

### BEFORE THE LAB

1. **This activity will require the equivalent of roughly two class periods. However, students may do much of the work at home or at the library. If the class can be taken to the library for their research, make the necessary arrangements with the librarian in advance.**
2. **Well in advance of this activity, start collecting pictures from old magazines and textbooks of different mammals representing the 14 major orders. You may be able to solicit student help by offering extra credit for each usable picture they bring in. The 14 orders that should be included are: Monotremata, Marsupalia, Insectivora, Chiroptera, Carnivora, Edentata, Proboscidea, Perissodactyla, Rodentia, Lagomorpha, Sirenia, Cetacea, Artiodactyla, and Primates.**
3. **At least one day before the lab, use a permanent marker to label each of the pictures with a number or letter for future identification. Prepare a key to the identification codes on the pictures so students may check their work. Be sure that sufficient white construction paper and tape is available for each student group. Groups of 2 to 3 students are appropriate, but larger groups may be necessary if not enough pictures are available.**

## LABORATORY ACTIVITY

### *Classifying and Comparing Mammals*

**Purpose**

In this activity, you will classify different mammals and compare their characteristics.

**Materials** *(per group)*

Reference materials that provide information on the 14 major orders of mammals
Pictures of mammals of different orders, cut from magazines and old texts
14 sheets of white construction paper
Tape

*TABLE 1*

| Identification Code | Characteristics | Predicted Order |
|---|---|---|
| | | |
| | | |
| | | |
| | | |

*TABLE 2*

| Identification Code | Name of Organism | Correct Order |
|---|---|---|
| | | |
| | | |
| | | |
| | | |

**Procedure**

1. Using reference books, make a list of the fourteen major orders of mammals and their distinguishing characteristics.
2. Fold each of 14 sheets of construction paper in half along its width.
3. Tape the sides of each sheet, forming a pocket envelope. Do not tape the top of the paper.
4. Study the pictures of mammals provided by your teacher. Each picture should have an identification code.
5. Label each pocket envelope with the name of an order.
6. Place each picture in the envelope you think represents that animal's order.
7. Copy Table 1 on a sheet of paper. Place the identification code from each of your pictures in the proper space.
8. In the second column of the table, fill in the characteristics of each of the animals based on your study of the pictures.
9. Fill in the order in which you classified each of the mammals.
10. Copy Table 2 on the same sheet of paper as Table 1. Check with your teacher to find out the correct order and the name of each mammal. Fill in this information in Table 2. Compare the correct order with your predicted order.

**Observations and Conclusions**

1. How do the monotremes differ from the marsupials?
2. What distinguishes the marsupials from the other mammals?
3. Which orders of mammals appear most similar? Explain your answer.
4. List the characteristics that all mammals have in common.

298

### PRE-LAB DISCUSSION

Have students read the complete laboratory procedure. Discuss the procedure by asking a few questions similar to the following:

- **What is the purpose of this laboratory activity?** (to classify mammals and compare their characteristics)
- **What is meant by "orders of mammals?"** (If students have forgotten classification groupings, it may be necessary to review Chapter 6.)

List the 14 major orders of mammals on the chalkboard, and tell students they are to look up characteristics of these orders in reference books.

- **What kinds of characteristics will be most useful for deciding what order a mammal belongs to?** (Accept all logical answers.)

### SKILLS DEVELOPMENT

Students will use the following skills while completing this activity.

1. Classifying
2. Comparing
3. Relating
4. Inferring
5. Applying
6. Recording

# CHAPTER REVIEW

## SUMMARY

**12-1 Characteristics of Mammals**

- Mammals are warmblooded vertebrates that have hair or fur.
- Female mammals feed their young milk from mammary glands. Most mammals give their young more care and protection than do other animals.
- Mammals have well-developed brains.
- Mammals can be placed into three basic groups depending upon how their young develop. These groups are the egg-laying mammals, the pouched mammals, and the placental mammals.

**12-2 Egg-Laying Mammals**

- Like birds and reptiles, monotremes are mammals that lay eggs.
- The spiny anteater and duckbilled platypus are examples of egg-laying mammals.

**12-3 Pouched Mammals**

- Marsupials are pouched mammals.
- The young of the marsupials are born only partially developed. They further develop in the pouch of their mother.
- The koala, kangaroo, and opossum are examples of marsupials.

**12-4 Placental Mammals**

- In placental mammals, the placenta provides food for the developing young inside the females.
- There are more than 20 groups of placental mammals.
- Insect-eating mammals include moles, hedgehogs, and shrews.
- Bats are the only flying mammals.
- Flesh-eating mammals, or carnivores, include sea-living animals such as walruses. The land-living carnivores include any members of the dog, cat, and bear families.
- Armadillos, anteaters, and sloths are "toothless" mammals.
- The only members of the trunk-nosed mammal group are the elephants.
- Hoofed mammals are divided into those with an even number of toes on each hoof and those with an odd number of toes. Even-toed mammals include pigs, camels, goats, cows, and giraffes. Odd-toed mammals include horses, rhinoceroses, and tapirs.
- Gnawing mammals, such as beavers, chipmunks, rats, mice, and porcupines, have chisel-like incisors for chewing.
- Rabbits, hares, and pikas are examples of rodentlike mammals.
- Mammals such as whales, dolphins, porpoises, dugongs, and manatees are water-dwelling mammals.
- Human beings, monkeys, and apes are known as primates.

## VOCABULARY

*Define each term in a complete sentence.*

**canine**
**carnivore**
**herbivore**
**incisor**
**mammary gland**
**marsupial**
**monotreme**
**placenta**



## OBSERVATIONS AND CONCLUSIONS

**1.** Monotremes lay eggs. Marsupials give birth to live young.
**2.** Marsupial offspring complete their development inside their mother's pouch. Most mammals complete their early development inside the mother's body.
**3.** Answers will vary. Students may place Artiodactyla, Perissodactyla, and Proboscidea together in one group, for example, and Insectivora, Rodentia, and Lagomorpha in another.
**4.** Mammals are warmblooded, produce milk in mammary glands, have hair, highly developed brains, and four-chambered hearts.

## GOING FURTHER: ENRICHMENT

**Part 1**

If students initially placed some of the pictures in the wrong envelopes, ask them to try to determine why they did so. Are some mammals difficult to classify from pictures alone? Ask students to list those characteristics that are especially difficult to determine from pictures.

**Part 2**

If time permits, have students gather information on some of the families of some large mammal orders like Carnivora, Perissodactyla, or Artiodactyla. Then have them take the pictures from the envelope and rearrange them in family groupings.

## TEACHING STRATEGY FOR LAB PROCEDURE

**1.** As students search through reference books for information, circulate among them and give assistance when needed. If students are working in teams, check to see that all are participating.
**2.** After students have finished classifying the pictured mammals, have them return the coded pictures for other classes to use. The same pictures can be kept on hand for use in future years.
**3.** Be sure that students copy and complete Tables 1 and 2. Tell them each table will need to be big enough for the pictures of the 14 orders of mammals in their envelopes.

## CHAPTER REVIEW

### MULTIPLE CHOICE

**1.** b **2.** a **3.** d **4.** c **5.** d **6.** b **7.** c **8.** c **9.** b **10.** c

### COMPLETION

**1.** warm
**2.** mammary glands
**3.** spiny anteater
**4.** placenta
**5.** carnivore
**6.** incisors
**7.** trunk-nosed
**8.** rodents
**9.** rodentlike
**10.** primates

### TRUE OR FALSE

**1.** F vertebrates
**2.** F warm-blooded
**3.** T
**4.** T
**5.** F bats
**6.** F incisors
**7.** F prey
**8.** T
**9.** T
**10.** F human being

### SKILL BUILDING

**1.** Answers will vary. Students will probably think that monotremes are the most similar to birds because they lay eggs. Others may think that bats are the most similar because they fly.
**2.** Check charts to make sure they reflect the information presented in the chapter.
**3.** Although they resemble placental bears, koalas are actually marsupials.
**4.** Whales come to the surface to breathe.
**5.** Hoofed mammals may feed in herds because they can ward off predators when in groups, or because an individual in a group is less likely to be preyed upon than one that is alone.
**6.** In general, animals that are more complex provide more care for their young than less complex animals. Students' examples will vary.
**7.** Experiments should be logical and well thought out. Answers will vary.

### CONTENT REVIEW: MULTIPLE CHOICE

*Choose the letter of the answer that best completes each statement.*

**1.** All mammals have
a. pouches. b. well-developed brains.
c. feathers. d. fins.

**2.** The kangaroo is a(n)
a. pouched mammal. b. egg-laying mammal.
c. placental mammal. d. gnawing mammal.

**3.** The only North American marsupial is the
a. kangaroo. b. koala. c. platypus. d. opossum.

**4.** Young mammals that develop totally within the female belong to the group called
a. egg-laying mammals. b. pouched mammals.
c. placental mammals. d. marsupial mammals.

**5.** Which is an insect-eating mammal?
a. whale b. elephant c. bear d. mole

**6.** Teeth used for tearing and shredding food are
a. carnivores. b. canines. c. incisors. d. all of these.

**7.** A "toothless" mammal is the
a. skunk. b. mole. c. armadillo. d. camel.

**8.** The largest land animal is the
a. blue whale. b. rhinoceros. c. elephant. d. giraffe.

**9.** An example of a water-dwelling mammal is the
a. spiny anteater. b. dolphin. c. shrew. d. elephant.

**10.** Human beings belong to a group of mammals called the
a. insect-eating mammals. b. rodents.
c. primates. d. carnivores.

### CONTENT REVIEW: COMPLETION

*Fill in the word or words that best complete each statement.*

**1.** Mammals are a group of ______________ blooded vertebrates.
**2.** Female mammals feed their young with milk from ______________.
**3.** An egg-laying mammal that is covered with spines is the ______________.
**4.** The internal structure through which the developing young of some mammals receive their food is the ______________.
**5.** Another name for flesh-eating mammals is ______________.
**6.** The teeth that are used for biting are the ______________.
**7.** Elephants are ______________ mammals.
**8.** Gnawing mammals are also known as ______________.
**9.** Rabbits are examples of ______________ mammals.
**10.** Mammals that have five fingers on each hand and five toes on each foot are called ______________.



### ESSAY

**1.** Monotremes: lay eggs. Marsupials: young develop in pouches. Placentals: young develop within the body of the mother and are fed through the placenta.
**2.** An anteater is a placental mammal; a spiny anteater is a monotreme.
**3.** Kangaroo, opossum, koala. The young are born in an immature state and complete their development inside the mother's pouch.
**4.** Carnivores often have keen senses of sight and smell to help them find their prey. They may be swift runners or have sharp claws for attacking prey. Mammals that are carnivores often have sharp teeth, particularly canines, for biting their prey.
**5.** Insect-eating, flying, flesh-eating, "toothless," trunk-nosed, hoofed, gnawing, rodentlike, water-dwelling, primates. Examples should be similar to those given in the text.

## CONTENT REVIEW: TRUE OR FALSE

*Determine whether each statement is true or false. If it is true, write "true." If it is false, change the underlined word or words to make the statement true.*

1. Mammals are invertebrates.
2. In coldblooded animals, the body temperature remains about the same all the time.
3. The duckbilled platypus is an example of a monotreme.
4. Animals that eat only plants are called herbivores.
5. The flying squirrels are the only examples of flying mammals.
6. Canines are teeth that are used for biting.
7. Organisms upon which other organisms feed are called predators.
8. A squirrel is an example of a gnawing mammal.
9. Rhinoceroses are examples of hoofed mammals.
10. The dolphin has the most well-developed brain of all the mammals.

## CONCEPT REVIEW: SKILL BUILDING

*Use the skills you have developed in the chapter to complete each activity.*

1. **Relating facts** Which group of mammals is most similar to birds? Explain your answer.
2. **Making charts** Prepare a chart with three columns. In the first column, list the ten groups of mammals that you learned about in this chapter. In the second column, list the characteristics of each group. And in the third column, list at least two examples of each group.
3. **Classifying animals** Although koalas are often called koala bears, they are not really bears. Explain this statement.
4. **Applying concepts** Why do whales usually come to the surface of the ocean several times an hour?
5. **Making inferences** Many species of hoofed mammals feed in large groups, or herds. What possible advantage could this behavior have for the survival of these mammals?
6. **Making generalizations** What is the relationship between how complex an animal is and the amount of care the animal gives to its young? Provide an example to support your answer.
7. **Designing an experiment** Design an experiment that will test the following hypothesis: Chimpanzees are able to understand the meaning of the spoken words for the numbers one through ten.

## CONCEPT REVIEW: ESSAY

*Discuss each of the following in a brief paragraph.*

1. What are the three groups of mammals? How do they differ from one another?
2. What is the difference between an anteater and a spiny anteater?
3. Name two pouched mammals. Explain how their young develop.
4. What features make carnivores good predators?
5. List ten groups of placental mammals and give an example of each.



## ISSUES IN SCIENCE

The following issue can be used as a springboard for discussion or given as a writing assignment.

At the beginning of the eighteenth century, there were about 9 million harp seals in the North Atlantic Ocean. Overhunting once drove the harp seal to the edge of extinction, but strict hunting laws allowed the population to recover. Today, harp seals have a population of about 2.6–3.8 million, which is growing by about 5% each year.

For the first two weeks of life, young harp seals have beautiful white coats. In the past, large numbers of seal pups have been killed for their fur, usually by clubbing them to death in the presence of their mothers. Conservation and humane groups strongly oppose the killing of the seals. They think that killing young seals for their fur is cruel and may have a negative effect on the seal population and the environment. Seal hunters, facing economic difficulty, want permission to kill greater numbers of young seals. They believe that the seal population is no longer in danger, and that the young seals can be killed humanely. Should seal hunters be permitted to harvest greater numbers of seal pups. What is your opinion and why?

## ADDITIONAL QUESTIONS AND TOPIC SUGGESTIONS

**1.** Look at the walruses in Figure 12-13. Why can walruses be active in this environment, while frogs cannot? (Walruses are warmblooded and their body temperature does not change with the surroundings. Frogs are coldblooded.)

**2.** Explain why a koala bear is more closely related to an opossum than it is to a grizzly bear. (The koala and opossum are both marsupials, but the grizzly bear is a placental mammal.)

**3.** Suppose a friend brings you the skull of a small mammal she has found. It has two long, chisel-shaped incisors in both the upper and lower jaws. There are no canine teeth in either jaw. Is this the jaw of a carnivore or a herbivore? (The skull is probably a herbivore. It is not likely to be that of a carnivore since it has no canine teeth.)

## Unit Two

## CLASSIFICATION OF LIVING THINGS

### ADVENTURES IN SCIENCE: JANE GOODALL AND THE CHIMPS OF GOMBE STREAM

### BACKGROUND INFORMATION

Jane Goodall is the first scientist to make daily observations of animals in their natural habitat over long periods of time. She is also the first to follow the behavior patterns and development of specific individuals, instead of observing the animals only as a group. Goodall's research has proved incorrect many previously held assumptions about primates, as well as certain ideas about what separates human beings from other animals.

The most significant aspect of Jane Goodall's work is her observations that chimpanzees make and use tools. These observations refute the previously accepted idea that people are the only animals capable of making and using tools. Also significant is Goodall's discovery that chimpanzees develop and behave as individuals. Prior to such observations, scientists tended to assume that, with the exception of human beings, all animals of a species are more or less alike. In fact, the ability to individuate was considered to be a trait that separated humans from other animals on the evolutionary tree.



Adventures in Science

# Jane Goodall and the Chimps of Gombe Stream

Deep in the heart of the African jungle, Jane Goodall has been doing research among the wild animals. She is studying one of our closest living relatives, the chimpanzee.

Goodall has always wanted to study the animals in Africa. When she was a young girl, she was inspired by Rudyard Kipling's jungle tales. Ever since reading Kipling's vivid descriptions of the jungle and its inhabitants, Goodall has known where she wanted to spend her time.

For more than 20 years, Goodall has lived at Gombe Stream, Tanzania's Game Preserve on Lake Tanganyika. She is both an ethologist, a person who studies animal behavior, and a primatologist, a person who studies primates. She hopes that her studies will help scientists better understand prehistoric and present-day humans.

In 1960, she stepped off a small boat and came face to face with the majestic mountains of Africa. Over the years, she has learned to live with blistering heat and drenching rain. She has learned how to travel in a country with deep valleys, high mountains, and steep gorges. She has adjusted to living among threatening animals such as cobras and leopards.

But for the first few months after her arrival, Goodall was in despair. Until she discovered wild pig and baboon trails, she had to fight her way through dense underbrush to do her research. Although she could hear the loud calls of the chimps and the rustling of the



### TEACHING STRATEGY: ADVENTURE

#### Motivation

- **How many of you have ever owned a dog or cat?** (Answers will vary.)
- **How many of you have owned, at the same time or at different times, more than one dog or cat?** (Answers will vary.)
- **For those of you who have owned more than one dog or cat, would you say that all the animals were alike in personality, or that each was distinctly different in personality?** (Accept all answers. Many will probably say that each animal was distinctly different in personality.)

#### Content Development

Use the previous discussion to stress an interesting aspect of Jane Goodall's research, such as her observations about the development and behavior of individual chimps. Point out that until Goodall's research, the study of animals in their natural habitat had been limited to observation of animals as a group.

Also emphasize to students another of Goodall's important findings: Chimpanzees are able to modify natural objects and turn them into tools. Prior to Goodall's research, scientists believed that human beings were the only animals capable of making and using tools.

- **How did Goodall's chimps make**

Chimps often chew leaves, which they then use as "sponges" to soak up drinking water from places such as hollows in logs.

leaves as the chimps moved, she rarely saw them. They would always run when they heard her approaching. After eight months, Goodall had not come within 500 yards of the chimps. She was about to give up!

Suddenly her luck changed. A chimp she had named David Greybeard came shyly to her camp one day. He continued coming to the camp every day. After two weeks of observing her camp, David Greybeard actually took a banana from Goodall's hand. The other chimps, after seeing their leader's bold move, gradually allowed Goodall to get close to them too.

At about the same time, Goodall began to make some amazing discoveries. One of the most important is that chimps use objects as tools. Once again, David Greybeard was the star of the show. Goodall saw David Greybeard and a female chimp she had named Flo strip the leaves from thin branches. They then went over to a termite mound. For protection, termites seal the tops of their tunnels. Chimps look for these protective coverings and peel them off. Goodall watched as David and Flo dipped their branches into the tunnels and pulled them out covered with termites. The chimps then ate the termites off the sticks, just as you might take a piece of meat from a fork. Goodall also noticed that the chimps would go out of their way to select sticks and carry them to termite nests that were far away.

This behavior seems to show that the chimps not only are able to use tools, but can think about their use ahead of time.

In time, Goodall was able to observe other chimp behavior. She saw another male, Evered, strip the leaves off a small branch and stuff them into his mouth. He chewed them for a short time and took them out of his mouth. He then dipped them, like a sponge, into a fallen log to get water. Goodall observed that the other chimps also used leaves to get water. So she collected some leaves and chewed them in order to understand what the chimps were doing. She found that chewing the leaves before dipping them helped to increase their absorbency.

When Goodall is not at Gombe Stream, she and her son live in Palo Alto, California, where she does laboratory research. However, she much prefers studying in the wild. Even in California, she keeps in touch with what is happening at Gombe by corresponding with her assistants there. She has lived among the wild chimps for more than twenty years. If asked why she chooses to live her life this way, Goodall replies that it is her "inborn love of the place and the beasts."

Jane Goodall (left) and companion observe the behavior of an adult and a young chimp.

## ADDITIONAL QUESTIONS AND TOPIC SUGGESTIONS

**1.** Have students locate the country of Tanzania on a map of Africa. Then have them locate Lake Tanganyika.
**2.** In addition to her work with chimpanzees, Jane Goodall has studied various other types of primates, as well as hyenas. Have students read about and report on some of Goodall's findings about these animals.
**3.** Challenge students to read and report on one of the following books: *Jane Goodall,* by Eleanor Coer, G. P. Putnam's Sons, New York, 1976; *My Friends the Wild Chimpanzees,* by Baroness Jane Van Lawick-Goodall, National Geo. Society, 1967; *In the Shadow of Man,* Jane Van Lawick-Goodall, Houghton Mifflin Co., Boston, 1971.

## CRITICAL THINKING QUESTIONS

**1.** How did Jane Goodall know that David Greybeard was a leader of the chimpanzees? (David was the first chimp to come to Goodall's camp. When he finally took a banana from her hand, the other chimps let Goodall get close to them also. In addition, David was among the chimps who demonstrated superior toolmaking skills.)
**2.** Why do you think toolmaking is such an important measure of an animal's intelligence and level of evolutionary development? (The ability to make a tool demonstrates an animal's capacity to think and plan ahead, and to see the relationship between a problem and its solution.)

**tools for eating termites?** (The chimps stripped leaves from branches to make sticks that could pull the tops off termite mounds; then they used the sticks as eating utensils to dip into the termites and eat them.)

- **What type of tool did the chimps make to help them get water?** (They chewed leaves, and then used the leaves to absorb water like a sponge.)

### Skills Development

***Skills: Making a model, making a diagram***

Have students work in small groups. Challenge each group to make a model or diagram of a chimpanzee tool that is described in the text. For example, students might show how chewed lettuce leaves (caution students to use only edible leaves) can be used to absorb water. Students might also make a diagram showing a chimp stripping leaves from a branch, removing the top off a termite mound, and then eating the termites off the branch.

### Teacher's Resource Book Reference

After students have read the Science Gazette article, you may want to hand out the reading skills worksheet based on the article in your Teacher's Resource Book.

## Unit Two

## CLASSIFICATION OF LIVING THINGS

### ISSUES IN SCIENCE: CONSERVATIONISTS TO THE RESCUE?

### BACKGROUND INFORMATION

Which plants and animals on earth are most valuable? Does the value of an organism depend on its beauty? On its usefulness to human beings? On its role in an ecological system? Or are all species of living things created equal?

These are some of the questions with which conservationists must wrestle as they attempt to determine which plants and animals are most worth saving from extinction. As one expert has explained, it is not possible to save all species from extinction because there are only enough resources available to save a small fraction of them. The question is, which species should be saved and which should be saved first.

For some conservationists, the issue is a moral one; they see all living things as having an intrinsic value of their own. A conservationist with this point of view would be less concerned about the usefulness or economic importance of a species, and more concerned about the right of any species to live and multiply.

Some conservationists are much more pragmatic and materialistic in their views. They feel that a species should be saved if it is economically valuable, or if it is useful to human beings. Lions in Kenya, for example, are economically valuable because they are a great tourist attraction; certain plants are useful because they can be used to produce drugs.

A third consideration is the effect of a certain species on ecosystems. A particular species may prey on another species, thus keeping its population under control; or a particular species may serve as food for another species. Thus, the balance of an ecosystem may depend at least in part on the preservation of a particular organism.

SCIENCE GAZETTE

Issues in Science

# Conservationists to the Rescue?

Bathed in early morning light, a herd of large, dark brown sambar deer grazes on a grassy hillside at the edge of a forest in southern Asia. Suddenly, one of the deer raises its head in alarm and sniffs the air. Soon the entire herd is wary and watchful—ready to flee in an instant.

The reason for their alarm quickly becomes obvious. Emerging from the woods a mere 225 meters away, a tiger strides up the hillside. The big cat's powerful muscles ripple as it moves. For a moment, the tiger stops and turns its head toward the deer. Then it continues on its way. This morning, at least, the tiger is not hungry. The deer are left to graze in peace.

Tigers are a vanishing species. They have disappeared from many parts of Asia where they were once common. In an effort to save them, conservation groups such as the World Wildlife Fund have raised more than one million dollars. Some of the money has been spent on research into tiger behavior. And the rest of the money has been used to establish preserves in the wild where the cats are protected.

To some extent, the effort has been successful. India now has 11 preserves for

304

### TEACHING STRATEGY: ISSUE

#### Motivation

Display pictures of various types of wild animals, such as tigers, lions, monkeys, zebras, giraffes, and so on.

- **If all of these species were to become extinct except one, which one would you most want to survive?** (Answers will vary.)
- **Why?** (Answers will vary.)

#### Content Development

Use the previous discussion to introduce the idea of the value of a species. Point out to students that by responding to the question above, they each were making a value judgment. For most students, this value judgment was probably based on personal preference—perhaps a liking for a certain animal's appearance, or an interest in studying a certain animal's behavior. But suppose a person

tigers. In some of these protected areas, the cat population is increasing once more. Conservationists are hopeful that tigers can continue to live in the wild.

But not everyone wants tigers living in the wild. In areas near some of India's preserves, tigers have killed both animals and people. Some Indians living near preserves oppose the government's attempts to save the tiger. They feel the tiger is not a species that should be saved, but a threat that should be removed.

## A Fight to the Finish

Each year, more plants and animals become endangered. As conservationists work to save them, questions often arise about whether all species of living things should be saved. And if not all, which ones should be saved?

The answers are not simple. Some situations involve human needs versus the needs of animals. Other situations require a decision about which species is worth more time, effort, and money. For example, should developers be prevented from building in the Pine Barrens forest of New Jersey if the construction endangers a type of moth that lives there? Or is it worth thousands of dollars to save a kind of sparrow that lives only in a small part of Florida near Cape Canaveral?

## Facing the Facts

To do their job, conservationists have to face tough questions like these. Some leading conservationists are very practical about the issues. Norman Meyers, for example, is a well-known environmental scientist. He notes that by the end of the 1980s, one species an hour, counting insects and other invertebrates, will disappear.

"Sad to say," Meyers writes, "the question is not how to save *all* these species; we just do not have the resources to rescue more than a small fraction."

Meyers suggests that first the value of each species be determined. The decision would be based on economics, the environment, and the survival of people. Then, the most valuable species should receive help first.

For example, in the African country of Kenya, lions are a big tourist attraction. Tourism brings lots of money to the country. Therefore, lions may be worth saving.

A tropical plant called the periwinkle is the source of two drugs used to treat cancer. And the venom of the Malayan pit viper, a cousin of the rattlesnake, is used to stop blood clots that cause heart attacks. In Meyers' view, the periwinkle and the pit viper should be among the first to receive help.

Environmental scientists predict that by the end of the 1980s, one species an hour will disappear.

really had the power to determine the survival of one species over another? On what basis would such a person choose? How would he or she determine which animal is most "valuable?"

Emphasize to students that this is exactly the kind of dilemma that is facing conservationists today. Extinction is a very real threat to many animals whose presence we take for granted. Can students imagine, for example, a world without tigers? Yet tigers are, as the text points out, a vanishing species. Tigers are surviving in certain parts of the world only because conservation groups have set up preserves in which tigers can be protected.

Extinction is also a very real threat to many plants and animals that are relatively unknown to the average person, yet have great value to human beings as sources of drugs and other products. Still other plants and animals play valuable roles in the ecosystems in which they live; their extinction would cause the lives of many other organisms to be disrupted.

### Skills Development

***Skill: Relating cause and effect***

The opening paragraph of this article describes an important relationship among wild animals—that of the predator and the one preyed upon. This relationship is often represented in food chains and food webs.

- **What do you think would happen if an animal or plant at the beginning of a food chain were to become extinct?** (Many of the animals in the chain would lose their food supply, and possibly become extinct themselves.)
- **Might some of the animals in the chain still be able to survive? If so, how?** (Yes, because they might have other sources of food besides those dependent on the extinct organism.)
- **What is the relationship called when several food chains are interconnected, making animals dependent on more than one chain?** (a food web)
- **Suppose that an animal at the end of a food chain were to become extinct. What might be the impact on**

## ADDITIONAL QUESTIONS AND TOPIC SUGGESTIONS

**1.** Have students research several types of plants and animals that have become extinct during the earth's history. Ask students to find out as much as they can about why the organisms became extinct.

**2.** Have students list some of the plants and microscopic organisms that are useful in fighting disease. Some students may wish to find out about organisms that are useful in producing modern "wonder" drugs such as antibiotics; other students may enjoy finding out about some of the plants and organic substances that have been used in the "folk" remedies of primitive peoples.

## CRITICAL THINKING QUESTIONS

**1.** Write a paragraph in which you either agree or disagree with the following statement: "People should let nature take its course. It is just as much of an interference to try and save animals from extinction as to destroy animals by hunting, trapping, and other activities." (Accept all logical answers.)

**2.** List the different ways in which a plant or animal can be "valuable." Then rank these considerations in order of what *you* consider most important and least important. (Answers will vary. Some different values include economic value; usefulness in industry; usefulness in fighting disease; natural beauty; importance in ecological balance; and the basic value of all living things.)

**3.** Do you think that human beings are as responsible for plants and animals that become extinct due to natural causes as they are for organisms that become extinct due to human activities? Why or why not? (Accept all logical answers.)

## CLASS DEBATE

Are humans the measure of all things? Do humans have the right to determine which species of living things are most valuable? Should usefulness to humans be the criteria by which plants and animals are either saved or abandoned? Use these questions as a springboard for a class debate.

This California condor is one of several species of animals that faces possible extinction.

### The Unknown Factor

It is clearly to our advantage to invest time and money in preserving such species of plants and animals. The trouble is, there are probably thousands of species that could be very valuable to people—only no one yet knows it!

Are we wiping out species that hold secrets to curing cancer or solving the energy crisis? Scientists and conservationists Paul and Anne Ehrlich of Stanford University think we might be. And they also see a further complication. When species disappear, the ecosystems, or environments, to which they belong are changed or even destroyed. All life on the earth depends on ecosystems. Ecosystems provide important services such as the maintenance of soils and the control of crop pests and transmitters of human disease. As the Ehrlichs write, "Humanity has no way of replacing these free services should they be lost . . . and civilization cannot persist without them."

The Ehrlichs cite yet another reason for saving as many species as possible—"plain old-fashioned compassion." For many conservationists, stopping the extinction of species is a moral issue.

### Concern Around the World

The International Union for the Conservation of Nature and Natural Resources is a worldwide group associated with the United Nations. The union's position is that we are "morally obliged to our descendants and to other creatures" to act wisely when it comes to conserving plants and animals.

Most conservationists would agree that each time a species vanishes, the world is a bit poorer. Consider just the sheer beauty of many of the endangered plants and animals. The main reason for saving tigers and condors, for example, is that they are among the earth's most magnificent creatures.

Perhaps sadly, people tend to feel more concern for species that are pretty or striking than for those that are plain. The black rhinoceros of Africa and the Higgin's eye mussel of midwestern rivers are both in danger. The world knows about the threat to the rhinoceros. But few people know about the Higgin's eye, even though it is closer to extinction than the rhino.

How can conservationists decide which endangered species to help or, at least, to help first? Norman Meyers has an interesting approach. When many people are injured in a big disaster, physicians often first treat victims who are badly hurt but who will recover if treated promptly. Meyers suggests that the same approach be used for species in danger. First help those species that are in the greatest danger and that have a good chance of surviving extinction. Still, says Meyers, it will not be easy to decide which species to aid and which to ignore. Making these decisions, he adds, "will cause us many a sleepless night."

306

## ISSUE (continued)

**the rest of the chain? Why?** (The population of animals near the top of the chain would tend to increase because they would not be preyed upon by the extinct organism.)

- **What do you think would happen if an animal in the middle of a food chain were to become extinct?** (The populations of organisms below the extinct organism would tend to increase, while the populations of organisms above the extinct organism would tend to decrease.)
- **Based on what you have concluded, can you explain how the extinction of an organism might affect the balance in an ecosystem?** (The balance of an ecosystem might be affected by the extinction of an organism in the following ways: If the extinct organism was essential to the food supply of other organisms, those organisms would decrease in population and possibly become extinct; if the extinct organism preyed upon other organisms, the populations of those organisms would increase.)

### Teacher's Resource Book Reference

After students have read the Science Gazette article, you may want to hand out the reading skills worksheet based on the article in your Teacher's Resource Book.

Futures in Science

The Corn Is as High as a Satellite's Eye
Farms of the future will move into space, and crops will be far out, too.


The old man held his grandaughter's hand as the high-speed elevator zoomed 180 floors to the top of the Triple Towers. Out on the observation deck, the young girl stared in wonder at the huge city that fanned out to meet the horizon in all directions. Far below, crowds of little dots moved along in neat columns. "Look at all those people!" the girl exclaimed. "There must be millions of them."

The old man nodded. "Twenty-two million, to be exact," he said softly.

"But Grandpa, how do all these people keep from starving? I read that years ago people in cities all over the world couldn't get enough to eat. And there are even more people now than there were back then."

The old man frowned as he remembered his own youth. "Yes, that's true, Lisa," he said with a sigh. "Twenty percent of the world was going hungry when I was a young man in 1985. The farmers could use only a quarter of the earth's land for farming."

"Why couldn't they use more of the land, Grandpa?" Lisa asked.

"Well," he answered, "the rest was either sizzling hot deserts, barren mountains, or frozen tundras. Of the available farmlands, some had to be used for nonfood crops like



## TEACHING STRATEGY: FUTURE

### Motivation

• **Which do you think would produce more fruit—a tree 2½ m tall, or a tree 1 m tall?** (Answers may vary, but most students will probably say the tree 2½ m tall.)

### Content Development

Use the previous question to introduce the idea of "dwarf" trees that can produce as much or more fruit than trees that are much taller.

• **Assuming that the quality of the fruit is the same, what would be the advantage of the smaller trees?** (Answers may vary. Guide students to understand that smaller trees take up less space. Also, though less important, smaller trees might be easier

## Unit Two

## CLASSIFICATION OF LIVING THINGS

## FUTURES IN SCIENCE: THE CORN IS AS HIGH AS A SATELLITE'S EYE

### BACKGROUND INFORMATION

A major concern of twentieth-century scientists is how to increase and improve food production so as to accommodate a growing world population. One of the greatest obstacles to farming is lack of adequate land. The article states that one-quarter of the earth's land is used for farming; the precise figure is about 30%. Of the land that is not suitable for farming, 20% is too dry; 20% is too mountainous; 20% is sea- or snow-covered; and 10% is land without topsoil.

Clearly, it is to the advantage of agriculturists to maximize the use of available farmland, and also to find alternative areas that can be used for farming. The article emphasizes the possibility of farming in outer space. While it is true that this idea is being developed, its application is aimed primarily at providing food for people in space colonies or for astronauts who spend extended periods (for example, two years) in space. Much more germane to life on earth are the techniques of genetic engineering that seek to maximize the yields of essential food crops. Also important are the use of alternative growing methods such as hydroponics (growing plants without soil) and aquaculture (farming in the ocean).

Several techniques of future farming not discussed in the article include the use of genetic engineering to improve the nutritional quality of food; the use of plant "vaccines" to immunize plants against certain diseases and other environmental hazards; and the genetic engineering of microorganisms that could be used to target insects, weeds, and other pests that are harmful to crops.

## ADDITIONAL QUESTIONS AND TOPIC SUGGESTIONS

**1.** Make a graph to show the percentage of the earth's land that is suitable for farming. For the land that is not suitable for farming, find out what percentages of land are too dry, too mountainous, lacking in topsoil, frozen, and underwater.
**2.** Farming in the ocean is called aquaculture. Find out where in the world aquaculture is being used, and what kind of crops are grown.
**3.** Write a science fiction adventure story about farming in space. You may wish to use the idea of a space colony in which people farm and carry out other occupations.

## CRITICAL THINKING QUESTIONS

**1.** How might farms in space extend and improve the science of space exploration? (Instead of having to carry with them the food needed for a space mission, astronauts would be able to grow their own food. This would certainly increase the amount of time that an astronaut could stay away from earth. In addition, the presence of farms in space might eventually make it possible to develop "space colonies"—groups of people who actually live in space.)
**2.** How do you think the occupation of farming might change as a result of the innovations discussed in this article? What type of person might be attracted to farming in the future? (Accept all answers. It is likely that



cotton. Less food was available and more and more people were being born."

"Are people still starving, Grandpa?" Lisa asked in a worried voice.

"Well, dear, some are, but because of techniques that were developed during the late twentieth century, farmers can now produce more food."

"Oh, Grandpa, that sounds interesting. Will you tell me how more food is made?" Lisa asked.

"Of course I will, Lisa. Let's sit down on that bench and I will tell you all about it," the old man said with a smile.

### *Space Crops*

"Look up there. What do you see?" the old man asked.

"Stars, Grandpa. Why?" the girl responded.

"Well, Lisa, much of the world's food is now being grown in space. We have space stations that contain rotating drums filled with plants. The rotation simulates the earth's gravity. The plants are bathed in fluorescent light, which simulates the sun. Crops that do not grow very tall or that grow in the ground, like lettuce, potatoes, radishes, and carrots, are grown in the drums," the old man stated proudly.

"Wow, that's neat, Grandpa. I never knew that salad came from outer space," Lisa said excitedly.

Laughing, the old man said, "Well, most of it does. We also have space stations where plants grow out of the sides of walls or are moved along conveyor belts through humid air."

### *Genetics Makes the Difference*

"Is all our food grown in outer space, Grandpa?" asked Lisa with a curious look on her face.

"Why, of course not. Methods of growing plants on Earth have been improved too. When I was young, Lisa, twenty percent of the crops that were grown on Earth died off. Now, because of changes scientists have made in the genetic material of plants, very few crops die. The genetic changes have resulted in plants with increased resistance to temperature changes, diseases, herbicides, and harsh environments. Plants have been made resistant by several methods. Some plants were altered by a process called protoplast fusion," the old man said.

Short food plants such as lettuce can be grown out of the sides of revolving drums. Such devices could be used in orbiting space stations to grow food crops in the future.



## FUTURE (continued)

and more efficient to harvest.)

Point out that the development of dwarf fruit trees is just one application of a technique called protoplast fusion, which is a form of genetic engineering. In protoplast fusion, the walls of cells from two different types of plants are dissolved by an enzyme. The contents of the cells are called protoplasts. Once the cell walls have been dissolved, the protoplasts can be combined, or fused, to bring together genetic material of species that could not otherwise be crossed. After the two protoplasts are fused, they are placed in a solution that stimulates the growth of a new cell wall.

### Content Development

You may wish to point out to students that a forerunner of protoplast fusion is a process called tissue culture. In this process, scientists use tiny pieces of plant tissue to reproduce cells that later develop into full-sized plants. Plants produced in this way have the advantage of being genetically uniform and of taking less time to reach maturity than plants grown from seeds. In addition, chemicals can be added to the cell solution in order to produce plants that are stronger and more resistant to disease.

Dwarf trees being grown right now may one day supply enough fruit to fill most of our needs.

"That sounds complicated," Lisa said.

"It's not really very complicated," he replied. "The plant cell walls are dissolved by enzymes. The part remaining is called a protoplast. This protoplast is then fused to another protoplast from a different strain of the same plant or from an entirely different plant. The cell wall is regrown around the fused protoplasts, and a new plant develops. We now have many varieties of such plants. And you might think it strange, but many are smaller than those that used to be grown."

"Smaller? But that doesn't make sense," Lisa protested.

"Well, the smaller plants can produce as much as or more than the big ones because they have been genetically changed to be highly productive," explained the old man. "Also, more of them can be grown in the same amount of space. Two examples of the smaller plants are dwarf rice plants and dwarf peach trees."

"Wow, that's neat. Tell me more," Lisa said.

"All right. We also have plants that can get the nitrogen they need directly from the air. Plants need nitrogen to build amino acids, proteins, and other cell chemicals. In the old days, most plants got their nitrogen from the soil. But this process was using up all the nitrogen in the soil, and the plants didn't grow well. So fertilizer had to be added to the soil. Soybeans, legumes, alfalfa, and clover have always been able to use nitrogen directly from the air with the help of nitrogen-fixing bacteria. Other plants that can use nitrogen directly from the air were developed by combining genes from these nitrogen-fixing bacteria with genes from plants that normally took their nitrogen from the soil. Now almost all the plants take their nitrogen directly from the endless supply in the air."

"I didn't know that. Are any other plants different from the plants you saw when you were young?" Lisa asked.

"Oh, yes," the old man declared. "Many of the crops are now considered to be high-yield. That means they can be harvested faster than they could before, so that more can be grown. By redesigning the plants genetically, scientists have shortened the plants' growing time."

"It seems that genetic changes have kept people from starving," Lisa said thoughtfully. "That and the growing of food in space or on land."

"That's true," said her grandfather. "But there's still a third place. Some of the crops are grown in the ocean. This leaves more room for crops that can be grown only on land."

"That's a neat story, Grandpa," Lisa exclaimed.

"It's not a story, Lisa. It's the truth. But scientists still have to keep working to improve crop production because the population continues to grow," the old man observed.



farming will require knowledge of genetically engineered plants and new growing techniques; some farmers may want to actually carry out processes such as protoplast fusion; still other farmers may travel to outer space! It is likely that a person with scientific curiosity and a taste for adventure and experimentation may be attracted to farming in the future.)

**3.** According to the article, can you expect to see these innovations in farming during your lifetime? How can you tell? (Yes. The story is being told by an older man who was a young man in 1985.)

## CLASS DEBATE

Have students debate the following question: Will the improved techniques of food production discussed in this article benefit those who need help most—people in underdeveloped, overpopulated nations—or will they just be more expensive "gadgetry" to make the United States get richer while the rest of the world remains poorer?

### Reinforcement

Review with students the four ways discussed in this article that scientists hope to improve food crops through the use of genetic engineering.

**1.** By developing smaller plants that can produce as much or more food in less space

**2.** By producing plants that are stronger, so that space is not wasted by plants that die

**3.** By producing plants that can use nitrogen directly from the air

**4.** By producing plants that require shorter growing time

### Teacher's Resource Book Reference

After students have read the Science Gazette article, you may want to hand out the reading skills worksheet based on the article in your Teacher's Resource Book.

### Enrichment

Another way in which scientists hope to improve crops is by genetically engineering new, environmentally safe microorganisms that can attack and destroy specific weeds, insects, and other pests. Have students find out how, in 1981, the microorganism bacillus thuringiensis was used to destroy gypsy moths in the northeastern United States.

### Enrichment

The growing of plants without soil is called *hydroponics*. Have students find out the techniques of hydroponics, where this method is being used, and how successful the method has been. Students may be able to gain information on this topic by contacting the Kraft Foods company, which is experimenting with hydroponics, or by contacting the EPCOT in Orlando, Florida.

## Unit Three

# MATTER

## UNIT OVERVIEW

In Unit Three, students are introduced to the nature of matter. They learn about the properties of matter, including mass and volume. They also explore the meaning of inertia, gravity, and weight, and gain experience with the concept of density.

Next, students study physical properties and changes, including phase changes. Finally, they gain an understanding of the nature of chemical properties and changes.

## UNIT OBJECTIVES

1. **Define *matter*.**
2. **Distinguish among mass, weight, and volume.**
3. **Define and carry out calculations involving density.**
4. **Explain and give examples of physical properties and physical changes.**
5. **Explain and give examples of chemical properties and chemical changes.**

## INTRODUCING UNIT THREE

Begin your teaching of the unit by having students examine the unit-opening paragraph. Before they read the unit introduction, ask them the following questions.

- **What do you see in the picture?** (Ice, snow, rock, water, and penguins are visible.)
- **Where do you think this photograph was taken? How can you tell?** (It must have been taken near Antarctica, since penguins are found naturally only in that region.)
- **What is the snowy form on which the penguins are standing?** (It is an iceberg.)
- **How are icebergs formed?** (They are formed generally as pieces of glacier break off into the ocean.)
- **Is most of the iceberg below the water's surface or above it?** (Most of it is below the surface.)

Now have students read the unit introduction and answer the following questions.

- **Of what are icebergs made?** (They are made of pure, not salt, water.)
- **What property of icebergs enables them to float in liquid water?** (Their density is less than that of liquid water.)
- **How would you define density?** (Density is mass per unit volume.)
- **How does the melting of an ice-**

# Matter

Whipped by angry waves, a huge white form sweeps across the Antarctic Ocean. Carried eastward by icy currents, this giant raft of ice will circle the Antarctic continent like a ghost ship. Now and again, penguins will hitch a ride and then plunge into the sea in search of fish. To a sailor, the iceberg rising 50 meters into the sky spells great danger. No ship that strikes such an object is likely to survive. But the sailor also knows that a ship must stay far from the iceberg's glistening sides, for there is danger beneath the waves as well. It is there that most of the iceberg hides: more than 400 meters of it deep beneath the smashing waves. Perhaps the sailor does not know why so much of the iceberg sinks below the ocean's surface. The danger is the sailor's only concern. But as you read on, you will learn why there is more to the iceberg than meets the eye!

CHAPTERS



*Adelie Penguins hitch a ride on an iceberg.*



## CHAPTER DESCRIPTIONS

**13 General Properties of Matter** In Chapter 13, the nature of matter is examined. Mass and inertia are explained. Gravity and weight are related. Volume is introduced. A verbal and mathematical treatment of the concept of density is presented.

**14 Physical and Chemical Changes** In Chapter 14, physical properties and changes are first contrasted. The idea of physical phase and phase change is introduced and related to changes in heat energy. The chapter concludes with an examination of chemical properties and chemical changes.

**berg differ from the burning of a piece of paper, in terms of how reversible the processes are?** (The melting can be reversed, so that the liquid water can be frozen to form ice. The burning cannot be reversed. The ashes and smoke cannot simply be turned back into paper.)

## Chapter 13

# GENERAL PROPERTIES OF MATTER

### CHAPTER OVERVIEW

Everything in the world is made up of matter. The earth itself and the things of the earth represent a bewildering variety of matter. Yet in all of this variety, there are general properties of matter that make all matter the same.

The general properties of matter are mass, weight, volume, and density. Using these properties, specific matter can be identified and its special features studied.

Mass and weight are not the same. Mass is the amount of matter in an object. Mass can be defined as a measure of inertia. Inertia is the property of a mass to resist changes in motion. The mass of an object doesn't change.

Weight is a force caused by the gravitational attraction between any two objects. On Earth, the weight of an object is a result of the gravitational pull between the planet and the object. The same mass has a different weight on a mountain, in a valley, or on the moon. Because it is a force, weight is measured in force units.

All matter takes up space. The amount of space a certain object takes up is called its volume. The volume and the mass of an object are related by the formula: density = mass/volume, D = M/V, or density is mass per unit volume. All matter has density. The density of a specific kind of matter is a property that helps to identify and distinguish it.

### INTRODUCING CHAPTER 13

Have students observe the photo on page 312.

• **What do you predict is happening in the picture?** (Students might suggest the diver is looking for treasure, living organisms, or a sunken ship.

Explain that the diver is examining the ruins of a ship called the *Mary Rose*. Have the students read the chapter introduction.

Point out that from 1545 to 1970, a period of 425 years, the *Mary Rose* rested on the ocean floor without anyone knowing what had happened. Tell the students that according to the article, it took divers five years to find the *Mary Rose*.

• **Why do you think it took the divers so long to find it?** (Accept all ánswers. Lead students to suggest that the ship was laying deep in the ocean.)

• **How did the divers stay down near the bottom of the ocean?** (Students should remember that the divers wore lead belts.)

• **What did the lead belts do for the divers?** (Accept all logical answers. Students should suggest that the belts gave the divers extra weight.)

• **Without the belts, what do you predict would have happened to the divers?** (Most students will say that the divers would not have been able

# 13 General Properties of Matter

**CHAPTER SECTIONS**



**CHAPTER OBJECTIVES**

*After completing this chapter, you will be able to:*

**13-1** Explain what is meant by the term matter.
**13-1** Identify the general properties of matter.
**13-2** Define mass.
**13-2** Relate mass and inertia.
**13-3** Define gravity and explain its relation to weight.
**13-4** Explain what is meant by the term volume.
**13-5** Provide a word definition and a mathematical formula for density.

As King Henry VIII of England watched proudly, his fighting ship the *Mary Rose* sailed slowly out of Portsmouth Harbor. The *Mary Rose* was among a fleet of English warships that set out on the morning of July 19, 1545, to battle an invading French fleet. In addition to a crew of 415 sailors, the ship carried 285 soldiers and a number of new, heavy bronze cannons.

But the *Mary Rose* never met the French ships. As the story goes, a sudden gust of wind sprang up, and in seconds the *Mary Rose* tipped over and sank to the bottom of the sea. Did this really happen? Or did the mighty ship actually sink at the hands of the French?

Partly to answer this question, teams of scuba divers began to search for the wreck of the *Mary Rose* in 1965. Some of the divers wore lead belts so that they could hover above the sandy ocean bottom in search of the *Mary Rose*.

In 1970, the divers found it. And with their discovery came the answer to its mysterious disappearance. The weight of the cannons had made the ship top-heavy. And when it tipped, water had flowed into the open spaces inside the ship, replacing the air. Without air inside it, the *Mary Rose* had sunk like a rock to the sea floor. It was as simple as that.

But perhaps there is still a mystery to solve. How can a diver wearing a lead belt hover in the sea, while a ship weighted with cannons and excess water plunges to the bottom? The solution to the mystery can be found as you read on.

*Examining ruins of the Mary Rose*



to reach the ocean floor or that they would have floated back to the surface of the ocean.)

• **What do you predict will happen to a person in ocean water?** (Accept all logical answers. Students are likely to say that the person will either float or sink below the surface, but not sink to the bottom.)

• **If the divers would not naturally sink to the bottom of the ocean, why did the ship sink to the bottom?** (Accept all logical answers. Students might refer to water replacing the air in the ship's interior, or they might say the ship was heavier.)

Point out that some objects sink while other objects float.

• **What do you think causes objects to float or to sink?** (Accept all logical answers. Allow students time to use descriptive words in describing objects and in explaining why they sink or float.)

## TEACHER DEMONSTRATION

Set a tall 500- to 1000-mL glass graduated cylinder on the demonstration table. Nearly fill the cylinder with clear, carbonated soda. Show the class a small pea-size stone.

• **What do you predict will happen if the stone is dropped into the liquid?** (It will sink.)

Drop the stone into the liquid.

• **What happened?** (The stone fell to the bottom.)

• **What do you predict will happen if a raisin or kernel of popcorn is dropped into the liquid?** (Students might guess that the raisin or corn kernel will float, sink completely, or sink only halfway to the bottom. The objects usually sink to the bottom, float upward, lose air bubbles, and then sink.)

Drop several raisins and/or popcorn kernels into the liquid.

• **What happened?** (They fell to the bottom and then floated up. Some only floated part of the way up.)

• **What change occurred in the objects to cause them to both float up and sink down?** (Students might suggest a change in weight, magic, or they might notice that the air bubbles have something to do with the movement.

Point out that students will be able to provide a more scientific explanation for their observations after completing the chapter.

## TEACHER RESOURCES

### Audiovisuals

*Investigations in Science: Properties of Matter Series,* film loops, BFA Educational Media

*Matter and Energy* (2nd ed.), film, Coronet

*Matter and Energy: Properties of Matter,* filmstrip, SVE

*Matter and Its Physical Condition,* film, Macmillan

### Books

Goodstein, D. L. *States of Matter,* Englewood Cliffs, N.J.: Prentice-Hall.

Solomon, J. *Structure of Matter,* New York: Halsted Press.

### Software

*Physical and Chemical Properties,* Prentice Hall

## 13-1 MATTER

### SECTION PREVIEW 13-1

Matter is all around you. You are matter. The water you drink is matter. The air you breathe is matter. Matter is in the form of solids, liquids, and gases. Matter has common characteristics. These characteristics are called the properties of matter.

### SECTION OBJECTIVES 13-1

1. **Explain what is meant by the term matter.**
2. **Describe some of the properties of matter.**

### SCIENCE TERMS 13-1

**matter p. 314 property p. 315**

### ANNOTATION KEY

**❶ Students may list rocks, water, trees, clouds, air, and bushes. Other answers are possible. (Making observations)**

**❷ No. The mass of each object is too small to produce the results shown. (Making comparisons)**

**❶ Thinking Skill: Making observations**

**❷ Thinking Skill: Making generalizations**

## 13-1 Matter

Stop what you are doing for a moment, and look around you. You are surrounded by a tremendous variety of materials. There are objects such as a desk, a chair, the book you are reading, and a pencil that you may be holding. Through the window you may see the sky and clouds. Observe your surroundings ❶ again. Then make a mental list of several of the objects you see. How would you describe them? Would you use such words as "large," "small," "red," "blue," "cold," "hot," "heavy," "light," "rough," "smooth"?

You see and touch hundreds of things each day. And although most of these things are probably very different from each other, they share one important quality. They are all forms of **matter.** Matter is what the world is made of. All materials consist of matter.

Through your senses of smell, sight, taste, and touch, you are familiar with matter. Some kinds of matter are easily recognized. Wood, water, salt, clay, glass, gold, plants, and animals are examples of matter that are easily observed. Oxygen, carbon dioxide, ammonia, and air are kinds of matter that are not as easily recognized. Are these different kinds of matter

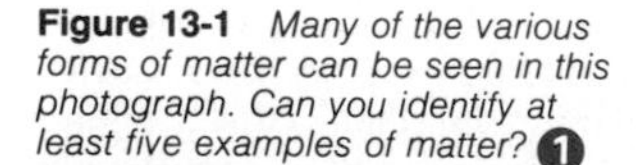

**Figure 13-1** *Many of the various forms of matter can be seen in this photograph. Can you identify at least five examples of matter?* ❶

### TEACHING STRATEGY 13-1

#### Motivation

Have students observe the photograph on page 314. Read the caption. Allow students enough time to determine that the matter in the photograph includes rocks, water, trees, clouds, air, and bushes.

- **What do all these things have in common?** (Students might suggest they are all "something.")

#### Content Development

Point out that matter is all around. Explain that the students themselves are matter. Have the students look around the classroom and list some things that are made of matter.

Tell the students that everything in the universe is made of matter. Explain that matter can be in the form of solids, liquids, and gases. Point out that when we talk about matter, we use words to describe that matter. We describe the characteristics of the matter that we are talking about. Point out that some of the characteristics are very specific, while others are very general.

- **What is a specific characteristic of this desk?** (wood)
- **What is a general characteristic of this desk?** (Accept all logical answers. Some students might point out that it is heavy.)

Explain that the characteristics of matter are called properties of matter. Point out that all matter has common characteristics, or properties. Some properties are specific; they distinguish one form of matter from another. Some properties are general; they describe how all matter is the same.

similar in some ways? Is salt anything like ammonia? Do water and glass have anything in common?

In order to answer these questions, you must know something about the **properties,** or characteristics, of matter. Properties describe an object. Color, odor, shape, texture, and hardness are properties of matter. They are very specific properties of matter, however. Specific properties make it easy to tell one kind of matter from another.

Some properties of matter are more general. Instead of describing the differences among forms of matter, general properties describe how all matter is the same. **General properties of matter include mass, weight, volume, and density.** ❷

### *SECTION REVIEW*

1. What is matter?
2. Name four general properties of matter.

## 13-2 Mass

The most important general property of matter is that it has **mass.** All of the objects you put on your mental list have mass. But what is mass?

Think of the materials you listed. They are all made up of a certain amount of matter. **Mass is the amount of matter in an object.** The mass of an object does not change unless some matter is either removed from the object or added to the object. Mass, then, does not change when you move an object from one location to another. You, for example, have the same mass whether you are on top of a mountain, at the bottom of a deep mine, or on the moon! Later in this chapter you will discover why this is a very important concept.

Suppose you were given the choice of pushing either an empty supermarket cart up a hill or a cart full of groceries up the hill. The full cart, of course, has more mass than the empty cart. And you know from experience that it will be easier for you to push the empty cart. Now suppose both carts are at the top of the hill and begin to roll down. Again, you

**Figure 13-2** *Because of its large mass, a bowling ball is able to produce the results shown in this photograph. Would a hollow bowling ball or a tennis ball do the same?* ❷

## 13-2 MASS

### SECTION PREVIEW 13-2

The most important general property of matter is mass. Mass is the amount of matter in an object. The mass of a specific object will not change. Mass is measured in units called grams (g) and kilograms (kg).

The more mass an object has, the less likely that it can be moved. Objects that have mass resist changes in motion. Newton's first law of motion states that a body at rest will remain at rest unless acted upon by an outside force. A body in motion will remain in motion in the same direction and at the same speed unless acted upon by an outside force. The property of a mass that resists changes in motion is called inertia. Mass is a measure of the inertia of an object.

### SECTION OBJECTIVES 13-2

1. **Explain the general property called mass.**
2. **Discuss the relationship between mass and inertia.**

### SCIENCE TERMS 13-2

**mass p. 315** **inertia p. 316**

### TEACHER DEMONSTRATION

Show the students two small pieces of typing paper.

- **What do these two pieces of paper have in common?** (Students might suggest color, shape, size, kind of material, or other properties.)

Use a match to light and burn one of the pieces of paper. As the paper is burning, place it in a beaker.

- **What is in the beaker?** (Most students will say ashes.)

Show the students the remaining piece of paper.

- **Now what do the pieces of paper have in common?** (Students might suggest that they have nothing in common, or that they might both still contain some tiny bits of the same kind of matter.)

### Skills Development

***Skill: Making generalizations***

Divide the class into teams of 4-6 students per team. Give each team a sealed bottle (half-filled with air and half-filled with water), block of wood, stone, sheet of paper, ball of plastic clay, piece of cloth, and inflated balloon. Have the teams examine the items and list all their general properties.

- **What are the general properties of all the items?** (Accept all logical answers. Students are likely to suggest size, shape, and weight.

Point out that the general properties of all matter are mass, weight, volume, and density.

### Section Review 13-1

1. Anything with mass and volume.
2. Mass, weight, volume, and density.

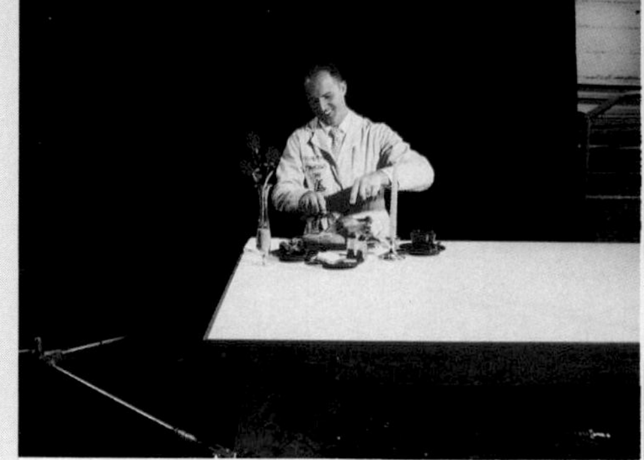

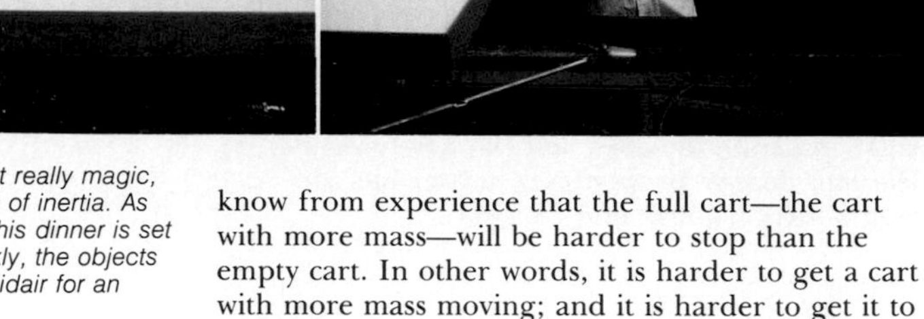

**Figure 13-3** *It's not really magic, just a demonstration of inertia. As the table on which this dinner is set* (left) *is moved quickly, the objects are suspended in midair for an instant* (right).

know from experience that the full cart—the cart with more mass—will be harder to stop than the empty cart. In other words, it is harder to get a cart with more mass moving; and it is harder to get it to stop again.

Scientists use this idea to define mass in another way. Objects that have mass resist changes in their motion. For example, if an object is at rest, a force must be used to make it move. If you move it, you notice that it resists your push or pull. In the same way, if an object is moving and you stop it, you see that it resists this effort too. The property of a mass to resist changes in motion is called **inertia** (in-ER-shuh). **Mass is a measure of the inertia of an object.**

The more mass an object has, the greater its inertia. So the force that must be exerted to overcome its inertia is also greater. That is why you must push or pull harder to speed up and slow down a supermarket cart if it is loaded with groceries than if it is empty.

Mass is measured in units called grams, g, and kilograms, kg. One kilogram is equal to 1000 grams. The mass of small objects is usually expressed in grams. A nickel, for example, has a mass of about 5 grams. The mass of this book is about 1600 grams, or 1.6 kilograms.

### Activity

*Demonstrating Inertia*

You can demonstrate that objects at rest tend to remain at rest by using a drinking glass, playing card, and coin.

**1.** Place the glass on a table.
**2.** Lay a flat playing card on top of the glass. Place the coin in the center of the card.
**3.** Using either a flicking motion or a pulling motion of your fingers, quickly remove the card so it flies out from under the coin. Can you remove the card fast enough so the coin lands in the glass? You might need to practice a few times.

How does this activity demonstrate inertia? What happens to the coin if you remove the card slowly? How does removing the card slowly demonstrate inertia?

### *SECTION REVIEW*

1. What is mass?
2. What is inertia?
3. How are mass and inertia related?

316

## Activity

**Demonstrating Inertia**
**Skills: Observing, relating**
**Level: Average**
**Type: Hands-on**
**Materials: Glass, playing card, coin**

Students are able to demonstrate and reinforce their understanding of inertia through this simple activity. Caution students not to flick the coin in such a way that it might strike another student.

Students should be able to pull the card out fast enough so that the coin drops into the glass. Make sure they can relate this result to the concept of inertia: The coin at rest resists a change in its motion and remains at rest.

When the card is pulled slowly, the coin will move along with the card and not fall into the glass.

## TEACHING STRATEGY 13-2

### Motivation

Divide the class into teams of 4–6 students per team. Have the teams find and list common classroom objects that have masses: (a) less than their science text; (b) greater than their science text. The teams might exchange lists and check to see if they agree with the estimates of mass made by the other teams.

### Content Development

Point out that the mass of an object is a most important property of matter. Explain that the mass of an object is the amount of matter that is in that object. Tell students that the mass of an object can only be changed by removing or adding matter to the object.

• **How is mass measured?** (Students may recall that a balance is used to measure mass and that the units of mass are grams and kilograms.)

### Content Development

Point out that the property of a mass to resist changes in motion is called *inertia.* Explain that an object with a small mass needs only a small force to set it in motion and only a small force to stop the motion. An object with a large mass needs a large force to set it in motion and a large force to stop the motion.

Explain that if an object is at rest, it takes a force to move it. Point out that the force it takes to move the object must be greater than the inertia of the object.

Explain that to stop a moving object, a force is needed. Point out that the force needed to stop a moving object must be greater than the inertia of the object. The inertia of an object is related to both the mass and speed of the object.

## 13-3 Weight

In addition to mass giving an object inertia, mass is also the reason an object has **weight.** Weight is another general property of matter. In order to understand what weight is, you must know something about gravity.

You have probably noticed that a ball thrown up in the air soon falls to the ground. And you know that an apple that drops off a tree falls down, not up. The ball and the apple fall to earth because of the earth's force of attraction for these, and all other, objects. The force of attraction between objects is called **gravity.**

Gravitational force is not a property of the earth alone. All objects exert a gravitational attraction on other objects. Indeed, your two hands attract each other, and you are attracted to books, papers, and chairs. Why, then, are you not pulled toward these objects as you are toward the earth? In fact, you are! But the attractions in these cases are just too weak

**Figure 13-4** *In his novel* From the Earth to the Moon, *author Jules Verne pictured this condition of weightlessness, or zero gravity, inside a spaceship* (left). *His prediction came true, as you can see from the orange-juice sphere and book floating in this astronaut's space capsule.*



## 13-3 WEIGHT

### SECTION PREVIEW 13-3

Weight is another general property of matter. An object has weight because its mass is attracted by the gravitational force of the earth. The force of attraction between objects is called *gravity.*

Gravitational force exists between all objects. The gravitational force between objects is directly proportional to the mass of the objects and inversely proportional to the square of the distance between them. Earth's gravity is great because it is so massive.

The pull of gravity on an object determines the object's weight. The weight of an object can change from place to place. The mass of an object remains constant.

### SECTION OBJECTIVES 13-3

1. **Define gravity.**
2. **Explain the relationship between gravity and mass that causes weight.**
3. **Explain the difference between weight and mass.**

### SCIENCE TERMS 13-3

**weight p. 317** **gravity p. 317**

### TEACHING STRATEGY 13-3

#### Motivation

Have students observe Figure 13-4 and read the caption.

• **Why are the people and objects floating?** (Because they are weightless.)

• **What makes them weightless?** (Some students might suggest the possibility of zero gravity.)

Point out that gravity is not pulling on the people or the objects in Figure 13-4.

• **What does gravity do?** (Students might suggest a variety of answers, such as holds us on Earth, makes things fall, etc.

#### Reinforcement

Encourage students to identify examples of inertia in everyday life.

• **What does a seat belt do for a passenger when a car stops suddenly?** (It prevents him from moving forward.)

• **Why would the passenger move forward without the restraining force of the belt?** (The passenger's inertia causes her to keep moving forward until acted on by a force.)

• **What would stop a passenger not wearing a seat belt?** (the windshield of the car)

#### Section Review 13-2

1. The amount of matter in an object.
2. The ability of a mass to resist changes in its motion.
3. The more mass, the more inertia.

## ANNOTATION KEY

❶ The gravity of Jupiter is 2.7 times the gravity of Earth. (Making comparisons)
❷ 2500 mL (Making calculations)
① Thinking Skill: Relating cause and effect.
② Thinking Skill: Making conversions

### Activity

**A Quick Weight Change**
**Skills: Computation, making inferences**
**Level: Remedial**
**Type: Computational**

This simple computational activity will help students relate the facts that mass is constant and weight changes. The inhabitant of Planet X should weigh 90 jupes on Earth (243/2.7). On the moon, the inhabitant should weigh 15 jupes. Remind students that the gravity of the moon is one-sixth that of Earth.

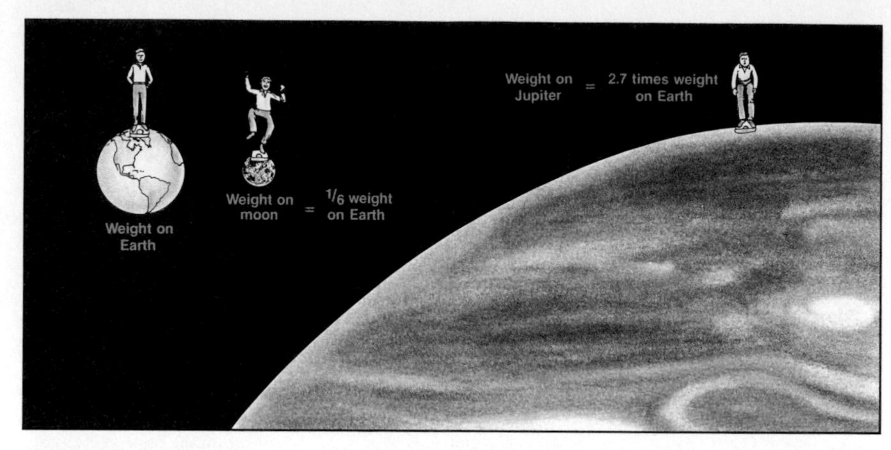


**Figure 13-5** *Because of differences in the gravity of the earth, the moon, and Jupiter, this boy's weight changes from one place to another. His mass, however, stays the same.*

for you to notice them. The earth's gravity, however, is great because the earth is so massive. In fact, the greater the mass of an object, the greater its gravitational force. ❶ How do you think the gravity of Jupiter compares with that of the earth?

**The pull of gravity on an object determines the object's weight.** On the earth, your weight is a direct measure of the planet's force pulling you toward the surface. The pull of gravity between objects weakens as the distance between the centers of the objects becomes greater. ① So at a high altitude, on top of a tall mountain for example, you actually weigh less than on the surface of the earth. Again, this is because you are farther from the center of the earth when you are on top of a mountain. And when an object is sent into space far from the earth, the object is said to be weightless. However—and this is a very important point—the object *does not* become massless. Mass, remember, does not change when location changes. So no matter what happens to the force of gravity, mass stays the same. Only weight changes.

### Activity

*A Quick Weight Change*

An inhabitant of Planet X weighs 243 "jupes" on his home planet. The gravity of Planet X is 2.7 times greater than that of the earth. How many "jupes" will he weigh on the earth? On the moon?

### SECTION REVIEW

1. What is gravity?
2. What is weight?
3. What is the basic difference between mass and weight?



## 13-3 (continued)

### Content Development

Point out that gravity is a force of attraction between all objects in the universe. On Earth we might call the attraction between Earth and an object "Earth force" or "Earth pull." Earth force is the pull of Earth on objects. We normally call this force our "weight." Thus, our weight is the force of attraction between the mass of our body and the mass of Earth. Explain that unless an object has mass, it cannot have weight.

Point out that all objects attract each other. Explain that all objects have a gravitational attraction for one another.

## Section Review 13-3

**1.** The force of attraction between objects.
**2.** A measure of the force of gravity on an object.
**3.** When the force of gravity changes, weight changes but mass does not.

## 13-4 Volume

Did you have this book on your mental list of examples of matter? Let's use the book to help discover another general property of matter. Suppose you could wrap a piece of paper around the entire book and then remove the book. How would you describe what is left inside the paper? You probably would use the word "space." For an important property of matter is that it takes up space. And when the book is occupying its space, nothing else can be in that same space. You might go ahead and prove this to yourself.

**The amount of space an object takes up is called its volume.** The units used to express **volume** are liters, L; milliliters, mL; and cubic centimeters, $cm^3$. ❷ One liter is equal to 1000 milliliters. How many milliliters are there in 2.5 liters? ❷

### Activity

*Volume of a Solid*

The volume of a liquid is easily measured by using a graduated cylinder. Can this method be used for a solid?

Fill a measuring cup with water. Note the volume of the water. Now place a small solid object in the measuring cup. You might choose a rock, a block of wood, or a bar of soap. If the object floats, use a piece of wire to push it under the water's surface. Note the volume again.

Subtract the original volume from the new volume. This difference in volume is the volume of the object.

### SECTION REVIEW

1. What is volume?
2. In what units is volume commonly measured?

### Career: *Scuba Diver*

**HELP WANTED: SCUBA DIVER** To perform tasks under water for limited periods of time. Requires scuba diver certification with advanced scuba training.

On a warm, muggy day in August, a salvage diving crew finally raised a 29-meter-long sailing vessel off the bottom of the Los Angeles harbor. The vessel had been at the bottom of the harbor for a year after mysteriously sinking. The crew consisted of specially trained **scuba divers** who earn their livings by completing underwater tasks with the use of scuba gear.

The tasks of scuba divers differ from job to job. Their work may include inspecting dams, pipelines, and cables. Or it may involve recovering sunken objects, or doing underwater repair and construction. They may work from a boat, off a dock, or from the shore.

Scuba divers must wear wet suits to protect them from the cold temperatures of deep water. They must also wear air tanks. Anyone interested in scuba diving as a career can obtain more information by writing to NASDS Headquarters, P.O. Box 17067, Long Beach, CA 90807.



## 13-4 VOLUME

### SECTION PREVIEW 13-4

Students will be introduced to the idea that volume measures the amount of space taken up by matter. Then they will learn the units in which volume is measured. At this point, students should realize that they know about three important general properties by which matter can be described.

### SECTION OBJECTIVES 13-4

**1. Explain what is meant by volume.**

### SCIENCE TERMS 13-4

**volume p. 319**

### Activity

**Volume of a Solid**
**Skills: Making measurements, manipulative**
**Level: Average**
**Type: Hands-on**
**Materials: Measuring cup, water, small solid object**

In this activity, students are introduced to a laboratory technique to find the volume of solid objects. Students should note that the volume of the object is the volume of the object in water minus the volume of the volume of the water alone.

## TEACHING STRATEGY 13-4

### Motivation

Divide the class into teams of 4–6 students per team. Give each team a 100-mL graduated cylinder and three small cups or bottles. (These containers should have a volume less than 100 mL.) Have the teams label the containers A, B, and C. Teams can then use water and the graduated cylinder to measure the volume of each container. Teams might trade containers with other teams to see if their measurements for the volume of each container are the same.

### Content Development

Show the class a box or cube that measures 10 cm by 10 cm by 10 cm. Tell the class that this amount of space is called a liter (L). Point out that one thousandth of a liter is called a milliliter (mL). Tell the class that volumes of liquids and gases are measured in liters and milliliters.

### Section Review 13-4

1. The amount of space an object takes up
2. Liters, L; milliliters, mL; and cubic centimeters, $cm^3$

## 13-5 DENSITY

### SECTION PREVIEW 13-5

Density is the amount of matter in a given space. More accurately, density is mass per unit volume. For instance, 1 mL of water has a mass of 1 g. Therefore the density of water is 1 g per mL. To find the density of a substance, the formula density = mass/volume is used.

### SECTION OBJECTIVES 13-5

1. **Define density.**
2. **Explain the mathematical formula for density.**

### SCIENCE TERMS 13-5

**density p. 320**

### Activity

**Describing Properties**
**Skills: Making observations, classifying objects, describing objects**
**Level: Remedial**
**Type: Hands-on, vocabulary/writing**
**Materials: Variety of ordinary objects such as rocks, wood, keys, coins, books, marbles, sand, colored water, etc.**

This activity permits students to improve their powers of observation, classification, and description. You may wish to set the activity up as a contest, using teams of students to guess the identities of the samples from the descriptions given by the other students.

### Activity

*Describing Properties*

Collect at least six different kinds of objects as samples of matter. You might include rocks, pieces of wood or metal, and objects made by people. Identify each sample by its general properties and its special properties. Now write a sentence that describes each sample. Be sure to use the following words in your descriptions.

color
density
hardness
mass
matter
property
shape
volume
weight

## 13-5 Density

Suppose you were asked to determine which is heavier, wood or lead. Could you answer this question right away? How would you go about it? Perhaps you would suggest comparing the masses of both on a balance. You are on the right track, but there is one problem with this solution. What size pieces of lead and wood should you use? After all, a small chip of lead does not have a greater mass than a baseball bat made of wood!

You probably realize now that in order to compare the masses of two objects, you need to use equal volumes of each. In other words, you need to compare the way masses of objects are related to their volumes. (1) When you do, you soon discover that a piece of lead has a greater mass than a piece of wood of the *same* size. A cubic centimeter of lead, for example, is heavier than a cubic centimeter of wood. An object's mass per unit volume is called its **density.** All matter has density. And the density of a specific kind of matter is a property that helps to identify it and distinguish it from other kinds of matter.

**Figure 13-6** *Salt water has a higher density than fresh water, so it's easy to float in the great Salt Lake, Utah.*

### TEACHING STRATEGY 13-5

#### Motivation

Place a large box on a table in front of the class. Fill the box with loosely inflated balloons. Count the balloons as you place them in the box.

- **How many balloons are in the box?** (Accept the answer.)
- **Could we place more balloons in the box?** (Some students may say yes; others may say no.)

Force several more balloons into the box.

- **What happened to the volume of the box when more balloons were added?** (Nothing—the volume is the same.)
- **What happened to the mass of the box when more balloons were added?** (The mass increased.)

#### Content Development

Remind students that all matter has specific properties. Show students a pencil, a piece of paper, a flask, and a metal bolt. Point out that all the objects are matter. Point out that all the objects have mass and volume.

- **If all the objects had the same mass, what would the size of each be?** (Accept all logical answers. Students should recognize that the metal bolt would be the smallest of the objects.)
- **If all the objects had the same**

**Since density is mass per unit volume, the formula used to calculate the density of an object is**

$$\text{density} = \frac{\text{mass}}{\text{volume}}$$

Density is often expressed in grams per milliliter or grams per cubic centimeter. The density of wood is about 0.8 g/cm$^3$. This means that any piece of wood 1 cubic centimeter in volume has a mass of about 0.8 gram. The density of lead is 11.3 g/cm$^3$. So a piece of lead has a mass about 14 times that of a piece of wood of the same size.

The density of water is 1 g/mL. Wood floats in water because its density is less than the density of water. What happens to a piece of lead when it is put in water? ❶

From the fact that ice floats, you now know that it too is less dense than liquid water. Actually, the density of ice is about 89 percent that of cold water. What this means is that only about 11 percent of a block of ice stays above the surface of the water. The ③

*DENSITIES OF SOME COMMON SUBSTANCES*

| Substance | Density *(g/cm³)* |
|---|---|
| Air | 0.0013 |
| Aluminum | 2.7 |
| Gasoline | 0.68 |
| Gold | 19.3 |
| Ice | 0.9 |
| Steel | 7.8 |
| Water (liquid) | 1.0 |

**Figure 13-7** *According to this chart, which substances will float on water?* ❷

**Figure 13-8** *Although this iceberg may appear harmless to passing ships, it is really quite dangerous. Only the tip of the iceberg is visible. Nearly 90 percent of it lurks beneath the water's surface!*

321

## ANNOTATION KEY

❶ **It sinks. (Making inferences)**
❷ **Air, gasoline, ice (Applying concepts)**
① **Thinking Skill: Making comparisons**
② **Thinking Skill: Applying formulas**
③ **Thinking Skill: Relating concepts**

## SCIENCE, TECHNOLOGY, AND SOCIETY

A group of American scientists is now trying to prove that the world's tallest mountain is not Nepal's Mt. Everest, but a mountain on the border of China and Pakistan known simply as K2 (its geologic survey number).

Scientists used a Doppler radar device placed about halfway up K2 to receive and record signals from a U.S. Navy satellite as it passed overhead. They found the height of K2 to be 248 m higher than previously thought, making its elevation of 8865 m above sea level—11 m greater than that of Mt. Everest.

Before K2 can be proclaimed as the world's tallest peak, more satellite readings will need to be taken to ensure the reliability of the new measurement. According to surveyors, even the most accurate satellite survey measurements have an approximate error of plus or minus 9.2 m—enough to eliminate the 11-m advantage of K2 over Mt. Everest calculated by this method. Also, the same satellite measure will have to be taken of Mt. Everest to make certain that its elevation has not also been underestimated by past surveyors.

**volume, how would their mass compare?** (Accept all logical answers. Students should say the metal bolt would be the heaviest.)

Point out that all matter has volume and mass. Explain that when equal volumes of matter are compared to their mass, the density of the matter is determined.

Tell students that the density of a specific kind of matter is a property that helps to identify it and distinguish it from other kinds of matter. Explain that the density of an object is the amount of mass in a given amount of volume.

Point out that when things are compared, it is important to compare similar volumes. Explain that the mass per unit volume of a substance is called density. Write:

density = mass/volume
density = grams/milliliters
density = g/mL

Show that if a substance has a mass of 220 g and a volume of 110 mL, its density would be:

density = mass/volume
density = 220 g/110 mL
density = 2 g/mL

Explain that 2 g/mL is read as "two grams per milliliter" and means that every 1 mL of the substance has a mass of 2 g.

## Activity

**Archimedes and the Crown**
**Skills: Applying concepts, designing an experiment**
**Level: Enriched**
**Type: Vocabulary/writing, library**

By doing this activity, students will learn about Archimedes' principle that an object immersed in a fluid undergoes an apparent weight loss (is buoyed up by an amount) that is equal to the weight of the fluid that it displaces.

Student answers should include the idea that because of their different compositions, pure gold and a mixture of gold and silver will not have the same density. Any means of determining density is acceptable, including the idea of weight of water displaced upon immersion.

## HISTORICAL NOTES

When the 46,328-ton "unsinkable" *Titanic* entered passenger-sea service in 1912, it was divided into 16 watertight compartments by 15 transverse watertight bulkheads. On most passenger ships, these bulkheads were 9 m above sea level. The *Titanic's* bulkheads were only 3 m above sea level. When the *Titanic* hit the iceberg on April 13, 1912, five of these compartments were penetrated, allowing water to enter the base of the ship. Thus the *Titanic* sunk about twenty minutes after midnight on April 14, 1912.

### Activity

*Archimedes and the Crown*

The famous Greek mathematician and scientist Archimedes was once faced with a most difficult task. He had to determine whether a crown received by King Hieron was made of gold or a mixture of gold and silver. And he had to accomplish this task without damaging the crown!

Pretend that you are Archimedes' assistant and describe an experiment that would help determine whether the crown is made of gold or a mixture of gold and silver. Hint: The concept of density is useful.

rest is below the surface. This fact is what makes icebergs, like the one pictured on page 321, so dangerous. What is visible is only the "tip" of the iceberg. The rest of it lurks silently below the water's surface!

Can you now solve the mystery posed at the beginning of this chapter? An object floats in water if its density is less than 1 g/mL. In order for the scuba diver to sink in the water, the diver's overall density has to be greater than 1 g/mL. So the diver wears a lead belt to increase mass.

Now, the density of water increases as the water gets colder. So below the surface, the density of water is greater than 1 g/mL. At a certain depth, the scuba diver's density becomes equal to the water's density. The diver will not sink further.

With its large volume partly filled with air, the *Mary Rose* was less dense than water, and so it floated. The air was able to balance the added mass of the cannons. However, when its volume was filled with water, the added mass of the cannons made the *Mary Rose*'s overall density greater than that of the surrounding water at any depth. Down, down went the *Mary Rose*!

**Figure 13-9** *The objects and liquids in this container have different densities. So some float while others sink.*

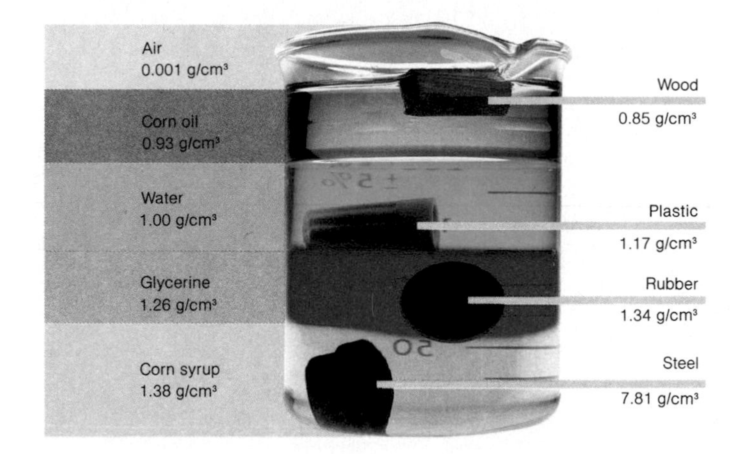


322

## 13-5 (continued)

### Motivation

Show the class an inexpensive piece of imitation gold jewelry, such as a bracelet or necklace.

- **How is it possible to find out if this jewelry is made of real gold without scratching or harming the jewelry?** (Because students are not yet likely to understand the idea of density that is developed in this section, accept all answers and suggest that after studying the text, they will be able to figure out how it could be done.)
- **What would you do to find out?** (Although students are not likely to know about density, they may suggest far out ideas such as using X-rays, lasers, and/or other sophisticated electronic devices.)

### Content Development

Have students observe Figure 13-9. Point out that water has a density of 1 g/cm$^3$. Remind students that a cubic centimeter (cm$^3$) is in fact the same as a milliliter (mL). Therefore, 1 g/cm$^3$ = 1 g/mL. Have students read and compare the densities of the other substances. Tell students that density distinguishes whether a substance is lighter or heavier than the same volume of water. Substances

**Figure 13-10** *It is the density of an object, not its size, that determines whether it will float or sink in water. This large ocean liner* (above) *has an overall density less than 1 g/mL, so it floats in water. The density of this coin* (left) *is greater than 1 g/mL,and so it sinks.*

***SECTION REVIEW***

1. What is density?
2. What determines whether an object floats or sinks in water?



## TEACHER DEMONSTRATION

Nearly fill a large liter beaker with water. Place the beaker on a tray. Show students a glass marble. Place the marble on top of the water.

• **What happened to the marble?** (It fell to the bottom.)

• **What happened to the level of the top of the water?** (It went up.)

• **Which substance has the greatest density?** (the marble)

• **How do you know that the density of the marble is greater than the density of water?** (The marble sinks in water.)

Place several small rocks into a small bottle (be sure the bottle is small enough to float in the beaker). Point out that the glass of bottle and marble have about the same density.

• **What do you predict will happen if this bottle with rocks is placed upright in the beaker of water?** (Students may guess incorrectly that the bottle will sink, or correctly that the bottle will float.)

Place the bottle upright in the beaker of water.

• **What happened?** (Accept all logical observations. The bottle floated. The water overflowed.)

• **Why did the glass bottle with rocks float?** (Students should suggest that the total density of the air, bottle, and rocks is less than one.)

## ANNOTATION KEY

❶ **Thinking Skill: Applying facts**

❷ **Thinking Skill: Relating cause and effect**

with a density less than one will float in water. Substances with a density greater than one will sink in water.

Have students observe Figure 13-10.

• **How can a ship of that size have a density less than 1 g/cm$^3$?** (Some students might suggest that the ship has airtight compartments and that air is not very dense.)

Allow students time to discuss how a ship is built. Explain that although steel is used in shipbuilding, the hull of the ship contains many "air-locked" compartments. Point out that air has a density of 0.001 g/cm$^3$. Explain that the air in the compartments allows the ship to float.

### Content Development

Point out that the density of the entire object (such as a ship) must be determined in order to know if the object is going to float or sink.

Point out that when water filled the air compartments of the *Mary Rose* the total density of the ship increased and the ship sank.

### Section Review 13-5

1. Mass per unit volume.
2. Whether its density is less than or greater than that of water (1g/mL).

# LABORATORY ACTIVITY INERTIA

## BEFORE THE LAB

1. **Divide the class into groups of 3–6 students per group.**
2. **Gather all materials at least one day prior to the investigation. You should have enough supplies to meet your class needs, assuming 3–6 students per group.**

## PRE-LAB DISCUSSION

Have students read the complete laboratory procedure.

- **What is the purpose of the laboratory investigation?** (to define mass in terms of inertia)
- **What is inertia?** (the resistance of a mass to changes in its motion)
- **Why is it important to be sure that each push by the "broom spring" has the same amount of force?** (Accept all logical answers. If this force is not the same for each trial, the results will not be accurate.)

## SKILLS DEVELOPMENT

Students will use the following skills while completing this investigation.

1. Manipulative
2. Safety
3. Observing
4. Comparing
5. Recording data
6. Concluding
7. Applying

## SAFETY TIPS

Remind students to be cautious about releasing the "broom spring" while others are very close by.



## *Inertia*

### Purpose

In this activity, you will define mass in terms of the behavior of objects, showing that the more resistance to change of position, or inertia, an object has, the more mass it has.

**Materials** (*per group*)
Several objects of various masses that will fit into shoe boxes
Smooth table surface
Household broom
Several shoe boxes of similar size
Metric ruler

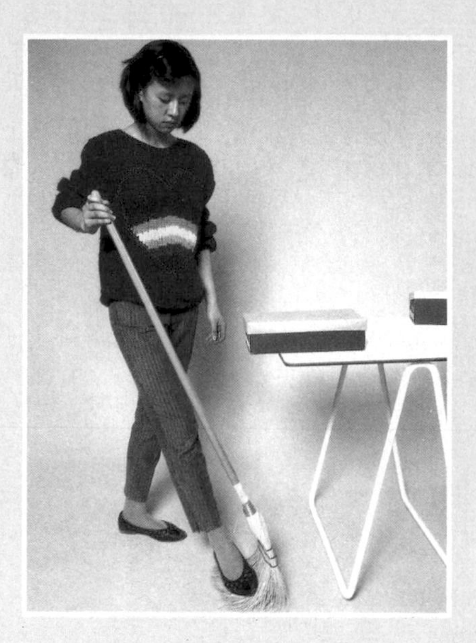

### Procedure

1. Place one object in each shoe box and replace the shoe box's lid.
2. Set the box so that it hangs over the edge of the table by 8 cm.
3. Stand the broom directly behind the table. Put your foot on the straw to hold it in place. The handle should be pointing up in the air directly behind the box.
4. Now slowly bring the handle back away from the box.
5. When you release the handle, it should spring forward, striking the middle of the end of the box.
6. Measure how far the box travels across the table owing to the force of the broom.
7. Repeat the procedure with each of the boxes.

### Observations and Conclusions

At this point, you will have noticed that when the same force—the impact of the broom handle—is delivered, the distance that the box and its contents move varies from box to box.

1. What part of the definition of inertia applies to your observations about the movements of the boxes?
2. Why do you think some boxes moved farther than others?
3. Open the boxes and examine the contents. What do you notice about the objects that moved the farthest from resting position? What do you notice about the objects that moved the least from resting position?
4. Mass can be determined on a balance, but do you now see another way to compare masses?



## TEACHING STRATEGY FOR LAB PROCEDURE

1. Have the students practice manipulating the "broom spring" several times and check that the same amount of force is applied to each box.
2. Have the teams follow the directions carefully as they work in the laboratory.
3. Discuss how the investigation relates to the chapter ideas by asking open questions similar to the following:

- **What did you find out?** (Accept all logical answers.)
- **What kinds of stationary objects have the least inertia?** (objects with smaller mass)
- **What kinds of stationary objects have the greatest inertia?** (objects with larger mass)
- **What do you predict about the inertia of an object on the moon as compared to on Earth?** (Even though the moon's gravity is less, the mass of

## CHAPTER REVIEW

### SUMMARY

**13-1 Matter**

- All objects are made up of matter.
- A quality or characteristic that describes an object is called a property.
- General properties tell how all matter is the same. Specific properties describe the differences among forms of matter.

**13-2 Mass**

- One property of matter is that it has mass. Mass is the amount of matter an object contains.
- The property of a mass to resist changes in motion is called inertia. Mass is a measure of the inertia of an object.
- Mass is commonly measured in grams and kilograms.

**13-3 Weight**

- The force of attraction between objects is called gravity.
- The pull of gravity on an object determines the object's weight.
- The pull of gravity between objects weakens as the distance between the centers of the objects becomes greater.
- The weight of an object can vary, but its mass remains unchanged.

**13-4 Volume**

- The amount of space an object takes up is called its volume.
- Volume is expressed in liters, milliliters, and cubic centimeters.

**13-5 Density**

- The density of an object is its mass per unit volume.
- The density of a specific kind of matter is a property that helps identify it.
- Objects with a density less than 1 gram per milliliter, which is the density of water, will float in water. Objects with a density greater than 1 gram per milliliter will sink.

### VOCABULARY

*Define each term in a complete sentence.*

**density**
**gravity**
**inertia**
**mass**
**matter**
**property**
**volume**
**weight**

### CONTENT REVIEW: MULTIPLE CHOICE

*Choose the letter of the answer that best completes each statement.*

1. Air, water, glass, and clay are examples of
 a. energy. b. matter. c. volume. d. properties.
2. Characteristics that tell how all matter is the same are called
 a. specific properties. b. universal differences.
 c. density numbers. d. general properties.



## GOING FURTHER: ENRICHMENT

**Part 1**

Have students devise other ways to produce a constant force. They might use a pendulum, ball rolling down a ramp, or some other method. Then have students repeat the activity using this different force. Students should then compare the data and conclusions.

**Part 2**

Students might explore the amount of mass needed to "just" stop a moving ball at different speeds. The speed of the ball could be varied by the height of a ramp. Students could then determine the inertia in terms of the amount of mass (in grams) required to stop the motion of the ball.

**Part 3**

Have students drop unequal (and nonbuoyant) masses through air to observe whether mass affects rate of fall. It turns out not to, as Galileo discovered. This actually indicates that the ratio of inertia to gravitational mass is equal for all objects.

the object stays the same. Therefore its inertia stays the same, regardless of the fact that its weight would be less.)

## OBSERVATIONS AND CONCLUSIONS

**1.** An object's mass causes it to resist changes in motion, or exhibit inertia.
**2–3.** The boxes that contained more mass had greater inertia and were not set in as rapid a motion as were the lighter boxes.
**4.** Use one object as a standard. Perform inertia (motion-resistance) tests on it and on other objects and assign mass values on this basis.

## CHAPTER REVIEW

### MULTIPLE CHOICE

1. b  2. d  3. a  4. c  5. b  6. d  7. b  8. a  9. c  10. b

### COMPLETION

1. property
2. mass
3. 1000
4. gravity
5. weight
6. decreases
7. volumes
8. density
9. sink
10. less

### TRUE OR FALSE

1. T
2. F Inertia
3. T
4. F 1000
5. F grams
6. F gravity
7. T
8. T
9. F stays constant
10. T

### SKILL BUILDING

**1a.** Volume of 20-g mass is smaller. **b.** Weight of 20-g mass is smaller. **c.** Density of both samples is the same because both samples are copper.
**2.** 0.78 $g/cm^3 \times 4.0\ cm^3 = 3.12$ g. This object will float in water.
**3.** Students may suggest determining the mass of air and showing that air has volume.
**4.** Weight could be reduced simply by changing position relative to the center of the earth, but such a change would not improve the fit of clothing. The statement should be changed to read: "I have to lose mass."
**5.** If water froze from the bottom of the lake up, the fish would die. However, since ice is less dense than liquid water, the ice rises to the top of the lake and the fish survive beneath the ice layer.
**6.** You are essentially weightless because there are no other objects exerting a gravitational force on you. Your mass is the amount of matter you contain. It is the same mass as you had on Earth. It would not be wise to kick the large boulder hurtling toward you! Although the boulder is essentially weightless, it still has mass. And its mass gives it inertia. Trying to kick it would probably prove rather painful, especially since its large mass and great speed would give it plenty of inertia!

**3.** The amount of matter an object possesses is called its
a. mass. b. volume. c. density. d. weight.
**4.** In describing mass, it is correct to say that
a. mass changes with altitude. b. mass changes with location.
c. mass remains unchanged. d. mass changes with weight.
**5.** An object's resistance to a change in motion is called its
a. density. b. inertia. c. mass. d. volume.
**6.** The force of attraction between objects is
a. inertia. b. weight. c. density. d. gravity.
**7.** As an object gets farther away from the earth,
a. its weight increases. b. its weight decreases.
c. its mass decreases. d. its weight remains the same.
**8.** The amount of space an object takes up is called its
a. volume. b. density. c. weight. d. graduated cylinder.
**9.** The formula for finding density is
a. volume/mass. b. volume × mass.
c. mass/volume. d. mass/weight.
**10.** General properties of matter include
a. mass, shape, and density. b. mass and volume.
c. weight, volume, and color. d. volume and density.

### CONTENT REVIEW: COMPLETION

*Fill in the word or words that best complete each statement.*

1. A quality or characteristic that describes an object is called a(n) ______________.
2. The amount of matter an object has is called its ______________.
3. One kilogram is equal to ______________ grams.
4. The force that pulls objects toward the earth is called ______________.
5. The earth's pull on an object determines the object's ______________.
6. As an object gets farther from the center of the earth, its weight ______________.
7. In order to compare the masses of two objects, you must use equal ______________.
8. An object's mass per unit volume is called its ______________.
9. An object with a mass of 10 g and a volume of 5 mL will ______________ in water.
10. An ice cube floats in water because it is ______________ dense than water.

### CONTENT REVIEW: TRUE OR FALSE

*Determine whether each statement is true or false. If it is true, write "true." If it is false, change the underlined word or words to make the statement true.*

1. All objects are made up of matter.
2. Volume is a measure of the resistance of an object to change in its motion.
3. An object at rest tends to remain at rest because of inertia.
4. One liter is equal to 100 milliliters.

326

### ESSAY

**1.** Aluminum's density is low, making it relatively light and therefore useful for airplane construction. Cast iron's relatively high density makes it more useful for heavy-machine construction.
**2.** Students may suggest doing various tests to determine its properties. These tests might include determining hardness and density.
**3.** The concrete boats will float only if they are less dense than water. To keep the density less than 1 g/mL (density of water), the boat's mass must be as low as possible and its volume as large as possible.

5. The mass of a small object is usually measured in liters.
6. The earth's force of attraction for all objects is called mass.
7. As the pull of gravity decreases, an object's weight decreases.
8. The amount of space an object takes up is called its volume and is measured in liters and milliliters.
9. As an object's weight decreases, its mass increases.
10. An object's mass per unit volume is called its density.

## CONCEPT REVIEW: SKILL BUILDING

*Use the skills you have developed in the chapter to complete each activity.*

1. **Making comparisons** You are given two samples of pure copper, one with a mass of 20 grams and the other with a mass of 100 grams. Compare the two samples in terms of (a) volume, (b) weight, and (c) density.
2. **Making calculations** If the density of a certain plastic used to make a bracelet is 0.78 $g/cm^3$, what mass would a bracelet of 4 $cm^3$ have? Would this bracelet sink or float in water?
3. **Designing an experiment** Air is matter, although it is less easily recognized than other kinds of matter. Using the general properties of matter, suggest two situations in which you could illustrate that air is matter.
4. **Applying concepts** "I have to lose weight" might be the reaction of a person who discovers that clothes fit too tightly. From a scientific point of view, discuss whether a weight loss will make the clothes fit better. What would be a more accurate way of stating the situation?
5. **Relating concepts** Explain why fish are able to survive in lakes during very cold winter months when the lakes freeze.
6. **Making inferences** Suppose you are an astronaut floating in space, far from any other object. What is your weight in space? What is your mass? Would you be willing to remove your space boot and with your bare foot kick a large boulder that comes hurtling past you? Explain your answer.

## CONCEPT REVIEW: ESSAY

*Discuss each of the following in a brief paragraph.*

1. Aluminum is used to make airplanes. Cast iron is used to make heavy machines. How do the densities of these metals make them useful for these purposes?
2. The common metal iron pyrite is often called "fool's gold" because it can be mistaken for gold. How would you go about determining whether a particular sample is iron pyrite or gold?
3. Each year some college students have a contest to build and race concrete boats. What advice would you give the students to make sure their boats float?

327

## ISSUES IN SCIENCE

The following issues can be used as springboards for discussion or given as writing assignments.

**1.** The United States is the only major industrialized nation that does not use the metric system of measurement for common manufactured goods such as screws, nuts, bolts, and other machine parts. How might this difference in the size of "standard" objects affect United States export sales to other countries?

**2.** The United States still uses the inch-pound system of measurement mainly because of tradition and the fear of change. The rest of the world has changed to the metric system because of simplicity, logic, and ease of communication and selling/buying. Should the United States change to the metric system? Why or why not?

## ADDITIONAL QUESTIONS AND TOPIC SUGGESTIONS

**1.** What could be done to increase the cargo capacity of a ship? (Increase the total volume of the hull with the same mass; make the ship out of less dense material; add extra air compartments.)

**2.** What is the displacement of a ship? (the amount of water that the ship will replace below the "safe" waterline of the ship)

**3.** Why is a ship rated in terms of displacement? (If the ship displaces more water than the rated displacement, it is likely to sink.)

**4.** How could you lose weight without losing mass? (Go to a higher elevation such as a mountain, or go to space, the moon, or a smaller planet.)

# Chapter 14

# PHYSICAL AND CHEMICAL CHANGES

## CHAPTER OVERVIEW

Everything in the universe that has volume (occupies space) and mass is matter. Matter can exist as a solid, liquid, gas, or plasma. These states are called *phases of matter*. Solids, liquids, and gases are familiar because they are the naturally occurring phases of matter on Earth. The fourth phase, plasma, exists mainly in stars. Plasma can be made on Earth, but only by using equipment that produces very high energy.

Solids, liquids, and gases can change from one phase to another. Energy must be added to or removed from a substance in order for a phase change to occur. During a phase change, the identity of the substance remains the same.

All matter has chemical properties as well as physical properties. During a chemical change a new substance is formed.

## INTRODUCING CHAPTER 14

Have the students observe the photo on page 328.

• **What happened to the oranges?** (Students might suggest that rain covered the oranges with water, which then froze when the temperature dropped.)

• **What do you predict will happen to the oranges?** (Most students will say the oranges will freeze and probably be ruined.)

Have the students read the chapter introduction.

• **Why do you think they sprayed the oranges with water to save them?** (to protect them from the cold)

• **How do you keep yourself warm on a cold day?** (by wearing sweaters, hats, mittens, jackets, and so on to keep warm)

• **What would you think if someone suggested that you keep yourself warm with a covering of ice?** (Answers will vary. Most students will probably say that the idea sounds crazy.)

• **How does freezing protect fruit from the cold?** (Students probably will not know the answer at this time, but someone may suggest: by releasing heat energy.)

• **Can you think of any other situations in which the process of freezing or cooling releases heat?** (Heat can be felt escaping from the back of a refrigerator or from the outdoor side of an air conditioner.)

• **Do you know why the process of freezing releases heat?** (Students probably will not know until they read the chapter; however, you can encourage them to speculate.)

Refer to the chapter introduction again and help students recognize

# 14 Physical and Chemical Changes

**CHAPTER SECTIONS**



**CHAPTER OBJECTIVES**

*After completing this chapter, you will be able to:*

**14-1** Distinguish between a physical property and a physical change.
**14-2** Classify matter according to phase.
**14-2** State Boyle's and Charles's laws.
**14-3** Identify the various phase changes.
**14-3** Relate phase changes to changes in heat energy.
**14-4** Distinguish between a chemical property and a chemical change.

All day long, the temperature had been steadily dropping. Only a few people working in the orange grove could remember a day as cold in that area of Florida. The branches of the orange trees were heavy with fruit that was not yet ripe enough for picking. If the temperature fell much lower, the juice in the oranges would freeze. The entire crop of fruit would be ruined.

Something had to be done quickly in order to save the orange crop. So workers lighted small fires in the groves. But they soon realized that this heat would not be enough to save the oranges. Suddenly, some of the workers carried large hoses into the grove. Racing against time and temperature, the workers sprayed the trees with water. As the temperature dropped, this water would freeze and turn into ice. Strange as it may seem, ice was being used to keep the oranges warm!

When the sun rose the next day and temperatures began to climb, the glistening ice that had covered the fruit trees melted away. The fruit was undamaged—cold but not frozen. The orange crop had been saved.

Was this some sort of magic? In a sense, yes. But it was magic that anyone who knows science can do. And as you read further, you will learn how freezing water can sometimes do a better job of keeping things warm than fire can.

*Keeping oranges warm with ice!*



## TEACHER DEMONSTRATION

Show the class a beaker of vinegar and a box of baking soda. Point out that the vinegar and baking soda are specific substances. Pour 50 mL of vinegar into a 1-L beaker. Add a spoonful of baking soda. Display the beaker.

- **What happened when the baking soda was added to the vinegar?** (Fizz and bubbles appeared.)
- **What was formed?** (gas bubbles)
- **Where did the gas bubbles come from?** (Answers will vary. Students might suggest a wide variety of explanations. At this time, accept their answers. Later in the chapter, they will learn that the bubbles are a new substance that formed from the chemical reaction of vinegar and baking soda.)

## TEACHER RESOURCES

### Audiovisuals

*Matter and Energy: Physical and Chemical Changes,* filmstrip, Singer Educational Division.
*Particles in Motion: States of Matter,* film, SFS, National Geographic
*Physical or Chemical: What Kind of Change,* filmstrip, CAR
*Solids, Liquids, and Gases,* film, McGraw-Hill

### Books

Booth, V. H., and M. Bloom, *Elements of Physical Science: The Nature of Matter and Energy,* New York: Macmillan.
Goodstein, David L., *States of Matter,* Prentice-Hall
Solomon, J., *Structure of Matter,* Halstead
Toulmin, S., and J. Goodfield, *The Architecture of Matter,* Chicago: University of Chicago Press.

### Software

*Boyle's and Charles's Laws,* Prentice Hall
*Physical and Chemical Properties,* Prentice Hall

that a rather obvious solution to the problem of protecting the fruit by heating did not work.

- **Why do you think the small fires in smokepots were not able to protect the fruit from the cold?** (The heat they provided was not very great, and what heat they did provide dissipated in the atmosphere.)
- **If you had been standing in the orange grove, what would you have thought of the idea of spraying the fruit with water?** (Answers will vary. Many students may say that they would have doubted that this procedure could save the fruit.)
- **Have you ever had a problem to solve in which an unusual and unexpected solution turned out to be better than an obvious solution?** (Encourage students to share experiences involving this type of a situation.)

## 14-1 PHYSICAL PROPERTIES AND CHANGES

### SECTION PREVIEW 14-1

The physical properties of matter are specific properties of the individual matter that distinguish it from other forms of matter. These physical properties can include color, shape, hardness, and density.

A change in which the physical properties of a substance are altered but the substance remains the same kind of matter is called a physical change.

### SECTION OBJECTIVES 14-1

1. **Define a physical property.**
2. **Describe a physical change.**

### SCIENCE TERMS 14-1

**physical property p. 330**
**physical change p. 331**

### ANNOTATION KEY

**❶ Wooden hockey stick: brown; long and thin with curved bottom; fairly hard but can be broken**
**Ice: solid; white; hard; turns to water; can be scraped**
**Cold air: gas; colorless; odorless; tasteless (Making observations)**

**❷ Some substances are poisonous. (Applying safety rules)**

**❸ Ice, liquid water, water vapor (Classifying matter)**

## 14-1 Physical Properties and Changes

How would you describe a copper penny? Or a wooden stick? Well, after reading Chapter 13, you would probably say both the penny and the stick are matter. And because they are matter, they have mass, weight, volume, and density. You are using general properties of matter to tell how the penny and the stick are alike.

What else could you say about them? You might say a copper penny is reddish-brown, round, hard to scratch, and dense enough to sink in water. In contrast, a wooden stick is brown, long and narrow, easy to scratch, and floats in water. Now you are using specific properties of matter to distinguish the penny from the wooden stick. Properties such as color, shape, hardness, and density are called **physical properties.**

Now suppose you take the wooden stick and break it into pieces. Are the pieces still wood? To answer that question, you must determine whether the pieces are still the same kind of matter. In this example, they are. You changed the shape of the wood, but it is still wood. In other words, the wood that makes up the stick has not changed into another substance. **Changes in which physical properties of**

**Figure 14-1** *Various types of matter, each having its characteristic properties, can be seen in this photograph. Can you identify three types and some of their properties? Remember, invisible gases are matter!* ❶

330

### TEACHING STRATEGY 14-1

#### Motivation

Show students a piece of paper and a pencil.

- **How would you describe the piece of paper?** (white, thin solid; has measurements; takes up space)
- **How would you describe the pencil?** (solid, takes up space, made of wood, long, thin, cylindrical or hexagonal, color, etc.)
- **How could we change them?** (They could be changed by breaking, tearing, cutting, and/or burning.)

#### Content Development

Have students describe other objects.

Point out that each object has a certain size, shape, color, hardness, and density. Tell students that when they describe an object or substance, they are describing the physical properties of the object or substance.

#### Skills Development

***Skill: Making observations***

Have students make a list of words describing the physical properties of objects in the classroom.

#### Content Development

Explain that some physical changes are difficult to recognize. Demonstrate stirring a lump of sugar into water until it dissolves.

Point out that both the sugar

**a substance are altered, but the substance remains the same kind of matter, are physical changes.**

Breaking an object into pieces is one example of a **physical change.** Another example is dissolving a substance. You can easily dissolve a cube of sugar in a glass of warm water. The sugar disappears from sight, and the liquid remains clear. You might be tempted to think that somehow the sugar has changed its identity—that it is no longer sugar. But the sugar is still there, as you can tell by the sweet taste of the liquid. Taste, then, is another physical property of matter. Although the sugar has lost its white color and its original shape, it is still the same kind of matter. It still has its physical property of sweet taste. This physical property makes it easy to identify. However, scientists rarely use taste to identify substances. Why do you think this is so? ❷

### *SECTION REVIEW*

1. Give three examples of physical properties.
2. Give an example of a physical change.

## 14-2 The Phases of Matter

Ice, liquid water, steam. Perhaps these three materials seem very different to you. Certainly you can use them in very different ways. You can cool a drink with ice, wash a car with liquid water, and cook vegetables in steam. And because they all look and feel different, you usually do not mix them up.

Ice, liquid water, and steam, however, are all made up of exactly the same substance in different states. These states are called **phases.** Phase is an important physical property of matter. **Matter can exist in four phases—solid, liquid, gas, and plasma.** Ice, liquid water, and steam are phases of water.

### *Solids*

Cubes of ice, cubes of sugar, metal coins, wooden sticks, rocks, and diamonds are several examples of **solids.** And although they look very different, as sol-

**Figure 14-2** *Even after this karate expert breaks the three boards, the wood remains the same kind of substance. Breaking an object into pieces is a physical change.*

**Figure 14-3** *Three familiar phases of the substance water are visible in this photograph. Can you identify them?* ❸

## 14-2 THE PHASES OF MATTER

### SECTION PREVIEW 14-2

Matter is classified according to its phase. The phases of matter are solid, liquid, gas, and plasma. As matter passes from one phase to another, its physical properties change.

In a solid, the tiny particles are packed very close together, giving a solid a definite shape. These particles cannot flow over or around one another. In a liquid, the particles can flow easily around one another, even though they are close together. A liquid takes the shape of its container. Gas particles are more free flowing than liquid particles. Like a liquid, a gas has no definite shape. Plasma contains very high energy particles. Matter in the plasma phase is not found naturally on Earth.

### SECTION OBJECTIVES 14-2

1. **Identify a phase change as an important physical property of matter.**
2. **Describe the four phases of matter.**
3. **State the Gas Laws.**

### SCIENCE TERMS 14-2

**solid p. 331**
**crystal p. 332**
**crystalline solid p. 332**
**amorphous solid p. 333**
**liquid p. 333**
**gas p. 334**
**Boyle's Law p. 335**
**Charles's Law p. 336**
**plasma p. 336**

and water have changed, but the beaker still contains water and sugar. Explain that the sugar could be separated from the water by evaporating the water. Place the beaker on a window ledge and observe the sugar that remains after the water has evaporated.

### Section Review 14-1

1. Color, shape, hardness, density.
2. Breaking an object into pieces, dissolving a substance.

### TEACHING STRATEGY 14-2

#### Motivation

Have students look up the word *phase* in the dictionary and read some of the definitions out loud. Discuss the definitions with the following questions in mind:

- **What are the key words used in each definition?** (part, change, cycle, etc.)
- **What do you predict a phase of something is?** (a change of appearance or behavior)

Point out that the word *phase* is used in science to describe changed forms or states. Tell students that there are phases of the moon, phases in electrical current, and many other phases in the universe.

## BACKGROUND INFORMATION

The particles of any substance are constantly in motion. The type and extent of the motion determines whether the substance is solid, liquid, or gas. Particles of a solid are in fixed positions. Their movements consist primarily of vibrations. Particles of a liquid are free to move from one place to another, but forces of attraction keep the particles close together. Particles of a gas are spread out, for there are essentially no forces of attraction between them. These particles are in constant random motion, and they collide with each other frequently.

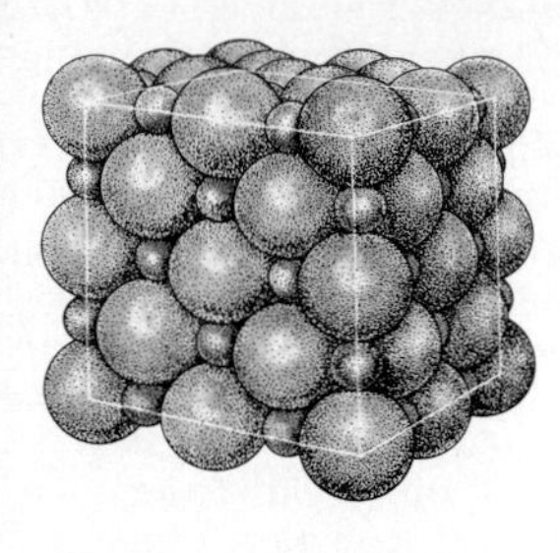

**Figure 14-4** *The white cubic crystals in this photograph* (left) *are sodium chloride, or table salt. They are growing on crystals of calcium sulfate, or gypsum. Crystals have a regular arrangement of particles, as shown in the drawing of the structure of sodium chloride* (right).

ids they share two basic characteristics. All solids have a definite shape and a definite volume.

1 All the tiny particles in a solid are packed very close together, so it keeps its shape. The particles cannot move far out of their places and, in most solids, the particles cannot flow over or around each other. In many solids, the particles are arranged in a regular, repeating pattern. Such a regular arrangement of particles is called a **crystal** (KRIS-tuhl). Solids made up of crystals are called **crystalline solids.** Common table salt is a good example of a crystalline solid. See Figure 14-4.

Crystals often have beautiful shapes that result from the arrangement of the particles within them. Snowflakes are crystals of water in the solid phase. If you look at them closely, you will see that all the

**Figure 14-5** *Snowflakes* (left) *are six-sided crystals of water in the solid phase. The repeating pattern of the particles in ice can be seen in this computer-generated drawing* (right) *of a portion of the crystal.*

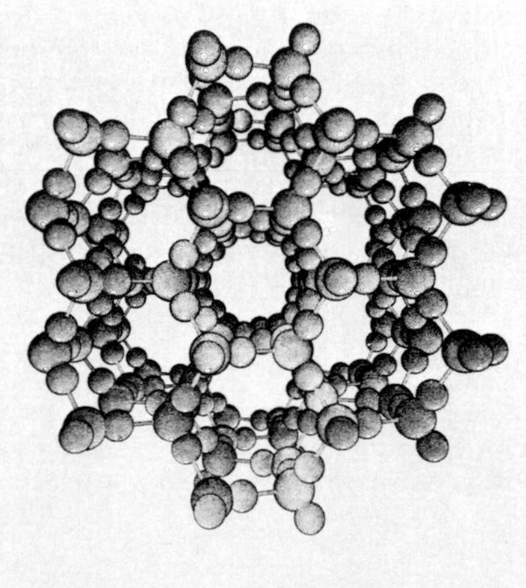

### 14-2 (continued)

### Content Development

Tell students that matter is classified by the phase in which it exists. Point out that matter can exist in any of four phases. Explain that matter is found naturally occurring on Earth as either a solid, liquid, or gas. Point out that the fourth phase of matter, plasma, does not occur naturally on Earth.

Write "solid," "liquid," and "gas" on the chalkboard. Explain that matter in the solid phase has less energy than matter in the liquid phase. Matter in the gas phase has more energy than matter in the liquid phase.

Write the following on the chalkboard:

ice ↔ water ↔ steam

Point out that phase changes are physical changes. Explain that one phase can change into another phase provided that the necessary amount of energy is gained or lost.

### Content Development

Point out that most liquids will become the corresponding solids if the temperature is lowered enough. Explain that the tiny particles in a solid are very close together and very slow moving. Tell students that the particles cannot flow over or around one another. Explain that the particles in a solid have little movement and only vibrate or jiggle.

### Content Development

Point out that when solids form, the particles may arrange themselves in two different ways, depending on conditions. Tell students that sometimes the particles are arranged in regular, repeating patterns called *crystals.* Have the students observe Figure 14-4. Point out that common table salt is the chemical compound sodium chloride.

- **What is the shape of the sodium chloride crystals?** (They look like squares or cubes.)
- **How many flat surfaces are on each cube?** (six)

Have students observe the structural drawing of sodium chloride in Figure 14-4. Point out that the small spheres represent sodium and the large spheres represent chlorine.

flakes have six sides. However, what is so amazing is that no two snowflakes in the world are exactly alike!

Have you ever played with sealing wax or silicone rubber? Both of these materials are solids. Yet from your experience, you may know that they do not keep their shapes permanently, as do crystalline solids. Left out on a table top for a long period of time, both of these solids will lose their shape and flatten out into a "puddle." Such solids that do not keep a definite shape are called **amorphous** (uh-MOR-fuhs) **solids.** Unlike crystals, the particles within amorphous solids are not arranged in a rigid way. So these particles can slowly flow around one another. ❷

Some scientists think of amorphous solids as slow-moving liquids. Tar, candle wax, and glass are examples. Are you surprised to learn that glass actually flows? You might be able to see this for yourself if you can look at windowpanes in very old houses. Such windowpanes are thicker at the bottom than at the top because the glass has flowed slowly downward. Given enough centuries, the glass might flow completely out of its frame! Glass is sometimes described as a supercooled liquid. It is formed when a material in the liquid phase is cooled to a rigid condition, but no crystals form.

**Figure 14-6** *Amorphous solids like this sealing wax lose their shape under certain conditions.*

### *Liquids*

When you put water into a glass, it takes on the same shape as the glass. But when you pour that water onto a table or floor, it takes on a different shape. This behavior is a property of another phase of matter—the **liquid** phase.

Liquids have no definite shape. They take the shape of the container into which they are placed. So liquid water in a square container is square. And liquid water in a round container is round. Liquids can take on different shapes because the particles in them can flow easily around one another, even though they are close together. ❸

Although liquids do not have a definite shape, they do have a definite volume. For example, pouring one liter of water into a two-liter bottle does not fill the bottle. The water does not spread out to fill the entire volume of the bottle. What happens if you try to pour two liters of water into a one-liter bottle? ❶

#### Activity

*Observing Viscosity*

Viscosity is the resistance of a liquid to flow.

**1.** Obtain samples of the following: catsup, corn syrup, milk, honey, maple syrup.

**2.** Cover a piece of cardboard with aluminum foil.

**3.** Place the cardboard on a plate or baking pan at about a 50–55-degree angle with the bottom of the plate or pan.

**4.** With four classmates helping you, pour a measured sample of each liquid from the top of the cardboard at a given signal.

**5.** Determine the order in which the liquids reach the bottom of the cardboard.

Which liquid is the most viscous? The least viscous?

333

## ANNOTATION KEY

❶ **The water will overflow. (Applying concepts)**

❶ **Thinking Skill: Identifying patterns**

❷ **Thinking Skill: Making comparisons**

❸ **Thinking Skill: Identifying patterns**

## Activity

**Observing Viscosity**
**Skills: Making observations, making comparisons**
**Level: Remedial**
**Type: Hands-on**
**Materials: Catsup, corn syrup, milk, honey, maple syrup, cardboard, aluminum foil, baking pan**

In this activity, students will observe the viscosities of various substances. Students should infer that the substance that flows the quickest is the least viscous. That substance will be the milk. The most viscous substances will flow the least. The single most viscous substance will depend on the substances students use.

- **How many chlorine spheres surround each sodium sphere within the cube?** (six)
- **How many sodium spheres surround each chlorine sphere within the cube?** (six)

## Content Development

Point out that the particles in some solids are not arranged in a rigid way. These particles are held together loosely in an irregular arrangement. When the particles are held together loosely, the substance does not keep the same shape. Explain that solids that have an irregular arrangement of particles are called *amorphous solids.* Point out that amorphous solids do not keep their shape permanently.

## Content Development

Fill a beaker or other cylindrical container with water. As students watch, pour the water into a rectangular container.

- **How do you know that this substance is a liquid?** (It pours; it changes shape; it stays together as it flows.)
- **Could the water be poured into any size container? Why not?** (No. It would overflow a container that is too small.)
- **Would the water fit any shape container provided the container was large enough? Why?** (Yes. A liquid takes the shape of its container.)
- **How would you predict the particles of a liquid are arranged?** (They are free to move around, but are still relatively close together.)

## BACKGROUND INFORMATION

Because the particles of a gas are in constant motion, they have kinetic energy. Increasing the temperature of a gas makes the particles move faster, thus increasing their kinetic energy. Collisions also increase kinetic energy as faster moving, high-energy particles transfer energy to slower moving particles.

## HISTORICAL NOTES

Robert Boyle (1627–1691), an English scientist, was the first person to investigate what he called the "spring of the air." His experiments showed that air had elasticity and explained the behavior of gases. He was a contemporary of Isaac Newton.

## ANNOTATION KEY

**❶ The container breaks or explodes. (Applying concepts)**

❶ Thinking Skill: Identifying patterns

❷ Thinking Skill: Identifying relationships

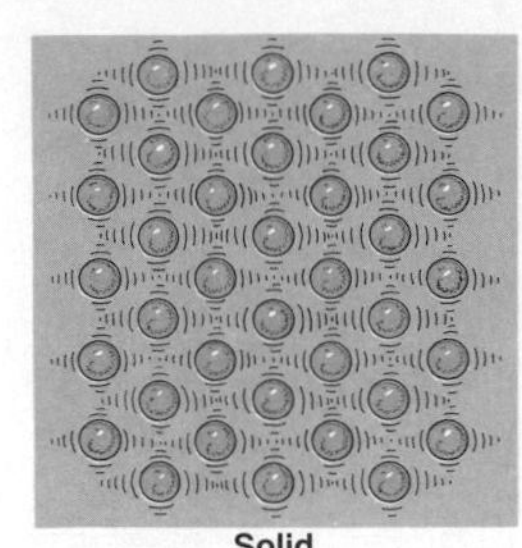

Solid

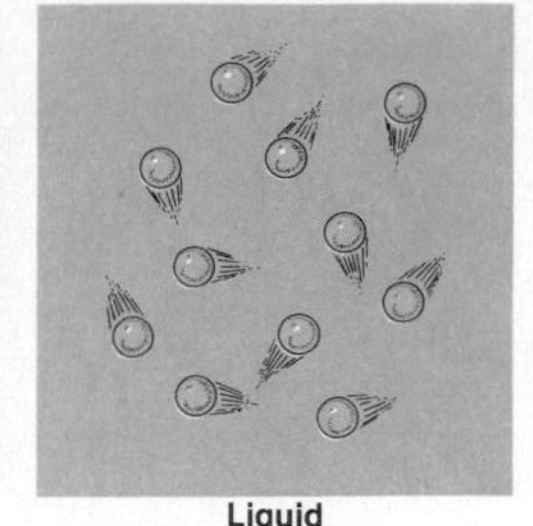

Liquid

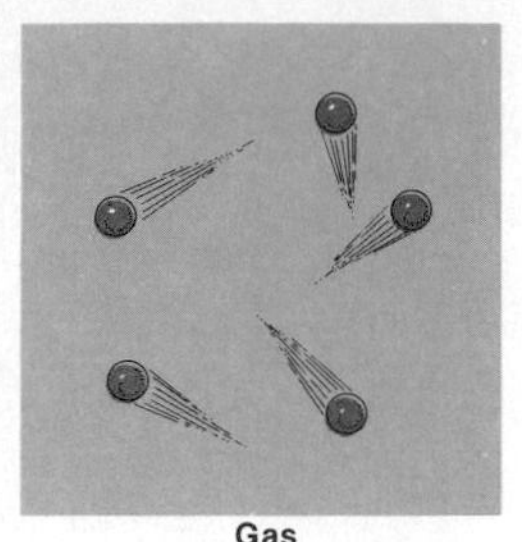

Gas

**Figure 14-7** *The arrangement and movement of particles vary in a solid, a liquid, and a gas.*

### Gases

Have you ever pumped air into a bicycle tire or blown up a balloon? If so, you may have observed an important property of another phase of matter—the **gas** phase. ❶ Gases have no definite volume. They always fill their container, regardless of the size or shape of the container.

When air is pumped into a bicycle tire or a balloon, a large amount of gas is being squeezed into a small volume. Fortunately, the particles in a gas can be pushed close together.

Just the opposite can also happen. A small amount of gas can spread out to fill a large volume. The smell of an apple pie baking in the oven comes to you because gases from the pie spread out to

**Figure 14-8** *A gas has no definite volume and will expand to fill its container. If allowed to, the gas will expand without limit, which is what has happened to the gas in Donald Duck's left hand. A hole in that part of the balloon has enabled the gas to escape into the air.*

## 14-2 (continued)

### Motivation

Show students an air-filled balloon.

- **What is inside this balloon?** (air)
- **What phase of matter is air?** (gas)
- **What shape is the air?** (the shape of the balloon)
- **What is the volume of the air?** (the same as the volume of the balloon)

Let the air out of the balloon.

- **What is the shape of the air now?** (It has combined with the air in the room and therefore has taken on the shape of the room.)
- **What does this tell you about the particles of a gas?** (They spread apart easily and do not seem to hold together.)

### Content Development

Point out that air is one example of a gas phase of matter. Matter in the gas phase has neither a definite shape nor a definite volume. Air is composed of a mixture of nitrogen, oxygen, carbon dioxide, water vapor, and other gases. Gas molecules move rapidly.

### Motivation

Have students imagine that they are standing in a crowded car of a train. Suddenly someone starts pushing and shoving.

- **What do you predict will happen?** (Someone may start pushing back; people will bump into one another; someone may fall down; some people will bang into the walls of the car, etc.)

Explain that the force of pushing people in a train car is similar to the pressure created within a closed container when gas molecules collide with one another.

every part of the room. In fact, if allowed to, gases will expand without limit. If not for the pull of gravity, all the gases making up the earth's atmosphere would do just that!

Like liquids, gases have no definite shape. They take the shape of their container. The particles that make up a gas are not arranged in any set pattern. So it is very easy for gas particles to move around, either spreading apart or moving close together.

### The Behavior of Gases

The world inside a container of gas particles is not as quiet as it may seem to you. Although you cannot see the particles of gas, they are in constant motion. Whizzing around at speeds of about 500 meters per second, these bulletlike particles are constantly hitting one another. In fact, each particle undergoes about 10 billion collisions per second! Added to that are the collisions the particles make with the walls of the container. The effect of all these collisions is an outward pressure, or push, by the gas. The pressure is what makes the gas expand to fill its container. What do you think happens to a container when the pressure of the gas becomes too great? ❶

**BOYLE'S LAW** Imagine you are holding an inflated balloon. If you press lightly on the outside of the balloon, you can feel the air inside pushing back. Now if you squeeze part of the balloon, what do you feel? You probably feel the air pressing against the walls of the balloon with even greater force.

This increase in pressure is due to a decrease in volume. By squeezing the balloon, you reduce the space the gas particles can occupy. As the particles are pushed a bit closer together, they collide with one another and the walls of the balloon even more. So the pressure from the moving gas particles increases. The relationship between volume and pressure is called **Boyle's Law:** The volume of a fixed amount of a gas varies *inversely* with the pressure of the gas. In other words, as one increases, the other decreases. If the volume increases, the pressure decreases. If the volume decreases, the pressure increases. ❷

#### Activity

*Determining Particle Space*

In this activity, you can determine how closely the particles of a solid and liquid are arranged.

**1.** Fill one 250-mL beaker with marbles, another with sand, and a third with water.

**2.** Describe the appearance of the beaker filled with marbles. Do the marbles occupy all the space in the beaker?

**3.** Carefully pour some sand from its beaker into the beaker of marbles. How much sand fits? Is all the space in the beaker now occupied by sand and marbles?

**4.** Carefully add some water from its beaker to the beaker of marbles and sand. How much water can you add?

Is there space between the particles of a solid or a liquid?

335

#### Activity

**Determining Particle Space**
**Skills: Making observations, manipulative**
**Level: Average**
**Type: Hands-on**
**Materials: 250-mL beaker, marbles, sand, water**

In this activity, students will note that there is space between the particles in a solid and a liquid. Students will note that the marbles do not fill all the available space in the beaker. Sand can be added, which will fill the spaces between the marbles. The amount of water that can be added will vary, but students should note that the water can also fill some of the space between the particles of sand and the marbles.

#### TIE-IN/ HEALTH

Students might do library research to find out how our body's breathing is an illustration of Boyle's Law. By increasing the chest cavity volume, the pressure inside the body is decreased below that of the outside air. This is accomplished by the diaphragm muscles. Then, with exhalation, the volume is decreased, and the pressure in the cavity becomes greater than the outside air.

#### Content Development

Point out that when the particles of a gas collide, the pressure increases. Explain that as the pressure increases, the gas particles move faster and faster, causing more and more collisions. Explain that anything that increases the number of particle collisions within a gas will increase the pressure.

Tell students to imagine two streets. One street is crowded with many cars; the other street has only two cars.

• **On which street do you predict it is more likely to have a collision? Why?** (The street crowded with cars. The cars have less room to move.)

• **Now picture a container full of gas particles. What will happen to the number of collisions if the container becomes smaller? Why?** (They will increase. Each particle has less space in which to move because the particles are closer to one another and to the sides of the container.)

• **Suppose the container becomes much larger. What will happen to the number of collisions? Why?** (They will decrease. Each particle has more space in which to move.)

Write on the chalkboard,
Volume up ↑, pressure down ↓
Volume down ↓, pressure up ↑

Point out that this relationship between volume and pressure is called *Boyle's Law.* Explain that Boyle's Law states that the volume of a fixed amount of a gas varies inversely with the pressure exerted on it, provided the temperature remains constant.

## ANNOTATION KEY

❶ **Its volume decreases. (Identifying relationships)**

❶ Thinking Skill: Identifying relationships

❷ Thinking Skill: Developing a model

❸ Thinking Skill: Relating facts

❹ Thinking Skill: Applying concepts

### Activity

**Charles's Law**
**Skills: Relating concepts, relating cause and effect, manipulative**
**Level: Enriched**
**Type: Hands-on**
**Materials: Balloon, string, metric tape measure, oven, freezer unit**

In this activity, students obtain a better understanding of Charles's Law by observing the effects of heat and cold on the diameter of a balloon. Students should note that the balloon increases in diameter when heated and decreases in diameter when cooled. You may want to work with students when placing the balloons in the heated oven. Instruct students to use heat-resistant gloves whenever working with heated objects.

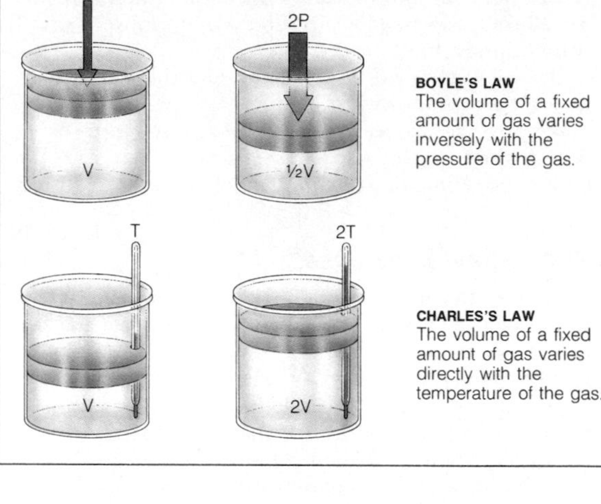

P
2P
V
½V
BOYLE'S LAW
The volume of a fixed amount of gas varies inversely with the pressure of the gas.
T
2T
V
2V
CHARLES'S LAW
The volume of a fixed amount of gas varies directly with the temperature of the gas.

**Figure 14-9** *You can see in this illustration that if the pressure of a fixed amount of gas increases, the volume of the gas decreases* (top). *This inverse proportion between pressure and volume is called Boyle's Law. According to Charles's Law, if the temperature of a fixed amount of gas increases, the volume of the gas increases* (bottom). *The relationship between temperature and volume is a direct proportion.*

### Activity

*Charles's Law*

**1.** Inflate a balloon, making sure it is not so large that it will break easily. Tie the end of the balloon so that air inside cannot escape.

**2.** Measure and record the diameter of the balloon.

**3.** Put the balloon in an oven set at a low temperature—not more than 150° F (65° C). Leave the balloon in the oven for about 15 minutes.

**4.** Remove the balloon and quickly measure its diameter.

**5.** Now place the balloon in a refrigerator for 15 minutes.

**6.** Remove the balloon and measure and record its diameter. What happens to its size at the different temperatures? Do your observations agree with Charles's Law?

**CHARLES'S LAW** Imagine you have that inflated balloon again. This time you heat it very gently. What do you think happens to its volume? As the temperature increases, the gas particles absorb more heat energy. They speed up and move further away from one another. So the increase in temperature results in an increase in volume. If the temperature had decreased, then the volume would have decreased. ❶ This relationship between temperature and volume of a gas is called **Charles's Law:** The volume of a fixed amount of gas varies *directly* with the temperature of the gas. What do you think happens as the temperature of a gas drops? Try to support your answer by putting an inflated balloon in your freezer. ❶

❷ Boyle's Law and Charles's Law together are called the Gas Laws. **The Gas Laws describe the behavior of gases with changes in pressure, temperature, and volume.**

### Plasma

The fourth phase of matter is quite rare on the earth. It is called the **plasma** phase. Matter in the plasma phase is very high in energy. In fact, the particles contain so much energy that they are

## 14-2 (continued)

### Content Development

Explain to students that when pressure is held constant, the volume of a gas increases as temperature increases and decreases as temperature decreases. This is Charles's Law. Point out that pressure remains constant during the volume change.

Have students observe Figure 14-9. Read the caption. Read the explanations accompanying the illustrations of Boyle's Law and Charles's Law.

- **What is the constant in Boyle's Law?** (temperature)
- **How does Charles's Law differ from Boyle's Law?** (Heat is added to the gas and pressure is kept constant.)

### Reinforcement

Have students determine which of the following statements represent an inverse relationship and which represent a direct relationship:

The greater the number of hours worked, the more money earned. (direct)

The greater a car's speed, the farther it travels in one hour. (direct)

The more hours you sleep in a day, the fewer hours you are awake. (inverse)

The greater the number of cars

dangerous to living things. Luckily, plasma is not found naturally on the earth. It is, however, common in stars, such as the sun. Plasma can only be made on the earth by using equipment that produces very high energy. But the plasma cannot be contained by the walls of ordinary matter, which it would immediately destroy. Instead, magnetic fields produced by powerful magnets keep the high-energy particles in a plasma from escaping. One day, producing plasmas on the earth may meet most of our energy needs.

### SECTION REVIEW

1. What are the four phases of matter?
2. How is a crystalline solid different from an amorphous solid?
3. State Boyle's Law and Charles's Law.

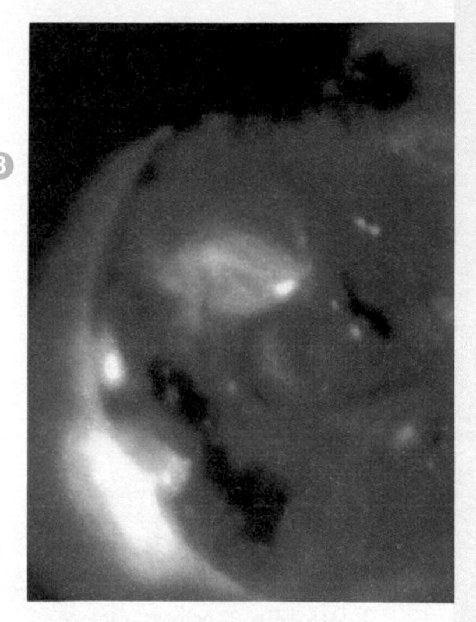

Figure 14-10 *These eruptions on the sun are examples of the high-energy plasma phase of matter.*

## 14-3 Phase Changes

Ice, liquid water, and steam are all the same substance. What, then, causes the particles of a substance to be in one particular phase rather than another? The answer has to do with energy—energy that can cause the particles to move faster and farther apart.

A solid substance tends to have less energy than that same substance in the liquid phase. A gas usually has more energy than the liquid phase of the same substance. So ice has less energy than liquid water, and steam has more energy than both ice and liquid water. The greater energy content of steam is what makes a burn caused by steam more serious than a burn caused by hot water!

Because energy content is responsible for the different phases of matter, substances can be made to change phase by adding or taking away energy. And the easiest way to do this is to heat or cool the substance, allowing heat energy to flow into or out of it. This idea should sound familiar to you since you frequently increase or decrease heat energy to produce phase changes in water. You put

Figure 14-11 *When heat is applied, ice changes to liquid water and then to water vapor.*

337

## 14-3 PHASE CHANGES

### SECTION PREVIEW 14-3

The five phase changes of matter are melting, freezing, vaporization, condensation, and sublimation. These phase changes are physical changes, and the identity of the substance involved remains the same. Phase changes are produced when energy is added to or taken away from a substance.

### SECTION OBJECTIVES 14-3

1. **Identify the phase changes in matter.**
2. **Explain how adding or taking away energy will produce a phase change.**
3. **Discuss the relationship between heat, energy, and phase change.**

### SCIENCE TERMS 14-3

**melting p. 338**
**melting point p. 338**
**freezing p. 339**
**freezing point p. 339**
**vaporization p. 339**
**evaporation p. 340**
**boiling p. 340**
**boiling point p. 340**
**condensation p. 341**
**sublimation p. 341**

on the road, the greater the chances of an accident. (direct)

The more people who enter the contest, the less chance I have of winning. (inverse)

### Section Review 14-2

1. Solid, liquid, gas, plasma.
2. Crystalline: regular arrangement of particles, keeps its shape; amorphous: no rigid arrangement of particles, can flow.
3. Boyle's: Volume of a fixed amount of gas varies inversely with pressure; Charles's: Volume of a fixed amount of gas varies directly with temperature.

### TEACHING STRATEGY 14-3

#### Motivation

Show students a block of ice or some ice cubes.

- **What phase of matter is the ice?** (solid)
- **What will happen if the ice sits at room temperature for a long time?** (It will melt.)
- **What phase of matter is melted ice?** (liquid)
- **Could the liquid phase change into the gas phase through a physical change? How?** (Yes. The liquid will eventually evaporate into the gas phase.)

Point out that the ice could easily go through three phases by starting out as a solid, then melting to a liquid, and then evaporating into a gas.

## Activity

**Melting Ice and Freezing Water**
**Skills: Making observations, making comparisons, manipulative**
**Level: Average**
**Type: Hands-on**
**Materials: Ice cubes, water, glass, Celsius thermometer, freezer unit**

Through this simple activity, the fact that the melting and freezing points of water are the same will be reinforced for students who have trouble grasping this concept. You may want to have interested students graph the results of the experiment.

## BACKGROUND INFORMATION

Most substances expand when heated, but water is an exception to the rule. Because the crystalline structure of ice involves large spaces between molecules, water expands as it freezes. Thus the density of ice is less than the density of liquid water. That is why ice floats in water.

A liquid boils when the vapor pressure of the liquid equals atmospheric pressure. It is the vapor pressure that gives the particles enough "push" to escape into the gas phase. As altitude increases, atmospheric pressure decreases. At high altitudes, the vapor pressure of a liquid does not need to be as high in order to equal atmospheric pressure. That is why water boils at a lower temperature on a mountain than at sea level.

### Activity

*Melting Ice and Freezing Water*

In this activity, you will discover an important relationship between the melting and freezing points of water.

**1.** Place several ice cubes, a little water, and a thermometer in a glass and wait several minutes before observing.

**2.** Observe and record the temperature as the ice melts. This temperature is the melting point of ice.

**3.** Place a glass of water with a thermometer in it in the freezer. Observe and record the temperature every 10 minutes. Record the temperature of the water when it begins to freeze. This temperature is the freezing point of water.

How does the melting point of ice compare with the freezing point of water?

liquid water in the freezer to remove heat and make ice. On a hot stove you add heat to make liquid water turn to steam. **The phase changes in matter are melting, freezing, vaporization, condensation, and sublimation.**

Changes of phase are examples of physical changes. In a physical change, a substance changes from one form to another, but it remains the same *kind* of substance. No new or different *kinds* of matter are formed, even though physical properties may change.

### Solid-Liquid Phase Changes

What happens to your popsicle on a very hot day if you do not eat it fast enough? Right—it begins to melt. **Melting** is the change of a solid to a liquid. Melting occurs when a substance absorbs heat energy. The rigid crystal structure of the particles breaks down, and the particles are free to flow around one another.

(1) The temperature at which a solid changes to a liquid is called the **melting point.** Most substances have a characteristic melting point. It is a physical property that helps to identify the substance. The melting point of ice is 0° C. The melting point of table salt is 801° C, while that of a diamond is 3700° C.

**Figure 14-12** *As some solids absorb heat energy, they begin to melt, or change to the liquid phase.*

338

## 14-3 (continued)

### Content Development

Ask students the following questions.

- **What are some examples of melting?** (snow melting on a warm winter day; ice cubes melting in a cold drink; frozen food thawing; wax melting on a candle; metal such as iron or lead becoming molten at very high temperatures)
- **What is added to cause the substance to melt?** (heat or heat energy)
- **What do you predict happens to the particles of a substance when the substance is heated and melts?** (They gain energy to move faster and more freely.)
- **Where does the energy come from?** (heat)

Point out that the temperature at which a solid changes to a liquid is called the *melting point.* Tell students that substances have their own individual melting points. Explain that ice has the melting point of 0°C. Write "Ice melts at 0°C" on the chalkboard.

- **What do you predict will happen to the particles of a liquid when heat is removed from the liquid?** (They will lose energy and slow down; eventually the substance will freeze, or turn into a solid.)

Point out that the temperature at which a liquid changes to a solid is

Figure 14-13 *The force of freezing water can cause a violent explosion. A cast-iron container about 0.6 centimeter in width is filled with water and placed in a beaker of dry ice and alcohol* (left). *As the water freezes and expands, a huge amount of energy is exerted against the walls of the container, causing an explosion* (right).

The opposite phase change, that of a liquid to a solid, is called **freezing.** Freezing occurs when a substance loses heat energy. The temperature at which a substance changes from a liquid to a solid is called the **freezing point.** Strangely enough, the freezing point of a substance is equal to its melting point. So water both melts and freezes at 0° C. ❷

Substances called alcohols have freezing points much lower than 0° C. These substances are used in automobile antifreeze because they can be cooled to low winter temperatures without freezing. One such substance, ethylene glycol, when mixed with water can lower the freezing point of the mixture to −49° C. ❸

The fact that freezing involves a loss of heat energy explains the "magic" worked by the orange growers you read about at the beginning of this chapter. The liquid water sprayed onto the trees released heat energy as it froze. Some of this heat energy was released into the oranges, preventing them from freezing and being destroyed.

### *Liquid-Gas Phase Changes*

Have you ever left a glass of water out overnight? If so, perhaps you noticed that the level of the water was lower the next morning. Some of the liquid changed phase and became a gas. The gas then escaped into the air.

The change of a substance from a liquid to a gas is called **vaporization** (vay-puhr-uh-ZAY-shuhn). During



## TEACHER DEMONSTRATION

Evaporation can be used to separate dissolved solids from the liquid solvent. As a class demonstration, leave a solution of salt and water exposed to the air until the water evaporates. (This may take several days.) You may want to use ocean water if it is available in your area. Have students develop logical hypotheses regarding their observations of this demonstration.

## HISTORICAL NOTES

People in hot areas such as deserts use bags of leather or canvas to keep their water cool. These containers are known as *evaporative coolers.* Water seeps through the leather or canvas to the outside of the container. The water on the surface of the container evaporates into the air, using some of the heat energy from the water in the bag. This leaves the water bag cooler than before. The process loses some water, but the water that remains is cool to drink.

## ANNOTATION KEY

❶ **Thinking Skill: Identifying relationships**

❷ **Thinking Skill: Identifying relationships**

❸ **Thinking Skill: Applying concepts**

called the *freezing point.* Tell students that most substances have their own individual freezing point. Explain that water freezes to ice at 0°C. Write "Water freezes at 0°C" on the chalkboard.

- **What do you notice about the melting point of ice and the freezing point of water?** (They are the same.)

Explain that the temperature at which a substance in the solid phase changes to the liquid phase, or its melting point, is the same temperature at which the liquid phase of the substance changes to the solid phase, or its freezing point. Emphasize that the melting point of a substance is equal to its freezing point.

### Skills Development

***Skill: Applying concepts***

Divide the class into teams of 4–6 students per team. Give each team a beaker, several ice cubes, and a Celsius thermometer. Have the team determine the temperature of the ice and the melted water.

- **What is the temperature of the ice before melting?** (0°C)
- **What is the temperature of the water from the melted ice?** (0°C)

### Content Development

Tell students that a change of phase from a liquid to a gas is called *vaporization.*

- **What must happen to the particles of a liquid in order for them to change to a gas?** (They must gain energy.)
- **What kind of energy?** (heat energy)

## BACKGROUND INFORMATION

Heat plays an important role in phase changes. Heat is energy that causes the particles of matter to move faster and farther apart. As the particles move faster, they leave one phase and pass into another.

The addition of heat to a substance is usually accompanied by a rise in temperature. But if a record were kept of the temperature and the heat energy involved in changing ice to steam, several interesting things would be observed. These observations can best be explained by constructing a phase-change diagram, which shows the relationship between heat energy and temperature during phase changes. This diagram clearly shows that phase changes are accompanied by increases in heat energy but not by increases in temperature. The heat energy that is absorbed is used to overcome forces that hold the particles of the substance together. Once the forces have been overcome and the substance has changed phase, added energy causes a rise in temperature.

It is important to remember that the gas phase consists of exactly the same particles of matter as the liquid phase and the solid phase. Phase changes produce changes in the physical properties of matter only. Regardless of its phase, it is still the same kind of matter.

**Figure 14-14** *How did the mist hanging over this forest form?*

**Figure 14-15** *During both evaporation and boiling, particles of a liquid absorb heat energy and change from the liquid phase to the gas phase.*

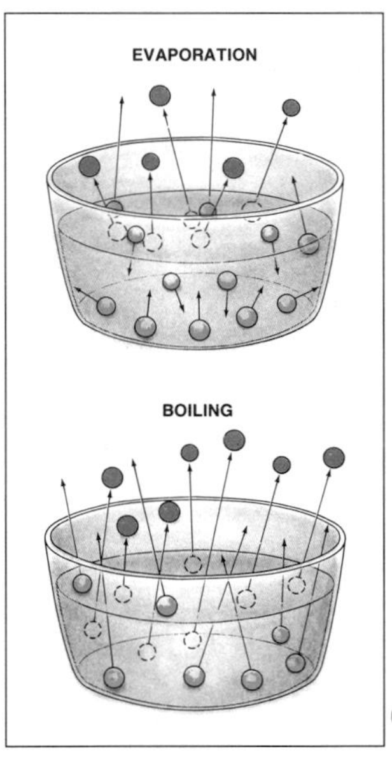


this process, particles in a liquid absorb enough heat energy to escape from the liquid phase. If vaporization takes place at the surface of the liquid, the process is called **evaporation** (ee-vap-uhr-AY-shuhn). (1) So some of the water in the glass left out overnight evaporated.

Evaporation is often thought of as a cooling process. Does this sound strange to you? Think for a moment of perspiration on the surface of your skin. As this water evaporates, it absorbs and carries away heat energy from your body. In this way, your body is cooled. Can you explain why it is important to sweat on a hot day? (1)

Vaporization does not occur *only* at the surface of a liquid. If enough energy is supplied, particles inside the liquid can change to gas. These particles travel to the surface of the liquid and then into the air. This process is called **boiling.** The temperature at which a liquid boils is called its **boiling point.** The boiling point of water at the earth's surface under normal conditions is 100° C. The boiling point of table salt is 1413° C, and that of a diamond is 4200° C!

The boiling point of a liquid is related to the pressure of the air above it. Since the gas particles must escape from the surface of the liquid, they need to have enough "push" to equal the "push" of the air pressing down. So the lower the air pressure, the more easily the bubbles of gas can form within the liquid and then escape. (2) Lowering the air pressure lowers the boiling point.

340

### 14-3 (continued)

### Skills Development

***Skill: Applying concepts***

Divide the class into teams of 4–6 students per team. Give each team a beaker, hot plate, and two Celsius thermometers. Have the team determine the temperature of boiling water and of the steam from the boiling water.

- **What is the temperature of the boiling water?** (It is about 100°C. *Note:* This temperature may be lower than 100°C if you are at an altitude above sea level. Higher altitudes have lower air pressure, and at a lower pressure water will boil at a temperature below 100°C.)
- **What is the temperature of the steam?** (It is about 100°C. Note: Here again, this temperature may be lower than 100°C if the altitude is above sea level. What should be observed is that the temperature of the boiling water is the same as that of the steam.)

### Content Development

Point out that if vaporization takes place on the surface of the liquid, it is called *evaporation.* Explain that evaporation is a cooling process. Tell students that the body maintains a constant temperature because of evaporation.

At high altitudes, air pressure is much lower and so the boiling point is reduced. If you could go many kilometers above the earth's surface, the pressure of the air would be so low that you could boil water at ordinary room temperature! However, this boiling water would be cool! Certainly you would not be able to cook anything in this water. For it is the heat in boiling water that cooks food, not simply the boiling process.

Gases can change phase too. If a substance in the gas phase loses heat energy, it changes into a liquid. Scientists call this change **condensation** (kahn-den-SAY-shuhn). You have probably noticed that cold objects, such as glasses of iced drinks, tend to become wet on the outside. Water vapor present in the surrounding air loses heat energy when it comes in contact with the cold glass. The water vapor condenses and becomes liquid drops on the glass.

### Solid-Gas Phase Changes

If you live in an area where winters are very cold, you may have noticed something unusual about fallen snow. Even when the temperature stays below the melting point of the water that makes up the snow, the fallen snow slowly disappears. What happens to it? The snow undergoes **sublimation** (suhb-luh-MAY-shuhn). When a solid sublimes, its surface

**Activity**

*Holding Particles Together*

The solid, liquid, and gas phases of matter can be explained in terms of the forces that hold the particles of each together. One of these forces is cohesion. ❸

Using books and reference materials in the library, find out what this force is. Then explain how this force affects the particles of each phase of matter, and what happens to this force when heat energy is added to or taken away from matter.

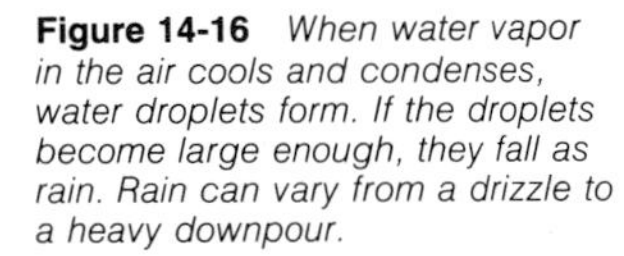

**Figure 14-16** *When water vapor in the air cools and condenses, water droplets form. If the droplets become large enough, they fall as rain. Rain can vary from a drizzle to a heavy downpour.*

**Activity**

**Holding Particles Together**
**Skills: Comparing, relating, applying**
**Level: Average**
**Type: Library**

By doing this activity, students will learn about cohesive forces that act between particles and about energy stored during phase changes. Have them report their findings to the class.

## ANNOTATION KEY

❶ **To cool the body, which otherwise would become overheated (Applying concepts)**

❶ **Thinking Skill: Identifying relationships**

❷ **Thinking Skill: Relating concepts**

❸ **Thinking Skill: Identifying relationships**

- **What happens to the particles of a gas as they change to the liquid phase?** (They lose energy.)
- **How is this energy loss accomplished?** (by removing heat)

### Content Development

Point out that when a substance in the gas phase loses energy, it changes into a liquid. Tell students that this process is called *condensation.*

- **What do you predict would happen to body temperature if body perspiration did not evaporate?** (Body temperature would go up.)

Point out that vaporization does not occur only at the surface of a liquid. Explain that if enough heat energy is supplied to a substance, the particles of the substance will travel to the surface of the liquid and then into the air as a gas. Tell students this is called *boiling*. Tell students that the temperature at which a liquid boils is called its *boiling point.*

### Content Development

Point out that gases also change to liquids.

- **What are some examples of a gas changing to a liquid?** (steam on windows during cool weather; dewdrops on the grass; steam condensing as it gives off heat in a steam engine or steam heating system)

### Content Development

Tell students that some solids do not pass through the liquid phase and then into the gas phase. Certain solids pass directly from a solid to a gas phase by a process called *sublimation.*

- **Can you think of any solids that go directly to the gas phases?** (Most students will not readily think of examples. You might suggest dry ice, paradichlorobenzene crystals [mothballs], and iodine crystals.)

## 14-4 CHEMICAL PROPERTIES AND CHANGES

### SECTION PREVIEW 14-4

Physical and chemical properties of matter are useful in determining the identity of a substance. The properties that distinguish one substance from another without changing the substance are called *physical properties.* The properties that describe how a substance changes into other new substances are called *chemical properties.*

Flammability, or the ability of a substance to burn, is a chemical property. The substance produced by the burning is a new substance that has been altered by a chemical change. Thus, flammability is a chemical property, and burning is a chemical change. Another name for a chemical change is a *chemical reaction.*

### SECTION OBJECTIVES 14-4

**1. Distinguish between physical and chemical properties of matter.**
**2. Explain how chemical properties are useful in identifying substances.**
**3. Define and discuss the chemical property of flammability.**
**4. Discuss the chemical property of the ability to support combustion.**
**5. Distinguish between a chemical property and a chemical change.**

### SCIENCE TERMS 14-4

**chemical property p. 343**
**flammability p. 343**
**chemical change p. 344**
**chemical reaction p. 345**

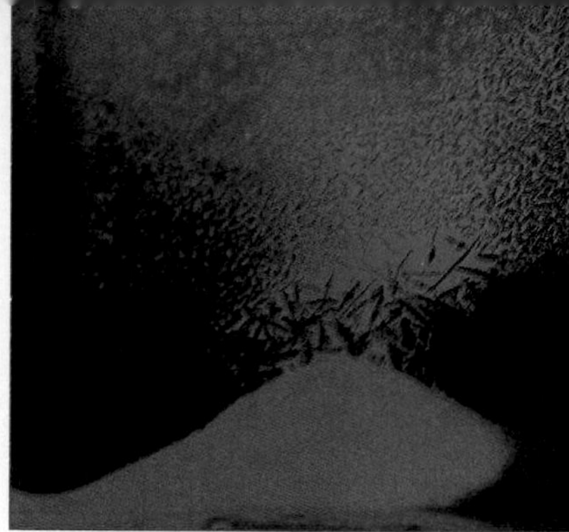

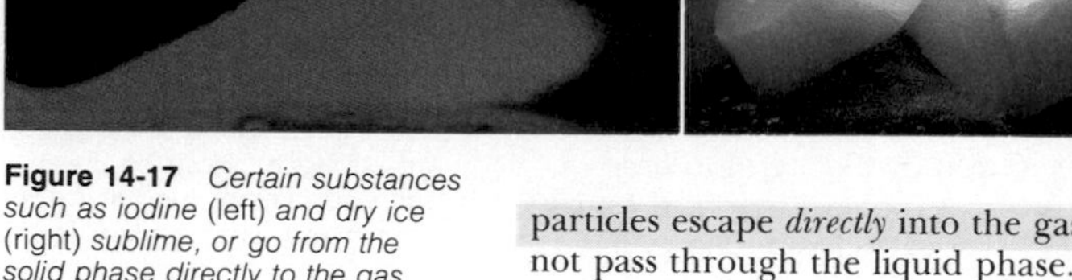

**Figure 14-17** *Certain substances such as iodine* (left) *and dry ice* (right) *sublime, or go from the solid phase directly to the gas phase.*

particles escape *directly* into the gas phase. They do not pass through the liquid phase.

A substance called dry ice is often used to keep other substances, such as ice cream, very cold. Dry ice is solid carbon dioxide. At ordinary pressures, it cannot exist in the liquid phase. So as it absorbs heat energy, it sublimes directly to the gas phase. By absorbing and carrying off heat energy as it sublimes, dry ice keeps materials that are near it cold and dry. Just think what would happen to an ice cream cake if it were packed in regular ice rather than dry ice!

**Activity**

*An Almost Ruined Day*

Using the following words, write a 250-word story about a day on which your birthday party was almost a disaster.

boiling, freezing, crystal, evaporation, liquid, melting, phase, physical change, solid, sublimation

#### *SECTION REVIEW*

1. Define melting and freezing.
2. Explain the difference between evaporation and condensation.
3. What is sublimation?

### 14-4 Chemical Properties and Changes

At the beginning of this chapter, you learned that you could easily tell a copper penny from a wooden stick by its physical properties. It was easy to see the differences in color, shape, hardness, and density. But now suppose you have to distinguish between two gases—oxygen and hydrogen. Both are colorless,



### 14-3 (continued)

#### Skills Development

***Skill: Applying concepts***

Place a piece of solid stick deodorant in a petri dish. Have students observe the deodorant every day for about a week.

- **What happened to the deodorant?** (Most of it disappeared. The smell of the deodorant could be detected in the air.)
- **What was the phase of the deodorant?** (solid)
- **What happened to the deodorant?** (It sublimed.)

#### Section Review 14-3

1. Change of solid to liquid; change of liquid to solid.
2. Evaporation: liquid-to-gas change at surface of liquid; condensation: gas-to-liquid change.
3. Change of solid directly to gas.

### TEACHING STRATEGY 14-4

#### Motivation

Show students a wooden splint.

- **What is this object made of?** (wood)
- **What phase is it in?** (solid)
- **Describe its physical properties.** (Answers will vary, but students should suggest its hardness, color, length, mass, weight, etc.)

odorless, and tasteless. Since they are gases, they have no definite shape or volume. And although each has a specific density, you cannot drop them into water to see what happens! So in this case, physical properties are not very helpful in identifying the gases.

Fortunately, physical properties are not the only way to identify a substance. Both oxygen and hydrogen can turn into other substances and take on new identities. And the way in which they do this can be useful in determining the gas. ❷ The properties that describe how a substance changes into other new substances are called **chemical properties.**

If you collected some hydrogen in a test tube and put a glowing wooden stick in it, you would hear a loud pop. The pop results when hydrogen combines with oxygen in the air. The hydrogen is burning. The ability to burn is called **flammability** (flam-uh-BIL-uh-tee). It is a chemical property. A new kind of

## Career: *Firefighter*

**HELP WANTED: FIREFIGHTER** High school diploma required. Must pass written test and physical examinations. Minimum age 18. Needed for firefighting team.

It was a dry October day when the field caught fire. Responding to the alarm sounded at their fire station, the men and women quickly jumped onto the trucks and raced to the burning field. Their first task was to keep the fire from spreading. The team began by clearing a strip of land a short distance from the approaching flames. They then set another fire between the cleared strip and the raging flames. This would keep the fire contained to a small area. Finally the fire was put out with water.

The brave individuals who often risk their lives in the line of duty are **firefighters.** It is their job to protect people and property from the thousands of fires that occur each year. In doing their job, they are sometimes exposed to explosive, flammable, or poisonous materials.

The duties of firefighters include driving emergency vehicles, hooking up hoses and pumps, setting up ladders, and rescuing victims. Some firefighters are trained to conduct fire safety checks in buildings and homes, while others investigate false alarms and suspicious fires. If you are interested in this career, write to the Department of Fire Protection and Safety Technology, Oklahoma State University, 303 Campus Fire Station, Stillwater, OK 74078.



## TIE-IN/ HOME ECONOMICS

Students might research the useful freezer life of normal food products.

### Activity

**An Almost-Ruined Day**
**Skills: Relating, applying, creative expression, writing**
**Level: Average**
**Type: Vocabulary/writing**

This activity gives students the opportunity to express themselves creatively and to reinforce their understanding of the scientific terms used to describe phases and phase changes. Encourage them to read their stories to their classmates.

## ANNOTATION KEY

❶ Thinking Skill: Applying facts
❷ Thinking Skill: Making generalizations

## BACKGROUND INFORMATION

The ability of hydrocarbons to burn is what makes these substances so useful as fuels. The burning of hydrocarbons, often called *combustion,* results in the production of carbon dioxide and water, plus lots of energy. Some examples of hydrocarbon fuels include gasoline, home heating oil, natural gas, and diesel fuel.

Light a match and set the splint on fire. Let it burn until at least half of it has become charred or has fallen off as ash. Place the splint in a beaker so the students can observe both the wood and the ashes.

• **What changes happened to the wood when it burned?** (The color changed to black; the burned part became brittle and crumbled easily; much of the wood seemed to disappear.)

• **Do you think that the identity of the original substance has changed?** (Answers may vary. Students are likely to suggest that the wood has changed.)

To support the idea that the substance has changed, try igniting the charred material.

• **What evidence supports the idea that the charred substance is not wood?** (It does not burn the way wood does.)

### Content Development

Point out that chemical properties describe the ways in which substances change into other substances.

• **What chemical property of wood did you observe in the demonstration with the wooden splint?** (the ability to burn)

## SCIENCE, TECHNOLOGY, AND SOCIETY

The pages of an average book printed today will last from 20 to 50 years before they start to turn brittle. The brittleness occurs when the chemical aluminum sulfate, found in the paper of a book, chemically reacts with moisture in the paper to form sulfuric acid. The acid breaks down the wood fibers in the paper, making the pages brittle.

Since the 1860s, aluminum sulfate has been used in the paper-making process to give paper a smooth finish and to prevent the running of ink. About 75 percent of the books published in the United States today contain aluminum sulfate.

To stop the further destruction of invaluable and irreplaceable books, scientists at the Library of Congress have developed a technique that neutralizes the acid in book pages. A book is treated by being placed in a vacuum chamber into which a gas known as diethyl zinc, or DEZ, is piped. The tiny DEZ molecules permeate the pages of the book, neutralizing the acid in the paper and preventing further destruction. The DEZ also reacts with moisture in the paper to produce zinc oxide, which will act to neutralize any acid that forms in the future.

**Figure 14-18** *Burning is a chemical change in which oxygen combines with another substance and produces heat and light.*

matter forms as the hydrogen burns. This substance is water—a combination of hydrogen and oxygen.

Oxygen is not a flammable gas. So you can distinguish it from hydrogen through the chemical property of flammability. However, although oxygen does not burn, it does support the burning of other substances. A glowing wooden splint placed in a test tube of oxygen will continue to burn until the oxygen is used up. This ability to support burning is another example of a chemical property.

The changes that substances undergo when they turn into other substances are called **chemical changes.** Chemical changes are closely related to chemical properties, but they are not the same. **A chemical property describes a substance's ability**

**Figure 14-19** *Rusting is a chemical change in which iron slowly combines with oxygen to form rust, or iron oxide.*

### 14-4 (continued)

#### Content Development

Point out that oxygen alone will not burn, nor rust items, nor produce fireworks. Explain that oxygen combined with other substances produces all of these processes, plus many others.

Emphasize that the ability to support burning or other chemical changes is a distinguishing property of a substance. Point out that this ability is a good example of a chemical property.

#### Skills Development

***Skill: Applying concepts***

Divide the class into teams of 4–6 students. Distribute a small candle (birthday type), jar lid, and matches to each team. Have the teams light the candle, set it on the lid to protect the desk, and then blow it out.

- **Why did the candle burn?** (It was lit with a match.)
- **What did the match do?** (heated the wick until it was hot enough to burn)
- **What combined with the wick and wax to cause combustion?** (oxygen)
- **Where did the oxygen come from?** (the air)

Distribute a 500-mL beaker to each team. Have the teams light the candle again, and then cover the can-

Figure 14-20 *This beautiful fireworks display is the result of a chemical change in which substances such as magnesium, phosphorus, and sulfur combine very rapidly with oxygen.*

**to change into a different substance; a chemical change is the process by which the substance changes.** For example, the ability of a substance to burn is a chemical property. However, the process of burning is a chemical change. Figures 14-18 through 14-20 show several chemical changes. ②

Another name for a chemical change is a **chemical reaction.** Chemical reactions often involve chemically combining different substances. For example, during the burning of coal, oxygen combines chemically with carbon—the substance that makes up most of the coal. This combining reaction produces a new substance—carbon dioxide. The carbon and oxygen have changed chemically. They no longer exist in their original forms.

The ability to use and control chemical reactions is an important skill. For chemical reactions produce a range of products from glass to pottery glaze to medicines. Your life is made easier and more enjoyable because of the products of chemical reactions: synthetic fibers such as nylon, plastics, soaps, building materials, even foods you eat. The next time you eat cheese or a slice of bread, remember that you are eating the product of a chemical reaction!

### SECTION REVIEW

1. Give two examples of chemical properties.
2. Give an example of a chemical change.

#### Activity

*Physical and Chemical Changes*

1. In a dry beaker, mix a teaspoon of citric acid crystals with a tablespoon of baking soda. Observe what happens.
2. Fill another beaker halfway with water. Pour the citric acid-baking soda mixture into the water. Observe what takes place.

What type of change took place in the first step of the procedure? In the second step? Did the water have an important role in the procedure? If so, what do you think was its purpose? Why are some substances marked "Store in a dry place only"?



## ANNOTATION KEY

① Thinking Skill: Identifying patterns
② Thinking Skill: Making generalizations

### Activity

**Physical and Chemical Changes**
**Skills: Making observations, making inferences**
**Level: Enriched**
**Type: Hands-on**
**Materials: Beakers, citric acid crystals, tablespoon, baking soda, water**

In this activity, students will observe a physical change and a chemical change. Combining citric acid and baking soda creates a physical change; the two substances merely mix without reacting. When water is added, a chemical change occurs. Evidence is the carbon dioxide bubbles that result. The water plays a vital role: it allows the two substances to dissolve and react chemically. Substances that may react with water are labeled "Store in a dry place only."

dle with the inverted beaker.
- **What happened?** (The flame went out.)
- **Why did the flame go out?** (The burning used up all the oxygen that was inside the beaker. Without oxygen, things will not burn.)

### Skills Development

***Skill: Applying concepts***

Have each student bring one or two pebbles or small rocks to class. Divide the class into teams of 4–6 students. Distribute a medicine dropper, beaker of vinegar, and paper towel to each team. Have the teams place several drops of vinegar onto each pebble.
- **Which rocks produced a chemical reaction with the vinegar?** (the rocks that fizzed)
- **How do you know a chemical reaction occurred?** (A new substance and gas bubbles were formed.)
- **Which rocks did not produce a chemical reaction with the vinegar?** (the ones that did not fizz)

### Section Review 14-4

1. Flammability, ability to support combustion.
2. Burning.

# LABORATORY ACTIVITY CONSERVATION OF MASS

## BEFORE THE LAB

1. **Divide the class into groups of 3–6 students.**
2. **Gather all materials at least one day prior to the activity. You should have enough supplies to meet your class needs, assuming 3–6 students per group.**

## PRE-LAB DISCUSSION

Have students read the complete laboratory procedure. Discuss the procedure by asking questions similar to the following:

- **What does "conservation of mass" mean?** (The amount of matter does not change even though the form of the matter may change.)
- **What is the purpose of the laboratory activity?** (To prove that changes in substances, whether chemical or physical, do not change the masses of the substances.)
- **What evidence will prove that the total mass is conserved during a chemical or physical change?** (The measured mass before, during, and after the change stays the same.)
- **Which substance should be put into the large outside test tube?** ($Fe[NO_3]_3$)
- **Which substance should be put into the small inside test tube?** (KSCN)
- **What is the purpose of the stopper?** (to cover the system so no substances [and thus no mass] are lost to the outside)

## SKILLS DEVELOPMENT

Students will use the following skills while completing this activity.

1. Manipulative
2. Safety
3. Observing
4. Measuring
5. Recording data
6. Comparing

# LABORATORY ACTIVITY

## *Conservation of Mass*

### Purpose

In this activity, you will prove that changes in the appearance of substances, whether chemical or physical, do not change the masses of the substances.

**Materials** *(per group)*

- 1 large test tube and stopper
- 1 small test tube
- 3 beakers
- Balance
- Potassium thiocyanate, KSCN
- Ferric nitrate, $Fe(NO_3)_3$
- Flask with stopper or peanut butter jar with screw cap
- Several chunks of rock salt
- Water

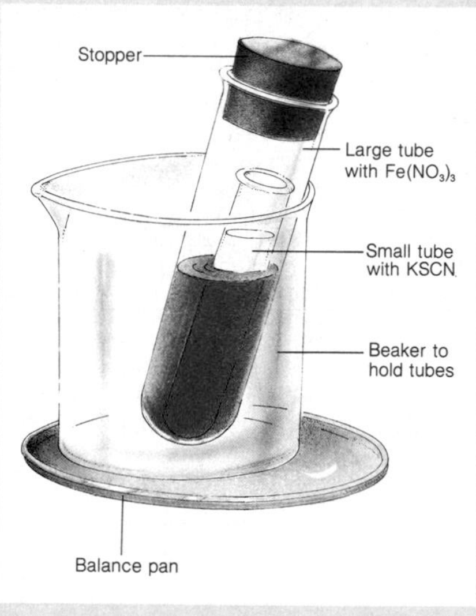


### Procedure

***A.*** *Chemical Change*

1. Fill a large beaker with 100 mL of water. Then add 1.0 gram of KSCN to the beaker.
2. Fill another beaker with 100 mL of water. To this beaker add 2.4 grams of $Fe(NO_3)_3$. **CAUTION:** *KSCN and $Fe(NO_3)_3$ are poisonous if swallowed.*
3. Fill half of the small test tube with the KSCN solution. Carefully put the small test tube into the large test tube.
4. Now put the $Fe(NO_3)_3$ solution in the large test tube so that it comes to a height no higher than the height of the solution in the small test tube. Put the stopper into the large test tube. Then place the test tubes into a beaker.
5. Determine the combined mass of the beaker and the test tubes.
6. Invert the large test tube to observe the reaction. A dark red material, $Fe(SCN)_3$, forms.
7. Redetermine the mass of the total system.

***B.*** *Physical Change*

1. Pour enough water into the glass container to make it half full. Add a few pieces of rock salt to the water and seal the container.
2. Use the balance to determine the mass of the container and its contents.
3. After some of the salt has dissolved, determine the mass of the container and contents again.

### Observations and Conclusions

1. Did the mass of the total system in procedure A change after you inverted the test tube?
2. In procedure B, was there a change in the mass of the container, water, and salt after some of the salt had dissolved?



## SAFETY TIPS

Alert students to be cautious about using any chemical in the laboratory. In addition to being poisonous if swallowed, many chemicals can be irritating or harmful to the eyes. Be sure to have students wear safety goggles while doing this laboratory activity. In addition, an approved eye-wash station and/or emergency shower should be readily available to all students. Demonstrate how to use the eye wash and emergency shower.

## TEACHING STRATEGY FOR LAB PROCEDURE

**1.** If your students are not skilled at using a metric balance, you might do a pre-lab activity to give them practice in measuring masses with precision.
**2.** Have the teams follow directions carefully as they work in the laboratory.
**3.** Discuss how the investigation relates to the chapter ideas by having students consider: why the large test tube was inverted (so the two chemicals

# CHAPTER REVIEW

## SUMMARY

### 14-1 Physical Properties and Changes

- Examples of physical properties of matter include color, shape, hardness, and density.
- Changes in which physical properties of a substance are altered but the substance remains the same kind of matter are called physical changes.

### 14-2 The Phases of Matter

- Matter can exist in any of four phases: solid, liquid, gas, and plasma.
- A solid has a definite shape and volume.
- A crystal is the regular, repeating pattern in which particles of a solid are arranged.
- Amorphous solids do not keep a definite shape because they do not form crystals.
- A liquid has a definite volume but not a definite shape. It takes the shape of its container.
- A gas has no definite volume or shape.
- Boyle's Law states that the volume of a fixed amount of gas varies inversely with the pressure. Charles's Law states that the volume of a fixed amount of gas varies directly with the temperature.
- Matter in the plasma state is very high in energy.

### 14-3 Phase Changes

- Phase changes are accompanied by either a loss or gain of heat energy.
- Melting is the change of a solid to a liquid at a temperature called the melting point. Freezing is the change of a liquid to a solid at the freezing point.
- Vaporization is the change of a liquid to a gas. Vaporization at the surface of a liquid is called evaporation. Vaporization throughout the liquid is called boiling.
- The boiling point of a liquid is related to the air pressure above the liquid.
- The change of a gas to a liquid is called condensation.
- Sublimation is the change of a solid directly to a gas.

### 14-4 Chemical Properties and Changes

- Chemical properties describe how a substance changes into other new substances.
- Flammability, the ability to burn, is a chemical property.
- When a substance undergoes a chemical change, or chemical reaction, it turns into a new and different substance.

## VOCABULARY

*Define each term in a complete sentence.*

**amorphous solid**
**boiling**
**boiling point**
**Boyle's Law**
**Charles's Law**
**chemical change**
**chemical property**
**chemical reaction**
**condensation**
**crystal**
**crystalline solid**
**evaporation**
**flammability**
**freezing**
**freezing point**
**gas**
**liquid**
**melting**
**melting point**
**phase**
**physical change**
**physical property**
**plasma**
**solid**
**sublimation**
**vaporization**

347

could mix); what type of a reaction occurred (chemical); what happens to mass during either a chemical or physical change (it is conserved).

## OBSERVATIONS AND CONCLUSIONS

1. The total mass did not change.
2. The total mass did not change.

## GOING FURTHER: ENRICHMENT

### Part 1

Students might verify the conservation of mass during chemical changes by using the same procedure and the following materials in the test tubes:

**1.** Dilute iodine solution in the small inside test tube and cornstarch powder or liquid in the large outside test tube. After the system is inverted, it should then be turned again to the upright position to allow the substances to mix.
**2.** Liquid bleach in the small inside test tube and bits of brown paper or colored cloth in the large outside test tube. After the system is inverted, it should then be turned again to the upright position to allow the substances to mix.
**3.** Vinegar in the small inside test tube and pieces of copper in the large outside test tube. After the system is inverted, it should then be turned again to the upright position to allow the substances to mix.

Discuss the difficulty of doing a conservation-of-mass experiment for chemicals that react to form a gas. The gas may be released with pressure and force the stopper off.

### Part 2

Have students demonstrate that mass does not change during a phase change. Have them find the mass of a beaker containing ice cubes, and then have them find the mass of the system again after the ice cubes have melted.

## CHAPTER REVIEW

### MULTIPLE CHOICE

| | | | | |
|---|---|---|---|---|
| **1.** c | **3.** b | **5.** d | **7.** b | **9.** b |
| **2.** a | **4.** b | **6.** d | **8.** c | **10.** a |

### COMPLETION

| | |
|---|---|
| **1.** liquid | **6.** 100°C |
| **2.** amorphous | **7.** decreases |
| **3.** decreases | **8.** condensation |
| **4.** plasma | **9.** sublimation |
| **5.** melting point | **10.** reaction |

### TRUE OR FALSE

| | |
|---|---|
| **1.** F physical | **6.** F Charles's Law |
| **2.** T | **7.** F loses |
| **3.** F crystalline | **8.** T |
| **4.** F gas | **9.** F condensed |
| **5.** T | **10.** F chemical |

### SKILL BUILDING

**1.** **(a)** Because boiling water contains less heat at high altitudes, frozen stringbeans would, in fact, require more boiling to be heated to the same temperature as stringbeans placed in boiling water at sea level. **(b)** Because of friction, the tire and the air inside have increased in temperature. According to the Gas Laws, an increase in temperature will cause an increase in the volume of the gas. But because the volume remains constant, the increase will be seen as an increase in pressure inside the tire.
**2.** **(a)** physical **(b)** chemical **(c)** physical **(d)** physical **(e)** chemical **(f)** physical
**3.** **(a)** chemical **(b)** chemical **(c)** physical **(d)** chemical **(e)** physical **(f)** physical **(g)** chemical **(h)** chemical
**4.** **(a)** 1/3 original **(b)** 1/2 original **(c)** 5 times original **(d)** 4 times original
**5.** Heat energy absorbed during a phase change is used to overcome forces that hold particles of the substance together. There is no increase in temperature because the particles are not moving faster.

### CONTENT REVIEW: MULTIPLE CHOICE

*Choose the letter of the answer that best completes each statement.*

**1.** Color, odor, and density are
a. chemical properties. b. general properties.
c. physical properties. d. solid properties.
**2.** The phase of matter characterized by definite shape and definite volume is
a. solid. b. liquid. c. gas. d. plasma.
**3.** A regular pattern of particles is found in
a. molecules. b. crystals. c. compressions. d. plasmas.
**4.** It is true that liquids have
a. definite shape and definite volume.
b. no definite shape but definite volume.
c. no definite shape and volume.
d. definite shape but no definite volume.
**5.** As the volume of a fixed amount of a gas decreases, the pressure of the gas
a. decreases. b. remains the same.
c. first increases and then decreases. d. increases.
**6.** As the temperature of a gas increases,
a. the volume decreases. b. the volume remains the same.
c. the volume increases and decreases. d. the volume increases.
**7.** The state of matter made up of very high-energy particles is
a. liquid. b. plasma. c. gas. d. solid.
**8.** A solid changes to a liquid by
a. evaporation. b. freezing. c. melting. d. sublimation.
**9.** Vaporization taking place at the surface of a liquid is called
a. boiling. b. evaporation. c. sublimation. d. condensation.
**10.** Flammability is an example of a
a. chemical property. b. physical property.
c. chemical change. d. physical change.

### CONTENT REVIEW: COMPLETION

*Fill in the word or words that best complete each statement.*

**1.** The phase of matter that has a definite volume but no definite shape is the ______ phase.
**2.** Solids that do not have a definite shape are called ______ solids.
**3.** If the volume of a certain amount of gas increases, the pressure ______.
**4.** High-energy particles that make up most of the matter on the sun are called the ______ phase of matter.
**5.** The temperature at which a solid changes to a liquid is called the ______.
**6.** The boiling point of water at the earth's surface is ______.
**7.** As air pressure decreases, the boiling point of water ______.
**8.** The process by which a gas changes into a liquid is called ______.
**9.** The process by which a solid changes directly to a gas is ______.
**10.** Another name for a chemical change is a chemical ______.

348

### ESSAY

**1.** A chemical reaction is one in which two substances are chemically combined.
**2.** Solid: definite shape, definite volume, particles tightly packed together and move through vibration. Liquid: definite shape, definite volume, particles farther apart than solids, particles flow around one another. Gas: no definite shape, particles spread to fill volume of container, particles very far apart, particles move in a random way with rapid motion.
**3.** The water on the clothes freezes. Then it sublimes from the solid phase directly to the gas phase. Clothes dry as moisture is removed during sublimation.
**4.** Evaporation is vaporization at the surface of a liquid. Boiling is vaporization throughout the liquid. Both processes require heat energy, but the amount of heat energy is less for evaporation than for boiling.

## CONTENT REVIEW: TRUE OR FALSE

*Determine whether each statement is true or false. If it is true, write "true." If it is false, change the underlined word or words to make the statement true.*

1. Hardness, shape, taste, and melting point are chemical properties of matter.
2. Particles that make up a solid are packed very close together.
3. Common table salt, which has a regular, repeating arrangement of particles, is called amorphous.
4. The particles of matter are spread farthest apart in a liquid.
5. The pressure a gas exerts on the walls of its container is due to collisions of the particles of the gas with each other and with the walls of the container.
6. The relationship between the temperature of a gas and the volume it occupies is described by Boyle's Law.
7. A liquid will freeze when it absorbs heat energy.
8. The process by which a liquid changes to a gas is called vaporization.
9. Drops of water on the outside of a cold glass are water vapor that has sublimed into a liquid.
10. New substances having different properties are formed as a result of physical changes.

## CONCEPT REVIEW: SKILL BUILDING

*Use the skills you have developed in this chapter to complete each activity.*

1. **Applying concepts** Explain the following statements:
   a. Frozen stringbeans have to be cooked for a longer time in Denver, Colorado, because of the city's high altitude.
   b. After traveling several kilometers on your bike, you notice that the tires feel hot and the pressure gauge indicates an increase in pressure.
2. **Classifying properties** Identify the following properties as either physical or chemical: (a) taste (b) combustibility (c) color (d) odor (e) flammability (f) ability to dissolve.
3. **Classifying changes** Identify the following changes as either physical or chemical: (a) burning coal (b) baking brownies (c) boiling water (d) digesting food (e) dissolving sugar (f) melting butter (g) exploding TNT (h) tarnishing silver.
4. **Relating cause and effect** Using the Gas Laws, predict what will happen to the volume of a gas if (a) the pressure triples (b) the temperature is halved (c) the pressure is decreased by a factor of five (d) the pressure is halved and the temperature is doubled.
5. **Making inferences** Explain why the temperature remains constant during a phase change even though the substance is absorbing heat.

## CONCEPT REVIEW: ESSAY

*Discuss each of the following in a brief paragraph.*

1. What is a chemical reaction?
2. Compare the solid, liquid, and gas phases of matter in terms of shape, volume, and arrangement and movement of particles.
3. Explain how wet clothes hung on a clothesline on a very cold day dry.
4. Explain how evaporation and boiling are similar. Different.

349

## ISSUES IN SCIENCE

The following issues can be used as a springboard for discussion or given as a writing assignment.

**1.** Improper use of household chemicals causes many accidents and deaths every year. Should state or local governments require greater controls on the sale and/or use of these chemicals? Why or why not?

**2.** Some acid rain occurs naturally and is helpful in breaking down rocks to form soil. Industrialized cities and states create excessive amounts of smoke and other pollutants that increase acid rain, causing harmful effects on the environment. How can we maintain our level of industry yet reduce the resulting destruction of the environment? Who should be responsible for the prevention and cleanup of pollution—the government, industry, or the taxpayer?

## ADDITIONAL QUESTIONS AND TOPIC SUGGESTIONS

**1.** Name three physical properties that apply to both solids and liquids. (They have a definite volume, a definite mass, and a definite density.)

**2.** Name three physical properties that apply to both liquids and gases. (They have no definite shape; the particles can move around one another; they have a definite density; they have a definite mass.)

**3.** Explain the statement, "Energy content is responsible for the different phases of matter." (Substances can be made to change phase by adding or taking away energy. The amount of energy in a substance determines the amount of movement of the particles of the substance.)

**4.** Using Charles's Law, explain why a balloon that is left in the sun will pop. (The balloon will pop because the sun heats the air inside the balloon, causing the volume of the air to increase.)

## Unit Three

## MATTER

### ADVENTURES IN SCIENCE: W. LINCOLN HAWKINS— HE SOLVED THE PUZZLE OF THE AGING WIRES

### BACKGROUND INFORMATION

At the time of his retirement in 1976, W. Lincoln Hawkins was assistant director of the chemical laboratories at Bell Laboratories in Murray Hill, New Jersey. Dr. Hawkins was in charge of 129 technical people, more than half of whom had Ph.D.'s in chemistry. It was Dr. Hawkins's job to organize, administer, and help direct the chemical research at Bell Laboratories. Dr. Hawkins has over 30 patents to his credit, he has written some 50 technical publications, and has contributed to several books.

As a research chemist, Dr. Hawkins's area of special interest was how to protect polymers—especially plastics—from the environment. Plastics are especially important in the telecommunications industry because they can protect the network components from damage induced by human and natural influences, and they can also act as insulators. Some of the polymers that are commonly used for these and other purposes include polyethylene, polyvinyl chloride, polycarbonates, polystryrene, polyurethane, and silicones.



Adventures in Science

# W. LINCOLN HAWKINS

# He Solved the Puzzle of the Aging Wires

Lincoln Hawkins examines a model of his "miracle" molecule. The molecule is that of a plastic used to protect wires from the weather.

When "Linc" Hawkins was twelve, his father and mother came home to find their son hard at work on a strange project. Linc was drawing plans for building a perpetual-motion machine. Such a machine never stops working. Linc thought that by laying out a series of tilted ramps, he would be able to keep a steel ball bearing rolling forever. What Linc did not realize was that friction and gravity would slow the ball down no matter how he tilted the ramps. But that did not stop Linc from trying to build his machine. After a few years of trying, however, Linc gave up the project. By then he had come to realize that the laws of physics make the invention of a perpetual-motion machine impossible.

Linc's impossible perpetual-motion machine was to be only the first in a series of challenges that awaited him. Although his first efforts had not been successful, they had convinced Linc that his future lay in science and engineering. Several years later, he attended Rensselaer Polytechnic Institute, from which he received a degree in chemical engineering.

Unfortunately for Linc, getting a degree was a lot easier than getting a job! For Linc graduated while the nation was experiencing the Great Depression of the 1930s. There were just no jobs available in engineering. But Linc Hawkins was not ready to call it quits. He went back to school and managed to earn a Master of Science degree from Howard University. A few years later, he received a higher degree, a Ph.D. from McGill University. And still he was not finished with his education. He continued his studies at Columbia University in New York. But was all this education leading anywhere?

Linc Hawkin's first break finally came in 1942. He was hired by the Bell System's Laboratories in New Jersey. There he worked on many chemical problems. But he did not find the problems challenging enough. So Linc convinced his supervisor to let him work with plastics. In this area, Linc believed, his training in chemistry and chemical engineering might be put to better use.



### TEACHING STRATEGY: ADVENTURE

#### Motivation

Have students observe the photograph of Dr. Hawkins on page 350. Point out that Dr. Hawkins is holding the model of a molecule.

- **How would you describe this molecule?** (Answers may vary. Students will probably notice the size and complexity of the molecule, and its rather unwieldy shape.)
- **What characteristics would you expect a molecule such as this one to have?** (Accept all answers.)

#### Content Development

Continue the above discussion by pointing out that Dr. Hawkins is holding a model of a polymer. A polymer is a giant molecule made up of many smaller molecules joined together. In fact, the term *polymer* comes from the Greek words *polus,* meaning "many" and *meros,* meaning "parts." The individual molecular units that form a polymer are called monomers. The type of monomers and the length and shape of the polymer chain determine the properties of a particular polymer.

Point out to students that many of the synthetic materials we use everyday are polymers. All plastics are

At this time, scientists were beginning to consider the use of plastics as insulators for electrical cables, including telephone cables. Insulators protect the cables. They keep the electricity in the wires. Up until then, lead had been used for this purpose. But lead was expensive, heavy, and in short supply.

Plastics known as polyethylenes, however, were lighter than lead. They were also inexpensive, flexible, strong, and water resistant. Some scientists felt that polyethylenes would make very good insulating coatings for electrical cables. But there was a problem. Tests showed that polyethylenes would wear out quickly when used outdoors. Exposure to light, heat, and moisture sometimes caused them to stiffen and crack. Was there a way to protect polyethylene against the weather?

This is where Linc Hawkins came in. He began to investigate how plastics become "old." And he also performed experiments to find ways of keeping the plastics "young." In other words, Hawkins was looking for a way to protect plastics from light, heat, and exposure to air. Each of these factors speeded up the aging of a plastic.

Hawkins soon realized that the key to keeping plastics young was to mix them with special chemicals. But which chemicals? The answer seemed simple. To keep a plastic safe from light, the plastic should be mixed with a chemical that did not break down when bathed in bright light for a long time. To protect a plastic against heat, the plastic should be mixed with a chemical that stood up to heat. And to keep a plastic from falling apart in air, the plastic should be mixed with a chemical that withstood long-time exposure to air.

Was there a single chemical with all these properties? And, if there were such a chemical, could it be mixed successfully with a plastic? Hawkins knew that such a single, miracle chemical probably did not exist. But there were different chemicals that were each strong in one way but not in others. So Hawkins began a search for a *combination* of chemicals. If he could find the right combination, he would be on the way to making the almost-perfect insulator.

Hawkins and an assistant prepare to test wires coated with protective plastic.

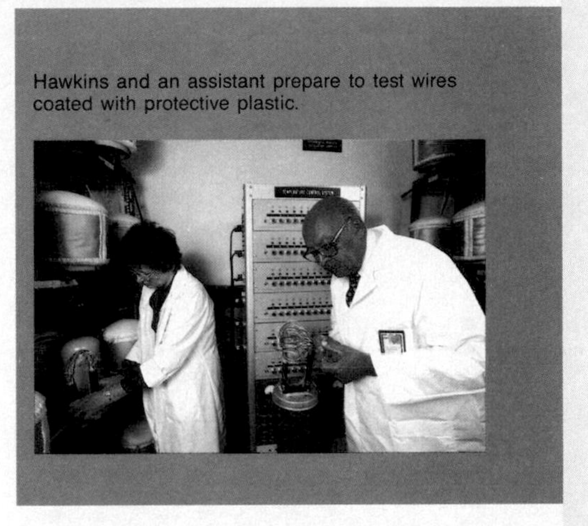

Days stretched into weeks, weeks into months, and months into years as Hawkins worked on his problem. Little by little he began to solve it. First he found a chemical that could protect plastics against oxygen in the air. Then he found a chemical that could protect plastics against light. And finally he found a chemical that helped plastics stand up to heat.

Today, these chemicals and others like them make up the plastic that covers billions of kilometers of electrical wires. So the next time you pick up the telephone, stop and think how W. Lincoln Hawkin's career has made it easier for your call to reach its destination. And how ability, education, and hard work, even under the most difficult circumstances, can achieve great things.

351

## ADDITIONAL QUESTIONS AND TOPIC SUGGESTIONS

**1.** Review Newton's law of motion. Which law was Linc Hawkins applying when he tried to build a perpetual motion machine? (Newton's first law, which states that an object at rest tends to stay at rest, and an object in motion tends to stay in motion unless acted on by an outside force)
**2.** What is the difference between applied research and pure research? Which type of research was Dr. Hawkins involved in? (Applied research is research undertaken to solve a specific practical problem, for example, how to make a plastic that can withstand high temperatures. Pure research is research that seeks to gather knowledge for its own sake, regardless of possible applications. Dr. Hawkins was involved in applied research.)
**3.** Look up the molecular structures for several familiar plastics. Then diagram the structures and indicate the different molecules that make up each polymer.

## CRITICAL THINKING QUESTIONS

**1.** What would be needed in order to make a perpetual motion machine that works? (a power source that could provide a force to overcome friction)
**2.** Why is plastic able to "keep electricity in wires"? (Plastic is a nonconductor, or insulator. Most nonmetals are nonconductors.)

polymers, and so are synthetic fabrics such as Dacron and polyester. Natural substances such as wool, silk, cotton, and natural rubber are also polymers.

### Content Development

Emphasize to students that Dr. Hawkins worked to change the properties of plastic polymers.

- **Why did he want to do this?** (in order to make the plastics more useful, especially for use outdoors)
- **What properties did he especially want to change?** (the ability of the plastics to withstand exposure to heat, light, and moisture)

Explain to students that part of Dr. Hawkins's task was to study the chemical processes that occur when plastic polymers are exposed to oxygen in air.

### Teacher's Resource Book Reference

After students have read the Science Gazette article, you may want to hand out the reading skills worksheet based on the article in your Teacher's Resource Book.

## Unit Three

## MATTER

### ISSUES IN SCIENCE: CAN THE ENERGY OF VOLCANOES BE HARNESSED?

### BACKGROUND INFORMATION

On May 18, 1980, Mount St. Helens in the state of Washington erupted with a tremendous explosion. A volcanologist named David Johnston, who was exploring in the area of the volcano, was killed by the blast.

Volcanologists study active volcanoes with the hope of learning how to predict volcanic eruptions. They also study volcanoes in order to learn more about the earth's interior. David Johnston's death illustrates how dangerous an occupation this can be. The practice of tapping volcanoes as an energy source would be just as risky.

Of course, many technologies involve taking risks. Perhaps the most obvious example is the technology of producing power in nuclear power plants. The question that scientists and the general public must ask themselves is whether the benefits of the technology make the risks worthwhile. A second question to consider is whether the risks can be reduced, or at least controlled, as the technology is improved.

This geothermal plant in Iceland helps bring heat and electricity to about 75 percent of homes in the country.

# Can the Energy of Volcanoes Be Harnessed?

For many years, people living in different parts of the world have used heat from within the earth as a source of energy. They have done this in many ways. For example, when underground water comes in contact with hot rocks, the water's temperature can rise to above its boiling point.

The water then begins to turn to steam. This causes a tremendous buildup of pressure. The pressure forces steam and hot water to come rushing out through weak spots in the earth's crust. This spouting, steaming water is called a hot spring.

Water from these natural hot springs has been used by people as a source of heat. In cases where the water does not come up naturally, wells have been drilled near hot

352

### TEACHING STRATEGY: ISSUE

#### Motivation

Display a photograph of a natural geyser such as Old Faithful.

- **Do you know what this natural phenomenon is called?** (a geyser or hot spring)
- **What is the water like that comes out of a geyser?** (The water is warm or hot.)
- **How do you think the water in a geyser or hot spring becomes so warm?** (Accept all answers.)

#### Content Development

Explain to students that in certain places on earth, called hot spots, water beneath the earth's surface comes into contact with hot rocks. The water becomes hot enough to change into steam. The pressure of steam builds up underground, and soon the water is forced up to the earth's surface through cracks in the earth's crust. The place where the hot water springs forth is called a hot spring or geyser.

Point out to students that heat within the earth—the same heat that warms water at a hot spot—is called geothermal energy. Geothermal energy is appealing as an energy source because, unlike many sources of energy, it is renewable.

springs. These wells bring underground steam and boiling water to the surface where they can be put to use.

Heat energy from deep within the earth is called *geothermal* energy. One of the most important uses of geothermal energy is producing electricity. Since 1904, steam from deep underground has been used to run machines that generate electricity. This has been done in Italy, New Zealand, and California. In a year, these power plants can produce the same amount of energy that is in 13 million barrels of oil. That's enough to meet the electricity needs of a city the size of San Francisco! And the energy is there for the taking.

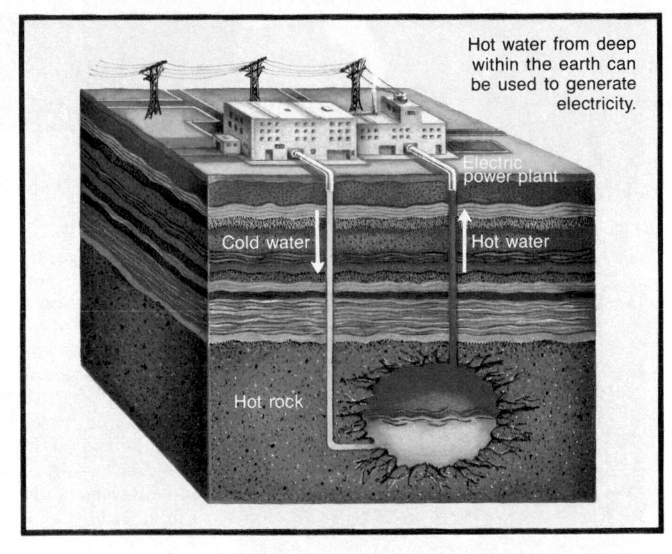

Hot water from deep within the earth can be used to generate electricity.

Up until now, most geothermal energy has come only from underground places near natural hot springs. But some engineers say that we could drill into underground hot spots far from hot springs.

If the hot spot turns out to be dry, cold water from the surface could be pumped down into the hot rocks. There the water would turn into steam. The steam would rush to the surface. And it could be used to generate electricity.

But several things could make this hard to do. Rocks hot enough to do the job might be much deeper underground than expected. Drilling very deep could cost a lot of money. Also, the steam rising from deep in the earth might cool on the way up. As a result, it would not provide enough pressure to run machines that produce electricity. Another problem might be that the deep rocks might have so many cracks and holes that drilling would cause them to break up. Then the ground would collapse, destroying an expensive well and perhaps triggering an earthquake!

## Taming a Volcano

Another way to use the earth's underground energy is to get it from an active volcano. This may seem like a crazy idea. After all, volcanoes are among the most powerful and "wild" natural furnaces on Earth. How can they be tamed? For one thing, the molten, or melted, rock of a volcano is often close to the surface of the earth. So there is no need to drill deep wells to get at the heat. Using the heat of volcanic rock is exactly what is done at the Eldfell volcano in Iceland.



## ADDITIONAL QUESTIONS AND TOPIC SUGGESTIONS

**1.** Find out how steam is used to produce electricity once it reaches an electric power plant.
**2.** Find out how geothermal energy is collected and used in the United States, Italy, Iceland, and New Zealand. Then compare and contrast the three countries.
**3.** Find out what chemicals are present in volcanic steam and hot water. Then find out some of the uses of these chemicals, and why they are considered valuable.

- **What do we mean when we say that an energy source is renewable?** (It is replenished by natural means; it cannot be used up.)
- **Can you think of energy sources that are renewable?** (Answers may vary; the most obvious renewable energy source is sunlight.)

### Reinforcement
Point out to students that the major limitation of geothermal energy is that hot spots exist only in certain locations. Also remind students that the required long-distance drilling to hot spots poses many problems. Among these are the possibility that the water might cool on the way up; the likelihood that deep drilling would be very expensive; and the possibility that drilling into deep rocks might cause the ground to collapse and might even trigger an earthquake.

### Content Development
Explain to students that the heat source for a hot spring or geyser is the same as the heart source for a volcano—hot molten rock beneath the earth's crust. The major difference between a hot spring and a volcano is that beneath a hot spring, the hot rock stays underground; only water heated by the rock is forced upward. Underneath a volcano, on the other hand, pressure forces the molten rock upward through openings in the earth's crust. When the hot rock reaches the earth's surface, it is called lava.

## CRITICAL THINKING QUESTIONS

**1.** How would you feel about being a worker asked to pump water or install pipes in the area of an active volcano? (Accept all answers.)
**2.** In what ways might it be easier to tap heat energy from a volcano than to tap heat energy from a hot spot? (Answers may vary. One possible answer is that at a volcano, the hot rock is already at the earth's surface, so drilling would not be necessary to reach the heat source.)
**3.** Not all volcanoes erupt violently; some volcanoes have lava flows that are quiet and gradual. Such volcanoes are called shield volcanoes. Find out more about shield volcanoes, and decide whether these might be a source of geothermal energy. What would be the advantage of tapping a shield volcano as a heat source? (There would be no danger from violent eruptions.)

## CLASS DEBATE

Use the questions raised in the background information section as a springboard for a class debate. Have students emphasize the pros and cons of tapping volcanic energy in terms of possible benefits versus possible risks.

Hot lava spurts freely from the ground in Iceland. However, in other parts of the country heat energy from inside the earth is trapped and used to warm homes and produce electricity. But there are risks in "taming" the earth's heat.

There spiral-shaped pipes are put into an area of hardening lava. Lava is molten rock that has reached the earth's surface. Water is then pumped through the pipes. The water is heated and then used to heat homes and do other jobs.

But there is a price to pay for getting heat from volcanoes. The pipes have to be made of a material that can withstand high temperatures. Unfortunately, making such pipes is very expensive. In addition, the water flowing through the pipes might not get hot enough to turn into high-pressure steam.

Of course, pipes don't have to be used. Water could be pumped directly into the cracks of a volcano crater. The steam, gases, and vapors that escape from the cracks could be passed through ordinary pipes to drive machines that produce electricity. But as one scientist said, "fooling around with a volcano involves risks."

Drilling holes and sending water into the cracks of a volcano could touch off a small but dangerous explosion. And a powerful volcanic eruption might be set off!

Does the need for energy make taking risks with volcanoes worthwhile? Many people say yes. They point out that electricity is not the only thing we could get by taming a volcano. For example, water heated up by a volcano could be used to heat homes, buildings, and greenhouses. But that is not all. Both steam and hot water could be used by industry to do a number of jobs. In addition, large amounts of valuable chemicals, particularly sulfur, are present in volcanic steam and hot water. Maybe these chemicals could be obtained from volcanoes. This would be a real bonus, point out many scientists.

Long ago, the people of Iceland learned to harness and use the energy from deep within the earth. Since then, Icelanders have used geothermal energy to improve their quality of living. Miles of greenhouses, warmed by the earth's own heat, provide large crops of fruits and vegetables for Icelanders. About 75 percent of their homes are heated by geothermal energy. Without this source of energy, Iceland would have to import millions of additional gallons of oil each year. Now the people are starting to use the earth's heat for fish farming, industry, and the generation of electricity. In time, Iceland may run entirely on geothermal power.

Is the greatest source of this power locked in volcanoes? Should we try to get at it? What do you think?



## ISSUE (continued)

### Reinforcement

Collect or have students collect newspaper and magazine articles that describe major volcanic eruptions in different parts of the world. If you live in the northwestern part of the United States, students may be especially interested in gathering photographs of the devastation caused by the eruption of Mount St. Helens.

### Teacher's Resource Book Reference

After students have read the Science Gazette article, you may want to hand out the reading skills worksheet based on the article in your Teacher's Resource Book.

Futures in Science

THE HOUSE OF THE FUTURE
WILL RUN ON ENERGY FROM THE SUN,
MAKE ITS OWN ELECTRICITY,
AND EVEN SELL SOME OF IT
TO THE ELECTRIC COMPANY.

It looks like an ordinary house, nestled among several others just like it. The lights are on in the kitchen. Good—dinner will be ready soon. Let's see . . . what was it you ordered for supper before you left the house this morning? Oh yes, a menu featuring your favorite Italian dishes. Mmm . . .

## POWER PANELS

The thought of collapsing into a comfortable chair in a cool room makes you walk even faster. Not such a good idea on a day that saw the temperature reach 35°C—for the sixth time in a row. The only advantage to this heat wave is that it will enable you to sell back lots of electricity to the power company. Yes indeed, those 64 solar panels rising up from the southern side of the roof really do their job. They produce enough electricity to run most of the major appliances in the house. The panels, covering a roof area of 5.5 square meters, are made up of *photovoltaic cells*. These cells change the energy in sunlight directly into electricity.



# Unit Three

# MATTER

## FUTURES IN SCIENCE: WIRED TO THE SUN

## BACKGROUND INFORMATION

Since the beginning of human civilization, people have used energy from the sun. Primitive peoples relied on the sun to dry goods and animal skins. Native Americans in the Western Hemisphere built adobe huts that absorbed the heat of the sun by day and radiated that heat indoors at night. The modern technology that is called solar energy is not a new idea—rather it is the development of new methods that make the use of solar energy more effective and reliable.

The major limitation of solar energy is that the sun does not always shine. Nighttime hours, rainy days, and cloud cover all inhibit the absorption of the sun's rays. Thus, a primary goal of modern technology is to find ways to more effectively collect and store solar energy.

An important use of solar energy is in home heating systems. Solar heating systems can be either passive or active. In a passive system, windows and glass panels are positioned to capture the maximum amount of sunlight. Special materials in the walls or floor of a house have the ability to effectively absorb sunlight, and then to radiate heat into the house. The trombe wall described in this article is an example of passive solar heating.

Active solar heating involves the use of one or more technological devices to collect, transfer, and store solar energy. Devices called solar collectors absorb energy from the sun. The heat absorbed by the collector is usually transferred by circulating water. Pumps, fans, and heat exchangers may aid in this process. A storage tank holds solar-heated water for future use.

## TEACHING STRATEGY: FUTURE

### Motivation

Begin by asking students the following question.

- **Suppose that a car with all of its windows closed has been sitting in the sun for several hours. How do you think the temperature inside the car will compare with the temperature outside the car?** (The temperature inside the car will be warmer.)
- **Why?** (Answers may vary. Guide students to understand that sunlight absorbed through the windows of the car changes into heat, and that the heat is then trapped inside the car by the glass.)

## ADDITIONAL QUESTIONS AND TOPIC SUGGESTIONS

**1.** Use the ideas discussed in this article, plus any additional information gained from reference sources, to design your own solar-powered home of the future. You may wish to join with several classmates and design a variety of houses and other buildings to form a solar-powered community.

**2.** Discuss some of the ways in which energy changes from one form into another in a solar-power house. (Radiant energy from the sun is absorbed by glass panels and changed into heat energy; radiant energy is also absorbed by photovoltaic cells and changed into electricity. The electricity is then changed into mechanical energy and light energy as it supplies power to appliances and lights.)

**3.** Write to the Public Information Department of the Georgia Power Company in Atlanta, Georgia. Ask for pictures and other information that describe the research projects that are developing homes like the one described in this article.



Now, of course, when there is no strong sunlight—like during the night and on cloudy days—the photovoltaic cells don't work. But you need not worry. At those times, your house is automatically switched to a local electric company's cables. You use its electricity and pay its prices. But so far this summer, the photovoltaic cells have produced more energy than the house needs. Some of the energy has been stored in the hot-water heater. The remainder has been sold back to the electric power company. The electricity is actually sent from the roof through cables to a nearby company station. Just think about it: The electricity you sell back is used to power one of those old-fashioned houses!

Each one of the 64 solar panels in the roof of this house is made up of photovoltaic cells. These cells capture the energy of the sun and change it directly into electricity.

### COMPUTER COMFORT

As you climb the stairs to the airlock entry, you happily notice that the outside window shutters have been automatically rolled down. These shutters reflect sunlight on hot days and help hold heat in the house on cold days. Shutters inside the windows have also been automatically lowered. Keeping direct sunlight out of the house on a day like today is very important.

Punching your code in the door keypad, you walk into the airlock entry. The airlock keeps hot air from entering the house in summer. In winter, it keeps the cold air out. The burglar-alarm system now shuts itself off. It feels a bit too cool in the house, so you signal the air-conditioning system to quit working so hard by punching a code into the thermostat.

As you pass from room to room, doors open automatically and lights switch on and off. The heat and motion sensors built into the floors, walls, and ceilings keep track of your path. And sure enough, as you enter the family room, your favorite music goes on.

With dinner cooking in the oven, which automatically went on when you walked in, you can sit down for awhile and relax. The dusting and vacuuming have all been done by the computer-driven robots. The kitchen computer keeps track of what food items you are getting low on. It will "call in" a list of groceries to the supermarket later.

So you're now free to sit and think about just how comfortable your home life is. The main computer of your house-management system takes care of almost everything—from controlling heating and cooling systems to providing news and sports information and educational courses. It opens locks, gives fire-alarm protection, cooks meals, turns lights on and off. It remembers when to turn certain appliances on or turn them off. It even takes



## FUTURE (continued)

### Content Development

Continue the discussion by explaining that solar heating is a lot like the warming of a closed car. Glass panels in a house or other building absorb solar energy, which is then converted into heat energy. The heat, which is trapped by the glass, radiates to the interior of the building.

Emphasize that the house described in this article uses not only a solar heating system, but a solar electrical system as well. Panels in the roof of the house contain photovoltaic cells, which are small "sandwiches" of silicon and metal. When sunlight strikes the surface of a photovoltaic cell, electrons flow across the layers of silicon and metal. Some students may have used calculators that contain photovoltaic cells. Instead of needing a battery for power, these calculators automatically "turn on" when exposed to sunlight.

### Content Development

Discuss with students some of the advantages of using sunlight as an energy source.

- **What advantages does solar energy have over an energy source such as oil or natural gas?** (Answer may vary. Students should recognize that solar energy is a renewable resource, while oil and natural gas are not; also, unlike oil and natural gas, the use of sunlight does not cause air pollution.)
- **What advantages does solar energy have over nuclear power?** (Answer may vary. An obvious consideration is safety, since nuclear

readings of the dust level on the solar panels and lets you know when they need washing! Maybe someday soon, the computer will take care of cleaning the photovoltaic cells too!

## SOLAR SERVICE

One amazing thing about your house is how well it uses energy. That's because it has several special features. These features are designed to get the most out of the sun's natural heating ability in winter. At the same time, they are designed not to add heat to the house in summer. How can this be done?

Both the trombe wall and the wall containing phasechange salts can be seen in this view of the house. These solar features, designed to get the most out of the sun's natural heating ability, are for winter use only.

The large, rounded shape in the front of the house is the airlock entry, which keeps hot air out of the house in summer and cold air in winter. The automatic shutters that reflect sunlight away in summer can be seen covering the upstairs windows.

That black wall covered with glass along the south side of the house is called a *trombe wall*. Its job is to collect the sun's rays. When these rays pass through the glass, they strike the trombe wall and are absorbed. The wall is painted black to make sure as much of the sun's energy as possible is absorbed. Dark colors absorb sunlight best. The wall heats up as it absorbs energy. Because the wall is very thick, a great deal of heat is stored. At night or on a cloudy day, that heat is slowly released into the house. The glass covering is about 15 centimeters from the wall and creates an air space. This space prevents the heat from escaping to the outside.

The trombe wall is a wintertime-only feature. During the summer months, the wall is shaded.

Phasechange salts are another amazing solar feature. Along the south side of the house, tubes no longer than 76 centimeters are built into the wall. These tubes are painted black on the outside to absorb the greatest amount of solar energy. Inside the tubes are special calcium compounds, which are solids at temperatures below 27.2° C, their melting point. During the day, the sun's radiation is absorbed by the salts. The salts melt if their temperature goes above 27.2° C. But during the time they melt and stay liquid, they store the sun's heat energy. Then at night, when the temperature drops below the melting point, the salts turn back to the solid phase. This phase change releases the stored heat to the house. However, like the trombe wall, the salt-containing tubes must be shaded during the summer months.

Well, these solar features have come a long way since they were first introduced back in the 1970s. Since then, they have been modified and improved. So now in 1997, they help keep your house warm as toast in winter and cool as a cucumber in summer. That thought reminds you your computer is calling—dinner is ready!



## CRITICAL THINKING QUESTIONS

**1.** Based on what you know about solar heating, would you say that glass is a conductor or a nonconductor of heat? Why? (Glass is a nonconductor. This is evident because heat produced by radiant energy does not pass back through glass to the outdoors.)

**2.** Do you think that solar heating would be more effective in some locations than others? Explain your answer. (Accept all logical answers. Solar heating is probably most effective in areas that tend to receive a fair amount of sunlight all year, such as the southern United States; solar heating also tends to be most effective in flat, open areas that are not shaded by mountains or trees.)

**3.** Could water be used instead of phasechange salts in the tubes along the south side of the house? Why or why not? (No, because the freezing point of water is too low—the temperature at night would have to go down to 0°C in order for the water to freeze. In addition, phase-change salts can hold much more heat per unit-of-volume than can water.)

power involves the risk of a nuclear accident; also, nuclear wastes tend to cause land, water, and air pollution.)

• **What are some disadvantages of using solar energy instead of other sources?** (Answer may vary. The main limitations of solar energy include high cost; the fact that the sun does not always shine; and its dependence on a geographic location that has a high exposure to sunlight year-round.)

### Reinforcement

Point out to students that scientists estimate that the amount of solar energy falling on a 200-square-km plot near the equator is enough to meet all of the world's energy needs, if somehow this solar energy could be collected and distributed.

### Content Development

Discuss with students how phase-change salts collect and give off heat. Point out that when a substance freezes, heat is given off; when a substance melts, heat is absorbed. The amount of heat per gram of substance that is released in freezing or absorbed in melting is called the heat of fusion for that substance. For example, the heat of fusion for water is 88 calories per gram.

### Teacher's Resource Book Reference

After students have read the Science Gazette article, you may want to hand out the reading skills worksheet based on the article in your Teacher's Resource Book.

# Unit Four

# STRUCTURE OF MATTER

## UNIT OVERVIEW

In Unit Four, students are introduced to the structure and basic classification of matter. They learn first about the atomic model of matter, as they study the nature of atoms and of the molecules the atoms form. Next they explore the differences between elements and compounds and gain practice in recognizing and writing chemical symbols, formulas, and equations. Finally, they learn about the properties and types of mixture and the components of solutions.

## UNIT OBJECTIVES

1. **Describe the structure of atoms.**
2. **Explain the nature of molecules.**
3. **Compare elements and compounds.**
4. **Recognize and write chemical symbols, formulas, and equations.**
5. **Describe the properties of mixtures, and distinguish between heterogeneous mixtures and homogeneous mixtures.**
6. **Describe the properties and components of suspensions, colloids, and solutions.**

358

## INTRODUCING UNIT FOUR

Begin your teaching of the unit by having students examine the unit-opener computer-generated image of a quasar. Before students read the unit introduction, you may wish to ask them the following questions.

- **What do you think the picture shows?** (Answers will probably vary widely. Some students may name very large, distant objects, such as galaxies or star clusters. Others may name very small objects, such as an atom, perhaps. Point out that the object is called a *quasar*.)
- **Do you know what the word *quasar* comes from?** (It comes from "quasi-stellar," which means starlike.)
- **How are quasars like stars?** (They give off light, are distant, and so on.)
- **How do you think they differ?** (Quasars produce vastly greater amounts of energy than typical stars.)

Now have students read the unit introduction, then ask them the following questions.

- **What is a light-year?** (It is the distance light travels in a year—nearly 10 trillion km.)
- **Why is looking at a quasar like looking back in time?** (The image we are seeing is formed by light that left the quasar very long ago.)
- **What do you think a quasar actually is?** (Answers will vary. Theories

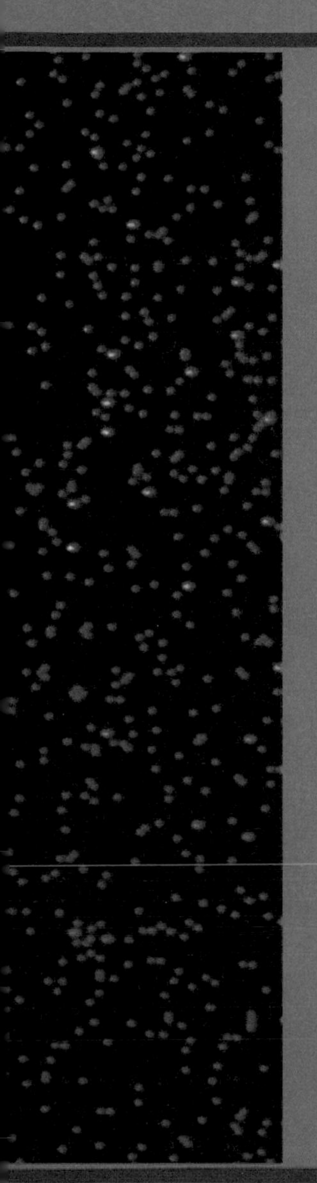

# Structure of Matter

Probing ever further into space, scientists have discovered more and more distant objects. Somewhere out there is the most distant object in the universe, a sphere of tightly packed particles shooting out incredible bursts of energy. This object is a quasar. Located perhaps 20 billion light-years from the earth, this quasar has existed for 20 billion years. And its light has been traveling for that long a time. Perhaps by viewing the light from this quasar, scientists can witness the very beginning of the universe. For in that flash of light is evidence of the birth of all the matter in space—galaxies, stars, planets, rocks, molecules, atoms, and even the atoms' tiniest parts. These small particles, set loose billions of years ago, hold many secrets about the world around you.

CHAPTERS



*Computer-generated image of Quasar 3C 273*



## CHAPTER DESCRIPTIONS

**15 Atoms and Molecules** Chapter 15 deals with the atomic model of matter. After a brief historical treatment of this topic, the principal subatomic particles—electrons, protons, and neutrons—are described, and the concepts of atomic number and of isotopes are explored. Finally, the modern atomic theory of the atom is sketched, and the nature of molecules is analyzed.

**16 Elements and Compounds** Chapter 16 focuses first on elements, or pure substances that contain only one kind of atom. Then it goes on to deal with compounds, pure substances that contain more than one kind of atom. Chemical formulas are presented and analyzed and, finally, the concept of balanced chemical equations is introduced.

**17 Mixtures and Solutions** The nature and properties of mixtures are treated first in Chapter 17. Then heterogeneous and homogeneous mixtures are contrasted, and examples of each are provided. The properties of suspensions and colloids are described. The chapter closes with a treatment of solutions and their two basic components, solutes and solvents.

differ among scientists. Many speculate that quasars are the beginning stages of formulation of other structures, such as galaxies, and that, at their centers, they contain a black hole.)

• **Do you think the matter that makes up a quasar is fundamentally different in kind from the matter that makes up the earth?** (Answers will vary. Most scientists would agree that it is not.)

## Chapter 15

# ATOMS AND MOLECULES

## CHAPTER OVERVIEW

All things in the universe are made of atoms. Atoms are the basic building blocks of all matter. Each atom has a structure and can be divided into smaller parts. These parts are called *subatomic particles*. The main subatomic particles of the atom are the protons, neutrons, and electrons. The nucleus or heart of an atom consists of positively charged protons and electrically neutral neutrons. The number of protons in the nucleus identifies the atom and gives the atom its atomic number. Surrounding the nucleus of an atom is an electron cloud. The electron cloud is space in which the electrons, or negatively charged subatomic particles, might be found. The location of an electron in the cloud depends upon how much energy the electron has. Electrons of the atom are arranged in energy levels.

Atoms can combine chemically with each other to produce new and different structures called *molecules*. A molecule is the smallest particle of a substance that has all the properties of that substance.

## INTRODUCING CHAPTER 15

Have the class observe the photograph on page 360. Point out that the photograph shows beads of mercury. Have students read the introduction to the chapter.

- **What is mercury?** (Students might suggest that mercury is the "liquid" in a thermometer. They may also suggest that it is matter.)
- **In your opinion, what causes some of the beads of mercury to be so small, while others are larger?** (They broke or were cut away from the original that way.)
- **What phase do you guess mercury to be? Why?** (Liquid, because it has no definite shape)
- **Is there some tiny bead of mercury in the picture that if sliced one more time would no longer be mercury?** (no)

Point out that mercury is a liquid and is made of one kind of particle. Explain that after observations, collecting facts, and experimenting, scientists developed a theory that some substances were made out of only one kind of particle. Tell them that first the scientists had to determine what the particle was.

- **What do you think one particle of mercury would be?** (mercury)
- **What would you do to identify the smallest single particle of any substance?** (name it)

Point out that scientists also needed to name the single smallest particle of matter. Have the students read the chapter title.

# 15 Atoms and Molecules

**CHAPTER SECTIONS**

**15-1 An Atomic Model of Matter**
**15-2 Structure of the Atom**
**15-3 The Molecule**

**CHAPTER OBJECTIVES**

*After completing this chapter, you will be able to:*

- **15-1** Discuss an early idea about the nature of matter.
- **15-1** Describe the models of the atom developed by Thomson and Rutherford.
- **15-2** Classify three subatomic particles according to their location, charge, and mass.
- **15-2** Define the terms atomic number, isotope, mass number, and atomic mass.
- **15-2** Describe the arrangement of electrons according to modern atomic theory.
- **15-3** Explain what a molecule is.

Beads of mercury gleam on a sheet of cloth. Some beads are small, others are large. But they are all still mercury, a pure substance. If you were to take the smallest bead of mercury and slice it in half once, twice, three times, even a thousand times, you would still be left with mercury—or would you?

Is there some incredibly tiny bead of mercury that if sliced one more time would no longer be mercury? It was just this kind of question that sparked the imagination and curiosity of early scientists. They hypothesized and argued as the years passed—for more than two thousand years, in fact.

Then, slowly, clues were found. Experiments were performed. New ideas were hatched. And finally an answer was developed. In many ways it was a very simple answer. To find it out, let's begin at the beginning.

Silvery beads of liquid mercury



## TEACHER DEMONSTRATION

Show the class a square sheet of aluminum foil. Point out that aluminum is a homogeneous pure substance. Tear the foil in half and give each half to a student. Have students repeat this procedure until each student has one small piece of foil. Have each student tear the piece of foil into the smallest possible pieces.

Tell students that if they could tear the foil into the smallest pieces that still had the properties of aluminum, each piece would be an aluminum atom.

• **Copper is an element that is a pure substance. What would we have if we cut a copper wire into the smallest pieces that still had the properties of copper?** (Students should respond by saying copper atoms.)

## TEACHER RESOURCES

### Audiovisuals

*Electrons and Protons in Chemical Change,* 2 filmstrips with 2 cassettes, PH Media

*Matter and Molecules: Into the Atom,* filmstrip with cassette, SVE

*Measuring Electron Charge and Mass,* 2 filmstrips with 2 cassettes, PH Media

*The Nucleus: Composition, Stability, and Decay,* filmstrip or slides with cassette, PH Media

### Books

Bohr, Niels, *Theory and the Description of Nature.* AMS Press

Condon, E. V., and H. Odabasi, *Atomic Structure,* Cambridge University Press

Conn, G. K., *Atoms and Their Structure,* Cambridge University Press

Koester, L., and A. Steyerl, *Neutron Physics,* Springer-Verlag

### Software

*The Atomic Nucleus,* Prentice Hall

*Atomic Structure,* Prentice Hall

• **What do you predict they named this single smallest particle of matter?** (an atom)

• **If an atom is the single smallest particle of matter, what do you predict is a molecule?** (Students might suggest two atoms or a combination of atoms.)

Explain that after scientists found this very tiny particle, they had to test the theory of the atom over and over again. Point out that a great many experiments over 2400 years changed and rechanged the idea of what an atom was.

• **Why would it take such a long time to research the atom?** (Students might suggest that the atom is too small to see, or that because it was a new idea other scientists were reluctant to accept it.)

Tell the students that the atom is the basic building block of all matter.

## 15-1 AN ATOMIC MODEL OF MATTER

### SECTION PREVIEW 15-1

We know that virtually everything in the world that occupies space and has weight exists as a solid, liquid, or gas and is called *matter.* In the search of a fuller description of matter, a Greek philosopher, Democritus, believed that matter was composed of tiny indivisible parts which he called atoms. Early in the 1800s, John Dalton discovered that gases combined as if they were made of atoms.

In 1897, while J. J. Thomson was studying the passage of an electric current through a gas, he reasoned that because some of the gas gave off rays made of negatively charged particles, the atom was divisible. These negatively charged particles are now known as *electrons.*

Later in 1908, Ernest Rutherford discovered the positively charged center of an atom, which he called the *nucleus.*

### SECTION OBJECTIVES 15-1

1. **Discuss an early idea about the nature of matter.**
2. **Describe the models of the atom developed by Thomson and Rutherford.**

### SCIENCE TERMS 15-1

**atom p. 362**
**electron p. 363**
**nucleus p. 364**

## 15-1 An Atomic Model of Matter

Can matter be divided into smaller and smaller pieces forever? This was one of the questions that puzzled Greek philosophers more than 2400 years ago. For they were trying to figure out what matter was made of. Some of their ideas were rather strange and were certainly incorrect. Yet others were steps along the trail that would finally lead to the truth.

### *Early Ideas About the Atom*

The search for a description of matter began with the Greek philosopher Democritus (de-MOK-rih-tuhs). He believed that an object could not be cut into smaller and smaller pieces forever. Eventually, the smallest possible piece would be obtained. And that piece could not be divided any further. It was indivisible. Democritus called this smallest piece of matter an **atom.** The word "atom" comes from the Greek word *atomos* meaning "indivisible."

What were these atoms? Democritus had no way of knowing. But he guessed they were small, hard particles that were all made of the *same* material but came in different shapes and sizes. Also, these atoms were infinite in number, always moving, and capable of joining together. Although Democritus was on the right trail, his theory of atoms was ignored and forgotten. In fact, it would be another 2100 years before others traveled down that trail.

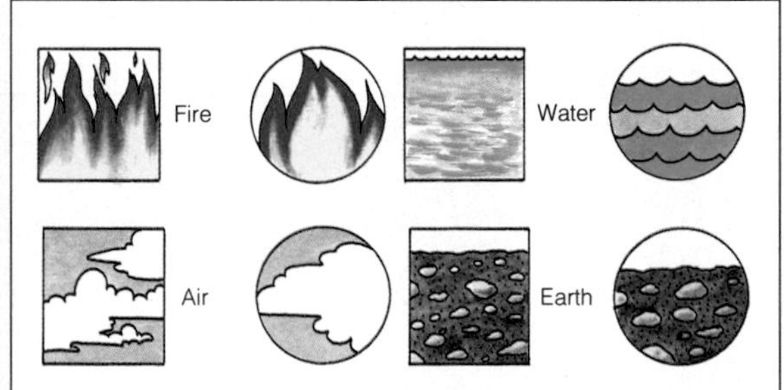


**Figure 15-1** *Ancient Greeks believed that all the matter on the earth was made of different combinations of four basic elements—fire, air, water, and earth* (right). *Alchemists, or early chemists, hoped to find a way to change ordinary metals into gold* (left).

362

### TEACHING STRATEGY 15-1

#### Motivation

Show students a beaker of water.

• **Can we divide this water?** (yes)

Fill a medicine dropper with the water. Slowly allow drops to fall from the dropper.

• **Have we divided the water?** (yes)

• **What did we divide the water into?** (drops)

• **How might we divide each drop in half?** (Answers will vary. Students might suggest using the dropper to "lift out" half the water.)

#### Content Development

Point out that a Greek philosopher named Democritus believed that an object could not be divided into smaller pieces forever. He called the smallest piece of matter an *atom.* Tell students that *atom* means indivisible. He guessed them to be solid and indivisible.

Point out that it was not until 1897 that a model of an atom was developed. Explain that J. J. Thomson found that negatively charged particles, which he called corpuscles, were part of an atom. These "corpuscles" of Thomson's are called *electrons* today.

Explain that this finding showed that atoms were neither solid nor in-

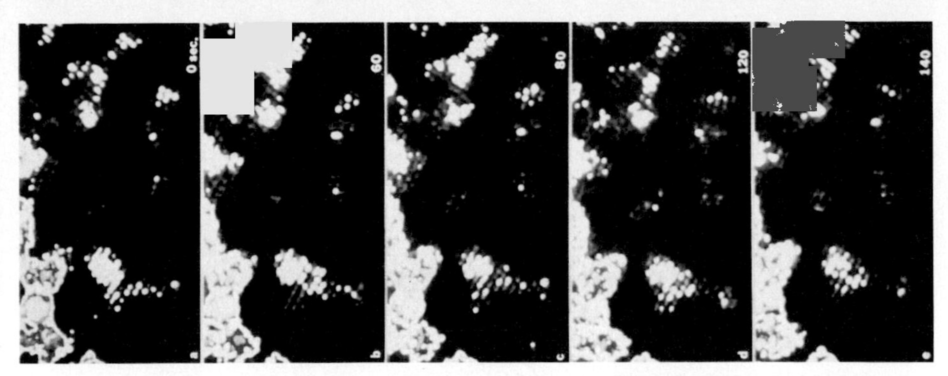

Figure 15-2 *Magnified more than 5 million times, the atoms that Democritus and other scientists labored to describe are clearly visible as blue, yellow, and red spots in these photographs of uranium atoms.*

### *Modern Ideas About the Atom*

In the early 1800s, the English chemist John Dalton did a number of experiments that led to an interesting conclusion. Dalton discovered that gases combined as if they were made of individual particles. These particles were the atoms of Democritus. For the first time, a theory of atomic structure was based on chemical experiments. Dalton had taken an important step along the atomic trail. But like Democritus, Dalton believed that atoms were "uncuttable." Was he right?

In 1897, the English scientist J. J. Thomson was studying the passage of an electric current through a gas. The gas gave off rays that Thomson showed were made of negatively charged particles. But the gas was known to be made of uncharged atoms. So where had the negatively charged particles come from? From within the atom, Thomson reasoned. A particle smaller than the atom had to exist. The atom was divisible! Thomson called the negatively charged particles "corpuscles." Today they are known as **electrons.**

Thomson's discovery of electrons presented him with a new problem to solve. If the atom as a whole was uncharged, yet it contained negatively charged particles, what balanced the negative charge? Thomson reasoned that the atom must also contain a positive charge. But try as he might, he was unable to find this positive charge.

So Thomson proposed a model of the atom that is sometimes called the "plum pudding" model. Thomson's proposed atomic model can be seen in Figure 15-3.

**Activity**

*A Mental Model*

Scientists have developed an atomic model based on many experiments. Because scientists cannot see the subatomic particles, they must rely on observing how atoms behave under certain conditions. In this activity, you will develop a mental model of something you cannot see.

Place three to five small items in a shoe box. Seal the box. Exchange boxes with a friend or classmate. Using a variety of tests on the box, describe what you think is inside it. You can shake, turn, weigh, and smell the box. You can also hold a magnet near it. But do not open the box or damage it.

After you have developed a mental model of the contents, open the box and see how close your model is to the actual contents.



## HISTORICAL NOTES

The Greek philosopher Aristotle believed that all matter was made of four fundamental "elements": earth, air, fire, and water. Although we know today that not one of these is a true element, it was Aristotle and other Greek philosophers who led the way in scientific investigations with their spirit of inquiry.

## TIE-IN/GEOGRAPHY

Show the class a road map of your state. Point out that the map does not show all the details about the highways and cities. Explain that a map is a model of reality that is helpful in understanding how to get to different places. A scientific model of the atom is helpful in understanding the behavior of matter.

### Activity

**A Mental Model**
**Skills: Inferring, relating, applying, hypothesizing, predicting**
**Level: Remedial/average**
**Type: Hands-on**
**Materials: Shoe box, several common items**

This activity will help students develop the ability to form hypotheses on the basis of indirect observation. The difficulty involved in doing this will help bring out the difficulties faced by atomic scientists, who cannot directly observe atoms.

divisible. They contain smaller parts and have a structure.

### Content Development

Point out that atoms are the building blocks of all matter. Explain that even though atoms are too small to be seen, the properties of atoms can be determined by indirect methods. The scientific model of what an atom is like has changed over the years. As each scientist made a discovery, the model was changed to fit the evidence.

### Reinforcement

Review the scientific method for the benefit of slower students, being sure to provide many examples. Have students identify the steps involved. Take extra care to make clear what is meant by indirect evidence, as this concept is somewhat difficult for slower students to grasp in the abstract. Once again, provide examples, making use of situations relating to the students' everyday lives.

### Motivation

Point out to students that the type of tube Thomson used in his experiments is a cathode ray tube. Such tubes are still used in television sets, in which a beam of electrons is directed in such a way that it makes a bright image appear on the screen.

## TEACHER DEMONSTRATION

Show the class a tennis ball. Toss the ball against the wall.

- **What happened to the ball?** (The ball bounced back.)
- **Why did the ball bounce back?** (It hit the hard wall.)
- **What would happen if the wall had many large holes similar to a wire fence?** (Sometimes the ball would bounce back and sometimes it would go through the holes.)

Point out that scientists used an experiment similar to tossing a tennis ball at a wall to develop ideas about the composition of matter.

## FACTS AND FIGURES

A mass of 12 g of carbon contains an unbelievable number of atoms. If every person on Earth were to help count these atoms at the rate of 1 atom per second, it would take 5 million years for all of the atoms to be counted!

## ANNOTATION KEY

❶ **Thinking Skill: Making inferences**
❷ **Thinking Skill: Making generalizations**

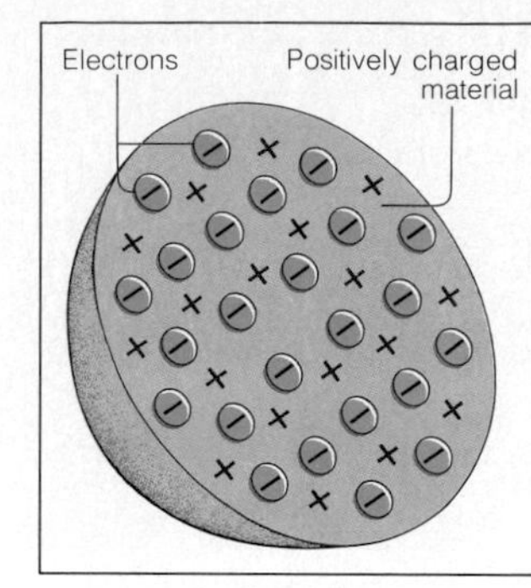


**Figure 15-3** *Thomson's model of the atom pictured a "pudding" of positively charged material throughout which were scattered negatively charged electrons.*

**According to Thomson's atomic model, the atom was made of a puddinglike positively charged material throughout which negatively charged electrons were scattered, like plums in a pudding.** Thomson's model was far from correct, but it was an important step toward understanding the atom.

In 1908, the British physicist Ernest Rutherford was hard at work on an experiment that seemed to have little to do with the mysteries of the atom. Rutherford was doing experiments in which he fired tiny, positively charged particles at thin sheets of gold. Rutherford discovered that most of these positively charged "bullets" passed right through the gold atoms without changing course at all. This could only mean that the gold atoms in the sheet were mostly empty space! The atoms were not a pudding filled with a positively charged material, as Thomson had thought.

Some of the "bullets," however, did bounce away from the gold sheet as if they had hit something solid. What could this mean? ❶ Rutherford knew that positive charges repel other positive charges. So he proposed that an atom had a small, positively charged center that repelled his positively charged "bullets." He called this center of the atom the **nucleus** (NOO-klee-uhs; plural: nuclei, NOO-klee-igh).

**In Rutherford's model, the atom looked like a tiny solar system in which electrons circled the nucleus in the same way that the planets circle the sun.** Useful in many ways, this model did not explain the arrangement of the electrons.

**Figure 15-4** *In Rutherford's experiment* (left), *most of the positively charged particles passed right through the gold sheet. A few were slightly deflected, while a very few bounced back. From these observations, Rutherford developed his model of the atom* (right) *as an object that was mostly space but had a dense, positively charged nucleus.*

EXPERIMENTAL SETUP

High-speed positively charged particles
Few particles
Very few particles
Screen
Lead box
Screen
Most particles
Particle emitter
Thin gold sheet

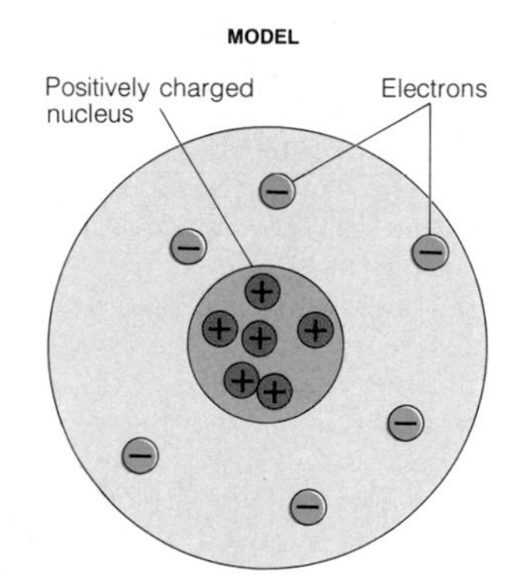


364

### 15-1 (continued)

### Motivation

Have students observe Figure 15-3. Discuss the figure, using questions similar to the following:

- **What kind of material was most of Thomson's model of the atom made of?** (positively charged material)
- **What do you predict might happen if a small bullet of positively charged material was shot at Thomson's atom?** (Students should suggest that the small bullets would bounce off.)
- **If you could actually shoot a small bullet of positively charged material at a real atom and the bullet bounced back, what would you conclude?** (Students might suggest a wide variety of answers. Lead students to suggest that if the bullet bounced back, then Thomson's model would be reasonable.)
- **If you could actually shoot a small bullet of positively charged material at a real atom and the bullet did not bounce back, what would you conclude?** (Students might suggest a wide variety of answers. Lead students to suggest that if the bullet went through the atom, then Thomson's model would be incorrect.)

### Content Development

Explain that because some of the bullets in the Rutherford experiment did bounce away, he concluded that an atom had a small, positively charged center. Point out that he called this center of an atom the *nucleus.*

### Section Review 15-1

1. Electron.
2. Nucleus; Rutherford.

It would be the job of future scientists to improve upon Rutherford's atomic model. And it would not be a simple task! For the diameter of the atom is only one ten-billionth of a meter. Examining its structure is like finding one particular grain of sand among all the grains of the earth.

### *SECTION REVIEW*

1. What atomic particle did J. J. Thomson discover?
2. What is the center of the atom called? Who discovered it?

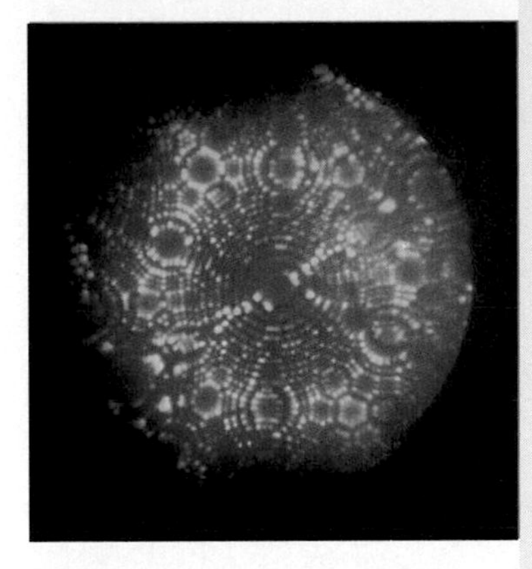

**Figure 15-5** *A crystal of tungsten, magnified 3,000,000 times, shows a regular arrangement of atoms.*

## 15-2 Structure of the Atom

When Thomson performed his experiments, he was hoping to find a single particle smaller than the atom. Certainly Thomson would be surprised to learn that today about 200 different kinds of such particles are known to exist! Because these particles are smaller than an atom, they are called **subatomic particles**.

**The three main subatomic particles are the proton, neutron, and electron.** As you read about these particles, note the location, mass, and charge of each. In this way, you will better understand the modern atomic theory. Let's begin with the nucleus, or center, of the atom.

### *The Nucleus*

The nucleus is the "heart" of the atom, the core in which 99.9 percent of the mass of the atom is located. Yet the nucleus is about a hundred thousand times smaller than the entire atom! In fact, the size of the nucleus compared to the entire atom is like the size of a bee compared to a football stadium!

**PROTONS** Those positively charged "bullets" that Rutherford fired at the gold sheets bounced back because of the **protons** in the nucleus of the gold atoms. Protons are positively charged particles found in the nucleus.

2 **Figure 15-6** *When subatomic particles collide, new and unusual particles may be produced. By studying the tracks made by these particles in a bubble chamber, scientists can learn more about the nature and interactions of subatomic particles.*

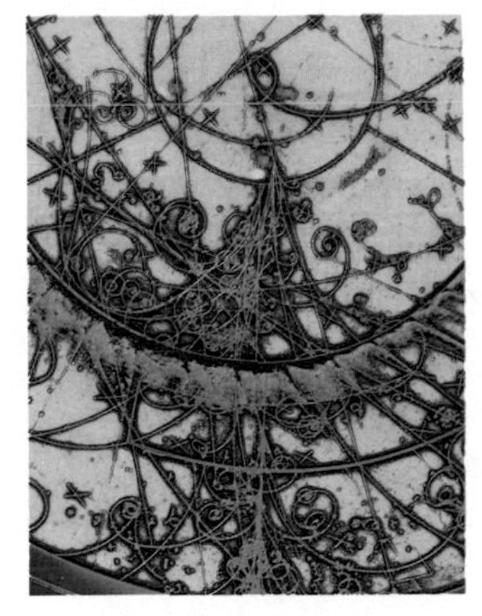

## 15-2 STRUCTURE OF THE ATOM

### SECTION PREVIEW 15-2

Each atom has a structure that can be divided into smaller parts. The positively charged nucleus contains protons and neutrons. The number of protons in the nucleus is called the *atomic number.* The atomic number of a substance never changes. Atoms that have the same number of protons but a different number of neutrons are called *isotopes.*

The positively charged nucleus is surrounded by negatively charged electrons. The number of negatively charged electrons is usually equal to the number of positively charged protons.

### SECTION OBJECTIVES 15-2

1. **Compare the proton, neutron, and electron in terms of charge, mass, and location within the atom.**
2. **Define atomic number and isotopes.**
3. **Describe the structure of the atom according to modern atomic theory.**

### SCIENCE TERMS 15-2

**subatomic particle p. 365**
**proton p. 365**
**atomic mass unit p. 366**
**neutron p. 366**
**atomic number p. 366**
**isotope p. 367**
**mass number p. 368**
**atomic mass p. 368**
**electron cloud p. 369**
**energy level p. 369**
**quark p. 370**

## TEACHING STRATEGY 15-2

### Motivation

Obtain two basketballs or volleyballs. Place one ball on the floor and step back about 3–6 m.

• **What do you predict will happen if I roll this ball and it hits the stationary ball?** (Students might suggest it will bounce back, glance off in a sideward direction, or some other response.)

Explain that scientists used evidence from common collisions to develop modes of matter and atoms.

### Content Development

Have students observe Figure 15-5. Point out that an atom has to be magnified over a million times in order to be seen. Explain that in the center of these magnified atoms we would find a small core. In it is concentrated nearly all the mass of the atom. Point out that this center core, or "heart", of the atom is called the *nucleus.* It contains one or more particles called *protons.* Explain that protons are positively charged particles. It is the protons that make atoms different from one another.

## BACKGROUND INFORMATION

The atomic mass unit is based on the assignment of a value of exactly 12.0000 amu to one atom of the isotope carbon-12, which has 6 protons and 6 neutrons. Until relatively recently, the amu was based upon oxygen-16.

## FACTS AND FIGURES

Several other subatomic particles are called baryons, mesons, leptons, and bosons.

### Activity

**Making Indirect Observations**
**Skills: Making observations; manipulative; making comparisons, making inferences**
**Level: Remedial**
**Type: Hands-on**
**Materials: 2 glasses, water, soap, flashlight**

In this activity, students use indirect evidence to infer that water particles collide with soap particles in order to keep the soap particles in suspension. Students may not be able to put their answer in this format, so accept all logical answers. To come to this conclusion, students first note that the beam of light passes directly through the water without soap (the control). The light, however, is scattered by the water with soap.

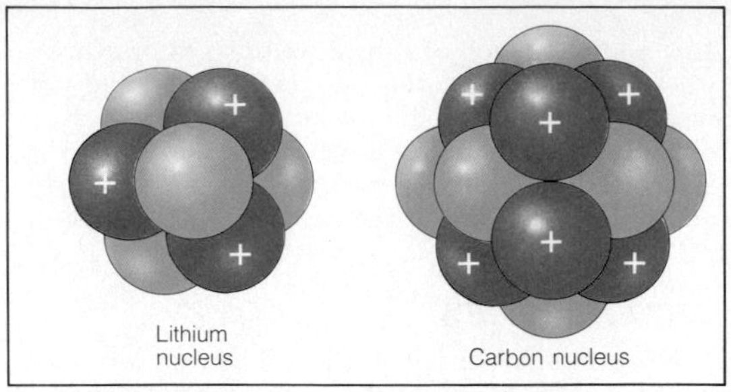

**Figure 15-7** *A lithium nucleus contains 3 protons and 4 neutrons. A carbon nucleus contains 6 protons and 6 neutrons.*

### Activity

*Making Indirect Observations*

Indirect evidence about an object is evidence you get without actually seeing or touching the object. As you gather indirect evidence, you can develop a model, or mental picture.

**1.** Fill two glasses almost completely full with water. Leave one glass as is. To the other glass add a piece of soap about half the size of a pea. Dissolve the soap by stirring the water.

**2.** Turn off all the lights in the room and make sure the room is completely dark.

**3.** Shine a flashlight beam horizontally from the side of the glass into the soapy water just under the surface. Repeat this procedure with the plain water. Observe and record the effect of the light beam in each glass of water.

What was the effect in the plain water? What was the effect in the soapy water? What caused the effect in the soapy water? What was the role of the plain water in this activity? (1)

Because the masses of subatomic particles are so small, scientists use a special unit to measure them. They call this unit an **atomic mass unit**, or amu. The mass of a proton is 1 amu. To get a better idea of how small a proton is, imagine the number 6 followed by 23 zeros. It would take that many protons to equal a mass of 1 gram!

**NEUTRONS** Sharing the nucleus with the protons are the electrically neutral **neutrons**. Neutrons have no charge. They do have mass, however. Each neutron has a mass of 1 amu.

### *Atomic Number*

You read before that atoms of different substances are different. What is it that accounts for this difference? The answer lies in the nucleus. For it is the number of protons in a nucleus that determines what the substance is. For example, a carbon atom has 6 protons in its nucleus. Carbon is a dark solid. There are 7 protons in the nucleus of a nitrogen atom. That is only 1 more than carbon. Yet nitrogen is a colorless gas.

The number of protons in the nucleus is called the **atomic number**. It identifies the kind of atom. All hydrogen atoms—and *only* hydrogen atoms—have 1 proton and an atomic number of 1. Helium atoms have an atomic number of 2. So there are 2 protons in the nucleus of every helium atom. Oxygen has an atomic number of 8, and 8 protons in the nucleus of each atom. How many protons does uranium, atomic number 92, have? (1)



## 15-2 (continued)

### Motivation

Have the class observe figure 15-9. Point out that all three drawings represent models of hydrogen atoms. Tell the class that the red spheres with +'s represent protons, the green spheres represent neutrons, and the blue spheres with −'s represent electrons. Discuss the diagram using questions similar to the following:

- **How are the three drawings similar?** (All three drawings show one proton and one electron.)
- **How are the three drawings different?** (They have different numbers of neutrons.)

Point out that the number of protons in the nucleus of an atom determines the kind of atom. Explain that similar atoms can have different numbers of neutrons. Each of the model drawings in Figure 15-9 represents an atom of hydrogen.

### Content Development

Point out that each nucleus contains both a proton and a neutron. Explain that a neutron is neutral and has no electrical charge.

Explain that the proton and the neutron make up nearly all the mass of the atom. Because the particles are so small, scientists have developed a special unit to measure them. This special unit is called the *atomic mass unit.* Write "amu" on the chalkboard.

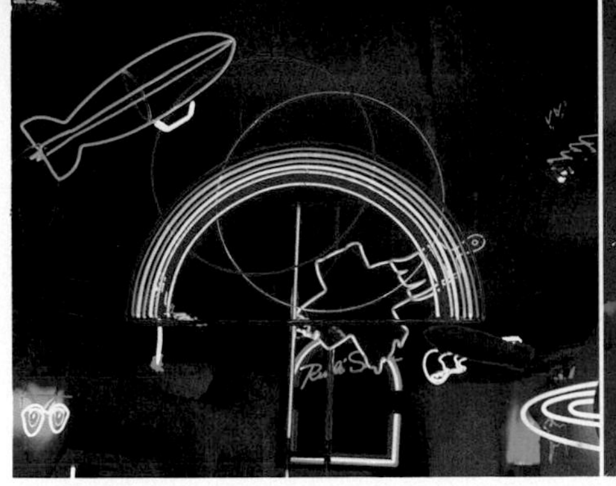

**Figure 15-8** *The number of protons in an atom's nucleus determines what the atom is. Neon atoms, which make up the gas often used in colored lights* (left), *have 10 protons. Helium atoms, which make up a colorless, odorless gas often used in balloons* (right), *have 2 protons.*

## *Isotopes*

The atomic number of a substance never changes. This means that there is always the same number of protons in the nucleus of every atom of that substance. But the number of neutrons is not so constant! Atoms of the same substance can have different numbers of neutrons.

Atoms that have the same number of protons but different numbers of neutrons are called **isotopes** (IGH-suh-tohps). Look at Figure 15-9. You will see three different isotopes of hydrogen. Notice that the number of protons does *not* change. Remember that the atomic number, or number of protons, identifies a substance. No matter how many neutrons there are in the nucleus, 1 proton always means the atom is hydrogen. ❷

**Figure 15-9** *The three isotopes of hydrogen. Notice that the number of protons remains the same; the number of neutrons changes.*

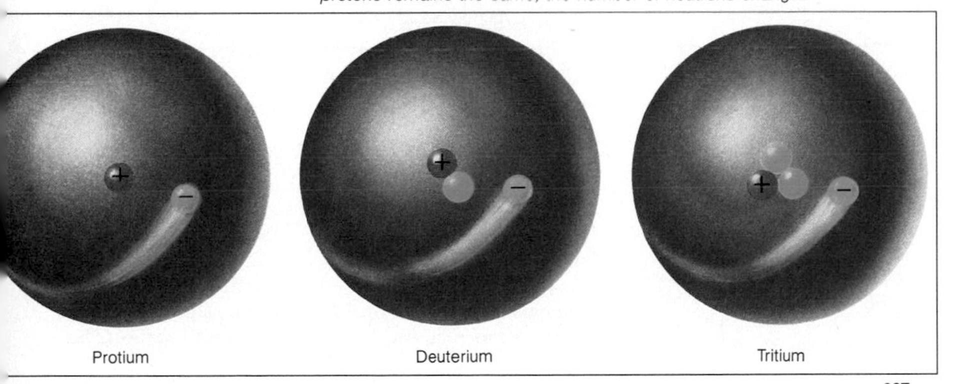




## HISTORICAL NOTES

Isotopes of a given element are nearly identical in their properties and thus went completely unrecognized as such until the present century. A mass spectroscope was used to demonstrate conclusively the existence of isotopes. The heavier the isotope of the given element, the farther the isotope traveled before being deflected sufficiently to leave a photographic record. The appearance on the mass spectrogram of different bands at different distances from the point of origin revealed the presence of isotopes.

## ANNOTATION KEY

❶ 92 (Applying concepts)

❶ Thinking Skill: Relating facts

❷ Thinking Skill: Making comparisons

Tell the students that the mass of a proton is 1 amu. The mass of a neutron is also 1 amu.

## Skills Development

***Skill: Developing a model***

Point out that the number of protons in the nucleus of hydrogen is 1, helium is 2, and lithium is 3. Divide the class into teams of 4–6 students. Give each team several toothpicks and a handful of pink and green miniature marshmallows. Have the teams use the marshmallows and small toothpick pieces as fasteners to construct models of the nucleus of the hydrogen, helium, and lithium atoms. Tell the teams to use pink marshmallows to represent protons and green marshmallows to represent neutrons. Remind students that the number of protons in each kind of atom must always be the same; the number of neutrons can vary. After the models have been completed, have the teams exchange models and determine the kind of atom represented by counting the number of protons.

- **How do you know if a model represents hydrogen?** (To represent hydrogen, the model must have only one proton.)
- **What are some neutron possibilities in the nucleus of hydrogen?** (none, 1, 2)
- **How do you know if a model represents helium?** (To represent helium, the model must have only 2 protons.)
- **What are some neutron possibilities in the nucleus of helium?** (none, 1, 2)
- **How do you know if a model represents lithium?** (To represent lithium, the model must have only 3 protons.)
- **What are some neutron possibilities in the nucleus of lithium?** (none, 1, 2, etc.)
- **What determines the kind of atom?** (the number of protons in the nucleus)

## TEACHER DEMONSTRATION

The concept of electron energy level is a difficult one for most students. The bookcase analogy is a useful one and can be demonstrated when the topic is introduced. If you wish, you can act it out in a sort of pantomime, asking students what you are doing. Using an empty narrow bookcase with several shelves, begin to fit books into the lowest shelf. When the shelf is full, attempt unsuccessfully to place another book onto it, and then place the book onto the second-lowest shelf, which you then fill in the same way. Do this until you have filled the three lowest shelves. Ask the following questions:

- **What am I attempting to do?** (You are attempting to fit the books into as low a position as possible.)
- **What limits my attempt?** (Each shelf can hold only a certain maximum number of books. When a shelf is filled, you must move up to a higher shelf.)

The shelves are analogous to energy levels, and the books are analogous to electrons, which tend to fill the lowest energy levels, each of which can hold only a limited number of them.

**Figure 15-10** *This chart shows the symbol, atomic number, and mass number for some common elements. Why is the mass number always a whole number while the atomic mass is usually not?* ❶

**COMMON ELEMENTS**

| Name | | Atomic Number | Mass Number |
|---|---|---|---|
| Hydrogen | H | 1 | 1 |
| Helium | He | 2 | 4 |
| Carbon | C | 6 | 12 |
| Nitrogen | N | 7 | 14 |
| Oxygen | O | 8 | 16 |
| Fluorine | F | 9 | 19 |
| Sodium | Na | 11 | 23 |
| Aluminum | Al | 13 | 27 |
| Sulfur | S | 16 | 32 |
| Chlorine | Cl | 17 | 35 |
| Calcium | Ca | 20 | 40 |
| Iron | Fe | 26 | 56 |
| Copper | Cu | 29 | 64 |
| Zinc | Zn | 30 | 65 |
| Silver | Ag | 47 | 108 |
| Gold | Au | 79 | 197 |
| Mercury | Hg | 80 | 201 |
| Lead | Pb | 82 | 207 |

### *Mass Number and Atomic Mass*

All atoms have a **mass number.** The mass number of an atom is the sum of the protons and neutrons in its nucleus. The mass number of the carbon isotope with 6 neutrons is 6 (protons) + 6 (neutrons), or 12. The mass number of the carbon isotope with 8 neutrons is 6 (protons) + 8 (neutrons), or 14. To distinguish one isotope from another, the mass number is given with the element's name.

Two common isotopes of the element uranium are uranium-235 and uranium-238. The atomic number, or number of protons, of uranium is 92. Since the mass number is equal to the number of protons plus the number of neutrons, the number of neutrons can easily be determined. The number of neutrons is determined by subtracting the atomic number from the mass number. How many neutrons are there in each uranium isotope? ❷

Any sample of an element as it occurs in nature will contain a mixture of isotopes. As a result, the **atomic mass** of the element will be the average of the masses of all the atoms in the sample. The atomic mass of an element refers to the average mass of all the isotopes of that element as they occur in nature. For this reason, the atomic mass of an element is not usually a whole number. The atomic mass of carbon is 12.011. This number indicates that in any sample of carbon there are more atoms of carbon-12 than there are of carbon-14.

### *The Electrons*

If you think protons and neutrons are small, picture this. Whirling around outside the nucleus are particles that are about 1/2000 the mass of either a proton or a neutron! These particles are electrons. Electrons have a negative charge and a mass of 1/1836 amu. In an atom, the number of negatively charged electrons is usually equal to the number of positively charged protons. So the total charge on the atom is zero. The atom is neutral.

Rutherford pictured electrons moving around the nucleus like planets orbiting the sun. But as scientists learned more about the atom, they had to reject this



## 15-2 (continued)

### Content Development

Review the concept of mass number until all students are clear about what this number signifies. Keep in mind that when students work with mass number, they often forget that this quantity is equal to the total number of protons and neutrons, not to the number of neutrons only. This error will affect their calculations involving numbers of subatomic particles present.

### Motivation

Attach a rubber band to a small rubber ball. Hold one end of the rubber band and move your hand to make the ball vibrate and circle your hand up/down, in/out, and around. Discuss the demonstration.

- **Where was the ball during this demonstration?** (Students might suggest a wide variety of answers, such as within a certain distance of your hand or within an imaginary sphere surrounding your hand.)

Point out that the ball could possibly be many different places at any single instant of time. Therefore it is difficult to say exactly where the ball might be during the next instant in time.

### Content Development

Explain that electrons are negatively charged particles that move around the nucleus of an atom. The number of electrons is usually equal to the number of protons in an atom.

Point out that the mass of an electron is 1/1836 amu. Point out that the mass of the proton and the neutron makes up most of the mass of the atom.

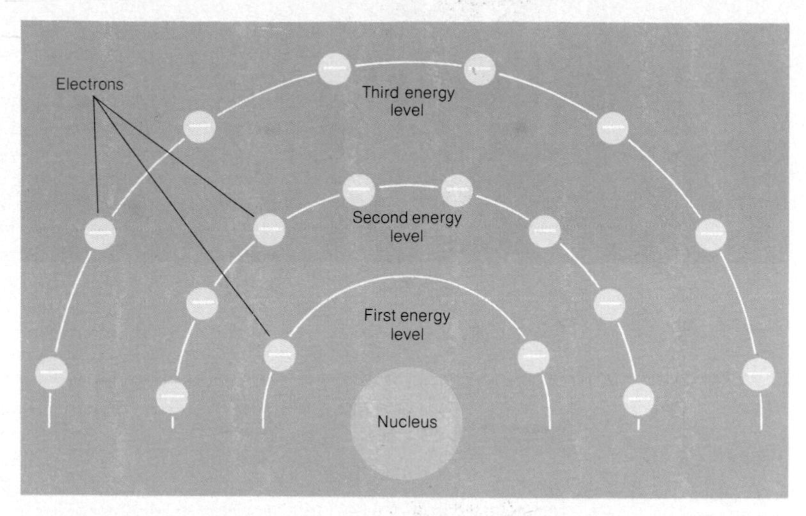


**Figure 15-11** *Each energy level in an atom can hold only a certain number of electrons. How many electrons are there in the first, second, and third energy levels shown here?* ❸

theory. Their experiments showed that electrons do not move in fixed orbits. In fact, the *exact* location of an electron cannot be known. Only the probability, or likelihood, of finding an electron at a particular place in an atom can be determined.

In fact, the entire space in which electrons are found is what scientists think of as the atom itself. Sometimes this space is called the **electron cloud.** But do not think of the atom as a solid center surrounded by a fuzzy, blurry cloud. For the electron cloud is a *space* in which electrons are likely to be found. It is somewhat like the area around a beehive in which the bees move. Sometimes the electrons are near the nucleus. Sometimes they are farther away from it. In a hydrogen atom, one electron "fills" the cloud. It fills the cloud in the sense that it can be found almost anywhere within the space.

Although the electrons whirl about the nucleus billions of times in one second, they do not do this in a random way. For each electron seems to be locked into a certain area in the electron cloud. The location of an electron in the cloud depends upon how much energy the electron has.

According to modern atomic theory, electrons are arranged in **energy levels.** Energy levels represent

### Activity

*Electron Arrangement*

Determine the electron arrangement for each of the following elements given their atomic numbers.

| | |
|---|---|
| sodium 11 | lithium 3 |
| nitrogen 7 | neon 10 |
| chlorine 17 | argon 18 |



## ANNOTATION KEY

❶ **Mass number is the number of protons and neutrons of the most common isotope. Atomic mass is the average mass of all the naturally occurring isotopes of an element. (Applying concepts)**

❷ **Uranium-235 = 143 neutrons; uranium-238 = 146 neutrons (Applying facts)**

❸ **2, 8, 8 (Interpreting diagrams)**

### Activity

**Electron Arrangement**
**Skills: Relating facts, applying concepts**
**Level: Enriched**
**Type: Computational**

Students will practice assigning electrons to energy levels and reinforce their understanding of atomic structure in this activity. The answers follow: sodium: 2 (electrons in first level), 8 (electrons in second level), 1 (electron in third level); nitrogen: 2, 5; chlorine, 2, 8, 7; lithium: 2, 1; neon: 2, 8; argon: 2, 8, 8.

Explain that the electron is found in a space known as the *electron cloud.* Relate to the text discussion and Motivation by asking questions like the following:

- **Is the electron always found in the same place?** (no)
- **What is meant when we say "an electron fills the cloud"?** (It means it can be anywhere within the space around the nucleus.)

## Skills Development

***Skill: Developing a model***

Have students draw a sketch of only the outside circle and inner proton of the protium hydrogen atom as shown in Figure 15-9. Tell the class that at any single instant the hydrogen electron could be anywhere *outside* the proton yet *inside* the outer circle. Have the students make 100 or more pencil dots within this region to represent random places that the electron *might* be at any instant. Discuss the activity using questions similar to the following:

- **Where might the hydrogen electron be at any single instant in time?** (anywhere *outside* the proton yet *inside* the outer circle)
- **Other than those dots you sketched, where else might the electron be located?** (between the dots, or anywhere outside the proton or within the outer circle)
- **Why do you think scientists refer to the space as an electron cloud?** (The model dots look like a cloud.)
- **The hydrogen atom only has 1 electron. What do you infer about the speed of that electron to be able to be almost anywhere within the outer circle?** (The electron must be moving extremely fast.)

## 15-3 THE MOLECULE

### SECTION PREVIEW 15-3

Everything we touch or see is made up of atoms or a combination of atoms. Atoms can combine chemically with each other to produce a new and different structure called a *molecule.* A molecule can be composed of a single type of atom or different types of atoms. For example, a molecule of oxygen is made up of 2 atoms. A molecule of water is composed of 2 atoms of hydrogen and 1 atom of oxygen.

A molecule is formed when 2 or more atoms chemically bind. The molecule is the smallest particle of a substance that has all the properties of that substance.

### SECTION OBJECTIVES 15-3

1. **Explain what a molecule is.**
2. **Give one example of a molecule.**

### SCIENCE TERMS 15-3

**molecule p. 370**

**Figure 15-12** *Not all molecules are as simple as water. Here you see computer images of the more complex molecules of a body fluid* (top) *and a virus* (bottom).

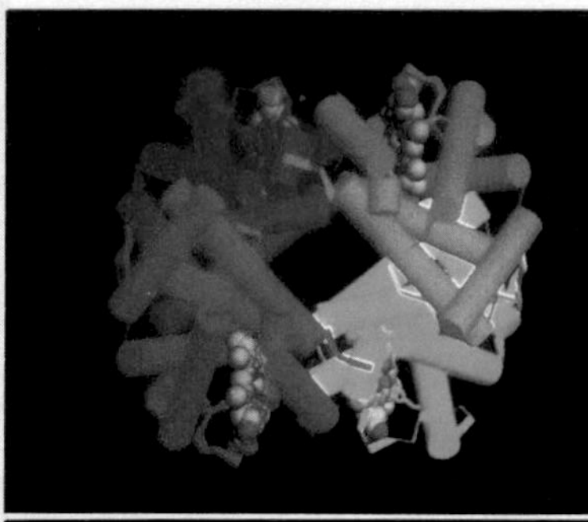

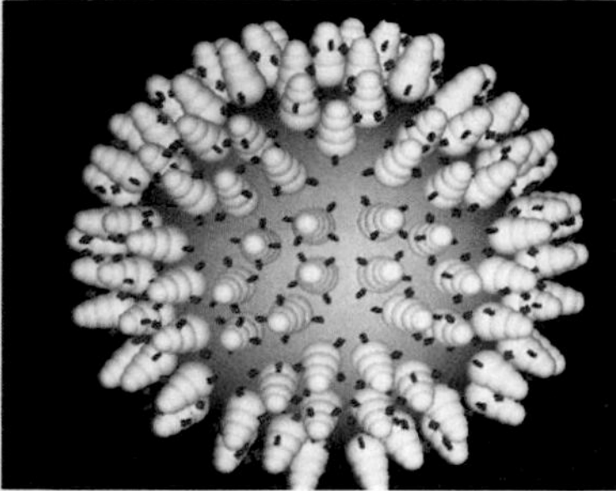

the *most likely* location in the electron cloud in which electrons can be found. Electrons with the lowest energy are found in the energy level closest to the nucleus. Those electrons with higher energy are found in energy levels farther from the nucleus.

Each energy level within an atom can hold only a certain number of electrons. The energy level closest to the nucleus—the lowest energy level—can hold at most 2 electrons. The second and third energy levels can each hold 8 electrons. See Figure 15-11. The properties of different kinds of atoms depend on how many electrons are in the various energy levels of the atoms. In fact, the electron arrangement in the electron cloud is what gives the atom its chemical properties.

Is the atom "cuttable"? The existence of protons, neutrons, and electrons proves it is. And these particles in turn can be separated into even smaller particles. It is now believed that a new kind of particle makes up all the other known particles in the nucleus. This particle is called the **quark** (kwahrk). There are a number of different kinds of quarks. All nuclear particles are thought to be combinations of three quarks. According to current theory, quarks have properties called "flavor" and "color." There are six different flavors and three different colors.

#### *SECTION REVIEW*

1. What two particles are found in the nucleus?
2. Define atomic number; isotopes; atomic mass.
3. Where are the electrons in an atom found?

### 15-3 The Molecule

Most atoms have a very important property. They can combine chemically with each other to produce new and different structures called **molecules** (MAH-luh-kyoolz). A molecule is made up of two or more atoms chemically bonded.

**A molecule is the smallest particle of a substance that has all the properties of that substance.** Suppose a sample of water was divided up into

370

### 15-2 (continued)

#### Enrichment

Although the third energy level can actually hold 18 electrons, 10 of the 18 have energies comparable to those of the electrons of the fourth energy level. These electrons are placed more "deeply" in the atom than are the fourth-level electrons. For this reason, the third energy level—and all higher energy levels—can never contain more than 8 *valence* electrons.

#### Section Review 15-2

1. Protons, neutrons.
2. Number of protons in the nucleus of atom; atoms with same number of protons but different numbers of neutrons.
3. In energy levels in the electron cloud.

### TEACHING STRATEGY 15-3

#### Motivation

Point out that when atoms chemically bind, a new substance is formed. Write "$H_2O$" on the chalkboard.

- **How were the atoms of hydrogen and oxygen combined to get water?** (Two atoms of hydrogen were chemically bonded to one atom of oxygen.)

#### Content Development

Point out that although there are only 92 different kinds of naturally occurring atoms, there are thousands of different kinds of substances. The smallest particle of a substance that has all the properties of that substance is called a *molecule.*

Some atoms such as hydrogen and oxygen combine easily to form the pure substance that we call water. The smallest possible particle of wa-

## Career: *Science Teacher*

**HELP WANTED: SCIENCE TEACHER** To teach physical science to junior high school students. Science program includes laboratory and field activities. Teaching certification in science required. Must be available to guide students in after-school activities such as science club.

Saturday afternoon provided ideal weather for some backyard gardening. But the teacher's mind was back in the classroom rather than on the sprouting bean plants. This week the unit on chemistry was to begin. Many students would have trouble understanding what could not be seen with the unaided eye. The teacher thought about the best way to begin the lesson.

**Science teachers** spend much of their time developing lesson plans for different topics. They create interesting ways to present information to their students. Activities such as lectures, demonstrations, laboratory work, and field trips are often used.

Science teachers have other duties as well. They must correct homework and test papers, make up tests and quizzes, record pupil attendance and progress, and issue reports to parents. They also attend meetings, conferences, and workshops. If you are interested in this career, you can learn more by writing to the American Federation of Teachers, 1012 Fourteenth Street N.W., Washington, DC 20005.

smaller and smaller samples. Eventually a particle that could not be further divided without losing its properties would be reached. This particle is the water molecule. It is made up of two atoms of hydrogen and one atom of oxygen chemically bonded.

Some atoms form bonds with other atoms very easily. Other atoms hardly ever form bonds. What accounts for the differences in bonding ability? The answer can be found in the arrangement of the electrons in an atom. The bonding ability of an atom is a chemical property. And it is the electron arrangement in the electron cloud that gives an atom its chemical properties.

**Figure 15-13** *This is the first photograph ever taken of atoms and their bonds. The bright round objects are single atoms. The fuzzy areas between atoms represent bonds.*

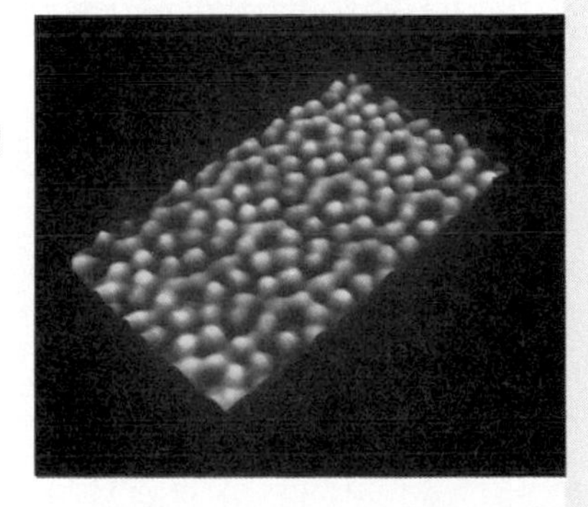

### *SECTION REVIEW*

1. What is a molecule?
2. Give one example of a molecule.

371

## SCIENCE, TECHNOLOGY, AND SOCIETY

A particle accelerator is a device used by nuclear scientists to probe the secrets of the atom's nucleus. A typical accelerator consists of a long tunnel in which beams of protons traveling at almost the speed of light are smashed into targets.

By using a particle accelerator, scientists are able to produce new elements as the nuclei in the targets absorb additional protons. Scientists are also able to analyze the tracks made as a result of particle collisions in an attempt to discover new subatomic particles. Two of the newest confirmed subatomic particles, the W and Z particles found in 1983, were discovered through the use of a particle accelerator and earned a Nobel prize for discoverer Carl Rubbia in 1984.

The United States is now planning to build the Superconducting Super Collider (SSC), an 85-km tunnel in which beams of protons will be smashed into each other at energies 20 times greater than what is now possible. The SSC will permit scientists to recreate the conditions believed to have existed 1/1000 of a trillionth of a second after the Big Bang creation of the universe. It is hoped that such experiments will give physicists some insight into the building blocks of all matter.

ter is a molecule made up of 2 atoms of hydrogen and 1 atom of oxygen that are chemically bonded together. Although at room temperature both hydrogen and oxygen are gases, the chemical produced is a liquid at normal room temperature.

The smallest possible particle of carbon is made of only a single atom of carbon. Carbon can combine with oxygen to form a molecule of the gas called carbon dioxide. Each molecule of carbon dioxide contains 1 atom of carbon and 2 atoms of oxygen.

The molecule of ordinary table sugar is a chemical combination of carbon, hydrogen, and oxygen atoms.

The arrangement of the electrons in atoms is responsible for the bonding that produces molecules of water, carbon dioxide, sugar, and many other pure substances.

Some atoms, such as helium and neon, hardly ever form bonds with other atoms.

### Section Review 15-3

1. Smallest particle of a substance that has all the properties of that substance.
2. Water molecule.

# LABORATORY ACTIVITY FLAME TESTS

## BEFORE THE LAB

1. **Divide the class into groups of 3–6 students.**
2. **Gather all materials at least one day prior to the activity. You should have enough supplies to meet your class needs, assuming 3–6 students per group.**

## PRE-LAB DISCUSSION

Have students read the complete laboratory procedure. Discuss the procedure by asking questions similar to the following:

- **Why is it important to wear a lab coat while doing this activity?** (6M HCl is a very strong acid and could destroy clothing.)
- **What should you do if acid is accidentally spilled on your lab coat or clothing?** (Immediately wash the area with water and then notify your teacher.)
- **What should you do if acid is accidentally spilled on your skin?** (Immediately wash the area with water and then notify your teacher.)
- **Why is it important to wear chemical safety goggles while doing this activity?** (6M HCl is a very strong acid and is very harmful to the eyes.)
- **What should you do if any chemical touches your eye or the inside of the safety goggles?** (Remove the goggles and use the emergency eye wash.)
- **What is the purpose of the laboratory activity?** (to use the flame tests to determine the color characteristics of various solutions caused by changes in electron energy levels due to heating)
- **Why is it important to clean the wire loop with acid and then heat it in a flame until the wire adds no color to the flame?** (This removes other chemicals that might be on the wire loop.)



### *Flame Tests*

**Purpose**

In this activity, you will observe the colors produced by excited atoms of different substances. When some substances are heated in a flame, their electrons are raised to higher energy levels by heat energy. When these electrons fall back into the lower energy levels, energy is released. The released energy produces a color characteristic of that substance.

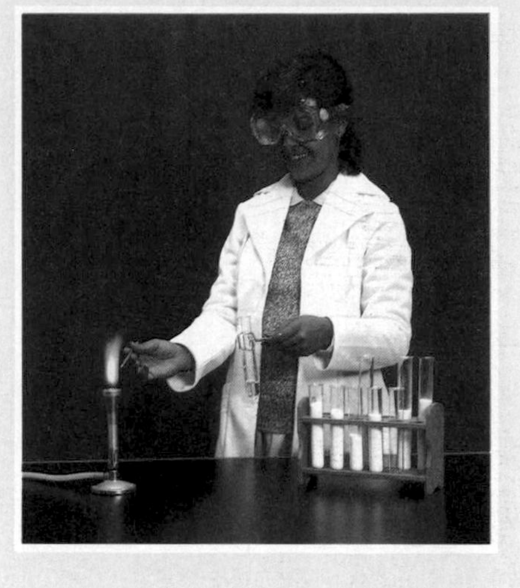

**Materials** *(per group)*

Bunsen burner
7 test tubes and test tube rack
Test tube clamp
Safety goggles
Wire loop with wooden handle
Hydrochloric acid, 6M HCl
.5M solutions of the nitrates of sodium, barium, calcium, lithium, potassium, strontium
Graduated cylinder
Test tube brush
Distilled water
Lab coat
Rubber gloves

**Procedure**   

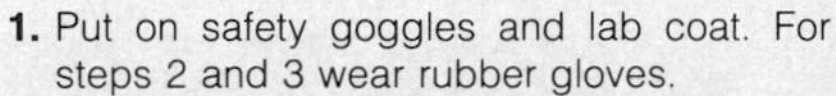

1. Put on safety goggles and lab coat. For steps 2 and 3 wear rubber gloves.
2. To clean each test tube thoroughly, hold it with the test tube clamp, place a few milliliters of 6M HCl into it and gently brush. Rinse the test tubes with tap water and then distilled water. **CAUTION:** *HCl burns. Avoid contact with skin, eyes, and clothes. If any should fall on you, immediately wash the area with water and then notify your teacher.*
3. Clean the wire loop by carefully dipping it first into some 6M HCl acid in a test tube and then holding it in the colorless flame of the Bunsen burner. Repeat until the wire adds no color to the flame. **CAUTION:** *Be careful when lighting and using the Bunsen burner.*
4. Pour 4 mL of the sodium nitrate solution into a clean test tube. Dip the tip of the clean wire loop into the solution, and then hold it in the flame just above the wire. Note: Heat only the tip of the wire.
5. Clean the wire loop as in Step 3.
6. Repeat Step 4, using in turn 4 mL of the solutions of the nitrates of barium, calcium, lithium, potassium, strontium. Clean the wire thoroughly after each test. Record the color of the flame for each substance.

**Observations and Conclusions**

1. How might scientists use a flame test to study unknown substances?



## SKILLS DEVELOPMENT

Students will use the following skills while completing this activity.

1. Manipulative
2. Observing
3. Comparing
4. Recording
5. Relating
6. Applying
7. Predicting
8. Safety

## SAFETY TIPS

Alert students to be cautious about using *any* chemical in the laboratory. Tell them that 6M HCl acid is very strong and is dangerous to the skin and eyes. In addition to being poisonous if swallowed, most chemicals can be irritating or harmful to the eyes. Be sure to have students wear chemical safety goggles *and* lab coats while doing this laboratory activity. In addition, an approved eye-wash station

# CHAPTER REVIEW

## SUMMARY

**15-1 An Atomic Model of Matter**

- More than 2000 years ago, Greek philosophers such as Democritus attempted to explain the nature of matter.
- J. J. Thomson's model of the atom pictured a positively charged material in which negatively charged electrons were scattered.
- Ernest Rutherford's discovery of the nucleus produced an atomic model in which the negatively charged electrons whirled around the positively charged nucleus.

**15-2 Structure of the Atom**

- The positively charged proton and the electrically neutral neutron make up the nucleus of an atom.
- The number of protons in the nucleus of an atom is the atomic number.
- Atoms that have the same number of protons but different numbers of neutrons are called isotopes.
- The mass number of an atom is the sum of the protons and neutrons in its nucleus.
- The atomic mass of an element is the average mass of all the naturally occurring isotopes of that element.
- The area in which electrons are likely to be found is called the electron cloud.
- Electrons are arranged in energy levels according to how much energy they have.
- All nuclear particles are thought to be made of smaller particles called quarks.

**15-3 The Molecule**

- Two or more atoms chemically bonded together form a molecule.

## VOCABULARY

*Define each term in a complete sentence.*

| | | | |
|---|---|---|---|
| **atom** | **electron** | **mass number** | **proton** |
| **atomic mass** | **electron cloud** | **molecule** | **quark** |
| **atomic mass unit** | **energy level** | **neutron** | **subatomic particle** |
| **atomic number** | **isotope** | **nucleus** | |

## CONTENT REVIEW: MULTIPLE CHOICE

*Choose the letter of the answer that best completes each statement.*

**1.** The name Democritus gave to the smallest possible piece of matter is
a. molecule. b. atom. c. electron. d. proton.

**2.** The scientist J. J. Thomson discovered the
a. proton. b. electron. c. neutron. d. nucleus.

**3.** The small, heavy center of an atom is the
a. neutron. b. proton. c. electron cloud. d. nucleus.

**4.** Ernest Rutherford's atomic model pictured a
a. negatively charged center with positively charged particles close to it.
b. positively charged material with negatively charged particles in it.
c. positively charged center with negatively charged particles around it.
d. positively charged center only.



and/or emergency shower should be readily available to all students. Demonstrate how to use the eye wash and emergency shower.

Baking soda is an excellent neutralizer for acid spills. Have several boxes of baking soda available around the lab at all times. Demonstrate how to neutralize acids by sprinkling baking soda over the spill.

## TEACHING STRATEGY FOR LAB PROCEDURE

**1.** Have the teams follow the directions carefully as they work.
**2.** Discuss how the investigation relates to the chapter ideas by asking open questions similar to the following:

- **What happened when the loop was dipped in a solution and then placed in the flame?** (The flame had different colors, depending on which solution was used.)
- **What causes the color?** (the energy released when the electrons fall back to lower energy levels)
- **Why do you think each solution has a different-colored flame?** (Answers will vary. Students might suggest that each chemical releases different amounts of energy and that amount of energy determines the color.)
- **What do you predict might happen if two of these solutions were mixed together and the mixture was tested in a flame?** (Answers will vary. Students might suggest that the colors might mix, or that one color might appear brighter than the other.)

## OBSERVATIONS AND CONCLUSIONS

**1.** Flame tests are often used to confirm the identity of an unknown metal since compounds of Group I and II metals impart characteristic colors to a Bunsen burner flame.

## GOING FURTHER: ENRICHMENT

Students might verify what happens when mixtures of the solutions are flame tested. Usually, the brightest color will dominate the other colors. Because a sodium flame is an intense yellow color, it will mask all other colors.

Students might heat thin strips of copper wire in the flame and notice emerald-green color.

Students might test common household chemicals and foods and find the flame. They are likely to notice that most common chemicals and foods contain some sodium. Thus, the intense yellow flame of sodium will mask others that may be present.

## CHAPTER REVIEW

### MULTIPLE CHOICE

1. b 2. b 3. d 4. c 5. d 6. a 7. d 8. a 9. d 10. c

### COMPLETION

1. atom
2. electrons
3. nucleus
4. protons
5. 18
6. isotopes
7. electron
8. energy levels
9. quark
10. molecule

### TRUE OR FALSE

1. T
2. F Dalton
3. F electrons
4. T
5. F nucleus
6. F neutrons
7. T
8. F isotopes
9. F closest to
10. T

### SKILL BUILDING

**1.** Atom X is sodium, atomic number 11.
**2.** Answers will vary, but should include the idea that probability represents the likelihood—or the most logical explanation—of a circumstance or event. Check that students' total probability equals 100%. Any change in a variable should affect results. The probability of an event can only be approximated; it can never be known for certain—whether it's an electron's location in an atom or a friend's whereabouts.
**3.** The atomic mass is exactly the same as the mass number since only one isotope exists. The atomic mass is an average of the masses of all naturally occurring isotopes.
**4.** Because atoms cannot actually be seen, scientists rely on indirect evidence, or observations of how matter behaves. From this indirect evidence, a model can be developed. A model is a mental picture that uses familiar ideas to explain unfamiliar or unobservable phenomena. A model can easily be changed as new information is gathered.

**5.** Particles smaller than the atom are
a. molecules. b. elements. c. compounds. d. subatomic particles.
**6.** The nucleus of an atom contains
a. protons and neutrons. b. protons and electrons. c. neutrons and electrons. d. protons, neutrons, and electrons.
**7.** The number of protons in an atom is called its
a. mass number. b. isotope number. c. quark number. d. atomic number.
**8.** An isotope of oxygen, atomic number 8, could have
a. 8 protons and 10 neutrons. b. 10 protons and 10 neutrons. c. 10 protons and 8 electrons. d. 6 protons and 8 neutrons.
**9.** All nuclear particles are thought to be made up of a combination of three
a. electrons. b. isotopes. c. molecules. d. quarks.
**10.** Two or more atoms chemically bonded together form a(n)
a. quark. b. isotope. c. molecule. d. nucleus.

### CONTENT REVIEW: COMPLETION

*Fill in the word or words that best complete each statement.*

**1.** The smallest particle of matter is a(n) ________.
**2.** The negatively charged particles that Thomson called "corpuscles" are today known as ________.
**3.** Rutherford is credited with the discovery of the ________, or the center of the atom.
**4.** The particles in an atom that have a positive charge and a mass of 1 amu are the ________.
**5.** An atom with an atomic number of 18 has ________ protons in its nucleus.
**6.** Atoms that have the same number of protons but different numbers of neutrons are called ________.
**7.** The subatomic particle that has a negative charge and a mass almost equal to zero is the ________.
**8.** The negative particles in an atom are arranged in ________, which represent the most likely places of finding them based on their energy content.
**9.** The particle that is now thought to make up all other nuclear particles is called the ________.
**10.** When two or more atoms combine chemically, they form a(n) ________.

### CONTENT REVIEW: TRUE OR FALSE

*Determine whether each statement is true or false. If it is true, write "true." If it is false, change the underlined word or words to make the statement true.*

**1.** The idea that matter was made up of indivisible particles called atoms was proposed by <u>Democritus</u>.
**2.** In the early 1800s, <u>Aristotle</u> developed a theory of atomic structure that was based on chemical experiments.
**3.** In Thomson's experiment, the gas in the tube gave off rays that were made of negatively charged particles called <u>neutrons</u>.
**4.** Rutherford's experiments detected the presence of a <u>positively</u> charged center in the atom.

374

### ESSAY

**1.** Sulfur: 2, 8, 6. Fluorine: 2, 7. Argon: 2, 8, 8. Lithium: 2, 1.
**2.** The atomic mass of this element would be closer to J than K because 82% of the sample contains the isotope of mass number J. The atomic mass of an element is the average mass of all the naturally occurring isotopes of that element.
**3.** The exact location of an electron cannot be determined. Only the probability of finding an electron in a certain location can be predicted.
**4.** Proton: nucleus, positively charged, 1 amu. Neutron: nucleus, neutral, 1 amu. Electron: electron cloud, negatively charged, 1/1836 amu.
**5.** Atomic number is the number of protons in the nucleus. The atomic number distinguishes one element from another.

5. Most of the mass of the atom is located in the electron cloud.
6. Subatomic particles with a mass of 1 amu and no electric charge are called protons.
7. Chlorine has an atomic number of 17. It has 17 protons in its nucleus.
8. Atoms having the same number of protons but different numbers of neutrons are called isomers.
9. Electrons having the least amount of energy are found farthest from the nucleus.
10. Two atoms of hydrogen and one atom of oxygen chemically bonded together form a water molecule.

## CONCEPT REVIEW: SKILL BUILDING

*Use the skills you have developed in the chapter to complete each activity.*

1. **Making calculations** You can calculate the atomic number of element X by doing the following arithmetic:
   a. Multiply the atomic number of hydrogen by the number of electrons in mercury, atomic number 80.
   b. Divide this number by the number of neutrons in helium, atomic number 2, mass number 4.
   c. Add the number of protons in potassium, atomic number 19.
   d. Add the mass number of the most common isotope of carbon.
   e. Subtract the number of neutrons in sulfur, atomic number 16, mass number 32.
   f. Divide by the number of electrons in boron, atomic number 5, mass number 11.

   Which of the following elements is X: fluorine, atomic number 9; neon, atomic number 10; sodium, atomic number 11?
2. **Developing a model** You are trying to locate a friend on a sunny Saturday afternoon. Although you cannot say with absolute certainty where your friend is, you can estimate the chances of finding your friend in various places. Your estimates are based on past experiences. Construct a table listing at least seven possible locations for your friend. Next to each location, give the probability, or likelihood, of finding your friend in that location in percent. Remember that your total probability should equal 100 percent. Would a change in the weather affect your results? How about a change in the day of the week? How does this activity relate to an electron's location in an atom?
3. **Identifying relationships** Sodium has only one naturally occurring isotope. Explain how the atomic mass of this isotope compares with the mass number.
4. **Making inferences** Why are models useful in the study of atomic theory?

## CONCEPT REVIEW: ESSAY

*Discuss each of the following in a brief paragraph.*

1. Describe the electron configuration of each element based on atomic number: sulfur, 16; fluorine, 9; argon, 18; lithium, 3.
2. A certain element contains 82 percent of an isotope of mass number J and 18 percent of an isotope of mass number K. Is the atomic mass of this element closer to J or to K? Explain your answer.
3. Why must scientists consider the concept of probability in describing the location of electrons?
4. Classify the three main subatomic particles according to location, charge, and atomic mass.
5. What is the atomic number of an element? What is its significance?

375

## ADDITIONAL QUESTIONS AND TOPIC SUGGESTIONS

1. Explain in your own words why J. J. Thomson proposed a model of the atom in which the puddinglike, positively charged material was scattered throughout with negatively charged electrons. (Thomson could not find the protons, or positively charged particles, in the atom. But he knew that the atom as a whole was neutral, or uncharged. Therefore there had to be positively charged material somewhere in the atom.)
2. Name the three principal subatomic particles of an atom. (protons, neutrons, electrons)
3. If you had 3 isotopes of the carbon atom, how many protons would each isotope have? (6)
4. What is an "electron cloud"? (The electron cloud is a space in which electrons are likely to be found arranged in energy levels.)
5. What causes a molecule to form? (A molecule forms when 2 or more atoms combine and chemically bond.)

## ISSUES IN SCIENCE

The following issues can be used as springboards for discussion or given as writing assignments.

1. In your opinion, why would the early scientists have believed that the atom had the same structure as the solar system.
2. Compare the experiments of Thomson and Rutherford. What are some of the similarities? Differences? In your opinion, would the results have remained the same if Rutherford had been using tiny negatively charged particles?

## Chapter 16

# ELEMENTS AND COMPOUNDS

## CHAPTER OVERVIEW

Matter is all around us. It makes up every living and nonliving thing we see or use. Early scientists thought matter could be broken up into smaller and smaller pieces until the pieces were so small that they no longer could be divided. Finally, in 1808, John Dalton, an English teacher and scientist, developed a theory about the structure of atoms. Dalton's theory led the way to classifying matter by grouping substances according to particular properties or makeup. Scientists use the properties of matter to help classify matter into groups. Two of these groups are elements and compounds. Elements and compounds are pure substances.

Elements are the simplest type of pure substances. A pure substance is made of only one kind of matter with a definite chemical composition. A pure substance is the same throughout. An element is composed of building blocks called atoms. An atom is the smallest particle of an element that has the properties of that element. All atoms of a particular element are alike. An element cannot be changed into simpler substances by heating or by any chemical process. Elements are represented by chemical symbols.

Compounds are pure substances that contain two or more elements chemically combined in a definite composition. A molecule of a compound can be a combination of atoms of different elements or atoms of the same element. A molecule is the smallest particle of a compound that has the properties of that compound.

Compounds are represented by chemical formulas, which are combinations of chemical symbols that show the kind of elements and number of atoms of each element in the compound. A description of a chemical reaction using symbols and formulas is called a *chemical equation.*

## INTRODUCING CHAPTER 16

Have students observe the photograph on page 376.

• **What is detective Sherlock Holmes doing?** (gathering clues)

Have students read the chapter introduction. Point out that Holmes's success had a simple, solid basis: logical thinking combined with a knowledge of chemistry.

• **How would a knowledge of chemistry help a detective?** (Students should suggest that by knowing chemistry the detective can analyze blood samples, liquids, or other forms of evidence found at the scene of the crime.)

• **How does knowing what a substance contains help the detective?** (The detective would know if the substance could have been used in the crime.)

Point out that all substances con-

# 16 Elements and Compounds

**CHAPTER SECTIONS**

**16-1 Elements**
**16-2 Compounds**

**CHAPTER OBJECTIVES**

*After completing this chapter, you will be able to:*

**16-1** Define an element and give several examples of elements.
**16-1** Give the chemical symbols for ten common elements.
**16-2** Distinguish between an element and a compound.
**16-2** Describe what information a chemical formula provides.
**16-2** Explain how a balanced chemical equation describes a chemical reaction.

A reddish stain on a scrap of fabric . . . some bits of dust gathered in the creases of a man's clothing . . . a seemingly unimportant clump of mud upon the floor . . . a few grains of tobacco let fall at the scene of a murder. . . . What could all these details mean? To detective Sherlock Holmes, the creation of British novelist Arthur Conan Doyle, they were clues to some of the most mysterious crimes imaginable. By paying close attention to the evidence, Holmes was able to solve many of the most perplexing crimes. And in so doing, he amazed not only the London police but also his own assistant, Dr. Watson.

Holmes's success had a simple, solid basis: logical thinking combined with a knowledge of chemistry. Using this knowledge, he was able to classify and analyze various substances that were clues to the mysteries. Holmes was a master at using scientific principles to solve crimes.

Criminology, of course, is not the only area in which scientific principles are applied. The whole world is a place of mystery, filled with puzzles and wonders that await the investigation of young detectives like you. But before you set out on your adventure, you will need to know how chemical substances are classified. And soon you will share the feelings of Holmes, who exclaimed at moments of discovery, "By Jove, Watson, I've got it!"

Detective Sherlock Holmes gathering clues



## TEACHER DEMONSTRATION

Before the demonstration, half fill three different flasks with water, clear vinegar, and dilute sulfuric acid.

Show students the flasks and tell them the following mystery story.

When the police arrived, they found a dead body on the floor of the apartment. On a table next to the body was a glass of clear liquid and these three flasks. The police assumed that one of these clear liquids killed the person.

• **How could they find out if their assumption was true?** (Most students will say the liquids have to analyzed.)

## TEACHER RESOURCES

### Audiovisuals

*A Look at Chemical Changes,* film, CRM/McGraw-Hill.

*Matter and Molecules: The Matter of Elements,* filmstrip, Singer Educational Division.

### Books

Ahrens, L. H., ed., *The Origin and Distribution of the Elements,* Elmsford, N.Y.: Pergamon

Donohue, J., *Structure of the Elements,* New York: Wiley

Schroeder, H. A., *Elements in Living Systems,* New York: Plenum

### Software

*Physical and Chemical Properties,* Prentice Hall

tain combinations of atoms or molecules.

• **Could the way certain atoms or molecules combine make a difference in the substance formed?** (yes)

• **Do you predict that when certain atoms or molecules combine they could be poisonous?** (yes)

Relate this answer back to the teacher demonstration. Show students the three liquid-filled flasks. Tell students one of the flasks contains water, another clear vinegar, and the other dilute sulfuric acid. Point out that two of the three liquids are perfectly harmless, but the third is deadly.

• **How could the flask holding the sulfuric acid be identified?** (by testing and analyzing the liquids)

Point out that atoms of hydrogen and oxygen combine to form water. Write $H_2O$ on the chalkboard. Explain that atoms of hydrogen, oxygen, and carbon combine to form vinegar. Write $HC_2H_3O_2$ on the chalkboard. Explain that atoms of hydrogen, oxygen, and sulfur combine to form sulfuric acid. Write $H_2SO_4$ on the chalkboard.

• **What two kinds of atoms are present in all three liquids?** (oxygen and hydrogen)

Point out that all matter is composed of different kinds of atoms. Explain that atoms are the building blocks of matter, and a molecule, which is made up of two or more atoms chemically bonded, is the smallest particle of a substance that has all the properties of that substance.

## 16-1 ELEMENTS

### SECTION PREVIEW 16-1

By applying the concepts of atoms and molecules, scientists can classify matter. Scientists divide matter into two groups: materials made up of parts that are all alike and materials made up of parts that are not alike.

Materials made up of parts that are all alike are called *pure substances.* Elements are the simplest type of pure substances. An element cannot be changed into simpler substances by heating or by any chemical process. An element is made up of only one kind of atom. An atom is the smallest particle of an element that has the properties of that element.

### SECTION OBJECTIVES 16-1

1. **Define an element and give several examples of elements.**
2. **Explain why elements are pure substances.**
3. **Give the chemical symbols for ten common elements.**

### SCIENCE TERMS 16-1

**pure substance p. 379**
**element p. 379**

### ANNOTATION KEY

**1 Thinking Skill: Applying definitions**
**2 Thinking Skill: Making comparisons**
**3 Thinking Skill: Classifying matter**

**Figure 16-1** *The gold in this rock* (left) *is an example of a pure substance. Gold has long been used in jewelry and coins. These gold bars* (right)*, called bullion, are part of a nation's money reserves.*

## 16-1 Elements

Suppose you were given a list of 100 different objects—objects common to your world. And you had to sort them out into different groups. How might you do it? Perhaps you would group them according to their color. Or their shape and size. Or their texture. Maybe you would sort them by their uses.

Well, scientists have faced just such a problem with the classification of matter. And by applying the concepts of atoms and molecules, they have been able to sort all forms of matter into logical groups. So a great deal of the work has already been done for you!

But before you discover what scientists have done, you should recall what you have learned about atoms and molecules. 1 An atom is a tiny building block of matter. And a molecule, which is made up of two or more atoms chemically bonded, is the smallest particle of a substance that has all the properties of that substance.

**Activity**

*Getting to Know the Elements*

Choose two elements that are commonly found in your surroundings. Mount a sample of each element on poster paper. Then illustrate the element's uses by cutting out and pasting pictures that show the various uses.

### *The Simplest Pure Substances*

The first thing scientists did was to divide matter into two groups: materials made up of parts that are all alike and materials made up of parts that are not



## TEACHING STRATEGY 16-1

### Motivation

Collect some samples of pure substances such as salt, sugar, iron, aluminum, copper, lead, silver, nickel, vinegar, and water. Be sure that some of the pure substances are elements (iron, copper, lead, silver, and nickel) but that not all of them are metals.

- **How would you group the objects?** (Students might suggest that the metals go together, the water and vinegar go together, and the salt goes by itself.)

### Content Development

Point out that scientists faced the same problem when they attempted to classify matter. Remind students that an atom is a tiny building block of matter. And a molecule, which is made up of two or more atoms chemically bonded, is the smallest particle of a substance that has all the properties of that substance. Tell students that using the concept of atoms and molecules, scientists classified matter into two groups: materials made up of parts that are all alike and materials made up of parts that are not alike.

### Motivation

Discuss the collection of pure substances with some of the following questions.

alike. For example, in a sample of water, all the parts making up the water are exactly alike. These parts are water molecules. Scientists call such a material a **pure substance**. A pure substance contains only one *kind* of molecule. The oxygen you breathe is another example of a pure substance. All the molecules in the gas are exactly the same. 2

Now suppose you are looking at a sample of a rock called granite, like the one in Figure 16-2. You might notice that scattered throughout the granite are different-colored particles. These particles are the minerals quartz, feldspar, and mica. So granite is not made up of only one kind of molecule. It is not a pure substance. You will learn exactly what granite is in Chapter 17.

Let's go back to oxygen for a moment. Every molecule of oxygen is made up of two atoms of oxygen bonded together. So each oxygen molecule contains only one *kind* of atom—the oxygen atom. **A pure substance made up of only one kind of atom is called an element.** Oxygen is an **element.** So is hydrogen. Carbon, aluminum, iron, mercury, silver, and uranium are also elements.

Elements are the simplest type of pure substance. 3 They cannot be changed into simpler substances by heating or by any chemical process. Suppose you melt a piece of iron by adding heat energy to it. You may think that you have changed the iron into a

**Figure 16-2** *Granite, which is not a pure substance, is made up of the minerals quartz, feldspar, and mica.*

**Figure 16-3** *The beautiful colors of fireworks* (left) *are produced by the rapid burning of elements such as magnesium, phosphorus, sulfur, barium, and strontium. The element silicon is used to make computer chips* (right).

## TEACHER DEMONSTRATION

Before the demonstration, collect several blocks of wood (or pieces of paper or dice) of equal size and color. Show the class a block of wood. Tell students to think of it as an atom.

- **What is an atom?** (the tiniest building block of matter)

Place another block on top of or to the side of the first block. Tell the class that the model is chemically combined.

- **What has just been made by adding another atom to the first atom?** (a molecule)

Add the remaining blocks.

- **Now what has been made?** (The model still represents a molecule because a molecule is made up of two or more atoms chemically bonded.)

Point out that all the atoms in the molecule are exactly the same.

### Activity

**Getting to Know the Elements**
**Skills: Observing, identifying, relating, applying**
**Level: Remedial**
**Type: Hands-on/field**

Students should choose solid elements, such as iron, copper, carbon, aluminum, and zinc, that are not dangerous. Students should indicate on their charts the symbols and sources of the elements, as well as their uses (in construction, coins, and so on).

- **What do you predict a "pure substance" is?** (a substance that is pure or the same throughout)

Point out that all of the items in the collection are pure substances.

### Content Development

Tell students that a pure substance contains only one kind of atom or one kind of molecule. Show students the items in the pure substance collection that are elements: iron, aluminum, copper, lead, silver, and nickel. Point out that an element contains only one kind of atom.

Refer to the demonstration and explain that a pure substance made up of only one kind of atom is called an *element*. Tell students that elements are the simplest type of pure substances. Point out that elements cannot be changed into simpler substances by heating or by any chemical process.

### Enrichment

Students might do research to find out how many elements have been discovered since the original 63 in Mendeleev's periodic table. They might present a report to the class on when these elements were discovered and by whom.

## FACTS AND FIGURES

In the earth's crust, oceans, and atmosphere are found 92 elements. These elements are known as the "natural" elements. Only eight of these elements are found in the earth's crust. Oxygen makes up nearly 50% and silicon about 25%. Aluminum, iron, calcium, magnesium, sodium, potassium, and titanium together make up 24%. Carbon, which is fundamental to all living things, is present in amounts less than 0.1%.

## ANNOTATION KEY

❶ **There are only 26 letters in the alphabet and 108 elements, so two letters must be used at times. (Relating facts)**

❷ **K; Pb (Applying facts)**

❶ **Thinking Skill: Identifying patterns**

**Figure 16-4** *These were the symbols used by the alchemists.*

simpler substance. But the liquid you now have still contains only iron atoms. True, the iron has changed phase—from solid to liquid. But it is still iron. No new or simpler substance has been formed.

### *Chemical Symbols*

For many years, scientists had to spell out the full names of elements when writing about them. As you can imagine, this practice was time-consuming. Then in 1813, a system of representing the elements with symbols was introduced. After all, why couldn't chemists do what mathematicians and musicians did?

❶ **Chemical symbols** are a shorthand way of representing the elements. Each symbol consists of one or two letters, usually taken from the element's name. The symbol for the element oxygen is O. The symbol for hydrogen is H; for carbon, C. The symbol for aluminum is Al; and for chlorine, Cl. You should note that when two letters are used in a symbol, the

### Career: *Melter Supervisor*

**HELP WANTED: MELTER SUPERVISOR, trainee** To train as supervisor of steel company crew. Must complete four- to five-year training program, which includes classroom instruction. High school diploma optional. Minimum age 18.

What do cars, toasters, and business machines have in common? All require steel in their manufacture—steel that must be supplied by one of about 200 companies in the United States. Steel companies use iron ore, iron scrap, carbon, manganese, and other substances to make steel. These materials are placed in furnaces in carefully measured amounts. The molten steel that is produced is poured into molds and then cooled. Finally, it is sent to the manufacturers of steel products.

The steelworker who oversees work crews during the operation of such a furnace is called the **melter supervisor.** As crew supervisor, the melter directs the various activities involved in making steel. These activities include adding materials to the furnace, "charging" the furnace with steel scrap, and adding molten iron from another furnace to the scrap. Lime is added to the mixture to carry off unwanted compounds.

The melter supervisor determines when the process is complete. If the temperature is over 1000° C, and the quality of the molten steel is acceptable, it is poured into molds.

If you think you might wish to pursue a career in steelwork, for further information write to the American Iron and Steel Institute, Communications and Education Services Department, 1000 Sixteenth Street N.W., Washington, DC 20036.

## 16-1 (continued)

### Motivation

Have students observe Figure 16-4. Explain that early scientists, known as alchemists, used symbols to represent the various elements.

- **Can you think of anything that is represented by a symbol today?** (Answers will vary: logos for manufacturing companies; broadcast channels; driving directions; laboratory safety precautions, etc.)

Have students make a bulletin-board display with some of the symbols and names they can identify. Point out that the use of a symbol is an easy way to recognize a product, company, direction, precaution, and so on.

### Content Development

Point out that the alchemists used symbols for the elements. Explain that these symbols helped the alchemists to recognize and represent the element. Point out that an element is made of only one kind of atom. Atoms of different elements are different. Tell students that like people, every element has its own special name.

- **What are some of the shorthand names (nicknames) of your friends?** (Accept all answers.)
- **What are their real names?** (Accept all answers.)

Explain that the names of all elements can be represented in a shorthand method. The shorthand representation of an element is called *symbol.* Lead students to infer shorthand names with the following questions.

- **What is an atom?** (the smallest particle of an element that has the properties of the element; the building blocks of matter)
- **What is a chemical symbol?** (a shorthand way of representing an element)
- **If the element's name is oxygen, what logical shorthand symbol would you use?** (Students should answer "O.")
- **If the element's name is nitrogen, what logical shorthand symbol**

**COMMON ELEMENTS**

| Name | Symbol | Name | Symbol | Name | Symbol |
|---|---|---|---|---|---|
| Aluminum | Al | Helium | He | Oxygen | O |
| Bromine | Br | Hydrogen | H | Potassium | K |
| Calcium | Ca | Iodine | I | Silver | Ag |
| Carbon | C | Iron | Fe | Sodium | Na |
| Chlorine | Cl | Lead | Pb | Sulfur | S |
| Copper | Cu | Magnesium | Mg | Tin | Sn |
| Fluorine | F | Mercury | Hg | Uranium | U |
| Gold | Au | Nitrogen | N | Zinc | Zn |

**Figure 16-5** *This table shows the chemical symbols for some of the most common elements. What is the symbol for potassium? For lead?* ❷

first letter is *always* capitalized, but the second letter is *never* capitalized. Do you know why two letters are sometimes needed for a symbol? Hint: Scientists have already identified more than 108 elements! ❶

What do you think the symbol for gold is? Is it G? Is it Go? The symbol for gold is Au. Does that surprise you? Gold is not spelled with an "a" or a "u." But the reason for the symbol is really not so strange. The Latin name for gold is *aurum.* Scientists often use the Latin name of an element to create its symbol. Here are some other examples. The symbol for silver is Ag, from the Latin word *argentum.* The Latin word for iron is *ferrum.* So the symbol for this element is Fe. Mercury's symbol is Hg, from the Latin name *hydrargyrum.* The table in Figure 16-5 lists some common elements and their symbols.

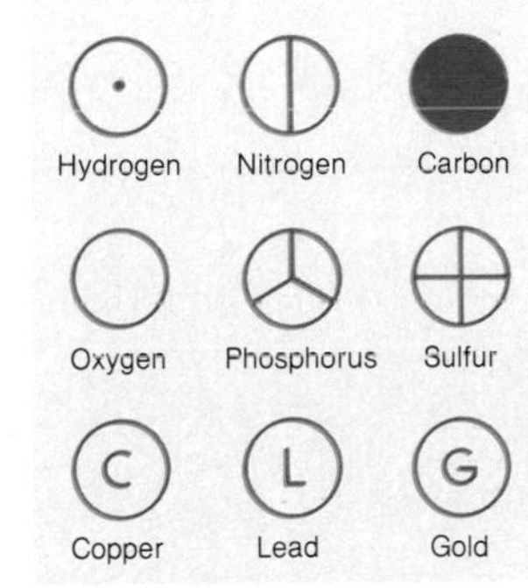


**Figure 16-6** *These symbols were part of Dalton's system for representing the elements.*

### SECTION REVIEW

1. What type of pure substance is made up of only one kind of atom?
2. What is the shorthand way of representing the elements?
3. What is the chemical symbol for: bromine; nitrogen; iron; calcium; oxygen; sodium?



## SCIENCE, TECHNOLOGY, AND SOCIETY

Lead is an element that even in very small amounts can cause tremendous damage if taken into the body. Some of the health problems that are caused by lead poisoning include: weight loss; dehydration; weakness; inability to sleep; crankiness; inability to absorb and use important substances such as iron, vitamin D, and calcium; and high blood pressure.

In an attempt to reduce the amount of lead found in the environment, the United States Government has ordered the production of unleaded gasoline for automobiles and has banned the manufacture of leaded paint. These measures have greatly reduced the problem of lead poisoning. But Americans are still exposed to lead through drinking water, soil, lead-containing fertilizers and pesticides, air pollution, and from the glaze on foreign ceramicware that has not been fired sufficiently.

Those who are most susceptible to lead poisoning are young people whose growth can be affected, pregnant women whose unborn children can be affected, and middle-aged men who battle heart disease and high blood pressure.

Individuals can limit their lead intake by replacing lead-based paint in their homes, checking plumbing lines for lead traces, carefully examining foreign ceramicware, and eating a balanced diet to prevent maximum absorption of lead by body tissues.

**would you use?** (Students should answer: N.)

Point out that scientists often use the Latin name of an element to create the symbol. Also point out that the symbols for some elements consist of two, not one, letters.

## Skills Development

***Skill: Identifying relationships***

Advance Preparation: Make a chalkboard list of the elements listed below, showing both the English word and the Latin word for the element (without the chemical symbol).

Have students predict the symbol for each element. After students have listed their symbols, write the correct symbols on the chalkboard.

O = oxygen
Ca = calcium
Cu = copper (cuprum)
C = carbon
H = hydrogen
Cl = chlorine
Al = aluminum
Ni = nickel
He = helium
Au = gold
Pb = Lead (plumbum)
K = potassium (kalium)
Ag = silver (argentum)
Fe = iron (ferrum)
Hg = mercury (hydrargyrum)

## Section Review 16-1

1. Element.
2. Chemical symbols.
3. Br; N; Fe; Ca; O; Na.

## 16-2 COMPOUNDS

### SECTION PREVIEW 16-2

Compounds are pure substances consisting of two or more elements chemically combined in a definite composition. When two or more atoms chemically combine, a molecule is formed. A molecule can be a combination of atoms of different elements or the same element. A molecule is the smallest particle of a compound that has the properties of that compound.

Names of compounds are represented by combinations of chemical symbols that show the kind and the number of each kind of atom in the compound. A description of a chemical reaction using symbols and formulas is called a *chemical equation.*

### SECTION OBJECTIVES 16-2

1. **Distinguish between an element and a compound.**
2. **Describe what information a chemical formula provides.**
3. **Explain how a balanced chemical equation describes a chemical reaction.**

### SCIENCE TERMS 16-2

**compound** p. 382
**chemical formula** p. 383
**subscript** p. 383
**chemical reaction** p. 384
**chemical equation** p. 385
**coefficient** p. 385

**Figure 16-7** *Sodium is a highly reactive element that must be stored in oil* (top). *Chlorine is a yellow-green, poisonous gas* (center). *The compound formed when these two elements chemically combine is sodium chloride, or common table salt* (bottom).

## 16-2 Compounds

Let's go back to the sample of water again. Water is a pure substance. It contains only *one kind of molecule.* That molecule is made up of two hydrogen atoms chemically bonded to one oxygen atom. So each molecule of water contains *more than one kind of atom.* Pure substances such as water are known as **compounds.** Compounds are made up of molecules that contain more than one kind of atom. **Compounds are two or more elements chemically combined.** Sugar is a compound. Each sugar molecule is made up of atoms of carbon, hydrogen, and oxygen. Baking soda is a compound you may be familiar with. Carbon dioxide, ammonia, and TNT are compounds. Can you name some other compounds? ❶

Unlike elements, compounds can be broken down into simpler substances. Heating is one way of separating some compounds into their elements. For example, the ore known as chalcocite is the compound copper sulfide. When heated to a high temperature, this ore breaks down into the elements copper and sulfur.

Sometimes the elements in a compound are so strongly bonded that heating cannot separate them. So some other form of energy must be used. Often electric energy is used to break down a compound. For example, by passing an electric current through water, this compound can be broken down into the two different elements, hydrogen and oxygen, that make it up.

The properties of the elements that make up a compound are often very different from the properties of the compound itself. Would you want to flavor your French fried potatoes with a poisonous gas and a highly active metal? Probably not! Yet, in a way, this is exactly what you are doing when you sprinkle salt on your potatoes. Chlorine is a yellow-green gas that is poisonous. Sodium is a silvery metal that explodes if placed in water. But when chemically combined, these elements produce a harmless white compound—sodium chloride—that has a tasty flavor!

### TEACHING STRATEGY 16-2

#### Motivation

Hold up a clean, shiny iron nail and a rusty iron nail. Conduct an open-ended discussion of the two nails.

- **How are these nails alike? Different?** (Answers will vary, but likenesses will probably include solid and iron, whereas differences will include shiny/dull, rusty/smooth.)
- **What is under the rust?** (Pure iron is under the rust.)
- **What do you think might have caused the rust?** (Answers will vary and may include materials or processes such as water, acid, other rust, chemicals, or weathering.)

#### Content Development

Relate the text discussion and the Motivation demonstration by asking the following questions:

- **What are the names of some compounds?** (Answers will vary and may include rust, sugar, carbon dioxide, baking soda, ammonia, TNT, and other compounds listed in the text.)
- **What is a compound?** (A compound is a pure substance made of two or more elements that are chemically combined.)

#### Skills Development

*Skill: Applying concepts*

- **Describe some compounds that have properties different from the properties of the elements in them.** (Accept all correct answers derived from the text or Motivation demon-

### Chemical Formulas

When you began to read, you probably started by learning the alphabet. Well, you can think of chemical symbols as the letters of a chemical alphabet. Just as you learned to put letters together to make words, chemical symbols can be put together to make chemical "words." Combinations of chemical symbols are called **chemical formulas.** These formulas are a shorthand for the names of chemical substances. ❷

Most chemical formulas represent compounds. For example, ammonia is a compound made up of the elements nitrogen and hydrogen. A molecule of ammonia has the formula $NH_3$. Sometimes, however, the formula represents a molecule of an element. For example, the symbol for the element oxygen is O. But oxygen occurs naturally as a molecule containing 2 atoms of oxygen bonded together. So the formula for a molecule of oxygen is $O_2$. Some other gases that exist only as pairs of atoms are hydrogen, $H_2$, nitrogen, $N_2$, fluorine, $F_2$, and chlorine, $Cl_2$. Remember, the *symbols* for the elements just listed are the letters only. The *formulas* are the letters with the small number 2 at the lower right.

Carbon dioxide is a compound of the elements carbon and oxygen. Its formula is $CO_2$. By looking at the formula, you can tell that every molecule is made up of 1 atom of carbon, C, and 2 atoms of oxygen, O. Water has the formula $H_2O$. How many hydrogen atoms and oxygen atoms are there in a molecule of water? ❷ ❸

When writing a chemical formula, you use the symbol of each element in the compound. You also use small numbers called **subscripts.** Subscripts are placed to the lower right of the symbols. A subscript gives the number of atoms of the element in the compound. When there is only 1 atom of an element, the subscript 1 is *not* written. It is understood to be 1.

Can you now see the advantages of using chemical formulas? Not only does a formula save space, but it tells a lot about the compound. It tells you the elements that make up the compound. And it tells you how many atoms of each element combine to form the compound. ❹

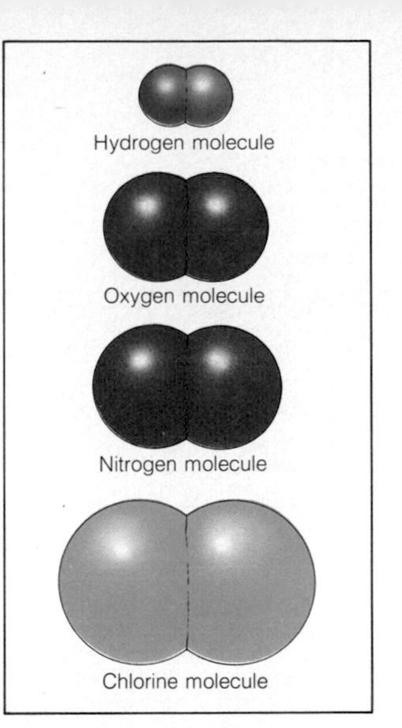


**Figure 16-8** *Hydrogen, oxygen, nitrogen, and chlorine exist as pairs of atoms chemically bonded to form molecules.*

**Figure 16-9** *A carbon dioxide molecule, $CO_2$, is made up of two atoms of oxygen bonded to one atom of carbon. A water molecule, $H_2O$, consists of two hydrogen atoms bonded to one oxygen atom.*

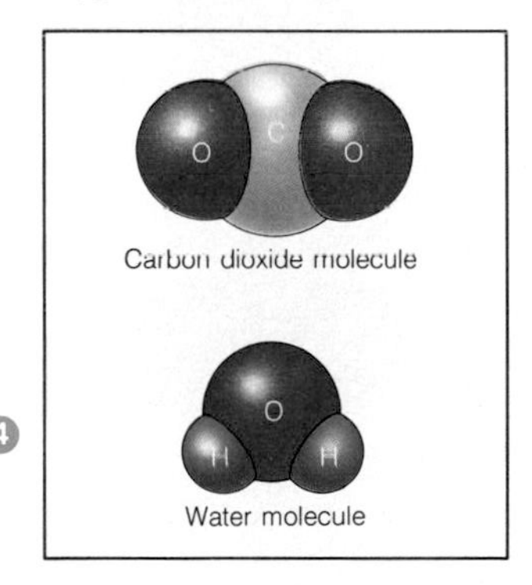




## ANNOTATION KEY

❶ **Answers will vary but may include carbon monoxide, salt, sugar, alcohol, butane, hydrogen peroxide, milk of magnesia, any acid or base. (Applying facts)**

❷ **Two H, one O (Interpreting formulas)**

❶ **Thinking Skill: Making generalizations**

❷ **Thinking Skill: Identifying patterns**

❸ **Thinking Skill: Applying formulas**

❹ **Thinking Skill: Relating facts**

## Motivation

Have students observe Figures 16-8 and 16-9.

- **How many atoms of hydrogen are in one hydrogen molecule?** (two)
- **How many atoms of oxygen are in one oxygen molecule?** (two)
- **What is the formula for one molecule of oxygen?** ($O_2$)
- **How many oxygen atoms are in a carbon dioxide molecule?** (two)
- **How many carbon atoms are in a carbon dioxide molecule?** (one)
- **What is the formula for the carbon dioxide molecule?** ($CO_2$)
- **What is the formula for the molecule of water?** ($H_2O$)

## Content Development

Relate the text ideas to the shiny and rusty nails. Tell the class that $Fe_2O_3$ is the chemical formula for red rust. Write $Fe_2O_3$ in the chalkboard. Tell students that Fe is iron and O is oxygen. Explain that $Fe_2O_3$, or red rust, is the compound formed when these two elements chemically combine.

- **How many atoms of iron are in the formula $Fe_2O_3$?** (two)
- **How many atoms of oxygen are in the formula $Fe_2O_3$?** (three)
- **What are the numbers 2 and 3 called?** (subscripts)

stration. Examples: Sugar is an edible white solid made of black carbon, colorless oxygen gas, and colorless hydrogen gas. Salt is an edible white solid made of explosive sodium metal and poisonous chlorine gas. Rust is a red solid made of solid iron metal and colorless oxygen gas.)

## Reinforcement

- **How are compounds different from elements?** (Compounds can be broken down into elements by chemical means. Elements cannot be broken down into simpler substances by chemical means. Compounds are made up of two or more elements.)
- **How do the properties of a compound compare to the properties of the elements in the compound?** (The properties of a compound are generally very different from the properties of the elements in it.)

## BACKGROUND INFORMATION

Oxygen is abundant in the atmosphere. Oxygen combines easily with many other elements. Many important chemical reactions involve the combination of a substance or substances with oxygen. Burning is one example of such a chemical reaction. Rusting is another example. Rusting, however, is a slower process. The processes of burning and rusting are called oxidation. Burning is rapid oxidation; rusting is slow oxidation.

**Figure 16-10** *The changing colors of autumn leaves* (left) *are caused by chemical reactions. Several chemical reactions are involved in the manufacture of plastic fibers* (right).

## ANNOTATION KEY

❶ **Thinking Skill: Writing chemical formulas and equations**

❷ **Thinking Skill: Balancing chemical formulas**

❸ **Thinking Skill: Applying concepts**

**Figure 16-11** *The formation of a carbon dioxide molecule involves the chemical bonding of one carbon atom to one oxygen molecule.*

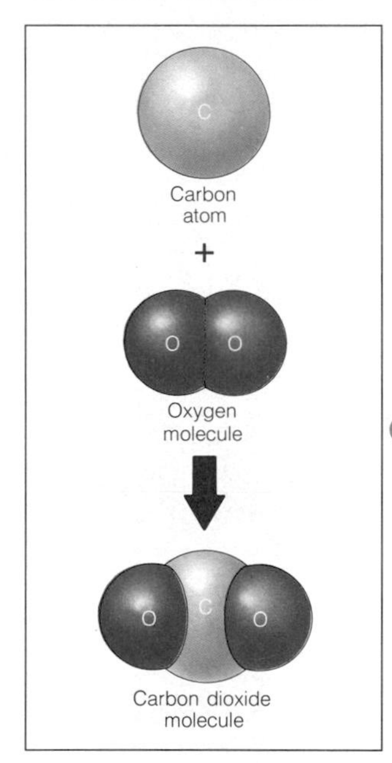


### *Chemical Equations*

If symbols are "letters" and formulas are "words," chemical "sentences" can be written. But what are chemical "sentences"? They are a way of describing a chemical process, or **chemical reaction.** In a chemical reaction, substances are changed into other substances through a rearrangement or new combination of their atoms. New chemical substances with new properties are formed.

Have you ever seen charcoal burning in a barbecue grill? If so, you were watching a chemical reaction. The carbon atoms in the charcoal were combining with the oxygen molecules in the air to form the gas carbon dioxide. This reaction could be written:

❶ carbon atoms plus oxygen molecules produce carbon dioxide molecules

But using symbols and formulas, this reaction can be written in a simpler way:

$$C + O_2 \rightarrow CO_2$$

The symbol C represents a carbon atom. The formula $O_2$ represents a molecule of oxygen. And the formula $CO_2$ represents a molecule of carbon dioxide. The arrow is read "yields," which is another way

## 16-2 (continued)

### Motivation

Place a tablespoon of baking soda in a glass. Add a teaspoon of vinegar. Ask the class the following questions:

- **What did you observe?** (The material fizzed.)
- **What was produced?** (a gas that bubbled)
- **How were the starting substances different from the material that was produced?** (The starting substances were a solid [baking soda] and a liquid [vinegar]. A gas was produced.)
- **What evidence is there that this was a chemical reaction?** (A new and different substance was produced.)

### Content Development

Discuss the idea of a chemical equation by asking the following questions:

- **What is a chemical equation?** (A chemical equation is a description of a chemical reaction using symbols and formulas.)
- **What is a coefficient?** (A coefficient is a number placed in front of a chemical symbol or formula in a chemical equation so that the equation is balanced.)
- **What does a coefficient indicate?** (A coefficient indicates the number of molecules of the element or compound that are needed to balance the equation.)

### Skills Development

***Skill: Interpreting diagrams***

Have students observe Figure 16-11.

- **How many molecules of carbon dioxide are produced by the chemical reaction?** (one)
- **How many molecules of carbon are needed to produce one molecule of carbon dioxide?** (one)
- **How many atoms of carbon are needed to produce one molecule of carbon dioxide?** (one)
- **How many atoms of oxygen are needed to produce one molecule of carbon dioxide?** (two)
- **How many molecules of oxygen are needed to produce one molecule of carbon dioxide?** (one)

of saying "produces." This description of a chemical reaction using symbols and formulas is called a **chemical equation.**

Here is the chemical equation for the formation of water from the elements hydrogen and oxygen:

$$H_2 + O_2 \rightarrow H_2O$$

Look closely at this equation. It tells you what elements are combining and what product is formed. But there is something wrong. Do you know what it is?

Look at the number of oxygen atoms on each side of the equation. Are they the same? On the left side of the equation, there are 2 oxygen atoms. On the right side, there is only 1 oxygen atom. Could 1 oxygen atom have disappeared? Scientists know that atoms are *never* created or destroyed in a chemical reaction. Atoms can only be rearranged. So there must be the same number of atoms of each element on each side of an equation. The equation must be balanced.

$$2H_2 + O_2 \rightarrow 2H_2O$$

Now count the atoms of each element on each side of the equation. You will find they are the same: 4 atoms of hydrogen on the left and on the right, and 2 atoms of oxygen on the left and on the right. The equation is correctly balanced.

An equation can be balanced by placing the appropriate number in front of the chemical formula. This number is called a **coefficient** (koh-uh-FI-shuhnt). The equation now tells you that 2 molecules of hydrogen combine with 1 molecule of oxygen to produce 2 molecules of water.

A balanced chemical equation is "evidence" of a chemical reaction. Using your knowledge of elements, compounds, symbols, and formulas, you can now solve the mysteries of chemical reactions. Like Sherlock Holmes, you too can say, "I've got it!"

### SECTION REVIEW

1. What is a compound? How is it different from an element?
2. What does a formula indicate about a compound?
3. Why must a chemical equation be balanced?

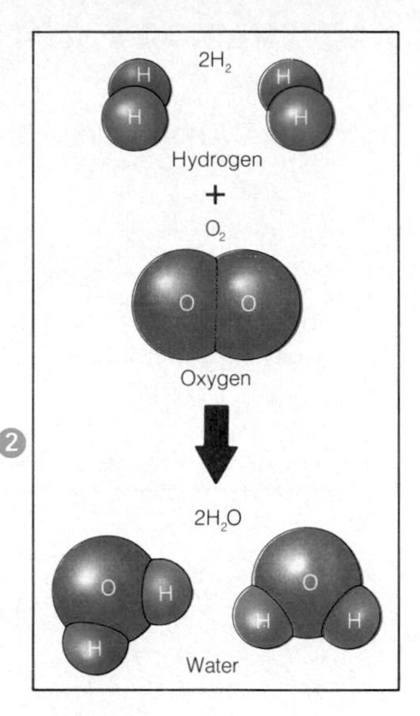


**Figure 16-12** *When two molecules of hydrogen chemically combine with one molecule of oxygen, two molecules of water are formed.*

#### Activity

*Naming and Counting Atoms*

The chemical formula for baking soda is $NaHCO_3$. The chemical formula for table sugar is $C_{12}H_{22}O_{11}$. What elements make up each of these compounds, and how many atoms of each element are there?



### Activity

**Naming and Counting Atoms**
**Skills: Applying facts, interpreting formulas, making calculations**
**Level: Remedial-average**
**Type: Computational**

This activity helps to reinforce understanding of chemical formulas and the subscripts they contain. $NaHCO_3$ contains 1 sodium atom, 1 hydrogen atom, 1 carbon atom, and 3 oxygen atoms; $C_{12}H_{22}O_{11}$ contains 12 carbon atoms, 22 hydrogen atoms, and 11 oxygen atoms.

### Reinforcement

Have students make a list of several chemical reactions. If some students suggest that melting ice or other physical changes belong on the list, review the nature of both physical and chemical changes.

### Enrichment

Have students make observations of chemical changes in their everyday life and report on them to the class. Observed chemical changes might include: combustion of gasoline in a car engine; tarnishing of silver; all types of burning; formation of acid rain; baking process; digestion.

### Skills Development

***Skill: Applying definitions***

Have students observe Figure 16-12.

- **How many molecules of water are produced by the chemical reaction?** (two)
- **How many molecules of hydrogen are needed to produce two molecules of water?** (two)
- **How many atoms of hydrogen are needed to produce two molecules of water?** (four)
- **How many atoms of oxygen are needed to produce two molecules of water?** (two)
- **How many molecules of oxygen are needed to produce two molecules of water?** (one)
- **What is the balanced chemical equation for the formation of water from hydrogen and oxygen?** ($2H_2 + O_2 \rightarrow 2H_2O$)

### Section Review 16-2

**1.** A pure substance made of more than one element. A compound can be broken down into simpler substances.
**2.** The elements and number of atoms of each element that make up the compound.
**3.** Atoms cannot be created or destroyed, so the same number of atoms of each element must appear on each side of an equation.

# LABORATORY ACTIVITY MARSHMALLOW MOLECULES

## BEFORE THE LAB

**1.** Gather all materials at least one day prior to the activity. You should have enough supplies to meet your class needs, assuming 6 students per group.
**2.** Plan to do Part A: Making Marshmallow Atoms during one class period, allowing the food coloring to dry overnight. Do "Part B: Assembling the Marshmallow Molecules" during the next class period.
**3.** Prepare 2 beakers (cups) of each color needed by adding 15 mL of food coloring to 250 mL of water (or 4 tsp of food coloring to 1 cup of water). Orange can be made by mixing equal parts of red and yellow. Purple can be made by mixing equal parts of red and blue.
**4.** Prepare two coloring stations for each color. Each station should have marshmallows, a beaker of color, paper towels, and toothpicks on a tray. Each team will need a tray covered with waxed paper.
**5.** Prepare a place for the trays of colored marshmallows to dry overnight. If you have several class sections, you may want to have each section do the coloring on a different day to be sure that you have enough storage space for the marshmallows to dry.
**6.** Some students are bound to eat the marshmallows. One way to avoid this problem is to sprinkle a small amount of powdered alum on the marshmallows. Alum has a very bitter taste and will make the mouth pucker. Powdered alum is available in the spice section of most grocery stores. It is commonly used in canning pickles. Be sure to warn students that the marshmallows have been contaminated with alum and are not edible! (A student will find this out very fast if the marshmallows are tasted.)

## PRE-LAB DISCUSSION

Before doing the coloring in Part A, discuss how to color a marshmallow by inserting a toothpick in it, dipping it into the color, and placing it (with pick) on the wax paper to dry. Tell each team to label their tray for easy retrieval.

Remind teams to keep each colored marshmallow separate from the others so the colors do not mix.

Remind students to wash their hands *thoroughly* with soap to remove all traces of the alum.

Before doing part B, tell the class that they will be making models to help them understand chemical reactions.

## SKILLS DEVELOPMENT

Students will use the following skills while completing this investigation.

1. Manipulative
2. Comparing
3. Relating
4. Applying
5. Inferring
6. Developing a model

## LABORATORY ACTIVITY

### *Marshmallow Molecules*

**Purpose**

In this activity, you will make model molecules and use them to learn why coefficients are necessary to balance chemical equations.

**Materials** *(per group)*
Toothpicks
Food coloring: red, yellow, green, blue, purple (red/blue), orange (yellow/red)
25 large marshmallows

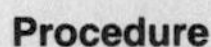

**Procedure**

*A. Making Marshmallow Atoms*

1. Set 25 marshmallows out overnight so that they dry out.
2. Prepare marshmallow atoms by applying food coloring as follows:
   N (nitrogen)—red (2)
   H (hydrogen)—blue (6)
   Cu (copper)—green (4)
   O (oxygen)—white (8)
   C (carbon)—yellow (1)
   K (potassium)—orange (2)
   Cl (chlorine)—purple (2)
3. Allow the marshmallows to dry for a few hours.

*B. Assembling the Marshmallow Molecules*

1. Using two red marshmallows and a toothpick, make a molecule of $N_2$. Then make a molecule of $H_2$ using blue marshmallows.
2. Ammonia, $NH_3$, is used in cleaning solutions and fertilizers. A molecule of ammonia contains 1 nitrogen atom and 3 hydrogen atoms. Using the marshmallow molecules that you made in step 1, produce an ammonia molecule of nitrogen and hydrogen. You may use as many nitrogen and hydrogen molecules as you need to make ammonia molecules as long as you do not leave any atoms over. Note: Hydrogen and nitrogen must start out as molecules consisting of two atoms each. Now balance the equation that produces ammonia:
   $$\_\_ N_2 + \_\_ H_2 \rightarrow \_\_ NH_3$$
3. Using two green marshmallows for copper and one white one for oxygen, prepare copper oxide, $Cu_2O$. Using a yellow marshmallow for carbon, manipulate the molecules to represent and balance this equation, which produces metallic copper:
   $$\_\_ Cu_2O + \_\_ C \rightarrow \_\_ Cu + \_\_ CO_2$$
4. Using orange for potassium, purple for chlorine, and white for oxygen, assemble $KClO_3$.
5. The decomposition of $KClO_3$ is a way to produce $O_2$. Take apart your $KClO_3$ to make KCl and $O_2$. You may need more than one molecule of $KClO_3$ to do this.

**Observations and Conclusions**

1. How many molecules of $N_2$ and $H_2$ are needed to produce two molecules of $NH_3$?
2. How many molecules of copper are produced from two molecules of $Cu_2O$?
3. How many molecules of $O_2$ are produced from two molecules of $KClO_3$?

# CHAPTER REVIEW

## SUMMARY

**16-1 Elements**

- A pure substance is made up of parts that are all alike.
- Elements are the simplest type of pure substance because they are made up of only one kind of atom. They cannot be broken down into simpler substances by chemical means.
- Examples of elements include oxygen, carbon, aluminum, iron, mercury, silver, uranium, magnesium, and silicon.
- Chemical symbols made up of one or two letters are used to represent the elements. When two letters are used in a symbol, the first letter is always capitalized, but the second letter is never capitalized.

**16-2 Compounds**

- Pure substances that contain more than one kind of atom are called compounds.
- Compounds are two or more elements chemically combined. Compounds can be broken down into simpler substances. The elements in a compound are most often separated by heating and by electricity.
- The properties of the elements that make up a compound are often very different from the properties of the compound itself.
- Combinations of chemical symbols, used to represent substances, are called formulas.
- Most chemical formulas represent compounds. Sometimes, however, a formula represents a molecule of an element. This is true for substances that exist only as pairs of atoms, such as oxygen, nitrogen, hydrogen, chlorine, and fluorine.
- Subscripts indicate the number of atoms of an element present in a compound.
- A chemical reaction is described by a chemical equation.
- In a chemical equation, the same number of atoms of each element must appear on each side of the equation. Therefore, a chemical equation must be balanced. Coefficients are used to balance an equation.
- A balanced chemical equation indicates what substances are combining, what products are formed, and how many atoms and molecules are involved in the reaction.

## VOCABULARY

*Define each term in a complete sentence.*

| | | | |
|---|---|---|---|
| **chemical equation** | **chemical symbol** | **compound** | **pure substance** |
| **chemical formula** | **coefficient** | **element** | **subscript** |
| **chemical reaction** | | | |

## CONTENT REVIEW: MULTIPLE CHOICE

*Choose the letter of the answer that best completes each statement.*

**1.** A material made up of parts that are all alike is called a(n)
a. atom. b. pure substance. c. mixture. d. granite.

**2.** The simplest form of matter made up of only one kind of atom is a(n)
a. element. b. mixture. c. compound. d. molecule.

**3.** Oxygen, iron, and silver are examples of
a. particles. b. formulas. c. compounds. d. elements.



$\underline{1}\ N_2 + \underline{3}\ H_2 \rightarrow \underline{2}\ NH_3$

$\underline{2}\ Cu_2O + 1\ C \rightarrow \underline{4}\ Cu + \underline{1}\ CO_2$

$\underline{2}\ KClO_3 \rightarrow \underline{2}\ KCl + \underline{3}\ O_2$

## OBSERVATIONS AND CONCLUSIONS

**1.** 1 molecule of $N_2$ and 3 molecules of $H_2$
**2.** 4 molecules of copper
**3.** 3 molecules of oxygen

## GOING FURTHER: ENRICHMENT

**Part 1**

You may want to have students construct additional models of the following molecules and then balance the equations.

$\underline{\ \ }\ N_2 + \underline{\ \ }\ O_2 \rightarrow \underline{\ \ }\ NO$
$(\underline{1}\ N_2 + \underline{1}\ O_2 \rightarrow \underline{2}\ NO)$

$\underline{\ \ }\ KCl \rightarrow \underline{\ \ }\ K + \underline{\ \ }\ Cl_2$
$(\underline{2}\ KCl \rightarrow \underline{2}\ K + \underline{1}\ Cl_2)$

$\underline{\ \ }\ CO + \underline{\ \ }\ O_2 \rightarrow \underline{\ \ }\ CO_2$
$(\underline{2}\ CO + \underline{1}\ O_2 \rightarrow \underline{2}\ CO_2)$

**Part 2**

Discuss how models help scientists and students understand nature.

- **How do models help you understand chemical reactions?** (Students might suggest that the marshmallow models helped them visualize what happened.)
- **How were the marshmallow models like real atoms?** (Each marshmallow model was a single atom and each color represented a different kind of atom.)
- **How were the marshmallow models different from real atoms?** (All the model marshmallows were the same size and composition.)
- **What is a model?** (a representation of something impossible to see or something requiring observation and analysis before going further)

## SAFETY TIPS

Alert students that although food coloring is safe, spills could cause color damage to clothing.

If you have sprinkled the marshmallows with alum, warn students not to eat the marshmallows because of alum contamination.

## TEACHING STRATEGY FOR LAB PROCEDURE

**1.** If you have sprinkled the marshmallows with alum, be sure to warn students that the marshmallows are not to be eaten because of alum contamination.
**2.** After the teams have completed the investigation, share and discuss the results. Have teams take turns showing and explaining one model and the balanced equation.

## CHAPTER REVIEW

### MULTIPLE CHOICE

1. b 3. d 5. d 7. c 9. c
2. a 4. a 6. b 8. b 10. b

### COMPLETION

1. pure substance
2. element
3. symbols
4. compound
5. compounds
6. heat
7. two
8. formulas
9. reaction
10. two

### TRUE OR FALSE

1. T
2. F compound
3. F pure
4. F elements
5. T
6. F symbols
7. T
8. F two
9. F subscripts
10. T

### SKILL BUILDING

**1.** Answers will vary. Sample criteria are the first letter of the name of month; number of letters in the name of a month; average temperature; hours of daylight; number of days; school and nonschool months.
**2.** The symbols and formulas are the same throughout the world. This permits scientists to communicate without misunderstanding or the need to translate symbols of one system to another.
**3.** Classifying matter according to makeup is more specific and avoids confusion.
**4.** $NaHCO_3$: 1 sodium, 1 hydrogen, 1 carbon, and 3 oxygen atoms
$C_2H_4O_2$: 2 carbon, 4 hydrogen, and 2 oxygen atoms
$Mg(OH)_2$: 1 magnesium, 2 oxygen, and 2 hydrogen atoms
$3H_3PO_4$: 9 hydrogen, 3 potassium, and 12 oxygen atoms
**5.** The substances that enter the reaction, the substances formed by the reaction, and the number of atoms and molecules of each substance.
**6.** Pass an electric current through water and collect the $O_2$ and $H_2$ into which it decomposes. Test each gas appropriately for identification. Evaporate the water, leaving behind the salt.
**7.** a. $2Mg + O_2 = 2MgO$
b. $2NaCl = 2Na + Cl_2$
c. $CH_4 + 2O_2 = CO_2 + 2H_2O$
d. $2H_2 + O_2 = 2H_2O$
**8.** The changing of subscripts would change the identity of the substances. The equation is supposed to represent the formation of water ($H_2O$), not hydrogen peroxide ($H_2O_2$). The balanced equation makes use of coefficients: $2H_2 + O_2 \rightarrow 2H_2O$.

### ESSAY

**1.** In a chemical reaction, substances are changed into other substances through a rearrangement or new combination of their atoms. Thus atoms in a chemical reaction get rearranged or combined differently than they were before.
**2.** Two or more atoms chemically bonded together. It is represented by a chemical formula.
**3.** An atom is the smallest particle of an element that has the properties of that element. It is the basic building



**4.** One or two letters used to represent the name of an element are called a chemical
a. symbol. b. formula. c. compound. d. equation.
**5.** Two or more elements chemically combined form a(n)
a. mixture. b. atom. c. symbol. d. compound.
**6.** Two methods of separating a compound into its elements are
a. heating and filtering. b. heating and electric current.
c. electric current and magnetizing. d. filtering and magnetizing.
**7.** A shorthand way of representing a chemical substance as a combination of symbols is a(n)
a. equation. b. reaction. c. formula. d. atom.
**8.** Numbers that indicate the number of atoms of an element in a certain compound are called
a. coefficients. b. subscripts. c. fractions. d. superscripts.
**9.** The process by which substances are changed into other substances as a result of rearranging atoms is called a(n)
a. chemical equation. b. chemical symbol.
c. chemical reaction. d. chemical formula.
**10.** A chemical equation can be balanced by using numbers called
a. subscripts. b. coefficients. c. superscripts. d. fractions.

#### CONTENT REVIEW: COMPLETION

*Fill in the word or words that best complete each statement.*

**1.** A material that contains only one kind of molecule is called a(n) ______.
**2.** Oxygen, which is made up of only one kind of atom, is called a(n) ______.
**3.** Chemical ______ are used to represent elements.
**4.** A(n) ______ is a pure substance made up of more than one kind of atom.
**5.** Carbon dioxide and water are examples of ______.
**6.** Two forms of energy often used to break down a compound into simpler substances are electricity and ______.
**7.** Molecules of gases such as chlorine and oxygen occur naturally as ______ atoms of the element bonded together.
**8.** Chemical ______ are used to represent compounds.
**9.** A chemical process in which substances are changed into other substances is called a chemical ______.
**10.** In order to balance the equation $MgO \rightarrow Mg + O_2$, the number ______ must be placed in front of MgO and Mg.

#### CONTENT REVIEW: TRUE OR FALSE

*Determine whether each statement is true or false. If it is true, write "true." If it is false, change the underlined word or words to make the statement true.*

**1.** The tiny building blocks of matter are called <u>atoms</u>.
**2.** Two or more atoms chemically bonded make a <u>mixture</u>.
**3.** A substance made up of parts that are all alike is called a <u>complex</u> substance.
**4.** Silver, gold, and <u>carbon</u> are examples of <u>compounds</u>.

5. Elements cannot be changed into simpler substances by heating them.
6. Ca and Fe are examples of formulas of chemical substances.
7. Two or more elements chemically combined form a compound.
8. One molecule of nitrogen, $N_2$, is made up of one atom of nitrogen.
9. Small numbers placed at the lower right of a chemical symbol in a formula are called superscripts.
10. The description of a chemical reaction using symbols and formulas is called a chemical equation.

## CONCEPT REVIEW: SKILL BUILDING

*Use the skills you have developed in the chapter to complete each activity.*

1. **Classifying data** Develop a classification system for the months of the year. State the property or properties you used to develop your system. Make your system as useful and specific as possible. Do *not* use the four seasons.
2. **Making inferences** The language of chemistry is a universal language. Scientists all over the world use the same representations for chemical substances. Explain why the system of symbols and formulas you learned about is so important to this idea.
3. **Relating concepts** Why is it more useful to classify matter according to makeup than according to phase?
4. **Making calculations** Calculate how many atoms of each element are present in the following compounds:

   $NaHCO_3$ $Mg(OH)_2$
   $C_2H_4O_2$ $3H_3PO_4$
5. **Making generalizations** What three things does a chemical equation indicate about a chemical reaction?
6. **Designing an experiment** Describe an experiment to demonstrate that water is a compound, not an element, and that salt water is not a pure substance.
7. **Applying facts** Balance the following equations:

   a. $Mg + O_2 \rightarrow MgO$
   b. $NaCl \rightarrow Na + Cl_2$
   c. $CH_4 + O_2 \rightarrow CO_2 + H_2O$
   d. $H_2 + O_2 \rightarrow H_2O$
8. **Applying concepts** The equation below is not balanced. Explain why it would be incorrect to balance it by changing $H_2O$ to $H_2O_2$. Hint: The name for $H_2O_2$ is hydrogen peroxide.

   $$H_2 + O_2 \rightarrow H_2O$$

## CONCEPT REVIEW: ESSAY

*Discuss each of the following in a brief paragraph.*

1. Explain what happens to atoms in a chemical reaction.
2. What is a molecule? How is a molecule of an element or a compound represented?
3. What is an atom? How do atoms of the same element compare? Atoms of different elements?
4. Write the chemical symbols for aluminum, calcium, iron, sulfur, sodium, and helium.
5. Why are elements and compounds considered pure substances? Do you think breakfast cereal with bananas and milk would be a pure substance? How about iron filings mixed with powdered sulfur? Explain your answers.
6. Explain the following statement: "A balanced chemical equation is evidence of a chemical reaction."

389

block of matter. Atoms of the same element are alike, while atoms of different elements are different.
**4.** Al, Ca, Fe, S, Na, He
**5.** Pure substances are substances whose parts are all exactly alike. Pure substances contain only one kind of atom (elements) or one kind of molecule (diatomic elements and compounds). Both of these examples are not pure substances. They are not exactly alike throughout; they do not contain only one kind of atom or molecule; and they can be separated by ordinary physical means. Pure substances either cannot be separated (elements) or can be separated only by chemical means (compounds).
**6.** A balanced equation indicates the substances entering into the reaction and in what amounts (numbers of atoms and molecules) and the substances formed by the reaction and in what amounts. If a balanced chemical equation for a reaction can be written, it generally indicates that the reaction can take place, provided the numbers of reacting atoms and molecules are present as indicated by the correctly balanced equation. A balanced equation shows that the law of conservation not being disobeyed.

## ADDITIONAL QUESTIONS AND TOPIC SUGGESTIONS

**1.** If a pure substance contains only one kind of molecule, why is an element considered the simplest type of pure substance? (An element is considered the simplest type of pure substance because it contains only one kind of atom in its molecule.)
**2.** What evidence is there that water, sugar, and salt have a definite composition? (Compounds have a definite composition. Water, sugar, and salt are compounds that have the following compositions: water is $H_2O$, sugar is $C_{12}H_{22}O_{11}$, and salt is NaCl.)
**3.** Would a molecule of an element be the same as a molecule of a compound? (A molecule of an element is either a single atom or two atoms of the same element. A molecule of a compound contains two or more atoms of different elements.)
**4.** Why is it important to use subscripts when writing a chemical formula? (Subscripts indicate the number of atoms of each element in the compound.)
**5.** What is the difference between a chemical equation and a chemical formula? (Chemical formulas, or combinations of chemical symbols, represent compounds. A chemical equation is a description of a chemical reaction using symbols and formulas.)

## ISSUES IN SCIENCE

The following issues can be used as springboards for discussion or given as writing assignments.
**1.** In your opinion, why were the alchemists' and Dalton's systems of element representation eliminated?
**2.** Predict future development or research in the field of discovering new elements and explaining the chemical compositions of compounds.

## Chapter 17

# MIXTURES AND SOLUTIONS

## CHAPTER OVERVIEW

Mixtures consist of two or more pure substances that are not chemically combined. Substances in a mixture can be present in any amount and can be separated by physical means. When mixtures are separated, the individual substances keep their own identity and most of their own properties. Mixtures are classified according to the size of the particles in them. The particles in a heterogeneous mixture are large and can be separated easily by physical means.

The particles in a homogeneous mixture are smaller than the particles in a heterogeneous mixture. The particles in a homogeneous mixture are too small to settle out when the mixture is allowed to stand. Colloids are homogeneous mixtures.

Solutions are a special type of mixture. A solution is a homogeneous mixture that has uniform composition and properties throughout the entire solution. A solution is formed when one substance dissolves in another. Although the solution as a whole is homogeneous, each substance in the solution retains individual properties and can be separated out. Solutions can be combinations of liquids, solids, or gases. All solutions have two important properties: the particles are evenly spread out; the particles are too small to be seen or to scatter light.

## INTRODUCING CHAPTER 17

Have students observe the photo on page 390. Point out that the superburger is a mixture of substances.

- **How many substances do you think are in the burger?** (Most students will say 20 to 30.)
- **How are the substances alike? Different?** (Students will probably make reference to a type of classification such as solid, liquid, vegetable, or meat.)
- **How would separate bites of the burger be alike? Different?** (Students will probably refer to the different types of ingredients on separate bites of the burger.)
- **Why do different parts look, taste, and feel different?** (Each part is a different substance or combination of substances and thus has different properties.)

Have students read the introduction. Refer to the text discussion and teacher demonstration.

- **Could you separate the superburger into substances as you could the cornflakes and raisins? How?** (Yes. The burger is made up of meat patties, bun, pickles, etc. All of the substances could be separated, although some might be more difficult than others.)

Point out that many of the sub-

# 17 Mixtures and Solutions

**CHAPTER SECTIONS**



**CHAPTER OBJECTIVES**

*After completing this chapter, you will be able to:*

**17-1** Describe a mixture.

**17-1** Identify three important properties of a mixture.

**17-2** Distinguish between heterogeneous mixtures and homogeneous mixtures.

**17-2** Describe a suspension and a colloid.

**17-3** Define the terms solute, solvent, and solution.

Hold the ketchup! Extra mustard, a dab of mayonnaise, sliced onion, pickle relish, and cheese! Don't forget the lettuce, but no tomato, please!

Is this the way you order your hamburger? Or do you prefer it plain? In either case, you are getting a combination of many ingredients. Substances such as fats, oils, proteins, and minerals are in the hamburger. Flour, salt, water, and other ingredients are mixed together in the bun. If you order a milkshake, you are drinking a blend of milk, sugar, ice cream, various flavorings, and air.

Making a list of all the chemical substances found in your burger and shake might surprise you. For your list would probably contain thousands of entries! After reading this chapter, why not try it?

*A dazzling combination of chemical substances*



## TEACHER DEMONSTRATION

Show students 2 cups of cornflakes in a wide-mouthed jar. Hold up a box or bag of raisins.

• **What kind of substances are cornflakes and raisins?** (solids)

Point out that a mixture is a combination of substances. Pour 1/4 cup of raisins onto the cornflakes. Cover the jar. Shake and turn the jar to mix the raisins and cornflakes. Hold the jar up for the class to observe.

• **What kind of matter is present now?** (Most students should respond by saying that it is a mixture.)

Explain that a mixture is a combination of two or more substances not chemically combined. Pour out the cornflake-and-raisin mixture onto a sheet of paper.

• **How could the cornflakes be separated from the raisins?** (Students will probably say that the raisins can be picked out of the mixture.)

## TEACHER RESOURCES

**Audiovisuals**

*Matter and Molecules: Clue-Compounds and Changes,* filmstrip, Singer Educational Division

*Naming Chemical Substances,* Parts 1 and 2, filmstrip, PH Media

**Books**

Stone, A. H. *The Chemistry of a Lemon,* Prentice-Hall

Stone, A. H., and D. Ingmanson. *Crystals from the Sea: A Look at Salt,* Prentice-Hall

Zubrowski, B. *Messing Around With Baking Chemistry: A Children's Museum Activity Book,* Little, Brown

stances that they can see in the photograph are themselves combinations of substances.

• **What substance is itself a combination of substances?** (Most students will identify the meat as a combination of beef and spices; pickle relish as a combination of cucumbers, peppers, and spices; ketchup as a combination of tomatoes, water, and spices.)

• **Could you take these combinations of substances apart as easily as you could the cornflakes and raisins?** (No. The particles in these substances are too small.)

• **Do you predict that the ketchup is the same kind of combination as the entire burger is?** (no)

Point out that mixtures are a combination of substances.

• **Would you predict different mixtures are made of different kinds of molecules?** (yes)

Remind students that in an element, all the atoms are the same, and that a compound is composed of two or more elements chemically combined (Chapter 16). Point out that in a mixture, more than one kind of substance is present. The ingredients or substances in a mixture are physically, rather than chemically, combined. Thus, mixtures can be separated by physical means.

## 17-1 PROPERTIES OF MIXTURES

### SECTION PREVIEW 17-1

Mixtures are composed of two or more pure substances that are mixed but not chemically combined. Mixtures can be combinations of elements or compounds. They can be in any of the four phases—such as solid, liquid, gas, or plasma. They also can be combinations of different phases.

Substances in a mixture can change in physical appearance, but they do not change in chemical composition. The substances in a mixture can vary in amounts and will always retain their original properties when combined.

### SECTION OBJECTIVES 17-1

1. **Describe three important properties of a mixture.**
2. **Describe how certain mixtures can be separated.**

### SCIENCE TERMS 17-1

**mixture p. 392**

### ANNOTATION KEY

**❶ Substances in a mixture can be present in any amount. In compounds, elements combine in exact amounts, or in definite proportions. (Applying definitions)**

**❶ Thinking Skill: Making generalizations**

**❷ Thinking Skill: Interpreting observations**

**Figure 17-1** *Granite is a mixture that contains crystals of the minerals quartz, feldspar, and mica.*

## 17-1 Properties of Mixtures

In Chapter 16, you learned that pure substances are made up of a single kind of particle. So it follows that any amount of a pure substance—a gram, a kilogram, or a ton—has the same properties as any other amount. You also learned that elements are the simplest type of pure substance. Compounds, which are chemical combinations of two or more elements, are also pure substances.

Do you remember the picture of a piece of granite? See Figure 17-1. Granite contains particles of different minerals. Looking closely at the rock in the photograph, you can see crystals of quartz, feldspar, and mica. Granite, then, is *not* made up of a single kind of particle. Such a material is called a **mixture.**

**Figure 17-2** *There are two different mixtures in this photo of a fish tank* (top). *The air above the water is a gaseous mixture made up mostly of oxygen and nitrogen molecules. The liquid mixture contains water molecules, nitrogen molecules, and oxygen molecules* (bottom).



### TEACHING STRATEGY 17-1

#### Motivation

Have students observe Figure 17-1 and read the caption. If possible, pass several pieces of granite around the classroom for students to observe.

- **What do you see in the granite?** (small pieces of different colored rock)

Point out that the lighter pieces of rock are probably quartz, the orange-colored pieces feldspar, and the darker pieces, mica. Each of these substances in the granite is a pure mineral.

Point out that if the piece of granite were separated, each would retain its individual properties.

#### Content Development

Explain that mixtures are made of materials that are combined physically. A mixture consists of two or more pure substances that are mixed, but not chemically combined. Each substance in a mixture retains most of its physical and chemical properties. No new substance is formed when a mixture is made.

The amounts of the different substances that make up a mixture are not fixed; any amount of one substance can be added to any amount of another substance.

### *Makeup of Mixtures*

**A mixture consists of two or more pure substances that are mixed together but not chemically combined.** Beach sand is a mixture, as is soil. Salad dressing, concrete, and sea water are other examples of mixtures. The pure substances that make up a mixture can be elements or compounds. For example, in a mixture of sugar and water, there are molecules of the compound sugar, $C_{12}H_{22}O_{11}$, and molecules of the compound water, $H_2O$. In the mixture known as air, there are molecules of the elements oxygen, $O_2$, and nitrogen, $N_2$. There are also other pure substances, such as molecules of the compound carbon dioxide, $CO_2$.

Because the substances in a mixture are not chemically combined, they keep their separate identities and most of their own properties. Think of the mixture of sugar and water you just read about. The water is still a colorless liquid. And the sugar, although dissolved in the water, still keeps its property of sweetness. Your sense of taste tells you this is so. The same molecules of water and sugar are present after the mixing as before it. No new chemical substances have been formed. Substances in a mixture may change in physical appearance, as when they dissolve. But they do not change in chemical composition.

If you eat cereal for breakfast, you are probably making a mixture. That is what you produce when you pour milk over the cereal. And if you put berries or raisins into your cereal, you make an even more complex mixture. But you do not use exactly the same amount of cereal, milk, and fruit each time. This illustrates another property of mixtures.

The substances that make up a mixture can be present in any amount. The amounts are not fixed. So you can make a mixture of a lot of cereal, a little milk, and loads of fruit. Or, if you prefer, a little cereal with lots of milk and fruit! This property illustrates an important difference between mixtures and compounds. Do you know what this difference is? You might remember that when elements combine chemically to form a compound, they do so in exact amounts. For example, 2 hydrogen atoms always combine with 1 oxygen atom—no more, no less—to form a molecule of water.

**Activity**

*A Mixture Collection*

You can find examples of mixtures all around you. Make a display of several mixtures from three of the places listed below. Include a description of the substances in each mixture.

- drugstore
- supermarket
- schoolyard
- park
- beach
- your home

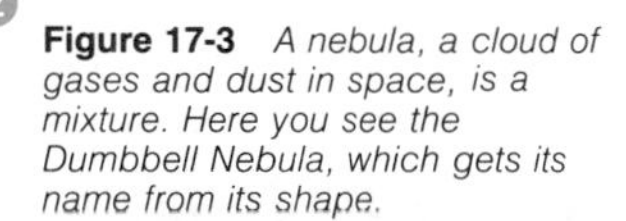

**Figure 17-3** *A nebula, a cloud of gases and dust in space, is a mixture. Here you see the Dumbbell Nebula, which gets its name from its shape.*

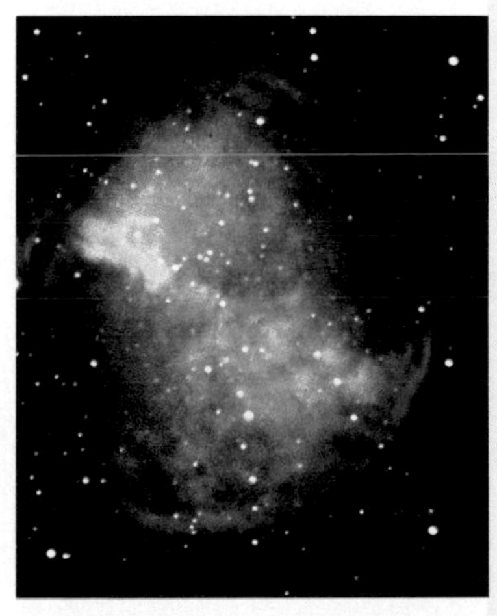

393

**Activity**

**A Mixture Collection**
**Skills: Writing, diagraming, relating, applying**
**Level: Remedial**
**Type: Hands-on**

It should be easy for students to find examples of mixtures and display them. Have them do library research to identify the chemical components of some of the mixtures they have chosen.

## SCIENCE, TECHNOLOGY, AND SOCIETY

It was in August 1986 when the volcanic Lake Nyos spewed large clouds of carbon dioxide into the air—clouds that killed 1700 residents of nearby villages. According to environmental researchers, Nyos and about 40 other lakes in Cameroon still remain a hazard. Lakes such as these release toxic gases.

A mass of magma more than 80 kilometers below the bottom of Lake Nyos released carbon dioxide into the lake through a volcanic feeder tube. The carbon dioxide then dissolved in the water at the bottom of the lake. A brief rainstorm, minor tremor, or landslide might have been the event that forced the deep water to the surface. As the pressure on the water dropped, about 1 cubic kilometer of gas bubbled out explosively and travelled quickly across the countryside. In another of Cameroon's lakes, Lake Monoun, a similar scenario had taken place in 1984, killing 40 people.

Scientists hope to monitor the percentage of carbon dioxide in Lake Nyos, allowing for evacuation of the area if levels become too high. Environmental scientists have suggested eliminating the carbon dioxide hazard by piping water from the bottom of the lake to the surface, thus allowing the gas to be released more slowly.

## Skills Development

***Skill: Relating facts***

Relate this activity to the text discussion on the mixture of sugar and water.

Divide the class into groups of 4–6 students. Give each group 250 mL of water and 30 mL of sugar. Have students dissolve 5 mL of sugar in the water at a time. Have them taste the liquid after each addition.

- **What did you observe?** (The mixtures were sweet and wet. They became sweeter as more sugar was added. Different amounts of sugar can be mixed with the water to form mixtures.)

## Enrichment

Students might like to separate a given amount of any fruit–cereal mixture. After sorting the different ingredients, they might determine the mass of the individual ingredients.

## TIE-IN/ART

Paint is a familiar fluid used by artists. Paint is usually a mixture of a liquid and finely powdered pigment.

Sometimes the liquid in paint is only a wetting agent (such as water or thinner) used to make the pigment brushable. The liquid evaporates, and the dried pigment remains as a pure substance on the painted surface. Watercolors and tempera are usually mixed only with a wetting agent. Although this kind of paint sometimes *looks* homogeneous, it is really a heterogeneous mixture.

Sometimes the liquid in paint is a wetting agent plus other liquid chemicals that cause the dry paint to have a hard, shiny surface. The liquid part of the paint is usually a homogeneous solution. When the pigment is added, the paint becomes a heterogeneous mixture that can look as if it is homogeneous. An example of this kind of paint is artists' oil paints, a combination of linseed oil and solvent in solution added to a powdered pigment. When the wetting agent evaporates, the surface contains a heterogeneous mixture of pigment and a dull or shiny hardening substance.

### ANNOTATION KEY

❶ **Magnetism (Inferring)**

❷ **Threads of different colors in a fabric, etc. (Applying concepts)**

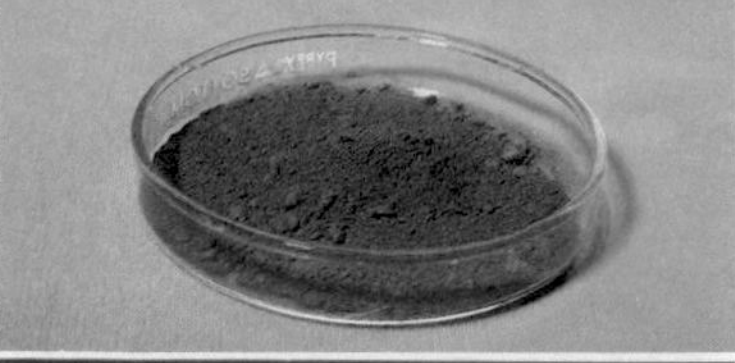

**Figure 17-4** *By combining powdered iron* (top) *with powdered sulfur* (center), *an iron-sulfur mixture is formed. What physical property of iron is being used to separate the mixture* (bottom)? ❶

### Separation of Mixtures

Look at the photographs in Figure 17-4. A mixture has been made by combining powdered iron with powdered sulfur. Iron is black and sulfur is yellow. The resulting mixture has a grayish color, although particles of iron and sulfur are clearly visible. This mixture illustrates the two properties of mixtures you have just learned: The substances in a mixture retain their original properties. And the substances can be present in varying amounts. Look again at the figure to discover another property.

Iron is attracted by a magnet. Sulfur is not. The powdered iron can be separated from the powdered sulfur by holding a strong magnet near this mixture. If the two elements were chemically combined in a compound, they could not be separated with a magnet. This happens to be the case for a compound called iron sulfide, FeS.

You can use a number of methods to separate substances in a mixture. The method you choose will depend on the type of mixture. See Figure 17-5. You should note that all the methods of separating mixtures are based on physical properties. No chemical reactions are involved.

### SECTION REVIEW

1. Explain why a mixture is a physical combination of substances.
2. Describe three properties of a mixture.
3. List three ways to separate a mixture.

**Figure 17-5** *Salt water is a mixture of various salts and water. When the water evaporates, which is a physical change, it leaves behind deposits of salt* (right). *Heavy pieces of gold can be separated from rock, sand, and dirt by shaking the mixture in a pan of water* (left). *The gold will settle to the bottom of the pan.*

### 17-1 (continued)

#### Content Development

Point out that all mixtures can be separated because they are two or more substances physically, not chemically, combined. Mixtures can be in any of the four phases. When a mixture is separated, each substance in the mixture retains its own properties.

Tell students the methods most commonly used for separation of mixtures are filtration and evaporation. Point out that if metallic iron is involved, separation can be done with a magnet.

#### Skills Development

***Skill: Applying concepts***

Divide the class into groups of 4–6 students. Give each group three small beakers, watch glasses, filters, filter paper, magnet, water, small amounts of sand, salt, and metal filings. Tell the groups to make the following mixtures: sand and water; salt and water; and sand, salt, and metal filings. Have them determine how to separate the mixtures. Discuss their findings.

- **What was the best method to separate the sand from the water?** (filtration)
- **How was the salt separated from the water?** (evaporation)
- **How were the sand, salt, and metal filings separated?** (A magnet was used to remove the filings; water

## 17-2 Types of Mixtures

Does it surprise you to learn that both concrete and stainless steel are mixtures? Concrete is a mixture of pieces of rock, sand, and cement. Stainless steel is a mixture of the elements chromium and iron. But in looking at each of these mixtures, you might describe the stainless steel as "better mixed" than the concrete because you cannot see individual bits of chromium and iron in the steel. **Mixtures are classified according to how "well mixed" they are as either homogeneous or heterogeneous.**

### *Heterogeneous Mixtures*

Looking closely at a piece of broken concrete, you will see particles of rock mixed in with sand and cement. You will also see that no two parts of the concrete piece appear exactly the same.

A mixture such as concrete or sand is said to be **heterogeneous** (het-uhr-uh-JEEN-ee-uhs). This means that no two parts of the mixture are identical. A heterogeneous mixture is the least "well mixed" of mixtures. The particles that make up the mixture are large enough to be seen. For this reason, they are easy to recognize and to separate from the mixture. A mixture of different-sized buttons or of different types of coins is an example of a heterogeneous mixture. Can you think of some other examples? ❷

Not all heterogeneous mixtures contain solid particles. Shake up some pebbles or sand in water to make a solid-liquid mixture. This mixture is easily separated just by letting it stand. Oil and vinegar make up a liquid-liquid heterogeneous mixture. When the mixture is well shaken, large drops of oil spread throughout the vinegar. This mixture, too, will separate when allowed to stand. Both these mixtures contain particles that are mixed together but not dissolved. Such mixtures are called **suspensions.**

### *Homogeneous Mixtures*

When the particles of a mixture are very small, are not easily recognized, and do not settle when the

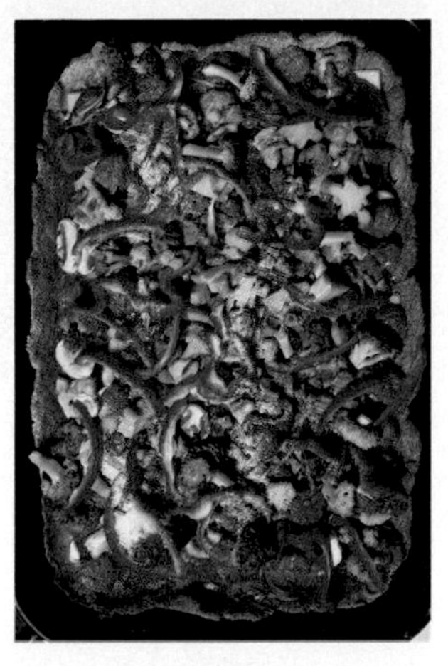

**Figure 17-6** *Pizza is a heterogeneous mixture in which the parts are easy to recognize and to separate from the mixture.*

**Figure 17-7** *This suspension of soil and water is a solid-liquid mixture. Notice how the particles separate when the mixture is allowed to stand.*

## 17-2 TYPES OF MIXTURES

### SECTION PREVIEW 17-2

Mixtures are classified by the size of the particles in them and by how "well mixed" they are. There are two general types of mixtures: heterogeneous and homogeneous.

The particles in a heterogeneous mixture are large enough to be seen. In a heterogeneous mixture, no two parts of the mixture are identical. A heterogeneous mixture is easily separated by allowing the particles to settle to the bottom.

The particles in a homogeneous mixture are very small and not easily recognizable. The particles are well blended, but not dissolved. Because of their small size, they do not settle out. Therefore, the parts of a homogeneous mixture appear to be identical. Colloids are homogeneous mixtures. Colloids contain particles or groupings of molecules larger than a molecule but smaller than the particles in a suspension.

### SECTION OBJECTIVES 17-2

1. **Distinguish between heterogeneous mixtures and homogeneous mixtures.**
2. **Give an example of a suspension and a colloid.**

### SCIENCE TERMS 17-2

| | |
|---|---|
| **heterogeneous p. 395** | **homogeneous p. 396** |
| **suspension p. 395** | **colloid p. 396** |

was added to allow the salt to mix with the water; the sand was filtered out; the salt was separated from the water by allowing the water to evaporate.)

### Enrichment

Students can make a list of all the mixtures they use in one day. The list can then be expanded by identifying the substances in each mixture.

### Section Review 17-1

**1.** The substances in a mixture do not combine chemically. They are merely "mixed" together.
**2.** Substances in a mixture keep their own identities and most of their own proportions; substances can be present in any amount; substances can be separated by physical means.
**3.** Using a magnet, using a filter device, allowing the liquid portion to evaporate.

### TEACHING STRATEGY 17-2

#### Motivation

Pour 1 cup of water into a beaker or flask. Add 1/4 cup of vegetable oil.

- **What happens to the oil?** (It sits on top of the water.)
- **How could the oil be blended into the water?** (Answers will vary, but most students will say by shaking or stirring the mixture.)

Shake or stir the water–oil mixture and allow it to settle.

- **What happens?** (After a while, the oil and water separate.)

## ANNOTATION KEY

❶ Emulsion; gel (Interpreting charts)

① Thinking Skill: Making comparisons

② Thinking Skill: Making generalizations

## TEACHER DEMONSTRATION

Pour 1/2 cup of water into a beaker. Add 1 tsp of dried milk powder. Stir to mix.

- **Is this a mixture?** (yes)
- **Can you separate it easily?** (no)

Tell students that milk is a homogeneous mixture.

### Activity

**Expressing Solubility**
**Skills: Applying definitions, writing**
**Level: Average**
**Type: Library/vocabulary**

This library activity gives students the opportunity to learn the meanings of certain terms used in discussing solutions. You may wish to have them actually prepare examples of dilute and concentrated solutions, or of saturated or super-saturated solutions, to illustrate these concepts.

**Figure 17-8** *Gelatin is a colloid containing liquid particles mixed in a solid, while whipped cream contains gas particles in a liquid* (left). *Fog is a type of colloid in which liquid particles are mixed in a gas* (right).

mixture is allowed to stand, the mixture is "well mixed." As a result, different parts seem to be identical. This type of mixture is said to be **homogeneous** (ho-muh-JEEN-ee-uhs). The stainless steel you read about is a homogeneous mixture.

① Although you may not be aware of it, many of the materials you use and eat each day are homogeneous mixtures. Milk, whipped cream, toothpaste, mayonnaise, and suntan lotion are just a few examples. In these homogeneous mixtures, the particles are mixed together but not dissolved. As a group, these mixtures are called **colloids** (KAHL-oyds). The particles in a colloid are larger than ordinary molecules but too small to be easily seen through a microscope. In fact, it is the large size of the particles that often makes a colloid appear cloudy. Have you ever wondered why milk has a white, cloudy appearance? The colloidal particles in milk are just large enough to scatter light in all directions. The result of this scattering is the cloudy white color you see.

Colloidal particles, however, are too small to settle when the mixture is allowed to stand. There are several different types of colloids, as you can see in the table in Figure 17-9.

**Figure 17-9** *Colloids include many commonly used materials, such as milk, toothpaste, liquid glue, jelly, and plastics. What type of colloid is mayonnaise? Butter?* ❶

*TYPES OF COLLOIDS*

| Name | Example |
|---|---|
| **Fog** (liquid in gas) | Clouds |
| **Smoke** (solid in gas) | Smoke |
| **Foam** (gas in liquid) | Whipped cream |
| **Emulsion** (liquid in liquid) | Mayonnaise |
| **Sol** (solid in liquid) | Paint |
| **Gel** (liquid in solid) | Butter |

### *SECTION REVIEW*

1. Based on how "well mixed" they are, what are the two types of mixtures?
2. What is the name given to mixtures such as milk and mayonnaise?

396

## 17-2 (continued)

### Content Development

Explain that mixtures are classified into two different types: those that separate easily and those that do not separate easily. Point out that mixtures separate differently because of the size of their particles. Tell students that the two types of mixtures are heterogeneous and homogeneous.

Have students observe Figures 17-6 and 17-7. Point out that each is a heterogeneous mixture. Both the pizza and the mixture of soil and water can be separated with relative ease.

- **What are the differences between the pizza as a heterogeneous mixture and the soil–water combination as a heterogeneous mixture?** (Most of the ingredients can easily be taken off the pizza, while some of the soil seems to be relatively well mixed with the water.)

Explain that when soil is mixed with water, the particles of soil slowly settle to the bottom. The particles of soil are made of groups of molecules. A mixture of this type in which the molecules slightly "hang" in the liquid but eventually settle out is known as a suspension. Explain that in a suspension, the particles or group of molecules of one of the substances are small enough to be held suspended by the other substance. Tell students that the group of molecules is not dissolved and will eventually settle out.

## 17-3 Solutions

When you are thirsty, you might drink a glass of lemonade. But did you know that you were drinking a homogeneous mixture called a **solution?** A solution is the "best mixed" of all mixtures. In fact, the substances making up the solution are not just mixed together. They are dissolved in one another.

### *Properties of Solutions*

By picturing a glass of lemonade, you can discover several important properties of a solution. First of all, how was the lemonade made? Lemon juice and sugar were probably added to water. They dissolved in the water. **In a solution, the substance that is dissolved is called the solute and the substance that does the dissolving is called the solvent.** 2 So the lemon juice and the sugar are **solutes** (SAHL-yoots). The water is the **solvent** (SAHL-vunt).

Looking at the glass of lemonade, you will notice that the particles are not large enough to be seen.

**Activity**

*Expressing Solubility*

Solubility can be described in several different ways. Using books in the library, find out what each of the following words means. Give an example of each term.

dilute
concentrated
unsaturated
saturated
supersaturated

**Career:** ***Perfumer***

**HELP WANTED: PERFUMER** To create fragrances for cosmetic company. High school diploma required. On-the-job training provided.

Perhaps you have noticed that the second-hand car your neighbor just bought smells brand new. Or that your friend's new plastic wallet smells like real leather. Companies known as fragrance houses create scents to make products more attractive.

Most of the scents made by fragrance houses are used in soaps, detergents, perfumes, and other grooming items. The people who create the fragrances are **perfumers.** They combine many ingredients, including oils of flowers, to make a desired fragrance.

There are more than 3000 natural and synthetic ingredients used in making fragrances. In a process that can take up to two years, perfumers mix solutions of ingredients, check to make sure the ingredients are properly balanced, and test the fragrances. For more information on this career, write to the Fragrance Foundation, 116 East Nineteenth Street, New York, NY 10003.



## 17-3 SOLUTIONS

### SECTION PREVIEW 17-3

Solutions are a special kind of homogeneous mixture that actually have the same makeup throughout the mixture. In solutions, one substance dissolves another substance. The substance that does the dissolving is called the solvent. The substance that is dissolved is called the solute. The substance in the greater amount is normally considered the solvent. When a solid and water combine, the water is considered the solvent.

The substances in a solution are not chemically combined but cannot be separated easily by simple physical means. The particles or groups of molecules in a solution are approximately the same size and mix uniformly. Solutions can be solids dissolved in liquids, liquids dissolved in liquids, gases dissolved in liquids, or gases dissolved in gases.

### SECTION OBJECTIVES 17-3

1. **Define the terms solution, solute, solvent.**
2. **Give examples of solutions.**

### SCIENCE TERMS 17-3

**solution** p. 397
**solute** p. 397
**solvent** p. 397
**soluble** p. 398
**solubility** p. 398
**insoluble** p. 398
**alloy** p. 399

### Section Review 17-2

1. Heterogeneous, homogeneous.
2. Colloids.

### TEACHING STRATEGY 17-3

#### Motivation

Open a bottle of carbonated water. Have students observe the gas bubbles that rise from the liquid.

• **What gives the carbonated water its fizz?** (the gas)

Point out that the carbonated water is a solution where gas was dissolved in water.

#### Content Development

Explain that a solution is the "best mixed" of all mixtures. One of the substances in a solution is dissolved in another substance. Point out that a solution can be a solid, such as sugar, dissolved in a liquid, such as water. It can be a gas dissolved in a liquid, a liquid dissolved in a liquid, or a gas dissolved in a gas.

Refer to the text discussion to point out that the substance that is dissolved is called the solute and the substance that does the dissolving is called the solvent. Explain that the substance that is in the greater amount is normally considered the solvent.

## ANNOTATION KEY

❶ Alloy (Applying concepts)

① Thinking Skill: Applying definitions

## Activity

**Water Dissolves Most Substances**

**Skills: Manipulative, observing, comparing, relating, applying, diagraming**

**Level: Remedial–average**

**Type: Hands-on**

**Materials: 7 glasses of water, sugar, starch, salt, flour, cooking oil, baking soda, cleaning powder**

Students can do this simple activity at home. They will learn that some household substances, such as sugar and salt, are very soluble in water, whereas others, such as flour and cooking oil, have low solubility.

*TYPES OF SOLUTIONS*

| Solute | Solvent | Example |
|---|---|---|
| Gas | Gas | Air (oxygen in nitrogen) |
| Gas | Liquid | Soda water (carbon dioxide in water) |
| Gas | Solid | Charcoal gas mask (poisonous gases on carbon) |
| Liquid | Gas | Humid air (water in air) |
| Liquid | Liquid | Antifreeze (ethylene glycol in water) |
| Liquid | Solid | Dental filling (mercury in silver) |
| Solid | Gas | Soot in air (carbon in air) |
| Solid | Liquid | Ocean water (salt in water) |
| Solid | Solid | Gold jewelry (copper in gold) |

**Figure 17-10** *Nine different types of solutions can be made from the three phases of matter. What are solutions of solids dissolved in solids called?* ❶

### Activity

*Water Dissolves Most Substances*

You can find out what substances will dissolve in water by performing the following activity. Fill seven glasses with water. Into each glass of water, add a small amount of one of the substances listed below. Stir and let stand for several minutes. Observe what happens. Make a chart of your observations. Include your conclusions about what substances dissolve in water.

sugar
starch
salt
flour
cooking oil
baking soda
cleaning powder

398

The particles in a solution are individual atoms or molecules. For this reason, most solutions cannot be easily separated by simple physical means. Unlike many colloids, liquid solutions appear clear and transparent. The particles in a liquid solution are too small to scatter light.

Tasting the lemonade illustrates another property of a solution. Every part of the solution tastes the same. This might lead you to believe that one property of a solution is that its particles are evenly spread out. And you would be right!

There are nine possible types of solutions, as you will see in Figure 17-10. Many liquid solutions contain water as the solvent. Ocean water is basically a water solution containing many salts. Body fluids are also water solutions. Because water can dissolve many substances, it is called the "universal solvent."

### *Solubility*

A substance that dissolves in water is said to be **soluble** (SAHL-yuh-bul). Salt and sugar are soluble substances. Mercury and oil do not dissolve in water. ① They are **insoluble.**

The amount of solute that will completely dissolve in a given amount of solvent *at a specific temperature* is called **solubility.** What is the relationship between temperature and the solubility of solid solutes? In general, as the temperature of a solvent

## 17-3 (continued)

### Motivation

Have students make a list of things that "mix" with water. Example: sugar, salt, shampoo, dish detergent, vinegar, alcohol, flour, etc. Then have students star (*) each item on their list that dissolved in the water.

### Content Development

Tell students that a substance that dissolves in another substance is said to be soluble. If a substance does not dissolve in another substance, it is said to be insoluble. Point out that water dissolves so many other substances that it is the most commonly used solvent and is thought of as the "universal solvent."

### Content Development

Tell students that solutions that have a small amount of solvent are said to be dilute. Solutions with a large amount of solute are known as concentrated. Point out that the amount of solute that will completely dissolve in a given amount of solvent at a specific temperature is called solubility. Explain that as the temperature of the solvent increases, the solubility of the solute increases.

### Skills Development

***Skill: Designing an experiment***

Divide the class into groups of 4–6 students. Give each group a 250-mL beaker, 2 boiling thermometers, heat source (Bunsen burner or hot plate), and 50 mL of salt. Water should be available.

Have the teams design an experiment to test what happens to the boiling point of 125 mL of water containing 50 mL of salt.

If necessary, guide the students to first find the normal boiling point of 125 mL of water. This should be near 100°C at sea level and normal air pressure. At higher elevations water will boil at lower temperatures.

**Figure 17-11** *In the process of making steel, a solution of scrap steel and iron is made in a blast furnace* (left). *Near the end of the process, alloying substances are dissolved in the steel. In a water-purification plant, air dissolves in water as the water flows from sprinklers* (right). *The air-water solution now contains an increased amount of oxygen.*

increases, the solubility of the solute increases. What about gaseous solutes? An increase in the temperature of the solvent usually decreases the solubility of a gaseous solute. This explains why soda that warms up goes flat. The "fizz" of soda is due to bubbles of carbon dioxide dissolved in the solution.

Some substances are not very soluble in water. But they do dissolve easily in other solvents. For example, one of the reasons you use soap to wash dirt and grease from your skin or clothing is that the soap dissolves these substances, while water alone does not. Soap is made up of long molecules, with one end dissolving in water and the other end serving as a solvent for grease. The soap dissolves the grease and then, along with the grease, is washed away by the water.

Not all solutions are liquid, as the table in Figure 17-10 indicates. Metal solutions called **alloys** are examples of solids dissolved in solids. Gold jewelry is actually a solid solution of gold and copper. Brass is an alloy of copper and zinc. Sterling silver contains small amounts of copper in solution with silver. And the stainless steel you read about before is an alloy of chromium and iron. You may find it interesting to learn about the makeup of other alloys such as pewter, bronze, and solder.

### SECTION REVIEW

1. What is a solute? A solvent?
2. What condition can increase or decrease the solubility of a substance?

### Activity

*Solubility of a Gas in a Liquid*

You can determine what conditions affect the solubility of a gas in a liquid.

1. Remove the cap from a bottle of soda.
2. Immediately fit the opening of a balloon over the top of the bottle. Shake the bottle several times. Note any changes in the balloon.
3. Heat the bottle of soda very gently by placing it in a pan of hot water. Note any further changes in the balloon.

What two conditions of solubility are being tested here? What general statement about the solubility of a gas in a liquid can you now make?



## TEACHER DEMONSTRATION

Show students 125 mL (1/2 cup) of water and 125 mL of salt.

• **Do you predict that 125 mL of salt will dissolve in 125 mL of water at room temperature?** (Students may guess either yes or no.)

Combine the salt and the water, stirring to mix as well as possible.

• **What happened?** (Some of the salt dissolved but some did not.)

• **How do you know some salt did not dissolve?** (Some salt is still on the bottom.)

### Activity

**Solubility of a Gas in a Liquid**
**Skills: Manipulative, observing, comparing, relating, hypothesizing, concluding**
**Level: Average**
**Type: Hands-on**
**Materials: Club soda in bottle, balloon, hot plate or Bunsen burner**

In this activity, students investigate the properties of a gas in a liquid. Students should find that the balloon inflates when the liquid is heated. The two conditions being tested are pressure and temperature. Students should conclude that the solubility of a gas in a liquid decreases with a decrease in pressure and decreases with an increase in temperature.

Next have the teams add 50 mL of salt to the 125 mL of water and heat until boiling. Students should find that the salt solution boils at a higher temperature than plain water.

### Content Development

Tell the students that there are other solvents besides water. Point out that grease does not dissolve in water, but when soap or laundry detergent, which does dissolve in water, is added, other molecules in the soap or detergent dissolve the grease. Gasoline enhanced with ethanol or methol will dissolve grease and the carbon build up in engines.

### Content Development

Have students observe Figure 17-11. Point out that the scrap steel and the iron were both heated to their melting point. Explain that a solution was made by mixing the melted scrap steel and the melted iron. Tell students that when the solution of scrap steel and iron cooled, it became a solid solution. Metal solutions are called alloys. Explain that alloys are examples of a solid being dissolved in a solid.

### Section Review 17-3

1. The substance that is dissolved; the substance that does the dissolving.
2. Temperature.

## LABORATORY ACTIVITY EXAMINATION OF FREEZING-POINT DEPRESSION

### BEFORE THE LAB

1. **Divide the class into groups of 3–6 students per team.**
2. **Gather all materials at least one day prior to the investigation. You should have enough supplies to meet your class needs, assuming 3–6 students per group.**

### PRE-LAB DISCUSSION

Have students read the complete laboratory procedure. Discuss the procedure.

- **What is the purpose of the laboratory activity?** (to observe the effects of various dissolved substances on the freezing point of water)
- **Why is it important to use 10 mL of water in each tub?** (to test the same amount of water each time)
- **What is the purpose of the control?** (to find the freezing point of tap water to which no substances have been added)
- **Why is it important to lower the test tube into the ice-salt bath at least to the level of the liquid?** (This allows the same amount of cooling each time.)
- **What should you do if most of the ice has melted before you complete testing all seven trials?** (Add more ice.)
- **When should you record the temperature of the liquid in each tube?** (when the liquid begins to freeze into ice)

## LABORATORY ACTIVITY

### *Examination of Freezing-Point Depression*

**Purpose**

In this activity, you will observe the effects of various dissolved substances on the freezing point of water.

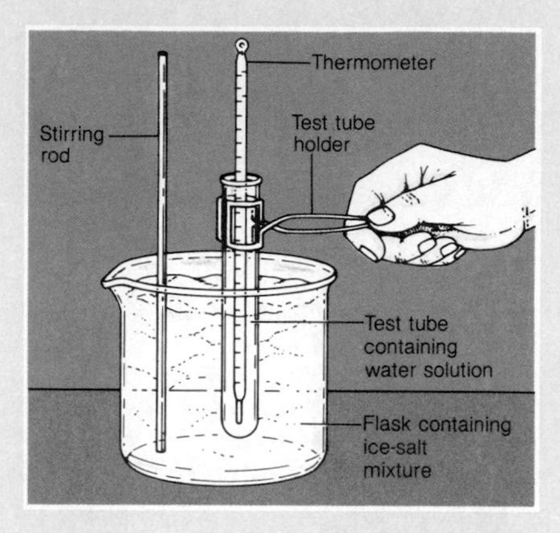


**Materials** *(per group)*

| | |
|---|---|
| Water | Glass-marking pencil |
| Glucose | Coarse salt |
| Sucrose (table sugar) | Crushed ice |
| Sodium chloride | Glass stirring rod |
| Test tube holder | 500-mL beaker |
| Test tube rack | Triple-beam balance |
| | Graduated cylinder |

7 large test tubes and stoppers
Thermometer (−10 ° to 100 °C)

**Procedure**

1. Label seven large test tubes with the numbers 1 through 7.
2. Place 10 mL of tap water in each of the test tubes.
3. Use the balance to pour out the following amounts of glucose ($C_6H_{12}O_6$), sucrose ($C_{12}H_{22}O_{11}$), and sodium chloride (NaCl).
   1. control
   2. 1.8 g glucose
   3. 3.6 g glucose
   4. 3.4 g sucrose
   5. 6.8 g sucrose
   6. 0.6 g NaCl
   7. 1.2 g NaCl
4. Add the measured substances to the corresponding numbered test tubes. Cover each test tube with a stopper. Shake until the substance is completely dissolved in the water.
5. Set up a low-temperature bath by mixing together 150 mL crushed ice, 100 mL coarse salt, and 100 mL water in a 500-mL beaker. Stir with a glass stirring rod for a few minutes. If all the ice should melt at any time during the activity, simply add more.
6. Place a thermometer in test tube 1. Using the test tube holder, lower the tube into the ice-salt bath until it is submerged at least up to the level of the liquid it contains. Stirring gently with the thermometer, record the temperature at which the water begins to freeze. Remove the tube and thermometer. Wipe the thermometer clean.
7. Repeat step 6 for tubes 2 through 7.

**Observations and Conclusions**

1. What effect does the presence of a dissolved substance have on the freezing point of water?
2. What effect does doubling the amount of a given dissolved substance have on the freezing-point?
3. Tubes 2 and 4 contained roughly equal numbers of molecules of dissolved substances. Does the kind of substance have much effect on freezing-point?



### SKILLS DEVELOPMENT

Students will use the following skills while completing this activity.

1. Manipulative
2. Safety
3. Observing
4. Measuring
5. Applying
6. Relating
7. Hypothesizing
8. Comparing

### SAFETY TIPS

Alert students to be cautious about putting the thermometer in a place where it won't roll or be knocked onto the floor.

### TEACHING STRATEGY FOR LAB PROCEDURE

1. Have the teams follow the directions carefully as they work in the laboratory.
2. Discuss how the investigation relates to the chapter ideas by asking the following questions:

- **What happens to the freezing point of water when the water contains other substances in solution?** (The freezing point is lowered.)

# CHAPTER REVIEW

## SUMMARY

### 17-1 Properties of Mixtures

- Mixtures are made up of different substances mixed together but not chemically combined. They can be separated by ordinary physical means.
- Substances in a mixture keep their properties and are present in varying amounts.

### 17-2 Types of Mixtures

- No two parts of a heterogeneous mixture are identical. The particles are large enough to be seen, and they settle when the mixture is allowed to stand.
- Homogeneous mixtures, which contain very small particles, have parts that seem to be identical.
- Colloids are homogeneous mixtures of very small particles. Colloids often appear cloudy.

### 17-3 Solutions

- A solution consists of a solute dissolved in a solvent. The particles of each are individual atoms or molecules.
- Solutions are transparent. They cannot be easily separated by simple physical means.
- Soluble substances dissolve in water. Insoluble substances do not.
- The amount of solute that completely dissolves in a given amount of solvent is called solubility.
- The solubility of a solute depends upon the temperature of the solvent. The solubility of a solid in a liquid usually increases as the temperature of the solvent increases. The solubility of a gas in a liquid decreases as the temperature of the solvent increases.
- Alloys are solid solutions.

## VOCABULARY

*Define each term in a complete sentence.*

| | | | |
|---|---|---|---|
| **alloy** | **homogeneous mixture** | **solubility** | **solution** |
| **colloid** | **insoluble** | **soluble** | **solvent** |
| **heterogeneous mixture** | **mixture** | **solute** | **suspension** |

## CONTENT REVIEW: MULTIPLE CHOICE

*Choose the letter of the answer that best completes each statement.*

**1.** Which of the following substances is *not* a pure substance?
a. oxygen b. sugar c. salt water d. carbon dioxide

**2.** The particles in a mixture
a. are always too small to be seen.
b. are identical throughout the mixture.
c. are chemically combined.
d. are not identical throughout the mixture.

**3.** In a mixture, substances are present in
a. varying amounts. b. fixed amounts.
c. a ratio of 2 to 1. d. the liquid phase only.



• **What difference did you notice when the amount of the dissolved substance was increased?** (The freezing point was lowered by an even greater amount.)

## OBSERVATIONS AND CONCLUSIONS

**1.** It lowers the freezing point.
**2.** It doubles or nearly doubles the amount of freezing-point depression (about 1.86°C for solutions 2 and 4; about 3.72°C for 3 and 5; about 3.72°C for solution 6; and about 7.44°C for solution 7).
**3.** The kind of substance does not seem to have much effect. (This is true for nonionic substances. The depression is about 1.86°C for a 1-molal, or 1 *m*, solution. Solutions 2 and 4 were both 1 *m*. However, solution 6 was of the same concentration—as you may wish to inform your students—and its freezing-point depression was about twice as great. This is because each NaCl unit dissociated into 2 ions, so a 1-*m* solution produced a 2-*m* solution of ions.)

## GOING FURTHER: ENRICHMENT

**Part 1**

Students might investigate the freezing point depression of other common liquid substances such as milk, orange juice, vinegar, window-cleaning solution, and other available solutions.

**Part 2**

Students might investigate the freezing point depression of water containing mixtures of glucose/sucrose, glucose/NaCl, sucrose/NaCl, and glucose/sucrose/NaCl.

## CHAPTER REVIEW

### MULTIPLE CHOICE

| | | | | |
|---|---|---|---|---|
| **1.** c | **3.** a | **5.** b | **7.** c | **9.** b |
| **2.** d | **4.** b | **6.** a | **8.** c | **10.** c |

### COMPLETION

| | |
|---|---|
| **1.** mixture | **6.** emulsion |
| **2.** physical | **7.** solvent |
| **3.** suspension | **8.** solubility |
| **4.** homogeneous | **9.** insoluble |
| **5.** colloid | **10.** alloys |

### TRUE OR FALSE

| | |
|---|---|
| **1.** F mixture | **6.** T |
| **2.** F retain | **7.** T |
| **3.** T | **8.** F water |
| **4.** F physical | **9.** F less |
| **5.** F homogeneous | **10.** T |

### SKILL BUILDING

**1.** Heterogeneous; sausage pizza, chocolate chip cookies; homogeneous: air in balloon, glass of water.

**2.** Substances keep their own identity: you can recognize the milk, blueberries, and cereal by both sight and taste. Substances can be present in any amount; you can use varying amounts of each substance and still have a mixture of the three. Substances are easily separated according to their physical properties: you can filter the cereal flakes and berries from the milk and then pick out the berries from the flakes.

**3.** The parts of the solution are not chemically combined. A solution is made up of two or more different kinds of particles. A solution can be separated by physical means. A solution has the same properties as its ingredients.

**4.** **a.** Evaporate water **b.** Use magnet for getting iron out **c.** Float in water **d.** By size

**5.** Treated so that all parts are identical throughout. Milk that is not homogenized separates into cream at top and milk below.

**6.** **a.** Temperature **b.** Warm bottle. Cold Bottle. As temperature increases, the solubility of a gas in a liquid decreases. The soda in a warm bottle bubbles rapidly because the solubility of the carbon dioxide gas is decreased.

**7.** Heterogeneous mixtures contain large clumps of particles that will not fit through the small openings of most filters. The particles in solutions are very well mixed or are mixed on the molecular level, and are small enough to pass through.

**8.** Answers will vary but should include the idea that if such a solvent existed, there would be no possible container to hold it because the solvent would dissolve any container.

**9.** The sugar would dissolve rapidly and completely in the hot tea. Dissolving would occur much more slowly in the cold tea.

**10.** The physical properties of particle size and separation of solute and solvent particles distinguish a solution from a suspension and a colloid. The particles in a solution are very small—so small, in fact, that they do not separate out upon standing (as do the particles of a suspension) nor do they scatter light (as do the particles of a colloid).

**4.** Ways of separating a mixture depend on
a. chemical properties. b. physical properties.
c. chemical reactions. d. nuclear reactions.

**5.** A mixture in which no two parts are identical is a
a. homogeneous mixture. b. heterogeneous mixture.
c. solution. d. solvent.

**6.** A mixture that contains large particles that settle when the mixture is allowed to stand is called a
a. suspension. b. solution. c. colloid. d. solvent.

**7.** Homogeneous mixtures such as milk, toothpaste, and mayonnaise are examples of
a. suspensions. b. solutions. c. colloids. d. solvents.

**8.** In a solution, the substance that is dissolved is called the
a. solvent. b. gel. c. solute. d. emulsion.

**9.** In a saltwater solution, the water is the
a. solute. b. solvent. c. alloy. d. colloid.

**10.** Metal solutions of solids in solids are
a. salts. b. colloids. c. alloys. d. suspensions.

### CONTENT REVIEW: COMPLETION

*Fill in the word or words that best complete each statement.*

**1.** Two or more pure substances mixed together but not chemically combined form a(n) ______.

**2.** Methods of separating mixtures are based on the ______ properties of the substances in the mixture.

**3.** Oil mixed with vinegar is an example of a mixture called a(n) ______.

**4.** Mixtures whose particles are very small and whose different parts seem to be identical are said to be ______.

**5.** A mixture that often appears cloudy, does not separate when allowed to stand, and contains very small particles is a(n)

**6.** Mayonnaise is a colloid known as a(n) ______.

**7.** A(n) ______ is the substance that does the dissolving in a solution.

**8.** ______ is the extent to which a substance dissolves in another substance at a given temperature.

**9.** Substances that do not dissolve in water are said to be ______.

**10.** Stainless steel, sterling silver, and brass are examples of solid solutions called ______.

### CONTENT REVIEW: TRUE OR FALSE

*Determine whether each statement is true or false. If it is true, write "true." If it is false, change the underlined word or words to make the statement true.*

**1.** The rock granite is a compound.

**2.** In forming a mixture, substances lose their original properties.

**3.** The substances that make up a mixture are present in varying amounts.

**4.** Separating mixtures is based on the chemical properties of the substances.

**5.** A mixture whose particles are very small and whose parts are identical throughout is said to be heterogeneous.

6. A suspension will separate when allowed to stand.
7. In a solution of sugar water, the sugar is the solute.
8. The universal solvent is alcohol.
9. A gas is usually more soluble in warm water than in cold.
10. Brass, a solution of copper and zinc, is an example of an alloy.

## CONCEPT REVIEW: SKILL BUILDING

*Use the skills you have developed in the chapter to complete each activity.*

1. **Classifying matter** Classify the following materials as either homogeneous or heterogeneous mixtures: sausage pizza, chocolate chip cookies, air inside a balloon, glass of water.
2. **Relating facts** You have learned that mixtures have three important properties. Using the example of breakfast cereal with milk and blueberries, how can you illustrate each of these properties?
3. **Relating concepts** Explain why a solution is classified as a mixture instead of as a compound.
4. **Applying concepts** Describe a method of separating the following mixtures:
   a. sugar and water
   b. powdered iron and powdered aluminum
   c. wood and gold
   d. nickels and dimes
5. **Applying definitions** Most milk sold in stores is homogenized. What do you think this means?
6. **Identifying cause and effect** The caps are removed from a warm bottle and a cold bottle of carbonated beverage. The soda in the cold bottle bubbles slightly. The soda in the warm bottle bubbles rapidly.
   a. What condition affecting solubility is present here?
   b. Which bottle is the variable? The control?
   c. Give an explanation for what happens.
7. **Making generalizations** Why can heterogeneous mixtures be separated by filtering but solutions cannot?
8. **Making inferences** Explain whether you believe there exists a true "universal solvent," capable of dissolving all other substances. Include a description of the kind of container you would put such a solvent in.
9. **Making predictions** Describe what would happen if you were to drop two lumps of sugar into a cup of hot tea; a cup of cold tea.
10. **Making comparisons** How is a solution different from a suspension and a colloid? What simple procedure could you perform to distinguish among the three mixtures?

## CONCEPT REVIEW: ESSAY

*Discuss each of the following in a brief paragraph.*

1. What two physical properties determine whether a mixture is a suspension or a colloid?
2. What is a solution? What are its two parts?
3. Describe three properties of a solution.
4. What is solubility? What factor affects the solubility of a solute?
5. Describe three types of colloids.
6. Describe three properties of a mixture.



**6.** The substances keep their separate identities and most of their own properties; the substances can be present in any amount; the substances can be separated by simple physical means.

## ADDITIONAL QUESTIONS AND TOPIC SUGGESTIONS

**1.** What is the difference between an element, compound, and mixture? (An element is a pure substance with atoms that are all the same; a compound is two or more elements chemically combined in definite proportions and is also a pure substance; a mixture is two or more pure substances mixed in varying proportions and not chemically combined.)
**2.** How do the particles in a heterogeneous mixture differ from those in a homogeneous mixture? (The particles in a heterogeneous mixture are larger than the particles in a homogeneous mixture.)
**3.** How do the particles in a colloid differ from the particles in a solution? (The particles in a colloid are larger than the particles in a solution.)
**4.** Why do some particles dissolve? (because the particles of the solvent attract the particles of the solute and they mix together uniformly)
**5.** Why is an alloy considered a solid solution? (because the particles or groups of molecules in each combining metal mix uniformly but do not bond chemically)

## ESSAY

**1.** Particle size and separation of solute and solvent particles upon standing. In a suspension, the solute particles are larger than atoms, ions, or molecules. A suspension separates out upon standing. In a colloid, the solute particles are larger than those of a solution, but smaller than those of a suspension. The solute particles are kept permanently suspended and do not settle out.
**2.** Homogeneous mixture in which one substance is dissolved in another; solute and solvent.
**3.** Contains solvent and solute; particles are individual atoms, ions, or molecules; most cannot be easily separated by simple physical means; particles are evenly spread.
**4.** A measure of how much solute can be dissolved in a given amount of solvent under certain conditions. Temperature.
**5.** Answers will vary, but any of those listed in Figure 17-9, page 396, are acceptable.

## ISSUES IN SCIENCE

The following issues can be used as a springboard for discussion or given as a writing assignment.
**1.** Some scientists think solutions should not exist in the general classification of mixtures. Other scientists believe that solutions are a special type of mixture. What is your opinion?
**2.** In your opinion, what is the benefit of knowing the differences that exist in mixtures?
**3.** Streams, rivers, and lakes become polluted because of unwanted substances that dissolve in the water. Although very costly, it is possible to remove these substances from streams, rivers, and lakes. How would you solve this problem?

## Unit Four

# STRUCTURE OF MATTER

## ADVENTURES IN SCIENCE: SHIRLEY ANN JACKSON: HELPING OTHERS THROUGH SCIENCE

### BACKGROUND INFORMATION

According to the information division at Bell Laboratories, Dr. Shirley Ann Jackson "is as intriguing and dynamic as the microscopic particles she studies." In addition to her work at Fermilab and Bell Laboratories, Dr. Jackson served as a visiting scientist in the theoretical division at C.E.R.N. (European Organization for Nuclear Research) in Geneva, Switzerland. Dr. Jackson has written numerous articles for leading physics journals and is a frequent speaker at scientific meetings.

Although Dr. Jackson was accepted to do graduate work at Harvard, Brown, and the University of Chicago, she chose to remain at M.I.T., where she had completed her undergraduate studies, because she wanted to encourage the enrollment of more black students there. She continues to maintain a connection with M.I.T. as a member of the board of trustees. Dr. Jackson is also involved in organizations dedicated to the cause of helping women gain education and employment in the sciences.

SCIENCE GAZETTE

Adventures in Science

SHIRLEY ANN JACKSON: HELPING OTHERS THROUGH SCIENCE

Imagine what it would be like to catch a glimpse of the universe as it was forming—to look back in time nearly 20 billion years! Of course, no one can really see the beginning of time. But physicists such as Shirley Ann Jackson believe that learning about the universe as it was in the past will help us understand the universe as it is now and as it will be in the future.

By unraveling some of the mysteries of the universe, Dr. Jackson hopes to fulfill a basic ambition: to enrich the lives of others and

404

### TEACHING STRATEGY: ADVENTURE

#### Motivation

Bring in photographs of telephones, radios, television sets, and computers. Display the photographs and explain that much of the research that makes the production of devices such as these possible is done at a place called Bell Laboratories in Murray Hill, NJ. Point out that in this Adventure lesson, students will read about a prominent physicist at Bell Laboratories, who is Dr. Shirley Ann Jackson.

#### Content Development

Have students read the article about Dr. Jackson. Point out that, like Dr. Stephen Hawkins, whom they read about in the last Adventure lesson, Dr. Jackson is an unusual person. As a black and as a woman, Dr. Jackson probably had to overcome certain obstacles in order to obtain her goal.

Explain to students that the area of physics that Dr. Jackson is currently involved in is called solid-, or condensed-state, physics. Solid-state physics is concerned with the physical properties of crystalline solids. Many students have probably heard the expression *solid-state* applied to various types of electronics equipment. This

make the world a better place in which to live. This contribution, Dr. Jackson believes, can be achieved through science.

Jackson was born and raised in Washington, D.C. After graduating from high school as valedictorian, she attended the Massachusetts Institute of Technology, M.I.T. There, her role as a leader in physics began to take root. Jackson became the first American black woman to receive a doctorate degree from M.I.T. She also achieved the distinction of being the first American black woman to receive a Ph.D. in physics in the United States.

After graduate school, Jackson began work as a research associate in high-energy physics at the Fermi National Accelerator Laboratory in Batavia, Illinois. This branch of physics studies the characteristics of subatomic particles—such as protons and electrons—as they interact at high energies.

Using devices at Fermilab called particle accelerators, physicists accelerate subatomic particles to speeds that approach the speed of light. The particles collide and produce new subatomic particles. By analyzing these subatomic particles, physicists are able to learn more about the structure of atoms and the nature of matter.

The experiments in which Jackson participated at Fermilab helped to prove the existence of certain subatomic particles whose identity had only been theorized. This information is important in understanding the nuclear reactions that are taking place at the center of the sun and other stars.

Jackson's research is not limited to the world of subatomic particles alone. Her work also includes the study of semiconductors—materials that conduct electricity better than insulators but not as well as metal conductors. Semiconductors have made possible the development of transistor radios, televisions, and computers.

Jackson's current work in physics at Bell Laboratories in Murray Hill, New Jersey, has brought her from the beginnings of the universe to the future of communication. This talented physicist is presently doing research in the area of optoelectronic materials. This branch of electronics—which deals with solid-state devices that produce, regulate, transmit, and detect electromagnetic radiation—is changing the way telephones, computers, radios, and televisions are made and used.

Shirley Ann Jackson, in her office at Bell Laboratories, is presently doing research in the field of optoelectronic materials used in communication devices.

Looking back on her past, Jackson feels fortunate to have been given so many opportunities at such a young age. And she is optimistic about the future. "Research is exciting," she says. Motivated by her research, Shirley Ann Jackson is happy to be performing a service to the public in the way she knows best—as a dedicated and determined scientist.

◂ This particle-accelerator generator at Fermilab is familiar equipment to Shirley Ann Jackson.



## ADDITIONAL QUESTIONS AND TOPIC SUGGESTIONS

**1.** What aspects of the scientific method are illustrated by Dr. Jackson's work at Fermilab? (Scientists had predicted the existence of certain subatomic particles. Dr. Jackson gathered information to test this idea by carrying out experiments. When the results of these experiments were analyzed, she concluded that these particles do indeed exist.)
**2.** According to the article, what is Dr. Jackson's chief motivation as a scientist? Does your own attitude toward science fit in with her viewpoint? (to help other people through science; answers will vary.)
**3.** In what ways is Dr. Jackson's story encouraging to women? To members of minorities? (She was the first American black woman to receive a doctorate from M.I.T.)

## CRITICAL THINKING QUESTIONS

**1.** Pure research involves the gathering of knowledge for its own sake. Applied research has practical goals in mind, often related to technology or medicine. Based on the article, which of these areas has Dr. Jackson been involved in? Explain your answer. (Both areas: The work at Fermilab was probably pure research, but the work at Bell Labs was applied to communications technology.)
**2.** Can you think of reasons why studying subatomic particles might lead to an understanding of the past and future of the universe? (One possible answer is that because atoms are the basis of all matter, they may hold the key to the changes that have taken and will take place in the universe.)

label refers to items based on or composed of transistors or related semi-conductor devices.

### Teacher's Resource Book Reference

After students have read the Science Gazette, you may want to hand out the reading skills worksheet based on the article in your Teacher's Resource Book.

## Unit Four

# STRUCTURE OF MATTER

## ISSUES IN SCIENCE: WASTING TIME: THE NUCLEAR CLOCK TICKS DOWN

### BACKGROUND INFORMATION

In the United States, radioactive wastes are generated primarily by 95 nuclear power plants located throughout the country. Each year, approximately 2000 tons of nuclear wastes are produced. Experts estimate that by the year 2020, there will have accumulated approximately 100,000 tons of nuclear garbage.

There are two main classifications of radioactive wastes: high-level and low-level. High-level radioactive waste consists of spent fuel from commercial nuclear power plants and the highly radioactive materials resulting from atomic-energy defense activities. According to the U.S. Nuclear Regulatory Commission, this waste must be permanently isolated.

Low-level radioactive wastes include products such as those used in medical diagnosis and treatment, as well as discarded filters and protective clothing from nuclear power plants and other industries. Most low-level wastes decay rather quickly. The disposal of low-level wastes is the responsibility of the states in which they are produced.

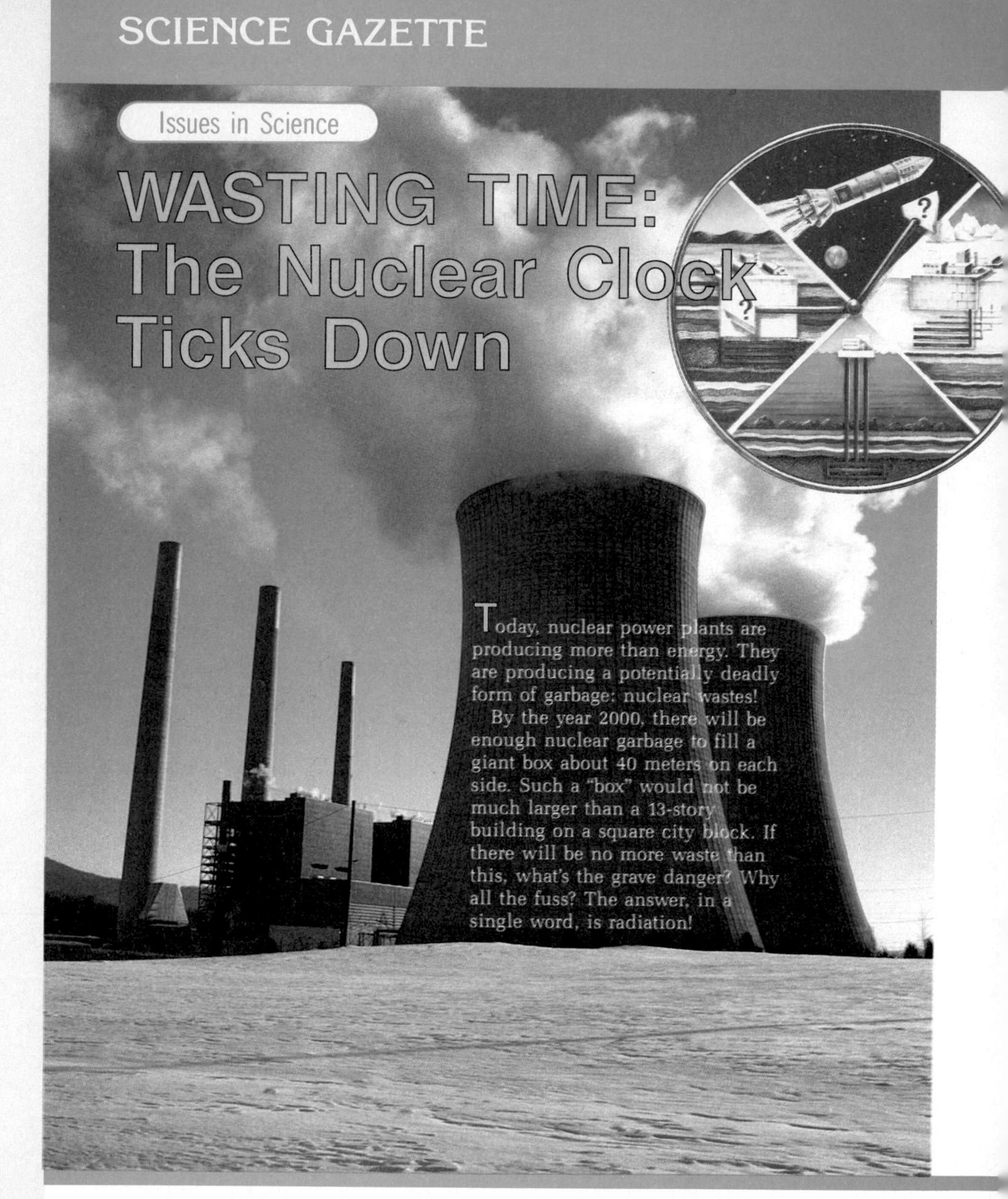

SCIENCE GAZETTE
Issues in Science
WASTING TIME:
The Nuclear Clock
Ticks Down
Today, nuclear power plants are producing more than energy. They are producing a potentially deadly form of garbage: nuclear wastes!
By the year 2000, there will be enough nuclear garbage to fill a giant box about 40 meters on each side. Such a "box" would not be much larger than a 13-story building on a square city block. If there will be no more waste than this, what's the grave danger? Why all the fuss? The answer, in a single word, is radiation!
406


### TEACHING STRATEGY: ISSUE

#### Motivation

Begin by asking the following questions:

- **What do you do with garbage at your house or apartment?** (Put it in trash cans or garbage disposals; put it out for garbage collection.)
- **Do you know what happens to household garbage once it is removed by garbage collectors?** (It usually goes to a local dump. Sometimes garbage that has been collected is burned or buried.)
- **What do you think would happen if no one came to take the garbage away?** (It would begin to smell; it would pile up and look bad; it would take up a lot of room; it could eventually cause disease and attract unwanted pests such as rodents.)

#### Content Development

Continue the discussion by pointing out that just as household activities generate garbage, so do the activities of a nuclear power plant. The problem with nuclear garbage, however, is that it has the potential to pollute the environment for thousands of years. In addition, nuclear wastes are potentially harmful to living things because they give off radiation.

### Invisible Danger

Radiation is invisible energy. And nuclear radiation is *powerful* invisible energy. It is so powerful, in fact, that even small doses over a period of time can permanently harm, or even kill, living things.

Substances such as uranium and plutonium are common fuels for nuclear power plants. Like other fuels, these substances leave behind waste materials when they are used. But these wastes are not at all like the ashes that are the waste products of burning wood. Ashes are harmless. Nuclear wastes are extremely hazardous. The deadly radiation they give off can penetrate most ordinary substances. And this radiation can last for hundreds of thousands of years. So the disposal of nuclear wastes, even the smallest amount, is a giant problem.

Obviously, nuclear wastes cannot be disposed of like ordinary garbage. And they cannot be kept in giant containers, either. So how and where can they be safely stored? Scientists and engineers are trying to answer this question. But they cannot take forever to find a solution. For the wastes are piling up. Over the next few years, special tanks located near nuclear power plants will serve as temporary storehouses. But what is needed is a permanent home for these hazardous materials.

### Space-Bound Garbage

If we cannot find a place on earth to get rid of nuclear garbage, why not send it into space? Rockets loaded with nuclear wastes could be launched into orbit between the earth and Venus. Traveling at the right speeds, the rockets could stay in orbit for a million years or more without bumping into either planet. By that time, the nuclear wastes would have become harmless.

Critics of this idea point to its cost and potential danger. An accident during rocket launch could harm thousands of people. These critics believe that the solution is not in the stars but on earth. Only where can these nuclear burying grounds be found?

### Down-to-Earth Alternatives

The Antarctic ice sheet is more than 2500 meters thick in some places. Could nuclear wastes be buried under this huge, frozen blanket? No, according to some critics of this idea. Not enough is known about the behavior of ice sheets. And what is known is not comforting. For example, ice sheets move rapidly about every 10,000 years. Their movement might allow the wastes to get loose. In addition, nuclear wastes produce a tremendous amount of heat—enough heat, in fact, to melt the ice. Where the ice melted, nuclear radiation might leak out into the oceans and air.

If not under the Antarctic ice sheet, then how about a nuclear cemetery under the ocean floor? Thick, smooth rock layers have been building up there for millions of years. Nuclear wastes deposited in these rock layers would probably remain there almost forever.

But as with other proposals, there are problems with this idea. At present, the technology to do the job does not exist. And not enough is known about the various forces to which such rock is exposed. For example, the force of hot currents might pull the stored nuclear wastes out of the rock.

### Rock Candidates

With ice sheet cemeteries and underwater graveyards all but impossible, one idea still remains. That idea is to put nuclear wastes in "rooms" dug out of underground rock.

In order to determine the best place to bury nuclear wastes, scientists must know all they



## ADDITIONAL QUESTIONS AND TOPIC SUGGESTIONS

**1.** France has found a method of disposing of wastes from nuclear power plants that seems to be working. Find out what this method is and whether the United States plans to try it. (In France, radioactive materials are converted into salts, mixed with molten glass, cast into ingots, and then buried in the ground. The United States plans to use this method to dispose of liquid radioactive wastes from the military.)

**2.** Find out what percentage of electric power in the United States is produced by nuclear energy. Then find out what percentage of electric power is produced by nuclear energy in France, Switzerland, Finland, Belgium, Sweden, the U.S.S.R., and Japan. Make a graph or chart to display and compare the information.

## CRITICAL THINKING QUESTIONS

**1.** Why is it important that rock used to make storehouses for radioactive wastes be waterproof? (If water were to seep into the rock, it would become contaminated with radiation. This water would eventually pass through the rock and contaminate the groundwater system in that area.)

**2.** The radioactive materials that make up nuclear wastes tend to have long half-lives—sometimes as long as a million years. Why does this make their disposal even more difficult? How might the disposal of nuclear wastes be easier if half-lives were much shorter?

Review with students the concepts of radioactive decay and half-life. Explain that certain elements are radioactive, and that the atoms of radioactive elements have unstable nuclei. As these unstable nuclei change and break down, matter and energy are released.

Radioactive elements break down at fixed rates. The half-life of an element is the amount of time it takes for half the atoms in a given sample of that element to decay. For example, suppose the half-life of radioactive element X is 20 years, and you have a 4-g sample of the element. After 20 years have passed, only 2 g of element X will remain.

- **Do you know what happens to the portion of a radioactive element that decays?** (It changes into other elements.)

Explain to students that a radioactive element will continue to break down until a stable, nonradioactive element is obtained. Some elements change only one or two times before reaching a stable form; others may go through numerous changes before a stable element is reached.

Emphasize that the major problem with radioactive materials is that radiation is released as they break down. Radiation can be in the form of alpha rays, beta rays, or gamma rays. Gamma rays, which are released

(Long half-lives make consistent monitoring of nuclear wastes nearly impossible. If half-lives were much shorter, it might be possible to contain the wastes and carefully watch them until nearly all of the radioactive element were to break down.)

**3.** The storage of radioactive wastes requires shipping the wastes to disposal sites, which may be many states away from a nuclear power plant. What factors do you think should be considered in choosing a route for transporting radioactive wastes? (Accept all logical answers. The main concern is to prevent an accident that could cause the radioactive materials to be released. Some considerations published by the U.S. Department of Transportation include the following: Use the most direct route possible; bypass cities and heavily populated areas when possible; use roads that are in excellent condition; and use highways that have the lowest frequency of accidents.)

## CLASS DEBATE

Have students imagine that they are citizens of a town that has been chosen as a site for a nuclear waste dump. Challenge students to dramatize a town council meeting in which the following "cast of characters" debates the issue: citizens who are against the proposed waste dump; the mayor of the town, who is in favor of the dump; a representative from the federal government who is trying to convince citizens to accept the dump; and a representative from the state environmental protection agency, who is opposed to the dump.



can about the rock. Here are the properties scientists have determined are best: The rock must be strong, heat-resistant, and waterproof. The rock must be at least 6100 meters deep. And the ground where the rock exists must be very dry and free of earthquakes.

As you can see from the map, such rock formations exist in the United States. Scientists are now studying many of these formations. They plan to choose the best location and build a nuclear-waste garbage dump there between the years 1998 and 2006. There are four kinds of rocks that scientists believe will be the best.

Basalt is a volcanic rock that is strong and waterproof. It does not lose its strength when heated. However, basalt formations usually contain seams. Some scientists believe that water, which might carry nuclear wastes, could flow along these seams.

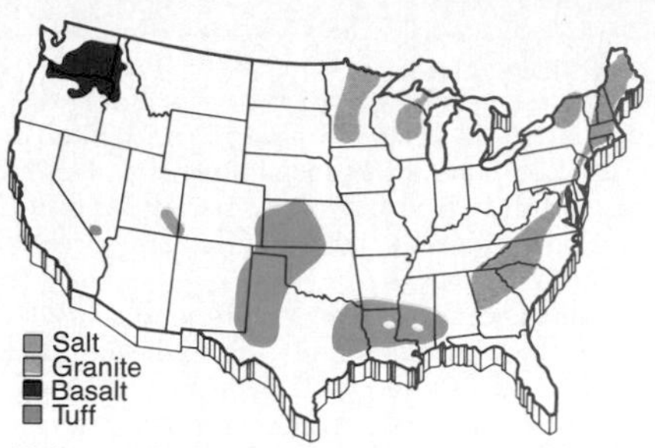


Various underground sites are being considered for the storage of nuclear waste material. There are a number of different kinds of rock in which wastes could be put. The map shows where some of these types of rock are located.

Tuff is another volcanic rock that has some of the properties of basalt. It is being studied in Nevada, where there is little underground water. Scientists point out that there is still much to be learned about tuff.

Granite is a very hard rock that resists heating. But water is found in many granite formations. And in addition, granite is very difficult to drill through. So building a nuclear-waste "room" in granite might be a very tough and costly job.

Salt is found in huge underground deposits. Although salt looks solid, it is really a flowing material. Scientists point out that this property of salt helps to seal cracks in salt formations. Unfortunately, salt is often not dry. And when heated, the salt will dissolve in the water and produce a saltwater solution called brine. The brine could carry nuclear wastes out of the salt formation.

So far, scientists have been unable to find the perfect graveyard for nuclear wastes. But the search goes on. Unfortunately, in the meantime the wastes pile up.

Storage of nuclear wastes requires a complex network of facilities. Here you see an artist's concept of the surface and underground features of a disposal site mined out of rock.

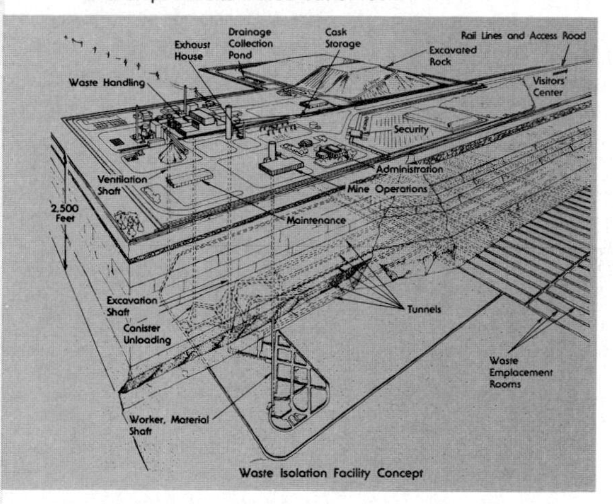




## ISSUE (continued)

along with alpha or beta rays, are by far the most harmful to human beings.

Radiation can contaminate soil, water, and air. For this reason, burying radioactive wastes in the ground or leaving them in open-air dumps does not solve the problem of disposal. Burning radioactive materials would be disastrous, since radiation would be released even more rapidly into the atmosphere. An added problem is that gamma rays can penetrate all but the most dense solids, so it is even difficult to place nuclear wastes in containers.

### Reinforcement

Review with students the possible methods for the disposal of radioactive wastes that have been proposed thus far: sending the wastes into outer space; burying them under the antarctic ice sheet; burying the wastes under the ocean floor; and storing the wastes in "rooms" built in underground rock. Have students discuss why the first three solutions have been rejected, and why the fourth solution remains as a possible option. Then have students recall the properties of rock that would be required in order to effectively contain radioactive wastes: The rock would have to

SCIENCE GAZETTE

THE RIGHT STUFF—

Futures in Science

# Plastics

*Cars, buildings, and appliances of the twenty-first century will be lighter, safer, and stronger than anything we have today.*

Standing apart from a group of her classmates, Jennifer bounced her glass ball on the sidewalk. She was not paying attention to the other members of her class or to what her teacher, Ms. Parker, was saying.

As the sparkling ball bounced, Jennifer was barely aware of the sidewalk changing color to reduce the glare of sunlight. She did not notice a young woman hurrying along the street, carrying an auto engine in one hand and groceries in the other. She did not look up as cars sped by towing boats and houses by plastic threads.

Jennifer and her Design for Living class were on a field trip to a brand new "house of the future." The house had just been completed that year—2086. It certainly made the students' homes seem old-fashioned.

For days before the field trip, Ms. Parker had talked about the house of the future. She had described how safety, security, cleaning, maintenance, entertainment, home education, and communications systems in the house were completely controlled by a central computer. She had showed the students how the combination of a waste-recycling unit, solar cells, and wind turbines met all of the house's energy needs. Ms. Parker had spent a long time talking about the new materials that made such a house possible. She had drawn pictures of organic and inorganic molecules on the classroom computer screen.

Even with all this preparation, Jennifer did not care about the house. She was interested only in the designs of the past. Jennifer's real love was for twentieth-century antiques. In her opinion, the appliances, furnishings, and materials of 2086 were boring and ugly.

"*Jennifer!*" Ms. Parker's voice cut through the group.

Jennifer scooped up the bouncing ball and thrust it into her shoulder bag.

"Yes, Ms. Parker?"

"Please come forward and give us the benefit of your views on home design."

As Jennifer moved to the front of the group, she knew she was on the spot. The class



## Unit Four

## STRUCTURE OF MATTER

## FUTURES IN SCIENCE: THE RIGHT STUFF— PLASTICS

### BACKGROUND INFORMATION

This fanciful article depicts life in the year 2086—just a little less than a hundred years from now. As a result of a "materials revolution," familiar objects such as cars, refrigerators, and furniture are made of strong, flexible, lightweight plastic instead of heavy and rigid materials such as metal and wood.

The article stresses that the plastics of the future are being developed today, in the 1980s. The plastics are being made from a new family of polymers that use silicon atoms instead of carbon atoms. Described in the article are applications of these materials that may come about within the next hundred years.

In addition to plastics, the article touches on some other materials and inventions that may be part of life in the late twenty-first century. These include the use of solar and wind energy for household power; laser brake systems in automobiles; homes controlled by central computer systems; elastic glass; and building materials that change color to adapt to sun glare.

be heat resistant, waterproof, and at least 6100 m deep; and the ground would have to be dry and free from earthquakes.

### Teacher's Resource Book Reference

After students have read the Science Gazette article, you may want to hand out the reading skills worksheet based on the article in your Teacher's Resource Book.

### TEACHING STRATEGY: FUTURE

### Motivation

Begin by asking students the following questions:

- **What do you think life would be like without plastic?** (Accept all answers.)
- **How many items have you used so far today that are made of plastic?** (Answers will vary.)
- **How many items can you think of in your home that are made of plastic?** (Answers will vary.)
- **What types of materials do you think were used to make these items before plastics were invented?** (Accept all answers.)

### Content Development

Use the previous discussion to introduce the idea that even though we tend to take plastics for granted, they

## ADDITIONAL QUESTIONS AND TOPIC SUGGESTIONS

**1.** Have you ever felt that you like old-fashioned things better than modern things, or things made of natural materials rather than synthetic ones? If so, give examples and explain why you feel the way you do.
**2.** Based on the article, do you think you would enjoy living in the year 2086? Why or why not? (Accept all answers.)
**3.** Imagine that you have been asked by a museum to create a "time capsule." The time capsule will be preserved to show future generations what life is like in the United States in the late twentieth century. Decide what items from your home, school, or personal possessions you would include to give an accurate picture of twentieth-century materials and technology.



was already aware of Jennifer's "good-old-days" views, which contrasted with Ms. Parker's enthusiasm for modern materials and designs.

Nervously, Jennifer started to talk. "Plastics, silicones, polymers . . . I don't think the scientists of the late twentieth-century did us a favor by developing all this phony stuff. I wish we could go back to using wood, steel, aluminum, copper, and real glass. Remember those beautiful steel-and-chrome cars and the polished wood furniture in the design museum?"

"Yes," Ms. Parker agreed, "many things produced in the last century were beautiful and well made. It's easy to admire them in museums. But I don't think we'd find them convenient to live with every day."

"I would," Jennifer insisted.

"All right," responded Ms. Parker. "Let's hop into a mental time machine and find out. What age should we go back to?"

"How about the 1980s?" Jeremy suggested.

"Perfect," said Ms. Parker. "The 1980s were just the beginning of the great materials revolution. Great changes in industry, electronics, and materials development began at that time. By the way, Jennifer, you'd better leave your bouncing ball behind. They didn't have elastic glass in the 1980s. Light-sensitive building materials that change color to reduce glare were also unheard of. However, the research to produce these materials was already under way."

### A Trip Back in Time

The house of the future was forgotten as Ms. Parker and the class moved to a nearby park to talk about materials of the past.

Ms. Parker began by asking the students, "What would be the first thing you'd notice in the world of the 1980s?"

"Heaviness. The great weight of almost everything," Jeremy volunteered.

"Go on," Ms. Parker said.

"Appliances like refrigerators, stoves, washing machines, and air conditioners were still made of metal at that time. They were so heavy that it was almost impossible for a single person to lift one of them. And a lot of everyday things were much heavier than they are now. Many food items in supermarkets still came in metal cans and glass bottles. A bagful of those containers could weigh a lot. It wasn't until the 1990s that lightweight, tough plastic containers had completely replaced them."

### A Safer Future

Turning to another student, Ms. Parker asked: "Carlee, what would you notice most about life in the 1980s?"

"That it wasn't safe," Carlee responded.

"Why?"

Carlee thought for a moment and then said, "I guess what I was thinking of was the danger of riding in a 1984 car. I've seen pictures of how those old metal cars hurt people in accidents. Sometimes the heavy metal engines and batteries in the front end were pushed back to where people were sitting. Sharp pieces of broken metal and glass were all over

This plastic automobile engine weighs much less than an all-metal engine. Plastic engines have already been used to power race cars, and it may not be long before they are used in passenger cars.

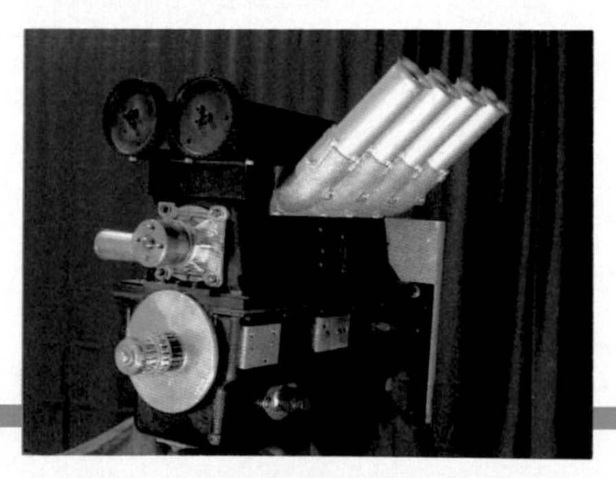



## FUTURE (continued)

are a rather recent invention. Plastics are made of polymers, and the first polymer was manufactured in 1909. Items that are today made of plastic were once made of materials such as wood, glass, china, and metal. Other synthetic materials that have been developed during the twentieth century include synthetic rubber, and synthetic fabrics such as nylon and polyester.

Once students recognize how much familiar appliances and other items have changed during the last hundred years, they may find it easier to imagine how much these same items might change during the next hundred years.

## Skills Development

***Skill: Relating concepts***

Have students think of examples of machines, appliances, and other familiar items that have changed in appearance and construction during the last century. Some of these items include automobiles, vacuum cleaners, radios, typewriters, computers, sewing machines, bicycles, and record players. Students may enjoy collecting pictures of these items as they have looked at different times in history and then comparing them to modern versions.

Discuss with students some of the reasons for the changes they see. Explain that items such as automobiles and record players have changed largely as a result of improved technology. Items such as vacuum cleaners have changed primarily because of new materials. Some old-fashioned vacuum cleaners were

Cars of the future may be made entirely of plastics and powered by solar batteries. Such a car would be lightweight, durable, and clean and inexpensive to operate.

the scene of an accident. Cars could blow up or catch fire."

"That can't happen now," Marian said. "Our cars, including the engines, are made of superplastics and silicones inside and out. These materials are very light and strong, and they bounce. Even if a laser brake system fails, no one can get seriously hurt."

"Cars not only are lighter and safer now," Jeremy added, "they require less energy to run. The changeover from metal to plastic engines in the 1990s led to great energy savings. And, the changeover since then to solar battery-powered cars has meant even greater savings. If we'd kept on using heavy metal cars, the world might have run out of oil and other materials by now."

"You're quite right, Jeremy," Ms. Parker agreed. "Let's sum up what's happened since the 1980s. A great materials revolution started at that time. Chemists discovered how to produce polymers, very long chains and loops of carbon, oxygen, hydrogen, and nitrogen atoms.

"Around the same time, other scientists created a new family of polymers. Silicon atoms were used instead of carbon atoms, so the materials were called silicones. The result of all this chemistry was a new range of super strong, light, cheap plastics.

*The result of all this chemistry was a new range of super strong, light, cheap plastics.*

"The new materials can be made into anything that was once made with metal, wood, glass, or ceramic. In fact, we can do many things with combinations of the new materials that we couldn't do with the old materials, like making a transparent bouncing ball." Ms. Parker smiled at Jennifer.

Jennifer smiled back, still not convinced of the advantages of living in 2086.



## CRITICAL THINKING QUESTIONS

**1.** Do you think that the lightweight plastic cars of 2086 would be as totally safe as the article describes? Support your answer. (Accept all logical answers. Students may wish to consider the possibility that a powerful impact at high speed might still be dangerous, even with light, strong materials that bounce.)

**2.** Why do you think a plastic engine would be especially useful in a racing car? (Answers may vary. A logical answer is that the lighter an object is, the less force and therefore energy it takes to accelerate the object. Since the primary purpose of a racing car is speed, a lightweight plastic engine would be extremely practical.)

**3.** What aspect of the article indicates that some of the modern inventions described are likely to become a reality in the near future? (The article indicates that the materials for these items began to be produced in the 1980s. This means that many of the materials for the items have already been invented.)

made of heavy metal, while most of today's vacuum cleaners are made of lightweight plastic. Computers have changed drastically because of both improved technology and new materials. The first computers were a mass of metal cranks and gears, and were so massive that they took up a whole room. Today's computers are so small and lightweight that they can be picked up and carried as easily as a briefcase.

### Reinforcement

Have students list the items described in the article that are clearly of the future. These items include Jennifer's ball made of elastic glass; an auto engine that can be carried in one hand, plastic threads that can tow boats and cars; a computer-controlled house; the use of solar cells and wind turbines for household power; and a sidewalk that changes color to reduce the glare of the sun.

### Teacher's Resource Book Reference

After students have read the Science Gazette article, you may want to hand out the reading skills worksheet based on the article in your Teacher's Resource Book.

## Unit Five

# COMPOSITION OF THE EARTH

## UNIT OVERVIEW

In Unit Five, students are introduced first to the nature and properties of minerals. They learn to identify minerals by various methods. They also study the uses of minerals and gems. Next, they are introduced to the study of rocks. They examine the nature and formation of igneous, sedimentary, and metamorphic rocks. Finally, they study how soil is formed. The composition and types of soil are also described.

## UNIT OBJECTIVES

1. **Identify and describe the uses of minerals.**
2. **Compare the properties and formation of igneous, sedimentary, and metamorphic rocks.**
3. **Describe the rock cycle.**
4. **Describe the different soil levels.**
5. **Compare the types of soil in the tropical rain forest, temperate forest, prairie grassland, desert, and tundra.**

412

## INTRODUCING UNIT FIVE

Begin your teaching of the unit by having students examine the opening photograph, which shows uncut amethyst crystals. Amethyst is a form of quartz, a mineral that is the most widely found in the earth's crust. Quartz is actually silicon dioxide ($SiO_2$). In amethysts, traces of iron give the quartz its purple color. Most gem-quality amethysts are mined in the United States, the Soviet Union, India, Australia, South Africa—and, in particular, in Brazil.

Have students read the unit introduction and ask the following questions.

- **What are some of the properties of the object shown in the photograph?** (It is a solid, transparent, crystalline substance that is pale purple in color. The substance is embedded in a grayish-whitish material.)
- **What do you think the purplish crystals are?** (They are amethyst crystals.)
- **What properties make them crystals?** (Their regular shape and internal lattice arrangement make them crystals.)
- **How do you think such crystals were formed?** (They were formed within the earth, as molten materials that slowly solidified.)
- **What conditions exist within the**

# Composition of the Earth

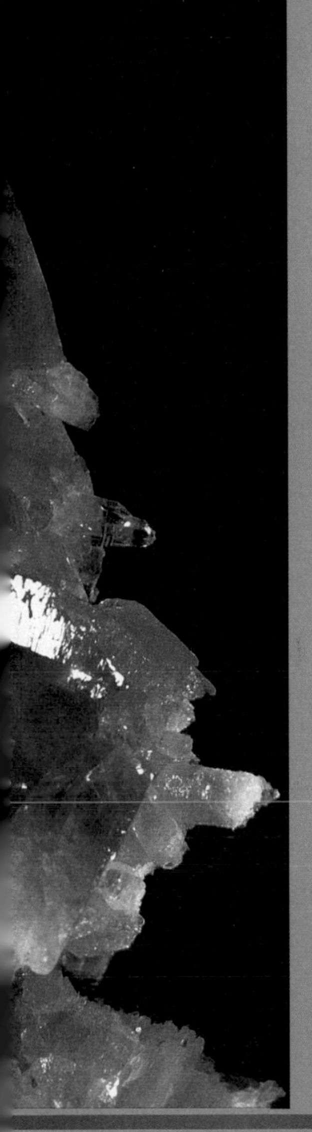

Imagine a place 160 kilometers below the surface of the earth. You are taking a journey deep into the earth in an unbelievably strong bubble—a kind of underground diving capsule. Rivers of orange, glowing rock ooze past your observation window. The temperature gauge on your instrument panel indicates that the outside temperature is a rock-melting 1500° C. You hammer your fist on the pressure gauge. Can it be right? It reads 70,000 kilograms per square centimeter.

Your instruments are accurate. They are giving you a picture of the inside of your planet—the birthplace of rocks and minerals. One day, perhaps millions of years into the future, the molten rocks around you will reach the earth's surface. If conditions are right, unusual and beautiful crystals could form in these rocks as they cool. And what was once formless, glowing ooze may become the rarest rocks on the earth—gemstones!

CHAPTERS



*Amethyst uncut*



## CHAPTER DESCRIPTIONS

**18 Minerals** In Chapter 18, the nature of minerals is explained. Minerals are identified according to physical properties and the Mohs hardness scale. Finally the uses of minerals in various practical devices and processes and in jewelry are described.

**19 Rocks** Chapter 19 opens with a treatment of igneous rocks and their formation. Sedimentary rocks and metamorphic rocks are discussed in turn. An explanation of the rock cycle concludes the chapter.

**20 Soils** The different levels of soil are discussed in Chapter 20. Next, the various types of soil are described. These include the soils in tropical rain forests, temperate forests, prairie grasslands, deserts, and tundras.

**earth?** (high temperatures and pressures)

- **How would you define a mineral?** (A mineral is a natural, nonliving substance that forms in the earth.)
- **What makes a mineral, like amethyst, desirable as a gemstone?** (Its color, clarity, hardness, and rarity make it desirable as a gemstone.)

# Chapter 18

# MINERALS

## CHAPTER OVERVIEW

A mineral is a naturally occurring solid substance with a definite chemical composition and atomic arrangement, found in the earth's surface. Minerals are all around us—from the salt (NaCl) we eat to the diamonds we wear. Every rock we see is made up of minerals. The inorganic portion of soil is mostly comprised of minerals. Our bodies require certain minerals for normal growth.

This chapter introduces students to the basic physical and chemical properties that can be used to identify minerals. The five characteristics of all minerals are discussed.

The various uses of minerals are also discussed. Students should be aware that they use many of the earth's minerals to meet their daily needs. A discussion of metals and nonmetals removed from ores and precious and semiprecious gems is presented in this chapter. The chapter also includes a discussion of the various tests that may be conducted to identify minerals.

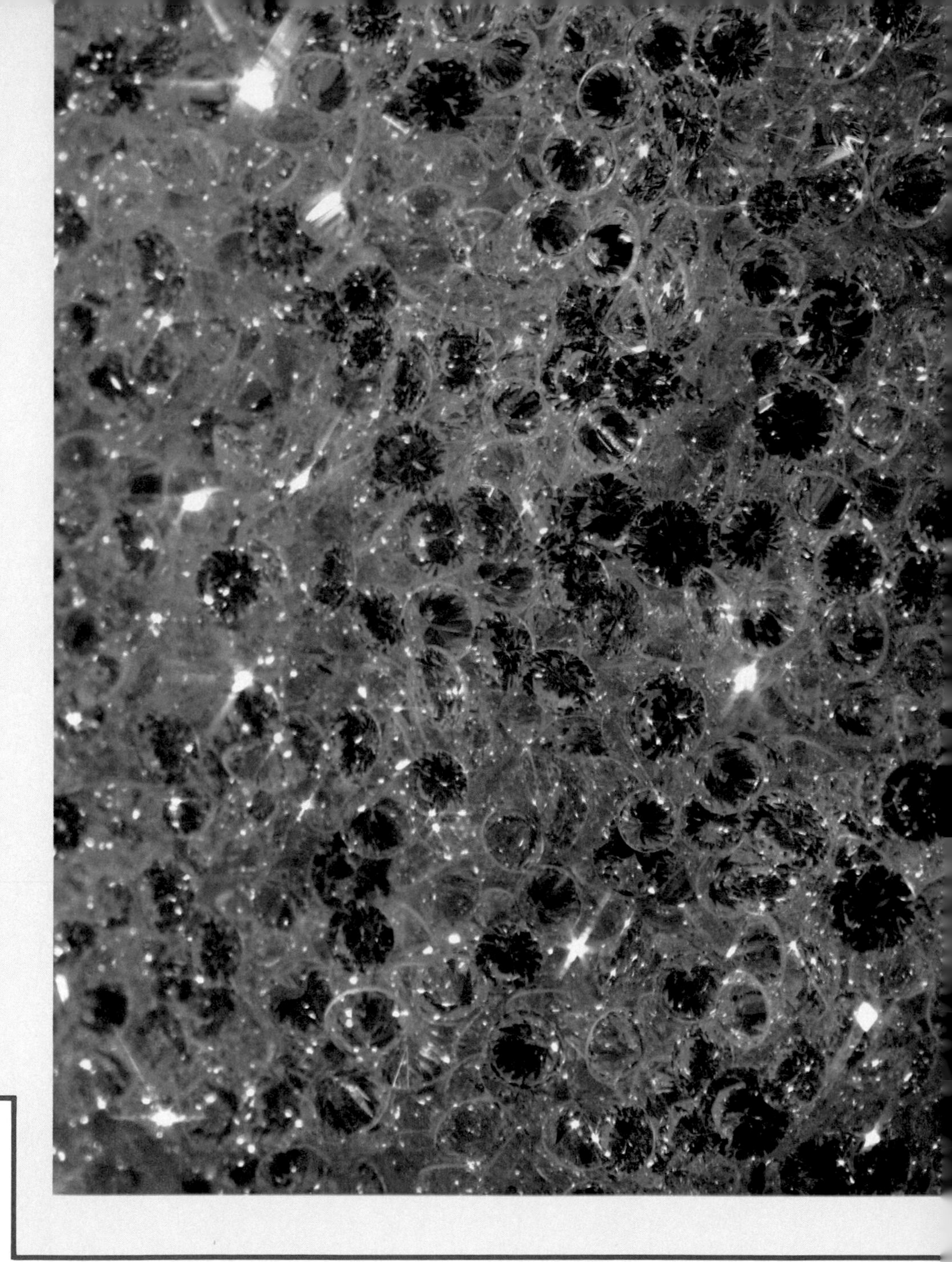

## INTRODUCING CHAPTER 18

Have students observe the photograph on page 414. Point out that it is a photograph of diamonds.

• **How do you know if a mineral is a diamond?** (Most students will suggest seeing if it cuts glass.)

Point out that it is true that diamonds will cut glass, but so will other substances: corundum, topaz, and artificial diamonds. Glass has a hardness of 7. Diamonds have a hardness of 10. Any substance with a hardness greater than 7 will scratch glass. There are tests that can be conducted by gemologist to determine if the substance in question is a diamond. These tests include tests for hardness, tests for refraction of light, and tests to determine the specific gravity of the substance.

• **What are diamonds used for?** (Accept all responses. Point out that one use of diamonds is as a gem—for jewelry and for investment and trading.)

Because diamonds are the hardest natural substance (10), they are used commercially as abrasives and cutting edges. Certain types of saw blades, drill bits, and sandpaper have small pieces of diamonds on their cutting edges.

Diamonds are also used for points on stereo needles because the diamond points' hardness does not easily wear down.

• **Where are most diamonds found?** (Most diamonds are found in South Africa. Although small deposits of diamonds have been found in Brazil,

# 18 Minerals

**CHAPTER SECTIONS**

**18-1 What Is a Mineral?**
**18-2 Uses of Minerals**

**CHAPTER OBJECTIVES**

*After completing this chapter, you will be able to:*

**18-1** Define the term mineral.
**18-1** Describe the properties of minerals.
**18-1** Identify minerals by their physical properties.
**18-2** Describe some uses of minerals.
**18-2** Compare metals and nonmetals.
**18-2** Compare ores and gems.

Throughout history, diamonds have been prized as a symbol of great wealth and power. Many of the world's largest diamonds decorate the crowns and jewelry of royalty. Take, for example, the Orloff diamond—if you can find it!

The story goes that this diamond was once in the eye of an idol in a Buddhist shrine in India. In the 1700s the diamond was stolen by a French soldier, who sold it to a British sea captain. After many thefts and some violent crimes, including murder, the diamond found its way to Russia. It was bought by Count Gregory Orloff from an Armenian merchant for about four million dollars. The Count gave it to Catherine II, Empress of Russia. Later, the diamond disappeared during the Russian Revolution.

Other diamonds also have disappeared before their owners' eyes—not through magic, but through ignorance. Early diamond hunters in South America thought they had a simple test for diamonds. They knew that diamonds are the hardest natural substance known. When the hunters found a stone that looked like a diamond, they hit it with a hammer. Usually, the stone shattered into bits and pieces. To the hunters, the shattering just proved the object was worthless. Actually, the stone may have been a diamond.

What the hunters did not know was that diamonds are brittle. Tap a diamond with a knife edge at exactly the right angle and it breaks cleanly and beautifully along an absolutely straight line. Tap it at the wrong angle and the diamond shatters.

Why do diamonds and other precious stones behave in this way? The answer is just a part of the fascinating story of minerals.

*Diamonds—one of the most precious minerals*



## TEACHER DEMONSTRATION

Place in the front of the room or list on the chalkboard some of the items in your classroom that are made of minerals. Then ask students if they can add to the list.

- **What kinds of minerals are found in these items?** (Answers will vary depending on items selected but may include: window glass—quartz; steel chair—hematite; pencil lead—graphite; trays—aluminum; pennies—copper; nickels—nickel.)
- **How would our lives be different if we had no minerals or mineral products?** (Accept all logical answers but lead students to the idea that minerals are extremely useful in our daily lives.)

## TEACHER RESOURCES

**Audiovisuals**

*Investigating Rocks: Recognizing Rock-Making Minerals,* filmstrip with cassette, EBE
*Minerals (Problem of Conservation series),* 16 mm film, EBE
*Rocks and Their Minerals,* 20 slides, SVE
*Rocks and Minerals,* 16 mm film, BFA

**Books**

Desautels, Paul E., *Rocks and Minerals,* Grosset and Dunlap
Dietrich, R. V., and B. J. Skinner, *Rocks and Rock Minerals,* Wiley
Frye, Keith, *Modern Mineralogy,* Prentice-Hall
Mason, Anita, *The World of Rocks and Minerals,* Larousse

India, and the U.S. [California and Arkansas], the largest producer of commercial and gem-grade diamonds is still South Africa.)

- **Is a diamond shiny when found in its natural setting?** (No. To get the beautiful, lustrous shine and shape we associate with diamonds, they must be carefully cut.)

Because diamonds, like all gems, break along cleavage lines, extreme caution must be taken while cutting them. For a large diamond, a gem cutter may spend several weeks determining where to make the first cut or "hit." Diamonds are cut by placing a very sharp knife along the angle of cleavage and tapping or hitting it with a small wooden mallet.

Although all diamonds are thought of as being shiny and clear, this is not true. The clarity, or clearness, of a diamond is reflected in its price or usage. Many cloudy or yellow diamonds are used for commercial purposes. Because some diamonds do not possess the clarity of others the size or weight does not necessarily determine the value. A clear, well-cut diamond may be twenty times more valuable than a cloudy or yellow diamond of equal weight. Therefore, it is imperative before purchasing a diamond as a gem to have it appraised by a qualified gemologist.

# 18-1 WHAT IS A MINERAL?

## SECTION PREVIEW 18-1

Minerals are composed of one or more elements. The large number of naturally occurring elements form more than 2000 different minerals. For a substance to be classified as a mineral, it must be an inorganic solid that occurs naturally in the earth. It must also have a definite chemical composition and an orderly atomic structure. Minerals are the building blocks of rocks on the earth. About 12 minerals make up most of the earth's crust.

The atoms of the elements that make up minerals combine in a fixed proportion and in a definite geometric pattern. The orderly arrangement of the atoms determines the crystal shape of a mineral. There are six basic crystal shapes: isometric, tetragonal, orthorhombic, monoclinic, triclinic, and hexagonal.

## SECTION OBJECTIVES 18-1

1. **Define the term mineral.**
2. **Identify minerals by their physical properties.**
3. **Describe some uses of minerals.**

## SCIENCE TERMS 18-1

**mineral p. 416**
**inorganic p. 416**
**crystal p. 416**
**magma p. 417**
**luster p. 418**
**hardness p. 419**
**Mohs hardness scale p. 419**
**streak p. 421**
**cleavage p. 422**
**fracture p. 422**

## 18-1 What Is a Mineral?

The substance you just read about—diamond—is a mineral. A **mineral** is a naturally occurring substance formed in the earth. A mineral may be made of a single element, such as copper, gold, or sulfur. Or a mineral may be made of two or more elements chemically combined to form a compound. For example, the mineral halite is the compound sodium chloride, which is made of the elements sodium and chlorine.

In order for a substance to be called a mineral, it must have five special properties. The first property of a mineral is that it is an **inorganic** substance. Inorganic substances are not formed from living things or the remains of living things. Calcite, a compound made of calcium, carbon, and oxygen, is a mineral. ① It is formed underground from water containing these dissolved elements. Coal and oil, although found underground, are not minerals because they are formed from decayed plant and animal life.

The second property of a mineral is that it occurs naturally in the earth. Steel and cement are manufactured substances. So they are not minerals. But gold, silver, and asbestos, all of which occur naturally, are minerals.

The third property of a mineral is that it is always a solid. The minerals you just read about—halite, calcite, gold, silver, and asbestos—are solids.

The fourth property of a mineral is that, whether it is made of a single element or a compound, it has a definite chemical composition. The mineral silver is ② made of only silver atoms. The mineral quartz is made entirely of a compound formed from the elements silicon and oxygen. So even though a sample of quartz may contain billions of atoms, its atoms can only be silicon and oxygen joined in a definite way.

The fifth property of a mineral is that its atoms are arranged in a definite pattern repeated over and ③ over again. This repeating pattern of atoms forms a solid called a **crystal.** A crystal has flat sides, or faces, that meet in sharp edges and corners. All minerals have a characteristic crystal shape.

**Figure 18-1** Chalk (top), *aragonite* (center), *and pearl* (bottom) *are all made of calcium carbonate or limestone. But only aragonite is a mineral. Why?* ①

416

## TEACHING STRATEGY 18-1

### Motivation

Display samples of such common minerals as salt, mica, sulfur, quartz, gold, silver, copper, and graphite. Have students observe the samples and note the characteristics of each one. Also have students note ways in which some of the substances are similar. For example, gold, silver, and copper are all metals; salt and sulfur are powdery; mica and quartz are glassy and have an obvious crystal structure.

### Content Development

Emphasize the definition of a mineral as a substance that is not live and was never part of a living thing. Some students may be familiar with the terms *organic* and *inorganic*. Explain that a mineral is an inorganic substance, because it was not formed from a living thing, or the remains of a living thing.

Point out that although certain types of minerals such as precious metals and gems have similar properties, the entire group of substances known as minerals is amazingly diverse. The diversity of minerals is evidenced by the fact that minerals are found in almost every aspect of our lives.

### Content Development

Review the five characteristics that all minerals must meet. Point out that substances such as coal and oil are

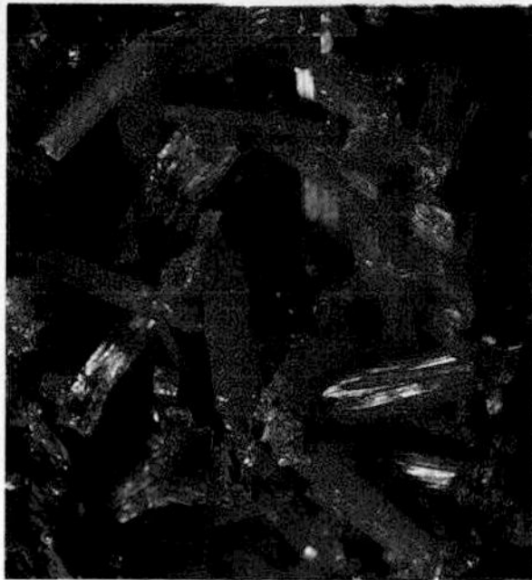

Figure 18-2 *Crystals are found in many different shapes, such as six-sided cubes of fluorite* (top right), *radiating crystals of wavelite* (top left), *and needlelike structures of croccolite* (bottom).

There are more than 2000 different kinds of minerals. But all minerals have the five special properties you just read about. Using these five properties, you can now define a mineral in a more scientific way. **A mineral is a naturally occurring, inorganic solid that has a definite chemical composition and crystal shape.**

### *Formation and Identification of Minerals*

Almost all minerals come from the material deep inside the earth. This material is hot liquid rock called **magma.** When magma cools, mineral crystals are formed. How magma cools and where it cools determine the size of the mineral crystals.

When magma cools slowly beneath the earth's crust, large crystals form. When magma cools rapidly beneath the earth's crust, small crystals form. Sometimes the magma reaches the surface of the earth and cools so quickly that crystals do not form at all. ❹

Crystals may also form from a mineral dissolved in a liquid. When the liquid evaporates, or changes to a gas, it leaves behind the mineral as crystals. Here too, the size of the crystals depends on the speed of evaporation.

Because there are so many different kinds of minerals, it is not an easy task to tell one from another. However, minerals have certain physical properties that can be used to identify them.

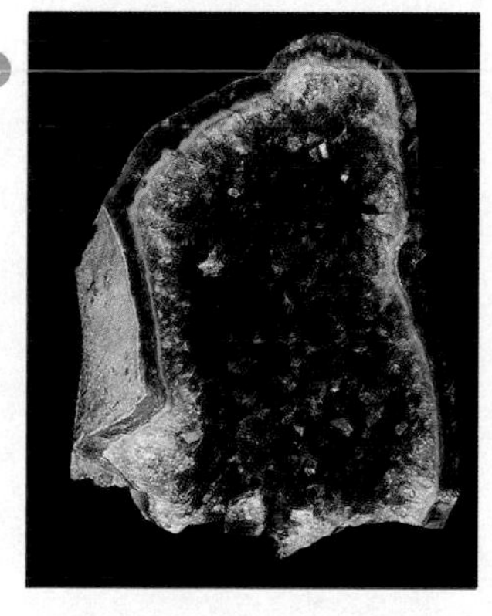

Figure 18-3 *A geode is a rock whose hollow interior is lined with mineral crystals, usually quartz, formed by evaporation.*



## ANNOTATION KEY

❶ **Chalk and pearl are formed from living things. A mineral, such as aragonite, must be inorganic. (Applying concepts)**

❶ Thinking Skill: Applying definitions

❷ Thinking Skill: Relating properties

❸ Thinking Skill: Identifying patterns

❹ Thinking Skill: Identifying processes

## TEACHER DEMONSTRATION

Heat enough water to fill 3 Styrofoam cups. Fill each cup with hot water and place a thermometer in each. Record the starting temperature. Pour the heated water in cup #1 into a flat tray. On cup #2, place a lid. On cup #3, place a lid and wrap it with a towel. Wait 10 minutes and check the temperatures. Have students predict the results.

- **Which cup will have cooled the most?** (cup #1)
- **Why did it cool the fastest?** (exposure to air)
- **How might this compare to minerals being formed in and on the earth's surface?** (The more insulation, the longer it takes to cool. The longer it takes to cool, the larger the crystal of the same mineral.)

found naturally occurring on the earth but that they are not minerals since they do not have a crystal shape and are organic substances made from the remains of once-living things.

### Content Development

Explain to students that most minerals come from hot molten rock called magma. Magma that makes it to the earth's surface is called lava. Not all magma makes it to the earth's surface—most remains below. Magma that remains deep within the earth's surface cools slowly, and this allows the crystal time to grow larger than it would if it cooled quickly.

### Motivation

Have students add salt to a beaker of water until no more salt will dissolve. Gently heat the solution or leave the beaker uncovered for a couple of days. As the water evaporates, crystals will form. Have students look at the crystal shape under a microscope or other magnifier.

### Reinforcement

Show students various types of minerals that exhibit crystal shapes. Review the basic characteristics of minerals as students examine your samples.

## BACKGROUND INFORMATION

One way to identify a mineral is by its color. However, color is one of the most variable characteristics of many minerals. Many minerals have several different colors. A mineral may also vary in color because of the impurities from other elements contained in the mineral.

Luster refers to the appearance of the unweathered surface of a mineral and the way in which it reflects light. Minerals may have metallic or nonmetallic lusters.

A mineral's hardness, or resistance to scratching, is another physical property that can be used for identification.

Another useful property is the streak of a mineral because the streak of a mineral is always the same. For example, hematite is always reddish-brown. Many minerals have a streak that is different from the color of the actual specimen.

Heft refers to the comparison of the density of two minerals by holding one mineral in each hand. The mineral that is more dense will seem heavier than the other. When minerals are similar in density, scientists use specific gravity to determine the actual density of the minerals.

The type of crystal formed by a mineral can also be used to identify a mineral. Other properties are cleavage and fracture.

**Figure 18-4** *The color of a mineral is not always a reliable way to identify it. Here you see the mineral quartz in three different varieties: citrine quartz* (top left), *smoky quartz* (top right), *and amethyst* (bottom). *Note that each variety has a characteristic color.*

### Color

The color of a mineral is an easily observed physical property. But color can be used to identify only those few minerals that always have their own characteristic color. The mineral malachite is always green. The mineral azurite is always blue. No other minerals look quite the same as these.

Many minerals, however, come in a variety of colors. The mineral quartz is usually colorless. But it may be purple, yellow, or pink. You would not want to rely on color alone to identify such minerals.

Color is not always a reliable way to identify minerals for another reason. The color of most minerals changes. For example, minerals such as silver and copper turn color when they tarnish. Tarnish forms when the surface of a mineral reacts chemically with oxygen in the air. Rain, heat, cold, and pollution can also change the color of a mineral.

### Luster

The **luster** of a mineral describes the way a mineral reflects light from its surface. Certain minerals reflect light the way highly polished metal does. Such minerals, which include silver, copper, gold, and graphite, have a metallic luster. The mineral in Figure 18-5 has a metallic luster.

**Figure 18-5** *These crystals of pyrite have a metallic luster. How would you describe this luster?* ❶

## 18-1 (continued)

### Motivation

Hold up several different minerals and ask students,

- **Are these minerals the same?** (no)
- **What are these minerals?** (Most students will not know the identity of the various minerals.)

Explain to students that all minerals have certain physical properties. Once these distinct properties are known they can be used to aid in identification. The most common physical properties used to identify minerals are:

1. Color
2. Luster
3. Hardness
4. Streak
5. Density
6. Crystal shape
7. Other special properties

These physical properties are observed and measured by certain tests. These tests will be explained throughout this section.

### Content Development

Explain to students that color can be used to identify only certain specific minerals such as malachite (green), azurite (deep blue), and realgar (red). Color is often caused by impurities in the mineral. The amount and type of impurities can cause distinct color differences. For example, quartz in its pure state is colorless but when iron is added, it may take on a deep-violet color. Manganese will turn quartz to a pink-red color. Other impurities can cause quartz to be blue, brown, yellow, and other col-

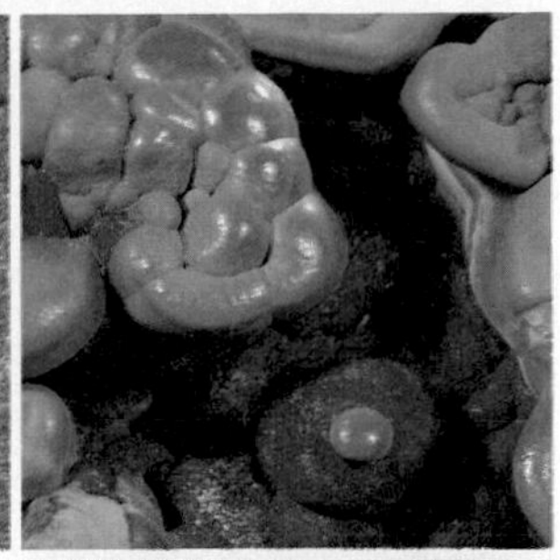

**Figure 18-6** *Some minerals have a nonmetallic luster. Tourmaline has a glassy luster* (top left). *Diamond has a brilliant luster* (top center). *Malachite is described as having a silky luster* (top right), *while mica has a pearly luster* (center). *Serpentine looks as if it is covered by a thin layer of oil. It has a greasy luster* (bottom).

Minerals that do not reflect much light often appear dull. Such minerals have a nonmetallic luster. Their appearance is often described as glassy, pearly, silky, greasy, or brilliant. Quartz and tourmaline have a glassy luster. They appear transparent or partly transparent. Mica has a pearly luster. Malachite has a silky luster. Serpentine looks as if it is covered with a thin layer of oil. It has a greasy luster. Diamond has a brilliant luster. As the rays of light are reflected from diamond, they break up into sparkles and flashes of color. ①

### *Hardness*

The ability of a mineral to resist being scratched is known as its **hardness.** Hardness is one of the most useful properties for identifying minerals. Friedrich Mohs, a German mineralogist, worked out a scale of hardness for minerals. He used ten common minerals and arranged them in order of increasing hardness. The number 1 is assigned to the softest mineral, talc. Diamond, the hardest of the ten minerals, is given the number 10. Each mineral will scratch any mineral with a lower number and will be scratched by any mineral with a higher number. Figure 18-7 shows the minerals of the **Mohs hardness scale** with their assigned numbers. ②



## ANNOTATION KEY

❶ **Reflects light the way highly polished metal does (Relating concepts)**

① **Thinking Skill: Identifying patterns**

② **Thinking Skill: Applying concepts**

## FACTS AND FIGURES

A few tiny zircon crystals found in western Australia are 4.2 billion years old. These are the oldest minerals ever found on the earth.

## SCIENCE, TECHNOLOGY AND SOCIETY

Minerals are important natural resources with a variety of uses. Unfortunately, minerals are nonrenewable and the demand for them increases every day. As a result, recycling has not only become a way of conserving minerals but a way of helping the economy.

In the past, aluminum soft-drink cans, which are an important source of the mineral bauxite, were discarded. Today many areas of the country are setting up recycling centers to conserve this mineral. In 1986, of the 69 billion aluminum soft-drink cans that were made, 33.3 billion cans were recycled. A large industrial company paid recyclers nearly $93 million for the recycled aluminum. In turn, this aluminum produced approximately 8 billion new soft-drink cans! In some areas, people receive 5 cents per soft-drink can. In other areas, a person can receive 61 to 88 cents for one kilogram. As you can see, recycling is beneficial to society and nature.

- **Why are companies paying people to recycle products?** (Answers will vary but should include the idea that society needs to be encouraged to participate in recycling efforts.)
- **How many students participate in recycling efforts?** (Answers will vary.)
- **What products in your home can be recycled?** (Answers will vary, but may include aluminum foil, tin cans, newspapers, plastic containers, and glass.)

ors. The amount of the impurities will also affect the shade of the color of quartz.

## Skills Development

***Skill: Hypothesizing***

- **Why is the color of a mineral not used as a single method of identification?** (Many minerals have several color varieties. Some minerals tarnish or change color when exposed to air.)

## Skills Development

***Skill: Observing and classifying***

Pass out various minerals such as galena, pyrite, quartz, chalcopyrite, hematite, biotite, and orthoclase to the students.

- **Describe the luster of the various minerals.** (Accept all possible answers.)

## BACKGROUND INFORMATION

Students sometimes confuse color and streak. Streak is the color of the powdered mark made by a mineral when it is rubbed against a hard surface such as porcelain. The color of the streak is not always the same as the external color of the mineral. For example, the outer color of chalcopyrite is gold, but it leaves a greenish-black streak.

| MOHS HARDNESS SCALE | |
|---|---|
| **Mineral** | **Hardness** |
| Talc | 1 |
| Gypsum | 2 |
| Calcite | 3 |
| Fluorite | 4 |
| Apatite | 5 |
| Feldspar | 6 |
| Quartz | 7 |
| Topaz | 8 |
| Corundum | 9 |
| Diamond | 10 |

**Figure 18-7** *The Mohs hardness scale is a list of ten minerals that represent different degrees of hardness. Each mineral on the scale is harder than the minerals it scratches and softer than the minerals that scratch it. Which mineral is the hardest? The softest?* ❶

To determine the hardness of an unknown mineral, the mineral is rubbed against the surface of each mineral in the hardness scale. ❶ If the unknown mineral is scratched by the known mineral, it is softer than the known mineral. If the unknown mineral scratches the known mineral, it is harder than that mineral. If two minerals do not scratch each other, they have the same hardness.

Suppose that you have a mineral sample that is scratched by fluorite, number 4, but not by calcite, number 3. The sample, however, scratches gypsum, ❷ number 2. Using the Mohs scale, what mineral could your sample be made of? You are right if you say calcite or any mineral that has a hardness of 3.

It is not always possible to have the minerals of the Mohs hardness scale with you. In such cases, a field scale is convenient to use. Although a field scale is not as exact as the Mohs scale, the materials it uses are easily obtained. Figure 18-8 shows a field hardness scale.

### Streak

The color of the powder left by a mineral when it is rubbed against a hard, rough surface is called its

**Figure 18-8** *A field hardness scale can be used when the minerals from the Mohs scale are not available. What mineral sample could be scratched by a penny but not by a fingernail?* ❷

| FIELD HARDNESS SCALE | |
|---|---|
| **Hardness** | **Common Tests** |
| 1 | Easily scratched with fingernail |
| 2 | Scratched by fingernail (2.5) |
| 3 | Scratched by a penny (3) |
| 4 | Scratched easily by a knife, but will not scratch glass |
| 5 | Difficult to scratch with a knife; barely scratches glass (5.5) |
| 6 | Scratched by a steel file (6.5); easily scratches glass |
| 7 | Scratches a steel file and glass |



## 18-1 (continued)

### Content Development

Point out that luster of a mineral is due to the way light is reflected from that mineral. Luster is different from color. Minerals of the same color may have different lusters. Common terms such as greasy, oily, waxy, or silky may be used to describe the luster of minerals. The two major categories of luster are metallic (like that of a shiny metal surface) or nonmetallic or dull luster. Other terms such as vitreous (glassy), adamantine (gemlike), resinous (resinlike), or earthy are used to describe the various lusters.

To use luster as an identification method, observe a freshly broken surface. Because, like color, the luster of a mineral is affected by tarnish.

### Motivation

Ask students to use the field hardness scale and explain why minerals other than diamonds scratch glass. (Glass has a hardness of 5.5–6.)

Have students list some common minerals that scratch glass. (quartz, topaz, vanadinite, garnet)

### Content Development

Point out that on the Mohs hardness scale the number 10 assigned to diamond does not mean it is 10 times harder than talc. It only means it is the hardest mineral.

### Skills Development

***Skill: Applying concepts***

To help students remember the order of the minerals on the hardness scale have them make up a sentence using the first letter of each mineral's name.

### Reinforcement

Have students work in pairs. For each pair of students, prepare two sets of 10 index cards. On one set, write the name of a mineral on the Mohs hardness scale on each card. On the second set, write a number from 1 to 10 on each card. Have student pairs see how quickly they can match the mineral names with the correct number on the Mohs hardness scale.

### Content Development

- **Why is a streak test a better method of identifying minerals than the**

**Figure 18-9** *Talc is a mineral that leaves a white streak* (left). *Graphite is a mineral that leaves a gray-black shiny streak* (right). *Graphite mixed with clay is the "lead" used in pencils.*

**streak.** Streak color can be an excellent clue to identifying minerals that have a characteristic streak color and are fairly soft. Even though the color of a mineral may vary, its streak is always the same. Yet this streak color is often different from the color of the mineral itself. See Figure 18-9. ❸

Streak color is determined by rubbing the mineral sample across a piece of unglazed porcelain. The back of a piece of bathroom tile or a streak plate is good to use. A streak plate has a hardness slightly less than 7. So a streak test is useful only with minerals whose hardness is less than 7. Harder minerals do not leave a streak.

Many minerals have white or colorless streaks. Talc, gypsum, and quartz are examples. Streak is not a useful physical property in identifying minerals such as these. Some other physical property must be used.

## Density

Every mineral has a property called density. Density is the amount of matter in a given space. Density can also be expressed as mass per unit volume. The density of a mineral is always the same, no matter what the size of the sample is. Because each mineral has a characteristic density, one mineral can easily be compared with any other mineral.

### Activity

*Mineral Hardness*

You can become familiar with the field hardness scale by doing this activity.

**1.** Obtain five different mineral samples, a penny, a penknife, a piece of glass, and a steel file.

**2.** Test the minerals to find their hardness according to the scale in Figure 18-8. Most minerals will probably fall between 3 and 7 on the scale.

**3.** Inside a shallow box, draw a chart listing the minerals by name in order of increasing hardness. Attach the minerals to the chart next to their names.

421

## ANNOTATION KEY

❶ Diamond, talc (Interpreting charts)
❷ Calcite (Interpreting charts)
❶ Thinking Skill: Applying concepts
❷ Thinking Skill: Relating concepts
❸ Thinking Skill: Making observations

## Activity

**Mineral Hardness**
**Skills: Observing, comparing, relating, applying, diagraming**
**Level: Enriched**
**Type: Hands-on**
**Materials: Five different mineral samples, penny, penknife, glass, steel file, shallow box**

If five different minerals cannot be found by students in your local area, supply students with various minerals obtained from science supply houses. Have each group of students compare their results with other groups to see if they agree on the hardness of each mineral specimen. Then have them compare the estimates with a mineral hardness chart.

**color of the mineral?** (The streak of a mineral is its color when made into a fine powder. Although the color of the mineral may vary, the streak will remain the same.)

Point out that streak tests work best on identification of dark colored minerals. The hardness of a mineral being tested should be less than 6.5 because the streak plate has a hardness of 7. Some streak colors of common minerals are pyrite—greenish-black, hematite—cherry-red, limonite—yellow-brown, sulfur—white, galena—lead-gray.

## Motivation

Have students compare the weight of approximately equal-sized samples of galena to quartz.

- **Which one feels heavier?** (galena)
- **Why do equal volumes of different minerals weigh differently?** (The density of each is different.)

## Content Development

Most nonmetal minerals are less dense than metallic minerals.

## Motivation

Have students compare the densities of several specimens by hefting them.

## Skills Development

***Skill: Making a model***

Have students work in small groups. Ask each group to prepare a demonstration of the property of heft, using everyday objects that are the same size and shape, but different weights. For example, a group might display a bowling ball and a balloon; a Ping-Pong ball and a golf ball; a brick and a Styrofoam block.

## Activity

**Crystal-System Models**
**Skills: Applying, relating, manipulative, making models**
**Level: Remedial**
**Type: Hands-on**
**Materials: Straws, note cards**

This activity will help reinforce students' understanding of the different crystal systems by having them construct various crystal-system models. You may also want to have students find a photograph of an example of each crystal system. They can then attach the photograph to the appropriate note card.

## BACKGROUND INFORMATION

The classification of crystals into six systems is based on the relative lengths of the three intersecting axes and the angle of intersection.

## HISTORICAL NOTES

It was not until the development of X-ray analysis in 1912 that the internal atomic structure of crystals could be studied.

### Activity

*Crystal-System Models*

Make the six crystal systems easier to understand by doing this activity.

**1.** With some straws, construct the six basic crystal systems. Use Figure 18-10 to help you.

**2.** Label the straw model of each system with a note card.

You can compare the densities of two minerals of about the same size by picking them up and hefting them. The denser mineral feels heavier.

### Crystal Shape

Most minerals form crystals, or solids that have a definite geometric shape. The shape of a crystal results from the way the atoms or molecules of a mineral come together as the mineral is forming. Each mineral has its own pattern of atoms or molecules. So each mineral has its own crystal shape.

There are six basic shapes of crystals, or crystal systems. Each shape has a number of faces, or flat surfaces, that meet at certain angles to form sharp edges and corners. See Figure 18-10.

### Cleavage and Fracture

The way a mineral breaks is called **cleavage** or **fracture.** When a mineral breaks along smooth, definite surfaces, cleavage occurs. Cleavage is a characteristic property of a mineral. Halite, for example, always cleaves in three directions. It breaks into small cubes. Mica cleaves along one surface, making layer after layer of very thin sheets.

When a mineral breaks unevenly, fracture occurs. The fracture surfaces are usually rough or jagged. Like cleavage, fracture is a property that helps identify a mineral.

**Figure 18-10** *Garnet, corundum, sulfur, crocoite, anatase, and kyanite illustrate the six basic crystal systems, which are due to the way the atoms in a crystal bond.*

CRYSTAL SYSTEMS

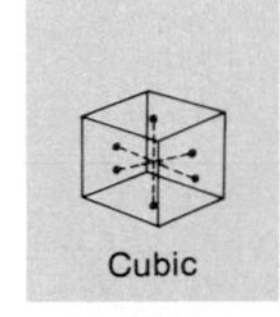

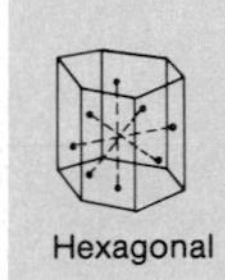

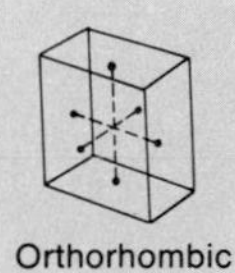

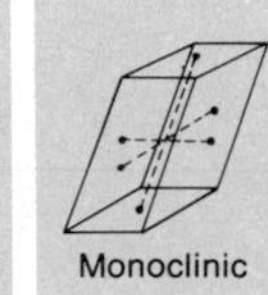

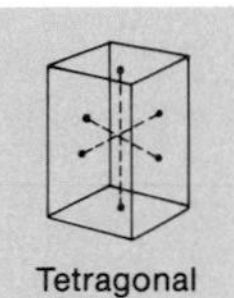

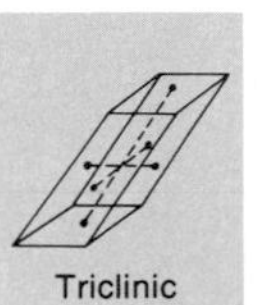


Garnet

Corundum

Sulfur

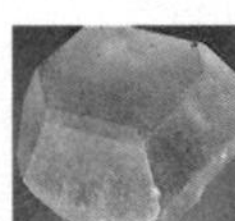

Crocoite

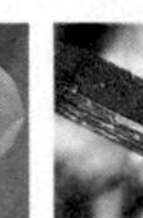

Anatase

Kyanite

Courtesy Carolina Biological Supply Company

## 18-1 (continued)

### Content Development

Crystal forms, or patterns, are formed by the internal atomic structure. This structure could not be studied until the early 1900s. At that time, early analysis of X-ray defraction was used. Now, more accurate equipment is being used. This internal pattern is responsible for the regular geometric shapes that crystals take. It is also responsible for other physical properties such as hardness, fracture, and cleavage. Because this internal structure is the same in all specimens of a mineral, the crystal shape remains the same.

### Skills Development

***Skill: Making observations***

Have students look at the mineral halite (table salt) under a hand lens. Also provide them with a quartz crystal.

- **To what crystal system would halite belong?** (cubic)
- **To what crystal system would quartz belong?** (hexagonal)

### Content Development

The way a mineral breaks is called cleavage, or fracture. Explain that a smooth uniform break is called a fracture. The break occurs between rows of the internal atomic structure. A good example of cleavage is mica, which cleaves in a single direction breaking into thin sheets. Galena, calcite, and halite cleave in three directions breaking into small cubes. Minerals such as sulfur and dolomite fracture. Some fractures are conchoidal (like a shell), hackly (jagged with sharp edges), or fibrous (like splinters). Explain to your students that

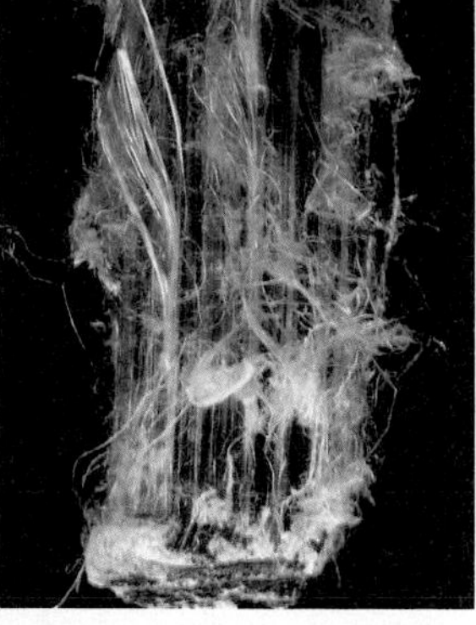

Figure 18-11 *Many minerals break into different shapes. Galena cleaves into cubes* (left). *Hematite cleaves into thin sheets* (center). *Asbestos fractures into fibers* (right).

### *Special Properties*

Some minerals can be identified by special properties. Magnetite, a mineral made of iron and oxygen, is naturally magnetic. Fluorite, made of calcium and fluorine, glows when put under ultraviolet light. Halite, common table salt, has a special taste. ❸ Sulfur has a distinct smell. Calcite fizzes when hydrochloric acid is added to it. And jade has a bell-like ring when tapped.

### Career: *Gem Cutter*

**HELP WANTED: GEM CUTTER** Technical school coursework in gem cutting and jewelry making required. High school graduate preferred. Experience helpful but on-the-job training available.

Among the most valuable treasures obtained from the earth are gems. These hard, beautiful stones include diamonds, rubies, and emeralds. However, rough gems do not look very much like the stones used in jewelry. To become sparkling gems they must be cut and polished. This work is done by **gem cutters.**

Gem cutting requires a great deal of patience and concentration. It also requires a great deal of skill. If a gem cutter makes a mistake, a stone that could have had a value of hundreds of thousands of dollars may suddenly be worthless.

People who want to become gem cutters begin as apprentices. They learn their trade through on-the-job training. At first, apprentices only cut inexpensive stones. As they become more skillful, they work on more valuable stones.

You can learn more about gem cutting by writing to the Gemological Institute of America, 1660 Stewart Street, Santa Monica, CA 90400.



fracture is an important identifying characteristic.

## Content Development

Many minerals have special properties. Halite, epsomite, and borax all have a special taste. To be identified by its taste, the mineral must be soluble in water. The use of taste should be discouraged because some minerals can be poisonous, such as chalcanthite (hydrous copper sulfate).

Certain minerals, such as pyrite and arsenopyrite, have characteristic smells. Some minerals react to acids. This is important in identifying minerals of the carbonate family. Other minerals, such as uranium and thorium, are radioactive.

Fluorescence is also a special property exhibited by some minerals. These minerals will glow red, blue, green, or orange when exposed to (U.V.) ultraviolet light.

## FACTS AND FIGURES

The largest recorded crystal of gem quality was a 520,000-carat aquamarine found in Brazil in 1910. A carat is a unit of weight for gems equal to 200 milligrams. This crystal produced more than 200,000 carats of cut gems. (A diamond engagement ring usually weighs one carat or less.)

## ANNOTATION KEY

❶ Thinking Skill: Classifying crystal shapes

❷ Thinking Skill: Making observations

❸ Thinking Skill: Identifying properties

## TEACHER DEMONSTRATION

Use ball-and-stick models to illustrate the six basic crystal shapes shown on page 422. Emphasize that the shape of a crystal results from the way atoms or molecules of a mineral are bonded together. You may want to have students look at the drawings in Figure 18-10 as you display the models. Point out that in an actual crystal, each structure would be tightly packed with atoms or molecules.

Have students describe, in their own words, each shape as you display the model.

- **What are the characteristics of a cubic crystal?** (All six faces are squares, and all six faces have equal dimensions.)
- **In a cube, are all the faces at right angles to one another?** (yes)
- **Are there some crystal shapes in which the faces are not at right angles to one another?** (yes)
- **What are these?** (monoclinic and triclinic)
- **How would you describe the shape of a hexagonal crystal?** (The top and bottom of the crystal are hexagons; the six sides are rectangles.)

## FACTS AND FIGURES

Most gems are minerals, but some—including amber, coral, and pearl—come from living or once-living things.

## ANNOTATION KEY

❶ **Graphite (Interpreting charts)**
❷ **Hematite, bauxite (Interpreting charts)**
① **Thinking Skill: Making comparisons**
② **Thinking Skill: Applying definitions**
③ **Thinking Skill: Applying technology**

## 18-1 (continued)

### Enrichment

Have students work in small groups. Challenge each group to create a detective story in which the mystery centers around the identity or authenticity of a mineral. Groups can present their stories in written form or as dramatizations.

### Section Review 18-1

**1.** A naturally occurring substance formed in the earth.
**2.** Color, luster, hardness, streak, density, crystal shape, cleavage and fracture.
**3.** Cleavage: breakage that occurs along smooth, definite surfaces. Fracture: breakage that occurs along uneven surfaces.

***CHARACTERISTICS OF COMMON MINERALS***

| Mineral | Color | Luster/ Hardness | Cleavage or Fracture | Uses |
|---|---|---|---|---|
| Jade | White or green or greyish green | Glassy to silky 6.5-7 | Fracture | To make jewelry, vases, and figurines |
| Calcite | White or colorless | Glassy 3 | Perfect cleavage or fracture | In medicine and toothpaste; also found in marble and limestone, which are used as building material |
| ① Graphite | Black to iron grey | Metallic 1 | Perfect cleavage | In pencils and as a lubricant in machinery, clocks, and locks |
| Malachite | Bright green | Silky 3.5-4 | Perfect to fair cleavage or fracture | In jewelry and for table tops |
| Silver | Silver white | Metallic 3 | Fracture | In electrical equipment, photographic chemicals, and jewelry |

**Figure 18-12** *This table gives characteristics of some common minerals. Which mineral is easily scratched with a fingernail?* ❶

424

## TEACHING STRATEGY 18-2

### Motivation

Direct students' attention to the list of mineral uses in Figure 18-13. Ask,

- **How many of the items under "Uses" have you used in the last several days?** (Answers will vary.)
- **Which of the items listed do you have in your home?** (Answers will vary.)
- **Look around our classroom. Can you identify some items that contain minerals?** (Answers will vary.)

### Content Development

Although this section emphasizes minerals that are used as gems, many minerals have practical uses that are important to everyday life. Emphasize to students the diversity of these uses—from table salt to watch crystals to building materials. Also emphasize that different minerals have very different characteristics, which make them useful for different purposes.

### Motivation

Begin this section with a discussion.

- **Where did the metal in your car come from?** (steel or iron—hematite, aluminum—bauxite)
- **What economic advantage is gained by a country with rich mineral resources?** (Answers will vary.)
- **What are some economic disadvantages of reliance on a single industry such as steel production?** (Talk about the problems in the

### SECTION REVIEW

1. What is a mineral?
2. List seven physical properties of minerals.
3. What is the difference between the cleavage and fracture of minerals?

## 18-2 Uses of Minerals

Throughout history, people have used minerals. At first, minerals were used just as they came from the earth. Later, people learned to combine and process the earth's minerals. **Today many of the earth's minerals are used to meet the everyday needs of people.** Minerals are raw materials for a wide variety of products from dyes to dishes and from table salt to televisions.

### Ores

Minerals from which metals and nonmetals can be removed in usable amounts are called **ores.** **Metals** are elements that have certain special properties. Metals have shiny surfaces and are able to conduct electricity and heat. Metals also have the property of **malleability** (mal-ee-uh-BIL-uh-tee). Malleability is the ability of a substance to be hammered into thin sheets without breaking. Another property of metals is **ductility** (duhk-TIL-uh-tee). Ductility is the ability of a substance to be pulled into thin strands without breaking. Iron, lead, aluminum, copper, silver, and gold are metals. 2

Most metals are found combined with other substances, or impurities, in ores. After the ores are removed from the earth by mining, the metals must be removed from the ores. During a process called smelting, the ore is heated in such a way that the metal can be separated from it. For example, iron can be obtained from the ores limonite and hematite. 3 Lead can be processed from the ore galena. And aluminum comes from the ore bauxite.

Metals are very useful. Probably the most useful metal is iron, which is used in making steel. Lead is a metal used in pipes. Copper is a metal used in pipes,

**COMMON MINERALS AND THEIR USES**

| Minerals | Uses |
|---|---|
| Alum | Used in cosmetics, dyes, and water purification |
| Bauxite | Source of aluminum |
| Corundum | Used to make emery boards and to grind and polish metals |
| Feldspar | Used to make pottery, china, and glass |
| Halite | Table salt |
| Hematite | Source of iron |
| Quartz | Used in radios, television, and radar instruments |
| Sulfur | Used to make matches, medicine, rubber, and gunpowder |

**Figure 18-13** *According to this chart of common minerals and their uses, what mineral is the source of iron? Of aluminum?* 2



## 18-2 USES OF MINERALS

### SECTION PREVIEW 18-2

Metals such as iron, lead, aluminum, copper, silver, or gold can be characterized by their shiny surfaces, hardness, ability to conduct electricity and heat, and their malleability. Nonmetallic minerals include a wide range of substances that have a large number of uses in chemical compounds for industry and as building materials. Clay, cement, building stone, crushed stone, sand and gravel, gypsum, sulfur, salt, fluorite, barite, and asbestos are just a few examples of nonmetallic minerals.

Precious and semiprecious stones called gems are also minerals. Gems are durable, beautiful, and very rare. Only about 80 of the 2000 minerals have the qualities necessary to be called gems. Diamonds, emeralds, rubies, and sapphires are the best-known precious gems.

### SECTION OBJECTIVES 18-2

1. **List some common uses of minerals.**
2. **Define what ore is.**
3. **Explain the terms malleability and ductility.**
4. **Describe and name some precious and semiprecious gems.**

### SCIENCE TERMS 18-2

**ore** p. 425
**metal** p. 425
**malleability** p. 425
**ductility** p. 425
**nonmetal** p. 426
**gem** p. 426

"steel towns" and the loss of jobs as production of steel is shifted to other geographic areas.)

### Content Development

Explain that metals are elements that conduct electricity and heat, have shiny surfaces, and are malleable.

Metals are found in nature as ores. To extract the metal from the ores, heating or smelting must occur. Various metallic ores yield various metals.

### Skills Development

*Skill: Inferring*

Challenge students to answer the following question:

- **Why were diamonds used to cut and polish other gems?** (A diamond is the hardest mineral, with a Mohs rating of 10. As a result, a diamond can scratch—or cut any other mineral.)

### Enrichment

Ores are obtained by mining, and valuable minerals are removed from ores by smelting. Have students research these two processes and report their findings to the class.

## Activity

**Mineral Deposits**
**Skills: Diagraming, researching data, relating, applying**
**Level: Remedial**
**Type: Hands-on/library**

This activity will help reinforce students' ability to use reference materials and present their findings in a map or chart form. Students' answers should include: uranium: United States, Canada, South America, Africa, Europe, India, and Australia; native sulfur: Japan and along the Gulf of Mexico from Alabama to Mexico; bauxite (aluminum ore): United States, Canada, South America, Africa, Australia, India, and the U.S.S.R.; halite: Europe, South America, Middle East, U.S.S.R., and in the area of the Gulf of Mexico in the United States; gold: South Africa, U.S.S.R., Canada, Western Australia, Philippines, Japan, and Alaska and South Dakota in the United States.

**Figure 18-14** *The mineral aluminum can be obtained from the ore bauxite.*

### Activity

*Mineral Deposits*

This activity will help you find out where some of the major mineral deposits in the world are located.

**1.** From the library, find a map of the world. Draw or trace the map on a piece of paper. Label Africa, Asia, Europe, North America, South America, Australia, and Antarctica.

**2.** Find out where uranium, sulfur, aluminum, iron, halite, and gold deposits are located.

**3.** Using a symbol to represent each mineral, show the locations of these deposits on the map.

**4.** Make a key by writing the name of each mineral next to its symbol. Make your map colorful and descriptive.

pennies, and electrical wire. Aluminum is a metal used in the production of cans, foil, lightweight motors, and airplanes. Silver and gold are metals used in fillings for teeth. Silver and gold are also used in decorative objects such as jewelry.

**Nonmetals** are minerals that are not shiny, are poor conductors of electricity and heat, and are not malleable or ductile. Sulfur, asbestos, and halite are nonmetals.

Some nonmetals are removed from the earth in usable form. Other nonmetals must be processed to separate them from the ores in which they are found. For example, halite can be found in large deposits in usable form. But asbestos must be separated from other minerals, such as serpentine.

Nonmetals are also useful. Sulfur is one of the most useful nonmetals. It is used to make matches, medicines, and fertilizers. It is also used in iron and steel production.

### Gems

Some minerals are called **gems.** Gems are rare minerals. They are also beautiful and durable, or lasting. Not very many minerals have all of these qualities. The rarest and most valuable gems are called precious stones. Diamonds and emeralds are precious stones. Other gems are called semiprecious stones. Semiprecious stones are not as rare or valuable as precious stones. Amethysts, zircons, garnets, turquoises, and opals are semiprecious stones. They are all beautiful and durable. But they are more common than precious stones.



## 18-2 (continued)

### Content Development

Explain that nonmetals are poor conductors of heat and electricity and are not malleable. One of the most useful nonmetals is sulfur.

### Skills Development

***Skill: Applying technology***

Have students research the uses of sulfur and how sulfur is mined.

### Motivation

Demonstrate how metals can be obtained from ores. To do this put a small amount of black copper oxide and powdered charcoal into a test tube. Gently heat the test tube for several minutes, then pour the mixture into a beaker of water. The copper will settle to the bottom of the beaker and the charcoal will remain floating on top of the water.

### Content Development

Explain to students that gems are minerals that are cut, shaped, and polished from gemstones. The rarer the gemstone.

- **Why do people own gems?** (Accept all logical answers.)
- **What gem would you want to own and why?** (Accept all logical answers.)
- **Why are some gemstones semiprecious?** (not as rare as precious stones)

### Skills Development

***Skills: Observing, comparing***

Have students observe the photographs of cut and uncut gems in Figure 18-15. Have them compare the cut and uncut gems shown for each type. Discuss with students how the crystal shape of the uncut stone may influence the shape and appearance of the cut stone.

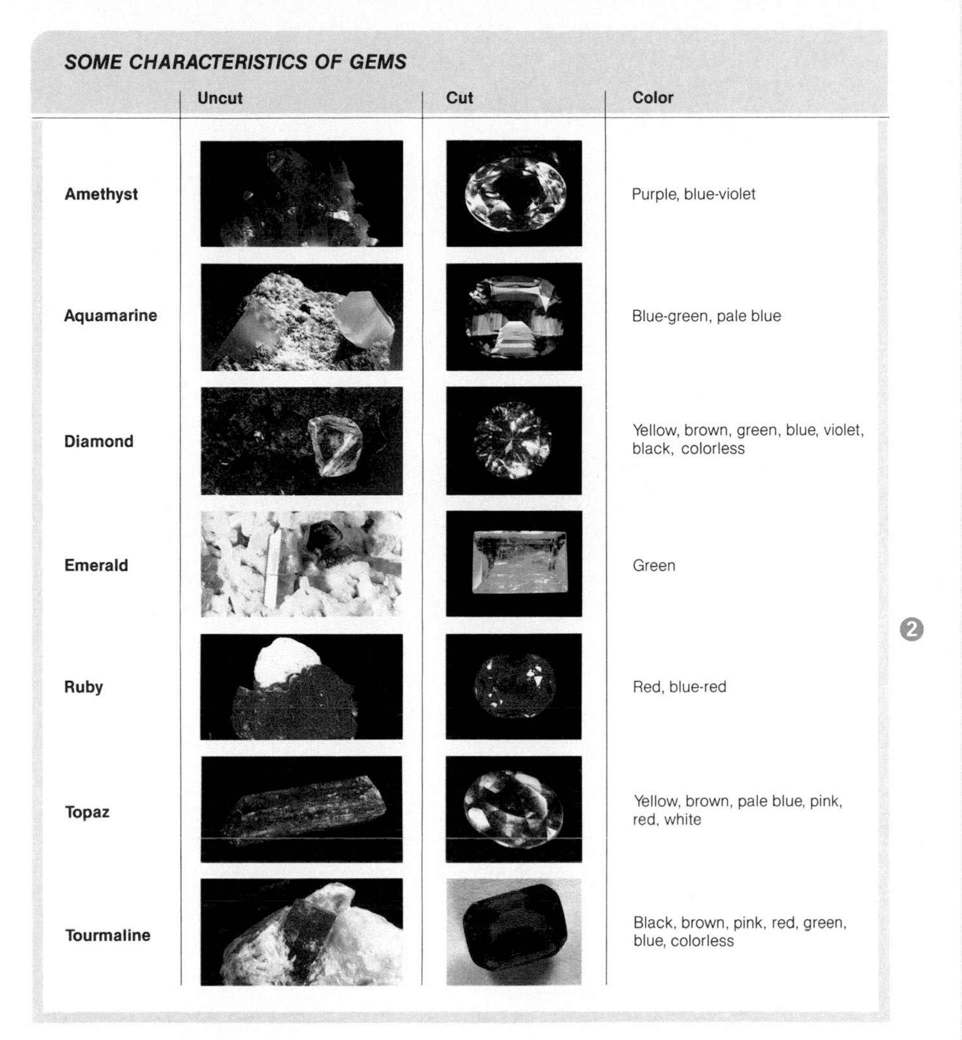

**SOME CHARACTERISTICS OF GEMS**

| | Uncut | Cut | Color |
|---|---|---|---|
| Amethyst | | | Purple, blue-violet |
| Aquamarine | | | Blue-green, pale blue |
| Diamond | | | Yellow, brown, green, blue, violet, black, colorless |
| Emerald | | | Green |
| Ruby | | | Red, blue-red |
| Topaz | | | Yellow, brown, pale blue, pink, red, white |
| Tourmaline | | | Black, brown, pink, red, green, blue, colorless |

❷

**Figure 18-15** *Very few minerals have the qualities that make them gems: beauty, rarity, and durability. Some gems in their uncut and cut forms are shown. Green is the characteristic color of what gem?* ❶

### *SECTION REVIEW*

1. What is an ore? A gem?
2. What is malleability? Ductility?
3. List some examples of precious and semiprecious stones.



## TEACHER DEMONSTRATION

Some mineral crystals form when a mineral is dissolved in a liquid, and then the liquid evaporates. You can demonstrate this concept to students by adding salt to a beaker of water until the solution is saturated. Then gently heat the solution until all of the water boils away. As the liquid changes into a gas, crystals will form. Have students take turns examining the crystals with a hand lens or under a microscope. You or your students may wish to experiment with different rates of boiling or evaporating the water to see what effect this has on the crystal size.

## ANNOTATION KEY

❶ **Emerald (Interpreting charts)**

❶ **Thinking Skill: Making comparisons**

❷ **Thinking Skill: Making comparisons**

Only about 20 mineral species qualify as gemstones. These account for about 35 of the most common gemstones. The value of gems is based on rarity and quality of the gemstone.

### Enrichment

Find out if a gem cutter is located in your area. If so, arrange for students to visit the shop and observe how stones are cut and polished. Students may also enjoy discussing with the gem cutter the career description found on the previous page, asking how it compares with his or her own experience.

### Enrichment

Many of the paints used by artists are made from minerals. Have students find out the names and colors of some pigments that contain minerals. If possible, contact the art department in your school or a local artist supply store to obtain samples of some of these paints.

### Content Development

Diamonds with a slight hint of yellow or brown are less valuable than those with no color. Diamonds with a blue or pink color are more valuable than those with no color at all.

### Reinforcement

Gemstones are minerals that are not beautiful. After they are cut, shaped, and polished, they are considered gems.

Gems are minerals that are rare, have durability, and possess beauty.

### Section Review 18-2

**1.** Mineral from which usable amounts of metals and/or nonmetals can be removed.
**2.** Ability of metals to be hammered into thin sheets without breaking; ability of metals to be pulled into thin strands.
**3.** Precious stones: diamonds, rubies, emeralds, sapphires. Semiprecious stones: opal, jade, turquoise, amethysts, malachite, zircon.

## LABORATORY ACTIVITY FORMING MINERAL CRYSTALS

## BEFORE THE LAB

1. **At least one day prior to the investigation, gather enough materials for your class, assuming six students per group.**
2. **Keep in mind that this activity will require two class periods about two days apart.**

## PRE-LAB DISCUSSION

Review with students the information in Section 18-1 about the crystal structure of minerals. Emphasize that each mineral has its own crystal shape, and that this is one way in which minerals are identified.

Discuss with students the use of the scientific method in this activity.

- **What factors are kept constant in this investigation?** (the amount of solid dissolved in the water, the amount of water, and the use of a petri dish and piece of dental floss)
- **What factor is the variable?** (the type of solid used)
- **How do you think the results for each type of solid will differ?** (Answers may vary; the differences will appear in the shape of each crystal type, and the time it takes for each type of crystal to form.)

## SAFETY TIPS

Caution students to be careful when handling glassware. Be sure that they do not handle any of the chemicals without the use of a spatula.

## TEACHING STRATEGY FOR LAB PROCEDURE

For #3 in Observations and Conclusions, you may wish to have students make a graph to compare the time of formation for each type of crystal.

## LABORATORY ACTIVITY

### *Forming Mineral Crystals*

**Purpose**

In this activity, you will grow different types of crystals.

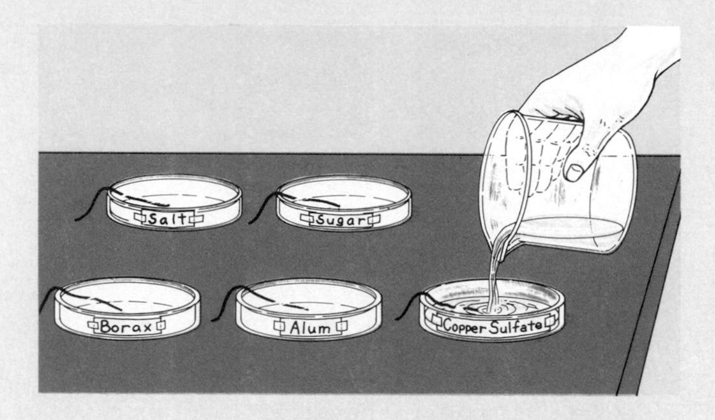


**Materials** *(per group)*

Table salt
Table sugar
Borax
Alum
Copper sulfate
250-mL beaker
Dental floss
Stirring rod
5 petri dishes
Magnifying glass or hand lens
Marking pencil

**Procedure**

1. Using a beaker and stirring rod, dissolve 25 grams of table salt in 200 milliliters of hot water.
2. Fill a petri dish with this solution. Use a marking pencil to label the dish.
3. Place a piece of dental floss in the solution and let it hang over the edge of the dish.
4. Repeat steps 1 to 3 four times, using sugar, borax, alum, and copper sulfate instead of salt.
5. Allow the solutions to evaporate slowly for a day or two. Note which crystals form quickly and which form slowly.
6. With a hand lens or magnifying glass, observe the crystals formed in the dish and along the dental floss. Note the difference in the crystal forms. Each substance has its own distinct crystal shape.

**Observations and Conclusions**

1. Write a brief statement describing the results of this activity.
2. Describe the appearance of each of the different crystals that you grew.
3. Draw a chart showing how long it took each of your crystals to grow.



## OBSERVATIONS AND CONCLUSIONS

1. Regularly shaped crystals grew on the dental floss and around the edge of the flat-bottomed dish.
2. Sugar: slanted rectangles, white crystals; alum: octahedrons, hexagons, or slanted rectangles and colorless crystals; borax: slanted rectangles, colorless crystals; copper sulfate: slanted rectangles, green or blue crystals.
3. Answers and charts will vary.

## GOING FURTHER: ENRICHMENT

**Part 1**

Have students grow larger crystals of alum by growing small seed crystals in the bottom of a dish and then suspending the crystals from dental floss in a saturated alum solution in a flask. The saturated solution should be permitted to evaporate slowly for about one week.

# CHAPTER REVIEW

## SUMMARY

**18-1 What Is a Mineral?**

- A mineral is a naturally occurring, inorganic solid that has a definite chemical composition and crystal shape.
- Almost all minerals come from magma, the hot liquid rock deep inside the earth. The rate of cooling of magma determines the size of the mineral crystals.
- When magma cools slowly beneath the earth's crust, large crystals form. When magma cools rapidly beneath the earth's crust, small crystals form.
- Crystals also form when a mineral that is dissolved in a liquid is left behind as the liquid evaporates.
- The following physical properties are used to identify minerals: color, luster, hardness, streak, density, crystal shape, and cleavage and fracture.
- The luster of a mineral describes the way a mineral reflects light from its surface. Minerals have a metallic or nonmetallic luster.
- The ability of a mineral to resist being scratched is known as its hardness.
- The color of the powder left by a mineral when it is rubbed against a hard, rough surface is called its streak.
- Hefting is a way of comparing the densities, or mass per unit volume, of two minerals.
- The shape of a crystal results from the way the atoms or molecules of a mineral come together as the mineral is forming.
- There are six basic shapes of crystals, or crystal systems.
- The way a mineral breaks is called cleavage or fracture.

**18-2 Uses of Minerals**

- Ores are minerals from which metals and nonmetals can be removed in usable amounts.
- The ability of a substance to be hammered into thin sheets without breaking is called malleability.
- Ductility is the ability of a substance to be pulled into thin strands without breaking.
- Metals and nonmetals are elements that have properties useful to people.
- Gems are rare minerals that are beautiful and durable.

## VOCABULARY

*Define each term in a complete sentence.*

| | | | |
|---|---|---|---|
| **cleavage** | **gem** | **magma** | **Mohs hardness scale** |
| **crystal** | **hardness** | **malleability** | **nonmetal** |
| **ductility** | **inorganic** | **metal** | **ore** |
| **fracture** | **luster** | **mineral** | **streak** |

## CONTENT REVIEW: MULTIPLE CHOICE

*Choose the letter of the answer that best completes each statement.*

**1.** Minerals are
a. solid. b. found in the earth. c. inorganic. d. all of these.



**Part 2**

An interesting crystalline structure is rock candy, which is made out of sugar. Obtain or have students obtain samples of rock candy to display to the class. Also have students find out how rock candy is made.

## CHAPTER REVIEW

### MULTIPLE CHOICE

1. d  2. a  3. d  4. b  5. c  6. b  7. d  8. a  9. b  10. a

### COMPLETION

1. crystal
2. magma
3. large
4. metallic
5. Mohs hardness scale
6. streak
7. density
8. ores
9. ductility
10. precious stones

### TRUE OR FALSE

1. F inorganic
2. T
3. F magma
4. F smaller
5. F physical
6. F luster
7. F Mohs
8. F cleavage
9. T
10. F gems

### SKILL BUILDING

**1.** Minerals must be inorganic and naturally occurring solids. They also must have a definite chemical composition and a crystal shape. A mineral, such as gold, is inorganic because it is not formed from living things. On the other hand, oil, a nonmineral, is formed from fossilized plant and animal life. Gold occurs naturally in the earth, while a nonmineral such as cement is a manufactured substance. Gold is a solid, while oil is not. Because gold is made up of gold atoms, it has a definite chemical composition. The nonmineral cement is a mixture of substances and does not have a definite chemical composition. Lastly, gold has a crystal shape while cement does not.

**2.** When magma, or the hot liquid rock inside the earth, cools, crystals form. Sometimes, the magma reaches the earth's surface and cools so quickly that crystals do not form at all, such as in obsidian.

**3. a.** Color is the least helpful in identifying different minerals. Factors, such as rain, heat, cold, and pollution, can change the color of a mineral. Also different minerals may have the same color. **b.** Hardness is the most helpful in identifying different minerals. The Mohs scale is used to determine the relative hardness of minerals. **c.** Hardness **d.** Hardness and luster

**4.** This skill building activity will provide good experience for students in the concept of classification because it makes students classify numbers. When the activity is completed, students should realize that classification systems provide a logical system for identifying relationships between groups of things.

**5.** Classification is a process of elimination. You have a mineral you want to identify. Suppose you first make a hardness test and find it has a hardness of 7. This eliminates all minerals with a hardness other than 7. You then make a streak test and find it is white. This eliminates all minerals that do not have a white streak. By making two tests, you have eliminated about 1900 minerals that do not have these properties. You can eliminate many more with a few more tests. Then, at last, you have a good idea of what you are holding. In other words, you determine the mineral by showing what it cannot be.

**6.** Charcoal, graphite, and diamonds each have very different properties.

---

2. Hot liquid rock deep inside the earth is called
   a. magma. b. plasma. c. plastic. d. mantle.
3. The physical property of a mineral that can be changed by rain, heat, cold, and pollution is
   a. ductility. b. cleavage. c. streak. d. color.
4. The way in which a mineral reflects light from its surface is called
   a. streak. b. luster. c. fracture. d. malleability.
5. The ability of a mineral to resist scratching is called
   a. ductility. b. malleability. c. hardness. d. durability.
6. The softest mineral is
   a. fluorite. b. talc. c. diamond. d. calcite.
7. The mass per unit volume of a mineral is called its
   a. ductility. b. malleability. c. streak. d. density.
8. The breaking of a mineral along smooth, definite surfaces is called
   a. cleavage. b. fracture. c. splintering. d. none of these.
9. The ability of a substance to be hammered into thin sheets is called
   a. ductility. b. malleability. c. hardness. d. durability.
10. Elements that have shiny surfaces and are able to conduct electricity and heat are called
    a. metals. b. nonmetals. c. ores. d. geodes.

#### CONTENT REVIEW: COMPLETION

*Fill in the word or words that best complete each statement.*

1. A solid in which the atoms are arranged in a definite and repeating pattern is called a(n) ______________.
2. Hot liquid rock beneath the earth's surface is called ______________.
3. When magma cools slowly beneath the earth's crust, the size of the crystals formed is ______________.
4. Minerals that reflect light the way highly polished metal does are described as having a(n) ______________ luster.
5. A commonly used scale that rates the hardness of minerals from 1 to 10 is called the ______________.
6. The color of the powder left by a mineral when it is rubbed against a hard surface is called its ______________.
7. The property of a mineral that can be expressed as mass per unit volume is called ______________.
8. Minerals from which metals and nonmetals can be removed in usable amounts are called ______________.
9. The ability of a substance to be pulled into thin strands without breaking is called ______________.
10. Gems such as diamonds and emeralds are called ______________.

#### CONTENT REVIEW: TRUE OR FALSE

*Determine whether each statement is true or false. If it is true, write "true." If it is false, change the underlined word or words to make the statement true.*

1. Substances not formed from living things or the remains of living things are called organic.
2. The repeating pattern of atoms in a solid forms a crystal.
3. Hot liquid rock is called plasma.

430

4. The faster magma cools, the larger the size of the mineral crystals that are formed.
5. The chemical properties of minerals are used to identify them.
6. The hardness of a mineral describes how the mineral reflects light from its surface.
7. The Dalton hardness scale gives the relative degrees of hardness of minerals.
8. When a mineral breaks along smooth, definite surfaces, fracture occurs.
9. Minerals from which metals and nonmetals can be removed in usable amounts are called ores.
10. Minerals that are rare, beautiful, and durable are called nonmetals.

## CONCEPT REVIEW: SKILL BUILDING

*Use the skills you have developed in the chapter to complete each activity.*

1. **Making comparisons** Using the five special properties of a mineral, compare a mineral with a nonmineral.
2. **Relating facts** The black, glassy rock obsidian has no mineral crystals. Obsidian comes from volcanoes. How does obsidian's formation account for the absence of crystals?
3. **Applying concepts** You are using the following physical properties to distinguish among several different minerals: color, luster, and hardness.
   a. Which physical property is the least helpful? Why?
   b. Which physical property is the most helpful? Why?
   c. What property can you use to distinguish galena from graphite?
   d. What two properties can you use to distinguish talc from galena?
4. **Developing a classification scheme** Across the top of a piece of paper, print the numbers 1 through 20. On the next line, separate the odd numbers into one-digit and two-digit numbers. Do the same with the even numbers.

   Look carefully at the numbers you have written. Some numbers are made from only straight lines. Other numbers are made from only curved lines. Still others are made from both straight and curved lines. Classify each number in the one-digit and two-digit groups according to the following three headings: straight lines, curved lines, straight and curved lines.

   Explain how classification systems are useful in relating similar things.
5. **Applying concepts** Explain the following statement. You can determine the identity of a mineral by showing what it cannot be. Use specific properties of a mineral in your explanation.
6. **Drawing a conclusion** Charcoal, graphite, and diamonds are all made of carbon. Yet they are not considered types of the same mineral. Rubies, sapphires, and corundum are all made of aluminum oxide. They are considered types of the same mineral. Explain why this is so.

## CONCEPT REVIEW: ESSAY

*Discuss each of the following in a brief paragraph.*

1. Describe six properties used to identify minerals. Which properties of a mineral can be tested without damaging the sample?
2. Compare metals and nonmetals.
3. Describe what your life would be like without minerals.
4. Why is density more useful than heft in identifying a mineral?



Rubies, sapphires, and corundum all have some very similar properties—hardness, crystal shape, cleavage, etc. They vary only in color and size.

## ESSAY

**1.** The physical properties used to identify minerals are hardness, the ability to scratch or be scratched by other materials; streak, the color of a mineral in powder form; heft, the apparent weight or heaviness; color; luster, the way the mineral reflects light; crystal shape; the way a mineral breaks (cleavage or fracture). The properties of a mineral that can be tested without damaging the sample are: hardness, streak, heft, color, luster, and crystal shape. For the most part, the tests for streak and hardness does not damage a sample, but in the strictest sense of the word, some damage may be made to the sample. Therefore, some students may omit these two tests in their answer because they cause some damage.
**2.** Metals are elements that have shiny surfaces, are good conductors of heat and electricity, and are malleable and ductile. Nonmetals, on the other hand, are not shiny, are poor conductors of heat and electricity and are not malleable or ductile.
**3.** Answers will vary but should include the idea that minerals are important to people because most common objects are made from minerals. Without minerals, these objects would not exist. Life would be very difficult.
**4.** Density is more useful than heft in identifying a mineral because the density of a mineral is always the same, no matter what the size of the sample is. By hefting two minerals of about the same size, you can determine which mineral is denser because it feels heavier. This method only works well for minerals with very different densities.

## ADDITIONAL QUESTIONS AND TOPIC SUGGESTIONS

**1.** Describe and discuss the different methods that can be used to identify minerals (color, luster, hardness, streak, density, crystal shape, various special properties, cleavage, and fracture).
**2.** Briefly explain how a mineral differs from a nonmineral. (Minerals, unlike nonminerals, are natural substances in the earth and composed of one or more elements. Minerals are not products of living things.)
**3.** Discuss five different types of nonmetallic lusters and give one or more mineral examples of each type. (Types [and examples] of nonmetallic lusters are glassy [quartz], pearly [mica], greasy [serpentine], brilliant [diamond], and silky [malachite].)

## ISSUES IN SCIENCE

The following issue can be used as a springboard for discussion or given as a writing assignment.

Many of our mineral resources are in danger of running out. Yet we use these mineral resources for a wide variety of things. Should the use of certain mineral be controlled? Should we find new materials that can be used in place of certain minerals? Or should we simply let future generations tackle this difficult and expensive problem?

# Chapter 19

# ROCKS

## CHAPTER OVERVIEW

This chapter summarizes some important and interesting information about the three types of rocks—igneous, sedimentary, and metamorphic. All rocks are composed of one or more minerals.

The characteristics of each type of rock is discussed. Students will learn the distinguishing characteristics of each type of rock and how each was formed. They will also learn some important things about each kind of rock group, as well as about individual kinds of igneous, sedimentary, and metamorphic rocks.

In the last section of the chapter, students will be introduced to the rock cycle. The rock cycle shows how rocks are continuously changed from one kind to another over long periods of time.

## INTRODUCING CHAPTER 19

Begin by having students observe the photograph on page 432 and read the chapter introductory material. The photograph shows the Giant's Causeway, which is composed of a type of rock called basalt. Basalt is one of an important group of rocks called igneous rocks.

- **What do you see in the photograph?** (a person at the top of huge steplike stones)
- **How does the size of the person compare with the size of the stones?** (The person appears very tiny compared to the stones.)
- **What is the name of the stones shown in the picture?** (the Giant's Causeway)
- **If you had not read the text, how might you think the stones had been placed?** (Answers may vary, but many students will probably say that the stones appear to have been placed there by people, in much the same way as a stadium or amphitheater is built out of stones.)
- **What makes the stones appear to have been placed by builders?** (the regular pattern that creates a series of steps)
- **According to the legend, how were the stones placed?** (The stones were placed by giants who wished to cross

# 19 Rocks

CHAPTER SECTIONS



CHAPTER OBJECTIVES

*After completing this chapter, you will be able to:*

**19-1** Explain how igneous rocks are formed.
**19-1** Compare extrusive and intrusive igneous rocks.
**19-2** Explain how sedimentary rocks are formed.
**19-2** Classify various types of sedimentary rocks.
**19-3** Explain how metamorphic rocks are formed.
**19-4** Describe the rock cycle.

On the Giant's Causeway, row after row of black, six-sided rock pillars rise from the shore. The thousands of rocky pillars form huge platforms that look like stepping stones for giants. And according to legend, that is how the Giant's Causeway got its name. As the legend goes, giants built the rocks so they could cross the sea from the northern coast of Ireland to Scotland.

If such giants had existed, they might have discovered another unusual rock formation south of Scotland. On the southeastern coast of England, white cliffs tower 114 meters above the town of Dover. These cliffs form a wall of bright, gleaming stone. But if you were to examine the stone through a microscope, you would discover countless tiny, beautiful shells. From these white cliffs, it is said, comes one of the ancient names for England—Albion, the "White Land."

These are but two examples of different and amazing forms of rock. There are many others. Each has its own history—a history hidden in the earth's past. This history is full of action and change. And although the next rock you see is not likely to move before your eyes, it got to be where it is by great movements on and within the earth—movements that built the Giant's Causeway and the white cliffs of Dover.

*The Giant's Causeway*



## TEACHER DEMONSTRATION

Obtain 5 to 10 different rock samples from areas near your school. Wash the rocks so that the colors and textures are clearly visible. Display the rocks and have students note the similarities and differences that exist among the samples.

- **What characteristics may help you classify these rocks?** (Answers may vary. Possible answers include color, texture, size, shape, luster.)
- **What causes a rock to have certain characteristics?** (The way the rock was formed; the particular materials that make up the rock; the effects of weathering and other environmental factors on the rock.)

## TEACHER RESOURCES

### Audiovisuals

*Earth Science: The Rock Cycle,* filmstrip with cassette, SVE
*Investigating Rocks,* 9 filmstrips with 9 cassettes, EBE
*Minerals and Rocks,* 16 mm film, EBE
*The Earth: Rocks and Minerals,* filmstrip with cassette, SVE

### Books

Blatt, Harvey, G. V. Middleton, and R. C. Murray, *Origin of Sedimentary Rocks,* Prentice-Hall
Fay, Gordon, *Rockhound's Manual,* Barnes & Noble
Firsoff, V. A., *The Rockhound's Handbook,* Arco
Shakley, M. L., *Rocks and Man,* St. Martin

the sea from Ireland to Scotland.)

- **What is a legend?** (a popular story handed down through history that serves to explain an event or natural phenomenon)
- **Are legends true?** (Not really, because they cannot be verified; however, they often are based to some extent on fact, and are usually believed by many people to be true.)
- **How do you think the stones in Giant's Causeway were actually placed?** (Answers may vary; the correct answer is that they were placed naturally as rocks formed on the surface of the earth.)

As you go through this chapter, emphasize the important connection between minerals (Chapter 18) and rocks to students. Point out that rocks are made up of minerals in various combinations, and that certain minerals characterize particular types of rock. Also remind students that minerals in usable amounts are found in rocks called ores.

## 19-1 ROCKS OF LIQUID AND FIRE: IGNEOUS ROCKS

### SECTION PREVIEW 19-1

In this section, students will learn about the formation and characteristics of igneous rocks. They will discover that igneous rocks form from hot molten rock called magma that exists deep within the earth.

When magma cools and hardens, crystals form. These crystals make up igneous rocks. The igneous rocks that form deep below the earth's surface are called intrusive rocks. Igneous rocks that form at or near the earth's surface are called extrusive rocks. When molten rock at the earth's surface cools and hardens very rapidly, the result is a glassy igneous rock that contains no crystals.

### SECTION OBJECTIVES 19-1

1. **Explain how igneous rocks are formed.**
2. **Compare extrusive and intrusive igneous rocks.**

### SCIENCE TERMS 19-1

**magma p. 434**
**lava p. 434**
**rock p. 434**
**extrusive rock p. 434**
**intrusive rock p. 435**

**Figure 19-1** *This photograph shows hot, molten lava erupting from a volcano on Hawaii.*

**Figure 19-2** *A pattern of roughly six-sided rock columns forms where mud flats dry out. This pattern is similar to the skin of solid rock that shrunk and split to form the Giant's Causeway.*

## 19-1 Rocks of Liquid and Fire: Igneous Rocks

The strange six-sided rocks of the Giant's Causeway were forged in fire. But not by giants, and not in an ordinary fire.

Deep within the earth it is so hot that some of the minerals that make up rocks melt and flow like liquid. The liquid rock is less dense than solid rock, so the liquid rock tends to rise—just as bubbles rise through water. This hot, molten rock deep inside the earth is called **magma.**

Sometimes the magma reaches the surface. It may erupt from a volcano. Or it may seep through weak spots in the ground. Magma that has reached the earth's surface is called **lava.**

At the site of the Giant's Causeway, melted rock seeped toward the surface, forming large pools of melted rock material. As the rock cooled and solidified, it shrank and cracked. The cracks formed a six-sided pattern. You can see a pattern something like this where mud flats dry out in places like the Painted Desert in Arizona. As the lower levels of the pool of melted rock material cooled and hardened, the pattern of cracks spread downward. Each pool broke up into rows of six-sided rock columns.

The rocks of the Giant's Causeway are basalt. They belong to one group of rocks called **igneous** (IG-nee-us) **rocks.** The word "igneous" comes from the Latin word "ignis," meaning coming from fire. **All rocks formed from the cooling and hardening of magma are igneous rocks.**

### *Extrusive Rocks*

Basalt is only one kind of igneous rock. It is a kind known as **extrusive** (ek-STROO-siv) **rock.** Extrusive rocks form from melted rock or lava that cools and hardens at or near the earth's surface. Some extrusive rocks originally were pushed slowly onto the earth's surface, much as toothpaste is pushed out of a tube. Other extrusive rocks formed as melted rock material erupted rapidly from places such as volcanoes.

434

### TEACHING STRATEGY: 19-1

#### Motivation

Begin by asking students,

- **Have you ever seen a hot liquid substance that cooled into a solid?** (Most students will probably say yes.)
- **What are some of the substances that you have seen?** (Answers will vary; possible answers include glass, fudge, hard candy, cake icing, metal heated as in welding.)

#### Content Development

Point out that igneous rocks form when hot liquid rock cools and hardens.

- **What causes rock to become liquid?** (high temperatures deep within the earth)
- **What is this liquid rock called?** (magma)
- **Where does magma cool and harden into igneous rocks?** (Some magma cools far below the earth's surface; some cools at or near the earth's surface.)

#### Skills Development

***Skill: Making a model***

Students can simulate the mud flats shown in Figure 19-2 by performing the following activity.

Have students work in small groups. Provide each group with a pie pan or shallow dish. Have students make mud by mixing soil and water. (The mud can be prepared beforehand if you wish to save time.) Have each group fill their pie pan

**Figure 19-3** *These photographs show that basaltic rock can have very different shapes. Devil's Postpile National Monument* (left) *is a spectacular mass of blue-gray basalt columns. Shiprock* (right) *is the remains of the basaltic inner core of a volcano.*

Obsidian is another kind of extrusive rock. Because it is glassy-looking, it is often referred to as volcanic glass. It is usually dark in color and forms from molten material that cools very rapidly on the earth's surface. Rapid cooling is the key to making all kinds of glass.

Sometimes igneous rocks form so quickly that bubbles of volcanic gases are trapped in them. Rocks formed in this way contain many holes and tiny needlelike slivers of volcanic glass. Pumice (PUH-mis) is an example of this kind of rock. Because pumice is filled with bubbles, it is a rock that actually floats on water! The particles of volcanic glass in pumice make it a good polishing agent. Have you had your teeth cleaned at a dentist's office? The gritty stuff in the cleaning powder was probably crushed pumice. 3

**Figure 19-4** *This obsidian dome is an example of an extrusive rock that cools rapidly on the earth's surface.*

### *Intrusive Rocks*

Another kind of igneous rock is granite, one of the earth's most common types of rock. Chemically, granite is similar to obsidian. However, granite is usually white or gray, rather than black. Also granite contains large crystals, while obsidian contains none.

Rocks that form from melted rock or magma that cools and hardens deep below the earth's surface are called **intrusive** (in-TROO-siv) **rocks.** They intrude, or push between surrounding rocks, while in a molten state. Huge masses of granite form within the earth



## BACKGROUND INFORMATION

Rocks are naturally occurring solid materials made of one or more minerals. The principal elements that make up these minerals are oxygen, silicon, aluminum, iron, calcium, sodium, potassium, and magnesium.

Rocks are classified according to how they are formed. Igneous rocks are formed from magma. Magma is hot liquid rock that contains dissolved gases. Magma originates in the lower part of the earth's crust and in the upper part of the mantle. Magma moves upward through the crust. When magma rises to the earth's surface and flows out of a volcano, it is called lava.

## TIE-IN/ LANGUAGE ARTS

The prefix *in-* (intrusive) means within. The prefix *ex-*(extrusive) means out of. The root *-trusive* comes from the Latin word trudere, meaning to thrust.

## ANNOTATION KEY

1 **Thinking Skill: Relating cause and effect**

2 **Thinking Skill: Classifying rocks**

3 **Thinking Skill: Identifying processes**

with mud, making sure that the mud is spread evenly. Then have students put their pans of mud in a warm, dry place for several days until the mud is completely dried out.

- **What changes do you observe in your pans of mud?** (Answers may vary, but students will probably note that the mud has cracked and become lighter in color.)
- **What caused these changes?** (The loss of moisture from the mud caused it to dry and crack.)
- **How does the appearance of your "mud flat" compare with that of the mud flat shown in Figure 19-2?** (Answers will vary.)
- **What type of rocks form in a way that is similar to the drying out of a mud flat?** (extrusive igneous rock such as the basalt)

### Content Development

Point out to students that there are two basic types of igneous rocks. Their classification is based on where the magma has cooled. Magma that has cooled within the surface of the earth is called intrusive. When magma reaches the surface of the earth and is cooled there, it is called extrusive. The size of the crystals found in igneous rock is determined by two factors: One is the chemical composition of the mineral and the other is the length of time the rock had to cool. Therefore, the longer a rock has to cool, the larger the crystals found within the rock. The size of these crystals is referred to as the texture of a rock. The texture and chemical composition are used in classifying igneous rocks.

## Activity

**Sedimentation**
**Skills: Drawing conclusions, manipulative, observing, applying, relating, inferring, hypothesizing**
**Level: Average**
**Type: Hands-on**
**Materials: 500-mL jar with cap, garden soil, potting soil, coarse-grained sand, fine-grained sand, coarse gravel, fine gravel, water**

In this activity, students will use hands-on skills to demonstrate how sediments settle out in standing water. Observations are recorded in tables. Then students are asked to interpret their observations and draw conclusions from the activity. Students will observe that the heavier gravel will settle out first. The finer grained sand and garden soil will take longer to settle out. It may take several hours for the water to clear completely.

## ANNOTATION KEY

❶ Thinking Skill: Making comparisons
❷ Thinking Skill: Identifying processes
❸ Thinking Skill: Applying definitions

**Figure 19-5** *This batholith is a huge body of igneous rock containing granite that formed underground and was eroded at the earth's surface.*

in this way. In some places, the masses are more than a thousand kilometers long and 200 kilometers wide. Scientists call these gigantic masses of rock batholiths (BA-thuh-liths), or deep stones. Batholiths form the stony base of the earth's great mountains.

Why do obsidian and granite look so different? The reason is because granite forms from molten rock that cools and hardens *within* the earth instead ❶ of at the earth's surface. As a result, granite cools and hardens much more slowly than the molten rock that becomes obsidian. Why? Think of a pot of soup on a stove. Exposed to air, it cools quickly. With the lid on the pot, the soup cools much more slowly. Granite forms "with the lid on." Obsidian forms "with the lid off."

**Figure 19-6** *The igneous rock scoria cools so quickly that gas is trapped in it, producing hundreds of holes.*

The amount of time it takes liquid rock to cool and harden affects the texture of the rock. Texture refers to the size and type of crystals, or grains, that make up that rock. The slower the rate of cooling, the larger the crystals. The faster the rate of cooling, the smaller the crystals.

❷ Most extrusive igneous rocks, such as basalt, have small crystals and are fine-grained because they cool and harden very quickly at the earth's surface.

Most intrusive igneous rocks, such as granite, have large crystals and are coarse-grained because they cool and harden very slowly beneath the earth's surface. Sometimes, magma cools and hardens so



## 19-1 (continued)

### Reinforcement

Use the information below to summarize important facts about the crystal sizes of igneous rocks.

**Large crystals:** Rock type—intrusive
texture—coarse-grained
cooling time—slow
cooling location—deep in the earth

**Small crystals:** rock type—extrusive
texture—fine-grained
cooling time—rapid
cooling location—at or near earth's surface

### Section Review 19-1

**1.** Magma: found deep inside the earth. Lava: found at the earth's surface.
**2.** Extrusive rock: cools and hardens at or near the earth's surface; obsidian. Intrusive rock: cools and hardens deep below the earth's surface; granite.
**3.** Time it takes liquid rock to cool and harden.

## TEACHING STRATEGY 19-2

### Motivation

Bring in some common sand and ask students to identify it. Then show

quickly that crystals do not form at all. Obsidian, or volcanic glass, is an extrusive rock that did not have time to form crystals.

### SECTION REVIEW

1. What is the difference between magma and lava?
2. What is the difference between extrusive and intrusive rocks? Give an example of each.
3. What determines the size and type of crystals in rocks?

## 19-2 Rocks in Layers: Sedimentary Rocks

The rocks in the limestone cliff in Arizona, shown in Figure 19-7, appear to be stacked very neatly. One layer is piled on top of another. The layers are almost as straight as if they had been drawn using a ruler. Layered rocks like these are called **sedimentary** (sed-uh-MEN-tuh-ree) **rocks.** ③

How did this layering happen? The answers are in the rocks themselves. The rocks of this cliff are made up largely of the same material as the white cliffs of Dover—the shells of small sea creatures.

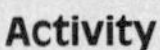

**Activity**

*Sedimentation*

1. In a 500-mL jar, put one layer each of garden soil, potting soil, coarse-grained sand, fine-grained sand, coarse gravel, and fine gravel. Each layer should be 2 cm deep.
2. Add enough water to fill the jar three-fourths full. Cap the jar and shake vigorously.
3. Carefully observe the various sediments as they settle out of the water. Which sediments settle out first? Last?
4. After all of the sediments have settled out, record the order in which they settled.

What can you tell about the way layers are formed in sedimentary rocks?

**Figure 19-7** *The sedimentary rocks of the limestone cliffs in Arizona* (left) *and New Zealand* (right) *are stacked neatly like piles of pancakes.*



## 19-2 ROCKS IN LAYERS: SEDIMENTARY ROCKS

### SECTION PREVIEW 19-2

Sedimentary rocks are formed by accumulation of layer upon layer of deposited materials. As the sediments are laid down, they form strata, or layers. Several strata make up a sedimentary rock and are sometimes called stratified rocks. Increasing pressure on the bottom strata by new layers causes them to become compacted and cemented.

There are many different types of sedimentary rocks. They can be grouped into three categories according to origin of the sediments and how the rocks were formed. The three categories are clastic, organic, and chemical rocks. Clastic rocks are made up of rock fragments mixed with sand, clay, and mud cemented together. Organic rocks are formed when the remains of once-living plants or animals are part of sediments. Another type of sedimentary rock is formed by chemical means, as when a sea dries up.

### SECTION OBJECTIVES 19-2

1. **Explain how sedimentary rocks are formed.**
2. **Describe the processes of compaction and cementation.**
3. **Compare clastic, organic, and chemical rocks.**

### SCIENCE TERMS 19-2

**sedimentary rock p. 437**
**clastic rock p. 438**
**organic rock p. 439**
**chemical rock p. 440**

students a sample of sandstone. Ask if they can guess what this rock is called. Then ask why. Explain that most surface rocks started out as another rock. As rocks erode and weather, the small particles are reformed into new rocks.

### Content Development

Sedimentary rocks are made up of various materials. The most common is other rock fragments. Materials such as shells from dead sea animals also can be part of a sedimentary rock.

There are two major ways in which material that forms sedimentary rock is moved from where it was eroded—wind and water. In both, the weight and size of the particles have a direct relationship to how long the particle can be carried until it is deposited. The heavier the particle, the less time it can be suspended. The speed or velocity of the wind or water also affects the distance a particle can be suspended. The greater the velocity, the greater the distance an object can be carried. Therefore slow-moving water can carry only very small particles until they settle out of the suspension.

## TEACHER DEMONSTRATION

The following demonstration will illustrate to students how crystals are formed in igneous rocks. For the demonstration you will need some sulfur, a Bunsen burner or hot plate, a small container for heating the sulfur, a small dish, and a small beaker filled with water.

Heat some sulfur in the small container until the sulfur melts. Pour the sulfur into the dish and let it cool slowly. Heat some more sulfur and pour it into the beaker of water. Have students observe and compare the different crystals that form in both cases.

- **How do you account for this difference?** (The sulfur in the dish cooled slowly, while the sulfur in the beaker cooled rapidly.)
- **Can you relate what you have seen to the formation of crystals in igneous rocks?** (Crystals in igneous rocks are larger when the rock cools slowly and smaller when the rock cools rapidly.)

**Figure 19-8** *Sedimentary limestone rock contains the shells of once-living marine animals.*

**Figure 19-9** *Conglomerates are made up of rock particles of different sizes, as well as sand and mud.*

Some 250 million years ago, this part of Arizona was believed to be at the bottom of a shallow sea. The sea stretched from what is now western Texas north to Canada. The sea was rich in life. Sharks and many kinds of fish swam in the waters. Most abundant of all were sea creatures that had shells. As these animals died, their remains settled to the sea floor. Steadily, more and more shells fell to the bottom. This process continued for millions of years.

Meanwhile, tropical forests of fernlike trees grew. Ancestors of the dinosaurs waddled on the shores of the sea. Rain and wind washed soil into rivers. The rivers flowed down to the sea, carrying muddy soil, pebbles, and rocks. These, too, built up in layers on the sea bottom.

Bones, shells, mud, and pebbles all settled to the sea floor as sediment. The sediment piled up in layers many hundreds of meters thick. Lower layers were pressed together, or compacted, more and more tightly under the weight of the layers above. Larger rock fragments became cemented together by dissolved minerals. **As a result of compaction and cementation, the sediments hardened into strata, or layers, of rock called sedimentary rock.**

Sedimentary rocks usually are formed in water. Most of the earth was under water at some time in the past. That is true now too—70 percent of the earth is covered by oceans. Consequently, sedimentary rocks are common all over the world. Sedimentary rocks often are rich in fossils—the remains or traces of animals or plants that lived in prehistoric times and were trapped in the layers of sediment.

### *Clastic Rocks*

Conglomerates, sandstones, and shales are all examples of **clastic rocks.** They are sedimentary rocks made up of rock fragments mixed with sand, clay, and mud cemented together.

**CONGLOMERATES** Pebbles and other rocks of different sizes often cement together gradually by sand, mud, and clay to form a single large rock. It is an odd-looking rock. The sand, mud, and clay form the cement surrounding the pebbles. You can see in Figure 19-9 why the scientists who first studied this kind



## 19-2 (continued)

### Content Development

Emphasize the idea that most sedimentary rocks are formed in water.

- **Why is water the most likely place for sedimentary rocks to form?** (Water acts as a medium to carry and collect the sediment that makes up sedimentary rock.)
- **Sedimentary rocks are found in many parts of our country that are not near water. How is this possible?** (Many, many years ago, these areas were covered by ocean.)

Stress the idea that sedimentary rocks are characterized by layers.

- **Based on the way in which sedimentary rocks form, can you explain why sedimentary rocks consist of layers?** (Sediment that collects in ocean water eventually settles to the bottom of the ocean. Because new sediment is constantly collecting in the ocean, new layers of sediment are continually being added to the ocean floor. The weight of new, top layers of sediment press down on the older bottom layers, causing the rocks to have a "pancake" appearance.)

### Motivation

Conglomerate rock was first called puddingstone because of the similar lumpy appearance of yorkshire pudding from Yorkshire, England. This pudding is a batter made of eggs, flour, and milk that is baked in the drippings of roast beef.

### Skills Development

***Skill: Relating concepts***

Point out to students that unlike igneous rocks, sedimentary rocks often contain the remains of once-living organisms.

- **Why do you think that igneous rocks do not contain the remains of living things?** (Igneous rocks are formed from magma that is extremely hot. This intense heat would burn up the remains of any living

Figure 19-10 *The unusual sandstone formations in Bryce Canyon, Utah were formed from cliffs made of layers of red sedimentary rocks.*

of rock called it "puddingstone." Later, scientists gave it another name—conglomerate rock. The name means the same thing. Conglomerate rock translates as "rock-made-up-of-many-things-put-together."

**SANDSTONE** The layered rocks of sandstone, shown in Figure 19-10, are located in Bryce Canyon National Park, Utah. As their name implies, they form from sand grains cemented together. This sandstone is colored red due to traces of iron. Sandstone is very resistant to wear and decay and is often used as a building stone.

**SHALE** Mud and clay that has hardened into layers of rock is called shale. Shales often form in quiet waters such as swamps and bogs. Small particles of clay and mud make up shales. These particles could only settle to the bottom in quiet waters. In fast-moving rivers, these small particles would be swept along by the currents.

### *Organic Rocks*

Not all sedimentary rocks are made from pieces of rock. There are two other kinds of sedimentary rock. Some sedimentary rocks, such as limestone, may be built up from the remains of living things. Such rocks are called **organic rocks.** The name for

Figure 19-11 *This large deposit of shale is made up of small particles of clay and mud.*

## ANNOTATION KEY

**❶ Thinking Skill: Identifying processes**

**❷ Thinking Skill: Comparing clastic rocks**

## FACTS AND FIGURES

Seventy-five percent of the rocks on the earth's surface are sedimentary rocks.

## FACTS AND FIGURES

Fragmental sedimentary rocks are also known as clastic rocks.

## SCIENCE, TECHNOLOGY AND SOCIETY

Technology has enabled scientists to study many natural processes. For example, the forces that can change existing rock into a metamorphic rock, primarily heat and pressure, can be reproduced in the laboratory. Using two gem-quality diamonds as anvils, minute rock samples are subjected to extreme pressure. A laser, directed through the transparent diamonds, heats the rock samples to corelike temperatures. Using a microscope, scientists can observe the changes that the minerals in the existing rock samples undergo as they change into metamorphic rock.

thing before it could become part of a rock.)

- **Why are the remains of living things likely to be found in sedimentary rocks?** (Sediments are soft and moist; they provide a burial for the remains of living things without destroying them.)

## Enrichment

Layers of sedimentary rocks can be seen in the Grand Canyon. Have students find photographs of the Grand Canyon in which these layers are clearly visible. Also challenge students to find out how the Grand Canyon and its rock structures were formed.

## Skills Development

***Skill: Observing***

Have students observe Figure 19-9.

- **How can you tell by looking at this rock that it is a sedimentary rock?** (The rock contains various rock particles cemented together.)
- **Why could this rock not be an igneous rock?** (Igneous rocks have uniform composition, and they are made up of crystals, not pieces of rock cemented together.)

## Reinforcement

Students may enjoy thinking of some everyday examples of conglomerates. For example: a pizza with anchovies, pepperoni, mushrooms, and extra cheese; a banana split with three flavors of ice cream and four different toppings; a piece of clothing made of cotton, satin, fake fur, and sequins; a collage that consists of paint, photographs, and bits of paper.

## Activity

**Coral Conversions**
**Skill: Making calculations**
**Level: Remedial**
**Type: Computational**

Students will need their metric conversion skills to complete this activity. Students should find that 2000 kilometers equals 2,000,000 meters and 200,000,000 centimeters. The length of the United States is approximately 2.25 times the length of the Great Barrier Reef.

## Activity

**Rock Quarries**
**Skill: Relating facts, researching, applying**
**Level: Average**
**Type: Library/writing**

**1.** Limestone, basalt, granite, marble, slate
**2.** Rocks are removed from a quarry by several methods, such as the "plug and feather" method.
**3.** Most of the rock materials taken from quarries are used in the construction of foundations, walls, steps; supports for bridges and piers; and the finishing and decorating of all types of structures.
**4.** Pennsylvania, Ohio, West Virginia, Illinois, Nebraska, and Kentucky.

**Figure 19-12** *As the water in Mono Lake, California dries up, it leaves behind unusual shaped formations of rock salt.*

these rocks comes from the word "organism," meaning living thing. Shells of animals such as clams and oysters sink to the ocean bottom and eventually form limestone. Limestone may also contain mud and sand. The hard outer coverings of corals are cemented together to become limestone reefs. Chalk is another organic sedimentary rock. It is made up of small pieces of animal shells and crystals of limestone cemented together.

### Activity

*Coral Conversions*

The largest coral reef is the Great Barrier Reef, which parallels the northeastern coast of Australia for a distance of about 2000 kilometers. How many meters long is the Coral Reef? How many centimeters? Compare this distance to the distance across the United States which is 4517 km from east to west.

### Chemical Rocks

1 Some sedimentary rocks are formed when a sea or lake dries up leaving large amounts of minerals that were dissolved in the water. These minerals may collect into large formations called **chemical rocks.** Examples of this type of sedimentary rock include rock salt and gypsum (JIP-sum).

### SECTION REVIEW

1. What are clastic rocks made of? What are two examples?
2. What are organic rocks made of? What are two examples?
3. How are chemical rocks made? What are two examples?



## 19-2 (continued)

### Motivation

Have students demonstrate the formation of stalactites. To do this, have them pour Epsom salts into two small containers until no more will dissolve in the water. Next have students place one end of a string in one container and the other end in the second container. Have them hang the string between the containers over a small dish. An Epsom salt "stalactite" will form on the string. Have students compare this to the formation of stalactites in caves.

### Section Review 19-2

**1.** Rock fragments mixed with sand, clay, and mud cemented together; conglomerates, sandstone, shale.
**2.** Remains of living things; limestone, chalk.
**3.** When a sea or lake dries up leaving large mineral deposits; rock salt, gypsum.

## TEACHING STRATEGY 19-3

### Motivation

Ask students to pick up familiar solid objects and hold them in their hands. Objects such as a baseball or tennis ball, a wood block, a chalkboard

## 19-3 Rocks That Change: Metamorphic Rocks

There is a very simple chemical test to determine if a rock is limestone. Add some strong vinegar to limestone and the rock will fizz and begin to dissolve. If you add strong vinegar to a bit of marble, the marble will also fizz and begin to dissolve. Why? Marble is chemically the same as limestone. However, the resemblance ends there.

Marble does not look like chalk, a kind of limestone. Chalk is soft and powdery, which is one reason you can use it to write on a chalkboard. But you cannot write with a piece of marble. Why? The reason is because marble is much harder and not powdery like chalk. Marble can be smoothed and polished until it gleams. And, unlike chalk, you will find no trace of fossil shells in marble. ❷

Yet all marble was once limestone, including organic limestone made of fossil shells. Any kind of limestone may become buried deep within the earth. Under very high temperatures and tremendous pressures, the limestone changed. For millions of years the limestone was bent, folded, twisted, squeezed, and changed by forces in the earth. A different kind of rock finally formed. In this case, the new rock was marble.

**Activity**

*Rock Quarries*

Quarrying is a method of taking solid blocks or smaller pieces of rock from the earth and preparing them for various uses. Using reference materials in the library, write a report about quarrying stone and include the answers to the following questions:

**1.** What kinds of rocks are usually taken from quarries?

**2.** Describe three methods used to remove rocks from a quarry.

**3.** What are some of the uses of rock material taken from quarries?

**4.** Where are some large rock quarries located in the United States?

**Figure 19-13** *The igneous rock granite can be changed by heat and pressure into the metamorphic rock gneiss. How would you describe the appearance of gneiss?* ❶

## 19-3 ROCKS THAT CHANGE: METAMORPHIC ROCKS

### SECTION PREVIEW 19-3

In this section, students will learn how existing rocks can be changed by high temperatures and tremendous pressure deep within the earth. The rocks that result from these changes are called metamorphic rocks.

Students will discover that most metamorphic rocks were once sedimentary or igneous rocks. They will also discover that metamorphic rocks tend to have different characteristics than the original rocks. For example, metamorphic rocks are usually harder and denser than the rocks from which they were formed.

### SECTION OBJECTIVES 19-3

1. **Describe the conditions under which metamorphic rocks form.**
2. **List some examples of metamorphic rocks.**

### SCIENCE TERMS 19-3

**metamorphic rock p. 442**

### ANNOTATION KEY

**❶ Alternating light and dark bands of minerals (Making observations)**

**❶ Thinking Skill: Identifying processes**

**❷ Thinking Skill: Making comparisons**

eraser, or a plastic or metal container work well.

• **What do you think would happen to these objects if you were strong enough to exert tons of pressure on them?** (Answers may vary; most likely, the objects would flatten, or become bent or twisted; a brittle object might shatter or crack.)

• **What do you think would happen to these objects if they were subjected to a temperature of many hundred degrees Celsius?** (Answers may vary; most likely, the objects would melt or change chemically in some way.)

### Content Development

Point out that when existing rock is placed under extreme heat and/or pressure, certain chemical and physical changes take place. These changes cause the rock to assume new textures and structures. This new rock quite often shows minerals in a distinct crystalline form.

• **Explain where the heat comes from to change the rocks.** (The exact reason for underground heat is not known. But many kilometers below the surface of the earth the temperatures are very high. Allow the student to suggest several reasons.)

• **Can metamorphic rocks become sedimentary rocks? Explain your answer.** (Yes. They can be exposed to the surface and then eroded, and the fragments can become sediments to form a new sedimentary rock.)

## BACKGROUND INFORMATION

Metamorphic rocks begin to form at depths of 12 to 16 km beneath the earth's surface and at temperatures of 100°C to 200°C. They continue to form at temperatures of up to 800°C. At these high temperatures, the rock becomes soft enough for minerals within it to undergo change. The mineral crystals may change their size or shape, or they may separate into layers. Chemical reactions involving the minerals may also occur. As a result, major changes in the rock's composition take place.

There are two basic types of metamorphism. The first type is called contact metamorphism. Contact metamorphism occurs when rocks are heated by contact with magma or lava. Existing rocks are changed by the heat, chemicals, and gases released by magma. Contact metamorphism can occur on the earth's surface as lava flows over rocks. The second type of metamorphism is called regional metamorphism. Regional metamorphism occurs over large areas when rocks, deep in the earth, are changed by increases in temperature and pressure.

There are two basic textures of metamorphic rocks: foliated and unfoliated. Rocks with a foliated texture have mineral crystals arranged in parallel layers, or bands. These rocks tend to break along these bands. Foliated rocks form when mineral crystals in the original rock recrystallize or flatten under pressure. Foliation can also take place when minerals of different densities separate into layers. Metamorphic rocks that do not have these bands of crystals are called unfoliated.

**Figure 19-14** *The sedimentary rock shale can be changed by heat and pressure into the metamorphic rock slate. What type of texture does slate have?* 1

Rocks made in this way are called **metamorphic** (met-uh-MAWR-fik) **rocks.** These rocks are usually the hardest and densest rocks. The word "metamorphic" means a rock that has been changed.

1 **Metamorphic rocks can be formed from igneous rocks, sedimentary rocks, and other metamorphic rocks that are exposed to tremendous heat, great pressure, and chemical reactions.** Slate is a metamorphic rock made from clay or from shale, a sedimentary rock. Shale is porous. But under high temperatures and pressures, shale turns to slate, which is waterproof. Slate is used to make roofs, pathways, and chalkboards.

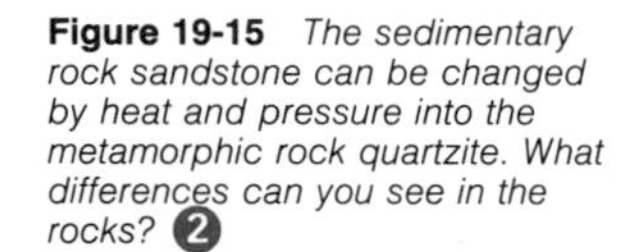

**Figure 19-15** *The sedimentary rock sandstone can be changed by heat and pressure into the metamorphic rock quartzite. What differences can you see in the rocks?* 2

442

## 19-3 (continued)

### Content Development

Students may be misled by the text into thinking that the chemical composition of a metamorphic rock is always the same as that of the original rock. Although this is true for some rocks, such as marble and limestone, it is not always the case. The tremendous heat and pressure involved in metamorphism can cause chemical reactions to occur in an igneous or sedimentary rock. When this happens, the resulting metamorphic rock will have a chemical composition that is different from that of the original rock.

### Skills Development

***Skill: Relating cause and effect***

- **Why do you think slate is waterproof, while shale is porous?** (When shale is changed into slate, great pressure is exerted on it. This pressure causes the pores in shale to close up. As a result, slate is nonporous.)
- **Why do you think quartzite is harder than sandstone?** (Tremendous heat and pressure push the grains of sandstone closer together. The result is a rock that is harder and denser than sandstone.)

### Enrichment

Have students research the various

Another example of metamorphic rock is quartzite (KWORT-zight). This rock is formed from the sedimentary rock sandstone. Under high temperature and pressure, the sand grains change shape and are packed closer together. The result is that quartzite is harder than sandstone.

### SECTION REVIEW

1. What are the two physical conditions needed for metamorphic rocks to form?
2. Metamorphic rock was once what kind of rock?
3. Name two metamorphic rocks. Name the rock from which each is formed.

#### Activity

*A Rock Walk*

Write a 500-word essay describing a walk through a rocky area. Make sure you use the following terms: igneous rock, sedimentary rock, metamorphic rock.

## CAREER: *Geochronologist*

**HELP WANTED: GEOCHRONOLOGIST** Bachelor's degree in geology or related field required. Graduate study or coursework in geochronology a plus. Training will be provided during fieldwork and laboratory research.

For a very long time scientists could only guess at the age of the earth and the rocks upon it. Scientists tried determining age by estimating the time it took for the ocean to gain its present saltiness. They also made guesses on the time it takes for sediments to deposit, and then they measured the thickness of all the sedimentary rocks. Unfortunately, the scientists came up with age estimates of the earth ranging from 3 million to 1600 million years with no precise answer.

In the early 1900s, scientists began determining the age of rocks by using the half-life of radioactive minerals. Today scientists who specialize in the study of geologic time are called **geochronologists.** They figure out the age of rocks and landforms by the radioactive decay of certain elements such as carbon and uranium. They test igneous, metamorphic, and sedimentary rocks that range in age from being just recently formed to being almost as old as the earth itself, about 4.6 billion years.

Geochronologists work to learn more about how the earth came to be as it is. Through the study of sediments and fossils, they try to provide pictures of the earth throughout its history. These scientists have contributed to knowledge about the development of life from the first single-celled organisms to prehistoric human beings.

Some qualities that lead to a career in geochronology are natural curiosity, problem-solving ability, and enjoyment of the outdoors. Physical stamina is also required to do the fieldwork. More information may be obtained by writing to the American Geological Institute, 5205 Leesburg Pike, Falls Church, VA 22041.



#### Activity

**A Rock Walk**
**Skills: Applying facts, observing, applying definitions**
**Level: Average**
**Type: Field/writing**

Check students' essays first for their use of the terms mentioned and second for their imagination and writing style. Students should be able to use each term in its correct context. You may want to have selected students read their essays to the class. You may also want to suggest that students actually take such a rock walk and write about the rocks they have seen.

## TIE-IN/ LANGUAGE ARTS

The word metamorphic comes from two Greek words: *meta-*, which means change or transformation; and *morphe,* which means form.

## ANNOTATION KEY

**❶ Mineral crystals arranged in parallel layers (Making observations)**

**❷ Quartz grains are different in size and shape. (Making observations)**

**❶ Thinking Skill: Applying definitions**

uses of slate. Students may be able to contact a roofing company or outdoor supply company to learn about the use of slate in buildings and landscaping. Students can also look for examples in their area of slate roofs and pathways

### Reinforcement

Have students work in small groups. Assign each group one of the three major rock types: igneous, sedimentary, or metamorphic. Challenge each group to answer the following questions about their rock type:

- **How is this type of rock formed?**
- **Where is it formed?**
- **What are its major characteristics?**
- **How is it different from other types of rock?**
- **Can you name at least three examples of this rock type?**

### Section Review 19-3

1. High temperatures and high pressures.
2. Igneous or sedimentary.
3. Marble from limestone; slate from shale; quartzite from sandstone.

## 19-4 THE ROCK CYCLE

### SECTION PREVIEW 19-4

In this section, students will be introduced to the rock cycle. The rock cycle is the continuous changing of rocks from one kind into another over long periods of time.

Students will learn that rocks are changed in many ways. Some of these changes occur on the surface of the earth as rocks are broken down by weathering and erosion. Other changes occur as existing rocks become buried deep beneath the earth's surface, and are subjected to high temperatures and great pressure.

Students will discover that as a result of the rock cycle, rocks can change from one major type into another. For example, an igneous rock can be weathered into sediments that eventually form sedimentary rock.

### SECTION OBJECTIVES 19-4

1. **Describe the rock cycle.**
2. **Explain how rocks are changed above and below the earth's surface.**

### SCIENCE TERMS 19-4

**rock cycle p. 445**

## 19-4 The Rock Cycle

Rocks go through many changes. Igneous and sedimentary rocks can change to metamorphic rocks. Metamorphic rocks, too, may be remelted and become igneous rocks again. ❶ **The continuous changing of rocks from one kind to another over long periods of time is called the rock cycle.**

**Figure 19-16** *This diagram of the rock cycle shows the many ways that rocks are changed. What changes sedimentary rocks to metamorphic?* ❶

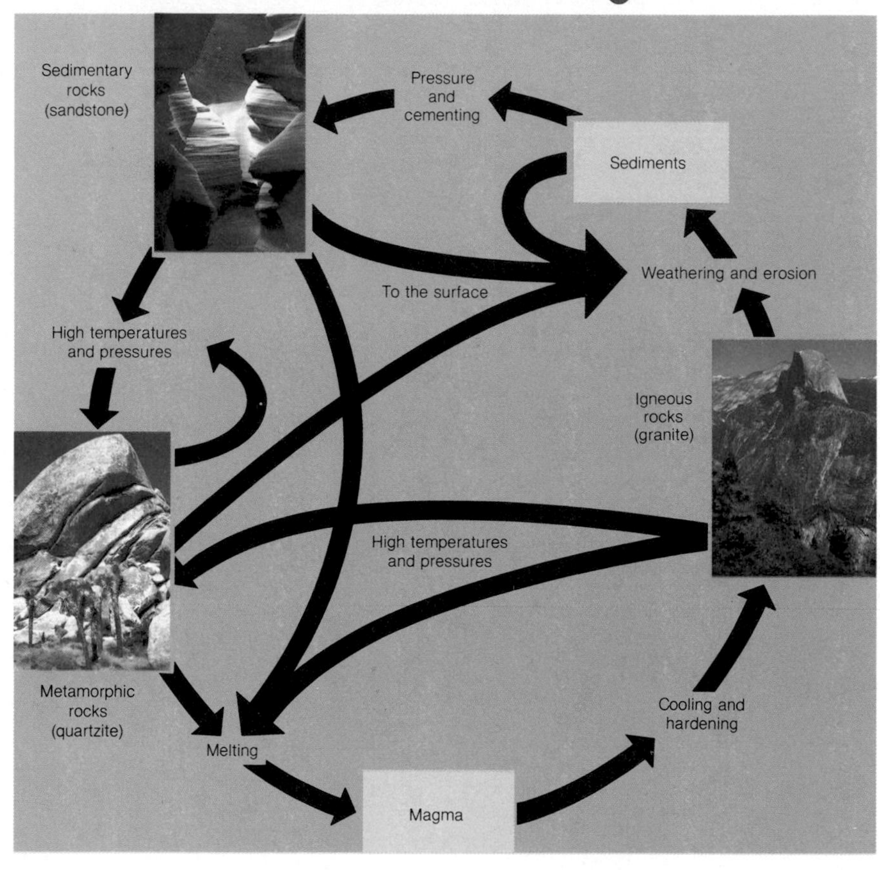


444

### TEACHING STRATEGY 19-4

#### Motivation

Have students observe Figure 19-16.

- **What types of rock are shown in the diagram?** (sedimentary, metamorphic, and igneous rocks)
- **What do these large black arrows tell you?** (Each rock type can change into another rock type.)

#### Content Development

Emphasize to students that existing rocks are constantly being changed, both on and below the earth's surface.

- **What are some factors that change rocks on the earth's surface?** (Wind, rain, changes in temperature, and chemical reactions. Point out that some chemical reactions occur as a result of air pollution, which introduces acids and other chemicals into the atmosphere.)
- **What are the factors that change rocks below the earth's surface?** (intense heat and pressure)
- **Can a sedimentary rock be changed into an igneous or metamorphic rock and then be changed again into a sedimentary rock?** (yes)

#### Content Development

The rock cycle explains the relationship between the three classes of rocks. All rock begins as magma. It is extruded onto the earth's surface or is cooled within the earth. Once it cools and solidifies, it becomes igneous rock. If parts of this igneous rock are eroded or broken away and then reconsolidated into rock, this new rock is called sedimentary. This rock can then be changed back to igneous rock through subduction—the process whereby the ocean's floor

Let's follow a rock on its ages-long journey through the **rock cycle.** See Figure 19-16. A huge dome of granite, an igneous rock, lies exposed to the wind and the rain, the cold of winter, and the heat of summer. Millions upon millions of years ago, this mass of granite was seething molten rock within the earth. It pushed upward against the solid rock above it, forming a dome. The dome hardened into solid rock underground. Slowly, the rock and soil that covered the dome were washed and ground away by wind and rain. Now the harder granite stands alone on the earth's surface.

But though the granite is hard, it also eventually will be worn down under the steady force of wind, water, and temperature changes. Bits of granite will flake off. Dragged along in rushing streams, these sediments may be reduced to powder. Much of this powder will be sand grains, which are made up mainly of the mineral quartz. The quartz will mix with water and other minerals, such as aluminum. The small particles that form are called clay. Some of this clay, as you will see in the next chapter, will become part of the soil.

But these sediments may take another path. They may be deposited on the sea floor. Under high pressure, the sediments may be cemented together in layers to form a sedimentary rock such as sandstone. These rocks may become buried deep under the earth's surface.

High pressure and high temperature may then change the sandstone into quartzite, a metamorphic rock. The quartzite may be exposed at the earth's surface and eventually become eroded to sediments again. Or the quartzite may become molten deep inside the earth. The magma that forms may cool and harden back into granite again. So the rock cycle goes on and on. ❷

### *SECTION REVIEW*

1. What are the three main groups of rocks? Give one example from each group of rocks.
2. How do wind, temperature, and water cause changes in rocks?
3. What is the rock cycle? What two factors in this cycle may change sandstone to quartzite?

**Activity**

*Rock Samples*

You can become more familiar with the various types of rocks by collecting and observing them.

**1.** Start a rock collection. To obtain a good sampling, try to collect rocks from several different locations. Note where you found each specimen.

**2.** Bring your rock collection to class. Using reference materials, identify each rock.

**3.** Tape each rock to a labeled note card or label the rock directly. Indicate the type of rock, its name, and the location in which it was found.

**4.** Organize your rock specimens according to the various classifications you have learned about in this chapter.



## Activity

**Rock Samples**
**Skills: Classifying, observing, comparing, applying, inferring**
**Level: Enriched**
**Type: Field/hands-on**
**Materials: Tape, note card, reference books**

This activity will force students to use field and observational skills as they build their rock collections. You can alter this activity by taking students on a guided walk around the school grounds. Have them locate samples of igneous, sedimentary, or metamorphic rocks and note the locations and environmental conditions of the areas where they find the different rock samples. Using a field guide or other reference book, help students identify the types of rocks they find.

## ANNOTATION KEY

❶ **High temperatures and pressures (Interpreting diagrams)**

❶ **Thinking Skill: Applying concepts**

❷ **Thinking Skill: Relating cause and effect**

## TIE-IN/PHYSICAL SCIENCE

The rock cycle illustrates the law of conservation of matter. The law of conservation of matter states that, under normal conditions, matter may change from one form into another, but the total amount of matter is neither created nor destroyed.

along trenches is pushed back into the interior of the earth and the rock is remelted.

The sedimentary rock could also change into metamorphic rock. This can occur when the sedimentary rock is exposed to extreme heat and pressure. The new metamorphic rock could then erode to form a new sedimentary rock. This continuous process is known as the *rock cycle.*

## Skills Development

***Skills: Interpreting diagrams, making diagrams***

To make the rock cycle more understandable to students, have each student choose one type of rock—sedimentary, igneous, or metamorphic—and diagram its possible pathways. Challenge students to show clearly those changes that occur on the earth's surface, and those that occur below the earth's surface. Have students share their diagrams with the class, and compare them to Figure 19-16.

## Section Review 19-4

**1.** Igneous, sedimentary, metamorphic; granite, sandstone, quartzite.
**2.** They wear down rocks.
**3.** The continuous changing of rocks from one kind to another. High pressure and high temperature.

## LABORATORY ACTIVITY MAKING A SEDIMENTARY ROCK

### BEFORE THE LAB

1. **Keep in mind that this activity requires one class period, followed by about 20 minutes of class time 1 to 3 days later.**
2. **At least one day prior to the investigation, gather enough materials for your class, assuming 6 students per group.**

### PRE-LAB DISCUSSION

Review the discussion of sedimentary rocks in Section 19-2. After students have reviewed this material and read the lab, ask,

- **What is the medium that carries and collects the sediments that form sedimentary rocks?** (water)
- **How is this medium going to be used in this activity?** (The substances that are to form the sedimentary rock will be dissolved in water.)

Discuss with students the idea of making a model.

- **Why is the making of a model useful?** (A model can be used to represent something that is difficult to observe or measure directly; it can also be a miniature representation of something.)
- **What does your model represent?** (a miniature version of a sedimentary rock)
- **Would it be possible to observe an actual sedimentary rock being formed?** (no, not really)
- **Why not?** (Sedimentary rocks take thousands, even millions of years to form.)

Point out that earth scientists often need to make miniature versions of natural phenomena in the laboratory, because natural processes may occur too slowly or over too large an area to be observed directly.

## LABORATORY ACTIVITY

### *Making a Sedimentary Rock*

**Purpose**

In this activity, you will make your own sedimentary rock.

**Materials** *(per group)*
25 g alum
Sand
Metric ruler
Plastic butter or margarine container
100-mL graduated cylinder
Stirring rod
250-mL beaker
Triple-beam balance

**Procedure**  

1. Use a stirring rod to dissolve 25 g of alum in 100 mL of warm water.
2. Fill the bottom of a plastic container to a depth of 2 cm with sand.
3. Pour the alum solution onto the sand in the container. Stir the mixture so that the alum solution is evenly distributed throughout the sand.
4. Pour off and throw away any excess solution. Gently tap the plastic container to help the sand grains settle together.
5. Place the container in a location where it can cool slowly and evaporate to dryness. This usually takes about a day.
6. Gently twist the sides of the container to free its contents. You have created a sedimentary rock.

**Observations and Conclusions**

1. Describe your rock in terms of color, density, and overall appearance.
2. How is your rock similar to a piece of real sandstone?
3. How is your rock different from a piece of real sandstone?

446

### SKILLS DEVELOPMENT

Students will use the following skills while completing this activity.

1. Observing
2. Comparing
3. Manipulative
4. Relating
5. Measuring
6. Safety
7. Applying

### SAFETY TIPS

Have students be careful when handling glassware and the alum.

## CHAPTER REVIEW

### SUMMARY

**19-1 Rocks of Liquid and Fire: Igneous Rocks**

▬ Deep within the earth it is so hot that rock-forming minerals exist only in a liquid form known as magma.

▬ If magma reaches the surface of the earth, it is known as lava.

▬ Igneous rocks formed from the cooling of molten rock at or near the earth's surface are called extrusive rocks. Examples include basalt, obsidian, and pumice.

▬ Igneous rocks formed from the cooling of molten rock deep beneath the earth's surface are called intrusive rocks. Granite is an example.

**19-2 Rocks in Layers: Sedimentary Rocks**

▬ The hardening of layers of sediment results in the formation of sedimentary rocks. These rocks usually are formed in water.

▬ Clastic sedimentary rocks are formed by the cementing together of rock fragments. Conglomerates, sandstones, and shales are examples of clastic rocks.

▬ Organic sedimentary rocks are formed from the cementing together of the remains of living things. Chalk is an example.

▬ Chemical sedimentary rocks are formed as large amounts of minerals—once dissolved in water—harden together.

**19-3 Rocks That Change: Metamorphic Rocks**

▬ Rocks that underwent high temperatures and high pressures without melting lost their original features and are called metamorphic rocks. Marble and slate are examples.

**19-4 The Rock Cycle**

▬ The continuous changing of rocks from one kind to another is called the rock cycle.

### VOCABULARY

*Define each term in a complete sentence.*

| | | |
|---|---|---|
| **chemical rock** | **intrusive rock** | **organic rock** |
| **clastic rock** | **lava** | **rock cycle** |
| **extrusive rock** | **magma** | **sedimentary rock** |
| **igneous rock** | **metamorphic rock** | |

### CONTENT REVIEW: MULTIPLE CHOICE

*Choose the letter of the answer that best completes each statement.*

**1.** Molten rocks that cool at the earth's surface are
a. igneous rocks. b. extrusive rocks.
c. volcanic glass. d. all of these.

**2.** A kind of igneous rock that has cooled so quickly that gas bubbles remain trapped inside is called
a. basalt. b. pumice. c. obsidian. d. granite.

**3.** Which of these is an example of an intrusive rock?
a. granite b. obsidian c. magma d. pumice



## TEACHING STRATEGY FOR LAB PROCEDURE

You may want to circulate through the room to ensure that students are performing the activity correctly. Provide a place for students to leave their plastic containers. For easy identification, have one member of each group place their names on the outside of their containers.

## OBSERVATIONS AND CONCLUSIONS

**1.** The color of the rock formed should not be much different from that of the original sand. The rock will probably resemble many kinds of sandstones in its general characteristics.
**2.** Small grains are cemented together, as in real sandstone.
**3.** It will be more brittle, tending to crumble between the fingers, and the cementing alum will dissolve in water, releasing the grains of sand. Its density will probably be lower than that of most sandstones.

## GOING FURTHER: ENRICHMENT

**Part 1**

Students can repeat the steps in this activity after their piece of "sandstone" has been made. In this way they can see how new layers are deposited on an existing stone, forming the strata that are characteristic of sedimentary rocks.

**Part 2**

Challenge students to design an experiment in which they investigate the rates of sedimentation of various materials. For example, students might mix gravel, sand, and garden soil, then determine which material settles first and which takes longest to settle.

## CHAPTER REVIEW

### MULTIPLE CHOICE

| | | | | |
|---|---|---|---|---|
| **1.** d | **3.** a | **5.** b | **7.** b | **9.** d |
| **2.** b | **4.** c | **6.** d | **8.** c | **10.** d |

### COMPLETION

**1.** lava
**2.** obsidian
**3.** batholiths
**4.** conglomerates
**5.** shale
**6.** organic
**7.** corals
**8.** chemical
**9.** slate
**10.** rock cycle

### TRUE OR FALSE

**1.** F magma
**2.** T
**3.** F intrusive rocks
**4.** T
**5.** T
**6.** F fragmental rock
**7.** T
**8.** F limestone
**9.** T
**10.** T

### SKILL BUILDING

**1.** Obsidian is glassy-looking and does not contain crystals, whereas granite is crystalline and grainy. Obsidian is extrusive and is formed by rapid cooling, whereas granite is intrusive and is formed by slow cooling, which gives large crystals a chance to form.
**2.** Sedimentary rocks form when layers of material are laid down over a period of time and are often cemented together. Such rocks are often formed in water and can contain the remains of once-living things that were deposited on the floor of the body of water.
**3.** Texture and chemical composition would be the first ways to identify the two rocks. Features such as hardness, ability to easily break into smaller pieces, presence of crystals, and presence of mica flakes would be helpful.
**4.** The sediments from which sedimentary rocks are formed can be deposited without disturbing dead organisms. The heat of the magma from which igneous rocks are formed would destroy most organisms. The heat and pressure needed to form metamorphic rocks would also destroy most organisms.
**5.** Metamorphic quartzite may become eroded at the earth's surface, or may be melted to magma inside the earth. After cooling and hardening, it forms granite. The granite may be worn down by weathering and erosion and become fragments that may be reduced to powdery sediments made up mainly of quartz sand grains. This may mix with water and minerals and form clay, which may in turn become part of the soil. The powdery sediments may, alternately, be deposited in water and cemented together to form sedimentary rocks, such as sandstone. After such rocks are deeply buried, high pressure and temperature may change the stone back into quartzite.
**6.** Help students identify the sedimentary, metamorphic, and igneous rocks they have collected. Work with students to help them complete their rock cycle.
**7.** The rock formed when the mineral crystals in the original rock flattened under pressure.

---

**4.** The amount of time it takes molten rock to cool and harden mainly affects the rock's
a. color. b. mass. c. texture. d. all of these.
**5.** Which of these igneous rocks cooled most slowly?
a. obsidian b. granite c. pumice d. volcanic rock
**6.** Which sedimentary rock is formed from the cementing together of sand grains?
a. granite b. shale c. limestone d. sandstone
**7.** Which of these is an example of a clastic rock?
a. basalt b. shale c. chalk d. obsidian
**8.** Organic rocks are formed from
a. the cooling of magma.
b. the cementing together of sand and clay.
c. the remains of living things.
d. all of these.
**9.** Which of these is an example of a metamorphic rock?
a. pumice b. shale c. chalk d. marble
**10.** In the rock cycle, weathering and erosion can change which rocks into sediments?
a. igneous b. sedimentary c. metamorphic d. all of these

#### CONTENT REVIEW: COMPLETION

*Fill in the word or words that best complete each statement.*

**1.** Molten rock that reaches the earth's surface is called ______.
**2.** ______ is often called volcanic glass.
**3.** Gigantic masses of granite called ______ form the bases of the earth's mountain ranges.
**4.** An example of a rock formed by the cementing together of rock fragments is ______.
**5.** ______ is a sedimentary rock formed from the hardening of layers of mud and clay.
**6.** Limestone is ______ sedimentary rock formed from the shells of ocean animals.
**7.** Limestone reefs form from the hard outer coverings of ______ that become cemented together.
**8.** Sedimentary rocks that form from large amounts of minerals that were once dissolved in water are called ______ rocks.
**9.** ______ is a metamorphic rock formed from clay or shale.
**10.** The continuous changing of rocks from one kind to another over long periods of time is called the ______.

#### CONTENT REVIEW: TRUE OR FALSE

*Determine whether each statement is true or false. If it is true, write "true." If it is false, change the underlined word or words to make the statement true.*

**1.** Hot, molten rock deep inside the earth is called lava.
**2.** Granite and obsidian are igneous rocks of similar chemical composition.

448

3. Molten rocks that cool and harden beneath the earth's surface are extrusive rocks.
4. Due to rapid cooling, the crystals of extrusive rocks are smaller than those of intrusive rocks.
5. Sedimentary rocks may be rich in fossils.
6. A conglomerate is an example of a(n) chemical rock.
7. Chalk is a sedimentary rock made of small pieces of animal shells and limestone.
8. Strong vinegar will react chemically with most sandstone.
9. Metamorphic rocks can change into igneous rocks.
10. Under high temperature and high pressure, sandstone, a sedimentary rock, may change into quartzite, a metamorphic rock.

## CONCEPT REVIEW: SKILL BUILDING

*Use the skills you have developed in the chapter to complete each activity.*

1. **Applying facts** Obsidian and diorite are both igneous rocks. However, they have different appearances. Obsidian is dark and glassy. Diorite is light-colored and coarse-grained. How do you account for the differences in these two rocks?
2. **Identifying patterns** Describe how and where sedimentary rocks form. Explain why scientists study sedimentary rocks to learn about prehistoric life.
3. **Classifying rocks** What information would you use to determine whether a rock sample is shale or slate?
4. **Relating cause and effect** Explain why fossils are present in sedimentary rocks but are not usually in metamorphic or igneous rocks.
5. **Sequencing events** Describe what may happen to a huge outcrop of quartzite as it goes through the rock cycle.
6. **Making diagrams** Obtain samples of sedimentary, metamorphic, and igneous rocks. Using Figure 19-16 on page 444, illustrate the rock cycle with the rocks you have collected.
7. **Making inferences** Suppose you find a metamorphic rock that breaks in layers. How do you think this rock formed? Hint: The minerals in the rock have different densities.

## CONCEPT REVIEW: ESSAY

*Discuss each of the following in a brief paragraph.*

1. Clastic rocks, organic rocks, and chemical rocks are all examples of sedimentary rocks. Describe the differences among these three types and list one example of each.
2. How does an igneous rock become part of a sedimentary rock?
3. Explain why 75 percent of the rocks on the earth's surface are sedimentary rocks.
4. Relate the cooling rate of magma to the crystal size in igneous rocks.
5. Compare extrusive and intrusive igneous rocks. Give an example of each.
6. How can the shell of a snail become part of a sedimentary rock?
7. What three factors are responsible for the formation of metamorphic rocks from other rocks?

449

## ESSAY

**1.** Clastic rocks, such as sandstone, are made up of rock fragments mixed with sand or clay and cemented together. Organic rocks, such as limestone, are built up from the remains of living things. Chemical rocks, such as rock salt, are formed by evaporation of water.
**2.** During the chemical and physical weathering process, small pieces of the igneous rock could be cemented together with other sediments to form a sedimentary rock.
**3.** Most of the earth was under water at one time in the past. Sedimentary rocks are formed from layers of sediments carried by wind and water.
**4.** Magma that cools quickly on the earth's surface forms igneous rocks with small crystals. Magma that cools slowly beneath the earth's surface forms igneous rock with large crystals.
**5.** Intrusive igneous rocks are formed deep within the earth; granite, diorite, gabbro, peridotite. Extrusive igneous rocks are formed at the earth's surface: rhyolite, andesite, pumice, basalt, obsidian.
**6.** When the snail dies, the shell could break apart or begin to decay. The sediments of the shell could then become part of a sedimentary rock.
**7.** Tremendous heat, pressure, and chemical reactions

## ADDITIONAL QUESTIONS AND TOPIC SUGGESTIONS

**1.** In the introduction to this chapter, you read about the legend of Giant's Causeway. Visit the library and find photographs and a description of the White Cliffs of Dover. Imagine that you are living in ancient times. Can you create your own "legend" to describe how the White Cliffs of Dover were formed?
**2.** A fossil of a fish is found on a mountaintop in South America. Based on your knowledge of rocks, how do you think the fossil got there? (Fossils are found in sedimentary rocks, and most sedimentary rocks form in water. The most logical explanation for the existence of the fossil is that this area of South America was once covered by ocean, and the fish fossil formed while the region was under water. When the ocean water receded, the fossil remained in its original location.)

## ISSUES IN SCIENCE

The following issue can be used as a springboard for discussion or given as a writing assignment.

A cultural exchange committee in the U.S. proposes that an Egyptian stone pyramid be brought to New York City and put on display in Central Park. Egyptian officials protest the idea, saying that in less than a year's time, the pyramid will be ruined. Based on your knowledge of rocks, do you think that the Egyptians' objections are valid? Explain. Can you think of a better place in the United States to display the pyramid?

# Chapter 20
# SOILS

## CHAPTER OVERVIEW

In this chapter, students will learn about soil. They will discover how soil is formed, and how different environmental factors produce various kinds of soil.

Students will learn that soil consists of distinct layers—topsoil, subsoil, and parent rock. They will also learn about the importance of humus, which is the rich, dark soil that is part of topsoil.

Students will be introduced to the major soil types found on Earth. These are tropical rain-forest soil, temperate forest soil, prairie grassland soil, desert soil, and tundra soil. Students will discover why certain soils are more fertile than others, and how leaching affects soil formation.

## INTRODUCING CHAPTER 20

Begin by having students observe the chapter-opener photograph and read the text.

• **What do you see in this photograph?** (a vast area of grass)
• **What is most striking about this land surface?** (It is very flat, with little variation in the land surface.)
• **Where do you think the land in this photograph is located?** (in the midwestern part of the United States)
• **To early settlers, how did this land differ from the land in the eastern part of the United States?** (The land in the East was like a great forest, full of tall trees.)
• **In what other important way was the land in the Midwest different from the land in the East?** (The soil was different.)
• **Do you think that the different soil accounts for the fact that large trees grew in the East, while grass grew in the Midwest?** (yes)
• **What did the early settlers not know about the soil in the Midwest?** (that it is some of the best farming soil in the world)
• **According to the text, what kind of land lay west of the prairie grasslands?** (desert)
• **What kind of soil is found in this region?** (Much of it is barren powdery rock.)
• **What do you think causes different soil types to form in the eastern, midwestern, and western parts of the United States?** (Answers will vary. Stress that as students study this chapter, they will learn more about what causes different soil types to form, and how these factors affect the plants and trees that grow in a region.)

As you discuss this chapter, keep in mind the many cause-and-effect relationships that exist with regard to soils. Emphasize the idea that environmental factors such as temperature and rainfall affect the soil type

# 20 Soils

**CHAPTER SECTIONS**



**CHAPTER OBJECTIVES**

*After completing this chapter, you will be able to:*

**20-1** Describe the different soil layers.

**20-1** Identify the environmental factors that cause weathering.

**20-2** Relate leaching to soil formation in tropical rain forests.

**20-3** Explain why temperate forest soil does not provide enough minerals for widespread plant growth.

**20-4** Explain how prairie grassland soil came to be so fertile.

**20-5** Explain why little humus forms in desert soil.

**20-6** Explain why tundra soil is different from most types of soil.

To the first settlers from Europe, much of North America looked like one vast forest. It stretched westward as far as the settlers could see.

Or so it seemed. The first settlers cleared some forest and built their towns along the Atlantic coast of the new continent. As more settlers arrived and families grew, the settlers slowly pushed westward. By the 1800s, they reached the edge of the forest—the Midwest.

It was like standing between two worlds. Behind the settlers, to the east, lay the forest—shaded, cool and damp. Under a thin layer of leaves, the soil was white and ashlike. To the west, in brilliant sunshine, a sea of grass rippled in the wind. Beneath the grass, though the settlers did not know it yet, lay some of the best farming soil in the world.

Beyond the western edges of the great grasslands, the grass grew scarcer and died away into desert land with cactus plants and shrubs. Here much of the soil was barren powdery rock.

The pioneers that settled on this new continent did not know why soils were different in forests, grasslands, and deserts. They may have guessed that factors in the environment were important. And they would have been right.

*The prairie grasslands*



that forms in a region. Also stress that the soil type found in a region influences the environment—first in terms of the type of plant growth, then in terms of the area's economic and social development.

As they study this chapter, students should become aware of how soil type has influenced the growth and development of their own region. If they live in an area where farming or forestry are important occupations, students are probably quite aware of the importance of soil and will easily understand how soil type has affected their region. If, however, they live in a highly industrialized or urban area, students may be somewhat oblivious to the importance of soil. If this is the case, they may want to obtain information about areas in their state where soil plays an important role in the economy and lifestyle.

## TEACHER DEMONSTRATION

This demonstration will illustrate to students the composition of a typical soil sample. For the demonstration, you will need a large sheet of paper and a bucket of soil from a forest area, grassy field, or other natural site.

Spread out the soil on a large sheet of paper. Have students assist you in separating out large rocks, small rocks, decaying organisms, and humus.

- **How is humus different from the rest of the soil** (It is darker and more dense; it may also be more moist than the rest of the soil.)
- **What special value do you think humus might have?** (Humus is rich in the nutrients needed for growing plants, and it is able to hold a great deal of water.)

Challenge students to identify, if they can, the decaying organisms found in the soil.

- **What do you think will happen to these decaying organisms if they remain in the soil?** (They will become part of the rich, dark soil called humus.)
- **What do you think will happen to the rocks if they remain in the soil?** (The rocks, if they remain near the surface of the soil, will be broken down by water, wind and other environmental factors. Eventually they will become part of the soil.)

## TEACHER RESOURCES

**Audiovisuals**

*Erosion,* 16 mm film, BFA
*Geological Weathering,* 20 slides, SVE
*Soil: Its Meaning for Man,* filmstrip with cassette, PH Media
*The Changing Land,* filmstrip with cassette, NGS

**Books**

Bloom, A. L., *The Surface of the Earth,* Prentice-Hall
Carroll, Dorothy, *Rock Weathering,* Plenum
Ollier, C. D., *Weathering,* Longman

## 20-1 SOIL

### SECTION PREVIEW 20-1

In this section, students will learn about the formation and structure of soil. They will also learn how different environmental factors produce various kinds of soil.

Students will learn that soil consists of three main layers: topsoil, subsoil, and parent rock. Topsoil contains the rich, dark soil called humus.

Students will discover that soil is produced by the weathering of rocks. They will learn that rocks are weathered by wind, water, chemicals, and changes in temperature.

### SECTION OBJECTIVES 20-1

1. **Identify the three main soil layers.**
2. **Explain the importance of humus.**
3. **Describe the process of weathering.**

### SCIENCE TERMS 20-1

**topsoil p. 452**
**humus p. 452**
**subsoil p. 453**
**parent rock p. 453**
**weathering p. 454**

## 20-1 Soil

Some people think of "soil" as being another word for dirt. And when people say something is "dirt cheap," they mean it is practically worthless. But soil is precious and vital to life on the earth. Without soil, most plants could not grow on the land. Without plants, no animals would survive. The earth's surface might look as barren as the moon.

**Figure 20-1** *The two main ingredients of soil are pieces of rock and organic material, which is material that was once living or was formed by living organisms. How can you tell that the soil in the top photograph is low in organic material and the soil in the bottom photograph is high?* 1

### *Soil Layers*

You can find out a lot about how a house is built by looking at what holds it up and what holds it together—its structure. But usually you cannot see much of the structure without taking the house apart. The framework of a house is hidden in the walls and floors. The foundation is underground.

The structure of soil is much the same. Soil is a covering over most of the earth's land surface. This soil covering can range in thickness from a few millimeters to a few meters. However, all you usually see of the soil covering is the very top.

Like a house with several floors, soil has different layers. **The three layers of soil are the topsoil, subsoil, and parent rock.** The uppermost layer is made up of **topsoil.** Topsoil is a mixture of small grains of rock and the decayed matter of plants and animals. 1 How did this mixture come about? You can see its beginnings on the forest floor. Year after year a carpet of wood, pine needles, leaves, and bits of twigs is laid down. A rich variety of animals lives on and under this carpet, including worms, snails, insects, and mice.

Year after year, the remains of plant and animal life decay to become a part of the topsoil. The decayed matter combines with rocky soil particles to form a rich, dark soil called **humus.** Humus supplies essential chemicals for growing plants. Because humus is spongy, it stores water. In fact, humus can hold up to 600 times its own weight in water. Humus soil also contains many air spaces, or pores. When mixed with rock grains, humus forms a soil that allows air and water to reach plant roots.



### TEACHING STRATEGY 20-1

#### Motivation

Obtain soil samples of varying qualities. Try to include one sample of very rich humus; one moderately rich sample of topsoil; one sample of dry, dusty soil; and one sample of soil containing mostly sand or gravel. (A nursery or farm in your area may be helpful in providing these samples.) Display the soil samples and have students observe the differences.

- **Which of these soils do you think would be best for growing plants?** (the humus and the topsoil)
- **Why?** (The richer soils contain more of the nutrients that plants need.)

#### Content Development

Use Figure 20-2 to emphasize the different soil layers. Use the drawing to point out how some plant roots penetrate well into the subsoil, while others extend only a short distance into the topsoil. This concept will be important for students' understanding of leaching.

#### Skills Development

*Skill: Inferring*

- **Which type of soil is more valuable for growing crops—soil in which the topsoil layer is a few centimeters thick, or soil in which the topsoil layers is more than a meter thick?** (soil in which the topsoil layer

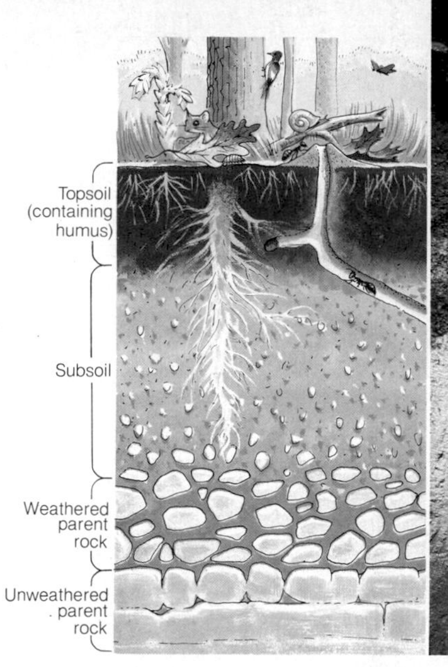

**Figure 20-2** *Soil has different layers, as shown in an artist's view* (left) *and in a photograph* (right).

The thickness of the topsoil may be anywhere from a few centimeters to a meter or more. Beneath it is the **subsoil** level. This level is made up of larger bits of rock and has little or no humus.

As you dig down further into the soil, the rock particles become larger and more numerous. Finally you reach solid rock, which is called the **parent rock.** Most minerals and other rocky particles in the soil first originated from this parent rock. But how could tiny rock particles come from solid parent rock?

### *Soil and the Environment*

If you place a rock on the ground and strike it with a hammer, the rock will break into pieces. Smash these pieces and they become even smaller. Continue smashing the pieces and they may even become part of the soil. Billions of years ago, before soil formed, the land surfaces of the earth were solid rock—parent rock. Naturally, there were no giant hammers crushing the parent rock into soil. But in a way environmental factors such as wind, water, heat,

#### Activity

*Humus*

**1.** Obtain some topsoil from a forest or grassy area or from a gardener who uses compost to enrich the soil.

**2.** Carefully sort through the soil or compost. Using a magnifying glass, separate the small particles of soil from the particles of decaying plants and animals.

What type of soil particles are in your topsoil? What does the soil look like?



## ANNOTATION KEY

❶ **The soil in the bottom photograph is high in organic material because it is darker in color than the top photograph. The bottom photograph also contains small pieces of plant matter. (Applying concepts)**

❶ **Thinking Skill: Classifying soil layers**

## Activity

**Humus**
**Skills: Classifying objects, comparing, relating, manipulative, observing**
**Level: Remedial**
**Type: Hands-on**
**Materials: Topsoil, magnifying glass**

This hands-on activity reinforces students' observational skills and calls upon them to describe and analyze their observations. The particles students find in their soil samples will vary depending on the locale and the soil sample being observed.

is more than a meter thick)

- **Why?** (Many more minerals are present in the thicker topsoil, and these minerals are more easily reached by plant roots. With very thin topsoil, most plant roots will extend beyond the topsoil layer.)

## Motivation

Begin by asking students to think about the food they eat every day.

- **How are you dependent upon soil for food?** (Answers will vary. Certain foods such as lettuce, carrots, potatoes, and beans grow in soil. Dairy products come from cows, who eat grass that grows in soil. Meat and poultry come from animals, who eat grains that grow in soil.)

Point out that plants depend upon soil *directly* for the nutrients they need. Our dependence on soil is *indirect.* We eat plants, and we also eat animals who eat plants.

## Content Development

Emphasize to students the idea that a layer of rock always exists beneath soil. Point out that this rock is called bedrock. Bedrock continues to weather and form more soil. Stress that this weathering process occurs faster when animals such as earthworms and ants burrow in the soil above the rock. This is because the holes created by these animals enable water to reach the underlying rock.

## FACTS AND FIGURES

Although horizon, or soil, depths vary from place to place, some average values for fertile soil found in the U.S. are listed below.

A horizon (topsoil): 0–30 cm
B horizon (subsoil): 30–90 cm
C horizon (parent rock): 90 + cm

### Activity

**Studying Soil Layers**
**Skills: Making observations, manipulative, comparing, relating, applying**
**Level: Enriched**
**Type: Hands-on**
**Materials: Shovel, soil sample**

Students should gain firsthand observational knowledge of soil profiles by examining soil from their local region. Students' answers to the questions will vary depending on the nature of the soil sample being examined. The depth of the various soil layers in the samples dug out by students will depend on various general and local factors. You may wish to have students make drawings of their findings and display them in the classroom. In certain regions, very little topsoil will be present. In such cases, you may wish to have students relate their findings to the environmental conditions.

**Figure 20-3** *This photograph shows the longest natural bridge in the world. Made of sandstone, it was formed by the process of weathering.*

### Activity

*Studying Soil Layers*

Using a shovel, dig down about 0.5 meter and obtain a soil sample from your neighborhood. Be sure not to disturb the soil sample or ground too much or you will not be able to see the different soil layers. From your observations, answer the following questions.

1. How deep is the topsoil layer? What color is this layer?
2. How deep is the subsoil layer? What color is this layer?
3. How do the soils in the two layers differ?
4. Did you find the layer of weathered parent rock? Describe this soil layer.

and cold acted much like a hammer. The process by which rocks are broken down by the environment is called **weathering.** Weathering helped produce soil when the earth was young, and weathering still is producing soil today.

How can weathering produce soil? There are many ways. Water, for example, may seep into cracks in the rock. In cold weather, the water freezes. As water freezes, it expands to take up more space than liquid water. The ice pushes against both sides of the cracks and splits the rock into smaller pieces. Small plants, such as lichens, may grow on these rocks. These plants produce acids that break down the rocks into even smaller pieces. Eventually, weathering breaks down parent rocks into mineral particles. Remains of plants and animals get mixed in with these mineral particles and soil is formed.

Running water can also break down rocks into smaller soil particles. Whipping winds can do the same. Environmental factors such as these not only help produce soil, they also distribute the soil—often far from the region where it formed. On the steep slope of a mountainside, for example, strong winds and heavy rains can break down rock, which may eventually become a part of the soil. But as fast as the soil forms, floods can sweep it away. The soil may eventually be deposited many kilometers away on a flat plain, forming the rich soil important to farmers.

But to say that soil is created in this way is only part of the story. There are many kinds of soil. Different soils are shaped by different environmental



## 20-1 (continued)

### Reinforcement

Emphasize to students that soil results from the breakdown of both inorganic (nonliving) and organic (living) materials. Soil particles result from the weathering of rocks, which is inorganic. Humus results from the decay of plants and animals, which is organic.

### Section Review 20-1

1. Topsoil, subsoil, parent rock.
2. Wind, water, heat, and cold.

## TEACHING STRATEGY 20-2

### Motivation

Have students observe the photographs in Figures 20-5 and 20-6.

- **Are these plants and animals familiar to you?** (no)
- **Where would you expect to see these types of plants and animals?** (in the jungle)

### Content Development

Emphasize to students that the major factors influencing the formation of tropical rain forest soil are warm temperatures and heavy rainfall.

- **How do warm temperatures influence soil formation?** (Plant and animal remains decay very quickly.)

factors, including the weather, the land, and the organisms that live where the soil was formed. At the same time, the kind of soil in a place has an important influence on the kinds of things that can live in that place. For example, plants such as rhododendrons and azaleas grow best in soils containing large amounts of iron. Other plants have special needs for various elements in the soil.

### *SECTION REVIEW*

1. Describe the three layers of soil.
2. What environmental factors cause weathering?

## 20-2 Tropical Rain Forest Soil

There are three major areas in the world where tropical rain forests grow: the Amazon area of South America, parts of Africa, and parts of Southeast Asia. This green belt of tropical rain forests is near the equator, where there is much rain and the temperatures are warm year round.

If you were to walk through a tropical rain forest, you might be amazed by the many different kinds of plants and animals. All around you tall trees, some 60 meters or more in height, soar toward the sky to capture sunlight. And soar they must, for the forest

**Figure 20-4** *A tropical rain forest has many kinds of trees that grow very close together.*

## 20-2 TROPICAL RAIN FOREST SOIL

### SECTION PREVIEW 20-2

In this section, students will be introduced to tropical rain forest soil. They will learn that heavy rainfall and warm temperatures characterize tropical rain forest regions, and that these areas are rich in plant and animal life.

Students will learn that heavy rainfall in the tropical rain forest causes important minerals to be washed downward into the soil. This process is called leaching. Because of leaching, soil in these regions has only a thin layer of humus. Students will discover, however, that in this hot, damp environment, plant and animal remains decay quickly. As a result, needed minerals are constantly returned to the soil.

### SECTION OBJECTIVES 20-2

1. **Describe the characteristics of tropical rain forest soil.**
2. **Explain how leaching affects the development of soil.**

### SCIENCE TERMS 20-2

**leaching p. 457**

- **How does heavy rainfall influence soil formation?** (Minerals are leached deep into the subsoil.)
- **How do warm temperatures and heavy rainfall make up for one another?** (Leaching removes minerals rapidly from the topsoil; however, the quick decay of plant and animal matter constantly restores minerals to the topsoil.)

### Skills Development

*Skill: Inferring*

- **Would you expect large trees in a tropical rain forest to have deep roots?** (no)
- **Explain.** (Minerals are leached too deep into the ground for roots systems to reach them.)
- **How do trees in the rain forest receive nourishment?** (They have large surface root systems that can take up nutrients from decaying plants and animals.)

### Enrichment

Have students work in small groups. Challenge each group to create a display entitled, "Jungle Safari." Each display should show the different animals, plants, and trees that thrive in the tropical rain forest.

## TEACHER DEMONSTRATION

This demonstration will illustrate to students the size of various particles found in soil. For the demonstration you will need several screens with openings of various sizes; a large sheet of paper; and a bucket of soil from a forest area, grassy field, or other natural site.

Spread out the soil on a large sheet of paper. Have students assist you in separating out large rocks, small rocks, decaying organisms, and humus. Then pass the soil through the different-sized screens, beginning with the screen that has the largest opening. If there is enough variation in the screen sizes, you should be able to separate particles of sand, silt, gravel, and clay.

## BACKGROUND INFORMATION

The characteristics of tropical soil illustrate the importance of the nutrient cycle. In the nutrient cycle, plants take up nutrients from the soil, then return nutrients to the soil when they die and decay.

Plants with high nutrient demands are especially valuable to soil fertility, because they prevent nutrients from being leached. By taking nutrients from the soil and using them for growth, these plants are able to return the same nutrients to the top layer of soil when they die.

## FACTS AND FIGURES

Tropical soil tends to be red in color because it contains iron oxide. A common name for iron oxide is rust.

**Figure 20-5** *Some plants grow piggyback on the trunks and limbs of trees in a tropical rain forest.*

floor is dim and shadowy. Even at noon only scattered flecks of sunlight shine through the ceiling of trees and reach the forest floor.

You might wander for over a kilometer without seeing the same kind of tree twice! From some trees hang strange vines that weave their way around the tree trunks, forming coils and loops of living matter. Other vines snake through the air, waving in the warm tropical breeze. Colorful plants such as orchids grow on the bark of some trees. Surprisingly, you have little trouble walking on the forest floor. Very few plants block your way because they cannot grow where the light is so dim.

The tropical rain forest is a biologist's delight. Animals of all kinds, ranging from huge tarantulas that capture small birds for prey to brightly feathered parrots munching on nuts, live in the forest. Some animals, such as hunting leopards, glide through the shadows of the forest floor in search of food. Others, such as lizards and snakes, climb up and down tree trunks. Still others, including over 150 species of bats, fly through the air. In fact, scientists estimate that two-thirds of all plant and animal species make their home in tropical rain forests.

As you continue on your walk, you notice that many plants have leaves with unusually long pointed ends. These ends, called "drip tips," function like rain gutters. They allow water to quickly drip off the leaves of these plants. Drip tips are a big advantage in a tropical rain forest in which rainfall averages more than 200 centimeters per year. If the rain

**Figure 20-6** *Many interesting animals, such as these colorful macaws* (left) *and this large orangutan* (right), *live in a tropical rain forest.*

### 20-2 (continued)

### Skills Development

***Skill: Making maps***

Have students find out the specific areas of South America, Africa, and Southeast Asia that have tropical rain forest soil. Then have students make maps to show the location of these ares.

### Skills Development

***Skill: Inferring***

• **Why do so many animals live in the tropical rain forest?** (Abundant food supply; warm temperatures all year; easy to find shelter.)

### Section Review 20-2

**1.** Washing of minerals out of the topsoil; spreads minerals throughout the subsoil.
**2.** Rain constantly washes it away.

## TEACHING STRATEGY 20-3

### Motivation

Have students recall the chapter-opener text.

• **What was the first type of land that early settlers saw when they came to America?** (forests)

• **Where were these forests located?** (along the eastern coast of the United States)

• **What types of trees are present in**

could not quickly flow off the leaves, the plants might become covered with mold and die.

The heavy rainfall in the tropical rain forest plays a role in soil formation. Heavy rains can wash minerals important for plant growth downward into the soil. This process of washing minerals out of the topsoil is called **leaching.** Leaching is important in the formation of soils. **Normally, leaching can spread important minerals throughout the subsoil where roots can easily reach them. But in the tropics, where there is often too much rain for this to happen, the minerals are leached so far into the ground that they are lost to the plants.** ❶

If leaching removes important minerals from the rain forest soil, why are so many different kinds of plants able to survive? The answer is right above you—all you need do is look up to see the forest's main source of minerals. Leaves, flowers, and other plant parts constantly fall to the forest floor. Added ❷ to these remains are the bodies of dead animals. Since the forest is hot and damp all year long, these remains decay very quickly and form a rich supply of food for the plants. Although rain constantly washes away this thin layer of humus, it is constantly replaced by a new layer. Also, the root systems of rain forest plants are so efficient that almost all the decaying matter is absorbed by living plants.

**Figure 20-7** *Many tropical rain forest trees have extensive surface root systems that are so efficient they can recycle nearly all the nutrients from decaying plants.*

### *SECTION REVIEW*

1. What is leaching? Why is leaching important in the formation of soils?
2. Tropical soil has a thin layer of humus. Why?

## 20-3 Temperate Forest Soil

"The American forests have generally one very interesting quality, that of being entirely free from underwood. This is owing to the extraordinary height and spreading tops of the trees; which thus prevent the sun from penetrating to the ground."

So wrote a traveler in Pennsylvania in 1806. Today only small areas of these temperate forests still



## 20-3 TEMPERATE FOREST SOIL

### SECTION PREVIEW 20-3

In this section, students will be introduced to temperate forest soil. Temperate forest soil is found in most parts of the eastern United States.

### SECTION OBJECTIVES 20-3

1. **Describe the characteristics of temperate forest soil.**
2. **Explain why temperate forest soil does not provide enough minerals for widespread plant growth.**

### ANNOTATION KEY

❶ Thinking Skill: relating cause and effect

❷ Thinking Skill: Identifying patterns

**these forests?** (very large trees; many are evergreen trees; others are deciduous trees)

### Content Development

Emphasize that the two main factors that influence temperate forest soil formation are cold temperatures in winter and the leaching of minerals into the subsoil.

- **Rainfall is not heavy in temperate forest regions, yet important minerals are often leached out of the topsoil. Why does this happen?** (The heavy tree cover prevents sunlight from reaching the soil in order for rainwater to evaporate.)
- **How do cold temperatures affect the formation of temperate forest soil?** (When temperatures are cold, decay of plant and animal remains stops. Thus, little humus forms at this time.)

Students will learn that temperate forest soil contains little humus and that important minerals are often leached into the subsoil. For this reason, temperature forest soil is better for growing large trees than small plants.

Students will discover that climate plays an important role in the formation of temperate forest soil. Because there are four different seasons in temperate forest regions, plant and animal decay does not continue all year. Little humus is produced in cold weather, adding few minerals to the soil during this time.

## SCIENCE, TECHNOLOGY AND SOCIETY

As the world's population reaches 5 billion, the agricultural industry must keep up with the continuing demand for food. To successfully supply the world with food, this industry must better manage its most important natural resource—soil. Poor management of soil in some areas of the world has resulted in the desertification of over 90 million hectares of grassland. When a grassland is overgrazed by cattle and/or sheep in an attempt to produce more food, it becomes a desert. The land's nutrients are no longer replaced; erosion increases; plants die; and animals and insects move elsewhere.

Ironically, the solution to this problem is grazing, but controlled grazing. In controlled grazing, large herds of cattle or sheep are allowed to graze in a pasture while another pasture is left untouched. This method has many beneficial effects: The animals' hooves break up the soil so that plants can take root; the animals' trampling breaks off dead plant growth and provides nutrients for seedlings; and areas that are grazed are then given time to recover. Applying controlled grazing can help meet the world's demand for food and prevent a grassland from becoming a desert.

exist in the United States. But they still take up a lot of ground. Temperate forests include such trees as beech, maple, oak, hemlock, wild cherry, and dogwood, as well as a variety of evergreens.

The traveler's description of the temperate forest sounds something like that of the tropical rain forest. Of course, there are important differences. The climate of the tropical forest is hot and damp. (1) The climate of the temperate forest ranges from cold and fairly dry in the north to warm and damp further south. There are four different seasons in all temperate forests. In the cold seasons, particularly in the northern forests, most decay of plant and animal remains stops. As you might expect, little humus is produced in the cold weather and few minerals are added to the soil. However, even in warm weather, the temperate forest floor cannot support a wide variety of plant life. Why?

### Career: *Soil Conservationist*

**HELP WANTED: SOIL CONSERVATIONIST** Bachelor's degree in soil conservation or natural resource sciences such as agronomy, forestry, or agriculture necessary.

The rain poured down hard and fast. Most of the valuable topsoil washed into the stream that ran across the farm. Enough soil filled the stream to cause flooding and destruction of the crops beside it. To obtain help, the farmer contacted the area's soil conservation service. A **soil conservationist** was immediately sent to inspect the situation.

Soil conservationists advise farmers and ranchers how best to water livestock or to prevent overgrazing. Other concerns include water conservation, sound land use, and environmental improvement. Soil conservationists prepare plans using soil conservation practices such as crop rotation, reforestation, terracing, and permanent vegetation. Usually in planning a soil program, soil conservationists work closely with government workers, farmers, ranchers, foresters, and urban planners.

Anyone who is interested in conserving and protecting our natural resources and who enjoys working with people might consider a career in soil conservation. In rural areas, positions arise in banks, insurance firms, and mortgage companies. Public utility, lumber, and paper companies also employ soil conservationists. Government positions are also available, such as in the Department of the Interior. For more information about this career, write to the United States Department of Agriculture, Soil Conservation Service, Room 5218—South, Fourteenth Street and Independence Avenue S.W., Washington, DC 20250.

### 20-3 (continued)

#### Skills Development

***Skill: Making maps***

Have students work in small groups. Challenge each group to find out where in the United States different soil types are found. Then have each group make a map to show this information.

#### Reinforcement

Emphasize to students that most minerals found in soil are the result of plant and animal decay. Thus, minerals appear first in the topsoil. Point out that heavy rainfall will "push" the minerals down into the subsoil.

- **What is this process called?** (leaching)
- **What happens when minerals are leached far down into the subsoil?** (Roots cannot reach them.)

#### Enrichment

Have students work in small groups. Have each group find out which plants and trees grow in northern temperature forest regions, and which grow in southern temperate forest regions. Then challenge each group to find a creative way to display this information.

**Figure 20-8** *Trees in a temperate forest grow many leaves during the summer* (left). *But these leaves do not grow during the winter. In the winter, organic decay and plant growth stop* (right).

In temperate forests, there is no heavy year-round rainfall. However, although the forest floor is matted with leaves and evergreen needles, rainwater often leaches out important minerals. **Unlike tropical rain forest soil, little decay of dead plant matter occurs in temperate forest soil before leaching carries minerals deep into the subsoil.** So little humus forms in temperate forest soil. ❷

Because the trees in temperate forests have deep roots, they can reach important minerals that have leached into the subsoil. However, smaller plants with shorter root systems cannot get to these minerals needed for plant growth. The lack of smaller plants is particularly true in northern temperate forests where almost all the plants are evergreen trees. The carpet of evergreen needles is especially poor in many important minerals. In more southern temperate forests, the leafy floor is richer in minerals and the warmer weather allows more decay to proceed at a faster pace. Why is this soil more fertile than soil in the most northern temperate forests?

### *SECTION REVIEW*

1. Would you expect the rate of decay of dead plant and animal matter to be faster in a tropical rain forest or in a temperate forest? Why?
2. List three trees, other than evergreens, that grow in temperate forests in the United States.

**Activity**

*Leaching*

In a 500-word essay describe how leaching affects each of the following soil types: tropical soil, temperate forest soil, prairie grassland soil, desert soil, and tundra.



## ANNOTATION KEY

❶ Thinking Skill: Relating cause and effect

❷ Thinking Skill: Making comparisons

## Activity

**Leaching**
**Skills: Comparing, relating, applying**
**Level: Remedial**
**Type: Writing**

Have students review Chapter 20 to write their essays on the effects of leaching on the various soil types. Students may want to write an overview based on the information given in the chapter, or they may be encouraged to concentrate on the leaching problems in one type of soil. In the latter case, students will have to supplement the text material with library research.

## HISTORICAL NOTES

Much of the knowledge that soil scientists have today comes from work done more than a hundred years ago by soil scientists in Russia. The Russian scientists noticed that over the vast area of their country, soil characteristics seemed closely associated with large-scale patterns of climate and vegetation. The studies of these scientists indicated that soil profiles show the influences of five factors: nature of the parent rock; climate; site; organisms; and time. Scientists today call these the factors of soil formation.

## Skills Development

***Skills: Identifying cause and effect; predicting***

Challenge students to consider the following question:

- **Suppose the weather in a temperature forest region were much warmer and wetter than usual. What effect would this have on the quality of the soil?** (Answers may vary. Above average rainfall would probably cause more leaching of minerals from the topsoil. Warmer temperatures would probably cause more rapid decay of plant and animal matter—thus improving the quality of the topsoil.)

## Section Review 20-3

**1.** Tropical rain forest. The warmer temperatures of the tropical rain forests help in the decay of plant and animal matter.
**2.** Oak, maple, beech.

## FACTS AND FIGURES

In general, the soil found in the southern United States is deeper than the soil found in the northern United States. This is because glacial sheets retreated from the northern U.S. only about 10,000 years ago, causing the soil to be younger.

## 20-4 PRAIRIE GRASSLAND SOIL

### SECTION PREVIEW 20-4

In this section, students will be introduced to prairie grassland soil. It is this type of soil that covers most of the midwestern United States.

Students will learn that prairie grassland soil is one of the best soil types for growing crops. Regions that have this type of soil are often called "breadbaskets of the world" because they grow so much of the world's wheat, corn, barley, and rye.

Students will discover that North America's prairie grassland soil formed as desert soil blew eastward. They will learn how the soil came to be extremely fertile as the growth and decay of grasses, heat and high winds all contributed to the formation of a deep, rich layer of humus.

### SECTION OBJECTIVES 20-4

1. **Describe the characteristics of prairie grassland soil.**
2. **Explain how prairie grassland soil became fertile.**

### ANNOTATION KEY

❶ Thinking Skill: Sequencing events

❷ Thinking Skill: Making comparisons

**Figure 20-9** *Darrel Coble* (far right) *was three years old in April 1936 when he and his family fled this dust storm in Oklahoma. During the 1930s, prairie grassland soils became useless for growing crops.*

## 20-4 Prairie Grassland Soil

To the first settlers, the great prairie grasslands of the midwestern United States appeared as a vast sea of grasses. It stretched from the eastern forest region to the Rocky Mountains.

At first, the settlers felt that these prairie grasslands were useless for growing crops. The settlers were used to planting crops in cleared forest land. They had never seen anything like this land and believed that where trees did not grow in large numbers, the ground could not be good for crops.

In time, the settlers discovered that the soil of these regions was some of the richest in the world. Today the Canadian and United States prairie grasslands are considered the "breadbaskets of the world," for here is where great oceans of wheat are grown. Wheat is the grain from which bread is made. There are similar regions in Argentina and the Ukraine region of the Soviet Union. Together, these regions produce nearly all of the grain crops grown in the world. Grain crops include corn, wheat, barley, and rye.

❶ The soils of North America's prairie grassland were not always so fertile. Millions of years ago most of the land west of the Rocky Mountains slowly turned to desert. As the soil dried up and crumbled to dust, winds carried it eastward and deposited the

### Activity

*The Use of Fertilizers*

Fertilizers are natural or artificial chemicals that take the place of humus. In the library, look up information about the many kinds of fertilizers used to improve the soil. What kinds of fertilizers are best suited for different kinds of soil?

Some fertilizers may have harmful as well as beneficial effects. What are some of these harmful effects? What factors should be considered before using certain fertilizers to improve the soil? Prepare a chart of the various fertilizers and their uses. In your chart include the advantages and disadvantages of each kind of fertilizer.



### TEACHING STRATEGY 20-4

#### Motivation

Bring in samples of foods made from corn, wheat, barley, and rye.

- **How many of these foods have you eaten in the past week?** (Answers will vary.)
- **Why are these foods important?** (Foods made from grains are important sources of carbohydrates in people's diets.)

#### Content Development

Stress that prairie grassland soil is especially good for growing grains because important minerals are contained in a rich layer of topsoil. Because grain plants have relatively short roots, they are able to take nourishment from the topsoil. Help students to understand why leaching is not a major factor in the formation of prairie soil.

- **Why are minerals not leached from the topsoil in prairie grassland soil?** (The action of the sun and wind cause the rainwater to evaporate quickly, before it can drive minerals down into the subsoil.)
- **How are the sun and wind able to reach the soil surface easily?** (There are no trees to block the sun or break the wind.)

#### Enrichment

One reason prairie soil is so fertile is

**Figure 20-10** *On the same land in Oklahoma 42 years later, Darrel Coble and his two sons walk through fertile fields of wheat. Changes in climate greatly influence soil and plant changes.*

soil over the prairie. Soon grasses began to grow in the newly laid soil. Year after year, thick mats of grass roots and stems decayed, building a deep layer of humus. Since the prairie, unlike the forests, was exposed to lots of sun and high winds, rainfall quickly evaporated. This prevented important minerals in the humus from leaching out of the topsoil. **Today prairie grassland soil is rich in humus, dark brown to black in color, and very fertile.**

### *SECTION REVIEW*

1. Why are the Canadian and United States prairies called the "breadbaskets of the world"?
2. Why is there less leaching in prairie soil than in forest soil?

## 20-5 Desert Soil

You probably imagine deserts as being very hot, dusty places. And, in fact, many deserts are quite hot—at least during the day. However, the main characteristic of all deserts is not heat but a lack of water. This lack of water is a problem for many plants and animals that make their home in the desert. But it also means there is very little leaching of



## BACKGROUND INFORMATION

Most grasslands are found between very dry lands and humid lands covered with forests. Some grasslands are found in humid climates.

The three types of grasslands are: prairies with tall grasses; steppes with short grasses; and tropical savannas with coarse grasses. Prairies are found in humid climates and often have patches of forests. Steppes are semidry grasslands in all parts of the world. Tropical savannas have a dry season in winter and a rainy summer.

### Activity

**The Use of Fertilizers**
**Skills: Comparing, relating, applying**
**Level: Average—enriched**
**Type: Library/Writing**

This activity can be carried out at varying levels of sophistication, depending on the background, ability level, and interest of the student. More advanced students may wish to concentrate on the chemical nature and aspects of fertilizers, and they may wish to learn more about the chemical composition of these substances and of the components of soil. Other students may limit themselves to a more qualitative and descriptive approach to fertilizer use. The issues surrounding such use are rather controversial, and may be dealt with in class discussion or debates.

that it is formed from a material called loess. Have students find out what loess is, what makes it so fertile, and how and when it was deposited in the midwestern United States.

### Section Review 20-4

**1.** Great amounts of wheat are grown in these regions.
**2.** Rainfall is quickly evaporated by the sun and high winds.

## TEACHING STRATEGY 20-5

### Motivation

Obtain photographs of deserts in various parts of the world. Have students observe the photographs and note the characteristics of desert land. Guide students to recognize the windswept, sandy soil that characterizes deserts.

### Content Development

Point out that deserts and prairie grasslands have one thing in common—important minerals are not leached from the topsoil. However, the reasons for this are quite different in each case.

- **Why are minerals not leached from the topsoil in the desert?** (There is very little rainfall.)
- **In prairie grasslands?** (In the prairie grasslands, the amount of rainfall is adequate for plant growth, but wind and sun evaporate excess water before it can cause leaching.)

## 20-5 DESERT SOIL

### SECTION PREVIEW 20-5

In this section, students will be introduced to desert soil. Desert soil is found in many parts of the western United States.

Students will learn that the main factor in the formation of desert soil is lack of water. As a result, there is little leaching of minerals into the subsoil. Students will discover, however, that there is little formation of humus and topsoil in desert regions. This is because desert plants are far apart and dry desert weather causes plant matter to decay very slowly.

### SECTION OBJECTIVES 20-5

1. **Describe the characteristics of desert soil.**
2. **Explain why little humus forms in desert soil.**

### ANNOTATION KEY

❶ **By irrigating the land (Inferring)**

❶ **Thinking Skill: Making comparisons**

**Figure 20-11** *In the spring, desert flowers bloom in a spectacular display. But the flowers soon vanish so the plants can complete their life cycles before the start of the hot, dry summer.*

**Figure 20-12** *These barrellike cacti act like living accordions by swelling up to store water and then contracting slowly as the water supply decreases. Cacti also have a waxy coating that keeps the stored water from evaporating.*

minerals from the desert soil. So you might expect desert soil to be rich in minerals—and it is.

Despite the minerals in the desert soil, there is little or no humus in the soil. Why? Desert plants often have root systems that spread out in a wide area around the plant. In this way, plants can take in as much water as possible whenever an infrequent rain shower occurs. However, because of their widespread root systems, there are often large spaces between desert plants. With so few plants in any one area, there is not a constant "rain" of plant matter onto the desert floor. **What little plant matter does fall to the desert floor decays so slowly in the dry desert heat that very little humus or topsoil can form.** How, then, do you think scientists might be able to grow plants and food crops in a desert? ❶

#### *SECTION REVIEW*

1. What is the main characteristic of all deserts?
2. Why is there little leaching in the desert?

## 20-6 Tundra Soil

Beyond the edge of the northern evergreen forests and in much of northern Alaska lies the arctic tundra. Tundra also is found in other arctic regions in the world. In the tundra, the climate is very cold most of the year, and strong winds blow constantly.

### 20-5 (continued)

#### Motivation

If students can obtain samples of the various types of soils, have them plant a typical plant from their local area and see how well the plant develops in the different types of soils. Make sure students recreate the conditions in each soil area. For example, they should not water desert soil more than once every two months.

#### Content Development

After students have read the description of various soil types, ask,

• **What two characteristics make grassland and fertile prairie soils fertile?** (They are rich in humus and they receive abundant rainfall.)

• **Why is fertilizer often necessary to make forest soil suitable for growing crops?** (The topsoil is very thin and heavy rainfall causes most minerals to be leached. The fertilizer replaces the lost mineral nutrients.)

• **Why is irrigation able to make desert soil suitable for farming?** (Desert soil is rich in minerals but it receives too little rainfall. Irrigation provides the necessary water.)

• **Which soil type is not suitable for growing crops?** (tundra)

• **Why is tropical soil able to support abundant plant life even though heavy rain causes leaching of most minerals?** (The warm, moist climate causes plant and animal matter to decay quickly and replace the lost nutrients to the soil.)

#### Section Review 20-5

1. Lack of water.
2. Lack of water.

There is little or no topsoil in the tundra and much of the soil just below the surface is permanently frozen. This frozen soil is called **permafrost.** As you might expect, few plants can grow in the harsh tundra, particularly during the long winter. Those plants that do grow are small plants, mainly reindeer moss, lichens, and dwarf trees. When these plants die, they decay very slowly due to the cold weather. So little humus forms in the tundra.

**Unlike most soils, tundra soil does not have different levels. There is a constant mixing of the soil due to freezing and thawing.** In summer, a thin layer at the top of the soil thaws. It stays wet because the water cannot go through the permanently frozen layers beneath. During this time, small lakes and marshes form on the tundra. Ducks of all kinds flock to the tundra lakes during the short summer months. When winter does return and the ground soil again freezes, the ducks and many other tundra animals will have already migrated to warmer areas farther south.

**Figure 20-13** *Another kind of soil is mountain soil. Soil on the slopes of mountains is made up of jagged pieces of weathered parent rock and very small amounts of clay and sand. There usually is no topsoil in these regions.*

**Figure 20-14** *Many small plants such as grasses, lichens, and reindeer moss grow on the tundra during summer* (left). *But during the winter the ground is snow-covered, and animals such as these caribou must keep traveling in search of food* (right).

### *SECTION REVIEW*

1. What is permafrost?
2. Tundra soil has no distinct layers. Why?



## 20-6 TUNDRA SOIL

### SECTION PREVIEW 20-6

In this section, students will learn about tundra soil, which is found in the world's arctic regions. Northern Alaska is the only part of the United States that has tundra soil.

Students will discover that tundra soil differs from other soil types in that it does not have distinct layers. Instead, there is a constant mixing of the soil caused by freezing and thawing.

Students also will discover that tundra soil just below the surface is permanently frozen. This frozen soil is called permafrost.

### SECTION OBJECTIVES 20-6

1. **Describe the characteristics of tundra soil.**
2. **Explain why tundra soil lacks definite layers.**
3. **Define permafrost.**

### SCIENCE TERMS 20-6

**permafrost p. 463**

## TEACHING STRATEGY 20-6

### Motivation

Have students observe the photograph of the snow-covered landscape in Figure 20-4.

• **What do you think life is like for plants and animals in an area such as this?** (Answers may vary; students should recognize that most plants would freeze or be buried by snow; animals would have to survive very cold temperatures and strong winds; because of the lack of plants, animals would have difficulty finding food.)

### Content Development

Emphasize to students that tundra soil is unlike any other soil type found on earth.

• **How is tundra soil different from other soil types?** (It has no distinct layers, and below the surface it remains frozen all year long.)

• **What factors cause tundra soil to be this way?** (Extreme cold is the main factor, as well as the thawing of a thin layer of surface soil in the summer.)

• **Why does little humus form in the tundra?** (The cold climate prevents plant growth and decay.)

### Content Development

Students should be made aware that the minerals found in soil come not only from the decay of plant and animal matter, but also from the parent rock. For example, limestone and basalt contain calcium, magnesium, potassium, and sodium. Soils that have these rocks as parent material will contain these minerals.

### Section Review 20-6

1. Permanently frozen soil.
2. There is a constant mixing of the soil due to freezing and thawing.

# LABORATORY ACTIVITY DETERMINING RATES OF WEATHERING

## BEFORE THE LAB

**At least one day prior to the activity, gather enough materials for your class, assuming 6 students per group.**

## PRE-LAB DISCUSSION

Discuss with students the various factors that affect the rate of weathering on rocks. One of these factors is the composition, or mineral content, of the rock. Some minerals are more resistant to weathering than others.

Another factor that affects the rate of weathering is climate. In general, rocks weather more rapidly in moist climates. Air pollution also can affect the rate at which rocks weather. Certain pollutants released into the air combine with moisture to produce acids. These acids tend to erode rock.

Time and surface area also determine the rate at which a rock weathers. Old rocks are less likely to weather than newly formed rocks. Rocks that are broken into small pieces, or that have many cracks and joints, will weather rapidly because a great deal of surface area is exposed.

After students have read the lab, ask,

- **Which factor that affects weathering is being tested in this activity?** (surface area)
- **What is the variable in this experiment?** (the amount of exposed surface area of the antacid tablet)
- **What factors are constant?** (the amount of water added to each tablet and the general environment—temperature, humidity, air quality—in which both tablets are being tested)

## SKILLS DEVELOPMENT

Students will use the following skills while completing this activity.

1. Observing
2. Comparing
3. Manipulative
4. Relating
5. Applying
6. Recording
7. Safety

# LABORATORY ACTIVITY

## *Determining Rates of Weathering*

### Purpose

In this activity, you will find out what effect particle size has on the speed, or rate, at which a rock reacts with its surroundings.

**Materials** *(per group)*

| | |
|---|---|
| 1 antacid or seltzer tablet | Paper |
| 4 100-mL beakers | Pen or pencil |
| Graduated cylinder | Water |

### Procedure

1. Break an antacid tablet in two.
2. Place one of the halves on a piece of paper, and fold the paper so that it covers the tablet.
3. With one end of your pencil or pen, gently tap the half tablet under the paper several times in order to break it into smaller pieces. Be careful not to crush the material to powder.
4. Place the other half of the tablet in one beaker. Place the crushed, smaller pieces in a second beaker.
5. Add 10 mL of water to each of the remaining beakers.
6. At the same time, pour the water from these two beakers into the beakers containing the seltzer tablets.
7. Observe both reactions until they end. Then record your observations. Be sure to describe the reactions you observed in both beakers and the time it took for the reactions to end.

### Observations and Conclusions

1. What difference in reaction rates did you observe in the two beakers?
2. When you broke the tablet into very small pieces, you increased the surface area of that part of the tablet. State the relationship between surface area and reaction rate in a chemical change.
3. Your observations in this activity can be used to make a general statement about weathering processes that occur in the natural environment. Draw a graph showing the relationship between the particle size of a substance and the rate at which it weathers.

464

## TEACHING STRATEGY FOR LAB PROCEDURE

This activity should be performed when Section 20-1 has been completed.

## SAFETY TIPS

Caution students to be careful when handling glassware.

## OBSERVATIONS AND CONCLUSIONS

**1.** The broken tablet reacted more quickly than the unbroken one.
**2.** In general, the greater the surface area, the higher the reaction rate.
**3.** The graph, which plots weathering rate ($y$-axis) versus particle size ($x$-axis), should show a curve that extends from upper left to lower right in the first quadrant, indicating that rate tends to decrease as particle size increases.

# CHAPTER REVIEW

## SUMMARY

### 20-1 Soil

- Soil is the thin layer of minerals, decayed plant and animal matter, water, and air that covers the earth's surface.
- Topsoil, the uppermost soil layer, is made up of small rock grains, remains of dead organisms, and humus.
- Subsoil, the soil layer beneath the topsoil, contains larger rock grains and little or no humus.
- Parent rock is the solid rock that lies beneath the subsoil.
- The environmental factors that affect soils include climate, mineral content of parent rock, and different types of plants and animals.

### 20-2 Tropical Rain Forest Soil

- Tropical soils are found in the rain forests of South America, Africa, and Asia.
- Huge amounts of rain keep the forest floor damp, which increases the decay of plant and animal matter.
- The washing of minerals downward out of the topsoil is called leaching. In tropical soils, leaching quickly washes minerals out of reach of plant roots.
- Decayed remains become part of the thin humus layer. The humus layer repeatedly washes away and reforms.
- Root systems of rain forest plants are so efficient that they absorb almost all of the decaying matter from plants and animals.

### 20-3 Temperate Forest Soil

- Unlike tropical rain forests, temperate forests have four seasons. In winter, most decay of plant and animal matter stops.
- The process of leaching removes many minerals from temperate forest soils.
- Southern temperate forests have more fertile soil than northern temperate forests.

### 20-4 Prairie Grassland Soil

- Prairie grassland soils are found in the grassland areas of the midwestern United States and in similar regions of the world.
- Prairie grassland soils are very fertile, rich in humus, and dark in color.

### 20-5 Desert Soil

- Desert soils are found in very dry regions.
- Because there is little rain, there is little leaching and therefore little washing away of minerals.
- Because there are very few plants and animals as a result of the lack of water, there is little humus and little topsoil.

### 20-6 Tundra Soil

- Tundra soil is found in the arctic regions of the world.
- Tundra topsoil is very thin. The permanently frozen subsoil is known as permafrost.
- Few plants grow in tundra topsoil, and it is very cold. So there is little organic decay.

## VOCABULARY

*Define each term in a complete sentence.*

**humus**
**leaching**
**parent rock**
**permafrost**
**subsoil**
**topsoil**
**weathering**



## GOING FURTHER: ENRICHMENT

### Part 1

Have students investigate other types of reactions in which reaction rate is affected by surface area. One easy-to-observe reaction is the dissolving of sugar. Have students compare the dissolving rates of a sugar cube, a sugar cube cut in half or in fourths, and an equivalent amount of granulated sugar.

### Part 2

Challenge students to design an experiment in which they test another factor that affects the rate at which rocks weather.

## CHAPTER REVIEW

### MULTIPLE CHOICE

1. d 2. d 3. c 4. b 5. d 6. d 7. a 8. c 9. a 10. c

### COMPLETION

1. centimeters
2. humus
3. topsoil
4. weathering
5. drip tips
6. leaching
7. topsoil
8. humus
9. tundra soil
10. permafrost

### TRUE OR FALSE

1. F topsoil
2. F plants
3. F many
4. T
5. T
6. F equator
7. T
8. F poor
9. T
10. T

### SKILL BUILDING

**1.** With soil, the earth is able to support an amazingly wide variety of living things that are adapted to particular environmental and soil conditions. If there were no soil, very few kinds of land organisms could exist.
**2.** The Antarctic has very little precipitation, which explains why it is considered to be a desert despite its coldness.
**3.** Cold, dry regions: gravity, abrasion, oxidation since all of these factors do not require water or high temperatures; hot, humid regions: temperature, organic activity, gravity, abrasion, water, oxidation, carbonation, sulfuric acid, plant acids since all these factors require either warmth or the presence of water.
**4.** Weathered rocks form soil, and all living things depend directly or indirectly on soil as a source of minerals and food. Plants obtain needed minerals and water from the soil. Animals eat plants or other animals to obtain the materials they need to live.
**5.** Students' graphs should have two lines labeled "With Humus" and "Without Humus." Both lines should indicate that the amount of water retained decreases as the particle size increases. The graph should also indicate that the soil with humus holds more water than the soil without humus when particle size is the same for both. Students should infer that the best soil to supply water to plant roots during a long period of little rainfall would be made of small particles and would contain humus.

### ESSAY

**1.** a. tropical rain forest—damp, contains thin layer of humus; minerals are leached deep into soil and abundant plant and animal life decay quickly, replacing needed minerals; b. temperate forest soil—fertile, contains little humus; important minerals leached into subsoil; c. prairie grassland soil—fertile, rich in humus; d. desert soil—dry, contains a large amount of minerals; little topsoil and humus; e. tundra soil—no distinct layers, permafrost.
**2.** By rock that is continuously broken down by weathering.
**3.** Humus is material formed from the decay of plants and animals. It supplies essential chemicals for growing plants.
**4.** Leaching is the process by which minerals are washed out of the topsoil.

#### CONTENT REVIEW: MULTIPLE CHOICE

*Choose the letter of the answer that best completes each statement.*

1. Soil contains
   a. rock particles. b. decayed plant and animal matter.
   c. air. d. all of these.
2. The color of the humus layer of the topsoil is normally
   a. reddish. b. reddish-yellow. c. whitish. d. black or brown.
3. The mixture of decayed plant and animal matter that combines with rocky particles to form a fertile soil is called
   a. permafrost. b. subsoil. c. humus. d. leaching.
4. Which of these is most closely related to parent rock?
   a. topsoil b. subsoil c. humus d. decayed plants
5. Climate is an important factor because it affects the kinds and amounts of
   a. plants. b. animals. c. soil. d. all of these.
6. Which of these environmental factors can contribute to soil formation?
   a. wind b. heat and cold
   c. running water d. all of these
7. In which type of soil is the humus layer constantly replaced?
   a. tropical soil b. forest soil c. desert soil d. tundra soil
8. Which area has the most fertile soil?
   a. tropical rain forests b. forests
   c. prairie grasslands d. deserts
9. The minerals and humus needed by plants remain in prairie grassland soils because there is little
   a. leaching. b. subsoil. c. sun and wind. d. permafrost.
10. The soil of very dry regions with little water and little leaching is called
    a. tropical soil. b. forest soil.
    c. desert soil. d. tundra soil.

#### CONTENT REVIEW: COMPLETION

*Fill in the word or words that best complete each statement.*

1. Topsoil can range in thickness from a few ____________ to a few meters.
2. Material that results from the decay of dead plants and animals is known as ____________.
3. The uppermost layer of soil is called ____________.
4. The breakdown of parent rock by environmental factors is called ____________.
5. The leaves of many tropical rain forest plants have ____________ that help to draw off water.
6. The washing of minerals out of the topsoil is called ____________.
7. Heavy rainfalls cause many minerals to leach from the ____________ layer in a tropical rain forest.
8. The topsoil of prairie grassland soil is dark in color because it contains a lot of ____________.
9. The type of soil characteristic of the Arctic is called ____________.
10. The permanently frozen subsoil of the Arctic is called ____________.

466

## CONTENT REVIEW: TRUE OR FALSE

*Determine whether each statement is true or false. If it is true, write "true." If it is false, change the underlined word or words to make the statement true.*

1. The uppermost layer of soil is subsoil.
2. Humus is used by animals as food.
3. Humus soil contains few air spaces.
4. The process in which rock is broken down into soil is called weathering.
5. If the leaves of plants in tropical rain forests do not dry quickly, molds may grow on them and kill the leaves.
6. Tropical rain forests are located near the earth's poles.
7. Rapid decaying and forming of humus occurs in tropical soils.
8. A mat of evergreen needles is rich in minerals needed for plant growth.
9. Plants and animals of the desert areas can survive because they are able to save and store water.
10. Constant mixing of the soil due to freezing and thawing prevents tundra soils from having different soil levels.

## CONCEPT REVIEW: SKILL BUILDING

*Use the skills you have developed in the chapter to complete each activity.*

1. **Making comparisons** Soil is vital to life. Compare what the earth is like with soil to what it would be like without soil.
2. **Applying concepts** Many scientists consider the Antarctic among the world's largest deserts. Explain how such a cold place can be a desert.
3. **Making inferences** What environmental factors of weathering have more impact on soil formation in cold, dry regions? In hot, humid regions? Explain your answers.
4. **Relating concepts** How is the weathering of rocks helpful to life on the earth?
5. **Analyzing data** In an experiment to measure soil's ability to hold water, particle size and amount of humus were tested. The results are shown below.

   Construct a graph that represents the relationship between the amount of water retained and the size of the soil particles. Your graph should have two lines. Label each line appropriately as either "Without Humus" or With Humus."

   Based on your graph, describe a type of soil that would supply water for plant roots during a period of little rainfall.

| | Small Particles | | Medium Particles | | Large Particles | |
|---|---|---|---|---|---|---|
| | *With humus* | *Without humus* | *With humus* | *Without humus* | *With humus* | *Without humus* |
| **Water retained by soil** | 50.0 mL | 20.8 ml | 44.6 mL | 13.6 mL | 39.8 mL | 10.2 mL |

## CONCEPT REVIEW: ESSAY

*Discuss each of the following in a brief paragraph.*

1. Describe the five different soil types.
2. How is soil formed?
3. What is humus? How is it formed?
4. Describe the process of leaching and explain why leaching is important in the formation of soil.

467

Leaching is important in the formation of soil because it can spread important minerals throughout the subsoil where roots can easily reach them.

## ADDITIONAL QUESTIONS AND TOPIC SUGGESTIONS

**1.** In general, which do you think is better for soil development—frequent light rains or heavy downpours? Explain your answer. (Frequent light rains would be better because most of the water would be absorbed by plants and there would be little leaching of minerals from the topsoil. A heavy downpour would tend to flood the soil with water and cause many minerals to be leached out.)

**2.** Suppose a forest in a temperate forest region is cut down. What do you think will happen to the soil quality? (Answers may vary. Probably the soil quality will get worse because there will be no more plant life to take up nutrients and then return the nutrients to the soil. Also, the roots of trees and other plants will no longer "hold" the soil in place. As a result, extensive leaching and erosion by wind and water are likely.)

**3.** Where in the United States would fertilizers be most useful to farmers? Why? (Fertilizers would be most useful in the Northeast and in the Southeast, because these areas have temperate forest soil. Temperate forest soil is characterized by a lack of important minerals in the topsoil. Fertilizer would add these minerals, which are vital to plant growth.)

## ISSUES IN SCIENCE

The following issues can be used as springboards for discussion or given as writing assignments.

**1.** Certain methods of farming, such as contour plowing, help to preserve the quality of soil. Some farmers, however, refuse to use these techniques, claiming that they are not profitable on a short-term basis. Find out more about soil-conserving methods of farming, including their advantages and disadvantages. Then offer your opinion as to whether farmers should use these methods.

**2.** There has been much publicity recently about some northern New Jersey soil that is contaminated with radon, a radioactive element. Find out how this soil became contaminated, and why it is dangerous to residents. Also find out what is being done to remove the soil, and why this procedure is causing controversy. Finally, express your opinion as to the severity of the problem, and whether you agree or disagree with the methods being used to overcome it.

## Unit Five

## COMPOSITION OF THE EARTH

### ADVENTURES IN SCIENCE: MARIA REICHE SOLVING THE MYSTERIES OF THE NAZCA LINES

### BACKGROUND INFORMATION

Experts judge that the Nazca markings were made sometime between 1000 B.C. and A.D. 1000, thus predating the Inca civilization. There are about 30 figures in all, including a monkey, a bird, a lizard, a fish, flowers, plants, hands, and an owllike man. Most of the figures are between 30 and 120 km long, although a bird measures more than 270 km.

One of the most interesting drawings is that of a monkey. This drawing is unique in that it is formed from a single, uninterrupted line that traces out various geometric shapes and also outlines the whole monkey. Sixteen parallel lines below the monkey may represent the rising and setting of a star in Ursa Major. A series of zigzag lines near the monkey may symbolize water.

Most of the drawings include geometric shapes such as quadrangles, triangles, and spirals, as well as long, incredibly straight lines. Scientists are trying to find out how the markings were made in order to determine how the Nazcas developed such a remarkable sense of abstract geometry and technical precision.

One thing that makes the origin of the lines difficult to determine is that the Nazca culture left no written records, and few other remains have been found. Maria Reiche has made an important breakthrough, however, in establishing how the figures were made. She discovered that the Nazcas used a unit of measurement 38 cm long, and that they drew all the curves in the figures from angles that are multiples of 15 degrees. Reiche believes that the Indians drew a small scale model of the figure first, then transposed it to the ground.



Adventures in Science

# Maria Reiche Solving the Mysteries of the Nazca Lines

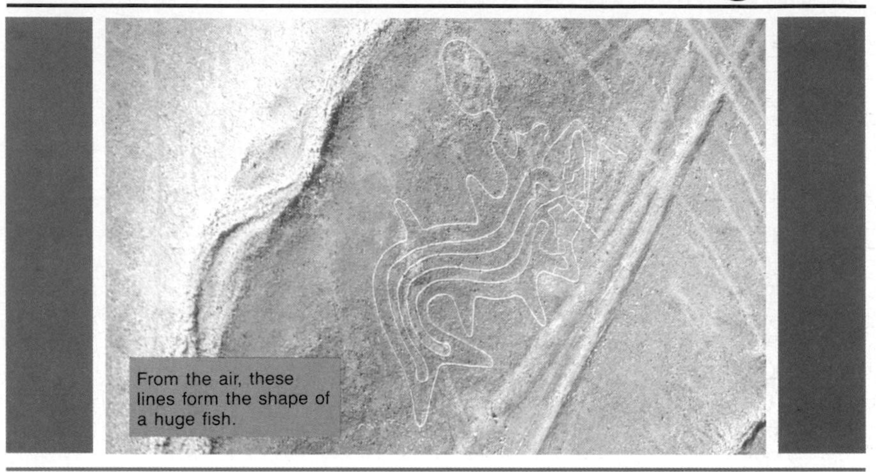

From the air, these lines form the shape of a huge fish.

One of the driest spots on the earth is the Nazca Desert of Peru. Here, huge drawings of animals have been scratched into the earth's surface. These drawings are so large that they can be recognized only from the air.

The figure of a monkey capers across 80 meters of flat, dusty landscape. Nearby, a gigantic bird spreads its wings over more than 270 meters of desert soil. These are just two of about 30 huge mysterious drawings that, from the ground, look simply like furrows in the dirt.

The drawings are called the Nazca Lines, after the desert on which they are drawn. They were made many centuries ago by Peruvian Indians who left behind almost nothing else of their culture. But why did the ancient Peruvians make the drawings? This question remains unanswered.

Many scientists, as well as interested visitors, have tried to unlock the secret of the Nazca Lines. But the efforts of one woman stand out. Her name is Maria Reiche, a woman on a lifelong quest to solve the



### TEACHING STRATEGY: ADVENTURE

#### Motivation

Begin by asking students the following question:

- **Suppose that you are walking along a road and you suddenly come upon some elaborate figures of animals and plants carved into the pavement. What questions might you ask about the drawings?** (Accept all answers. Possible questions include the following: Who made the drawings? Why were they made? Do they have special meanings?)

#### Content Development

Continue the discussion by explaining that the Nazca lines were first discovered by an American archaeologist named Paul Kosok. When Maria Reiche met Kosok and found out

mystery of the lines. Maria Reiche is an 80-year-old mathematician and geographer. For more than 20 years she has lived in the desert studying the Nazca Lines and making careful measurements.

Speaking of her search for the answer to the mystery of the Nazca Lines, Maria Reiche says, "We have to penetrate into the minds of the Indians who drew these lines and find the way they worked to find the answer."

Reiche came to Peru from Germany in the 1930s. Her fascination with the desert drawings began in 1940, when she met an American archeologist named Paul Kosok. Kosok was in Peru studying ancient systems of watering crop lands. He had charted one of the winding desert markings thinking that it might be part of such a watering system. Instead, he found that the line formed the huge figure of a bird. It was the first drawing to be discovered. Reiche has since devoted herself to the drawings.

From her many years in the desert, Reiche discovered how the ancient Peruvians carved the lines in the soil. They created the 7- to 8-centimeter deep furrows by removing the surface of brown stones to reveal the yellow soil beneath. The yellow soil—a mixture of sand, clay, and calcite—contrasts sharply with the stones.

Maria Reiche seeks to unlock the secrets of mysterious lines carved into the ground of the Nazca Desert of Peru.

The Indians could not have chosen a better spot for their artwork. The Nazca Desert receives less than one-quarter of a centimeter of rain each year. It is the dryness that has preserved the lines for the one or two thousand years since they were made. A cushion of warm air hovers over the desert surface and helps protect the drawings from being covered by blowing sand.

## A Recent Breakthrough

Some scientists, including Reiche, believe that the drawings had something to do with religious ceremonies. But Reiche feels that religious ceremony was not their only function. She has been trying to prove that the lines were a kind of calendar that was used to chart the movement of the sun, moon, and stars. Many ancient civilizations had such calendars and used them to set dates for planting, harvesting, and festivals.

To test her hypothesis, Reiche must find a relationship between the lines and the positions of the sun and moon at special times of the year. Such times are the solstices, which mark the beginning of summer and winter, and the equinoxes, which mark the beginning of spring and fall.

In the early morning, before the desert heat becomes too intense, Maria Reiche can be seen shuffling across the desert. Even before the sun has risen, she is noting the position of the lines in relation to the sunrise. In this way, she has found that some of the straight lines in the figures were solstice lines. They pointed to where the sun rose and set on June 21 and December 21 hundreds of years ago.

Still, many scientists are not convinced that the Nazca Lines were drawn as a calendar. Although Reiche has proven to her own satisfaction that some lines mark ancient solstices and equinoxes, too many lines seemingly point to nothing at all. Also, scientists argue, the animal drawings remain a mystery.

Maria Reiche is not about to give up, however. The lines, she says, "can't be senseless. There is too much work in them."



## ADDITIONAL QUESTIONS AND TOPIC SUGGESTIONS

1. Use reference sources to find additional pictures of some of the figures portrayed by the Nazca lines. Then make large-scale drawings of the figures to display on the walls of your classroom.
2. Read about other ancient phenomena that have proved mysterious in their origin and purpose. For example, Stonehenge in England is an intriguing grouping of stones that may have served as an astronomical observatory.
3. Present an oral report that describes some of the outdoor calendars that have been used by ancient civilizations. One of the most famous of these is the Aztec calendar stone.

## CRITICAL THINKING QUESTIONS

1. Write a short paper discussing what you think might be the meaning of the Nazca lines and why.
2. Do you think that each figure in the Nazca lines was drawn by one person or by more than one person? Why? (Accept all answers. Students should consider that the size of the drawing makes it likely that each figure was drawn by more than one person.)

about the lines, she was so fascinated that she wanted to answer questions such as how, why, and by whom the lines were made.

Point out that the lines are of interest to scientists not only because they contain beautiful figures, but because they display a high level of mathematical and technical skill. Most of the figures include perfectly drawn geometric shapes, and lines that are so straight that experts from the Smithsonian Institution could find only a slight deviation in an entire kilometer. Also significant is Reiche's belief that the figures were drawn to scale from much smaller, original drawings. This would indicate that the Nazcas had a working knowledge of ratio and proportion.

Emphasize to students that the lines have been preserved due to the natural conditions of the Nazca Desert. The processes of erosion by water and wind are inhibited by the extreme dryness of the region and a protective blanket of warm air that tends to block the wind.

### Teacher's Resource Book Reference

After students have read the Science Gazette article, you may want to hand out the reading skills worksheet based on the article in your Teacher's Resource Book.

## Unit Five

# COMPOSITION OF THE EARTH

## ISSUES IN SCIENCE: DIVING FOR NATURAL TREASURES: A RISKY BUSINESS?

### BACKGROUND INFORMATION

Not too long ago, it was an accepted fact that beneath the ocean's surface lay nothing but dark, still water and perhaps some stagnant wastes. It is only in the last century that scientists have learned of life and other resources at the bottom of the sea.

Interest in exploring the bottom of the ocean began in the late 1800s. In 1872, the British ship *Challenger* undertook a three-and-a-half-year voyage that uncovered minerals and strange organisms in deep ocean waters. Despite the obvious success of the *Challenger* expedition, interest in exploring the ocean did not surface again until the 1950s. At this time, satellite technology was just beginning, and in time satellite mapping gave an extraordinary view of the ocean floor. In the 1970s, scientists who descended to the bottom of the ocean to investigate the topography of the ocean floor made an amazing discovery—not only is the ocean floor loaded with minerals, but it is also teeming with life around strange openings in the ocean floor called deep-sea vents.

Deep-sea vents are similar to geysers and hot springs on the earth's surface. Water from the ocean seeps



Issues in Science

# DIVING FOR NATURAL TREASURES:

## A Risky Business?

Edith Widder, with the Wasp diving suit in the background, prepares for a dive in California's Santa Barbara Channel.

Six hundred meters below the ocean's surface, Edith Widder looks out on a fascinating world. The powerful beam from her lamp illuminates oddly shaped fish. Glowing masses of plankton, or microscopic plants and animals, float within reach of her outstretched arms.

Edith Widder is one of five pioneer scientists who participated in a program of deep-ocean exploration conducted by the University of California at Santa Barbara in the summer of 1984. The program had two purposes. One was to study the strange and beautiful animals that live in the ocean depths. The other was to test a diving suit called the Wasp. As its name suggests, this special new suit makes the wearer look like a wasp, with a big head and a tapered body.

The scientists made a series of dives down to 600 meters in the deep Santa Barbara Channel off the coast of California. Despite the crushing water pressure, each scientist was able to spend long periods of time moving around on the ocean floor. Few divers have ever ventured to such a depth before. Those that have were not able to stay down for long. In fact, most deep-ocean exploration has been done by scientists in submarines and diving bells or by tanklike robots.

The Santa Barbara Channel dives have opened up a whole new way to explore the ocean depths. The Wasp provides deep-ocean protection and life support while giving the wearer remarkable freedom of movement. Some people have dubbed it the "body submarine" because it combines the best features of a submarine and a diving suit. Such a combination is vital for future exploration of the ocean depths and for future use of the ocean's wealth.

## TEACHING STRATEGY: ISSUE

### Motivation

Begin by asking students the following questions:

- **If you were to descend to the bottom of the ocean, what do you think you would find?** (Accept all answers.)
- **What problems do you think you would encounter in this journey?** (Accept all answers.)

### Content Development

Continue the discussion by explaining that one of the most difficult areas to explore is the ocean depths. The United States was able to land two men on the moon before it was able to send human beings to the bottom of the sea.

Emphasize that the main deterrent to under-ocean exploration is intense pressure. Pressure below the ocean's surface increases at a rate of 1 atmosphere per 10 m. (One atmosphere is the standard pressure of air at sea level.) At a depth of 1 km—which is less than one-seventh the depth of the deepest part of the ocean—the pressure of water on an underwater vessel is about 7 tons.

Continue by pointing out that until the last hundred years or so, scientists did not believe that any significant resources existed beneath the ocean's surface. The lumps of min-

## BURIED TREASURE

The treasures that lie beneath the waves are many and varied. People eat some of these treasures, such as fish, sea animals, and sea plants. Scientists have also discovered that many ocean plants and animals are a source of valuable medicines. Vast oil fields lie beneath the ocean floor. Some are being tapped now. Many more lie waiting to be tapped in the future. Large deposits of hard minerals have been discovered on the ocean floor as well. These minerals include manganese, tungsten, vanadium, nickel, cobalt, tin, titanium, silver, platinum, and gold.

Some of the minerals can be found just below the surface of the ocean floor. Others are a bit deeper in the sediment layers and would have to be mined. But perhaps the most interesting minerals are those that are just sitting on the ocean floor waiting to be scooped up. Trillions of tons of these minerals in the form of potato-sized lumps, or nodules, cover the ocean bottom in many places.

A number of American companies have designed equipment to gather these nodules from the ocean floor. One craft, a tractorlike robot, would roam the ocean floor scooping up minerals. The craft would then wash and crush them and pipe them to the surface for further processing.

Trillions of tons of valuable metal ores lie on the ocean floor. These potato-sized rocks hold magnesium.

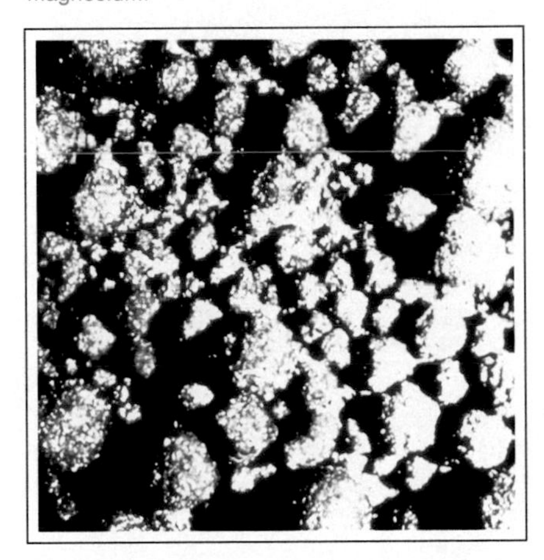

## QUESTIONS RISE TO THE SURFACE

The harvesting of minerals raises some difficult questions. Will ocean mining pollute the water? Will mining in one area hurt

Before the Wasp was invented, deep ocean exploration had to be done in vehicles such as the one shown in this photo.

into cracks in the ocean floor and comes into contact with hot rock—mostly basalt—in the earth's mantle. As the water become hot and changes into steam, it is forced back up into the ocean through the "vents." After its encounter with hot basalt, the water is rich in hydrogen sulfide. Bacteria in the bents use the sulfides in much the same way that green plants use sunlight to produce food. The rest of the organisms around the vents depend on the bacteria for nourishment. Small animals eat the bacteria, and bigger animals eat the smaller ones. Life around the vents is unique in that, unlike other ecological communities, these groups of organisms do not in any way depend on sunlight.

## ADDITIONAL QUESTIONS AND TOPIC SUGGESTIONS

**1.** Find out about the underwater discoveries made by crew members of the submersible *Alvin,* which first descended to the bottom of the ocean in the 1970s. Report your findings in the style of a short newspaper or magazine article.

**2.** Many ocean organisms can be classified according to the depth at which they live. Make a chart to show some of the different living things that can be found at various depths.

erals called nodules were first discovered in 1872. These potato-sized objects are often referred to as "manganese nodules" because they contain about 35% manganese. However, they also contain large amounts of iron, copper, nickel, and cobalt. The extent of this resource is enormous; scientists estimate that on the floor of the Pacific basin alone there are over 1.5 trillion tons of manganese nodules.

Emphasize that the most recent discovery at the bottom of the ocean is that of life around deep-sea vents. Deep-sea vents are like hot springs on the earth's surface. Water that seeps into cracks on the ocean floor becomes superheated when it comes into contact with molten rock beneath the earth's crust. The water and steam that are forced back up into the ocean act like a miniature furnace, heating the surrounding water to temperatures of 300°C or more. In the 1970s, the submersible *Alvin* descended to a depth of 7500 m to discover the most amazing creatures living around deep-sea vents. Clams, tube-worms, sea anemones, and mussels were among the creatures that the scientists found. The crew members of the *Alvin* were so ecstatic about their discovery that they quickly dubbed five locations on the ocean floor Clambake 1, Clambake 2, Dandelion Patch, Oyster Bed, and the Garden of Eden.

### Skills Development

***Skill: Comparing***

Point out to students that until the invention of the Wasp, divers could

## CRITICAL THINKING QUESTIONS

**1.** What are some of the advantages of the Wasp compared with a deep-ocean vessel such as the one pictured on page 471? (Answers may vary. Students should recognize the obvious advantage of the "hands-on" exploration that is possible with the Wasp, as well as the Wasp's ability to move freely under human power.)

**2.** What do you think will be the impact of the United States' refusal to sign the Law of the Sea treaty? (Accept all logical answers. It is likely that the treaty would be considerably weakened by the United States' action, since the United States is a major economic power and a leader in underwater technology. However, it is also possible that the United States could find itself at a disadvantage if all other major nations were to agree to the treaty.)

**3.** Why do you think the Law of the Sea treaty states that all mining technology must be shared with the Seabed Authority? (Answers may vary. One consideration is that in this way, no one nation will rule the ocean floor simply because of superior technology.)

## CLASS DEBATE

Have students choose one of the following statements as a basis for a class debate: (1) The United States has behaved irresponsibly by not signing the Law of the Sea treaty, or (2) The Law of the Sea treaty is unrealistic in terms of the normal economic and political aspirations of nations.



fishing in nearby areas? Will mining wipe out rare fish and other sea life?

Studies such as the one conducted in the Santa Barbara Channel may help to answer these questions. Divers wearing the Wasp would be able to keep a close watch on mining and drilling operations on the ocean floor. This kind of monitoring might reduce the chances of oil leaks and other such accidents. Deep-ocean divers could also keep track of the effects of mining and drilling operations on marine life.

International political questions also arise. Who has the right to mine the ocean depths? Where should this mining be allowed? Who should set the rules? And who should benefit from ocean mining?

These questions are difficult to answer. Since 1959, representatives from many nations have been meeting to try to create a Law of the Sea. This law would regulate ocean mining, oil drilling, fishing, energy usage, dumping of wastes, exploration, and research. Finally, in 1982, 119 nations signed a Law of the Sea treaty. But the United States was not one of them.

### THE LAW OF THE SEA

Under provisions of the treaty, each coastal nation is given an exclusive economic zone 200 nautical miles from its shore. Within that zone, a nation controls all natural resources, dumping, economic use, and scientific research. Where economic zones of different nations overlap, the nations must work out agreements.

Outside of the economic zones, no single nation controls the ocean floor. The International Seabed Authority administers this vast ocean floor region, called the International Seabed Area. The Authority sets the rules for mining in the International Seabed Area. And further, all mining technology must be shared with the Authority.

The United States could not accept these provisions and, therefore, did not sign the Law of the Sea treaty. On March 10, 1983, the United States declared its own economic zone extending 200 nautical miles off the coasts of the United States and all its territories. Since then, a number of American companies have announced that they plan to drill for oil and mine hard minerals in the zone.

Conflict over the zone has already arisen. And it is likely that many disputes will come up when people start mining the sea floor for its mineral wealth.

Offshore oil rigs are already tapping vast oil reserves that lie beneath the ocean floor. Divers wearing the Wasp may be able to repair cracks in the rig and prevent underwater oil leaks.

## ISSUE (continued)

descend to depths of only about 600 m. Scuba diving equipment could accommodate a diver up to a depth of 100 m; pressure-controlled diving suits could allow divers to descend as far as 600 m. Diving machines called bathyscaphs could enable scientists to approach the deepest parts of the ocean.

Challenge students to use reference sources to find out the characteristics of scuba equipment, pressure-controlled diving suits, and bathyscaphs. Then ask students to compare and contrast these three types of underwater equipment. Finally, have students compare these methods with the Wasp.

### Reinforcement

Emphasize to students that the three major resources at the bottom of the ocean are minerals, oil, and ocean organisms. Have students discuss ways in which each of these resources is valuable.

### Teacher's Resource Book Reference

After students have read the Science Gazette article, you may want to hand out the reading skills worksheet based on the gazette in your Teacher's Resource Book.

*KANSAS CITY STAR: JANUARY 12, 2021*

*WANTED: Moon Miners and Engineers to provide new space colony with building materials. We're looking for people who can turn moon rocks and lunar topsoil into usable metals such as aluminum, magnesium, and titanium. These metals will be used to make tools and to build support structures for the colony. We also want workers who can extract silicon from the moon's silicates. Silicon is needed to make solar cells and computer chips. Glass that will be used for windows in space colony homes also comes from lunar silicates. And iron and carbon mined from the moon need to be turned into steel.*

*In addition, we're looking for lunar gold miners to search the moon's surface for deposits of gold, as well as nickel and platinum. And oxygen, which will be used for both life-support systems and rocket fuel, needs to be extracted from moon rocks.*

Terry had answered the advertisement. Now, in the year 2024, she was hard at work at Moon Mine Alpha.

Terry expertly plunged the heavy shovel of her bulldozer into the soil to pick up another load of valuable material. When she raised the shovel, it was filled with moon rocks and lunar dust. Terry dumped her cargo into her lunar hauler. "My last truckload of the day," she thought, jumping down from the bulldozer. "Now I can go watch the mass driver in action."

Terry swung into the driver's seat of the lunar hauler. Soon she was bumping along the moon highway, a dusty path her hauler and others like it had carved out of the lunar topsoil over the past few months.



## Unit Five

# COMPOSITION OF THE EARTH

## FUTURES IN SCIENCE: WANTED! SPACE PIONEERS

### BACKGROUND INFORMATION

NASA is currently involved in research on the use of the Space Shuttle to build and supply orbiting manufacturing centers. These centers would use materials mined in space as well as materials brought from Earth.

NASA has recently undertaken a major sales effort to interest private companies in space mining and manufacturing. Among the companies that have decided to move ahead with experiments aboard the Space Shuttle is McDonnell Douglas.

One of the most promising aspects of space mining is the ease of materials processing due to weightlessness. The molecules of certain materials that are difficult or impossible to separate on Earth can be easily separated in the zero-gravity environment of space. Since some of the substances in this category are used to make pharmaceuticals, this industry stands to gain a great deal from the new technology.

Despite the appeal of space mining, however, private industry on the whole has not been quick to make commitments. As a *New York Times* writer explained, the rewards of space ventures may be dozens of years away, and most companies work on a three- to five-year plan. Everyone is waiting for someone else to prove and perfect the technology.

Space mining would probably become more attractive to industry if an inexpensive and reliable method of space travel could be developed. At the present time, the space shuttle is just too expensive to make space manufacturing centers economically practical. As an official at NASA's Johnson Space Center pointed out, a Space Shuttle, "like the first train west out of Saint Louis a century ago," must come first if space mining and manufacturing are to become a reality.

### TEACHING STRATEGY: FUTURE

#### Motivation

Have students read the "want ad" on the opening page of this article.

- **Does this sound like the typical want ad you would find in a local newspaper?** (Answers may vary. Obviously, the ad describes a job that is located on the moon, and one that involves some very futuristic things such as a space colony, moon mines, and homes in outer space.)
- **Where might an ad like this appear?** (Answers may vary. Most likely, this ad would appear in a newspaper or other publication at some time in the future, when moon mining and space colonies are a reality.)
- **How would you feel about answering an ad like this?** (Accept all answers.)

## ADDITIONAL QUESTIONS AND TOPIC SUGGESTIONS

**1.** The moon contains significant quantities of aluminum, iron, silver, and oxygen. Find out some of the important uses of these elements, and then make a chart to display your findings.
**2.** How does soil on the moon differ from soil on Earth? (The moon's soil is made of rocks and dust. Unlike soil on Earth, it contains no water or living things. In some ways, the soil resembles that of the most barren deserts on Earth. Since there is no wind and no water, there is no soil erosion on the moon.)
**3.** What type of education and background do you think a person would need to qualify for the jobs of moon or asteroid miner and engineer? (Accept all logical answers. Students should consider that a person would need education and training similar to that of an astronaut, as well as training in engineering and/or mining.)

A typical lunar colony with all the comforts of home!

Fifteen minutes later, she spotted the huge piece of machinery known as the mass driver. She saw the "flying buckets" of the mass driver suspended magnetically above special tracks. The buckets, filled with packages of lunar rocks, soon would be sent speeding along above these tracks by powerful magnetic forces. When the buckets reached high enough speed, they would fling their contents into space at 2.4 kilometers per second. At this speed, an object can escape the moon's gravity. Hundreds or thousands of kilometers away, a mass catcher would be waiting to grab the lunar cargo. The lunar materials would then be turned into fuel or building materials for a new space colony. Surely the mass driver, developed in the 1970s at the Massachusetts Institute of Technology, had proved to be a valuable tool for space colonization.

What will this space colony be like? Stationed nearly 400,000 kilometers from the earth, a huge sphere more than 1.5 kilometers in diameter will rotate in space. The rotation will create artificial gravity on the inner surface of the sphere. Here 10,000 people or more will live and work inside an earthlike environment powered by the sun. The Space Colony will be constructed of materials mined almost entirely on the moon.

### Resources in Space

As many scientists see it, our growing needs for raw materials, energy, and jumping-off places for journeys to the planets and stars make us look into space. Where else is there a free, continuous supply of solar energy, uninterrupted by darkness or weather? Where else does weightlessness, which will aid in the construction of huge structures, exist? Where else is there an untapped source of minerals?

A wealth of energy and materials is available in space. Let's start with the moon, a mere 356,000 kilometers away. This natural satellite could be an important source of aluminum, iron, silicon, and oxygen. A permanent base established on the moon could supply all the resources needed to support space settlement and exploration. Although these resources are abundant on the earth, bringing millions of tons of materials into space is out of the question!

The moon could become a gigantic "supply station" in the sky. Metals and lunar soil could be mined to build huge structures inside of which comfortable, earthlike homes would be

474

## FUTURE (continued)

### Content Development

Emphasize to students that the possibility of space mining offers two major advantages. First, space mining would provide more resources for use on Earth—a benefit that is particularly important in light of dwindling supplies of certain elements and minerals. Second, space mining could provide the raw materials needed to support colonies in space. Space colonies may become increasingly important as research in space extends to other planets and as space travel becomes more common place. These colonies would not be able to exist if all necessary resources had to be "imported" from Earth.

### Content Development

Point out that part of the task of mining asteroids may involve "capturing" a nearby asteroid and bringing it closer to Earth. This proposed idea has generated considerable controversy. For one thing, some experts worry about what will happen if scientists "miss" when they try to harness the asteroid—the asteroid could come crashing to Earth with the force of a nuclear explosion. In addition, the capture of an asteroid raises some thorny political questions. For example, who would own this new satellite? Would it be the sole property of the nation that brought it to Earth? Or would any nation with sufficient technology have the right to land on the asteroid and extract its mineral riches?

Mining the asteroids yields precious metals, minerals, and water.

constructed. Since the moon's gravitational force is one sixth the strength of the earth's, it would be cheaper and easier to build such space factories and colonies on the moon. These buildings could become part of the permanent moon base.

Almost half of the moon is made up of oxygen. This oxygen could be used to make rocket fuel. Liquid hydrogen mixed with liquid oxygen is a basic rocket fuel. Rockets bound for the outer plants could be launched more easily from the moon than from Earth, where the pull of gravity is six times greater.

## A New Frontier

The moon is not the only source of natural materials in near space. Asteroids are also vast treasure houses of minerals. They contain metals such as nickel, iron, cobalt, magnesium, and aluminum. Phosphorus, carbon, and sulfur are also present on asteroids. And they may contain the precious metals gold, silver, and platinum. Asteroids are also important sources of water. One small asteroid can perhaps yield between one and ten billion tons of water.

"Hey, Terry, how's it going?" The voice belonged to Bill, one of the workers who ran the mass driver.

"Oh, I still like being a moon miner," Terry answered, "but a few years from now, I hope to be mining the asteroids instead. It should be a challenge trying to capture a small asteroid or land on a big one."

"I hope you like traveling, Terry," Bill said with a worried look. "The trip could take months or years."

"It would be worth it," said Terry as she waved to Bill and headed to her two-room apartment under the plastic dome of Hadleyville.

Turning on her TV set to watch live coverage from Earth of the 2024 Summer Olympics, Terry first tuned in the *Moon Miner's Daily Herald*, a TV "newspaper." Suddenly, an advertisement caught her eye.

WANTED: *Asteroid Miners and Engineers to capture small asteroids and collect samples from larger asteroids. Workers must be willing to spend long periods of time far from home. Travel to the asteroid belt, which lies between the orbits of Mars and Jupiter about 160 to 300 million kilometers away, is required.*

"Why not?" Terry thought as she began to type out a reply on her computer keyboard.



## CRITICAL THINKING QUESTIONS

**1.** Observe the photograph on page 474 and read the caption. Do you agree that this typical lunar colony has "all the comforts of home"? Explain your answer. (Accept all answers. Encourage students as they answer the question to imagine what life would be like in a colony such as this one.)

**2.** What are some of the advantages of developing human colonies in space? (Answers may vary. Possible advantages include relieving some of the overpopulation on Earth; obtaining valuable resources such as those described in the article; increasing human knowledge about the universe; and providing people with a new and unusual life style.)

**3.** Why would it be easier and cheaper to construct a large building on the moon than on Earth? (Gravity on the moon is only one-sixth that of gravity on Earth. This means that building materials would weigh less. It would take less manpower to move and handle the materials, less energy to transport the materials; and less human energy to construct the building. These factors would result in reduced costs.)

## Skills Development

***Skill: Forming an opinion***

Have students work in small groups. Challenge each group to imagine that they are board members of a corporation that is considering building a manufacturing center in space. Have students list the pros and cons of such a venture. Then have each group present to the class a dramatization of a board meeting in which the proposal is debated. Based on the information presented, have the class decide whether they think the corporation should go ahead with the project.

## Enrichment

Challenge students to imagine that they are real estate sales people attempting to sell the latest space colony home or apartment. Have students compose newspaper ads for the futuristic dwelling, using the real estate section of a local newspaper as a model for style. Encourage students to emphasize not only the indoor features of the house or apartment but the advantages of location as well. For example, an ad might read, "just a few blocks away from the nearest space shuttle" or "a quick commute to Moon Mine Alpha."

## Teacher's Resource Book Reference

After students have read the Science Gazette article, you may want to hand out the reading skills worksheet based on the article in your Teacher's Resource Book.

## Unit Six

# STRUCTURE OF THE EARTH

### UNIT OVERVIEW

In Unit Six, students are introduced first to the internal structure of the earth. The layers of the earth are described, and the phenomenon of earthquakes is explained. Next, the students learn about the earth's landmasses and the various types of landscapes. Then they read about saltwater and freshwater sources and the water cycle. They also study and analyze the structure of the earth's atmosphere. They then explore the various layers of the atmosphere and the role of ozone. Finally, they learn about the earth's magnetism and magnetosphere.

### UNIT OBJECTIVES

1. **Describe two kinds of earthquake waves.**
2. **Name and describe the layers of the earth.**
3. **Decribe the three major kinds of landmasses.**
4. **Identify the major sources of salt water and fresh water, and explain the water cycle.**
5. **Describe the layers of the earth's atmosphere.**
6. **Describe the earth's magnetosphere.**

### INTRODUCING UNIT SIX

Begin your teaching of the unit by having students examine the unit-opening photograph. You may wish to see whether students can infer the answers to the following questions before they read the unit introduction.

- **What do you see in the photograph?** (Dark-colored mountains, snow, clouds, and sky are visible.)
- **What kinds of mountains do you think these are?** (They are volcanic mountains.)
- **What do you think makes up the very dark material on the mountainsides?** (Volcanic stone formed from cooled lava makes up this material.)
- **How do you account for the depressed areas at the top of each peak?** (The volcanic craters, or openings, are located there.)

Now have students read the unit introduction, and then ask them the following questions:

- **If Mauna Kea is actually the world's tallest mountain, why is Mount Everest often classified as the tallest?** (Mount Everest rises highest above sea level. Much of Mauna Kea lies below the surface of the ocean.)
- **The slopes of Mauna Kea are more gentle than are those of many other volcanoes. How might this be explained on the basis of the type of lava this volcano releases?** (Mauna

# Structure of the Earth

Rising from the ocean floor up through the clouds, Hawaii's Mauna Kea from bottom to top is the tallest mountain in the world. Its 10,203-meter height is greater than that of Mount Everest by 1355 meters. Although the skiers who glide down the mountain's gentle slopes may not know it, Mauna Kea is the product of violent forces trapped deep beneath the earth's surface. For Mauna Kea is a volcano, one of many that have molded the Hawaiian Islands. The great eruptions of such volcanoes around the world have done more than change the earth's surface. Clouds of dust and gases shooting into the sky from these thundering mountains have dramatically changed the earth's atmosphere—sometimes for a short while, sometimes permanently.

CHAPTERS



*Hawaii's Mauna Kea*



## CHAPTER DESCRIPTIONS

**21 Internal Structure of the Earth**
In Chapter 21, the kinds of earthquake waves are described first. Next the earth's core and mantle are detailed. Finally, the structure of the earth's crust and the floating of the continents are explained.

**22 Surface Features of the Earth**
Chapter 22 begins with a treatment of the earth's major landmasses. The earth's mountain, plateau, and plain landscapes are then described. An explanation of saltwater and freshwater sources and the water cycle ends the chapter.

**23 Structure of the Atmosphere**
The major layers of the earth's atmosphere—the troposphere, stratosphere, mesosphere, and thermosphere—are described in Chapter 23. The roles of ozone and of the jet stream are also explained. The chapter closes with an examination of the earth's magnetosphere.

Kea—a shield volcano—releases a very runny kind of lava that can flow and spread quickly, producing a broader, gently sloping mountain.)

- **How can volcanoes change the earth's atmosphere?** (Answers will vary. The dust the volcanoes release can prevent light from passing through and can result in cooling. Also, the kinds of gases released can change the chemical composition of the atmosphere.)

## Chapter 21

# INTERNAL STRUCTURE OF THE EARTH

## CHAPTER OVERVIEW

In this chapter, students will be introduced to the internal structure of the earth. They will also learn about the earth's crust and the floating continents.

Students will discover that due to high temperatures and great pressure, the interior of the earth cannot be explored directly. Students will read how scientists have studied earthquake waves in order to learn about the interior of the earth. Students will be introduced to the different kinds of earthquake waves. They will also learn how earthquake waves are recorded on a seismograph.

The structure and composition of the four main layers of the earth will be discussed. These layers are the inner core, outer core, mantle, and crust. Students will discover that the crust consists of about 20 floating parts called plates. They will learn how the motion of these plate causes movements of the continents and the ocean floor.

## INTRODUCING CHAPTER 21

Have students observe the chapter-opener photograph and read the chapter introduction.

• **How would you describe this photograph?** (Answers may vary. Encourage students to comment on the photograph using such statements as: "It reminds me of . . . " or "It looks like . . . .")

Explain to students that this is a photograph of the Incredible Pit, a 134-m natural shaft in Ellison's Cave in Georgia.

• **How do you think the depth of this cave compares with the distance to the center of the earth?** (It is much less. Inform students that the depth of the cave is 134/6,450,000, or about 1/50,000, the distance to the center of the earth.)

• **According to the text, has anyone ever visited the center of the earth? Explain.** (No. intense heat and pressure make it impossible for anyone to reach the center of the earth.)

• **What do you think causes such great heat and pressure?** (Answers will vary.)

• **Why is it easier for scientists to reach far into outer space than to reach deep into the earth?** (Satellites and spaceships can be built to withstand the conditions in outer space,

# 21 Internal Structure of the Earth

**CHAPTER SECTIONS**

**21-1 The Earth's Core and Earthquake Waves**
**21-2 Exploring the Earth's Mantle**
**21-3 The Earth's Crust and Floating Continents**

**CHAPTER OBJECTIVES**

*After completing this chapter, you will be able to:*

**21-1** Relate seismic wave movements to the composition of the earth's core.
**21-1** Describe the characteristics of the inner core and the outer core.
**21-2** Describe the properties and composition of the mantle.
**21-2** Define the term Moho.
**21-2** Explain how mantle rocks form new ocean floor.
**21-3** Describe the earth's crust.
**21-3** Explain how continents can float.

Would you like to visit the center of the earth? It is not far away—no farther than the distance from New York City to Berlin, Germany. That comes to about 6450 kilometers.

But in some ways, the earth's center is harder to reach than planets that are millions of kilometers away. Scientists are sending probes into outer space that are traveling beyond the solar system. But no mechanical probe into inner space that scientists are able to build today can survive the enormous heat and pressure at the earth's center.

Yet there *are* probes that can reach the center of the earth and return to tell their story. As long ago as 1906, the Irish geologist Richard Dixon Oldham said that such a probe "enables us to see into the earth . . . as if we could drive a tunnel through it."

Oldham was a geologist who studied earthquakes. Great earthquakes topple buildings, shake the ground, and can change the shape of entire mountains. They also send shock waves through the earth. These shock waves penetrate the depths of the earth and return to the surface. Detected by special instruments, the shock waves can be used by scientists to "see" into the inside of the earth.

These shock waves were the probes that Oldham spoke of. Let's take a look at what they let Oldham, and the scientists who followed him, see!

*Descending into the earth*



but no scientific device can withstand the conditions in the interior of the earth.)

• **According to the text, what is the only way that scientists have been able to learn about the earth's interior?** (by studying earthquakes)

• **Why are earthquakes able to provide useful information about the earth's interior?** (The shock waves produced by earthquakes travel through the interior of the earth. By studying these waves, scientists can learn what the interior of the earth is like.)

As you continue with this chapter, keep in mind that the study of the earth's interior is an excellent example of the use of indirect evidence. Guide students to recognize that gathering indirect evidence is an important part of the scientific method when something cannot be observed or measured directly.

## TEACHER DEMONSTRATION

Use plastic putty to demonstrate the property of plasticity. Before showing the putty to the class, roll the putty into a smooth ball or egg shape.

• **Is this substance a liquid or a solid?** (Most students will answer solid.)

• **Why do you say solid?** (It has a definite shape; it is not runny or wet.)

Begin to pull and stretch the putty, then mold it into a very different shape.

• **Do you still think this substance is a solid?** (Answers will vary.)

• **How is its behavior different from that of most solids?** (It can be stretched, bent, and easily molded into different shapes.)

Explain that this special quality of the putty is what is called plasticity. Although considered a solid, a plastic material has the ability to flow.

If a refrigerator is available in or near your classroom, you may want to leave the putty in the freezer overnight and then let the class see what happens to the putty's plasticity. (The putty should lose all or most of its plasticity, become brittle, and break easily.)

• **Can you determine a relationship between temperature and plasticity?** (Plasticity decreases as temperature decreases and increases as temperature increases.)

## TEACHER RESOURCES

**Audiovisuals**

*Earth in Change—The Earth's Crust,* 16 mm film EBE
*Strata—The Earth's Changing Crust,* 16 mm film BFA
*The Earth's Crust,* filmstrip with cassette, SVE
*What Moved the Mountains? What Shaped the Seas?,* 16 mm film UEVA

**Books**

Sawkins, Frederick J. et al., *The Evolving Earth* (2nd ed.), Macmillan
Sparks, John and Arthur Bourne, *Planet Earth: Earth's Atmosphere and Crust,* Doubleday

**Software**

Earthquakes, Prentice Hall

## 21-1 THE EARTH'S CORE AND EARTHQUAKE WAVES

### SECTION PREVIEW 21-1

In this section, students will learn about two different kinds of earthquake waves—primary waves and secondary waves. Students will learn that primary, or P, waves travel faster than secondary, or S, waves. P waves can travel through solids, liquids, and gases, whereas S waves can travel only through solids.

Students will discover that the different properties of P and S waves enabled scientists to determine the structure of the earth's core. By noting where P and S waves disappeared and reappeared on the earth's surface, scientists were able to conclude that the outer core is composed of hot churning liquid, while the inner core is a solid under great pressure.

### SECTION OBJECTIVES 21-1

1. **Describe two kinds of earthquake waves.**
2. **Explain why earthquake waves are not detected in the shadow zone.**
3. **Describe the structure of the earth's core based on the study of earthquake waves.**

### SCIENCE TERMS 21-1

**seismograph p. 480**
**core p. 480**
**seismic wave p. 481**
**focus p. 481**
**primary wave (P wave) p. 481**
**secondary wave (S wave) p. 482**
**shadow zone p. 482**

## 21-1 The Earth's Core and Earthquake Waves

In Tokyo, Japan, at a little past two in the morning, a strong earthquake occurs more than a hundred kilometers beneath the ground. The date is April 18, 1889. The earthquake is so deep that few people are aware of it and little damage is done. However, in Tokyo, a special instrument called a ❶ **seismograph** (SIGHZ-muh-graf) records a strange message from the earthquake. The message is written as a pattern of squiggly lines.

In Potsdam, Germany, on the other side of the world, it is just past seven on the evening of April 17. There the seismograph is quiet—but not for long. One hour and four minutes after the earthquake in Tokyo, the Potsdam seismograph makes a pattern of squiggles like those in Figure 21-1.

Scientists quickly realized that the message recorded in Potsdam was really due to shock waves from the earthquake in Tokyo. And these waves had traveled through the earth's **core,** or center region, to reach Potsdam! This was the first time one of these messages from an earthquake was recorded by a seismograph on the opposite side of the world.

**Figure 21-1** *A seismograph* (left) *detects and records earthquake waves, or seismic waves. A typical pattern of seismic waves is shown* (right).

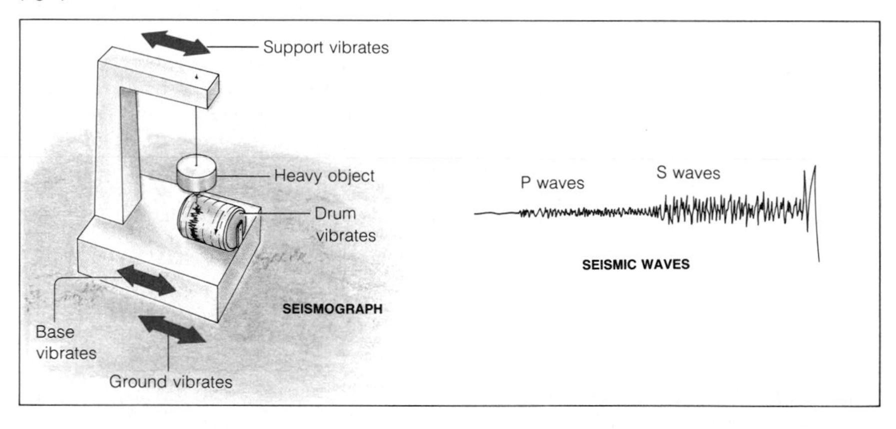


480

### TEACHING STRATEGY 21-1

#### Motivation

Obtain two shoe boxes with lids. Fill one shoe box with tissue paper and the other with a heavier material such as sand. Tape the lids securely on both boxes so they cannot be opened. Display the boxes and ask,

• **If you are not allowed to open the boxes, how might you go about finding out what is in each box?** (Accept all logical answers.)

Allow students to try some of the methods they have described. They will probably want to shake the boxes, compare the weight, and put their ears to the boxes.

• **What do you think is in each box?** (Answers will vary.)

Open the boxes and have students observe the contents. Discuss the accuracy or lack of accuracy of their guesses. If many students did not guess correctly, ask,

• **What additional information might have helped you to determine the contents of each box more accurately?** (Answers will vary.)

#### Content Development

Using the Motivation activity, introduce the idea of indirect evidence. Point out that scientists often use in-

Figure 21-2 *A powerful earthquake in Alaska ripped a hole in the ground near this house.*

Here was a chance, scientists believed, to look deep inside the earth. A worldwide network of seismograph stations was set up to record future earthquakes. But how can earthquake waves provide a picture of the earth's interior? To find out, let's take a look at how these waves are produced.

### *Kinds of Earthquake Waves*

Major earthquakes begin with a massive breaking and sliding of rocks underground. Shock waves called **seismic** (SIGHZ-mik) **waves** travel in every direction from the underground point of origin of the breakage. The point beneath the earth's surface where the rocks break and move is called the **focus** of the earthquake. ❷ Some seismic waves pass through the earth and come to the surface at points far from the earthquake's focus. Seismographs detect and record these waves, measure their strength, and record their time of arrival.

All earthquakes produce at least two kinds of shock waves at the same time. **Primary waves,** or **P waves,** result from a back-and-forth vibration of



## ANNOTATION KEY

❶ **Thinking Skill: Applying technology**

❷ **Thinking Skill: Sequencing events**

## TIE-IN/LANGUAGE ARTS

The word "seismic" comes from the Greek word *seismos,* which means earthquake or shock.

direct evidence to learn about something that cannot be observed or measured directly.

Emphasize to students that most of the earth's interior can be studied only by indirect methods.

- **Why must indirect methods be used to learn about the earth's interior?** (The conditions of intense heat and pressure make it impossible to study the inner earth directly.)
- **What indirect method have scientists used to learn about the earth's interior?** (Scientists have studied earthquake waves that have been recorded on seismographs around the world.)
- **What initially prompted scientists to study earthquakes to learn about the inner earth?** (About a hundred years ago, shock waves from an earthquake in Tokyo, Japan, were recorded on the opposite side of the world in Germany. Scientists knew that these waves had to have traveled through the center of the earth.)

## Skills Development

***Skill: Interpreting diagrams***

Have students observe the diagram of a seismograph in Figure 21-1.

- **What is the first stimulus that activates a seismograph?** (the vibration of the ground)
- **How is this stimulus translated into a record of seismic waves?** (The vibration of the ground causes the base and support of the seismograph to vibrate. Finally the drum vibrates, and this causes the waves to be recorded.)

## Motivation

Have students observe the photograph in Figure 21-2.

- **What caused the hole to open in the earth?** (an earthquake)
- **Have any of you ever experienced an earthquake?** (Answers will vary; have students share any experiences they may have had and any results of an earthquake that they may have seen.)

Point out that while the photograph shows a very obvious result of an earthquake, some effects of an earthquake are not so obvious. The movement of P waves and S waves do not cause holes to open up in the ground, but they do show up on sensitive instruments called seismographs. It is the movement of these waves that has taught scientists most of what they know about the earth's inner and outer core.

## TEACHER DEMONSTRATION

Hold a Slinky with one end in each hand so that the Slinky is in a horizontal position. Illustrate to students how the Slinky can stretch and compress. Then, begin moving the Slinky to the left or to the right while continuing small stretching and compressing movements. Try to make the stretching and compressing movements the same size each time and in a regular rhythm. Explain that the stretching and compressing movements represent vibrations.

- **In what direction is the Slinky vibrating?** (back and forth along a horizontal line)
- **In what direction is the Slinky traveling?** (sideways along a horizontal line)

Explain to students that this is the way a P wave vibrates—along the same line that the wave is traveling.

Next, have a student volunteer assist you by holding one end of a 3 m rope while you hold the other end. Make sure both ends of the rope are the same height off the ground. While the student holds his or her end steady, vibrate your end of the rope up and down so that a regular pulse is produced in the rope. Together with the student, slowly begin walking to one side of the room so that the rope is travelling sideways. Keep the up-and-down pulse going in the rope.

- **In what direction is the rope traveling?** (sideways, or horizontally)
- **In what direction is the rope vibrating?** (up and down)

Explain to students that this is the way an S waves vibrates—up and down, while the wave travels horizontally.

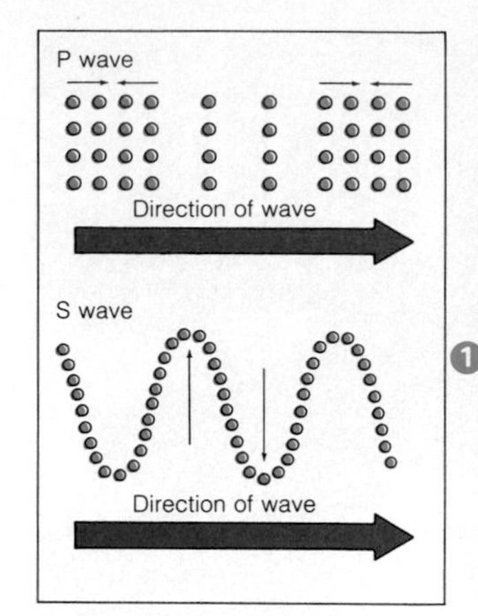


**Figure 21-3** *P waves push together and pull apart rock particles in the direction of the wave movement. The slower S waves move rock particles from side to side at right angles to the wave movement.*

rock particles. These particles vibrate in the same direction as the path of the wave. See Figure 21-3. P waves move rapidly and can travel through both solids and liquids. **Secondary waves,** or **S waves,** result from an up-and-down or side-to-side vibration of rock particles. These particles vibrate at right angles to the path of the wave. See Figure 21-3. S waves are slower than P waves and can pass only through solids. **After observing the speeds of P waves and S waves, scientists have concluded that the earth's core, or center, is actually made of two very different layers.**

### The Mystery of the Shadow Zones

By 1900, a worldwide network of seismographs was in place. And very soon, scientists noticed a regular but mysterious pattern to the P and S waves produced by earthquakes.

Suppose, for example, an earthquake strikes under the North Pole. Seismograph stations around the northern part of the earth record both P and S waves, but from northern Canada to southern Mexico there is a **shadow zone.** See Figure 21-4. In a

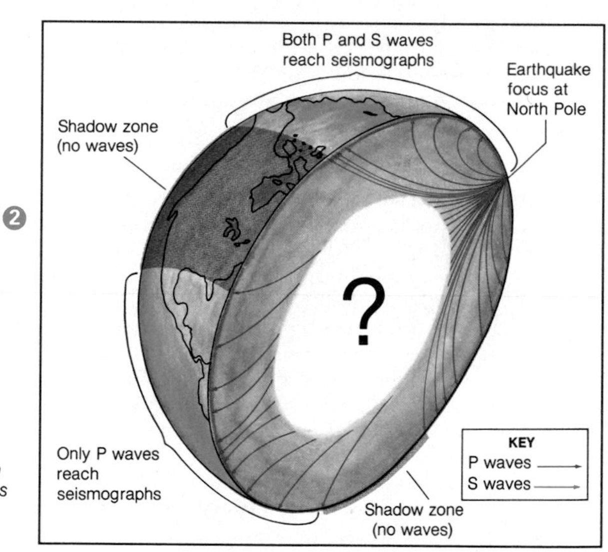


**Figure 21-4** *There are no earthquake waves in a shadow zone. Something inside the earth blocks S waves and concentrates P waves on the other side of the world.*

482

## 21-1 (continued)

### Skills Development

***Skill: Identifying cause and effect***

Begin by asking students,

- **What happens to P waves as they pass through a liquid?** (They slow down considerably.)
- **What happens to S waves as they encounter a liquid?** (They cannot pass through liquids.)
- **What behavior of these waves convinced scientists that the earth's outer core is liquid?** (S waves stopped completely at the outer core, and P waves slowed down.)
- **What convinced scientists that the earth's inner core is solid?** (The P waves, which had been slowed down as they passed through the outer core, began to speed up.)

### Content Development

Explain to students that a wave is a rhythmic disturbance that transmits energy from on place to another.

- **In the case of earthquake waves, where does the energy come from?** (from the breaking and sliding of rocks under the earth)
- **Where does this energy travel?** (From the focus of the earthquake outward; some of the waves travel upward to the surface of the earth; others travel downward to the earth's interior.)

shadow zone, no earthquake waves are detected by seismographs. Why are there shadow zones? At first scientists could not solve this mystery.

Adding to the mystery is the fact that south of the shadow zone, very strong P waves *are* detected. But *no* S waves are found in this area. The P waves are strongest around the South Pole—the one place on the earth directly opposite the North Pole! Scientists needed to find out how this information was related to the mystery of the shadow zones.

Scientists discovered that the pattern is the same for every earthquake. Both P and S waves are detected up to a little more than a quarter of the way around the world from the focus of the earthquake. Beyond that lies a shadow zone. Then a strong concentration of P waves appears on the opposite side of the earth from the focus. S waves do not appear.

Scientists reached two conclusions. The first conclusion was that something inside the earth was blocking S waves and preventing them from getting through to the other side of the earth. The second conclusion was that something inside the earth was concentrating P waves on the other side of the world. The mystery had to be solved.

### Seeing the Earth's Core

What did this pattern of shadow zones and concentration of P waves mean? Scientists found out that this pattern is a message from the depths of the earth that actually answers two questions. The answers to these questions are coded into the wiggly lines that record earthquake waves.

**THE FIRST QUESTION** What happened to the missing S waves? These waves disappear at the edge of the shadow zone and are not found on the opposite side of the world.

Scientists believed that something was deflecting the S waves. And since S waves pass through solids but not liquids, that something must be liquid. The earth's core, scientists concluded, is at least partly liquid. Using data from seismographs, the geologist Richard Dixon Oldham was able to map the boundary of the earth's liquid core at about 2900 kilometers beneath the earth's surface.

**Activity**

*Speed of Seismic Waves*

Seismic waves travel at a speed 24 times the speed of sound. The speed of sound is 1250 km/hr. How fast do seismic waves travel?

❸ **Figure 21-5** *This photograph shows a seismologist at the National Earthquake Information Service in Colorado. Each drum records the results from a different seismograph station.*

**Activity**

**Speed of Seismic Waves**
**Skills: Making calculations, making comparisons**
**Level: Remedial**
**Type: Computational**

This activity will help reinforce students' computational skills. Students will find that seismic waves travel at a speed of 30,000 kph.

**ANNOTATION KEY**

❶ **Thinking Skill: Making inferences**
❷ **Thinking Skill: Interpreting a diagram**
❸ **Thinking Skill: Drawing a conclusion**

**TIE-IN/PHYSICAL SCIENCE**

A wave that vibrates in the same direction as the path of the wave is called a longitudinal wave. Sound waves are examples of longitudinal waves. A wave that vibrates at right angles to the path of the wave is called a transverse wave. Light waves are examples of transverse waves.

**Reinforcement**

Reinforce the idea that P waves are not visible in the shadow zone by using the following analogy. A person running down a road in the bright sunlight can be easily seen. If, however, the person ran into a tunnel the person would seem to disappear for a short time. Once the person reaches the end of the tunnel, he or she would be visible again. Explain the movement of P waves through the earth is similar to the movement of the runner. At the surface of the earth, the waves are easy to detect, but as the waves travel through the interior of the earth, they seem to disappear. As the waves emerge from the interior of Earth, they can be easily detected again.

**Enrichment**

Earthquakes produce a third kind of wave called L waves. Have students find out the properties of L waves, and why they are important to people who live in the earthquake area.

## BACKGROUND INFORMATION

One of the greatest problems with regard to earthquakes is that it has been nearly impossible to predict them. By the time seismic waves are recorded, an earthquake is in progress and extensive damage to the earth's surface already has occurred.

Scientists have been working to find ways that will help them predict earthquakes accurately. This information would help in the preparation of evacuations and emergency procedures.

In March 1986, Japanese scientists set up an earthquake prediction center. The center, which is run by the Japan Meteorological Agency, consists of a computerized fast-response system that receives and processes data from 133 different instruments. These instruments record phenomena such as seismic waves, changes in land tilt, warping of land due to stress, changes in tidal levels, and electromagnetic changes in the earth's crust. The center records about 40 earthquakes a day; most of these are minor.

Japanese experts feel confident that, in most cases, they will be able to predict a major earthquake a few hours in advance. Some experts believe that a major earthquake may be predicted as much as two days in advance.

Earthquake prediction is difficult because there is not one signal that indicates the onset of an earthquake. Instead, there tends to be a variety of signals that occur any time from one month to one day before a major earthquake.

## TIE-IN/ PHYSICAL SCIENCE

The bending of a wave as it passes from one medium into another is called refraction. Refraction occurs because waves travel at different speeds through different mediums. The angle at which a wave bends as it enters a new medium is called the angle of refraction. The greater the angle of refraction, the greater the change in speed of the wave.

**Figure 21-6** *The bending of light makes this straw appear broken. Earthquake waves, like light waves, change speed and bend when they enter a different substance.*

As you see in Figure 21-4, the earth's core deflects S waves, forming a shadow zone where no S waves appear. This deflection of S waves also explains why there are no S waves on the opposite side of the world from the earthquake's focus. Part of the mystery of the shadow zones had been solved.

**THE SECOND QUESTION** Why is there a high concentration of P waves on the opposite side of the world from the earthquake's focus? The answer was decoded from earthquake waves by Inge Lehmann, who was chief of the Royal Danish Seismological Department from 1928 to 1953.

To see how she did it, let's look at a straw resting in a glass of water. See Figure 21-6. The straw seems to be broken right at the surface of the water—the boundary between air and water. Actually, the straw only makes visible what happens to light waves as they cross such a boundary—they bend because their speed changes. This bending of light causes the straight straw to appear broken.

Lehmann discovered that what is true of light waves is also true of earthquake waves speeding through the earth. The earthquake's shock waves bend at boundaries between different substances—boundaries that mark changes in the inner structure of the earth. Lehmann suggested that the earth's core is made up of two parts, a liquid outer core and a solid inner core. The answer to the mystery of the S waves had already shown that the liquid outer core boundary begins 2900 kilometers beneath the earth's surface. She suggested that the inner core is like a solid ball within the outer core. Lehmann estimated the radius, or distance from the edge to the center, of the inner core to be about 1200 kilometers.

Now Lehmann could explain why P waves are concentrated on the side of the world opposite the earthquake's focus. The waves would be bent four times—when entering the outer core, when entering the inner core, when leaving the inner core, and when leaving the outer core. The result of all this bending is that the waves are concentrated around the area opposite the earthquake's focus. See Figure 21-7. As you can see, P waves are also bent away from the shadow zones. The mystery of the shadow zones had been solved.

### 21-1 (continued)

#### Reinforcement

Reinforce the idea that S waves are not found in the shadow zone by using the analogy of a ball bouncing off a brick wall. Because the ball cannot pass through the wall, it is deflected, or bounced back from, the wall. Remind students that S waves cannot pass through liquid—so when S waves reach the liquid outer core, they bounce off the liquid and are deflected away from the shadow zone.

#### Skills Development

***Skills: Making a model, predicting*** Challenge students to use a globe and Figure 21-7 to predict where earthquake waves would be recorded from an earthquake occurring at: (1) Cairo, Egypt; (2) Seattle, Washington; (3) Reykjavik, Iceland; (4)

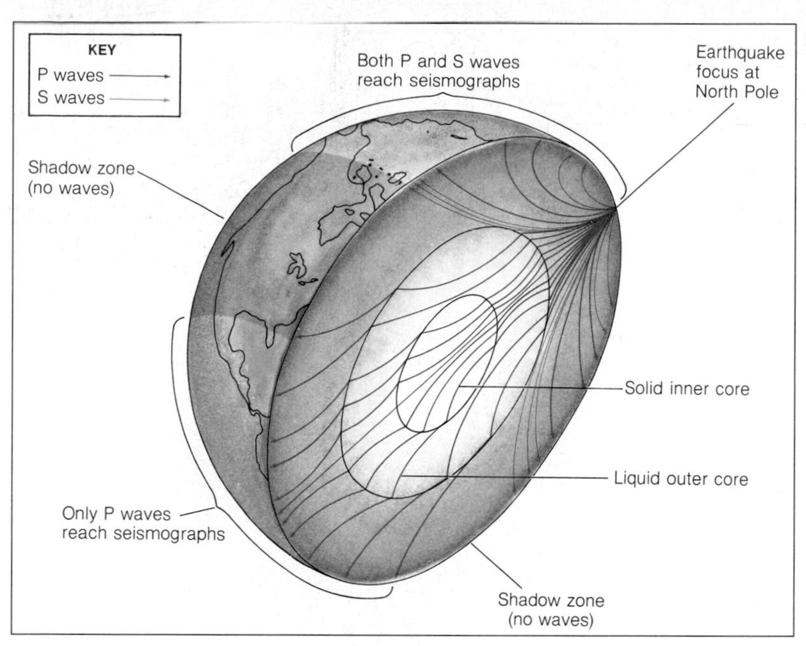


**Figure 21-7** *Inge Lehmann solved the mystery of the shadow zones when she explained how P waves bend and change speeds as they move through the earth's outer and inner cores.*

Lehmann wrote about her discovery in 1936. Since then, other scientists using more sensitive instruments have developed an even clearer picture of the earth's core.

Lehmann's estimate of the size of the inner core remains amazingly accurate. Scientists now calculate that the radius of the inner core is about 1300 kilometers. An outer core about 2250 kilometers thick surrounds the inner core. The outside boundary of the outer core begins about 2900 kilometers beneath the earth's surface. Add these three numbers together and you get the approximate distance from the earth's surface to the actual center of the earth—6450 kilometers. ❷

From studies of earthquake waves, scientists have learned much more about the earth's core. For example, they have learned that the hot outer core is probably a region of constantly churning liquid,

### Activity

*How Many Earths?*

The distance from the earth's surface to the center of the earth is about 6450 kilometers. The average distance from the earth to the sun is 150 million kilometers. How many earths lined up in a row are needed to reach the sun?



## ANNOTATION KEY

❶ **Thinking Skill: Applying concepts**
❷ **Thinking Skill: Making calculations**

## Activity

**How Many Earths?**
**Skills: Making calculations, making comparisons**
**Level: Average**
**Type: Computational**

This activity involves a computation, by division, that helps not only to reinforce mathematical skills but also to reveal the enormity of astronomical distances, even when compared with very large geologic distances. The answer to the question is about 11,628 earths. In order to arrive at the right answer, students must divide 150 million km by 12,900 km, the diameter—not the radius—of the earth.

Mexico City, Mexico; (5) Prague, Czechoslovakia.

Also challenge students to estimate the location of the shadow zone for each earthquake.

### Enrichment

Earthquakes tend to occur frequently in the same areas of the world. These areas also tend to have frequent volcanic eruptions. Challenge students to find out where the world's major earthquake and volcano zones are located. Then have students work in groups to make maps that show these zones.

### Enrichment

The strength of an earthquake is measured according to the Richter scale. Have students find out the origin of the Richter scale and how it is calibrated.

## ANNOTATION KEY

❶ Thinking Skill: Making comparisons

❷ Thinking Skill: Interpreting a diagram

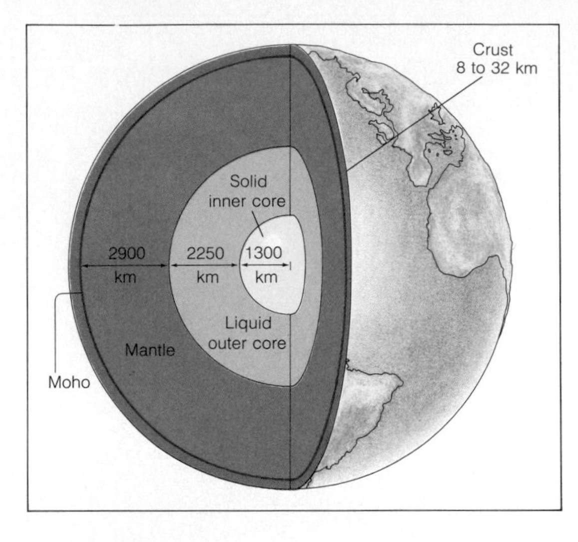


**Figure 21-8** *The earth is made up of different layers. This drawing shows the distances you would have to travel through these layers from the surface to the earth's center.*

mostly iron and nickel. The very hot inner core is made up mostly of solid iron and nickel under tre- ❶ mendous pressure. Temperatures probably range from about 2200° C in the upper part of the outer core to about 5000° C in the inner core.

### SECTION REVIEW

1. Name and describe two kinds of earthquake waves.
2. What is a shadow zone?
3. What is the outer core made of? What is the inner core made of?

### Activity

*Model of the Earth's Interior*

1. Obtain a Styrofoam ball 15 cm or more in diameter.
2. Carefully cut out a wedge from the ball so that the ball looks similar to Figure 21-8.
3. Draw lines on the inside of the ball and of the wedge to represent the four layers of the earth.
4. Label and color each layer on the ball and wedge.

## 21-2 Exploring the Earth's Mantle

The distance from the edge of the liquid outer core to the earth's surface is about 2900 kilometers. Most of this distance makes up the earth's **mantle.** Exactly what is the mantle? How is it different from the earth's core? Part of that question was answered in 1909. On October 8, an earthquake shook Croatia, which is now part of the country of Yugoslavia. In

486

### Activity

**Model of the Earth's Interior**
**Skills: Making models, applying, relating**
**Level: Average**
**Type: Hands-on**

This activity provides students with the opportunity to take information from the chapter to construct a stratigraphic model of Earth's interior. Students should use the following scale to make their model: inner core from the center out to 1.5 cm, outer core from 1.5 cm to 4 cm, mantle from 4 cm to the edge of the ball, crust is the surface of the ball.

## 21-1 (continued)

### Section Review 21-1

1. P, or primary, waves result from a back-and-forth vibration of rock particles which move rapidly and can travel through both solids and liquids. S, or secondary, waves result from an up-and-down or side-to-side vibration of rock particles. S waves are slower than P waves and can only pass through solids.
2. A zone where no earthquake waves are detected.
3. Liquid iron and nickel; solid iron and nickel.

## TEACHING STRATEGY 21-2

### Motivation

Have students, either orally or in writing, imagine what it would be like to actually be the first explorers to travel to the center of the earth. Allow any imaginative, fanciful stories students may come up with, despite their scientific inaccuracy. After completing this chapter, have students refer to their original stories about what the trip would be like now that they have more information about the center of the earth.

### Content Development

Point out that Mohorovicic was puzzled because different seismographic

Zagreb, the capital of Croatia, Yugoslavian geologist and seismologist Andrija Mohorovicic (moh-hoh-ROH-vuh-chich) studied seismograph records from many stations.

Mohorovicic noticed something odd about the seismograph recordings. Stations within about 160 kilometers of the earthquake's center showed a pattern of squiggles that indicated a strong earthquake followed by a weak one. But at stations more than 160 kilometers away from the earthquake the pattern of squiggles indicated a weak earthquake followed by a strong one. The second message was the reverse of the first message. But both messages had been sent out by the same earthquake at the same time! Here was a mystery to be solved.

Mohorovicic thought he must be misreading the messages. He knew there was only one earthquake. The stronger signals came from shock waves that had traveled a shorter distance. The weaker signals from the same earthquake came from shock waves that had traveled a longer distance and had lost some strength along the way. That seemed to explain the first set of messages from seismographs within 160 kilometers of the earthquake. The weaker waves traveled a longer distance and came in second. See Figure 21-9.

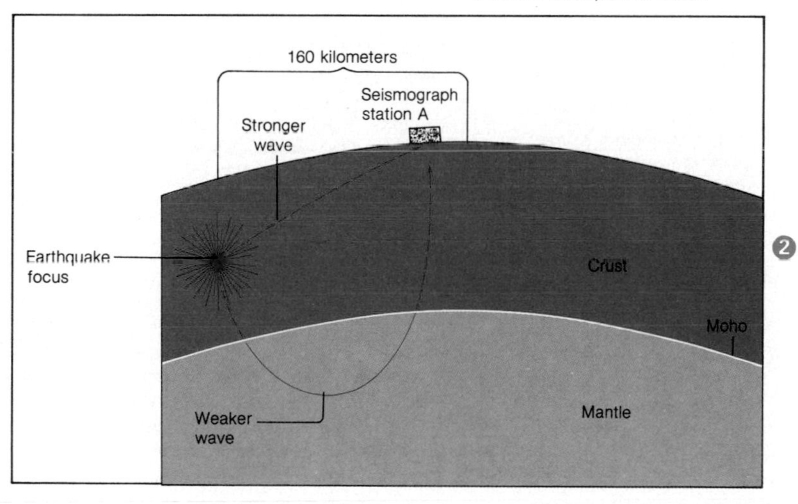


**Figure 21-9** *The stronger earthquake waves arrive before the weaker waves at a seismograph station less than 160 kilometers from the earthquake's focus.*



## 21-2 EXPLORING THE EARTH'S MANTLE

### SECTION PREVIEW 21-2

In this section, students will be introduced to features of the earth's mantle. Students will learn about the boundary, called the Moho, that separates the crust from the mantle.

Students will learn that the location and structure of the mantle were first determined in 1909 by Yugoslavian scientist Andrija Mohorovicic. Mohorovicic learned about the mantle by studying seismic waves.

Students will discover that in the 1950s, scientists found a doorway to the mantle in the ocean floor. Scientists discovered that molten rock from the mantle flows upward through cracks, or rifts, in the ocean floor. The molten rock then hardens into solid rock. This rock becomes a new part of the ocean floor.

### SECTION OBJECTIVES 21-2

1. **Describe the features of the mantle.**
2. **Explain how the location and structure of the mantle were determined by studying seismic waves.**
3. **Describe the Moho.**
4. **Describe how mantle rocks form new ocean floor.**

### SCIENCE TERMS 21-2

**mantle p. 486**
**Moho p. 488**
**Mid-Atlantic Rift Valley p. 489**
**Mid-Ocean Ridge p. 489**

stations reported exactly opposite versions of the same earthquake.

- **At first, what did Mohorovicic think had happened?** (He had misread the messages.)
- **What possible explanation or hypothesis did Mohorovicic think of later?** (Something must have speeded up the weaker waves by the time they were 160 k from the earthquake.)
- **What did he conclude this "something" was?** (The weaker waves must have passed through a deeper portion of the earth, through which waves can travel faster.)
- **What did he conclude about this portion of the earth in which waves can travel faster?** (The structure of this portion must be different from the structure of the crust because waves travel at different speeds through different materials.)

### Enrichment

Some scientists believe that the composition of meteorites lends support to the idea that the earth's core consists of iron. Have students find out reasons why scientists believe this. What do other scientists believe?

## ANNOTATION KEY

❶ Thinking Skill: Interpreting a diagram
❷ Thinking Skill: Applying concepts
❸ Thinking Skill: Making inferences

❶

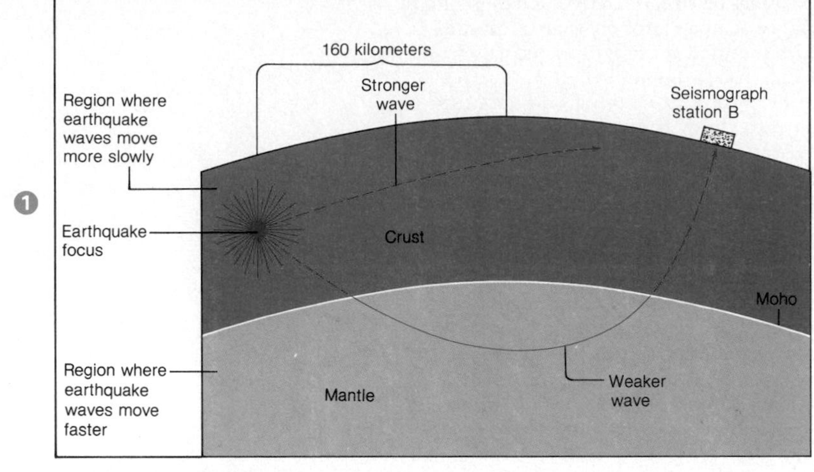


**Figure 21-10** *The weaker earthquake waves arrive before the stronger waves at a seismograph station more than 160 kilometers from the earthquake's focus.*

But what about those stations farther from the earthquake's center than 160 kilometers? There the weaker waves came in first, although they had traveled farther than the stronger signals. Mohorovicic reasoned that there could be only one answer. Something had speeded up the weaker waves so that they outraced the stronger waves to the station. Mohorovicic calculated that the stronger waves traveled near the surface of the earth. The weaker waves had traveled through a deeper region well beneath the earth's surface. ❷ In this region, Mohorovicic thought, the earth's structure must be different than on the surface. The weaker waves speeded up in this lower region and were bent back toward the surface. Although these weaker waves traveled a longer distance, they came in first. See Figure 21-10.

You know that earthquake waves bend. They also change speed. The bending and the change in speed occur as the waves travel through the boundary between different layers of the earth. Mohorovicic calculated that the boundary between the different layers was about 32 kilometers beneath the earth's land surface. He had discovered the boundary between the earth's crust, or outermost layer, and the mantle. This boundary is now called the **Moho.**

### Activity

*Visiting the Earth's Core*

Read *A Journey to the Center of the Earth.* Most libraries will have a copy. It was written by the French author Jules Verne more than a hundred years ago. At that time, scientists did not have any accurate knowledge of what lay just a kilometer beneath their feet—and it is about 6450 kilometers to the center of the earth. Compare Verne's description of what is inside the earth to what you have learned in this chapter.



### Activity

**Visiting the Earth's Core**
**Skills: Making comparisons, relating, applying**
**Level: Average**
**Type: Library/writing**

Jules Verne's novel *A Journey to the Center of the Earth* makes delightful reading. Students will enjoy contrasting Verne's imaginative account with their own newly gained knowledge of the earth's structure.

## 21-2 (continued)

### Skills Development

***Skill: Interpreting diagrams***

Have students observe the diagram in Figure 21-10.

- **In which layer of the earth do earthquake waves travel faster?** (in the mantle)
- **In which layer do they travel more slowly?** (in the crust)
- **Which wave arrived at station B first—the stronger wave or the weaker wave?** (the weaker wave)
- **Which wave had to travel a longer distance to station B?** (the weaker wave)
- **Can you explain why this wave arrived first, even though it is weaker and had to travel a longer distance?** (A large portion of the weaker wave's path included the mantle. As it passed through the mantle, the weaker wave was able to speed up enough to beat the stronger wave.)

### Reinforcement

Reinforce the idea that a rift is a crack in the earth's crust. Point out that the crack and the area on either side of it form a rift valley. Stress that molten rock from the mantle flows upward through a rift. This molten rock then hardens and becomes part of the earth's crust.

### *Doorway Into the Mantle*

Over many years, scientists discovered that the Moho extends around the world. Beneath it, the mantle goes down about 2900 kilometers into the earth until it reaches the edge of the outer core.

The distance from the earth's surface to the Moho varies. It is a long distance from the surface of the land to the Moho. In comparison, it is a very short distance from the ocean floor to the Moho. In fact, on the floor of the Atlantic Ocean, scientists have found a doorway into the mantle.

In the 1950s, scientists mapped the floor of the Atlantic Ocean. Ships sailed back and forth, bouncing thousands of sound signals over the entire bottom of the Atlantic Ocean. From these sound signals, Marie Tharp, a draftswoman working for the Lamont-Doherty Geological Observatory in New York, made the first map of the Atlantic Ocean's bottom. Her map showed a great crack, or rift, in the crust that ran down the middle of a massive underwater mountain range in the Atlantic Ocean. This crack and the area on either side of it are called the **Mid-Atlantic Rift Valley.** Later, scientists found that the Mid-Atlantic Rift Valley is part of a much larger system of rift valleys and mountain ranges—a system called the **Mid-Ocean Ridge** that extends for about 74,000 kilometers under all the world's oceans.

Partially molten, or melted, rock from the mantle flows upward through a rift valley. It spreads outward on both sides of the valley. As it spreads, the partially molten rock cools quickly and hardens into solid rock. This rock becomes a new part of the ocean floor. Because it was "frozen" in the act of flowing, the solid rock shows a pattern of twisted ridges. See Figure 21-11.

This rock came from the earth's mantle. Now, for the first time, scientists could hold a piece of the mantle in their hands. **After studying rock samples, scientists have determined that the mantle is made mostly of the elements silicon, oxygen, iron, and magnesium.** So the unknown substance of the mantle is made up of the same ordinary materials found in rocks on the earth's surface.

But the mantle rocks were not acting like ordinary rocks. Are these rocks liquid? Evidence from

**Figure 21-11** *Partially molten rock has been "frozen" while flowing from a rift valley in the ocean floor. Notice the ridges that formed when the rock hardened.*

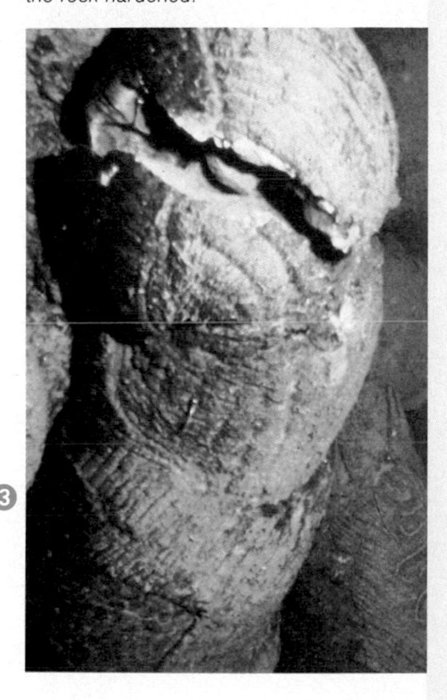



## SCIENCE, TECHNOLOGY, AND SOCIETY

Earth scientists have developed a more detailed model of the earth's interior by using an instrument similar to a CT (computerized tomography) that is used in medical technology. In a CT, X-rays penetrate the human body from all directions and provide doctors with a three-dimensional image of the body. Bones, tumors, and organs are easily identified because they have different densities and therefore absorb X-rays differently.

Similarly, in seismic tomography, X-rays are used. These X-rays replace seismic waves. As you know, seismic waves travel at different speeds in the earth depending on the temperature and density of the rock through which they travel. By examining seismic tomograms, scientists have learned that the layers of the earth are not uniform. The tomograms also have enabled scientists to map the location and shape of the hot, less dense; and cool, more dense regions of the earth's interior.

- **What are some other ways in which scientists might use seismic tomograms?** (Accept all reasonable answers, such as, to learn about the internal structure of other planets.)
- **What other types of scientific technology have been applied in new ways?** (Answer might include: space technology spinoffs and the use of seismographs in detecting nuclear explosions.)

## FACTS AND FIGURES

The mantle makes up 80% of the earth's volume and 68% of the earth's mass.

## TIE-IN/ASTRONOMY

Scientists have studied the moon and other planets in our solar system in order to learn more about the earth's inner structure. Many scientists believe that the earth and the moon formed together, which means that they should have similar composition. There is doubt, however, that the moon has an iron inner core.

## Enrichment

Some important examples of rift valleys include: Death Valley in California, the Valley of the Dead Sea between Israel and Jordan, the Great Rift Valley of East Africa, and the Rhine Valley in Germany. Have students choose one of these valleys and prepare a report. The report should include the valley's location, size, physical features, geographic importance, and history of formation.

## BACKGROUND INFORMATION

Many geologists believe that the earth was once molten. The heat that caused this molten state is believed to have come from the decay of radioactive elements. Radioactive elements are elements that break down into other elements, giving off energy in the process.

Large amounts of radioactive elements such as thorium and uranium were probably found among the matter that formed the earth. As these elements broke down into other elements, energy in the form of heat was released. This heat became trapped in the earth, eventually causing much of the rock that formed the earth to melt.

Since the earth is no longer molten, scientists believe that there has been a significant decrease in radioactive heating. However, there is still evidence of radioactive heating in the high temperature found in the earth's interior.

## ANNOTATION KEY

1. Thinking Skill: Relating cause and effect
2. Thinking Skill: Relating cause and effect
3. Thinking Skill: Interpreting a diagram

**Figure 21-12** *A research vessel travels through a rift valley of the Mid-Atlantic Ridge System. The partially molten rock near the rift's center will eventually become new ocean floor.*

earthquake waves suggests that the temperature in the upper part of the mantle is about 900 to 1000°C. This is more than hot enough to melt the mantle rocks completely. However, these rocks are not completely melted.

Are these rocks solid? Scientists found that S waves can travel through the mantle. And you know that S waves travel through solids but not liquids. But how can solid rock flow? How could a rock act like a solid and like a liquid at the same time? Here was another mystery to be solved.

### *Flexible Rocks and Conveyor Belts*

Scientists found that the answer to this mystery is pressure. Pressure is the amount of force over a certain area. (1) Mantle rocks are under great pressure from the weight of all the rocks above them. Under

490

### 21-2 (continued)

#### Enrichment

In addition to the ocean floor, volcanoes have provided scientists with information about the mantle. In fact, sometimes volcanoes are referred to as "windows to the earth's interior." Have students find out what scientists have learned about the inner earth from the study of volcanoes. Also challenge students to find out how volcanoes are formed, and how they change the earth's surface.

#### Content Development

Remind students of the demonstration on plasticity. Point out that the putty had properties of both a solid and a liquid.

- **How does the term plasticity apply to the earth?** (The rocks have properties of both a solid and a liquid.)
- **In what way does mantle rock behave like a solid?** (S waves can travel through the mantle, so it must be solid.)

#### Reinforcement

Have students outline how Mohorovicic's work demonstrates the steps of the scientific method. Some students may enjoy working in small groups and dramatizing Mohorovicic's discovery.

great pressure, the solid rocks flow very slowly. The mantle rocks do not melt under this great pressure but become a flexible solid that flows slowly like a gooey liquid—like very thick molasses. S waves can travel through this thick liquid.

But why does mantle rock flow steadily upward through the rift valleys? The answer is that the great heat from the earth's core sets up churning motions in the solid-liquid rocks of the mantle. They are much like the churning motions you can see in a pot of boiling soup on a stove. In this case, the soup acts like rocks in the mantle, and the stove provides heat as does the earth's core. The hottest soup, nearest the heat from the stove, rises to the top of the pot. There the soup cools and sinks back to the bottom of the pot. This soup is reheated and again moves upward to the surface.

Scientists reason that material in the mantle must move in the same way. These churning motions keep molten rock pouring out of the rift valleys. This molten rock becomes a new part of the ocean floor. As the flow out of a rift valley goes on, the older parts of the ocean floor are pushed farther away from the rift valley on either side.

Think of a conveyor belt like the ones that move groceries at supermarket counters. Then imagine a

### Activity

*Simulating Plasticity*

The ability of a solid to flow is called plasticity.

**1.** Put 15 g of cornstarch into a beaker. Use a medicine dropper to add one drop of water at a time to the cornstarch. Stir after each addition.

**2.** Stop adding water when the mixture becomes difficult to stir.

**3.** Try to pour the mixture into your hand. Try to roll it into a ball.

How is the mixture similar to the earth's mantle? Different?

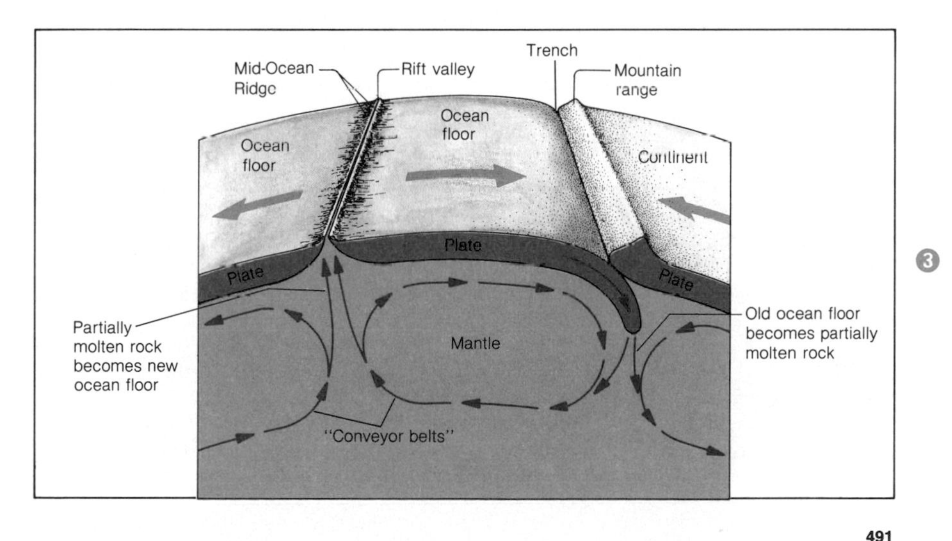


**Figure 21-13** *The continents and ocean floor ride on plates that may move like "conveyor belts."*



## Activity

**Simulating Plasticity**
**Skills: Making observations, comparing, relating, measuring, applying, manipulative, hypothesizing**
**Level: Remedial**
**Type: Hands-on**
**Materials: Cornstarch, beaker, medicine dropper, water, stirring rod**

Students will be able to observe that cornstarch mixed with water behaves like a solid and a liquid, depending on how much pressure is applied to the mixture. When they squeeze it in their hands or roll it into a ball with constant pressure, it has the characteristics of a solid. When they release the pressure, the mixture flows like a liquid. The trick to achieving this effect is in adding just the right amount of water to the cornstarch.

The mixture has the property of plasticity, which is characteristic of the mantle. Students can see this happening when they roll the mixture into a ball and press it. However, the mixture is not under intense heat and pressure as the mantle is.

## 21-3 THE EARTH'S CORE AND FLOATING CONTINENTS

### SECTION PREVIEW 21-3

In this section, students will learn about the thin outer layer of the earth that is called the crust. Students will also learn how the crust is divided into plates that carry the continents and the ocean floor.

Students will learn how new crust is formed as molten rock from the mantle flows upward at rift valleys, and older crust moves back down into the mantle at trenches. Students will discover that the trenches and rift valleys divide the earth's crust into floating parts called plates. Students also will discover that as the plates float on the solid-liquid rocks of the mantle, the continents and ocean floor also move.

### SECTION OBJECTIVES 21-3

1. **Describe the features of the earth's crust.**
2. **Describe the formation of new crust and the movement of older crust.**
3. **Define plate.**
4. **Explain how continents float.**

### SCIENCE TERMS 21-3

**crust p. 492**

**Figure 21-14** *The elements that make up the earth's crust are listed in this chart. What two elements are the most abundant?* ❶

***ELEMENTS IN THE EARTH'S CRUST***

| Element | Percentage in Crust |
|---|---|
| Oxygen | 46.60 |
| Silicon | 27.72 |
| Aluminum | 8.13 |
| Iron | 5.00 |
| Calcium | 3.63 |
| Sodium | 2.83 |
| Potassium | 2.59 |
| Magnesium | 2.09 |
| Titanium | 0.40 |
| Hydrogen | 0.14 |
| Total | 99.13 |

second conveyor belt at the end of the first one. The second belt is moving in the opposite direction. It is as if two such "belts" made of partially molten rock are rolling up out of a rift valley and spreading in opposite directions. See Figure 21-13.

These belts move much more slowly than conveyor belts at the supermarket. They add new floor to the ocean at the rate of only a few centimeters a year. But over millions of years, this adds up to hundreds of kilometers. Over such a length of time, the Atlantic Ocean has become hundreds of kilometers wider.

If you think about it, that is like saying that the continents, or large landmasses, on either side of the Atlantic have moved farther apart. In fact, scientists now have good evidence that all the continents on the earth's surface are slowly moving. And this is part of the story of the earth's crust.

### *SECTION REVIEW*

1. What two changes happen to earthquake waves when they enter a different substance?
2. Where did scientists find a doorway into the mantle?
3. What causes mantle rock to become a flexible solid that flows like thick molasses?

## 21-3 The Earth's Crust and Floating Continents

You now have traveled 6450 kilometers from the center of the earth's core to the earth's surface, or **crust**—the place on which you live. **The crust is the thin outer layer of the earth and is made up mainly of solid rocks.** These rocks contain mostly two elements—silicon and oxygen. Figure 21-14 provides a list of the most abundant elements in the crust of the earth.

The thickness of the crust varies. The crust beneath the continents has an average thickness of about 32 kilometers. But beneath the ocean floor,



### 21-2 (continued)

### Section Review 21-2

1. They bend and change speeds.
2. On the floor of the Atlantic Ocean.
3. Pressure.

### TEACHING STRATEGY 21-3

#### Motivation

Display an apple and ask,

• **How would you describe the thickness of the peel of this apple as compared to the whole apple?** (The peel is only a very thin covering compared to the whole apple.)

Point out that the thickness of the earth's crust is like the peel on this apple as compared to the entire earth. The crust is only a very thin covering. However, all life on earth exists on or within a few hundred meters above the earth's crust.

#### Content Development

Remind students of their study of rocks in Chapter 19.

• **What three types of rocks make up the earth's crust?** (igneous, sedimentary, and metamorphic rocks)

Have students observe the diagram in Figure 21-15. Explain that the lighter rocks that make up the continents are primarily granite, whereas the heavier rocks are primarily basalt.

• **Based on the diagram, what type of rock must make up the ocean floor?** (basalt)

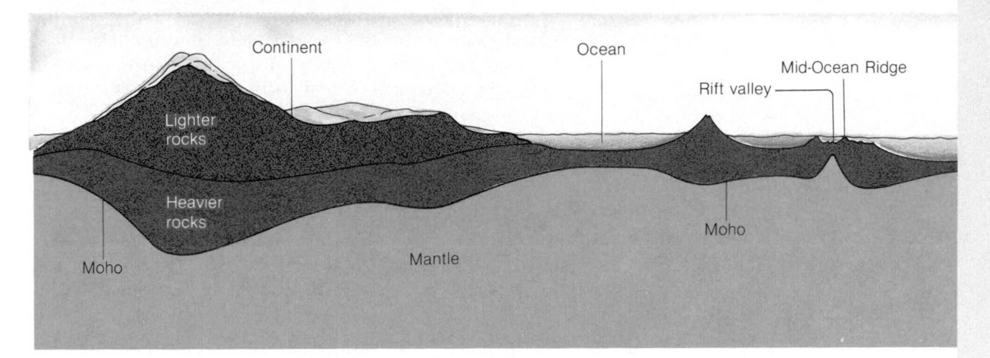


**Figure 21-15** *The earth's crust has two layers. The upper layer is made of lighter rocks and is found only under the continents. The lower layer is made of heavier rocks. It is found both under the continents and under the oceans, where it forms the ocean floor.*

the crust has an average thickness of only about 8 kilometers. You learned earlier that new ocean floor is being added on either side of the rift valleys. If this is so, then why doesn't the ocean floor pile up higher and higher?

Let's go back to our picture of the supermarket conveyor belts. Although the belts come up to the counter in one place, they also go back under the counter at another. The "belts" that move the continents are like this too.

At a rift valley, the molten rock flows up from inside the earth, forming new crust. At other places, called trenches, older crust moves back down into the mantle to again form part of its solid-liquid rock. See Figure 21-13. After many millions of years the slowly moving mantle rocks once again emerge at a rift valley—just as the supermarket conveyor belt travels part of the time *under* the counter and then comes up again.

The trenches and rift valleys of the earth divide the crust into about 20 floating parts like the pieces of a jigsaw puzzle. Scientists call these parts plates. Plates usually move away from rift valleys and toward trenches. Each plate is like a package riding on the upper, visible part of a supermarket conveyor belt. The motion of these belts moves the continents

### Activity

*The Earth's Crust*

Use sand, gravel, clay soil, water, and an empty aquarium to construct a model of a cross section of the earth's crust. Your model should look similar to Figure 21-15. Line the "ocean" with clear plastic before adding water to your model.



## ANNOTATION KEY

**❶ Oxygen and silicon (Interpreting charts)**

## Activity

**The Earth's Crust**
**Skills: Making observations, comparing, manipulative, relating, applying, designing**
**Level: Remedial**
**Type: Hands-on**
**Materials: Sand, gravel, clay, soil, water, empty aquarium, clear plastic**

This activity will help students understand the difference between the two layers of the earth's crust. After checking students' models, have students print each of the labels in Figure 21-15 on small pieces of pieces of paper. Then have the students attach the labels to the outside of the aquarium with tape. Make sure labels are in their proper positions.

## Enrichment

The earth can be divided into layers according to the rigidity, or stiffness, of the rock. Have students find out the meaning of the terms lithosphere, asthenosphere, and mesosphere.

## Reinforcement

Reinforce the idea that new crust is formed at rift valleys and old crust returns to the mantle at trenches.

Also emphasize that the movement of rock is upward from the mantle to the surface at rift valleys and downward from the surface to the mantle at trenches. Students should be able to make their own simple diagrams to show this cyclic process.

## Skills Development

***Skills: Comparing, inferring***

Have students study the chart in Figure 21-14.

- **How does the composition of the crust compare with the composition of the mantle?** (The composition of the crust is more complex, because many more different elements are present in the crust.)
- **Are all of the elements found in the mantle also found in the crust?** (yes)
- **What are these elements?** (oxygen, silicon, iron, and magnesium)
- **Why do you think the elements found in the mantle also would be found in the crust?** (Crust is formed as mantle rock pushes upward to the earth's surface at rift valleys.)

## TEACHER DEMONSTRATION

This demonstration will illustrate to students how melting ice during the end of an Ice Age could result in the rising of the land that was once covered by ice. For the demonstration, you will need a block of wood, a small basin or medium-sized bowl, several ice cubes, water, and a waterproof marker.

Fill the basin about three-fourths full of water and place the wooden block in the water. Place several ice cubes on the block. Have a student volunteer mark the level of the water on the block.

Allow the ice cubes to melt. Then have a student again mark the level of the water on the block. Remove the block from the water and have students observe the water-level marks.

- **What happened to the wooden block when the ice cubes melted?** (The wooden block floated higher on the water.)
- **Why did the block float higher after the ice melted?** (The ice added extra weight to the block. Once the ice melted, the mass was removed and the block rose.)

and ocean floor, which rest on the upper part of the plates. The plates float on the solid-liquid rocks of the earth's mantle. You can also think of the continents and ocean floor as floating ships slowly riding on the plates.

Balance is the key to the floating continents, as it is to floating ships. When a ship takes on cargo, it rides deeper in the water. When the cargo is unloaded, the ship rises in the water. Continents also rise higher or sink deeper into the mantle. Over the last ten thousand years, for example, parts of northern Europe have risen as much as 250 meters—higher than a 60-story building. Why? Ten thousand years ago, this land was covered with sheets of ice

### Career: *Cave Guide*

**HELP WANTED: CAVE GUIDES** To conduct and instruct groups of all ages and backgrounds through local natural caves. High school or college students preferred. Courses or interest in geology or public speaking desirable. Summer or part-time work available.

There are more than 50,000 naturally formed caves in the United States. Caves are found at or near the surface of the earth's crust. The deepest cave in the world extends only about 1.5 kilometers into the crust. New caves are discovered, explored, surveyed, and named each year. One cave in Texas, for example, was discovered by highway engineers while building an interstate roadway. Visitors to this cave are shown special formations such as "icicles" made of stone. Fossils of extinct animals have also been discovered in this cave.

Workers who lead groups of visitors safely through a cave are the **cave guides.** Cave guides lecture about a cave's size, history, and how it was formed. They point out features of special interest and answer questions. They are concerned with cave conservation, preservation, and safety.

Each cave is different, so guides must spend time learning about their particular cave before they begin work. Visitors on cave tours

walk or ride in vehicles such as jeeps. In caves with underground lakes or rivers, visitors may travel by boat. Besides explaining geologic formations to visitors, guides discuss topics such as underground waterfalls, fossils, Indian artifacts, cave life, and the constant temperature maintained in caves.

If you are interested in experience as a cave guide or would like to know more about the caves in your area, write to the National Caves Association, Route 9, Box 106, McMinnville, TN 37110.



### 21-3 (continued)

#### Enrichment

Have students find out the elevation and depth of major land and ocean features, such as Mt. Everest, Mariana Trench, Mt. McKinley, and Death Valley. Then have students work in small groups to make charts or diagrams that display this information.

#### Skills Development

***Skill: Making graphs***

Have students use the information in this chapter and other reference sources to make a graph showing how the temperature of the earth changes with depth. More advanced students may also enjoy finding out how pressure changes with depth, and making a graph of this information.

#### Enrichment

Scientists have made some interesting studies about the structure of the moon compared with the structure of the earth. Have students find out more about this topic and present their findings to the class. Encourage students to include in their presentations a diagram of what is currently believed to be the structure of the moon.

**Figure 21-16** *The plate that carries Egypt* (right) *is separating from the plate that carries Saudi Arabia* (left). *The Red Sea has become wider as a result of this separation.*

several kilometers thick. This ice was like a cargo of additional material. As the ice melted, the continent rose like a lightened ship.

You now have come to the end of the journey from the center of the earth to the surface. That familiar land on which you live and walk, with its rocks and water and soil is just the top of the earth's crust. And the crust is afloat, sinking and rising on the mantle and powered by heat from the very center of the earth.

### SECTION REVIEW

1. What is the average thickness of the crust beneath the continents? Beneath the ocean floor?
2. How is new crust formed?
3. Approximately how many floating plates make up the crust?

#### Activity

*Mohorovicic's Discovery*

The boundary between the crust and the mantle is called the Moho after Andrija Mohorovicic. Write a 300 word essay explaining how Mohorovicic discovered this boundary. Your essay should emphasize all of the following vocabulary words:

| | |
|---|---|
| crust | focus |
| mantle | Moho |
| P waves | S waves |
| seismograph | |



## FACTS AND FIGURES

The internal temperature of the earth rises 2–3°C for every 100 m below the earth's surface.

## FACTS AND FIGURES

The densities of various layers of the earth are:

Inner core—10.7 g/cm$^3$
Mantle—4.5 g/cm$^3$
Crust—2.7–4.3 g/cm$^3$

The average density of the earth is 5.5 g/cm$^3$

## Activity

**Mohorovicic's Discovery**
**Skills: Report writing, vocabulary comprehension**
**Level: Enriched**
**Type: Library/vocabulary**

This activity not only helps provide students with more information on the mantle–crust boundary but also serves to reinforce vocabulary and composition skills.

## Section Review 21-3

**1.** 32 k; 8 k.
**2.** Molten rock flows upward from inside the earth and forms new crust.
**3.** 20.

# LABORATORY ACTIVITY
# BUILDING AN ACTIVE GEYSER

## BEFORE THE LAB

1. **At least one day prior to the activity, gather enough materials for your class, assuming 6 students per group.**
2. **Make sure that *each* student has a pair of safety goggles in wearable condition.**

## PRE-LAB DISCUSSION

Review with students the idea that the temperature of the earth increases drastically beneath the earth's surface. Ask

- **What evidence is there of this intense heat?** (Molten lava flowing out of volcanoes; molten rock that has "frozen" on the ocean floor; igneous and metamorphic rocks that have been formed in intense heat; the existence of hot springs and geysers in various parts of the world.)
- **What do you think causes this great heat in the inner earth?** (Answers may vary; most scientists believe that it is energy from the breakdown of radioactive elements that causes the inner earth to be so hot.)

Point out to students that the heat energy of the earth's interior is often referred to as geothermal energy. Geothermal energy from hot springs and geysers is being used in places such as Iceland to heat homes and power electric generators. Environmentalists are enthusiastic about the use of geothermal energy because it is a clean and renewable energy source.

## SKILLS DEVELOPMENT

Students will use the following skills while completing this activity.

1. Observing
2. Relating
3. Safety
4. Applying
5. Inferring
6. Hypothesizing



### *Building an Active Geyser*

**Purpose**

In this activity, you will build a model geyser and observe it in action.

**Materials** *(per group)*
250-mL beaker
Bunsen burner
Glass funnel
Tripod
Wire gauze
Safety goggles
Water

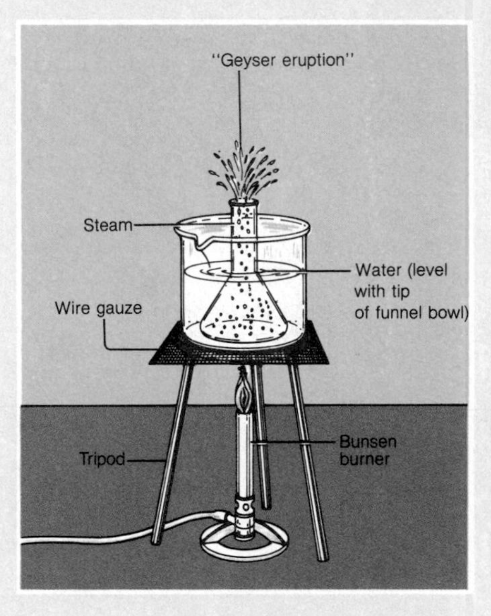


**Procedure**

1. Place a glass funnel upside-down in a beaker.
2. Add water until the bowl of the funnel is covered.
3. Place the beaker, water, and funnel on a piece of wire gauze resting on a tripod.
4. Put the Bunsen burner beneath the tripod and light it. **CAUTION:** *Be very careful in lighting the burner. Avoid putting your face or hands near the setup when the geyser becomes active. Be sure to wear your safety goggles.*
5. Allow time for the water to boil. Notice the bubbles of steam that push the hot water up the funnel stem. As the steam bubbles expand, you should be able to observe water spurting from the funnel stem, much as water acts in a coffee percolator.

**Observations and Conclusions**

1. Describe the behavior of your model geyser while it was erupting.
2. A Bunsen burner provided the heat source for your model geyser. What is the source of energy that heats the water in a real geyser?
3. According to your observations of the model geyser, what is the relationship between geyser activity and available heat energy?
4. If a natural geyser stopped erupting, what might have been the cause?
5. "Old Faithful" is a well-known geyser in Yellowstone National Park. It erupts on a very regular basis, just about every 66 minutes! What do you think must be true of the heat source and water supply that keep this geyser operating?



## SAFETY TIPS

1. Read aloud to students the **CAUTION** in step 4 of the procedure. Remind them that steam or boiling water can cause a very bad burn.
2. Point out to students the need to be careful when handling the glass funnel. If a funnel becomes cracked or broken, students should notify you immediately.

## OBSERVATIONS AND CONCLUSIONS

1. Water should spurt periodically from the funnel stem.
2. Heated rock below the surface of the earth provides the heat energy.
3. Increased heat energy would cause more frequent or more vigorous eruption, assuming an unlimited water supply.

# CHAPTER REVIEW

## SUMMARY

**21-1 The Earth's Core and Earthquake Waves**

- Seismographs detect and record the shock waves from earthquakes.
- Shock waves, or seismic waves, originate at the focus of an earthquake and travel in every direction.
- Earthquakes produce at least two kinds of waves: primary or P waves and secondary or S waves.
- Primary waves vibrate in the direction of the path of an earthquake wave. They travel through both solids and liquids.
- Secondary waves vibrate at right angles to the path of an earthquake wave. They travel only through solids.
- Shadow zones are areas on the surface of the earth where seismographs cannot detect earthquake waves.
- The earth's solid inner core has a radius of about 1300 kilometers.

**21-2 Exploring the Earth's Mantle**

- The earth's mantle is about 2900 kilometers thick and surrounds the earth's outer core.
- By knowing that earthquake waves bend and travel at different speeds through different substances, Andrija Mohorovicic was able to calculate how far below the earth's surface the boundary is located that separates the mantle from the crust. This boundary was named the Moho in his honor.
- The Mid-Atlantic Rift Valley is a crack that runs down the middle of an underwater mountain range in the Atlantic Ocean. The Mid-Ocean Ridge is part of a mountain range system that extends under all the world's oceans.
- Mantle rock is a flexible solid that flows like a thick gooey liquid.
- It is believed that heat from the earth's core sets up churning motions that cause mantle rock to move. You can think of this motion as being like very slow conveyor belts in the mantle that bring molten rock up through rift valleys to form new ocean floor.

**21-3 The Earth's Crust and Floating Continents**

- The crust is the thin, outermost layer of the earth.
- At rift valleys, mantle rocks form new crust. At trenches, old crust becomes mantle rock.
- The crust is divided into about 20 pieces called plates. The continents and ocean floor ride on the upper part of the plates.
- Plates usually move away from rift valleys and toward trenches.
- Floating continents rise higher or sink deeper into the mantle depending on the amount of material the continents have.

## VOCABULARY

*Define each term in a complete sentence.*

| | | |
|---|---|---|
| **core** | **Mid-Atlantic Rift Valley** | **secondary waves (S waves)** |
| **crust** | **Mid-Ocean Ridge** | **seismic waves** |
| **focus** | **Moho** | **seismograph** |
| **mantle** | **primary waves (P waves)** | **shadow zone** |



**4.** A reduction in the amount of either heat or water could cause the stoppage.
**5.** The heat source and water supply must be fairly constant.

## GOING FURTHER: ENRICHMENT

**Part 1**

Have students experiment with water levels that are lower than the tip of the funnel bowl; also have them experiment with a lower flame. Have students relate their results to question 4 in the Observations and Conclusions.

**Part 2**

Have students work in small groups. Challenge each group to prepare a report or other type of project that describes the use of geothermal energy in various parts of the world. Encourage students to discuss the advantages and disadvantages of this energy source, and whether scientists expect it to be used more extensively in the future.

## CHAPTER REVIEW

### MULTIPLE CHOICE

| | | | | |
|---|---|---|---|---|
| **1.** d | **3.** b | **5.** a | **7.** b | **9.** b |
| **2.** c | **4.** d | **6.** d | **8.** d | **10.** a |

### COMPLETION

**1.** seismograph
**2.** core
**3.** focus
**4.** S, or secondary, waves
**5.** mantle
**6.** pressure
**7.** Mid-Ocean Ridge
**8.** Moho
**9.** crust
**10.** silicon

### TRUE OR FALSE

**1.** T
**2.** F S waves
**3.** T
**4.** F crust
**5.** T
**6.** T
**7.** F is
**8.** T
**9.** F thinner
**10.** F plates

### SKILL BUILDING

**1.** The S waves did not arrive in Moscow at all. Because Moscow is on the other side the world, the earthquake waves have to travel through the earth's core. S waves cannot travel through the earth's liquid outer core.
**2.** The reason the inner core is a solid at such high temperature is due to the extreme pressures found within the inner core.
**3.** Students' graphs should accurately depict the data given in Figure 21-14. The most general conclusion they will arrive at is that the elements oxygen and silicon are far more abundant than other elements found in the earth's crust.

### ESSAY

**1.** S, or secondary, waves result from an up-and-down or side-to-side vibration of rock particles. S waves are slower than P waves and can pass only through solids. P, or primary, waves result from a back-and-fourth vibration of rock particles. P waves move rapidly and can travel through both solids and liquids.
**2.** Heat from the earth's core sets up churning motions in the solid-liquid rocks of the mantle. As in a pot of soup on a stove, the hottest material, nearest the source of heat, rises to the top, where it cools and sinks back down, only to be reheated.
**3.** Continents may sink when, for example, large sheets of ice are deposited on them. They may rise when such ice melts away, just as a cargo ship may sink or rise in the water as the ship becomes heavier or lighter by addition or removal of cargo.
**4.** P waves pass slowly through liquids. S waves are stopped completely. Thus scientists have been able to determine the liquid nature of the outer core surrounding the solid inner core.
**5.** Moho is the boundary between the earth's crust and the mantle. As earthquake waves travel through the Moho, they bend and undergo a change in speed.

### CONTENT REVIEW: MULTIPLE CHOICE

*Choose the letter of the answer that best completes each statement.*

**1.** Seismographs
a. detect earthquake waves. b. record earthquake waves.
c. measure the strength of earthquake waves. d. do all of these.
**2.** Primary waves are earthquake waves that can travel through
a. liquids. b. solids. c. liquids and solids. d. none of these.
**3.** Areas in which seismographs cannot detect earthquake waves are called
a. Mohos. b. shadow zones. c. rifts. d. seismozones.
**4.** The distance in kilometers from the earth's surface to the earth's center is
a. 1300. b. 2900. c. 5000. d. 6450.
**5.** Geologists now know that the earth's core is
a. a solid center surrounded by a liquid.
b. a liquid center surrounded by a solid.
c. completely solid.
d. completely liquid.
**6.** The distance from the earth's surface to the Moho
a. is 2900 kilometers. b. is longer than the radius of the earth's core.
c. is 1300 kilometers. d. varies between 8 and 32 kilometers.
**7.** The crack in the Atlantic Ocean crust through which mantle rock flows to form new ocean floor is called the
a. Moho. b. Mid-Atlantic Rift Valley.
c. Mid-Ocean Ridge. d. Mid-Atlantic Trench.
**8.** Rocks on the earth's surface are made mostly of
a. iron and silicon. b. oxygen and magnesium.
c. iron and magnesium. d. oxygen and silicon.
**9.** Which is the thinnest layer of the earth?
a. mantle b. crust c. outer core d. inner core
**10.** The earth's crust
a. is least thick beneath the ocean floor. b. is least thick beneath the continents.
c. is equally thick around the world. d. is none of these.

### CONTENT REVIEW: COMPLETION

*Fill in the word or words that best complete each statement.*

**1.** A(n) ______________ is an instrument that records shock waves from an earthquake.
**2.** The center region of the earth is known as the ______________.
**3.** The underground point of origin, or center, of an earthquake is the ______________.
**4.** Earthquake waves that travel through solids and are blocked by liquids are called ______________.
**5.** The ______________ is the region between the earth's outer core and the earth's crust.
**6.** Mantle rocks flow like a thick liquid because of the great ______________ from the weight of all the rocks above them.
**7.** The mountain range system that runs under the world's oceans is called the ______________.

8. The ______________ is the boundary between the earth's crust and the mantle.
9. The thin, solid outer layer of the earth is called the ______________.
10. About 74 percent of the earth's crust is made of the elements ______________ and oxygen.

## CONTENT REVIEW: TRUE OR FALSE

*Determine whether each statement is true or false. If it is true, write "true." If it is false, change the underlined word or words to make the statement true.*

1. Earthquakes produce both primary and secondary shock waves.
2. There are no P waves in the shadow zones because they are deflected by the earth's core.
3. Each time earthquake waves enter and leave the earth's outer and inner cores, they bend.
4. The boundary between the earth's mantle and outer core is called the Moho.
5. Earthquake waves change speeds as they travel through different substances.
6. Scientists are able to analyze mantle rock because mantle material flows up through rift valleys.
7. The temperature of the mantle is not hot enough to melt rocks.
8. Older crust moves back into the mantle through trenches.
9. The crust is thicker beneath the ocean floor than beneath the continents.
10. Continents and ocean floors ride on about 20 floating parts called ridges.

## CONCEPT REVIEW: SKILL BUILDING

*Use the skills you have developed in the chapter to complete each activity.*

1. **Applying concepts** Suppose a mild earthquake occured in Washington, D.C., at 9:00 A.M. Earthquake waves were felt 8040 kilometers away in Moscow, U.S.S.R. According to seismographic readings, P waves arrived in Moscow at about 9:11 (Washington time). Did the S waves arrive before or after 9:11? Explain your answer.
2. **Relating facts** The temperature of the inner core reaches about 5000° C. The temperature of the outer core begins at about 2200° C. Yet the outer core is liquid while the inner core is solid. Explain how this can be so.
3. **Making graphs** Use the data in Figure 21-14 to construct a graph. Plot each element in the earth's crust on the horizontal axis. Plot the percentage of each element on the vertical axis. What general conclusions can you draw from your graph?

## CONCEPT REVIEW: ESSAY

*Discuss each of the following in a brief paragraph.*

1. Compare S waves and P waves.
2. Explain how the movement of mantle rock is like a pot of soup boiling on a stove.
3. Compare the rising and sinking of floating ships to floating continents.
4. What is the importance of P waves and S waves in learning about the structure of the earth's interior?
5. What is the Moho? How does it affect earthquake waves?

499

## ADDITIONAL QUESTIONS AND TOPIC SUGGESTIONS

**1.** Why do you think the crustal plates tend to move away from rift valleys and toward trenches? (New crust is formed at rift valleys, and older crust returns to the mantle at trenches. As new crust forms, it spreads out; eventually it reaches a trench.)
**2.** A worker at a research center in St. Louis, Missouri makes the following statement: "We rarely have earthquakes in this part of the world. I see no reason for St. Louis to have all this elaborate equipment to detect seismic waves." Do you agree with this statement? Explain your answer. (Disagree. Because seismic waves can be detected all the way on the other side of the world, these distant recordings have often given scientists more knowledge about the inner earth.)
**3.** Suppose three bullets are fired: one into the air, one into wood, and one into lead. How will the movement of the bullets differ? Relate what you have described to the movement of seismic waves through the movement of seismic waves through the earth. (The bullet will travel fastest through air, and more slowly through wood. The bullet will probably not be able to pass through the lead. Instead, the bullet will either penetrate a short distance and stop, or be deflected. This is similar to seismic waves in that waves, like bullets, travel at different speeds through different media. Some waves, like some bullets, cannot penetrate certain materials, and are deflected.)

## ISSUES IN SCIENCE

The following issues can be used as springboards for discussion or given as writing assignments.
**1.** For many years, scientists have believed that the earth's magnetic field is due to the presence of iron in the earth's core. Recently, however, new theories have been proposed that refute this idea. Find out about these alternative theories, and how widely accepted they are. Then express your own opinion as to the causes of the earth's magnetic field.
**2.** In March 1986, the Japanese government set up an earthquake prediction center in Tokyo. At the center, scientists are learning how to predict major earthquakes a few hours in advance so that emergency measures can be taken. Earthquake specialists have urged that comparable evaluation and warning systems be installed in 13 regions of the United States that are particularly vulnerable to earthquakes. To date, there has been no response to this suggestion. Find out more about earthquake prediction, particularly as it is being done in Japan. Then express you opinion as to whether earthquake prediction centers should be set up in the United States.

## Chapter 22

# SURFACE FEATURES OF THE EARTH

## CHAPTER OVERVIEW

In this chapter, students will be introduced to the surface features of the earth. They will learn about the earth's landmasses and the earth's water. Students will also learn about the water cycle, which replenishes the earth's water supply.

Students will learn about the different physical features that make up the earth's land. Areas that have certain physical features are called landscapes. The three major landscape regions found on the earth are plains, plateaus, and mountains.

Students will learn that approximately 70% of the earth's surface is covered by water. Most of this water is salt water, which is found in the oceans. The rest of the earth's water, which is called fresh water, is found in lakes, rivers, streams, and ponds. In addition, a significant amount of fresh water is frozen in ice sheets around the North and South poles.

## INTRODUCING CHAPTER 22

Begin by having students observe the chapter-opener photograph.

- **What do you see in this photograph?** (a man standing in front of a small wooden structure, some defoliated trees, and hills)
- **What is coming out of one of the hills?** (smoke)
- **What do you think the hill is?** (the volcano Paricutín)
- **Do you think Paricutín may have had an effect on the appearance of the landscape? Explain.** (Yes. The eruptions of Paricutín spewed lava, rocks, and ash on the land, which may have destroyed buildings or vegetation, making the landscape barren.)

Have students read the chapter-opener text. Tie in Pulido's discovery of a crack in the earth with the discussion of rift valleys in the previous chapter.

- **What are cracks and areas around cracks in the earth's surface called?** (Rifts are cracks in the earth crust; the crack and the area around the crack is called a rift valley.)
- **What materials flow out of rifts?** (hot, liquid rock from the mantle)
- **What did Pulido see in his cornfield that is similar to a rift?** (the crack he found in his land)
- **What was flowing out of the crack in Pulido's cornfield** (hot rocks and ash)
- **Where do you think the hot rock originated?** (from the mantle)

Remind students that in Chapter 21, they read about hot, liquid rock flowing out of cracks in the earth and "freezing" to form new crust on

# 22 Surface Features of the Earth

**CHAPTER SECTIONS**

**22-1 The Earth's Surface: Landmasses**

**22-2 The Earth's Surface: Water**

**CHAPTER OBJECTIVES**

*After completing this chapter, you will be able to:*

**22-1** Describe the three major kinds of landscapes.

**22-1** Explain how mountains are formed.

**22-1** Compare plateaus and plains.

**22-2** Identify the major sources of salt water and fresh water.

**22-2** Explain how the water cycle works.

On February 20, 1943, a Mexican farmer, Dionisio Pulido, went to work as usual in his cornfield. But his cornfield had changed overnight. He stared in disbelief at a 24-meter-long crack in his land. Hot rocks and ash oozed from the crack. Pulido ran to the nearby village of Paricutín to tell what he had seen.

Pulido had witnessed a dramatic change in the earth's surface—the birth of a volcano! The volcano, called Paricutín after Pulido's village, grew rapidly. It was about 10 meters high by the next morning, and about 168 meters high after the first week. The volcano was about 335 meters high after the first year. By then, great amounts of ash had covered the land around for about 50 square kilometers. On February 25, 1952, Paricutín reached its final height of about 410 meters above the surrounding land.

Nine years and 12 days after its birth, Paricutín became inactive. During its active years it had thrown out about 3.6 billion tons of lava, rocks, and ash. Pulido had lost his cornfield, and about 4500 people in two villages had lost their homes. The surface of the land was completely different.

Volcanoes like Paricutín are among the great forces that shape and reshape the earth's surface. More slowly, folding of the earth's crust forms huge mountain ranges. And even more slowly, the mountains are worn down. The flow of water also changes the shape of the land. This process has been going on for billions of years. It is still going on today.

*Paricutín erupting*



the ocean floor.

• **What changes in the earth's surface are described in this story?** (the gradual building up of a volcano to a height of 410 m; the covering of 50 km of land with lava and ash)

• **Do you think that volcanoes tend to change the earth's surface rapidly or slowly?** (quite rapidly, particularly with regard to violent eruptions)

• **Does the text suggest that there are other factors that reshape the earth's surface more slowly?** (yes)

• **What are some of these factors?** (folding of the earth's crust; wearing down of mountains; changing of land by running water)

• **Do you think that the earth's surface will ever stop changing? Explain.** (Probably not—the forces that change the earth's surface are always at work. As long as the earth exists, it will continue to change.)

## TEACHER DEMONSTRATION

This demonstration will illustrate to students the difference between measuring the height of a mountain relative to sea level and measuring the height of a mountain relative to the surrounding land. For the demonstration, you will need a desk or table, a few books, and a meterstick.

Clear the top of the desk or table and position it so that all students can see it clearly. Place the books on top of each other so they reach a height of approximately 1 m.

• **What is the height of this stack of books?** (Answers will vary. Many students will estimate that the stack of books is approximately 1 m high, but some students may consider the height of the books from the floor, or they may ask you which height you mean.)

Point out that in order to be accurate, it is necessary to specify the height of an object *relative* to another object. In this case, the height of the books can be determined relative to the floor, or relative to the table top. Have a student volunteer measure the height of the books relative to the tabletop. Record the measurement on the chalkboard. Then have another student measure the height of the books relative to the floor. Record this measurement on the chalkboard.

• **Can the height of the table be determined from these measurements?** (yes)

• **How can this be done?** (by subtracting the height of the books relative to the table from the height of the books relative to the floor)

## TEACHER RESOURCES

**Audiovisuals**

*Landforms on the Earth's Crust,* 20 slides, SVE

*The Earth and Its Wonders,* 2 filmstrips with 2 cassettes, EBE

*The Earth: Oceans,* filmstrip, SVE

**Books**

Bradshaw, Michael J. et al., *The Earth's Changing Surface,* Halsted

Hart, John F., *The Look of the Land,* Prentice-Hall

Wegener, A., *Origin of Continents and Oceans,* Gordon Press

## 22-1 THE EARTH'S SURFACE: LANDMASSES

### SECTION PREVIEW 22-1

In this section, students will be introduced to the earth's landmasses. They will learn that most of the visible land on earth rests on the seven continents.

Students will discover that there are many different surface features on the earth's continents. They will also learn that areas with certain physical features are called landscapes.

Students will be introduced to the three major landscape types that are found on the earth's surface. These landscapes are mountains, plateaus, and plains. Students will learn about the characteristics of each landscape, and where in the United States each landscape is found.

### SECTION OBJECTIVES 22-1

1. **Describe the three major types of landscapes.**
2. **Explain how mountains are formed.**
3. **Describe the characteristics of plateaus, mountains, and plains.**
4. **Compare coastal plains and inland plains.**

### SCIENCE TERMS 22-1

**mountain landscape p. 503**
**mountain range p. 506**
**mountain belt p. 506**
**plateau landscape p. 507**
**plains landscape p. 509**
**coastal plains p. 509**
**inland plains p. 510**

## 22-1 The Earth's Surface: Landmasses

Nearly all the visible land on our planet rests on seven great landmasses—the continents. There are many different surface features on these continents. For example, Asia holds the highest place on earth—Mount Everest, the crown of the mighty Himalaya mountain range. Everest's peak is 8848 meters above sea level. Mount Everest is in Nepal, a small country north of India. Thousands of kilometers to the west, but still within Asia, lies the lowest place on earth's land surface—the shore of the Dead Sea in Israel, 392 meters below sea level.

Asia, the largest continent, stretches from the frozen Arctic tundra in the north to the steaming tropical rain forests and rice paddies of places such as Vietnam and Cambodia in the south. Asia also stretches from the parched Arabian Desert in the west to rich farmlands on the Pacific coasts of China and the Soviet Union.

But even within the smallest continent, Australia, the land takes many shapes. Along the western coast, the land is low and flat. Looking eastward, you can see what appears to be a long line of mountains. As you approach them, you realize that they are not

**Figure 22-1** *Mount Everest* (left) *is the highest place on the earth's surface, and the shore of the Dead Sea* (right) *is the lowest place.*

### TEACHING STRATEGY 22-1

#### Motivation

Display a globe or world map and have students locate the seven continents: Europe, Asia, Africa, North America, South America, Antarctica, and Australia. Also have students locate large islands, such as Iceland, Greenland, New Zealand, Hawaii, and Cuba. Point out that the continents of North America and South America are joined by a tiny piece of land to make up one large landmass; also point out that Africa is joined to Asia to make up the world's largest landmass, which consists of Europe, Asia, and Africa.

#### Content Development

Emphasize to students that three major landscape types are found on the earth's surface. These are mountains, which may be low or high; plains, which may be inland or coastal; and plateaus. Also point out that the continent of Antarctica has surface features that are different from those found on the rest of the continents.

#### Skills Development

***Skill: Interpreting maps***

Have students observe the map in Figure 22-2.

- **What types of landscape regions**

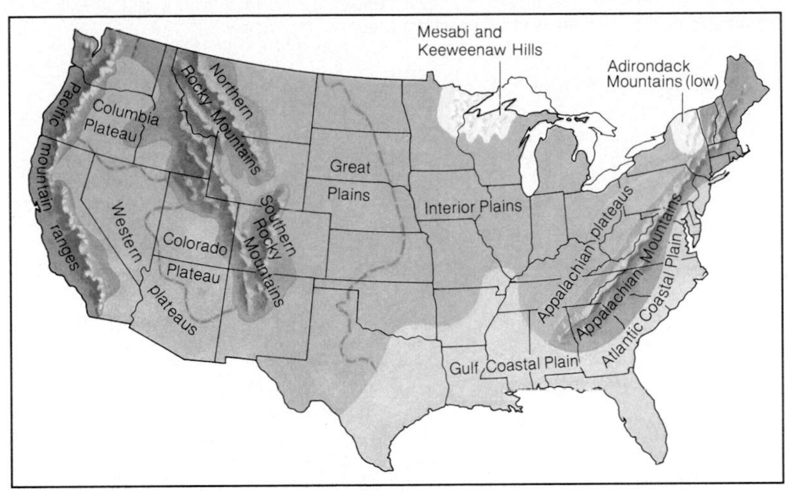


**Figure 22-2** *This map shows the major landscape regions of the continental United States. What type of landscape region covers most of the United States?* ❶

mountains but steep cliffs. A climber reaching the tops of these cliffs finds a great area of roughly level land. This highland region covers most of western Australia. It is like an enormous table whose edges are cliffs. In fact, such highlands are often called tablelands.

Mountains, highlands, and lowlands are different kinds of physical features on the earth's land surface. Such areas that have certain physical features are called landscapes. **There are three main types of landscape regions on the earth's surface: mountains, plains, and plateaus.** Each type has different characteristics. Figure 22-2 shows the major landscape regions of the United States. In which landscape region do you live? ❷

### *Mountains*

You are walking through a region of tall, jagged peaks, long slopes, and high, narrow ridges. You can see deep valleys, forests, and small, rushing streams. Where are you? The region you are visiting is an example of a typical **mountain landscape.**

What is the difference between a mountain and a hill? Most geologists agree that a mountainous area rises at least 610 meters above the surrounding land.



## BACKGROUND INFORMATION

Although the earth's land is divided into seven continents, there are actually only four major landmasses on earth. The largest landmass consists of the continents of Europe, Asia, and Africa. The second largest landmass consists of the continents of North America and South America. A much smaller landmass is Antarctica, and the smallest landmass is Australia.

## ANNOTATION KEY

❶ Plains (Interpreting maps)

❷ Answers will vary. (Interpreting maps)

- **What two types of mountains are shown on the map?** (high mountains and low mountains, or hills)
- **Where are low mountains located?** (near the Great Lakes and in the northern part of New York)
- **What section of the country consists only of mountains and plateaus?** (the western states)

### Enrichment

Students may enjoy doing research on the exploration of the continent of Antarctica. Challenge them to find out how and by whom this exploration was carried out, as well as some of the problems encountered by the explorers in this very cold region.

### Enrichment

Early in the earth's history, the continents were very different from what they are today. Have students find out how the earth's landmasses have moved and split apart. Also challenge them to discuss the factors that have caused these changes.

### Reinforcement

Have students use reference materials to prepare reports on the geography and geology of Antarctica. A particularly helpful pamphlet called: *The Antarctic and Its Geology* (1978-261-226/50) can be ordered from the U.S. Government Printing Office, Washington, DC 20402.

**are found in the United States:** (plateaus, mountains, hills, interior plains, coastal plains)

- **What types of landscape regions are found in our state?** (Answers will vary.)
- **What type of landscape region makes up the area in which we live?** (Answers will vary.)
- **Which states have all three major landscape regions—plains, mountains, and plateaus?** (New York, Pennsylvania, New Mexico, Alabama, Tennessee)
- **What type of landscape is found in Northern New England?** (mountains)
- **Where are the major mountain ranges located?** (along the west coast; in the interior western states; and inland on the east coast)
- **What landscape feature is located between the plateaus in the western states?** (mountain)

## BACKGROUND INFORMATION

The earth's crust under the continents extends considerably deeper than the crust under the oceans—on the average about 40 km as compared to about 10 km. The deepest part of the continental crust is under mountain ranges. Mountain crust extends about 70 km downward, forming "roots" that keep the continents in equilibrium.

## ANNOTATION KEY

❶ Thinking Skill: Interpreting charts
❷ Thinking Skill: Making comparisons
❸ Thinking Skill: Development a model

**Figure 22-3** *Some of the world's mountains are described below.*

❶

***SOME OF THE WORLD'S MOST FAMOUS MOUNTAINS***

| Name | Height Above Sea Level (meters) | Location | Interesting Facts |
|---|---|---|---|
| **Aconcagua** | 6959 | Andes in Argentina | Highest mountain in the Western Hemisphere |
| **Cotopaxi** | 5897 | Andes in Ecuador | Highest active volcano in the world |
| **Elbert** | 4399 | Colorado | Highest mountain of Rockies |
| **Everest** | 8848 | Himalayas on Nepal-Tibet border | Highest mountain in the world |
| **K2** | 8611 | Kashmir | Second highest mountain in the world |
| **Kanchenjunga** | 8598 | Himalayas on Nepal-India border | Third highest mountain in the world |
| **Kilimanjaro** | 5895 | Tanzania | Highest mountain in Africa |
| **Logan** | 5950 | Yukon | Highest mountain in Canada |
| **Mauna Kea** | 4205 | On volcanic island in Hawaii | Highest island mountain in the world |
| **Mauna Loa** | 4169 | On volcanic island in Hawaii | Famous volcanic mountain |
| **McKinley** | 6194 | Alaska | Highest mountain in North America |
| **Mitchell** | 2037 | North Carolina | Highest mountain in the Appalachians |
| **Mont Blanc** | 4807 | France | Highest mountain in the Alps |
| **Mount St. Helens** | 2549 | Cascades in Washington | Recent active volcano in the United States |
| **Pikes Peak** | 4301 | Colorado | Most famous of the Rocky Mountains |
| **Rainier** | 4392 | Cascades in Washington | Highest mountain in Washington |
| **Vesuvius** | 1277 | Italy | Only active volcano on the mainland of Europe |
| **Whitney** | 4418 | Sierra Nevadas in California | Highest mountain in California |



## 22-1 (continued)

### Motivation

Collect and display photographs that show the various landscape regions in the United States. An excellent source of such photos would be the tourist information centers of states in each region. Have students observe the photographs and discuss the characteristics of each different landscape.

### Content Development

Some students may have the mistaken idea that relief refers to the elevation of a region, rather than to the differences in elevation within a region. Emphasize the idea that both a plateau and a coastal plain have low relief, even though a plateau is high above sea level while a coastal plain in close to sea level.

### Skills Development

***Skill: Making maps***

Have students choose an area of the world other than the United States and find out what major types of landscapes are found in that area. Then have students make a map of the area similar to the map shown in Figure 22-2.

Remind students of the demonstration in which the height of a stack of books was measure relative to the top of a table, and then relative to the floor. Point out that Pikes Peak in Colorado rises about 2700 m above the surrounding land, but that the height of the mountain relative to sea level is 4301 m.

- **What is the height of the surrounding land?** (1601 m)
- **How did you arrive at this an-**

But the actual height of a mountain is given as its height above sea level. For example, Pikes Peak in Colorado rises about 2700 meters above the surrounding land. But its actual height above sea level is 4301 meters.

The height of a mountain on an island is also measured from sea level. For example, Hawaii's Mauna Kea is listed in Figure 22-3 as 4205 meters above sea level. Actually, Mauna Kea is the world's tallest mountain, rising 10,203 meters from the sea floor to its peak.

Most of the world's mountains have been built up by the slow folding and wrinkling of the earth's crust. See Figure 22-4. These mountains include the Appalachians of the eastern United States, the Himalayas of Asia, the Andes of South America, the Rockies of the western United States, the Urals and Caucasus of the Soviet Union, and the Alps of Europe.

You can make a model of this kind of folding with a large piece of cloth. Lay it flat on a table. Put your hands on opposite ends of the cloth and push them slowly together. Wave-shaped wrinkles form in the cloth, making miniature mountains and valleys. You may notice that the wrinkles tend to form parallel rows at right angles to the direction of your push. If you flew over the Alps of Europe, you would see a large-scale version of your cloth model. There are row after parallel row of snow-capped, jagged peaks with steep, narrow valleys in between.

Mountains are built very slowly. It is thought that the Rocky Mountains began to form about 65 million years ago. It took about 10 million years for these mountains to reach their maximum height. It will take much longer for them to wear down. But wear down they will. Someday, many millions of years in the future, the majestic sharp-peaked Rockies will be worn down by the action of wind and moving water. On that day, the Rockies will be little more than low rounded hills.

Some mountains are formed when the earth's crust breaks into huge blocks. See Figure 22-4. The Sierra Nevada Mountains of northern and central California were formed in this way.

Sometimes the crust cracks, and lava and ashes come through the cracks and pile up, layer on layer.

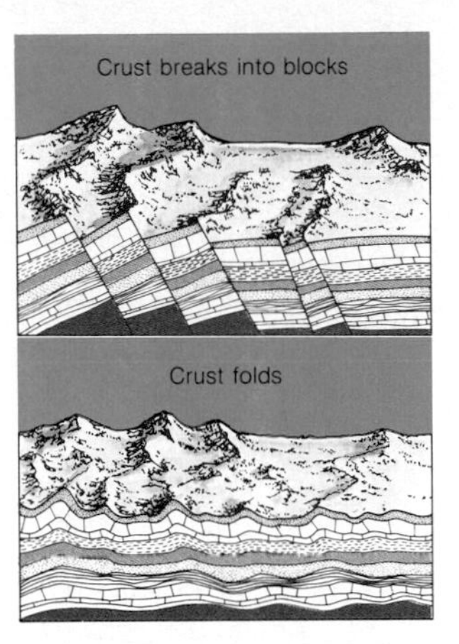


**Figure 22-4** *Mountains may form when the earth's crust breaks into great blocks that are then tilted or lifted* (top). *Folded mountains form when layers of the earth's crust wrinkle into wavelike folds* (bottom).

**Figure 22-5** *The Alps are an example of folded mountains. These mountains show parallel rows of snow-capped, jagged peaks with steep, narrow valleys in between.*

## TIE-IN/SPORTS

One of the most rugged and demanding outdoor sports is mountain climbing. Some mountain climbers enjoy scaling a major peak, such as Mt. McKinley. This type of climb usually requires a team effort and tremendous physical endurance. Other people enjoy hikes, which are often called treks, through a range of mountains such as the Alps or the Himalayas. Students can obtain interesting information on mountain climbing and trekking from the Appalachian Mountain Club (AMC), Pinkham Notch, Gorham, NH.

## FACTS AND FIGURES

Mountains cover about a fifth of the earth's land surface.

**swer?** (by subtracting the height of the mountain relative to the surrounding land from the height of the mountain relative to sea level)

## Enrichment

Various major mountain ranges formed at different times during the earth's history. Have students find out when mountains such as the Andes, Alps, Caledonian Mountains of Scandinavia, Appalachian Mountains, Himalayas, Rocky Mountains, Ural Mountains of Russia were formed. Then have students work in groups to create time lines to show this information.

## Skills Development

***Skills: Interpreting charts, drawing a conclusion***

Have students study the chart in Figure 22-3.

- **What is the highest mountain in the United States?** (Mt. McKinley)
- **What is the highest mountain in the world?** (Mt. Everest)
- **Where is Mt. Rainier located?** (in the Cascades in the State of Washington)
- **What is the highest mountain in the Rockies?** (Mt. Elbert)

## Skills Development

***Skill: Making a model***

Have students work with a partner or in small groups to create a model of the folding of the earth's crust as described in the text. Students may enjoy experimenting with ways to keep the folds in the cloth to create a model landscape. For example, the cloth can be coated with spray starch or a paste solution before it is pushed together; then, when the solution dries, the folds will have hardened into place.

## Activity

**Earth's Mountains**
**Skill: Making calculations**
**Level: Average**
**Type: Computational**

This computational activity will provide students with a reference base as to the amount of land that is covered by mountains. Students must multiply the total land area given by $^{1}/_{2}$ to find the area covered by mountains. This area is 29,660,000 $km^2$.

## ANNOTATION KEY

❶ **Circum-Pacific belt (Interpreting maps)**
❶ **Thinking Skill: Applying definitions**
❷ **Thinking Skill: Relating facts**

### Activity

*Earth's Mountains*

Mountains cover about one fifth of the earth's land surface. The total land area on the earth's surface is about 148,300,000 square kilometers. How much of the land surface is covered by mountains?

Such pile-ups may form volcanic mountains, such as the Cascade Mountains of Washington and Oregon. Many individual mountains are volcanoes. For example, Paricutín in Mexico, Vesuvius in Italy, and Kilimanjaro in Africa are all volcanoes.

Most mountains are part of a group of mountains. A group of mountains is called a **mountain range.** ❶ Examples of mountain ranges are the Alps, Andes, Himalayas, Rockies, and Appalachians.

Most mountain ranges are part of a larger group of mountains called a **mountain belt.** The Rocky Mountains, for example, are just one part of a huge chain of mountains that extends along the Pacific Coast of North and South America from Alaska to the southern tip of South America. This chain is one of the two major mountain belts on the earth's surface. Figure 22-6 shows the pattern of these two mountain belts. One is called the circum-Pacific belt. This belt roughly circles the Pacific Ocean. The other belt is the Eurasian-Melanesian belt, which extends across northern Africa, southern Europe, and southern Asia. The two belts meet in Indonesia.

The great mountain belts on the continents are rivaled by the undersea mountains of the worldwide Mid-Ocean Ridge system discussed in Chapter 21. This system extends through the Atlantic, Pacific, Indian, and Antarctic oceans. On either side of the rift valleys of the Mid-Ocean Ridge system are belts of

**Figure 22-6** *Most of the mountains on the earth's surface are located in the two major mountain belts shown on this map. Which major mountain belt runs through the United States?* ❶

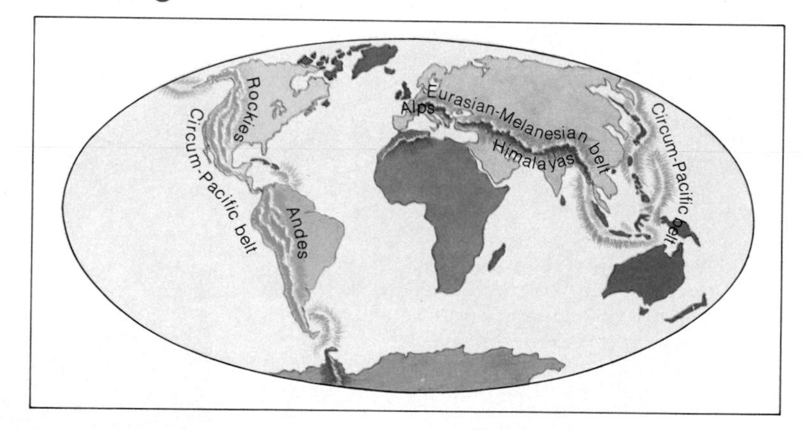




### 22-1 (continued)

### Reinforcement

Reinforce the concept that most mountains are part of a mountain range, which in turn is part of a mountain system, which in turn is part of a mountain belt. You may wish to have students make diagrams in which they show the relationships between a mountain, a mountain range, a mountains system, and a mountain belt.

### Content Development

Challenge students to picture the earth's surface with all of the ocean water removed.

- **What do you think the earth's surface would look like?** (Answers may vary. Many will probably say that there would be huge depressions in the earth's surface where the ocean waters had been.)
- **What do you think the ocean bottom would look like?** (Answers may vary.)

Point out that large portions of the ocean floor are covered by mountain ranges and rift valleys. If ocean waters were to be removed from the earth's surface, an impressive system of mountains and valleys would be revealed.

### Skills Development

***Skills: Relating concepts, relating cause and effect***

Have students look at Figure 22-2 again. Reinforce the idea that plateaus tend to be located adjacent to mountains.

Explain to students that weather systems in the United States usually move from west to east. As a result, mountains to the west of a plateau may block precipitation from falling

Figure 22-7 *In this photograph of the Colorado Plateau and Grand Canyon, a barren, dry landscape is revealed.*

high undersea mountains. These mountains were built by volcanic action. Many of the mountains are still active volcanoes.

### Plateaus

Now you are walking through a different landscape. You are in a highland area dotted with enormous flat-topped hills. This region is dry, almost a desert. But the same kind of landscape at another place on the earth may be damp and covered with forests. No matter whether it is dry or wet, warm or cold, this type of region is a **plateau landscape,** or simply a plateau.

A plateau begins as flat land that is gradually lifted by movements of the earth's crust. A plateau is usually more than 600 meters above sea level. Some plateaus reach heights of more than 1500 meters. 2

The plateau upon which you are standing lies west of the Rocky Mountains, among some of the wildest and most beautiful scenery on the earth. It is the Colorado Plateau, which includes the Grand Canyon. The Colorado Plateau is roughly centered on the point where the boundaries of Colorado, New Mexico, Arizona, and Utah meet.

Plateaus usually form near mountain ranges. In fact, the Colorado Plateau began rising as the Rocky Mountains were forming some 65 million years ago.



## BACKGROUND INFORMATION

Mountains that form when the earth's crust breaks into huge blocks are called fault-block mountains. A fault is a break or crack in the earth's crust along which rocks move. When the rock above a fault moves downward, the fault is called a normal fault. Fault-block mountains are most likely to form in areas where there are many normal faults.

Fault-block mountains form when a block of rocks between normal faults is pushed up. Fault-block mountains are common in western North America. The Sierra Nevada Mountains in the western United States are fault-block mountains.

## BACKGROUND INFORMATION

Mountain ranges on the ocean floor form differently from mountain ranges on land. Mountain ranges on land are formed by the folding and faulting of the earth's crust. Mountain ranges on the ocean floor form when molten rock from the mantle flows upward through rifts. The buildup of molten rock or lava does form volcanic mountains on land, but these mountains are usually single mountains, rather than mountain ranges.

on the plateau.

• **Based on this information, what would cause a plateau to have a very dry climate?** (a mountain range lying to the west of the plateau)

• **What would cause a plateau to have a moist climate?** (an adjacent mountain range lying to the east of the plateau)

Have students observe the map in Figure 22-2.

• **Would you expect the Western, Colorado, and Columbia Plateaus to be moist or dry? Explain.** (Very dry—the plateaus are bordered by mountain ranges to the west. In addition, mountains to the east would block any precipitation that might come from that direction.)

• **Would you expect the Appalachian Plateaus to be moist or dry? Explain.** (moist—there are no mountains to the west of the plateau, only to the east)

## Skills Development

***Skill: Interpreting maps***

Have students observe the map in Figure 22-6.

• **In which mountain belt are the Rocky Mountains located?** (the Circum-Pacific belt)

• **Which other group of mountains in the United States is located in this belt?** (the Pacific Mountain ranges)

• **Which mountain range in the United States is located in neither of the two major mountain belts?** (the Appalachian Mountains)

• **The major mountain belts run primarily through which continents?** (North and South America; Europe and Asia)

• **In which continent is there no major mountain belt?** (Australia)

## ANNOTATION KEY

❶ Thinking Skill: Comparing plateaus
❷ Thinking Skill: Sequencing events

### Activity

**Continental Sizes**
**Skills: Making calculations, sequencing, manipulative, applying, relating, observing**
**Level: Enriched**
**Type: Hands-on/computational**

This activity will not only reinforce students' computational skills and graph paper skills but it will also serve as an excellent tool in showing students how to compare regions with completely different shapes in terms of their relative sizes. As such, students gain specific information on the areas of the earth's major landmasses through this exercise. They also learn about a simple geometric technique used to estimate area. Be sure that the grid size of the graph paper they use is appropriate. If the squares are too large, the estimate will not be accurate; if they are too small, it will take too long to count them. You may wish to have students calculate the mathematical ratios for the continental areas. All students should obtain the same values, regardless of grid size.

**Figure 22-8** *Here you can see the forests and rounded slopes of the Appalachian Plateau.*

### Activity

*Continental Sizes*

Find out how the continents compare in size.

**1.** From a globe, trace the outline of each of the seven continents. Cut out the outlines. Trace each cutout on a piece of graph paper. Shade in the outlined continents.

**2.** Consider each square on the graph paper as an area unit of 1. Calculate the area units for each of the seven continents to the nearest whole unit. For example, suppose a continent covers all of 45 units, about one-half of 20 units, and about one-fourth of 16 units. The total area units the continent covers will be 45 + 10 + 4, or 59.

List the continents from largest to smallest.

But a plateau forms without the sideways folding and buckling of the crust that forms mountains. So its top stays more or less flat. In its youth, the Colorado Plateau was a single broad structure, towering 1500 meters and more above the surrounding land, but lower than the Rockies to the east.

Today, however, the Colorado Plateau is made of many plateaus. As you look around, you see an awesome landscape of huge flat-topped tables of rock. They are separated by steep clifflike canyon walls. Along the cliffsides, layers of brightly colored rock glow in the sun. The canyon walls are often terraced—almost as if a giant hand had carved a series of ledges there to make convenient steps.

Actually, the carving was done by the Colorado River and its branches over millions of years. They sliced deeper and deeper into the rock, cutting it up into separate plateaus. The steps of the terraces are layers of more water-resistant rock. The streams cut around this rock.

The Colorado River has carved the mighty Grand Canyon out of the hard rocks of the Colorado Plateau. It is a huge canyon about 2 kilometers deep, up to 29 kilometers wide, and 349 kilometers long. Today, the Colorado River continues to carve out more of this canyon.

❶ By contrast, the Appalachian Plateau, west of the Blue Ridge Mountains, has more rounded slopes and forests. The Appalachian Plateau gets lots of rain. This plateau is many millions of years older than the



## 22-1 (continued)

### Enrichment

An important volcano in the United States is Mt. St. Helens in Washington. Challenge students to find pictures of Mt. St. Helens and the surrounding land. Also challenge students to find newspaper or magazine articles that describe some recent eruptions of the volcano.

### Content Development

Define the term *plain* and help students identify the processes that are involved in the formation of a plain. Point out that coastal plains can have very different characteristics from inland plains, particularly in terms of climate and weather patterns.

### Reinforcement

Have students collect and display photographs that show various landscape regions of the United States. Possible sources of such photographs would be tourist information centers in various states, or a local travel agency.

### Enrichment

Many different crops are grown in coastal plains regions. Have each student choose a state that is all or partially covered by coastal plains. Then have students find out the types of

Colorado Plateau. So there has been much more time for rain, streams, and wind to smooth and round its slopes.

### Plains

Now you are visiting another of the major landscape types. Here the land is mostly flat and of about equal height, like plateaus. But, unlike plateaus, this landscape is lower than the surrounding land. You are in a lowland area that is not far above sea level. You see broad rivers or streams, and the vegetation is mainly grass. The region you are visiting is a **plains landscape.**

The two major types of plains areas are determined by location and height above sea level. They are either along a coast or farther inland. Let's visit these two types of plains in the United States.

**COASTAL PLAINS** Low, flat regions called **coastal plains** spread gently upward from the Atlantic and Gulf coasts of the United States. These plains extend along the Atlantic coast from Cape Cod, Massachusetts, to Florida and from the Gulf coast of Florida to Texas. They average about 150 meters above sea level.

Many crops can be grown in coastal plain soil. Why? The reason is that the soil is very rich in substances needed for plant growth. In the United States, citrus crops, cotton, tobacco, and many different kinds of vegetables can be grown on the rich coastal plains.

Where does the rich soil come from? The soil is brought to the coastal plains by rivers. Rivers can break down rocks into sand and mud. 2

Rivers flow from high mountains and plateaus down to low-lying plains and out to sea. These rivers move fastest at the start of their journey. A fast-moving river is a powerful tool for wearing down rock. Because it is fast-moving, it can carry a large load of sand and mud.

As the river hits the gentler slopes of the lowlands, it slows down. And the more slowly it flows, the more sand and mud drops out of the flow. So, year after year, the rivers help to wear down the highlands and build up the lowland plains.

**Figure 22-9** *In this plains landscape, the flat lowlands are not far above sea level. Such landscapes may include broad rivers or streams. The natural vegetation is mainly grass.*

## BACKGROUND INFORMATION

Plateaus are formed by some of the same forces that cause mountains to rise. Strong forces horizontally push on an area of rock, causing a plateau to be uplifted. Unlike mountains, however, plateaus are not faulted or folded. Instead, the raised rock layers remain flat.

Plateaus can also be formed by a series of lava flows. Flowing lava that pushes upward through cracks in the earth's surface spreads out over a large area and hardens onto sheets. The hardened sheets eventually pile up to form a raised plateau. The Columbia Plateau in the northwestern United States was formed in this way.

Another uplifted area that is formed by the movement of molten rock is a dome. The rock that forms a dome, however, never reaches the earth's surface. Instead, it pushes upward against surface layers of rock, causing them to swell. The result is a dome-shaped hill that looks roughly like the top half of a sphere. The Black Hills of South Dakota and Wyoming are domes.

## BACKGROUND INFORMATION

Coastal plains are formed as soil and silt are deposited on the edge of a continent. Some of the sediments are deposited by rivers and streams. Other sediments are left behind when shallow oceans that once covered an area recede. Interior plains, such as the Great Plains of the United States, are formed as mountains and hills are worn down by wind, streams, and glaciers. The main difference between these two types of formations is that coastal plains are formed by the building up of sediments, whereas interior plains are formed by the breaking down of existing rock.

crops that are raised in the state they have chosen. Also discuss with students how crops vary according to climate.

## Skills Development

***Skill: Making diagrams***

Have students make diagrams to show the shape, elevation, and relief of mountains, plains, and plateaus. Have them use the following figures in their diagrams:

Mountains: 600 m–8 km
Plateaus: 600–1500 m
Interior plains—450 m above sea level
Coastal plains—150 m above sea level

## Reinforcement

Draw an outline on the chalkboard that shows a mountain, plateau, interior plain, and coastal plain. Have students identify each landscape feature and its approximate elevation.

## TIE-IN/SOCIAL STUDIES

The landscape of an area greatly affects its economic and social development. Have students choose different regions in the United States that have various landscapes. Then have them research the life style and economics of each region.

## BACKGROUND INFORMATION

The difference in elevations within an area is called relief. Areas that are basically flat have low relief. Plains and plateaus are landscapes that have low relief. Areas that are hilly or mountainous have high relief; thus mountain landscapes have high relief. The relief of an area is not to be confused with its height above sea level. Plateaus are high above sea level, but because they are flat, they have low relief.

### Activity

**Surface Features**
**Skills: Applying facts, relating information, comparing**
**Level: Remedial**
**Type: Writing/vocabulary**

Students' essays should be well written and use all of the vocabulary terms listed. Have the best essays read to the class.

**Figure 22-10** *The flat land of the inland plains has very fertile soil that is good for growing crops such as wheat and corn.*

### Activity

*Surface Features*

Suppose you take a trip across the earth's land surface. You will see many different kinds of features. Write a 200 word essay about what you see using the following vocabulary terms: mountain landscape, mountain range, plateau landscape, coastal plains, and inland plains. Your essay should describe the major similarities and differences between these features.

There is another reason that the coastal plains have rich soil. Long ago, shallow oceans covered these areas. As the shallow seas disappeared, they left behind rich deposits of soil that formed the coastal plains.

But not all coastal plains are available for farming. Some of the most fertile soil in coastal plains is not usable because it is found in swamps and marshes. This land is below river level and sometimes even below sea level. So the land usually floods with water.

**INLAND PLAINS** The central part of the United States is divided into two very large **inland plains**—the Interior Plains and the Great Plains. See Figure 22-2. Inland plains average about 450 meters above sea level. How were these plains formed? Over millions of years, mountains and hills were worn down by wind, rivers, and glaciers, leaving plains that have become the heartland of the United States farming country.

Why are these plains considered among the world's best for farming? There are three reasons. Many parts of these plains receive very rich soil deposits from such rivers as the Mississippi, Ohio, Missouri, Arkansas, and Red. More than half the material that built up the 65 million-year-old Rocky Mountains has been spread over these plains by rivers. Another reason is that plenty of rain falls on much of the plains during the growing season. A third reason is that these northern and western plains are much higher above sea level than are the coastal plains. So flooding, which could carry away rich soil, tends not to occur there.

Even where there is less water and rainfall, as in the western part of the Interior Plains and in the Great Plains, crops such as corn and wheat cover the land. Only within the "rain shadow" of the Rockies does farmland give way to semidesert and desert. Within a rain shadow, the mountains cut off the rains moving in from the west.

Water is very important to all living things. As you read in this section, too much or too little water can affect the growth of crops and the value of land to farmers. The next section will discuss the other major surface feature of the earth—water.



## 22-1 (continued)

### Skills Development

***Skill: Applying concepts***

• **Why do you think that the Great Plains in the United States are higher in elevation than the coastal plains?** (The Great Plains were formed as mountains and hills were worn down by erosion. The coastal plains were formed by the deposition of sediments. It is reasonable to assume that the wearing down of mountains would produce higher elevations than the building up of sediments.)

### Motivation

Have students use clay or plaster of Paris to make models of the three main landscape regions discussed in the book. The models should be scaled to conform to the elevation and relief descriptions in the text for mountains, plains, and plateaus.

### Section Review 22-1

**1.** Mountains, plateaus, and plains.
**2.** Slow folding and wrinking of the earth's crust; the earth's crust breaks into huge blocks; volcanic pileups.
**3.** Coastal and inland.

## TEACHING STRATEGY 22-2

### Motivation

Obtain a copy of a NASA photograph of the earth taken from the Apollo spacecraft. Display the photograph and ask,

• **What characteristic of the earth's**

## Career: *Surveyor*

**HELP WANTED: SURVEYOR** Bachelor's degree in surveying required. Must also be licensed by the state. Involves outdoor work for road and bridge construction. Reply to the Army Corps of Engineers.

In ancient Egypt, farmers placed boundary stones around their land to mark off property lines. But the annual flooding of the Nile River moved the stones or washed them away. So the Egyptians, who were excellent mathematicians, learned to survey their land. Thus they could replace any lost boundary markers accurately.

Today boundaries and areas of property on the earth's surface are determined by scientists called **surveyors.** They use an instrument called a transit to measure and plot angles and distances. Surveyors can measure the exact location of points and elevations on the earth's surface. The results of their survey are recorded, the data are checked, and sketches, maps, and reports are prepared. This information is used for construction projects, mapmaking, land division, mining, and deciding property ownership.

Surveyors must make very precise measurements. Their maps and land descriptions are recorded in the courthouse for anyone to check. A state licensing exam is taken after several years of work experience.

A person interested in surveying must have good judgment, enjoy problem solving, be able to communicate, and be attentive to detail. For more information, write to the American Congress on Surveying and Mapping, 210 Little Falls Street, Falls Church, VA 22046.

### *SECTION REVIEW*

1. What are the three main types of landscapes on the earth's surface?
2. What are three ways that mountains can form?
3. What are the two major types of plains?

## 22-2 The Earth's Surface: Water

The earth is a watery planet. Oceans cover about 70 percent of its surface. And about 97 percent of all the water on our world is in the oceans. This is salt water. Every hundred kilograms of sea water contains about 3.5 kilograms of ordinary salt, as well as a number of other salts dissolved in the water.



## 22-2 THE EARTH'S SURFACE: WATER

### SECTION PREVIEW 22-2

In this section, students will explore the 70% of the earth's surface that is covered by water. They will discover most of this water, approximately 97%, is salt water.

Students will learn that the earth's supply of usable fresh water is continually being replenished by the water cycle. In the water cycle, water on the earth's surface evaporates into the atmosphere. Eventually this water vapor forms clouds, which in turn produce precipitation.

Students will learn how some of the water that falls to the earth forms streams and rivers, while some collects in lakes and ponds.

### SECTION OBJECTIVES 22-2

1. **Identify the major sources of salt and fresh water.**
2. **Explain how the water cycle works.**
3. **Explain how streams, rivers, and lakes form.**
4. **Describe the formation of ground water and discuss the importance of the water table.**

### SCIENCE TERMS 22-2

**evaporation p. 513**
**condensation p. 513**
**precipitation p. 513**
**ground water p. 515**
**water table p. 515**
**continental glacier p. 516**
**valley glacier p. 516**
**iceberg p. 517**

**surface is most apparent in this photograph?** (Answers may very, but most students will probably notice the abundance of swirling blue water on the water earth's surface.) Point out to students that the earth has sometimes been referred to as the "blue planet" or "Planet Ocean." These names are not really inaccurate, because more than half (approximately 70%) of the earth's surface is covered by water.

### Content Development

Point out to students that all of the earth's oceans are connected to form one continuous body of water. You may wish to demonstrate this to students by having them observe the oceans on a globe.

Emphasize to students that 97% of the earth's water is found in the oceans, and that this water is salt water. Point out that living things cannot use water containing salt. Therefore, all of the earth's life is dependent upon the 3% of the earth's total water supply that is fresh water.

### Motivation

As a class exercise, have students make a list of all the activities in a typical week that require water. Place a star next to those activities that are essential to a person's well-being or survival. Then discuss the difficulties that would result if little or no water were available for these activities.

## TEACHER DEMONSTRATION

For this demonstration, you will need a quart-sized glass jar, a pie tin, hot water, and a few ice cubes.

Add enough hot water to the glass jar so that it is one-fourth full. Place a few ice cubes in the pie tin and then place it on top of the jar. Have students observe the setup.

- **What is forming on the underside of the pie tin?** (Water is collecting on the underside of the pie tin.)
- **Which action in this demonstration is similar to precipitation?** (the falling water droplets)
- **Which action in this demonstration is similar to evaporation?** (the rising hot water vapor)
- **Where is condensation occurring?** (on the underside of the pie tin)
- **Why is condensation occurring here?** (As the moisture-laden hot air nears the cold pie tin, the air is cooled. Cool air cannot hold as much water vapor as hot air, so some of the water vapor is changed into a liquid.)

### Activity

**Reservoirs**
**Skills: Relating, applying**
**Level: Average**
**Type: Library/writing**

This activity will help students understand the importance of reservoirs. If you wish, you may encourage students to bring in photographs of some major reservoirs.

**Figure 22-11** *Oceans cover most of the earth's surface.*

### Activity

*Reservoirs*

The most frequently used sources of fresh water are artificial lakes called reservoirs. Using reference materials in your library, write a short report about reservoirs and include answers to the following questions:

**1.** How are reservoirs built?
**2.** How are reservoirs important during periods of heavy rain? During periods of drought?
**3.** How are some reservoirs used in the generation of electrical power?
**4.** What are the names and locations of some major reservoirs in the United States? In the rest of the world?

Where do these salts come from and how do they get into sea water? Ocean salts are found in undersea volcanoes, in rivers, and along shores. Ocean water dissolves salts in volcanic rocks. Rivers dissolve salts in rocks and soil on land. The rivers then carry these salts to the oceans. As waves constantly pound on shores, the erosion from waves helps to dissolve the salts in rocks.

What about the other 3 percent of the earth's water? Most of it is locked up in the ice packs of the Arctic Ocean, and in the great sheets of ice that cover a large part of the continent of Antarctica. This is fresh water. The rest of the fresh water is in the ground and in lakes, ponds, streams, and rivers. Fresh water does not contain salt, though it does have small amounts of other dissolved minerals. Fresh water is drinking water. It is essential for all plants and animals that live on land.

Where does fresh water come from? The answer is part of the story of the water cycle. **In the water cycle, water circulates continually between the atmosphere and the surface of the earth.**

### *The Water Cycle*

There is an old saying that every drop of water from the ocean rains five times before it returns to the ocean. Although this is not a very scientific statement, it does have a kind of scientific basis. Here is the reason.



## 22-2 (continued)

### Enrichment

Have students work in small groups. Challenge each group to find out what types of features are present on the ocean floor. (Major features include abyssal plains, seamounts, guyots, trenches, ridges, rift valleys, and reefs.) Then ask each group to prepare a creative presentation that describes the topographic features of the ocean floor. Discuss with students how the surface of the ocean floor compares with the earth's land surface.

### Enrichment

Have students find out about the scientific satellite SEASAT that was launched in 1978. Encourage students to find copies of maps and photographs that have been obtained from SEASAT. Also challenge students to discuss how SEASAT has been able to provide scientists with previously unknown information about the ocean.

### Content Development

When explaining the water cycle, it is important to point out that when water evaporates from the earth's surface, precipitation does not necessarily return it to exactly the same place. In fact, scientists have con-

Every day, the heat of the sun causes water on the earth's surface to change to a gas. This process is called **evaporation** (i-va-puh-RA-shun). Enormous amounts of water are evaporated from the oceans. Water also evaporates from fresh-water sources on land. See Figure 22-12.

When water evaporates, it gets into the air as an invisible gas—water vapor. There is always some water vapor in the air, even over the driest deserts.

Air can hold only a certain amount of water vapor. Warm air can hold more water vapor than can cold air. As warm air, loaded with water vapor, rises, it cools. The cooler air can no longer hold all of its water vapor. Some of the water vapor changes back into a liquid. This process is called **condensation** (kahn-den-SAY-shun). Next, clouds form. They are made up of billions of tiny water droplets so light that they can float in the air. Sooner or later, this water falls to the earth as rain, sleet, hail, or snow. This process is called **precipitation** (pree-sip-uh-TAY-shun). The water that falls is fresh water because when ocean water evaporates, dissolved solids are left behind. The cycle then begins again.

**Activity**

*Simulating the Water Cycle*

**1.** Stir salt into a small jar filled with water until no more salt will dissolve. Pour a 1-cm deep layer of the salt water into a large, wide-mouthed jar.

**2.** Place a paper cup half filled with sand in the jar.

**3.** Cover the jar's mouth with plastic wrap secured with a rubber band. Place a small rock on the plastic wrap above the paper cup.

**4.** Place the jar in direct sunlight. Observe for several hours.

What processes of the water cycle can you identify?

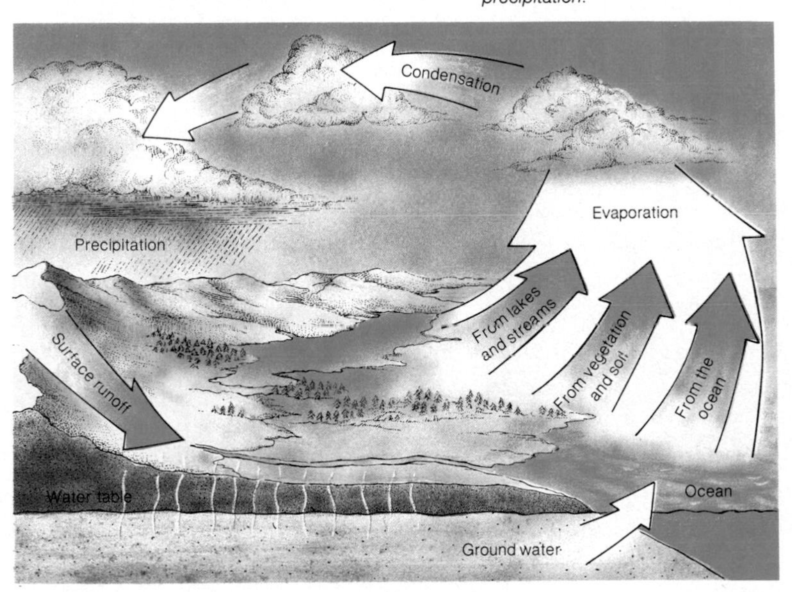


**Figure 22-12** *The water cycle is made up of the processes of evaporation, condensation, and precipitation.*

## Activity

**Simulating the Water Cycle**
**Skills: Manipulative, observing, inferring, relating, applying**
**Level: Average**
**Type: Hands-on**
**Materials: Small jar, water, salt, wide-mouthed jar, paper cup, sand, plastic wrap, rubber band, small rock**

In this activity, students observe the evaporation and then condensation of salt water. They should note that the purpose of sealing the jar is twofold. First, the seal allows students to collect and taste the water that condenses on the plastic wrap. Second, the seal simulates the atmosphere of the earth. Students should note that the water on the wrap is not salty. They should infer that the salt has been left behind when the water evaporated, as is the case in the water cycle.

## FACTS AND FIGURES

The average cloud droplet has a diameter of about 10 micrometers. It takes approximately 1 million cloud droplets to make a rain droplet.

## FACTS AND FIGURES

The three major oceans of the world are the Atlantic, the Pacific, and the Indian. The Pacific is the largest and also the deepest. The Atlantic is the second largest, but the Indian is the second deepest.

cluded that more water evaporates from the oceans than is returned to the oceans, and more water is returned to the land than evaporates from land. The balance of loss and gain is maintained by the flow of water from rivers, streams, and ground water sources into the oceans.

### Reinforcement

Have students work in groups of three. Challenge each group to present in a creative way the three main steps of the water cycle. Students should include in their demonstrations how the first step follows the last step to begin the cycle again.

### Skills Development

***Skill: Interpreting diagrams***
Have students observe the diagram of the water cycle in Figure 22-12.

- **From what sources does water evaporate into the atmosphere?** (from the oceans, from lakes and streams, from vegetation and soil)
- **According to the diagram, what happens to surface runoff?** (It flows into rivers and streams, which eventually flow into the ocean.)
- **What eventually happens to ground water?** (It flows to the ocean.)

Point out that the water cycle includes the movement of water from the land to the oceans, then back to the land.

## BACKGROUND INFORMATION

The land area in which surface runoff flows into rivers and streams is called a watershed. A watershed is a natural flow of water, not something made or controlled by human beings. Because of this, it is important that the environmental conditions that created a watershed by preserved.

## FACTS AND FIGURES

Ground water in any location can usually be reached by digging or drilling a well within 100 m of the earth's surface.

## ANNOTATION KEY

1. Thinking Skill: Sequencing events
2. Thinking Skill: Relating cause and effect

## BACKGROUND INFORMATION

One problem with the earth's water supply is that it is not evenly distributed. Part of this unequal distribution results from droughts or heavy rains in certain areas. Unequal distribution also results from the fact that some areas of the world have many more natural water sources than do other areas.

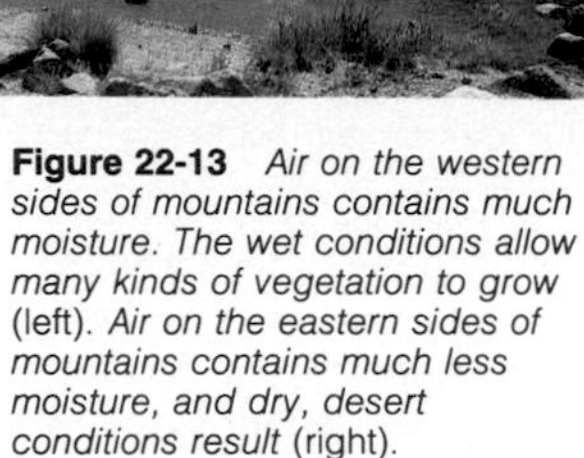

**Figure 22-13** *Air on the western sides of mountains contains much moisture. The wet conditions allow many kinds of vegetation to grow* (left). *Air on the eastern sides of mountains contains much less moisture, and dry, desert conditions result* (right).

**Figure 22-14** *The grooves in these mountains show how running water can cut deeply into mountains over millions of years.*

In some places, high mountain ranges generally prevent moisture from reaching the land, producing a desert climate. For example, moisture-laden air from the Pacific flows eastward over the California coast. Plenty of rain falls on the farms of the Great Central Valley in California. The air reaches the Sierra Nevada Mountains inland and is forced to rise. As it rises it cools. Lots of rain falls on the western slopes of the mountains. But by the time the air crosses the Sierras, it no longer contains enough moisture. So little rain falls on the desert lands east of the Sierras, including Death Valley. For the same reason, the eastern side of the Rockies is mostly desert land.

### *Running Water*

Some of the water that falls to the earth evaporates again. But about a third of the water remains on the surface. It begins moving downhill and flows into rivers and streams. (1) The water entering a river or stream is called surface runoff. In the spring, melting snow and ice add to the surface runoff.

From high mountains and steep plateau cliffs, water flows downward. It collects in basins, grooves, and channels worn in the rock by earlier runoffs. Streams like these come together to become the sources of the world's rivers. The rivers return to the oceans. Water evaporates from the oceans, and the cycle repeats endlessly.

Some rain water soaks into the ground. The ground takes on water like a sponge. This water is



## 22-2 (continued)

### Content Development

The water table generally follows the shape of the land. Where the water table meets the land's surface, swamps, ponds, and springs can form.

### Skills Development

***Skill: Making observations***

Have students study the permeability of certain soil samples. To do this, have them puncture a small hole in the bottom of several paper cups, then fill the cups with a variety of soil samples such as gravel, clay, or sand. Compare the permeability of the soils by timing how long it takes a given volume of water to flow through the soil and out the hole.

### Skills Development

***Skill: Making maps***

Have students work in small groups. Ask each group to find out the sources of running water and standing water in the county or state in which they live. Then challenge each group to make a map showing these freshwater sources.

### Reinforcement

Many students in your class may have

called **ground water.** About 0.6 percent of the earth's fresh water is in ground water. Even in the driest deserts, there is water underground. There is a place under the desert sands where the ground is soaked with water. Water fills spaces between the grains of soil. Eventually, ground water flows underground to the oceans. This water can then become part of the water cycle again.

The top level to which ground water rises when underground is called the **water table.** In the desert, the water table may be hundreds of meters underground. In places with more rainfall, the water table lies nearer the surface. There, water seeps up from the water table, keeping the roots of plants moist and supplying water for wells. See Figure 22-16. In some places, such as coastal plains, the land lies slightly below the level of the water table. Such land is usually swampy or marshy. When there is a long period without rain—a drought—the water table sinks deeper into the ground. Ponds begin to dry up. Wells go dry, the shallowest wells first.

On the other hand, after unusually heavy rains through much of the year, the water table may rise toward the surface. Normally dry ground becomes muddy. Puddles form everywhere. The ground is waterlogged right up to its top. Plants may be killed because their roots cannot get enough air.

### Standing Water

Unlike rivers and streams, the water in lakes and ponds usually stays in one location. Both can be important sources of water. Moosehead Lake, in Maine, for example, is a natural source of fresh water. It is 56 kilometers long and 3 to 16 kilometers wide. The pine-forested shores of the lake can hold enormous amounts of water from rains and melting snow. This water is released slowly to the lake, making flooding less likely. During times of drought, the lake holds a huge supply of water in reserve.

### Frozen Water

Aside from the oceans, the most enormous store of water on the earth lies in the ice sheets around the North and South poles. About 2 percent of the

**Figure 22-15** *An oasis can form in a desert when ground water rises to the surface.*

**Figure 22-16** *These diagrams show the changes in a water table after a long period of rain and after a long period without rain.*

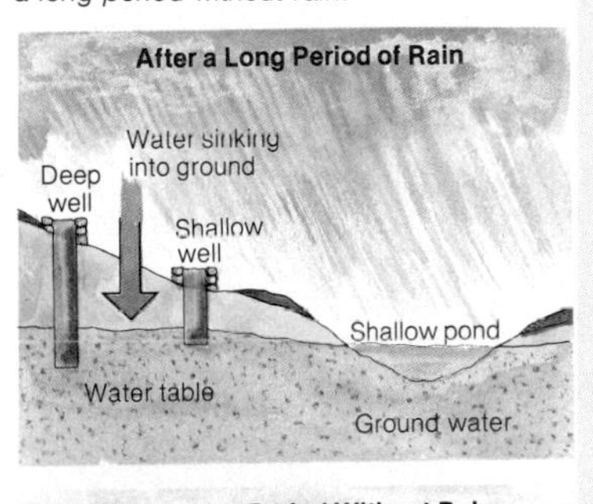


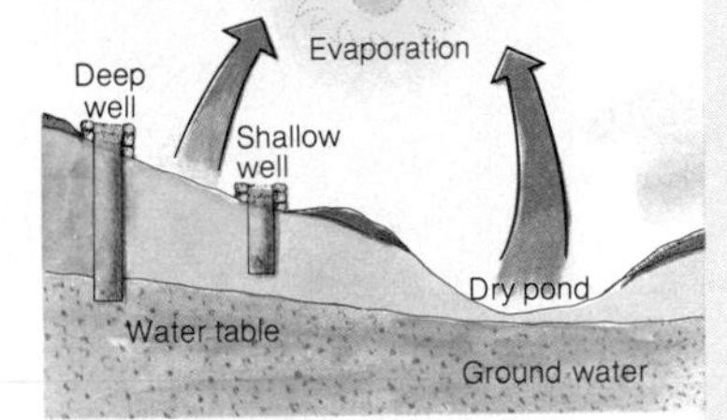


experienced water shortages, droughts, or floods. Have students share any experiences they may have had. Ask students if they can recall what factors caused these events, and what measures were taken to alleviate the problems that resulted.

**Content Development**
In addition to lakes and ponds, reservoirs are important sources of fresh water. Reservoirs are artificial lakes that are usually built by damming a stream or a river in a low-lying area.

## SCIENCE, TECHNOLOGY, AND SOCIETY

Approximately half of all Americans depend on ground water for cooking, drinking, bathing, and other household needs. Ground water is also important in sustaining the flow of streams and wetland ecosystems.

In the last few decades, the quality of ground water has been threatened by the improper disposal of hazardous chemical wastes by industry; leaks in underground storage tanks, sewer lines, and septic tanks; solid-waste disposal; and agricultural runoff containing pesticides. Another concern is the amount of ground water that will be available in the future. Increasing populations have placed a great demand on ground water. As ground water is removed for community use, the land above it begins to sink. In some coastal areas the land has sunk approximately 1 m because of the extraction of ground water.

Water is a natural resource that we cannot live without. As a result, the conservation of ground water, is important economically, scientifically, and politically.

- **Why is the protection of ground water important?**
  - a. **economically?** (Other sources of fresh water are expensive.)
  - b. **scientifically?** (maintenance of stream and river flow, wetland ecosystem; prevention of subsidence)
  - c. **politically?** (Accept all reasonable answers.)

- **From what source do you obtain your local drinking water?** (A call to the city water department will answer this question.)
- **What factors might endanger your drinking water?** (Accept all reasonable answers.)

## Activity

**Drought and the Water Table**
**Skills: Making observations, manipulative, relating, applying, hypothesizing, predicting**
**Level: Remedial**
**Type: Hands-on**
**Materials: clear baking dish, sand, water**

This simple hands-on activity allows students through their own observations to draw conclusions about the effects of drought on the water table. The activity shows students a representation of how the water table would look above and below the earth's surface. Students should notice that the water level drops over a period of time. This drop represents what would happen to an area in which there is a drought or lack of rainfall. Students should predict that a lowering of the water table would result in some shallow wells going dry and a reduced amount of water available to certain areas affected by the reduced rainfall.

## ANNOTATION KEY

1. Thinking Skill: Comparing glaciers
2. Thinking Skill: Relating cause and effect

### Activity

*Drought and the Water Table*

**1.** Fill the bottom of a deep, clear baking dish about halfway with sand.

**2.** Slowly add enough water so that the sand becomes saturated and about 1 cm of water is visible above the surface of the sand.

**3.** Add more sand on top of the water in *only one-half* of the baking dish. You can now see how the water table looks above and below the "earth's" surface after a period of rain.

**4.** Observe the water level above and below the sand's surface during the next few days.

What changes do you notice in the water level?

earth's fresh water is frozen in ice sheets. If all this ice melted, the oceans would rise about 50 meters above their present level. New York, New Orleans, Houston, and Seattle would be under water. Memphis, Tennessee, now 560 kilometers from the sea, would become a major seaport. In Washington, D.C., the top floors of the Capitol Building and the tip of the Washington Monument would be above water. But the Pentagon would be completely covered.

The very thick ice sheets that cover most of the polar regions are called **continental glaciers.** Glaciers are huge, slow-moving masses of ice. During the Ice Ages, when the earth's climate was much colder, the ice piled up several kilometers thick around both poles. The pressure of the piled-up ice caused it to flow outward away from the poles. At times, these glaciers covered a third of all the land on the earth.

(1) **Valley glaciers** are long, narrow glaciers. They move downhill between the steep sides of mountain valleys. Usually they follow channels worn by running water in the past.

As the glacier slides downward, it tears rock fragments from the mountainside. These fragments become frozen in the glacier. They cut deep grooves in the valley walls. Finer bits of rock sandpaper the walls, smoothing them. Glaciers often turn a steep, V-shaped valley into a U-shaped valley with smooth sides. There are many such valleys in the Alps.

**Figure 22-17** *This valley glacier is slowly moving downhill between the steep sides of a mountain valley.*

## 22-2 (continued)

### Reinforcement

As a class exercise, have students make a list of all the activities in a typical week that require water. Place a star next to those activities that are essential to a person's well-being or survival. Then discuss the difficulties that would result if little or no water were available for these activities.

### Content Development

Explain to students that glaciers form because of pressure. Glaciers form from snow, which is normally light and fluffy with considerable air space between crystals. As new snow piles on top of old snow, however, individual snowflakes on the bottom layer are pushed together. The amount of air space between crystals decreases, and a dense, closely packed form of snow called firn results. This process usually takes about 1 year. Then, as decades or even centuries pass, pressure causes the air spaces in the firn to disappear. The result is glacial ice.

If you live in an area that has snow, students can make a model of the formation of glacial ice by making a snowball and then holding it very tightly in their hands. The pressure from the hands will cause the snow to compress and become icy.

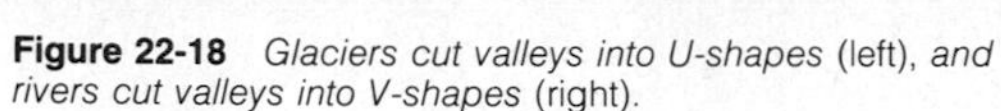

**Figure 22-18** *Glaciers cut valleys into U-shapes* (left), *and rivers cut valleys into V-shapes* (right).

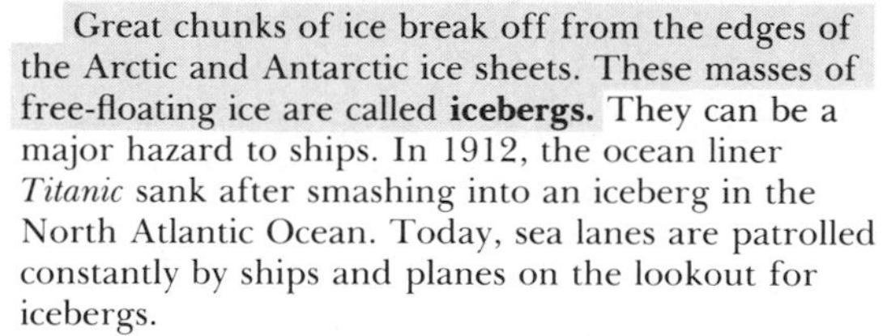

Great chunks of ice break off from the edges of the Arctic and Antarctic ice sheets. These masses of free-floating ice are called **icebergs.** They can be a major hazard to ships. In 1912, the ocean liner *Titanic* sank after smashing into an iceberg in the North Atlantic Ocean. Today, sea lanes are patrolled constantly by ships and planes on the lookout for icebergs.

Around the Great Lakes are masses of boulders, rocks, and sand, heaped up into ridges and hills. They are deposits left by glaciers that covered this area thousands of years ago. As the climate grew warmer, the glaciers melted and dropped their cargo of rocks in piles. These piles mark the farthest advance of the glaciers. 2

You now have explored the earth's interior and the earth's surface. In the next chapter, you will explore the earth's atmosphere.

### *SECTION REVIEW*

1. What percentage of the earth's water is fresh water? Where is most of it found?
2. Describe the water cycle.
3. What is the water table?
4. What is the difference between a valley glacier and a continental glacier?

517

## BACKGROUND INFORMATION

Glaciers are formed above the annual snow line. The snow line is the elevation above which some winter snow lasts through the summer and does not melt. Snow line varies according to the amount of winter snowfall and the temperatures during the summer.

In polar regions, the snow line is at sea level. Thus these areas are always snow covered, and ice sheets are continually forming.

Snow lines reach the highest elevations—about 6000 m—at latitudes 20° north and south of the equator. Near the equator, snow lines are about 1000 m lower due to greater precipitation and more cloudiness during the summer.

## FACTS AND FIGURES

Ten percent of the earth's land surface is covered by ice.

## FACTS AND FIGURES

The amount of fresh water locked up in glaciers is equal to 60 years' worth of rainfall and snowfall over the entire earth.

## Skills Development

***Skill: Inferring***

• **What danger could result if a glacier or iceberg began to melt?** (The release of large amounts of water could cause flooding.)

## Enrichment

Have students choose a mountainous area of the United States and research the annual snow line for the region. Have them find out what factors cause the snow line to vary from place to place or from year to year.

## Content Development

There are two types of glaciers—valley glaciers and continental glaciers. Valley glaciers are formed in the valleys of mountains and they flow downward. Continental glaciers cover nearly 80–90% of Greenland and Antarctica. They are nearly 3200 m thick at the center.

## Section Review 22-2

1. Three percent; in the ice packs of the Arctic Ocean and in the great sheets of ice that cover a large part of Antarctica.
2. Water evaporates from the oceans and freshwater sources. As it cools, it condenses back to liquid and falls as precipitation. The cycle then begins again.
3. The top level to which ground water rises when it is underground.
4. Valley glaciers are long, narrow glaciers that move downhill between the steep sides of mountain valleys. Continental glaciers are very thick sheets of ice that cover most of the polar regions.

## LABORATORY ACTIVITY EXAMINING DIFFERENCES BETWEEN FRESH AND SALT WATER

### BEFORE THE LAB

1. **Prepare solutions A, B, and C as follows. For solution C, make a saturated sodium chloride solution. Pour equal amounts of solution C into two beakers. Label one beaker C. Dilute the solution in the other beaker by one half. This is solution B. Pour equal amounts of solution B into two beakers. Label one beaker B. Dilute the solution in the other beaker by one half. Label this solution A. Be sure to prepare a sufficient amount of each solution for your class.**
2. **At least one day prior to the activity, gather enough material for your class, assuming 6 students per group.**

### PRE-LAB DISCUSSION

Discuss with students the physical property of density. Point out that density is a measure of the mass per unit volume of a substance. Use a familiar example such as a golf ball compared to a Ping-Pong ball to illustrate how objects of the same size can have different masses.

- **Which object, the Ping-Pong ball or the golf ball, has the greater density?** (the golf ball)
- **Why?** (Although the golf ball has the same volume as the Ping-Pong ball, it has more mass.)

Explain to students that a hydrometer is a device used to measure the density of a liquid. Guide students to understand that in this activity, they will construct their own hydrometers out of a drinking straw, steel BBs, and clay.

## LABORATORY ACTIVITY

### *Examining Differences Between Fresh and Salt Water*

#### Purpose

In this activity, you will examine some of the differences in the physical characteristics of salt water and fresh water.

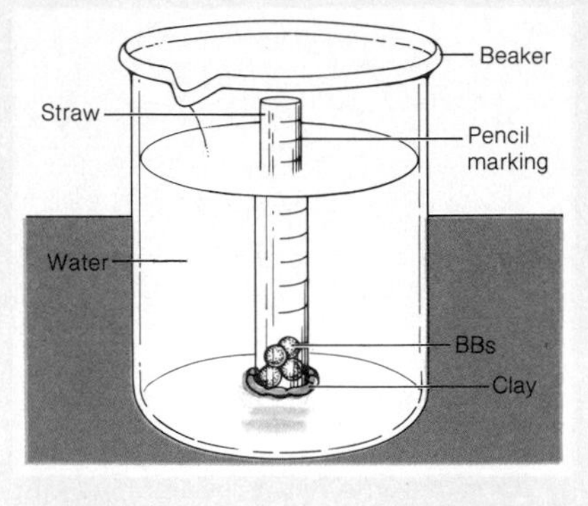


**Materials** *(per group)*
1 plastic drinking straw
Small piece of clay
4 to 6 steel BBs or ball bearings
250-mL beaker
Fresh water
Pencil
Metric ruler
3 samples of liquids

#### Procedure

1. Cut a straw in half and plug one end with a small piece of clay.
2. With a pencil, make a series of marks ½ cm apart along the entire length of the straw.
3. Add water to the beaker or glass container until it is about three-quarters full.
4. Carefully drop two or three BBs into the open end of the straw and let them roll down to the clay at the bottom.
5. With the clay end down, gently place the straw into the water. It should float. Add as many BBs as are necessary to cause your straw hydrometer to float very low in the water. Only two or three pencil lines on the hydrometer should show above the water's surface. Note exactly the level at which it floats. Your hydrometer is now ready to be used to compare densities of liquids. The higher it floats, the more dense (salty) the liquid will be.
6. Examine the three beakers containing sample fluids that have been provided by your teacher. They are labeled A, B, and C. Each sample has had different amounts of salt added. As a result, each will have a different density.
7. Gently place your hydrometer into the fluid of the first container. Carefully note and record the level at which it floats. Repeat the procedure for each of the two remaining samples.

#### Observations and Conclusions

1. List the order of the samples tested, from the one that was least dense to the one that was most dense. Include the sample of fresh water in your list.
2. How would the hydrometer float in a fluid that was less dense than the fresh water?
3. How would the hydrometer float in a fluid that was more dense than any of those tested?
4. How would the level of any floating object in fresh water compare to the floating level of that same object in salt water?
5. What do you think happens to the density of sea water at the surface of the ocean whenever it rains for a long period of time?

518

### SKILLS DEVELOPMENT

Students will use the following skills while completing this activity.

1. Observing
2. Comparing
3. Manipulative
4. Relating
5. Measuring
6. Inferring
7. Recording
8. Applying
9. Hypothesizing
10. Predicting

### TEACHING STRATEGY FOR LAB PROCEDURE

**1.** Students may find it helpful if they number the markings on the drinking straw every centimeter.
**2.** Remind students to be sure to record the level of the hydrometer in fresh water (step 5).

# CHAPTER REVIEW

## SUMMARY

**22-1 The Earth's Surface: Landmasses**

- Mountains, plateaus, and plains are the three main types of landscapes on the earth.
- Most mountains have been formed by the slow folding and wrinkling of the earth's crust.
- Some mountains are formed when the earth's crust breaks into huge blocks. Volcanic material also can form mountains.
- Mountains are usually parts of larger groups called ranges and belts.
- Plateau landscapes are usually areas of flat land of about equal height that are more than 600 meters above sea level and are higher than the surrounding land.
- Plains landscapes are areas of flat land of about equal height that are lower than the surrounding land. Plains are lowland areas that are usually not far above sea level.
- Coastal plains are low, flat regions that extend along coasts and usually average about 150 meters above sea level. Rich soil is brought to the coastal plains by rivers. Flooding is a problem in these low-lying plains.
- Inland plains are low, flat regions that average about 450 meters above sea level. Rich soil is brought to such plains by rivers.

**22-2 The Earth's Surface: Water**

- About 97 percent of the water on the earth is in the oceans. This ocean water is salt water. The other 3 percent is fresh water found in glaciers, ground water, lakes, and rivers.
- The water cycle is a continual process. It involves the processes of evaporation, condensation, and precipitation.
- Fresh water from streams, rivers, and ground water eventually returns to the oceans.
- A water table is the top level to which ground water rises.
- Fresh water from lakes and ponds usually stays in one location. About 2 percent of all fresh water is frozen in glaciers.

## VOCABULARY

*Define each term in a complete sentence.*

| | | | |
|---|---|---|---|
| **coastal plains** | **ground water** | **mountain landscape** | **precipitation** |
| **condensation** | **iceberg** | **mountain range** | **valley glacier** |
| **continental glacier** | **inland plain** | **plains landscape** | **water table** |
| **evaporation** | **mountain belt** | **plateau landscape** | |

## CONTENT REVIEW: MULTIPLE CHOICE

*Choose the letter of the answer that best completes each statement.*

1. The highest place on the earth's surface is
   a. Mauna Kea in Hawaii. b. Pikes Peak in Colorado.
   c. Mount Everest in Nepal. d. Vesuvius in Italy.
2. Most of the world's mountains have been built up by
   a. river deposits. b. folding and blocking of the earth's crust.
   c. earthquakes. d. all of these.



## GOING FURTHER: ENRICHMENT

**Part 1**

Have students carry out the same activity using a substance other than salt dissolved in water—for example, sugar. Then have students try the same activity using a solvent other than water—for example, ethyl alcohol.

**Part 2**

Have students find out how differences in densities of ocean water cause deep ocean currents to form.

## OBSERVATIONS AND CONCLUSIONS

**1.** The fresh water was least dense, the solution with the most dissolved salt was most dense.
**2.** It would be more deeply submerged.
**3.** It would float higher.
**4.** It would float higher in salt water than in fresh water.
**5.** The density becomes lower.

## CHAPTER REVIEW

### MULTIPLE CHOICE

**1.** c **3.** d **5.** b **7.** b **9.** d
**2.** b **4.** b **6.** c **8.** b **10.** b

### COMPLETION

**1.** continents
**2.** mountain belt
**3.** plateau
**4.** coastal plains
**5.** evaporation
**6.** condensation
**7.** precipitation
**8.** water table
**9.** continental glaciers
**10.** fresh

### TRUE OR FALSE

**1.** F plateaus
**2.** T
**3.** F belts
**4.** F plateaus
**5.** F plains
**6.** F lower
**7.** T
**8.** F condensation
**9.** F running
**10.** T

### SKILL BUILDING

**1.** New sources of water and conservation are necessary because all living things need to have an adequate supply of fresh water.
**2.** In a year, each person will need 912.5 L (2.5L × 365 days) of water. To determine the amount of water needed per day by the class, multiply 2.5L by the number of students in the class. Then to determine the amount of water needed by the class in a year, multiply the last answer by 365 days.
**3.** Because industry uses a great deal of water, the water table would become lower.
**4.** Chemicals dumped on the land are washed into rivers, lakes, and streams by rain.
**5.** The coastal regions would be most affected because they are low-lying areas located next to major bodies of water. These areas would become flooded quite easily.
**6.** When water evaporates from the ocean, the salt is left behind. This water returns to the earth as fresh water in some form of precipitation. It then falls into your source of fresh water. And the cycle begins again.
**7.** Student's answers should be presented in a laboratory report format, with a detailed problem described, a number procedure, and a list of all materials needed. Students should also include any observations and conclusions they can draw from the experiment. Basically, students' experiments should include the fact that they first taste the salty water to determine that it is in fact salty. Then the water should be boiled using a hot plate. The steam that rises from the water must be passed through a glass tubing of some sort and collected in a beaker. If the beaker sits in an ice bath, the steam will readily condense into the beaker. When students taste the water from the beaker, they should note that it is no longer salty.

### ESSAY

**1.** Mountain landscapes contain individual features that rise at least 610 m above the surrounding land. Plateaus are raised flat regions. Plains are lower flat regions and can occur along coasts or inland.

**3.** A mountain belt is a
a. group of volcanic mountains.
b. group of underwater mountains.
c. group of mountains such as the Appalachians.
d. group of mountain ranges.

**4.** Plateaus
a. have jagged peaks.
b. are sometimes called highlands.
c. are lower than the surrounding land.
d. are always found in dry, desert areas.

**5.** On the earth's surface flooding is a major problem for
a. inland plains. b. coastal plains.
c. plateaus. d. mountains.

**6.** Some of the best farmland is found
a. in marshes. b. on mountain slopes.
c. on inland plains. d. on plateaus.

**7.** An example of standing water is
a. rivers. b. lakes. c. glaciers. d. streams.

**8.** The process by which the sun's heat causes water on the earth's surface to change to a gas is called
a. precipitation. b. evaporation.
c. condensation. d. the water cycle.

**9.** Fresh water is not found in
a. lakes. b. ground water. c. glaciers. d. oceans.

**10.** Long, narrow masses of ice that move downhill along the sides of mountains are called
a. icebergs. b. valley glaciers.
c. polar ice sheets. d. continental glaciers.

### CONTENT REVIEW: COMPLETION

*Fill in the word or words that best complete each statement.*

**1.** There are seven large landmasses on the earth called ______.
**2.** Most mountain ranges are part of a larger group of mountains that is called a(n) ______.
**3.** A(n) ______ is a major landscape feature that forms when a flat land surface is uplifted by movements from within the earth's crust.
**4.** Lowland areas that are not far above sea level are called ______.
**5.** The process in the water cycle by which the sun changes surface water to water vapor is called ______.
**6.** The process in the water cycle by which water vapor changes to liquid water is called ______.
**7.** The process in the water cycle by which water falls to the earth as rain, sleet, hail, or snow is ______.
**8.** The top level at which ground water is found is called the ______.
**9.** The thick ice sheets that cover most of the earth's polar regions are known as ______.
**10.** Most of the ______ water on the earth is found frozen in ice sheets.

520

## CONTENT REVIEW: TRUE OR FALSE

*Determine whether each statement is true of false. If it is true, write "true." If it is false, change the underlined word or words to make the statement true.*

1. Three main types of landscapes on the earth's surface are mountains, hills, and plains.
2. Most of the world's mountains were built by the slow folding and wrinkling of the earth's crust.
3. There are two major mountain ranges on the earth's surface.
4. Plains are highland areas that look like enormous tabletops.
5. Rivers bring rich soil to plateau regions.
6. Coastal plains are higher above sea level than inland plains.
7. About 97 percent of all the water on the earth's surface is salt water.
8. Evaporation is a process in the water cycle by which water vapor changes into liquid water.
9. Rivers and streams are examples of standing water.
10. Most fresh water is frozen in glaciers.

## CONCEPT REVIEW: SKILL BUILDING

*Use the skills you have developed in the chapter to complete each activity.*

1. **Making generalizations** Why is it important to find new sources of fresh water and to conserve the sources now available?
2. **Making calculations** An average person needs about 2.5 liters of water a day to live. How much water will each person need in a year? How much water will your class need in a day to live? In a year?
3. **Making predictions** What would you predict might happen to the water table if a large industry using a well were built in the area?
4. **Applying concepts** Sometimes dangerous chemicals that have been dumped on land are found in drinking water. Explain how this is possible.
5. **Making inferences** Most of the earth's ice is found in or around Antarctica. Suppose the temperature of the South Pole were to rise high enough to melt all of Antarctica's ice. Which landscape region would be most affected? Why?
6. **Relating concepts** The water you drink today may have once been part of the Atlantic Ocean. Explain this statement.
7. **Designing an experiment** Clouds are not salty. The salt from the oceans is left behind when the water evaporates. Devise an experiment to illustrate this fact. Describe the problem, materials, procedure, expected observations, and conclusions of your experiment.

## CONCEPT REVIEW: ESSAY

*Discuss each of the following in a brief paragraph.*

1. Briefly describe the three major kinds of landscapes.
2. Discuss three reasons why the inland plains of the United States are among the best farming land in the world.
3. Describe three ways in which mountains are formed.
4. Compare inland plains and coastal plains.
5. Describe the processes of the water cycle and explain how it is a continual process.



**2.** Their rich soil deposited by rivers, their plentiful rainfall, and the lack of flooding problems account for their high quality as farming land.
**3.** Mountains can form by slow folding and wrinkling of the earth's crust, by the breaking of the crust into blocks, and by lava and ashes that rise through cracks in the crust.
**4.** Coastal and inland plains are both flat and low-lying and tend to have rich soil. The soil of coastal plains comes from soil-carrying rivers and from shallow oceans, and it tends to build up the plains. The soil of inland plains comes from rivers, and the plains themselves are formed by the wearing down of mountains and hills
**5.** Water evaporates from the oceans and freshwater sources. As it cools, it condenses back to liquid and falls as precipitation. The cycle then begins again.

## ADDITIONAL QUESTIONS AND TOPIC SUGGESTIONS

**1.** Many paintings depict landscape regions of the earth's surface. Have students go to a museum or to the art section of a library and find examples of art from different parts of the world that show mountains, deserts, coastal plains, and interior plains.
**2.** Why does running water tend to move from areas of higher elevation to areas of lower elevation? (The force of gravity pulls water downward; thus it flows naturally from areas of higher elevation to areas of lower elevation.)
**3.** Why do you think lakes often form in areas where glaciers were once present? (As glaciers move, they tear away rock and soil. This carving of the earth's surface often results in depressions that can easily fill with water to become lakes.)

## ISSUES IN SCIENCE

The following issues can be used as springboards for discussion or given as writing assignments.
**1.** Some scientists believe that the earth could experience another Ice Age within the next few thousand years. These scientists claim that climate conditions today resemble those conditions that preceded the last Ice Age. Other scientists scoff at this idea. They claim that the earth's climate is getting warmer, and that the real danger is that someday the polar icecaps will melt and flood the earth! Find out more about both sides of this issue. Then express your opinion about whether the earth is likely to experience another Ice Age in the near future.
**2.** A technological process that removes salts from ocean water is called desalination. Some experts feel that desalination could greatly reduce freshwater shortages in many parts of the world. At the present time, however, desalination is very expensive. Find out more about desalination and where it is being used. Also find out some of the problems associated with this process. Then express your opinion as to whether desalination is a practical way to obtain fresh water.

## Chapter 23

# STRUCTURE OF THE ATMOSPHERE

## CHAPTER OVERVIEW

In this chapter, students will learn about the earth's atmosphere. They will discover that the atmosphere consists of four distinct layers: the troposphere, the stratosphere, the mesosphere, and the thermosphere.

Students will learn that the atmosphere is divided into layers according to the way temperature changes with altitude. For example, in the troposphere, which is the layer closest to the earth, temperature decreases with altitude.

Students will discover that temperature in the stratosphere stays about the same as altitude increases. This constant temperature is due primarily to the presence of ozone.

Students will discover that the coldest layer of the atmosphere is the mesosphere, where temperatures decrease with altitude. In contrast, the warmest layer of the atmosphere is the thermosphere, where temperatures rise to more than 1480°C.

In the last part of the chapter, students will learn about the magnetosphere. The magnetosphere, which is located beyond the earth's atmosphere, is the magnetic field that surrounds the earth.

## INTRODUCING CHAPTER 23

Today's students are probably more aware of the earth's atmosphere than were students of previous generations, due to recent emphasis on space exploration. However, some students may still have the misconception that the atmosphere is just a vast undifferentiated mass of air. It is important to emphasize that the atmosphere has a definite structure, just as does the interior of the earth.

Begin by having students observe the cloud photograph that opens the chapter.

- **What do you see in this picture?** (billowing masses of clouds)
- **Are clouds all the same or do they differ?** (Most students will have noticed that clouds come in a variety of shapes and are found at different altitudes. Some will point out that rain clouds appear darker.)
- **Have any of you ever ridden above the clouds in an airplane?** (Answer will vary.)
- **Did the clouds you saw resemble the clouds in the photograph?** (Answers will vary.)
- **Did you find that the airplane ride tended to be bumpy or smooth as you flew over clouds?** (Answers may vary, but a flight above the clouds is usually smooth.)

# 23 Structure of the Atmosphere

**CHAPTER SECTIONS**

**23-1 The Layers of the Atmosphere**

**23-2 The Magnetosphere**

**CHAPTER OBJECTIVES**

*After completing this chapter, you will be able to:*

**23-1** Describe the layers of the atmosphere.

**23-1** Explain the relationship between temperature and the layers of the atmosphere.

**23-1** Relate the density of air at various altitudes to temperature and pressure.

**23-2** Describe the magnetosphere.

**23-2** Identify the probable causes of the magnetosphere.

"I caught my breath as I looked out. The scene as we topped 100,000 feet was utterly magnificent. For long moments, we drank in the beauty of earth, sea, and sky. Our horizon lay some 400 miles away. A narrow lower segment of the sky appeared bright and whitish-blue. I recognized it as the troposphere, the layer of the atmosphere closest to the earth. Above came another layer, a much richer, deeper, cleaner blue. Above that and over our heads, the blue darkened to a blue-black towards the void of space." That was the view from the pressurized cabin of a balloon soaring over the Gulf of Mexico.

Today, research balloons commonly explore the lower parts of the atmosphere. Rockets go up even farther, crossing that vague boundary where the last bit of earth's air fades into outer space. And down below, people travel through the lowest parts of the atmosphere—by foot, in elevators and cable cars, and in airplanes. These trips are not often scientific or mysterious. Yet there are mysteries in the air and scientists to solve them.

For example, a flight up through the clouds is often bumpy. But, above the clouds, the flight is smooth. There seems to be an invisible wall above which clouds and bumpy weather cannot rise. What is this wall? You will find the answer to this question and learn more about the atmosphere as you read the following pages.

*Clouds in the lower atmosphere*



## TEACHER DEMONSTRATION

Make a circle graph to demonstrate the composition of the earth's atmosphere. Use the following percentages: 78%, nitrogen; 21%, oxygen; 1% carbon dioxide, water vapor, argon, and trace gases (neon, helium, krypton, xenon, methane, hydrogen, ozone).

Display the graph on the chalkboard, on posterboard, or by using an overhead projector. You may want to compare your graph to Fig. 23-3 on page 526. Explain to students that air is a mixture of gases. (In fact, air is so well mixed that it is often referred to as a solution, which is the "best mixed" of all mixtures.)

- **What is the most abundant gas in the earth's atmosphere?** (nitrogen)
- **Do you know why nitrogen is important to living things?** (Answers may vary; the correct answer is that plants use nitrogen to synthesize proteins. Animals then eat plants and produce animal proteins. Thus, without nitrogen, there would be no food for living things.)
- **What is the second most abundant gas in the atmosphere?** (oxygen)
- **Why is oxygen important to living things?** (It is essential for respiration.)
- **Are you surprised to see that some familiar gases make up such a small percentage of the atmosphere?** (Answers may vary; some students will probably be surprised to learn that there are such small percentages of carbon dioxide and hydrogen in the atmosphere.)

Continue by having students read the chapter-opener text.

- **In the first paragraph of the text, who is describing the earth's atmosphere?** (a person riding in a balloon)

Explain that, based on the description, the balloon is probably in the second layer of the atmosphere (the stratosphere) at an altitude of about 33 km.

- **According to the person in the balloon, how does the stratosphere look different from the troposphere?** (The stratosphere is a much richer, deeper, cleaner blue.)
- **Where do you think the earth's atmosphere ends?** (Answers may vary; the text implies that there is no definite boundary—the air just gets thinner and thinner until it fades away.)

## TEACHER RESOURCES

### Audiovisuals

*Meteorology: The Atmosphere,* filmstrip with cassette, Eye Gate

*The Air Around Us,* filmstrip with cassette, CRM/McGraw-Hill

*The Atmosphere and Its Effect,* filmstrip with cassette, Eye Gate

*The Atmosphere in Motion,* 16 mm film, EBE

### Books

Goody, Richard, and James C. Walker, *Atmospheres,* Prentice-Hall

Riehl, H., *Introduction to the Atmosphere,* McGraw-Hill

## 23-1 LAYERS OF THE ATMOSPHERE

### SECTION PREVIEW 23-1

In this section, students will be introduced to the four layers of the atmosphere: troposphere, stratosphere, mesosphere, and thermosphere.

Students will realize that in the troposphere, air becomes colder and less dense as altitude increases. It is in the troposphere that almost all of the earth's weather occurs. Students will learn that in the stratosphere, the presence of ozone keeps the temperature relatively constant. Students will discover that the mesosphere is the coldest layer of the atmosphere, while the thermosphere is the warmest layer.

### SECTION OBJECTIVES 23-1

1. **Identify and describe the four layers of the atmosphere.**
2. **Explain how temperature, pressure, and density of air change with altitude.**
3. **Discuss the importance of the ozone layer.**

### SCIENCE TERMS 23-1

**air pressure p. 526**
**troposphere p. 527**
**stratosphere p. 529**
**ozone p. 529**
**jet stream p. 529**
**mesosphere p. 532**
**meteoroid p. 532**
**meteor p. 533**
**thermosphere p. 533**
**ion p. 534**
**ionosphere p. 534**
**aurora p. 534**
**magnetosphere p. 536**

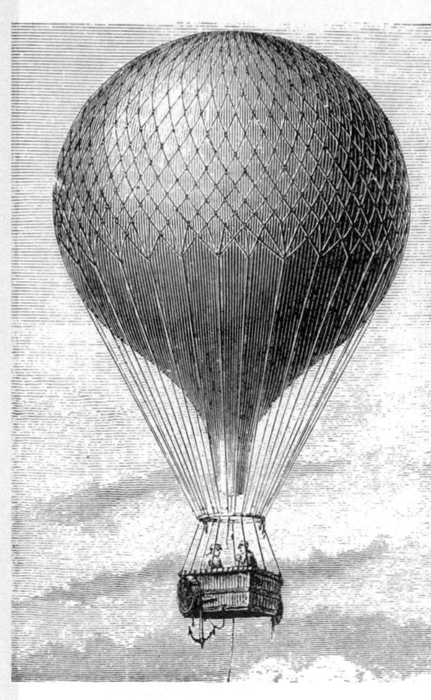

**Figure 23-1** *In the last century, scientists studied the atmosphere from wicker baskets attached to balloons* (left). *Today, scientists use high-altitude balloons carrying instruments to study the atmosphere* (right).

## 23-1 The Layers of the Atmosphere

In the nineteenth century, some weather scientists began exploring the atmosphere. Their ships of exploration were wicker baskets! Baskets? Yes, that's right. The scientists soared into the sky in huge baskets hanging from hydrogen-filled balloons.

James Glaisher, an English scientist, was one of these early explorers. On a warm day in September 1862, he and another scientist, Henry Coxwell, began a balloon trip. They soon found themselves in serious danger.

Twenty-five minutes after leaving the ground, they reached an altitude of 4800 meters. They were nearly as high as the tallest mountain in Europe, Mont Blanc. The temperature had dropped to −11° C. Ice coated the basket and the rigging of the balloon. The explorers could have returned to the ground by slowly releasing hydrogen gas and deflating the balloon. They decided to go on instead.

The temperature continued to drop as they rose. At 5000 meters it was −12.5° C. At 10,000 meters it



### TEACHING STRATEGY 23-1

#### Motivation

Have students observe the drawing of the balloon in Figure 23-1.

• **What do you think it would be like to explore the atmosphere in this balloon?** (Answers will vary.)

• **What factors might make this balloon a difficult vehicle in which to travel?** (no protection from wind, cold, and weather; rather delicate construction; limited means of steering or controlling the speed of the balloon)

Explain to students that the balloon is filled with hydrogen.

• **Why do you think hydrogen was chosen to fill the balloon?** (It is much lighter than air.)

• **Why do you think that wicker was chosen for the basket?** (It is very light.)

You may wish to discuss with students the characteristics of modern airplanes and space ships—how power and durability are combined with lightweight construction.

#### Content Development

Use the story of Glaisher and Coxwell to introduce to students the idea that the temperature and density of air change with altitude. Point out that Glaisher and Coxwell were trav-

was −45° C. Finally they reached 11,000 meters, well over 2000 meters higher than the top of Mount Everest. Though warmly dressed and in bright sunshine, the two men were numbed with cold. It was −52° C.

And something else was wrong. The scientists could move only very slowly. It was almost as if they were paralyzed. The bright sun turned dim as their eyesight failed. Glaisher lost consciousness. Coxwell tried to open the gas valves to get them down. But the valve cord was tangled up with the ropes that held the basket under the balloon. Although frostbitten, Coxwell somehow climbed up the ropes. He was surrounded by long icicles hanging from the balloon.

At last he was able to untangle the valve cord. His hands were completely numb. He climbed back down into the basket by clinging to the ropes with his elbows. Coxwell knew they would both soon die if he could not bring the balloon down. He grabbed the valve cord with his teeth and tugged the valve open. They were on their way to safety.

The two explorers lived to tell their story—and to make other balloon flights. Some other daring scientists were not so lucky. But around the world, scientists continued to explore the atmosphere in balloons. Everywhere, the picture of the atmosphere was the same. The temperature, scientists discovered, drops at an average rate of 6.5° C per kilometer of altitude.

The air also gets "thinner," or less dense, with increasing altitude. It becomes more spread out. So there is less of it in a given space at a higher altitude. This is why Glaisher and Coxwell moved so slowly and why Glaisher became unconscious. They did not have enough oxygen to breathe. In fact, at 11,000 meters, they had to take about four breaths to get the same amount of oxygen that is in one breath at the surface of the earth.

What causes air to be less dense at higher altitudes? Suppose you take a couple of handfuls of cotton and squeeze them into a small ball. Obviously, the ball has the same amount of cotton in it that once filled up both your hands. But it is squeezed into a much smaller space.

In much the same way, the air around you is pressed together by the weight of all the air above it.

**Figure 23-2** *Air becomes colder as altitude increases. That is why climbers need heavy clothing on a high mountain. Air also contains less oxygen at higher altitudes, so pilots must wear oxygen masks.*



## TEACHER DEMONSTRATION

The following demonstration will help students understand why cold air near the earth's surface is more dense than warm air. For the demonstration you will need 40 marbles (beans or jelly beans may be substituted), a small saucer, and a large plate or tray.

Divide the marbles into two groups of 20 each. Ask students to imagine that each marble represents an air particle.

Place one group of 20 marbles in the small saucer. Have the marbles grouped as close together as possible. Then place the other 20 marbles on the large plate or tray. Spread them out so that there is as much space between the marbles as possible.

- **In which container is the density of marbles greater?** (In the small saucer)
- **Why?** (There are many more marbles per unit of of space.)

Explain that when air particles become warm, they gain energy and spread out, like the marbles on the large plate. However, when air particles are cold, they stay close together, like the marbles in the small saucer. Emphasize that the number of particles does not change when air becomes warm; the particles simply spread out to occupy more space.

## FACTS AND FIGURES

At an altitude of 5.5 km, there is only half as much oxygen available as there is at the earth's surface.

eling in the troposphere—the layer of the atmosphere closest to earth. In this region, air becomes colder as one travels higher. Also, the air becomes thinner, or less dense. That is why the two explorers could not get enough oxygen; there was just not enough air present.

### Content Development

Have students observe the photograph in Figure 23-2.

- **How are the people in this photograph dressed?** (very warm clothing and oxygen masks)
- **What do you notice about the surface of the mountain in the background?** (It is snow-covered.)
- **Would you believe it if someone told you that this scene is taking place in the middle of July?** (Answers may vary, but students should be guided to recognize that high above the earth's surface, the temperature can be below freezing even if the temperature at the foot of the mountain is very warm.)

### Enrichment

Mountain climbers experience some of the same difficulties that Glaisher and Coxwell experienced. Have students find out how people who have climbed mountains such as Mt. McKinley and Mt. Everest have coped with these difficulties. Then have them share their findings with the class.

## BACKGROUND INFORMATION

Without the gases, temperature control, and moisture that the atmosphere provides, human beings could not live on Earth. The atmosphere provides protection for life on Earth. It also provides the materials necessary to sustain life. One of the reasons that life as we know it does not exist on other planets is that these planets lack an atmosphere like ours.

## TIE-IN/ PHYSICAL SCIENCE

The currents that form as warm air near the earth's surface, which rise and are replaced by colder air, are called convection currents. These convection currents help keep the earth at a comfortable temperature. Similar convection currents form in a room that is warmed by baseboard heating or by a stove or fireplace: warm air near the floor rises and is replaced by colder air from above.

## ANNOTATION KEY

❶ **Nitrogen (78%) and oxygen (21%) (Interpreting graphs)**

❷ **Air pressure decreases (Interpreting charts)**

① **Thinking Skill: Relating facts**

② **Thinking Skill: Relating concepts**

③ **Thinking Skill: Classifying atmospheric layers**

## 23-1 (continued)

### Skills Development

***Skill: Relating concepts, making calculations***

Have students choose several of the mountains listed in Fig. 22-3 of the previous chapter. Using the information that the temperature of air decreases 6.5°C per kilometer of altitude, have students calculate the temperature at the top of each mountain if the ground temperature is 20°C. (Example: height of Mt. Everest = 8.8 km. Temperature ar the top of Mt. Everest would be 20°C − [6.5 × 8.8] = 20°C − 57.2° = −37.2°C.)

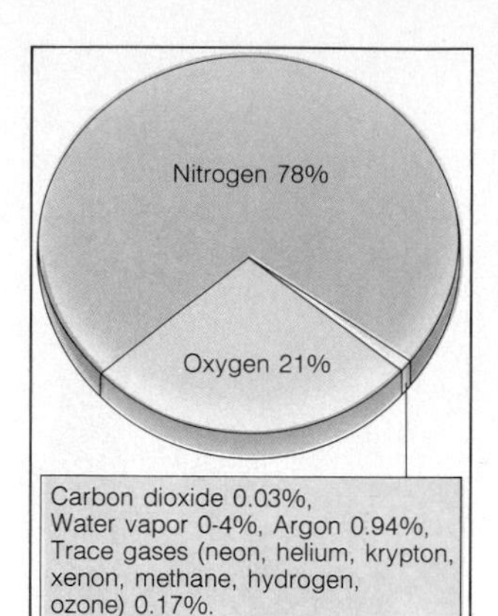


**Figure 23-3** *The atmosphere is a mixture of many gases. Which two gases make up most of the earth's atmosphere?* ❶

**Figure 23-4** *According to this chart, how does air pressure change as you go higher into the atmosphere?* ❷

***AIR PRESSURE AND ALTITUDE***

| Altitude (meters) | Air Pressure (g/cm²) |
|---|---|
| Sea level | 1034 |
| 3000 | 717 |
| 6000 | 450 |
| 9000 | 302 |
| 12,000 | 190 |
| 15,000 | 112 |

This pressing force due to the weight of the air above you is called **air pressure.** Air pressure at sea level is about 1000 grams on each square centimeter. The air pressure on your body is around 12 tons. That would be enough to flatten you into a very thin pancake—except that there is also air inside your body under the same pressure, balancing the outside pressure.

As you go upward into the atmosphere, whether by climbing a mountain or flying in an airplane or even riding in an elevator, the air pressure decreases. See Figure 23-4. That's because there is less air above you, so there is less weight and less air pressure. The air pressure inside your body decreases too, but a little more slowly. That is why you sometimes sense the change in pressure in your ears.

Air is like a spring. Under pressure, it is squeezed into a smaller space. When the pressure is lower, the air expands, taking up more space. This not only explains why the air is thinner at higher altitudes. It ① also explains why the air gets colder there.

Think of the stream of air coming out of a tire when you open the valve. The tire may be warm, and the day may be warm—but the air stream is cold. The air in the tire is under higher pressure than the air around it. When you open the valve, air rushes out and expands. When air expands, it cools.

As air rises from the earth's surface, it expands because the air pressure becomes less and less. The higher it rises, the more it expands. And the more it ② expands, the colder it gets up to a certain height.

Air is always rising from the earth's surface. As you know, the earth's surface is heated by the sun. So air near the surface is warmed by that heat. Warm air is less dense than cold air. So the warm air heated at the surface floats up through the colder air above it. Cold air sinks to take its place.

So in the lower atmosphere, the air is constantly in motion. These churning motions or currents caused by heat are like the currents in the earth's mantle that you read about in Chapter 21.

Scientists who study the earth's atmosphere have divided it into four layers. **The four layers of the atmosphere are classified according to temperature changes into the troposphere, the stratosphere, the ③ mesosphere, and the thermosphere.**



### Content Development

Explain to students that the division of the atmosphere into layers is based on *how temperature changes with altitude* within each layer. In the troposphere, temperature decreases as altitude increases. In the stratosphere, temperature remains fairly constant as altitude increases. In the mesosphere, temperature decreases as altitude increases. In the thermosphere, temperature increases with altitude.

### Skills Development

***Skill: Making graphs***

Have students graph the data shown in Figure 23-4. Let the x-axis indicate altitude and the y-axis indicate air pressure. When students have completed the graphs, ask,

- **What is the shape of the graph?** (a curved line that slopes downward)
- **What does this tell you?** (that air pressure decreases as altitude increases)

Figure 23-5 *In this photograph, you can see fog covering the Golden Gate Bridge in San Francisco. Fog is simply low-lying clouds that form when water vapor in cool, moist air condenses near the earth's surface.*

### *The Troposphere*

The lowest layer of the atmosphere is called the **troposphere** (TRAH-puh-sfeer). This name means "turning over." Turning gives you the idea of constant change, and that is what the troposphere is all about. Almost all of the earth's weather takes place in the troposphere. Ninety-nine percent of the clouds, water vapor, dust, and pollution in the atmosphere is located in the troposphere. Many airplane flights take advantage of the air currents characteristic of this layer. These currents also keep the air in the

Figure 23-6 *Winds and thunderstorms occur in the troposphere.*

**Activity**

*Air Pressure*

Air exerts a pressure of about 1000 grams per square centimeter at sea level. Use the following steps to demonstrate the effects of air pressure.

**1.** Obtain a glass juice or milk bottle, a hard-boiled egg with the shell removed, scrap paper, a match, and a pencil.

**2.** Stuff most of the paper into the bottle.

**3.** Have *your teacher* light the rest of the paper and push it into the bottle with a pencil.

**4.** Immediately place the egg on the mouth of the open bottle. **CAUTION:** *Be careful when working near any flame.* What happens to the egg? Why?



## Activity

**Air Pressure**
**Skills: Making observations, making inferences, relating concepts**
**Level: Average**
**Type: Hands-on**
**Materials: Glass bottle, hard-boiled egg, matches**

You may wish to do this activity yourself as a demonstration. If you have students do it, caution them to be very careful when using matches and to have a pan of water and a fire extinguisher at hand.

The hard-boiled egg will be drawn into the bottle. The reason is that the pressure inside the bottle decreases as a result of the combustion, and the outside pressure, now no longer balanced by pressure within, can push the egg inside.

## FACTS AND FIGURES

The boundary that divides the troposphere from the stratosphere is called the tropopause.

## BACKGROUND INFORMATION

In large quantities, ozone is a poisonous gas with a sharp, disagreeable odor. So it is vital that the ozone "umbrella" be high above the earth. When small quantities of ozone form on the earth during a thunderstorm or near electric motors, the sharp odor is sometimes mistaken for "clean air"–which is why certain towns and neighborhoods have names such as "Ozone Park."

Ozone is more reactive than oxygen, and it has the ability to break down many compounds. For this reason, ozone is valuable in water and sewage treatment, where it is used to destroy bacteria.

## Skills Development

***Skills: Relating cause and effect, predicting***

Have students observe Figure 23-5.

• **Often fog forms in the morning. What do you think will happen to fog if the sun comes out warm and strong in the afternoon?** (It will disappear.)

• **Why?** (The condensed water vapor that formed the fog will evaporate once again into the atmosphere.)

## Enrichment

A barometer is an instrument used to measure air pressure. Have students find out about different types of barometers and how they work. If possible, contact the science department of your local high school or college and obtain a barometer to display to the class.

## Activity

**Temperature Changes in the Troposphere**
**Skills: Making graphs, observing, manipulative, relating, applying, recording, measuring**
**Level: Average**
**Type: Hands-on/computational**
**Material: Celsius thermometer, graph paper**

Students should find that the temperature at the 1.25-m mark changes more rapidly and by a greater amount over time. There are many reasons for this, but the one students will most likely point to is the fact that land absorbs and retains heat longer than air. During the day the land absorbs heat and is warmer than the air above it. During the night the land releases heat, but the air above the land does not retain this heat and the temperature changes in the air at 1.25 m will be greatest. Another fact students may point out is that the winds are stronger farther above the land. If students have trouble with this concept, ask them where the temperature stays most stable on a sunny day at the beach.

## ANNOTATION KEY

1 Thinking Skill: Applying concepts
2 Thinking Skill: Relating concepts

### Activity

*Temperature Changes in the Troposphere*

**1.** At three times during both the day and evening, use an outdoor thermometer to measure air temperature 1 cm above the ground and 1.25 m above the ground. You may have to leave the thermometer in place for a few minutes.

**2.** On a chart, record the time of day and the temperature for both locations.

**3.** On a piece of graph paper, plot time versus temperature for each thermometer location. Label both graphs.

In which area did temperature change more rapidly? By a greater amount over the entire time period? Why?

troposphere well mixed. And, most important, the troposphere is the layer in which you live.

The most important single fact about the troposphere is that it gets steadily colder from bottom to top. This is what makes the churning of its air possible. Suppose the troposphere were nearly the same temperature throughout. Then there would be no rising of warm air, no settling of cold air—no currents at all.

For a long time, scientists thought that the atmosphere just kept getting colder all the way up to the edge of space. By 1902, a French scientist, Léon Philippe Teisserenc de Bort, had shown this was not so. He had checked the temperatures instruments recorded from 236 balloon flights. The balloons rose from altitudes between about 9.5 kilometers and 14.5 kilometers. The instruments showed that at an altitude of about 12 kilometers, the temperature stopped falling. There seemed to be a "roof" over the troposphere!

The troposphere actually is like a sloping roof. On the average, the troposphere's roof is lowest at the poles and slopes upward toward the equator.

**Figure 23-7** *Rainbows* (left) *and clouds* (right) *are caused by weather conditions in the troposphere.*

## 23-1 (continued)

### Reinforcement

Reinforce the idea that air in the stratosphere is calm, while air in the troposphere tends to be turbulent.

• **What causes air currents to form in the troposphere?** (Warm air near the earth's surface rises and is replaced by colder air from above.)

• **Why are there no air currents in the stratosphere?** (Air in the stratosphere is all about the same temperature; thus, there is no significant movement of cold air replacing warm air.)

### Content Development

Discuss with students the formation of convection currents as warm air rises and is replaced by cooler air.

• **Why do you often feel a draft in a room where there is an open fire?** (Convection currents are set up as warm air from the fire rises and is replaced by cooler air.)

### Enrichment

Have students observe Figure 23-7.

• **Do you know how rainbows form?** (Answers may vary; the correct answer is that sunlight is scattered as it passes through water droplets.)

### Reinforcement

Use the text discussion to review the

Over the United States, which is about halfway between the poles and the equator, the roof of the troposphere is 11 kilometers up. The temperature of the roof of the troposphere over the United States is about −55° C. About 76 percent of the earth's air is found in the troposphere.

### *The Stratosphere*

The layer of the earth's atmosphere above the troposphere is called the **stratosphere** (STRAT-uh-sfeer). The stratosphere extends from about 16 kilometers to about 48 kilometers above the earth's surface. About 24 percent of the air lies in the stratosphere.

From about 12 kilometers high in the troposphere layer up to about 24 kilometers in the stratosphere layer, the temperature usually stays the same. The temperature remains about −55° C. Above that height, the air in the stratosphere actually warms up slowly with increasing height. At the top of the stratosphere, the temperature normally reaches between −2° C and 0° C.

Why doesn't the stratosphere get colder with the increasing height? Because **ozone** is present. Ozone is a high-energy form of oxygen. It has a sharp smell, something like the smell given off by sparking electric wires. Oxygen in the stratosphere reacts with sunlight to form ozone. The ozone forms a layer in the stratosphere. This ozone layer absorbs powerful ultraviolet rays from the sun. These rays warm the stratosphere. The ozone layer also acts as a screen, shielding the earth's surface from these harmful ultraviolet rays. Without this shield, most living things on the earth would be killed. The ozone layer ends at the top of the stratosphere.

In the troposphere, air rises when there are layers of colder air above it. This does not happen in the stratosphere. The temperature is constant or rises with increasing height. So there are no currents. There are no storms. This is why airliners fly in the stratosphere when they can.

Another reason for flying in the stratosphere is the **jet stream.** The jet stream is a narrow, fast-moving current of air that forms along the troposphere-stratosphere boundary.

**Figure 23-8** *Jets fly in the stratosphere to avoid the storms and strong winds of the troposphere.*



## HISTORICAL NOTES

The earth's atmosphere was not always the same as it is today. In fact, when the earth first formed over 4 billion years ago, the atmosphere consisted primarily of two deadly gases: methane ($CH_3$) and ammonia ($NH_3$). The atmosphere also contained some water vapor.

The atmosphere began to change when sunlight triggered chemical reactions among these three substances. As a result of these reactions, new gases formed. One of these gases was nitrogen, now the most abundant gas in our atmosphere. Hydrogen and carbon dioxide also formed at this time.

Because hydrogen gas is so light, it escaped from the atmosphere and disappeared into space. Sunlight began breaking down the water vapor into hydrogen and oxygen. Once again, the hydrogen escaped, but the oxygen remained. The atoms of oxygen began combining to form ozone. This ozone eventually became the ozone layer that now occupies much of the stratosphere.

Once the ozone layer formed, green plants began to grow on earth. These plants used the energy from sunlight to combine carbon dioxide and water to produce food. A by-product of this process was oxygen. As plants continued to grow, the oxygen content in the atmosphere increased. By about 600 million years ago, the composition of the atmosphere was about the same as it is today.

## TIE-IN/PHYSICAL SCIENCE

Ozone is an allotropic form of oxygen. An allotrope is one of two or more forms of an element existing in the same physical state but having different molecular structures. A molecule of ordinary oxygen gas consists of 2 atoms of oxygen bonded together ($O_2$); a molecule of ozone gas consists of 3 atoms of oxygen bonded together ($O_3$).

Ozone is formed when oxygen reacts in the presence of an electric spark or other form of energy. During the formation of the earth's atmosphere, it was radiant energy from the sun that caused the production of ozone.

basic characteristics of the troposphere and the stratosphere. Make sure that students fully understand that virtually all weather takes place within the troposphere. Review with students that the single most important fact about the troposphere is that temperature decreases steadily from bottom to top. Remind students of the mountain climbers on page 525. One major difference between the stratosphere and the troposphere is that above a height of about 24 km, temperature in the stratosphere actually increases slowly with increasing height. The reason that the stratosphere does not get increasingly colder with height is the presence of ozone.

## BACKGROUND INFORMATION

Jet streams were discovered in the stratosphere in the 1940s, when military aircraft were developed that could fly at altitudes higher than 10 km. Jet streams are high-velocity streams of air that are hundreds of kilometers wide and several kilometers thick. Jet streams gain much of their energy from the difference in temperature between cold polar air and warm tropical air. Because jet streams blow from west to east, they enable eastbound airplanes to shorten their travel time.

Each hemisphere has, on the average, two jet streams. These are the subtropical jets, which are associated with subtropical highs, and the polar front jets, which are usually located above the polar fronts.

Jet streams follow paths that resemble sine-curves around the earth. Although jet streams usually travel in predictable "waves," sometimes the patterns shift. These pattern shifts can cause weather under the displaced waves to become colder or warmer or wetter or drier than normal, depending on which way the air is curving.

## FACTS AND FIGURES

A jet stream moving at 652.8 km per hour at a height of about 47 km above the earth's surface was recorded over Scotland on December 13, 1967.

1

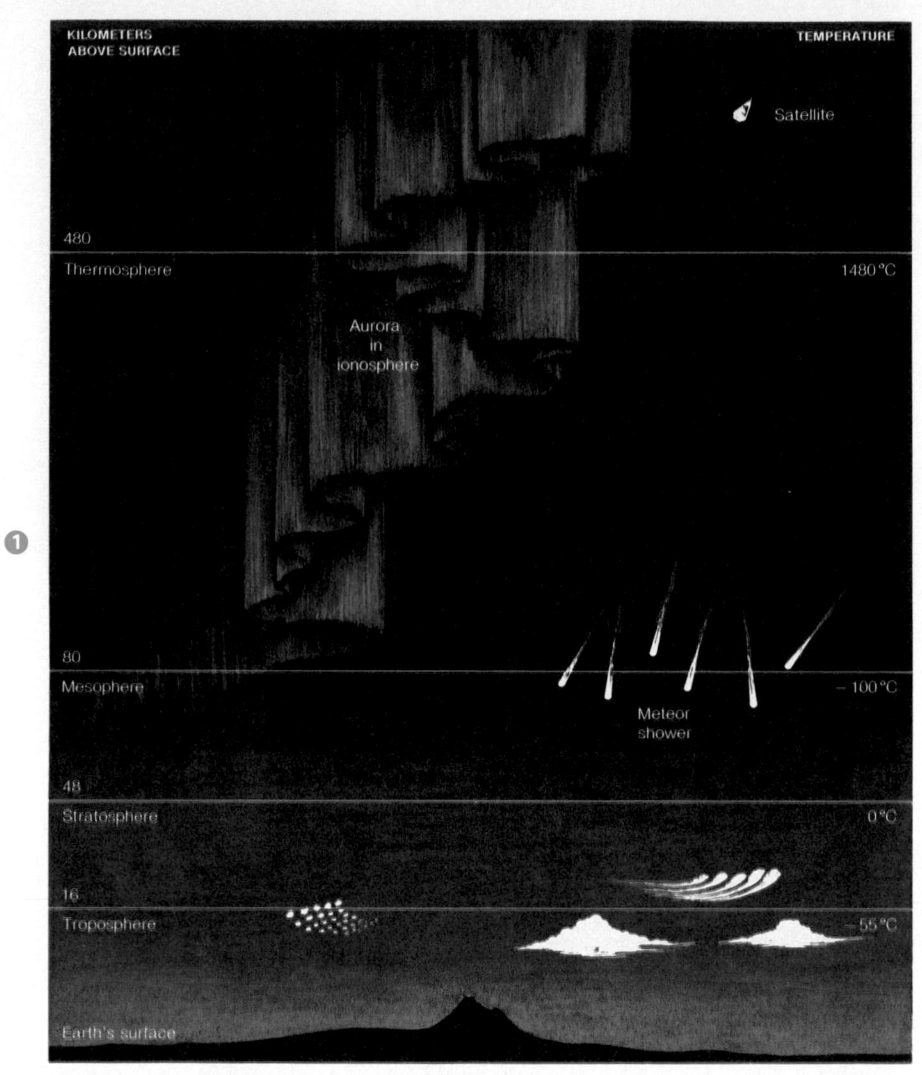


**Figure 23-9** *The layers of the earth's atmosphere, from lowest to highest altitude, are the troposphere, the stratosphere, the mesosphere, and the thermosphere. How do scientists classify the layers of the atmosphere?* 1

530

### 23-1 (continued)

### Skills Development

***Skill: Relating cause and effect***

Ask students to consider the following question:

- **Why do eastbound flights across the United States take about an hour less than westbound flights?** (The jet streams blow from west to east. Thus the winds help eastbound planes make better time than westbound planes.)

### Enrichment

Challenge students to find out what patterns are followed by the jet streams as they blow around the earth. Then have students make maps to show the patterns.

### Skills Development

***Skill: Interpreting diagrams***

Have students observe Figure 23-9.

- **According to the diagram, what characterizes the mesosphere?** (meteor showers)
- **What characterizes the thermosphere?** (the presence of the aurora)
- **Where do artificial satellites orbit the earth?** (above the atmosphere, in what is often referred to as "outer space")
- **In which layer of the atmosphere are most clouds found?** (in the troposphere)
- **Are any clouds found in other layers?** (yes)
- **In which layer(s)?** (in the lower part of the stratosphere)
- **Can you tell from the picture what type of clouds form in the stratosphere?** (thin, wispy clouds; probably cirrus clouds)

In 1922, a weather balloon was sent aloft at Hampshire, England. It crossed the boundary between the troposphere and the stratosphere. Suddenly, it started sailing eastward at great speed. It came down four hours later in Leipzig, Germany, 900 kilometers east of its starting point. The balloon had been carried by a jet stream blowing at more than 200 kilometers an hour. ❷

A jet stream usually forms where cold air from the poles meets warmer air from the equator. So there are two major jet streams that circle the earth. One jet stream is roughly halfway between the North Pole and the equator. The other is roughly halfway between the South Pole and the equator. Each usually blows in an easterly direction at more than 105

## Career: *Physical Meteorologist*

**HELP WANTED: PHYSICAL METEOROLOGIST** Bachelor's degree in meteorology required. Master's degree preferred. Courses in air pollution and hydrology helpful. Apply to National Weather Service.

As you read these words, it is possible that somewhere in the world a volcano is "blowing its top." Almost countless tons of rock, dust, and gases are exploding into the earth's atmosphere. Wind will carry this volcanic "junk" around the world—perhaps to the sky above your head. At the same time, factories near your home and cars on your street will puff more dust and gases into the atmosphere. Not so strangely, all of this material can affect the weather where you live. How? That is a question that **physical meteorologists** try to answer.

These scientists study the effects of gases and other particles in the air on weather and on other important things that affect the lives of people. The work of physical meteorologists can aid in the fight against air and water pollution.

Physical meteorologists study both the electrical and the chemical properties of the atmosphere. They observe the processes that produce winds and those that produce clouds, rain, snow, and hail. They do research to learn how the atmosphere affects light, sound, and radio waves. Meteorologists work for the government, universities, airlines, aerospace companies, and the radio and television industry.

If you are a curious person who enjoys detailed work in a laboratory and outdoors, physical meteorology may be a career for you. High school courses in physics, chemistry, and mathematics are very helpful. To learn more, write to the American Meteorological Society, 45 Beacon Street, Boston, MA 02108.



## Skills Development

***Skill: Making a model***

Have students make scale drawings showing the relative size of the earth compared with the thickness of the atmosphere surrounding the earth. Have students use 6450 km as the radius of the earth and 480 km as the height of the upper atmosphere above the earth's surface. Once students have made their drawings, ask,

- **How does the thickness of the atmosphere compare with the thickness of the whole earth?** (The thickness of the atmosphere is about 1/16 the radius of the earth.)

Remind students that when they studied the structure of the earth, the thickness of the earth's crust was likened to the peel of an apple. Point out that the relative thickness of the atmosphere is somewhat greater, rather like the skin on a very thick-skinned orange.

## FACTS AND FIGURES

About 300 km above the earth's surface, or halfway into the thermosphere, an air particle would have to travel on the average more than 12 km before hitting another air particle. At the top of the thermosphere, the distance between air particles would be about 6400 km.

## ANNOTATION KEY

❶ **According to major changes in temperature (Applying definitions)**

❶ Thinking Skill: Interpreting diagrams

❷ Thinking Skill: Relating cause and effect

## FACTS AND FIGURES

The lowest temperature ever recorded in the atmosphere is −107.4°C at a height of about 80 to 96 km above Sweden in 1963.

## SCIENCE, TECHNOLOGY, AND SOCIETY

Fast Food, Trash and Skin Cancer?

Every time we drive through a fast-food restaurant and pick up a hamburger in a plastic foam container, we are increasing our chances of getting a severe sunburn, and possibly skin cancer!

As they slowly break down, the plastic foam packaging used by fast-food restaurants give off chlorofluorocarbons (CFC). CFCs are man-made chemicals that attack the ozone layer. Bombardment of CFCs by ultraviolet radiation releases chlorine. Each chlorine atom then goes on to attack and destroy 100,000 ozone molecules. Our fast-food, fast-paced society is upsetting an atmospheric balance that has protected us from ultraviolet radiation for millions of years.

Brainstorm other activities, both human and natural, that might contribute to the destruction of the earth's atmosphere and climate. (Answers might include industrial pollution, burning of fossil fuels, deforestation, and volcanic activity.)

## TIE-IN/ MATH

When calculating the effect of the jet stream on an eastbound plane flight, the speed of the plane must include the positive effect of the jet stream wind. When calculating a westbound flight, the speed of the plane must include the negative effect of the jet stream. For example, if a plane flies 600 km/hr in still air, and the jet stream is blowing at 180 km/hr, the eastbound speed of the plane is (600 + 180) km/hr, while the westbound speed of the plane is (600 – 180) km/hr.

### Activity

**Meteor Hunt**
**Skill: Making observations**
**Level: Average**
**Type: Computational/field**

This field activity, which can only be done in areas in which the sky is clear and there is no light pollution from nearby cities, will enable students to practice their observational skills. Students should find that meteors can be seen on most nights. You may wish to tell them that at certain times during the year the earth passes through areas during its orbit around the sun that have increased meteor activity. For example, meteor activity during the first 2 weeks in August provides an impressive show of meteors with a "shooting star" occurring about once every 60 seconds.

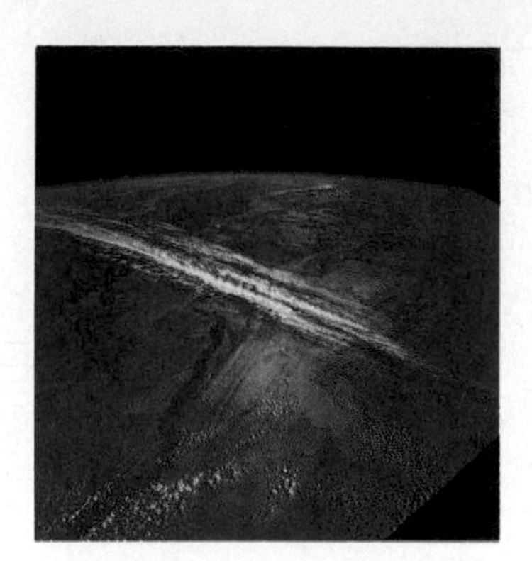

**Figure 23-10** *A jet stream forms where cold air from the poles meets warmer air from the equator. This high-altitude jet stream is moving over the Nile Valley and the Red Sea.*

kilometers per hour. However, speeds may exceed 320 kilometers per hour. Airplanes flying eastward use the jet stream to make a faster trip and to save fuel. Airplanes traveling westward climb above the jet stream to avoid a head wind that would slow them down. There, of course, they do not get a push from the jet stream. That's why an airline flight schedule will show a shorter time for a flight from San Francisco to New York than a return flight, for example.

Jet streams also can steer large storms. You may have seen a television weather forecaster showing a jet stream on a map of the United States. Usually the jet stream is shown as a sort of twisted, moving belt snaking its way east across the country. As the jet stream changes its path, usually from day to day, the track of big storms changes with it.

### The Mesosphere

1 The layer of the earth's atmosphere above the stratosphere is called the **mesosphere** (MES-uh-sfeer). This layer extends from about 48 kilometers above the earth to about 80 kilometers high. The air in the mesosphere is many thousands of times less dense than the air at sea level. Here again, the temperature begins to drop. At the top of the mesosphere, the temperature has dropped to about – 100° C. This is the coldest part of the atmosphere.

Why does the temperature drop? Because there is no more ozone. Above the stratosphere, there is not enough oxygen left to form an ozone layer.

With very little air in the mesosphere, you might think that nothing ever happens there. But something does happen and you may see it when you look up into the sky at night. For it is in the mesosphere that most **meteoroids** (MEET-ee-uh-roidz) burn up. A meteoroid is a chunk of rocklike matter from outer space.

At speeds of 160,000 kilometers an hour or more, a meteoroid may strike the atmosphere. As it rubs against the air, the meteoroid is quickly heated and glows white hot.

This is what happens to the heat shield of a spacecraft as it reenters the atmosphere. But meteoroids do not have heat shields to protect them. If

#### Activity

*Meteor Hunt*

Have you ever seen a meteor in the night sky?

**1.** On a clear night, go outside and observe the sky. If you live in an area that has considerable light pollution, find a location away from most lights.

**2.** Observe the stars and record the number of meteors you see. Do this for several nights.

**3.** Determine whether there is any increase in meteor activity.



## 23-1 (continued)

### Enrichment

Students may be interested to learn about the atmosphere of other planets in our solar system. Have students find out what gases these atmospheres are composed of and how the environment of each planet is affected by its type of atmosphere.

### Reinforcement

Have students work in pairs. Provide each pair with four index cards, each labeled with one of the following words: troposphere, stratosphere, mesosphere, and thermosphere. Have each pair arrange the cards vertically in the correct order. Then have the two students take turns describing the characteristics of each atmospheric layer.

### Content Development

Explain to students that the streak of light, or meteor, talked about in this section, is sometimes referred to as a shooting star.

### Enrichment

Two popular songs that students might know are "Up, Up, and Away" and "On a Clear Day, You Can See Forever." These songs use imagery of the atmosphere to express feelings of

Figure 23-11 *A heat shield protects the Space Shuttle from burning up in the earth's atmosphere.*

you are looking at the sky at night, you may see a streak of light, or **meteor** (MEE-tee-er). That streak of light is a meteoroid burning up in the mesosphere. ❷

Figure 23-12 *Most artificial satellites orbit above the thermosphere layer of the atmosphere.*

### The Thermosphere

The upper layer of the atmosphere is called the **thermosphere** (THER-muh-sfeer). This layer extends upward from about 80 kilometers to between 480 kilometers and 600 kilometers. At the top of the thermosphere is the beginning of space. But there is no sharp line where one ends and the other begins. As you travel near the top of the thermosphere, you are in the neighborhood of that invisible boundary.

"Thermosphere" means "heat sphere" or "warm layer." And in the thermosphere the temperature rises to more than 1480° C. ❸ But don't be fooled about the high temperatures in the thermosphere. This is a very different kind of heat. On the earth's surface, such temperatures would melt many metals. However, a thermometer placed in the thermosphere would register far below zero!



## FACTS AND FIGURES

Of the total mass of the earth's atmosphere, 99% is below an altitude of 32 km. Only 1% of the total mass of air exists from 32 km to 600 km.

## BACKGROUND INFORMATION

Temperature is a measure of the average kinetic energy of molecules. The liquid in a thermometer must be bombarded by high-energy molecules in order to gain enough heat to expand. When few air molecules are present, little energy is transferred to the thermometer—even if each individual molecule is very hot and possesses a great deal of energy. That is why an ordinary thermometer does not register high temperatures in the upper portion of the atmosphere.

## FACTS AND FIGURES

In the thermosphere, the density of air and air pressure are only about one ten-millionth of what they are at the earth's surface.

## ANNOTATION KEY

❶ **Thinking Skill: Classifying layers of the atmosphere**

❷ **Thinking Skill: Relating facts**

❸ **Thinking Skill: Applying definitions**

freedom and vision. Try to obtain records or tapes of these songs and play them for the class. Then have students use words or drawings to express their reactions to the music.

### Skills Development

***Skill: Relating concepts***

Ask students to consider the following question:

- **If the mesosphere is the coldest part of the atmosphere, how it is that meteoroids burn up as they pass through the mesosphere?** (It is the friction between a meteoroid and the gas particles in the atmosphere that creates enough heat to burn up the meteoroid.)
- **Can you think of an everyday situation in which heat is generated by friction?** (Answers may vary, but an excellent example is the way the blade of a skate cuts a figure on ice. The ice is very cold but the friction of the blade creates enough heat to melt the ice under the blade and make a cut.)

## HISTORICAL NOTES

Radio waves, which are a very low-frequency form of electromagnetic radiation, were discovered in 1888 by German physicist Heinrich Rudolf Hertz. About a decade prior to this discovery, the existence of radio waves was predicted, though not proven, by Scottish physicist James Clerk Maxwell.

## BACKGROUND INFORMATION

Nitrogen, oxygen, and the inert gases are present in the same relative proportions everywhere in the atmosphere. Relative amounts of the other gases, including carbon dioxide and water vapor, vary according to location.

## ANNOTATION KEY

1 Thinking Skill: Applying concepts

2 Thinking Skill: Relating facts

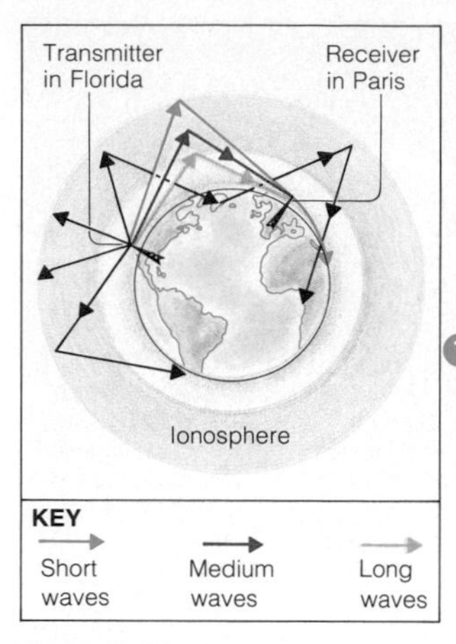


**Figure 23-13** *Radio waves are bounced off the ionosphere to transmit radio messages overseas or across continents. There are three types of waves and each travels to a different height in the ionosphere.*

**Figure 23-14** *Weather satellites orbit the earth in the upper part of the thermosphere. They transmit information used by scientists to track weather patterns. This photograph was taken by a weather satellite.*

Why? Temperature is a measure of how fast particles in air move. The faster the air particles travel, the higher the temperature. On the earth, particles in the air are packed so tightly together that a particle travels a very short distance before hitting another particle—about 0.06 microns. A micron is a millionth of a meter. In the thermosphere, however, the particles are few and far between. About 99.999 percent of the earth's atmosphere lies below the (1) thermosphere. So the particles travel a long distance before they hit other particles.

A thermometer on the earth's surface is bombarded by countless billions of air particles. But in the thermosphere, few particles ever strike the thermometer. These particles are moving very fast and are far apart. They are very hot, but there would not be enough of them to warm a thermometer. However, special instruments can measure the temperature of particles in the thermosphere.

The particles are moving very fast because they are absorbing energy from the sun and outer space. This energy causes many of the particles to become charged with electricity. These charged particles are called **ions.**

Ions are found mainly in the lower part of the thermosphere and extend up for several hundred kilometers. This part of the thermosphere is a kind of sphere within a sphere. Scientists call it the **ionosphere** (igh-AHN-uh-sfeer).

There are a number of layers of ions in the ionosphere. They act like a mirror, reflecting radio waves back to the earth. Radio waves, like light waves, travel in straight lines. But when reflected by the ionosphere, they can travel long distances around the earth. These radio waves can be received thousands of kilometers from their source.

Sometimes streams of fast-moving ions from the sun streak through the ionosphere. They collide with other particles in the air, which causes them to glow. This is much the same thing that happens in a neon sign. A stream of electrically charged particles is passed through a small amount of neon gas in the sign. The neon glows brightly.

These sky glows are called the northern and southern lights or **auroras** (aw-RAW-ruhz). They are most brilliant in the polar regions. Why? As you will



## 23-1 (continued)

### Skills Development

***Skill: Interpreting diagrams***

Have students observe the diagram in Figure 23-13.

- **From where are radio waves being sent?** (from New York)
- **What happens to the waves once they are sent out?** (Some of them pass into space. The rest hit the edge of the ionosphere and bounce back to earth.)
- **What is a more scientific way of saying that the waves bounce back?** (We say that the waves are reflected.)
- **According to the diagram, where do the reflected waves go?** (Some reach Paris; others reach Greenland; still others reach parts of Africa and parts of South America.)
- **What do you notice about the angle with which the radio waves strike the ionosphere and the angle with which the waves bounce back?** (The angles are the same; the angle of incidence equals the angle of reflection.)

### Enrichment

Have students find out what solar flares are and how they can disturb radio communication.

**Figure 23-15** *Auroras are bright bands of light in the upper atmosphere caused by electrically charged particles.*

see in the next section, the earth is like a giant magnet. The ends of this magnet are in the polar regions. Electrically charged particles are attracted toward the ends of this magnet. ❷

### *SECTION REVIEW*

1. In what layer does weather take place?
2. List two reasons that airplanes fly in the stratosphere.
3. Which is the coldest layer of the atmosphere?
4. In what layer are ions and auroras found?

## 23-2 The Magnetosphere

Place a pane of thin glass over a bar magnet. Sprinkle some iron filings over the top of the glass. Tap it gently. The tapping moves the filings and

### Activity

*Listening to the Radio*

To find out from how far away radio waves can be received during the day and night, try the following: Listen to an AM radio during the day and late at night. Record the number of stations you can receive during the day and at night. Record what city these stations are broadcasting from. Plot the location of these stations on a map.

Answer the following questions:

**1.** Which of the radio stations you were able to receive during the day was the farthest away? At night?

**2.** How many kilometers away is the farthest station you heard during the day? During the night?

**3.** On the average, the ionosphere at night is higher above the earth's surface. What happens to radio signals when the ionosphere is higher?



### Activity

**Listening to the Radio**
**Skills: Observing, comparing, inferring, relating**
**Level: Average**
**Type: Field**

Students will find that they can receive signals from more stations and stations farther away at night. The increased altitude of the ionosphere at night enables radio waves to travel farther before being reflected, increasing the total efficient distance they tend to travel.

### FACTS AND FIGURES

The rate at which the sun's rays strike the top of the earth's atmosphere is known as the solar constant. The average value for the solar constant is 1.94 cal/cm$^2$/min. During a year, the solar constant fluctuates least near the equator and most near the poles. On the average, location at or near the equator receive two and half times more solar radiation per year than do locations near the poles.

### Section Review 23-1

1. Troposphere.
2. No air currents and storms; jet stream.
3. Mesosphere.
4. Thermosphere.

## TEACHING STRATEGY 23-2

### Motivation

As described in the text, use a bar magnet and iron fillings to demonstrate to students how magnetic force lines surround a magnet. (A piece of paper can be substituted for the pane of thin glass.)

- **Can you describe the pattern of force lines that surround the magnet?** (The lines of force spread outward from the north and south poles of the magnet, then form two large semicircles between the north and south poles.)
- **Look at Figure 23-16 on page 536. How do the lines of force around the bar magnet compare with the lines of magnetic force around the earth?** (They are very similar.)
- **What does this tell you about the earth?** (The earth acts like a magnet; therefore the earth must have magnetic properties.)

## 23-2 THE MAGNETOSPHERE

### SECTION PREVIEW 23-2

In this section, students will learn about the magnetosphere. The magnetosphere, which is located beyond the earth's atmosphere, is the magnetic field that surrounds the earth.

Students will discover that magnetic lines of force surround the earth, just as magnetic lines of force surround a bar magnet. In fact, the earth acts like a giant bar magnet, with its north magnetic pole near, but not at, the geographic North Pole, and its south magnetic pole near the geographic South Pole.

### SECTION OBJECTIVES 23-2

1. **Describe the location of the magnetosphere.**
2. **Explain how the earth acts like a magnet.**

### SCIENCE TERMS 23-2

**magnetosphere p. 536**

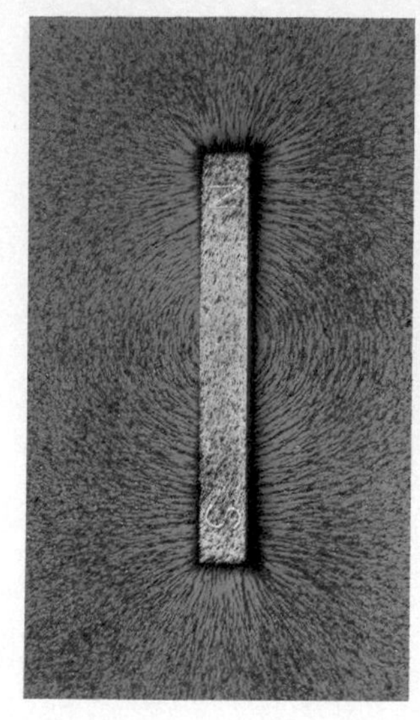

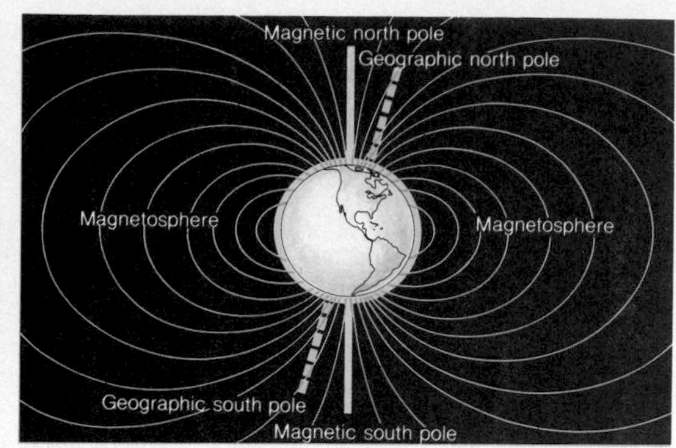


**Figure 23-16** *The earth acts like a giant bar magnet* (right) *whose lines of force produce the same pattern as a small bar magnet* (left).

they form a pattern around the magnet, as shown in Figure 23-16.

The pattern of the iron filings shows you the invisible lines of force that are between the north and south poles of a bar magnet. The earth acts like a giant bar magnet whose lines of force produce the same pattern.

You can see in Figure 23-16 that the magnetic poles of the earth are in a different location from the geographic poles. The magnetic north pole is near Bathurst Island in northern Canada, about 1600 kilometers from the geographic north pole. The magnetic south pole is in Wilkes Land, a part of Antarctica, about 2570 kilometers from the geographic south pole.

The magnetic field around the earth extends beyond the earth's atmosphere. This field is called the **magnetosphere** (mag-NEE-toh-sfeer). The magnetosphere begins at an altitude of about 1000 kilometers and extends out into space about 64,000 kilometers on the side of the earth facing the sun. This magnetic field extends even farther out into space on the other side of the earth. Why? The difference is caused by streams of electrically charged ions from the sun. These ions form a kind of solar wind. This wind pushes the magnetosphere farther into space on the side of the earth away from the sun.



### 23-2 (continued)

#### Content Development

Use the Teacher Demonstration and the Motivation demonstration to introduce the concept of the earth's magnetic field. Point out that although the magnetic force of the earth operates thousands of kilometers above the earth's surface, the force can still be felt by a tiny compass needle here on earth.

It should be emphasized that although the magnetic field of the earth is strongest around the earth's magnetic poles, these poles are not located precisely at the earth's geographic poles.

- **If this is the case, what must you do if you want to use a compass to find direction?** (You must take into account the extent to which the compass is not point to true north.)

Explain to students that this deflection of a compass from true north is called magnetic declination. Anyone who uses a compass for navigation must adjust the measurements according to magnetic declination.

#### Enrichment

Have students find out about the Van Allen radiation belts. Also have them find out about James Van Allen, the scientist whose work led to the discovery of the radiation belts.

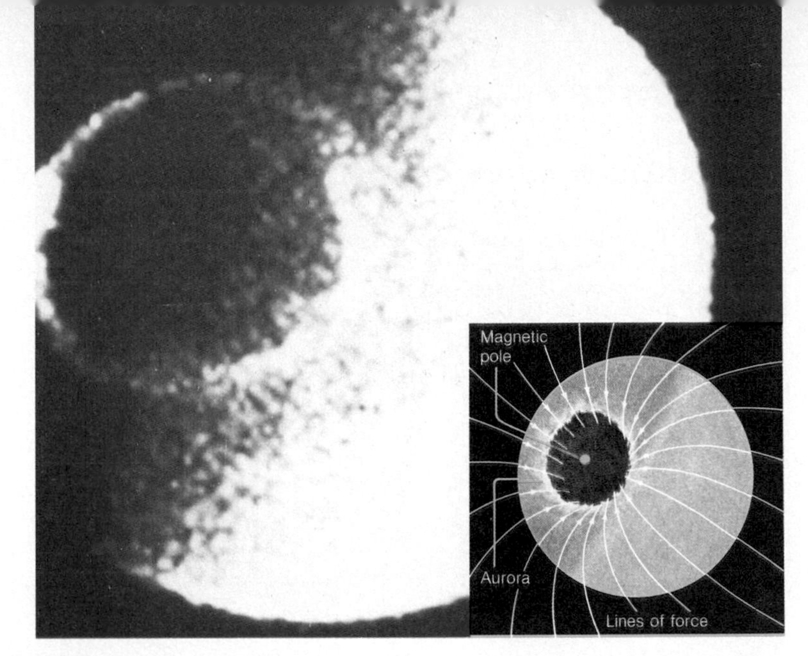


**Figure 23-17** *An entire aurora was photographed for the first time by a satellite approximately 22,000 kilometers above the earth's magnetic north pole on September 15, 1981. An artist's drawing shows the magnetic lines of force that give the aurora its shape* (lower right).

What causes the earth's magnetic field? **Most scientists believe that the magnetosphere probably is caused by a combination of the earth's rotation and electric currents in the earth's core.** As the earth spins, these currents produce the magnetosphere.

By studying the magnetic direction of iron particles found in Earth's rocks, scientists such as Allan Cox and Richard Doell of the U.S. Geological Survey have discovered that Earth's magnetic poles have reversed many times. When poles reverse, magnetic north becomes magnetic south, and magnetic south becomes magnetic north. It is estimated that the magnetic poles have reversed 171 times in the last 76 million years. How? Scientists do not yet know the answer.

### *SECTION REVIEW*

1. What is the magnetosphere?
2. What may cause the earth's magnetic field?
3. How do scientists know that magnetic poles have reversed?



## Section Review 23-2

**1.** The magnetic field around the earth.
**2.** Combination of earth's rotation and electric currents in the core.
**3.** By studying the magnetic direction of iron particles found in the earth's rocks.

## TEACHER DEMONSTRATION

As a class demonstration, make a compass out of needle and a piece of cork. For this demonstration, you will need a magnet, a compass, a large needle, a cork that has been cut in half to expose a flat surface, a shallow dish, and some water.

Use the magnet to magnetize the needle. Fill the dish with water, place the cork in the water, and balance the magnetized needle on the cork. The needle should quickly orient itself in a north–south direction.

- **Ask a student volunteer to hold the compass next to the cork and needle and compare the way the two needles are pointing.** (They should be pointing in the same direction.)
- **Do you know in what direction these needles are pointing?** (north)
- **Do you know why they point north?** (Answers may vary; correct answer is that the magnetized needles are attracted to the north magnetic pole of the earth.)

## HISTORICAL NOTES

The magnetic properties of the earth were first observed by the ancient Greeks. The Greeks discovered that if a piece of lodestone, which is a natural magnet, were suspended from a string, the lodestone would always align itself in a north–south direction.

The earliest compasses consisted of magnetized needles that floated in water on a piece of cork. Later on, the needles were mounted on a card and the various directions were marked off about the rim of a circle. Because the directions encompassed the rim of the card, the magnetized needle came to be called a compass.

The first use of a compass to make an ocean voyage was recorded by Europeans in the twelfth century. Using a compass to find direction was very useful, for up until this time only the positions of the sun and North Star could be used for navigation—and these methods depended upon clear weather. It is unlikely that Columbus or other European explorers would have been able to undertake voyages to the New World without the aid of a compass.

## LABORATORY ACTIVITY USING ATMOSPHERIC PRESSURE TO CRUSH A CAN

### BEFORE THE LAB

1. **At least one day prior to the activity, gather enough materials for you class, assuming 6 students per group.**
2. **Check to see that all bunsen burners and pieces of wire gauze are in good condition.**

### PRE-LAB DISCUSSION

Explain to students that this activity will not only demonstrate the presence of atmospheric pressure; it will also explore the relationship between temperature and pressure. This is important since both temperature and pressure are properties of air that vary from one layer of the atmosphere to another.

• **How do you think temperature and pressure are related?** (Answers may vary; the correct answer is that pressure decreases as temperature decreases.)

• **Can you think of some everyday examples of this?** (Tires look "lower" in the winter than in the summer; bottles of carbonated beverage may explode if they get too hot.)

Point out to students that this laboratory activity is somewhat similar to the Activity entitled Air Pressure on page 527.

• **In the Activity on page 527, what caused the egg to be pulled into the bottle?** (a decrease of pressure inside the bottle)

• **What force pushed the egg into the bottle?** (atmospheric pressure)

• **How was atmospheric pressure inside the bottle reduced?** (by consuming the oxygen in the bottle)

• **How is this Laboratory Activity different from the Activity on page 526?** (In this Laboratory Activity, pressure inside of a can will be reduced by changing the temperature.)

Point out to students that in both activities, an imbalance in pressure is created—the pressure inside the container becomes less than the pressure outside the container. As a result, the force of the larger pressure creates a visible effect.

### SKILLS DEVELOPMENT

Students will use the following skills while completing this activity.

1. Observing
2. Comparing
3. Relating
4. Inferring
5. Safety
6. Applying
7. Hypothesizing
8. Predicting

## LABORATORY ACTIVITY

### *Using Atmospheric Pressure to Crush a Can*

**Purpose**

In this activity, you will demonstrate the presence of atmospheric pressure by using the force caused by the weight of the air to crush a metal can.

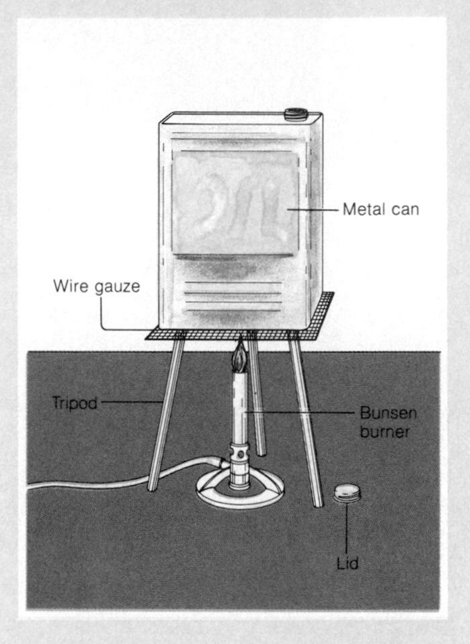


**Materials** *(per group)*

Safety goggles
A metal can that can be securely sealed
Graduated cylinder
Tripod
Bunsen burner
Oven mitten
Wire gauze

**Procedure**  

1. Put on your safety goggles and wear them during the entire activity.
2. Add 200 mL of water to a metal can. Place the unsealed can on a tripod. Position a Bunsen burner under the can and light it. **CAUTION:** *Be careful when lighting the Bunsen burner.*
3. Heat the water to boiling and allow it to boil for a few minutes after you have seen condensed water droplets leaving the opening of the can. Heating the liquid water will cause it to change to water vapor, which will force the air out of the can.
4. Turn off the Bunsen burner. Using the oven mitten, hold the heated can and carefully place the top on the can. Be sure it is on securely. **CAUTION:** *Be very careful; the can will be extremely hot.*
5. At the sink, carefully pour cold water over the hot can to cool it.

**Observations and Conclusions**

1. What happened when the heated can was sealed and allowed to cool? Why?
2. Describe a procedure to restore the can to its original state.

# CHAPTER REVIEW

## SUMMARY

**23-1 The Layers of the Atmosphere**

- Air pressure will decrease as altitude increases.
- Warm air is less dense than cold air, and so it rises.
- The atmosphere is divided into four layers based on temperature changes.
- The troposphere contains most of the earth's air. It is the weather layer. The temperature in the troposphere decreases with increasing altitude but then levels off.
- The stratosphere contains ozone, which absorbs ultraviolet rays from the sun. This accounts for the gradual warming within the stratosphere.
- The jet stream is the narrow, fast-moving current of air found between the troposphere and the stratosphere.
- The mesosphere is the coldest part of the atmosphere.
- Meteoroids, chunks of rocklike matter from outer space, are found in the mesosphere. A meteoroid burning up in the mesosphere is called a meteor.
- The thermosphere, the outer limit of the atmosphere, is heated by the sun's energy. It contains electrically charged particles called ions in an area called the ionosphere. A stream of these ions causes auroras.

**23-2 The Magnetosphere**

- The earth acts like a giant bar magnet surrounded by a magnetic field called the magnetosphere. This magnetic field is probably the result of the combination of the earth's rotation and electric currents in its core.

## VOCABULARY

*Define each term in a complete sentence.*

**air pressure**
**aurora**
**ion**
**ionosphere**
**jet stream**
**magnetosphere**
**mesosphere**
**meteor**
**meteoroid**
**ozone**
**stratosphere**
**thermosphere**
**troposphere**

## CONTENT REVIEW: MULTIPLE CHOICE

*Choose the letter of the answer that best completes each statement.*

**1.** The force pressing down on the earth's surface due to the weight of the air is called air
a. density. b. mass. c. temperature. d. pressure.

**2.** Generally, as air rises it expands, and its temperature as recorded by an ordinary thermometer
a. decreases. b. increases.
c. remains the same. d. first increases, then decreases.



(If students actually attempt to do this, they must be very careful not to overheat the can or to have their faces or hands near the jet of escaping steam.)

## GOING FURTHER: ENRICHMENT

**Part 1**

Challenge students to restore the can to its original shape. To do this, heat the can until its sides are pushed out to their normal position, then carefully open the can. (Be careful not to overheat the can.) For the sake of safety, you may wish to have one group perform this activity as a class demonstration. That way, you can adequately supervise the process.

**Part 2**

Challenge students to obtain and learn about various devices that measure or make use of pressure. Some possible devices include: gauge used to check tire pressure; pressure cooker, aneroid barometer; sphygmomanometer (the arm cuff use to measure blood pressure); mercury barometer.

**Part 3**

Given that average atmospheric pressure amounts to about 1 $kg/cm^2$, students can calculate the total atmospheric "pressure" (figured here in equivalents of mass rather than force) pushing against the entire surface of the can. They should measure the can and calculate its surface area, in $cm^2$, and then multiply by 1 $kg/cm^2$. The result will be surprisingly large.

## SAFETY TIPS

**1.** Have a student read aloud the **CAUTIONS** in steps 2 and 4.
**2.** Review the rules of laboratory safety for heat safety, fire safety, and eye and face safety.
**3.** Caution students not to put their faces close to the boiling water or hot can.

## OBSERVATIONS AND CONCLUSIONS

**1.** The can was crushed inward. The water vapor inside condensed to liquid, resulting in reduced gas pressure inside. The unbalanced outside air pressure pushed in the can.
**2.** The can may be heated again, vaporizing the water. It can then be reopened when its sides have been pushed out to their normal position.

## CHAPTER REVIEW

### MULTIPLE CHOICE

1. d 3. c 5. d 7. d 9. b
2. a 4. a 6. b 8. a 10. c

### COMPLETION

1. increases
2. troposphere
3. stratosphere
4. ozone
5. jet stream
6. mesosphere
7. meteoroids
8. ions
9. aurora
10. magnetosphere

### TRUE OR FALSE

1. T
2. F troposphere
3. F ozone
4. T
5. F meteor
6. F ionosphere
7. T
8. T
9. F away from
10. F rotations

### SKILL BUILDING

**1.** The earth is like a magnet in that the magnetic lines of force that form around a magnet are similar to the magnetic lines of force that form around the earth. The magnetic field around the earth is called the magnetosphere.
**2.** Troposphere 16 km, stratosphere 32 km, mesosphere 32 km, and thermosphere 400 km.
**3.** At this elevated altitude, the air is less dense. It therefore offers less resistance to the passage of objects through it. Beamon was able to jump farther because of this lowered air resistance.
**4.** An airplane flying eastward may use the jet stream, which flows from west to east, as a tail wind to help speed the plane in its motion. An airplane flying westward cannot make use of the jet stream and thus tends to move more slowly.
**5.** Graphs should show that the temperature drops from about 20°C to −55°C, then rises to 0°C at the top of the stratosphere. Between the stratosphere and the mesopause the temperature drops to about −100°C. From the mesosphere through the thermosphere and beyond, the temperature continues to rise as high as 2000°C.

**3.** The division of the earth's atmosphere into four layers is based on
a. surface pressure. b. wind currents.
c. temperature. d. density.
**4.** The weather layer of the atmosphere is the
a. troposphere. b. stratosphere.
c. mesosphere. d. thermosphere.
**5.** As the altitude increases within the troposphere, the temperature
a. increases continuously.
b. increases and then decreases.
c. decreases and then increases.
d. decreases and then levels off.
**6.** The layer of the atmosphere that extends from about 16 kilometers to about 48 kilometers above the earth's surface is called the
a. mesosphere. b. stratosphere.
c. troposphere. d. thermosphere.
**7.** A high-energy form of oxygen produced when oxygen reacts with sunlight is
a. hydrogen. b. helium. c. water vapor. d. ozone.
**8.** The coldest layer of the atmosphere is the
a. mesosphere. b. stratosphere. c. ionosphere. d. troposphere.
**9.** The upper layer of the atmosphere is the
a. mesosphere. b. thermosphere.
c. stratosphere. d. troposphere.
**10.** Scientists believe that the earth's magnetic field is caused by
a. a combination of the earth's revolution and rotation.
b. a combination of the earth's revolution and electric currents in its core.
c. a combination of the earth's rotation and electric currents in its core.
d. a combination of the magnetic poles and geographic poles.

#### CONTENT REVIEW: COMPLETION

*Fill in the word or words that best complete each statement.*

**1.** Within the atmosphere, as the altitude decreases, the air pressure ______________.
**2.** Clouds, dust, and pollution are characteristics of the layer of the atmosphere called the ______________.
**3.** Because there are no air currents or storms, pilots prefer to fly in the layer of the atmosphere called the ______________.
**4.** The ______________ layer in the stratosphere shields the earth from harmful ultraviolet rays.
**5.** The narrow, fast-moving current of air that forms at the boundary of the troposphere and the stratosphere is known as the ______________.
**6.** The layer of the atmosphere characterized by the lowest ordinary temperatures and very little oxygen is the ______________.
**7.** Chunks of rocklike matter from outer space are called ______________.
**8.** Electrically charged particles found in the thermosphere are ______________.
**9.** The glowing of the sky due to a stream of electrically charged particles is called a(n) ______________.
**10.** The area beyond the earth's atmosphere that is a magnetic field is known as the ______________.

540

### ESSAY

**1.** Auroras appear near magnetic poles because electrically charged ions are attracted to magnetic poles. Auroras are actually fast-moving streams of ions.
**2.** Ozone acts as a shield for the earth's surface. If the ozone did not absorb most of the ultraviolet energy, people's skin would be burned and their eyes blinded.
**3.** In the troposphere, temperature decreases steadily with altitude. In the stratosphere, temperature stays about the same from about 12 km high to about 24 km. Above that altitude, temperature in the stratosphere actually warms up slowly with increasing height. In the mesosphere, temperature again decreases with altitude. And, in the thermosphere, temperature is very high, as high as 2000°C. Temperature in the thermosphere

## CONTENT REVIEW: TRUE OR FALSE

*Determine whether each statement is true or false. If it is true, write "true." If it is false, change the underlined word or words to make the statement true.*

1. Warm air is less dense than cold air.
2. Most of the air is in the stratosphere.
3. Sunlight and oxygen react in the stratosphere to form the gas peroxide.
4. The coldest part of the atmosphere is the mesosphere.
5. A meteoroid burning up in the mesosphere is called a northern light.
6. The part of the thermosphere in which charged particles can be used to transmit radio waves is the mesosphere.
7. Magnetic lines of force are the strongest at the poles of a magnet.
8. The earth's magnetic poles are not in the same place as the earth's geographic poles.
9. The magnetic field around the earth extends farthest into space on the side of the earth facing toward the sun.
10. Most scientists believe that the earth's revolution is partly responsible for its magnetic field.

## CONCEPT REVIEW: SKILL BUILDING

*Use the skills you have developed in the chapter to complete each activity.*

1. **Making comparisons** Compare the earth to a magnet.
2. **Making calculations** Figure 23-9 shows the layers of the earth's atmosphere and the altitudes at which they begin and end. Use this information to calculate the average thickness of each layer.
3. **Relating concepts** In the 1968 Summer Olympic Games held in Mexico City, Mexico, long jumper Bob Beamon set a world record of 8.90 meters. The altitude of Mexico City is 2309 meters. Relate this to the density of air and the resistance the air offers a jumper. Then explain Beamon's extraordinary jump.
4. **Applying concepts** Airline guides give flying time from Seattle, Washington, to New York, New York, as 5 hours, 5 minutes. The reverse trip from New York to Seattle takes 5 hours, 35 minutes. Explain this difference in travel time.
5. **Making graphs** Compare the temperatures found in the four main layers of the atmosphere by drawing a line graph of the data. Plot the altitude of the layers on the X-axis and the temperatures on the Y-axis. Use the average winter or summer temperature in your area as the beginning of the troposphere. Then explain why your graph is *not* a straight line.

## CONCEPT REVIEW: ESSAY

*Discuss each of the following in a brief paragraph.*

1. Why do auroras appear near magnetic poles?
2. How might life on the earth be affected if the ozone layer were destroyed by the use of certain chemicals?
3. Describe the temperature changes in the four main layers of the atmosphere.
4. What is the magnetosphere? What are its possible causes?
5. Explain why air pressure decreases with altitude. What effect does this decrease have on air temperature?
6. Describe the ionosphere. Explain why it is important in transmitting radio waves.



must be measured with special instruments.

**4.** The magnetosphere is the magnetic field around the earth. Most scientists believe that it is caused by a combination of the earth's rotation and electric currents in the earth's core.

**5.** The layers of air that surround the earth are held close to it by gravity. The layers of air push down on the earth's surface. This is called air pressure. The upper layers of air also push down on the lower layers. Because of this, air pressure is greater near the surface of the earth than it is farther away from it. Decreasing air pressure causes a decrease in temperature. This is because as air pressure decreases air expands, and when air expands, it cools.

**6.** The ionosphere is a sphere within the thermosphere. There are layers of ions in the ionosphere which reflect radio waves back to earth. When reflected by the ionosphere, these radio waves can travel long distances around the earth.

## ADDITIONAL QUESTIONS AND TOPIC SUGGESTIONS

**1.** In what ways does an astronaut's space suit act like the atmosphere? (Like the atmosphere, the spacesuit provides a comfortable temperature and protection from the sun's ultraviolet rays; the suit also provides oxygen and moisture.)

**2.** How does the force of gravity affect the composition and structure of the atmosphere? (The force of gravity is strongest close to the earth; therefore air is most concentrated in the lower portion of the atmosphere. In addition, the heavier gases have tended to stay close to earth, while gases such as hydrogen have escaped into space.)

## ISSUES IN SCIENCE

The following issues can be used as springboards for discussion or given as writing assignments.

**1.** Many scientists are concerned about what they call a "hole" in the ozone layer over Antarctica. They claim that in years to come there will be a dramatic increase in skin cancer because of this problem. No one seems to know exactly why the hole has formed, nor why it is located over Antarctica. Find out more about this issue; then decide whether you think the ozone hole poses a threat to life on Earth.

**2.** The burning of fuels such as gasoline releases carbon dioxide into the atmosphere. As a result, the amount of carbon dioxide in our atmosphere is steadily increasing. Many scientists are worried about this because of what is called the greenhouse effect. Find out what is meant by the greenhouse effect and why scientists are concerned about it. Then offer your opinion as to whether an increase in carbon dioxide poses a threat to the earth's environment.

## Unit Six

# STRUCTURE OF THE EARTH

## ADVENTURES IN SCIENCE: WILLIAM HAXBY MAPS THE INVISIBLE OCEAN FLOOR

### BACKGROUND INFORMATION

Although it may seem incongruous, it was by measuring the height of the ocean's surface that *Seasat* was able to provide information for a detailed map of the ocean floor. This was possible because the pull of gravity on ocean water is affected by various ocean floor features. In fact, scientists have found that the ocean's surface tends to "parallel" the topography of the ocean floor, with water piling up above a large underwater object and sinking above an underwater valley or crevasse.

The instrument used aboard *Seasat* to record differences in the height of the ocean's surface was an altimeter. An altimeter is similar to an echo sounder, except that an echo sounder uses sound waves to measure the depth of the ocean floor, while an altimeter uses electronic pulses to measure the height of the ocean's surface. The satellite sends out a fine stream of electronic pulses straight down to the surface of the water. The altimeter then measures the time it takes for the pulses to reach the ocean's surface and bounce back.

SCIENCE GAZETTE

Adventures in Science

# William Haxby Maps the Invisible Ocean Floor

William Haxby's undersea maps provided a view of the ocean floor never before seen by people. In this photo, parts of Africa and South America are shown in gray. Red dots show locations of underwater earthquakes. Dark blue indicates areas of the ocean that have a greater depth than light-colored areas.

Vast canyons and craggy mountains make the ocean floor as mysterious as the surface of a far-off planet. It may be centuries before the bottom of the world's seas are fully explored. Yet using information from a space satellite, a 33-year-old scientist has created a startling map of the undersea landscape. This map is almost as detailed as if the water had been drained out of the seas and a man had walked over the land, making a map as he went.

That man is William Haxby of the Lamont Geological Observatory in Palisades, New York. Haxby fed readings from the satellite *Seasat* into a computer. Using computer graphics, he produced a three-dimensional map in vivid colors. This map provided a view of the ocean floor never before seen by people. Cracks in the sea bottom, underwater volcanoes, and other features of the ocean floor popped up on Haxby's map. These features provided new evidence about some of the most important earth science theories.

One of these theories, continental drift, suggests that all continents were once part of a single, large landmass. This landmass gradually broke up into the fragments now called continents. And the continents drifted to their present positions. Another theory, called plate tectonics, states that the earth's crust is made up of a number of very large plates. Heat and motion deep within the earth cause the plates to move. The movement of plates triggers earthquakes, thrusts up mountain ranges, and cuts deep ridges.

Haxby's map shows many signs of plate movement along the sea floor. One deep crack under the Indian Ocean may have been made when India drifted away from Antarctica and headed for Asia millions of years ago. Geologists believe a twisting ridge on the ocean floor off the southern tip of Africa was also formed millions of years ago, when Africa, South America, and Antarctica separated. The ridge, concealed under layers of sediment, was detected by Haxby's computer.



## TEACHING STRATEGY: ADVENTURE

### Motivation

Begin by asking students the following questions:

- **Suppose that you wanted to make a map of the ocean floor. How would you go about doing it?** (Accept all answers.)
- **What difficulties might you encounter as you try to make this map?** (Accept all answers.)
- **Suppose that all the water were drained out of the oceans, and you were free to go hiking on the ocean floor. What type of "land" surface do you think you would discover on your journey?** (Accept all answers.)
- **Do you think the task of mapping the ocean floor would be easier now, with the water removed? Why?** (Accept all answers.).

### Content Development

Point out to students that the task of mapping the ocean floor is one that must be accomplished by indirect evidence. Indirect evidence is information obtained about something that cannot be observed or measured directly.

- **Can you think of other tasks in science that must be accomplished by indirect evidence?** (Answers may vary. Some obvious examples include

## Mapping the Ocean Floor

Haxby started his mapping project, which took 18 months, in 1981. His work was based on *Seasat's* measurements of height differences on the sea's surface. Even if there were no waves or wind, Haxby points out that the surface of the ocean would not be perfectly flat. The height of the sea's surface varies by dozens of meters from one place to another. The reason for the variations in the sea's surface is the gravitational pull of structures on the bottom. Structures with large mass, such as undersea mountains, pull on the water with more force than those with less mass, such as canyons. The stronger the pull, the more water is attracted to a place above the structure. "As a result," says Haxby, "water piles up and there is a bump in the sea over a big object." So the sea surface imitates the sea bottom.

Measurements of differences in the sea's surface were recorded on *Seasat* by an instrument called an altimeter. It measured distances between the satellite and the surface of the sea to within a few centimeters. The satellite sent out 1000 electronic pulses a second. The altimeter measured the time it took for the pulses to hit the sea's surface and bounce back. Launched in 1978 for a five-year orbit, *Seasat* became silent after three months due to a short circuit in its electrical system. But the eight billion readings it had radioed back to tracking stations on the earth were enough for Haxby to start his computer work.

Tall and slender, Haxby has been interested in geology since his boyhood, when he was a "rockhound." At the University of Minnesota, he became interested in continental drift and plate tectonics. He eventually did graduate work in geophysics, leading to a doctoral degree from Cornell University. When he first began computer analysis of the satellite information, however, he did not intend to map the entire ocean bottom. All he wanted to do was chart some small areas by matching sea surface heights with the gravitational forces that created them. This would enable him to figure out the mass of objects on the sea floor. His first maps were so detailed, however, that Haxby decided to go further.

William Haxby is an expert on continental drift and plate tectonics. His maps of the sea floor have helped scientists all over the world solve riddles about the earth.

## Colorful Results

Sitting at his computer, often working late into the night, Haxby gradually expanded his map to include all of the ocean floor. He assigned different colors—ranging from blue to pink—to various sea levels. More than 250 different color intensities helped him create his three-dimensional map. Ridges, mountains, trenches, and canyons stand out by color. The detail—Haxby can pinpoint objects as small as 30 kilometers across—and colors reveal structures that are not on other large maps of the sea bottom. In fact, the images produced by Haxby's computer look so real that people looking at the images feel as if they are standing on the bottom of the sea. And around the globe, scientists now use Haxby's maps to help solve riddles about the earth.



## ADDITIONAL QUESTIONS AND TOPIC SUGGESTIONS

**1.** Early measurements of the ocean floor were made in the 1970s by the crew of the British ship *Challenger.* Find out how these measurements were made and some of the information that they provided.
**2.** Scientists believe that the ocean floor is built up by a process called ocean-floor spreading. Use reference sources to find out how this process takes place, and how it fits in with the theory of continental drift.
**3.** One of the most important discoveries made by *Seasat* is the presence of mountain ranges on the ocean floor. Find out where these underwater mountains (called mid-ocean ridges) are located. Then make a map to display the information.

## CRITICAL THINKING QUESTIONS

**1.** What do you think would happen to the level of the ocean's surface if a large area of dense iron ore was present beneath a level ocean floor? Why? (The water level would rise at that place because of the strong gravitational pull on the iron ore.)
**2.** Observe the map on page 542. What do you notice about the locations of underwater earthquakes? (Answers may vary. Guide students to recognize that underwater earthquake locations form a pattern with locations of earthquakes on land. The earthquake locations do not occur randomly, but as part of a long, continuous line.)

determining the structure of the inner earth; determining the structure of the atom; and calculating the distance from Earth to distant stars and galaxies.)

Explain that Haxby used indirect measurements made by *Seasat* to form a detailed map of the ocean floor. This technique is called "satellite mapping." In satellite mapping, a satellite takes readings which are relayed back to earth. These readings are then fed into a computer which "translates" them into a map. Emphasize that maps such as those made by Haxby utilize two types of modern technology—scientific satellites and computers.

### Teacher's Resource Book Reference

After students have read the Science Gazette article, you may want to hand out the reading skills worksheet based on the article in your Teacher's Resource Book.

## Unit Six

# STRUCTURE OF THE EARTH

## ISSUES IN SCIENCE: SHOULD PEOPLE BUILD LAKES?

### BACKGROUND INFORMATION

The earth's supply of fresh water includes surface water and ground water. An important form of surface water is standing water, which includes lakes and ponds.

Lakes form naturally when a mature river meanders or when a river or stream is obstructed in some way or when a glacier carves a hollow in the earth's crust. Large lakes can be extremely beneficial to an area. The shores of a lake can often hold large amounts of water from rains and melting snow. This water is released slowly into the lake, reducing the possibility of flooding. In times of drought, a lake forms a natural reserve water supply. Plants and wildlife tend to flourish in the lush area surrounding a lake.

Artificial lakes are often created as a byproduct of the construction of dams. The dams are built in order to obtain hydroelectric power, which is the using of energy from a flowing river to generate electricity. Sometimes artificial lakes are created deliberately either by damming the flow of a river or by pumping water from underground springs.

When a lake forms naturally, the process is quite gradual and the environment tends to adjust itself. The formation of an artificial lake, however, can be very disruptive to the environment. The most obvious disruption is that a certain portion of land must be flooded in order to make room for the lake. Forests and other forms of vegetation may be arbitrarily eliminated. Wildlife that has made its home in a reasonably dry environment suddenly finds itself in a sink-or-swim situation. The animals that cannot flee or adapt die; other may die later because their food supply in the form of plants or smaller animals has been cut off.

Issues in Science

Lakes affect the land around them. Runoff from a lake feeds surrounding trees and plants. More plant life attracts birds and animals to the area. But artificial lakes, if not well planned, can harm the environment.

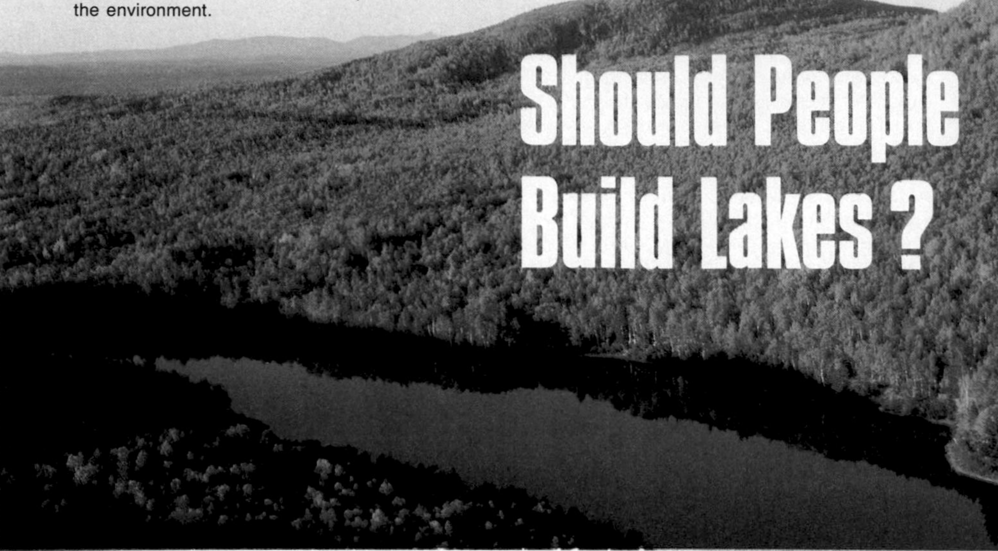
Should People
Build Lakes?

### The Benefits May Be Obvious, But What Are the Drawbacks?

Although it was a tiring trip, I looked forward to seeing my friend Jeff again. But I didn't look forward to leaving the air-conditioned comfort of the bus.

All around me, the landscape was dry, dusty, and sun-baked. There were few trees. I knew the temperature outside must be about 38° C.

Moments later, as we pulled into town, I spotted Jeff's car. And there was Jeff, looking tanned and happy.

I was neither tanned nor happy by the time we reached Jeff's house about four miles out of town. In fact, I was hot, dusty, and sweaty. So, Jeff's suggestion that we go for a swim at a nearby lake was very welcome. The suggestion was also surprising.

"I don't remember any lakes in this area," I remarked.

Jeff replied, "It's brand new. It was built since the last time you were here."

I had always thought that lakes were created only by nature way back when the earth was young. Then I realized that Jeff's artificial lake was probably no more than a muddy pond.

Did I get a pleasant surprise when we arrived at the lake! A great expanse of beau-

### TEACHING STRATEGY: ISSUE

#### Motivation

Obtain from travel brochures or other sources attractive photographs of several large lakes. Display the photographs and ask students the following questions:

- **What do these pictures make you think of?** (Accept all answers.)
- **Do any of these pictures make you think of endangered wildlife, land and water pollution, or other damage to the environment?** (Accept all answers.)

#### Content Development

Continue the discussion by pointing out that when we hear the words "pollution" or "danger to the environment" we tend to think of obviously negative things such as air

tiful, clean water sparkled before us. It was surrounded by green grass, bushes, and young trees. There were sandy beaches at several spots along the winding shore. People were swimming and fishing in the clear water. A few boats bobbed on the lake's surface.

## Lakes Influence Climate

"This is great," I said. "Even the air here feels fresher and cooler."

"Lakes influence local climate and vegetation," Jeff responded. "Water vapor rising from the lake surface makes the air moist and cool. Wait until you come back in a few years. The trees will be taller and the greenery will have spread. It will be a brand new ecosystem."

"But where does the water for this new lake come from?" I asked.

"The water for this lake comes from an underground spring," Jeff replied. "It's part of the groundwater system. There's lots of water down there."

"Do all artificial lakes get their water from underground springs?"

"Oh, no! Most of these lakes are created by damming rivers or streams. A few have their water piped in. If necessary, we could pipe water all the way down from Canada."

Over the following two weeks, I became very fond of Jeff's lake. In a dry, hot, dusty world, it was a delightful change. I was convinced that it would be a good idea to build lakes in all the dry regions of the country.

## Life and Death of Lakes

By the time I got home, I was fascinated by the subject of lakes. But as I learned more and more about lakes, I realized that artificial lakes could have serious drawbacks as well as benefits. The drawbacks are caused by the very nature of lakes.

For example, all lakes eventually die. But artificial lakes tend to have much shorter life spans than natural lakes.

Lakes die in various ways. For one thing, water evaporates from the surface. If the loss of water through evaporation is greater than the incoming flow of water, the level of the lake will go down rapidly.

Artificial lakes tend to have shorter lives than do natural lakes. In each case, old age often turns a lake into a swamp.



Proponents of artificial lakes argue that once a lake is formed, it benefits the environment in the same ways as a natural lake by providing a guard against flooding and drought, by moderating climate, and by supporting wildlife and various kinds of fish. Even this assertion is open to question, however, for artificial lakes tend to be short-lived compared with natural lakes—and a dead lake is likely to turn into a muddy swamp that benefits no one.

## ADDITIONAL QUESTIONS AND TOPIC SUGGESTIONS

**1.** Find out if there are any artificial lakes in your state or in a nearby state. Then find out about the condition of these lakes and how they benefit or perhaps endanger the environment.
**2.** Research newspaper and magazine articles that describe Brazil's project to dam the Amazon River for hydroelectric power. Report your findings to the class, including the effect that this project will have on the environment.
**3.** The National Wildlife Federation has been active in the fight to place environmental safeguards on dam projects. Find out more about the work of this federation and its effectiveness in preserving the natural habitat of wildlife.

pollution from factories, nuclear wastes, exhaust from automobiles, and so on. Sometimes, however, something that is beneficial—such as a beautiful lake—can also be hazardous to the environment, particularly if it has been formed by artificial means.

### Reinforcement

Review with students the factors that can make an artificial lake hazardous to the environment. These include the following: A dying lake may turn into a swamp; the lake may become polluted if water input does not balance evaporation; and taking too much water from the earth to fill a lake may cause the ground to collapse.

Emphasize that environmental damage is also likely to occur when the lake is being built. Point out that a certain amount of land must be flooded in order to make room for the lake. This flooding can disrupt the lives of plants and animals, destroy the lives of plants and animals, destroy forests, or eliminate land that could be used for farming.

### Content Development

Stress that most artificial lakes are formed by damming rivers and streams. Some of these lakes are formed as byproducts when dams are built to obtain hydroelectric power.

• **What is hydroelectric power?** (The harnessing of energy from a flowing river to generate electricity.)

Point out that when a river is dammed in order to obtain hydro-

## CRITICAL THINKING QUESTIONS

**1.** Do you think that possible ground collapse due to overpumping of underground water is a valid argument against building artificial lakes? Why or why not? (Accept all logical answers. One point to consider is that most artificial lakes are not created by pumping ground water but by damming rivers.)

**2.** Why is an artificial lake likely to be more shallow than a natural lake? (Answers may vary. A logical answer is that builders of a lake will not want to spend extra money and energy making the lake any deeper than necessary, nor will they want the difficult task of finding enough water to fill a deep lake.)

**3.** In what ways is wildlife endangered by the building of an artificial lake? (In order to build a lake, a portion of land must be flooded. Animals and plants living on this land will have their natural habitat disrupted. Some plants and animals may die; some animals may flee to drier ground. Food chains will be disrupted, and some species may decrease or die out due to the loss of their food sources.)

## CLASS DEBATE

Challenge students to imagine that they are members of a state environmental protection agency. Have them debate the pros and cons of the proposed construction of a large artificial lake in the state. Encourage students to include in their debate the purpose of the lake, and how its purpose affects the validity of the lake's construction.



Lake Meade in Nevada is an artificial lake that is a recreation spot for many people.

This is a problem especially in hot, dry areas. So unless the flow of water from the underground spring into Jeff's lake is greater than the evaporation from its surface, that lake might disappear in just a few years.

Evaporation is not the only thing that kills a lake. Water flowing into a lake is likely to carry with it large quantities of soil, sand, salts, and minerals. This material settles on the bottom of a lake to form a layer of sediment. It may be joined by leaves, branches, and other decaying vegetation.

Layer by layer, the sediment builds up. And if this process is combined with evaporation, a once clear lake might quickly turn into a muddy swamp.

This is more likely to happen to an artificial lake than to a natural lake. The reason is that artificial lakes are usually shallow and built in areas where water is scarce to begin with. Also, soil and sand are usually more easily washed into a lake in those areas.

A number of lakes that were built for resort communities over the past 30 years have turned into swamps. Others have become bitter or salty as a result of the seepage of fertilizers from surrounding farms and from the fall of acid rain—rain polluted by smoke and gases from industry.

A big drawback of creating lakes that might soon die is that water that could be used for other purposes is wasted. Also, taking great amounts of water from the ground may injure the earth. In Arizona, for example, so much water has been pumped out of the ground that great cracks have opened in the earth's surface. As ground water is removed, empty spaces form deep under ground. These spaces can collapse, creating depressions called sinkholes that can swallow homes and people.

In order for a lake to survive over a long period, it should be deep. Its water output should not exceed its input. And the water in the lake should be able to circulate. Care should be taken to prevent the buildup of soil, sand, salts, acids, fertilizers, and other pollutants. If these things are not considered, it might be better not to build any lakes at all. What do you think?



## ISSUE (continued)

electric power, people must weigh the need for energy versus the possible damage to the environment caused by the dam. Hydroelectric power tends to be cheap and reliable, and in many parts of the country it is a necessary source of energy. Also, some alternative energy sources, such as nuclear power, also damage the environment.

## Skills Development

***Skill: expressing an opinion***

- **Do you think that building a dam is more justified when the purpose is to obtain hydroelectric power rather than to create an artificial lake for recreational purposes? Support your answer.** (Accept all logical answers.)

## Teacher's Resource Book Reference

After students have read the Science Gazette article, you may want to hand out the reading skills worksheet based on the article in your Teacher's Resource Book.

Futures in Science

# the longest winter

The team of Survivors shivered as the bouncing raft made its way across the river. An icy July wind whipped up the water all around the raft. The shore to which the Survivors were heading was covered with a thick coat of new snow. Although it was noon, the day was strangely dark.

Jonah, the Team Leader, peered into the strange darkness ahead. Members of the team lowered long poles into the water and pushed the raft toward the shore.

"There," Jonah said, pointing to a good landing spot. "Head for that flat beach."

Within seconds, the team members were ashore. Soon after, they had built a fire and cooked a meal of canned foods. And then they had settled down for Jonah's afternoon "story" about the past.

## TEACHING STRATEGY: FUTURE

### Motivation

Begin by having students read only the first paragraph of the story.

- **What information given in this paragraph lets you know that there is something strange about this story?** (the fact that there is snow and an icy wind in July; and the fact that it is dark at noon)
- **Can you think of possible explanations for these phenomena?** (Answers may vary. The cold July weather could indicate that the story takes place in the Southern Hemisphere or at the North Pole; it is also possible that this weather is the result of a dramatic climate change or abnormal weather patterns. The darkness at noon could be the result of a solar eclipse or a stormy day; it could also indicate that the story takes place

## Unit Six

## STRUCTURE OF THE EARTH

## FUTURES IN SCIENCE: THE LONGEST WINTER

### BACKGROUND INFORMATION

It is the atmosphere that makes life possible on Earth. The earth's atmosphere provides the gases necessary to sustain living things. The atmosphere also provides the earth with a comfortable temperature, moisture in the form of water vapor and precipitation, and protection against overexposure to the sun's ultraviolet rays.

Any serious damage to the atmosphere would have a profound affect on life as we know it. Such damage could even spell the end of life on Earth.

A tremendous threat to the atmosphere would be a nuclear war. Millions of tons of smoke would rise into the air, creating a smoke screen that would block out much of the sun's radiant energy. The result would be an endless winter with temperatures well below freezing. Many plants and animals would die of the cold. Even if humans managed to survive a nuclear holocaust, it is questionable whether they could survive the climatic changes and the loss of plant and animal life.

The story told in this gazette presents the possibility of such an event. The story is told by the leader of an outdoor club to his followers (the group is on a winter hiking trip in Argentina). The leader, whose name is Jonah, remembers life in the 1980s, when scientists were concerned about the possible consequences of a nuclear holocaust. The story keeps the reader in suspense until the very last moment, for until the time it seems as if the members of the outdoor club are indeed survivors of a nuclear war. By the end of the story, however, it becomes clear that the members of the outdoor club are living at a time in the future when nuclear war is no longer a threat. How this enviable state of international existence was arrived at is left to the reader to ponder.

## ADDITIONAL QUESTIONS AND TOPIC SUGGESTIONS

**1.** Make a diagram to show the effect that a nuclear soot cloud would have on the heating of the earth by the sun.
**2.** Why do you think that soot particles absorb sunlight better than other kinds of particles? (Answers may vary. Probably the black color of the particles is important, since black absorbs more radiant energy than any other color. It is also possible that the chemical makeup of the particles makes them readily absorb sunlight.)
**3.** Find out what is being done by the United States and other leading nations in the field of disarmament. Do you think that a world in which nuclear war is no longer a threat could become a reality in your lifetime? (Accept all answers.)



"Well, my friends," Jonah began, "this day makes me think of a frightening story that began long, long ago. If I remember correctly, the time was winter, 1983. The football season was in full swing. The holidays were approaching. Everything was kind of normal and happy. Of course, people were worried about such things as taxes and war. But that wasn't too unusual for those days. Then something happened that scared a lot of people. A group of some of the finest scientists in the United States made a report.

The scientists—there were 20 of them—had been trying to figure out what might happen if there were a nuclear war. And the report described what they had discovered."

Jonah paused and looked up at the dark sky as if searching for the July sun. But all he could see was an unbroken gray cloud that seemed to stretch forever. Jonah sighed and went on.

"To get an idea of what the scientists concluded, you must first know some basic facts about this planet of ours. Our atmosphere is very special. For one thing, as one of the scientists said, our atmosphere normally 'acts as a window for sunlight, but as a blanket for heat'."

"What does that cause," one of the Survivors asked as she tugged a blanket tightly around her shoulders.

"Normally, it causes the earth to stay warm. You see, the light from the sun is a form of energy. When this energy hits the earth, a new kind of energy is produced. It is called heat. Our atmosphere normally traps a lot of this heat. The heat keeps us warm. It keeps our oceans and lakes from freezing. And it makes plant and animal life possible on the earth. Now what do you think would happen if the 'window for sunlight' had a shade pulled down over it?"

A young Survivor, busily rubbing his hands together to keep them warm, answered. "The light from the sun would not reach the earth. So heat energy would not be produced. And the earth would cool down."

"Right," said Jonah. "And that's exactly what the scientists said could happen if there were a nuclear war. The scientists figured out that such a war would set millions of fires. Tons of black, sooty smoke would rise into the atmosphere. This smoke, they said, would have a mass of more than 100 million tons!

"The particles of soot would rise very high into the atmosphere, the scientists calculated. As high, maybe, as 20 kilometers. There the winds of the atmosphere would spread the soot over the entire earth.

"Now there's something very special about these two things—the soot and the height to which it would rise. Experiments showed that small particles of black soot absorb sunlight better than other kinds of particles. But that's not all. Researchers also found that the higher in the air particles are found, the longer they are going to stay there."

"That means that the soot closes the

## FUTURE (continued)

near the South Pole, where the earth experiences nearly 24 hours of darkness in winter.)

### Content Development

• **What was the cause of the cold July weather and darkness at noon?** (The story takes place in Argentina, which is in the Southern Hemisphere. The Southern Hemisphere experiences winter in July and the hours of daylight are few. Also, weather on this particular day is apparently overcast.)

Display a world map or globe and have students locate Argentina. Point out that the southern part of Argentina is not far from Antarctica.

• **As you were reading the story, did you think that there might be another reason for the weather conditions?** (Answers may vary. Many students may say that they thought a nuclear war had occurred.)

• **Were you surprised by the ending of the story? Why or why not?** (Accept all answers.

Some students may have thought that the survivors were indeed survivors of a nuclear war, and that they were experiencing the changes in the earth's climate that Jonah was describing.)

'windows for sunlight.' And if the soot is high in the air, the 'window' stays closed for a long time," said one of the Survivors.

"Sadly, that's true," Jonah replied in a whisper. "Using the very best computers of the time, the scientists figured out that about 95 percent of the sun's light would not reach the ground. In their report, the scientists said that in some places the brightness at noon 'could be as low as that of a moonlit night'."

"Just like it is now," another Survivor said softly.

Jonah sighed. Telling this story wasn't easy, he thought to himself.

"So what would happen then?" asked another Survivor.

"Again, the computer came up with some answers. And they were chilling. Temperatures over land areas like North America, Europe, and Asia would suddenly drop about 40°C! That means that the average temperature in these places would be about −25°C."

"Why that's way below the freezing point of water!" exclaimed a Survivor.

"I'm afraid so," said Jonah. "And these freezing temperatures would last for months, said the scientists. The scientists called it a 'nuclear winter.' And the winter would cover the whole world."

"But that would kill off all sorts of plants, and animals that live on plants," the Survivor continued. "Why there would be no trees, no grass, no cows and sheep that feed on grass, no corn or wheat, no food for us."

"How about fish in lakes and streams?" another Survivor asked.

"Frozen solid," said Jonah about these waters. "Life in the water would die out."

"But wouldn't living things survive in the oceans and on the coasts of oceans? After all, ocean water holds a lot of heat for a long time. It might keep the land nearby warm enough to grow crops," suggested a Survivor.

"The scientists thought of that too," replied Jonah. "But the very cold air over the land meeting the warm air over the sea would cause terrible storms. Farming in areas struck by constant hurricanes would be impossible."

Jonah stopped speaking. Again he looked up at the sky, searching for a ray of sunshine. But there was none to be seen. The Survivors sat very quietly. Finally, one of them spoke.

"If that had happened, Jonah, the whole world would have looked and felt like this day. Everywhere it would be winter in July—not just here in Argentina and other southern countries. And we would be real survivors instead of members of an outdoor club."

"That's right," said Jonah, who now smiled. "But leaders of our country and of the other countries of the world saw to it that there was no nuclear war. So we still have our trees, and our grass, our forests, our animals . . . and our wilderness to explore. So let's pack up our gear and be on our way. We've got a long hike ahead of us."

## CRITICAL THINKING QUESTIONS

**1.** Why do you think Jonah was telling this story to his followers? (Answers may vary. Probably, he was telling the story to his followers so that they would appreciate living at a time when nuclear war is no longer a threat. He may also have wanted to impress upon them the idea that no nuclear war should ever be allowed to occur in the future.)

**2.** Why is the story called "The Longest Winter"? (because in the event of atmospheric damage due to nuclear war, winter would last forever)

**3.** What major effect of nuclear war is not discussed in this story? (the effect of radiation on living things and the environment)

### Content Development

Emphasize to students that the heating of the earth depends on two processes. First, radiant energy from the sun must strike the earth and be changed into heat. Second, heat radiated back into space from the earth's surface must be trapped by the atmosphere.

- **Why would a soot cloud such as the one described in the story interfere with the heating of the earth?** (It would absorb sunlight before it could reach the earth.)
- **Why would the soot cloud not keep the earth warm, if the particles of soot absorb sunlight?** (The soot particles are too high in the atmosphere to warm the earth. At a height of 20 km, the soot would be in the stratosphere. Gases such as carbon dioxide, which trap heat from the earth's surface, are located nearer to the earth in the troposphere.)

### Teacher's Resource Book Reference

After students have read the Science Gazette article, you may want to hand out the reading skills worksheet based on the article in your Teacher's Resource Book.

# For Further Reading

If you have an interest in a specific area of General Science or simply want to know more about the topics you are studying, one of the following books may open the door to an exciting learning adventure.

**Chapter 1:** Exploring Living Things

National Geographic Society. *Hidden Worlds.* Washington, DC: National Geographic Society.

Smith, N. F. *How Fast Do Your Oysters Grow? Investigate and Discover Through Science Projects.* New York: Messner.

**Chapter 2:** The Nature of Life

Adler, I. *How Life Began.* New York: John Day.

Silver, D. *Life on Earth.* New York: Random House.

**Chapter 3:** Cells

Cobb, V. *Cells: The Basic Structure of Life.* New York: Watts.

Silverstein, A., and V. B. Silverstein. *Cells: Building Blocks of Life.* Englewood Cliffs, NJ: Prentice-Hall.

**Chapter 4:** Tissues, Organs, and Organ Systems

Kelly, P. M. *The Mighty Human Cell.* New York: John Day.

Milne, L., and M. Milne. *The How and Why of Growing.* New York: Atheneum.

**Chapter 5:** Interactions Among Living Things

Carson, R. *The Edge of the Sea.* Boston: Houghton Mifflin.

Mabey, R. *Oak & Company.* New York: Greenwillow.

**Chapter 6:** Classification

Simon, S. *Strange Creatures.* New York: Four Winds/Scholastic.

Venino, S. *Amazing Animal Groups.* Washington, DC: National Geographic Society.

**Chapter 7:** Viruses, Bacteria, and Protists

Asimov, I. *How Did We Find Out About Germs?* New York: Walker.

Curtis, H. *The Marvelous Animals: An Introduction to the Protozoa.* New York: Natural History Press.

**Chapter 8:** Nonvascular Plants and Plantlike Organisms

Johnson, S. A. *Mushrooms.* Minneapolis: Lerner.

Shuttleworth, F. S., and H. S. Zim. *Non-Flowering Plants.* New York: Golden Press.

**Chapter 9:** Vascular Plants

Bauman, R. P. *Plants as Pets.* New York: Dodd, Mead.

Varnard, P. *Don't Tickle the Elephant Tree: Sensitive Plants.* New York: Messner.

**Chapter 10:** Animals: Invertebrates

Dallinger, J. *Grasshoppers.* Minneapolis: Lerner.

Johnson, S. A. *Snails.* Minneapolis: Lerner.

**Chapter 11:** Animals: Vertebrates

Rowland-Entwistle, T. *Illustrated Facts and Records Book of Animals.* New York: Arco.

Sadoway, M. *Owls: Hunters of the Night.* Minneapolis: Lerner.

**Chapter 12:** Mammals

Boitani, L. *Simon and Schuster's Guide to Mammals.* New York: Simon & Schuster.

Patent, D. *Whales: Giants of the Deep.* New York: Holiday House.

**Chapter 13:** General Properties of Matter

Asimov, I. *A Short History of Chemistry.* Garden City, NY: Anchor Books.

Lapp, R. E., and the editors of Time-Life Books. *Matter.* New York: Time-Life Books, Inc.

**Chapter 14:** Physical and Chemical Changes

Adler, I., *The Wonders of Physics: An Introduction to the Physical World.* New York: Golden Press.

Cobb, V. *Gobs of Goo.* New York: Lippincott.

**Chapter 15:** Atoms and Molecules

Asimov, I. *How Did We Find Out About Atoms?* New York: Walker.

Frisch, O. J. *Working With Atoms.* New York: Basic Books.

**Chapter 16:** Elements and Compounds

Drummond, A. H. *Molecules in the Service of Man.* Philadelphia: Lippincott.

Hyde, M. O. *Molecules Today and Tomorrow.* New York: McGraw-Hill.

**Chapter 17:** Mixtures and Solutions

Dickinson, E. *Colloids in Foods.* New York: Elsevier.

Stone, A. H. *The Chemistry of a Lemon.* Englewood Cliffs, NJ: Prentice-Hall.

**Chapter 18:** Minerals

Gilbert, M. *The Science Hobby Book of Rocks and Minerals.* Minneapolis: Lerner.

O'Neil, P., and the editors of Time-Life Books. *Planet Earth: Gemstones.* Alexandria, VA: Time-Life Books, Inc.

**Chapter 19:** Rocks

Gallant, R. A., and C. J. Schubeth. *Discovering Rocks and Minerals: A Nature and Science Guide to Their Collection and Identification.* New York: Natural History Press.

Shepherd, W. *Wealth from the Ground.* New York: John Day.

**Chapter 20:** Soils

Graham, A., and F. Graham. *The Changing Desert.* New York: Scribner's.

Shimer, J. A. *The Changing Earth: An Introduction to Geology.* New York: Barnes & Noble.

**Chapter 21:** Internal Structure of the Earth

Adler, I., and R. Adler. *The Earth's Crust.* New York: John Day.

Matthews, W. H. *Introducing the Earth.* New York: Dodd, Mead.

**Chapter 22:** Surface Features of the Earth

Bauer, E. *Wonders of the Earth.* New York: Watts.

Jacobs, L., Jr. *The Shapes of Our Land.* New York: Putnam.

**Chapter 23:** Structure of the Atmosphere

Chandler, T. J. *The Air Around Us: Man Looks at His Atmosphere.* New York: Natural History Press.

Weiss, M. *Storms: From the Inside Out.* New York: Julian Messner.

# Appendix A THE METRIC SYSTEM

The metric system of measurement is used by scientists throughout the world. It is based on units of ten. Each unit is ten times larger or ten times smaller than the next unit. The most commonly used units of the metric system are given below. After you have finished reading about the metric system, try to put it to use. How tall are you in metrics? What is your mass? What is your body temperature in degrees Celsius?

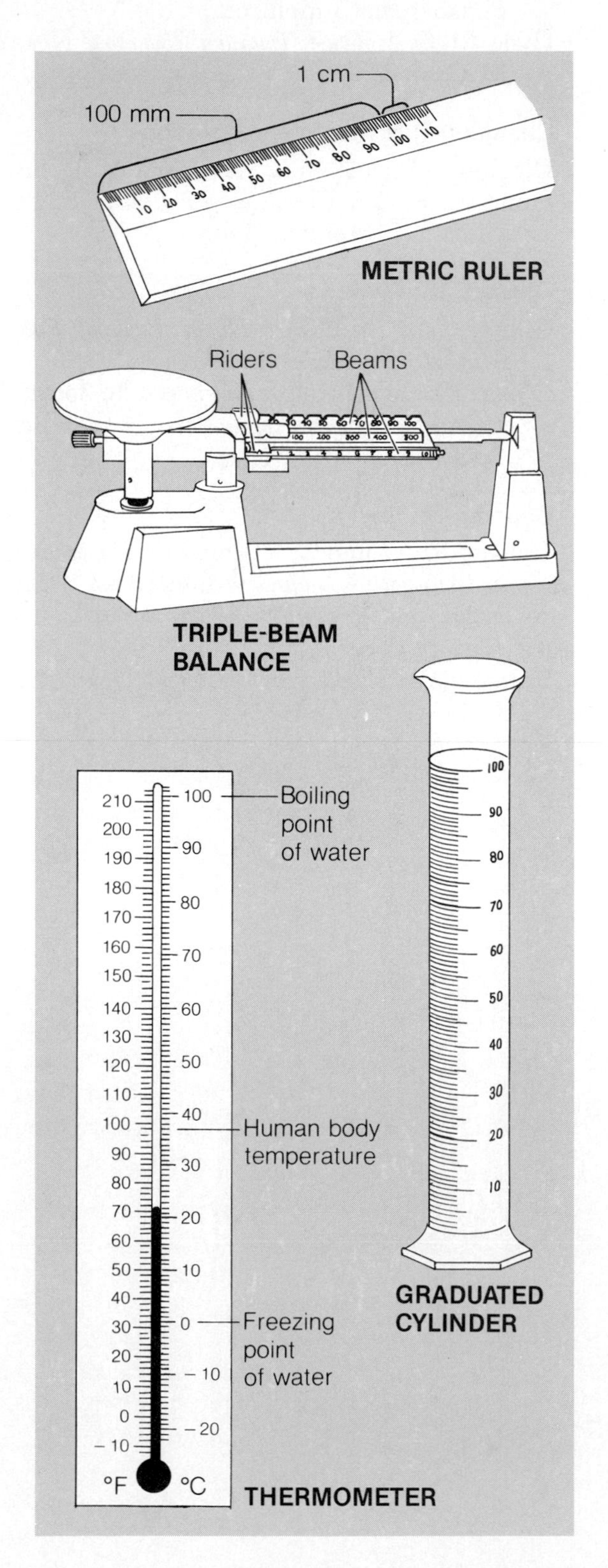


## COMMONLY USED METRIC UNITS

**Length** The distance from one point to another

| | |
|---|---|
| meter (m) | 1 meter = 1000 millimeters (mm) |
| (a meter is slightly longer than a yard) | 1 meter = 100 centimeters (cm) |
| | 1000 meters = 1 kilometer (km) |

**Volume** The amount of space an object takes up

| | |
|---|---|
| liter (L) | 1 liter = 1000 milliliters (mL) |
| (a liter is slightly larger than a quart) | |

**Mass** The amount of matter in an object

| | |
|---|---|
| gram (g) | 1000 grams = 1 kilogram (kg) |
| (a gram has a mass equal to about one paper clip) | |

**Temperature** The measure of hotness or coldness

| | |
|---|---|
| degrees Celsius (°C) | 0 °C = freezing point of water |
| | 100 °C = boiling point of water |

## METRIC–ENGLISH EQUIVALENTS

2.54 centimeters (cm) = 1 inch (in.)
1 meter (m) = 39.37 inches (in.)
1 kilometer (km) = 0.62 miles (mi)
1 liter (L) = 1.06 quarts (qt)
250 milliliters (mL) = 1 cup (c)
1 kilogram (kg) = 2.2 pounds (lb)
28.3 grams (g) = 1 ounce (oz)
°C = 5/9 x (°F − 32)

# Appendix B THE LABORATORY BALANCE

The laboratory balance is an important tool in scientific investigations. You can use the balance to determine the mass of materials that you study or experiment with in the laboratory.

Different kinds of balances are used in the laboratory. One kind of balance is the double-pan balance. Another kind of balance is the triple-beam balance. The balance that you may use in your science class is probably similar to one of the balances illustrated in this appendix. To use the balance properly, you should learn the name, function, and location of each part of the balance you are using.

## THE DOUBLE-PAN BALANCE

The double-pan balance shown in this appendix has two beams. Some double-pan balances have only one beam. The beams are calibrated, or marked, in grams. The upper beam is divided into 10 major units of one gram each. Each of these units is further divided into units of 1/10 of a gram. The lower beam is divided into 20 units, and each unit is equal to 10 grams. The lower beam can be used to find the masses of objects up to 200 grams. Each beam has a rider that is moved to the right along the beam. The rider indicates the grams used to balance the object in the left pan.

Before you begin using the balance, you should be sure that both riders are pointing to zero grams on their beams and that the pans are empty. The balance should be on a flat, level surface. The pointer should be at the zero point. If your pointer does not read zero, slowly turn the adjustment knob until it does.

The following procedure can be used to find the mass of an object with a double-pan balance:

1. Place the object whose mass is to be determined on the left pan.
2. Move the rider on the lower beam to the 10-gram notch.
3. If the pointer moves to the right of the zero point on the scale, the object has a mass less than 10 grams. Return the rider on the lower beam to zero. Slowly move the rider on the upper beam until the pointer is at zero. The reading on the beam is the mass of the object.
4. If the pointer did not move to the right of the zero, move the rider on the lower beam notch by notch until it does. Move the rider back one notch. Then move the rider on the upper beam until the pointer is at zero. The sum of the readings on both beams is the mass of the object.
5. If the two riders are moved completely to the right side of the beams and the pointer remains to the left of the zero point, the object has a mass greater than the total mass that the balance can measure.

The total mass that most double-pan balances can measure is 210 grams. If an object has a mass greater than 210 grams, return the riders to zero.

## PARTS OF A BALANCE AND THEIR FUNCTIONS

### DOUBLE-PAN BALANCE

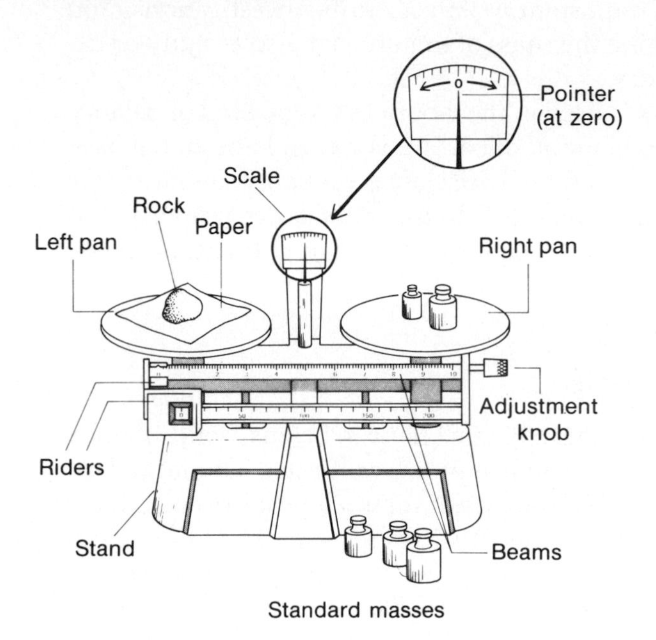


**Scale** Graduated instrument along which the pointer moves to show if the balance is balanced

**Pointer** Marker that indicates on the scale if the balance is balanced

**Zero point** Center line of the scale to which the pointer moves when the balance is balanced

**Adjustment knob** Knob used to balance the empty balance

**Left pan** Platform on which an object whose mass is to be determined is placed

**Right pan** Platform on which standard masses are placed

**Beams** Scales calibrated in grams

**Riders** Moveable markers that indicate the number of grams needed to balance an object

**Stand** Support for the balance

The following procedure can be used to find the mass of an object greater than 210 grams:

1. Place the standard masses on the right pan one at a time, starting with the largest, until the pointer remains to the right of the zero point.

2. Remove one of the large standard masses and replace it with a smaller one. Continue replacing the standard masses with smaller ones until the pointer remains to the left of the zero point. When the pointer remains to the left of the zero point, the mass of the object on the left pan is greater than the total mass of the standard masses on the right pan.

3. Move the rider on the lower beam and then the rider on the upper beam until the pointer stops at the zero point on the scale. The mass of the object is equal to the sum of the readings on the beams plus the mass of the standard masses.

## THE TRIPLE-BEAM BALANCE

The triple-beam balance is a single-pan balance with three beams calibrated in grams. The back, or 100-gram, beam is divided into 10 units of 10 grams each. The middle, or 500-gram, beam is divided into 5 units of 100 grams each. The front, or 10-gram, beam is divided into 10 major units of 1 gram each. Each of these units is further divided into units of 1/10 of a gram.

## TRIPLE-BEAM BALANCE

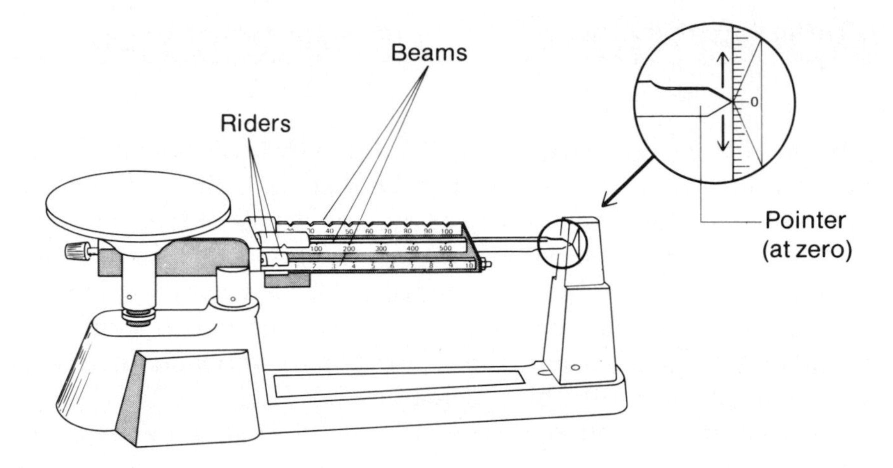


The following procedure can be used to find the mass of an object with a triple-beam balance:

1. Place the object whose mass is to be determined on the pan.
2. Move the rider on the middle beam notch by notch until the horizontal pointer drops below zero. Move the rider back one notch.
3. Move the rider on the front beam notch by notch until the pointer again drops below zero. Move the rider back one notch.
4. Slowly slide the rider along the back beam until the pointer stops at the zero point.
5. The mass of the object is equal to the sum of the readings on the three beams.

# Appendix C THE MICROSCOPE

The microscope is an essential tool in the study of life science. It enables you to see things that are too small to be seen with the unaided eye. It also allows you to look more closely at the fine details of larger things.

The microscope you will use in your science class is probably similar to the one illustrated on the following page. This is a compound microscope. It is called compound because it has more than one lens. A simple microscope would only contain one lens. The lenses of the compound microscope are the parts that magnify the object being viewed.

Typically, a compound microscope has one lens in the eyepiece, the part you look through. The eyepiece lens usually has a magnification power of 10 ×. That is, if you were to look through the eyepiece alone, the object you were viewing would appear 10 times larger than it is.

The compound microscope may contain one or two other lenses. These two lenses are called the low- and high-power objective lenses. The low-power objective lens usually has a magnification of 10 ×. The high-power objective lens usually has a magnification of 40 ×. To figure out what the total magnification of your microscope is when using the eyepiece and an objective lens, multiply the powers of the lenses you are using. For example, eyepiece magnification (10 ×) multiplied by low-power objective lens magnification (10 ×) = 100 × total magnification. What is the total magnification of your microscope using the eyepiece and the high-power objective lens?

To use the microscope properly, it is important to learn the name of each part, its function, and its location on your microscope. Keep the following procedures in mind when using the microscope:

1. Always carry the microscope with both hands. One hand should grasp the arm, and the other should support the base.
2. Place the microscope on the table with the arm toward you. The stage should be facing a light source.
3. Raise the body tube by turning the coarse adjustment knob.
4. Revolve the nosepiece so that the low-power objective lens (10 ×) is directly in line with the body tube. Click it into place. The low-power lens should be directly over the opening in the stage.
5. While looking through the eyepiece, adjust the diaphragm and the mirror so that the greatest amount of light is coming through the opening in the stage.
6. Place the slide to be viewed on the stage. Center the specimen to be viewed over the hole in the stage. Use the stage clips to hold the slide in position.
7. Look at the microscope from the side rather than through the eyepiece. In this way, you can watch as you use the coarse adjustment

## MICROSCOPE PARTS AND THEIR FUNCTION

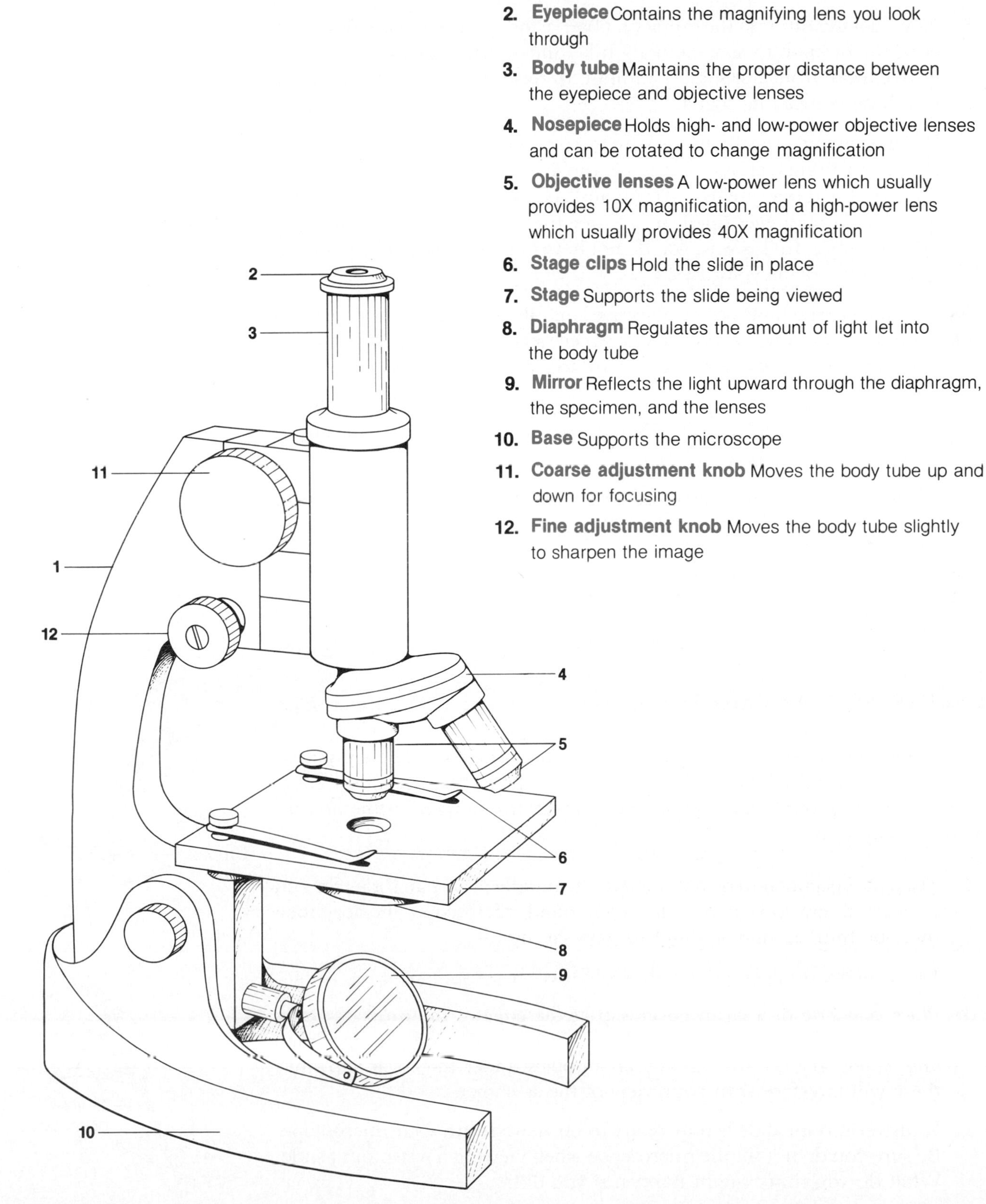


1. **Arm** Supports the body tube
2. **Eyepiece** Contains the magnifying lens you look through
3. **Body tube** Maintains the proper distance between the eyepiece and objective lenses
4. **Nosepiece** Holds high- and low-power objective lenses and can be rotated to change magnification
5. **Objective lenses** A low-power lens which usually provides 10X magnification, and a high-power lens which usually provides 40X magnification
6. **Stage clips** Hold the slide in place
7. **Stage** Supports the slide being viewed
8. **Diaphragm** Regulates the amount of light let into the body tube
9. **Mirror** Reflects the light upward through the diaphragm, the specimen, and the lenses
10. **Base** Supports the microscope
11. **Coarse adjustment knob** Moves the body tube up and down for focusing
12. **Fine adjustment knob** Moves the body tube slightly to sharpen the image

knob to lower the body tube until the low-power objective *almost* touches the slide. Do this slowly so you do not break the slide or damage the lens.

8. Now, looking through the eyepiece, observe the specimen. Use the coarse adjustment knob to *raise* the body tube, thus raising the low-power objective away from the slide. Continue to raise the body tube until the specimen comes into focus.

9. When viewing a specimen, be sure to keep both eyes open. Though this may seem strange at first, it is really much easier on your eyes. Keeping one eye closed may create a strain, and you might get a headache. Also, if you keep both eyes open, it is easier to draw diagrams of what you are observing. In this way, you do not have to turn your head away from the microscope as you draw.

10. To switch to the high-power objective lens (40 ×), look at the microscope from the side. Now, revolve the nosepiece so that the high-power objective lens clicks into place. Make sure the lens does not hit the slide.

11. Looking through the eyepiece, use only the fine adjustment knob to bring the specimen into focus. Why should you not use the coarse adjustment knob with the high-power objective?

12. Clean the microscope stage and lens when you are finished. To clean the lenses, use lens paper only. Other types of paper may scratch the lenses.

## PREPARING A WET-MOUNT SLIDE

Most specimens to be observed with the compound microscope are placed on a slide in a liquid solution. This preparation is know as a wet-mount slide. To make a wet-mount slide, follow these directions along with the accompanying diagrams.

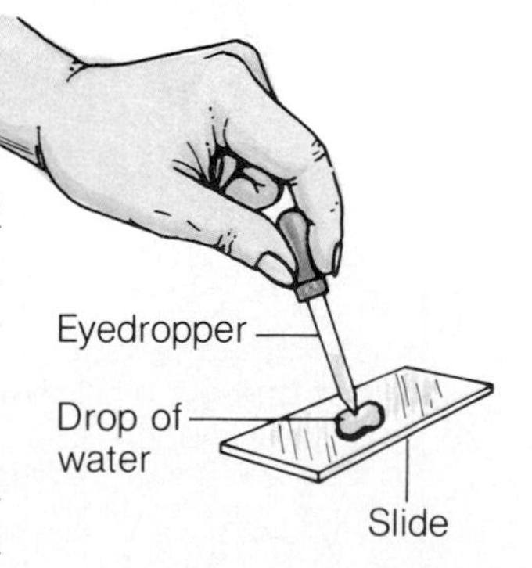


1. Place the specimen to be observed in the middle of a clean slide. For you to observe any specimen with a compound microscope, the specimen must be thin enough for light to pass through it.

2. Using an eye dropper, place a drop of water on the specimen.

3. Place one side of a clean cover slip at the edge of the drop of water. Using a needle or probe, slowly lower the cover slip over the specimen and water. Try not to trap any air bubbles under the cover slip since these will interfere with your view of the specimen.

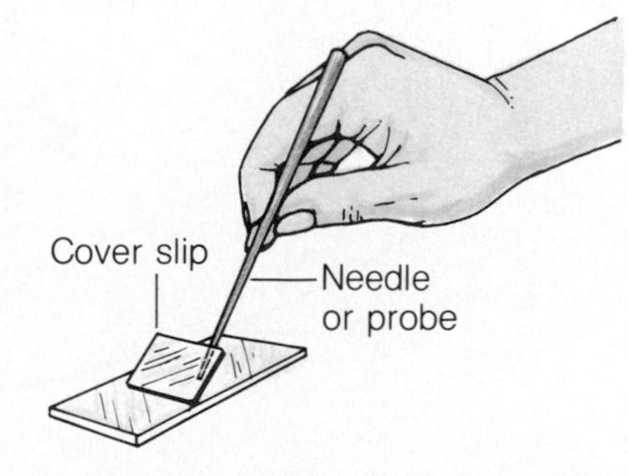


4. Your wet-mount slide is now ready to be viewed with your microscope. Be sure you do not tilt the microscope when viewing a wet-mount slide. What do you think might happen if you did?

# Glossary

**air bladder:** tiny grape-shaped structure that acts like an inflatable life-preserver in brown algae
**air pressure:** pressing force due to the weight of air
**alga** (AL-guh; plural: algae, AL-gee): member of the largest group of nonvascular plants
**alloy:** solid metal dissolved in another solid metal
**amino acid:** building block of protein
**amorphous** (uh-MOR-fuhs) **solid:** solid that does not keep a definite shape
**anal pore:** tiny opening through which a paramecium eliminates waste
**angiosperm** (AN-jee-oh-sperm): type of seed plant whose seeds are covered by a protective wall
**annual ring:** one year's growth of xylem cells
**antibiotic:** substance produced by helpful bacteria that destroys or weakens disease-causing bacteria
**asexual reproduction:** formation of an organism from a single parent
**astronomy:** study of planets, stars, and other objects beyond the earth
**atom:** tiny particle of matter consisting of a nucleus that contains protons and neutrons and an electron cloud that contains electrons
**atomic mass:** average of the masses of all the existing isotopes of an element
**atomic mass unit** (amu): unit used to measure the masses of subatomic particles; a proton has a mass of one amu
**atomic number:** number of protons in the nucleus of an atom; number that identifies the kind of atom
**aurora** (aw-RAW-ruh): bands of light in the upper atmosphere caused by electrically charged particles
**autotroph** (AWT-uh-trohf): organism that makes its own food from simple substances

**bacillus** (buh-SIL-uhs; plural: bacilli): rod-shaped bacterium
**bacteriophage** (bak-TEE-ree-uh-fayj): virus that infects bacterial cells
**bacterium** (plural: bacteria): unicellular microorganism
**binary fission:** reproductive process in which a cell divides into two cells
**binomial nomenclature** (bigh-NOH-mee-uhl NOH-muhn-klay-cher): naming system in which organisms are given two names: a genus and a species
**bioluminescence** (bigh-oh-loo-muh-NE-suhns): firefly-like glow produced by a kind of fire algae
**boiling:** process in which particles of a liquid change to gas, travel to the surface of the liquid, and then into the air
**boiling point:** temperature at which a liquid boils
**botany:** study of plants
**Boyle's Law:** law stating that the volume of a fixed amount of gas varies indirectly with the presure of the gas
**budding:** reproductive process in yeast, in which a new yeast cell is formed from a tiny bud

**cambium** (KAM-bee-uhm): growth tissue of the stem where xylem and phloem cells are produced
**canine** (KAY-nighn): pointed tooth next to the incisor
**cap:** fruiting body of a mushroom
**carbohydrate:** organic compound made up of the elements carbon, hydrogen, and oxygen, and a main source of energy for living things
**carnivore:** (KAR-ni-vor): flesh-eating animal
**cartilage:** flexible tissue that gives support and shape to body parts
**catalyst:** substance that speeds up or slows down chemical reactions, but is not itself changed by the reaction
**cell:** basic unit of structure and function in a living thing; building block of life
**cell membrane:** thin, flexible envelope of protoplasm that forms the outer covering of an animal cell and that is inside the cell wall of a plant cell; controls movement of materials into and out of the cell
**cell wall:** outermost boundary of plant and bacterial cells that is made of cellulose
**Celsius:** temperature scale used in the metric system at which water freezes at 0° and boils at 100°
**Charles's Law:** law stating that the volume of a fixed amount of gas varies directly with the temperature of the gas
**chemical change:** change a substance undergoes when it turns into another substance
**chemical equation:** description of a chemical reaction using symbols and formulas
**chemical formula:** combination of chemical symbols used as a shorthand for the names of chemical substances
**chemical property:** property that describes how a substance changes into another new substance

**chemical reaction:** process by which a substance is changed into another substance through a rearrangement or new combination of its atoms
**chemical rock:** sedimentary rock formed from large amounts of minerals when a body of water dries up
**chemical symbol:** shorthand way of representing the elements; usually consists of one or two letters, the first of which is always capitalized, but the second is never capitalized
**chemistry:** study of what substances are made of and how they change and combine
**chlorophyll:** green substance, needed for photosynthesis, found in green plant cells
**chloroplast:** large, irregularly shaped structure that contains the green pigment chlorophyll; food-making site in green plants
**chromosome:** thick, rodlike object found in the nucleus that directs the activities of the cell and passes on the traits of the cell to new cells
**cilium** (SIHL-ee-uhm; plural: cilia): small, hairlike projection on the outside of a cell
**class:** classification group between phylum and order
**clastic rock:** sedimentary rocks made up of rock fragments mixed with sand, clay, and mud
**cleavage:** breakage of a mineral along definite lines or smooth, flat surfaces
**coastal plains:** low, flat areas that average about 150 meters above sea level
**coccus** (KAHK-suhs; plural, cocci): sphere-shaped bacterium
**coefficient** (koh-uh-FI-shuhnt): number placed in front of a chemical formula in a chemical equation so that the equation is balanced
**coldblooded:** having a body temperature that can change somewhat with changes in the temperature of the environment
**colloid** (KAHL-oyd): homogeneous mixture in which the particles are mixed together but not dissolved
**commensalism:** symbiotic relationship in which one organism benefits and the other is neither helped nor harmed
**community:** the living part of an ecosystem; all the different organisms that live together in an area
**competition:** struggle among living things to get the proper amount of food, water, and energy from the environment
**compound:** substance made up of molecules that contain more than one kind of atom; two or more elements chemically combined
**compound light microscope:** microscope containing more than one lens that uses light to make an object look larger
**condensation** (kahn-den-SAY-shun): process in which a gas loses heat energy and changes into a liquid
**conjugation** (kahn-joo-GAY-shun): type of sexual reproduction in which hereditary material is exchanged
**consumer:** organism that feeds directly or indirectly on producers; heterotroph
**continental glacier:** very thick ice sheet that covers most of both polar regions
**contour feather:** largest bird feather needed for flight
**control:** experiment run exactly the same way as the experiment with the variable, but the variable is left out
**core:** center region of the earth
**cotyledon** (kaht-uh-LEE-duhn): leaflike structure of an embryo plant that stores food
**crop:** saclike organ that stores food in an earthworm
**crust:** thin outer layer on the earth's surface made up mainly of solid rocks
**crystal** (KRIS-tuhl): regular, repeating pattern of particles in a solid
**crystalline solid:** solid made up of crystals
**cytoplasm:** jellylike substance outside the nucleus

**data:** recorded observations and measurements
**decomposer:** organism that breaks down dead plants and animals into simpler substances
**density:** mass per unit volume, or how much mass is contained in a given volume of an object
**diffusion:** process by which food, oxygen, water, and other materials enter and leave a cell through openings in the cell membrane
**digestion:** process by which food is broken down into simpler substances
**division of labor:** division of the work that keeps an organism alive among the different parts of its body
**DNA:** (deoxyribonucleic acid) nucleic acid that stores the information needed to build proteins and carries genetic information about an organism
**down feather:** short, fuzzy bird feather that acts as insulation
**ductility** (duhk-TIHL-uh-tee): ability of a substance to be pulled into thin strands without breaking

**ecology:** study of the relationships and interactions of living things with one another and with their environment
**ecosystem:** group of organisms in an area that interact with one another and with their nonliving environment
**electron:** negatively charged particle that moves around the nucleus of an atom in a region called the electron cloud
**electron cloud:** space in which electrons are most likely to be found
**electron microscope:** microscope that uses a beam of electrons to magnify the image of an object
**element** (EL-uh-ment): pure substance made up of only one kind of atom that cannot be broken down into simpler substances by ordinary means
**endoplasmic reticulum** (en-doh-PLAZ-mik ri-TIK-yuh-luhm): maze of clear tubular passageways that leads out from the nuclear membrane; involved in the manufacture and transport of proteins
**endospore:** oval-shaped structure that protects a bacterium
**energy level:** most likely location in the electron cloud in which an electron can be found
**environment:** all the living and nonliving things with which an organism may interact
**enzyme:** special type of protein that regulates chemical activities within the body
**epidermis** (e-puh-DER-mis): outer protective layer of the leaf
**evaporation** (ee-va-puh-RA-shun): process in which a liquid changes into a gas by absorbing heat energy

**excretion:** process of getting rid of waste materials
**exoskeleton:** rigid, outer covering of an organism
**extrusive** (ek-STROO-sive) **rock:** igneous rock formed from melted rock or lava that cools and hardens at or near the earth's surface
**eyespot:** light-sensitive organ in lower organisms

**family:** classification group between order and genus
**fat:** energy-rich organic compound made up of carbon, hydrogen, and oxygen, that is solid at room temperature
**fermentation:** energy-releasing process in which sugars and starches are changed into alcohol and carbon dioxide
**fertilization:** joining of egg and sperm
**flagellum** (fla-JEL-um; plural, flagella): whiplike structure that propels some one-celled organisms
**flammability** (flam-uh-BIL-uh-tee): ability to burn
**flower:** structure containing the reproductive organs of the angiosperm
**focus:** underground center of an earthquake
**food chain:** illustration of how groups of organisms within an ecosystem get their food and energy
**food web:** all the food chains in an ecosystem
**fracture:** mineral breakage that forms an irregular surface that may be rough or jagged
**freezing:** change of a liquid to a solid
**freezing point:** temperature at which a liquid changes to a solid
**frond:** leaf of a fern plant
**fruit:** ripened ovary of an angiosperm
**fruiting body:** in a fungus, the spore-containing structure
**fungus** (FUHNG-guhs; plural: fungi): group of simple organisms that have no chlorophyll and must get food from other organisms

**gas:** phase in which matter has no definite shape or volume
**gem:** rare, very valuable, and beautiful mineral
**genus** (plural: genera): group of organisms that are closely related; classification group between family and species
**germination** (jer-muh-NA-shuhn): early growth stage of an embryo plant; also called sprouting
**gill:** structure through which fish and other aquatic animals breathe; in a mushroom, the spore-producing structure
**gizzard:** in an earthworm, the organ that grinds up food
**glucose:** simple sugar into which all carbohydrates are broken down in the body to produce energy
**gravity:** force of attraction between objects
**ground water:** water that has soaked into the ground
**guard cell:** sausage-shaped cell that regulates the opening and closing of stomata
**gullet:** funnel-shaped structure in the paramecium extending from the oral groove to the food vacuole
**gymnosperm** (JIM-nuh-sperm): type of seed plant whose seeds are not covered by a protective wall

**habitat:** place in which an organism lives
**hardness:** ability of a mineral to resist being scratched
**herbaceous** (her-BAY-shuhs) **stem:** green, soft stem
**herbivore** (HER-bi-vor): animal that eats only plants
**heterogeneous** (het-uhr-uh-JEEN-ee-uhs) **mixture:** mixture in which no two parts are identical
**heterotroph** (HET-uhr-u-trohf): organism that is not able to make its own food and thus feeds on other organisms
**hibernation:** winter sleep during which all body activities slow down
**homeostasis** (ho-mee-o-STAY-sis): ability of an organism to keep conditions inside its body the same even though conditions in its external environment change
**homogeneous** (ho-muh-JEEN-ee-uhs) mixture: mixture in which different parts seem to be identical
**host:** organism in which a parasite lives
**humus:** dark, rich soil made of decayed plant and animal matter combined with rocky soil particles
**hypha** (HIGH-fuh; plural: hyphae, HIGH-fee): threadlike structure in a fungus that produces digestive enzymes
**hypothesis** (high-PAH-thuh-sis): suggested solution to a scientific problem

**iceberg:** mass of free-floating ice that has broken off from the edges of the Arctic or Antarctic ice sheets
**igneous** (IG-nee-us) **rock:** rock formed from cooled and hardened magma or lava
**incisor** (in-SIGH-zer): one of the four front teeth used for biting
**inertia** (in-ER-shuh): property of a mass to resist changes in motion
**ingestion:** eating
**inland plains:** low, flat areas that average about 450 meters above sea level and are confined to the interior or central part of a country or region
**inorganic:** composed of material that is not and never was living
**inorganic compound:** compound that does not usually contain the element carbon
**insoluble:** cannot be dissolved in water
**intrusive** (in-TROO-siv) **rock:** rock that forms from melted rock or magma that cools and hardens deep below the earth's surface
**invertebrate:** animal without a backbone
**ion:** electrically charged particles
**ionosphere** (i-AH-nuh-sfeer): lower part of the thermosphere that contains ions
**isotope** (IGH-suh-tohp): atom of a substance that has the same number of protons but a different number of neutrons as another atom of the same substance

**jet stream:** narrow, fast-moving current of air that forms along the troposphere-stratosphere boundary

**kilogram:** basic unit of mass in the metric system
**kingdom:** largest classification grouping

**larva:** stage of insect that emerges from the egg
**lava:** magma that has reached the earth's surface
**leaching:** process in which rain washes minerals out of the top soil

**lens:** curved pieces of glass that bends light rays as they pass through it
**lichen** (LIGH-kuhn): organism made up of a fungus and an alga that live together
**life span:** maximum length of time a particular organism can be expected to live
**liquid:** phase in which matter has no definite shape but does have a definite volume
**liter:** basic unit of volume in the metric system
**liverwort:** nonvascular, green leaflike plant; similar to moss but smaller
**luster:** way a mineral reflects light
**lysosome:** small, round structure involved with the digestive activities of a cell

**magma:** hot, molten rock deep inside the earth
**magnetosphere** (mag-NEE-toh-sfeer): magnetic field of the earth
**malleability** (mal-ee-uh-BIHL-uh-tee): ability of a substance to be hammered into thin sheets without breaking
**mammary gland:** milk-producing gland in mammals
**mantle:** part of mollusk that produces material that makes up the hard shell; layer that extends from the edge of the liquid outer core to where the earth's crust begins
**marsupial** (mahr-SOO-ip-uhl): mammal that has a pouch
**mass:** amount of matter in an object
**mass number:** sum of the protons and the neutrons in the nucleus of an atom
**matter:** what all materials are made of
**meiosis** (migh-OH-sis): form of cell division that halves the number of chromosomes in a male and female sex cell as they form
**melting:** change of a solid to a liquid
**melting point:** temperature at which a solid changes to a liquid
**mesosphere** (MES-uh-sfeer): coldest layer of the atmosphere
**metabolism** (muh-TA-buh-li-zuhm): all chemical activities in an organims essential to life
**metal:** element that is a good conductor of heat and electricity, is shiny, has a high melting point, is ductile and malleable, and forms positive ions
**metamorphic** (met-uh-MAWR-fik) **rock:** rock changed by heat, pressure, or chemical action
**metamorphosis** (met-uh-MAWR-fuh-sis): change in appearance due to development
**meteor** (MEE-tee-er): bright trail or streak of light that results when a meteoroid burns up in the mesosphere
**meteoroid** (MEE-tu-uh-roid): chunk of rocklike matter from outer space
**meter:** basic unit of length in the metric system
**metric system:** universal system of measurement
**microbiology:** science of microorganisms
**microscope:** instrument that produces an enlarged image of an object
**Mid-Atlantic Rift Valley:** great crack, or rift, and the area on either side of it in the earth's crust that runs down the middle of a massive underwater mountain range in the Atlantic Ocean
**Mid-Ocean Ridge:** system of rift valleys and mountain ranges that extends for 74,000 kilometers under all the world's oceans
**migrate:** to move to a new environment during the course of a year
**mineral:** natural substance that forms in the earth
**mitochondrion** (plural: mitochondria): rod-shaped structure that is referred to as one of the powerhouses of a cell
**mitosis** (migh-TOH-sis): duplication and division of the nucleus and of the chromosomes during cell reproduction
**mixture:** two or more pure substances that are mixed but not chemically combined
**Moho:** boundary between earth's crust and mantle
**Mohs** (mohz) **hardness scale:** scale used to determine the hardness of a mineral
**molecule** (MAH-luh-kyool): smallest particle of a substance that has all the properties of that substance; made up to two or more atoms that are chemically bonded
**monotreme** (MAHN-uh-treem): egg-laying mammal
**moss:** nonvascular plant that contains chlorophyll
**mountain belt:** group of mountain ranges
**mountain landscape:** area of tall, jagged peaks, long slopes, and high, narrow ridges
**mountain range:** group of mountains
**multicellular:** having many cells
**mutualism:** symbiotic relationship in which both organisms benefit

**nematocyst** (NEM-uh-toh-sist): stinging cell around the mouth of a coelenterate
**neutron:** electrically neutral particle found in the nucleus of an atom
**niche:** everything an organism does and everything an organism needs within it habitat
**nonmetal:** element that is a poor conductor of heat and electricity, has a dull surface, low melting point, is brittle, breaks easily, and forms negative ions
**nonvascular plant:** plant that does not have transportation tubes to carry water and food
**nuclear membrane:** thin membrane that separates the nucleus from the protoplasm of the cell
**nucleic acid:** large organic compound that stores information that helps the body make the proteins it needs
**nucleolus** (noo-KLEE-uh-luhs): cell structure located in the nucleus and made up of RNA and protein; may play important role in making proteins for the cell
**nucleus** (NOO-klee-uhs; plural: nuclei, NOO-klee-igh): spherical cellular structure that directs all activities of the cell; positively charged center of an atom

**oil:** energy-rich organic compound made up of carbon, hydrogen, and oxygen that is liquid at room temperature
**oral groove:** mouthlike indentation in the paramecium
**order:** classification group between class and family
**ore:** rock from which metals and other minerals can be removed in usable amounts
**organ:** group of different tissues working together
**organic compound:** compound that contains the element carbon
**organic rock:** sedimentary rock built up from the remains of living things
**organism:** entire living thing that carries out all the basic life functions

**osmosis:** special type of diffusion by which water passes through the cell membrane
**ovule** (OH-vyool): structure that contains the female sex cells of a plant
**ozone:** high-energy form of oxygen that forms a layer mainly in the stratosphere

**parasite** (PA-ruh-sight): organism that feeds on other living organisms
**parasitism:** symbiotic relationship in which one organism benefits and the other is harmed
**parent rock:** solid rock found where soil is formed
**permafrost:** permanently frozen soil just below the surface
**petal:** colorful leaflike structure that surrounds the male and female reproductive organs in a flower
**phase:** state in which matter can exist; solid, liquid, gas, plasma are the phases of matter
**pheromone** (FER-uh-mohn): chemical substance given off by insects and other animals to attract a mate
**phloem** (FLOH-uhm): tubelike plant tissue that carries food down the plant
**photosynthesis** (foh-tuh-SIN-thuh-sis): process by which green plants make glucose for food by combining carbon dioxide and water by using the sun's energy
**phylum** (FY-luhm; plural: phyla): second largest classification group; between kingdom and class
**physical change:** change in which physical properties of a substance are altered, but the substance remains the same kind of matter
**physical property:** characteristic that distinguishes one type of matter from another and can be observed without changing the identity of the substance
**physics:** study of different forms of energy
**pistil:** female reproductive organ of the flower
**placenta** (pluh-SEN-tuh): structure through which developing young receive food and oxygen while in the mother
**plains landscape:** broad, flat lowland area divided into coastal or inland plains that are not far above sea level
**plasma:** phase in which matter is very high in energy and cannot be contained by the walls of ordinary matter; very rare on Earth
**plateau landscape:** broad, flat highland area usually more than 600 meters above sea level
**pollen:** particle containing the male sex cells of a plant
**pollination:** transfer of pollen to the stigma of a flower
**population:** group of the same type of organism living together in the same area
**pore:** opening on the outer surface of an animal through which materials enter and leave
**precipitation** (pree-sip-uh-TA-shun): water falling to the earth as rain, sleet, hail, or snow
**primary waves** (P waves): shock waves produced by a back-and-forth vibration of rock particles during an earthquake
**producer:** green plant that makes its own food; autotroph
**property:** quality or characteristic that describes an object
**protein:** organic compound made up of carbon, hydrogen, oxygen, nitrogen, and sometimes sulfur and phosphorus, necessary for the growth and repair of body structures
**proton:** positively charged particle found in the nucleus of an atom
**protoplasm:** all the living material found in both plant and animal cells
**protozoan:** unicellular animal-like organism
**pseudopod:** "false foot" in amoeba
**pupa** (PYOO-puh): stage in an insect's life cycle when it is wrapped in a cocoon
**pure substance:** substance that contains only one kind of molecule

**quark** (kwark): particle that makes up all subatomic nuclear particles

**radula:** filelike structure in the mouth of a univalve used to scrape food from an object
**reproduction:** process by which living things give rise to the same type of living thing
**respiration:** process by which living things take in oxygen and use it to produce energy
**response:** some action or movement of the organism brought on by a stimulus
**rhizoid** (RIGH-zoid): rootlike structure through which mosses absorb water
**rhizome** (RIGH-zohm): stem growing along or under the ground
**ribosome:** grainlike body made up of RNA and attached to the inner surface of an endoplasmic passageway; a protein-making site of the cell
**RNA** (ribonucleic acid): nucleic acid that "reads" the genetic information carried by DNA and guides the protein-making process
**rock cycle:** continuous changing of rocks from one kind to another

**saprophyte** (SA-pruh-fight): organism that feeds on dead things
**scavenger:** animal consumer that feeds on the bodies of dead animals
**scientific method:** basic steps that scientists follow in uncovering facts and solving scientific problems
**secondary waves** (S waves): shock waves produced by an up-and-down or side-to-side vibration of rock particles during an earthquake
**sedimentary** (sed-uh-MEN-tuh-ree) **rock:** rock formed in layers
**seed:** structure from which a plant grows; contains a young plant, stored food, and a seed coat
**seismic** (SIGHZ-mihk) **wave:** shock wave produced by an earthquake
**seismograph** (SIGHZ-muh-graf): instrument that records shock waves due to earthquakes
**sepal** (SEE-puhl): leaflike structure enclosing a flower when it is still a bud
**sexual reproduction:** formation of an organism from the uniting of two different sex cells
**shadow zone:** area in which no earthquake waves are detected by seismographs
**slime mold:** bloblike organism found on decaying material
**solid:** phase in which matter has a definite shape and volume

**solubility:** amount of solute that will completely dissolve in a given amount of solvent at a specific temperature
**soluble** (SAHL-yuh-bul): can be dissolved in water
**solute** (SAHL-yoot): in a solution, substance that is dissolved
**solution:** homogeneous mixture in which particles are dissolved in one another
**solvent** (SAHL-vunt): in a solution, substance that does the dissolving
**species** (SPEE-sheez): group of organisms that are able to interbreed, or produce young
**spirillum** (spigh-RIL-uhm); plural: spirilla): spiral-shaped bacterium
**spontaneous generation:** theory that states that life can spring from nonliving matter
**stalk:** in a mushroom, a stemlike structure with a ring near its top
**stamen** (STAH-muhn): male reproductive organ of the flower
**stimulus:** signal to which an organism reacts
**stoma** (STOH-muh; plural: stomata): opening in the surface of a leaf through which carbon dioxide enters and oxygen and water vapor pass out
**stratosphere** (STRAT-uh-sfeer): layer of the atmosphere where the ozone layer and jet streams are found
**streak:** color of a mineral in powder form
**subatomic particle:** particle that is smaller than an atom
**sublimation** (suhb-luh-MAY-shuhn): process in which the surface particles of a solid change directly into gas
**subscript:** number placed to the lower right of a chemical symbol to indicate the number of atoms of the element in the compound
**subsoil:** level beneath topsoil that is made up of larger bits of rock with little or no humus
**suspension:** heterogenious mixture containing particles that are mixed together but not dissolved
**swim bladder:** sac in a bony fish that enables the fish to rise or sink in water
**symbiosis:** (sim-bigh-OH-sis): relationship in which one organism lives on, near, or even inside another organism
**system:** group of organs that work together to perform certain functions

**talon:** sharp claw on the toe of birds of prey
**taxonomy:** (tak-SAH-nuh-mee): science of classification
**thermosphere** (THER-muh-sfeer): upper layer of the atmosphere where the ionosphere and auroras are found
**tissue:** group of cells similar in structure and joined together to perform a special function
**topsoil:** uppermost level of soil that is made of a mixture of small grains of rock and decayed matter of plants and animals
**toxin:** poison produced by bacteria
**transpiration:** process for regulating water loss through the leaves of a plant
**troposphere:** (TRAH-puh-sfeer): lowest layer of the atmosphere where all the earth's weather occurs

**unicellular:** having one cell

**vacuole** (VA-kyoo-ohl): large, round sac floating in the cytoplasm of a cell in which water, food, enzymes, and other materials are stored
**valley glacier:** long, narrow glacier that moves down hill between the steep sides of mountain valleys
**vaporization** (vay-puhr-uh-ZAY-shuhn): change of a liquid to a gas
**variable:** factor being tested in an experiment
**vascular plant:** plant that contains transporting tubes that carry material throughout the plant
**venom:** poison produced in special glands by snakes
**vertebra** (plural: vertebrae): one of the bones that make up a vertebrate's backbone
**vertebrate** (VER-tuh-brit): animal with a backbone
**virus:** tiny particle that contains hereditary material
**volume:** amount of space an object takes up

**warmblooded:** able to maintain a constant body temperature
**water table:** top level to which ground water rises when underground
**weathering:** process by which rocks are broken down by such factors as wind, water, heat, and cold
**weight:** measure of the force of attraction between objects due to gravity
**woody stem:** rigid stem containing woody xylem tissue

**xylem** (ZIGH-luhm): tubelike plant tissue that carries water and minerals through the plant

**zoology:** study of animals

# Index

## E



## F

## G



## H



## I



## J



## K



## L

## M



## N

## O



## P



## Q



## R

## S

## Photograph Credits

**1,** Rick Smolan/Contact/*Woodfin Camp;* **2,** top, David O. Houston/*Bruce Coleman;* bottom left, Ed Cooper/*H. Armstrong Roberts;* bottom right, Wesley Frank/*Woodfin Camp;* **5,** top, NIH Science Source/*Photo Researchers;* bottom, T. Daniel/*Bruce Coleman;* **6,** top, Manfred Kage/*Peter Arnold;* bottom, Fred Bavendam/*Peter Arnold;* **7,** top, Taronga Zoo, Sydney-Tom McHugh/*Photo Researchers;* bottom, O.S. Pettingill/*Photo Researchers;* **8,** top, Werner Muller/*Peter Arnold;* bottom, Vulcain-Explorer/*Photo Researchers;* **9,** top, Steve Vidler/*Leo de Wys;* bottom, Margot Conte/© *Earth Scenes;* **14,** © Tom McHugh/*Photo Researchers;* **16,** Peter B. Kaplan; **17,** Peter B. Kaplan; **19,** top, *dpi;* bottom left, NASA; bottom right, Wally McNamee/*Woodfin Camp & Associates;* **21,** top left, Ken Karp; top right, Ken Karp; bottom left, Ken Karp; bottom right, Ken Karp; **22,** Jerry Wachter/*Focus on Sports;* **23,** NASA; **24,** Kim Taylor/*Bruce Coleman;* **25,** top, J.A.L. Cooke/*Animals Animals;* center, David Scharf/*Peter Arnold;* bottom, Manfred Kage/*Peter Arnold;* **26,** Dr. E.R. Degginger/*Bruce Coleman;* **27,** top, Howard Sochurek/*Woodfin Camp & Associates;* bottom, NIH/Science Source/*Photo Researchers;* **28,** left, Dan McCoy/*Rainbow;* right, Howard Sochurek/*Woodfin Camp & Associates;* **36,** top left, M.P. Kahl/*Bruce Coleman;* bottom left, Breck P. Kent; bottom right, Jen & Des Bartlett/*Bruce Coleman;* top right, Peter Ward/*Bruce Coleman;* **38,** right, Chris Newbert/*Bruce Coleman;* left, Wayne Lankinen/*Bruce Coleman;* **39,** T. Daniel/*Bruce Coleman;* **40,** left, Tom McHugh/*Photo Researchers;* right, Jen & Des Bartlett/*Bruce Coleman;* **41,** left, Harry Rogers/*dpi;* right, Harry Rogers/*dpi;* **42,** top left, John Shaw/*Bruce Coleman;* top right, Grant Heilman; bottom, Shelley Rotner/*OPC;* **43,** Runk/Schoenberger/*Grant Heilman;* **44,** top, R. Mariscal/*Bruce Coleman;* bottom left, Kjell B. Sandved/*Bruce Coleman;* bottom right, Bill Bridge/*dpi;* **45,** left, Miguel Castro/*Photo Researchers;* right, W. H. Hodge/*Peter Arnold;* **46,** David C. Fritts/*Animals Animals;* **47,** left, Manuel Rodriguez; right, Dr. E. R. Degginger; **49,** top, Dr. E. R. Degginger; bottom, Dr. E. R. Degginger; **50,** top, Barry L. Runk/*Grant Heilman;* bottom, Bill Stanton/*International Stock Photo;* **51,** Tripos Associates/*Peter Arnold;* **56,** Lennart Nilsson/The Incredible Machine/*The National Geographic Society;* **60,** Jack McConnell, *McConnell McNamara Assoc./dpi;* **61,** Fawcett/*Photo Researchers;* **62,** Ken Karp; **63,** Omikron/*Taurus Photos;* **64,** K. R. Porter/*Photo Researchers;* **67,** Manfred Kage/*Peter Arnold;* **70,** left, M. Sheetz/*University of Conn. Health Center/J. Cell. Biol.;* center, M. Sheetz/*Univ. of Conn. Health Center/J. Cell. Biol.;* right, M. Sheetz/*Univ. of Conn. Health Center/J. Cell. Biol.;* **71,** left, Runk/Schoenberger/*Grant Heilman;* right, Grant Heilman; **72,** Carolina Biological Supply Company; **73,** left, Sven Olaf Lindblad/*dpi;* right, Craig Aurness/*West Light;* **78,** Peter Davey/*Bruce Coleman;* **80,** top, © Hans Pfletschinger/*Peter Arnold;* bottom, Dr. Merlin D. Tuttle; **82,** top left, Manfred Kage/*Peter Arnold;* top center, Manfred Kage/*Peter Arnold;* top right, Manfred Kage/*Peter Arnold, Inc.,* bottom, © Hans Pfletschinger/*Peter Arnold;* **83,** G. Ziesler/*Peter Arnold;* **85,** left, Rod Allin/*Bruce Coleman;* right, Eric Crichton/*Bruce Coleman;* bottom, Wil Blanche/*dpi;* **87,** Charles G. Summers, Jr./*dpi;* **88,** Ken Karp; **92,** Raymond A. Mendez/*Animals Animals;* **95,** *dpi;* **96,** Kim Taylor/*Bruce Coleman;* **97,** J. Alex Langley/*dpi;* **98,** C. A. Morgan/*Peter Arnold;* **99,** Des and Jen Bartlett/*Bruce Coleman;* **100,** © Jonathan T. Wright/*Bruce Coleman;* **101,** Fred Bavendam/*Peter Arnold;* **103,** top left, Jane Burton/*Bruce Coleman;* top right, G. Ziesler/*Peter Arnold;* bottom, Lovett E. Williams, Jr./*dpi;* **104,** top left, Walker/*Photo Researchers;* second left, Michael Abbey/*Photo Researchers;* third left, © George Holton/*Photo Researchers;* bottom left, Chris Bry/*dpi;* bottom right, Jen & Des Bartlett/*Bruce Coleman;* **107,** R. Andrew Odum/*Peter Arnold;* **108,** Charlie Ott/*dpi;* **109,** left; Charlie Ott/*dpi;* right, J.M. Barr/*dpi;* **110,** left, Lee Lyon/*Bruce Coleman;* right, Peter Ward/*Bruce Coleman;* **111,** Leonard Lee Rue III/*dpi;* **112,** top, Mike Price/*Bruce Coleman;* bottom, Wilt Hodge/*Peter Arnold;* **113,** top left, John Bechteler/*Shostal;* top right, Kevin Schafer/*Tom Stack & Associates;* center right, Phil Degginger/*Bruce Coleman;* bottom right, Roger Werth/*Woodfin Camp and Associates;* **114,** left, John Gerlach/*Earth Scenes;* right, Liane Enkelis/*Mono Lake Committee;* **116,** Ken Karp; **120,** Laboratory of Ornithology/Photo: *Joyce Poole;* Jen & Des Bartlett/*Bruce Coleman;* **122,** David Scharf/*Peter Arnold;* **123,** Eric Kroll/*Taurus;* **124,** Michael Habicht/*Animals Animals;* **126,** left, Smithsonian Institution; center, Ken Karp/*OPC;* right, Joel Greenstein/*OPC;* **128,** Bryon Crader/*Tom Stack;* **130,** Peter Scoones/*Seaphoto/Colorific;* **132,** Mazonowicz/*Monkmeyer Press;* **133,** right, Robert C. Simpson/*Tom Stack;* left, John Pawloski/*Tom Stack;* top, © Boyd Norton/*Peter Arnold;* **136,** Dr. E.R. Degginger; **137,** G.B. Schaller/*Bruce Coleman;* **139,** first row, left to right, Phil Dotson/*dpi;* Jack Dermid, Charles G. Summers, Jr./*dpi;* Theodore Zywotko/*dpi;* second row, left to right, Dr. E. R. Degginger/*Bruce Coleman;* David M. Stone; Barbara K. Deans/*dpi;* third row, left to right, Jerry Frank/*dpi;* J. Alex Langley/*dpi;* fourth row, left to right, Phil Dotson/*dpi;* Phil Dotson/*dpi;* Francisco Erize/*Bruce Coleman;* fifth row, left to right, Lois and George Cox/*Bruce Coleman;* James Theologos/*Monkmeyer Press;* Mimi Forsyth/*Monkmeyer Press;* sixth row, left to right, Kenneth W. Fink/*Bruce Coleman;* Wil Blanche/*dpi;* Phil Dotson/*dpi;* bottom, Jon A. Hull/*Bruce Coleman;* **140,** Stephen Dalton/*Animals Animals;* **141,** Eric Grave/*Phototake;* **146,** Martin Rotker/*Taurus Photos;* **148,** left, Wendell Metzen/*Bruce Coleman;* right, *Grant Heilman;* **149,** Biology Media/*Photo Researchers;* **150,** E. H. Cook/*Photo Researchers;* **151,** Lee D. Simon/*Photo Researchers;* **153,** left, Manfred Kage/*Peter Arnold;* center, Manfred Kage/*Peter Arnold;* right, Dr. E. R. Degginger; **154,** top, *L. V. Bergman & Associates;* bottom, D. Jorgenson/*Tom Stack & Associates;* **155,** top, William E. Ferguson; botton, Manfred Kage/*Peter Arnold;* **156,** Martin M. Rotker/*Taurus Photos;* **158,** top, William E. Ferguson; botton, Manfred Kage/*Peter Arnold;* **159,** Jonathan D. Eisenback/*Phototake;* **161,** Michael Abbey/*Photo Researchers;* **162,** top, Manfred Kage/*Peter Arnold;* bottom, *L. V. Bergman & Associates;* **168,** Rick Smolan/*Contact/Woodfin Camp;* **170,** *Grant Heilman;* **171,** top, *Doug Wechsler;* bottom, *Bruce Coleman;* **172,** *Grant Heilman;* **173,** left, David C. Fritts/*Earth Scenes;* right, F. G. Love; **174,** left, L. S. Stepanowicz/*Photo Researchers;* right, Kim Taylor/*Bruce Coleman;* **175,** left, Dr. I. Metzner/*Peter Arnold;* right, Manfred Kage/*Peter Arnold;* **176,** Runk/Schoenberger/*Grant Heilman;* **177,** top, Breck P. Kent; bottom, Gordon Leedale/*Photo Researchers;* **178,** Oxford Scientific Films/*Animals Animals;* **179,** top left, Robert P. Carr/*Bruce Coleman;* top right, Heather Angel/*Biofotos;* bottom, Hal McKusick/*dpi;* **180,** top left; Manuel Rodriguez; top right, Cal Harbert/*dpi;* bottom left, John H. Gerard/*dpi;* bottom right, Manuel Rodriguez; **182,** top, Heather Angel/*Biofotos;* bottom, Eric V. Gravé/*Phototake;* **183,** Runk/Schoenberger/*Grant Heilman;* **184,** Cary Wolinsky/*Stock Boston;* **185,** Charles Ott/*dpi;* **186,** Dr. E. R. Degginger; **187,** top, David M. Stone; bottom, John H. Gerard/*dpi;* **188,** Ken Karp; **192,** Fred Bavendam/*Peter Arnold;* **194,** left, J. M. Barrs/*dpi;* right, *Field Museum of Natural History, Chicago;* **195,** top, Wendy Neefus/*Earth Scenes;* bottom, Adrienne T. Gibson/*Earth Scenes;* **196,** Jerome Wexler/*dpi;* **197,** Wendy Neefus/*Earth Scenes;* **198,** top left, Robert L. Dunne/*Bruce Coleman;* top right, Robert L. Dunne/*Bruce Coleman;* bottom right, Jerry Howard/*Photo Researchers;* **199,** Richard Kolar/*Earth Scenes;* **200,** top, *Ardea Photographs: London;* bottom, Manuel Rodriguez; **201,** left, *Peter Arnold;* right, Breck P. Kent/*Earth Scenes;* **202,** left, David M. Stone; right, *dpi;* **204,** top, Robert Weinreb/*Bruce Coleman;* bottom, W. H. Hodge/*Peter Arnold;* **205,** top, Norman O. Tomalin/*Bruce Coleman;* bottom, Charlie Ott/*dpi;* **206,** Darwin Van Campen/*dpi;* **207,** top, Manuel Rodriguez; bottom, Manuel Rodriguez; **208,** Patti Murray/*Animals Animals;* **209,** Elliott Varner Smith/*International Stock Photos;* **210,** Oxford Scientific Films/G. I. Bernard/*Earth Scenes;* **211,** left, Ann Hagen Griffith/*OPC;* center, Breck P. Kent/*Earth Scenes;* right, Marcia W. Griffen/*Earth Scenes;* **216,** Stephen Dalton/*Animals Animals;* **218,** top left, Stephen Dalton/*Animals Animals;* top right, Dr. E. R. Degginger/*Animals Animals;* bottom left, Tim Rock/*Animals Animals;* bottom right, Dr. E. R. Degginger/*Animals Animals;* **220,** left, Mike Schick/*Animals Animals;* top right, Carl Roessler/*Animals Animals;* bottom right, Carl Roessler/*Animals Animals;* **221,** top, Oxford Scientific Films/G. I. Bernard/*Animals Animals;* bottom, Steve Earley; **222,** top, Runk/Schoenberger/*Grant Heilman;* bottom left, Carl Roessler/*Animals Animals;* bottom right, Z. Leszczynski/*Animals Animals;* **223,** top, Jeff Rotman; bottom, Tim Rock/*Animals Animals;* **224,** Oxford Scientifc Films/*Animals Animals;* **225,** left, Phil Degginger/*Bruce Coleman;* right, G. I. Bernard/Oxford Scientific Films/*Animals Animals;* **226,** top, Runk/Schoenberger/*Grant Heilman;* bottom, *L. V. Bergman;* **228,** Hans Pfletschinger/*Peter Arnold;* **229,** *Culver Pictures;* **230,** Harry Hartman/*Bruce Coleman;* **231,** left, Hans Pfletschinger/*Peter Arnold;* right, Phil Degginger/*Bruce Coleman;* **232,** top, Bill Wood/*Bruce Coleman;* bottom left, Douglas Faulkner/*Photo Researchers;* bottom right, Jack Dermid; **233,** top, *Grant Heilman;* bottom, Steinhart Aquarium/Tom McHugh/*Photo Researchers;* **234,** top, Jeff Rotman; bottom, Z. Leszczynski/*Animals Animals;* **235,** left, Jeff Foott/*Bruce Coleman;* right, Phil Dotson/*dpi;* **236,** top, Fred Bavendam/*Peter Arnold;* bottom, Steinhart Aquarium/Tom McHugh/*Photo Researchers;* **237,** Jeff Rotman; **238,** top, Steve Martin/*Tom Stack;* bottom, Richard Kolar; **239,** top, W. Bayer/*Bruce Coleman;* bottom, John Shaw; **240,** top left, James H. Carmichael, Jr./*Bruce Coleman;* top right, John Shaw; bottom, M. P. L. Fogden/*Bruce Coleman;* **241,** left, Hans Pfletschinger/*Peter Arnold;* right, Runk/Schoenberger/*Grant Heilman;* **242,** Raymond A. Mendez/*Animals Animals;* **243,** Manuel Rodriguez; **244,** top and bottom left, John H. Gerard/*dpi;* top right, John H. Gerard/*dpi;* bottom, Verna R. Johnson/*dpi;* **245,** top, Stephen Dalton/*Animals Animals;* bottom, Phil Dotson/*dpi;* **246,** Leeanne Schmidt/*dpi;* **247,** left, Oxford Scientific Films/Stephen Dalton/*Animals Animals; bottom right, Jack Dermid; top right, Peter Ward/Bruce Coleman;* center right, Thomas Eisner/Daniel Aneshansley/*Discover* magazine; **255,** top, Carl Roessler/*Bruce Coleman;* bottom, Fred Bavendam/*Peter Arnold;* **256,** left, Roessler/*Animals Animals;* right, C. C. Lockwood/*Animals Animals;* bottom, Heather Angel/*Biofotos;* **257,** Valerie Taylor/*Ardea: London: **258,** top left, Valerie Taylor/Ardea: London;* top right, Bob Evans/*Peter Arnold;* bottom, Bill Wood/*Bruce Coleman;* **259,** left, Dr. E. R. Degginger; right, Oxford Scientific Films/*Animals Animals;* **260,** top, Andrew Gifford/*dpi;* bottom, Michael Fogden/*Bruce Coleman;* **261,** left, Brian Rogers/*Biofotos;* right, Hans Pfletschinger/*Peter Arnold;* **262,** left, John Shaw/*Bruce Coleman;* right, Robert L. Dunne/*Bruce Coleman;* **263,** top, Runk/Schoenberger/*Grant Heilman;* bottom, Dr. E. R. Degginger; **264,** Miguel Castro/*Photo Researchers;* **265,** Dr. E. R. Degginger; **266,** top left, E. Hanumantha Raa/*Photo Researchers;* top right, Kjell B. Sandved/*Photo Researchers;* bottom, Susan Pierres/*Peter Arnold;* **267,** left, Tom Brakefield/*Bruce Coleman;* right, John H. Pontier/*Animals Animals;* **268,** Phil Degginger; **269,** left, Carol Hughes/*Bruce Coleman;* right, G. Ziesler/*Peter Arnold;* **270,** left, Richard Kolar/*Animals Animals;* right, Marty Stouffer/© *Animals Animals;* **271,** Doug Allan/Oxford Scientific Films/*Animals Animals;* **272,** Dr. E. R. Degginger; **273,** top left, James R. Leard/*dpi;* bottom left, John E. Swedberg/*Bruce Coleman;* right, Panuska/*dpi;* **274,** top, John L. Pontier/*Animals Animals;* bottom, R. W. Young/*dpi;* **275,** Dr. E. R. Degginger/*Animals Animals;* **280,** Taronga Zoo, Sydney—Tom McHugh/*Photo Researchers;* **282,** F. Sohier/*Ardea: London;* **283,** Jerry Frank/*dpi;* **284,** Jen & Des Bartlett/*Bruce Coleman;* **285,** left, Jean-Paul Ferrero/*Ardea: London;* right, Ken Stepnell/*Taurus Photos;* **286,** top, Tom McHugh/*Photo Researchers;* bottom, Rob Chabot/*dpi;* **287,** top, Leonard Lee Rue III/*Photo Researchers;* bottom, Jen & Des Bartlett/*Bruce Coleman;* **288,** top, Erwin and Peggy Bauer/*Bruce Coleman;* bottom, John Serao/*Photo Researchers;* **289,** top left, Leonard Lee Rue III/*dpi;* top right, Stouffer Productions/*Animals Animals;* bottom, Dr. Melvin D. Tuttle; **290,** Al Giddings/*Bruce Coleman;* **291,** John Dominis/*Life* magazine © 1967, Time, Inc.; Barry L. Runk/*Grant Heilman;* **292,** left, Phil Dotson/*dpi;* right, Gunter Ziesler/*Peter Arnold;* **293,** left, J. Alex Langley/*dpi;* right, Leonard Lee Rue III/*dpi;* **294,** left, J. Alex Langley/*dpi;* top right, Dr. E.R. Degginger; bottom right, Norman Owen Tomalin/*Bruce Coleman;* **294,** top right, Dr. E.R. Degginger; **295,** top left, Stephen Dalton/*Photo Researchers;* top right, C. Robinson & J.A. Grant/*Bruce Coleman;* bottom, Charlie Ott/*dpi;* **296,** right, Jeff Foott/*Bruce Coleman;* left, M. Timothy O'Keefe/*Tom Stack & Associates;* **297,** top left, Halperin/*Animals Animals;* top right, Robert W. Hernandez/*Photo Researchers;* bottom, Norman O. Tomalin/*Bruce Coleman;* **303,** top, Wrangham/*Anthro Photo;* bottom, Wrangham/*Anthro Photo;* **306,** Jeff Foott/*Bruce Coleman;* **308,** Merle H. Jensen; **309,** Runk/Schoenberger/*Grant Heilman;* **310,** O.S. Pettingill, Jr./*Photo Researchers;* **312,** © William Curtsinger/*Photo Researchers;* **314,** © Porterfield/Chickering/*Photo Researchers;* **315,** Paul Kennedy/*Leo de Wys;* **316,** left, Loomis Dean/*Life* magazine © 1954, Time, Inc.; **317,** left, The Bettmann Archive; right, NASA; **319,** Ron Sefton/*dpi;* **320,** Margaret Durrance/*Photo Researchers;* **321,** © Tatarsky/*dpi;* **322,** Ken Karp; **323,** top, J. Alex Langley/*dpi;* bottom, Runk/Schoenberger/*Grant Heilman;* **324,** Ken Karp; **328,** Red Huber/*The Orlando Sentinel;* **330,** © Stan Pantovic/*Photo Researchers;* **331,** top, B. Benedict/*H.*

*Armstrong Roberts;* bottom, © Bill Ross/*West Light;* **332,** top, © Vulcain-Explorer/*Photo Researchers;* bottom left, Clyde H. Smith/*Peter Arnold;* bottom right, Crystal lattice of ice by John W. Moore, William G. Davies, and Robert G. Williams, Eastern Michigan University; reprinted by permission of the authors. **333,** R. Megna/ *Fundamental Photographs;* **334,** *dpi;* **337,** top, NASA; bottom, Ken Kay/*dpi;* **338,** © Patricia Agre/*Photo Researchers;* **339,** left and right, Life Science Library/WATER photograph by Ken Kay, Time–Life Books, Inc. Publisher © 1966 Time, Inc.; **340,** © Renate Jope/*Photo Researchers;* **341,** © Frank Schreiber/*Photo Researchers;* **342,** left, Fundamental Photographs; right, Ken Karp; **343,** © Kent & Donna Dannen/ *Photo Researchers;* **344,** top, Paul Stephanus/*dpi;* bottom, L. Jacobs/*H. Armstrong Roberts;* **345,** © Werner H. Muller/*Peter Arnold;* **350,** Ken Karp; **351,** Ken Karp; **352,** K.D. Franke/*Peter Arnold;* **354,** Robert Rose/Picture Group Charles Belinsky/*Photo Researchers;* **356,** Georgia Power Co.; **357,** bottom, Georgia Power Co.; top, Georgia Power Co.; **358,** Courtesy of Harvey Tananbaum, Harvard/Smithsonian Center for Astrophysics; **360,** Fritz Goro/*Life* magazine, © 1949 Time, Inc.; **362,** The Bettmann Archive, Inc.; **363,** Prof. Albert V. Crewe/*The University of Chicago/Fermi Institute;* **365,** top, Dr. G. Cranstoun/*Photo Researchers;* bottom, Patrice Loiez/*Photo Researchers;* **367,** left, Ken Karp; right, © James Lester/*Photo Researchers;* **370,** left, Laboratory of Molecular Biology/MRC/Science Photo Library/*Photo Researchers;* right, Dan McCoy/*Rainbow;* **371,** Scott Thode/*International Stock Photo;* **372,** Rodney Jones; **376,** Movie Still Archives; **378,** left, © Lowell Georgia/*Photo Researchers;* right, © Tom McHugh/*Photo Researchers;* **379,** top, © Educational Dimensions Group; bottom left, Werner Muller/*Peter Arnold;* bottom right, Photo courtesy of Sperry Corp; **380,** Cary Wolinsky/*Stock Boston;* **382,** top, © Dave Schaefer/*The Picture Cube;* center, Fundamental Photographs, Inc.; bottom, Barry L. Runk/*Grant Heilman;* **384,** left, Gerhard G. Scheidle/ *Peter Arnold;* right, J. Alex Langley/*dpi;* **386,** Rodney Jones; **390,** R. Dias/*H. Armstrong Roberts;* **392,** top, Educational Dimensions Group; bottom, Runk/ Schoenberger/*Grant Heilman;* **393,** Lick Observatory Photograph; **394,** top left, Dr. E. R. Degginger, FPSA; second left, Dr. E. R. Degginger; third left, Dr. E. R. Degginger; bottom right, Alan Pitcairn/ *Grant Heilman;* bottom left, © Van Bucher/*Photo Researchers;* **395,** top, Fredrik D. Bodin/*Stock, Boston;* bottom, Ken Karp; **396,** left, Ken Karp; right, *dpi;* **397,** Bob Stern/*International Stock Photo;* **399,** left, Farrell Grehan/*Photo Researchers;* right, Runk/Schoenberger/*Grant Heilman;* **404,** Fermilab; **405,** AT&T Bell Labs; **406,** © 1980 Roy H. Blanchard/ *dpi;* **408,** top, DOE; bottom, DOE; **410,** Amoco Chemical Co.; **412,** Breck P. Kent/© *Earth Scenes;* **414,** Fred Ward/*Black Star;* **416,** top, Dr. E. R. Degginger; center, Dr. E. R. Degginger; bottom, Stephanie S. Ferguson, William E. Ferguson; **417,** top right, Breck Kent/*Earth Scenes;* top left, Breck Kent/*Earth Scenes;* top bottom, Jane Burton/*Bruce Coleman;* bottom bottom, Charles R. Belinsky/*Photo Researchers;* **418,** top right, Breck P. Kent; top left, Dr. E. R. Degginger; top bottom, Runk/ Schoenberger/*Grant Heilman;* bottom bottom, Brian Parker/*Tom Stack;* **419,** top left, Breck P. Kent; top center, Dr. E. R. Degginger; top right, Breck P. Kent; center, Dr. E. R. Degginger; bottom, Breck P. Kent/*Earth Scenes;* **421,** Ken Karp; **423,** left, Breck P. Kent/*Earth Scenes;* center, Breck P. Kent; right, John Caucalos/*Tom Stack;* bottom, Steve Vidler; **424,** top to bottom, Charles R. Belinsky/*Photo Researchers;* Breck P. Kent/ *Earth Scenes;* Breck P. Kent; Breck P. Kent; Breck P. Kent/*Earth Scenes;* **426,** left, Dr. E. R. Degginger; right, Arthur Levine/Leo de Wys; **427,** top left to bottom left, Breck P. Kent/© *Earth Scenes;* © Breck P. Kent, Dr. E. R. Degginger; © Breck P. Kent; Dr. E. R. Degginger; Dr. E. R. Degginger; Dr. E. R. Degginger; top right to bottom right, © Russ Kinne/*Photo Researchers;* © Russ Kinne/*Photo Researchers;* Dr. E. R. Degginger; © Russ Kinne/*Photo Researchers;* © Russ Kinne/*Photo Researchers;* © Russ Kinne/*Photo Researchers;* Ward's Natural Science Establishment; **432,** Jack Fields/*Photo Researchers;* **434,** top, Dr. E. R. Degginger; bottom, Mickey Gibson/*Earth Scenes;* **435,** left, John Gerlach/*Tom Stack;* top right, © William E. Ferguson; bottom right, Photo by Stephanie S. Ferguson © William E. Ferguson; **436,** top, Dr. E. R. Degginger; bottom, G. R. Roberts; **437,** left, G. R. Roberts; right, G. R. Roberts; **438,** top, Dr. E. R. Degginger; bottom, Breck P. Kent/© *Earth Scenes;* **439,** top, Spencer Swanger/*Tom Stack;* bottom, © Phil Degginger; **440,** John Gerlach/*Tom Stack;* **441,** left and right, Hubbard Scientific Company; **442,** top left, top right, bottom left, and bottom right, from Viewfile "Rocks," Courtesy Hubbard Scientific, Northbrook, Illinois; **443,** Wil Blanche/*dpi;* **444,** top, Tom Bean/*Tom Stack;* bottom left, Dr. E. R. Degginger; bottom right, Dr. E. R. Degginger; **446,** Mickey Gibson/*Earth Scenes;* **450,** SUVA/*Omni-Photo Communications;* **452,** USDA Photo; **453,** USDA Photo; **454,** © Richard W. Tolbert/*dpi;* **455,** © Jacques Jangoux/ *Peter Arnold;* **456,** top, W. H. Hodge/*Peter Arnold;* bottom left, Henry Ausioos/ *Animals Animals;* bottom right, © Mickey Gibson/*Animals Animals;* **457,** © Jacques Jangoux/© *Peter Arnold;* **458,** USDA Photo; **459,** left, Dr. E. R. Degginger/*Earth Scenes;* right, Dr. E. R. Degginger/*Earth Scenes;* **460,** Arthur Rothstein/© Culver Pictures; **461,** © Arthur Rothstein; **462,** top, Stephen J. Krasemann/© *Peter Arnold;* bottom, © Mickey Gibson/*Earth Scenes;* **463,** top, Johnston/*dpi;* bottom left, Richard Farnell/*Earth Scenes;* bottom right, Richard Farnell/*Animals Animals;* **464,** Ken Karp; **468,** Peru National Tourist Office; **469,** Dr. Georg Gerster/*Photo Researchers;* **470,** Univ. of California; **471,** left, Dr. Bruce Heezen/Lamont-Doherty; right, Dave Woodward/*Taurus;* **472,** © C. B. Jones 1982/*Taurus;* **476,** Nicholas Devore/*dpi;* **478,** © Michael Nichols/ *Magnum Photos;* **481,** Eros Data Center/U.S.G.S.; **483,** © Eric Kroll 1981/*Taurus Photos;* **484,** Ken Karp; **489,** Frederick Grassle/*Woods Hole Oceanographic Institution;* **494,** National Caves Association; **495,** NASA; **500,** U.S. Geological Survey © Wilcox, R.E.; **502,** left, Sue and Jon Hacking/© *Earth Images;* right, *dpi;* **505,** Edward Johnson/ *Leo de Wys;* **507,** Steve Vidler/*Leo de Wys;* **508,** Jay Lurie Photography/*Bruce Coleman;* **509,** G. R. Roberts; **510,** Brian Parker/*Tom Stack;* **511,** © Pam Taylor/*Bruce Coleman;* **512,** *dpi;* **514,** top left, Stuart L. Craig/*Bruce Coleman;* top right, Keith Gunnar/*Bruce Coleman;* bottom, William E. Ferguson; **515,** Hackenberg/*Leo de Wys;* **516,** G. R. Roberts/ *OPC;* **517,** left, F. G. Love; right, G. R. Roberts; **522,** Margot Conte/© *Earth Scenes;* **524,** left, The Bettman Archive, Inc.; right, National Center for Atmospheric Research/National Science Foundation; **525,** bottom, Novack/*Leo de Wys;* top, Keith Gunnar/*Bruce Coleman;* **527,** top, J. Alex Langley/*dpi;* bottom, Monty Monsees/*dpi;* **528,** left, Breck P. Kent; right, Keith Gunnar/*Bruce Coleman;* **529,** Jerome Wyckoff; **531,** © Pat Lanza Field/ *Bruce Coleman;* **533,** both, NASA; **535,** Don Benson/*Tom Stack;* **536,** William E. Ferguson; **537,** NASA; **542,** William Haxby; **543,** Anita Brosius; **544,** © Dan Budnik 1982/*Woodfin Camp;* **545,** Alfred Owczarzak/*Taurus;* **546,** Chuck O'Rear/*West Light*

Text illustrations by Lee Ames & Zak Ltd. with special acknowledgement to David Christensen. Gazette art by Phil Carver & Friends, Inc.